Animal Biotechnology

Models in Discovery and Translation

Animal Biotechnology

Models in Discovery and Translation

Edited by

Ashish S. Verma
Amity Institute of Biotechnology
Amity University Uttar Pradesh,
NOIDA (UP), India

Anchal Singh
Amity Institute of Biotechnology
Amity University Uttar Pradesh,
NOIDA (UP), India

AMSTERDAM • BOSTON • HEIDELBERG • LONDON • NEW YORK • OXFORD • PARIS
SAN DIEGO • SAN FRANCISCO • SINGAPORE • SYDNEY • TOKYO

Academic Press is an imprint of Elsevier

Academic Press is an imprint of Elsevier
The Boulevard, Langford Lane, Kidlington, Oxford, OX5 1GB, UK
225 Wyman Street, Waltham, MA 02451, USA

First published 2014

Notices
Knowledge and best practice in this field are constantly changing. As new research and experience broaden our understanding, changes in research methods, professional practices, or medical treatment may become necessary.

Practitioners and researchers must always rely on their own experience and knowledge in evaluating and using any information, methods, compounds, or experiments described herein. In using such information or methods they should be mindful of their own safety and the safety of others, including parties for whom they have a professional responsibility.

To the fullest extent of the law, neither the Publisher nor the authors, contributors, or editors, assume any liability for any injury and/or damage to persons or property as a matter of products liability, negligence or otherwise, or from any use or operation of any methods, products, instructions, or ideas contained in the material herein.

British Library Cataloguing in Publication Data
A catalogue record for this book is available from the British Library

Library of Congress Cataloguing in Publication Data
A catalogue record for this book is available from the Library of Congress

ISBN: 978-0-12-416002-6

For information on all Academic Press publications
visit our website at store.elsevier.com

Dedication

This book is dedicated in fond memories of
Dr. Har Swarup Verma

(1941–1995)
Loving Father,
Admirable Professor,
&
Compassionate Physician.

Ashish (son) & Anchal

Contents

Section II
Animal Biotechnology: Tools and Techniques

Section III
Animal Biotechnology: Applications and Concerns

Contributors

Ashok K. Adya BIONTHE (Bio- and Nano-technologies for Health & Environment) Centre, Division of Biotechnology & Forensic Sciences, School of Contemporary Sciences, University of Abertay, Dundee, Scotland, UK

Neelesh Agarwal Epidemiology and Genomics Research Program, Division of Cancer Control and Population Sciences, National Cancer Institute, National Institutes of Health (NIH), Bethesda, MD, USA

Ashima Agarwal Amity Institute of Biotechnology, Amity University Uttar Pradesh, NOIDA (UP), India

Udita Agrawal Department of Pharmaceutical Sciences, Dr. Hari Singh Gour University, Sagar, M.P., India

Vasco Azevedo Laboratorio de Genetica Celular eMolecular, Departmento de Biologia Geral, Instituto de Ciencias Biologics, Universidade Federal de Minas Gerais Belo Horizonte, Minas Gerais, Brazil

Ruby Bansal Crosslay Wellness Program, Pushpanjali Crosslay Hospital, Ghaziabad (UP), India

Debmalya Barh Centre for Genomics and Applied Gene Technology, Institute of Integrative Omics and Applied Biotechnology (IIOAB), Nonakuri, Purba Medinipur, India

Mausumi Bharadwaj Division of Molecular Genetics & Biochemistry, Institute of Cytology & Preventive Oncology (ICMR), Noida, Uttar Pradesh, India

Anant Narayan Bhatt Institute of Nuclear Medicine and Allied Sciences, Defense Research Developmental Organization (DRDO), Ministry of Defense, Delhi, India

Manish Biyani Department of Biotechnology, Biyani Group of Colleges, Jaipur, India, Department of Bioengineering, The University of Tokyo, Tokyo, Japan

Madhu Biyani Department of Biotechnology, Biyani Group of Colleges, Jaipur, India, Department of Functional Materials Science, Saitama University, Saitama, Japan

Elisabetta Canetta Cardiff School of Biosciences, Cardiff University, Cardiff, Wales, UK

Vijender Chaitankar Department of Computer Science, Virginia Commonwealth University, Richmond, VA, USA

Rakhi Chaturvedi Department of Biotechnology, Indian Institute of Technology-Guwahati, Guwahati, Assam, India

Sushmita Chaudhary Amity Institute of Biotechnology, Amity University Uttar Pradesh, NOIDA (UP), India

Sudhir Chowbina Advanced Biomedical Computing Center, SAIC-Frederick, Inc., Frederick National Laboratory for Cancer Research, National Cancer Institute, Frederick, MD, USA

D. Kar Chowdhuri Embryotoxicology, CSIR-Indian Institute of Toxicology Research, Mahatma Gandhi Marg, Lucknow, Uttar Pradesh, India

Madhumita Roy Chowdhury Genetic Unit, Department of Pediatrics, All India Institute of Medical Sciences, New Delhi, India

Bhudev C. Das Laboratory of Molecular Oncology, Dr. B. R. Ambedkar Center for Biomedical Research (ACBR), University of Delhi, Delhi, India

Mukul Das Food, Drug and Chemical Toxicology Group, CSIR-Indian Institute of Toxicology Research, Lucknow, U.P., India

Surajit Das Department of Life Science, National Institute of Technology, Rourkela, Odisha, India

Amrita Datta Department of Pharmacology, Tulane University Medical Center, New Orleans, LA, USA

Gautami Devi Department of Biotechnology, Indian Institute of Technology-Guwahati, Guwahati, Assam, India

Devyani Dube ISF College of Pharmacy, Moga, Punjab, India

Sudhisha Dubey Department of Genetic Medicine, Sir Ganga Ram Hospital, New Delhi, India

Bilikere S. Dwarakanath Institute of Nuclear Medicine and Allied Sciences, Defense Research Developmental Organization (DRDO), Ministry of Defense, Delhi, India

Premendra D. Dwivedi Food, Drug and Chemical Toxicology Group, CSIR-Indian Institute of Toxicology Research, Lucknow, U.P., India

Rasha El Baz Department of Microbiology and Immunology, Drexel Institute for Biotechnology and Virology Research, Drexel University College of Medicine, Doylestown, PA, USA

R.K. Gaur Department of Science, Faculty of Arts, Science and Commerce, Mody Institute of Technology and Science, Rajasthan, India

Preetam Ghosh Department of Computer Science, Virginia Commonwealth University, Richmond, VA, USA

Madhu Gupta Department of Pharmaceutical Sciences, Dr. Hari Singh Gour University, Sagar, M.P., India

Anuj Kumar Gupta C-11/Y-1, C-Block, Dilshad Garden, Delhi, India

U.D. Gupta National JALMA Institute for Leprosy & Other Mycobacterial Diseases (ICMR), Agra, UP, India

Abhik Gupta Department of Ecology & Environmental Science, Assam University, Silchar, India

Showket Hussain Division of Molecular Genetics & Biochemistry, Institute of Cytology & Preventive Oncology (ICMR), Noida, Uttar Pradesh, India

Pooja Jain Department of Microbiology and Immunology, Drexel Institute for Biotechnology and Virology Research, Drexel University College of Medicine, Doylestown, PA, USA

Anurag Jyoti Nanotherapeutics & Nanomaterial Toxicology Group, CSIR-Indian Institute of Toxicology Research, Lucknow, U.P., India

S.K. Kashyap Department of Vet Microbiology & Biotechnology, Rajasthan University of Veterinary & Animal Sciences, Bikaner, Rajasthan, India

Zafar K. Khan Department of Microbiology and Immunology, Drexel Institute for Biotechnology and Virology Research, Drexel University College of Medicine, Doylestown, PA, USA

Fahim Halim Khan Department of Biochemistry, Faculty of Life Sciences, Aligarh Muslim University, Aligarh, India

Suchit Khanna Institute of Nuclear Medicine and Allied Sciences, Defense Research Developmental Organization (DRDO), Ministry of Defense, Delhi, India

Mohammad Reza Khorramizadeh Endocrinology and Metabolic Research Institute, Tehran University of Medical Sciences, Tehran, Iran and Department of Medical Biotechnology, School of Advanced Technologies in Medicine, Tehran University of Medical Sciences, Tehran, Iran

Naveen Kumar Central Institute for Research on Goats, Indian Council of Agricultural Research, Makhdoom, District-Mathura, UP, India

Sandeep Kumar Food, Drug and Chemical Toxicology Group, CSIR-Indian Institute of Toxicology Research, Lucknow, U.P., India

Satyendra Mohan Paul Khurana Amity Institute of Biotechnology, Amity University, Haryana, India

Aruna Kumar Amity Institute of Biotechnology, Amity University, Noida, Uttar Pradesh, India

S. Maherchandani Department of Vet Microbiology & Biotechnology, Rajasthan University of Veterinary & Animal Sciences, Bikaner, Rajasthan, India

Avinash Marwal Department of Science, Faculty of Arts, Science and Commerce, Mody Institute of Technology and Science, Rajasthan, India

Shet Masih Department of Microbiology and Immunology, Drexel Institute for Biotechnology and Virology Research, Drexel University College of Medicine, Doylestown, PA, USA

Pawan Kumar Maurya Center for Reproductive Medicine, College of Medicine, Taipei Medical University, Taipei, Taiwan

Ravi Mehrotra Division of Cytopathology, Institute of Cytology & Preventive Oncology (ICMR), Noida, Uttar Pradesh, India

Debasis Mondal Department of Pharmacology, Tulane University Medical Center, New Orleans, LA, USA

Koichi Nishigaki Department of Functional Materials Science, Saitama University, Saitama, Japan

Pravinkumar Purushothaman Department of Microbiology & Immunology, University of Nevada, Reno, School of Medicine, Center for Molecular Medicine, Reno, NV, USA

K. Ravi Ram Embryotoxicology, CSIR-Indian Institute of Toxicology Research, Mahatma Gandhi Marg, Lucknow, Uttar Pradesh, India

Farshid Saadat Department of Immunology, School of Medicine, Guilan University of Medical Sciences, Rasht, Iran

Anurag Kumar Sahu Department of Science, Faculty of Arts, Science and Commerce, Mody Institute of Technology and Science, Rajasthan, India

Emmanuel O. Salawu Institute of Bioinformatics and Structural Biology, National Tsing Hua University, Hsinchu, Taiwan; PhD informatics Program, Taiwan International Graduate Program, Academia Sinica, Taipei, Taiwan; and Institute of Information Science, Academia Sinica, Taipei, Taiwan

Rishi Shanker Nanotherapeutics & Nanomaterial Toxicology Group, CSIR-Indian Institute of Toxicology Research, Lucknow, U.P., India

Anchal Singh Amity Institute of Biotechnology, Amity University Uttar Pradesh, NOIDA (UP), India

Gulshan Singh Nanotherapeutics & Nanomaterial Toxicology Group, CSIR-Indian Institute of Toxicology Research, Lucknow, U.P., India

Iqram Govind Singh Amity Institute of Biotechnology, Amity University Uttar Pradesh, NOIDA (UP), India.

Surinder Pal Singh CSIR-National Physical Laboratory, New Delhi, India

Mithilesh Singh G. B. Plant Institute of Himalayan Environment and Development, Sikkim Unit, Pangthang, Gangtok, Sikkim, India

Neha Singh Department of Biotechnology, Panjab University, Chandigarh, India

Priyanka Srivastava Department of Biotechnology, Indian Institute of Technology-Guwahati, Guwahati, Assam, India.

Richa Tripathi Division of Molecular Genetics & Biochemistry, Institute of Cytology & Preventive Oncology (ICMR), Noida, Uttar Pradesh, India

Hitomi Tsuiji Laboratory for Motor Neuron Disease, RIKEN Brain Science Institute, Saitama, Japan

Abhishek Tyagi Laboratory of Molecular Oncology, Dr. B. R. Ambedkar Center for Biomedical Research (ACBR), University of Delhi, Delhi, India

Kailash C. Upadhyaya Amity Institute of Biotechnology, Amity University, Noida, Uttar Pradesh, India

Mukesh Verma Epidemiology and Genomics Research Program, Division of Cancer Control and Population Sciences, National Cancer Institute, National Institutes of Health (NIH), Bethesda, MD, USA

Mudit Verma Epidemiology and Genomics Research Program, Division of Cancer Control and Population Sciences, National Cancer Institute, National Institutes of Health (NIH), Bethesda, MD, USA

Subhash Chandra Verma Department of Microbiology & Immunology, University of Nevada, Reno, School of Medicine, Center for Molecular Medicine, Reno, NV, USA

Ashish S. Verma Amity Institute of Biotechnology, Amity University Uttar Pradesh, NOIDA (UP), India

Anju Verma University of Missouri, Columbia, MO, USA

Poonam Verma Department of Biochemistry, All India Institute of Medical Sciences, New Delhi, India

Vipin Verma Corning Life Sciences, Gurgaon, India

Alok Kumar Verma Food, Drug and Chemical Toxicology Group, CSIR-Indian Institute of Toxicology Research, Lucknow, India

Denys V. Volgin Department of Animal Biology, School of Veterinary Medicine, University of Pennsylvania, Philadelphia, PA, USA

Suresh P. Vyas Department of Pharmaceutical Sciences, Dr. Hari Singh Gour University, Sagar, India

Dinesh K. Yadav Amity Institute of Biotechnology, Amity University, Gurgaon, Haryana, India

Neelam Yadav Amity Institute of Biotechnology, Amity University, Gurgaon, Haryana, India

Koji Yamanaka Laboratory for Motor Neuron Disease, RIKEN Brain Science Institute, Saitama, Japan; Research Institute of Environmental Medicine, Nagoya University, Nagoya, Japan

Eugenia Ch Yiannakopoulou Department of Basic Medical Lessons Faculty of Health and Caring Professions, Technological Educational Institute of Athens, Athens, Greece

Foreword

Animal biotechnology is one of the eight disciplines – along with environmental, food, plant, aquaculture, industrial, molecular, and medical studies – of biotechnology. This volume, drawn together by Professor Ashish Verma and Dr. Anchal Singh, is a comprehensive overview of animal biotechnology from a diverse set of perspectives. The volume is comprised of 32 chapters divided into three main sections: (1) *in vivo* and *in vitro* models of human disease, (2) tools and techniques, and (3) applications and concerns.

The term animal biotechnology is broadly applied when the production or the processing of products derived from animals or aquatic species is subjected to a particular set of scientific and engineering principles in order to enhance accessibility and services. Some classic examples are the development of transgenic animals or aquatic species, the use of cloning techniques to generate nearly identical animals, and various gene knockout strategies. Transgenic animals, including cattle, pigs, and poultry, have been developed to enhance the production of human pharmaceuticals and proteins such as enzymes, antibodies, clotting factors, and albumin. Somatic cell nuclear transfer has been used to clone several important mammalian species, including sheep, pigs, goats, cattle, rats, and mice. Because success rates for implanted embryos are often quite low, this offers opportunities for research and development. It is critical that this stimulating and wide-ranging progress be assembled, assessed, and considered in a timely manner, for the development of future initiatives, and to provide appropriate and accessible background for agricultural and health regulators. This treatise does just that. From a societal perspective, there are two main questions: (1) How is animal biotechnology addressing the needs of human agriculture and health? (2) Are products from the technology safe for human consumption and not detrimental to the environment? This volume does not shy away from those tough discussions, and anchors the responses in science.

Section I of the volume offers 10 chapters on *in vivo* and *in vitro* model systems that have been developed for animal biotechnology research. It includes discussions on the applications of *Drosophila*, and the use of animal models for tuberculosis, human neurodegenerative diseases, and aging, as well as work on cancer, HIV and other antiretrovirals, HPV diagnosis, and DNA tumor viruses.

Section II assembles 11 chapters on the basic tools and techniques that are being used in contemporary animal biotechnology. These include the use of multicellular spheroids in cancer research, animal tissue culture and tissue engineering, and the applications of nanotechnology, antibodies, and molecular markers. The techniques and uses of gene expression and ribotyping are discussed, and the future of sequencing strategies presented. Finally, the importance of biomolecular displays and *in silico* modeling of networks and complex diseases in contemporary research are delineated.

Section III, which also consists of 11 chapters, focuses on applications and societal concerns. It provides summaries of the development and applications of transgenic animals, the saga of stem cells in medical research and therapy, the role of cytogenetics in medicine, and the applications of antibodies and vaccines. The importance of safety assessment of crop-derived foods is presented, together with the use of nanotechnology for the detection of pathogens, the development of marine animal biotechnology, and discussions on how the phytochemistry and pharmacology of herbal medicine biotechnology are linked to animal health. Finally, there are two chapters that provide an overview of the human genome and its relationship to animal biotechnology, and a consideration of the ethical issues that are fundamental to many aspects for the future evolution of animal biotechnology.

This volume makes clear both the vibrant diversity of the field of animal biotechnology, and the ethical and societal concerns that must be addressed. It is therefore an important volume for a wide audience, including researchers, veterinarians, physicians, agricultural and developmental economists, and policy regulators. The next few years are likely to see major breakthroughs in this field, which will be necessary to meet the nutritional and health care needs of a burgeoning global society.

Geoffrey A. Cordell, Ph.D.
Professor Emeritus, University of Illinois at Chicago
Adjunct Professor, University of Florida
President, Natural Products Inc.

Preface

Lately, "biotechnology" has become a buzz-word in both the academic arena and in day-to-day life. It is still debatable as to when and where the term originated. Who is its originator? Was biotechnology always known to the world in its present form? The answers to these questions are not known. The scientific literature tells us that Karl Erkey, a Hungarian Engineer, coined the term biotechnology in 1919. The next question is, did nature sire biotechnology or is it human beings that have created it in its present form? Again, it is difficult to come to any conclusion about the current state of knowledge. Let us go back and review the evolution of life from the most primitive form of organisms (i.e. viruses) to the most evolved form of life (i.e. human beings).

Certainly, one of the most important and advanced aspects of biotechnology and biotechnological tools is the manipulation of the genome of an organism. These manipulations can have either good or bad implications, but the answer lies in the final outcome. The most primitive form of life (i.e. viruses: bacteriophages) infects bacteria and replicates in bacterial hosts due to the integration of the viral genome into the bacterial genome. Is it Nature's biotechnological experiment to integrate genomes of two entirely different organisms? It is probably a natural need of life to compete and evolve with selection of better traits to survive against adversaries. It can be easily concluded that the present state of biotechnology has evolved due to the in-depth understanding of some of these natural processes and biological phenomenon.

There is no doubt that the life sciences have seen tremendous improvements by virtue of keen observations and discoveries made by numerous great scientists. Antibiotics and vaccinations are two of the most pronounced examples. During previous years, knowledge gained through various branches of science, namely biochemistry, molecular biology, virology, and recombinant DNA technology, etc., has tempted scientists to imitate Nature's experiments in laboratories. For successful and useful manipulations, there are three essential requirements: (1) to understand the mechanism of the biological process, (2) to replicate the same process exactly in an experimental model, and (3) to have a logical hypothesis. If these manipulations are successful, we may be able to find solutions to many prevailing and unresolved problems, namely famine, malnutrition, infectious diseases, new and emerging infections, genetic disorders, aging, debilitating diseases, etc. No doubt advancements in biotechnology, with reference to the animal sciences, have already provided solutions for some of these issues. Some issues are even partially resolved, while others are still in experimental stages.

The explosion in the knowledge of biotechnology is attributed to two important discoveries: (1) the structure of DNA, and (2) the Polymerase Chain Reaction (PCR). Advancements and applications of biotechnology have become so fascinating that it is almost difficult to confine it to the domain of scientists and high-end laboratories. This information has to be passed to the general public in order to increase awareness and to reap the benefits of these discoveries. With the explosion of biotechnology, numerous large and small companies dealing with the production and commercialization of biotechnology products have come into existence. To survive and thrive in the biotech market, companies are in a perdurable search for trained manpower.

That's how biotechnology as an educational course found its niche in the university curricula. The demand for trained biotechnologists led to the development of undergraduate and postgraduate courses in biotechnology at various universities and academic institutions. Realizing the needs of industry, some institutions developed management courses pertaining to biotechnology. In the last couple of decades it was realized that biotechnology education had to be imparted even to younger students, and that is the reason biotechnology was also included in the curricula of 10th and 12th Standard. Biotechnology itself is an amalgamation of various disciplines in the life sciences. Some of these disciplines are well evolved and have numerous good books to cater to the needs of audiences: biochemistry, molecular biology, genetics, microbiology, etc. However, animal biotechnology as a subject is still in its infancy, and has yet to develop and evolve as a full discipline in academic departments at universities. As such, it is difficult to find books in animal biotechnology that can fulfill the need of biotechnology students.

We teach animal biotechnology to undergraduate and postgraduate students. We have had a tough time teaching this course because of major limitations like an ever-evolving curricula and unavailability of reasonable textbooks on the subject. The only available resources are research publications and books semi-related to research topics. On the one hand it's hard for students to find a place to start when learning the subject, and on the other hand instructors have a difficult time locating and organizing materials and resources for the classroom. The ultimate resource for instructors and students is the World Wide Web (WWW). In our teaching experience, we come across curious students who ask numerous intelligent questions almost every day. Their quest for information and knowledge remains insatiable due to the limitation of consolidated sources of information. Not only this, but we routinely face questions from students about where they can get more information on a specific subject or topic, and to their utter disappointment, it's hard for us to pinpoint one book or a good resource to answer all their questions. We frequently discuss the issue of the lack of applicable literature, almost every day over coffee with our colleagues. Discussing various options and trying to narrow down our search to fill this void of content in the area of animal biotechnology was not getting us anywhere.

After numerous deliberations, it was Dr. Anchal Singh who came up with the idea to explore the possibility of developing a book on animal biotechnology to partially (if not completely) fill this void. Then we deliberated on our *modus operandi* to develop this book. Finally, we decided to develop a book by inviting chapters from experts in the field who have relevant research experience and an understanding of the intricacies of the subject. We had in mind a book that would help to alleviate most of the worries of both students and instructors. We discussed, argued, and disagreed until we came up with the thought that a resource book would be a reasonable format, as it could provide sufficient information and literature for instructors to teach the subject, while providing students with ample information to gain better insight about the subject. Once we formulated these thoughts to develop a resource book, the ball started rolling, and we identified various experts and convinced them to contribute chapters.

Bringing this book to completion was a joint effort. We could not possibly assemble all subjects together in one book, therefore we tried to bring together some of the important topics that usually interest students and instructors of animal biotechnology. The subject matter of this book varies from the basics of animal biotechnology, to animal tissue culturing, to the production of antibodies against infectious agents like HIV. Included are chapters dealing with animal models of important diseases like cancer and tuberculosis, and also *in silico* models, to emphasize their importance in understanding disease pathogenesis. An attempt was made to include the latest tools and technology related to the subject, namely, ribotyping, epigenetics, cytogenetics, bimolecular display technologies, next generation sequencing, and many more such topics not listed here.

This is our maiden effort to produce a book to help students and instructors of animal biotechnology. We hope that we will get support from the readers of this book. We are always open to criticism, suggestions, and recommendations that can help to improve the content and presentation of the book. Your suggestions and criticisms will give us an opportunity to explore other aspects of animal biotechnology in our future ventures and endeavors.

Ashish S. Verma
Anchal Singh

Acknowledgments

We are grateful to The GOD, because of whom we exist. God has gifted us (Human Beings) with a brain to hypothesize and analyze, courage to dream, and motivation to achieve.

Then we would like to thank Prof. Geoffrey A. Cordell, who agreed to write a Foreword for our book. This turned out to be a power boost for the editors.

Anchal would like to thank her dad, Mr. Kanhaiya Ji Singh, her mom, Ms. Mohini Singh, her brother, Abhisar, and his wife, Meenakshi, for their support, love, and help. Anchal's eight-year-old son, Aviral, was a stress buster whose unstoppable questions and witty answers alleviated the stress and pressure of editing this book.

I (Ashish) would like to express my indebtedness to my mother, Ms. Sushma Saxena. I do exist due to her great efforts to raise and groom me. She has always been the person in my life whom I can bank upon for anything, anytime. My brother, Mr. Saumya Swarup, his wife, Ms. Nimisha Swarup, and their kids, Utkarsh and Shreeparna, have also supported me as and when, I needed them. Similarly, my sisters and their family members always encouraged me to remain focused on this book. It is not only family members who inspired me during the development of this book, but also Anchal's son, Aviral with his inquisitiveness and unending innocent queries, which kept me refreshed. We must admit that it is young kids who are our prime stress relievers.

We are thankful to Dr. A. K. Chauhan, Founder President for his support and encouragement. Our special thanks go to Prof. Ajit Varma, who was always with us when we needed him, with his excellent and practical advices. We would not do justice to this project if we do not acknowledge the role of Prof. Soom Nath Raina, one of our colleagues. For us, Prof. Raina is more than a colleague. His affection, care, and concern made him a part of our extended family. We will always remain thankful for his untiring support, which kept us motivated in this long and sometimes clumsy journey of book editing.

Our students – Priyadarshini Mallick, Shruti Rastogi, Shishir Agrahari, Sneha Saran, Deepak Kushwaha, and Ajay Yadav – contributed both directly and indirectly towards the development of this book. We remain thankful for their support. Some of them provided help and support to organize us better, some of them offered their viewpoints, and some of them did not forget to offer us their critiques. We admire all of them for what they have contributed to this book. We have discovered that the biggest motivations for teachers are always their students and students' needs.

We are deeply indebted to Mr. Dinesh Kumar, who has worked with us since we joined this organization and has always provided crucial secretarial assistance. To Mr. Yogendra Singh, who has worked for a long time as a member of our group, and is always there with freshly brewed coffee to fulfill our caffeine requirements. Mr. Sandeep Kumar, who, though recently joined our group, also contributed with his efforts to this project.

Our special thanks go to Ms. Chirstine A. Minihane, who helped us initiate and develop this book. Last but not least, this book could never have been completed without constant support from dedicated persons at Elsevier: Mr. Unni Kannan (Technical Assessor), Ms. Catherine A. Mullane (Editorial Project Manager), Graham Nisbet (Acquisitions Editor), and Edward Taylor (Production Manager). They all provided the support and motivation to push us through to the completion of this project.

As editors, we would like to express our gratitude and thanks to all the contributing authors who shared their expertise and experience by writing chapters in their respective fields. Finally, as the editors, we would like to convey our heartfelt thanks to everyone who has contributed directly or indirectly towards this book.

Ashish S.Verma
Anchal Singh

Section I

Human Diseases: In Vivo and In Vitro Models

Chapter 1

Drosophila: A Model for Biotechnologists

K. Ravi Ram and D. Kar Chowdhuri
Embryotoxicology, CSIR-Indian Institute of Toxicology Research, Mahatma Gandhi Marg, Lucknow 226001, Uttar Pradesh, India

Chapter Outline

SUMMARY

Drosophila offers a miniature yet versatile and manipulable model to address basic biological questions with potential implications plus applications to other metazoans. In this chapter, we emphasize contributions of Drosophila to genetics and biotechnology and the translational versatility of this model along with associated ethical issues and available resources.

WHAT YOU CAN EXPECT TO KNOW

The initial contents of this chapter facilitate learning of basic concepts of Drosophila that evolved this organism as a model for genetics and development. The historical perspective helps the reader to assimilate the contributions of Drosophila research findings to genetics and biotechnology. Drosophila models for human diseases exemplify the ultimate power of biotechnological tools available in Drosophila and their translational significance.

HISTORY AND METHODS

INTRODUCTION

Innovative genetic technologies facilitated the advancement of science. For example, cloning and manipulation of gene sequences armed the generation of transgenics towards a substantial understanding of biological concepts and also for the betterment of life. However, the pivotal role played by model organisms for biotechnologists to achieve these challenges is absolutely phenomenal. One such model

Animal Biotechnology. http://dx.doi.org/10.1016/B978-0-12-416002-6.00001-8

organism that helped biotechnologists to realize their dreams is Drosophila. With significant contributions to genetics, and development over 100 years, Drosophila continues to inspire the creativity of biotechnologists. Drosophila is so amenable to genetic manipulation that it is constrained only by the imagination of biotechnologists.

Drosophila is a tiny fly also known as the vinegar-loving fly. The term "*Drosophila,*" meaning "dew-loving," is a modern scientific Latin adaptation of the Greek words *drósos* ("dew") and *phílos* ("loving") combined with the Latin feminine suffix "*-a*." It belongs to the Phylum Arthropoda, class Insecta, and order Diptera, and the famous family of Drosophilidae. *Drosophila* is a small fly, typically pale yellow to reddish brown or black, with red eyes. The plumose (feathery) arista, bristling of the head and thorax, and wing venation are characters used to identify the family. Most are small, about 2–4 mm long, but some, especially many of the Hawaiian species, are larger than a house fly. The genus *Drosophila* is found all around the world, from deserts to tropical rainforests to cities to alpine zones. Most species breed in various kinds of decaying plant and fungal material including fruit, bark, slime fluxes, flowers, and mushrooms.

Of the various species of Drosophila, *Drosophila melanogaster* offers several advantages as a model for molecular studies. Being small, these flies are extremely simple to handle. The sexual dimorphism (males and females are different) permits easy differentiation of the sexes. Further, these flies are non-pathogenic, have a short-generation lifetime (10–12 days), and can be cultured at a low cost in a limited space. In addition, Drosophila offers various molecular and genetic tools that a biotechnologist can only dream of, and the fully sequenced genome coupled with bioinformatics tools enhance the translational utility of this model.

In this chapter, we initially describe the classical aspects of Drosophila such as life cycle, cytology, and development. Subsequently, we provide a historical perspective of the research hallmarks that led to the utility of Drosophila as a model for molecular studies. In addition, we discuss the translational significance of Drosophila by emphasizing Drosophila models available for human diseases. Finally, we discuss the ethical issues and concerns associated with this model.

Classical Aspects of *Drosophila melanogaster*

Physical Appearance

The body of Drosophila, like that of any other insect (and typical of Arthropoda), is segmented. The body plan typically consists of head, thorax, and abdomen. While multiple segments give rise to the head, three constitute the thorax, and eight segments form the abdomen. The head consists of antennae whereas legs and wings arise from thoracic segments. *D. melanogaster* have transverse black rings across their abdomen. Males are easily distinguishable from females by the presence of a distinct black patch at the abdomen that is absent in females. Males also have sex combs, a row of dark bristles on the tarsus of the first leg that are absent in females. Furthermore, males have a cluster of spiky hairs (called claspers) that surrounds the anus and genitals and is used to attach to the female during mating.

Life Cycle

D. melanogaster is a popular experimental animal because it is easily cultured in mass out of the wild, has a short generation time, and mutant animals are readily obtainable. Typically, in a laboratory, *D. melanogaster* is grown on a cornmeal–yeast–fruit juice mixture at 25°C. The life cycle of this organism consists of a number of stages: embryogenesis, three larval stages, a pupal stage, and the adult stage. The development period for *D. melanogaster* varies with temperature. The time required for complete development at 25°C is 8 to 9 days. Females lay some 400 eggs, about five at a time, on overripe fruit or on other suitable material. The eggs, which are about 0.5 mm long, hatch after 20–22 hr at 25°C. The resulting larvae grow for about 3 days while molting twice into 2nd- and 3rd-instar larvae at about 24 hr and 48 hr, respectively, after eclosion. The larva then encapsulates in the puparium that is immobile and undergoes a four-day-long metamorphosis at 25°C. A fly finally emerges from a puparium after metamorphosis, a process referred to as eclosion. The life cycle of Drosophila is schematically depicted in Figure 1.1.

Drosophila Development

Drosophila contributed to most of our existing knowledge on the mechanisms of the development of organisms. In Drosophila, the complex adult body plan is realized from the fertilized embryo through developmental processes. Development in Drosophila is holometabolous, which involves developing stages morphologically distinct from adults (1993). Early processes of development occur in the fertilized egg laid by the female to give rise to the larva. The larva subsequently gives rise to the pupa after undergoing a series of modifications and two moltings driven by hormonal titers and molecular signals. During the pupal stage, many larval structures are broken down, and adult structures undergo rapid development. In this section, we will describe the key aspects of Drosophila development: embryogenesis, pattern formation, and homeotic genes.

Embryogenesis in Drosophila

The early development of Drosophila begins with the formation of oocytes through oogenesis in the ovary (Hartenstein

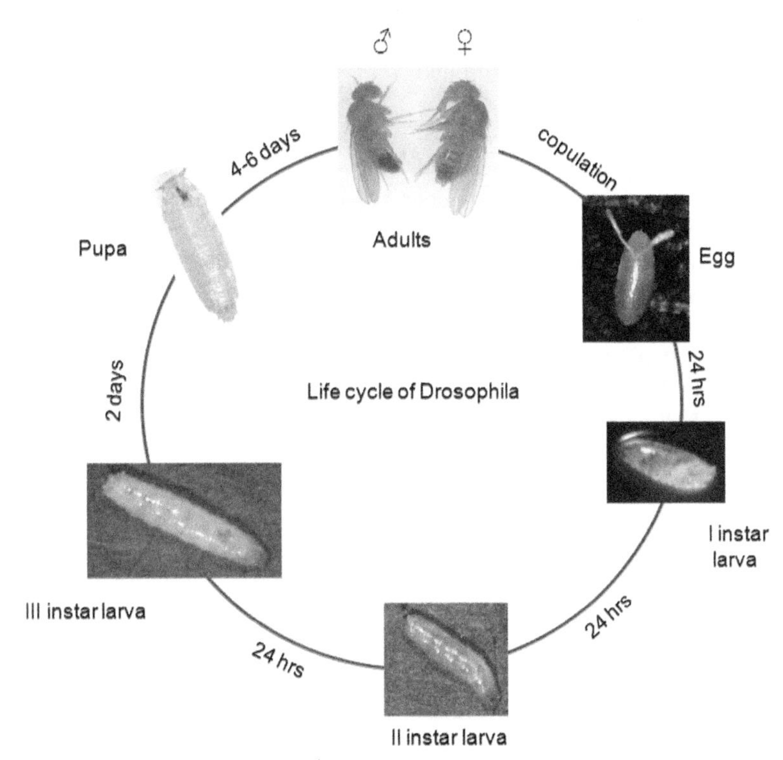

FIGURE 1.1 Life cycle stages of *Drosophila melanogaster*.

1993). These oocytes are packed with maternal RNA, protein, ribosomes and mitochondria that assist in the egg-to-embryo transition. Fertilization (union of these oocytes and sperm) triggers mitosis in the embryo. Several nuclear divisions without cytokinesis (division of cytoplasm) occur in the early embryo, which results in a cell with many nuclei in the cytoplasm. At the tenth nuclear division, these nuclei migrate towards the surface of the embryo, which results in the formation of the syncytial blastoderm. At the thirteenth nuclear division, membrane invaginations enclose the nuclei and lead to cellularization and the formation of the cellular blastoderm. At the time of cellularization, the major body axes and segment boundaries are determined. After cellularization, the embryo proceeds through gastrulation: cells from the ventral surface invaginate to create the ventral furrow. The ventral furrow is critical for the formation of the mesoderm. Subsequently, at the anterior and posterior ends of the ventral furrow, the invagination of the prospective endoderm occurs. Gastrulation is followed by the convergence of certain ectodermal cells on the surface with the mesoderm and their migration towards the ventral midline to form the germ band, a collection of cells that will form the trunk of the embryo. The germ band extends posteriorly and wraps around the dorsal surface of the embryo. At this extended state, several morphogenetic processes occur: organogenesis, segmentation, and segregation of imaginal discs that will unfold during metamorphosis to form adult fly structures such as antennae, legs, and wings (Campos-Ortega and Hartenstein 1985; Martinez Arias 1993). Interestingly, at all the stages of development, the general body plan remains the same. The generalized body plan consists of a segmented region sandwiched between a distinct head or anterior region and the tail or posterior region. Drosophila, with its versatile genetic tools, has led to the understanding of pattern formation and differentiation during early animal development. Three scientists, namely, Ed Lewis, Christiane Nusslein-Volhard, and Eric Wieschaus pioneered our understanding of pattern formation and differentiation. These scientists not only laid the platform for our understanding of development, but also deciphered the underlying dynamics. Nusslein-Volhard and Wieschaus (1980) focused their studies on understanding early embryogenesis while Ed Lewis concentrated on late embryogenesis.

Pattern Formation in Drosophila

The metamorphosis of a simple egg into an adult with a complex body plan requires three classes of genes. These classes comprise the maternal genes, the segmentation genes, and the homeotic genes. Nusslein-Volhard and Wieschaus (1980) were the first to report the key contributions of each gene that regulate a particular pattern formation event, the segmentation of embryo. They looked for recessive embryonic lethal mutations in a systematic genetic screen encompassing the whole genome to identify genes critical for embryonic development. Subsequently, they analyzed the phenotypes of dead embryos and classified these genes according to their phenotype before death. Based on their phenotypic analyses, Nusslein-Volhard and Wieschaus (1980) identified three categories of mutations. The first category comprised mutations that resulted in the loss of multiple adjacent segments (gap genes). The second category included mutations that cause alternate segment size units to go missing (pair-rule genes). The third category of mutations triggered the loss of part of each segment and duplication of the remaining part of the segment (segment polarity genes). In view of their findings, Nusslein-Volhard and Wieschaus (1980) proposed that gap genes, pair-rule genes, and segment polarity genes (together called segmentation genes) are critical for sub-dividing the embryo and for segment formation. Another class of genes, including homeotic genes, defines the identity of the segment. However, to put these genetic cascades into motion, maternal components are essential.

Maternal components play a critical role in determining embryonic patterning (Hartenstein 1993). As discussed above, even before fertilization, Drosophila eggs are loaded with regulatory molecules that determine the antero-posterior axis of the egg and development of the organism. Eggs are preloaded with bicoid and nanos mRNA, and these are translated upon fertilization (Hartenstein 1993). Of these two, bicoid is essential for the formation of the head: females carrying mutant alleles of bicoid give rise to offspring with defects in head development. Prior to cellularization, these proteins form concentration gradients in the embryo. At the anterior end, bicoid is at a higher concentration whereas at the posterior end, nanos is abundant. These concentration gradients are critical for regulating segmentation genes (see above), which define the segmentation pattern. Temporal and spatial activation of gene cascades is the hallmark of development. During the initial phase of development, bicoid and nanos differentially regulate the gap genes. At the anterior end, bicoid triggers the transcription of a gap gene, *hunchback*. However, at the posterior end, nanos inhibits *hunchback* RNA from being translated, thereby forming a hunchback protein gradient in the embryo (Wreden *et al.* 1997). This hunchback protein triggers, in a concentration-dependent manner, the transcription of other gap genes (like *Kruppel*, etc.) (Schulz and Tautz 1994; Zuo *et al.* 1991)), which in turn define large areas surrounding the antero-posterior axis (Gilbert *et al.* 2003). The gap genes encode transcription factors that regulate the expression of certain pair-rule genes, which in turn regulate other pair-rule genes. These pair-rule genes, which are expressed along the stripes of the embryo, divide the embryo into pairs of segments. Pair-rule genes encode transcription factors that regulate segment polarity genes. These segment polarity genes define the antero-posterior axis of each of the segments. Once the pattern of segmentation is established, the segments achieve unique identities through homeotic genes.

Homeotic Genes in Drosophila

The term homeosis represents the transformation of one structure of the body into the homologous structure of another body segment. In Drosophila adults, as mentioned earlier, structures like legs and wings develop from the thoracic segments whereas antennae appear on the head. The segment-specific development of these structures requires the action of homeotic genes. Homeotic genes are a group of genes that regulate pattern formation. These genes, although they do not specify the elements of the pattern, do indeed assign identities to these elements. Mutations in these genes result in development of the elements of the specified pattern with inappropriate identities. The best example to describe these genes is Antennapedia (ANTP). As the name suggests, a dominant mutation in this gene transforms the antennal structures on the head into an additional second leg. Generally, normal ANTP is required in the second thoracic segment to initiate the cascade of events that lead to the development of the leg. Perhaps, this involves the regulation of several genes. Hence, the homeotic genes are considered master controllers of developmental programming (Abbott and Kaufman 1986).

Most of our knowledge on homeotics is due to the pioneering works of Ed Lewis on bi-thorax complex (BX-C) (Lewis 1978). The homeotic genes consist of a 180 nucleotide consensus sequence called the homeobox. The homeobox corresponds to a 60 amino acid domain, namely, the homeodomain, which is involved in DNA binding. These homeobox-containing (or HOX) genes are not limited to Drosophila. They have also been found and studied in many other organisms ranging from invertebrates to vertebrates, including mammals (Ruddle *et al.* 1994; Santini *et al.* 2003). HOX genes occur in clusters, and interestingly, not only these genes, but also the synteny (relative gene order) within the cluster is conserved (Ruddle *et al.* 1994). Moreover, the order of HOX-genes on the chromosome is the same as the order of the segments that they affect along the anterio-posterior axis. Genetic analysis revealed the existence of posterior dominance: genes acting at the anterior are regulated by their posterior neighbors. For example, BX-C comprises of regions encoding three homeodomain genes called Ultrabithorax (Ubx), Abdominal-A (AbdA), and Abdominal-B (AbdB), and also a non-coding RNA, iab-8-ncRNA (Gummalla *et al.* 2012; Lewis 1978). In this case, iab-8-ncRNA represses AbdA (Gummalla *et al.* 2012) and both AbdA and AbdB repress Ubx (Lewis 1978). In addition to intrahomeotic regulation, these homeotic genes are also regulated by segmentation genes. For example, hunchback (gap gene) is known to limit Ubx expression (Wu *et al.* 2001) and mutations in pair-rule genes influence the expression of homeotic genes. At present, however, knowledge on how segmentation genes regulate homeotic gene expression/repression is limited. Nevertheless, as discussed

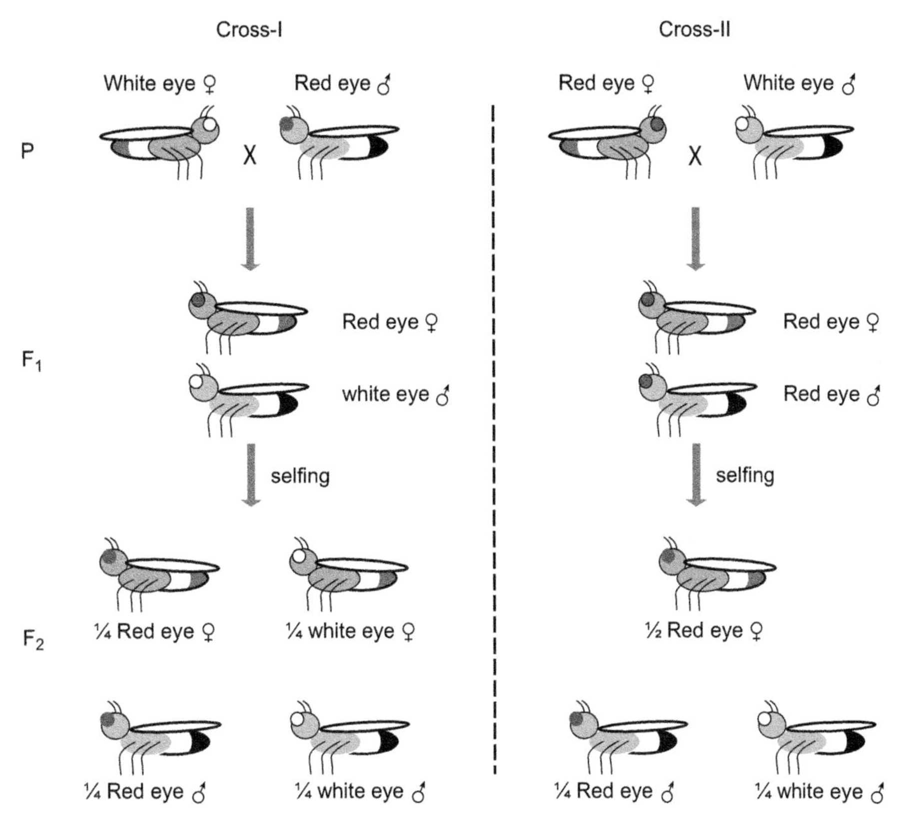

FIGURE 1.2 Schematic diagram to show discovery of sex-linked inheritance in *Drosophila melanogaster* in Morgan's classic experiment.

so far, studies on Drosophila have tremendously contributed to the understanding of development in metazoan animals. The events associated with pattern formation in Drosophila are schematically depicted in Figure 1.2.

Drosophila Genome

D. melanogaster has four pairs of chromosomes: an X/Y pair and three autosomes labeled 2, 3, and 4. The fourth chromosome is quite tiny. The size of the genome is 165 million base pairs and contains an estimated 14,000 genes (Adams *et al.* 2000). By comparison, the human genome has 3,400 million base pairs and may have about 25,000 genes (International human genome consortium, 2004). The *Drosophila* genome contains a considerable amount of non-protein-coding DNA sequences that are involved in the control of gene expression. Determination of sex in *Drosophila* occurs by the ratio of X chromosomes to autosomes.

HISTORY

The significance of Drosophila as a model to the advancement of genetics and the current biotechnology dates back to almost 1900. After the rediscovery of Mendelian principles (using pea plants), when plant models were dominating the research within the field of genetics, T. H. Morgan pioneered the use of Drosophila as a model in 1908. The findings on Drosophila very quickly established its superiority as a model for genetic studies. Here we provide some of the landmarks in Drosophila research that led to major conceptual and technical breakthroughs in our manipulation of genomes using biotechnology.

Drosophila was first introduced by Prof. Charles W. Woodworth, who was credited with the first quantity breeding of Drosophila. He proposed to Prof. W.E. Castle that Drosophila be used for studying genetics. It was his collaborator, E.B. Lutz, who introduced this tiny creature to T. H. Morgan. In 1910, T. H. Morgan, in his pursuit to prove mutation theory with his experimental heredity work, discovered the first of many mutants, a white-eyed fly. Normally, *D. melanogaster* (both males and females) have red eyes. Because the trait first seemed to occur only in males, Morgan referred to it as a "sex-limited" trait. However, after the first cross, he mated the original male with some of the F_1 red-eyed females and obtained approximately equal numbers of red- and white-eyed males among the progeny. Thus, the trait proved to be sex-related, not sex-limited. The pattern of sex-linked inheritance is schematically depicted in Figure 1.3. Subsequently, a number of mutants that were linked to the X chromosome were isolated. Using simple crosses, within a span of five years, Morgan (together with his three students A. H. Sturtevant, C. B. Bridges, and H. J. Muller) proposed a revolutionary chromosome theory of heredity. In 1913, A. H. Sturtevant constructed the first genetic map and showed that genes were arranged in a linear fashion. In 1916, Bridges, exploiting chromosome nondisjunction in XXY females, provided the elegant first proof that chromosomes harbor genes. In 1918, Muller introduced the use of balancer chromosomes for the stable maintenance of lethal mutations as heterozygotes. These accomplishments led to Morgan being awarded the Nobel Prize in 1933.

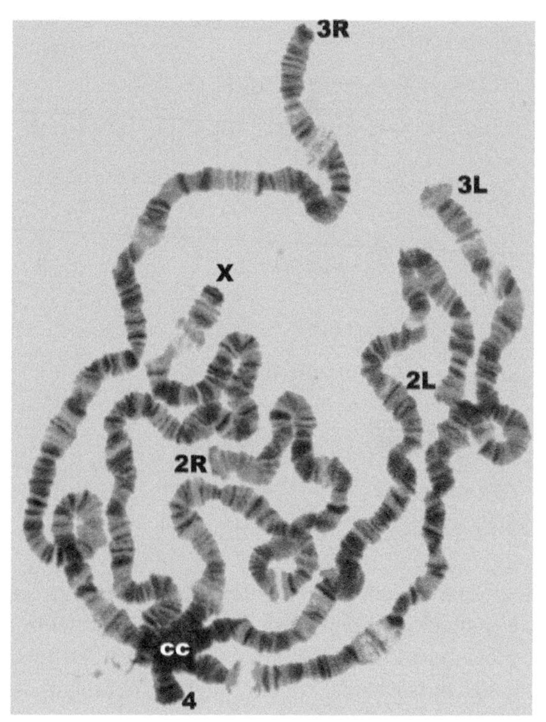

FIGURE 1.3 Polytene chromosome preparation from salivary gland of third instar larva of *Drosophila melanogaster*. Each arm of the chromosome is marked. For example, 2L and 2R represent the left and right arm of the second chromosome. Centromeres of all the four chromosomes converge to form the chromocenter. *(Kindly provided by Anand K Singh, Cytogenetics Laboratory, Banaras Hindu University, Varanasi, India)*

In 1934, following the discovery of salivary gland polytene chromosomes in the fly *Bibio hortulanus* by Heitz and Bauer, T. S. Painter published the first drawings of *D. melanogaster* polytene chromosomes. Polytene chromosomes are specific interphase chromosomes consisting of thousands of deoxyribonucleic acid (DNA) strands arranged side by side, thereby displaying a characteristic band–interband morphology (Figure 1.4). They appear very large. When there is a need for rapid development of an organ/tissue within a short span of development, polyteny arises in tissues and organs at developmental stages without altering the level of function. Thus, organs/tissues containing cells with polytene chromosomes are involved in intense secretory functions accomplished during a short time against a background of rapid growth. By 1938, Bridges meticulously sketched and published polytene maps (cytogenetic maps), which included the chromosomal localization of several genes. The accuracy of these localizations is exemplified by their relevance and usage even today in this era of molecular biology. These maps are so accurate that molecular biology techniques like *in situ* hybridization corroborate the placement of genes within intervals of less than 100 kb. In the early 1950s it was first suggested that puffs (swollen regions in the polytene chromosomes) in Drosophila and Balbiani rings in Chironomus are chromosome regions that represent an active state of gene transcription (Beermann 1952; Mechelke 1953; Pavan and Breuer 1952). At a given stage of development, the spectrum of puffs and Balbiani rings is strictly specific to each tissue. In 1970 Ashburner provided a detailed timetable of changes in the activity of the various puffs. Puffs are indeed sites of active gene expression, as was eloquently demonstrated by Charles Ritossa in 1962 when he showed that larvae of *Drosophila busckii* exposed to high temperature displayed a specific number of puffs on their polytene chromosomes, which were absent in the unexposed larvae. Based on these observations, he called them heat shock puffing, which later became an active area of research. He also demonstrated that

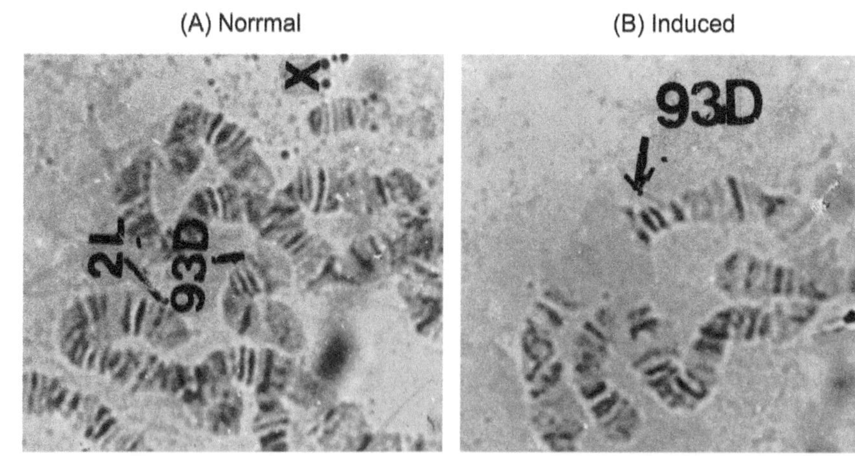

FIGURE 1.4 Puff on polytene chromosome as the site for inducible gene activity. Chemical-induced puffing at 93D locus on the right arm of third chromosome from *Drosophila melanogaster*.

these puffs appear if poisons that blocked oxidative phosphorylation were used. Somewhat later it was shown that ecdysone in Chironomus and Drosophila triggers the activation of a whole cascade of interrelated loci, with products of the activities of the puffs induced earlier being needed for the induction of puffs arising later. Therefore, during the pre-molecular biology era or at the onset of the molecular biology regime, puffing on polytene chromosomes of Drosophila and Chironomus remained an important tool to study gene expression under the light microscope (Figure 1.5).

Drosophila owes its success to the vast number of mutants available for studies. However, dependence solely on the spontaneous mutations would have limited the success of this model. In this context, the accomplishments of Mueller assume significance. In 1927, Muller showed that mutations, including chromosomal rearrangements, could be induced by exposure of Drosophila to X-rays (ionizing radiation). Fittingly, this finding brought him a Nobel Prize in 1946. By the 1930s, a generation of deficiencies and duplications by combining X-ray-induced chromosomal aberrations with closely spaced break points was feasible. Later, Lindsley and his co-workers exploited these methods in 1970 to generate an ordered set of duplications and deletions spanning the major autosomes. This work opened the avenue for whole-genome screens for phenotypic perturbations. Such a rich resource is typical to Drosophila and most likely can be duplicated only in *C. elegans*. In an elegant study, Hoskins *et al.* (2001) demonstrated systematic characterization of a dense set of molecular markers in *Drosophila* by using a sequence-tagged, site-based physical map of the genome. For this study, a set of P-element strains that facilitate high-resolution mapping was used, and subsequently,

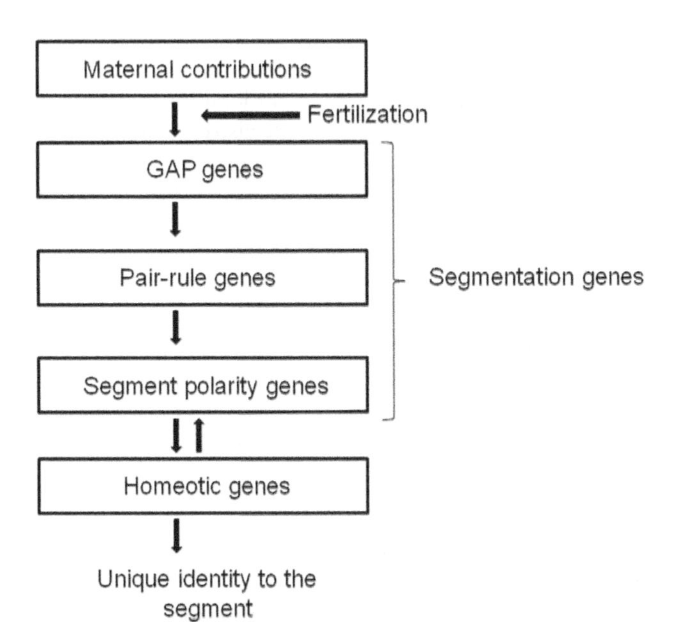

FIGURE 1.5 Schematic depiction of classes of genes associated with pattern formation in *Drosophila melanogaster*.

application of the new markers in a simple set of crosses to map a mutation in the *hedgehog* gene to an interval of < 1 Mb was successfully done. This new map resource significantly increased the efficiency and resolution of recombination mapping and was of immense value to the *Drosophila* research community (Hoskins *et al.* 2001).

The seeds for modern biotechnology research in Drosophila were sown by the 1970s. In 1974, random clones for Drosophila, the first for any organism, were generated in the D. S. Hogness laboratory at Stanford University. By early 1975, clone libraries representing the entire genome were generated and screens for clones carrying specific sequences with the newly developed method of colony hybridization were in place. In early 1979, cloning of a gene, ultrabiothorax, was achieved for the first time. By late 1980, many mutant alleles had been cloned and shown to be the consequence of chromosomal breakage or transposable element insertion. Subsequently, the availability of transposable element vectors only added to the growth of Drosophila as a model in the field of biotechnology. The use of transposable elements for generating transgenic flies has revolutionized gene manipulation in Drosophila and pioneered the development of a powerful array of techniques in Drosophila, many of which were ultimately adapted to other metazoans. These methods ranged from enhancer traps (1987), large-scale insertional mutagenesis (1988), and site-specific recombination for generating chromosomal rearrangements (1989), to the highly popular binary systems for controlling ectopic gene expression (1993). By 1999, over 1,300 genes were cloned and sequenced, and functions were characterized using loss of function phenotypes. These enormous tools were so meaningful that most researchers did not even consider whole-genome sequencing of Drosophila. Ultimately, when the Drosophila genome was sequenced in 2000, it gave another value addition to the field of biotechnology, the whole-genome shotgun approach, for genome sequencing. Subsequently, the vast resources made available in recent years by the research community for genome-wise ectopic expression and knockdown (RNAi), both *in vitro* and *in vivo*, have enhanced the utility of the Drosophila model. The fully sequenced genome coupled with these various genetic and molecular tools led to the upsurge of Drosophila as a model for basic as well as translational research.

PRINCIPLE

Drosophila is a well-studied and highly tractable genetic model system to decipher the molecular mechanisms underlying various biological processes. The completion of genome sequencing and annotation discovered the high degree of conservation of fundamental biological processes between Drosophila and mammals. This has prompted biotechnologists to utilize Drosophila to understand the molecular basis of human diseases. The ease with which Drosophila transgenics can be created was also been instrumental in the success of this model for understanding human diseases. Using a plethora of molecular tools available for Drosophila, biotechnologists genetically manipulated Drosophila by either inserting human genes in the fly genome or by modifying the function of human disease orthologs in Drosophila and sensibly developed Drosophila-based models for human diseases.

METHODOLOGY

Given the focus of this chapter, here we describe only those Drosophila-based methods essential for germ-line transformation of Drosophila to generate transgenics.

Culturing of Drosophila

Drosophila melanogaster is reared on standard Drosophila food medium at 22 ± 1°C. Stocks are usually maintained in vials (up to 20–30 flies/vial) and experimental cultures are maintained in bottles (as they permit the growth of flies in large numbers).

Preparation of Drosophila Food Medium

Materials Required

Glass or plastic vials (70–90 mm Height with 25–30 mm outer diameter)

Round flat bottom half-pint glass or plastic bottles

Agar-Agar, Maize powder, Sugar, Yeast, Methyl paraben, and Propionic acid

The Recipe to prepare one liter of fly food is as follows:

Agar-Agar	8 g
Maize powder	15 g
Sucrose	100g
Bakery dry Yeast	100g
10% Benzoic Acid	5 ml
Propionic acid	8 ml
Water	1000 ml

One litre of water is added to a 2 L (glass or stainless steel) beaker and the same is kept on a hot plate. Sugar is added slowly to the water, the beaker is covered with a glass plate, and the water is heated. In the mean time, the solid ingredients (agar-agar, maize powder, and dry yeast) should be mixed and added to the water once it starts to boil, with constant stirring. The contents are boiled for 15–20 minutes. Subsequently, measured quantities (as above) of 10% benzoic acid and propionic acid are added with thorough stirring. The heater is turned off and the food is brought to the table and can be poured (3–5 ml) into the vial depending upon the requirement. The food should be allowed to cool and solidify before plugging the vials with cotton. A couple of yeast granules should be added to these vials and they need to be left overnight. Now these food vials are ready for use. The same is the case with bottles except that the quantity of food will be proportionally higher.

Handling of Flies

Drosophila, which belongs to the class of insects, tends to fly. Therefore, these flies need to be put to sleep for sexing of males and females and to setup/carry out experiments, depending upon requirements. Several methods are available to put flies to sleep. These include exposure of flies to ether, chilling (or cooling), CO_2, or nitrogen, the latter three being least harmful. Of these three choices, cooling is a little bit messy, but is the simplest because it does not require any sophisticated equipment and needs only ice and petri dishes. In addition, it is the only method that will not affect fly neurology. The remaining two methods require commercially available gas cylinders and controllers to provide a regulated supply of the gas to incapacitate the flies.

Fly Disposal

This is a very essential step when using flies. A bottle or beaker with (new/used) oil (whose density is heavier than water, such as mineral oil), referred to as a fly morgue, is generally used. The anesthetized (but unused flies) should be dumped directly into the fly morgue. Generally, they drown at the bottom, but this needs to be ensured (especially in the case of an old morgue). Discarding of flies in the morgue is aimed at minimization of stock contamination and keeping the lab environment fly free. The old vials and bottles containing flies should be autoclaved to kill the flies prior to discarding them.

Egg Collection

Abundant batches of eggs synchronized in age are required for experiments. In general, 200–300 adults from fresh cultures should be transferred into bottles or collection chambers containing tiny petri plates containing fly food (or grape juice–agar food). To optimize egg collections, flies are starved for 4–6 hours under light in empty bottles. Subsequently, flies are transferred to collection chambers and kept in the dark. The first hour's collection should be discarded to avoid those eggs retained by females in anticipation of fresh food. Thereafter, egg collection plates can be removed and replaced with new ones at 30 min intervals.

Dechorination of Eggs

To prepare the eggs for microinjection, the outer covering of the egg (namely, chorion) should be removed. For chemical dechorination, eggs are washed from the egg collection plate to a netwell using egg wash buffer (0.03% Triton X-100, 0.4% NaCl) and the netwell containing the eggs is placed into 2% sodium hypochlorite (bleach) for 2 min. Thereafter, the eggs are thoroughly washed with egg wash buffer and collected onto an agar bed or onto a petri plate. Using a fine brush, the eggs are arranged in a vertically linear fashion and transferred to a slide using a double stick tape (Scotch 665). After optimal dehydration, the eggs are covered with a layer of halocarbon oil. Later, DNA is injected (through a fine glass needle) into these eggs under the inverted phase contrast microscope using a micromanipulator.

Preparation of DNA for Injection

To generate the transgenic, the gene of interest should first be cloned into a vector used for germ-line transformation. For Drosophila, P-element based vectors (for example, pPUAST, pPCASPER, etc.) that can integrate the gene of interest into the fly genome are quite commonly used. Depending upon the requirement, the complete gene or the coding region can be cloned into the P-element vectors. Once the gene is cloned, the plasmid DNA is extracted. The successful transformation requires DNA at an optimal concentration of 1 μg/μl and should be endotoxin free. Thereafter, the plasmid DNA is mixed with a helper plasmid DNA at a ratio of 3:1. The helper encodes for transposase required for the transposition of the transgene into the fly genome. Subsequently, the mixed DNA is precipitated using ethanol and resuspended in the injection buffer (5 mM KCL; 0.1mM $NaPO_4$ buffer pH 7.5).

PROTOCOLS

Protocol for Germ-Line Transformation in Drosophila

The protocol given below is adapted from those of Prof. John Belote, Syracuse University, USA and Prof. Heifetz, University of Jerusalem, Israel. This is schematically represented in the Flow Chart 1.1.

Materials Required

Plasmid purification kit; helper plasmid (generally, Δ 2–3); injection needle: glass with diameter of tip opening of approximately 0.5 uM; Fly strain: white eye of *Drosophila melanogaster* (w^{1118}); Petri dishes; grape juice; agar; potassium chloride; sodium phosphate (monobasic and dibasic); double stick tape (Scotch 665); tweezers (fine needle, sharp, Sigma); fine needles; Halocarbon oil; permitted food color (red or green); net wells.

Procedure

A. Preparation of DNA

(i) Prepare the endotoxin free DNA of the gene clone using a plasmid maxi-prep kit. A wide variety of kits is commercially available; follow the manufacturer's instructions to prepare the DNA.

(ii) In a 0.6 ml eppendorf tube, mix 15 μg of plasmid of interest with 5 μg of helper plasmid, ethanol precipitate the mixture using 1/10 volume of 3M NAOAC and two times volume of 100% ethanol. Leave tubes overnight at −20°C. The next morning, centrifuge the tubes at 14,000 rpm for 30 min, wash thrice with 70% ethanol (each for 3 min at 13,000 rpm) and ultimately resuspend the pellet in 50 μl injection buffer (5 mM KCL; 0.1mM $NaPO_4$ buffer at pH 7.5). This can be stored at −20°C until further use.

B. Collection of fly embryos

(i) Use 4–6 days old adult flies from fresh cultures for embryo collection.

(ii) Place 200–300 w^{1118} adults in bottles or collection chambers containing tiny petri dishes with grape juice-agar medium.

(iii) Place the bottle in dark.

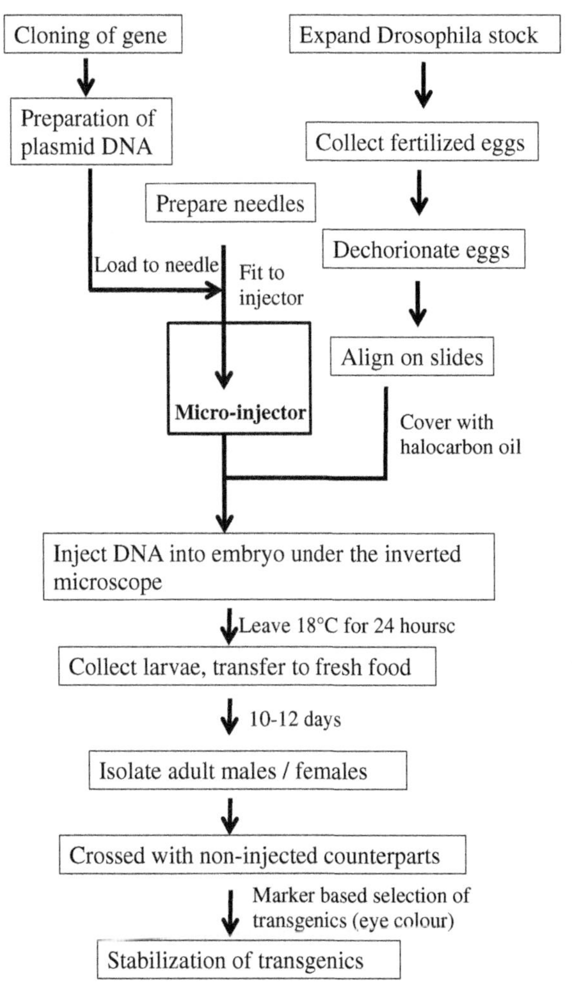

FLOW CHART 1.1 Flow chart of germ-line transformation in Drosophila.

(iv) Discard the first hour's collection plate and replace with a fresh one.
(v) Replace egg collection plates with new ones at 30 min intervals and use the embryos for injection.

C. Removal of chorion from embryos
(i) Wash the embryos from the egg collection plate to a netwell using egg wash buffer (0.03% Triton X-100, 0.4% NaCl).
(ii) Place the netwell containing the embryos in 2% sodium hypochlorite (bleach) for 2 min.
(iii) Wash the embryos thoroughly with egg wash buffer and collect onto an agar bed or petri dish.

D. Preparation of embryos for injection
(i) Cut a small strip of double stick tape (Scotch 665) and stick it in the center of a slide.
(ii) Using a fine brush, arrange the embryos in a vertically linear fashion on the agar bed.
(iii) Transfer the embryos to the slide by reversing the agar bed onto the edge of the double stick tape. Final orientation of the embryos should be such that the posterior end of the embryo hangs off the tape.
(iv) Observe the quality and developmental stage of the embryo. Discard those embryos with poor quality or aged embryos (visualization of syncytial blastoderm).
(v) Place the slide in a petri dish containing the desiccant for 4–5 min (optimal dehydration times are to be determined depending upon the relative humidity of the injection room).
(vi) After optimal dehydration, cover the embryos with a layer of halocarbon oil using a glass/plastic pipette.

E. Preparation of needle for injection
(i) Depending upon the microinjector being used, either pull a glass needle using a needle puller or buy commercially available glass needles meant for Drosophila injections. In both cases, the diameter of the tip opening is expected to be approximately 0.5 μM.
(ii) Mix 15 μl of DNA prepared for injection with 4 μl of food color (green or red, depending upon availability).
(iii) Load the needle with 2–5 μl of DNA either by using a pipette or through capillary action (particularly in case of pulled needles).
(iv) Connect the loaded needle to the holder connected to a micromanipulator or a syringe. In the case of pulled needles, be sure to break the tip of the needle by rubbing the same against the corner of the cover glass, placed on a slide.
(v) Place a slide containing a drop of halocarbon oil and ensure that the needle is in the same vertical focal plane as the embryos.

F. Injection of embryos
(i) Lift the needle vertically and replace the slide containing the halocarbon oil with the slide containing the embryos.
(ii) Position the needle slightly away from the middle of the posterior tip of the embryo and penetrate the embryo. After penetration, draw the needle as far back as possible, without leaving the embryo, and deposit the DNA in the posterior-most region of the embryo cytoplasm.
(iii) Inject all the embryos on the slide in a similar manner.

G. Post-injection care of embryos
(i) Remove the uninjected embryos to avoid false negatives.
(ii) Place the slide with the injected embryos on a petri plate containing the moist tissue paper.
(iii) After 24–28 hrs, transfer the hatched larvae using a fine needle to a vial containing fresh food.
(iv) Monitor the development of these larvae and cross each eclosed adult to the opposite sex partner from w^{1118} stock.
(v) After 10–12 days, screen for the red/orange/yellow eyed flies, which are transformants.

H. Generation of stable transgenic flies
(i) Collect unmated transformants and cross to the opposite sex partner from w^{1118} stock.
(ii) Collect unmated heterozygous males and females resulting from this cross and place the same in fresh vials.
(iii) In the resultant progeny, isolate unmated homozygous males and females (will have dark eye color) from heterozygous flies (eye color lighter than homozygotes) and set up a cross of homozygotes.
(iv) Cross-check the resultant progeny for similar eye color in all flies to ensure that the line is stable.

ETHICAL ISSUES

Biological research involving organisms requires ethical clearance. The rationale of introducing ethical issues lies majorly on the judicious use of the animals and their humane treatment before, during, and after the experiments. Therefore, it is now mandatory in a research institution or university where biological research is pursued to have an animal ethics and human ethics committee in place. These committees look into the details of the program that will be undertaken by the researchers for the possible use of laboratory animals (essentially mammals like mice, rats, guinea pigs, etc.) and for human samples. No research data on animals/humans is published nowadays in journals if the respective researchers do not provide the details of the ethical clearance data/number.

While in toxicology and other related fields researchers depend on the data generated on laboratory animals for extrapolation to humans, several factors that remain as hurdles for the genuine extrapolation are intra-species variations and compounding effects resulting from one experiment to another.

Smaller animals have played pivotal roles in our present-day understanding of fundamental facets of biology. In parallel, information generated in these organisms due to ease of handling, their isogenic conditions negating intra-species variation, simple yet functionally homologous tissue architecture to higher mammals, and more importantly in the post genomic era, the gene homology existing between the smaller organisms like Drosophila, *C. elegans*, etc. to mammals and humans, have led to important discoveries that have relevance to humans.

In this context, limited ethical concerns are raised for the use of smaller organisms, especially invertebrates and lower vertebrates, in biological experiments and testing. The European Centre for the Validation of Alternative Methods has recommended several such organisms to promote the basic principles of the 3Rs (refine, replacement, and reduce) in biological research and testing (Balls *et al.* 1995).

TRANSLATIONAL SIGNIFICANCE

How does Drosophila research over the years made significant inroads in the area of biotechnology? This is a pertinent question. Over the years, several significant bits of information pertaining to genetics and development have been generated using Drosophila that have advanced our knowledge. During the last decade, Drosophila has been a model for a number of human diseases.

DROSOPHILA MODELS OF HUMAN DISEASE

Model organisms have made invaluable contributions for studies of human disease mechanisms due to the high degree of conservation of fundamental biological processes throughout the animal kingdom. Completion of the *Drosophila* genome sequence and subsequent annotation (Adams *et al.* 2000) paved the way for the utilization of *Drosophila melanogaster* as a model for studying the molecular basis of human disease. Comparison of human and Drosophila genomes suggested the presence of orthologs in Drosophila for about 75% of human disease-related genes. In view of these advances, researchers have exploited the versatile molecular tools available for Drosophila and modeled several human diseases in Drosophila (reviewed in Lu and Vogel 2009). In this context it is very important to remember that human organization is much more complex than that of Drosophila. Despite sharing fundamental processes, Drosophila and humans might still differ in mechanisms through which these processes are regulated/implemented. Consequently, researchers sensibly developed Drosophila-based models for human diseases associated with mutations in one or two genes, or for those diseases where the essential pathways are conserved. These models are excellent for understanding disease progression, pathogenesis, and underlying mechanisms. Further, these models are also useful in drug discovery, although caution should be exercised before extrapolating the dose data and/or delivery route to human use. Nevertheless, Drosophila based models for human disease present a quick and inexpensive approach for first-tier screening of a large number of molecules during the drug discovery process. Below, we have given a few examples of Drosophila-based models for human diseases.

Drosophila-Based Models for Understanding Human Neurodegenerative Diseases

Neurodegenerative diseases (ND), as the name suggests, arise due to progressive loss of specific neuronal populations. Human genetic studies led to the identification of genes associated with certain ND, but mechanistic understanding of pathways and processes underlying the disease remained incomplete due to ethical and/or technical constraints. In this context, Drosophila, with its complex behavior, including learning and memory, and driven by a sophisticated brain and nervous system, provided excellent models for deciphering signaling pathways and understanding the cellular processes defining human ND (reviewed in Hirth 2010; Lu and Vogel 2009). For example, Drosophila-based models are available for Alzheimer's disease (AD), the most common neurodegenerative disease, characterized by age-dependent gradual impairment in memory and cognitive abilities. In humans, selective atrophy of the hippocampus and the frontal cerebral cortex, Amyloid β (Aβ) plaque deposition, and aggregations of the microtubule-associated protein tau, are the hallmarks of AD pathology. The Aβ plaque is mainly composed of Aβ-40 and Aβ-42 peptides, the products of endoproteolysis of the amyloid precursor protein (APP) through secretases. Autosomal-dominant mutations and defective trafficking of APP affect the onset and/or progression of AD; both circumstances promote the generation of amyloid Aβ peptides. The Drosophila genome contains homologs for human AD-related genes: APP, presenilin, and tau. Therefore, researchers employed Drosophila as a model system for AD research. Drosophila models of APP-mediated AD do simulate certain characteristics of human AD pathology, including Aβ plaque deposition. In addition to providing insights into the pathology, these models are also useful in identifying/targeting the modulators of AD pathology through genetic or pharmacological interference. Likewise, Drosophila models are available for Parkinson's disease.

Parkinson's disease (PD) is the second most common neurodegenerative disorder, and is characterized by impaired motor skills: uncontrollable tremor, imbalance, slow movement, and muscle rigidity. PD pathology is associated with progressive loss of dopamine neurons in the substantia nigra pars compacta of the ventral midbrain. Formation of Lewy bodies, mainly composed of synuclein and ubiquitin (among other proteins), is the pathological hallmark of PD. Several genes have been associated with PD in humans: alpha-synuclein, parkin, ubiquitin carboxy-terminal hydrolase L1 (UCHL1), phosphatase and tensin homolog (PTEN)-induced kinase 1, (PINK1), DJ-1, leucine-rich-repeat kinase 2 (LRRK2), high-temperature requirement protein A2 (HTRA2), glucocerebrosidase, and polymerase gamma and tau. Mutations in alpha-synuclein and Parkin have been associated with PD in humans. Except for alpha-synuclein, Drosophila carries the homologs of the remaining PD-associated genes. Accordingly, analysis of Drosophila mutations/transgenes of PD-associated gene homologs provided valuable insights into PD pathogenic mechanisms and targets of PD-related genes. Lack of an alpha-synuclein homolog did not deter researchers from using Drosophila to study the mechanism underlying alpha-synuclein-related PD and its targets. By utilizing the powerful genetic tools available with Drosophila, researchers expressed human alpha-synuclein in Drosophila to investigate its functional properties in PD and to elucidate its targets.

Drosophila as a Model for Understanding Human Metabolic Disorders

Diabetes and obesity are the most common metabolic disorders encountered in present-day lifestyle. To date, obesity concerns are limited mainly to developed nations, although developing countries are catching up quickly. Further, obesity in the majority of cases leads to diabetes. Diabetes, arising mainly from errors in glucose metabolism, is the most common metabolic disorder with a worldwide distribution. The World Health Organization (WHO) already declared diabetes as an epidemic because the number of diabetic patients or prevalence of diabetes has gone up dramatically over the last few decades. Diabetes has multiple consequences in that it causes 2–4 times more heart disease and stroke, 60% of amputations, and 44% of kidney diseases. Worldwide, diabetics are projected by the International Diabetes Federation to grow by over 54% from 246 M in 2007 to about 380 M by 2025. WHO predicted that developing countries will be at a higher risk of this epidemic in the 21st century.

There are two types of diabetes, namely Type-I and Type-II, depending upon the status of insulin, a peptide hormone, produced by the islets of Langerhans present in the pancreas. In humans, insulin plays a key role in glucose homeostasis along with glucagon (another peptide hormone), which is produced by the alpha cells of the pancreas. Insulin influences the cellular uptake of glucose from the blood whereas glucagon signals the release of glucose stored in liver cells into the blood, thereby maintaining healthy blood glucose levels. In Type-I diabetes, previously named juvenile onset diabetes, insulin production is affected due to autoimmune destruction of pancreatic cells. Consequently, there is a rise in the circulating sugar levels due to failure of insulin-dependent glucose transport, triggering a compensatory starvation response breakdown of glycogen and fat to produce energy since the body fails to generate energy from ingested food resources. Type-2 diabetes, previously known as adult-onset diabetes, differs from Type-1 by having insulin but not being able to utilize the same efficiently, probably due to defects arising in insulin receptors or due to a complicated medical condition called "insulin resistance." The conservation of the insulin/insulin-like growth factor-1 signaling (IIS) pathway between Drosophila and mammals (Brogiolo *et al.* 2001) and recent understanding of metabolic processes, glucose homeostasis, and endocrinology, have led to Drosophila-based models relevant to human metabolic disorders and diabetes.

Drosophila has seven insulin-like peptides and five of these show significant homology to human insulin. These peptides are produced by neurosecretory cells in the pars intercerebralis (PI) region of the protocerebrum in both larvae and adults. Drosophila insulin-like peptides (dilps) affect the growth of the organism and energy homeostasis. In addition to dilps, Drosophila also contains the adipokinetic hormone (AKH), further illustrating the conservation of endocrine mechanisms regulating circulating glucose levels. In Drosophila, ablation of AKH (synthesized and secreted by cells in the corpus cardiacum) results in decreased circulating carbohydrate levels. This antagonistic relationship between dilps and AKH is reminiscent of that between insulin and glucagon in mammals (Bharucha *et al.* 2008). To model the loss of insulin as observed in Type-1 diabetes, Zhang and his colleagues (2009) used genetic approaches and deleted dilps 1–5 in Drosophila. Drosophila lacking dilps 1–5 recapitulated symptoms that appear similar to human diabetes: knockout flies had elevated sugar levels in their circulation and despite feeding at normal levels, they generated energy from fat, which is generally a hallmark of starvation. Type-2 or insulin-resistant diabetes is caused by multiple factors. These include genetic factors, diet, and most obviously, obesity. Further, Type-2 diabetes due to a single mutation is rare in humans, and therefore, models based on single-gene manipulations would limit the understanding of the mechanism underlying Type-2 diabetes. To model the Type-2 diabetes resulting from diet/obesity, Musselman and his colleagues (2011) fed Drosophila a high-sugar diet and induced characteristics representing Type-2 diabetes: hyperglycemia, insulin resistance, and increased levels of triglycerides and free fatty acids. Though these flies are yet to be exploited for drug discovery, the generated models help to understand not only the mechanism underlying human diabetes, but also the downstream signaling components regulating glycogenolysis (Haselton and Fridell 2010). In addition, the power of Drosophila genetic screens would help to unravel candidates critical for carbohydrate metabolism and energy homeostasis, thereby providing a window for approaches leading to pharmacological intervention.

Drosophila as a Model for Understanding Nephrolithiasis (Kidney Stones)

An efficient kidney is critical for the filtration and elimination of waste from the blood. Given these roles, analysis of the renal system has the potential to provide indications of immune, toxic, and other insults to the body. Stone formation is the second most encountered medical complication of the kidney. The cause of stone formation is multifactorial and attempts of the prevention of the same have been complicated due to difficulties in finding a "model" system. Vertebrate kidneys are glomerular and structurally complex. On the contrary, the Drosophila renal system has aglomerular tubules and is simple. However, there exist certain structural similarities in the renal systems of insects and vertebrates. The malphigian tubule in Drosophila is the recognizable counterpart of the vertebrate renal tubule. Primarily, these tubules generate urine and may selectively reabsorb some solutes. In addition to tubules, Drosophila, like other insects, also possess nephrocytes having roles in detoxification of filtration and endocytosis, followed by metabolism. Taking these things into account, Dow and Romero (2010) exposed Drosophila to certain dietary foods to rapidly and reliably develop kidney stones. Drosophila is quite amenable for the study kidney stones because these stones are hardly lethal to the simple tubule architecture of the insect. Indeed, two classes of stone can be constitutively produced in insects without being harmful to the organism. In the tubule lumen, calcium phosphate is stored as concentric spherites. The accumulation of uric acid crystals in the tubule lumen can potentially provide a natural model for urate nephrolithiasis. Apart from this, calcium oxalate (CaOx) stone formation, which is the predominant type of kidney stone disease, can also be modeled in Drosophila. Towards this direction, Chen and his co-workers (2011) have shown that feeding Drosophila lithogenic agents such as ethylene glycol, hydroxyl-l-proline, or sodium oxalate results in the formation of calcium oxalate crystals within the lumen of the tubule in a dose-dependent manner. However, the extent to which the formation of stones observed in Drosophila mimics that of humans remains to be understood. Further studies on these models have the potential to provide insight into the pathophysiology underlying kidney stones, and would also be useful in developing new therapeutic approaches.

Drosophila-Based Model for Understanding HIV Pathology

Acquired immunodeficiency syndrome (AIDS) is a disease of the immune system caused by infection with the human immunodeficiency virus (HIV). The pathophysiology of AIDS is quite complex. After gaining entry to the body, HIV replicates to the level of several millions per ml of blood. This replication is associated with depletion of CD4+T cells, thereby weakening the immune system and leading to secondary infections. To facilitate the replication of the viral genome, HIV encodes a trans-activator of transcription (Tat), which functions as an activator of transcription by interacting with components from the transcription complex. Recent evidence points to the involvement of a host–cytoskeleton in HIV pathogenesis. To test if Tat is involved in the cytoskeleton organization, Battaglia and his co-workers (2001) generated Tat transgenic Drosophila lines and evaluated the effect of Tat by expressing it during oogenesis. These HIV transgenic flies helped to understand the action of viral gene products not just in a single cell, but within a defined territory. The principle behind these studies is that any expansion or restriction of the territory in which the gene is expressed results in mutated phenotypes. Studies on these transgenic flies have shown that Battaglia and his co-workers have demonstrated that Tat can interact with tubulin to alter the MT polymerization rate in HIV-infected cells, thereby furthering our understanding of the molecular mechanism underlying Tat-mediated pathogenesis.

DROSOPHILA-BASED THERAPEUTIC PEPTIDE PRODUCTION

Drosophila is not only useful for understanding the processes underlying diseases, but would also be useful for producing peptides of therapeutic importance. By exploiting the conserved nature of peptide processing across Metazoa, and the versatility as well as the scope of Drosophila genetics, Park and his colleagues (2012) have successfully produced functional human insulin *in vivo* in Drosophila. This innovative metabolic engineering in Drosophila, however, requires several adjustments and improvements to make it cost-effective and commercially feasible. Nevertheless, coupling of this novel method with the power of Drosophila genetics would immensely help to meet the increasing demand for therapeutic bioactive peptides.

We have discussed here only a few of the numerous human diseases. Drosophila is being used to study other human diseases, including oxidative stress-based disorders, tau neuropathies, Poly-Q-related neurodegeneration, cancers, epilepsy, and muscular dystrophy. Details pertaining to these and other diseases for which Drosophila can be a potential model are available on the Homophila website that lists the Drosophila counterparts of several human diseases.

TABLE 1.1 Network Resources Available for Drosophila Researchers and Their Utility

Name of Resource	Web Address	Purpose/Utility
FlyBase	http://flybase.org/	Gateway to *Drosophila melanogaster* genome
Berkely Drosophila genome project (BDGP)	http://www.fruitfly.org/	
Drosophila Genomics Resource Center (DGRC)	https://dgrc.cgb.indiana.edu/	
Drosophila 12 species genome Assembly/Annotation	http://rana.lbl.gov/drosophila/	Sequences, assemblies, annotations and analyses of the genomes of 12 members of the genus Drosophila
FlyAtlas	http://flyatlas.org/	Road map to Drosophila gene expression
FlyRNAi	www.flyrnai.org	Drosophila RNAi screening Center database
TRiP	http://www.flyrnai.org/TRiP-HOME.html	Transgenic RNAi project
Flymine	www.flymine.org/	An integrated database of gene expression and protein data for Drosophila and Anopheles
Flymove	http://flymove.uni-muenster.de/	Resource for university students and teachers studying Drosophila developmental biology
Flynet	https://www.cistrack.org/flynet/	Genomic resource for *Drosophila melanogaster* transcriptional regulatory networks
FlyEx	http://urchin.spbcas.ru/flyex/	Database of Segmentation Gene Expression in Drosophila
FlySNP	http://flysnp.imp.ac.at/	High-density genome-wide map of single nucleotide polymorphisms (SNPs) and SNP genotyping
Flybrain	http://web.neurobio.arizona.edu/	Database of Drosophila nervous system
Homophila	http://superfly.ucsd.edu/homophila/	Human disease to Drosophila gene database
BioGrid	http://thebiogrid.org/	Protein and genetic interaction database
Drosophila proteome atlas	http://www.dgrc.kit.jp/~jdd/proteome_atlas/index.html	Proteomes of adult Brain, eye, and reproductive system
Drosophila RNAi Screening Center (DRSC)	http://www.flyrnai.org/cgi-bin/DRSC_MinoTar.pl	To predict miRNA targets in Drosophila sequence
TargetScanFly	http://www.targetscan.org/fly_12/	
FlyTF	http://www.flytf.org/	Drosophila Transcription Factor database
TaxoDros	http://www.taxodros.uzh.ch/	Database on Taxonomy of Drosophilidae
FLIGHT	http://flight.icr.ac.uk/	Fly database for the Integration of Genomic and High-Throughput data
Drosophila population genomics project (DPGP)	http://www.dpgp.org/	Drosophila population genomics
Drosophila polymorphism database	http://dpdb.uab.cat/dpdb/	
FlyTrap	http://flytrap.med.yale.edu/	GFP protein trap database
DrosDel	http://www.drosdel.org.uk/	Isogenic deficiency kit for *Drosophila melanogaster*
FlyPNS	http://www.normalesup.org/~vorgogoz/FlyPNS/page1.html	*Drosophila melanogaster* embryonic and larval peripheral nervous system

TABLE 1.2 Physical Resources Available for Drosophila Researchers and Their Utility Stocks and Reagents

Name of Resource	Web Address	Utility
Bloomington Stock Center	http://flystocks.bio.indiana.edu/	Various fly stocks (including mutants and transgenics); also provides resources/recipes for fly work
NIG-FLY, Japan	http://www.shigen.nig.ac.jp/fly/nigfly/index.jsp	Mutants and RNAi stocks of Drosophila
Drosophila Genomics Resource Center (DGRC), USA/Japan	https://dgrc.cgb.indiana.edu/ http://kyotofly.kit.jp/cgi-bin/stocks/index.cgi	Stocks and reagents for Drosophila research
VDRC stock Center, Austria	http://stockcenter.vdrc.at/control/main	Drosophila RNAi lines
UC San Diego Drosophila stock center, USA	https://stockcenter.ucsd.edu/info/welcome.php	Source for Drosophila species stocks and also Genomic DNA and BAC libraries
FLY-TILL	http://tilling.fhcrc.org/fly/	Delivers EMS-induced mutations in requested genes

In addition, Drosophila is also being used in the field of toxicology. Transgenic Drosophila containing LacZ under the control of hsp70 promoter has extensively contributed to the understanding of the stress–response mechanism underlying heavy metal and/or pesticide toxicity (reviewed in Gupta *et al.* 2010). Given the power of Drosophila genetics/tools, we will be learning much more about our health using Drosophila in the future.

WORLD WIDE WEB RESOURCES

Resource availability is vital for the development of any community. Thanks to the cooperative work culture among Drosophila researchers both in the pre- and post-genome sequencing era, the Drosophila research community is very well equipped with versatile physical and network resources for knowledge generation and dissemination of the same. The physical resources range from stock centers that provide the required wild type stocks/mutants/transgenic flies to the specialized centers that cater to the biochemical reagent needs of Drosophila researchers. Virtual resources literally assist the quick assembly/annotation of genomes across various species of Drosophila, and also assist in functional genomics/proteomics. Here we list a few commonly used network resources and the resources for stocks and reagents along with their utility (Tables 1.1 and 1.2). More details regarding the listed resources can be obtained by visiting the sites using the web address given. There are many more than those listed here and they can be accessed at http://flybase.org/static_pages/allied-data/external_resources5.html

ACKNOWLEDGMENTS

This is IITR communication no. 3044. We are thankful to the Director, CSIR-IITR, for his constant support and encouragement, all the members of Embryotoxicology lab for their suggestions and support while preparing this manuscript, and Ms. Snigdha Misra for relevant graphics work. We are grateful to Prof. S. C. Lakhotia and Dr. Anand K. Singh, Department of Zoology, Banaras Hindu University, India, for providing us with the image of polytene chromosomes.

REFERENCES

Abbott, M. K., & Kaufman, T. C. (1986). The relationship between the functional complexity and the molecular organization of the Antennapedia locus of *Drosophila melanogaster*. *Genetics*, *114*, 919–942.

Adams, M. D., Celniker, S. E., Holt, R. A., Evans, C. A., Gocayne, J. D., et al. (2000). The genome sequence of *Drosophila melanogaster*. *Science*, *287*, 2185–2195.

Balls, M., Goldberg, A. M., Fentem, J. H., Broadhead, C. L., Burch, R. L., et al. (1995). The three Rs: the way forward: the report and recommendations of ECVAM Workshop 11. *Alternate to Lab Animals*, *23*, 838–866.

Battaglia, P. A., Zito, S., Macchini, A., & Gigliani, F. (2001). A Drosophila model of HIV-Tat-related pathogenicity. *Journal of Cell Science*, *114*, 2787–2794.

Beermann, W. (1952). Chromomore constancy and specific modifications of the chromosome structure in development and organ differentiation of *Chironomus tentans*. *Chromosoma*, *5*, 139–198.

Bharucha, K. N., Tarr, P., & Zipursky, S. L. (2008). A glucagon-like endocrine pathway in Drosophila modulates both lipid and carbohydrate homeostasis. *Journal of Experimental Biology*, *211*, 3103–3110.

Brogiolo, W., Stocker, H., Ikeya, T., Rintelen, F., Fernandez, R., et al. (2001). An evolutionarily conserved function of the Drosophila insulin receptor and insulin-like peptides in growth control. *Current Biology*, *11*, 213–221.

Campos-Ortega, J. A., & Hartenstein, V. (1985). *The embryonic development of Drosophila melanogaster*. Berlin ; New York: Springer-Verlag.

Chen, Y. H., Liu, H. P., Chen, H. Y., Tsai, F. J., Chang, C. H., et al. (2011). Ethylene glycol induces calcium oxalate crystal deposition in Malpighian tubules: a Drosophila model for nephrolithiasis/urolithiasis. *Kidney International*, *80*, 369–377.

Dow, J. A., & Romero, M. F. (2010). Drosophila provides rapid modeling of renal development, function, and disease. *American Journal of Physiology.Renal Physiology*, *299*, F1237–1244.

Gilbert, S. F., Singer, S. R., Tyler, M. S., & Kozlowski, R. N. (2003). *Developmental biology*. Sunderland, MA: Sinauer Associates.

Gummalla, M., Maeda, R. K., Castro Alvarez, J. J., Gyurkovics, H., Singari, S., et al. (2012). abd-A Regulation by the iab-8 Noncoding RNA. *Plos Genetics*, *8*, e1002720.

Gupta, S. C., Sharma, A., Mishra, M., Mishra, R. K., & Chowdhuri, D. K. (2010). Heat shock proteins in toxicology: How close and how far? *Life Sciences*, *86*, 377–384.

Hartenstein, V. (1993). *Atlas of Drosophila development*. Plainview, NY: Cold Spring Harbor Laboratory Press.

Haselton, A. T., & Fridell, Y. W. (2010). Adult Drosophila melanogaster as a model for the study of glucose homeostasis. *Aging (Albany NY)*, *2*, 523–526.

Hirth, F. (2010). Drosophila melanogaster in the study of human neurodegeneration. *CNS Neurol Disord Drug Targets*, *9*, 504–523.

Hoskins, R. A., Phan, A. C., Naeemuddin, M., Mapa, F. A., Ruddy, D. A., et al. (2001). Single nucleotide polymorphism markers for genetic mapping in Drosophila melanogaster. *Genome Research*, *11*, 1100–1113.

Lewis, E. B. (1978). A gene complex controlling segmentation in Drosophila. *Nature*, *276*, 565–570.

Lu, B., & Vogel, H. (2009). Drosophila models of neurodegenerative diseases. *Annual Review of Pathology*, *4*, 315–342.

Martinez Arias, A. (1993). Development and patterning of the larval epidermis of Drosophila. In M. Bate & V. Hartenstein (Eds.), *The development of Drosophila melanogaster* (pp. 517–608). Long Island, NY: Cold Spring Harbor Laboratories.

Mechelke, F. (1953). Reversible structural modifications of the chromosomes in salivary glands of Acricotopus lucidus. *Chromosoma*, *5*, 511–543.

Musselman, L. P., Fink, J. L., Narzinski, K., Ramachandran, P. V., Hathiramani, S. S., et al. (2011). A high-sugar diet produces obesity and insulin resistance in wild-type Drosophila. *Disease Models and Mechanisms*, *4*, 842–849.

Nusslein-Volhard, C., & Wieschaus, E. (1980). Mutations affecting segment number and polarity in Drosophila. *Nature*, *287*, 795–801.

Park, D., Hou, X., Sweedler, J. V., & Taghert, P. H. (2012). Therapeutic peptide production in Drosophila. *Peptides*, *36*, 251–256.

Pavan, C., & Breuer, M. E. (1952). Polytene chromosomes in different tissues of Rhynchosciara. *Journal of Heredity*, *43*, 151–157.

Ruddle, F. H., Bartels, J. L., Bentley, K. L., Kappen, C., Murtha, M. T., et al. (1994). Evolution of Hox genes. *Annual Review of Genetics*, *28*, 423–442.

Santini, S., Boore, J. L., & Meyer, A. (2003). Evolutionary conservation of regulatory elements in vertebrate Hox gene clusters. *Genome Research*, *13*, 1111–1122.

Schulz, C., & Tautz, D. (1994). Autonomous concentration-dependent activation and repression of Kruppel by hunchback in the Drosophila embryo. *Development*, *120*, 3043–3049.

Wreden, C., Verrotti, A. C., Schisa, J. A., Lieberfarb, M. E., & Strickland, S. (1997). Nanos and pumilio establish embryonic polarity in Drosophila by promoting posterior deadenylation of hunchback mRNA. *Development*, *124*, 3015–3023.

Wu, X., Vasisht, V., Kosman, D., Reinitz, J., & Small, S. (2001). Thoracic patterning by the Drosophila gap gene hunchback. *Developmental Biology*, *237*, 79–92.

Zhang, H., Liu, J., Li, C. R., Momen, B., Kohanski, R. A., et al. (2009). Deletion of Drosophila insulin-like peptides causes growth defects and metabolic abnormalities. *Proceedings of National Academy of Science (USA)*, *106*, 19617–19622.

Zuo, P., Stanojevic, D., Colgan, J., Han, K., Levine, M., et al. (1991). Activation and repression of transcription by the gap proteins hunchback and Kruppel in cultured Drosophila cells. *Genes and Development*, *5*, 254–264.

FURTHER READING

Sullivan, W., Ashburner, M., & Hawley, R. S. (Eds.), (2000). *Drosophila protocols*. Cold Spring Harbor, NY, USA: published by Cold Spring Harbor laboratory press.

Dahmann, C. (Ed.), (2007). *Drosophila methods and protocols*. Seacaucus, NJ, USA: published by Springer science and business media.

Ashburner, M. (Ed.), (1989). *Drosophila: A laboratory hand book*. Cold Spring Harbor, NY, USA: published by Cold Spring Harbor laboratory press.

Kohler, R. E. (Ed.), (1994). *Lords of the fly: Drosophila genetics and experimental life*. Chicago, IL, USA: published by The University of Chicago press.

Roberts, D. B. (Ed.), (1998). *Drosophila: A practical approach*. Oxford, UK: published by IRL press at Oxford University Press.

GLOSSARY

Allele An alternate form of a gene. A gene may have many different alleles that differ from each other by as little as a single base or by the complete absence of a sequence.

Centimorgan (cM) The metric used to describe the distance between two genes, which is determined by using the frequency of recombination between these genes. For example, a recombination frequency of 10 amounts to 10 cM. This term is named in honor of Thomas Hunt Morgan, who first conceptualized linkage while working with Drosophila.

Genotype The set of alleles for a given character. A genotype can be either homozygous (with two identical alleles) or heterozygous (with two different alleles) or hemizygous (in the case of sex-linked alleles).

Homolog A gene whose sequence is similar to a greater extent to a gene from another species and has commonalities in origin and function.

Inbred Organisms that result from the process of brother–sister matings for multiple generations. This process is called inbreeding.

Gene linkage Presence of loci so close to each other on a chromosome that they tend to be inherited together such that recombination between them is reduced to a level significantly less than 50%.

Locus Any genomic site mapped to a chromosome through formal genetic analysis.

Mutation A heritable variation in the sequence of a gene that alters the amino acid sequence of its protein. These mutations can influence the production, structure, and function of proteins.

Phenotype The physical manifestation of a genotype within an organism. For example, hair color in humans is a phenotype.

Transgene A fragment of foreign DNA incorporated into the genome through the manipulation of embryos. For example, insertion of the human insulin gene into Drosophila by manipulating their embryos.

Transgenic Refers to organisms containing a transgene or genes that are foreign. For example, cotton seeds developed by Monsanto are transgenic.

Isogenic Characterized by essentially identical genes. For example, identical twins (monzygotic) are isogenic.

Eclosion The emergence of an adult insect from its pupal case. For example, emergence of silk moth from its coccoon.

Transposable element A genetic element capable of moving from one chromosome to another or within the same chromosome. These elements can potentially disrupt the function of other genes.

Homeobox A regulatory DNA sequence present in the genes that controls pattern formation in organisms during development.

Ortholog Genes in different species with a similarity to each other due to their common ancestral origin.

Balancer chromosome Chromosome comprising inversions that facilitate stable maintenance of lethal mutations as heterozygotes in a manner that does not require selection.

Neurodegeneration This refers to progressive loss and/or death of structure and/or function of neurons. For example, Parkinson's disease is a result of neurodegeneration

Neurodegenerative disorder A disease condition that involves or causes neurodegeneration.

ABBREVIATIONS

ND Neurodegenerative diseases
AD Alzheimer's disease
Aβ Amyloid β
APP Amyloid Precursor Protein
PD Parkinson's disease
UCHL1 Ubiquitin carboxy-terminal hydrolase L1
PTEN Phosphatase and tensin homolog
PINK-1 PTEN-induced kinase 1
LRRK2 Leucine-rich-repeat kinase 2
HTRA2 High-temperature requirement protein A2
WHO World Health Organization
RNA Ribonucleic Acid
RNAi RNA interference
AntP Antennapedia
BX-C Bithorax Complex
Hox Homeobox-containing
Ubx Ultrabithorax
AbdA Abdominal A
AbdB Abdominal B
AKH Adipokinetic hormone
DILP Drosophila insulin-like peptides
PI Pars intercerebralis
AIDS Acquired immunodeficiency syndrome
HIV Human immunodeficiency virus
Tat Trans-activator of transcription
Hsp Heat shock protein

LONG ANSWER QUESTIONS

(1) Describe sex-linked inheritance.
(2) Describe the molecular events underlying the development of metazoans.
(3) Describe germ-line transformation in Drosophila.
(4) What is the translational significance of Drosophila biology?
(5) Explain the contributions of bridges for the development of chromosome maps of Drosophila.

SHORT ANSWER QUESTIONS

(1) State the different stages in the life cycle of Drosophila.
(2) What is holometabolous development?
(3) Provide two examples of Drosophila models for human diseases.
(4) How conserved are the genomes between Drosophila and humans?
(5) What is germ-line transformation?
(6) Can Drosophila be a model for therapeutic peptide production?

ANSWERS TO SHORT ANSWER QUESTIONS

(1) Egg, larva (three instars), pupa, and imago (adult).
(2) A specific type of insect development involving complete metamorphosis involving different stages of development to give rise to an imago that appears entirely different from the developmental stages.
(3) Diabetes and Parkinson's diseases.
(4) As many as 75% of human disease genes are conserved in *Drosophila.*
(5) Germ-line transformation is a method through which DNA is incorporated into the germ line of the individual for its faithful inheritance to subsequent generations of transgenic organisms.
(6) By exploiting the conserved nature of peptide processing across Metazoa and the versatility as well as the scope of Drosophila genetics, Park and his colleagues (2012) have successfully produced functional human insulin *in vivo* in Drosophila. Coupling of this novel method with the power of Drosophila genetics would immensely help to meet the increasing demand for therapeutic bioactive peptides.

Chapter 2

Animal Models of Tuberculosis

Devyani Dube*, Madhu Gupta†, Udita Agrawal† and Suresh P. Vyas†

*ISF College of Pharmacy, Moga, Punjab, India, 142001, †Department of Pharmaceutical Sciences, Dr. Hari Singh Gour University, Sagar, M.P., India, 470003

Chapter Outline

SUMMARY

Animal models for tuberculosis research provide valuable and specific information about the nature of the disease (pathology and the immune response). Refinement of animal models may pave the way to new information of great importance. The choice of model is mainly dependent on cost, availability, and space, as well as biosafety requirements.

WHAT YOU CAN EXPECT TO KNOW

There are no naturally occurring animal reservoirs for *Mycobacterium tuberculosis*; many different animal species are susceptible to infection with this organism. It has been observed that there is extreme variation in the pattern of pathological reactions between different species. Although different animal models for tuberculosis research have been in use for a long time, and have provided valuable information about the nature of the disease (including specific information about the disease pathology and the immune response to the infection), none completely mimic the human model. However, refinement of animal models may pave the way to new information of great importance. No single model is good enough for evaluation. The choice of model is mainly dependent on cost, availability, space, and biosafety requirements. The most commonly used experimental animal models of TB include the mouse, rabbit, and guinea pig. There are established protocols for infecting animals with TB and further analysis. These protocols are summarized at the end of the chapter and provide valuable information regarding the course of infection, the basic immune response and the extent of lung pathology of experimental pulmonary tuberculosis.

Animal Biotechnology. http://dx.doi.org/10.1016/B978-0-12-416002-6.00002-X

HISTORY AND METHODS

INTRODUCTION

Tuberculosis (TB) still remains to be one of the focal public health priorities for many of the developing countries of the world. According to WHO, globally, there were an estimated 8.87 million new incidents of TB in 2011. Most of the cases occurred in the South-East Asian (55%) and African (30%) regions. The five countries with the largest numbers of cases include India, China, South Africa, Nigeria and Indonesia. Of the 8.87 million new TB cases in 2011, about 15% were HIV positive; 78% of these HIV-positive cases were in the Africans and 13% were in the South-East Asia regions. Identification of multidrug resistant (MDR) strains (defined as mycobacteria resistance to at least rifampicin and isoniazid, two first line anti-TB drugs) and extensively drug resistant (XDR) strains (defined as MDR mycobacteria with additional resistance to fluoroquinolones and at least one of the injectable second line antituberculosis drugs) has worsened the condition. Notably MDR- and XDR-TB have been recognized by the WHO as a major challenge to be addressed in the fight against TB (Dube et al., 2012).

When humans are infected with *M. tuberculosis*, they may develop primary active TB, latent TB, chronic active TB, or reactivation disease. Ten percent of non-immunosuppressed individuals progress from latent to reactivation TB over their lifetimes, while HIV-infected individuals have a 10% annual risk of reactivating latent disease, suggesting that not all above-mentioned manifestations are mutually exclusive. The individual outcome is determined by various factors such as immunosuppression, HIV infection, and nutritional status, intensity of exposure, BCG vaccination, and age. Re-exposure to TB and re-infection play a role in the risk of developing the disease, but are less commonly reported. However, it attracts an increased importance. Each of these stages of infection in humans can be approached by the use of one or more of the animal models that are discussed further in this chapter.

For more than a hundred years, animals have indeed taught humans a great deal about TB, and they promise to have potential as increasingly useful tools in studying immunologic, genetic, molecular, and pharmacologic characteristics of bioactives, infection and related pathogenesis. Animal models have become standard tools for the study of a wide array of human infectious diseases. Many animal models of TB have been developed. Given the complexity of human TB, animal models of TB offer a vast resource to study a multitude of unresolved questions: the genetics of host defense, microbial virulence, latency, reactivation, reinfection, drug therapy, and immunization, just to name a few. Researchers are fortunate to have many well-developed experimental animal models from which exhaustive knowledge can be attained. The most commonly used experimental animal models of TB include the mouse, rabbit, and guinea pig. Although, substantial differences in TB susceptibility and disease manifestations exist between these species, they have contributed significantly to understanding various aspects of TB. Current concepts in TB pathogenesis have also been derived from animal studies involving experimentally induced infections with related mycobacteria (e.g., *M. bovis*). The manifestations in select animal hosts may mimic the etiology of tuberculosis in human TB.

COMPARATIVE PATHOLOGY OF TB IN HUMANS AND ANIMALS

Robert Koch recognized and reported the spectrum of pathology of TB in different animal species based on his seminal studies on TB. The examination of clinical specimens from infected humans, cattle, deer, badgers, and possums confirmed the extreme variation in the pattern of pathological reactions between different species (Table 2.1). There is also an associated spectrum of resistance to infection. Guinea pigs, ferrets, possums and badgers are innately susceptible to TB, while humans, rabbits, mice, cattle and deer express varying levels of resistance, depending on their genotype.

Studies in laboratory animals such as guinea pigs and rabbits have significantly enhanced the understanding of the etiology and pathogenesis of TB. Humans, cattle, deer, guinea pigs and rabbits have similar pathology, however differ in susceptibility to TB. As compared to humans and ruminants that are relatively resistant, fewer than five virulent organisms introduced by the aerosol route into guinea pigs consistently produce lung lesions, bacteremia and fatal disease. Studies in guinea pigs and rabbits have made an important contribution in the understanding relating to the virulence and pathogenesis of TB, but they have limited use in the study of the protective immune response (Smith and Wiegeshaus, 1989).

Characteristics of a model for TB with respect to infection and pathogenesis:

- Experimental infection mimics natural disease
- Infection results from low-dose challenge
- Route of exposure simulates natural exposure
- Pathology present in relevant target organs
- Lesions analogous to those found in naturally infected host
- Spectrum of disease equivalent to that in the naturally susceptible host

PATHOGEN DIVERSITY: CROSSING SPECIES BARRIERS

In 1865, Villemin proved that TB was an infectious disease by inoculating laboratory rabbits with mycobacterium isolated from infected humans and cattle, and in 1882, Robert Koch defined the etiology of TB using pure cultures in experimental animal models, particularly the highly susceptible guinea pigs. At that time, Koch regarded mycobacterial isolates from humans and cattle as interchangeable, but after

TABLE 2.1 Pathogenesis Found in Different Animal Species Infected with Virulent *M. bovis*

		Pathology				
Animal Species	Predominant Site for Lesions	Caseation Fibrosis	Langbans Giant Cells	Acid Fast Bacilli	DTH	Antibody (IgG)
Guinea Pig	Lung, liver, lymph node, spleen	++	–	+	++++	+
Ferret	Lung, liver, lymph node, spleen, kidney	–	–	+++	–	++
Rabbit	Lung, liver, lymph node, spleen, kidney	+	++	+	+	++
Mouse	Lung, liver, spleen, lymph node	–	–	++	+/–	+/–
Possum	Lung, liver, spleen, kidney, lymph node	–	–	+++	+/–	+/–
Badger	Lung, kidney, lymph node	–	–	+++	–	+/–
Cattle	Lung, lymph node	++	+++	+	++	+/–
Deer	Lung, lymph node	++	+++	++	++	++
Human	Lung, lymph node	++	+++	++	+++	++

following Theobald Smith's distinction of *M. bovis* from *M. tuberculosis* in 1896, Koch argued strongly that the bovine tubercle bacillus presented minimal health risk to humans. Emil von Behring, a junior colleague of Koch, took a different view, and advocated vaccination with the human tubercle bacillus, which he considered to present minimal health risk to cattle. The host-restriction of mycobacterial strains remains imprecise. "*M. tuberculosis*" is generally associated with human disease, but can be found in cattle; "*M. bovis*" is generally associated with animal disease, but can be found in humans as well. This increased precision was due to the use of genetic analysis (Hershberg et al., 2008).

Now we recognize seven major lineages as members of the "*M. tuberculosis* complex"; six *M. tuberculosis* lineages differentially distributed amongst different human populations and one *M. bovis* lineage that includes multiple "ecotypes" differentially distributed amongst different mammalian species (Smith et al., 2006). Some of the major questions which are still to be answered include whether these distributions are simply the products of history and geography, or did each of the variants uniquely adapt to different species or ethnic groups? Should we consider pathogen genotype when selecting host–pathogen combinations for experimental models? Major differences in disease progression were seen with different isolates in animal models. They certainly warrant further analysis and subsequently von Behring observed that repeated passage (in culture and in animals) has a significant and differential impact on the ability of isolates to cause a disease in different hosts.

HOST DIVERSITY: FUNDAMENTAL PROCESSES AND FINE TUNING

There are distinct differences in the pathological manifestations of TB in different mammalian species in terms of the patterns of cellular aggregates (more commonly known as granulomas) surrounding infected foci (Basaraba, 2008). Although human TB is associated with a diverse range of lesion types, caseous necrosis – comprising a well-structured ring of lymphocytes surrounding the remnants of dead cells – is regarded as the hallmark of human pathology. Physiological roles that have been assigned to these lesions include a positive contribution to containment of the infection, as well as detrimental contributions associated with sequestration of the bacteria from drugs and provision of a hypoxic microenvironment that produces a non-replicating, drug-tolerant state of bacterial persistence (Via et al., 2008). Breakdown of caseous necrotic lesions results in cavities that are capable of supporting the extensive bacterial replication required for subsequent transmission. In contrast, TB in the lungs of mice is generally associated with more diffuse cellular infiltration, lacking the structural organization and the necrotic foci that are the characteristics of the human disease. Caseous necrosis is seen in guinea pigs and in rabbits, while TB in cynomolgus macaques produces the full repertoire of lesions similar to human beings (Lin et al., 2006). This difference in disease pathology is central to the argument that the mouse is not a good model of TB. Although mice provide a model to explore the general effect of drugs on bacterial growth during infection, they are inevitably unreliable in the case of drugs customized for activity against lesion-specific bacterial sub-populations.

Approaching the issue of host diversity in terms of immunology lends a different perspective. The extensive range of immune reagents and recombinant inbred strains has allowed very detailed analysis of the immune response to mycobacterial infection and vaccination in mice (North and Jung, 2004). In naïve mice there is an initial period during which cells of the innate immune system – macrophages, neutrophils and NK cells are engaged but largely fail to contain the infection. A few weeks after infection, recruitment of antigen-specific T cells and stimulation antimicrobial function of macrophages restricts further bacterial growth but fails to eliminate the existing infection. Chronic infection persists, leading to death a year or so later. By establishing a population of primed T cells, vaccination has the effect of initiating and accelerating the involvement and engagement of the adaptive immune response, thus lowering the bacterial "set point" in the chronic phase; as a result delaying, but not preventing, death. IFN-γ and TNF-α are crucial mediators in the containment of infection.

Although opportunities for detailed immunological analysis are limited in other species, the basic features of the mouse model would appear to hold true for all mammals, including humans. For control of human TB, the critical role of T cells is evident from the dramatic increase in risk of disease for individuals co-infected with HIV; the role of TNF- α is shown by TB reactivation during anti- TNF- α therapy and the central role of IFN-γ was demonstrated by hyper-susceptibility to mycobacterial infection in individuals with rare mutations which affect IFN-γ signaling. This last example is interesting in that the IFN-γ effect is generally not reproduced in the common tissue culture model using monocyte-derived macrophages from human peripheral blood.

A general conclusion is that mice provide a robust and predictive model for studying fundamental features of the immune response to TB, a conclusion that is integral to the triage system that has led to selection of the front-line TB vaccine candidates currently moving into clinical trials. However, it is important to consider whether effective vaccination against reinfection/reactivation of TB in adults will require modification of core immune mechanisms, or whether the fine-tuning of immune regulation will perhaps require manipulation of species-specific processes that contribute to differences in lesion architecture discussed above. To address this, Apt and Kramnik (2009) put forward a compelling case that, rather than discarding the mouse and its immune opportunities, we should take advantage of inbred and recombinant genetics to select mice that recapitulate the relevant aspects of human pathology.

ANIMAL MODELS OF TB: LIMITS AND LESSONS

Artificially infected guinea pigs, mice, and rabbits have served as indispensable tools through which transmission, immunopathogenesis, tuberculin response, vaccine and antimicrobial efficacy, genetic resistance, and many other important facets of TB have been studied. Results, however, are usually not entirely reflective of TB infection and disease in humans. Substantial differences in TB susceptibility, disease patterns, and temporal course exist among species (see Table 2.2). The extent of organ involvement, immune response to aerosol or parenteral infection, and histopathology also vary considerably from species to species. In addition, a variety of clinical and laboratory strains of *M. tuberculosis* exist to infect animals experimentally, and these mycobacterial strains often differ greatly in infectivity, virulence, and immunogenicity in different animal models (Gagneux and Small, 2007). Well-defined host and pathogen variability allows researchers to control these factors, selecting those combinations needed to create animal models suited to the purpose being solved. Although infection by inhalation is the most relevant model for human infection, animal infections are also produced by parenteral inoculation. Like humans, TB in animal models also is treated with antimicrobials given orally (by gavage) or by parenteral routes.

In addition to these considerations, animal species vary based on size, laboratory space requirements, rearing costs, and ability to approximate the disease process in humans. In subsequent sections, the key features of the well-developed animal models of TB, as well as less commonly used animal species, will be discussed. Despite several important differences outlined in the sections that follow, the murine, rabbit, and guinea pig models have emerged at the forefront of TB research because

(1) Infection can occur with inhalation,

(2) Animals manifest an innate and acquired immune response,

TABLE 2.2 Common Animal Models of TB – A Comparative Study

Model	Histopathology			Relative Susceptibility to *M. tuberculosis*	Immunologic Reagents Available	Laboratory Space and Cost
	Necrosis	*Caseation*	*Cavitation*			
Mouse	Minimal	Usually not	No	Low	Extensive	Relatively small
Rabbit	Yes	Yes	Yes	Very low	Moderate	Relatively large
Guinea Pig	Yes	Yes	Infrequent	Very high	Relatively few	Moderate
Non-human primate	Yes	Yes	Yes	High	Extensive	Large

(3) Animals often initially control bacillary growth in the lung, and
(4) They ultimately succumb to the disease.

VARIOUS ANIMAL MODELS

Mouse Model

Robert Koch discovered the very first time that experimental mice can be used as an animal model for TB infection. The inoculation with *M. tuberculosis* induced lesions is similar to those seen in the case of the natural disease in humans. Subsequent work established the pattern of disease in the more resistant mouse model. The strong immune response of this model for TB infection has emerged slowly. Mice are generally more resistant models for TB infection as compared with rabbits, guinea pigs and even humans, as evidenced by their ability to tolerate relatively large bacillary numbers within their lungs and other organs without signs of illness. They develop non-caseating granulomata in response to infection, and generally manifest a chronic phase of disease that represents the immune-mediated tissue destruction on a background of slowly progressive bacterial growth, ultimately resulting in death. This persistent stage of infection might be due to the chronic exposure of TB antigens on T-cell function, signifying the role of CD4/CD8 T-cell responses in the mouse remains robust over time. Previous reports addressed that murine macrophages use Toll-like receptors on their cell surfaces to identify the mycobacterial antigens and ultimately trigger cytokine production, which are responsible for granulomatous response (Ito et al., 2007). The mouse model also possesses the T-cell independent (natural killer cell) production of IFN-γ (a cytokine), which is crucial to the host immunologic response against TB. Several available options such as inbred and genetically knockout strains of mice have contributed in the understanding of the role of many specific cytokines, cells and cell surface markers in containing bacillary growth.

The mouse models were successfully infected through a pulmonary route using aerosolized microorganisms by means of nose-only exposure chambers to easily attain an easy infection with relatively low doses (~50 colony-forming units [CFUs]) of *M. tuberculosis*. The growth of *M. tuberculosis* is logarithmic, and then plateaus after around 10^6 organisms have arrived in the lungs. Cell-mediated immunity (CMI) develops during the first 4 weeks. The plateau phase is an indicator of the persistent stage of infection, in which the TB antigen can be more metabolically quiescent within macrophages. As reported earlier, this model is not capable of replicating paucibacillary, latent human TB infection. TB-induced mice presented little necrosis before the last stages of the disease, and the animals developed only weaker sensitivity to tuberculin. Mice, that inhaled the virulent bacilli, died of a successive enlargement of pulmonary granulomatous tubercles. In addition, the virulence of a given strain of *M. tuberculosis* depends on several factors, indicating route of infection, the manner of preparation of the suspension, and the dispersion as well as size of the suspension. Earlier studies reported that the high-dose intravenous challenge model in mice (which have low protein levels) showed T-cell defects with a loss of control of virulent infection and impaired granuloma formation. But, it was also observed that mouse models could recover from TB infection, if they are allowed feeding with an adequate diet. The loss of TB resistance in this model was mainly due to the decreased nitric oxide production by activated macrophages that occurred secondary to an IFN-γ defect, especially in malnourished animals (Chan et al., 1996).

The course of disease can be also influenced by the genetic variation among various inbred strains of mice. The number of viable H37Rv bacilli in the lungs of BALB/c mice was 2 log less than their number in DBA/2 mice during the stationary phase after intravenous injection. The mice of BALB/c and C57BL/6 strains survived 2 times longer than DBA/2 and C3H/HeJ strains. The knockout mice (lacking genes for acquired immune response) also showed higher numbers of bacillary titres during the stationary phase.

In Webster–Swiss male mice, the TB was induced by H37Rv strain of *M. tuberculosis* administered intravenously (McCune et al., 1966). The treatment was accomplished by using oral INH and pyrazinamide (PZA) for 12 weeks after bacterial inoculation. The drug based therapy might be successfully employed for reducing the number of bacilli in mice tissues for up to 3 months after termination of INH/PZA, to the extent that mycobacteria could not be cultured or otherwise isolated from lung or other tissue homogenates (e.g. essentially "sterilized"). If one carefully observed for longer periods of time, almost one-third of similarly treated animals spontaneously developed reactivation of TB, which is characterized by a recrudescence of the bacterial burden in their tissues. All the "apparently sterile" animals that harbor dormant or latent organisms could be reactivated at or after three months. The study concluded that appropriately timed steroids led to reactivation of TB in most of these presumably "sterile" mice, owing to residual viable organisms. Although this latent TB model is imperfect, it could be an important adjunct to the conventional chronic TB disease model. Other researchers have manipulated a number of factors – such as the dose/duration of antibiotics, the time interval between antibiotic use and immunosuppression, and type of immunosuppression – to map specific cellular and cytokine mechanisms operative in the reactivation process.

The mouse is a very cost-effective model for evaluation of drugs and it was more accurate as a measure of drug efficacy than other models, such as the guinea pig. Before entering the clinical phase, the anti- tuberculous drug has to undergo *in vitro* testing to measure its growth inhibitory potential, followed by pharmacological activity for sterilizing activity in animals. The experimental studies investigated the sterilizing activity of effective drugs, as well as the efficacy of shorter treatment durations using combinations of new and existing drugs. Recent studies concluded that various effective regimens can substitute the existing moiety such as PA-824 for INH and rifampin, or rifapentine as an alternate of RIF. Rifapentine presented longer half-life and finally greater area under the curve (AUC), especially when administered daily or thrice weekly. Lenaerts et al., 2008, evaluated a newer moiety within a subcategory of quinolones, the 2-pyridones. This compound was found to be more effective than INH, but not superior to moxifloxacin.

Mice are a particularly good option for studying the immunology of mycobacterial infections and have contributed significantly to understanding the roles of various immunological mechanisms of resistance. From an experimental standpoint, mice are generally easier to maintain in BSL3 facilities and suggest an affordable, high-yield means to study vaccines, antitubercular drugs, immune mechanisms, host genetics and the contribution of host and pathogen strain differences leading to infection. The major disadvantage associated with the mouse model is the development of resistance for classic TB disease; as a result mice tend to be non-respondent to therapy.

Guinea Pig Model

Guinea pigs models have been used to create models of TB transmission due to their exceptional susceptibility to infection with a few inhaled mycobacteria. Guinea pigs are used as models in various forms of TB infection such as childhood TB and TB in immunosuppressed hosts, followed by the granulomata formation, primary and hematogenous pulmonary lesions, dissemination and caseation necrosis (McMurray, 1994). In guinea pigs, lesions developed after low-dose aerosol treatment with *M. tuberculosis* and had similarities to natural infections in humans – all this makes them attractive animal models to study bacterial persistence. Similar to humans, naive guinea pigs firstly develop primary lesions that differ in their morphology as compared to secondary lesions. Primary lesions originated after initial exposure, while secondary lesions from hematogenous dissemination developed after activation of acquired immunity (Lenaerts et al., 2007). It was previously reported that TB is an airborne infection; hence, this model can easily be experimentally induced with living air samples and airborne tubercle bacilli generated by TB patients. The infection was confirmed by both mycobacterial culture and histologic examination of various tissues namely lungs, spleens, and lymph nodes. To induce TB infection in guinea pigs, a limited number of animals were required as compared to mice, since to they have larger minute ventilation (Nardell, 1999). The course of infection firstly enters the logarithmic phase of bacillary multiplication in the lungs over 2 to 4 weeks. After that, it enters a stationary phase in lungs, whereas other organs show hematogenous dissemination.

The guinea pig model is the worst among the above-mentioned animal models in the context of the availability of immunologic reagents for studying guinea pig host immune responses. Moreover, a number of reagents are available to study cytokines and other inflammatory cells involved in pathogen recognition. Recent research has explored new avenues in the cloning of guinea pig cytokine and chemokine genes and their expression in recombinant guinea pigs as well as reagents which are used in the study of the response of guinea pigs to infection with virulent *M. tuberculosis* (Jeevan et al., 2006; Ly et al., 2007). (Figure 2.1)

Rabbit Model

Rabbits show higher resistance with *M. tuberculosis* infection and occupy an important position, as humans are also relatively resistant. Only 10% of infected individuals allow progression of their TB infection. The rabbit-based model enables us to understand the pathology of TB infection by *M. bovis* infection. However, previous experiments demonstrated that rabbits are generally resistant to infection with airborne *M. tuberculosis*. The pulmonary infections with *M. tuberculosis* formed the cavity, which ultimately may regress and heal. The rabbits are infected with bovine mycobacterial infection, i.e. *M. bovis*; their pulmonary pathology to inhaled bovine tubercle infection is more similar to human *M. tuberculosis* infection than those recorded in the case of other models, such as mice and guinea pigs. In rabbits, the pulmonary cavity is first developed, followed by the bronchial spread of microorganisms. Subsequently, it was presented that delayed type hypersensitivity (DTH) and CMI are major contributory factors for developing the cavity in the rabbit-based model. The cavity formation is mainly due to pulmonary cavities, which possess large populations of bacilli reaching to the bronchial tree, and also due to the degree of sputum culture positivity, showing high bacillary burden. Rabbits, which are known to produce cavitation, provide major perspectives in the study of disease transmission as compared to other animal models. Rabbits are also employed as models for studying the latent or paucibacillary TB states in humans. Rabbits easily get the paucibacillary state through their own immune systems' involvement in the control of infection. Moreover, the experimentally immunosuppressed animals achieve the latent stage of infection easily. However, the rabbit model lacks the immunological reagents. Rabbits are more costly as compared to mice and guinea pigs and require a large laboratory space.

Non-Human Primate Model

The non-human primates are also successfully employed as models for latent TB infection. This latent state is metastable

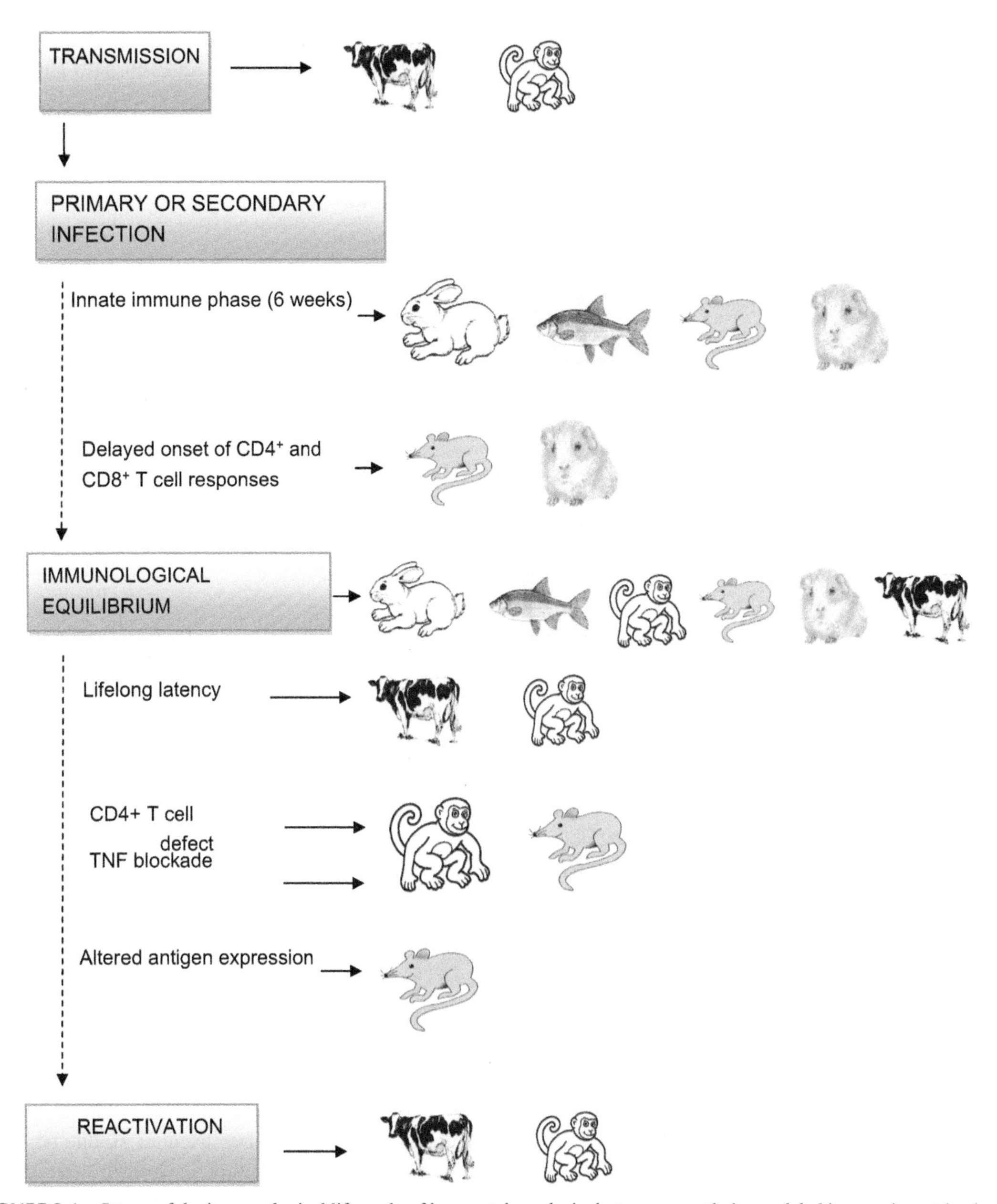

FIGURE 2.1 Stages of the immunological life cycle of human tuberculosis that can currently be modeled in experimental animals.

and can be reactivated easily when the animals are stressed, or possibly even without any exogenous immunosuppression administration. The main reason may be because they are closely related to humans, but owing to their higher cost, they have not been utilized on a large level. The other animal models, however, fail to mimic the human disease entirely; therefore research has been shifted and focused on the non-human primate model for TB infection. The Philippine cynomolgus monkey (*Macaca fasicularis)* developed acute symptoms for rapidly progressive, highly fatal multilobar pneumonia, when administered a high dose of *M. tuberculosis* (10^5 or 10^4 CFU) intratracheally. On the other hand, a lower dose of *M. tuberculosis* (10^3 CFU) caused a chronic, slowly progressive, and localized form of pulmonary TB. The significant proportion of monkeys might be challenged with 10^2 or 10^1 CFU to induce the infection in a subclinical state. In non-human primates, the cynomolgus macaque has been effectively used to simulate TB infection and has also been infected with low-dose aerosol concentrations (~25 CFUs). They manifested dormant infection and reactivation as it appears in humans. Capuano et al., 2003 concluded that a cohort of macaques might be challenged with low-dose aerosol. Indications for TB infection were provided by tuberculin skin test or lymphocyte proliferation assays to PPD; however, only 60% subsequently developed active TB. This model may be an attractive option to study

TABLE 2.3 Summary of Current Experimental Models of TB and the Particular Aspects Associated with These Models

Model	Pathology	Immunology	Genetics	Drugs	Vaccines
Zebra fish	Excellent for imaging of early pathology	Good for innate immunity	Good for making mutants		
Mice	Loosely organized granulomas	Extensive range of immunology reagents and recombinant animals	Diverse genetic backgrounds, wide range of mutants	Routinely used model, works for current drugs	Routinely used model
Guinea pigs	Well-structured granulomas, caseous necrosis	Limited range of immunology reagents		Routinely used model	Routinely used model, large "window" for BCG
Rabbits	Well-structured granulomas, caseous necrosis			Useful for assessing lesion specific Activities	
Cattle		Moderate range of immunology reagents	Well-defined lineages (Bos taurus, Bos indicus), cross-breeds and inbred herds		Experimental challenge and natural transmission models
Non-human Primates	Range of lesions parallel those in humans	Most human immunology reagents can be used		Expensive; use for proof-of-concept	Mimics human response (highly diverse); use for proof-of-concept

the interaction between simian viral immunodeficiency (SIV) and TB as a model for human HIV/TB co-infection. Moreover, their antigens cross-react with immunologic reagents developed for human cells and tissue as well as macaque-specific reagents and allow for immunohistochemical examination to understand the mechanisms of disease. However, non-human primates are restrictive due to higher cost, handling difficulties, and space requirements with BSL3 facilities, as well as higher susceptibility to TB infection and capable to horizontally transmit the disease.

Cattle Model

In the cattle model, the *M. bovis* based infections have been successfully established and used to understand the molecular mechanisms of TB infection. In context with the pathology, bovine TB appears to be similar to human TB in regard to granulomatous reactions and CMI, but differs in regard to cavitation. Cattle are a natural host for TB and the disease induces comparable pathological and immune responses to those that are seen in humans. The infection in cattle is mainly localized to the respiratory tract and clinical disease may take years to develop. Several immunological agents are also available to study infection in this species. The human *M. tuberculosis* based infection was diagnosed by new IFN-γ release assays; originally it was developed to diagnose TB in cattle (Vordermeier et al., 1999). Moreover, the cattle model is an eminent option for the secondary screening of TB vaccines, as there is more similarity between the disease in cattle and humans, and also outbred animals could be used (Table 2.3).

ETHICAL ISSUES

Infectious diseases such as TB continue to cause substantial morbidity and mortality. Continued research is critical to finding safe and effective ways to prevent and treat infectious diseases. The challenge experiment is an important method that is sometimes used to study the pathogenesis of infectious diseases and, especially, to evaluate initial efficacy of vaccines before large-scale field tests are conducted. In challenge experiments, infections are deliberately induced under carefully controlled and monitored conditions, usually in inpatient settings. Research volunteers are exposed to bacteria, viruses, or parasites. Induced infections are usually either self-limiting or can be fully treated within a relatively short period of time.

Experiments conducted by physician investigators designed to cause infections that have uncomfortable symptoms in human subjects are likely to evoke serious moral concern. The Three Rs principle (Replacement of animal experiments with alternative approaches, Reduction of animal numbers and Refinement to improve animal welfare) emerged as a way for scientists to ease this dilemma by developing research methods that decrease pain and distress. Nevertheless, the use of animals in research is still

TABLE 2.4 Potential Causes of Pain and Distress in Studies on Experimental Infection with *M. tuberculosis*

Infection Route	
Intratracheal instillation	Surgical procedure under general anesthesia: Inoculum delivery through an incision in the trachea, that heals in 2–3 days.
Intraperitoneal injection	This injection method offers no possibility to visually confirm correct delivery, and accidental penetration of the bladder, intestine, muscular or fatty tissue may occur.
Treatment Administration	
Intraperitoneal injection	This injection method offers no possibility to visually confirm correct delivery, and accidental penetration of the bladder, intestine, muscular or fatty tissue may occur.
Repeated oral gavage	Difficult procedure with risk of fluid aspiration by the lungs or perforation of oesophagic or gastric wall. Irritation, swelling and ulceration of the oesophagus from repeated dosing. Unexpected deaths as well as inappetence and weight loss reported in experimental infection studies. Reports of increased TB susceptibility due to gavage-induced stress.
Immunization	
Footpad immunization	Immune reaction to antigen, causing swelling and inflammation *in situ*, potentially causing pain and lameness.
Intramuscular immunization	Painful injection that may cause mechanical trauma and potential nerve damage; immune reaction may lead to painful swelling.
Health Status	
Signs of disease	Respiratory distress, hunched posture, lack of grooming; failure to eat or drink, fever, severe cachexia. Increasingly severe clinical signs, progressing to a hypokinetic irresponsive ("moribund") state, culminating in death.

controversial, with recent voices also questioning the translational validity into humans.

Franco et al., 2012 identified the main sources of animal distress and assessed the possible implementation of refinement measures in experimental infection research, using mouse models of TB as a case study. Table 2.4 briefly describes the most relevant experimental procedures with an impact on animal welfare, as well as the main welfare issues raised by the manifestation of active disease.

Literature published between 1997 and 2009 was analyzed, focusing on the welfare impact on animals used and the implementation of refinement measures to reduce this impact. The number of articles per year increased almost fivefold between 1997 and 2009. Regarding genetic status of the animals, the majority of studies (71% overall for all years) used non-genetically modified inbred strains. Information on the sex of animals used was not available in 34% of articles. The proportion of articles reporting the induction of experimental *M. tuberculosis* infection through aerosol exposure rose significantly between 1997 and 2009. The use of the intravenous route – originally the most recurrent method – decreased, whereas the use of the intratracheal route remained relatively stable throughout the analysis period. The intraperitoneal route was the least chosen.

Information on important research parameters, such as methods for euthanasia or sex of the animals, was absent in a substantial number of papers. Eighty percent of the articles omitted this information, with no significant differences between years. Moreover, when information on euthanasia was given, it was often incomplete and therefore difficult to interpret. For example, anesthetic overdose was often reported without indicating the route, compound or dose, and exsanguination was frequently referred to with no indication whether it was under anesthesia or not.

In this 12-year period, it was observed that a rise in reports of ethical approval of experiments had taken place. The proportion of studies classified into the most severe category did however not change significantly over the studied period. The majority of the studies analyzed were terminated before infected animals reached very severe morbidity, while the remaining mice were allowed to reach terminal stages, with no significant variation of this proportion across years.

Overall, the study showed that progress has been made in the application of humane endpoints in TB research, but that a considerable potential for improvement remains. Of course, such measures should not be taken at the expense of research quality and relevance. The best way to avoid a conflict between ethical constraints and scientific motivations is probably for scientists to be proactive and initiate a critical discussion within their own field, rather than awaiting limitations imposed from outside. In the field of experimental studies of important infections such as TB, a reassessment of the need for such a large proportion of studies to involve end-stages of the disease seems particularly pertinent.

TRANSLATIONAL SIGNIFICANCE

Animal models have and will continue to aid in early discovery as well as the pre-clinical testing phase of new drugs for efficacy and toxicity. The goal in modeling TB in animals is to mimic as closely as possible the pathology and clinical progression of the naturally occurring disease.

For practical or economic reasons, some species are more widely used for efficacy studies, while others are preferred for pharmacokinetic and toxicity studies. For example, despite the documented differences in the immune response between mice and humans, mice are still the most widely used animal model for studying the immunological responses to *M. tuberculosis* infection and TB vaccines. However, due to species-specific differences in disease progression and lesion morphology, responses to drug therapy in mice may or may not reflect the desired effects in people.

The differences in lesion morphology among the different animal species infected with *M. tuberculosis* provide different levels of stringency for testing new drugs. Mouse strains that develop only solid lesions are best suited for discovery and early testing of drugs for *in vivo* effects and toxicity. Certain highly susceptible mouse strains have the added benefit of not only developing necrotic lesions but also having a more rapid disease progression, thus shortening the *in vivo* testing intervals. Species such as guinea pigs and cotton rats provide a wider variety of lesion types that include necrotic and mineralized lesions for a higher level of *in vivo* testing stringency and to test adjunct therapies against novel therapeutic targets. The non-human primate and rabbit models develop an even wider variety of lesion types and are the most appropriate models to test drugs specially designed to treat cavitary lesions.

Since experimental *M. tuberculosis* infections are progressive in the majority of model species, all are suitable for testing the effects of drugs on extra-pulmonary lesions.

WORLD WIDE WEB RESOURCES

The sixteenth global report on tuberculosis (TB) published in 2011 by WHO is in a series that started in 1997. It provides a comprehensive and up-to-date assessment of the TB epidemic and progress in implementing and financing TB prevention, care, and control at global, regional, and country levels using data reported by 204 countries and territories that account for over 99% of the world's TB cases (Global Tuberculosis Control 2011: http://www.who.int/tb/publications/global_report/2011/en/).

The official website of the Nobel Prize can be assessed for exploring the work of Emil von Behring who got the Nobel Prize in Physiology or Medicine in 1901 for his work on serum therapy, especially its application against diphtheria, by which he opened a new road in the domain of medical science and thereby placed in the hands of the physician a victorious weapon against illness and death (http:// nobelprize.org/nobel_prizes/medicine/ laureates/1901/behring-lecture.html).

The Laboratory Biosafety Manual published by WHO provides practical guidance on biosafety techniques for use in laboratories at all levels. The manual covers risk assessment and safe use of recombinant DNA technology, and provides guidelines for the commissioning and certification of laboratories. Laboratory biosecurity concepts are introduced in the manual, and the latest regulations for the transport of infectious substances are reflected. Material on safety in health-care laboratories, previously published elsewhere by WHO, has also been incorporated (http://www.who.int/csr/resources/publications/biosafety/WHO_CDS_CSR_LYO_2004_11/en/). Furthermore, WHO also provides Tuberculosis Laboratory Biosafety Manual (http://apps.who.int/iris/bitstream/10665/77949/1/9789241504638_eng.pdf) where the recommendations are based on assessments of the risks associated with different technical procedures performed in different types of TB laboratories; the manual describes the basic requirements for facilities and practices, which can be adapted to follow local or national regulations or as the result of a risk assessment.

PROTOCOLS

This section of the chapter deals with describing the established protocols for infected animal models of TB, subsequent processing for analysis, and studying various aspects of biosafety that must be observed. All the protocols are illustrated by means of flow charts.

Preparing *M. tuberculosis* Inoculum for Aerosol Exposure

From the known concentration of *M. tuberculosis* stock solution, dilutions are prepared to infect mice or pigs by aerosols. To verify the amount of inoculums used in the nebulizer, a diluted aliquot of bacterial suspension is plated on 7H11 agar. Care should be taken during serial dilutions to prevent the generation of aerosol. All pipette tips, spills or drops should be disinfected with 5 % Lysol (Flow Chart 2.1).

Aerosol Infection of Mice Using the Middlebrook Apparatus

The Middlebrook airborne infection apparatus is the most widely used aerosol generation device. It is used to establish an animal model of pulmonary tuberculosis to resemble normal route and site of infection in humans. This instrument

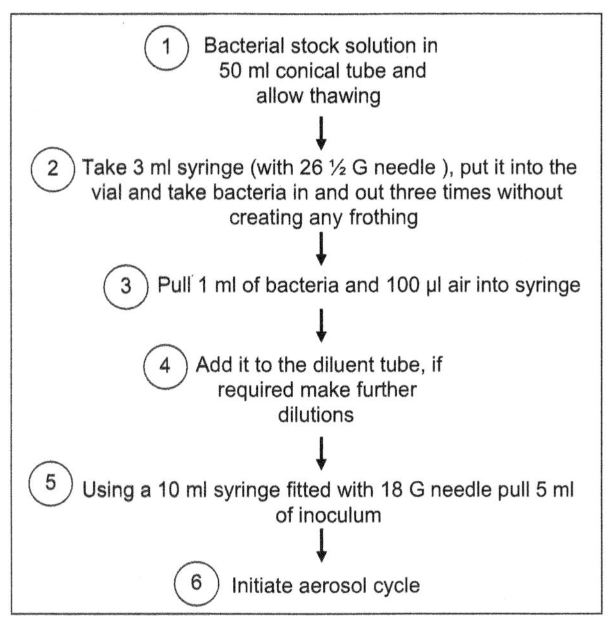

1 Bacterial stock solution in 50 ml conical tube and allow thawing
2 Take 3 ml syringe (with 26 ½ G needle), put it into the vial and take bacteria in and out three times without creating any frothing
3 Pull 1 ml of bacteria and 100 µl air into syringe
4 Add it to the diluent tube, if required make further dilutions
5 Using a 10 ml syringe fitted with 18 G needle pull 5 ml of inoculum
6 Initiate aerosol cycle

FLOW CHART 2.1

consists of an aerosol chamber that contains a basket or cage with five compartments that can accommodate 25 mice. Compressed air flows through the nebulizer and produces a fine mist of bacterial suspension carried into the aerosol chamber, which is inhaled by the animals. Finally, a UV lamp is used to decontaminate the surface (Flow Chart 2.2, Figure 2.2).

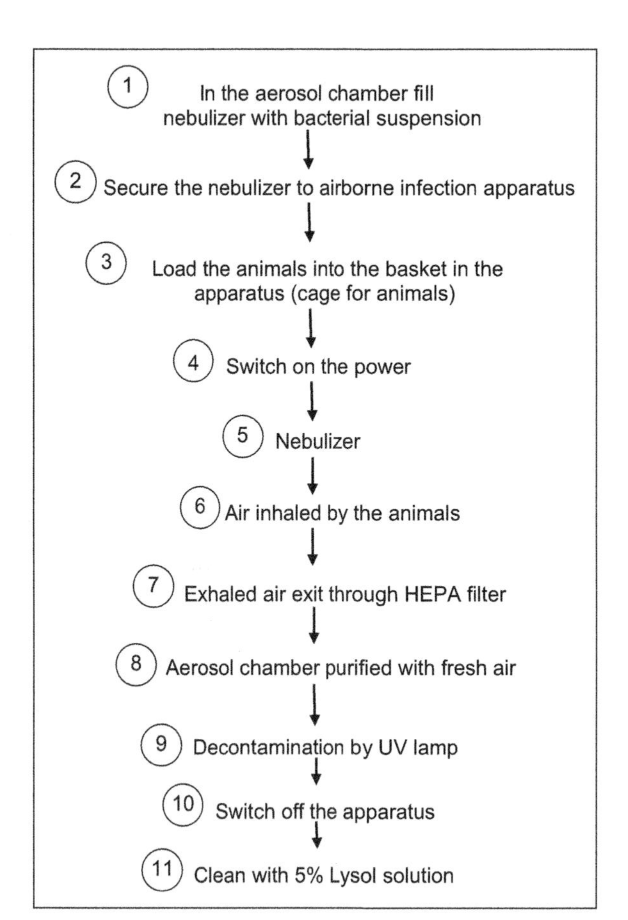

1 In the aerosol chamber fill nebulizer with bacterial suspension
2 Secure the nebulizer to airborne infection apparatus
3 Load the animals into the basket in the apparatus (cage for animals)
4 Switch on the power
5 Nebulizer
6 Air inhaled by the animals
7 Exhaled air exit through HEPA filter
8 Aerosol chamber purified with fresh air
9 Decontamination by UV lamp
10 Switch off the apparatus
11 Clean with 5% Lysol solution

FLOW CHART 2.2

Aerosol Infection of Guinea Pigs Using a Madison Chamber

A Madison chamber is an aerosol generation device used for guinea pigs because of their larger size. (Figures 2.3 and 2.4) The surfaces of the Biosafety Class II Cabinet should be sterilized with 5% Lysol and 70% ethanol.

A test run to ensure adequate pressure driving is shown in Flow Chart 2.3.

Bacterial Loading

Before bacterial loading, each of the corresponding guinea pig cages should be labeled by writing "aerosolized with *M. tuberculosis* on date/initials." A stainless steel container filled with 5% Lysol solution should also be placed next to the infection chamber (Flow Chart 2.4).

Intravenous Infection of Mice with *M. Tuberculosis*

Intravenous infection requires absolute concentration, and good eyesight with no distraction. The mouse is placed in a restraint device so that the tail can be immobilized and then the injectate is injected into the lateral vein (Flow Chart 2.5).

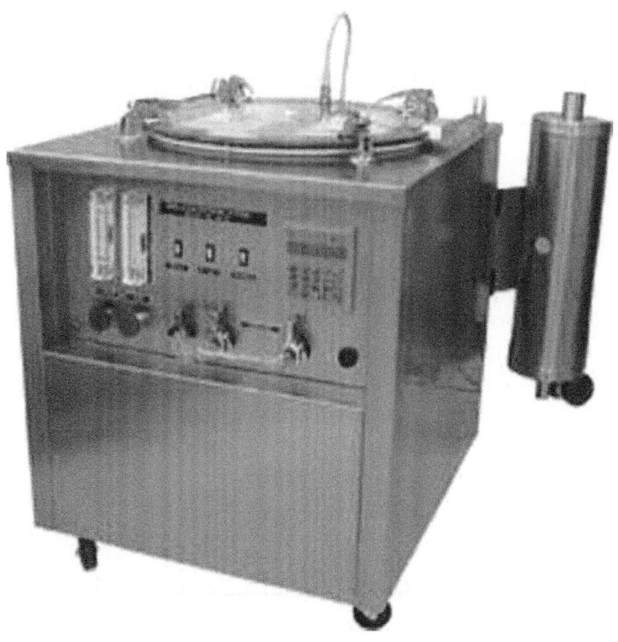

FIGURE 2.2 Glass-Col apparatus. This instrument is the advancement of earlier models developed by Middlebrook. 80 to 100 mice can be exposed to aerosol at one time.

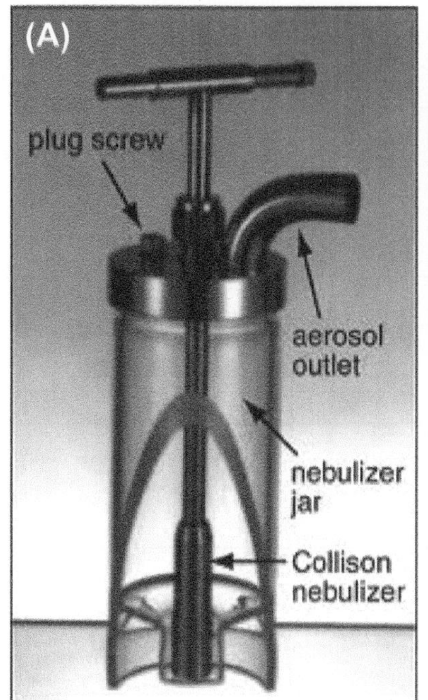

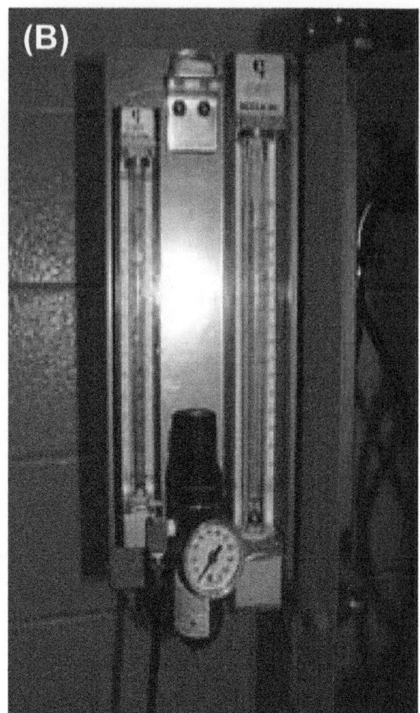

FIGURE 2.3 (A) Collison nebulizer unit, complete with the surrounding glass nebulizer jar, for the Madison guinea pig aerosol chamber, (B) Photohelic meter for Madison chamber.

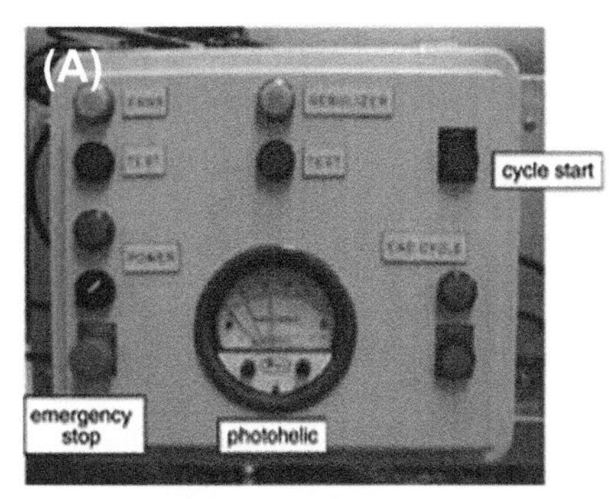

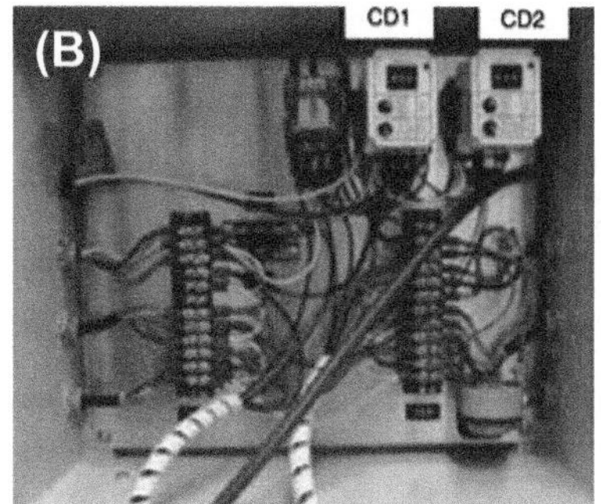

FIGURE 2.4 The exterior (A) and interior (B) of the Madison chamber control box.

Isolation of Samples for Determining *M. tuberculosis* Load by RT-PCR

Expression of the multiple genes during infection of *M. tuberculosis* can be measured by isolation of total RNA and then running a PCR assay. During the PCR assay, contamination of test samples with RNAase and extraneous sources of DNA should be prevented. Samples containing RNA and cDNA should be dealt in separate working areas. The protocol describes sample collection from the infected tissues to be processed for the PCR analysis (Flow Chart 2.6).

Determination of Bacterial Loads in Target Organs

Processing of infected animals should be done in a biosafety cabinet. All the necessary items should be kept in the biosafety hood without restricting the airflow. Animals should be euthanized according to the guidelines provided by the IACUC. Due to the chronic nature of the infection, the design of the study depends on the time course to be studied. Number of animals per time point should also be considered, which can be determined through statistical power calculation. For the bacterial load curve, a power of > 0.8 can be achieved using four to five mice (Flow Chart 2.7).

The process includes:

1. Preparation of agar plates
2. Bacterial count set-up
3. Necropsy
4. Homogenization of tissues
5. Plating
6. Bacterial count colonies

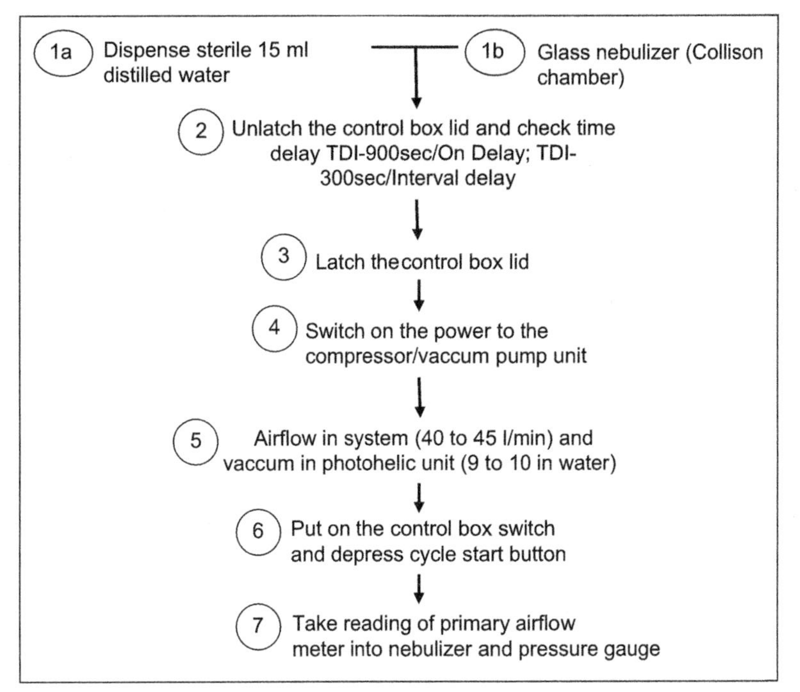

FLOW CHART 2.3

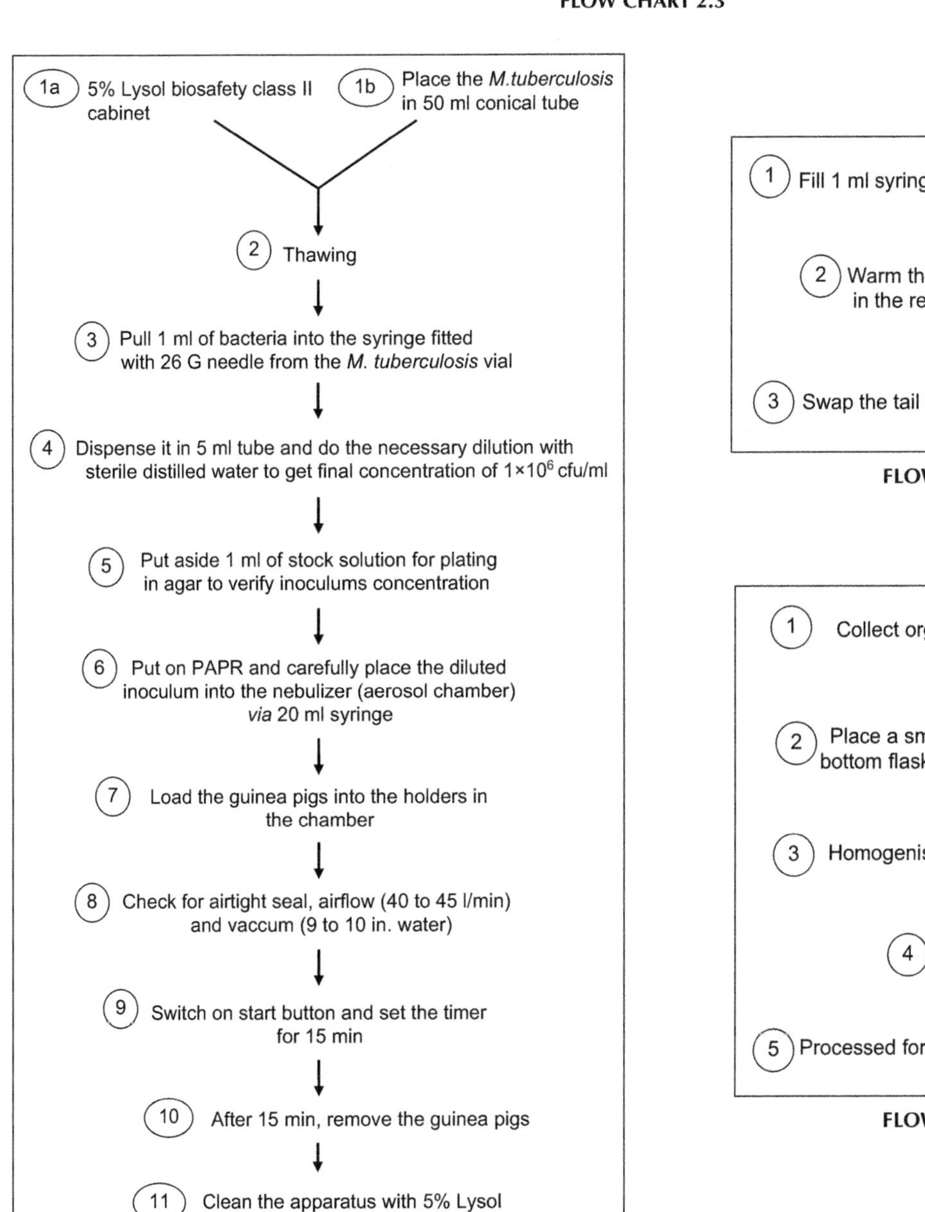

1a 5% Lysol biosafety class II cabinet
1b Place the *M.tuberculosis* in 50 ml conical tube
2 Thawing
3 Pull 1 ml of bacteria into the syringe fitted with 26 G needle from the *M. tuberculosis* vial
4 Dispense it in 5 ml tube and do the necessary dilution with sterile distilled water to get final concentration of 1×10^6 cfu/ml
5 Put aside 1 ml of stock solution for plating in agar to verify inoculums concentration
6 Put on PAPR and carefully place the diluted inoculum into the nebulizer (aerosol chamber) *via* 20 ml syringe
7 Load the guinea pigs into the holders in the chamber
8 Check for airtight seal, airflow (40 to 45 l/min) and vaccum (9 to 10 in. water)
9 Switch on start button and set the timer for 15 min
10 After 15 min, remove the guinea pigs
11 Clean the apparatus with 5% Lysol

FLOW CHART 2.4

1 Fill 1 ml syringe with the injectate without air bubbles
2 Warm the tail of the mouse kept in the restraint device with the lamp
3 Swap the tail with 70% ethanol and inject the injectate

FLOW CHART 2.5

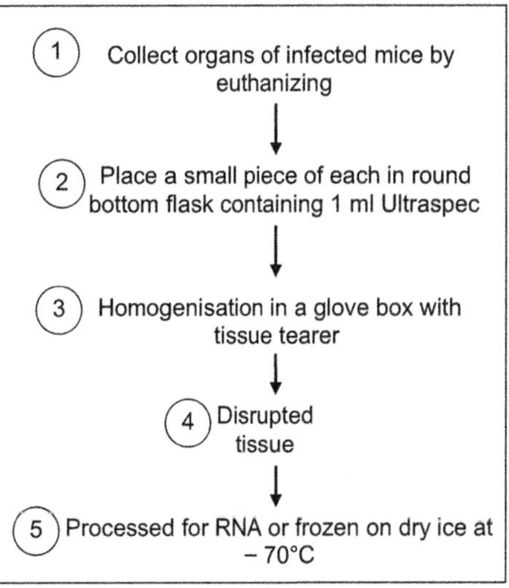

1 Collect organs of infected mice by euthanizing
2 Place a small piece of each in round bottom flask containing 1 ml Ultraspec
3 Homogenisation in a glove box with tissue tearer
4 Disrupted tissue
5 Processed for RNA or frozen on dry ice at − 70°C

FLOW CHART 2.6

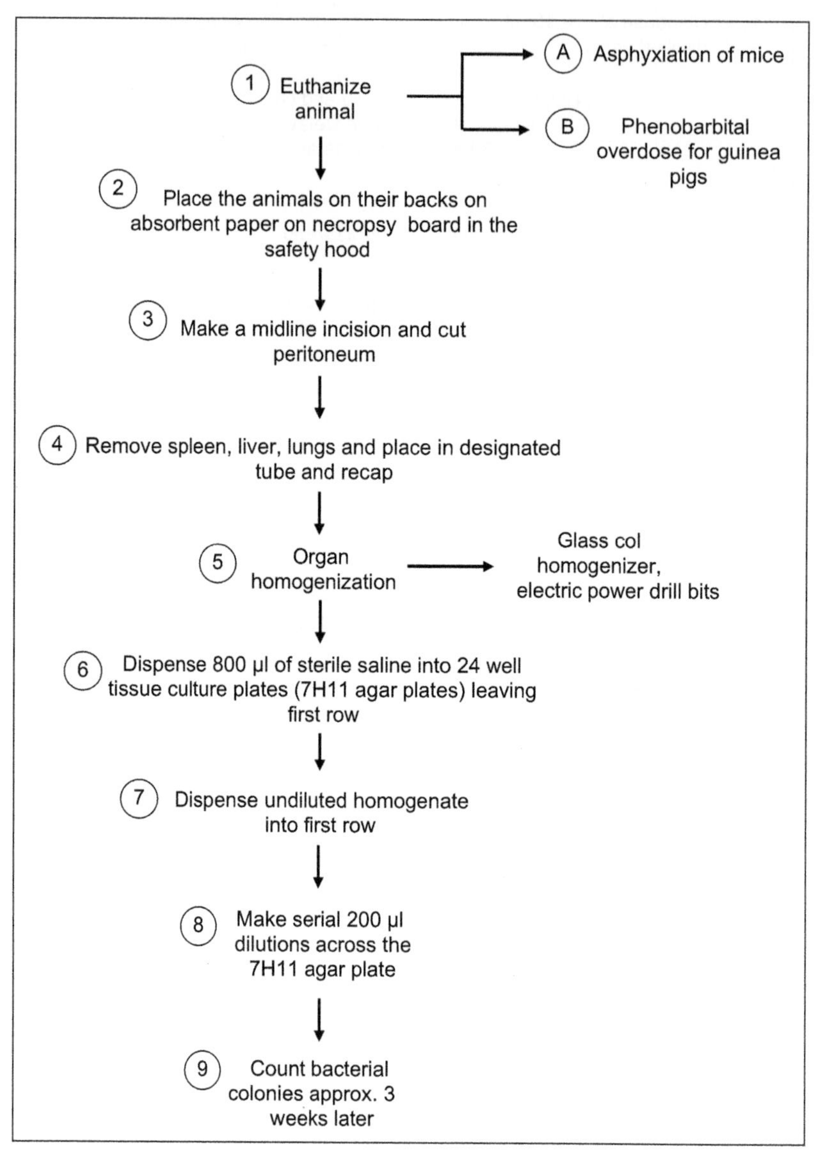

FLOW CHART 2.7

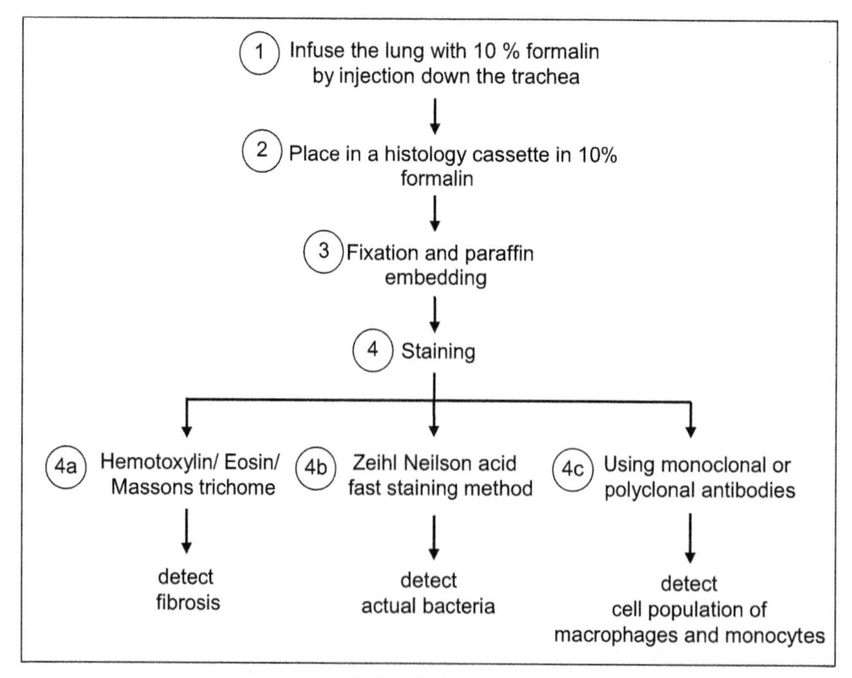

FLOW CHART 2.8

Preparation of Lungs or Other Tissues for Histology

Useful information about the host response can be acquired by the histological examination of organs from infected animals. The size of granulomas, their make-up in terms of lymphocytes and macrophages, the degree of lung tissue they are consolidating, the development of necrosis, and so forth, can be invaluable information. It is important that the specimens are frozen as soon as they are collected to preserve their morphology and integrity of the antigens. Sections are stained with hematoxylin and eosin after fixation and paraffin embedding (Flow Chart 2.8).

Preparation of Lung Cell Suspension

Cells suspensions from naïve and *M. tuberculosis* infected mice are used for single cell analysis studies by flow cytometry by cell culture. Cells from the culture of infected animals provide information such as cell activation, proliferation and cytokines as well as chemokine production (Flow Chart 2.9).

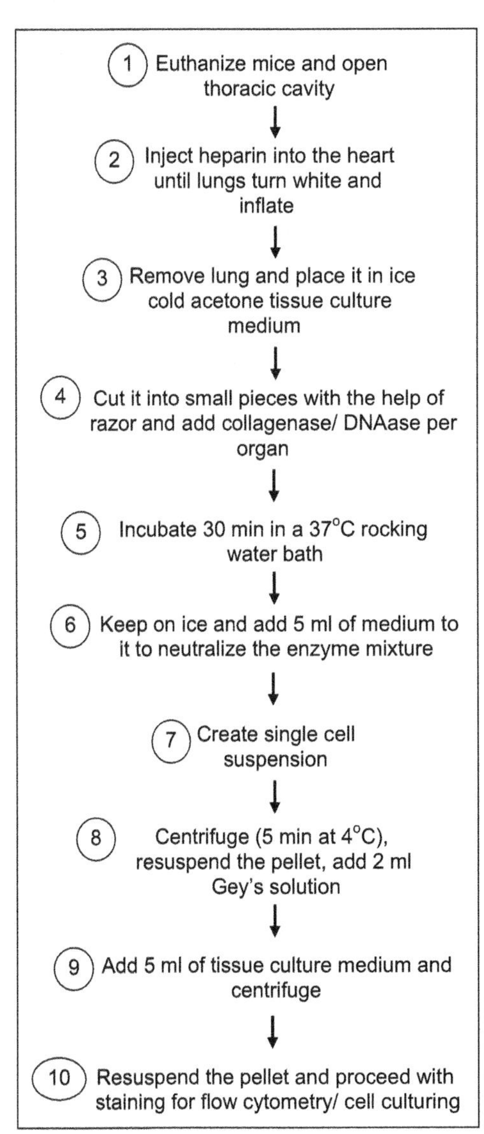

1 Euthanize mice and open thoracic cavity
2 Inject heparin into the heart until lungs turn white and inflate
3 Remove lung and place it in ice cold acetone tissue culture medium
4 Cut it into small pieces with the help of razor and add collagenase/ DNAase per organ
5 Incubate 30 min in a 37°C rocking water bath
6 Keep on ice and add 5 ml of medium to it to neutralize the enzyme mixture
7 Create single cell suspension
8 Centrifuge (5 min at 4°C), resuspend the pellet, add 2 ml Gey's solution
9 Add 5 ml of tissue culture medium and centrifuge
10 Resuspend the pellet and proceed with staining for flow cytometry/ cell culturing

FLOW CHART 2.9

REFERENCES

Apt, A., & Kramnik, I. (2009). Man and Mouse TB: Contradictions and Solutions. *Tuberculosis (Edinb.), 89*, 195–198.

Basaraba, R. J. (2008). Experimental Tuberculosis: The Role of Comparative Pathology in the Discovery of Improved Tuberculosis Treatment Strategies. *Tuberculosis (Edinb.), 88*(1), S35–S47.

Capuano, S. V., Croix, D. A., Pawar, S., Zinovik, A., Myers, A., Lin, P. L., Bissel, S., Fuhrman, C., Klein, E., & Flynn, J. L. (2003). Experimental *Mycobacterium tuberculosis* Infection of Cynomolgus Macaques Closely Resembles the Various Manifestations of Human *M. tuberculosis* Infection. *Infection Immunity, 71*(10), 5831–5844.

Chan, J., Tian, Y., & Tanaka, K. E. (1996). Effects of protein calorie malnutrition on tuberculosis in mice. *Proceedings of the National Academy of Sciences USA, 93*(25), 14857–14861.

Dube, D., Agrawal, G. P., & Vyas, S. P. (2012). Tuberculosis: From Molecular Pathogenesis to Effective Drug Carrier Design. *Drug Discovery Today, 17*, 13–14.

Franco, N. H., Correia-Neves, M., & Olsson, I. A. S. (2012). Animal Welfare in Studies on Murine Tuberculosis: Assessing Progress over a 12-Year Period and the Need for Further Improvement. *Plos One, 7*(10), e47723.

Gagneux, S., & Small, P. M. (2007). Global Phylogeography of *Mycobacterium tuberculosis* and Implications for Tuberculosis Product Development. *The Lancet Infectious Disease, 7*(5), 328–337.

Hershberg, R., Lipatov, M., Small, P. M., Sheffer, H., Niemann, S., Homolka, S., Roach, J. C., Kremer, K., Petrov, D. A., Feldman, M. W., & Gagneux, S. (2008). High Functional Diversity in *Mycobacterium tuberculosis* Driven by Genetic Drift and Human Demography. *Plos Biology, 6*(12), e311.

Ito, T., Schaller, M., Hogaboam, C. M., Standiford, T. J., Chensue, S. W., & Kunkel, S. L. (2007). TLR9 Activation is a Key Event for the Maintenance of a Mycobacterial Antigen-Elicited Pulmonary Granulomatous Response. *European Journal of Immunology, 37*(10), 2847–2855.

Jeevan, A., McFarland, C. T., Yoshimura, T., Skwor, T., Cho, H., Lasco, T., & McMurray, D. N. (2006). Production and Characterization of Guinea Pig Recombinant Gamma Interferon and its Effect on Macrophage Activation. *Infection and Immunity, 74*(1), 213–224.

Lenaerts, A. J., Bitting, C., Woolhiser, L., Gruppo, V., Marietta, K. S., Johnson, C. M., & Orme, I. M. (2008). Evaluation of a 2-pyridone, KRQ-10018, Against *Mycobacterium tuberculosis in vitro* and *in vivo*. *Antimicrobial Agents and Chemotherapy, 52*(4), 1513–1515.

Lenaerts, A. J., Hoff, D., Sahar, A., Stefan, E., Koen, A., Luis, C., Orme, I. M., & Basaraba, R. J. (2007). Location of Persisting Mycobacteria in a Guinea Pig Model of Tuberculosis Revealed by R207910. *Antimicrobial Agents and Chemotherapy, 51*(9), 3338–3345.

Lin, P. L., Pawar, S., Myers, A., Pegu, A., Fuhrman, C., Reinhart, T. A., Capuano, S. V., Klein, E., & Flynn, J. L. (2006). Early Events in *Mycobacterium tuberculosis* Infection in Cynomolgus Macaques. *Infection and Immunity, 74*(4), 3790–3803.

Ly, L. H., Russell, M. I., & McMurray, D. N. (2007). Microdissection of the Cytokine Milieu of Pulmonary Granulomas from Tuberculous Guinea Pigs. *Cell Microbiology, 9*(5), 1127–1136.

McCune, R. M., Feldmann, F. M., Lambert, H. P., & McDermott, W. (1966). Microbial Persistence. I. The Capacity of Tubercle Bacilli to Survive Sterilization in Mouse Tissues. *The Journal of Experimental Medicine, 123*(3), 445–468.

McMurray, D. N. (1994). Guinea Pig Model of Tuberculosis. In B. R. Bloom (Ed.), *Tuberculosis: Pathogenesis, Protection and Control* (1st ed., pp. 135–147). Washington, DC: American Society for Microbiology.

Nardell, E. A. (1999). Air Sampling for Tuberculosis: Homage to the Lowly Guinea Pig. *Chest, 16*(4), 1143–1145.

North, R. J., & Jung, Y. J. (2004). Immunity to Tuberculosis. *Annual Review of Immunology, 22*, 599–623.

Smith, D. W., & Wiegeshaus, E. H. (1989). What Animal Models Can Teach Us About the Pathogenesis of Tuberculosis in Humans. *Reviews of Infectious Diseases, 11*(2), S385–S393.

Smith, N. H., Kremer, K., Inwald, J., Dale, J., Driscoll, J. R., Gordon, S. V., van Soolingen, D., Hewinson, R. G., & Smith, J. M. (2006). Ecotypes of the *Mycobacterium tuberculosis* Complex. *Journal of Theortical Biology, 239*(2), 220–225.

Via, L. E., Lin, P. L., Ray, S. M., Carrillo, J., Allen, S. S., Eum, S. Y., Taylor, K., et al. (2008). Tuberculous Granulomas are Hypoxic in Guinea Pigs, Rabbits, and Nonhuman Primates. *Infection and Immunity, 76*, 2333–2340.

Vordermeier, H. M., Cockle, P. C., Whelan, A., Rhodes, S., Palmer, N., Bakker, D., & Hewinson, R. G. (1999). Development of Diagnostic Reagents to Differentiate Between *Mycobacterium bovis* BCG Vaccination and *M. bovis* Infection in Cattle. *Clinical and Diagnostic Laboratory Immunology, 6*(5), 675–682.

FURTHER READING

Bloom, B. R. (1994). Tuberculosis: Pathogenesis, Protection and Control. *American Society for Microbiology*.

Current Protocols in Microbiology, Wiley online library.

Dannenberg, A. M. (2006). *Pathogenesis of Human Pulmonary Tuberculosis. Insights from the Rabbit Model*. Washington, DC: ASM Press.

Kaufmann, S. H. E. (2008). *Helden P van, Rubin E, Britton WJ. Handbook of Tuberculosis*. Wiley-Blackwell.

WHO Tuberculosis Laboratory Biosafety Manual.

GLOSSARY

Caseous necrosis A form of cell death in which the tissue maintains a cheese-like appearance. The dead tissue appears as a soft and white proteinaceous dead cell mass.

Cavitation The formation and then immediate implosion of cavities in a liquid – i.e. small liquid-free zones ("bubbles") – that are the consequence of forces acting upon the liquid.

Ecotype A population of a species that survives as a distinct group through environmental selection and isolation and that is comparable with a taxonomic subspecies.

Gavage Forced feeding by means of a tube inserted into the stomach through the mouth.

Paribacillary Having or made up of few bacilli.

ABBREVIATIONS

AUC Area Under the Curve
BCG Bacillus Calmette–Guérin
BSL Biosafety Level
CFUs Colony-Forming Units
CMI Cell-Mediated Immunity
DTH Delayed Type Hypersensitivity
HIV Human Immunodeficiency Virus
IFN-γ Interferon-gamma
INH Isoniazid
MDR Multi Drug Resistant
NK Cells Natural Killer Cells
PPD Purified Protein Derivative
PZA Pyrazinamide
RIF Rifampicin
SIV Simian Viral Immunodeficiency
TB Tuberculosis
TNF-α Tumor Necrosis Factor-alpha
WHO World Health Organization
XRD Extensively Drug Resistant

LONG ANSWER QUESTIONS

1. Discuss in detail the pathogenesis of tuberculosis and pathogenic diversity in different animal species.
2. What do different animal models for tuberculosis teach us about the disease?
3. Write a detailed description of the mouse model of tuberculosis.
4. Give a comparative outline of different animal models of tuberculosis.
5. Discuss in detail the established protocols of aerosolized infections of mouse and guinea pig using different apparatus.

SHORT ANSWER QUESTIONS

1. With the help of a flow chart, discuss how mice can be infected with *M. tuberculosis* intravenously.
2. Briefly summarize the initial findings of Robert Koch.
3. Summarize the lessons learned from animal models of tuberculosis.
4. Briefly summarize the importance of rabbit as a model for tuberculosis.
5. What are the characteristics of models of tuberculosis with respect to infection and pathogenesis?

ANSWERS TO SHORT ANSWER QUESTIONS

1. Intravenous infection of the mouse is not by itself a difficult procedure, but requires the investigator's expertise. The procedure is as follows (Flow Chart 2.10):
2. Robert Koch recognized and reported the spectrum of pathology of TB in different animal species based on his seminal studies on TB. Through experiments, he described TB as an infectious disease caused by *M. tuberculosis*. He was one of the first people to

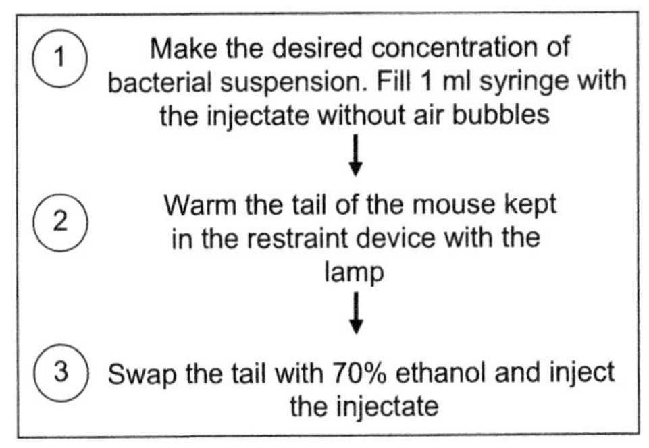

1
Make the desired concentration of bacterial suspension. Fill 1 ml syringe with the injectate without air bubbles
2
Warm the tail of the mouse kept in the restraint device with the lamp
3
Swap the tail with 70% ethanol and inject the injectate

FLOW CHART 2.10

envisage a vaccine for the control of TB. After 8 years since discovering the etiologic agent of tuberculosis, he announced the means of curing this disease. He suggested a vaccine for both the prevention and treatment of TB. Despite his first reports of a remedy for tuberculosis (studied in the guinea pig model), clinical trials soon demonstrated the ineffectiveness of his therapy.

3. Artificially infected guinea pigs, mice, and rabbits have served as indispensable tools through which important facets of TB have been studied. Results, however, are usually not entirely reflective of TB infection and disease in humans. Substantial differences in TB susceptibility, disease patterns, and temporal course exist among species. In addition to these considerations, animal species vary based on size, laboratory space requirements, rearing costs, and ability to approximate the disease process in humans. Despite several important differences, the murine, rabbit, and guinea pig models have emerged at the forefront of TB research.
4. Rabbits showed higher resistance with *M. tuberculosis* infection. The rabbits are infected with bovine mycobacterial infection i.e. *M. bovis*; their pulmonary pathology to inhaled bovine tubercle infection is more similar to human *M. tuberculosis* infection than those recorded in the case of other models such as mice and guinea pigs. In rabbits, the pulmonary cavity is first developed, followed by bronchial spread of microorganisms. Rabbits are also employed as models for studying the latent or paucibacillary TB states in humans.
5. The characteristics of different models are summarized in tabular form as follows (Table 2.5):

TABLE 2.5 Characteristics of Tuberculosis Models

	Histopathology					
Model	Necrosis	Caseation	Cavitation	Relative Susceptibility to *M. tuberculosis*	Immunologic Reagents Available	Laboratory Space and Cost
Mouse	Minimal	Usually not	No	Low	Extensive	Relatively small
Rabbit	Yes	Yes	Yes	Very low	Moderate	Relatively large
Guinea Pig	Yes	Yes	Infrequent	Very high	Relatively few	Moderate
Non-human primate	Yes	Yes	Yes	High	Extensive	Large

faces of TB have been studied. [illegible] however, [illegible] not entirely [illegible] of TB [illegible] disease in humans, substantial differences in TB susceptibility, disease [illegible] and [illegible] exist among species. [illegible] [illegible] [illegible] costs, [illegible] in humans. Despite [illegible] the current model [illegible] at the forefront of TB research.

[illegible]

Make the desired concentration of bacterial suspension. Fill 1 ml syringe with the inoculum without air bubbles

↓

Attach the [illegible] in the [illegible] device [illegible]

↓

[illegible]

[illegible]

Chapter 3

Animal Models for Neurodegenerative Disorders

Hitomi Tsuiji* and Koji Yamanaka*†

*Laboratory for Motor Neuron Disease, RIKEN Brain Science Institute, Saitama, Japan; †Research Institute of Environmental Medicine, Nagoya University, Nagoya, Japan

Chapter Outline

Animal Biotechnology. http://dx.doi.org/10.1016/B978-0-12-416002-6.00003-1

SUMMARY

Recent advances in neurodegenerative disease research have been made through animal models recapitulating human genetic mutations. We introduce neurodegenerative diseases with a focus on the motor neuron diseases. The clinical and genetic information, methodologies of engineering mouse models, and the research for disease mechanisms are provided.

WHAT YOU CAN EXPECT TO KNOW

In this chapter, you expect to know the clinical and genetic overview of neurodegenerative diseases, with a focus on motor neuron diseases, such as amyotrophic lateral sclerosis (ALS), spinal muscular atrophy (SMA), and spinal and bulbar muscular atrophy (SBMA), as a basis for generating animal models. For modeling motor neuron diseases in mice, we provide practical information including construction of transgenic mouse and Cre-loxP technology for conditional gene deletion or expression in mice. Finally, we review recent research progress through establishment and analysis of the animal models for neurodegenerative diseases.

HISTORY AND METHODS

INTRODUCTION

Neurodegenerative diseases are progressive neurological disorders characterized by death of specific nerve cells, excluding conditions such as ischemia, infection, intoxication, and malignant tumors. Representative examples include Alzheimer's disease, which is the most common cause of dementia and compromises cognitive and memory functions of patients, and Parkinson's disease, which is a progressive movement disorder exhibiting symptoms such as tremors, increased muscle tone, and slow movements. Many neurodegenerative diseases are genetically inherited. Through recent advances in genetics, many causative genes for neurodegenerative diseases have been identified. The breakthrough discovery of dominant mutations in amyloid precursor protein (APP) genes as causative for the familial Alzheimer's disease in 1991 led researchers to study molecular mechanisms of neurodegenerative disease. Subsequently, the causative genes enable us to further investigate the pathomechanism of diseases by generating animal models recapitulating human neurodegenerative diseases. In this chapter, we first provide an overview of neurodegenerative diseases, and then describe recent advances in neurodegenerative disease research, particularly focusing on motor neuron diseases through establishment and analysis of genetically modified mouse models.

Neurodegenerative Diseases

Representative neurodegenerative diseases include Alzheimer's disease (AD), Parkinson's disease (PD), Frontotemporal lobar degeneration (FTLD), Huntington's disease (HD), spinocerebellar degeneration (SCD), and amyotrophic lateral sclerosis (ALS). One of the characteristics of neurodegenerative diseases is death of specific neurons. In AD patients, the hippocampus, which is crucial for memory and learning, is affected at an early stage, and the disease spreads to other brain lesions as the disease progresses. In PD, FTLD, HD, SCD, or ALS patients, dopaminergic neurons in the midbrain, neurons in frontal and temporal lobes of cerebral cortex, neurons in basal ganglia, cerebellar neurons, or motor neurons are specifically affected, respectively. Here, characteristics of well-known neurodegenerative diseases are summarized (Table 3.1). Some diseases (such as Huntington's disease) are all genetically inherited, while others (such as Alzheimer's disease, Parkinson's disease, and ALS) are not. Animal models have been generated for almost all diseases listed here based on the causative genes that have been identified.

The motor neuron diseases make up the group of neurodegenerative diseases that is characterized by selective death of the neurons controlling motor functions, without affecting other neural systems, such as sensory and cognition. The motor neuron diseases include ALS, spinal muscular atrophy (SMA), spinal and bulbar muscular atrophy (SBMA), and other miscellaneous disease conditions. In the motor pathway, the signals from nerve cells in the primary motor cortex of the cerebrum (which are called "upper motor" neurons) are transmitted to the nerve cells in the brain stem and spinal cord (which are called "lower motor" neurons). The signals from lower motor neurons then reach the muscles innervated by the particular motor neurons. Both upper and lower motor neurons are affected in ALS, while only lower neurons are affected in SMA or SBMA. Clinical information and overview of ALS, SMA, and SBMA are provided below as an introduction of this chapter.

Amyotrophic Lateral Sclerosis (ALS)

Amyotrophic lateral sclerosis (ALS) is the most common form of adult motor neuron disease. As first described by the French neurologist, Charcot, in 1869, the primary symptom of disease is linked to the premature death of upper and lower motor neurons starting in adulthood. Neuronal death results in progressive paralysis of muscle in limbs and swallowing muscles, which typically is fatal in 2–5 years after the onset, due to respiratory muscle paralysis. The disease prevalence is roughly same worldwide (6 in 100,000 population). The disease onset is usually middle to late in life (50–70 years old), but 10% of cases begin before age 40.

TABLE 3.1 Representative Neurodegenerative Diseases

Name of Disease	Prevalence	Percent of Familial Cases	Main Clinical Symptoms	Affected Brain Area	Availability of Animal Models
Alzheimer's disease (AD)	16% of age 65 and older (USA)	1%	Progressive disturbance of memory, learning, and cognition	Hippocampus, cerebral cortex	Yes
Parkinson's disease (PD)	1–2% of age 60 and older	5–10 %	Involuntary movement (tremor), clumsiness, rigidity	Midbrain (dopaminergic neurons)	Yes
Frontotemporal lobar degeneration (FTLD)	15–20/100,000 of age 45–64	≈30%	Progressive disturbance of cognition, abnormal behavior, or a loss of language	Frontal and temporal lobe of cerebral cortex, hippocampus	Yes
Spinocerebellar degeneration (SCD)	10–20/100,000 populations	30–40 %	Progressive loss of balance and co-ordination of walking, speech	Cerebellum	Yes
Huntington's disease (HD)	0.3 (Asia), 3–7 (US, Europe) / 100,000 populations	100% (autosomal dominant)	Progressive chorea, cognitive impairment, psychiatric problem	Striatum	Yes
Amyotrophic lateral sclerosis (ALS)	6/100,000 populations	≈10%	Progressive motor weakness and paralysis	Spinal cord and cerebral cortex (motor neurons)	Yes
Spinal and bulbar muscular atrophy (SBMA)	1–2/100,000 males	100% (X-linked recessive)	Slowly progressive proximal motor weakness and atrophy, infertility	Spinal cord (motor neurons)	Yes

A list of adult-onset neurodegenerative diseases with prevalence, percent of familial cases, main clinical symptoms, affected brain area, availability of animal models is provided.

Males are slightly more affected than females (incidence of male patients is 1.5–1.8 times higher than female patients). Often the hands or feet (distal part of limbs) are affected first, while swallowing muscles are first affected in 10–20 % of cases. Sensation and bladder functions are spared. Among the motor functions, eye muscles are not usually affected, providing a clue to understand the mechanism of selective motor neurodegeneration. In many cases, intelligence is not affected, but 10–30% of ALS patients show cognitive decline.

Although most of ALS cases are sporadic, 10 percent of ALS cases are inherited. To date, researchers have identified approximately 18 ALS causative genes, the investigation of which lead to many hypotheses to explain the pathomechanisms of ALS (Table 3.2). Despite many studies, however, the causes of motor neuron death have not yet been clarified. There are no effective therapies for ALS, since the one prescribed drug (riluzole) approved in many countries prolongs life by only 2–3 months (Mitchell and Borasio, 2007).

Spinal Muscular Atrophy (SMA)

Spinal muscular atrophy (SMA) is an autosomal recessive neurological disorder mainly affecting children. The reduced amount of survival of motor neuron (SMN) protein due to the recessive mutations in the gene for *survival of motor neuron 1* (*SMN1*) is generally regarded as causative for SMA. SMA only affects lower motor neurons, producing muscle weakness in limbs and the body trunk. The incidence of SMA is about 1 case in 6,000–20,000 live births, but it varies among ethnicities. The male to female ratio is about 2:1. SMA in childhood is classified into three types by the age of onset and severity of disease. SMA type 1, also known as Werdnig–Hoffmann syndrome, is the most severe form of the disease and is evident at birth or before the age of 6 months. Proximal muscles are more affected than distal ones, with progressive weakness, resulting in complete paralysis of all limbs and respiratory failure, and 85% of children die before the age of two years. SMA type 2, the intermediate form, usually becomes symptomatic from 6 month to 1.5 years old. The patients may be able to sit but unable to walk without support. The rate of disease progression is variable; however, some patients show respiratory failure upon infection. SMA type 3, also known as Kugerberg–Welander disease, is a mild form of disease. Onset of the disease is usually in late childhood to adolescence (between 1.5 and 17 years old). Children often achieve an independent gait,

TABLE 3.2 ALS Causative Genes

Notation	Inheritance	Chromosome	Gene (Year of Identification)	Clinical Features and Comments
Genes Identified				
ALS1	AD	21q22	SOD1 (1993)	Typical ALS (adult), 20% of inherited ALS
ALS2	AR	2q33	ALS2 (Alsin) (2001)	Infantile onset, slowly progressive
ALS4	AD	9q34	Senataxin (SETX) (2004)	Juvenile onset, slowly progressive
ALS5	AR	15q15	Spatacsin (alleic to SPG11)	Juvenile onset, slowly progressive
ALS6	AD	16q12	FUS/TLS (Fused in Sacroma) (2009)	Typical ALS
ALS8	AD	20q13.3	VAPB (2004)	Typical and slowly progressive, heterogeneous
ALS9	AD	14q11	Angiogenin (ANG) (2006)	Typical ALS (adult)
ALS10	AD	1p36.22	TDP-43 (TARDBP) (2008)	Typical ALS (adult)
ALS11	AR	6q21	FIG4 (2009)	
HMN7B	AD	2p13	Dynactin p150 subunit (2003)	Slowly progressive, lower motor neuron disease
ALS12	AR	10p14-15	Optineurin (OPTN) (2010)	Adult onset
ALS13	AD	12q24	Ataxin-2 (ATXN2) (2010)	Susceptible gene (27–39 CAG repeats)
ALS14	AD	9p13.3	VCP (2010)	
ALS15	X-linked	Xp11.21	Ubiquilin 2 (UBQLN2) (2011)	X-linked dominant, affecting male and female
ALS16	AR	9p13.3	Sigma-1 receptor (SIGMAR1) (2011)	Juvenile onset, slowly progressive
ALS-FTD	AD	9q21-22	C9orf72 (2011)	Accompanied with FTD, 20–40% of inherited ALS
ALS17	AD?	3p11.2	CHMP2B (2010)	Lower motor neurons are predominantly affected
ALS18	AD	17p13.2	Profilin 1 (PFN1) (2012)	
Loci Identified				
ALS3	AD	18q21	Unknown	Typical ALS
ALS7	AD	20p13	Unknown	Typical ALS

however, exhibit slowly progressive muscle weakness of proximal limbs. The disease course is mostly benign, and many children are quite intelligent and live a normal life span (Lunn and Wang, 2008).

As explained in the following section, the human is the only species that has two highly conserved *SMN* gene copies, *SMN1* and *SMN2*. Although the classification of SMA type 1, 2, and 3 is defined by clinical information, the earlier disease onset and disease severity among SMA patients is generally correlated with lower expression level of SMN protein derived from *SMN2,* copy numbers of which have large variation.

Spinal and Bulbar Muscular Atrophy (SBMA)

Spinal and bulbar muscular atrophy (SBMA; also known as Kennedy–Alter–Sung disease) is an adult-onset slowly progressive motor neuron disease affecting lower motor neurons. SBMA is a X-linked recessive inheritance form of spinal muscular atrophy, mainly affects men, and is caused by the abnormal expansion of a CAG trinucleotide repeat in the exon 1 of androgen receptor gene, the mutation of which was identified by La Spada and colleagues in 1991. Female heterozygous and homozygous carriers are usually asymptomatic, although some have subclinical phenotypes only evident in the electrophysiological examination of muscle and blood tests. The disease prevalence is 1–2 in 100,000 population. SBMA usually affects after the age of 40 with slowly progressive weakness of limb and swallowing muscles; patients are dependent on wheelchairs and susceptible to pneumonia in 10–15 years. Due to the deficient androgen receptor, testicular atrophy and decreased fertility are observed. Patients with SBMA do not show cognitive impairments. Like ALS patients, the brainstem motor neurons to control eye movements are not affected in SBMA.

The recent studies indicate that SBMA is caused by a gain of toxicity from abnormal androgen receptor protein with polyglutamine expansion to neurons. Very importantly, toxicity requires a male sex hormone, an androgen (a ligand for the androgen receptor), which explains why SBMA mostly affects male patients (Adachi et al., 2007).

PRINCIPLES

To engineer the animal models recapitulating genetic mutations observed in human neurodegenerative disease patients, it is very important to understand the genetics of neurodegenerative diseases. In this section, an overview of genetics in ALS, SMA, and SBMA is provided as a principle to design the animal models recapitulating motor neuron diseases.

Genetics of ALS

To date, 18 genes have been identified as causative for familial ALS. The list of causative genes for ALS is provided in Table 3.2.

SOD1-ALS

Twenty percent of inherited ALS cases are caused by dominant mutations in the gene encoding for superoxide dismutase 1 (SOD1). SOD1, a ubiquitously expressed enzyme, consisting of 153 amino acids, catalytically converts reactive superoxide to oxygen and hydrogen peroxide. Since it was first identified as an ALS gene in 1993, more than 140 different mutations in SOD1 genes have been identified to date (see Glossary). Most of them are missense mutations, while some are the frameshift mutations that result in truncated protein products of mutant SOD1. Many SOD1-mutated ALS cases show a milder disease progression rate than that of sporadic ALS, with a mean disease duration of SOD1-ALS being about 5–10 years. Studies from ALS patients with SOD1 mutations revealed that there are no correlations between enzymatic activities of mutants and the clinical course. In addition, some disease-causing SOD1 mutations retain full enzymatic activity. Therefore, it is now generally recognized that all different mutations of the SOD1 gene (both enzymatically active and inactive mutants) uniformly cause neurodegeneration, not by a loss of enzymatic activity, but rather, a gain of toxicity. This gain of toxicity hypothesis is established by studies using SOD1-ALS mice as described in the later sections (Bruijn et al., 2004).

ALS: Other Genes Implicated in RNA Metabolism

TDP-43

About one to three percent of each inherited ALS case is caused by dominant mutations in the gene encoding for TAR-DNA binding protein 43 (TDP-43). TDP-43 is an RNA- and DNA-binding protein consisting of 414 amino acids with two RNA recognition motifs (RRM) and a C-terminal glycine-rich region. More than 30 mutations have been found in familial and sporadic ALS cases to date, and all of them are found in the C-terminal glycine-rich region, except for one mutation. TDP-43 is predominantly localized in the nucleus, shuttles between nucleus and cytoplasm, and regulates RNA splicing and mRNA stability. Examples of the well-known functions of TDP-43 in RNA metabolism are the splicing regulation of *CFTR* (a gene mutated in the inherited disease, cystic fibrosis) to enhance an exon skipping through binding to the UG-repeat of intron, the splicing regulation of *SMN* and *ApoAII*, and regulating mRNA stability of TDP-43 itself via direct protein binding to 3′-untranslated region of its own mRNA.

Importantly, in sporadic ALS cases (that are approximately 90% of total ALS cases), TDP-43 is lost from the nucleus and forms abnormal protein aggregates in cytoplasm of the affected motor neurons of spinal cords. This abnormal localization of TDP-43 is also seen in familial ALS caused by TDP-43 mutations. To date, it remains unknown whether the neurodegeneration in ALS patients' motor neurons are provoked by loss-of-function of nuclear TDP-43 functions, gain-of-function of abnormal cytoplasmic aggregates, or both of them. Nevertheless, elucidating the mechanisms through which abnormal TDP-43 leads to neurodegeneration will reveal the pathomechanism of sporadic ALS in which abnormal TDP-43 is accumulated (Da Cruz and Cleveland, 2011).

FUS/TLS

Another five percent of inherited ALS cases are caused by dominant mutations in the gene that encode for fused-in sarcoma (FUS, also known as translocated in liposarcoma protein (TLS)). FUS is also an RNA- and DNA-binding protein consisting of two arginine/glycine rich (RGG) domains, a Zn-finger domain, RRM, a glycin-rich region, and Q/G/S/Y-rich region. Most of the disease-causing mutations are located in the C-terminal region, with a nuclear localization signal, but some mutations are in the glycine-rich region. Abnormal aggregates of FUS protein in the cytoplasm or nucleus are found in affected motor neurons of spinal cords. FUS also shuttles between nucleus and cytoplasm and regulates transcription and RNA splicing. Considering the similarity between TDP-43 and FUS function, and the domain structure of the protein, defects in RNA metabolism could be one of the major pathways leading to neurodegeneration of ALS motor neurons (Da Cruz and Cleveland, 2011).

C9orf72

A large hexa-nucleotide GGGGCC repeat extension within intron or 5′UTR of *C9orf72* gene on chromosomal 9q21 also causes dominantly inherited ALS.(DeJesus-Hernandez et al., 2011; Renton et al., 2011) In Finland, this repeat expansion within *C9orf72* accounts for approximately 46% of familial ALS. In North America and Europe (except Finland), abnormal repeat expansion of *C9orf72* is responsible for about 20–40% of familial ALS, however, it has barely been found in Asian ALS patients so far. All patients with the repeat expansion had the founder haplotype, suggesting a one-off expansion event occurred about 1500 years ago (Majounie et al., 2012). Expansion of the nucleotide repeat in the non-coding region often results in a gain of toxicity derived from abnormal RNA. Moreover, abnormal nuclear aggregation containing RNA with the GGGGCC repeat was found in ALS patients, suggesting that a gain of toxicity of C9orf72 transcripts with the GGGGCC repeat may cause motor neuron degeneration of ALS patients.

From the viewpoint of ALS causative genes, ALS is highly related to defects in RNA metabolism. Other ALS causative genes implicated in RNA metabolism include *senataxin* and *ataxin-2* (see Table 3.2). However, ALS-causative genes for approximately 50 % of inherited ALS cases remain unknown.

Genetics of SMA

SMA is caused by a reduced amount of survival of motor neuron (SMN) proteins. This alteration is caused by the complex genetic molecular basis of SMA. Due to an interchromatin duplication on chromosome 5q13, humans possess two copies of the gene encoding SMN, *SMN1* and *SMN2*. Deletion or gene conversion events render SMA patients homozygous null for *SMN1* gene, whereas they maintain a variable copy number of *SMN2* gene. A critical C-to-T transition at position six of exon 7 causes aberrant splicing of 85–90% of *SMN2* transcripts without exon 7, which encodes unstable truncated SMN protein. A small percentage of transcripts encodes full-length active SMN protein, and the SMN protein level correlates with disease severity (Figure 3.1A–C) (Burghes and Beattie, 2009). Mouse studies demonstrated that complete loss of SMN protein causes embryonic lethality (Schrank et al., 1997.)

SMN is widely and constitutively expressed, and has been implicated in a wide range of cellular process, among which small nuclear ribonucleoprotein (snRNP) assembly is the best characterized. U-rich snRNP is the major component of the spliceosome, which carries out pre-mRNA splicing. SMN protein complex, consisting of several gemin family proteins, assists U-rich small nuclear RNA (U snRNA) and seven Sm core proteins to form U snRNP. SMN is also implicated in mRNA transport in the axon of the nerve, the disturbance of which might explain vulnerability of motor neurons in SMA (Burghes and Beattie, 2009). The gene encoding SMN is evolutionally conserved, the effects of SMN protein reduction have been modeled in diverse organisms, including the invertebrates *Caenorhabditis elegans* and *Drosophila melanogaster*, and the vertebrates *Danio rerio* and *Mus musculus*. Zebrafish and the invertebrate models are well studied to large-scale screening of drugs or genetic knockdown libraries prior to validation in mammals.

Genetics of SBMA

Abnormal expansion of a CAG trinucleotide repeat in exon 1 of the androgen receptor (AR) gene localized in X chromosomes was identified as causative for SBMA. Since a CAG trinucleotide repeat encodes a stretch of glutamine (so called polyglutamine), mutant AR protein encoded by the AR gene with longer CAG repeats contains an abnormal glutamine stretch. Normally, the number of CAG repeats in the AR gene ranges between 14 to 32, while it ranges between 40 and 62 in SMBA patients (Adachi et al., 2007). An inverse correlation has been reported between the age of disease onset and the number of CAG repeats, suggesting that the expansion of polyglutamine repeats in the AR exhibits a gain of toxicity in motor neurons. This inverse correlation is also seen in the other polyglutamine diseases, such as Huntington's disease, indicating that abnormal expansion of polyglutamine causes neurodegeneration through a gain of toxic mechanism.

As revealed by the analysis of SBMA model mice, SBMA is not caused by a loss of AR function, but is rather caused by a gain of toxicity from abnormal androgen receptor proteins with polyglutamine expansion to neurons. The pathomechanism unique to SBMA is that the mutant AR proteins require androgen, a male sex hormone, to provoke toxicities to motor neurons, which is linked to androgen-dependent translocation of mutant AR to the nucleus.

METHODOLOGY

Genetically modified mice for modeling human diseases include transgenic mice, knockout mice, conditional-knockout, and knock-in mice. In this section, a methodology for generating transgenic mice is provided. In addition, principle and technical details for Cre-loxP technology are discussed. These methods are used to generate the mouse models described in the "Examples and their Applications" section. The list of the mice described here includes: $SOD1^{G37R}$, $SOD1^{G93A}$, and $SOD1^{WT}$ transgenic mice as models for ALS, SMA model mice, and

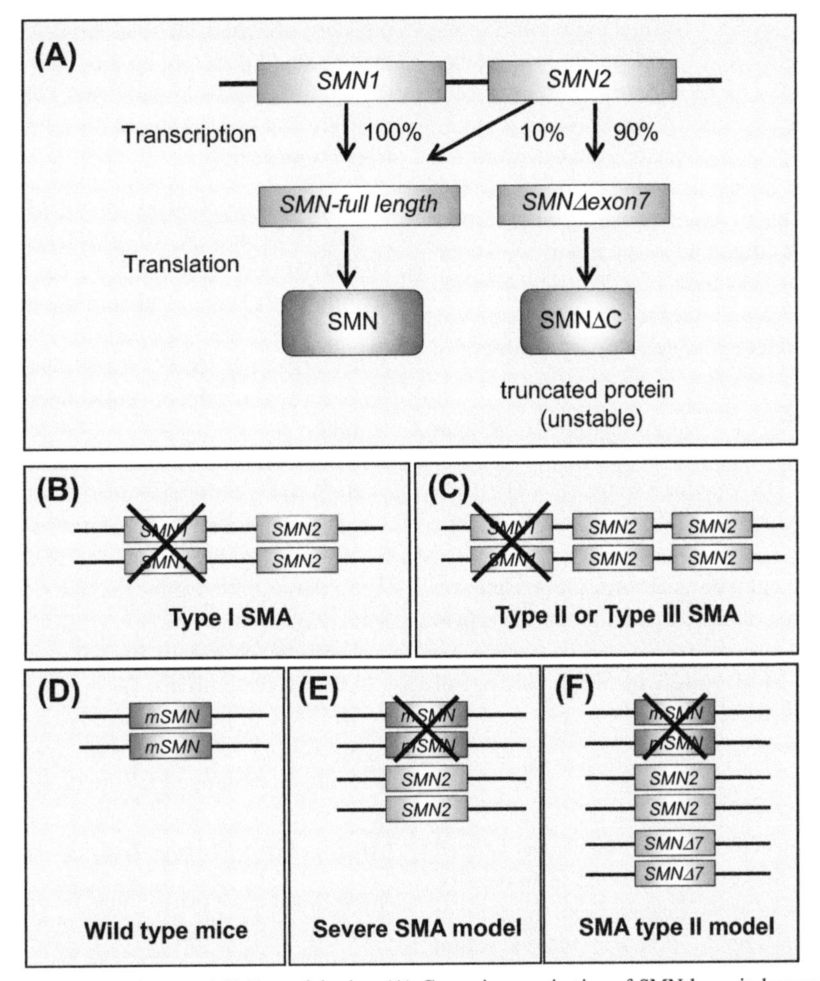

FIGURE 3.1 Genetic bases of SMA patients and SMA model mice. (A) Genomic organization of SMN locus in human healthy individual. (B, C) Genomic organization of SMN locus in SMA patients. (D–F) Genomic organization of wild type mice and SMA model mice.

AR-97Q and AR-24Q transgenic mice as models for SBMA.

Generation of Transgenic Mice

The brief protocol for generating transgenic mice is provided in this section according to the one used in our institute. The protocol consists of 6 basic steps: (1) preparation and purification of transgenic construct, (2) harvesting donor eggs, (3) microinjection of transgene to the fertilized egg, (4) implantation of microinjected egg to the pseudo-pregnant female mice, (5) screening of founder mice for transgene expression, and (6) establishing a stable transgenic line. The timeline is shown in Flow Chart 3.1. In order to generate transgenic mice, researchers should obtain approval from institutional committees for animal use and care, and biosafety for genetically modified organisms in advance (see Ethical Issues).

Preparation and Purification of Transgenic Construct (Step 1)

Designing the transgene construct is the most important step. First, fuse cDNA (coding region of the gene) with the

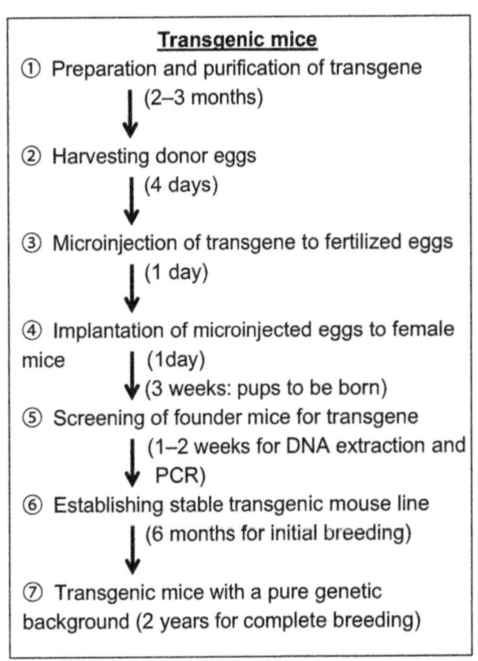

Transgenic mice
① Preparation and purification of transgene
(2–3 months)
② Harvesting donor eggs
(4 days)
③ Microinjection of transgene to fertilized eggs
(1 day)
④ Implantation of microinjected eggs to female mice
(1day)
(3 weeks: pups to be born)
⑤ Screening of founder mice for transgene
(1–2 weeks for DNA extraction and PCR)
⑥ Establishing stable transgenic mouse line
(6 months for initial breeding)
⑦ Transgenic mice with a pure genetic background (2 years for complete breeding)

FLOW CHART 3.1

appropriate promoter cassette for optimal expression in the cell types of interest. Alternatively, the genomic sequence of the gene of interest may be used. In this case, the gene will be transcribed under the control of its own promoter. The transgene DNA fragment, consisting of the promoter and the coding region of the gene, should be purified from the plasmid vector sequence. After digestion with restriction enzymes, the DNA fragment is separated with agarose gel electrophoresis, and then purified. The method to purify DNA either uses a commercial kit or a sucrose gradient centrifugation. Here is a step-by-step protocol to purify DNA.

1. Run the digested DNA transgene on 0.8–1.0% agarose gel.
2. Excise the transgene DNA under a long-wave UV light with a razor blade. (Minimize UV exposure to avoid the damage to the DNA.)
3. Purify DNA from the gel with QIAquick Gel Extraction kit (Qiagen, USA) according to the manufacturer's protocol.
4. Prepare DNA with 20–50 ng/μL in 50 μL TE buffer (10 mM Tris pH 8.0/0.1mM EDTA).

Harvesting Donor Eggs (Step 2)

The technical supports for step 2, 3, and 4 are usually offered by the animal facility of the research institute or university. Here, the outline for these steps is described. Egg-donor female mice receive injection of pregnant mare's serum gonadotropin (PMSG) and human chorionic gonadotropin (HCG) for superovulation. Approximately 400 fertilized eggs (zygotes) are obtained from 20 super-ovulated female mice.

Microinjection of Transgene to Fertilized Egg (Step 3)

Transgene DNA is microinjected to the pronucleus of a fertilized egg (zygote). Normally, DNA is diluted to 2 ng/μL, and approximately 2 pL of DNA solution is injected per zygote. Microinjected zygotes are subsequently incubated until 2-cell stage for 18 hours, and then 200 zygotes that are in good condition are used for implantation.

Implantation of Microinjected Egg to Pseudo Pregnant Female Mice (Step 4)

Typically 20–25 zygotes (per mice) are implanted to pseudo-pregnant female recipient mice.

Screening of Founder Mice for Expression of Transgene (Step 5)

Most recipient female mice are pregnant and deliver the pups after 20 days of embryo transfer. Normally, around 40 pups are obtained for screening. At 2–3 weeks of age, we make a small puncture on the ear for identification and take 0.5 cm of tail to isolate genomic DNA for genotyping.

Here is a brief protocol for extraction of genomic DNA from a mouse tail.

1. The tail is incubated with 500 μL of lysis buffer (0.4 mg/mL Proteinase K, 10 mM Tris, 100 mM NaCl, 10 mM EDTA, 0.5% SDS) at 55°C for 16 hours.
2. Add 500 μL of 1:1 phenol/chloroform solution and mix vigorously.
3. Centrifuge at 15,000 × g for 5 min. at room temperature.
4. Carefully transfer aqueous supernatant to new Eppendorf tube.
5. Add 40 μL of 3 M sodium acetate and 1 mL 100% ethanol, mix gently, and centrifuge at 16,000 × g, 4°C to pellet genomic DNA. Remove supernatant.
6. Wash DNA with 1 mL of 70% ethanol, centrifuge at 16,000 × g for 1 min, remove supernatant, repeat step 6, and air dry for 1 hr.
7. Dissolve DNA with 50–100 μL of TE buffer (pH 8). The DNA is ready to use for genotyping by PCR. Typically, 10–25 % of pups are positive for transgene. These pups are further screened for the transgene copy number by using quantitative PCR methods.

Establishing Stable Transgenic Line (Step 6)

A few transgene-positive pups (normally called "founders") with appropriate transgene copy number will be further bred with wild-type C57BL/6 mice to obtain the first generation (F1). Since transgene can be integrated to multiple sites of chromosomes, these mating steps (usually two to three generations) are necessary to establish a transgenic line with stable expression. In addition, to obtain the transgenic line with a pure genetic background (i.e. C57BL/6), it is necessary to breed mice for 7–10 generations with wild-type C57BL/6 mice.

Cre-loxP Technology

Cre-loxP system allows us to control the gene expression in a tissue-specific manner. Cre recombinase is a bacteria-derived enzyme that recognizes a portion of the specific DNA sequence (loxP) and deletes the DNA sequence between two loxP sites from the genome. LoxP sites are specific 34-base pair sequences (see Glossary). When two loxP sites are oriented in the same direction on the chromosome, Cre recombinase catalyzes the deletion of the DNA sequence between two loxP sites (Figure 3.2A), while Cre mediates inversion of DNA sequence when two loxP sites are oriented in the opposite directions (Figure 3.2B). The most popular application of Cre-loxP technology is a conditional knockout strategy to eliminate the gene of interest in the cell-type or tissue-specific manner, allowing examination

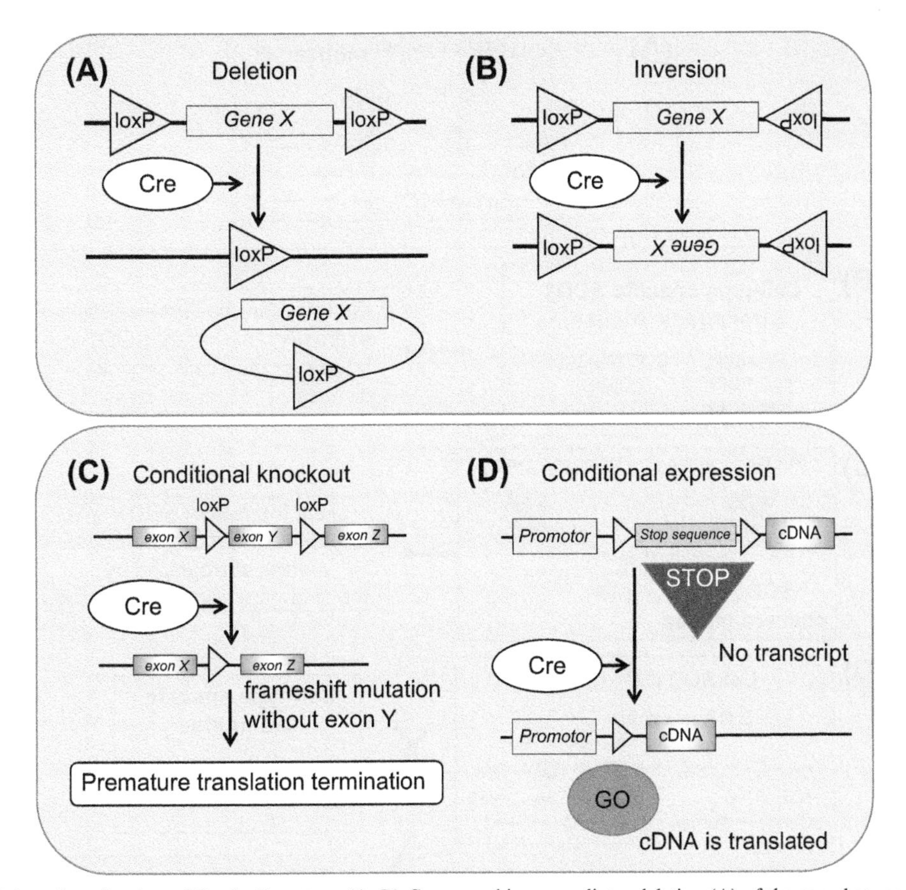

FIGURE 3.2 Principle and application of Cre-loxP system. (A, B) Cre recombinase mediates deletion (A) of the gene between two loxP sequences when loxP sites are oriented in the same direction, while it induces inversion (B) when two loxP sequences are oriented in the opposite directions. (C, D) Application of Cre-loxP system to gene manipulation in the mouse models. (C) Conditional knockout strategy. The mouse strain designed for conditional knockout carrying the loxP-flanked exon critical for a target gene. When mated with Cre-expressing mice, the loxP-flanked exon is deleted from genome, resulting in the premature termination of translation due to the frameshift mutation caused by the deletion of exon Y. (D) Conditional expression of the transgene. The mouse strain carrying a transgene incorporated whose promoter elements are separated from the coding region with an intervening stop sequence (with loxP sites). The Cre-mediated recombination to excise the stop sequence, occurs only in those cells expressing Cre. Therefore, the transgene can be transcribed in this specific cell type, whereas the transgene remains inactive in other Cre non-expressing cells.

of the gene function in the specific cell-types in the mice (Figure 3.2C). Another application is to express the gene of interest in the Cre-dependent, cell-type specific manner. For this purpose, you need to generate a transgenic or knock-in mouse carrying the conditional expression cassette, which consists of a promoter, an intervening stop sequence with two loxP sites at the each end, the coding sequence of interest, and a polyadenylation signal. The mouse expresses the transgene only when the intervening sequence is deleted by the action of Cre recombinase derived from the mating with a Cre transgenic mouse (Figure 3.2D). For this purpose, many Cre transgenic mice with a cell-type specific promoter have been generated to date.

ALS Models

SOD1^{G37R} Transgenic Mice

To generate SOD1 transgenic mice, a 12 kb genomic DNA fragment encoding the human SOD1 gene under an endogenous SOD1 promoter, with the mutation to convert codon 37 from glycine to arginine, was microinjected into hybrid (C57BL/6J x C3H/HeJ) F2 mouse embryos (Figure 3.3A) (Wong et al., 1995). From 128 initial pups, 23 founders were identified with the G37R mutant. Protein levels were measured for all founders by immunoblotting of whole blood using an SOD1 antibody. Lines were established from 8 founders expressing the highest levels of the G37R mutant. Quantitative immunoblotting coupled with SOD1 enzyme activity assays revealed that the mutant proteins accumulated between 5 to 12 times endogenous SOD1 in the spinal cord in the four G37R lines expressing the highest SOD1 level.

After backcrossing with C57/BL6 mice, two lines for SOD1^{G37R} transgenic mice were further characterized and frequently used for ALS research. SOD1^{G37R} (line 42) and SOD1^{G37R} (line 29) developed clinical signs of motor neuron disease around 4 and 10 months of age, and died around 6 and 13 months of age, respectively. SOD1^{G37R} (line

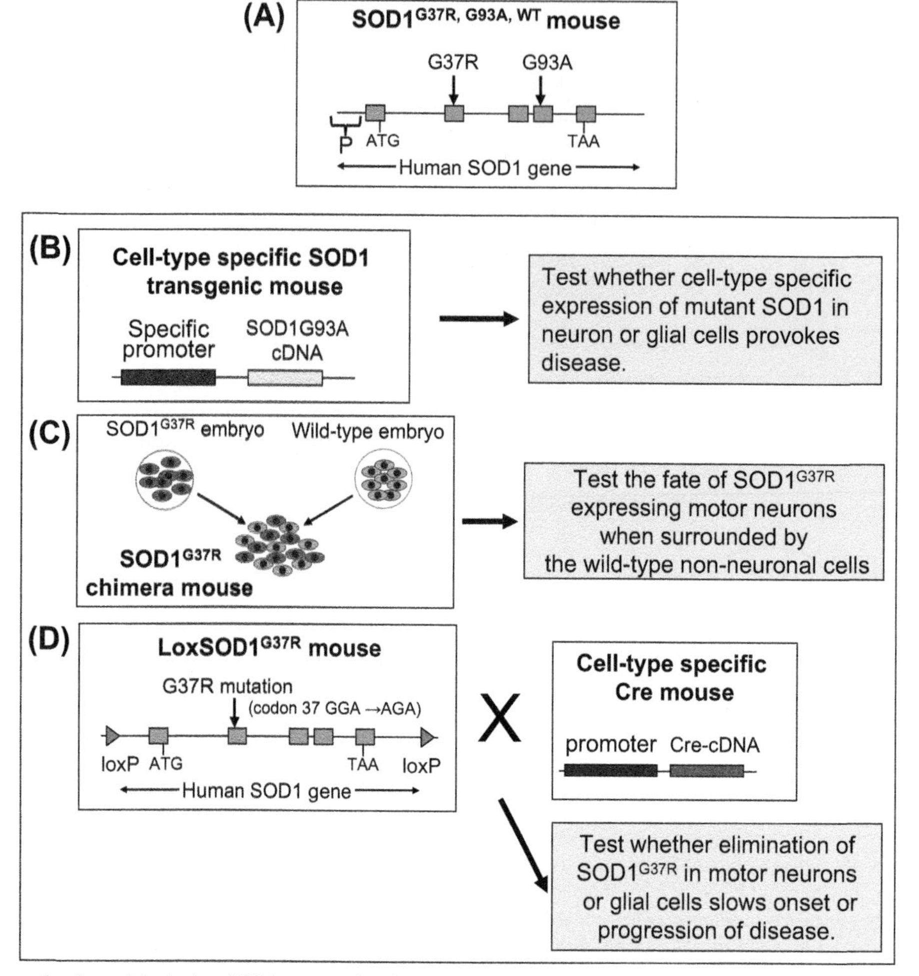

FIGURE 3.3 A schematic view of designing SOD1 transgenic mice. (A) A 12 kb human SOD1 gene carrying either G37R (glycine to arginine at codon 37) or G93A (glycine to alanine at codon 93) mutation is microinjected to generate SOD1^{G37R} or SOD1^{G93A} mice. SOD1^{WT} mice is generated from a wild-type human SOD1 gene. "P" indicates endogenous SOD1 promoter region. (B-D) The design to construct SOD1-ALS model mice to test the involvement of each central nervous system cell type. (B) SOD1 cDNA was expressed under the cell-type specific promoter (neuron or astrocyte). (C) Chimeric mouse approach to test the fate of mutant SOD1 expressing motor neurons in the environment of wild-type non-neuronal cells. (D) Schematic view of Lox SOD1^{G37R} transgene. To allow Cre-mediated gene excision, a pair of 34 base loxP sequence was added to each end of human SOD1^{G37R} gene. Mating with Cre expressing mouse will remove SOD1^{G37R} transgene in a cell-type specific manner.

42) expresses a higher number of transgene copies than SOD1^{G37R} (line 29), indicating that the disease severity is correlated with an expression level of mutant SOD1 (Wong et al., 1995).

Subsequently, to test the role of individual CNS cell types in ALS, loxSOD1^{G37R} mice expressing deletable SOD1^{G37R} transgene by using Cre-loxP technology were established. Addition of a pair of 34 base loxP sequences to each end of the SOD1 transgene allows Cre-mediated deletion of the transgene. While loxSOD1^{G37R} mice express mutant SOD1 in all the cell types under the authentic human SOD1 promoter, the mutant SOD1 transgene is excised in the specific cell type when the mice are mated with Cre expressing mice under the cell-type specific promoter (Figure 3.3D). The average life span of loxSOD1^{G37R} is 10–13 months (Boillee et al., 2006).

SOD1^{G93A} Transgenic Mice

A 12 kb human SOD1 gene carrying a disease-causing mutation at codon 93 from glycine to alanine was injected into a hybrid (C57BL6 × SJL) F1 embryo (Figure 3.3A) (Gurney et al., 1994). The transgenic mice with the highest copy number of SOD1 transgene with G93A mutation (24 copies) showed paralysis in limbs and died around 4.5 months of age. The SOD1^{G93A} line is most frequently used for ALS research. A mean survival time for the original line with mixed genetic background (C57BL6/SJL) or the line with pure C57BL/6 background is around 130 days or 155–160 days, respectively, indicating that the genetic background of ALS mice affects the disease onset and survival time.

$SOD1^{WT}$ Transgenic Mice

$SOD1^{WT}$ transgenic mice were generated in parallel to $SOD1^{G93A}$ or $SOD1^{G37R}$ mice using the human SOD1 gene (Figure 3.3A). They express SOD1 proteins at the level similar to that of the mutant mice, therefore are used as controls. In contrast to mutant SOD1 transgenic mice, the mice expressing high levels of wild-type SOD1 do not exhibit any clinical or pathological evidence of motor neuron disease (Wong et al., 1995; Gurney et al., 1994).

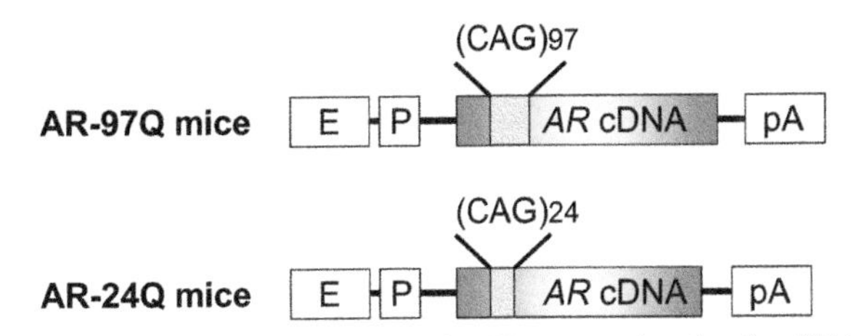

FIGURE 3.4 Schematic view of the AR transgenic mice for SBMA model. The microinjected fragment was composed of a cytomegarovirus enhancer (E), a chicken β-actin promoter (P), a human androgen receptor (AR) cDNA containing 24 or 79 CAG repeats with and a rabbit β-globin polyadenylation signal sequence (pA).

SMA Models

Severe SMA Mice ($mSMN^{-/-};SMN2^{+/+}$)

SMN knockout mice ($mSMN^{-/-}$ mice) were generated by the conventional gene targeting method to disrupt exon 2 of the mouse SMN gene by inserting the E. Coli LacZ gene (Schrank et al., 1997). SMN knockout mice ($mSMN^{-/-}$) die at the early embryonic stage, while heterozygote $mSMN^{+/-}$ mice do not show any detectable abnormalities. To rescue severe phenotype SMN knockout mice, the entire human *SMN2* gene including its promoter with a length of 35.5 kb was injected into embryos to generate SMN2 transgenic mice (Monani et al., 2000). Four founders were obtained. Two lines with 1 or 8 copies of *SMN2* genes were bred to C57BL/6J $mSMN^{+/-}$ mice to produce $mSMN^{+/-};SMN2$ progeny. The introduction of one copy of SMN2 gene rescued embryonic lethality from mSMN deficiency, resulting in severe SMA phenotype (Figure 3.1E). The $mSMN^{-/-};SMN2^{+/+}$ mice (called as severe SMA mice) can survive 8 days, but most of them die between 4–6 days. SMN knockout mice carrying higher copy number SMN2 transgene did not show any obvious phenotypes, indicating introduction of 8 copies of *SMN2* is sufficient to rescue the SMA phenotype.

SMA Type II Mice ($mSMN^{-/-};SMN2^{+/+};SMN\Delta7^{+/+}$)

To obtain SMNΔ7 transgenic mice, an SMN cDNA lacking exon 7 was introduced under the control of a 3.4 kb SMN promoter fragment (SMNΔ7) and microinjected into FVB/N mouse embryos (Le et al., 2005). Three founder lines carrying 2, 6, and 17 copies of SMNΔ7 transgenes were obtained. The integration of the transgene for all lines was a single tandem integration and followed Mendelian inheritance consistent with a single locus for genomic integration site of the transgene. The SMNΔ7 transgenic mice were crossed with mice expressing SMN2 and a mouse SMN knockout allele to obtain double transgenic mice $mSMN^{+/-};SMN2;SMN\Delta7$ on a FVB/N genetic background. These double transgenic mice were interbred to obtain $mSMN^{-/-};SMN2;SMN\Delta7$ mice. The SMNΔ7 transgenic mice with 6 copies of SMNΔ7 gave the highest levels of SMNΔ7 mRNA and protein, and gave the most improvement in life span of severe SMA. The $mSMN^{-/-};SMN2^{+/+};SMN\Delta7^{+/+}$ mice survived for a maximum 17 days with a mean of 13.3 days (Le et al., 2005). These mice are also called SMA type II model mice, and have become the most widely used SMA models (Figure 3.1F).

SBMA Models

AR-97Q and AR-24Q Transgenic Mice

To generate SBMA model mice, the full-length human androgen receptor (AR) cDNA harboring 24 or 97 CAG repeats with a chicken β-actin promoter and a cytomegarovirus enhancer was microinjected into BDF1 mouse embryos (Figure 3.4). Three founders with AR-24Q and five founders with AR-97Q were obtained, and these mice were backcrossed to C57BL/6J mice. Three lines with 1 to 5 copy numbers of AR-24Q were established, but none of them showed any manifested phenotypes. Three lines with 1 to 3 copy numbers of AR-97Q exhibited progressive motor impairment. All three lines of AR-97Q showed small body size, short life span, progressive muscle atrophy and weakness, as well as reduced cage activity, all of which were markedly pronounced and accelerated in the male AR-97Q mice, but not observed or far less severe in the female AR-97Q mice, regardless of the line (Katsuno et al., 2002).

EXAMPLES AND THEIR APPLICATIONS

The pathomechanisms of motor neuron diseases have been elucidated through the construction and analyses of mouse models recapitulating mutations of causative genes. Here, an example for evaluating the phenotypes and clinical course of SOD1-ALS model mouse is provided. Then, the representative examples of their applications, the research to understand the pathomechanisms for motor neuron diseases by using SOD1-linked ALS, other ALS, SMA, and SBMA model mice, are discussed.

SOD1-Linked ALS

Gain of Toxicity from Mutant SOD1 Established as Pathomechanisms through Engineering Mutant SOD1 Mice

The discovery of SOD1 mutations in familial ALS led researchers to develop animal models that recapitulate ALS-like disease. SOD1 is a metalloenzyme that coordinates one copper and one zinc in the protein, and forms a dimer to exert full enzymatic activity to convert superoxide to oxygen and hydrogen peroxide, thus eliminating reactive oxygen species harmful to the cells. Initially, ALS researchers hypothesized that a loss of enzymatic activity of ALS-causing mutant SOD1 proteins is causative for motor neuron degeneration. However, SOD1 knockout mice do not show overt phenotypes, while the mice show increased vulnerability to motor neurons only after axonal injury. In contrast, the transgenic mice overexpressing human SOD1 gene carrying a patient-derived mutation uniformly show a progressive neurodegenerative disease that closely resembles human pathology with a selective motor neuron death and gliosis accompanied by accumulation of misfolded proteins, while overexpressing wild-type SOD1 proteins in mice do not cause neurodegeneration. (Wong et al., 1995; Gurney et al., 1994) These results led to the consensus that all different mutations of the SOD1 gene (both enzymatically active and inactive mutants) uniformly cause toxicity in cells not by loss, but rather by gain, of function mechanisms. However, the exact mechanism and nature of toxicity are still unknown. Currently, numerous mechanisms of toxicity have been proposed that could mediate pathology in mutant SOD1-mediated ALS. The most important mechanisms are thought to be excitotoxicity from glutamate, ER stress, damage to mitochondria, neuroinflammation by secretion of superoxide and proinflammatory cytokines, axonal transport disruption, and spinal capillary microhemorrhages (Bruijn et al., 2004). All of these mechanisms are considered as the key processes for motor neurodegeneration and convergence of all of these events contributes to the development of ALS pathology.

Evaluating Phenotype and Clinical Course of Mutant SOD1 Transgenic Mice

$SOD1^{G93A}$ mice, widely used model for ALS research, (Gurney et al., 1994) show an observable phenotype of motor decline at around 90–100 days with a hind-limb tremor when hung by a tail. Subsequently, $SOD1^{G93A}$ mice show a walking abnormality, such as a waddling gait, due to the weakness of hind-limbs. A progressive loss of body weight is the objective measure of disease onset, which reflects a loss of muscle volume. As a measure of onset, researchers use a time for peak body weight (a time to start body weight decline) in combination with the neurological score. The hind-limb paralysis becomes clear at the age of 140–150 days (Figure 3.5). The endpoint is normally determined as a time when the mice are unable to right themselves within 15–30 seconds if laid on either side. The mean survival time of $SOD1^{G93A}$ mice is 155–160 days in C57BL6 genetic background, while approximately 130 days in the mix genetic background (C57BL6/SJL) of the originally constructed mice.

Toxicity from Mis-folded Mutant SOD1 Protein

A molecular characteristic common to almost all mutant SOD1 proteins is protein mis-folding: abnormally folded "misshaped" proteins that show altered biochemical properties, such as decreased solubility to detergents and increased aggregation propensities. Deposition of "misfolded proteins" is also common to neuropathology in many neurodegenerative diseases, including Alzheimer's disease (amyloid beta and tau proteins), Parkinson's disease (alpha-synuclein). Mis-folding of mutant proteins is linked to the above-mentioned mechanisms for motor neuron degeneration. One of the plausible hypotheses is mitochondrial dysfunction. Morphological abnormalities of mitochondria have been observed in the spinal cord lesion of mutant SOD1 mice (Wong et al., 1995). More importantly, the enzymatic activities of mitochondria are decreased rather selectively in the spinal cord, but not affected in the other brain regions of SOD1 mice. Mis-folded mutant SOD1 proteins are accumulated on the mitochondrial outer membrane and inter-membrane space by the analysis of biochemically isolated mitochondria from the spinal cord of mutant SOD1 transgenic mouse or rats, suggesting the active role of misfolded mutant proteins in mitochondrial dysfunction (Liu et al., 2004). Recently, *in vivo* evidence of the pathogenic role of mutant SOD1 in mitochondria was demonstrated by mice expressing $SOD1^{G93A}$ mutant selectively in the inter-membrane space of mitochondria of the central nervous system, which showed neurodegeneration, although it was not sufficient to recapitulate a full aspect of ALS phenotype (Igoudjil et al., 2011).

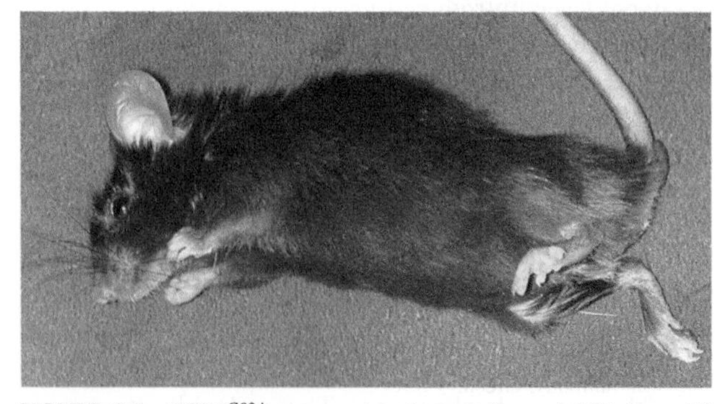

FIGURE 3.5 $SOD1^{G93A}$ mouse at end stage (around 160 days old) showing hindlimb paralysis.

Non-Cell Autonomous Neurodegeneration Demonstrated by SOD1 Mouse Models

The selective death of motor neurons initially led researchers to believe that cell autonomous mechanisms were central to pathogenesis. However, genetic and chimeric mouse studies indicate that non-cell autonomous processes might underlie motor neuron loss in these rodent models. Immuno-histological studies show glial cell involvement in ALS pathology where extensive activation and proliferation of glial cells (also called gliosis), such as astrocytes and microglia, is observed in motor neuron diseases (Figure 3.6). Astrocytes are glial cells that support the function of neurons by secreting neurotrophic factors and controlling neuronal functions by controlling ion balance, recycling the neurotransmitter glutamate, and vasculatures. Microglia are central to immune defense from scavenging pathogens and damaged neurons (see Glossary).

First, researchers tested the role of each cell type (motor neurons or glial cells) in mutant SOD1 mediated toxicities by generating transgenic mice to express mutant SOD1 in the specific cells in the central nervous system using a cell-type specific promoter (Figure 3.3B). When expression of SOD1 mutations was restricted to either neurons or astrocytes, but not both simultaneously, it did not lead to the development of ALS. In an alternate approach, chimeric mice, which are a mixture of cells derived from mutant SOD1 transgenic mice and the wild-type cells, were constructed to test whether mutant SOD1 expressing motor neurons are damaged through a cell-autonomous fashion (Figure 3.3C). In these experiments, mutant SOD1-expressing motor neurons, in chimeric mice with both wild type and mutant SOD1-expressing cells, escaped from degeneration when surrounded by wild-type glial cells, indicating that mutant SOD1-mediated ALS is provoked in a non-cell autonomous fashion (Clement et al., 2003; Yamanaka et al., 2008). Finally, to test the role of individual CNS cell types in ALS, researchers designed the loxSOD1^{G37R} transgenic mice expressing deletable mutant SOD1 gene by using Cre-loxP technology (Figure 3.3D). Reducing mutant SOD1 expression in either astrocytes or microglia using floxed SOD1 gene excised by Cre recombinase slowed the disease progression and extended lifespan (Boillee et al., 2006; Yamanaka et al., 2008). A series of these studies established the active role of glial cells in ALS pathogenesis through genetically engineering the ALS model mice.

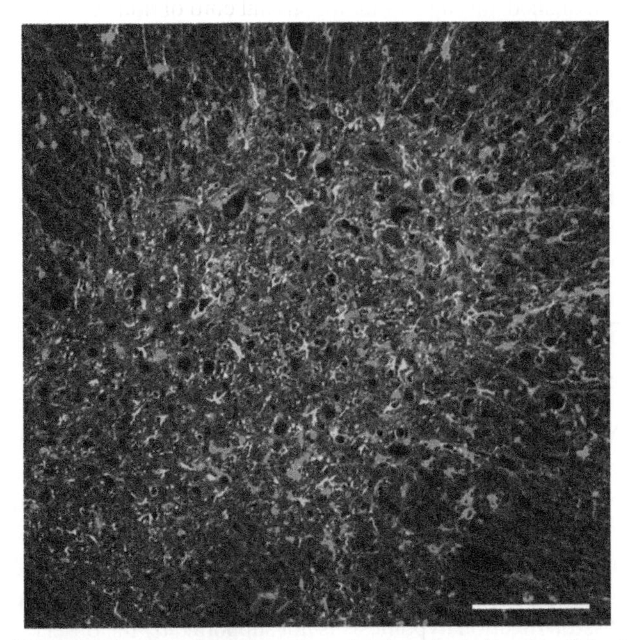

FIGURE 3.6 Activated Microglia and Astrocytes in Lumbar Spinal Cord of Symptomatic Mutant SOD1 Mice. Red: microglia stained with anti-Mac2 antibody, Green: astrocytes stained with anti-GFAP antibody, Blue: motor neurons stained with anti-neurofilament H antibody. Bar: 100 μm.

Stem Cell-Derived Motor Neurons Established from Mutant SOD1 Mice

These ALS model mice are also useful for modeling neurodegenerative diseases *in vitro*. Researchers established the embryonic stem (ES) cells from mutant SOD1 transgenic mice. The protocol to differentiate ES cell from motor neurons is available so that it is now feasible to model the motor neuron diseases *in vitro* to elucidate the detailed mechanisms underlying neurodegeneration (Wichterle and Peljto, 2008). Recent studies demonstrated that astrocytes expressing mutant SOD1 are toxic to ES cell-derived motor neurons, supporting the "non-cell autonomous" neurodegeneration hypothesis of mutant SOD1-linked ALS (Nagai et al., 2007). Moreover, ES cell-derived motor neurons carrying ALS causative genes are also useful for drug screening, as well as neurons derived from human inducible pluripotent stem cells (iPS cells).

Other ALS

Researchers have tried to develop ALS models recapitulating TDP-43 or FUS abnormalities, and many transgenic animals over-expressing human TDP-43 wild-type or ALS-linked mutants were generated; however, none of them represent the ALS-phenotype well. Both wild-type and ALS-linked mutant TDP-43 transgenic mice, when expressed at high level, showed neurological phenotypes, such as decline of motor performance and neurodegeneration. In contrast to SOD1-ALS models, cell-type specificity and mutant-specificity for neurodegeneration were not well achieved in TDP-43-ALS models. This may be because TDP-43 is essential to all cell types, and the amount of TDP-43 is tightly regulated by autoregulation of TDP-43 itself. Since TDP-43 downregulates its own mRNA level, a small amount of increase of TDP-43 does not affect the total amount of TDP-43. In contrast, the overexpression of a certain level of TDP-43 causes death in many types of cells, resulting in difficulty in recapitulating region-specific neurodegeneration. Currently, the mechanism underlying neurodegeneration in sporadic ALS is not well known, it may be mediated by a combination of both loss-of-function of normal RNA-binding protein and gain of toxic function. The alternative approaches to develop ALS models have been in progress, including the Cre-loxP system to express the target protein in the specific cell-types, virus-mediated expression of target proteins in the spinal cord of adult animals, and conditional deletion of target genes specific to motor neurons (Da Cruz and Cleveland, 2011).

SMA

Mice possess a single SMN gene, which has 82% amino acid identity with the human homolog; mouse and human SMN genes show a similar expression pattern. Homozygous deletion of the SMN gene results in massive embryonic cell death before implantation, as would be expected given the housekeeping functions of SMN (Schrank et al., 1997). Heterozygous SMN-null mice lack a marked clinical phenotype, and parental carriers of SMA-related mutations are phenotypically normal, indicating that presumably only approximately 50% of the wild-type SMN level is required for normal function in cells. When the level of SMN is substantially reduced to lower than ~20% of the normal level, most or all cells die. Therefore, there seems to be a critical level of SMN at which many cell types are relatively unaffected, but a few cell types, such as motor neurons and possibly muscle cells, are compromised. The selective vulnerability of motor neurons for the low levels of SMN is not well understood.

Human SMN2 Transgenic Mice

Since more than 95% of SMA patients show homozygous deletion of *SMN1* and varied expression of *SMN2* (Figure 3.1A–C), mice deficient for mouse *SMN* but expressing human *SMN2* seem to be an ideal model for SMA. Introduction of one or two copies of human *SMN2* genes to SMN knockout mice rescues the embryonic lethal phenotype, resulting in mice with severe SMA (*mSMN*$^{-/-}$;*SMN2*$^{+/+}$), that are indistinguishable from controls at birth, but die before postnatal day 7 (Monani et al., 2000; Hsieh-Li et al., 2000). Introduction of 8–16 copies of human *SMN2* gene completely rescued the phenotypes, (Monani et al., 2000) indicating that modestly enhanced expression of full-length SMN protein from *SMN2* can prevent SMA, confirming the idea that enhancing *SMN2* transcription is a potential therapeutic strategy (Burghes and Beattie, 2009).

Since the phenotype of mSMN$^{-/-}$ mice carrying two copies of SMN2 is severe, researchers tried to reduce the severity in many ways. For example, introduction of a second transgene containing human *SMNΔ7* into severe SMA mice (*mSMN*$^{-/-}$;*SMN2*$^{+/+}$) extends the lifespan from 6 to 13 days (Le et al., 2005). This mouse (*mSMN*$^{-/-}$;*SMN2*$^{+/+}$;*SMNΔ7*$^{+/+}$), also called SMA type II model, has become the most widely used of SMA models. Introduction of other transgenes containing patient-derived point mutations in *SMN1* (A2G, A111G) have also resulted in increased life span. However, none of these mutants alone rescue the embryonic lethality caused by *SMN* depletion, indicating the importance of retaining at least some full-length SMN in order to function.

Using SMA type II models, tissue specific abnormality of small nuclear RNA repertoires and profound pre-mRNA splicing defects were elucidated *in vivo* (Zhang et al., 2008). The work confirmed a key function of SMN complex in RNA metabolism and splicing regulation *in vivo*. The abnormality of small nuclear RNA repertories and splicing defects in SMA mice are not limited to the spinal cord, but are observed in many organs, and remain an unsolved question regarding the vulnerability of motor neurons in SMA.

Neuron-Specific Deletion of SMN in Mice Using Cre-loxP Systems

The Cre-loxP system allows us to modify the gene expression in a tissue specific manner. A mouse line carrying two loxP sequences flanking *SMN* exon 7 (*SmnF7*) has been established through homologous recombination. Crossing these mice with tissue-specific Cre mouse lines effectively produces complete deficiency of SMN in target tissues. Neuron-specific ablation of mSMN results in motor neuron loss and premature death of mice (Burghes and Beattie, 2009; Frugier et al., 2000). The complete absence of functional SMN, which never occurs in patients, suggests that these models provide limited insights into the pathogenesis of SMA.

SBMA

AR-97Q Mice as SBMA Model

After identification of expansion of polyglutamine in androgen receptor (AR) in SBMA patients, two hypotheses have been proposed for the role of polyglutamine-expanded mutant AR in the pathogenesis of SBMA: (1) Mutant AR acquires a toxic property for motor neurons, or (2) A loss of normal AR function causes motor neuron degeneration. The loss-of-function hypothesis is not supported by the observation that patients with testicular feminization lacking AR function and AR gene knockout mice do show motor neuron diseases. In contrast, the transgenic mice expressing a full-length human AR gene with expanded CAG repeats under the control of an ubiquitous promoter (AR-97Q mice) recapitulated motor neuron disease, while the mice with normal CAG repeats (24 repeats) did not show any abnormality (Katsuno et al., 2002). More importantly, AR-97Q mice showed that motor neuron disease is dependent on gender. The disease phenotype was seen only in male AR-97Q mice, not female, recapitulating a key phenotype in SBMA. This gender effect and CAG repeat-dependent neurodegeneration are also observed in the other transgenic mice expressing full-length androgen receptors with expansion of CAG repeats, establishing the successful SBMA model mice (Adachi et al., 2007).

Androgen Hormone and Mutant AR Receptor Central to SBMA Pathogenesis

This significant gender difference in the phenotype of AR-97Q mice led researchers to test whether motor neuron degeneration requires the ligand of AR, testosterone. AR-97Q male mice that had been castrated (surgically removal of testes) or pharmacologically altered to antagonize the effect of testosterone escaped from motor neuron disease and pathology; AR-97Q female mice that were administered testosterone did develop motor neuron disease (Katsuno et al., 2002). This ligand-dependent neurodegeneration is also observed in a fruit fly model of SBMA (Takeyama et al., 2002).

In contrast to the other polyglutamine diseases such as Huntington's disease, SBMA requires two elements (abnormal polyglutamine carrying mutant protein and its ligand, androgen) to develop motor neuron diseases. Based on the studies, it is now established that mutant AR proteins exhibit toxicity when translocated into a nucleus in the presence of androgens likely to provoke transcriptional dysregulation. Mutant AR proteins extracted from the affected tissue of male SBMA mice are mis-folded, and prone to form protein aggregates. Immunohistological studies revealed that extensive accumulation of AR proteins in the nucleus were observed in the male SMBA mice, despite that there was no significant difference in the AR mRNA levels between the male and female SBMA mice. In support of the hypothesis in which nuclear translocation of mutant AR is a key process to provoke neurotoxicity, neuronal dysfunction was halted by genetic manipulation to prevent nuclear import of pathogenic polyglutamine protein in the mouse model of SBMA (Montie et al., 2009).

ETHICAL ISSUES

Nowadays, animal experiments are widely used for understanding disease mechanisms, for developing new medicines, and for testing the safety of drug candidates. As many animal experiments cause pain and compromise the animals' quality of life and lifespan, researchers should be aware of the ethical problems associated with using laboratory animals.

The "3Rs" are the principles for researchers to follow in order to reduce the negative impact on animal research and to maintain accountability to society for the use of laboratory animals. The 3Rs are: Reduction, Refinement, and Replacement. Reduction means to reduce the number of animals used in experiments by improving techniques or methods. Refinement means to refine the experimental methods to reduce the suffering of animals by the use of anesthesia, or less invasive methods. Replacement means to replace the animal experiments with alternatives such as cell cultures or others. In practice, when mouse tissues are collected for research, we deeply anesthetize the mouse by the inhalational anesthetic drug isoflurane, then euthanize the animals. Similarly, when the animal models reach the end-stage of the disease, the animals should be euthanized to reduce pain and suffering from their disabilities due to the inability to move.

All experimental protocols using animals should be approved by the animal care and use committee in each research institute or university. In addition, to maximize safety during distribution of genetically engineered living organisms, more than 90 countries have ratified "the Cartagena Protocol on Biosafety," which has been in effect since 2003. This is an international agreement that aims to ensure the safe handling, transport, and use of living modified organism (LMOs) resulting from modern biotechnology that may have adverse effects on biological diversity, considering the potential risks to human health. When researchers create and use transgenic or knockout rodent models as discussed in this chapter, they should follow this protocol.

TRANSLATIONAL SIGNIFICANCE

The animal models for motor neuron diseases discussed here have been used to test candidate drugs for the therapy. SOD1-ALS models were used for many preclinical studies to test candidate drugs including antibiotics, minocycline and the cyclooxygenase 2 inhibitor, Celecoxib, on their disease courses. To date, more than 20 clinical trials on sporadic ALS patients have been conducted, based on the studies using SOD1-ALS rodent models. However, almost all clinical trials did not show efficacies (Benatar, 2007). Failure to translate results of rodent models to sporadic human patients was attributed to several reasons. First, in many preclinical studies, the drugs were administered to animals before onset. However, this is not the case for sporadic ALS patients, since human patients are treated after diagnosis. Second, in many cases the drug effects aiming to extend the survival time of mice were modest, with cohort sizes that were not sufficient. Adequate cohort size (more than 15 animals per genotype) as well as the timing of initiating drug treatment should be carefully considered in the rodent studies. Third, the disease mechanism of mutant SOD1-mediated familial ALS could be different from sporadic ALS. Recent discovery of new genes, TDP-43, FUS, responsible for ALS has provided new opportunity for the development of new animal models (Da Cruz and Cleveland, 2011). New rodent models useful for testing new candidate drugs are awaited.

In contrast, the translational efforts using SBMA animal models have been successful, partly because SBMA is a monogenic disease and the pathomechanisms are better understood. Leuprorelin, a potent luteinizing hormone-releasing hormone (LHRH) analog, is an approved drug for treating prostate cancer. Leuprorelin suppresses the release of the gonadotrophins, luteinizing hormone, and follicle-stimulating hormone. With subcutaneous administration of Leuprorelin, male AR-97Q mice improved motor function and showed significant extension of their survival. Based on the results from SBMA mice, the clinical trial of Leuprorelin for SBMA patients has been carried out. While the efficacy did not reach statistical significance, subgroup analysis demonstrated the efficacy of the drug to slow disease progression, resulting in a partial success (Katsuno et al., 2010).

In summary, the mouse models for neurodegenerative diseases (especially for motor neuron diseases) are very useful tools to evaluate the efficacy of drug candidates, when the protocol is appropriately designed. We hope this chapter provides a better understanding of the applications of animal models for neurodegenerative diseases to research for elucidating pathomechanisms, as well as developing therapies.

WORLD WIDE WEB RESOURCES

Gene Reviews: http://www.ncbi.nlm.nih.gov/sites/GeneTests/review

This open website provides detailed information of human inherited diseases and the research for genetic diseases, including motor neuron diseases.

OMIM (Online Mendelian Inheritance in Man): http://omim.org/

This website provides a comprehensive, authoritative compendium of human genes and genetic phenotypes that is freely available and updated daily. It is maintained by Johns Hopkins University, in collaboration with the National Institute of Health, United Status of America.

International Mouse Strain Resource (IMSR): http://www.findmice.org

The IMSR is a searchable online database of mouse strains, stocks, and mutant ES cell lines available worldwide, including inbred, mutant, and genetically engineered strains. The goal of the IMSR is to assist the international scientific community in locating and obtaining mouse resources for research.

ALS association: http://www.alsa.org/

The website introduces a non-profit orgaization in the United States of America that promotes research on ALS and provides support to ALS patients and the community. The detailed information of ALS diseases and research are found.

Families of SMA: http://www.fsma.org/

Like the above-mentioned organization, this non-profit organization provides information about spinal muscular atrophy.

Convention on Biological Diversity: http://bch.cbd.int/protocol/

The website provides information on "The Cartagena Protocol on Biosafety." Researchers who use genetically modified animals should follow these principles.

ACKNOWLEDGMENTS

The authors thank Drs. E. Takahashi, T. Arai, and C. Itakura (RIKEN Brain Science Institute) for help in describing the protocol for transgenic mice.

REFERENCES

Adachi, H., Waza, M., Katsuno, M., et al. (2007). Pathogenesis and molecular targeted therapy of spinal and bulbar muscular atrophy. *Neuropathology and Applied Neurobiology, 33*, 135–151.

Benatar, M. (2007). Lost in translation: treatment trials in the SOD1 mouse and in human ALS. *Neurobiology of Disease, 26*, 1–13.

Boillee, S., Yamanaka, K., Lobsiger, C. S., et al. (2006). Onset and progression in inherited ALS determined by motor neurons and microglia. *Science, 312*, 1389–1392.

Bruijn, L. I., Miller, T. M., & Cleveland, D. W. (2004). Unraveling the mechanisms involved in motor neuron degeneration in ALS. *Annual Review of Neuroscience, 27*, 723–749.

Burghes, A. H., & Beattie, C. E. (2009). Spinal muscular atrophy: why do low levels of survival motor neuron protein make motor neurons sick? *Nature Reviews Neuroscience, 10*, 597–609.

Clement, A. M., Nguyen, M. D., Roberts, E. A., et al. (2003). Wild-type nonneuronal cells extend survival of SOD1 mutant motor neurons in ALS mice. *Science, 302*, 113–117.

Da Cruz, S., & Cleveland, D. W. (2011). Understanding the role of TDP-43 and FUS/TLS in ALS and beyond. *Current Opinion in Neurobiology, 21*, 904–919.

DeJesus-Hernandez, M., Mackenzie, I. R., Boeve, B. F., et al. (2011). Expanded GGGGCC hexanucleotide repeat in noncoding region of C9ORF72 causes chromosome 9p-linked FTD and ALS. *Neuron, 72*, 245–256.

Frugier, T., Tiziano, F. D., Cifuentes-Diaz, C., et al. (2000). Nuclear targeting defect of SMN lacking the C-terminus in a mouse model of spinal muscular atrophy. *Human Molecular Genetics, 9*, 849–858.

Gurney, M. E., Pu, H., Chiu, A. Y., et al. (1994). Motor neuron degeneration in mice that express a human Cu, Zn superoxide dismutase mutation. *Science, 264*, 1772–1775.

Hsieh-Li, H. M., Chang, J. G., Jong, Y. J., et al. (2000). A mouse model for spinal muscular atrophy. *Nature Genetics, 24*, 66–70.

Igoudjil, A., Magrane, J., Fischer, L. R., et al. (2011). In vivo pathogenic role of mutant SOD1 localized in the mitochondrial intermembrane space. *The Journal of Neuroscience, 31*, 15826–15837.

Katsuno, M., Adachi, H., Kume, A., et al. (2002). Testosterone reduction prevents phenotypic expression in a transgenic mouse model of spinal and bulbar muscular atrophy. *Neuron, 35*, 843–854.

Katsuno, M., Banno, H., Suzuki, K., et al. (2010). Efficacy and safety of leuprorelin in patients with spinal and bulbar muscular atrophy (JASMITT study): a multicentre, randomised, double-blind, placebo-controlled trial. *Lancet Neurology, 9*, 875–884.

Le, T. T., Pham, L. T., Butchbach, M. E., et al. (2005). SMNDelta7, the major product of the centromeric survival motor neuron (SMN2) gene, extends survival in mice with spinal muscular atrophy and associates with full-length SMN. *Human Molecular Genetics, 14*, 845–857.

Liu, J., Lillo, C., Jonsson, P. A., et al. (2004). Toxicity of familial ALS-linked SOD1 mutants from selective recruitment to spinal mitochondria. *Neuron, 43*, 5–17.

Lunn, M. R., & Wang, C. H. (2008). Spinal muscular atrophy. *Lancet, 371*, 2120–2133.

Majounie, E., Renton, A. E., Mok, K., et al. (2012). Frequency of the C9orf72 hexanucleotide repeat expansion in patients with amyotrophic lateral sclerosis and frontotemporal dementia: a cross-sectional study. *Lancet Neurology, 11*, 323–330.

Mitchell, J. D., & Borasio, G. D. (2007). Amyotrophic lateral sclerosis. *Lancet, 369*, 2031–2041.

Monani, U. R., Sendtner, M., Coovert, D. D., et al. (2000). The human centromeric survival motor neuron gene (SMN2) rescues embryonic lethality in Smn(-/-) mice and results in a mouse with spinal muscular atrophy. *Human Molecular Genetics, 9*, 333–339.

Montie, H. L., Cho, M. S., Holder, L., et al. (2009). Cytoplasmic retention of polyglutamine-expanded androgen receptor ameliorates disease via autophagy in a mouse model of spinal and bulbar muscular atrophy. *Human Molecular Genetics, 18*, 1937–1950.

Nagai, M., Re, D. B., Nagata, T., et al. (2007). Astrocytes expressing ALS-linked mutated SOD1 release factors selectively toxic to motor neurons. *Nature Neuroscience, 10*, 615–622.

Renton, A. E., Majounie, E., Waite, A., et al. (2011). A hexanucleotide repeat expansion in C9ORF72 is the cause of chromosome 9p21-linked ALS-FTD. *Neuron, 72*, 257–268.

Schrank, B., Gotz, R., Gunnersen, J. M., et al. (1997). Inactivation of the survival motor neuron gene, a candidate gene for human spinal muscular atrophy, leads to massive cell death in early mouse embryos. *Proceedings of the National Academy of Sciences of the United States of America, 94*, 9920–9925.

Takeyama, K., Ito, S., Yamamoto, A., et al. (2002). Androgen-dependent neurodegeneration by polyglutamine-expanded human androgen receptor in Drosophila. *Neuron, 35*, 855–864.

Wichterle, H., & Peljto, M. (2008). Differentiation of mouse embryonic stem cells to spinal motor neurons. Chapter 1:Unit 1H.1. 1-1H. 1. 9, *Current protocols in stem cell biology*.

Wong, P. C., Pardo, C. A., Borchelt, D. R., et al. (1995). An adverse property of a familial ALS-linked SOD1 mutation causes motor neuron disease characterized by vacuolar degeneration of mitochondria. *Neuron, 14*, 1105–1116.

Yamanaka, K., Boillee, S., Roberts, E. A., et al. (2008). Mutant SOD1 in cell types other than motor neurons and oligodendrocytes accelerates onset of disease in ALS mice. *Proceedings of the National Academy of Sciences of the United States of America, 105*, 7594–7599.

Yamanaka, K., Chun, S. J., Boillee, S., et al. (2008). Astrocytes as determinants of disease progression in inherited amyotrophic lateral sclerosis. *Nature Neuroscience, 11*, 251–253.

Zhang, Z., Lotti, F., Dittmar, K., et al. (2008). SMN deficiency causes tissue-specific perturbations in the repertoire of snRNAs and widespread defects in splicing. *Cell, 133*, 585–600.

FURTHER READING

Cho, A., Haruyama, N., & Kulkarni, A. B. (2009). Generation of transgenic mice. *Curr Protoc Cell Biol*. Chapter 19: Unit 19.11.

Dion, P. A., Daoud, H., & Rouleau, G. A. (2009). Genetics of motor neuron disorders: new insights into pathogenic mechanisms. *Nature Reviews Genetics, 10*, 769–782.

Festing, S., & Wilkinson, R. (2007). The ethics of animal research: Talking point of the use of animals in scientific research. *EMBO Reports, 8*, 526–530.

Manfredi, G., & Kawamata, H. (2011). *Neurodegeneration: methods and protocols*. New York, USA: Humana Press.

Valentine, J. S., Doucette, P. A., & Zittin Potter, S. (2005). Copper-zinc superoxide dismutase and amyotrophic lateral sclerosis. *Annual Review of Biochemistry, 74*, 563–593.

GLOSSARY

SOD1: (Cu/ Zn superoxide dismutase) Copper/Zinc Superoxide dismutase is a ubiquitously expressed, major antioxidant enzyme. Subcellular localization is predominantly cytosol. It consists of 153 amino acids, normally exists as a 32 kDa homodimer, and contains one copper and zinc per protein for catalyzing reactive superoxide to oxygen and hydrogen peroxide, resulting in reducing the level of superoxide in the cells.

Microglia and Astrocytes Non-neuronal, glial cells in the central nervous system that support the functions of the neurons. Microglia are immune cells that eliminate pathogens and damaged cells at the sites of neuronal damage or inflammation. They have migratory and phagocytic properties, and secrete many proinflammatory molecules to provoke neuroinflammation when activated. Microglia have similar functions to peripheral monocytes and macrophages. Astrocytes support the function of neurons by secreting neurotrophic factors and controlling neuronal functions through controlling ion balance, recycling neurotransmitter glutamate, and vasculatures. Glial fibrillary acidic protein (GFAP), a structure protein exclusively expressed in the adult astrocytes, is often used to mark astrocytes by the immunohistochemistry.

Cre and loxP Cre protein is a bacterially derived enzyme, initially identified in P1 bacteriophage for its DNA replication. The loxP site is a specific 34-base pair (bp) sequence, consisting of an eight bp sequence in the middle, where recombination takes place, and two flanking 13-bp inverted repeats (example: ATAACTTCGTATAGCATACATTATACGAAGTTAT, the central 8-bp sequence is underlined). Cre protein recognizes loxP sequences for DNA recombination.

ABBREVIATIONS

AD Alzheimer's Disease
ALS Amyotrophic Lateral Sclerosis
AR Androgen Receptor
FUS Fusion in Sarcoma
FTLD Fronto-temporal lobar degeneration
GFAP Glial Fibrillary Acidic Protein

HD Huntington's Disease
PCR Polymerase Chain Reaction
PD Parkinson's Disease
SBMA Spinal and Bulbar Muscular Atrophy
SCD Spinocerebellar Degeneration
SMA Spinal Muscular Atrophy
SMN Survival of Motor Neuron
snRNA Small Nuclear RNA
snRNP Small Nuclear Ribonucleoprotein
SOD1 Copper/Zinc Superoxide Dismutase
TDP-43 TAR DNA Binding Protein-43

LONG ANSWER QUESTIONS

1. Neurodegenerative diseases are defined by many elements, and contain many diseases. (A) Provide a definition of neurodegenerative disease, and (B) Provide several examples.
2. There are many approaches for modeling mouse models for inherited neurodegenerative disease. If the mutation of the gene provokes toxicity in neurons, and mutations are transmitted as autosomal dominant inheritance, which type of genetic engineering of the mouse is appropriate for modeling? Explain your answer.
3. Some inherited disease is mediated by a loss of gene function. (A) Describe the most likely type of inheritance pattern. (B) Which type of genetic engineering of the mouse is appropriate?
4. You would like to disrupt the gene of interest from the mouse. However, you cannot obtain the gene-deficient mouse, because the systemic disruption of the gene of interest causes early embryonic lethality. In such a case, what kind of alternative approaches can you take?
5. When the model mouse for motor neuron disease is used for translational research to develop a medicine, the drugs effective for the mouse models are not always successful for the clinical trial for human patients. Explain the possible causes of failure in translation.

SHORT ANSWER QUESTIONS

1. Name a neurodegenerative disease affecting midbrain dopaminergic neurons, and describe its inheritance pattern.
2. Describe three representative motor neuron diseases, and their genetic inheritance.
3. Why are women carrying mutations of Androgen receptor not susceptible for SBMA (Spinal and Bulbar Muscular Atrophy)?
4. Which technique is used for conditional gene knockout (cell-type specific gene disruption)?
5. In spinal muscular atrophy (SMA), what type of therapeutic approach is likely to be promising, based on research of animal models?

ANSWERS TO SHORT ANSWER QUESTIONS

1. Parkinson's disease (most cases are not genetically determined, while about 5–10 % of cases are inherited).
2. ALS (10% of cases are inherited and about 18 ALS genes are identified), SMA (autosomal recessive inheritance), SBMA (X-linked recessive)
3. Male sex hormone, androgen is required for mutant androgen receptor to provoke toxicity.
4. Cre-loxP system
5. Increasing expression of functional SMN protein (derived from *SMN2* gene).

Chapter 4

Epigenetics and Animal Models: Applications in Cancer Control and Treatment

Mukesh Verma, Neelesh Agarwal and Mudit Verma

Epidemiology and Genomics Research Program, Division of Cancer Control and Population Sciences, National Cancer Institute, National Instititues of Health (NIH), Bethesda, Maryland

Chapter Outline

SUMMARY

Genetically engineered tumor-prone mouse models have proven to be powerful tools in understanding many aspects of carcinogenesis, including epigenetics. This article focuses on cancer because, compared to other diseases, cancer epigenetics has been studied extensively, from diagnostics to prognosis to therapy to survival. Translation of epigenetic information in a clinical setting is also included.

WHAT YOU CAN EXPECT TO KNOW

Tumorigenesis is a multistep process and epigenetics plays an important role in disease initiation and progression, especially by inactivating tumor suppressor genes and activating oncogenes. Epigenetic changes can be reversed by natural nutrients and biological food components (Verma and Srivastava, 2002; Dunn *et al.*, 2003; Verma *et al.*, 2003; Verma *et al.*, 2004; Kumar and Verma, 2009; Khare and Verma, 2012; Mishra and Verma, 2012). Their implication in preventing and treating diseases like cancer is tremendous. This article will provide advantages of using animal models to understand and implement epigenetic approaches in cancer control and treatment. A discussion about the strengths and weaknesses of various epigenetic technologies is also included.

Animal Biotechnology. http://dx.doi.org/10.1016/B978-0-12-416002-6.00004-3

HISTORY AND METHODS

INTRODUCTION

Cancer is a genetic and epigenetic disease. A number of animal models have been created based on animal biotechnology knowledge. The most useful models are the mouse models where epigenetic approaches have been applied. These models are useful because genome-wide methylation patterns and chromatin modifications are dynamic can be monitored in animal models. They do not persist throughout life but undergo precise, coordinated changes at defined stages of development. Epigenetically altered patterns are involved in the environmentally triggered phenotypes. Some of the phenotypes correlate with disease progression. Some sections below cover the general epigenetic regulation and animal models in different cancers. Ethical issues, protocols involved, and general questions are also discussed.

Animal models are substantially better than cell line models to understand disease process and develop treatment strategies, but they will never completely capitulate the intricacies of the human organ systems. Knowledge gained from animal model systems should be validated in human patients. It sounds simple, but involves some difficult, if not impossible, steps regarding ethical concerns, safety issues, and technical limitations. In ideal situation, disease progress (here carcinogenesis) could be studied in human patients without negatively impacting their safety, maintaining all ethical standards and care, with hope that treatment will be favorable, and disease will be managed. Progress in the field of animal models in understanding carcinogenesis in different organ sites is discussed.

HISTORY: LANDSCAPE OF EPIGENETICS

Both genetic and epigenetic changes occur simultaneously in an organism. Genetic alterations include mutations and single nucleotide changes, deletions, insertions, changes in copy number, and translocations. Epigenetic alterations (methylation, histone and selected non-histone protein alterations, non-coding RNA alterations, imprinting, and chromatin remodeling) broadly include non-genetic alterations that are capable of being transmitted from one cell generation to another. The epigenome is significantly perturbed in most cancers, raising questions about which epigenetic alterations are functionally important in cancer. The current focus is on identifying differentially expressed marks associated with cancer so that they can be used as screening tools to identify high-risk populations.

DNA methylation also can occur in other parts of genes. Because DNA methyl transfer reactions occur on one strand at a time, *de novo* methylation leading to full double-stranded DNA methylation can be thought of as two sequential one-stranded reactions. DNA methylation plays a key role in DNA repair, genome instability, and regulation of chromatin structure (Figure 4.1). DNA methylation can occur as hypermethylation or hypomethylation of CpG islands, resulting in gene inactivation or activation. Generally, gene-specific methylation is observed in the promoter region, and global methylation is observed in repeat sequences and transposable elements (i.e. Short Interspersed Nuclear Elements or SINE, and Long Interspersed Nuclear Elements or LINE) that contribute to genomic instability and altered gene expression, leading to disease development. Methylation involves covalent addition of a methyl

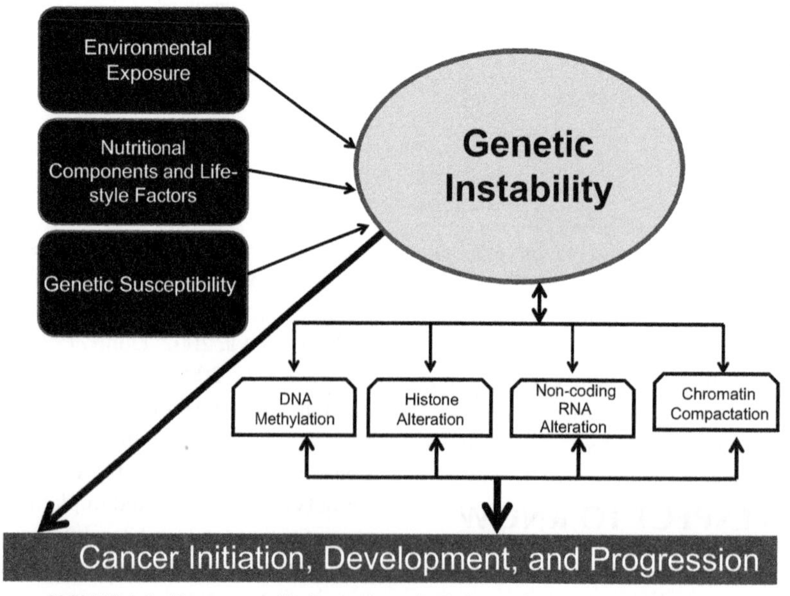

FIGURE 4.1 Factors contributing to the epigenetic regulation of gene expression.

group at position 5 of cytosine. DNA methylation mainly occurs in CpG islands. These regions are rich in phosphate-linked pairs of cytosine and guanine residues.

For normal function of a cell or organ, epigenetic regulation is needed; however, this regulation is disturbed during disease initiation and progression. Thus our genome is the "hardware" and our epigenome is the "software" of the body. Genetic information is static whereas epigenetic information is dynamic and transient. In the body, all the cells have the same genome but each cell has a different epigenome. The phenotype of a cell is determined by its epigenome.

Gene transcription depends strongly on chromatin structure. Euchromatin, with an open chromatin structure, indicates active transcription; heterochromatin, with a tight chromatin structure, represents transcription repression. Epigenetic mechanisms have evolved to regulate the structure of chromatin and, as a consequence, access to DNA for transcription. These modifications include the post-translational modifications of histones, remodeling of chromatin, polycomb suppressor complexes, and DNA modifications. Active chromatin complex formation involves energy-dependent reactions, specific histone variations, and chromatin remodeling. Few nucleosome remodeling factors, such as SWI/SNF, ISWI, CHD, and IN080, have been characterized. We also come across of the "histone code" hypothesis, which refers to an epigenetic marking system using different combination systems of histone modification patterns to regulate specific and distinct functional outcomes of the genome. The hypothesis involves several layers of gene regulation interpretations. One refers to the establishment of homeostasis of a combinatorial pattern of histone modifications in a given cellular and developmental context conducting "writing" and "erasing" messages via histone modifying enzymes. (Key examples of these enzymes are histone methyltransferase, which adds methyl group to histones, and demethylase, which removes the methyl group). Another interpretation involves interaction of modified histone with a different factor "effector," which results in transcription activation or suppression. Sometimes, neutralization of the charges on lysine residues by the acetylation step is considered an additional interpretation. All these regulatory mechanisms function broadly and set up an epigenetic environment that decides the fate of cell(s) for development or differentiation. Along with methylation and acetylation of histones, other modifications have also been reported, such as biotinylation, phosphorylation, and ubiquitination.

Packaged DNA in nucleosomes, each consisting of a histone octamer arranged as two H3-H4 tetramers and two H2A-H2B dimmers, is very stable in normal growth but gets unstable in a diseased state. The amino terminal part of histones protrudes out and becomes susceptible to enzymatic modifications, specifically at lysine residues, but also at other amino acids. More than 100 histone modifications of amino acids have been reported. The dimeric H3 and H4 form a tetramer, while H2A and H2B remain dimers. Post-translational modifications of histones provide the basis of gene expression and regulation in normal and disease states and DNA repair. Histone mark mis-reading, mis-erasing and mis-rewriting contribute to cancer development. At the same time, these modifications provide us clues to identify drug targets for cancer treatment. Methylation of histones can occur at both lysine and arginine residues and these modifications are reversible. During the process, different enzymes are involved, commonly called "writers" and "erasers." These enzymes are very specific and work with distinct factors. H3K4 methylation is established, for example, by the SET1 and mixed-lineage leukemia family of histone methyl transferases and is removed by the lysine-specific histone demethylase 1 and jumonji AT-rich interactive domain (JARID1) family of histone demethylases (HDMs). H3 methylation occurs at different lysine sites (H3K4, H3K9, H3K27, H3K36, H3K79). Addition of methylation can be mono-, di-, or tri-methylation. The distribution of these three methylated forms depends on the state of the cancer progression. H3K4me3 is generally associated with gene activation, and H3K27me3 with gene silencing. Pluripotent stem cells have almost equal distribution of these two forms, while differentiated cells have higher levels of H3K27me3 compared to H3K4me3.

Mixed lineage leukemia (MLL) catalyzes methylation of H3K4 whereas enhancer of zeste 2 (EZH2) catalyzes methylation of H3K27. MLL rearrangement and EZH2 disregulation are the most common mutations in leukemia and solid tumors. Global histone modifications are shown to be associated with cancer survival and recurrence.

Modified histones were observed in more than 400 small cell lung cancer patients compared to their matched controls. Liver cancer patients were followed for their histone modifications (methylation and acetylation) at different amino acid residues on H3 and prediction of prognosis came true. In gastric cancer, H3 and H4 modifications correlated well with the recurrence of cancer.

Throughout life, equilibrium is needed in the methylation state of the whole epigenome. Enzymes involved in methylation are called methyl transfereases; these enzymes either initiate or maintain methylation. Proteins that bind to methylated DNA have also been identified and characterized and are referred to as methylated binding proteins (MBP).

Another aspect of epigenetic regulation is non-coding RNAs, especially microRNAs (miRNAs). Genetic and epigenetic alterations in miRNA processing contribute to cancer development. MicroRNAs are highly conserved 19–25 nucleotide long non-protein coding RNAs involved in development and differentiation. These miRNAs bind to mRNAs and degrade them. RNA polymerase II codes miRNAs as a long precursor that is processed from nucleus to cytoplasm with the help of protein complexes. miRNAs are involved in cancer initiation and development. miRNAs are more stable than mRNAs, probably because of their small size. Molecular profiling of miRNAs helps in disease stratification, especially in identifying the stages of disease. The therapeutic potential of miRNAs also is being explored by several investigators.

One additional types of gene regulation is gene imprinting, which is paternal, or maternal allele-specific expression of a limited number of genes (50 to 80). Without proper imprinting control, abnormal growth occurs. Examples of diseases regulated epigentically via imprinting are Blackwith-Wiedemann syndrome (BWS), Silver–Russell Syndrome (SRS), and X-chromosome inactivation. Methylation of DNA occurs in the imprinting loci called imprinting control regions (ICR). Loss of imprinting (LOI) of IGF2 has been proposed in stem cell proliferation and cancer.

The phenotype of a cell is determined by its expression profiles based on the environmental changes of the epigenetic machinery. Epigenetics provides stability and diversity to this phenotype mediated by chromatin marks that affect local transcriptional potential. These marks are preserved or regenerated during cell division. CpG in the promoter represents the transcriptional suppressive mark. DNA methylation information is erased by standard molecular biology techniques, such as PCR, but it cannot be revealed by hybridization because the methyl group is located in the major groove of DNA rather than at the hydrogen bonds. That is the basic reason that methylated DNA is modified first before its analysis. Initially these technologies were applicable to the local regions that were rich in methylated DNA, but now these techniques can be applied to modify DNA of the whole genome (termed as genome-wide methylation).

Interaction of environmental factors with epigenetic regulation of genes is very important. Even transient exposures to hazardous substances can have persistent life-long phenotypic effects, and epigenetic equilibrium can be altered by exposure to toxic substances. During toxicology analysis, careful attention should be paid to the contribution of epigenetic factors. The agent under study should be toxic and should be capable of altering an individual's epigenetic profile; this should be recorded and stored for further analysis. Individual variations are contributed as a result of exposure to different behavioral conditions, chemicals, diet, radiation throughout life, and genetic background. In the case of cancer, such factors include toxic substances, radiation, infectious agents, specific dietary components, tobacco, alcohol, and environmental factors. It makes sense to identify epigenetic markers in normal and exposed populations. The advantages of identifying epigenetic biomarkers of exposure include improved exposure assessment, documentation of early alterations preceding cancer development, and identification of high-risk populations. Discovery-oriented as well as hypothesis-driven toxico-epigenomic studies can be conducted using factors related to these and additional information and resources. A few examples are discussed below.

Exposure to metal carcinogens such as nickel, chromate, arsenite, and cadmium has increased recently because of occupational exposures, the massive growth of manufacturing activities in developed and industrialized countries, increased consumption of nonferrous materials, and the disposal of waste products. DNA methyl transferaes (DNMTs) are the enzymes that transfer the methyl group to the cytosine. DNMT activity is very high during hypermethylation. DNMT1 is activated by toxins. According to recent theories, Polycomb-group proteins also are involved in activating DNMT1.

Metals are weak carcinogens and do not damage DNA directly (as does radiation); but they exert their carcinogenic effects by epigenetic mechanisms, especially after chronic exposure. A few more examples of toxic substances and their involvement in epigenetics is described below. Arsenic in drinking water is a health hazard. It causes genetic and epigenetic changes and contributes to the development of lung cancer. Arsenic induces global and gene-specific methylation changes and histone modifications in different populations. MiRNA profiling also changes with arsenic exposure. In addition, mitochondrial functions (oxidative stress) are affected by arsenic. Furthermore, bimethylated arsenic may cause oxidative damage. Arsenic exposure may lead to bladder, lung, and liver cancers. Benzene and its metabolite hydroquinone recently were reported to induce global hypomethylation and cytogenetic changes leading to leukemia. This study established a correlation between genetic and epigenetic changes that can lead to cancer. The investigators now are studying the degree of DNA breakage and adduct formation that occurs during benzene exposure. Cadmium exposure occurs in factories that produce certain batteries, in the electroplating process, and in metal industries, and has been linked to pulmonary diseases including lung cancer. Cadmium exposure induces hypermethylation in tumor suppressor gene *p16* and DNA repair gene *hMLH1*. It has been investigated that chromium toxicity and its contribution to lung cancer occur via methylation. Results described methylation in the *APC* (86%), *MGMT* (20%), and *hMKH1* (28%) genes compared to unexposed controls. This suggests the possibility of genomic instability resulting from chromium-induced epigenetic changes. Nickel, a nonessential metal, is used in the production of jewelry, coins, stainless steel, batteries, and medical devices. Prolonged exposure to nickel may change chromatin to heterochromatin and acetylate histones. Perfluorooctane sulfonate (PFOS) exposure and its effects on *GSTP1* methylation, global methylation, and hypomethylation of the Long Integrated Nuclear Element (LINE) and Short Integrated Nuclear Element (SINE) regulatory sequences have been investigated. PFOS affects gene regulation prenatally, as well as after birth, and contributes to carcinogenesis. Depleted uranium, a dense metal, is used in military applications. It emits alpha particles that can damage cellular processes. Along with genetic effects, uranium contributes to the causation of cancer by epigenetic mechanisms, mainly methylation. The involvement of uranium in leukemia via hypomethylation has been documented.

PRINCIPLE

Epigenetics reveals many molecular events of cancer biology. During epigenetic regulation due to external environmental, dietary, life-style, and behavior changes, key components of chromatin, histones, non-histone proteins and selected factors, DNA in the promoter region, and miRNA expression change and initiate the carcinogenesis process. Technologies exist to measure these changes. By following epigenetic changes cancer can be detected and intervention and therapeutic approaches can be applied.

Use of Mouse Models in the Epigenetics of Cancer

Two main approaches are being used in mouse models of epigenetic changes in cancer: (a) models that are genetically manipulated to overexpress or to lack specific genes that are direct regulators of DNA methylation and methylation-related gene expression; and (b) nutrition-, genetic-, or carcinogen-induced mouse models of cancer with assessment of somatic epigenetic alterations that arise in tumors (Mathers *et al.*, 2010). In the first approach, haplo-insufficient mice such as *Dnmt1*$^{wt/-}$ mice or mice that have tissue-specific gene alterations are used, which makes interpretation of results easy. For example, the *Apc*$^{Min/wt}$ mouse model system for intestinal cancer used mice with hypomorphic expression of *Dnmt3b* (DNA methyltransferase 3b) or *Mbd2* (methyl CpG domain-binding protein 2). In this model, *APC* allele modifications contribute to intestinal cancer. To elaborate, in the *Apc*$^{Min/wt}$ system, the double-mutant mice lack either the ability to establish new methylation patterns (*Dnmt3b* hypomorph) or the ability to repress genes that are associated with methylated DNA (*Mbd2* hypomorph). This provides an excellent system to directly demonstrate the role of methylation in cancer progression. This model also established the fact that *de novo* methylation and maintenance of methylation are required for disease development. Depletion of DNA methylation caused by a partial loss of the maintenance methyltransferase Dnmt1 leads to aggressive T-cell lymphomas (Gaudet *et al.*, 2003). In another kind of mouse model, a candidate gene approach was adopted. Tumor suppressor genes were identified in specific cancers. These genes are knocked down, and their effects on the development of animals and cancer are investigated. In the following section, selected cancer types and their models are described. The criteria used in selecting these cancers were their high incidence, prevalence, and mortality rates, as well as their complex biology.

Examples with Applications

Brain Cancer

A medulloblastoma is a pediatric brain tumor. Frequently in this tumor type, Kuppel-like factor-4 (*KLF4*) acquires homozygous deletions and occasional single nucleotide polymorphisms (SNPs). Functional analysis indicated inactivation of *KLF4* expression at the transcriptional and translational levels. When cells were treated with the demethylating agent 5-azacytidine, the *KLF4* gene was reactivated, which suggests the presence of CpG islands in the promoter region of the gene and epigenetic regulation of the gene. This study integrated genetically and epigenetically mediated gene regulation in cancer. Because *KLF4* also is targeted in other neoplasms, this study may provide the fundamental mechanism and model system to understand the underlying mechanism. In another study that was based on genome-wide methylation and copy number analyses, Hong et al. demonstrated suppression of growth suppressor *SLC5A8* in glioma and its reactivation by demethylating agents (Hong *et al.*, 2005).

Breast Cancer

Based on the genome-wide methylation analysis, Demircan et al. demonstrated similarity in the mouse and human methylomes in breast cancer models (Demircan *et al.*, 2009). In these experiments, human orthologs of *ATP1B2, FOXJ1*, and *SMPD3* were aberrantly hypermethylated in the human disease, whereas *DUSP2* was not hypermethylated in primary breast tumors. In human breast cancer, *BRCA1, CDKN2A, SFN, CDH1*, and *CST6* are regulated epigenetically, but not many mouse genes are regulated epigenetically. Studies by Demircan et al. identified the *Timp3, Rprm, Smpd3, Dusp2, Atp1B2*, and *FoxJ1* genes as being regulated epigenetically (Demircan *et al.*, 2009). Of all genes studied, only Wif1 and Dapk1 were found to not be silenced in either species. The protein encoded by *ATP1B2* is a member of the Na^+/K^+ ATPase beta chain protein family, which is an integral membrane protein responsible for establishing and maintaining the electrochemical gradients of Na^+ and K^+ ions across the plasma membrane. *FOXJ1* is a member of the forkhead gene family of transcription factors. Other genes discussed above are tumor suppressor genes.

Colorectal Cancer

The epigenetic regulation of colorectal cancer has been studied extensively. The epigenetic agents folic acid (FA) and sodium butyrate (NaBU) have shown potential to prevent colorectal cancer. In one experimental study, colorectal cancer was induced by 1,2-dimethylhydrazine (DMH) in ICR mice; FA and NaBU were administered after colorectal cancer developed. A lack of FA in the diet led to the hypomethylation of the total genomic DNA (global DNA) or proto-oncogenes. This also resulted in the development of tumors. Methylation profiling and H3 histone immunoprecipitation were measured in control and treated animals, and the protective effect of these agents was observed. In this study, investigators demonstrated that drug combinations enhanced tumor prevention and, when NaBu and FA were administered concomitantly, the beneficial effect was greater than that observed for either agent administered alone.

Esophageal Cancer

The methylene tetrahydrofolate reductase (*MTHFR*) gene is polymorphic, and the *p16, MGMT*, and *MLH1* genes are methylated during esophageal squamous cell carcinoma (ESCC). A diet rich in FA supplies the methyl group and influences the methylation process. Clinically, dietary FA has been reported to be beneficial for ESCC patients. Barrett's esophagus is the precancerous form of esophageal adenoma carcinoma, which is a metaplastic condition in which the normal squamous epithelial cells of the lower esophagus are replaced by small intestine-like columnar linings. In a large clinical study, surveillance and early detection of methylation biomarkers (a set of eight genes that included *p16, HPP1, RUNX3, CDH13, TAC1, NELL1, AKAP12*, and *SST*) was found to be very useful in reducing the incidence rate of this cancer. These studies established the stratification strategy for esophageal cancer progression, which might lead to better prognosis of the disease in the future.

Gastric Cancer

Kikuchi *et al.* demonstrated that *cycloxygenase-2* (*COX-2*) plays an important role in gastric cancer development (Kikuchi *et al.*, 2002). Gastric cancer is very common in Asian countries. It also is well established that *COX-2* plays a major role in inflammation and prostaglandin synthesis via the arachidonic acid pathway. In one model system, hypermethylation of *COX-2* resulted in its inactivation; epigenetic inhibitors reactivated *COX-2*, which suggests the involvement of methylation and histone alterations as key components of gastric carcinogenesis. Gene-encoding protease activated receptor 4 (PAR4) is involved in colon and prostate cancer, but its role in gastric cancer was not known until Zhang *et al.* demonstrated its regulation by epigenetics in a model system as well as in human tissues (Zhang *et al.*, 2011). PAR4s are G protein-coupled receptors that are involved in proteolysis. The involvement of PAR4 in gastric aggression also has been proposed. It also has been suggested that the epigenetic regulation of host genes is affected by *Helicobacter pylori* infection, thereby suggesting the involvement of *H. pylori* in the initiation and progression of gastric cancer. The involvement of cytidine deaminase has also been proposed by these investigators.

Head and Neck Cancer

Sun et al. demonstrated the role of transkelolase-like 1 (*TKTL1*) gene in head and neck squamous cell carcinoma (HNSCC) (Sun *et al.*, 2010). In cell-line and xenograft models, functional characterization of *TKTL1* was achieved by following mRNA and protein expression of fructose-6-phosphate, glyceraldehyde-3-phosphate, pyruvate, lactate, and the levels of HIF1α protein and its downstream glycolytic targets. Hypermethylation of *TKTL1* in HNSCC confirmed that epigenetic regulation is an integral part of HNSCC. A combination of protease inhibitor and histone deacetylase therapeutic agents showed promising results (reactivation of transcriptionally inactive genes), which further confirmed epigenetically mediated regulation of HNCCC. Note that proteasome inhibitor PS-341 has emerged as a novel therapeutic agent that works well in combination with other agents as a cancer inhibitor.

Lung Cancer

For lung cancer, more information has come from mutations and SNPs than from alterations in epigenetics and proteomics. Lung cancer mortality rates are high because the disease spreads to other organs, and little is known about the underlying mechanisms. The mouse is a good model for use in evaluating the efficacy of chemopreventive agents for lung cancer. Gene silencing by promoter hypermethylation is a critical component in the development and progression of lung cancer and an emerging target for preventive intervention by demethylating agents. Yue *et al.* demonstrated that fibulin-5, a vascular ligand for integrin receptors, worked as a suppressor of lung cancer invasion and metastasis by promoter hypermethylation (Yue *et al.*, 2009). It inhibited matrix metalloproteinase-7 (MMP-7), a marker of cell invasion. Results were confirmed by classical knockdown approaches. Further research demonstrated that epigenetic silencing of fibulin-5 promoted lung cancer invasion and that metastasis occurred by activating MMP-7 expression through the ERK pathway, which is an integral part of lung cancer development. Belinsky's group demonstrated that methylation events that were observed frequently in human lung cancer could be followed in mouse models (Vuillemenot *et al.*, 2006). The researchers identified four potential biomarkers for assessing intervention approaches for reversing epigenetically mediated gene silencing.

Lymphoma and Leukemia

In a mouse model of leukemia, transgenes were implicated in driving the proliferation of T cells. Some of these cells eventually acquired enough secondary mutations to give rise to either T-cell lymphomas or acute lymphoblastic leukemias (ALL). The methylation profile at the genomic level was determined to identify both hyper- and hypomethylation events in the induced benign proliferative state that preceded the formation of the tumors, as well as events in the established tumors.

In hematologic malignancies, epigenetic inhibitors can be used for the treatment of cancer. In one study, zebularine, a demethylating agent, was administered in mice where radiation was used to induce lymphogenesis. Unradiated mice of the matching weight and age were used as controls. Results indicated hypomethylation in lymphoma-associated genes and suppression of lymphoma in zebularine-treated animals.

To determine the early stages in the development of chronic lymphocytic leukemia (CLL), Chen *et al.* used a well-established mouse model for CLL and followed expression of human *TCL1*, a known CLL oncogene in murine B-cells, which led to the development of mature CD19+/CD5+/IgM+ clonal leukemia with a disease phenotype similar to that seen in human CLL (Chen *et al.*, 2009). Results indicated that the mouse model recapitulated the epigenetic events that were reported for human CLL. These events were detected as early as three months after birth, which was close to a year before disease manifestation. Accumulated epigenetic alterations during CLL pathogenesis occurred as a consequence of *TCL1* gene silencing and synthesis of the NFkβ repressor complex. This suggests that NFkβ may be used as a therapeutic target in CLL. The role of miRNA in leukemia also has been explored.

Another well-studied model is an epithelial multistep tumorigenesis model of squamous cell cancer formation. In this model, tumors are induced and promoted by chemical exposure, resulting in different progression stages of cancer development. This helps in the molecular classification of tumors. Fraga's group completed the methylation profiling in normal tissue and tumors and identified benign papilloma and invasive carcinoma-related markers, mostly tumor suppressor genes, and established an association between disease and the activation or inactivation of these tumor suppressor genes (Fraga *et al.*, 2004). Surprisingly, several tumor suppressor genes, including *Cdkn2a*, were found to be consistently methylated very early in disease progression, suggesting that methylation of these genes may be crucial to the formation of tumors in this system. Among early expression genes was *Cdkn2a*, which is a tumor suppressor gene; a few new genes also were identified.

Prostate Cancer

The protective role of FA is well established in colorectal cancer, and the underlying epigenetic mechanism also has been characterized extensively. Such studies have not been conducted in prostate cancer. Recent research demonstrated that similar protection could be obtained in prostate cancer. The transgenic adenoma of the mouse prostate (TRAMP) mouse model was used for this study because the model is very appropriate to follow the aggressiveness of the disease. Generally, prostate cancer takes 15 to 20 years to develop, but in a few cases this cancer is very aggressive. It is clinically significant to distinguish aggressive cancer from normally developing cancer, and the TRAMP model might be very useful in such situations.

Liver Cancer

Methylation sources in the body and one-carbon metabolism play important roles in epigenetically mediated gene regulation, and any abnormalities in the process may contribute to cancer initiation and progression. In a recent study of hepatocellular carcinoma (HCC), Teng *et al.* reported HCC development and fatty liver due to disturbed choline and one-carbon metabolism by betaine homocysteine-S-methyltransferase (BHMT) in a mouse model (Teng *et al.*, 2011). This study also demonstrated the integration of genetic and epigenetic alterations that are known to influence each other and contribute to cancer development. The function of BHMT is to transfer the methyl group from betaine to homocysteine, thereby forming dimethylcysteine and methionine. Mutations in the *BHMT* gene have been reported in breast cancer. To study the effect of mutations in *BHMT*, knockout mice of this gene were generated and functionally characterized. Higher BHMT activity was found in the liver and kidney compared to other organs. Because methionine is the precursor of S-adenosylmethionine, which is the major source of the methyl donor, it plays an important role in epigenetic regulation. The role of long non-coding RNA (lncRNA) and microRNA (miR-29) in HCC was demonstrated recently by Braconi *et al.* (2011). These studies were based on the microarray analysis of more than 23,000 lnc RNAs (long noncoding RNAs) and miRNAs. About 3 percent of RNAs were found to be down-regulated in HCC. Furthermore, heterochromatin histone modifications were reported in liver cancer.

Other Approaches

Along with the characterization of methylation patterns, proteins binding to methylated regions also have been characterized. These proteins are identified by methylated DNA immunoprecipitation (methyl DIP), which involves the hybridization of immunoprecipitated methylated DNA to microarrays or deep sequencing of the DNA in the immunoprecipitated DNA complex to assess the pattern at the genome level. However, improvements are required to adapt this process on a large scale to address such problems as low resolution when using microarrays, difficulty in obtaining sufficient coverage when deep sequencing is used, and high false-discovery rates. Thus, the technique used to assess the cancer methylome in these model systems is one crucial factor to include when interpreting the results of studies of epigenetic alterations in mouse models.

METHODOLOGY

Methodologies described below are suitable to assay epigenetic components in animal tissues. In the following section, specific methodologies are described.

Methylation Profiling

MethylLight technology, pyrosequencing and ChIP-on-Chip are the key technologies to measure epigenetic alterations in cancer. For methylation profiling, quantitative methylation specific polymerase chain reaction (QMSP) assays followed by pyrosequencing (for confirmation) are performed. Sodium bisulfite treatment followed by alkali treatment is the key for all assays. Bisulfite reacts with unmethylated cytosines and converts them to thymidine. In the PCR reaction all converted Cs behave like Ts. Methylated cytosines and other bases are not affected by bisulfate treatments. MethylLight is the most common method to determine methylation profile in real time. This is a high throughput quantitative methylation assay that utilizes fluorescence based real time PCR technology and does not require any manipulation after the PCR reaction. The technology can detect methylation allele in the presence of 1,000 unmethylated alleles. The fluorescence signal reflects the amount of PCR amplified DNA.

For the purpose of the background, initial forays into broad DNA methylation profiling were made with two-dimensional gene electrophoresis. However, the advancement of microarray technologies and sequencing technologies opened the door to single-base-pair resolution whole-genome-methylation analysis. Several high-resolution genome-wide profiling techniques were pioneered in model organisms with small genome size, but soon they were adopted for the mammalian genome successfully. The only thing that should be carefully considered is the objective of a project in terms of how much resolution is needed. In the past few years, the presence of 5-hydroxymethyl cytosine has been reported. Technologies exist that can distinguish 5-methyl cytosine from 5-hydroxymethyl cytosine in the same sample.

DNA methylation profiling methods vary in their accuracy, types of biases, ability to detect in cis-methylation patterns, extent of which they are influenced by target sequence density (or DNA methylation density), and reproducibility of results. For low CpG dense regions, methylation specific restriction enzyme based technologies are appropriate. In sequencing based technologies, where libraries of different size restriction fragments are used, the efficacy of library construction depends on the size of the fragments (smaller fragments are better than larger fragments).

Bisulfite treatment is an integral part of methylation analysis. Treatment of DNA with bisulfite converts cytosine residues to uracil, but leaves 5-methylcytosine residues unaffected. Thus bisulfite treatment introduces specific changes in the DNA sequence, which depends on the methylation status of individual cytosine residues, yielding single nucleotide resolution information about the methylation status of DNA. This treatment should not be done for a long time; it not only degrades DNA but can also lead to an increased incidence of methylated cytosines converting to thymine residues (resulting in underreporting of DNA methylation). Completion of the bisulfite treatment can be monitored with assays that are specifically designed to detect incomplete conversion by monitoring the conversion of spiked DNA controls, or by retention of non-target sequence dinucleotide cytosines (such as CpH) in most mammals. Incomplete bisulfite conversion can arise from incomplete denaturation before treatment or reannealing during the bisulfate conversion.

Histone Profiling

ChIP-on-chip is the standard method for determining alterations in histone. Monoclonal antibodies against histone modifications are used in binding to fragmented DNA. The most common histone modifications in cancer are H3K4me1 (monomethylated), H3K4me3 (trimethylated), H3K9/16Ac, and H3K27me3; monoclonal antibodies are available commercially against these histone-modified proteins. For high throughput, the ChIP-Seq technique is applied. High-throughput profiling of miRNA can be achieved by Illumina sequencing.

Nucleosome Mapping

The positioning of nucleosomes in the promoter dictates disease initiation and progression of disease and also responsiveness to treatment. Nuclease I digestion of chromatin and analysis by gel analysis is the most common method for nucleosome mapping.

Data Storage and Processing

Recent advances data storage and processing technology have created a huge amount of high-quality, readily accessible data. Researchers are able to quickly acquire data to investigate biological system to understand the underlying mechanisms. Examples include the 1,000 genome project (http://www.1000genomes.org), which provides public access to the fully ordered nucleotide base sequences in 1,000 healthy individuals (pathologically disease-free); PubMed http://www.ncbi.nlm.nih.gov/pubmed), which provides access to the abstract to biomedical papers; and the Gene Expression Omnibus database (http://www.ncbi.nlm.nih.gov/geo), which provides Omics information for more than half a million samples. These data are very helpful in understanding abnormal gene expression during diseases initiation, development, and progression.

PROTOCOLS

Four main protocols are used in epigenetic regulation in animal models.

- **Extraction and sodium bisulfate treatment of genomic DNA**. DNA from tissues is isolated using Qiagen Blood and Cell Culture DNA Isolation Kit (Qiagen, VA) and stored at −20°C before use.
- DNA is modified with EZ DNA Methylation Kit (Zymo Research, CA) using protocols supplied with the kit.
- For sodium bisulfate treatment, 1 μg DNA was incubated with sodium bisulfate at 50°C for 16 hours in the dark (solution of sodium sulfite is supplied with the kit).
- Using columns supplied with the kit, the excess bisulfate is removed and treated DNA is eluted with the elution buffer in a small volume (20 μL). One microliter is sufficient for PCR amplification for the next step.
- **Methylation specific PCR.** Primers are selected based on the gene of interest and PCR amplification is done using Qiagen HotStart Taq DNA Polymerase kit in the amplification buffer.
- Amplified fragments are analyzed by gel electrophoresis.
- **Pyrosequencing.** Small amount of amplified DNA (2–10 μL) is mixed with streptavidin-coated Sepharose beads, based on instructions provided with the lit (GE Healthcare Biosciences AB, Uppsala, Sweden).
- Beads are shaken for 10 minutes at room temperature.
- After centrifugation the supernatant is used to do sequencing using a PyroMark MD System (Biotage, AP, Uppsala, Sweden).
- Analysis of sequencing data is done using the software supplied by PyroMark.
- **Analysis of Histone (Acetylated) Content.** Cells are isolated from animal tissue, treated with micrococcal nuclease and fixed for 15 minutes in formaldehyde in PBS on ice.
- Cells are fixed one more time in 70% ethanol. Due to large amount of HDAC proteins present in the cell, deacetylation activity is very high.
- All washings are conducted in a refrigerated centrifuge using cold PBS, and finally cells are fixed by keeping them on ice.
- Fixed cells are washed with PBS buffer with 15 BSA and permealized using 0.1% Triton-X100 in PBS for 10 minutes at room temperature.
- After washing with PBS, samples are incubated with 500 μL of 20% normal goat serum in PBS for 20 minutes at room temperature.
- Histone acetylation is detected using monoclonal antibodies anti tetra acetylated histone H4 (Upstate Biochemicals, VA).
- Analysis of results is done by gel electrophoresis and Western Analysis.
- Detection of antibodies is done with FITC-conjugated, affinity purified Fab2 fragment of goat anti-mouse IgG at room temperature in dark.

Some Useful Points to Consider in a Project Involving Epigenetic Profiling

- Expression profiling of cells treated with epigenetic inhibitors (for methyl transferase of acetyltransferase) has also been used as a discovery tool for epigenetically silenced genes (associated with disease development). However, it is prone to false-positive and false-negative results and is not considered a reliable approach of DNA methylation at a given locus.
- For initial screening of a sample with an unknown methylation pattern, methylation specific restriction enzymes, such as MspI, can be used to obtain an idea of how many methylation sites exist. This method is still applicable for some locus-specific studies that require linkage of DNA-methylation-specific information across multiple kilobases, either between CpGs or between a CpG and a genetic polymorphism (could be disease-associated). Although this is a very sensitive technique, it is extremely prone to false-positive results caused by incomplete digestion for reasons other than DNA methylation.
- Restriction Landmark Genomic Screening (RLGS) is a technique for broad DNA methylation profiling. The differentially expressed restriction pattern of DNA fragments is based on profiling using a panel of restriction enzymes with methylation specific recognition sites. These fragments can be separated by two-dimensional gel electrophoresis. RLGS is used frequently to identify imprinted loci and sites that are methylated in cancer.
- Array analyses couple enzymatic methods to array-based analyses. The first one is methylated CpG island amplification (MCA), which is based on the differential methylation sensitivities as well as cutting behavior of two restriction enzymes, SmaI ans XmaI. The second one is called representational differential analysis (RDA) combined with array hybridization. The disadvantage of this technology is its poor sensitivity. An alternative approach is differential methylation hybridization. In this approach, one pool of genomic DNA is digested with methylation specific enzyme and the other one with mock enzyme digestion. Two fluorescent dyes are used to label genomic pools. Both pools are cut with MseI, which has a high frequency of cutting DNA. Then both pools are hybridized. Since MseI does not cut methylated DNA (it recognizes

AATT sequence), the locus specific methylated regions can be identified. One more modification of this approach is MethylScope, which shows better sensitivity and specificity.

- For greater accuracy and faster speed, next-generation-sequencing based techniques are used. This approach provides allele-specific methylation and does not require digestion-based arrays.
- Methyl-seq is sequencing by synthesis of libraries constructed from size-fractionated enzyme digests (HpaII or MspI) that are compared with randomly sheared DNA fragments.
- ChIP-chip and ChIP-seq are two other approaches. Chromatin immunoprecipitation followed by microarray hybridization is called ChIP-chip (used for histone characterization). Chromatin immunoprecioitation followed by next generation sequencing is called ChIP-seq, which is generally used for genome-wide histone analysis. An alternative approach is affinity-purified antibodies against 5-methyl cytosine, which can attract sheared genomic DNA with histone binding site. This complex is analyzed by peptide digestion and next generation sequencing.
- Combined bisulfate restriction analysis (COBRA) determines the sensitive quantification of DNA methylation levels at a specific genomic locus on a DNA sequence in a small sample of genomic DNA. IN COBRA, DNA methylation levels are easily and quickly measured without the need for laborious sub-cloning and sequencing. This method is considered simple, fast, and inexpensive. The weaknesses include that the assay is limited to using existing restriction sites in the region of interest, and methylation that does not occur in the context of a specific restriction site cannot be assayed.
- MeDIp is enrichment of methylated regions by immunoprecipitaion of denatured genomic DNA with an antibody specific for methylated cytosine, followed by hybridization to either a tiling array or CpG island array. Very frequently this approach is applied for determining mouse methylome.
- One should consider the capability of resolution of different technologies. For example, bisulfate Sanger sequencing can assess approximately 40,000 different CpGs located on 2,524 different amplicons. On the other hand, Illumina Infinitium DNA methylation platform can analyze 27,578 different CpGs located at 14,495 different gene promoters.
- In affinity or enzyme pretreatment, short read sequencing is used to determine the prevalence of different regions in the enriched sample by counting the number of reads that can be aligned to the genome easily. On the other hand, the bisulfite sequencing approach extracts the alignment and DNA methylation information from the sequence itself. A general convention among scientists is that the read count methods are prone to sources of bias (GC content, fragment size, and copy number variation) that influences the chances that a particular region is included in the sequenced fragments. Although bisulfite-based methods are also subject to these effects, they do not influence the DNA methylation quantitation (measurement) itself, because this information is derived from the sequence.
- Pyrosequencing has also been used to analyze bisulfite treated DNA. Following PCR amplification of the region of interest, this technology is used to determine the bisulfite-converted sequence of specific CpG sites in the region. The main principle is based on the ratio of C to T at individual sites that can be determined quantitatively based on the amount of C and T incorporation during the sequencing extension step. Pyrosequencing is a costly technology but its specificity is high, a large number of samples can be analyzed by this technology, and it gives fewer false-positive results compared to other technologies.
- A few investigators target methyl CpG binding domain 2 (MCBD2) and its multimerized form to improve affinity of the antibody and retrieving methylated region. This approach is called methylated CpG island recovery assay (MIRA).
- Chip with an antibody specific for native-methylation-binding-domain proteins has been used as an indirect measure of the distribution of methylcytosines present in a specific genome.
- For most methods, bioinformatic adjustments are needed for measuring varying CpG density at different regions of the genome.
- RLGS requires large amount of highly purified DNA (minimum 2 microgram), whereas Illumina Goldengate technology requires smaller DNA quantitites and works with degraded DNA or formalin fixed sample DNAs. Illumina Infinitium platform utilizes whole genome amplification (WGA) after bisulfate conversion and the system is capable of profiling 27,000 CpG sites with 1 microgram DNA.
- The main restriction of enzyme-based assays is that the number of sites recognized by the enzyme will determine the resolution. Such problem does not exist in sequencing-based technologies. However, the base composition and fragment size biases of sequencing platforms can introduce new coverage limitations along with measurement errors.
- Affinity based technologies do not require highly purified DNAs. However, these technologies are labor intensive and costly. On the other hand, sequencing based technologies on platforms such as Sequenom's EpiTYPER and Illumina Golden Gate are not labor intensive and can handle large number of samples with low cost.
- The ultimate comprehensive single base-pair resolution DNA methylation analysis technique is whole-genome

bisulfate sequencing. For smaller genomes, whole genome shotgun bisulfate sequencing has been achieved on Illumina genome Analyzer platform. One should keep in mind that approximately a tenth of the CpG dinucleotides in the mammalian genome remain refractory to alignment of bisulfite converted reads.

- For candidate gene approach, matrix-assisted laser desorption ionization time-of-flight (MALDI-TOF) mass spectrometry can be used. The advantage of this approach is that automation can be achieved on a platform, such as EpiTYPER by Sequenom.
- For all the technologies used in epigenetics, a direct head-to-head comparison is very difficult because the procedures are complex. Many technologies have competing strengths and weaknesses. Main factors for selecting techniques depend on quantity of available sample, extent of resolution required, number of samples to be analyzed, and region of the genome to be covered in a specific study.
- Methylation-sensitive single-strand conformational analysis (MS-SSCA) is suitable to assess all CpG sites as a whole in the region of interest rather than individual methylation sites. This technology differentiates between single-stranded DNA fragments of identical size, but the distinct sequence is based on differential migration in non-denaturating electrophoresis.
- As a rule of thumb, enzyme based techniques require double stranded purified DNA, while sequencing based technologies can use denatured and single stranded DNA.
- Purification of DNA is required whenever bisulfate treatment is done. Also, this treatment should be optimized because over-treatment may degrade DNA or make it unsuitable for further amplification. One should also be aware of the deaminases that convert unmethylated cytosine to uracil and can affect results.

ETHICAL ISSUES

For the successful development of intervention and treatment drugs, cell lines and animals are used at the initial stages. Animal models are also useful to understand disease biology. Proper handling and treatment of animals is of prime importance in these types of research. Questions such as "Are humans morally more important than all animals?", "Is there a sliding scale with humans at the top and the simplest animals at the bottom?", and "are humans and animals morally equal?" have been debated a number of times in past years. From time to time, these issues have been raised by different institutes and organizations, resulting in policies made by major agencies such as the National Institutes of Health (http://grants.nih.gov/grants/olaw/olaw.htm) and the United States Department of Agriculture (http://www.aphis.usda.gov/animal_welfare/). The number of animals used for research is estimated to be 35 million per year (12 million in the U.S. alone). The most frequent species among animals used for research is mice because considerable knowledge about its biology and genetics is known and there exists high genetic homology between mice and human. General guidelines for ethical issues are similar in different countries; however, each country has its own requirements and policies. The countries where specific guidelines and ethical issues are well documented include the U.S., Canada, all countries in the European Union, Australia and New Zealand. To follow proper ethical issues and policies, legislation requires adequate housing conditions, controls on animal pain, and critical review and approval by the institute's review board. The review board members should be experienced researchers and policy makers in animal model sciences. The institute may ask ethical justification for the use of animals in harmful experiments. The Public Health Services (PHS) of the U.S., which regulates NIH as well, developed guidelines to have an oversight system of Institutional Animal Care and Use Committees (IACUCs), to review and evaluate animal research protocols and care programs, assess and evaluate laboratory technicians and other staff, and investigate complaints about misuse or improper use of animals. The challenge is how to regulate private industries and pharmaceutical companies. The government does not provide funds to these organizations and hence cannot implement the same rules and regulations that are applied to investigators and institutes supported by federal money. The three "Rs" have also been proposed (replacement, refinement, and reduction) to improve ethical issues.

TRANSLATIONAL SIGNIFICANCE

Clinical implications of epigenetic inhibitors have been demonstrated by a number of investigators. These inhibitors activate tumor suppressor genes and/or inactivate oncogenes. A few examples of epigenetic inhibitors are shown in Table 4.1. Trichostatin A and suberoylanilide hydroxamic acid (SAHA) (histone deacetylase inhibitor), have been used for lung cancer treatment. Thilandapsins are another group of compounds in the histone inhibitor category with anti-ovarian cancer effect. Some cytotoxic issues are being addressed for this group of inhibitors. Romidepsins, panobinostat, varinostat and valproic acid are being tested in xenograft models of pancreatic cancer. 5-Azacytidine, a methylation inhibitor, was reported to increase regulatory T cells in ovalbumin-sensitized mice. Adenocortical carcinoma cells could be killed by the demethylating agent, decitabine. 5-Aza-2-deoxycytidine was effective against endometrial cancer due to its demethylating properties. A derivative of azacytidine,

TABLE 4.1 Epigenetic Inhibitors

Inhibitors	Other Names	Comments
DNA Methylatransferase Inhibitors		
5-Azacytidine	Vidaza	FDA approved (Vidaza by Celgene was approved for treatment of myeloplastic syndrome)
Isothiocynate		Preclinical
Zebularine		Preclinical
Decitabine (5-aza-2′-deoxycytidine)		FDA approved
Arabinosyl-5-azacytidine	Fazarabine	In Phase I/II trial
5-6-Dihydro-5-azacytidine	DHAC	In Phase I/II trial
Hydrazine		In Phase I trial
Histone Deacetylase Inhibitors		
Vorinostat	Zolinza, suberoylanilide hydroxamic acid (SAHA)	FDA approved for treatment of cutaneous T-cell lymphoma (SAHA is sold by Merck pharmaceutical company)
Butyrates (Phenylbutyrate)		For urea cycle disorder
Valproic Acid		FDA approved
FK228	Depsipeptide	In Phase I/II trial
Panobinostat	LBH589	In Phase I/II trial
Belinostat	PXD101	In Phase I/II trial

S110, was better tolerated in human xenograft models in mice. DZNep selectively inhibited trimethylation of H3K27, and trimethylation of H4k20. A combination of epigenetic inhibitors and other agents were also successful in reducing melanoma. Depsipeptide showed demethylation of DNA and histones.

Huge amounts of data are generated in the analysis of epigenetic results. During the last decade, major advances in biology, coupled with innovations in information technology, have led to an explosive growth of biological information. From the genomic revolution and many of its manifestations to recent developments in high-throughput, high-content screening, biomedical scientists have never before been exposed to research data that are so systematically collected, so rich, so complex, and so massive. This situation provides exciting, unprecedented opportunities for rapid discovery. At the same time, it presents investigators with the challenge of sifting or separating through mountains of seemingly orthogonal data and transforming them into new knowledge and clinical practice. This challenge is particularly evident in the field of cancer research, where the complexity and heterogeneity of the disease translates to more complex data generation conditions and higher data management and analysis overhead, creating a significant barrier to knowledge discovery and dissemination.

The fast growing field of biomedical informatics offers potential solutions to the "big data" problem. At the intersection of biology, medicine, computer science, and information technology, biomedical informatics concerns the development and application of computational tools to the organization and understanding of biomedical information so that new insights and knowledge can be discerned. The development of computational methods and tools in the form of computer software is a major component of biomedical informatics that is needed now. Application of existing software tools in biomedical research, such as computational data analysis and mathematical modeling, also should be improved. A single, preferably nationwide, coordination center should be formed so that data can be stored and disseminated to interested investigators.

The National Institutes of Health have been initiated the Epigenomics Roadmap Program which is comprised of five major initiatives: Reference Epigenome Mapping Centers, Epigenomic Data Analysis and Coordinating Centers, Technology Development in Epigenetics, Discovery of Novel Epigenetic Marks, and Epigenomics of Human Health and Diseases. This program proposes that the origins of health and susceptibility to disease are the result of epigenetic regulation of the genetic information. Specifically, epigenetic mechanisms that control stem cell differentiation and organogenesis contribute to the biological response to environmental and other factors in the form of stimuli that contribute in disease development. To accomplish this, the Roadmap Epigenomics Program plans to develop standardized platforms, procedures, and reagents for epigenomics research; conduct demonstration projects to evaluate how epigenomes change; develop new technologies for single cell epigenomic analysis and *in vivo* imaging of epigenetic activity; and create a public data resource to accelerate the application of epigenomics approaches. This program will transform biomedical research by developing comprehensive reference epigenome maps, developing new technologies for comprehensive epigenomic analyses and providing novel strategies for disease detection, diagnosis, treatment and prognosis.

Along with environmental factors, dietary components have a major role in both disease prevention and development. The folate pathway has been studied as a candidate biochemical and metabolic pathway for colon cancer. This pathway has been conserved among species, indicating its significance. Genetic variants in relevant genes have shown associations with diseases such as cancer, heart disease, and neural tube defects. In colon cancer adenomas, dietary folic acid supplementation has a protective effect, whereas either no effects or adverse effects have been observed in relation to colon cancer recurrence. Genetic explanations alone cannot explain these observations; therefore, attempts are being made to understand these associations by alternative mechanisms, such as epigenetics. Although reports indicate that nutrition plays a role in disease prevention, especially cancer, interactions among dietary bioactive food compounds and food combinations remain understudied. Colon cancer is one of the few areas of nutritional epigenetics that has been well studied. Folic acid is a well-known methyl donor and several foods are fortified with folic acid. Folic acid one carbon metabolism (FOCM) is an excellent example of a complex pathway with interconnected sub-pathways for folic acid and methionine metabolism, which in turn have their own feedback loops. Furthermore, folate biochemistry is well defined and enzymes involved in the metabolism of folate, whether they exist in the cytoplasm or mitochondria, are well characterized. Since methyl groups are the key component in CpG methylation, their levels influence gene expression. Alterations in homocysteine levels, DNA methylation, purine and thymidylate synthesis, and incorporation of uracil into DNA (misincorporation) occur simultaneously in the cells and contribute to DNA damage and repair pathways.

Genetic models of cancer are useful, but they can provide only limited information such as gains, losses, mutations, and rearrangements of chromosomal regions. An integrated genetic and epigenetic approach is necessary to determine the full complement of genes involved in tumorigenesis and to expose hidden therapeutic targets (Roukos, 2011). Gene silencing by aberrant epigenetic chromatin alteration is well recognized as an event that contributes to tumorigenesis. Although a number of examples of animal biotechnology in cancer epigenetics were presented in this article, more work remains to be done. The use of mouse models to assess DNA methylation in epigenetics still is in its early stages, but the approach is showing considerable promise for providing insights into this complex disease. Improvements are needed in areas such as the identification of appropriate mouse models of disease and identification of the most robust techniques for assessing DNA methylation, particularly in relation to technical reproducibility and coverage of the methylome. The use of mouse models is highly effective in clinical applications, especially in determining the toxicity and efficacy of methylation and histone deacetylase inhibitors (Roukos, 2011). Based on animal model results, large human trials can be planned. In the near future, we expect to see more anticancer drugs that may improve survival of cancer patients and their quality of life. A better understanding of the epigenetic mechanisms may open new avenues for disease prevention and treatment.

WORLD WIDE WEB RESOURCES

A very common and useful resource of information is the World Wide Web (WWW). However, no regulatory agency exists that checks the authenticity of the information provided. Our suggestion is to use www for a general background but refer to an agency or peer reviewed information (journals and books) for the correctness of the information. Wikepedia is a very popular source of information. On their home page, Wikepedia has indicated that the reliability of material posted on Wikepedia is assessed in many ways including comparative review, analysis of the historical patterns, and strengths and weaknesses inherent in the editing process unique to this site. The journal "Nature" acknowledged the quality and reliability of the material posted on Wikepedia and found it comparable to Encyclopedia Britannica (Giles, 2005). Material related with pathology, toxicology, oncology, pharmaceuticals, and psychiatry was found comparable to peer-reviewed sources (Clauson *et al.*, 2008; Leithner *et al.*, 2010; Wood and Struthers, 2010; Reavley *et al.*, 2012). However, some concerns were also raised by a scientific journal (Rajagopalan *et al.*, 2011). Generally, errors are detected early and corrected.

Based on the technologies used in animals to understand epigenetic regulated gene expression in normal and disease states, the following flow charts can be used (Flow Charts 4.1 and 4.2).

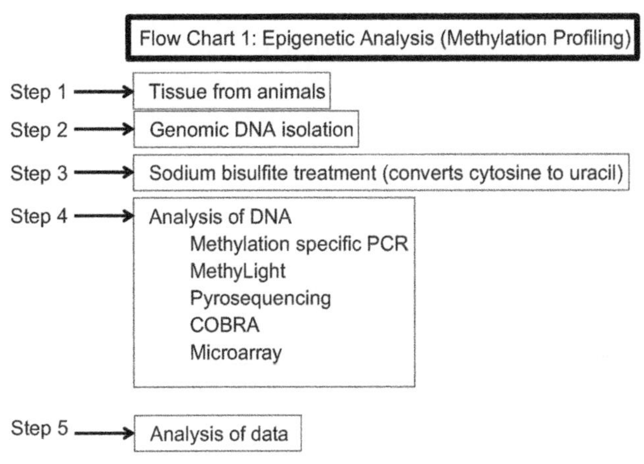

FLOW CHART 4.1 Epigenetic analysis (methylation profiling).

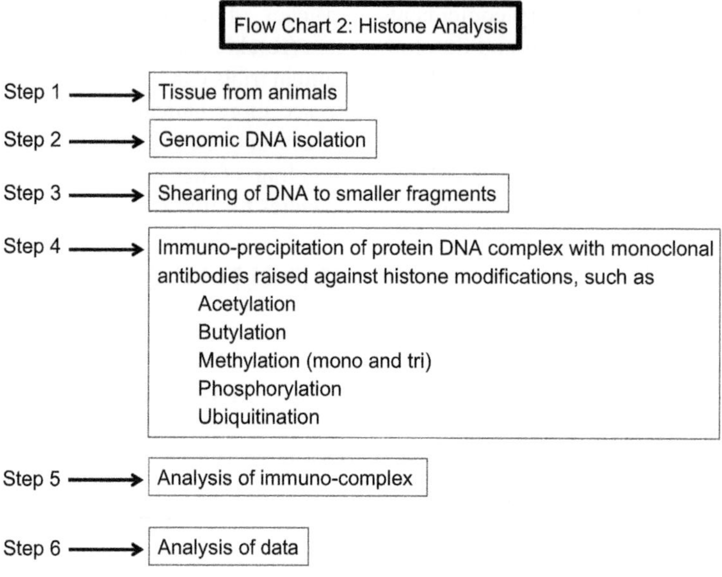

FLOW CHART 4.2 Histone analysis.

REFERENCES

Braconi, C., Kogure, T., Valerie, N., et al. (2011). Micro RNA 29 can regulate expression of the long non-coding RNA gene MEG-3 in hepatocellular cancer. *Oncogene*, *30*, 4750–4756.

Chen, S. S., Sherman, M. H., Hertlein, E., Johnson, A. J., Teitell, M. A., Byrd, J. C., & Plass, C. (2009). Epigenetic alterations in a murine model for chronic lymphocytic leukemia. *Cell Cycle*, *8*, 3663–3667.

Clauson, K. A., Polen, H. H., Boulos, M. N., & Dzenowagis, J. H. (2008). Scope, completeness, and accuracy of drug information in Wikipedia. *The Annals of Pharmacotherapy*, *42*, 1814–1821.

Demircan, B., Dyer, L. M., Gerace, M., Lobenhofer, E. K., Robertson, K. D., & Brown, K. D. (2009). Comparative epigenomics of human and mouse mammary tumors. *Genes, Chromosomes & Cancer*, *48*, 83–97.

Dunn, B. K., Verma, M., & Umar, A. (2003). Epigenetics in cancer prevention: early detection and risk assessment: introduction. *Annals of the New York Academy of Sciences*, *983*, 1–4.

Fraga, M. F., Herranz, M., Espada, J., et al. (2004). A mouse skin multistage carcinogenesis model reflects the aberrant DNA methylation patterns of human tumors. *Cancer Research*, *64*, 5527–5534.

Gaudet, F., Hodgson, J. G., Eden, A., Jackson-Grusby, L., Dausman, J., Gray, J. W., Leonhardt, H., & Jaenisch, R. (2003). Induction of tumors in mice by genomic hypomethylation. *Science*, *300*, 489–492.

Giles, J. (2005). Internet encyclopaedias go head to head. *Nature*, *438*, 900–901.

Hong, C., Maunakea, A., Jun, P., Bollen, A. W., Hodgson, J. G., Goldenberg, D. D., Weiss, W. A., & Costello, J. F. (2005). Shared epigenetic mechanisms in human and mouse gliomas inactivate expression of the growth suppressor SLC5A8. *Cancer Research*, *65*, 3617–3623.

Khare, S., & Verma, M. (2012). Epigenetics of colon cancer. *Methods in Molecular Biology (Clifton, NJ)*, *863*, 177–185.

Kikuchi, T., Itoh, F., Toyota, M., Suzuki, H., Yamamoto, H., Fujita, M., Hosokawa, M., & Imai, K. (2002). Aberrant methylation and histone deacetylation of cyclooxygenase 2 in gastric cancer. *Journal International Du Cancer*, *97*, 272–277.

Kumar, D., & Verma, M. (2009). Methods in cancer epigenetics and epidemiology. *Methods in Molecular Biology (Clifton, NJ)*, *471*, 273–288.

Leithner, A., Maurer-Ertl, W., Glehr, M., Friesenbichler, J., Leithner, K., & Windhager, R. (2010). Wikipedia and osteosarcoma: a trustworthy patients' information? *Journal of the American Medical Informatics Association: JAMIA*, *17*, 373–374.

Mathers, J. C., Strathdee, G., & Relton, C. L. (2010). Induction of epigenetic alterations by dietary and other environmental factors. *Advances in Genetics*, *71*, 3–39.

Mishra, A., & Verma, M. (2012). Epigenetics of solid cancer stem cells. *Methods Mol Biol*, *863*, 15–31.

Rajagopalan, M. S., Khanna, V. K., Leiter, Y., Stott, M., Showalter, T. N., Dicker, A. P., & Lawrence, Y. R. (2011). Patient-oriented cancer information on the internet: a comparison of Wikipedia and a professionally maintained database. *Journal of Oncology Practice / American Society of Clinical Oncology*, *7*, 319–323.

Reavley, N. J., Mackinnon, A. J., Morgan, A. J., Alvarez-Jimenez, M., Hetrick, S. E., Killackey, E., Nelson, B., Purcell, R., Yap, M. B., & Jorm, A. F. (2012). Quality of information sources about mental disorders: a comparison of Wikipedia with centrally controlled web and printed sources. *Psychological medicine*, *42*, 1753–1762.

Roukos, D. H. (2011). Novel cancer drugs based on epigenetics, miRNAs and their interactions. *Epigenomics*, *3*, 675–678.

Sun, W., Liu, Y., Glazer, C. A., Shao, C., Bhan, S., Demokan, S., Zhao, M., Rudek, M. A., Ha, P. K., & Califano, J. A. (2010). TKTL1 is activated by promoter hypomethylation and contributes to head and neck squamous cell carcinoma carcinogenesis through increased aerobic glycolysis and HIF1alpha stabilization. *Clinical Cancer Research: An Official Journal of the American Association for Cancer Research*, *16*, 857–866.

Teng, Y. W., Mehedint, M. G., Garrow, T. A., & Zeisel, S. H. (2011). Deletion of betaine-homocysteine S-methyltransferase in mice perturbs choline and 1-carbon metabolism, resulting in fatty liver and hepatocellular carcinomas. *The Journal of Biological Chemistry*, *286*, 36258–36267.

Verma, M., Dunn, B. K., Ross, S., Jain, P., Wang, W., Hayes, R., & Umar, A. (2003). Early detection and risk assessment: proceedings and recommendations from the Workshop on Epigenetics in Cancer Prevention. *Annals of the New York Academy of Sciences*, *983*, 298–319.

Verma, M., Maruvada, P., & Srivastava, S. (2004). Epigenetics and cancer. *Critical Reviews in Clinical Laboratory Sciences*, *41*, 585–607.

Verma, M., & Srivastava, S. (2002). Epigenetics in cancer: implications for early detection and prevention. *The Lancet Oncology*, *3*, 755–763.

Vuillemenot, B. R., Hutt, J. A., & Belinsky, S. A. (2006). Gene promoter hypermethylation in mouse lung tumors. *Molecular Cancer Research: MCR*, *4*, 267–273.

Wood, A., & Struthers, K. (2010). Pathology education, Wikipedia and the net generation. *Medical Teacher*, *32*, 618.

Yue, W., Sun, Q., Landreneau, R., Wu, C., Siegfried, J. M., Yu, J., & Zhang, L. (2009). Fibulin-5 suppresses lung cancer invasion by inhibiting matrix metalloproteinase-7 expression. *Cancer Research*, *69*, 6339–6346.

Zhang, Y., Yu, G., Jiang, P., Xiang, Y., Li, W., & Lee, W. (2011). Decreased expression of protease-activated receptor 4 in human gastric cancer. *The International Journal of Biochemistry & Cell Biology*, *43*, 1277–1283.

FURTHER READING

The following books may help in understanding epigenetic regulation in animal models and their implications in human system.

- Cancer Epigenetics (2012) (Book). Editors: Ramona G. Dumitrescu and Mukesh Verma. Springer Publications.
- Handbook of Epigenetics: The New Molecular and Medicine Genetics (2010) (Book). Editor: Trygve Tolletsbol. Amazon Publications.
- Epigenetics Protocols (Methods in Molecular Biology). (2011). (Book). Editor: Trygve Tolletsbol. Amazon Publications.
- DNA Methylation Inhibitors and Epigenetic Regulation of MicroRNA Expression. (2008). (Book). Editor: Jody Chouying Chuang. Amazon Publications.
- Epigenetics (2007). (Book). Editor: C. David Ellis. Barnes and Noble Publishers.
- Epigenetics in cancer prevention. (2003). (Book). Editors: Mukesh Verma, Barbara Dunn, and Asad Umar. Publisher: Annals of New York Academy of Sciences. NY.

GLOSSARY

Bisulfite treatment Bisulfite treatment of DNA converts cytosine to uracil, but leaves 5-methyl cytosines intact. Thus, 5-methyl cytosine patterns can be mapped by subsequent sequencing. This modified DNA is suitable for methylation analysis on different platforms (Illumina, Affymetrix).

Epigenetics Epigenetics is the stable, yet reversible chemical alterations to the genome that affects gene expression and genome function. These encompass mechanisms such as DNA methylation, histone modifications, non-coding RNA and other non-genetic methods of gene regulation. Mammalian DNA is packaged into chromosomes by wrapping the DNA around nucleosomes made up of an octamer of histone proteins. Each of the different histone types has N-terminal protein tails that extend outside of the nucleosome and can be modified by several large families of enzymes. Epigenetic changes affect gene expression without altering the genome and these changes are somatically heritable.

Cancer Cancer is a disease in which cells grow without any control and lose gene regulation.

Chromatin Chromatin describes the combination of DNA and proteins in the nucleus of a cell.

CpG Islands CpG rich regions of DNA that are often associated with the transcription start sites of genes that are also found in gene bodies and intergenic regions. These CpG islands are located in more than half of genes in the upstream promoter before the exon 1. In the normal cells, these CpG islands of the active genes are usually unmethylated, allowing gene expression. However, in cancer, regional hypermethylation of the CpG islands of several tumor suppressor genes was observed.

Histones Histones are present in the nucleus and they are needed for genomic stability and neutralizing the charge of DNA. The histones that form nucleosomes undergo numerous covalent modifications including acetylation, biotinylaion, methylation, ubiquitnylation, phosphorylation and sumoylation. Among all these modifications, histone acetylation and methylation have been found to be relatively stable and have been suggested as potential marks for carrying epigenetic information through cell divisions. All these changes influence the chromatin structure, keeping it in the active or repressed state and thus initiating or inhibiting gene expression.

Epigenotype The specific pattern of DNA methylation, histone modification, and miRNA expression at any given time is called epigenotype. Different epigenetic changes during disease development are called epimutations. Epigenotype is the intermediate state between genotype and phenotype. Programmed changes in the epigenotype represent a major candidate molecular mechanism by which early environmental exposure (radiation, toxins, smoking) can lead to permanent phenotypic changes. Furthermore, the identification of disease-predictive epigenotypes would indicate disease susceptibility and offer opportunities in intervention approaches.

Histone code All covalent alterations represent the histone code that can be "written" by the enzymes involved in these modifications. Histone acetyltransferases (HATs), histone deacetylases (HDACs), histone methyltransferases (HMTs) and histone demethylases (HDMTs) have been found to be part of the nucleosomal remodeling complexes, critical for changing the chromatin structure and controlling gene expression.

Hypomethylation Normal cells have repetitive genomic sequences heavily methylated, and the maintenance of this methylation plays an important role in preventing different chromosomal rearrangements (including translocations and gene disruption) through the reactivation of transposable elements. In cancer cells, global DNA hypomethylation has been suggested to contribute to structural changes in chromosomes, loss of imprinting, microsatellites and chromosomes instability through abnormal activation of proto-oncogenes and aberrant recombination events and increased mutagenesis.

Methylation Methylation of CpG islands in a gene promoter can occur during disease progression. The enzyme involved in the process is called methyl transferase. DNA methylation is heritable through mitosis and is copied to the new strand during DNA replication, and represents a mechanism of cellular memory of gene expression states in previous parental cells. This cellular memory is particularly important in development for tissue-specific and cell-lineage specific expression patterns.

MicroRNAs MiRNAs are 19 to 24 nucleotides long, transcribed as a precursor, and interfere in post-transcriptional processes. Due to their sequence specificity and complementary sequence to mRNAs, miRNAs can be made target specific and used for therapeutic purposes. In recent years, several cancer-specific miRNAs have been identified.

Polycomb proteins Polycomb proteins participate in the silencing of genes by mechanisms that do not involve DNA methylation. They often silence genes that are key regulators of differentiation.

ABBREVIATIONS

BHMT Betaine Homocysteine-S-Methyl Transferase
COX-2 Cycloxygenase-2
DNMT DNA Methyl Transferase
HAT Histone Acetyl Transferase
ESCC Esophageal Squamous Cell Carcinoma
HDAC Histone Deacetylase
HNSCC Head and Neck Squamous Cell Carcinoma
LINE Long Interspersed Nuclear Elements
lncRNA Long Non-Coding RNAs
Mbd2 Methyl CpG Domain Binding Protein 2
Methyl DIP Methylated DNA Immunoprecipitation
MCA Methylated CpG island Amplification
MTHFR Methylene Tetrahydrofolate Reductase
miRNA Micro RNA
PAR4 Protease Activated Receptor 4
RDA Representational Differential Analysis
SAHA Suberoylanilide Hydroxamic Acid
SINE Small Interspersed Nuclear Elements
TKTL1 Transkelolase-like 1

LONG ANSWER QUESTIONS

1. Why is it important to study epigenetics and how does it regulate gene expression?
2. What are the advantages of using epigenetic vs. genetic approaches in treatment of cancer?
3. Which events occur first, genetic or epigenetic, during development?
4. How can you use epigenetic approaches in risk assessment, disease diagnosis, and treatment?
5. Can you study epigenetics and genetics in the same animal model? If so, how?
6. Why is it important to focus on a non-coding region the genome?
7. What kinds of samples are appropriate for determining various components of epigenetics in an animal model, and which samples should be selected for cancer diagnosis and prognosis?
8. What is the transgenerational inheritance of epigenetics in mammals?
9. What is the critical time during life when exposure to pollutants and environmental agents can cause alterations in the epigenome?
10. Define the genomic regions most sensitive to epigenetic modifications that are transmitted across the generations.
11. Why is it important to study epigenetics in monozygotic twins?
12. How will you integrate information from genome-wide association studies (GWAS) with epigenome-wide association studies (EWAS) and use the information to identify high-risk populations?
13. What is the role of epigenetics in integration of RNA and DNA viruses in the host genome?

SHORT ANSWER QUESTIONS

1. What is epigenetics and how is it different from genetics?
2. What are the main components of epigenetics?
3. Is it necessary to study epigenetic regulation of cancer in cell-line models or in animal models?
4. Is cancer the only disease that is regulated epigenetically?
5. Can epigenetic drugs be used for cancer treatment?

ANSWERS TO SHORT ANSWER QUESTIONS

1. Epigenetics is the study of alterations in gene expression without changes in the primary nucleotide sequences. The main feature of genetics is a change in the primary sequence of a gene. Both processes are needed for proper gene expression.
2. The major components are DNA methylation, histone modifications, non-coding RNA expression, and chromatin compaction and relaxation.
3. Epigenetic regulation can be studied directly in humans, but cell-line and animal models are useful in identifying the functions of affected genes and pathways involved in the process.
4. No, cancer is one of many diseases that are regulated epigenetically. Other diseases in which epigenetic regulation has been reported are diabetes, cardiovascular diseases, neurological disorders, vision-related diseases, and infectious diseases.
5. Yes. Four epigenetic drugs have been approved by the U.S. Food and Administration and have been used successfully in treating blood, lung, colon, breast, prostate, and ovarian cancers in clinical trials.

Chapter 5

Development of Mouse Models for Cancer Research

Amrita Datta and Debasis Mondal
Department of Pharmacology, Tulane University Medical Center, New Orleans, Louisiana

Chapter Outline

SUMMARY

Mouse models are essential in cancer research. They are used to understand the genetic basis of tumor development and cancer progression. They can also be used to test the efficacies of different anti-cancer agents. This chapter will help students understand the applications of different mouse models in cancer research and provide them the knowledge necessary to carry out *in vivo* studies.

WHAT YOU CAN EXPECT TO KNOW

Numerous mouse models of human cancers have been developed to facilitate our understanding of the key processes in tumor development, and as pre-clinical models to test the efficacies of anti-cancer agents. These models recapitulate the crucial stages of tumor initiation from normal cells and their progression to aggressive tumors. These models have aided in the delineation of factors involved in tumor metastasis, tumor recurrence and therapeutic resistance. Pre-clinical studies in mouse models have also assisted in *in vivo* pharmacokinetics, toxicity and anti-tumor efficacy monitoring of numerous chemotherapeutic agents. In the genetically engineered mouse (GEM) models, alternations in genes believed to be responsible for human malignancies are mutated, deleted or overexpressed, and tumor development is monitored. In this model, tumorigenesis and chemotherapeutic responses can be studied in the presence of a normal immune system. However, since mouse tumor cells may not represent the tumorigenic process in humans, another widely used model is human tumor xenografts in

Animal Biotechnology. http://dx.doi.org/10.1016/B978-0-12-416002-6.00005-5

immunocompromised mice. In these models, human tumors are generated in different tissues using tumor specimen, cell lines, or cancer stem cells ($CD44^{+}/CD24^{-}$). However, investigations with tumor xenografts have been unable to address the role of tumor microenvironments, tumor-stroma, and the normal immune surveillance of tumors. Therefore, several new strains of humanized mouse models have been developed recently. This chapter provides several protocols and guidelines for the development of mouse models of cancer, which will help the student in initiation and successful completion of these *in vivo* experiments.

HISTORY AND METHODS

INTRODUCTION

To appreciate the successes and failures of clinical investigations using anti-cancer drugs, one needs to properly understand the implications of modern animal biotechnology as preclinical models for cancer (Cheon and Orsulic, 2011). It is well accepted that tumorigenesis is a complex process involving the accumulation of multiple genetic aberrations that transform normal cells, allowing for their abnormal growth, proliferation, and metastasis (Hanahan and Weinberg, 2000 Jan 7; Müller et al., 2001 Mar 1; Shacter and Weitzman, 2002 Feb). In order to fully understand the development and spread of cancer in the body and to elucidate the crucial role of tumor-associated stroma and tumor-angiogenesis, it is imperative to carry out research on tumor growth in live animals. To discover novel anti-cancer agents, it is important to first delineate the biology and genetics of a specific tumor and identify their relationship to tumor initiation, tumor promotion, progression, and metastasis, and lastly tumor recurrence and drug resistance development. This raises the need for *in vivo* models that resemble the human organ systems and portray the systemic physiology, and closely recapitulates the human disease. These animal models have enabled the testing of new approaches to cancer prevention and treatment, identification of early diagnostic markers, and novel therapeutic targets to prevent, treat and eliminate the cancerous growth. Unlike the *in vitro* cell culture systems, the use of *in vivo* models have been instrumental in obtaining both pharmacokinetics (PK) and pharmacodynamics (PD) data relevant to the successful clinical development of numerous chemotherapeutic drugs. The crucial role of both drug-transporters and drug-metabolizing enzymes in dictating the PK and PD of chemotherapeutic agents is also becoming very apparent, and how polymorphisms in these genes can regulate inter-individual variability in anti-cancer drug efficacies form the basis of a new field called pharmacogenomics (PG) (Mondal et al., 2012). Therefore, an appropriate animal model system can show the potential relationships among dose, concentration, efficacy, drug-drug interactions and/or toxicity in humans, and are a mainstay of modern cancer research investigations. More than any other animal model systems, mice have revolutionized our ability to study cancer. Recent advances in developing inbred mouse strains, transgenics and immunocompromised mice, have enabled the generation of clinically relevant mouse models which often recapitulate the human disease progression and response to therapy. Therefore, it is very important that one understands the rules and regulations, proper protocols and guidelines, as wells as the time and costs that are associated with *in vivo* studies (Table 5.1). Below is a comprehensive list of the advantages and limitations of the mouse models of cancer.

Although there is no single model that meets the criteria for all different cancers, a combination of the existing

TABLE 5.1 Advantages and Limitations of Mouse Models in Cancer Research

Advantages	Limitations
Small size, easy to handle and take care of, and short tumor generation time.	Mice are very different, in terms of size, life span, organ morphology and physiology, and thus differ in drug PK and PD from human beings.
Cheaper than other animal models of cancer, allowing the use of large numbers for statistical measurements.	One critical difference in the mouse is the activity of telomerase enzyme, which is largely inactive in adult human cells. (Mouse cells transform more readily and thus require fewer genetic alterations for malignant transformation.)
High tumor incidence and relatively rapid tumor growth.	Mouse models tend to develop relatively few metastases or display metastases with different tissue specificity as compared to human tumors.
Many mice can be treated at the same time to observe dose responses.	Differences in metabolic rate and pathways might result in a different drug response in mouse models (e.g. the cytochrome P450 pathway for drug metabolism).
Genetically, the best characterized of all mammals used in cancer research.	Due to a limited number of initiating genetic alterations, mouse tumors are typically more homogeneous and this can be an obstacle to modeling the heterogeneity of human cancers.

models may mimic the clinical features of particular human cancers. It is therefore imperative that one first understands how to decide on the optimum mouse model to be used, either to address specific questions regarding tumor development and/or to determine the therapeutic effects of antitumor agents *in vivo*.

HISTORY

A brief history of the use of mice in cancer research is provided below. As early as 1664, the eminent Microbiologist, Dr. Robert Hooke had used mice in his studies on infectious diseases. Around 1902, the era of modern mouse genetics began when a Harvard researcher, William Castle, began studying genetic inheritance in different strains of mice. Around the same time, Abbie Lathrop, a mouse breeder and entrepreneur, was generating colonies for mouse hobbyists that were later used in experiments by different researchers. These early studies included inbred mouse strains that were observed to develop tumors frequently. Ms. Lathrop eventually teamed with Dr. Leo Loeb at the University of Pennsylvania, and they authored numerous papers on their investigations using these mouse strains. The laboratory mice used by Castle and Lathrop are the ancestors of most of the strains that are routinely used by researchers nowadays.

Starting in the late 1980s, the advent of transgenic technologies enabled manipulation of mouse germ line DNA in embryonic stem (ES) cells, which brought about the development of numerous strains of genetically engineered mouse (GEM) models. The early 1990s brought about tremendous advances in gene-targeting approaches which allowed the development of mouse strains with "knockouts" of tumor suppressor genes or "knock-ins" of oncogenes which spontaneously developed tumors (Hanahan et al., 2007; Cardiff et al., 2006). However, early studies showed that many of these mutations manifested embryonic lethality in homozygous mice. Furthermore, variability in the time required for tumor formation in different mouse strains posed significant problems. Mutations carried in all somatic cells also confounded the organ specific tumor development, which is seen in humans. In the late 1990s, the ability to use the 'CRE-Lox recombinase' system allowed the development of conditional and inducible systems, which addressed the above drawback of GEM models. However, due to species-specific differences in tumor cells, the limitations of GEM in modeling human cancers were clearly evident. In this respect, the human tumor xenografts, either with actual human tumor tissue or with human cell lines, showed key advantages over the GEM models. In these xenografts, the tumor develops in a matter of few weeks to months and orthotopic tumor xenografts can be appropriately placed to reproduce the tissue environment in which the tumor grows in humans. One big challenge in using tumor xenografts was the lack of a functional immune system against tumor cells. In recent years, this problem has been partially solved using immunodeficient mice that have been 'humanized' by injection of human peripheral blood or bone marrow cells which allowed for an almost complete reconstitution of the immune response to the grafted tumors.

PRINCIPLE

Before embarking on studies using mice, it is imperative to develop procedures that reduce, replace or refine the animal studies. It is important to tailor the experimental designs and procedures, and follow the most ethical ways to treat the animals. Housing and feeding are the most important issues to address during the initial stages of *in vivo* experiment development. Appropriately sized cages, proper nutrition, and environments (temperature, opportunities for socialization, lighting, or water quality) that closely mimic their natural habitat can go a long way in reducing animal stress and increasing chances for protocol approval by the Institutional Animal Care and Use Committee (IACUC).

IACUC Approval

All organisms go through stress in many forms, but are normally able to adapt and recover. Thus, laboratory mice are allowed to acclimatize for at least a week before the start of experiments. However, unlike stress, distress is a negative state wherein the animal fails to adapt and return to its physiological or psychological homeostasis. Age, gender, genetic traits (including transgenic modifications) rearing and postnatal separation, psychological state, and housing conditions can all affect the animal in different ways. Therefore, care should be taken to try to relieve stress and eliminate distress as much as possible. In cases where alternatives to animal use are not available, it is important that researchers recognize and alleviate distress in laboratory animals. In this respect, it is important to differentiate between stress and distress. Thus, the following three Rs (Refine, Reduce, Replace) are important considerations and criteria when writing an IACUC protocol:

1. *Refine:* This involves careful consideration of the aspects of research methodology and animal maintenance. The *in vivo* techniques should be thoroughly reviewed so that they cause minimal amount of pain, stress, or suffering. If it is not possible to alleviate pain by giving medications then a humane endpoint should be defined so that the animal may go through as little suffering as possible. It is also important to have a clearly defined protocol of when to treat or euthanize the animals to reduce their suffering, and a better understanding of distress mechanisms may help to define earlier end points in the experiments.

2. *Reduce:* This aims at limiting the amount of animals needed for a study. Pilot studies to optimize techniques, treatments and conditions are ways of doing so, and the use of good statistical methods may help assess the accurate number of specimens required to get relevant results and limit unnecessary use of large numbers of animals. The development of new techniques can further reduce, or eliminate, the number of animals necessary.
3. *Replace:* This term takes into account that new technologies are developing that provide an alternative to animal testing, or at least minimize their use. The use of appropriate cell cultures can eliminate or minimize the use of whole organisms and still give required data. As an example, to test the anti-tumor activity of a drug, tumor cells can be grown *in vitro* for the majority of tests towards optimizing dosing and treatment regimen, instead of trying to optimize these *in vivo.* Also initial toxicology studies should be carried out using cultured liver cells or other normal cells lines such that supportive data is gained to test the efficacy and toxicity prior to animal testing.

IACUC Guidelines

Before any animal testing can be initiated, the protocol has to be first approved by the IACUC, which consists of at least five members appointed by the institution. The IACUC reviews the research protocols and conducts evaluations of the institution's animal care. The IACUC is required to ensure that the proposed work falls within the Animal Welfare Assurance. Furthermore, the approved animal use protocol must be reviewed by the IACUC every three years. To obtain an IACUC approval, the protocol needs to address the following points:

1. Identification of mouse strain and approximate number of mice to be used.
2. Rationale for using mice and the appropriateness of the numbers used.
3. A complete description of the proposed use of mice.
4. A description of anesthetic procedures to minimize discomfort and injury to animals.
5. Animals that experience severe or chronic pain that cannot be relieved will be painlessly killed at the end of the procedure or, if appropriate, during the procedure.
6. A description of appropriate living conditions that contribute to health and comfort.
7. A description of housing, feeding, and nonmedical care of the animals, which will be directed by a veterinarian or other trained scientists.
8. Personnel conducting procedures will be appropriately qualified and trained.
9. Euthanasia used will be consistent with the American Veterinary Medical Association (AVMA) guidelines.

METHODOLOGY

Inbred Mice

In scientific experiments homogeneity allows for controlled and reproducible experiments that necessitated the development of genetically identical mice by inbreeding. The inbred mice provide the advantage of data reproducibility. The first use of inbreeding in science can be traced back to laboratories of researcher Clarence C. Little, an American geneticist who was exploring coat color inheritance in his studies with the first inbred stock of laboratory mice, the DBA strain (Crow, 2002). Offspring of these ancestral DBA (Dilute, Brown and non-Agouti) mice are still available to researchers today. Today mice are considered to be stably inbred after 20 generations of the appropriate pairings. The utility of inbred mice as a model system lies in the fact that all the mice have an identical genetic background and are expected to generate similar responses to treatments that allows researchers to accumulate data in standardized collections.

Many scientists use a single inbred strain or F1 hybrid in their research because it has a repeatable, standardized, uniform genotype supported by a substantial body of previous research. In fact, the more that is known about a strain, the more valuable it becomes, and being of a single genotype, it is relatively easier to "know" about inbred strains. Another advantage of inbred strains is that sample size can be reduced in comparison with the use of out-bred stocks. An inbred strain may also be cross-bred with another strain to develop new models. For example, recombinant inbred strains, that have been out-crossed from two separate inbred strains, and then maintained for generations, are utilized for mapping traits and are in wide use (Casellas, 2011). Over 20,000 scientific papers used BALB/c mice between 2001 and 2005, and a further 10,000 used C57BL/6, both are well-defined inbred strains. Additionally, the work on the BALB/c inbred mouse strain led to the development of monoclonal antibodies and the development of embryonic stem (ES) cells used in developing transgenic mice (Figure 5.1).

Depending upon the type of cancer being investigated or the application of tumor model to be tested (either toward basic or translational research), a number of criteria can aid in the decision-making. A brief list of these criteria on different mouse model system to be chosen for cancer investigations is provided in Table 5.2.

Examples where genetic variability within inbred strains have often been used to measure cancer occurrence is "Consomic" strains; wherein an entire chromosome from one inbred strain replaces the corresponding chromosome in another inbred strain. Furthermore, researchers can rapidly breed "Congenic" strains from these consomic strains, such that the mice now carry only a small segment that differs between animals. This allows for narrowing of

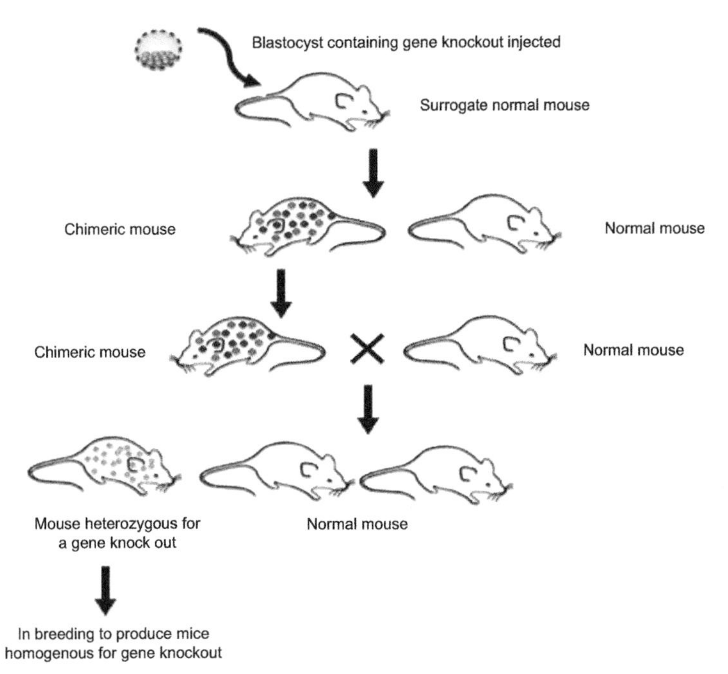

FIGURE 5.1 Generation of inbred strains of transgenic mice.

TABLE 5.2 Criteria for Choosing the Optimum Inbred Mouse Model

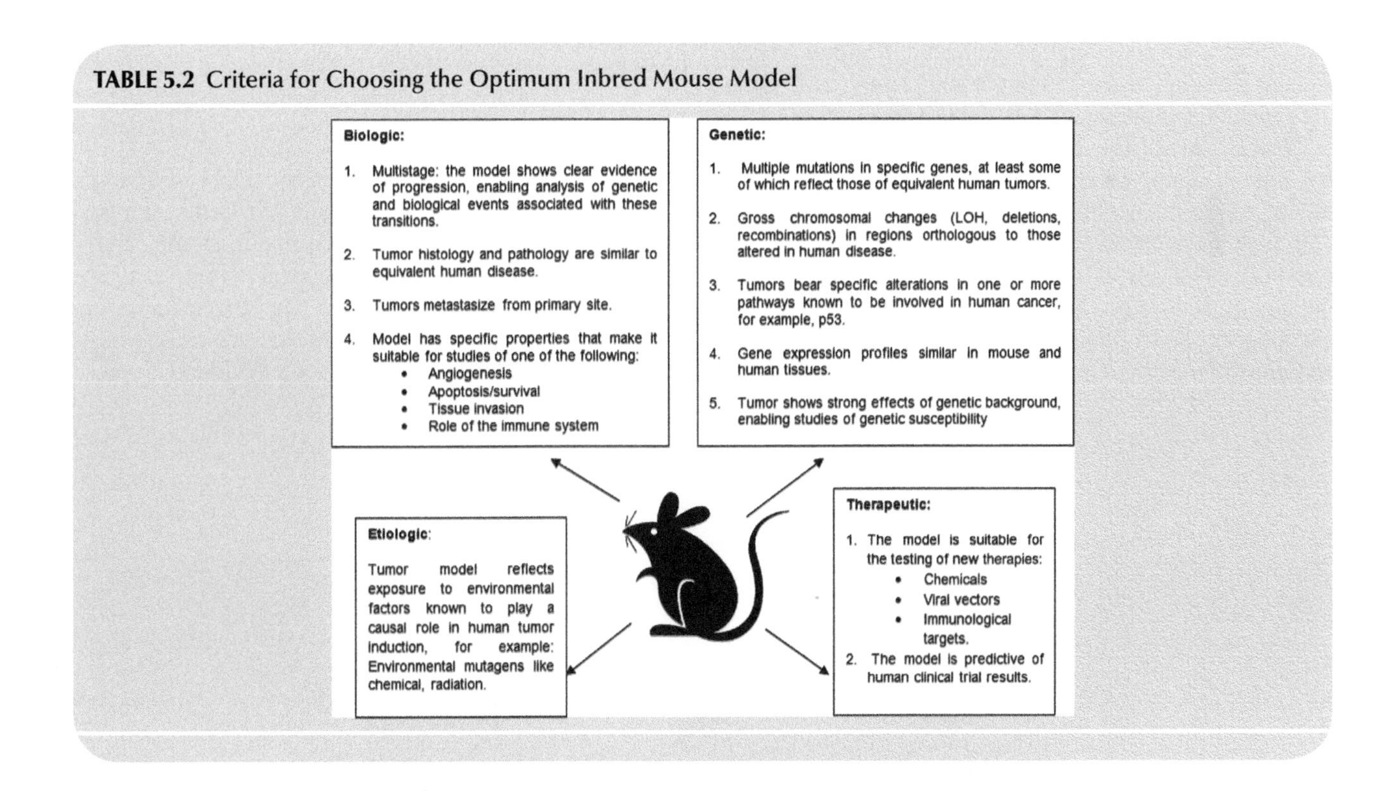

the region on a chromosome that can be targeted for gene identification related to oncogenesis. However, the main disadvantage of using a single strain is that it is not representative of the heterogeneity and polymorphisms observed in humans (Pal and Hurria, 2010; Tian et al., 2011). In the following sections, we are discussing more at length each of the different mouse models of cancer along with a short synopsis from a published study using each of these models. The following sections provide some examples for each of these mouse cancer models.

EXAMPLES WITH APPLICATIONS

Immunocompetent Mice

Spontaneous Tumor Models

Mice that are genetically engineered to carry pre-disposed mutations serve as very important tools in the study of cancer initiation and progression following exposure to carcinogens (Figure 5.2). The role of tumor promoters and or tumor inducing/silencing agents with specific characteristics can be studied during tumor development. Using these models, the effects of natural and environmental estrogens on normal mammary gland development and on carcinogenesis has been recently reviewed by Pelekanou et al. (2011) (Pelekanou and Leclercq, 2011). This review focused on the role of estrogen receptors (i.e. ERα and ERβ) in regulating growth, apoptosis and differentiation of mammary epithelial cells, and the bidirectional coordination between stroma and cancer cells in maintenance of tumor growth. Cravero et al. (2010), also wrote a recent review on new rodent systems that recapitulate both genetic and cellular lesions that lead to the development of pancreatic cancer (Ding et al., 2010). Interestingly, mice with mutant K-Ras oncogene (G12D) spontaneously develop tumors in the pancreas that are non-metastatic in nature. However, simultaneous mutation of the p53 tumor suprresor gene leads to the generation of metastatic pancreatic cancers.

There is also large variability in both the susceptibility and incidence of spontaneous lung tumors between mouse-inbred strains (Piegari et al., 2011). Typically strains with high spontaneous lung tumor incidence are also very responsive to chemical induction of lung tumors, for example, exposure to cigarette smoke, tar, or chemically pure carcinogens. Some strains are more resistant than others and this difference can often be attributed to an underlying genetic variance. For example, a *Cdkn2a* polymorphism was found between the intermediate resistant BALB/cJ and susceptible A/J strains. Chemical induction of lung tumors with carcinogens, such as polycyclic aromatic hydrocarbons ethyl carbamate (urethane), is very reproducible and invariably results in pulmonary adenoma and adenocarcinomas. In breast cancer research, one of the most commonly used models for the study of cancer preventive agents is the mouse model that develops tumors, either spontaneously or after carcinogen treatment. This model has been successfully used to demonstrate the chemopreventive activity of many agents, including selective estrogen receptor modulators (SERMs) such as Tamoxifen, Raloxifene and Idoxifene, and retinoid compounds. In DMBA-induced mammary tumors, Aryl hydrocarbon receptor (AHR) activation was shown to delay the development of tumors (Wang et al., 2011). It has been particularly useful for agents for the prevention and/or treatment of ER-positive mammary cancers. In addition to the effects of genetic mutations on cancer susceptibility, the carcinogen-induced tumor models are also useful in understanding the cancers caused by exposure to physical agonists (like Asbestos fibers), chemical (like Cadmium, Arsenic) and environmental (UV radiation) carcinogens, and also aid in the development of their effective treatments.

The GEM Mouse Models

To generate a more reproducible cancer model based on loss or gain of specific genes, scientists in 1987 utilized a procedure called homologous recombination in mouse embryonic stem cells (ES cells), allowing them to remove a gene (knockout), or replace it (knock in). This enabled a variety of alterations of mouse DNA segments that closely recapitulated human tumors. These genetically altered mouse

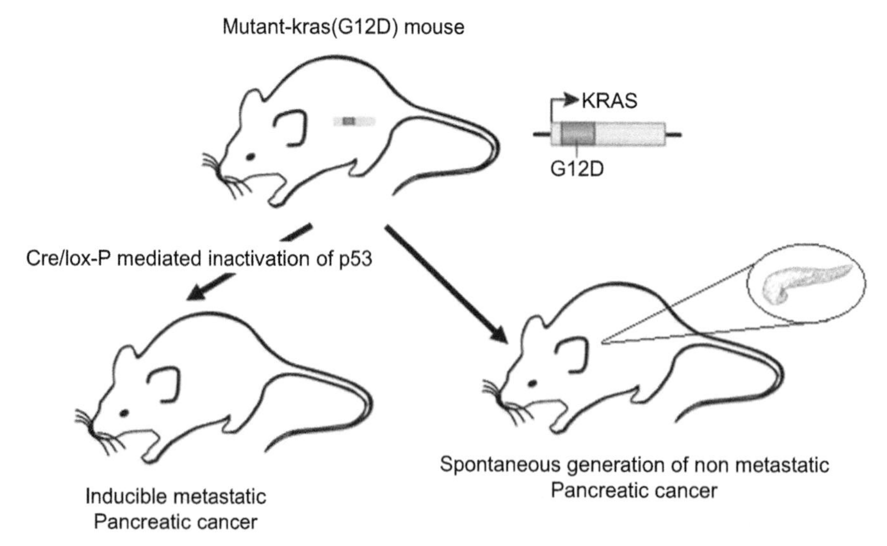

FIGURE 5.2 Generation of spontaneous tumor models for carcinogen studies.

models allow cancer researchers to test in the human context, as to how, when, where, and in which combinations particular gene alterations may be involved in the initiation and progression of cancer in immunocompetent animals (Figure 5.3).

The discovery of oncogenes and their association with the development of human tumors led to the first transgenic cancer models. The "knock-in" models of cancer included the over-expression of many oncogenes such as c-myc, v-Ha-ras or SV40 T-antigen, etc., which clearly indicated their roles in the development of different types of tumors, e.g. myelomas, lymphomas, carcinomas and sarcomas. For example, in a model of brain tumor, derived by delivering the viral oncogene SV40 T-antigen into mouse eggs, the targeting of CD8+ T cells was shown to be sufficient for tumor elimination (Tatum et al., 2008). The SV11 mouse line which expressed the SV40 T Ag (T Ag) as a transgene from the SV40 enhancer/promoter, lead to T Ag expression in the thymus as well as the choroid plexus tissue within the brain, which promoted the appearance of small papillomas by 35 days of age, and progressive tumor growth resulted in mortality by 104 days of age. Using this GEM model, investigators showed that immunotherapeutic approaches that target the recruitment of tumor-reactive CD8+ T cells were effective against well-established tumors. Furthermore, using donor lymphocytes derived from transgenic mice expressing a T-cell Ag (epitope IV) specific T-cells enabled rapid regression of established tumors.

Another example where GEM mice developing breast cancers were produced by delivering a mutant human oncogene called c-Myc by MMTV, a mouse virus that infects mouse mammary tissue. These mice developed mammary tumors that resembled human breast tumors. In these models, the role of tumor suppressor genes such as Rb, p53, and Brca1, was also implicated in the development of cancers. Disruption of tumor-suppressor genes can clearly demonstrate their role in neoplastic transformation and can aid in novel drug development (Singh and Johnson, 2006). Targeted activation of the K-RAS proto-oncogene and simultaneous inactivation of Rb and p53 in the mouse lung have resulted in GEM models that recapitulate many of the characteristics of lung cancers (both NSCLC and SCLC). The GEM models for breast cancers involve overexpression of oncogenes such as c-Myc, cyclin D1, Her2, and Wnt-1. Of particular importance are tumors overexpressing Wnt-1 which seem to show heterogenous ER status, making the model very relevant to the study of both ER-positive and ER-negative cancers.

Several important mouse models of gastrointestinal cancer have also been developed based on genetic alterations that are known to affect these pathways. These models clearly showed the process of multi-step carcinogenesis where a progressive series of mutations occur during colon cancer development, involving genes such as Wnt-1, APC, RAS, p53, and TGF-β (Ramanathan et al., 2012). The $Apc^{min/+}$ mouse, developed by Moser and colleagues in 1990, was the first mouse model of intestinal tumorigenesis to be generated by mutational inactivation of the adenoma polyposis coli (APC) gene through random chemical carcinogenesis. Thus, APC was classified as a tumor suppressor gene. The $APC^{min/+}$ mice and other *APC* mutant mice develop multiple intestinal adenomas and similar mutations in the *APC* gene was also found in patients that develop familial adenomatous polyposis. However, this GEM model has some limitations since the location of the polyps is predominantly restricted to the small intestine and some rare but evident occurrence of malignant progression to adenocarcinoma. Furthermore, most *APC* mutant strains live only

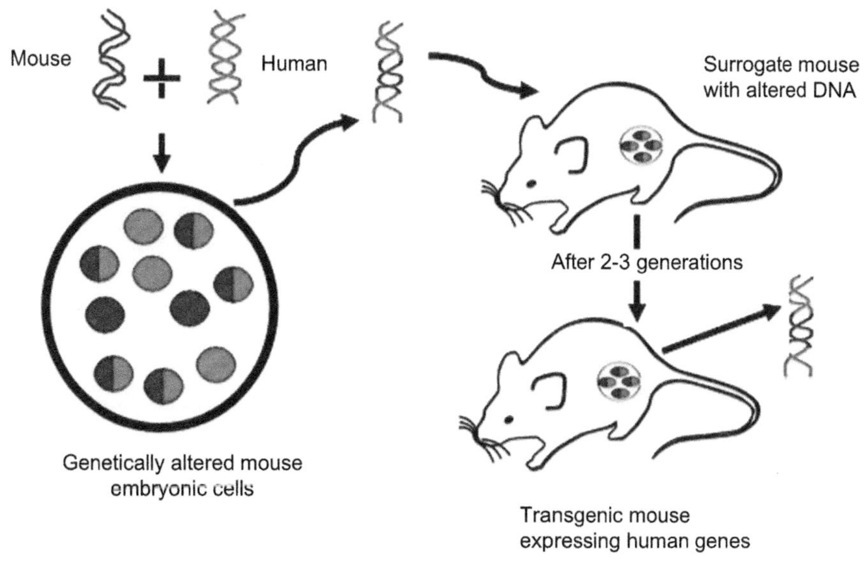

FIGURE 5.3 Generation of GEM models in immunocompetent mice.

to about 14 to 16 weeks of age (due to anemia from multiple polyp bleeds). This short lifespan may not be sufficient for the accumulation of additional genetic lesions that cooperate in promoting later-stage disease and thus do not represent the disease. Despite these limitations, the $APC^{min/+}$ mice model continues to be one of the most used models particularly for chemopreventive studies that involve the gastro-intestinal tract.

The Cre/Lox System: A Superior GEM Model

It has been almost 15 years since the discovery of the Cre/lox system, which is now frequently used as a way to artificially control gene expression (Sauer, 1998). This system has allowed researchers to create a variety of genetically modified animals where the gene of choice can be externally regulated. The Cre/lox system provides a method to produce a mouse that no longer has a target gene in only one cell type (Figure 5.4).

In this system, the induction of Cre-recombinase enzyme mediates the site-directed DNA recombination between two 34-base pair loxP sequences. To achieve this in mice, transgenic mice containing a gene surrounded by loxP sites are mated with transgenic mice that have the *cre* gene expressed in specific cell types. In tissues with no *cre* gene the target gene with function normally; however, in cells where *cre* is expressed, the target gene is deleted. The correct placement of Lox sequences around a gene of interest may allow genes to be activated, repressed, or exchanged for other genes. Furthermore, the activity of the *Cre* enzyme can be controlled so that it is expressed in a particular cell type or triggered by external stimulus like chemical signals or heat shock. This enables the induction of somatic mutations in a time-controlled and tissue-specific manner.

Since a majority of human colorectal cancers exhibit constitutive Wnt activity due to mutations in the APC or β-catenin genes, Wnt-targeted therapeutic strategies in which expression and activity of this gene product is altered specifically in cancer cells, has been successfully demonstrated using the Cre/Lox system (Bordonaro, 2009). By using this system, a mammary-specific deletion on mouse chromosome 11 could be achieved which accelerated Brca1-associated mammary tumorigenesis in comparison to Brca1 conditional knockout mice (Triplett et al., 2008). Similar to human BRCA1-associated breast cancers, these mouse carcinomas were ERα-negative and were of basal epithelial origin. Furthermore, to test the role of p53 in Brca1-associated tumorigenesis, a p53-null allele was introduced into mice with mammary epithelium-specific inactivation of Brca1. The loss of p53 accelerated the formation of mammary tumors in these females. Since the Brca1 knockout mice were found to die during embryonic development, the Brca1 conditional mice expressing *Cre* in the mammary gland epithelia enabled the development of mammary tumors only during adulthood. Therefore, this newly developed homologous recombination system closely models carcinogenesis in humans where tumors evolve from somatic gene mutations in normal cells in a time-controlled and tissue-specific fashion *in vivo*.

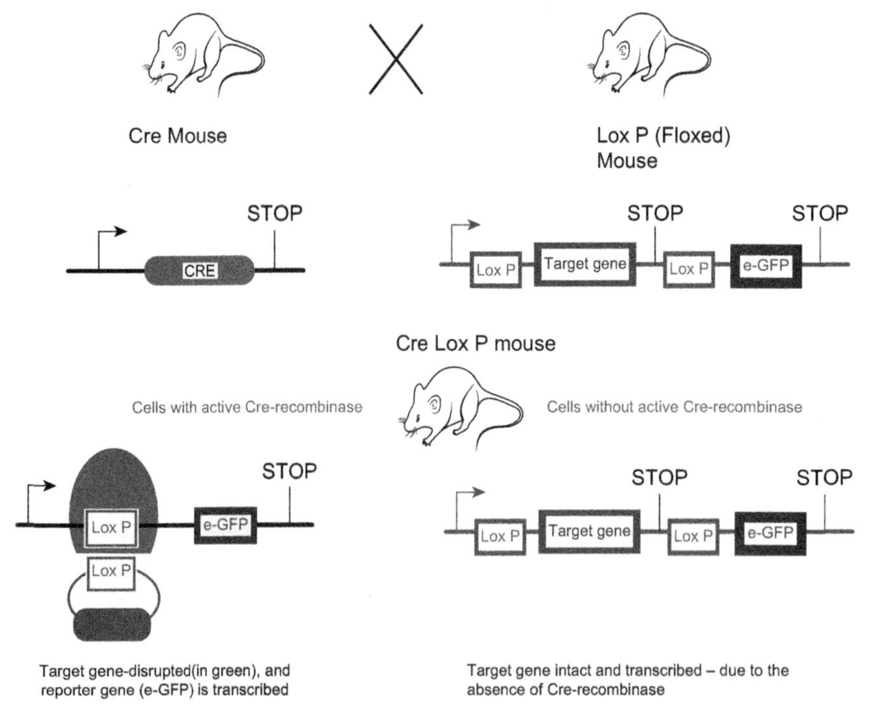

FIGURE 5.4 Generation of the Cre/Lox mouse model.

Immunodeficient Mice

In order to know whether a patient's tumor will respond to a specific therapeutic regime, it is essential to examine the anti-cancer response to the human tumor and not a mouse tumor. This is where the human tumor xenograft on athymic nude mice, SCID mice, or non-obese diabetic (NOD)/SCID humanized mice can be helpful. The availability of tumor grafts has made transplanted tumors the test models of choice to investigate anti-cancer therapeutics (Richmond & Su, 2008 Sep-Oct). The transplantation models allow us to propagate tumor tissues *in vivo*. Thus, transplanted tumors have contributed in a major way to our understanding of cancer biology, which would be unrealizable with the GEM models. Orthotopic xenografts can reproduce the organ environment in which the tumor grows to mimic the effects of tumor microenvironment. In addition, stromal cells can be included in the xenograft to more completely mimic the human tumor microenvironment; and xenografts using NOD/SCID mice that have been "humanized" by injection of peripheral blood or bone marrow cells, allow for an almost complete reconstitution of the immune response to the tumor (Figure 5.5). However, considering the benefits and limitations of these model systems, it is important to corroborate data in humans when interpreting the results in transplanted tumors (Richmond and Su, 2008; Kerbel, 2003).

Both retrospective and prospective studies reveal that human tumor xenografts can be remarkably predictive of cytotoxic chemotherapeutic drugs that have activity in humans, especially when the drugs are tested in mice using clinically equivalent or "rational" drug doses. However, the magnitude of benefit observed in mice, both in terms of the degree of tumor responses and overall survival, may be different as compared to the clinical activity of the drug observed in humans. Furthermore, since transplanted tumors allow large quantities of tissue of uniform character, both pathologic and molecular studies can be easily carried out. The choice of tumor and the site of transplantation must also be validated in advance, so that it is comparable with the normal tissue from which the tumor was derived. A variety of sites has been used for transplantation to mimic that observed in human, but the most commonly used are the subcutaneous (s.c.), intraperitoneal (i.p.), and intramuscular (i.m.) sites. Some tumors will grow as a dispersal of cells, free of a supporting stroma or vasculature, when innoculated i.p. This offers a particularly convenient since it permits precise cellular quantitation of inoculums of relatively "pure culture" of tumor cells. Furthermore, in contrast to the "spontaneous" cancers, transplanted tumors are most widely used for assaying the therapeutic effectiveness of a large array of anti-cancer compounds. Two different tumor transplantation models are currently used in different laboratories: allograft transplants and xenograft transplants.

Allograft Transplants

In the allograft (or syngeneic) transplantation models, the cancerous cells or solid tumors are itself of mouse origin, and are transplanted into another host mouse containing a specific genetic trait (e.g. GEM mice). Owing to the fact that recipient and the cancer cells have the same origin, the transplant is not rejected by mice with an intact immune system. Results of therapeutic interventions can thus be analyzed in an environment that closely recapitulates the real scenario where the tumor grows in an immunocompetent environment. A disadvantage, however, is that transplanted mouse tissue may not represent in whole the complexity of human tumors in clinical situations (Suthar and Javia, 2009). This transplantation strategy circumvents some of the difficulties observed in GEM models complicated breeding schemes, variable tumor latency and

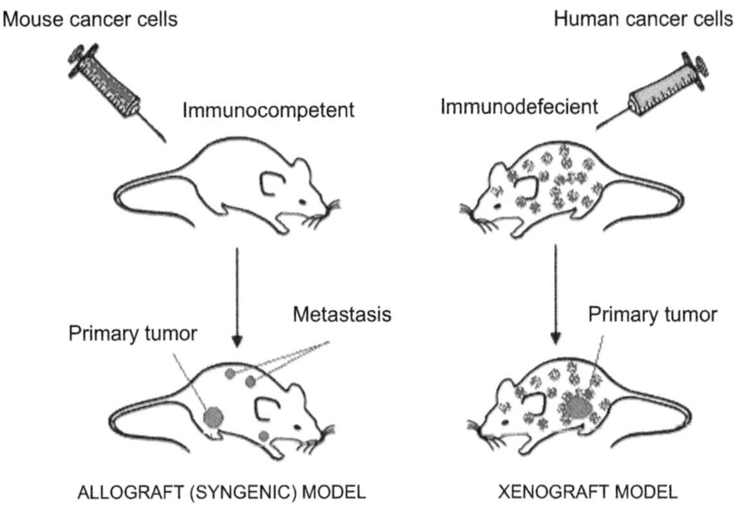

FIGURE 5.5 Generation of allograft and xenograft tumor models.

chemotherapeutic drug inefficacy. Tumor fragments from tumor-bearing MMTV-PyMT or cell suspensions from MMTV-PyMT, -Her2/neu, -Wnt1, -Wnt1/p53+/–, BRCA1/p53+/–, and C3(1)T-Ag mice can be transplanted into the mammary fat pad or s.c. into naïve syngeneic or immunosuppressed mice. Tumor development can be monitored and tissues processed for histopathology and gene expression profiling and metastasis can be scored at 40–60 days after removal of original tumors.

In a recent study by Prosperi et al. (2011), (Prosperi et al., 2011). APC mutation was shown to enhance PyMT-induced mammary tumorigenesis. In serial passages, regardless of the site of implantation, several other investigators showed that PyMT tumors from anterior glands grow faster than posterior gland derived tumors. Microarray analysis also revealed genetic differences between these glandular tumors. The differences in transplantation were reproducible using anterior tumors from multiple GEM and tumor growth rates correlated with the number of transplanted cells. Similar morphologic appearances were also observed in both original and transplanted tumors. Metastasis developed in >90% of mice transplanted with PyMT, 40% with BRCA1/p53+/– and wnt1/p53+/–, and 15% with Her2/neu tumors. Interestingly however, expansion of PyMT and wnt1 tumors by serial transplantation for two passages did not lead to significant changes in gene expression. Furthermore, PyMT-transplanted tumors and anterior tumors of transgenic mice showed similar sensitivities to cyclophosphamide and paclitaxel. Thus, allograft transplantation of GEM tumors can provide a large cohort of mice bearing mammary tumors at the same stage of tumor development and with defined frequency of metastasis in a well-characterized molecular and genetic background.

Xenograft Transplants

In the xenograft model, human tumor cells are transplanted (either under the skin or into the organ type (orthotopic) in which the tumor originated) into immunocompromised mice like athymic nude mice or severely compromised immunodeficient (SCID) mice, so that the human cells are not rejected. In this model human cancer cells or solid tumors are transplanted into a host mouse. In order to prevent rejection of the cells by the host's immune system, the host mice have impaired immune systems. These transplants may be orthotopic (meaning that the tumor is placed in the site it would be expected to arise naturally in the host, for example, human breast cancer cells placed in the mouse's mammary pads) or they may be subcutaneous or placed just beneath the host's skin. Because the cancer xenograft is of human origin it represents the properties and to an extent the complexities of the human cancer. On the other hand, the compromised host immune system is not truly representative of actual patients.

Tumors obtained from patients can be kept viable in the frozen state for prolonged periods. The usual method for preparing solid tumors for freezing is to immerse small pieces in media such as glycerol-glucose, in sealed sterile ampoules. The tumor-medium mixture should be slowly cooled to the final storage temperature and rapidly thawed (at 37°C) on removal, and after thawing, tumors should be immediately inoculated in mice. Depending upon the number of cells injected, once the tumor develops to an appropriate size the response to therapeutic regimes can be studied *in vivo*.

Using subcutaneous prostate tumor xenografts of both androgen-dependent (LNCaP) and castration-resistant (HR-LNCaP) prostate cancer cells, Schayowitz et al. (2010) showed that dual inhibition of AR and mTOR can prolong the hormone sensitivity of prostate cancers (Schayowitz et al., 2010). Male SCID mice 4–6 weeks of age (from the National Cancer Institute-Frederick Cancer Research Center) were inoculated with the LNCaP or HR-LNCaP cells (2.0×10^7) along with Matrigel (10 mg/mL) in 100 μL of cell suspension. The mice were then treated with a combination of inhibitors to block the AR and mTOR activation when tumors reached 500 mm^3, for 3 to 7 weeks. The addition of everolimus (androgen synthesis inhibitor) to bicalutamide (anti-androgen) in the treatment of resistant tumors significantly reduced tumor growth rates and tumor volumes and decreased serum prostate specific antigen (PSA) levels.

In another study by Kataoka et al. (2012), the efficacy of chemoendocrine therapy in both pre-menopausal and post-menopausal models with ER-positive human breast cancer xenografts were studied (Kataoka et al., 2012). Female 4–5-week-old BALB/c-nu/nu mice were inoculated with a suspension of MCF-7 breast cancer cells (5×10^6 cells/mouse) subcutaneously into the right flank. For the pre-menopausal breast cancer model, mice were subcutaneously implanted with slow-release estrogen pellets (0.25 mg/pellet 17β-estradiol) the day before tumor cell inoculation. After several weeks, mice bearing a tumors (~200–400 mm^3) were randomly allocated to control and treatment groups (8 mice each) and received 6-week oral therapy with 359 or 539 mg/kg/day capecitabine and/or tamoxifen (30 or 100 mg/kg/day). For the post-menopausal breast cancer model, mice were ovariectomized and subcutaneously implanted with slow-release androstenedione pellets (1.5 mg/pellet) the day before tumor cell inoculation. Mice bearing tumors (~200–400 mm^3 in volume) were treated for 6-week oral administration of capecitabine (359 mg/kg/day) and/or letrozole (0.1 mg/kg/day). In each model, control mice received vehicle alone. Tumor volumes and body weights were monitored two or three times a week starting from the first day of the treatment. The combination of 5′-deoxy-5-fluorouridine (5′-DFUR; an intermediate of capecitabine) with 4-hydroxytamoxifen (4-OHT; an active form of tamoxifen) or letrozole (aromatase inhibitor) decreased the number of estrogen-responding cells and size of breast tumor xenografts in both pre-menopausal and post-menopausal models.

Tumor xenografts have been used in the preclinical and clinical development of anti-cancer therapeutics. In human breast cancer xenografts, several investigators showed that herceptin (Trastuzumab™) can enhance the anti-tumor activity of paclitaxel and doxorubicin against HER2/neu-overexpressing cells and this led to its subsequent use successful clinical trials. In addition, herceptin is now a standard drug used in the treatment of Human Epidermal growth factor Receptor 2-positive (HER2+) early stage breast cancer. Multiple Myeloma has remained an aggressive and incurable cancer, and while melphalan and predinisone provide symptomatic relief, survival rates were only 3 years. However in the past decade, the combination of bortezomib and melphalen was demonstrated as effective for treatment of multiple myeloma, first in preclinical xenograft trials, and then in patients. This has led to the new standard of clinical care for multiple myeloma patients over 65 years of age or those with pre-existing conditions to whom the option of high-dose chemotherapy followed by stem-cell transplantation is an unavailable option. Therefore, xenograft models are useful for toxicity studies from targeted therapies, and in many cases to predict biomarkers of target modulation. However, several ethical issues need to be first addressed before the testing of toxic chemotherapeutic agents in mouse cancer models.

CHECKLIST FOR A SUCCESSFUL *IN VIVO* EXPERIMENT

Prior to initiating the studies, a successful *in vivo* experiment involves critical planning and addressing of multiple factors. It is very important to make sure that everything is ready and the guidelines and protocols have been well thought out and that the IUCAC approved animal protocol is in place.

1. *Identification of the Experimental Goals.* It is important to first identify what one plans to answer by using the animal experiments and justify the utility of mouse models. This could be based on either the role of a certain gene or the effect of a certain anti-cancer therapy. It is also important to pinpoint whether one aims to study tumor initiation, tumor progression, tumor metastasis or tumor recurrence/resistance. For this purpose, it is imperative that one chooses the best mouse model to be tested and decide on when to initiate the tumor transplant or drug treatment. At this stage, the investigator needs to choose whether an immunocompetent or immunodeficient model would be optimal and whether an easier sub-cutaneous tumor model or a more difficult orthotopic tumor model would be needed.
2. *Finalization of the Experimental Setup and Budget.* The next most important criteria would be to design an experimental setup based on the budget. Since animal experiments can be very costly, it is important to plan the number of animals available for the studies. Some immunocompetent mouse strains are relatively inexpensive ($10–30 per mouse); however, most of the transgenic strains and especially the immunocompromised (SCID or nu/nu) mice are much more expensive ($60–$100 per mouse). Multiple animals (at least 6–10 per group) would be necessary to generate statistically significant effects, and the use of multiple groups to assess drug treatments can become very expensive at the end. Furthermore, right from the beginning, the costs towards animal housing and care in the vivarium should be taken into account. Vivarium charges for per day and per animal can add up to be significantly expensive if the experiments are for long-term tumor growth and anti-cancer efficacy measurements. The investigators should be well aware of the time, efforts and costs associated with the *in vivo* experiment themselves, and should be able to analyze the validity of the data obtained from these *in vivo* studies. Therefore, time management is very crucial and enough time for performing the *in vivo* experiment and dosing of multiple animals will need to be coordinated. Rushing the surgical procedures and drug treatments may ruin the entire experiment.
3. *Training of Personnel.* Before starting an experiment, it is important that animal handling workshop/certifications are obtained for all personnel involved. It is crucial that each researcher understands the ethics involved in animal research, which will enable them to make important decisions about the protocol, the number of animals required, the end points needed for animal data collection and the method and timeliness of animal euthanization. Some experiment may take several hours to do, especially with large number of replicates and multiple treatment groups. Thus, for accurate and efficient progress it is advisable to have experienced assistance.
4. *Ordering of Animals.* Once the experiments are finalized, IACUC approval is obtained and trained personnel are available, it is advisable to order the animals at least two weeks in advance of initiating the experiments. Some experiments may need specific sex, such as prostate cancer model, breast cancer model etc. The age of the mice is also an important consideration because 6–8 weeks old mice may not respond to treatment the same way as mice that are 4 months old. Also, different strains respond to treatment differently, so one treatment that works in one strain may not work in another and *vice versa.* Therefore, make the choice of strain, sex, and age

according to the type of cancer and the application of experiment.

a. There are several different vendors that sell different mouse strains for researchers and the costs may vary significantly, e.g. Harlan Laboratories, Charles River Laboratories, etc. It is advisable to shop around for the best prices for the animals, but also be aware of previous publications using the mouse strains from different vendors since minute differences in the genetic makeups of different strains can significantly alter the end results.
b. Before ordering the animals, one needs to contact the vivarial staff in order to secure designated space for the animals, the specifics of animal housing, and treatments. In case vivarial staff, e.g. veterinarians, surgery specialists are needed, it would be important to schedule and coordinate the experiments in advance.

5. *Preparation of Animals for Anesthesia.* Mice should be healthy and well prepared for the experiments. All mice should be housed in a pathogen-free environment under controlled conditions (temperature 20–26°C, humidity 40–70%, light/dark cycle 12 h/12 h). All mice should be allowed to acclimatize and recover from shipping-related stress for at least 1 week prior to the study. The body fat, or lack thereof, age, sex, and strain can all impact a mouse's response to anesthetic agents. Pre-anesthetic fasting is not usually necessary in mice, however, if fasting is employed it should be limited to no more than 2–3 hours prior to the procedure. Because mice have a greater body surface area to body mass ratio than larger animals, thermal support is critical to their survival and successful anesthetic recovery. Parenteral anesthesia may be administered to mice via intraperitoneal, subcutaneous, or intravenous injection. Inhalation anesthesia may be delivered by chamber or facemask. Animal anesthesia can be achieved with isoflurane without the use of specialized anesthesia equipment. Alternative methods of anesthesia include injectable anesthesia that is achieved by Avertin (200 mg/kg) or ketamine/xylazine (100/10 mg/kg) i.p. injections. Irrespective of the method used, the mouse should be monitored to avoid excess cardiac and respiratory depression, and insufficient anesthesia.
6. *Preparation of Human Tumor Cells for Xenografts.* For some studies, it is advisable to use labeled tumor cells to be able to measure tumor growth and metastasis without invasive surgery. Therefore, fluorescent labeled cells (GFP or Luciferase) will need to be prepared in advance. Healthy tumor cells in logarithmic phase of growth are the key to a successful *in vivo* transplantation experiment. Cells should not be over confluent nor should they be too sparse, during passage of cells. Three days before the *in vivo* experiment, seed the cells into a new culture dish or flask, one day before the experiment, change to fresh medium, and when harvesting the cells, the cells must be 70–80% confluent (for monolayer culture). These cells should then be used to prepare single cell suspension. If cells are clumped, disrupt the suspension via aspiration or gentle vortexing. However, only one person should do all the cell injections to mice to minimize experimental errors. The quicker one can inject the cells into animal after harvesting, the better, and the more uniform will be the tumor growth.
7. *Preparation of Surgical Instruments and Cells.* Depending upon the experimental strategy, the site of injection may be orthotopic, subcutaneous, or directly under the skin. Sterile syringes and needles are required for the procedure, along with alcohol swabs. In many cases, matrigel or basement membrane-like matrix is mixed with the human cancer cells in a 1:1 ratio before injecting into mice. Matrigel provides a natural environment for the cancer cells to grow.
8. *Tumor Monitoring.* Human tumor usually takes 1–2 weeks to grow in a nude mouse depending upon cancer cell type and number of cells injected. It is advisable to inject not more than 100 μL of 1–5 million cells per mouse for a xenograft mouse model. Nutritional support is critical during post-procedure recovery periods and moistened rodent chow is recommended to encourage animals to eat following anesthetic events. The health of the mice should be monitored by daily observation. Regular animal monitoring and scheduling of sampling need to be coordinated with all of the personnel involved. When measuring tumor growth using the bioluminescence method, injection of D-luciferin (15 mg/mL) to a final 150 mg D-luciferin/kg body weight is used. At 12 min after D-luciferin injection, the animals can be imaged in the dorsal position for 2–5 s at field of view. The tumor regions of interest and tissues around the tumor sites are both imaged, and emitted signals are quantified as total photons/second using Living Image Software. Upon reaching a relevant tumor size/volume, mice are randomly separated into treatment groups, and treatment ensues. At the end of the study period, the primary tumors are surgically removed from the animals, isolated from surrounding tissues, and weighed. At this stage other organs, such as lung, liver, bone and lymph nodes, can also be dissected to measure metastasis. The mice should be euthanized if there is more than 20% weight loss, or the tumors are too big and impede movement, or the tumor burdens or treatment results in sickness. Scheduling of euthanization needs to be made and the vivarial staff should be informed in advance.
9. *Analysis of Data.* Once promising data with tumor growth or anti-tumor effects of the drug of interest

in obtained in the first set of mice, it is important to repeat a similar experiment in another set of mice. This will facilitate the statistical data analysis and manuscript preparation. Perform statistical analysis utilizing repeated measures analysis of variance (ANOVA) to evaluate the treatment effects on tumor growth. Compare tumor weight measurements between treatment groups utilizing one-way or two-way ANOVA. All data are presented as the mean standard error (SE). Statistical differences are evaluated by Student's t-test and ANOVA, followed by Dunnett's *post hoc* test. The criterion for statistical significance will need to set at $P < 0.05$. At this stage the *in vivo* data would be ready for publication. A flow chart (Flow Chart 5.1) is provided here to help the researcher to plan ahead. In addition, a detailed checklist that will help the animal researcher initiate their studies is provided below.

PROTOCOLS

The above checklist on mouse models of cancer will demonstrate the successes of using the xenografted tumors in many types of human cancers. Similar studies clearly illustrated that information learned from these studies can be translated into successful clinical trials for drugs. Thus, the human tumor xenograft models are the most-often utilized system to determine anti-cancer drug efficacies. In the following sections we include two examples of protocols using these models where each of the experimental steps are more thoroughly addressed.

An Orthotopic Mouse Model of Colorectal Cancer

The traditional subcutaneous tumor model is not ideal for studying colorectal cancer and does not replicate the human disease. Therefore, the orthotopic mouse model of

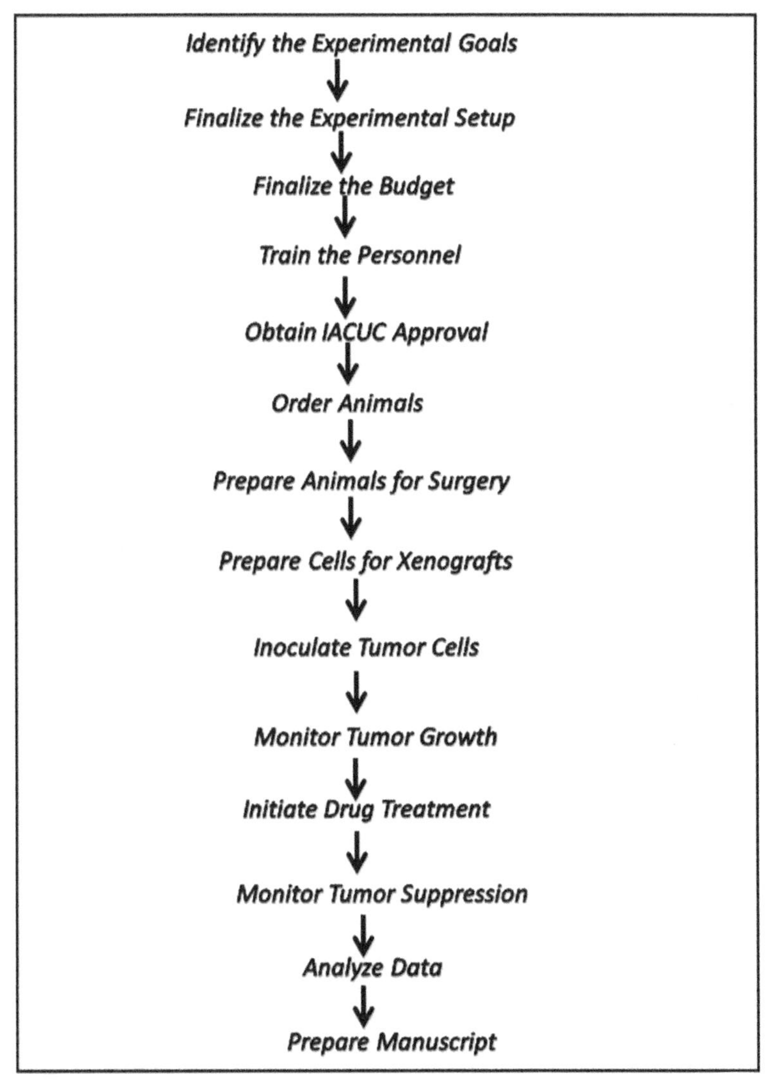

FLOW CHART 5.1 Guidelines for a successful mouse experiment.

colorectal cancer has been very useful for studying both the natural progression of cancer and testing of new therapeutic agents. In a recent publication by Bhattacharya et al. (2011), the effects of the natural antioxidant supplement selenium on tumor progression and angiogenesis in an orthotopic model of human colon cancer are discussed (Bhattacharya et al., 2011). Real-time imaging using both fluorescent protein imaging (FPI) and magnetic resonance imaging (MRI) were utilized to demonstrate tumor progression and angiogenesis. In this model, human colon carcinoma cells (GEO) were fluorescent labeled with GFP and implanted orthotopically into the colon of athymic nude mice. Beginning at five days post implantation, whole-body FPI was performed to monitor tumor growth *in vivo*. Tumor bearing animals were treated with daily oral administration of methyl-selenocysteine (0.2 mg/day) for five weeks. Dynamic contrast-enhanced MRI was performed to examine the change in tumor blood volume following treatment, and CD31 immunostaining of tumor sections was also performed to quantify microvessel density (MVD). Selenium treatment resulted in a significant reduction in blood volume and microvessel density of GEO-derived tumors and highlighted the usefulness of multimodal imaging approaches to demonstrate antitumor and anti-angiogenic therapies against colon cancers. Below are two protocols that have been used to establish the orthotopic model of colorectal cancer. The first involves injection of a colorectal cancer cell suspension into the cecal wall and the second technique involves transplantation of a piece of subcutaneous tumor, obtained from a different mouse, into the cecum.

Design and Execution

A. Cell Preparation/Tumor Preparation
Colorectal cancer cells (e.g. GEO or HT-29) are grown in culture and harvested when sub-confluent, and single cell suspension is prepared in phosphate buffered saline (PBS). Alternatively, a mouse with a previously established subcutaneous colorectal tumor is euthanized, the tumor is removed and divided into 2–3 mm pieces, and kept in PBS on ice.

B. Mouse Preparation

1. Use inhaled isoflurane (2-chloro-2-(difluoromethoxy)-1,1,1-trifluoro-ethane). Alternatively, use ketamine–xylazine–acepromazine (KXA) combination.
2. The depth of anesthesia is assessed using toe pinch and the absence of withdrawal reflex.
3. The anesthetized mouse is shaved at the site of injection, properly positioned, and the abdomen is prepped with a sterile betadine solution.

C. Laparotomy
A small nick is made in the skin and the abdominal wall musculature is grasped and lifted up to expose the abdominal cavity. The abdominal cavity is entered with a single blade of scissors, which is used to push the intra-abdominal contents away from the cecum. At this stage, the abdominal incision can be extended to 2–3 cm to gain better access.

D. Exposure of the Cecum
The cecum and its pouch are identified and exteriorized. The cecum is isolated from the rest of the mouse and placed on a pre-cut, sterile gauze. Warm saline is continuously used to keep the cecum moist.

E. Injection of Cells or Tumor Pieces

1. Injection of cells into the cecal wall.
   - **(a)** A 27 G needle is used to inject a 50 μL volume of cells ($2–5 \times 10^6$) into the cecal wall.
   - **(b)** The needle is removed, the injection site is inspected to ensure no leakage, and the cecum is returned to the abdominal cavity.
2. Transplantation of tumors into the Cecum
   - **(a)** A "figure 8" stitch is placed onto the cecum using a 6 or 7 sized suture. The cecal wall is lightly damaged and the tumor piece is positioned.
   - **(b)** The stitch is tied down and the cecum is returned to the abdominal cavity.

F. Mouse Abdominal Wall Closure

1. The abdominal wall is closed using three interrupted stitches using a 3 or 4 sized suture (alternatively, one can use a simple running stitch). Post-operative analgesics and a fluid bolus may be given at this point.
2. The mouse is allowed to recover from anesthesia.

G. Monitoring Tumor Development

1. Tumor formation may vary by cell line, amount of cells inoculated or size of tumor strips used, but palpable tumors (~ 100 mm^3) are seen within 4–6 weeks.
2. When animals develop visible tumors at the site of cell implantation (approximately 12 days after cell implantation), obtain images of the animals (ventral position).
3. Remove outlier animals with tumors larger or smaller compared to the majority of mice and randomize the remaining mice into different treatment groups.

H. Monitoring Drug Effects

1. Most drug treatments are started at this stage and treatment ended at about 8–10 weeks or when tumors reach a volume of ~ 500 mm^3.
2. For each treatment arm, mice are grouped in sets of 5 or 10 with equal tumor volumes. Tumor size is measured twice weekly with calipers, and tumor volume is calculated by the formula $4/3\pi r_1^2 \times r_2$, where r1 is the smaller radius.
3. At the end of treatments, mice are euthanized and tumors are excised, cleaned, weighed, and stored in −80°C for additional analysis.

Interpretation of Results

The time to developing primary tumors and liver metastases may vary depending on the technique, cell line, and mouse species used. Tumor response to anti-cancer therapy can vary dramatically depending on whether cancer cells are implanted in an ectopic (subcutaneous) versus orthotopic location. However, the orthotopic models of colorectal cancer replicate human disease with high fidelity. The two techniques have unique advantages and disadvantages. The use of colorectal cancer cell lines that have been growing *in vitro* allows for the inoculation of homogeneous cells. However, these cells may have reduced invasive or metastatic potential after several passages in culture. Transplantation of a piece of subcutaneous tumor introduces a more heterogeneous population of cancer cells that have been established *in vivo*. However, since tumors contain stromal cells, this may affect the consistency of the tumors in different mice. Primarily, invasive rectal cancers can develop in mice as early as one week after injection; however, mice do not usually develop metastatic lesions.

A Xenograft Model of Prostate Cancer Metastasis

To better understand the process involved in cancer metastasis and the effects of drugs in suppressing breast cancer metastasis, the design, experimental setup and analysis of tumor-bearing mice, are important aspects to consider. While prostate cancer metastasizes to lymph nodes, lung, and liver, the predominant site of metastatic prostate cancer is bone; approximately 80–90% of patients with advanced prostate cancer have bone metastases. We are presenting a recent study that determined the effects of Angiotensin-(1-7) in attenuating metastatic prostate cancer in mice (Krishnan et al., 2012). Angiotensin-(1-7) [Ang-(1-7)] is an endogenous, heptapeptide hormone with anti-proliferative and anti-angiogenic properties. *In vitro* studies were performed to establish the role of Ang-(1-7) in as an anti-metastatic agent against advanced prostate cancer. For example, in the *in vitro* migration studies, Ang-(1-7)-mediated a decrease in migration of both PC3 and DU145 cell lines. This suppression of cell migration was blocked by a specific Ang-(1-7) receptor antagonist, D-alanine^{7}-Ang-(1-7). This suggested that Ang-(1-7) may reduce the *in vivo* metastatis of prostate cancer cells. To demonstrate this in a xenograft mouse model, these investigators used luciferase expressing PC3`cells injected into the middle of the tibial plateau in 5–6 week old male SCID mice. The SCID mouse does not have an active immune system (as that would lead to clearance of tumor cells), and thus has a good tumor intake. The ability to make T or B lymphocytes, or activate some components of the complement system is impaired in SCID mice and they also cannot efficiently fight infections, nor reject tumors and transplants.

Design and Execution

A. Cell Preparation
1. Grow PC3 cells (CRL-1435; American Tissue Culture Collection) in RPMI1640 medium (HyClone) with 10% fetal bovine serum (FBS) and antibiotics.
2. Infect PC3 cells with the UBC-GFPLuc lentivirus construct (expressing a GFP-Firefly luciferase fusion protein behind the human ubiquitin C promoter) and select with 10 mg/mL blasticidin for 2 weeks to establish a stable cell line.

Note: The percentage of GFPLuc-positive cells can be determined by flowcytometry or via the expression of luciferase using a Luciferase Assay kit (Promega).

B. Mouse Preparation

Model 1: *In Vivo* Model of Prostate Cancer Metastasis Injection of tumor cells into the left ventricle is a commonly used method to mimic extravasation of tumor cells from the circulation into metastatic sites.
1. Anesthetize animals with 1% isoflurane. Make a midline incision over the trachea and throat of the mouse. Isolate the carotid artery and insert a catheter (14 mm long) into the artery. Insure that the tip of the catheter extends down the carotid artery to the aortic arch.
2. Inject one million (1×10^6) $PC3^{LUC}$ cells in 100 mL of Hank's buffered saline solution (HBBS) into the aortic arch catheter using a 28-gauge needle.

Model 2: Orthotopic Intra-Tibial Model
1. Anesthetize animals with 1% isoflurane. Make a 2–3 mm longitudinal incision over the mid-patellar region of the right hind limb. Make a percutaneous interosseal bore in the tibia using a 27-gauge needle.
2. Inject twenty thousand (2×10^4) $PC3^{LUC}$ cells into the middle of the tibial plateau.

Note: Hold a cotton swab over the injection site to prevent leakage of cells.

C. Mouse Closure

Perform skin closure with a 5-0 coated vicryl suture.

D. Ang(1-7) Treatment

Model 1
1. Two days prior to injection of $PC3^{LUC}$ cells into the aorta, implant a subcutaneous osmotic minipump to infuse 24 mg/kg/hr of Ang-(1-7) in SCID mice or control mice.

Model 2
1. Two weeks post-injection of $PC3^{LUC}$ cells into the bone, implant a subcutaneous osmotic minipump

to infuse 24 mg/kg/hr of Ang-(1-7) in SCID mice or control mice.
2. Treat mice for 5 weeks with Ang-(1-7).

E. Monitor Tumor Development
1. Monitor tumor metastasis in different organs at six weeks after treatment with Ang-(1-7) by bioluminescent imaging of the whole mouse.
2. Measure tumor growth by bioluminescence image analysis and determine tumor volumes in the bone by MRI.
3. At the end of treatments, mice are euthanized and tumors are excised, cleaned, weighed, and stored in −80°C for additional analysis.

Interpretation of Results

In model-1 (metastasis from aorta), even at six weeks after treatment, no detectable tumors were observed in mice treated with the Ang-(1-7), whereas 83% of the control mice developed metastatic lesions in either the tibia, mandible, or spine. In model-2 (orthotopic tumors) the 5-week regimen of Ang-(1-7) attenuated intra-tibial tumor growth. Circulating vascular endothelial growth factor (VEGF) levels were significantly higher in control mice compared to mice administered Ang-(1-7), clearly indicating its role as an anti-angiogenic factor. Furthermore, osteoclastogenesis in the bone was reduced by 50% in the presence of Ang-(1-7), suggesting its role in preventing the formation of osteolytic lesions to reduce tumor survival in the bone microenvironment. These findings using two different prostate tumor xenograft models clearly showed that Ang-(1-7) may serve as an anti-angiogenic and anti-metastatic agent for advanced prostate cancer and may provide effective therapy for bone metastasis produced from primary tumors sites.

ETHICAL ISSUES

Many believe that *in vivo* experimentations are unacceptable because they cause suffering to animals and the resulting benefits of such experiments are not clearly proven in human beings. However, another school of thought justifies that animal experiments are necessary for successful drug development, and should be acceptable if adverse physical and emotional effects on the animals are minimized. Indeed, animal studies should be avoided wherever alternative testing methods that produce equally valid results, and should only be conducted after *in vitro* studies have been successfully performed on suitable cell lines. It is, however, widely agreed upon that since all animal experiments cannot be immediately replaced, it is important to uphold the highest standards of welfare for the care and use of animals (Workman et al., 2010).

TRANSLATIONAL SIGNIFICANCE

Translational research on anti-cancer drug development is evolving at a very fast pace towards the discovery of novel therapeutics. The *in vitro* screens are primarily aimed at identifying targets or pathways of interest, then defining the effect a compound on the target(s) or pathway(s). Traditionally, lead compounds are selected for *in vivo* study based on their *in vitro* evidence of cytotoxicity and targeted effects on cancer cells. The use of *in vivo* models to obtain vast quantities of pharmacokinetic (PK) and pharmacodynamics (PD) data is a well-established pre-clinical approach. Before any clinical testing can be initiated in humans, it is important to compare the PK and PD properties of candidate molecules; model potential relationships among dose, concentration, efficacy, and/or toxicity in appropriate animal model systems. Indeed, the *in vivo* evaluations in numerous mouse cancer models have enabled the success of these screening paradigms, which regularly identify diverse types of lead compounds, and their progress towards promising anticancer therapeutics in the clinical setting.

In the past few years, the *in vivo* models have been able to capitalize on the explosion of new "data mining" technologies to cell-based *in vitro* assays which has also facilitated the identification of anti-cancer agents with non-classical endpoints such as those which suppress tumor angiogenesis or tumor invasion. The ultimate goal of *in vivo* model studies is to form a strictly pragmatic standpoint, and demonstration of unbiased and well-understood *in vivo* activity. Therefore, there is a critical need to maximize the utility of the animal model information in selecting agents for translational studies in humans. Many lead compounds that show compromised potency *in vitro* can turn out to be more effective *in vivo* because of their favorable pharmacokinetics, e.g. greater absorption, better distribution and stability, etc. Studies are first initiated in small animal models, e.g. mouse, rats, rabbits, to test for acute, sub-chronic, and then chronic toxicity. In acute toxicity tests, one administration of the drug or chemical is given to each animal in order to generate a safe and effective dose–response curve. Appropriate pharmacological testing in disease models is carried out to determine 50% effective dose (ED_{50}). Following acute administration, analytical methods are developed for determination of absorption, distribution, metabolism and excretion (ADME) of the drug. The sub-chronic toxicity tests usually involve animals exposed to the drug for 60–90 days' duration. Both multiple administrations and/or continuous exposure via food or water to one dose level of a chemical per animal, is/are carried out to measure drug accumulation and possible toxicities. Both genetically altered mice and immune-compromised mice are providing powerful tools for the preclinical evaluation of numerous pharmacological agents. Mutant mice, engineered to mirror the genetic alterations characteristic of human tumors, are also facilitating the identification and validation of biomarkers for early detection of cancer.

However, there are several challenging issues with interpretation of data obtained in small animal models and to prioritize the lead compounds towards human therapeutics. These are the intrinsic differences in pharmacology and drug metabolism observed between rodents and humans. Although several algorithms exist to predict the drug efficacy in humans, such as the extrapolation of cytochrome P450 (Cyp-450) metabolism or bioavailability features in small animals, even minor differences from the human can translate into decreased relevance for murine dosing and efficacy information as predicting clinical value. This is especially relevant since molecules that are tested for cancer treatments are usually exploited at close to their maximum tolerated dose. Indeed, in rodents, the pharmacokinetic parameters such as absorption, plasma protein binding, clearance mechanisms, and intrinsic susceptibility of the host tissues, will need to be determined precisely in order to extrapolate a safe and effective dose in humans. It is therefore essential to demonstrate that logarithmic dose increments of the agent can produce significant parallel reduction in tumor mass. It is becoming clear that in order to manifest highest anti-tumor efficacies and ultimately tumor-free survival, agents have to be administered in successive "cycles." Another paradigm that addresses the relationship between tumor cell inoculum to curability is that when anti-tumor efficacy is not achieved at a constant dose then "adjuvant" treatment programs using multiple drugs that target parallel pathways need to be implemented.

Efficient translational steps taken towards a successful anti-cancer agent development in mouse models should have a more early integration of pharmacological information, both kinetic and dynamic. A short list of currently available mouse cancer models used for drug discovery against different types of human cancers is provided (Table 5.3). In order to be efficacious, the *in vivo* model should also reflect pharmacological action at a distance and should be able to function across physiological and anatomical barriers. This should be evident at an acceptable therapeutic index and at the clinical proposed dose range and schedule. Thus, the ideal *in vivo* model use should be able to guide therapeutics development, the design and ultimate interpretation of the initial human clinical trial. Therefore, from an ethical standpoint, clear demonstration of *in vivo* activity in small animal models should form the basis for potentially justifying patient participation in the clinical study. In this way *in vivo* animal models can contribute not only to the initial qualification of a compound for human use, but also to a more refined way of advancing it to having its best chance for positive later-stage clinical trial efforts.

TABLE 5.3 A Short List of Available Mouse Cancer Models

Available Mouse Cancer Models

- Spontaneous
- Virus-induced
  - Classic models of virus-induced leukemia:
    - Rauscher
    - Moloney
    - LP-BMS
    - Friend
  - Non-leukemic tumors including:
    - Mammary tumors due to MMTV
    - Thymomas of AKR mice
- Transgenic
  - GEM mice (Knock-out/in)
    - Lung Adenocarcinoma (K)
    - Breast ductal carcinoma (Brca2; Trp53)
    - Prostate carcinoma (Pten; Pten:Nkx.1; Rb1:Trp53)
    - Liver carcinoma (Apc, Myc:Trp53, Myc: TGFA)
    - Many others available through commercial sources
    - http://emice.nci.nih.gov/emice/mouse_models
- Induced/carcinogens
    - DMBA [initiator mutagen] followed by TPA [pro-inflammatory]
    - 1,2 Dimethylhydrazine-2-HCl
    - Azoxymethane
    - Nitrosamine 4-(methyl-nitrosamine)-1-(3-pyridyl)-1-butanone
    - Methylcholanthrene
  - Epithelial tumorigenesis
  - GI tumorigenesis
  - Sarcoma induction
  - Lung tumorigenesis
- Transplanted
  - Syngeneic
    - B16 tumors in C57Bl/6 mice
  - Allogeneic – same species, different strain
    - M5076 sarcomas in athymic mice
  - Xenogeneic – different species
    - Human tumors growth in immunocompromised mice
  - Implant Site
    - Orthotopic
    - Heterotopic

WORLD WIDE WEB RESOURCES

Mouse models that mimic human diseases play a vital role in understanding the etiology (cause and origin) of cancer. Results of mouse model studies lend evidence toward the next step in biomedical research that leads to early detection of cancer, new cancer drugs, new combinations of treatments, or new methods such as gene therapy. There are a number of informative web resources that provide guidance and more thorough discussions on the choice of mouse strains to be used in cancer research. Some of these links are provided below.

1. http://www.ncbi.nlm.nih.gov/sites/entrez
   - Inactivating mutations in the p53 gene occur frequently in a variety of human cancers. For more information on the study, applications of the human p53 knock-in mouse model for human carcinogen testing are provided in the above website.
2. http://emice.nci.nih.gov/aam/mouse/inbred-mice-1
   - http://emice.nci.nih.gov/
   - The *electronic Models Information, Communication, and Education* (eMICE) database is maintained by the US National Cancer Institute (NCI) and provides information about a wide variety of animal models of cancer, including mice.
3. http://iospress.metapress.com/content/4806g13r56726741/
   - Genetically modified mice are prone to develop specific cancer types. Therefore, evaluation of imaging strategies could be undertaken at various stages during the course of cancer progression. Similarly, body fluids could be collected at early to late time points during the course of cancer progression. To learn more about this work, "Cancer Biomarkers, NCI Early Detection Research Network: 5th Scientific Workshop," visit the above website
4. http://www.informatics.jax.org/mgihome/other/mgi_people.shtml
   - The *Origins of Inbred Mice* is available online from the Mouse Genome Informatics (MGI) website. This website is maintained by The Jackson Laboratory and integrates access to several databases providing genetic, genomic and biological data on the laboratory mouse to aid its use as a model of human diseases.
5. http://tumour.informatics.jax.org/mtbwi/index.do
   - The *Mouse Tumor Biology Database* (MTBD) is part of the MGI database. It integrates data on tumor frequency, incidence, genetics, and pathology in mice to support the use of the mouse as a cancer model.
6. http://phenome.jax.org/
   - The *Mouse Phenome Database* is also maintained by The Jackson Laboratory and contains strain characterization data (phenotype and genotype) for the laboratory mouse, to facilitate translational research.

REFERENCES

Bhattacharya, A., Turowski, S. G., San Martin, I. D., Rustum, Y. M., Hoffman, R. M., & Seshadri, M. (2011). Magnetic resonance and fluorescence-protein imaging of the anti-angiogenic and anti-tumor efficacy of selenium in an orthotopic model of human colon cancer. *Anticancer Res.*, *31*(2), 387–393.

Bordonaro, M. (2009). Modular Cre/lox system and genetic therapeutics for colorectal cancer. *J Biomed Biotechnol*, *2009*, 358230. Review.

Cardiff, R. D., Anver, M. R., Boivin, G. P., Bosenberg, M. W., Maronpot, R. R., Molinolo, A. A., Nikitin, A. Y., Rehg, J. E., Thomas, G. V., Russell, R. G., & Ward, J. M. (2006). Precancer in mice: animal models used to understand, prevent, and treat human precancers. *Toxicol Pathol*, *34*(6), 699–707.

Casellas, J. (2011). Inbred mouse strains and genetic stability: a review. *Animal*, *5*(1), 1–7.

Cheon, D. J., & Orsulic, S. (2011). Mouse models of cancer. *Annu Rev Pathol*, *6*, 95–119. Review.

Crow, J. F. C.C. (2002). Little, cancer and inbred mice. *Genetics*, *161*(4), 1357–1361.

Ding, Y., Cravero, J. D., Adrian, K., & Grippo, P. (2010). Modeling pancreatic cancer in vivo: from xenograft and carcinogen-induced systems to genetically engineered mice. *Pancreas*, *39*(3), 283–292. Review.

Hanahan, D., & Weinberg, R. A. (2000). The hallmarks of cancer. *Cell.*, *100*(1), 57–70. Review.

Hanahan, D., Wagner, E. F., & Palmiter, R. D. (2007). The origins of oncomice: a history of the first transgenic mice genetically engineered to develop cancer. *Genes Dev.*, *21*(18), 2258–2270. Review.

Huang, B., Qu, Z., Ong, C. W., Tsang, Y. H., Xiao, G., et al. (2012). RUNX3 acts as a tumor suppressor in breast cancer by targeting estrogen receptor α. *Oncogene*, *31*(4), 527–534.

Kataoka, M., Yamaguchi, Y., Moriya, Y., Sawada, N., Yasuno, H., Kondoh, K., Evans, D. B., Mori, K., & Hayashi, S. (2012). Antitumor activity of chemoendocrine therapy in premenopausal and postmenopausal models with human breast cancer xenografts. *Oncol Rep*, *27*(2), 303–310.

Kerbel, R. S. (2003). Human tumor xenografts as predictive preclinical models for anticancer drug activity in humans: better than commonly perceived – but they can be improved. *Cancer Biol Ther*, *2*(4 Suppl. 1), S134–S139.

Krishnan, B., Smith, T. L., Dubey, P., Zapadka, M. E., Torti, F. M., Willingham, M. C., Tallant, E. A., & Gallagher, P. E. (2012). Angiotensin-(1–7) attenuates metastatic prostate cancer and reduces osteoclastogenesis. *Prostate*. http://dx.doi.org/10.1002/pros.22542. [Epub ahead of print].

Mondal, D., Gerlach, S. L., Datta, A., Chakravarty, G., & Abdel-Mageed, B. (2012). Readings in Advanced Pharmacokinetics – Theory, Methods and Applications. *Chapter Title: Pharmacogenomics dictate Pharmacokinetics: polymorphisms in drug-metabolizing enzymes and drug-transporters*. USA.: Intech Open Access Publications, Book Title. ISBN 979-953-307-315-5.

Moser, A. R., Pitot, H. C., & Dove, W. P. (1990). A dominant mutation that predisposes to multiple intestinal neoplasm in the mouse. *Science*, *247*, 322–324.

Müller, A., Homey, B., Soto, H., Ge, N., Catron, D., Buchanan, M. E., McClanahan, T., Murphy, E., Yuan, W., Wagner, S. N., Barrera, J. L., Mohar, A., Verástegui, E., & Zlotnik, A. (2001). Involvement of chemokine receptors in breast cancer metastasis. *Nature*, *410*(6824), 50–56.

Pal, S. K., & Hurria, A. (2010). Impact of age, sex, and comorbidity on cancer therapy and disease progression. *J Clin Oncol*, *28*(26), 4086–4093. Epub 2010 Jul 19. Review.

Pelekanou, V., & Leclercq, G. (2011). Recent insights into the effect of natural and environmental estrogens on mammary development and carcinogenesis. *Int J Dev Biol.*, *55*(7-9), 869–878.

Piegari, M., Díaz Mdel, P., Eynard, A. R., & Valentich, M. A. (2011). Characterization of a murine lung adenocarcinoma (LAC1), a useful experimental model to study progression of lung cancer. *J Exp Ther Oncol*, *9*(3), 231–239.

Prosperi, J. R., Khramtsov, A. I., Khramtsova, G. F., & Goss, K. H. (2011). Apc mutation enhances PyMT-induced mammary tumorigenesis. *PLoS One*, *6*(12), e29339. Epub 2011 Dec22.

Ramanathan, V., Jin, G., Westphalen, C. B., Whelan, A., Dubeykovskiy, A., Takaishi, S., & Wang, T. C. (2012). P53 gene mutation increases progastrin dependent colonic proliferation and colon cancer formation in mice. *Cancer Invest*, *30*(4), 275–286.

Richmond, A., & Su, Y. (2008). Mouse xenograft models vs GEM models for human cancer therapeutics. *Dis Model Mech*, *1*(2-3), 78–82.

Sauer, B. (1998 Apr). Inducible gene targeting in mice using the Cre/lox system. *Methods*, *14*(4), 381–392. Review.

Schayowitz, A., Sabnis, G., Goloubeva, O., Njar, V. C., & Brodie, A. M. (2010). Prolonging hormone sensitivity in prostate cancer xenografts through dual inhibition of AR and mTOR. *Br J Cancer*, *103*(7), 1001–1007.

Shacter, E., & Weitzman, S. A. (2002). Chronic inflammation and cancer. *Oncology*, *16*(2), 217–226. Review.

Singh, M., & Johnson, L. (2006). Using genetically engineered mouse models of cancer to aid drug development: an industry perspective. *Clin Cancer Res.*, *12*(18), 5312–5328. Review.

Suthar, Maulik P., & Javia., Ankur (2009). Xenograft cancer mice models in cancer drug Discovery. *Pharma Times -, Vol. 41*(No. 1).

Tatum, A. M., Mylin, L. M., Bender, S. J., Fischer, M. A., Vigliotti, B. A., Tevethia, M. J., Tevethia, S. S., & Schell, T. D. (2008). CD8+ T cells targeting a single immunodominant epitope are sufficient for elimination of established SV40 T antigen-induced brain tumors. *J Immunol*, *181*(6), 4406–4417.

Tian, T., Olson, S., Whitacre, J. M., & Harding, A. (2011). The origins of cancer robustness and evolvability. *Integr Biol (Camb)*, *3*(1), 17–30. Review.

Triplett, A. A., Montagna, C., & Wagner, K. U. (2008). A mammary-specific, long-range deletion on mouse chromosome 11 accelerates Brca1-associated mammary tumorigenesis. Neoplasia, *10*(12), 1325–1334.

Wang, T., Gavin, H. M., Arlt, V. M., Lawrence, B. P., Fenton, S. E., Medina, D., & Vorderstrasse, B. A. (2011). Aryl hydrocarbon receptor activation during pregnancy, and in adult nulliparous mice, delays the subsequent development of DMBA-induced mammary tumors. *Int J Cancer*, *128*(7), 1509–1523.

Workman, P., Aboagye, E. O., Balkwill, F., Balmain, A., Bruder, G., Chaplin, D. J., Double, J. A., Everitt, J., Farningham, D. A., Glennie, M. J., Kelland, L. R., Robinson, V., Stratford, I. J., Tozer, G. M., Watson, S., Wedge, S. R., & Eccles, S. A. (2010). Committee of the National Cancer Research Institute. Guidelines for the welfare and use of animals in cancer research. *Br J Cancer*, *102*(11), 1555–1577.

FURTHER READING

Abate-Shen, C. (2006). A new generation of mouse models of cancer for translational research. *Clin Cancer Res*, *12*(18), 5274–5276.

Frese, K. K., & Tuveson, D. A. (2007). Maximizing mouse cancer models. *Nat Rev Cancer*, *7*(9), 645–658.

Hunter, K. W. (2012). Mouse models of cancer: does the strain matter? *Nat Rev Cancer*, *12*(2), 144–149.

Klausner, R. D. (1999). Studying cancer in the mouse. *Oncogene*, *18*(38), 5249–5252. Review.

Politi, K., & Pao, W. (2011). How genetically engineered mouse tumor models provide insights into human cancers. *J Clin Oncol*, *29*(16), 2273–2281.

Walrath, J. C., Hawes, J. J., Van Dyke, T., & Reilly, K. M. (2010). Genetically engineered mouse models in cancer research. *Adv Cancer Res*, *106*, 113–164. http://dx.doi.org/10.1016/S0065-230X(10)06004-5.

GLOSSARY

Allograft A graft between individuals of the same species, but of different genotypes.

Angiogenesis Blood vessel formation. Angiogenesis that occurs in cancer is the growth of blood vessels from surrounding tissue to a solid tumor.

Benign A swelling or growth that is not cancerous and does not metastasize.

Cancer Stem Cells A small population of cells inside tumors that have the ability to self-renew while giving rise to different types of cells. Cancer stem cells might be resistant to many cancer drugs and reconstitute a tumor after chemotherapy.

Congenic Relating to a strain of animals developed from an inbred (isogenic) strain by repeated matings with animals from another stock that have a foreign gene, the final congenic strain then presumably differing from the original inbred strain by the presence of this gene.

Consomic An inbred strain with one of its chromosomes replaced by the homologous chromosome of another inbred strain via a series of marker-assisted backcrosses.

DCIS Ductal carcinoma *in situ*. A precancerous condition characterized by the clonal proliferation of malignant-looking cells in the lining of a breast duct without evidence of spread outside the duct to other tissues in the breast or outside the breast.

Epigenetic Having to do with the chemical attachments to DNA or the histone proteins. Epigenetic marks change the pattern of genes expressed in a given cell or tissue by amplifying or mutating the effect of a gene.

GEM Model Genetically Engineered Mouse model based on the loss or gain of specific genes, to study a disease like cancer in immunocompetent animals.

Germ Line Genetic material that is passed down through the gametes (sperm and egg).

Hazard Ratio A summary of the difference between two survival curves, representing the reduction in the risk of death on treatment compared to control.

HER2 (Human Epidermal Growth Factor Receptor-2) The HER2 gene is responsible for making HER2 protein, which plays an important role in normal cell growth and development.

Hyperplasia An overgrowth of cells.

Institutional Review Board (IRB) A board designed to oversee the research process in order to protect participant safety.

Karyotype A photomicrograph of an individual's complete set of chromosomes arranged in homologous pairs and ordered by size. Karyotypes show the number, size, shape and banding pattern of each chromosome type.

Leukemia Cancer of white blood cells.

Mitosis The process of cell division, resulting in formation of two daughter cells that are genetically identical to the parent cell.

Neoadjuvant Initial treatment, which is not the primary therapy (for instance, chemotherapy or radiation, prior to surgery).

Neoplasm An abnormal new growth of tissues or cells. Neoplasms can be benign or malignant.

Oncogene A mutated proto-oncogene that is locked into an active state and continuously stimulates unregulated cell growth and proliferation that leads to tumor development. The normal allele of an oncogene is called a proto-oncogene.

Palliative Treatment Treatment aimed at the relief of pain and symptoms of disease, but is not intended to cure the disease.

Recurrence The return of cancer, at the same site as the original (primary) tumor, or in another location, after the tumor had disappeared.

Remission A decrease in or disappearance of signs and symptoms of cancer. In partial remission, some, but not all, signs and symptoms of cancer have disappeared. In complete remission, all signs and symptoms of cancer have disappeared, although there still may be cancer in the body.

Telomeres Special DNA sequences at the ends of each chromosome that grow shorter each time a cell divides.

Telomerase An enzyme that rebuilds telomeres. Telomerase is overexpressed in many cancer cells, and it contributes to their immortality, or ability to divide endlessly.

VEGF (Vascular Endothelial Growth Factor) A protein that is secreted by oxygen-deprived cells, such as cancerous cells. VEGF stimulates new blood vessel formation, or angiogenesis, by binding to specific receptors on nearby blood vessels, encouraging new blood vessels to form.

Xenograft A graft of tissue transplanted between animals of different species.

ABBREVIATIONS

4-OHT 4-Hydroxytamoxifen
ADME Absorption Distribution Metabolism Excretion
AHR Aryl Hydrocarbon Receptor
Ang-(1-7) Angiotensin-(1-7)
ANOVA Analysis of Variance
APC Adenomatous Polyposis Coli
AR Androgen Receptor
ATCC American Tissue Culture Collection
AVMA Animal Veterinary Medical Association
Cyp-450 Cytochrome P450
DBA strain Dilute, Brown and Non-Agouti Strain
DMBA 7,12-Dimethylbenz(α)anthracene
DNA Deoxyribonucleic acid
ED_{50} 50% Effective Dose
eMICE Electronic Models Information, Communication, and Education Database
ERα Estrogen Receptor α
ERβ Estrogen Receptor β
ES cells Embryonic Stem Cells
FBS Fetal Bovine Serum
FPI Fluorescent Protein Imaging
GEM Genetically Engineered Mouse
GFP Green Fluorescent Protein
HER2 Human Epidermal Growth Factor Receptor 2
HBBS Hank's Buffered Saline Solution
IACUC International Animal Core and Use Committee
i.m. Intramuscular
i.p. Intraperitoneal
KXA Ketamine–Xylazine–Acepromazine
K-Ras Kirsten Rat Sarcoma Viral Oncogene Homolog
MRI Magnetic Resonance Imaging
mTOR Mammalian Target of Rapamycin
MVD Microvessel Density
MGI Mouse Genome Informatics Website
MMTV Mouse Mammary Tumor Virus
MTBD Mouse Tumor Biology Database
NCI National Cancer Institute
NOD Non-Obese Diabetic
NSCLC Non-Small Cell Lung Cancer
PK Pharmacokinetics
PD Pharmacodynamics
PG Pharmacogenomics
PBS Phosphate Buffered Saline
PSA Prostate Specific Antigen
PyMT Polyoma Virus, Medium T antigen
Rb Retinoblastoma Protein
RNA Ribonucleic Acid
s.c. Subcutaneous
SCLC Non-Small Cell Lung Cancer
SCID Severe Combined Immunodeficiency
SE Standard Error
SERM Selective Estrogen Receptor Modulators
SNP Single Nucleotide Polymorphism
SV40 Simian Virus 40
UV Ultraviolet
VEGF Vascular Endothelial Growth Factor

LONG ANSWER QUESTIONS

1. Discuss the different applications of Genetically Engineered Mouse (GEM) models used in cancer research.
2. Explain how different recombinant DNA vectors can be used to generate the (i) Constitutive and (ii) Inducible models of transgenic mice used in cancer research.
3. Explain the concept of "humanized mice" and provide an example of how these mice can be used to discover new anti-cancer agents.
4. Describe in detail how pre-clinical drug development has led to clinical approval of numerous chemotherapeutic agents, e.g. target identification, library screening, *in vitro* studies, lead candidates, and mouse PK and PD studies *in vivo*.
5. You have generated a prostate cancer (PC) cell line which grows in the absence of androgen, suggesting that it would be a good model to identify drugs against castration resistant prostate cancers (CRPC). Using this cell line, how will you study CRPC tumor development and screen for drugs *in vivo*?

SHORT ANSWER QUESTIONS

1. What are the advantages and disadvantages of mouse models in cancer research?
2. What points will need to be addressed to obtain an IACUC approval?

3. Provide an example where GEM models were used to test for tumor-suppressor genes.
4. What are the differences between xenograft and allograft models in cancer research?
5. Discuss the *Cre-lox* model of transgenic mouse development for cancer research.

ANSWERS TO SHORT ANSWER QUESTONS

1. The small size and short tumor generation time in mice make them ideal models of cancer studies. They are relatively inexpensive compared to larger animal models, which enable the use of large numbers for statistical measurements. Some inbred strains of mice are genetically well characterized and several transgenic and knock-out mouse models are available for cancer research. However, since the metabolic rates in mice are very different from humans, there are considerable differences in drug efficacies and toxicities, which need to be taken into account. Furthermore, most mouse models develop fewer metastases or display metastases with different tissue specificity as compared to human tumors. Due to a limited number of initiating genetic alterations, mouse tumors are typically more homogeneous and this can be an obstacle to modeling the heterogeneity of human cancers.
2. The 3Rs (Refine, Reduce, Replace) are important considerations and criteria when writing an IACUC protocol. Refinement involves careful consideration of the aspects of research methodology and animal maintenance. The *in vivo* techniques should be thoroughly reviewed so that they cause minimal amount of pain, stress, or suffering. Reduction aims at limiting the amount of animals needed for a study. Replace takes into account that new technologies are developing that provides an alternative to animal testing, or at least minimize their use. The initial toxicology studies should be carried out using cultured liver cells or other normal cells lines such that supportive data is gained to test the efficacy and toxicity prior to animal testing. The protocol needs to first identify the mouse strain, sex and age, and approximate number of mice that will be used. A thorough rationale for using mice and the appropriateness and a complete description of the proposed use of mice has to be justified. A description of anesthetic and euthanasia procedures, a description of appropriate living conditions such as housing, feeding, and nonmedical care of the animals will also need to be included.
3. In a recent publication, Huang et al. (2012), showed that RUNX3 acts as a tumor suppressor in breast cancer by targeting estrogen receptor α (Huang et al., 2012). *RUNX3*, a runt-related gene family protein, is known to act as a tumor suppressor in breast cancer. To demonstrate the role of *RUNX3* as a tumor suppressor, these investigators used heterozygous female mice with *RUNX3* mutation (*Runx3*$^{+/-}$) and compared mammary tumor development with control mice. In different age mice, they isolated mammary gland specimens from wild-type and *Runx3*$^{+/-}$ mice and evaluated RUNX3-immunohistochemical (IHC) staining. Tumor sections were stained with hematoxylin and eosin (HE) and evaluated by a pathologist to designate tumor cells. About one-fifth of the *Runx3*$^{+/-}$ female mice developed mammary tumors, whereas none of the wild-type mice developed mammary tumors. The *RUNX3* gene were overexpressed in human breast cancer cells (MCF-7) and their ability to form tumors in mouse xenografts were also studied. Both vector-MCF-7 (control) or RUNX3-MCF-7 cells were subcutaneously implanted in SCID mice (C.B-17/IcrCrl-scidBR). Compared with vector-MCF-7 cells, RUNX3-MCF-7 cells produced <80% smaller tumors. These results, using both GEM and xenograft models, indicated that RUNX3 acts as a tumor suppressor.
4. In the allograft (or syngeneic) transplantation models the cancerous cells or solid tumors are itself of mouse origin, and are transplanted into another host mouse containing a specific genetic trait (e.g. GEM mice). Owing to the fact that the recipient and the cancer cells have the same origin, the transplant is not rejected by mice with an intact immune system. Results of therapeutic interventions can thus be analyzed in an environment that closely recapitulates the real scenario where the tumor grows in an immunocompetent environment. A disadvantage, however, is that transplanted mouse tissue may not represent in whole the complexity of human tumors in clinical situations. In the xenograft model, human tumor cells are transplanted, either under the skin or into the organ type (orthotopic) in which the tumor originated, into immunocompromised mice like athymic nude mice (*nu/nu*) or severely compromised immunodeficient (SCID) mice, so that the human cells are not rejected. In this model, human cancer cells or solid tumors are transplanted into a host mouse. Because the cancer xenograft is of human origin, it represents the properties and to an extent the complexities of the human cancer. On the other hand, the compromised host immune system is not truly representative of actual patients. Tumor xenografts have been used in the preclinical and clinical development of anti-cancer therapeutics, and xenograft models are useful for toxicity studies from targeted therapies, and in many cases to predict biomarkers of target modulation.
5. The *Cre/lox* system is frequently used as a way to artificially control gene expression in transgenic mice. In this system, the induction of Cre-recombinase enzyme

mediates the site directed DNA recombination between two 34–base pair loxP sequences. To achieve this in mice, transgenic mice containing a gene surrounded by "lox-P" sites are mated with transgenic mice that have the *cre* gene expressed in specific cell types. In tissues with no *cre* gene the target gene will function normally; however, in cells where *cre* is expressed, the target gene is deleted. Therefore, the correct placement of *Lox* sequences around a gene of interest may allow genes to be activated, repressed, or exchanged for other genes. Furthermore, the activity of the *Cre* enzyme can be controlled so that it is expressed in a particular cell type or triggered by external stimulus like chemical signals or heat shock. This enables the induction of somatic mutations in a time-controlled and tissue-specific manner. Therefore, this homologous recombination system closely models carcinogenesis in humans where tumors evolve from somatic gene mutations in normal cells in a time-controlled and tissue-specific fashion *in vivo*.

Chapter 6

Human Papillomavirus (HPV): Diagnosis and Treatment

Mausumi Bharadwaj, Dr[*], Showket Hussain, Dr[*], Richa Tripathi, Dr[*], Neha Singh, Dr[†] and Ravi Mehrotra, Prof.[**]

*Division of Molecular Genetics & Biochemistry, Institute of Cytology & Preventive Oncology (ICMR), NOIDA, India, †Department of Biotechnology, Panjab University, Chandigarh, India, **Division of Cytopathology, Institute of Cytology & Preventive Oncology (ICMR), NOIDA, India

Chapter Outline

Animal Biotechnology. http://dx.doi.org/10.1016/B978-0-12-416002-6.00006-7

SUMMARY

Cervical carcinoma is a lethal and prevalent cancer in women. Among its various etiological factors, infection with high-risk HPV types is implicated with this cancer. This chapter presents the latest knowledge about cervical carcinoma and associated clinicomolecular approaches. The intended audience is undergraduate and post-graduate students, clinicians, cytologists, and public health specialists.

WHAT YOU CAN EXPECT TO KNOW

Cancer development is a multistep process characterized by uncontrolled cell growth and invasion of nearby tissue and distant sites that leads to alterations in genetic and molecular pathways of cells. Uterine cervical cancer is the third most common cancer among women worldwide. A large number of risk factors contribute to its high incidence, but the most important factor is infection with human papillomavirus (HPV). HPVs are small DNA viruses that are epitheliotropic (infect epithelial cells) in nature and cause a variety of benign epithelial lesions such as warts, condyloma acuminate, and neoplasias of the lower genital tract. More than 100 genotypes of HPV have been described to date, and are either high- or low-risk types. The oncogenic potential of HPV is attributed to its E6 and E7 genes. The products of these two genes stimulate cell proliferation by activating cell cycle-specific proteins and interfering with the functions of cellular growth regulatory proteins such as p53 and pRb.

It is imperative to mention that the incidence of cervical cancer has dropped due to awareness, vaccines, and regular Pap screening tests. Initiatives have been taken for the development of therapeutic vaccines against HPV, but most are in their infancy and will take time to become a clinical reality.

Wherever possible, in this chapter an attempt has been made to present the clinical, molecular, and epidemiological aspects of cervical cancer with special emphasis on the diagnostic/ prognostic applications for cervical cancer management.

HISTORY AND METHODS

CANCER OVERVIEW

Cancer refers to a class of disease wherein a cell or a group of cells divides and replicates uncontrollably due to accumulation of both genetic and/or epigenetic changes that occur in a multistep manner. This leads to unregulated cell proliferation, intrusion into adjacent cells and tissues (invasion), and ultimately, spread to other parts of the body (metastasis). These cells continue to grow despite restriction of space, nutrients, and initiating stimulus, with a tendency to invade or spread into adjoining and/or distant tissues (Parkin et al., 2005; Hanahan and Weinberg, 2000). Cancer development includes six essential alterations in cell physiology that include malignant growth, self sufficiency in growth signals, insensitivity to growth–inhibitory signals, evasion of programmed cell death (apoptosis), limitless replicative potential, sustained angiogenesis, and tissue invasion and metastasis (Kaivosoja, 2008).

CLASSIFICATION OF CANCER

Cancers may be classified according to their primary site of origin or by their histological grade or tissue types. The primary site of cancer origin, however, may be specific, such as uterine cancer, cervical cancer, breast cancer, oral cancer, esophageal cancer, lung cancer, prostate cancer, liver cancer, etc. Cancer may be classified into six major categories:

1. **Carcinoma**: Originates in epithelial tissue (i.e. tissue that lines organs and tubes).
2. **Sarcoma**: Originates in connective or supportive tissue (e.g. bone, cartilage, muscle).
3. **Myeloma**: Originates in bone marrow.
4. **Leukemia**: Originates in tissues that form blood cells.
5. **Lymphoma**: Originates in lymphatic tissue.
6. **Mixed**: Mixed mesodermal tumor, carcinosarcoma, adenosquamous carcinoma, teratocarcinoma, blastomas.

Carcinoma

This type of cancer originates from the epithelial layer of cells that form the lining of external parts of the body or the internal linings of organs within the body. This type of cancer accounts for 80 to 90 percent of all cancer cases since epithelial tissues are abundantly found in the body; they are present everywhere from the skin to the covering and lining of organs (e.g. the gastrointestinal tract). Carcinomas usually affect organs or glands capable of secretion, including breast, lungs, bladder, colon, and prostate. Carcinomas are of two types: adenocarcinoma and squamous cell carcinoma. Adenocarcinoma develops in an organ or gland, whereas squamous cell carcinoma originates in the squamous epithelium. Adenocarcinomas may involve mucus membranes, and are first seen as a thickened plaque-like white mucosa. These are rapidly spreading cancers with poor prognoses.

Sarcoma

Sarcomas originate in connective and supportive tissues including muscles, bones, cartilage, and fat. The classic examples of sarcomas are osteosarcoma (bone) and chondrosarcoma (cartilage). Other examples include fibrosarcoma (fibrous tissue), angiosarcoma or hemangioendothelioma (blood vessels), leiomyosarcoma (smooth muscles), rhabdomyosarcoma (skeletal muscles), mesothelial sarcoma or mesothelioma (membranous lining of body cavities), liposarcoma (adipose or fatty tissue), glioma or astrocytoma (neurogenic connective tissue found in the brain), myxosarcoma (primitive embryonic connective tissue), and mesenchymous or mixed mesodermal tumor (mixed connective tissue types).

Myeloma

These originate in the plasma cells of bone marrow. Plasma cells are capable of producing various antibodies in response to infections. Myeloma is a type of blood cancer.

Leukemia

These cancers are grouped within blood-related cancers. They affect bone marrow, which is the primary site for blood cell production. Types of leukemia include:

- **Acute Myelocytic Leukemia (AML):** These are malignancies of the myeloid and granulocytic white blood cell series common in children.
- **Chronic Myelocytic Leukemia (CML):** This is seen in adults.
- **Acute Lymphatic, Lymphocytic, or Lymphoblastic Leukemia (ALL):** These are malignancies of the lymphoid and lymphocytic blood cell series, and are common in both children and young adults.
- **Chronic Lymphatic, Lymphocytic, or Lymphoblastic Leukemia (CLL):** This is seen in the elderly.
- **Polycythemia Vera or Erythremia:** This is cancer of various blood cell products, with a predominance in red blood cells.

Lymphoma

These are cancers of the lymphatic system. Unlike leukemias, which affect the blood and are thus called "liquid cancers," lymphomas are "solid cancers." These may affect lymph nodes at specific sites like the stomach, brain, intestines, etc. These lymphomas are referred to as extranodal lymphomas. Generally, lymphomas are of two types: Hodgkin's and non-Hodgkin's. In Hodgkin's lymphoma, there is a characteristic presence of Reed–Sternberg cells in the tissue samples; these are are not present in non-Hodgkin's lymphoma.

Mixed Types

These have two or more components of a cancer. Some examples include mixed mesodermal tumor, carcinosarcoma, adenosquamous carcinoma, and teratocarcinoma. Blastomas are another type that involve embryonic tissues.

Tumor Grading

Grading involves examining tumor cells that have been obtained through tissue biopsy sections under a microscope. The abnormality of the cells determines the grade of the cancer. Increasing abnormality increases the grade (from 1 to 4). Cells that are well differentiated closely resemble mature, specialized cells. Cells that are undifferentiated are highly abnormal, that is, immature and primitive:

Grade 1: Cells that are slightly abnormal and well differentiated.
Grade 2: Cells that are more abnormal and moderately differentiated.
Grade 3: Cells that are very abnormal and poorly differentiated.
Grade 4: Cells that are immature and undifferentiated.

Cancer Staging

Staging is the classification of the extent of the disease. There are several types of staging methods. The tumor, node, metastases (TNM) system classifies cancer by tumor size (T), the degree of regional spread or node involvement (N), and distant metastasis (M).

Tumor (T)

T0: No evidence of tumor.
Tis: Carcinoma *in situ* (limited to surface cells).
T1–4: Increasing tumor size and involvement.

Node (N)

N0: No lymph node involvement.
N1–4: Increasing degrees of lymph node involvement.
Nx: Lymph node involvement cannot be assessed.

Metastases (M)

M0: No evidence of distant metastases.
M1: Evidence of distant metastases.

A numerical system also is used to classify the extent of disease.

Stage 0: Cancer *in situ* (limited to surface cells).
Stage I: Cancer limited to the tissue of origin; evidence of tumor growth.
Stage II: Limited local spread of cancerous cells.
Stage III: Extensive local and regional spread.
Stage IV: Distant metastasis.

CANCER-CAUSING AGENTS

Cancer-causing agents can be categorized into three groups: oncogenic viruses, chemicals, and radiation. All three either independently or in combination drive the carcinogenic cycle.

Oncogenic Viruses

Oncogenic viruses cause different types of cancers in humans. The potential of viruses/bacteria and their associated cancers is summarized in Table 6.1.

Chemicals

Numerous chemicals are known to cause cancer in humans. Many of these chemicals carry out their effects only on specific organs. The effect of chemical carcinogens and their interaction for cancer initiation and progression are divided into the following two categories: tumor initiators and promoters.

TABLE 6.1 Biological Agents That Cause Cancer

Virus/Bacteria	Type	Associated Cancers
Papillomavirus	HPV	Cervical Cancer
Hepatitis Virus	HBV, HCV	Liver Cancer
Herpes Virus	EBV, HHB	Nasopharyngeal carcinoma
Retrovirus	HTLV-1, Adult T-cell Leukemia	Lymphoma
HIV/AIDS	HIV-1/-2	Kaposi sarcoma
Helicobactor pylori	-	Stomach Cancer

Tumor Initiators

For tumor progression to occur, initiation must be followed by exposure to chemicals capable of promoting tumor development. Promoters do not cause heritable damage to the DNA, and thus on their own cannot generate tumors. Tumors ensue only when exposure to a promoter follows exposure to an initiator.

The effect of initiators is irreversible, whereas the changes brought about by promoters are reversible. Many chemicals, known as complete carcinogens, can both initiate and promote a tumor; others, called incomplete carcinogens, are capable only of initiation.

Proto-oncogenes and tumor suppressor genes are two critical targets of chemical carcinogens. When an interaction between a chemical carcinogen and DNA results in a mutation; the chemical is said to be a mutagen. Because most known tumor initiators are mutagens, potential initiators can be tested by assessing their ability to induce mutations in a bacterium (*Salmonella typhimurium*). This test, called the Ames test, has been used to detect the majority of known carcinogens.

Some of the most potent carcinogens for humans are the **polycyclic aromatic hydrocarbons**, which require metabolic activation for becoming reactive. Polycyclic hydrocarbons affect many target organs and usually produce cancers at the site of exposure. These chemical substances are produced through the combustion of tobacco, especially in cigarette smoking, and also can be derived from animal fats during the broiling of meats. They are also found in smoked fish and meat. The carcinogenic effects of several of these compounds have been detected through cancers that develop in industrial workers. For example, individuals working in the aniline dye and rubber industries have had up to a 50-fold increase in incidence of urinary bladder cancer that has been traced to exposure to heavy doses of aromatic amine compounds. Workers exposed to high levels of vinyl chloride, a hydrocarbon compound from which the widely used plastic polyvinyl chloride is synthesized, have relatively high rates of a rare form of liver cancer called angiosarcoma.

There are also chemical carcinogens that occur naturally in the environment. One of the most important of these substances is **aflatoxin B1**; this toxin is produced by the fungi Aspergillus flavusand *A. parasiticus*, which grow on improperly stored grains and peanuts. Aflatoxin B is one of the most potent liver carcinogens known. Many cases of liver cancer in Africa and East Asia have been linked to dietary exposure to this chemical.

Promoters

The initial chemical reaction that produces a mutation does not in itself suffice to initiate the carcinogenic process in a cell. For the change to be effective, it must become permanent. Fixation of the mutation occurs through cell proliferation before the cell has time to repair its damaged DNA. In this way, the genetic damage is passed on to future generations of cells and becomes permanent. Because many carcinogens are also toxic and kill cells, they provide a stimulus for the remaining cells to grow in an attempt to repair the damage. This cell growth contributes to the fixation of the genotoxic damage. The major effect of tumor promoters is the stimulation of cell proliferation. Sustained cell proliferation is often observed to be a factor in the pathogenesis of human tumors. This is because continuous growth and division increases the risk that the DNA will accumulate and pass on new mutations. Evidence for the role of promoters in the cause of human cancer is limited to a handful of compounds. The promoter best studied in the laboratory is tetradecanoyl phorbol acetate (TPA), a phorbol ester that activates enzymes involved in transmitting signals that trigger cell division. Some of the most powerful promoting agents are hormones, which stimulate the replication of cells in target organs. Prolonged use of the hormone diethylstilbestrol (DES) has been implicated in the production of post-menopausal endometrial carcinoma, and it is known to cause vaginal cancer in young women who exposed to the hormone while in the womb. Fats too may act as promoters of carcinogenesis, which possibly explains why high levels of saturated fat in the diet are associated with an increased risk of colon cancer.

Radiation

Among the physical agents that give rise to cancer, radiant energy is the main tumor-inducing agent in animals, including humans.

Ultraviolet Radiation

Ultraviolet (UV) rays in sunlight give rise to basal cell carcinoma, squamous cell carcinoma, and malignant melanoma of the skin. The carcinogenic activity of UV radiation is attributable to the formation of pyrimidine dimers in DNA. Pyrimidine dimers are structures that form between two of four nucleotide bases that make up DNA: the nucleotides cytosine and thymine, which are members of the chemical family called pyrimidines. If a pyrimidine dimer in a growth regulatory gene is not immediately repaired, it can contribute to tumor development.

The molecular basis of cancer is DNA repair defects. The risk of developing UV-induced cancer depends on the type of UV rays to which one is exposed (UV-B rays are thought to be the most dangerous), the intensity of the exposure, and the quantity of protection that the skin cells are afforded by the natural pigment melanin. Fair-skinned persons exposed to the sun have the highest incidence of melanoma because they have the least amount of protective melanin. It is likely that UV radiation is a complete carcinogen, that is, it can initiate and promote tumor growth, just as some chemicals do.

Ioinizing Radiation

Ionizing radiation (both electromagnetic and particulate) is a powerful carcinogen, although several years can elapse between exposure and the appearance of a tumor. The contribution of radiation to the total number of human cancers is probably small compared with the impact of chemicals, but the long latency of radiation-induced tumors and the cumulative effect of repeated small doses make precise calculation of its significance difficult. The carcinogenic effects of ionizing radiation first became apparent at the turn of the 20th century with reports of skin cancer in scientists and physicians who pioneered the use of X-rays and radium. Some medical practices that used X-rays as therapeutic agents were abandoned because of the high increase in the risk of leukemia. The atomic explosions in Japan at Hiroshima and Nagasaki in 1945 provided dramatic examples of radiation carcinogenesis: after an average latency period of seven years, there was a marked increase in leukemia, followed by an increase in solid tumors of the breast, lung, and thyroid. A similar increase in the same types of tumors was observed in areas exposed to high levels of radiation after the Chernobyl disaster in the Ukraine in 1986. Electromagnetic radiation is also responsible for cases of lung cancer in uranium miners in central Europe and the Rocky Mountains of North America.

CERVICAL CANCER

Anatomy of the Female Pelvis

The cervix (from the Latin for "neck") connects the upper body of the uterus to the vagina. The cervix is the lower one-third of the uterus, and is composed of dense, fibromuscular tissue lined by two types of epithelium: squamous epithelium and columnar epithelium. It is about 3 cm in length and 2.5 cm in diameter. The endocervix (upper part close to the uterus) is covered by glandular cells, and the ectocervix (lower part close to the vagina) is covered by squamous cells. The stratified squamous epithelium covers most of the ectocervix and vagina. It's lowest (basal) layer, composed of rounded cells, is attached to the basement membrane, which separates the epithelium from the underlying fibromuscular stroma. The columnar epithelium lines the cervical canal and extends outwards to a variable portion of the ectocervix. The transformation zone refers to the place where these two regions of the cervix meet. The original squamocolumnar junction (SCJ) appears as a sharp line, with a step produced by the different thicknesses of the columnar and squamous epithelia. The cervix produces cervical mucus that changes in consistency during the menstrual cycle to prevent or promote pregnancy. During childbirth, the cervix dilates to allow the baby to pass through. During menstruation, the cervix opens a small amount to permit passage of menstrual flow.

Cervical cancer forms in the cervix, the lower end of the uterus, and is preceded by a precancerous condition called cervical intraepithelial neoplasia (CIN), which may or may not develop into cancer (Das et al., 2008). The progression of cervical cancer is a multi-step process. Initially, normal cells undergo precancerous changes and ultimately develop into cancer cells. These precancerous conditions include cervical intraepithelial neoplasia (CIN), squamous intraepithelial lesion (SIL), and dysplasia. It takes several years to develop into an invasive cancer, and is preventable if detected early (Figure 6.1). Thus, cervical cancer provides an excellent human model for studying the process of carcinogenesis *in vivo*.

Historical Overview

Cervical cancer is a major health concern for all women. In most developing countries, cervical cancer is the leading female malignancy and a common cause of death among middle-aged women. In developed populations with good awareness and screening options, invasive cervical cancer is a relatively rare condition, whereas its precursors and the equivocal cytological results represent a major health burden (zur Hausen, 1982; zur Hausen, 2002; Parkin and Bray, 2006; Heins Jr. et al., 1958; Gasparini and Panatto, 2009).

The disease has been known since ancient times, however, the cause was unknown. In 400 BCE, the Greek physician Hippocrates wrote about the disease and even attempted to treat the cancer with a procedure known as the trachelectomy, but could not eradicate the cancer.

Mistaken Theories of Causation

Epidemiologists working in the early 20th century noted that cervical cancer behaved like a sexually transmitted disease and summarized it as follows:

1. Cervical cancer was common in female sex workers.
2. For centuries, doctors were confused by the cause of cervical cancer. The first theory rose to prominence in 1842 in Florence, when a doctor noticed that married women and prostitutes were susceptible to cervical cancer, but nuns had a very low incidence of the cancer (Rigoni in 1841). However, because nuns did suffer from breast cancer, it was incorrectly determined that the cause of both diseases was tight corsets.
3. It was more common in the second wives of men whose first wives had died from cervical cancer.
4. It was rare in Jewish women.
5. In 1935, Syverton and Berry discovered a relationship between rabbit papillomavirus (RPV) and skin cancer

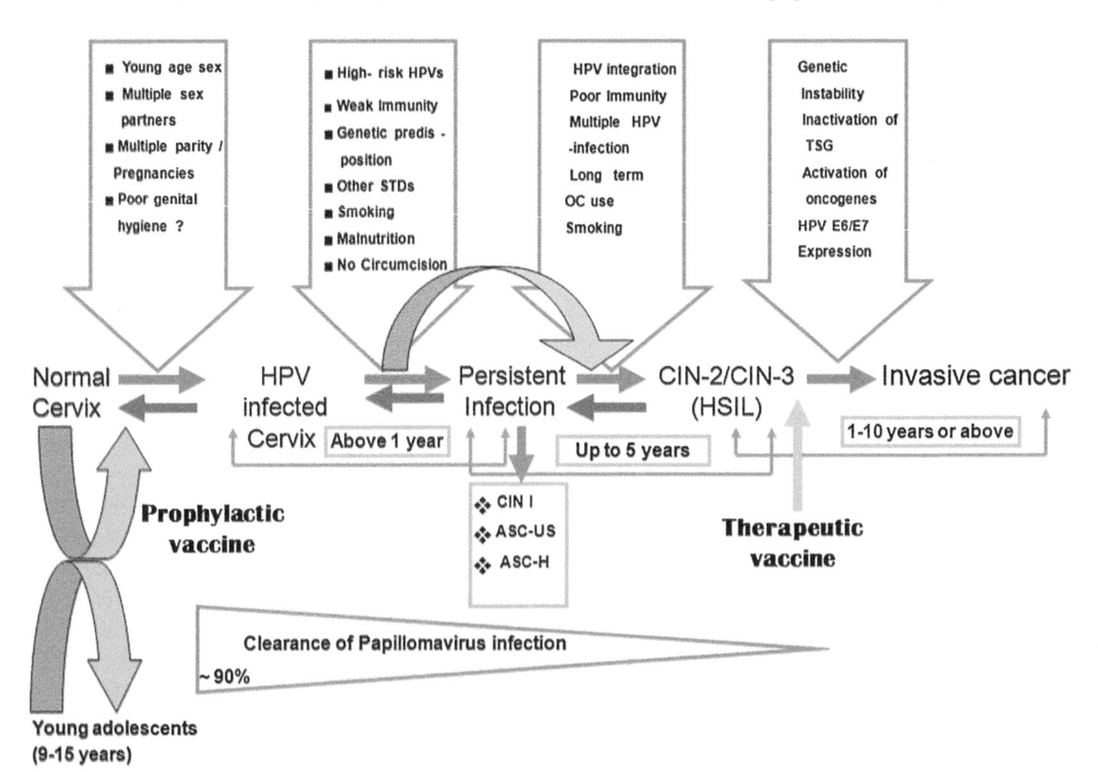

FIGURE 6.1 Biological behavior of HPV infection and development of cervical cancer.

in rabbits (HPV is species-specific, and therefore cannot be transmitted to rabbits).

This led to the suspicion that cervical cancer could be caused by a sexually transmitted agent. Initial research in the 1940s and 1950s put the blame on smegma. For example, (Heins Jr. et al., 1958) during the 1960s and 1970s it was suspected that infection with herpes simplex virus was the cause of the disease. HSV was seen as a likely cause because it was known to survive in the female reproductive tract, and to be transmitted sexually in a way compatible with known risk factors, such as promiscuity and low socioeconomic status. Herpes viruses were also implicated in other malignant diseases, including Burkitt's lymphoma, nasopharyngeal carcinoma, Marek's disease, and Lucké renal adenocarcinoma. HSV was recovered from cervical tumor cells. In the 1970s, the prevailing thought in American medicine was that cervical cancer was linked to herpes, which was also incorrect (Heins Jr. et al., 1958; Gasparini and Panatto, 2009; Scotto and Bailar 3rd, 1969; Syverton and Berry, 1935).

The First Breakthrough

While the majority of doctors were completely in the dark, in the 1930s, Dr. Richard Shope of the Rockefeller University studied wild rabbits that had developed "horns," which upon further analysis, was caused by a virus that could be transmitted. This research eventually led to the discovery that cervical cancer was caused by a papillomavirus.

Dr. George Papanicolaou's famous smear test was developed in the United States and was introduced into practice in the 1940s. He developed his cervical cytology research into an effective screening test, it entered widespread use in the UK as early as the 1950s, and a national cervical screening program was introduced in the UK in 1988; this effectively solved the screening problem. In 1951, the first successful *in vitro* cell line (HeLa) was derived from biopsies of the cervical cancer of Henrietta Lacks (Gasparini and Panatto, 2009; Syverton and Berry, 1935; Walboomers et al., 1999).

Nobel Prize for Discovering HPV

Dr. Shope paved the way for Dr. Harald zur Hausen's work in the 1980s. The link between genital HPV infections and cervical cancer was first demonstrated in the early 1980s by Harlad zur Hausen, a German virologist who cloned two of the most important high-risk HPV types (16 and 18) and showed the association between HPV infection and cervical cancer (Gasparini and Panatto, 2009). He did tremendous research on cancer of the uterine cervix (zur Hausen, 2002) and received the Nobel Prize in Physiology or Medicine (2008) for his discovery of HPV. This association is now well established by a large number of clinicoepidemiological, molecular, and experimental studies on HPV.

In 1988, the Bethesda System for reporting Pap results was developed, and in 2006 the first HPV vaccine was approved by the FDA (Das et al., 2008; Donnelly et al., 2005; Frazer, 2004; McMurray et al., 2001).

Cancer research is becoming multidisciplinary. Complex structural and therapeutic problems require synergistic approaches employing an assortment of molecular cancer biologies that synthesize the findings of three decades of recent cancer research and propose a conceptual framework that illuminates the conclusions of these discoveries. In the following sections, this chapter will continue to provide a detailed overview of various processes, and will cover the recent trends in the field of cervical cancer management, including the technologies available for early detection and cures. It also describes HPV biology, including the role of tumor suppressor genes and oncogenes used in the diagnosis and prognosis determination of cervical cancer. It will update various methods of cytology screening, including visual inspection of the cervix, the Pap test, colposcopy, detection of high-risk HPV16 and -18, and treatment with ultimate success in reducing cervical cancer mortality (zur Hausen, 2002; Hussain et al., 2012). Finally, development of the HPV vaccine is outlined.

Most importantly, the world has witnessed decreased incidence rates for cervical cancer in countries with organized screening. However, increased rates of cervical cancer have been reported in several populations due to failure to implement cytology-based screening programs. Therefore, complete awareness and knowledge of current diagnostic approaches of cervical cancer are important for tackling this fatal disease. Keeping this in mind, the following sections will certainly help clinicians, virologists, cytologists, epidemiologists, and public health specialists in understanding all clinical, molecular, and epidemiological aspects of cervical cancer. Offered are many pedagogical features such as cervical cancer epidemiology, screening, HPV biology, and vaccine development, which should help readers hone their analytical abilities and think objectively about a number of complex biological processes.

PREVALENCE AND EPIDEMIOLOGY OF CERVICAL CANCER

Global Scenario

Cervical cancer is a major reproductive health problem for women globally, and is the most common cause of cancer-related female mortality in developing countries. According to GLOBOCAN, 2008 (Ferlay et al., 2010) it is the third most frequent malignancy in women worldwide, with an estimated 530,232 new cases and 275,008 deaths every year, with nearly 80% in developing countries (a risk of 1.5% before the age of 65). In developed countries, cervical cancer accounts for only 3.6% of new cancers, with a cumulative risk (0 to 64) of 0.8%. The highest incidence rates are observed in sub-Saharan Africa, Melanesia, Latin America, the Caribbean, south central Asia, and southeast Asia. For example, incidences are 38 per 100,000 in the Second National Cancer Survey of the United States. Very low rates are also observed in China (6.8 per 100,000) and western Asia (5.8 per 100,000), and the lowest recorded rate is 0.4 per 100,000 in Ardabil, northwest Iran (Ferlay et al., 2010).

Symptoms of Cervical Cancer

Precancerous changes and early cancers of the cervix generally do not cause pain or other symptoms. When the disease gets worse, women may notice one or more of the following symptoms:

- Abnormal vaginal bleeding.
- Bleeding that occurs between regular menstrual periods.
- Bleeding after sexual intercourse, douching, or a pelvic exam.
- Menstrual periods that last longer and are heavier than before.
- Bleeding after menopause.
- Increased vaginal discharge.
- Pelvic pain.

Types of Cervical Cancer

There are several types of cervical cancer, classified on the basis of where they develop in the cervix. Cancer that develops in the ectocervix is called **squamous cell carcinoma**; around 80–90% of cervical cancer cases belong to this category, and arise from the metaplastic squamous epithelium of the transformation zone. The development of cancer in the endocervix is called **adenocarcinoma** (10%), and arises from the columnar epithelium of the endocervix. In addition, a small percentage of cervical cancer cases are mixed versions of the above two, and are called adeno–squamous carcinomas or mixed carcinomas.

Risk Factors for Cervical Cancer

A risk factor is anything that increases the chance of getting a disease such as cancer. They can be divided into two categories: genetic factors and non-genetic factors (Table 6.2).

TABLE 6.2 Risk Factors of Cervical Cancer

	Risk Factors
Non-genetic Factors	Lower socioeconomic status and lack of regular Pap tests Poor genital/sexual hygiene Multiple sexual partners or promiscuity Early age of first sexual intercourse below 18 years Oral contraceptives use and smoking Multiple pregnancies and parity. Socio-economic status Dietary factors Religion and ethnicity
Genetic Factors	High-risk type HPV (Human Papillomavirus) Multiple HPV infections Viral load (severity of viral infection) HPV variants Genetic predisposition Infections of other STDs like HIV Weakened immune system

Pre-Cancer Classification

WHO Classification

According to this classification, cervical lesions can be classified into dysplasia, carcinoma *in situ* (CIS), and invasive cancer. Cervical dysplasia can be further graded into mild, moderate, and severe dysplasia based on the degree of involvement (the ratio between the thicknesses of abnormal cervical epithelium to its complete thickness).

CIN Classification

a. **Richart** (1968) introduced the term Cervical Intraepithelial Neoplasia (CIN), which was used to distinguish severe dysplasia from carcinoma *in situ*. CIN is further classified into three groups, namely CIN1, CIN2, CIN3, that are compatible with the WHO classification in which mild dysplasia corresponds to CIN1, moderate to CIN2, and CIN3 both severe dysplasia and carcinoma *in situ*.
b. **Reagen and associates** promulgated the use of the term "dysplasia" to designate these intraepithelial lesions, and classified them as: CIN III (severe dysplasia), CIN II (moderate dysplasia), and CIN 1 (mild dysplasia). Cervical cancer follows a progressive course from epithelial dysplasia to carcinoma *in situ* to invasive cancer. It may take years for dysplasia to turn into carcinoma *in situ* or microinvasive cancer, but once this process occurs, the cancer can quickly become invasive and spread deeper into lymph nodes, nearby tissues, or other organs such as the bladder, intestines, liver, or lungs.

Bethesda Classification

In December 1988, the National Cancer Institute sponsored a workshop to develop a uniform reporting system for cervicovaginal cytology. It became known as "**The Bethesda System.**" The Bethesda System, 2001 classifies squamous cell abnormalities into the following categories (Apgar et al., 2003):

1. Atypical Squamous Cell (ASC), including lesions that have several abnormalities suggestive of Squamous Intraepithelial Lesions (SIL).
2. Low Squamous Intraepithelial Lesions (LSIL), encompassing cellular changes associated with HPVs and mild dysplasia or CIN1.
3. High Squamous Intraepithelial Lesions (HSIL) includes moderate dysplasia or CIN2, severe dysplasia, and CIS as CIN3.

Subsequently, modifications were made after a second meeting convened in 1991. The Bethesda System has the following advantages:

- A uniform diagnostic terminology to improve communication both among cytopathologists and between cytopathologists and health care providers.
- A descriptive diagnosis of atypical squamous cells of undetermined significance (ASCUS) and of atypical glandular cells of undetermined significance (AGUS), which refers to glandular cell nuclear enlargement, hyperchromasia and/or architechtural abnormalities.
- Inclusion of changes associated with human papillomavirus (HPV) such as koilocyctosis, along with cervical intraepithelial neoplasia (CIN) within the category of low-grade squamous intraepithelial lesion (LGSIL). In other words, the use of terminology that reflects the current understanding of the pathogenesis and biology of cervical neoplasias.
- Evaluation of specimen adequacy as an integral part of the report.

Cancer Classification

Cancer staging is one of the fundamental activities in oncology, and is of pivotal importance to the modern management of cancer patients. Tumor classification is generally conceived so that the clinical and/or pathological spread is stratified into four stages: Stage I refers to a tumor strictly confined to the organ of origin, hence of relatively small size; Stage II describes a disease that has extended locally beyond the site of origin to involve adjacent organs or structures; Stage III represents more extensive involvement (i.e. wide infiltration that reaches neighboring organs); and Stage IV represents clearly distant metastatic disease. These four basic stages are then classified into sub-stages that reflect specific clinical, pathological, or biological prognostic factors within a given stage.

Invasive cervical cancer can be classified by the **FIGO system** (International Federation of Gynaecology and Obstetrics, Montreal, 1994, Table 6.3) (Quinn et al., 2006).

TABLE 6.3 Figo Staging in Cervical Cancer (Quinn et al., 2006)

Stage 0	**Carcinoma *in situ*; it is found only in the top layer of cells in the tissue that lines the cervix.**
Stage I	The carcinoma is strictly confined to the cervix (extension to the corpus would be disregarded). **Ia:** Invasive carcinoma which can be diagnosed only by microscopy. **Ib:** All macroscopically visible lesions – even with superficial invasion – are allotted to Stage Ib carcinomas. Invasion is limited to a measured stromal invasion with a maximal depth of 5 mm and a horizontal extension not wider than 7 mm. Depth of invasion should not be more than 5 mm taken from the base of the epithelium of the original tissue (should not change the stage allotment). **Ia1:** Measured stromal invasion of not more than 3 mm in depth and extension of 7 mm. **Ia2:** Measured stromal invasion of more than 3 mm and not more than 5 mm with an extension of 7 mm. **Ib** Clinically visible lesions limited to the cervix uteri or preclinical cancers greater than Stage Ia. **Ib1:** Clinically visible lesions not greater than 4 cm. **Ib2:** Clinically visible lesions 4.0 cm.
Stage II	Cervical carcinoma invades beyond uterus, but not to the pelvic wall or to the lower third of vagina. **IIa:** No obvious parametrial involvement. **IIb:** Obvious parametrial involvement.
Stage III	The carcinoma has extended to the pelvic wall. On rectal examination, there is no cancer-free space between the tumor and the pelvic wall. The tumor involves the lower third of the vagina. All cases with hydronephrosis or non-functioning kidney are included, unless they are known to be due to other causes. **IIIa:** Tumor involves lower third of the vagina, with no extension to the pelvic wall. **IIIb:** Extension to the pelvic wall and/or hydronephrosis or non-functioning kidney.
Stage IV	The carcinoma has extended beyond the true pelvis or has involved (biopsy proven) the mucosa of the bladder or rectum. A bullous edema (as such) does not permit a case to be allotted to Stage IV. **IVa:** Spread of the growth to adjacent organs. **IVb**: Spread to distant organs.

Stage I

Stage I is carcinoma strictly confined to the cervix; extension to the uterine corpus should be disregarded. The diagnosis of both Stages IA1 and IA2 should be based on microscopic examination of removed tissue, preferably a cone, which must include the entire lesion.

Stage IA: Invasive cancer identified only microscopically. Invasion is limited to measured stromal invasion with a maximum depth of 5 mm and no wider than 7 mm.

Stage IA1: Measured invasion of the stroma no greater than 3 mm in depth and no wider than 7 mm in diameter.
Stage IA2: Measured invasion of stroma greater than 3 mm, but no greater than 5 mm in depth and no wider than 7 mm in diameter.
Stage IB: Clinical lesions confined to the cervix or preclinical lesions greater than Stage IA. All gross lesions even with superficial invasion are Stage IB cancers.
Stage IB1: Clinical lesions no greater than 4 cm in size.
Stage IB2: Clinical lesions greater than 4 cm in size.

Stage II

Stage II is carcinoma that extends beyond the cervix, but does not extend into the pelvic wall. The carcinoma involves the vagina, but not as far as the lower third.

Stage IIA: No obvious parametrial involvement. Involvement of up to the upper two-thirds of the vagina.
Stage IIB: Obvious parametrial involvement, but not into the pelvic sidewall.

Stage III

Stage III is carcinoma that has extended into the pelvic sidewall. On rectal examination, there is no cancer-free space between the tumor and the pelvic sidewall. The tumor involves the lower third of the vagina. All cases with hydronephrosis or a non-functioning kidney are Stage III cancers.

Stage IIIA: No extension into the pelvic sidewall, but involvement of the lower third of the vagina.
Stage IIIB: Extension into the pelvic sidewall or hydronephrosis or non-functioning kidney.

Stage IV

Stage IV is carcinoma that has extended beyond the true pelvis, or has clinically involved the mucosa of the bladder and/or rectum.

Stage IVA: Spread of the tumor into adjacent pelvic organs.
Stage IVB: Spread to distant organs

Human Papillomaviruses (HPVs)

Human papillomavirus (HPV) infection is a major etiological factor in the development of normal cervical epithelium into cancer (Hussain et al., 2012; Onon, 2011). More than 100 genotypes have been described to date: 15 types are categorized as high-risk (HR-HPVs) types (HPV16, 18, 31, 33. 35, 39, 45, 51, 52, 56, 58, 59, 68, 73, and 82) that are associated with genital and other epithelial cancers, and 12 low-risk (LR-HPVs) types (HPV 6, 11, 40, 42, 43, 44, 54, 61, 70, 72, 81, CP6108), which are responsible for benign tumors and genital warts. Table 6.4 shows the effects and diseases caused by different types of HPVs.

HPV types 16 and 18 are considered the most prevalent "high risk" types for cervical cancer, while types 6 and 11 are considered the most prevalent low-risk ones, and are associated with benign lesions and genital warts. Together, HPV16 and 18 are estimated to account for more than 80% of invasive cervical cancers. HPV induces hyperplastic, papillomatous, and verruicous squamous cell lesions in the skin and at various mucosal sites in a wide range of hosts, including humans. HPV infections have been reported in a number of body sites, including the anogenital tract, urethra, skin, larynx, tracheobronchial mucosa, nasal cavity, conjunctiva, and esophagus. HPVs can be passed from person to person through sexual contact. Most adults have been infected with HPV at some time in their lives (Onon, 2011; zur Hausen, 1991; Sankaranarayanan et al., 2008).

TABLE 6.4 Association of Various HPV Types with Different Disease Outcomes

Disease	HPV Type(s) Associated
Benign and Pre-Cancerous Lesions	
Plantar warts	1, 2, 4, 63
Common warts	1, 2, 3, 4, 7, 10, 26, 27, 28, 29, 41, 57, 65, 77,
Flat warts	3, 10, 26, 27, 28, 38, 41, 49, 75, 76
Other cutaneous lesions (e.g., epidermoid cysts)	6, 11, 16, 30, 33, 36, 37, 38, 41, 48, 60, 72, 73
Epidermodysplasia verruciformis	2, 3, 5, 8, 9, 10,12, 14, 15, 17, 19, 20, 21, 22, 23, 24, 25, 36, 37, 38, 47, 50
Recurrent respiratory papillomatosis	6, 11
Laryngeal papillomatosis	6,10 , 11, 16
Focal epithelial hyperplasia of Heck	13, 32
Conjunctival papillomas	6, 11, 16
Condyloma Acuminata (genital warts)	6, 11, 30, 42, 43, 45, 51, 54, 55, 70
Cervical Cancer	
High Risk	16, 18, 45,31, 33,52, 58, 35,59, 56,51, 39, 68,73 & 82.

Genomic Organization of Human Papillomavirus (HPV)

Human papillomaviruses are ubiquitous DNA viruses belonging to Family Papillomaviridae. They are small, non-enveloped DNA viruses with a circular, double-stranded DNA genome of approximately 7,200-8,000 base pairs (bp). The HPV particles are about 55 nm in diameter and consist of a 72-capsomere capsid containing the viral genome. A three-dimensional model of HPV is shown in Figure 6.1. Capsomeres are composed of two structural proteins: the 57 kDa late protein L1, which accounts for 80% of the viral particle, and the 43–53 kDa minor capsid protein L2. The genome can be divided into three regions: the long control region (LCR) without coding potential, the region of early proteins (E1–E8), and the region of late proteins (L1 and L2). HPVs have further been classified into subtypes: variants when they have 90 to 98% sequence similarity to the corresponding type, and variants when they show no more than 98% sequence homology to the prototype. Some naturally occurring variants have different biological and biochemical properties important in cancer risk. All the putative protein coding sequences that call open reading frames (ORFs) are restricted to one strand (zur Hausen, 1982; zur Hausen, 1991; Klug and Finch, 1965).

Upstream Regulatory Region/Long Control Region: The upstream regulatory region constitutes about 10% of the viral genome, varying between 800 bp and 900 bp. It is the non-coding region, but contains the origin of replication, viral promoter, and enhancer sequences. Viral gene expression is generally regulated by several viral and host-cell transcription factors, which bind to the URR (Munoz et al., 2003).

Early Region: This constitutes nearly 45–50% of the viral genome lying downstream of the URR, and consists of eight open reading frames (i.e. E1–E8). It encodes regulatory proteins. It is engaged in genomic persistence, DNA replication, and activation of the lytic cycle. E6 functions to activate telomerase and the SRC kinases, and to inhibit p53 and BAK. E7 inhibits RB, which releases E2F and results in the up-regulation of INK4A, but E7 also inactivates INK4A.

Late Region: This spans nearly 40%, and is responsible for encoding structural proteins for the production of the viral particles. It contains two open reading frames, L1 and L2.

Long Control Region: The LCR (500–1000 bp) is the origin of replication and regulation for HPV gene expression.

Non-Coding Elements: A short non-coding region (SNR) commonly exists between the E5 and L2 open reading frames of human papillomaviruses (HPVs). This region contains the cis-elements necessary for replication and transcription of the viral genome. The sizes and functions of papillomavirus proteins are given below:

Early Proteins

E1 (68–85 kDa): Helicase function; essential for viral replication and control of gene transcription; similar among types.

E2 (48 kDa): Viral transcription factor; essential for viral replication and control of gene transcription; genome segregation and encapsidation.

E3 (UNKNOWN): Function not known; only present in a few HPVs.

E1^E4 (10–44 kDa): Binding to cytoskeletal protein.

E5 (14Kd): Interaction with EGF/PDGF-receptors.

E6 (16–18 kDa): Interaction with several cellular proteins; degradation of p53 and activation of telomerase.

E7 (~10 kDa): Interaction with several cellular proteins; interaction with pRB and transactivation of E2F-dependent promoters.

E8^–E2C (20 kDa): New E2 protein, consisting of a fusion of the product of the small E8 ORF with part of the E2 protein, has been described. This fusion protein is able to repress viral DNA replication as well as transcription, and is therefore believed to play a major role in the maintenance of viral latency observed in the basal cells of infected epithelium.

Late Proteins

L1 (57 kDa): Major capsid protein expressed in upper epithelial layers allowing assembly of full virus.

L2 (43–53 kDa): Minor capsid protein expressed in upper epithelial layers allowing assembly of full virus; in its absence, virus-like particles are formed instead.

Among all, E6 and E7 are the most important oncogenic proteins. Transcription of the E6 and E7 genes was observed to always to occur in cervical carcinomas, and this was the first indication of an important role for these genes in HPV-associated tumorigenesis.

Transcriptional Regulation of Human Papillomavirus

Although E6 and E7 themselves possess intrinsic transactivation capacity on their homologous promoter, constitutive expression of E6 and E7 in immortalized or malignantly transformed human keratinocytes is mainly dependent on the availability of a defined set of transcription factors derived from the infected host cell. HPV 16 E6/E7 transcription is regulated by *cis*-acting elements contained within the Upstream Regulatory Region (URR) (Munoz et al., 2003).

HPV has a circular DNA genome in which the viral late and early genes are separated by the transcriptional control region called URR or LCR. Functionally, the 850 bp HPV 16 URR can be divided into three parts:

1. A 5′-terminal portion of unknown function, which only marginally contributes to the activity of the E6/E7 promoter.

2. A central 400 bp constitutive enhancer that is essential for E6/E7 promoter activity.
3. A promoter proximal region containing E6/E7 promoter p97 at its 3′-end.

The complete URR, as well as the constitutive enhancer regions of the virus, have been shown to exhibit a tissue preference for epithelial cells in their transcriptional activity.

Replication of HPV

The HPV life cycle is tightly linked to its host cell biology. Normal squamous epithelial cells grow as stratified epithelium, with those in the basal layers dividing as stem cells or transit amplifying cells. After division, one of the daughter cells migrates upward and begins to undergo terminal differentiation while the other remains in the basal layer as a slow-cycling, self-renewing population. HPV virions initially infect the basal layers of the epithelium, probably through microwounds, and enter cells via interaction with receptors such as α-6-integrin for HPV16. In infected cells at the basal layer, low levels of viral DNA are synthesized to an episomal copy number of approximately 50–100 genomes per cell. The early HPV genes *E1* and *E2* support viral DNA replication and its segregation so that the infected stem cells can be maintained in the lesion for a long period. As infected daughter cells migrate to the upper layers of the epithelium, viral late gene products are produced to initiate the vegetative phase of the HPV life cycle, resulting in high-level amplification of the viral genome.

As the viral DNA replication almost totally depends on host replication factors, except for viral helicase E1, other early genes *E5*, *E6* and *E7*, are considered to coordinate a host cell suitable for viral DNA replication, which sometimes induces host cellular DNA synthesis and prevents apoptosis. In the outer layers of the epithelium, viral DNA is packaged into capsids and progeny virions are released to re-initiate infection. Because the highly immunogenic virions are synthesized at the upper layers of stratified squamous epithelia, they undergo only relatively limited surveillance by cells of the immune system. In addition, E6 and E7 inactivate interferon (IFN) regulatory factor (IRF) so that HPV viruses can remain as persistent, asymptomatic infections.

High-risk HPV types can be distinguished from other HPV types largely by the structure and function of the E6 and E7 gene products. In benign lesions caused by HPV, viral DNA is located extra chromosomally in the nucleus. In high-grade intraepithelial neoplasias and cancers, HPV DNA is generally integrated into the host genome. In some cases, episomal and integrated HPV DNAs are carried simultaneously in the host cell. Integration of HPV DNA specifically disrupts or deletes the E2 ORF, which results in loss of its expression. This interferes with the function of E2, which normally down-regulates the transcription of the E6 and E7 genes, and leads to an increased expression of E6 and E7. In high-risk HPV types, the E6 and E7 proteins have a high affinity for tumor suppressor genes p53 and pRB, resulting in an increased proliferation rate and genomic instability. As a consequence, the host cell accumulates more and more damaged DNA that cannot be repaired. Efficient immortalization of keratinocytes requires the cooperation of the E6 and E7 gene proteins; however, the E7 gene product alone at high levels can immortalize host cells. Eventually, mutations accumulate that lead to fully transformed cancerous cells. In addition to the effects of activated oncogenes and chromosome instability, potential mechanisms contributing to transformation include methylation of viral and cellular DNA, telomerase activation, and hormonal and immunogenetic factors. Progression to cancer generally takes place over a period of 10 to 20 years (Doorbar et al., 1991; Thierry, 2009; Sellers and Kaelin Jr, 1997).

Functions of HPV Onco-Proteins E6 and E7

Both E6 and E7 proteins are essential to induce and maintain cellular transformation due to their interference with cell-cycle control and apoptosis. The HPV viral oncogenes, E6 and E7, have been shown to be the main contributors to the development of HPV-induced cervical cancer and increased expression, probably due to integration of the viral DNA in the host cell genome; they have been detected in invasive cancers and a subset of high-grade lesions.

Inactivation and Degradation of p53 through the E6/E6AP Complex

The most important function of the E6 protein is to promote the degradation of p53 through its interaction with a cellular protein, E6-associated protein (E6AP), an E3 ubiquitin ligase. The affinity of E6AP for p53 is likely to be modified by the association with E6. The *p53* tumor suppressor gene itself regulates growth arrest and apoptosis after DNA damage. When DNA damage is moderate, a prolonged p53-dependent arrest and DNA repair are induced, but when the damage is severe, apoptosis is provoked. Although aberrant inactivation of pRb family members would also normally induce apoptosis through p53, HPV-infected cells avoid such cell death by E6 inactivation of p53. In addition, E6 interferes with other pro-apoptotic proteins (Bak, FADD, and procaspase 8) to comprehensively prevent apoptosis. Alternatively, the susceptibility of E6-induced degradation of p53 has been suggested to link the polymorphisms in codon 72 of *p53*.

Inactivation of pRb: E7 is a small nuclear phosphoprotein separated into three conserved regions denoted in an analogous fashion to adenovirus E1A as CR1, CR2, and CR3. E7 is known to bind to the retinoblastoma tumor suppressor gene product, pRb, and its family members, p107 and p130, via a LXCXE (where X represents any amino acid)-binding motif conserved in its CR2 region. In the hypophosphorylated state, pRb family proteins can bind to transcription factors such as E2F family members and repress the transcription of particular genes involved in DNA synthesis and cell-cycle progression. Phosphorylation of pRb by G1 cyclin-dependent kinases releases E2F, leading to cell-cycle progression into the S phase. Because E7 is able to bind to unphosphorylated pRb, it may prematurely induce cells to enter the S phase by disrupting pRb–E2F complexes. Most recently, it was found that E7 promotes *C*-terminal cleavage of pRb by the calcium-activated cysteine protease calpain, and that this cleavage is required before E7 can promote the proteasomal degradation of pRb (Doorbar et al., 1991; Dyson et al., 1989). The E7 protein function enables HPV replication in the upper layers of the epithelium where uninfected daughter cells normally differentiate and completely exit the cell cycle. One cyclin-dependent kinase inhibitor, p16INK4a, which prevents the phosphorylation of pRb family members, is reported to be over-expressed when pRb is inactivated by HPV E7. Normally, over-expression of p16INK4a results in cell-cycle arrest, but with E7 expression, this is overcome. Thus over-expression of p16INK4a is suggested to be a useful biomarker for evaluating HPV pathogenic activity in cervical lesions (Burd, 2003; Butz and Hoppe-Seyler, 1993).

Diagnostic Methodologies of Cervical Cancer

Visual Methods

Two visual methods are available:

- Visual inspection with acetic acid (VIA).
- Visual inspection with Lugol's iodine (VILI).

Abnormalities are identified by inspection of the cervix without magnification, after application of dilute acetic acid (vinegar, in VIA) or Lugol's iodine (in VILI). When vinegar is applied to abnormal cervical tissue, it temporarily turns white (acetowhite), allowing the provider to make an immediate assessment of a positive (abnormal) or negative (normal) result. If iodine is applied to the cervix, precancerous and cancerous lesions appear well defined, thick, and mustard or saffron-yellow in color, while squamous epithelium stains brown or black; columnar epithelium retains its normal pink color. VIA and VILI are promising alternatives to cytology where resources are limited. They are currently being tested in large, cross-sectional, randomized controlled trials in developing countries. In research settings, VIA has been shown to have an average sensitivity for detection of pre-cancer and cancer of almost 77%, and a range of 56% to 94%. The specificity ranges from 74% to 94%, with an average of 86%. They are both short procedures, less costly, and cause no pain. Assessment is immediate, and no specimen is required.

Indications

VIA and VILI are indicated for all women in the target age group specified in national guidelines, provided that:

- They are pre-menopausal. Visual methods are not recommended for post-menopausal women because the transition zone in these women is most often inside the endocervical canal and not visible on speculum inspection.
- Both squamocolumnar junctions (i.e. the entire transformation zone) are visible. If the patient does not meet the above indications, and no alternative screening method is available in the particular clinical setting, she should be referred for a Pap smear.

Pap Test (or Pap Smear)

The Pap smear test is routinely used for early cancer detection worldwide. It is recommended for all women of reproductive age. The doctor uses a plastic or metal instrument (speculum) to widen the vagina in order to reach the cervix. It is most important that an adequate sample (cells) be taken from the squamo–columnar junction (the transformation zone), transferred to the cytology slide, and immediately fixed with a commercial fixative. The location of the squamo–columnar junction can be identified by a change in color and texture between the squamous and columnar epithelia. The squamous epithelium appears pale pink, shiny, and smooth. The columnar epithelium appears reddish with a granular surface. The Pap smear test is a rapid method for detecting cervical dysplasia and *in situ* cancer, as well as invasive cancer. A Pap smear evaluates cells harvested from the ectocervix and endocervix for abnormal changes associated with the development of cervical cancer.

Thin Prep Pap test: The Thin Prep Pap Test (FDA-approved) has been described as more effective than a conventional Pap smear, and is a liquid-based test that employs a fluid medium to collect and preserve cervical cells.

Speculoscopy: Recently approved by the U.S. Food and Drug Administration (FDA), this procedure involves the use of a magnifier and a special wavelength of light. Speculoscopy allows the physician to see cervical abnormalities that would otherwise be undetectable when performing a standard Pap test.

Colposcopy and Biopsy Colposcopy is the procedure of viewing the cervix, vagina, and vulva with a magnifying lens (colposcope) to identify abnormal epithelial patterns. The colposcopy procedure begins by wiping away cervical mucus with normal saline. Inspection of the cervix is done with a colposcope that magnifies the tissue with filtered and unfiltered light. A 3–5% acetic acid solution is then applied to the cervix and upper vagina. If an epithelial abnormality is identified, a biopsy is done. If the entire lesion is not visualized, an endocervical curettage is indicated. During pregnancy, colposcopy is done in order to exclude the presence of invasive cancer and to reassure the woman that her pregnancy will not be affected by the presence of an abnormal Pap test.

New Technologies

New technologies provide adjuncts to cervical cancer diagnosis:

1. **Autopap** provides automated cytology scanning with interpretation. It is primarily used as a secondary screening to enhance laboratory quality control. Autopap rescreens all satisfactory slides read as "within normal limits," and selects 10% of the slides most likely to be false negatives for review by cytotechnologists.
2. **Papnet** provides computerized selective rescreening of all slides read as "within normal limits," creates a digitized image of the entire slide, and selects 128 of the most questionable fields of each slide for review by a cytotechnologist or pathologist.
3. **Cervicography** is a visual adjunct to cervical screening. Because one limitation of the Pap smear is the false-negative rate, cervicography is being considered for screening and/or secondary triage. Advantages include the following: it is simple to perform, less expensive, and noninvasive, compared to colposcopy and biopsy. If cervicography is performed concomitantly with a screening Pap smear, it may improve the false-negative rate. If it is used to distinguish women with an abnormal Pap who should be referred for immediate colposcopy from those for whom follow-up with repeat Pap smears is appropriate, it may also effectively decrease the false-negative rate. Cervicography is not currently FDA-approved for primary screening.

DNA Cytometry

Dysplastic lesions progress to more severe lesions or regress to normalcy via an unknown series of changes in the abnormal epithelium. Microspectrophotometric determination of nuclear DNA content has contributed valuable information to understanding pathogenesis of cervical dysplasia. An increased aneuploidy rate with an increase in the grade of CIN has been observed by a number of investigators, and aneuploidy has been regarded as a malignancy-specific marker. Quantitative cytochemical DNA measurements in individual cells show a correlation between progressive morphological alterations with a progressive increase in nuclear DNA content. Hence, aneuploid DNA has been considered a risk indicator for the malignant potential of dysplastic tissues.

HPV DNA-Based Diagnostic Methods

Reliable diagnosis of HPV infection, particularly the "high-risk" types (16/18), may facilitate early identification of "high-risk" populations for developing cervical cancer, and may augment the sensitivity and specificity of primary cervical cancer screening programs by complementing the conventional Pap test. HPV is not generally used on its own as the primary screening test. It is mainly used in combination with cytology to improve the sensitivity of the screening, or as a triage tool to assess women with borderline indications. New screening procedures are based on the detection of high-risk HPV DNA in vaginal or cervical smears or tissue biopsies (Flow Chart 6.1).

Urine-Based Non-Invasive HPV DNA Detection Method

Conventional testing for genital HPV infections requires the collection of smears, scrapes, or tissue biopsies from the cervico-vaginal region, which involves pelvic examination and invasive procedures in a gynecological/cancer clinic. Such invasive methods are not only unsuitable for large-scale population screening, but are strictly prohibited for

HPV DNA Testing

1) Collection of cervical specimens from women aged 25-65 years

↓

2) Cervical biopsies or scrapes in 1X cold PBS solution.

↓

3) DNA isolation (phenol-chloroform method)

↓

4) DNA quantification in ethidium bromide stained 1% agarose gel or spectrophotometer

↓

5) Initial HPV detection by L1 region consensus sequence primers (MY09/11; 450 bp, GP 5+/6+ 150bp, SPF 1/2; 65 bp)

↓

6) Type specific PCRs for detection of HPV High –risk (16, 18 etc) or low-risk type (6, 11 etc)

↓

7) PCR product check in 2-3% ethidium bromide stained agarose gels

↓

8) Further confirmation can be done by direct DNA sequencing

FLOW CHART 6.1 HPV DNA testing.

adolescent or unmarried girls due to strong socio-cultural and religious reasons in developing countries. The use of non-invasive urine sampling for detection of various genital infections (including HPV) has developed as a useful technique, and has been validated over and over again to confirm its utility, sensitivity, and reliability by comparing the results via urine with those of biopsy specimens or cervical scrapes/swabs of the same patients. Exfoliated cells containing HPV DNA or virions that have been shed into the urine from epithelial lesions of the cervix could be detected by a highly sensitive technique such as PCR (Hussain et al., 2012).

Simple "Paper Smear" Method for Rapid Detection of HPV Infection

A simple "paper smear" method has been developed for dry collection, transport, and storage of cervical smears/scrapes at room temperature for subsequent detection of HPV DNA by a simple PCR assay. This method requires several types of biological specimens, such as imprint biopsies, blood, and fine-needle aspirates. It is simple, rapid, and cost effective, and can be effectively employed for large-scale population screening, especially for regions where the specimens are to be transported from distant places to the laboratory.

Detection of HPV by Multiplex PCR and RFLP

A low-cost method was developed for the detection of HPV types 6, 11, 16, 18, and 33, including co-infections from the cervical swabs of the females attending gynecological outpatient departments and cancer clinics. The method detects the five most prevalent HPV types commonly associated with cervical abnormalities. The technique involves RFLP of the approximately 450 bp amplicon, obtained after the amplification of the L1 region of the HPV genome by MY09/11 consensus primers (Hussain et al., 2012). MY09/11 primers are used routinely for HPV detection covering a broad spectrum of HPV types as compared to general primers GP5+/GP6+. Ninety percent of the cervical carcinomas (which contain some high-risk HPV types: 16, 18, 33, and a few others) are associated with CIN and cervical cancer, whereas HPV types 6 and 11 are associated with genital warts (condyloma accuminata and flat genital warts). Hence, the detection of co-infection is equally important to understanding the biological behavior of HPVs. The method detects the above five HPV types by digesting the PCR product of MY09/11 primers with Rsa-1 and resolving on 8% non-denaturing polyacrylamide gel. This method has an advantage over other conventional methods of HPV typing as it is cheaper and saves time for the second PCR by type-specific primers. Most of the PCR–RFLP studies show use of multiple restriction enzymes with two rounds of PCR. Hence, it was found to be less cumbersome, lower cost, and more user-friendly for the detection of HPV DNA from cervical swabs, both at the clinical and research levels.

ETHICAL ISSUES

PUBLIC HEALTH CONCERNS WITH SCREENING IMPLEMENTATION

In recent years, medical ethics has become an undisputed part of medical studies. Cervical screening programs do have the potential to save lives with minimum risk, but at considerable cost. One of the biggest limitations at present is that few of the cervical screening programs seem to address these ethical imperatives, and many cause more harm than good. Every patient has a right to full and accurate information about his or her medical condition. Hence, in planning a screening program, physicians must also think about the ethical responsibilities involved. Ethical principles to be followed in cancer screening programs are intended mainly to minimize unnecessary harm to the participating individuals. The dilemma facing physicians today is to decide whether the current method of cervical screening is justified in light of increasing evidence that these programs can be damaging to patients.

Screening is a public health intervention used on a population at risk, or a target-population. Screening can detect abnormalities in cells before they become cancer. Also, if cancer itself is detected early, it can be cured with proper treatment. Cervical cancer screening aims to test the largest possible number of at-risk women, and to ensure appropriate follow-up for those who have a positive or abnormal test result. Such women will need diagnostic testing and follow-up or treatment. Colposcopy and biopsy are often used to reach a specific diagnosis of the extent of the abnormality in women with a positive screening test. Decisions on the target age group and frequency of screening are usually made at the national level, on the basis of local prevalence and incidence of cervical cancer-related factors such as HPV, HIV prevalence, and availability of resources and infrastructure.

Risk

There is now growing evidence to suggest that cervical screening programs cause psychological harm to patients, particularly increased anxiety, embarrassment, and fear of outcome. False-positive results can lead to considerable distress, and perhaps unnecessary treatment. Negative results can produce false reassurance. These negative effects of screening are probably quite common, and in some cases can continue to cause long-term anxiety and distress. In order to choose between screening and not screening, a physician has to establish a balance to ensure that the benefit of screening is maximized and the risks minimized. This can be accomplished in several ways, such as rethinking how the program is organized, providing effective training of operators to ensure high-quality smears, and maintaining an effective quality control system. One of the most crucial ways of reducing anxiety is to ensure that before, during, and after the screening process, the patient is fully informed and thus involved in the decision-making process. It is, therefore, ethically imperative that all screening programs attend to these details.

Benefit vs. Cost

Socioeconomic status, access to care, and lack of health insurance coverage correlate with a delay in diagnosis, advanced-stage disease care, and impaired survival. It has also been calculated that it takes 40,000 smears and 200 excision biopsies to prevent one death from cervical cancer. This has been calculated to be the equivalent of $600,000 per life saved. However, the return is not worth the effort, especially as that effort is not directed to the population of women at greatest risk, in other words, those women who habitually fail to respond to screening invitations.

Patient Autonomy and Coercion

"Autonomy" means leaving the decision to have a cervical screening entirely up to the patient. Screening is effective only if a high proportion of women are screened, and such a proportion can be achieved only by infringing upon patient autonomy and by coercion. In this case, certain dangers have to be considered, as some patients may visit too often, use resources too frequently, and perhaps receive false reassurances. Some will not come at all, and they are most at risk. So, a patient should be properly informed prior to screening.

PUBLIC HEALTH CONCERNS ABOUT VACCINE IMPLEMENTATION

In a survey of parents of school girls in developing countries, it was discovered that the majority of parents are unaware of HPV, and perceived that their children were not at a risk of acquiring sexually transmitted HPV infection as they have good family backgrounds and the children are not allowed to be involved in premarital sexual activity (Das et al., 2008; Hussain et al., 2012). Some parents have diverse opinions that HPV vaccines would make sex safe, leading to freedom for promiscuity and risky sexual behavior, which is not very common in this region of the globe due to socio-cultural factors. They also suspected that the vaccine itself might cause infection in children. Therefore, it is extremely important to raise general awareness about HPV, to de-stigmatize HPV infection, and to subsequently gain acceptance for a mass vaccination program for pre-adolescent and adolescent girls in India. Potential strategies may therefore include vaccination of schoolgirls (which may miss the more vulnerable girls not attending school) through mother–daughter initiatives or other existing community outreach programs.

TREATMENT

If a biopsy shows either persistent LSIL or HSIL, or if there is a question of invasive cancer, further evaluation and/or treatment are indicated. Choice of treatment modality depends upon several factors, namely age of the patient, type of lesion, experience of the attending surgeon, facilities available at the hospital, etc. Other possible therapeutic modalities are both ablative and excisional. The ablative techniques include laser and cryosurgery. The excisional modalities include cold-knife conization, laser conization, loop electrocautery excision, and hysterectomy.

Ablative Techniques

- **Laser Ablation:** "Laser" stands for "light amplification by stimulated emission of radiation." Laser ablation is a procedure in which a carbon dioxide laser directed through a microscope is used to vaporize the cervical transformation zone. The procedure is done under local anesthetic, and takes 15 to 20 minutes to perform. Post-therapy, the patient may experience pain, some uterine cramping, and bloody discharge/vaginal spotting.
- **Cryotherapy:** Among ablative techniques, cryotherapy is a procedure performed under direct visualization in which a probe is placed against the cervix. The probe freezes the affected tissue, which in turn results in the destruction and sloughing of cervical cells. The entire procedure takes about 15 minutes. During the procedure, the patient may experience uterine cramping. Subsequently, the patient may expect to have a profuse watery discharge for 7 to 10 days. Recurrence rate is low, and four months after cryotherapy, the reevaluation via Pap smear is done.

Excisional Techniques

- **Electrosurgical Excision (LEEP):** Electrosurgical excision of the transformation zone is a procedure in which an electrical current generating a radio frequency is passed through a wire loop that excises the tissue and cauterizes the base. The procedure can usually be performed in an outpatient setting with the use of local anesthetic. Depending upon the size of the loop and the lesion, either the transformation zone or a "cone-like" specimen can be obtained. The patient must be grounded for safety purposes, because an electrical current is used. The electrical current also generates heat, which can cause distortion of the surgical margins, thus making accurate interpretation of the regions difficult, if not impossible. Risks of all excisional procedures include bleeding, cervical stenosis, cervical incompetence, decrease in cervical mucus, and possible infertility.
- **Laser Conization:** Laser conization is an operative procedure requiring an anesthetic in which a carbon

dioxide laser is used as a knife to generate the same type of specimen obtained with cold-knife conization. Post-operatively, the patient will experience cramping and some bleeding.

- **Cold-Knife Conization:** A cold-knife conization (CKC) is an operative procedure requiring either a regional or general anesthetic. During the procedure, the involved ectocervix and endocervix are excised using a circumferential excision. Post-operatively, the patient may have cramping and bleeding. Post-operative infection requiring antibiotic therapy also may occur.

Follow-Up for Excisional/Ablative Treatment

Following excisional/ablative treatment, a woman will need a follow-up Pap smear in three to four months. If the cytology report is normal at subsequent visits, the Pap smear should be repeated every six months for the first two years, and then annually thereafter, as long as the cervical smear is normal. If, however, a cervical smear is abnormal (based upon the Bethesda Classification), the patient should be reintroduced into the observation/treatment protocol.

Hysterectomy

Historically, a hysterectomy was performed either vaginally or abdominally for CIN III of the cervix. Currently, with the advent of colposcopy and good excisional therapy, a hysterectomy is indicated in only 5–10% of patients. There are still situations where a hysterectomy is acceptable management of the patient. In the presence of the more aggressive recurrent high-grade squamous intraepithelial lesions (HGSIL) in women who have a coexisting gynecologic disease and have completed childbearing, a hysterectomy may be an option. If a hysterectomy is performed for cancerous lesions, the patient's post-operative follow-up care should include a vaginal smear of the upper one-third of the vagina every six months for two years, and then annually. Once the patient has had her first post-operative return visit (which should occur within six weeks of treatment), she should follow an observation or follow-up protocol because of her increased risk of vaginal neoplasia. The role of HPV testing in this situation requires further investigation.

Therapeutic Interventions

Stage-Wise Management of Cervical Cancer

1. Treatment of microinvasive carcinoma is shown in Flow Chart 6.2.
2. Treatment of early invasive cancer is shown in Flow Chart 6.3.

When the tumor is more extensive, but predominantly situated in the cervix, possibly with some vaginal

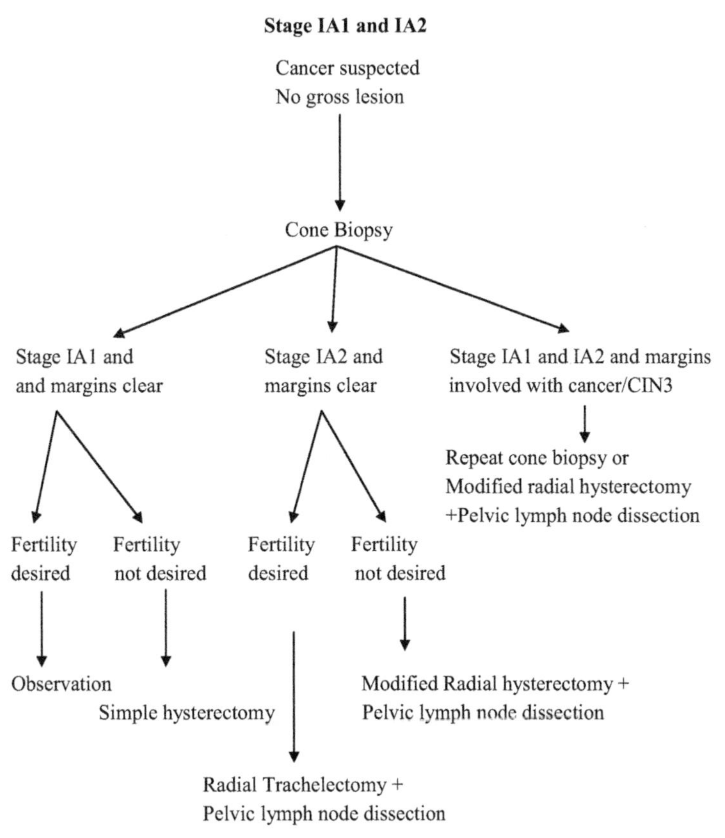

FLOW CHART 6.2 Treatment of microinvasive carcinoma.

STAGE IB1 AND IIA < 4 CM

Medically fit → Radial hysterectomy/pelvic lymphadenectomy
- → Negative nodes → Observe
- → Positive nodes → Pelvic teletherapy ± Brachytherapy ± Chemotherapy (cisplatin, 30–40 mg/m^2 per week)

Medically unfit → Treat with radiotherapy as for early bulky disease

FLOW CHART 6.3 Treatment of early invasive cancer.

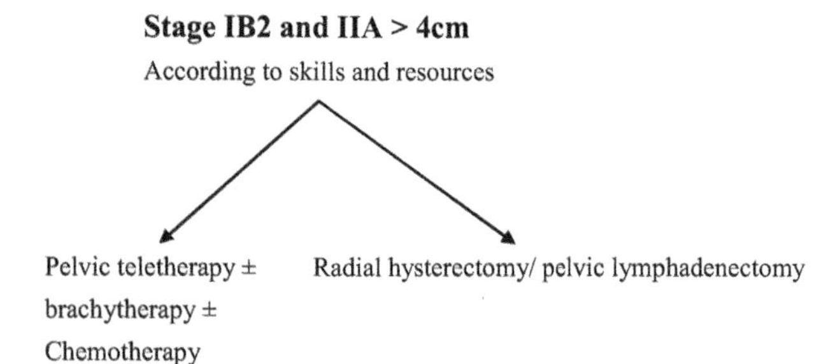

FLOW CHART 6.4 Treatment of early bulky disease (Stage IB2 and IIA > 4cm).

Stages IIB-IIIB

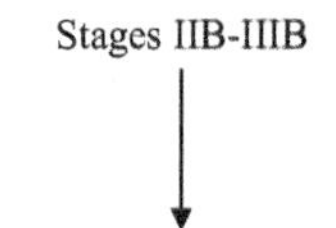

Pelvic teletherapy ± Brachytherapy ± Chemotherapy

FLOW CHART 6.5 Treatment of extensive disease: Stages IIB–IIIB.

Stages IVA

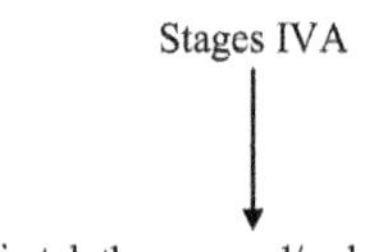

Pelvic teletherapy and/or brachytherapy

FLOW CHART 6.6 Treatment of Stage IVA.

involvement, surgical removal is preferred, except in the unfit patient.

3. Treatment of early bulky disease (Stage IB2 and IIA > 4 cm) is shown in Flow Chart 6.4.
4. Treatment of extensive disease (Stages IIB–IIIB) is shown in Flow Chart 6.5.

These patients are managed by radical (curative intent) radiotherapy, comprising teletherapy and brachytherapy. The role of chemotherapy has not yet been proven in a developing-country setting.

5. Treatment of Stage IVA is shown in Flow Chart 6.6.

Stage IVB (5% of cases) indicates the presence of distant hematogenous metastases, and is incurable by any currently known means.

6. Treatment of Stage IVB is shown in Flow Chart 6.7.

Preventive Measures

Primary approaches to prevent HPV infection include both risk reduction and development of HPV vaccines. Use of latex condoms and a spermicide may decrease the risk of contracting HPV. Condoms, however, are not totally reliable, since HPV may be contracted by contact with other parts of the body, such as the labia, scrotum, or anus, which are not protected by a condom.

TRANSLATIONAL SIGNIFICANCE

PROPHYLACTIC HPV VACCINES

Currently, two successful prophylactic HPV vaccines – quadrivalent "Gardasil" (HPV 16/18/6/11) developed by Merck, and bivalent "Cervarix" (HPV 16/18) by GlaxoSmithKline (GSK) – are recommended for vaccination of young adolescent girls at or before onset of puberty. In these vaccines, viral capsid proteins are present in the form of spontaneously reassembled virus-like-particles (VLPs) expressed either in yeast (for "Gardasil") or in baculovirus (for "Cervarix"). These two vaccines protect from infection by two of the most common cancer-causing HPV types (16 and 18); more than 70% of cervical cancer cases are associated with these two HPV types. Both the vaccines were found to be highly immunogenic, safe, well-tolerated, and effective in preventing incident and persistent HPV infections, including developing pre-cancerous lesions (Figure 6.1) (Das et al., 2008; Donnelly et al., 2005; Frazer, 2004).

Although these vaccines are expected to provide protection against other malignancies such as vaginal, anal, vulvar,

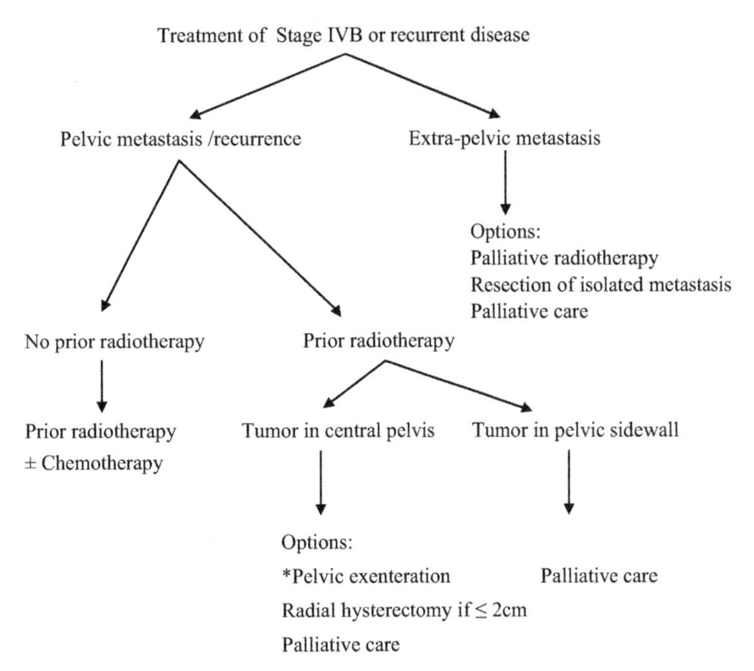

*Pelvic exenteration is infrequently used as it has major sequelae of urinary and colonic diversions, both of which is difficult to care for in developing countries, and are unacceptable to many patients when it is not possible to offer a cure.

FLOW CHART 6.7 Treatment of Stage IVB or recurrent disease. *Pelvic exenteration is infrequently used as it has major sequelae of urinary and colonic diversions, both of which are difficult to care for in developing countries; they are unacceptable to many patients when it is not possible to offer a cure.

oral, esophageal, and laryngeal papillomatosis (which are associated with these HPV types), they will certainly not provide protection against about 10–30% of the cervical cancers that arise due to infection of other high-risk HPV types (31, 35, 39, 45, 51, 52, 56, 58, etc). It is suspected that some of these HPV types may take the lead because of a change in the microenvironment (Das et al., 2008).

THERAPEUTIC HPV VACCINES

Although prophylactic vaccines appear to be successful, it would take decades to perceive the benefits because it takes 10–20 years to develop invasive cervical cancer. Therapeutic vaccines are used to bridge the temporal deficit by attacking already persistent HPV infections and to treat cervical cancer in women. Several animal studies showed promising results and indicated that therapeutic HPV vaccine may regress disease progression. As a result, several therapeutic HPV vaccines are in Phase I and Phase II clinical trials (Table 6.5) (Das et al., 2008; Frazer, 2004). Most efforts have been directed towards the early proteins, HPV E6 and E7, or small peptides derived from them, mainly because these are the major transforming viral proteins that are invariably retained and expressed throughout the full spectrum of HPV-related disease progression and cervical carcinogenesis. Financial concerns are compounded by health care structures that can impede a woman's timely progress to the appropriate provider, and may limit her access to the vaccine.

GENETIC-BASED DNA VACCINES

- No risk of infection.
- Cheap to produce.
- Stable product.
- High antigenicity.
- All proteins present in best effectiveness.

ISSUES/UNANSWERED QUESTIONS ASSOCIATED WITH HPV VACCINES

Vaccine Efficacy

- Will the HPV vaccine be effective lifelong?
- How long it will last?
- Are booster shots needed?
- How long will it protect from HPV infection?
- What is the end-point of the vaccine trial?

Vaccine Protection

- Does the HPV vaccine ultimately prevent development of cervical cancer?
- Does the vaccine require long-term monitoring of vaccinated woman (35–45 years) to prove that it prevents HPV-related cancers?

TABLE 6.5 Current Status of the Selective Human Papillomavirus Vaccines (Das et al., 2008)

Antigen Used	Type of Vaccine	Nature of Vaccine	Current Status	Mode of Administration
Quadrivalent HPV types HPV(16/18/6/11)L1 (Gardasil)	Prophylactic	Virus-Like Particle (VLP), protein	USA-FDA approved	Injectable
HPV 16/18 L1 (Cervarix)	Prophylactic	Virus-Like Particle (VLP), protein	Applied for FDA approval	Injectable
HPV 16 E6/E7	Therapeutic	Fusion protein	PhaseI Clinical trial	Injectable
HPV16 E7	Therapeutic	Peptide	PhaseI Clinical trial	Injectable
HPV16/18 E6/E7	Therapeutic	Recombinant Vaccinia virus	Late stage cervical cancer	Injectable
Second Generation vaccine Other high-risk (HPV31,45,33)L1	Prophylactic	Virus-Like Particle (VLP), protein		Injectable
HPV16 L1	Prophylactic	Capsomeres (pentameric) protein	Animal model	Intranasal delivery
HPV16 L1 in plant	Prophylactic	Virus-Like Particle (VLP), protein	Animal model	Oral delivery
HPV16 L1 as recombinant bacteria	Prophylactic	VLP produced in recombinant Lactobacillus, protein	Animal model	Mucosal
HPV16 L1	Prophylactic	DNA-based	Animal model	Parentaral, oral
HPV16 E7	Therapeutic	DNA-based	Clinical trial	Injectable with Microparticles (ZYC101)
HPV16 E7 Detox(Sig/E7 detox/HSP70)	Therapeutic	DNA-based	Phase I	Injectable
HPV16 L1/L2-E7	Chimeric (Prophylactic/Therapeutic)	Fusion protein	Animal model	Injectable
HPV16L2E7E6	Chimeric (Prophylactic/Therapeutic)	Fusion protein	Phase I & II Clinical trials	Injectable

Who Should Be Vaccinated?

- What age group of females should be vaccinated?
- Should vaccinations be given during the childhood years (1–5)?
- Young adolescent (9–19 years)?
- Immediately after marriage (18–25 years)?
- Should men also to be vaccinated?

FUTURE PROSPECTS OF HPV WITH RESPECT TO BIOTECHNOLOGY

It is an established fact that HPV infection is necessary but insufficient to cause malignancy. Furthermore, persistence of HPV-16 or -18 in women does not necessarily result in cancer. Persistence indicates the importance of other factors for malignant conversion of high-grade HPV infection. Progression of an HPV-infected cell to a malignant phenotype involves further modification of host gene expression and/or mutations. The appearance of chromosomal aberrations can lead to mutational inactivation or loss of tumor suppressor genes; activation and amplification of oncogenes play a central role in cancer progression.

The multi-step cervical carcinogenesis process is amendable to molecular therapeutics such as therapeutic nucleic acids (TNAs). TNA-based therapies for cervical carcinoma include ribozymes, antisense oligonucleotides (AS-ODNs), and small interfering RNAs (siRNAs). *In vitro* experiments with TNAs have successfully inhibited E6/E7 expression, and have caused induction of apoptosis and/or senescence in cervical carcinoma cells. Early ribozyme and AS-ODN approaches showed promise as therapeutic moieties for cervical cancer. Viral early genes E6 and E7 from high-risk HPV types are responsible for the transformation of epithelial cells, and their continuous expression is essential for ongoing cervical cancer cell survival, as they function as oncogenes. Therefore, E6 and E7 are ideal targets for RNAi therapy. In recent years, there have been a number of publications showing the potential use of RNAi as a treatment for cervical cancer.

The process of gene expression in response to physiological and environmental entities is mainly regulated at the transcriptional level where transcription factors bind to *cis*-regulatory DNA sequences. Transcription factors are proteins involved in the regulation of gene expression that bind to the promoter elements upstream of genes and either facilitate or inhibit transcription. They control and regulate gene expression through this process. Transcription factors are composed of two essential functional regions: a DNA-binding domain and an activator domain. The DNA-binding domain consists of amino acids that recognize specific DNA bases near the start of transcription. The activator domains of transcription factors interact with the components of the transcriptional apparatus (RNA polymerase) and with other regulatory proteins, thereby affecting the efficiency of DNA binding.

HPV16 E6/E7 transcription is regulated by *cis*-acting elements contained within the Upstream Regulatory Region (URR). URR is a transcriptional control region in HPV. It is an 850 bp region that is functionally divided into three parts: a 5′-terminal portion of unknown function, a central 400 bp constitutive enhancer for E6/E7 promoter activity, and a terminal 3′-region containing the E6/E7 promoter. All the transcription factors are known to bind the URR region (Thierry, 2009). These transcription factors do not function in isolation, but form regulatory networks in which several factors interact with them at a DNA-binding domain (also called a transactivation domain) that mediates the interaction between the host and the environment including viruses. A number of oncogenic signaling pathways all seem to converge on a limited set of nuclear transcription factors. These transcription factors are the final "switches" that activate the gene-expression patterns that ultimately lead to malignancy. Targeting a single transcription factor can block the effects of a multitude of upstream genetic aberrations that cause persistent activation. Certain transcription factors like Signal Transducer and Activator of Transcription (STAT), Activator Protein-1 (AP-1), and Nuclear Factor-kB (NF-kB) have been identified as important components of signal transduction pathways, leading to pathological outcomes such as inflammation and tumorigenesis, and are current molecular targets for cancer therapy. These transcription factors are normally modulated at the level of expression and/or their activation. Host transcription factors in association with viral factors are likely to dictate viral latency, vegetative replication, or oncogenic transcription during HPV infection.

Cancer stem cells (CSCs) within a tumor possess the capacity to self-renew and to cause the heterogeneous lineages of cancer cells that comprise the tumor. Cancer stem cell populations that fuel tumor growth have emerged as a new therapeutic intervention in cervical cancer progression. Cervical cancer contains a heterogeneous population of cancer cells. Several investigations have identified putative stem cells from solid tumors and cancer cell lines via the capacity to self-renew and drive tumor formation. Cervical cancer stem cells have a 10-fold higher binding capacity for papillomavirus-like particles than any other cervical epithelial subpopulation, suggesting that HPVs might indeed preferentially bind and infect cervical CSCs *in vivo*. Cancer stem cells are responsible for tumor initiation and maintenance, and are attractive targets for advanced cancer therapy. Sphere-forming cells (SFCs) from cervical cancer cell lines HeLa and SiHa have been isolated. These cells showed an expression pattern of CD44high/CD24low that resembles the CSC surface biomarker of breast cancer. HeLa-SFCs expressed a higher level (6.9-fold) of the HPV oncogene E6, compared with that of parental HeLa cells. Silencing oncogene expression in cervical cancer stem-like cells inhibits their cell growth and self-renewal ability.

CONCLUSION

Cervical cancer provides a unique window to study the deregulation of important molecular gate-keepers that are important for normal cell-cycle events. However, tumorigenic transformation of cervical epithelial cells takes 10–15 years to develop histopathologically well-characterized precursors and cancerous lesions. So the question arises, why does it take so long to develop cervical cancer compared to other cancers that have poor prognosis? Cervical intraepithelial lesions are a common pre-cancerous condition among HPV-infected cases, and are known to progress to invasive cancer. To intensify this scenario, until date, there has been no standard therapeutic modality available that can cure these viral infections. Therefore, for effective therapeutic intervention of HPV, and to prevent cervical cancer development at an early stage, it is important to develop an understanding of the molecular mechanisms involved in HPV-mediated cervical carcinogenesis. Therefore, it is mandatory to address important questions that would eventually help in understanding the molecular basis of cervical cancer.

KEY POINTS

HPV Infections:

- HPV infections are the cause of almost all cases of cervical cancer, and are also associated with the development of carcinomas of other organ sites.
- HPV is one of the most common sexually transmitted infections (STIs), but may not be an indicator of sexual practice or promiscuity.

- HPV infections are often asymptomatic, but can still transmit infection to sexual partners.
- The majority (~90%) of HPV infectious resolve spontaneously.
- Persistent infection of HPV is essential before initiation of events in host cells.
- Most predominant is the high-risk HPV type 16; HPV16 and -18 together account for more than 80% cervical cancers.
- High-risk HPV types 16/18 can immortalize human squamous epithelial cells *in vitro*.
- Systemic immunization with HPV-VLP can confer protection against HPV infection.

Cervical Cancer Screening Programs in Low-Resource Settings:

- Cervical cancer can be prevented if precancerous lesions are identified early through screening.
- Every woman should undergo cervical Pap smear screening or HPV DNA testing at least once in her lifetime.
- The optimal age for cervical cancer screening (to achieve maximum effect) is 26 to 40 years.
- Visual inspection with an acetic acid (VIA) and cytology-based Pap smear test should continue for screening, but HPV DNA testing must be incorporated for confirmation, particularly in unequivocal cases until an affordable, cost-effective, and reliable HPV test is available.

WORLD WIDE WEB RESOURCES

The American Cancer Society (ACS) Southwest Division

Tel: (505) 260-2105

http://www.cancer.org

5800 Lomas Blvd., NE, Albuquerque, NM 87110

The ACS is a nonprofit, nationwide organization that supports research, conducts educational programs, and offers a variety of services to people with cancer (and to their families). ACS helps women with cancer through various patient services and support groups.

CancerNet

E-mail: Cancernet@icic.nci.nih.gov

CancerNet is a way to obtain PDQ information summaries and other NCI information via the Internet and selected electronic information services. To use CancerNet, send a mail message to the address above. Enter the word "HELP" as the text of the message to receive materials in English; enter "SPANISH" to receive the information in Spanish.

People Living Through Cancer (PLTC)

Tel: (505) 242-3263

Fax: (505) 242-6756

323 Eighth Street SW, Albuquerque, NM 87102

PLTC was founded by and for those coping with a cancer diagnosis; it also serves those who have a friend or loved one with cancer. PLTC has cervical cancer resources.

American Institute for Cancer Research

http://www.aicr.org

Provides information on cancer and nutrition. Publishes a newsletter, cookbooks, and diet/nutrition brochures. AICR also has a hotline for nutrition-related cancer inquiries where callers are connected with a registered dietitian.

American Society of Plastic and Reconstructive Surgeons

http://www.plasticsurgery.org

For referrals to a plastic surgeon for corrective or reconstructive procedures, contact the American Society of Plastic and Reconstructive Surgeons for a list of local board-certified plastic surgeons.

Asian & Pacific Islander American Health Forum

http://www.apiahf.org/

The Asian & Pacific Islander American Health Forum is a national advocacy organization dedicated to promoting policy, program, and research efforts for the improvement of health status of all Asian-American and Pacific Islander Communities. The Women's Health Information Network strives to increase the public understanding of Asian and Pacific Islander women's health status, including cancer. Translations include Chinese, Tagolog, Vietnamese, Korean, Hindi, Gujurati, Urdu, Farsi, Thai, and Cambodian.

Avon's Breast Cancer Awareness Crusade

http://www.avoncrusade.com/

Avon's Breast Cancer Awareness Crusade, a national initiative of Avon Products, Inc., provides women, particularly low-income, minority, and older women, with direct access to a full range of breast cancer education and early-detection services.

Cancer Information Service

http://www.cis.nci.nih.gov/

CIS interprets and explains research findings to the public in a clear and understandable manner. The Northwest Regional Cancer Information Service serves Alaska. The CPCD provides bibliographic citations, abstracts of journal articles, book chapters, technical reports, papers, materials, curricula, and descriptions of cancer-prevention programs and risk-reduction activities at national, state, and local levels. It is available in English, Spanish, or on TTY equipment.

Cancer Mail

cancermail@icicc.nci.nih.gov

National Cancer Institute information about cancer treatment, screening, prevention, and supportive care. To obtain a contents list, send e-mail to the address with the word "help" in the body of the message.

CancerNet™

http://www.meb.uni-bonn.de/cancer.gov/index.html

CancerNet™ contains material for health professionals, patients, and the public, including information from PDQ about cancer treatment, screening, prevention, supportive care, clinical trials, and CANCERLIT (a bibliographic database).

Cancer Patient Education Database

http://chid.nih.gov

Provides information on cancer patient education resources for cancer patients, their family members, and health professions.

Cancer Research Foundation of America

http://www.preventcancer.org/

Information on women's health, ages 18–27, 28–39, 40–49, and 50+ (breast cancer and cervical cancer).

http://www.cancercare.org

Cancer Types: Cervical Cancer

Provides information and resources related specifically to cervical cancer.

Division of Cancer Studies, University of Birmingham

http://www.birmingham.ac.uk/schools/cancer/index.aspx

The School of Cancer Sciences is one of the world's premier translational cancer research institutes. Provides excellent basic science information, and works with clinical partners to improve the outlook for cancer patients.

REFERENCES

Apgar, B. S., Zoschnick, L., & Wright, T. C., Jr (2003). The 2001 Bethesda System terminology. *American Family Physician, 68*, 1992–1998.

Burd, E. M. (2003). Human papillomavirus and cervical cancer. *Clinical Microbiology Reviews, 16*, 1–17.

Butz, K., & Hoppe-Seyler, F. (1993). Transcriptional control of human papillomavirus (HPV) oncogene expression: composition of the HPV type 18 upstream regulatory region. *Journal of Virology, 67*, 6476–6486.

Das, B. C., Hussain, S., Nasare, V., & Bharadwaj, M. (2008). Prospects and prejudices of human papillomavirus vaccines in India. *Vaccine, 26*, 2669–2679.

Donnelly, J. J., Wahren, B., & Liu, M. A. (2005). DNA vaccines: progress and challenges. *Journal Immunology, 175*, 633–639.

Doorbar, J., Ely, S., Sterling, J., McLean, C., & Crawford, L. (1991). Specific interaction between HPV-16 E1-E4 and cytokeratins results in collapse of the epithelial cell intermediate filament network. *Nature, 352*, 824–827.

Dyson, N., Howley, P. M., Munger, K., & Harlow, E. (1989). The human papilloma virus-16 E7 oncoprotein is able to bind to the retinoblastoma gene product. *Science, 243*, 934–937.

Ferlay, J., Shin, H. R., Bray, F., Forman, D., Mathers, C., et al. (2010). Estimates of worldwide burden of cancer in 2008: GLOBOCAN 2008. *International Journal of Cancer, 127*, 2893–2917.

Frazer, I. H. (2004). Prevention of cervical cancer through papillomavirus vaccination. *Nature Reviews. Immunology, 4*, 46–54.

Gasparini, R., & Panatto, D. (2009). Cervical cancer: from Hippocrates through Rigoni-Stern to zur Hausen. *Vaccine, 1*(Suppl. 27), A4–5.

Hanahan, D., & Weinberg, R. A. (2000). The hallmarks of cancer. *Cell, 100*, 57–70.

Heins, H. C., Jr., Dennis, E. J., & Pratthomas, H. R. (1958). The possible role of smegma in carcinoma of the cervix. *American Journal of Obstetrics and Gynecology, 76*, 726–733. discussion 733–725.

Hussain, S., Bharadwaj, M., Nasare, V., Kumari, M., Sharma, S., et al. (2012). Human papillomavirus infection among young adolescents in India: impact of vaccination. *Journal of Medical Virology, 84*, 298–305.

Kaivosoja. (2008). *Helsinki Institute of Technology*. What is Cancer.

Klug, A., & Finch, J. T. (1965). Structure of viruses of the papilloma polyoma type. I. Human wart virus. *Journal of Molecular Biology, 11*, 403–423.

McMurray, H. R., Nguyen, D., Westbrook, T. F., & McAnce, D. J. (2001). Biology of human papillomaviruses. *International Journal of Experimental Pathology, 82*, 15–33.

Munoz, N., Bosch, F. X., de Sanjose, S., Herrero, R., Castellsague, X., et al. (2003). Epidemiologic classification of human papillomavirus types associated with cervical cancer. *The New England Journal of Medicine, 348*, 518–527.

Onon, T. S. (2011). History of human papillomavirus, warts and cancer: what do we know today? *Best Practice & Research. Clinical Obstetrics & Gynaecology, 25*, 565–574.

Parkin, D. M., & Bray, F. (2006). The burden of HPV-related cancers. *Vaccine, 3*(24 Suppl), S3/11–25. Chapter 2.

Parkin, D. M., Bray, F., Ferlay, J., & Pisani, P. (2005). Global cancer statistics, 2002. *CA: A Cancer Journal for Clinicians, 55*, 74–108.

Quinn, M. A., Benedet, J. L., Odicino, F., Maisonneuve, P., Beller, U., et al. (2006). Carcinoma of the cervix uteri. FIGO 26th Annual Report on the Results of Treatment in Gynecological Cancer. *International Journal of Gynaecology and Obstetrics, 1*(Suppl. 95), S43–103.

Sankaranarayanan, R., Bhatla, N., Gravitt, P. E., Basu, P., Esmy, P. O., et al. (2008). Human papillomavirus infection and cervical cancer prevention in India, Bangladesh, Sri Lanka and Nepal. *Vaccine, 12* (Suppl. 26), M43–52.

Scotto, J., & Bailar, J. C., 3rd (1969). Rigoni-Stern and medical statistics. A nineteenth-century approach to cancer research. *Journal of the History of Medicine and Allied Sciences, 24*, 65–75.

Sellers, W. R., & Kaelin, W. G., Jr (1997). Role of the retinoblastoma protein in the pathogenesis of human cancer. *Japanese Journal of Clinical Oncology, 15*, 3301–3312.

Syverton, J. T., & Berry, G. P. (1935). The cultivation of the virus of St. Louis encephalitis. *Science, 82*, 596–597.

Thierry, F. (2009). Transcriptional regulation of the papillomavirus oncogenes by cellular and viral transcription factors in cervical carcinoma. *Virology, 384*, 375–379.

Walboomers, J. M., Jacobs, M. V., Manos, M. M., Bosch, F. X., Kummer, J. A., et al. (1999). Human papillomavirus is a necessary cause of invasive cervical cancer worldwide. *The Journal of Pathology, 189*, 12–19.

zur Hausen, H. (1982). Human genital cancer: synergism between two virus infections or synergism between a virus infection and initiating events? *Lancet, 2*, 1370–1372.

zur Hausen, H. (1991). Viruses in human cancers. *Science, 254*, 1167–1173.

zur Hausen, H. (2002). Papillomaviruses and cancer: from basic studies to clinical application. *Nature Reviews. Cancer, 2*, 342–350.

FURTHER READING

Gynecological Cytopathology CERVIX with clinical aspects, (2007) (2nd ed.). Suresh Bhambhani, Mehta publishers.

The Health Professionals HPV Handbook (HPV and cervical cancer). Walter Prendiville, Philip Davies, Taylor and Francis.

Albuquerque, N. M. (1997). Cervical cancer in New Mexico. *A handbook for health care providers.*

Cancer, What's It Doing in My Life? A Personal Journal of the First Two Years of Chemotherapy in the Career of a Cancer Patient (1997) (2nd ed.). Mary Alice Geier Paperback, (128 Pages) Published by Hope Pub House.

Cancer and Its Management. Robert Souhami, Jeffrey Tobias (2005) (5th ed.). Paperback, (533 pages) Published by Blackwell Science Inc.

GLOSSARY

Cancer Cancer refers to a class of disease wherein a cell or a group of cells divide and replicate uncontrollably due to accumulation of both genetic and/or epigenetic changes that occur in a multi-step manner. This leads to unregulated cell proliferation, intrusion into adjacent cells and tissues (invasion), and ultimately, spread to other parts of the body than the location at which they arose (metastasis). Different types of cancer include cervical lung, breast, oral, colon, prostate, and ovarian cancers.

Care Taker Genes Genes involved in repair or prevention of DNA damage that arise by deletion, point mutation, or methylation (mutation is generally recessive). Their inactivation results in genetic instabilities that cause an increased mutation rate that affects all genes, including DNA repair genes (XPD, XRCC).

Cervical Cancer Cervical cancer results from the abnormal growth and division of cells at the opening of the uterus or womb (the area known as the cervix). The progression of cervical cancer is a multi-step process. Initially, normal cells undergo precancerous changes, and ultimately develop into cancer cells. These pre-cancerous conditions include cervical intraepithelial neoplasia (CIN), squamous intraepithelial lesion (SIL), and dysplasia. It takes several years to develop into an invasive cancer.

Human Papillomavirus (HPV) Human papillomaviruses are ubiquitous DNA viruses belonging to Family Papillomaviridae. They are small, non-enveloped DNA viruses with a circular, double-stranded DNA genome of approximately 7,200–8,000 base pairs (bp). Human apillomavirus (HPV) infection is a major cause of uterine cervical cancer and benign epithelial lesions such as warts and condyloma acuminate in the lower genital tract in humans. More than 100 genotypes have been described to date; 15 types are categorized as high-risk (HR-HPVs) types (HPV16, 18, 31, 33, 35, 39, 45, 51, 52, 56, 58, 59, 68, 73, and 82) that are associated with genital and other epithelial cancers; and 12 low-risk (LR-HPVs) types (HPV6, 11, 40, 42, 43, 44, 54, 61, 70, 72, 81, and CP6108) that are responsible for benign tumors and genital warts.

Proto-Oncogene and Oncogene A proto-oncogene is a normal gene that can become an oncogene due to mutations or increased expression. The resultant protein may be called an oncoprotein. Proto-oncogenes code for proteins that help to regulate cell growth and differentiation. Proto-oncogenes are often involved in signal transduction and execution of mitogenic signals, usually through their protein products. Upon *activation*, a proto-oncogene (or its product) becomes a tumor-inducing agent, an oncogene that has the potential to cause cancer. Examples of proto-oncogenes include RAS, WNT, MYC, ERK, and TRK. The MYC gene is implicated in Burkitt's Lymphoma. In tumor cells, they are often mutated or expressed at high levels. Since the 1970s, dozens of oncogenes have been identified in human cancer. Many cancer drugs target the proteins encoded by oncogenes.

Screening Screening is a public health intervention used on a population at risk, or target-population. Screening is not undertaken to diagnose a disease, but to identify individuals with a high probability of having or developing a disease. Screening can detect abnormalities before they become cancer. Also, if cancer itself is detected early, it can be cured with proper treatment. Example: Cervical cancer screening aims to test the largest possible proportion of women at risk and to ensure appropriate follow-up for those who have a positive or abnormal test result. Such women will need diagnostic testing and follow-up treatment. Methodologies include colposcopy and biopsy.

Tumor Suppressor Gene These genes normally function to inhibit cell growth/division and prevent cancer. If a tumor supressor gene is deleted or becomes mutated, this can contribute to cancer development. Examples include the BRCA1 gene, which is mutated in some breast cancers, and RB1, which is mutated in retinoblastoma. A tumor suppressor gene, or anti-oncogene, is a gene that protects a cell from being cancerous and encodes the proteins, which either have a dampening or repressive effect on the regulation of the cell cycle or promote apoptosis (and sometimes both). When this gene is mutated to cause a loss or reduction in its function, the cell can progress to cancer, usually in combination with other genetic changes (e.g. p53 tumor-suppressor protein encoded by the TP53 gene). Homozygous loss of p53 is found in 70% of colon cancers, 30–50% of breast cancers, and 50% of lung cancers. Mutated p53 is also involved in the pathophysiology of leukemias, lymphomas, sarcomas, and neurogenic tumors.

ABBREVIATIONS

Abs Antibodies
ACS Atypical Squamous Cell
ACS American Cancer Society
ALL Acute Lymphatic, Lymphocytic, or Lymphoblastic Leukemia
AML Acute Myelocytic Leukemia
AP-1 Activator Protein 1
ATF Activating Transcription Factor
Bcl-2 B-cell CLL/Lymphoma 2
Bcl_{XL} Member of Bcl-2 Family
BD Basic Domain
bp Base Pair
BSAP B-Cell Specific Activator Protein
bZIP Basic Region-Leucine Zipper
CDKs Cyclin-Dependent Kinases (Protein Required for Cell)
CIN Cervical Intraepithelial Neoplasia
CIS Carcinoma *In Situ*
CKC Cold-Knife Conization
CLL Chronic Lymphatic, Lymphocytic, or Lymphoblastic Leukemia
CML Chronic Myelocytic Leukemia
c-Myc Cytoplasmic Myelocytomayosis Oncogene
CSCs Cancer Stem Cells
DES Diethylstilbestrol
DNA Deoxyribonucleic Acid
dNTPs Deoxyribonucleoside Triphosphates
EBV Epstein–Barr Virus
E6AP E6-Associated Protein
EGF Epidermal Growth Factor
FDA Food and Drug Administration
FIGO International Federation of Gynecology and Obstetrics
HeLa Cell Line from Henrietta Lacks
HPV Human Papilloma Virus
HSIL High Squamous Intraepithelial Lesions
HTLV-1 Human T-Cell Leukemia Virus Type-1

IFN Interferon
IHC Immunohistochemistry
IKK Iκb Kinase
IL Interleukin
Ile Isoleucine
ISH *In Situ* Hybridization
Jak Janus Kinase
JIP Jun-Interacting Protein
JNK c-Jun N-terminal Kinase
Kb Kilobase
kDa Kilodalton
KGF Keratinocyte Growth
LEEP Electrosurgical Excision
LGSIL Low-Grade Squamous Intraepithelial Lesion
LSIL Low Squamous Intraepithelial Lesions
MAPK Mitogen-Activated Protein Kinase
MAPKK MAPK kinase
Mm Millimeter
NF-κB Nuclear Factor-Kappa B
ng Nanogram
°C Degrees Centigrade/Celsius
ORFs Open Reading Frames
PBS Phosphate-Buffered Saline
PCR Polymerase Chain Reaction
PLTC People Living Through Cancer
PNAs Peptide Nucleic Acids
RANK Receptor Activator of Nuclear Factor-κB
RAR Retinoic-Acid Receptor
RARβ2 Retenoic Acid Receptor β2
Rb Retinoblastoma Gene
RFLP Restriction Fragment Length Polymorphism
RHD Rel Homology Domain
RNA Ribonucleic Acid
rpm Revolutions per Minute
RSK Ribosomal S6 Kinase
RTK Receptor Tyrosine Kinase
SCJ Squamocolumnar Junction
Ser Serine
SFCs Sphere-Forming Cells
SIL Squamous Intraepithelial Lesion
SREs Serum Responsive Elements
STAT Signal Transducer and Activator of Transcription
TAD Transactivation Domain
Taq *Thermus aquatics* DNA Polymerase
TNAs Therapeutic Nucleic Acids
TNF Tumor Necrosis Factor
TNFR Tumor Necrosis Factor Receptor
TNM Tumor, Node, Metastases
TPA Tetradecanoyl Phorbol Acetate
TPA 12-O-Tetradecanoylphorbyl-13-Acetate
TSG Tumor Suppressor Gene
U Units
URR Upstream Regulatory Region
UV Ultraviolet
VIA Visual Inspection with Acetic Acid
VILI Visual Inspection with Lugol's Iodine
VLP Virus-Like-Particles
WDSCC Well-Differentiated Sqamous Cell Carcinoma
YY-1 Ying-Yang 1
α Alpha
β Beta
μg Microgram
μL Microlitre
μM Micromolar

LONG ANSWER QUESTIONS

1. Describe replication of HPV and its role in cervical cancer?

Normal squamous epithelial cells grow as stratified epithelium, with those in the basal layers dividing as stem cells and after division, one of the daughter cells migrates upward and begins to undergo terminal differentiation while the other remains in the basal layer as a slow-cycling, self-renewing population. HPV virions initially infect the basal layers of the epithelium, probably through microwounds and enter cells via interaction with receptors such as α-6 integrin for HPV16. In infected cells at the basal layer, low levels of viral DNA are synthesized to an episomal copy number of approximately 50 -100 genomes per cell. The early HPV genes *E1* and *E2* support viral DNA replication and its segregation so that the infected stem cells can be maintained in the lesion for a long period. As infected daughter cells migrate to the upper layers of the epithelium, viral late gene products are produced to initiate the vegetative phase of the HPV life cycle, resulting in high-level amplification of the viral genome.

Early genes *E5*, *E6* and *E7* are considered to coordinate a host cell suitable for viral DNA replication, which sometimes induces host cellular DNA synthesis and prevents apoptosis. In the outer layers of the epithelium, viral DNA is packaged into capsids and progeny virions are released to re-initiate infection. Because the highly immunogenic virions are synthesized at the upper layers of stratified squamous epithelia they undergo only relatively limited surveillance by cells of the immune system. In addition, E6 and E7 inactivate interferon (IFN) regulatory factor (IRF) so that HPV viruses can remain as persistent, asymptomatic infections.

High-risk HPV types can be distinguished from other HPV types largely by the structure and function of the E6 and E7 gene products. In benign lesions caused by HPV, viral DNA is located extra chromosomally in the nucleus. In high-grade intraepithelial neoplasias and cancers, HPV DNA is generally integrated into the host genome. In some cases, episomal and integrated HPV DNAs are carried simultaneously in the host cell. Integration of HPV DNA specifically disrupts or deletes the E2 ORF, which results in loss of its expression. This interferes with the function of E2, which normally down-regulates the transcription of the E6 and E7 genes and leads to an increased expression of E6 and E7. In high-risk HPV types, the E6 and E7 proteins have a high affinity for tumor suppressor genes p53 and pRB, resulting in increased proliferation rate and genomic instability. As a consequence, the host cell accumulates more and more damaged DNA that cannot be repaired. Efficient immortalization of keratinocytes requires the cooperation of the E6 and E7 gene proteins; however, the E7 gene product alone at high levels can immortalize host cells. Eventually, mutations accumulate that lead to fully transformed cancerous cells. In addition to the effects of activated oncogenes and chromosome instability, potential mechanisms contributing to transformation include methylation of viral and cellular DNA, telomerase activation, and hormonal and immunogenetic factors. Progression to cancer generally takes place over a period of 10 to 20 years.

SHORT ANSWER QUESTIONS

1. What is the status of cervical cancer incidences worldwide?
2. What are the symptoms and risk factors of cervical cancer?
3. What oncogenes in the HPV genome are involved in cervical cancer, and what are their functions?
4. What is a Pap test?
5. What are the commercially available HPV vaccines? Are they therapeutic or prophylactic in nature?

ANSWERS TO SHORT ANSWER QUESTIONS

1. According to Globocan 2008, it is the third most frequent malignancy in women worldwide, with an estimated 530,232 new cases and 275,008 deaths every year, nearly 80% in developing countries. Incidence rates are now generally low in developed countries.
2. Women may notice one or more of the following symptoms: abnormal vaginal bleeding, bleeding that occurs between regular menstrual periods, bleeding after sexual intercourse, douching, menstrual periods that last longer and are heavier than before, bleeding after menopause, increased vaginal discharge, pelvic pain. Risk factors fall into two categories:
   a. **Non-Genetic Factors**: Lower socio-economic status and lack of regular Pap tests, poor genital/sexual hygiene, multiple sexual partners or promiscuity, early age of first sexual intercourse below 18 years, oral contraceptives use, smoking, multiple pregnancies and parity, socio-economic status, dietary factors, religion and ethnicity.
   b. **Genetic Factors**: High-risk type HPV, viral load (severity of a viral infection), HPV variants, genetic predisposition, infections from other STDs like HIV, and weakened immune system.
3. Both E6 and E7 proteins are essential to induce and maintain cellular transformation due to their interference with cell-cycle control and apoptosis. The HPV viral oncogenes, E6 and E7, have been shown to be the main contributors to the development of HPV-induced cervical cancer and increased expression, probably due to integration of the viral DNA in the host-cell genome; they have been detected in invasive cancers and a subset of high-grade lesions. Both E6 and E7 HPV oncogenes interact with and inhibit the activities of tumor suppressors p53 and/or retinoblastoma protein (pRb), which is a common event for the carcinogenesis of human cells. This appears to result from deregulation of Plk1 by the loss of p53 through E6, and pRb family members by E7, overcoming the safeguard arrest response. Acute loss of pRb family members by E7 has also been shown to induce centrosome amplification and aneuploidy. In addition, E6 and E7 cause deregulation of cellular genes controlling the G2/M phase transition and progression through mitosis, such as the genes controlling centrosome homeostasis.
4. The Pap test looks for pre-cancers, cell changes on the cervix that might become cervical cancer if they are not treated appropriately. An adequate sample should be taken from the squamo-columnar junction (the transformation zone), transferred to the cytology slide, and immediately fixed with a commercial fixative. Pap smear screening is a rapid method for detecting cervical dysplasia and *in situ* cancer, as well as invasive cancer. A Pap smear evaluates cells harvested from the ectocervix and endocervix for abnormal changes associated with the development of cervical cancer.
5. HPV vaccines are of two types: prophylactic and therapeutic. Two successful **prophylactic vaccines** for HPV – quadrivalent "Gardasil" (HPV 16/18/6/11) developed by Merck, and bivalent "Cervarix" (HPV 16/18) by Glaxo SmithKline (GSK) – are recommended for vaccination of young adolescent girls at or before the onset of puberty. In these vaccines, viral capsid proteins are present in the form of spontaneously reassembled virus-like particles (VLPs) expressed either in yeast (for "Gardasil") or in baculoviruses (for "Cervarix"). These two vaccines protect from infection from two of the most common cancer-causing HPV types (16 and 18); more than 70% of cervical cancer cases are associated with these two HPV types. **Therapeutic vaccines** are used to bridge the temporal deficit by attacking already persistent HPV infections and to treat cervical cancer in women. Several animal studies have shown promising results, and indicate that therapeutic HPV vaccine may regress disease progression. As a result, several therapeutic HPV vaccines are in Phase I and Phase II clinical trials (Table 6.2). Most efforts have been directed towards the early proteins, HPV E6 and E7, or small peptides derived from them, mainly because these are the major transforming viral proteins that are invariably retained and expressed throughout the full spectrum of HPV-related disease progression and cervical carcinogenesis.

Chapter 7

Human DNA Tumor Viruses and Oncogenesis

Pravinkumar Purushothaman and Subhash Chandra Verma

Department of Microbiology & Immunology, University of Nevada, Reno, School of Medicine, Center for Molecular Medicine, Reno, Nevada

Chapter Outline

SUMMARY

A normal cell is transformed into a cancerous cell when a virus infects and persists in the infected cells. Tumor viruses are either DNA viruses, which include Epstein–Barr virus (EBV), Kaposi's sarcoma-associated herpesvirus (KSHV), human papillomavirus (HPV), hepatitis B virus (HBV), and Merkel cell polyomavirus (MCPyV); or RNA viruses such as hepatitis C virus (HCV) and human T lymphotrophic virus (HTLV-1).

WHAT YOU CAN EXPECT TO KNOW

Cancer involves the deregulation of multiple cell-signaling pathways that govern fundamental cellular processes such as cell death, proliferation, differentiation, and migration (Saha et al., 2010). The biologic pathways that lead to cancer are more complex and intertwined (Hanahan and Weinberg, 2000). Globally, it is estimated that 20% of all cancers are linked to oncogenic viruses (Parkin, 2006). However, most viral infections do not lead to tumor formation as several other factors influence the progression from viral infection to cancer development. Some of these factors include the host's genetic makeup, mutation occurrence, exposure to cancer-causing agents, and immune impairment. Initially viruses were believed to be the causative agents of cancers only in animals. It was almost half a century before the first human tumor virus, Epstein–Barr virus (EBV), was identified in 1964 (Moore and Chang, 2010). Subsequently, several human tumor viruses were identified. Tumor viruses are sub-categorized as either DNA viruses, which include EBV, Kaposi sarcoma-associated herpesvirus (KSHV), human papillomaviruses (HPV), hepatitis B virus (HBV), and Merkel cell polyomavirus (MCPyV); or RNA viruses such as hepatitis C virus (HCV) and human T lymphotropic virus (HTLV-1) (Saha et al., 2010). The normal cell is transformed into a cancer cell when a virus infects and persists in the infected cells either by integrating or retaining its genome as an extra-chromosomal entity. The

Animal Biotechnology. http://dx.doi.org/10.1016/B978-0-12-416002-6.00007-9

infected cells are regulated by the viral genes, which have the ability to drive abnormal growth. The virally infected cells are either eliminated via cell-mediated apoptosis or they persist in a state of chronic infection. Importantly, the chronic persistence of infection by tumor viruses can lead to oncogenesis (Pipas and a., 2009). This chapter will specifically focus on the major tumor DNA viruses associated with human cancer and their mechanisms of oncogenesis.

HISTORY AND METHODS

INTRODUCTION

Carcinogenesis, the development of cancer, begins with the accumulation of disruptions in several normal cellular activities that can eventually transform normal cells into cancer cells. These disruptions upset the normal balance between cell proliferation and death, allowing cells to acquire certain capabilities essential to malignant growth and spread. These general hallmarks of cancer include self-sufficient growth, insensitivity to anti-growth signaling, evasion of apoptosis, limitless replication, tissue invasion/metastasis, and angiogenesis (Hanahan and Weinberg, 2000). The progression stage of carcinogenesis occurs when cells acquire a combination of these abilities, which allows conversion of a normal cell into a cancer cell (Hahn et al., 1999). As the repertoire of capabilities continues to build, the next stage of carcinogenesis is observed when cells acquire the ability to degrade the local basement membrane, allowing them to spread and invade the surrounding tissues. In the case of solid tumors, these now invasive cancer cells can acquire the ability to induce blood vessel growth (blood supply) from pre-existing vessels via angiogenesis (Folkman, 2002). This provides the cancer with nutrients and oxygen needed to further grow and spread. At this point the cancer cells may concurrently acquire the ability to metastasize to distant sites (such as other organs) as tumor-mediated angiogenesis can provide primary tumor cells with a mode of transport to metastasize. This metastatic process includes the successful intravasation of cancer cells into blood/lymphatic vessels, transiting out of blood vessels, and finally, establishment of a secondary site of tumor growth. These cancer cells can then either lie dormant or can aggressively propagate into a secondary tumor. Each of these successive phases of carcinogenesis increases the likelihood of cancer-related morbidity and mortality, the metastasis stage representing the largest contributor. Additionally, promoting a permissive microenvironment is crucial to the progression of carcinogenesis, one major example being the modulation of immune responses (Hanahan and Weinberg, 2000). Similar to environmental and host-related oncogenic events, human tumor-associated viruses can lead to malignancies by providing viral mechanisms that promote one or more general hallmarks of cancer (Pipas and a., 2009). It is furthermore recognized that chronic inflammation and immunosuppression provide a microenvironment more conducive to the progression of carcinogenesis (Goedert, 2001). This chapter will focus specifically on DNA tumor viruses known to associate with various cancers, and will highlight viral mechanisms of oncogenesis.

TRANSFORMATION AND ONCOGENESIS

Studies on DNA tumor viruses have been instrumental to our understanding of basic cell biology and how the perturbations of cellular pathways contribute to the initiation and maintenance of cancer. DNA tumor virus infection leads to immortalization of the infected cell through deregulation of multiple cellular pathways via expression of many potent oncoproteins (Saha et al., 2010). as shown in Figure 7.1. Research on various viral oncoproteins has revealed many of their novel cellular targets that are directly associated with cellular signaling, cell-cycle control, and the host's defense system (Saha et al., 2010; Stevenson, 2004). (Table 7.1). Tumor viruses reprogram the host quiescent, G0 cell into the S phase of the cell cycle, allowing viral access to the nucleotide pools and cellular machinery that are required for replication and transmission. The host cellular innate immune responses respond to viral infection by activating tumor suppressor proteins, pRB1 and p53, to induce cell death. However, the tumor viruses have evolved the means to inactivate these signaling pathways for their own benefit (Bouvard et al., 2009; Goedert, 2001; Stevenson, 2004). Herpesvirus family members, EBV- and KSHV-encoded oncoproteins, have been shown to manipulate p53 and pRb functional activity to block apoptosis. The EBV-encoded proteins, EBNA3C and LMP1, modulate p53 function either by repressing its transcriptional activity or by blocking p53-mediated apoptosis (Saha et al., 2010). Additionally EBNA3C has also been shown to induce pRb degradation, thus leading to an establishment of latent infection (Moore and Chang, 2010). Similarly, KSHV-encoded LANA and ORF K8 proteins, block p53-mediated host cell death through their interaction with p53 (Saha et al., 2010; Verma and Robertson, 2007). The KSHV-encoded LANA can also directly interact with pRb and enhance E2F-dependent transactivation activity and contribute to KSHV-induced oncogenesis by targeting the pRb-E2F regulatory pathway (Cai et al., 2010; Mesri et al., 2010). Likewise, other DNA tumor virus-encoded oncoproteins also target tumor suppressor proteins; HPV-encoded E6 protein has been shown to bind and degrade p53 through the ubiquitin-proteasome pathway (Moore and Chang, 2010). In addition, HPV E7 oncoprotein bypasses cell cycle arrest through binding to the hypophosphorylated form of pRb, thereby inducing the degradation of pRb through a proteasome-mediated pathway (Saha et al., 2010; Damania, 2007). Also, HBV-encoded HBx interacts

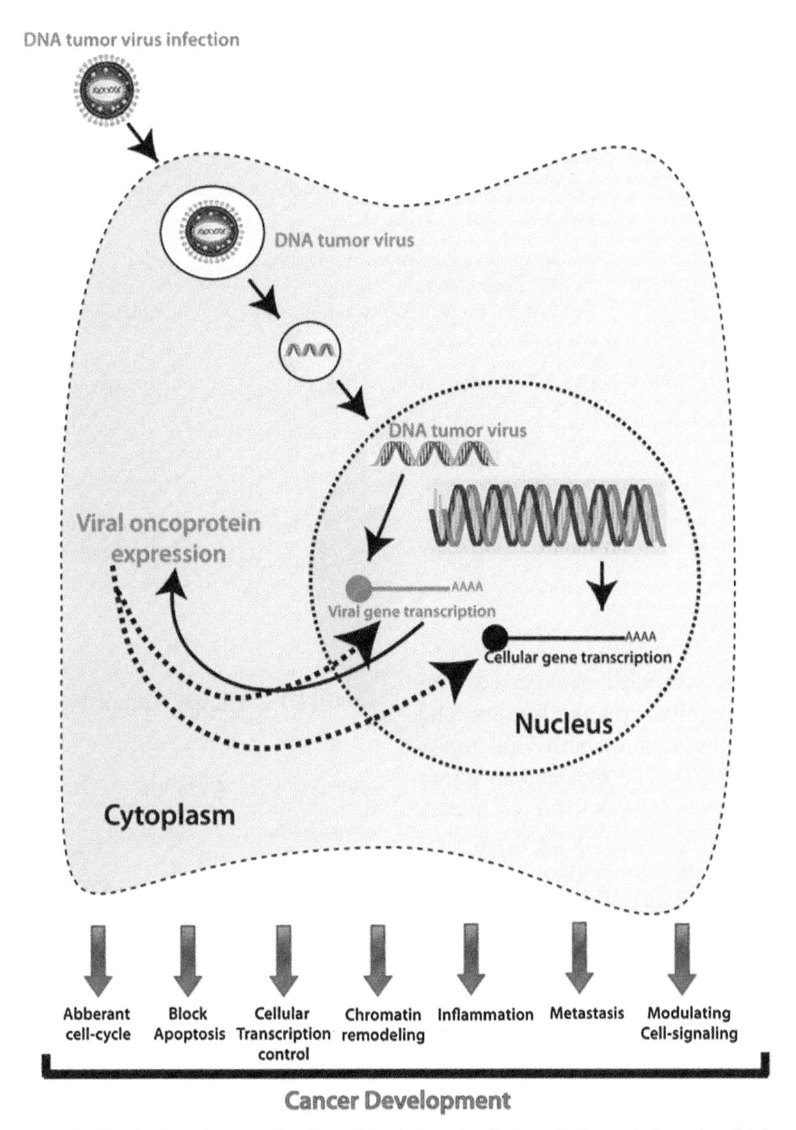

FIGURE 7.1 DNA tumor virus infection leads to immortalization of the infected cell through deregulation of multiple cellular pathways involved in cellular signaling, cell-cycle control, and defense systems via expression of many potent oncoproteins.

with p53 to inhibit its functional activity, which leads to the development of human hepatocellular carcinoma (HCC). In addition, the HBV-encoded HBx oncoprotein destabilizes pRb by up-regulating E2F1 promoter activity (Damania, 2007; Shepard et al., 2006). Recent studies show that KSHV-encoded LANA and HBV-encoded HBx can down-regulate von Hippel–Lindau (VHL), a tumor-suppressor gene, along with p53 (Saha et al., 2010). Association between viral oncoproteins and cell cycle regulatory factors cyclin/CDK complexes also play a crucial role in viral transformation. The EBV-encoded latent antigen, EBNA2, transactivates cyclin D2 through the activation of c-Myc in EBV-associated lymphomas. Similarly, KSHV-encoded LANA stabilizes β-catenin, resulting in increased expression of both β-catenin and cyclin D1 in KSHV tumors (Saha et al., 2010). Viral oncoproteins also deregulate various cellular signaling pathways that are directly linked to development of oncogenesis, such as Notch signaling, JAK/STAT, JNK, Wnt, IRF, the ubiquitin–proteasome system, TNF, and NFκB-signaling pathways, to facilitate their survival and evade host–immune response (Saha et al., 2010; Moore and Chang, 2010).

HISTORY OF HUMAN DNA TUMOR VIRUSES AND CANCER

The International Agency for Research on Cancer (Ignatovich et al., 2002). estimates that one in five cancer cases worldwide are caused by infection, mostly with viruses (Parkin, 2006). Viruses have played a central role in modern cancer research and have been providing profound insights into both infectious and non-infectious cancer causes.

TABLE 7.1 Viral Oncoproteins and Their Targets

Tumor Virus	Viral Oncoproteins	Important Cellular Binding Partners	Deregulated Signaling-Pathways
EBV	LMP1	p53, Mdm2, pRb, p300, Chk2, c-Myc, HDAC1, SUMO-1, SUMO-3, Cyclin A, E and D1, TRAFs, TRADD, JAK	Cellular transcription, cell-cycle, metastasis, ub-proteasome, apoptosis, inflammation, chromatin remodeling, cellular signaling
KSHV	vGPCR, vIL-6, vBcl2, vCyclin LANA and vFLIP	p53, pRb, c-Myc, core histones, apoptosis, TRAF2, transcriptional activators (Sp1, AP-1), transcriptional inhibitors (HP1, mSin3)	Cellular transcription, cell-cycle, apoptosis, ub-proteasome, chromatin remodeling, cellular signaling
HPV	E6 E7	p53, p73, c-Myc , pRb, p21CIP1, p27KIP1, IRF-1, cyclin A and E	cell-cycle, ub-proteasome
MCPyV	LT	p53, pRb	Cell-cycle
HBV	HBx	NFκB, p53, c-jun, c-fos, PKC, c-myc	Cell-cycle, apoptosis, cellular transcription, cellular signaling, metastasis

This diverse group of viruses revealed unexpected connections between innate immunity, immune sensors, and tumor suppressor signaling that control both viral infection and cancer (Damania, 2007). The first human tumor virus, Epstein–Barr virus (EBV; also known as human herpesvirus 4) was identified by Anthony Epstein, Bert Achong, and Yvonne Barr in 1964. Since then, four more human DNA viruses have been discovered (Table 7.2) that are now widely accepted as causes of invasive human tumors. These include Kaposi's sarcoma-associated herpesvirus (KSHV; also known as human herpesvirus 8, HHV8), human papillomavirus (HPV), hepatitis B virus (HBV), and the recently identified Merkel cell polyomavirus (MCPyV) (Moore and Chang, 2010). Infectious cancer agents have been divided into two broad categories: direct carcinogens, which express viral oncogenes that directly contribute to cell transformation into cancer cells, and indirect carcinogens that presumably cause cancer through chronic infection and inflammation, which in turn eventually leads to carcinogenic mutations in host cells.(Hausen, 2001). By definition, a direct viral carcinogen is present in each cancer cell and expresses at least one transcript to maintain the transformed tumor cell phenotype, as occurs with HPV, MCPyV, EBV, and KSHV-related cancers. Evidence supporting this comes from knockdown studies in which the loss of viral proteins results in the loss of host cancer viability (Moore and Chang, 2010). Even in these cases, external factors such as immunity and exposure to other infectious agents directly affect carcinogenesis. Indirect carcinogens could potentially also include "hit-and-run" viruses in which the viral genes are lost as the tumor begins to mature. Several agents such as HBV, HCV, and HTLV-I, which are involved in hepatocellular carcinoma (HCC), partly fit into this category (Moore and Chang, 2010). The oncogenic viruses are generally characterized by prolonged and often life-long latency and evasion of the host immune surveillance as only a small subset of viral genes are normally expressed that may elicit an immune

TABLE 7.2 Human Tumor Viruses

Virus	Genome	Notable Cancers	Year Identified
Epstein–Barr Virus (EBV)	174 kb linear dsDNA	Burkitt's lymphoma, nasopharyngeal carcinoma, Hodgkin's disease, infectious mononucleosis, X-linked lymphoproliferative disorders	1964
Kaposi's Sarcoma Herpesvirus (KSHV)	137 kb linear dsDNA	Kaposi's sarcoma, primary effusion lymphoma, multicentric Castleman's disease	1994
Human Papillomaviruses (HPV)	8 kb circular dsDNA	Most cervical cancer, penile cancers, anogenital, head & neck cancers	1983
Merkel Cell Polyomavirus (MCV)	5 kb circular dsDNA	Merkel cell carcinoma	2008
Hepatitis B Virus (HBV)	3 to 3.3 kb of relaxed circular, partially duplex DNA	Hepatocellular carcinoma	1965

response (Moore and Chang, 2010). Among DNA tumor viruses, latency is well studied in herpesvirus family members EBV and KSHV. During latency, the virus is tumorigenic and persists as episomes in a highly ordered chromatin structure whose propagation is dependent on their ability to hijack the replicative machinery of the host (Arvin et al., 2007; Cai et al., 2010; Mesri et al., 2010). The most likely explanation for the connection between virus latency and tumorigenesis is that productively replicating viruses initiate cell death, which has long been known to virologists as the cytopathic effect (CPE) (Moore and Chang, 2010). Small DNA tumor viruses (HPV and MCPyV) do not encode their own replication proteins, while large DNA tumor viruses (EBV, KSHV and HBV) do encode their own viral DNA polymerase, but still require components of cellular replicative machinery for efficient viral DNA synthesis. A contrasting feature between small and large DNA tumor viruses is that small DNA tumor viruses can integrate into the host chromosomal DNA during or before cancer progression (Damania, 2007; Moore and Chang, 2010).

Epstein–Barr Virus

Epstein–Barr virus (EBV) is a double-stranded DNA virus that belongs to the genus lymphocryptovirus of the human γ-herpesvirus family. EBV infects more than 90% of the world's adult population and is the causative agent of numerous acute diseases and cancers, such as infectious mononucleosis (IM), Burkitt's lymphoma, nasopharyngeal carcinoma, natural killer cell lymphoma, Hodgkin's disease, and X-linked lymphoproliferative disease (Arvin et al., 2007). Immunocompromised patients, including AIDS or post-organ transplant patients, have a high probability of getting EBV-associated lymphomas. The virus only infects cells expressing the receptor for complement C3d component (CR2 or CD21); these cells include epithelial cells and B-lymphocytes. EBV has powerful transforming potential for B-lymphocytes and establishes a lifelong latent infection in the B-cell of the infected host. The EBV genome is 174 Kb, which is maintained in the nucleus as an episome via tethering to the host chromosome (Arvin et al., 2007). EBV encodes several viral proteins that have transforming potential. EBV latency proteins (LMP1, EBNA2, EBNA3A and EBNA3C) have been shown to express in AIDS-associated lymphoma, post-transplant lymphoma patients and lymphoblastoid cell lines (LCLs) generated from *in vitro* EBV infection of primary B-cells (Saha et al., 2010). LMP1 and EBNA2 are essential for the ability of EBV to immortalize B-cells because deletion of LMP1 from EBV renders the virus non-transforming (Damania, 2007). EBV-encoded latent protein EBNA-LP functions like a co-stimulator of EBNA2-mediated transactivation of many cellular and viral genes shown to be critical for B-cell immortalization (Saha et al., 2010). EBNA1 is essential for the maintenance and segregation of the EBV genome. EBNA3A and 3C are also critical for B-cell immortalization, while EBNA3B enhances the survival of cells. All three EBNA3 proteins are shown to bind with RBP-Jκ/CBF1 and to regulate cellular gene transcription important for transforming B-cells into immortalized LCLs (Saha et al., 2010).

Kaposi's Sarcoma-Associated Herpesvirus (KSHV)

Kaposi's sarcoma-associated herpesvirus (KSHV), also called human herpesvirus 8 (HHV8), is a member of the gamma-herpesvirus family, which is tightly associated with human cancers, including Kaposi's sarcoma, primary effusion lymphoma (PEL), and multi-centric Castleman's disease (MCD) (Arvin et al., 2007). KSHV remains asymptomatic in healthy individuals and establishes lifelong persistence in B-lymphocytes after primary infection, similar to EBV (Blake, 2010). Kaposi's sarcoma, originally called "idiopathic multiple pigmented sarcoma of the skin," was first described by Moritz Kaposi, a Hungarian dermatologist (Arvin et al., 2007). KS was initially thought to be an uncommon tumor of Mediterranean populations until it was identified throughout sub-Saharan Africa and later became more widely known as one of the AIDS-defining illnesses in the 1980s (Mesri et al., 2010). The viral cause for this cancer was found by Yuan Chang and Patrick Moore in 1994 by isolation of DNA fragments of a herpesvirus from a Kaposi's sarcoma (KS) tumor in an AIDS patient (Chang et al., 1994). Similar to EBV, the KSHV has a linear double-stranded DNA genome of approximately 137 kb in size and codes for more than 90 open reading frames (ORFs); however, only a small subset of these genes is expressed during latency, which includes latency-associated nuclear antigen (LANA), vCyclin, vFLIP/K13, K12/Kaposin, and an miRNA cluster (Cai et al., 2010; Mesri et al., 2010). LANA, encoded by ORF73, is a multifunctional nuclear antigen and functional homolog to the EBV EBNA1 protein that plays a central role in deregulating various cellular functions, including maintenance of the viral episome, (Verma and Robertson, 2007). degradation of the p53 and pRb tumor suppressors, transactivation of the telomerase reverse-transcriptase promoter, promotion of chromosome instability in KSHV-infected B cells (Verma and Robertson, 2007). and accumulation of the intracellular domain of Notch in KSHV-mediated tumoriegenesis (Feng et al., 2008; Saha et al., 2010). LANA also plays a crucial role in maintaining latency by regulating the expression of RTA, another critical viral-encoded transcriptional activator required to switch from the latent to the lytic cycle (Verma and Robertson, 2007). During latency, LANA tethers the viral episomal DNA to the host chromosomes, which helps in the efficient partitioning of the viral DNA in the daughter cells after cell division. Disruptions of LANA expression lead to reduction in episomal copies, suggesting the importance of LANA in KSHV-mediated pathogenesis (Saha et al., 2010).

Kaposi's Sarcoma (KS)

Kaposi's sarcoma (KS) is a multifocal vascular tumor of mixed cellular composition that develops from the cells that line lymph or blood vessels, and is most often seen as a cutaneous lesion. The abnormal cells of KS form purple, red, or brown blotches or tumors on the skin called lesions (Arvin et al., 2007). KSHV is always found in the spindle cells of the lesion, which are thought to be of endothelial origin. In some cases, the disease causes painful swelling, especially in the legs, groin area, or skin around the eyes. KS can cause serious problems or even become life threatening when the lesions are in the lungs, liver, or digestive tract (Pipas and a., 2009). KS in the digestive tract, for example, can cause bleeding, while tumors in the lungs can cause difficulty in breathing. There are four epidemiological forms of KS:

1. Classic KS is seen in elderly men of Mediterranean, eastern European, and Middle Eastern heritage. Classic KS is more common in men than in women. Patients typically have one or more lesions on the legs, ankles, or the soles of the feet.
2. Endemic KS is a second type of KS, which is seen in Africa and affects HIV-positive and HIV-negative individuals, and even children (Mesri et al., 2010). Rarely, a more aggressive form of endemic KS is seen in children before puberty; it usually affects the lymph nodes and other organs and can lead to death within a year. There could be other factors in Africa (such as environmental cofactors and genetic predispositions in the population) that could contribute to the development of aggressive KS since the disease affects a broader group of people that includes children and women.
3. The third type of KS lesion is an iatrogenic form of KS that develops in post-transplant patients receiving immunosuppressive therapy (Damania, 2007; Mesri et al., 2010; Mitxelena et al., 2003). Greater than 95% of all KS lesions, regardless of type, have been shown to contain KSHV viral DNA, thereby indicating a strong epidemiological link between KSHV infection and KS (Damania, 2007; Mesri et al., 2010).
4. AIDS-associated epidemic KS is a highly aggressive tumor and is primarily detected in HIV-infected individuals whose immune systems are severely damaged. In these individuals the KS lesion is not restricted to the skin, and often disseminates to the liver, spleen, gastrointestinal tract, and lungs. KS is the most frequently detected tumor in AIDS patients.

Primary Effusion Lymphoma (PEL)

Primary effusion lymphoma (also called body cavity lymphoma) is a rare and aggressive B-cell lymphoma that arises as body cavity effusions in 5 to 20% of HIV-positive patients and is considered an AIDS-defining illness (Mesri et al., 2010; Chen et al., 2007). PEL has a unique clinical presentation in that it has a predilection for arising in body cavities such as the pleural space, pericardium, and peritoneum (Chen et al., 2007). PEL cells are morphologically variable, with a null lymphocyte immunophenotype with evidence of human herpesvirus (HHV)-8 infection. All PELs are KSHV-positive, indicating a strong epidemiological link between the presence of KSHV and the induction of PEL; however, 90% of these lymphoma cells often contain EBV as well. PELs are observed in both HIV-positive and HIV-negative individuals, with both types of PELs invariably containing KSHV viral DNA (Damania, 2007; Chen et al., 2007).

Multicentric Castleman's Disease (MCD)

Castleman's disease, also called angiofollicular or giant lymph node hyperplasia, is a clinically heterogeneous entity that can be either localized (unicentric) or multicentric. MCD is an atypical lymphoproliferative disorder of a plasma cell type, and is related to immune dysfunction (Cai et al., 2010; Mesri et al., 2010). There are two types of MCD, a hyaline vascular form, which presents as a solid mass, and a plasma-cell variant, which is associated with lymphadenopathy. Sometimes a mixture of both hyaline-vascular and plasma-cell variants can also be found (Damania, 2007). Nearly 100% of AIDS-associated MCD is positive for KSHV, whereas less than 50% of non-AIDS-associated MCD contains KSHV viral DNA (Damania, 2007). Patients with AIDS-associated MCD often develop malignancies like Kaposi's sarcoma and non-Hodgkin's lymphoma. More than 5% of all cancers worldwide are caused by persistent infection with this virus (Cai et al., 2010; Damania, 2007; Mesri et al., 2010).

Human Papillomavirus (HPV)

Infection by the human papilloma virus is the most common sexually transmitted disease, afflicting 50–80% of the world's population (Pipas and a., 2009). An association between human papillomavirus (HPV) infection and the development of cervical cancer was initially reported in 1983 by Harald zur Hausen (Hausen, 2001). HPV belongs to the papovaviridae family and has a genome size of about 8 kilobases (Damania, 2007). The majority of the known types of HPV are asymptomatic, but some types may cause warts, while others can lead to cancers of the cervix, vulva, vagina, and anus (in women), or cancers of the anus and penis (in men) (Damania, 2007; Narisawa-Saito, 2007). About 50% of sexually active men and women get HPV infection at some point in their lives. Most HPV infections in young females are temporary and have little long-term significance. Seventy percent of infections are gone in 1 year and ninety percent in 2 years. However, when the infection persists, in 5% to 10% of infected women, there is high risk of developing precancerous lesions of the cervix, which can progress to invasive cervical cancer (Narisawa-Saito, 2007). Cervical screening using a Papanicolaou (Pap) test or liquid-based cytology is used to detect abnormal cells that may develop into cancer (Pipas and a., 2009). So far, more than 160 HPV types have been identified and subsequently classified into low- or high-risk groups according to their potential for causing cervical cancer (Moore and Chang, 2010; Narisawa-Saito, 2007). About a dozen HPV types (including types 16, 18, 31, and 45) are called "high-risk" types because they can lead to cervical as well as anal, vulvar, vaginal, and penile cancer (Damania, 2007; Moore and Chang, 2010; Narisawa-Saito, 2007; Saha et al., 2010). Several types of HPV (type 16 in particular) have been found to be associated with HPV-positive oropharyngeal cancer (OSCC), a form of head and neck cancer (Damania, 2007; Narisawa-Saito, 2007). HPV-induced cancers often have viral sequences integrated into the cellular DNA (Damania, 2007). Low-risk viruses are associated with benign lesions such as condyloma accuminata or with basal cell and squamous-cell carcinomas of the skin (Damania, 2007). Another very rare inherited disease associated with HPV is epidermodysplasia verruciformis (EV), which is caused by an autosomal recessive mutation that leads to abnormal, uncontrolled papilloma virus replication (Damania, 2007; Narisawa-Saito, 2007). This results in the growth of scaly macules and papules on many parts of the body, but especially on the hands and feet. EV, which is associated with a high risk of skin carcinoma, is typically associated with HPV types 5 and 8 (other types may also be involved (Pipas and a., 2009).

Primary HPV infection occurs in the basal stem cells of the epithelium. The virus cannot bind to live tissue; instead, it infects epithelial tissues through micro-abrasions or other epithelial trauma as would occur during sexual intercourse or after minor skin abrasions that expose segments of the basal membrane (Narisawa-Saito, 2007; Pipas and a., 2009). The virus then traverses upwards and replicates in the terminally differentiated keratinocytes and is shed from the stratum corneum (Damania, 2007). HPV genes encode proteins responsible for replication, cellular transformation, control of viral transcription, and those necessary for the generation of viral progeny (Damania, 2007; Saha et al., 2010). The E6 and E7 proteins of high-risk HPV strains have strong transforming abilities. HPV oncoproteins E6 and E7 have been shown to act synergistically to immortalize cells in vitro and induce skin tumors in transgenic animals (Damania, 2007). The HPV viral proteins target tumor suppressors, oncoprotein E6 binds to p53, and E7 binds to the retinoblastoma (Rb) family of proteins, which induces their degradation through ubiquitin–proteasome-mediated degradation and leads to deregulation of the cell cycle and the inhibition of apoptosis (Cai et al., 2010; Narisawa-Saito, 2007; Saha et al., 2010).

Hepatitis B Virus (HBV)

Hepatitis B virus (HBV) is a member of the Hepadnavirus family that causes an infectious inflammatory disease of the liver. Humans are the only known natural host for HBV (Martin & Gutkind, 2008; Pipas and a., 2009). It is estimated that about 360 million individuals worldwide are infected with the virus at one point in their lives, and many of them are chronic carriers (Kao, 2011; Parkin, 2006). without any identifiable risk factor (Shepard et al., 2006). Most individuals with chronic hepatitis B remain asymptomatic for many years or decades. Patients with HIV infection or post-organ transplant patients taking immunosuppressive drugs are at higher risk of developing chronic infection (Damania, 2007; Shepard et al., 2006). However, patients with chronic hepatitis B are at risk of developing hepatocellular carcinoma (HCC), the fifth most common cancer and the third leading cause of cancer death worldwide (Damania, 2007; Kao, 2011; Shepard et al., 2006). Infection with hepatitis B virus (HBV) and hepatitis C virus (HCV) are considered the major contributors to HCC development, accounting for over 80% of all HCC globally (Moore and Chang, 2010; Saha et al., 2010).

Dr. Baruch Blumberg discovered the viral cause of HBV in 1965. The genome of HBV is made of circular DNA of about 3.3 kb full-length negative-sense strand and a 2.8 kb short-length sense-strand (Damania, 2007; Moore and Chang, 2010). One end of the full-length strand is covalently linked to the viral DNA polymerase and the other short strand has an RNA oligonucleotide at its 5′ end. Thus, neither DNA strand is closed, and the circularity is maintained by cohesive ends (Damania, 2007). Viral DNA enters the nucleus and integrates into the host DNA immediately after infection. HBV replication initiates increased DNA synthesis and interferes with normal cellular detoxification and repair functions, causing chronic liver cell injury that leads to HCC (Kao, 2011; Shepard et al., 2006). There are three different types of hepatitis B antigens encoded by the HBV genome. These include hepatitis B surface antigens (HBsAg, MHBsAg, and LHBsAg), hepatitis B core antigen (HBcAg), and hepatitis B early antigen (HBeAg). The hepatitis B surface antigen (HBsAg) is most frequently used for diagnosis and is the first detectable viral antigen to appear during infection. There is a strong correlation between HBsAg chronic carriers and the incidence of HCC (Kao, 2011; Shepard et al., 2006). It has been shown that HBsAg carriers have a risk of HCC that is 217 times that of a non-carrier, and 51% of the deaths of HBsAg carriers are caused by liver cirrhosis or HCC compared to 2% of the general population (Kao, 2011). The virus is classified into four major serotypes (adr, adw, ayr, ayw) based on antigenic epitopes presented on its envelope proteins, and into eight genotypes (Vaughn and Elenitoba-Johnson, 2001). according to overall genomic variation (Kao, 2011). The genotypes show a distinct geographical distribution and disease severity, and are used in tracing the evolution and transmission of the virus (Kao, 2011). HBV is transmitted by exposure to infectious blood or body fluids through sexual contact, blood transfusions, reuse of contaminated needles and syringes, and vertical transmission from mother to child (MTCT) during childbirth (Damania, 2007; Kao, 2011; Shepard et al., 2006). Without intervention, a mother who is positive for HBsAg confers a 20% risk of passing the infection to her offspring at the time of birth (Shepard et al., 2006). This risk is as high as 90% if the mother is also positive for HBeAg (Shepard et al., 2006). Holding hands, sharing utensils or glasses, kissing, hugging, coughing, sneezing, or breastfeeding, however, cannot spread hepatitis B viruses (Shepard et al., 2006).

Approximately 80% of HBV-related HCC shows integrated HBV sequences, however, HBV can also be found integrated in non-HCC tissues (Damania, 2007; Martin and Gutkind, 2008; Saha et al., 2010). Integration of HBV leads to severe mutagenic consequences such as large inverted duplications, deletions, amplifications, and translocations that result in chromosomal instability (Saha et al., 2010). HBV-encoded oncogene HBx is most commonly found to be integrated in patients with HBV-related cirrhosis and dysplasia (Moore and Chang, 2010; Saha et al., 2010). HBx has been shown to interact with EGF receptor, c-myc, c-jun, c-fos, p53, AP-1, NFκB, and SP1 in multiple ways to contribute to the molecular pathogenesis of human HCC (Martin and Gutkind, 2008; Moore and Chang, 2010; Saha et al., 2010).

Human Polyomaviruses

Polyomaviruses are double-stranded DNA viruses of the family Polyomaviridae. Ludwik Gross discovered the first murine polyomavirus in 1953 (Pipas and a., 2009). Subsequently, many polyomaviruses have been found to infect birds and mammals. Polyomaviruses were so named because they are potentially oncogenic and cause a wide range of tumors in a number of animal species. Polyomaviruses are icosahedral in shape, with a small circular DNA genome of approximately 5,000 base pairs (Damania, 2007). For several years after the discovery of leukoencephalopathy (JC virus) and nephropathy (BK virus) in 1971, only these two polyomaviruses were known to infect humans. Nonetheless, in 2008, using digital transcriptome subtraction (DTS) and high-throughput genome sequencing techniques, Yuan Chang and Patrick S. Moore and their colleagues identified the presence of Merkel cell polyomavirus (MCPyV) in Merkel cell carcinoma (Feng et al., 2008). Polyomavirus infections via respiratory tract are highly common in childhood and young adult infections. Most of these infections are asymptomatic, though the virus persists life-long in almost all adults (Silva . d, 2011). Clinical evidence of disease caused by human polyomavirus infection is most common among persons who become immunosuppressed by AIDS, who are of old age, or persons taking immunosuppressive drugs after organ transplantation. BK virus was identified from the urine of a kidney transplant patient, and JC virus was identified from the brain of a Hodgkin's lymphoma patient who developed progressive multifocal leukoencephalopathy (PML). The human JC virus (JCV) and BK virus (BKV) have been linked to several different human cancers. However, whether the role of the virus is causal or incidental has been the subject of much debate (Damania, 2007; Silva, 2011). JCV DNA has been identified from brain tumors found in patients with or without progressive multifocal leukoencephalopathy (PML) and with glial tumors and pediatric medulloblastomas (Damania, 2007). JCV has also been associated with colon cancer and central nervous system (CNS) lymphoma (Damania, 2007). Similarly, BK virus is associated with polyomavirus nephropathy (PVN), a form of acute interstitial nephritis, (Silva, 2011). and BK viral DNA has also been found in pancreatic islet tumors and brain tumors (Damania, 2007). This indicates that BKV virus is newly emerged as an opportunistic CNS infectious agent in AIDS and transplant patients, particularly those with a coexistent urologic disease and neurological decline (Silva, 2011).

Merkel Cell Polyomavirus (MCPyV)

Merkel cell polyomavirus (MCPyV) is associated with Merkel cell carcinoma (MCC). Also known as trabecular carcinoma of the skin, (Bhatia et al., 2011). it is a rare and highly aggressive cancer with a high mortality rate in patients with malignant cancer. It develops in hair follicles or on (or beneath) the skin (Bhatia et al., 2011). Merkel cell carcinoma (MCC) occurs most often on the sun-exposed face, head, and neck. Persons with AIDS, older people with extensive prior sun exposure, or persons taking immunosuppressive drugs after organ transplantation are at higher risk to develop MCC (Bhatia et al., 2011). MCPyV appears to be widely prevalent among healthy individuals. It has been shown that the prevalence of MCPyV seropositivity was 0% in infants, 43% among children aged 2–5 years old, and increased to 80% among adults older than 50 years (Bhatia et al., 2011). Additionally, MCPyV DNA was detected in cutaneous swabs from clinically healthy individuals with a prevalence of 40–100%. Although widely prevalent, active MCPyV infection appears to be asymptomatic in healthy individuals, with the exception of MCC. This virus has not yet been convincingly linked with any other human disease (Bhatia et al., 2011; Moore and Chang, 2010). The exact mode of transmission remains to be elucidated, and could involve cutaneous, fecal–oral, mucosal, or respiratory routes. MCPyV viral DNA has been detected in lower frequencies among respiratory secretions, on oral and anogenital mucosa, and in the digestive tract. Importantly, it appears that the virus is being shed chronically from clinically normal skin in the form of assembled virions (Bhatia et al., 2011) The MCPyV LT-antigen transcripts and oncoproteins are expressed in most MCC tumors and the LT-antigen appears to retain the major conserved features of other polyomavirus LT-antigens, such as Rb-binding motif and helicase/ATPase domains (Bhatia et al., 2011; Moore and Chang, 2010). MCPyV is the first polyomavirus that has been shown to integrate into human genomic DNA (Bhatia et al., 2011). The virus then undergoes at least two mutations, (Moore and Chang, 2010). the first being a non-homologous recombination with the host chromosome, and then a sequential large T (LT)-antigen truncation mutation that eliminates its viral replication functions but spares its Rb-targeting domain (Moore and Chang, 2010). These sequential mutational events result in persistent T-Ag expression, which play a key role in turning asymptomatic viral infection into an aggressive Merkel cell carcinoma (Bhatia et al., 2011).

Principle

DNA tumor viruses encode proteins with oncogenic potential, which alter the normal growth of cells by deregulating various cellular pathways. Here we discuss the roles of viral proteins and RNA in tumoriegenesis by taking the example of Epstein–Barr Virus (EBV). A diagrammatic representation of the EBV life cycle is depicted in Flow Chart 7.1: (1) First step in EBV primary infection is the viral entry through the buccal cavity. (2) EBV shows higher affinity to B cells in primary infection of naive B cells that occurs in the oropharyngeal mucosa. (3) Upon entry into the target cells, EBV enters into a short burst of lytic proliferation followed by a well-defined latency program. (4) EBV persists in the B cells in the peripheral circulation and this latent infection of B cells leads to various types of lymphomas. (5) Latently infected B cells mature into plasma cells. (6) Plasma cells undergo lytic reactivation releasing infectious EBV particles. (7) EBV produced thus could reinfect fresh naive B cells or epithelial cells to cause nasopharyngeal carcinoma.

EBV Genome Structure

A mature EBV viral particle is approximately 120 to 180 nm in diameter. It is composed of a double-stranded, linear DNA genome enclosed by a protein capsid surrounded by a protein tegument, which in turn is surrounded by a lipid envelope. The EBV genome is approximately 174 kilobase pairs (kb) in length, containing 0.5 kb terminal repeats (TRs), and internal repeat sequences (IRs) (Robertson, 2005). EBV strain B95-5 (derived from infectious mononucleosis) was the first herpesvirus to have its genome completely cloned and sequenced. The virus encodes for approximately 80 proteins, however not all of them have been fully characterized. The EBV genome was sequenced from a viral DNA BamHI fragment cloned library, hence open reading frames (ORFs), genes, and sites for transcription or RNA processing are frequently referenced to specific BamHI fragments, from A to Z, in descending order of fragment size (Arvin et al., 2007; Robertson, 2005).

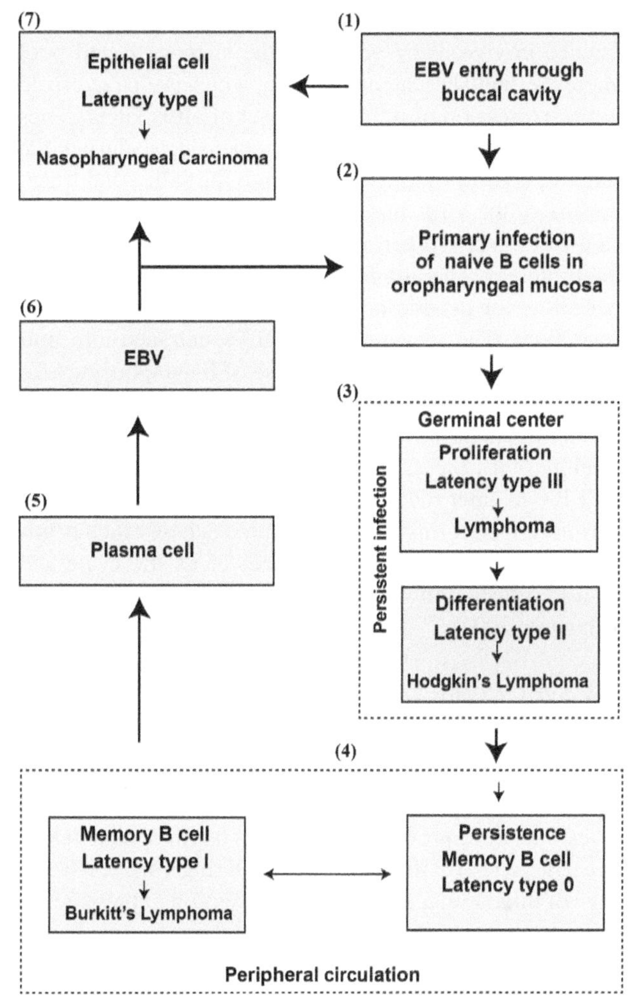

FLOW CHART 7.1 Various stages of the EBV lifecycle.

Entry into the Cell

EBV can infect both B-cells and epithelial cells. During primary infection, EBV infects B-lymphocytes through interaction of the glycoprotein gp350/220 with the complement receptor CD21. Following primary infection, EBV persists in the infected host as a lifelong asymptomatic infection (Arvin et al., 2007). To achieve long-term persistence in vivo, EBV colonizes the memory B-cell pool where it establishes latent infection, which is characterized by the expression of a limited subset of virus latent genes (Epstein, 1981). A low level of reactivation during the lytic cycle allows viral shedding into the saliva and transmission of the virus in vivo (Arvin et al., 2007). In vitro, EBV can transform peripheral human B-lymphocytes into indefinitely proliferating lymphoblastoid cell lines (LCLs) that allow for genetic manipulation of the virus (Robertson, 2005). The mechanisms of EBV entrance into epithelial cells are different from those of B-lymphocytes. To enter epithelial cells, viral protein BMRF-2 interacts with cellular β1 integrins that in turn triggers fusion of the viral envelope with the epithelial cell membrane, thus allowing EBV to enter the cell (Blake, 2010). There are several well-described forms of EBV latency, each of which is utilized by the virus at different stages of its life cycle and are also reflected in various EBV-associated malignancies (Damania, 2007).

EBV Lytic Replication

It has been shown that in newly infected cells, before the establishment of latency, EBV undergoes a short burst of lytic DNA replication. EBV expresses a small set of viral genes that were previously classified as immediate–early or early genes of the lytic cycle. During this pre-latent stage of infection, the immediate expression of these genes activates resting B cells and protects them from immediate activation of apoptosis (Blake, 2010). The cascade of events in the lytic phase of the EBV life cycle is divided into three phases of regulated gene expression: immediate–early, early, and late. The immediate–early gene products are transactivator proteins that trigger the expression of the early genes, the products of which include enzymes that are required for viral DNA replication (Blake, 2010). In turn, amplification of EBV DNA defines the boundary between early and late gene expression. During the late phase, viral structural proteins are expressed and assembled into virus particles into which the DNA is packaged prior to release of infectious virions. The principal switch from latency to productive infection involves activation of the immediate–early genes BZLF1 (Zta) and BRLF1 (Rta). These two proteins can be expressed from a major 2.9 kb and a minor 3.8 kb bicistronic R-Z RNA transcribed from the R-promoter (Rp) (Arvin et al., 2007). BZLF1, a viral transactivator protein, is involved in triggering the expression of the lytic genes and down-regulation of latent genes, culminating in cell death and the release of infectious virions. These proteins up-regulate the expression of other immediate early genes as well as their own expression. Both Zp and Rp are activated by ZEBRA, whereas Rta can up-regulate Zp and autoactivate its own synthesis. However, synergistic effects of ZEBRA and Rta induce maximum activation of the upstream Rp promoter. This synergism suggests that low levels of the two proteins are sufficient to trigger the lytic cascade. These immediate–early gene expressions in turn up-regulate the expression of early genes such as viral DNA polymerase (BALF5) and thymidine kinase. The major proteins of the lytic phase are the EBV DNA polymerase, BALF5, and the late lytic cascade major capsid protein, BcLF1 (Arvin et al., 2007). The lytic DNA replication of virion DNA starts from the lytic origin of replication (ori-lyt), which is distinct from the plasmid DNA replication origin (ori-P) that is used to maintain the episomal virus during latency. EBV ori-lyt lies within the BamHI H region of EBV DNA and contains two essential *cis*-acting regions, the BHLF1 promoter and a 0.5 kb distant region required for replication (Arvin et al., 2007). The ZREs present within the BHLF1 promoter are essential for ori-lyt-directed replication. The six-core replication protein together with ZEBRA is absolutely required for the amplification and replication of the viral genome from ori-lyt. ZEBRA is a member of the basic zipper family of transcription factors and binds as a homodimer to ZEBRA response elements (ZREs) within early gene promoters (Arvin et al., 2007). Moreover, ZEBRA is an essential component of lytic DNA replication, and its association with the EBV helicase targets ZEBRA to the viral DNA replication compartments within the nucleus. Generally, herpesviruses replicate their DNA in the G1-phase of the host cell cycle, which has been suggested to be advantageous to the virus due to the lack of competition with cellular DNA replication. As the key regulatory element, ZEBRA plays a critical role in the EBV lytic cycle by interacting with C/EBPα, which leads to an accumulation of p21CIP-1and G1 cell cycle arrest. Further studies suggest that apart from inhibiting cellular DNA synthesis, EBV induces an S-phase-like cellular environment during lytic replication. Latent EBV in B-cells spontaneously reactivates to switch to lytic replication, but the precise trigger for the induction is unknown. Many changes in physiological conditions or other non-related infections have been attributed to the triggering of spontaneous reactivation (Arvin et al., 2007). In vitro, latent EBV in B-cells can be reactivated by stimulating the B-cell receptor or by treating the cells with sodium butyrate or phorbol esters. EBV lytic replication does not inevitably lead to lysis of the host cell because the virions are produced by budding from the infected cell (Robertson, 2005).

EBV Latency

Latently infected B-cells maintain EBV genomes tethered to the host chromosome as 174 kb circular plasmids referred to as episomes, and express only a limited number of viral latent gene products (such as EBNA1, EBNA2, EBNA3A, EBNA3B, EBNA3C, and EBNA-LP) and three latency-associated membrane proteins (LMP1, LMP2A and LMP2B) (Blake, 2010). Transcripts referred to as BARTs (BamHI A rightward transcripts) from the BamHI A region (Bam A) of the EBV genome, as well as small non-polyadenylated RNAs, EBERs 1 and 2, are abundantly expressed during latency (Blake, 2010). Four patterns of EBV latency programs are recognized at present. In type I latency, represented mainly in Burkitt lymphoma (BL) cells, viral gene expression is restricted to the two EBER genes, the BART transcripts, and EBNA1 (EBV nuclear antigen 1). In latency II, additional expression of three latent-membrane proteins (LMP-1, LMP-2A, and LMP-2B) is observed and is most frequently seen in Hodgkin's lymphoma. Latency III is seen in lymphoproliferative diseases developed in immunocompromised individuals and EBV-transformed lymphoblastoid cell lines. In this group all six EBNAs, all three LMPs, and the two EBERs are expressed. Type IV latency is less strictly defined and pertains to infectious mononucleosis patients and to post-transplant lymphoproliferative disease. Some individuals also present with putative latency program (latency 0), which shows no detectable level of latent gene. The principal mediators of EBV-induced growth and cellular transformation of B-lymphocytes in vitro include EBNA2, EBNA3A, 3C, and LMP1 proteins (Arvin et al., 2007; Blake, 2010). The EBNA genes are important for transformation of primary B-lymphocytes, leading to transactivation and regulation of other cellular and viral genes. These proteins are involved in augmentation of the expression of genes coding for CD21, CD23, LMP1, and LMP2 proteins in B-lymphocytes. Growing evidence shows that EBV primarily persists in B-lymphocytes. Studies of EBV strains in donor–recipient pairs before and after bone marrow transplantation (BMT) have shown that the recipient's strain disappears from the oropharynx and is replaced by the donor's strain, thus indicating that the bone marrow B cells harbor EBV. Furthermore, patients with X-linked agammaglobulinemia (XLA), who are deficient in mature B cells, are found to be free of EBV infection, suggesting they are not able to maintain a persistent infection. Although much of the evidence described above implicates B cells as the site of persistence, a role for infection of squamous epithelial cells is also suggested by the detection of EBV in oral hairy leukoplakia (Arvin et al., 2007).

EBV Latent Genes

An understanding of EBV latent gene function is significant both to the factors contributing to the establishment of persistent infection and to the role of the EBV oncogenesis. Recent research on EBV unraveled the essential roles of EBNA2 and LMP1 in an in vitro transformation of B-cells, and highlighted roles for EBNA-LP, EBNA3A, EBNA3C, and LMP2A in this process. These studies confirm that EBV-induced B-cell transformation requires the cooperative effect of several latent genes. A brief description of EBV latent gene function involved in virus persistence and cellular transformation is as follows (Arvin et al., 2007).

EBNA1

EBNA1 is a DNA-binding protein that is required for the replication and maintenance of the episomal EBV genome. EBNA1 binds to ori-P, the episomal origin of viral replication. EBNA1 is a transcriptional transactivator and up-regulates Cp and the LMP1 promoter. The Gly–Gly–Ala repeat domain of EBNA1 is a *cis*-acting inhibitor of MHC Class I-restricted presentation, and regulates antigen processing via the ubiquitin–proteosome pathway. Targeting expression of EBNA1 to B cells in transgenic mice results in B-cell lymphomas, suggesting that EBNA1 might also have a direct role in oncogenesis. EBNA1 acts as a transcriptional activator of several viral and cellular genes (Arvin et al., 2007).

EBNA2

EBNA2 is a transcriptional activator of both cellular and viral genes, and up-regulates the expression of various B-cell antigens, including CD21 and CD23, as well as LMP1 and LMP2. EBNA2 also transactivates the Cp promoter, and thereby induces the switch from Wp to Cp detected early in B-cell infection. EBNA2 also transactivates c-myc oncogene, which is essential for EBV-induced B-cell proliferation and transformation (Arvin et al., 2007).

EBNA3 Family

EBNA3A and EBNA3C are essential for B-cell transformation in vitro, whereas EBNA3B is dispensable, EBNA3C has been shown to induce the up-regulation of both cellular (CD21) and viral (LMP1) gene expression. EBNA2 and the EBNA3 proteins work together to precisely control RBP-Jκ activity, thereby regulating the expression of cellular and viral promoters containing the RBP-Jκ cognate sequence. EBNA3C has been shown to interact with human histone deacetylase 1, which contributes to the transcriptional repression of Cp promoter. EBNA3C has also been shown to repress the Cp promoter and interacts with the retinoblastoma protein, pRb, to promote cell transformation (Arvin et al., 2007).

EBNA-LP

EBNA-LP is encoded as a variable size protein depending on the number of BamHI W repeats contained in a particular EBV isolate. The role of EBNA-LP in an in vitro B-cell transformation is not clear, but EBNA-LP is required for an efficient outgrowth of LCLs. EBNA-LP binds with pRb in LCLs and with both pRb and p53 in in vitro assays, but its expression appears to have no effect on the regulation of the pRb and p53 pathways (Arvin et al., 2007).

LMP1

LMP1 is the major transforming protein of EBV that behaves as a classical oncogene in rodent fibroblast transformation and is essential for EBV-induced B-cell transformation in vitro. LMP1 has pleiotropic effects when expressed in cells, resulting in induction of cell surface adhesion molecules and activation antigens, up-regulation of anti-apoptotic proteins (Bcl-2, A20), and stimulation of cytokine production (IL-6, IL-8). Recent studies have demonstrated that LMP1 functions as a constitutively activated member of the tumor necrosis factor receptor (TNFR) superfamily, activating a number of signaling pathways in a ligand-independent manner (Arvin et al., 2007).

LMP2

The gene encoding LMP2 yields two distinct proteins, LMP2A and LMP2B. Neither LMP2A nor LMP2B are essential for B-cell transformation. The LMP2A amino-terminal domain contains eight tyrosine residues, two of which (Y74 and Y85) form an immuno-receptor tyrosine-based activation motif (Matsuo & Itami, 2002). ITAM phosphorylation in the B-cell receptor (BCR) plays a central role in mediating lymphocyte proliferation and differentiation by the recruitment and activation of the src family of protein tyrosine kinases (PTKs). Expression of LMP2A in the B cells of transgenic mice abrogates normal B-cell development, thus allowing immunoglobulin-negative cells to colonize peripherally to lymphoid organs. This indicates that LMP2A can drive the proliferation and survival of B cells in the absence of signaling through the BCR. Together this suggests that LMP2 can modify the normal program of B-cell development to favor the maintenance of EBV latency and to prevent activation of the EBV lytic cycle (Arvin et al., 2007).

EBERs

Two small non-polyadenylated (non-coding) RNAs, EBERs 1 and 2, are mostly expressed in all forms of latency. However, the EBERs are not essential for EBV-induced transformations of primary B-lymphocytes. The EBERs assemble into stable ribonucleoprotein particles and bind the interferon-inducible, double-stranded RNA-activated protein kinase, PKR. EBER-mediated inhibition of PKR function is important for viral persistence, perhaps by protecting cells from interferon-induced apoptosis (Arvin et al., 2007).

BARTs

BARTs (BamHI A rightward transcripts) are derived from the BamHI A region (Bam A) of the EBV genome. BARTs were first identified in NPC tissue and subsequently in other EBV-associated malignancies such as Burkett's lymphoma and nasal T-cell lymphoma. BARTs encode a number of potential open reading frames (ORFs), including BARF0, RK-BARF0, A73, and RPMS1; however, protein products of these ORFs have not been identified (Arvin et al., 2007).

MicroRNAs

MicroRNAs (miRs) are small, noncoding RNA molecules of only 21–24 nucleotides in length, and have been shown to play a role in the posttranscriptional down-regulation of target mRNAs. The EBV miRs are arranged in two clusters within the viral genome, the BHRF1 cluster and the BART cluster that comprises the remaining 20 miRs located in the introns of the BART transcripts. These two clusters of EBV-encoded miRs are differentially expressed in cells exhibiting different forms of EBV latency. The BART cluster is expressed predominantly in latency I or II, whereas the BHRF1 miRs are associated with latency III. The expression levels of several miRs from both clusters are enhanced following induction of lytic infection. However, the precise function of these miRs remains unclear (Arvin et al., 2007).

Clinical Symptoms and Diagnosis of EBV

Clinical symptoms and diagnostic approaches differ according to the immune status of the patient. In healthy individuals, primary infection of EBV is most often asymptomatic. However, in immunocompromised individuals, EBV is associated with disorders with high rates of morbidity and mortality (Robertson, 2005). The spectrum ranges from benign B-cell hyperplasia resembling infectious mononucleosis (IM) to more classic malignant lymphomas. Molecular diagnostics is increasingly important for diagnosis and monitoring of patients affected by EBV-related diseases. As virus-specific treatments continue to be investigated, it becomes even more important to recognize these EBV-associated diseases so that proper clinical management decisions can be made. New molecular tests (qRT-PCR or EBV-encoded RNA (EBER)-RNA *in situ* hybridization) combined with traditional serological (heterophile antibody testing) or histochemical assays (immunofluorescence assay, *in situ* hybridization, or Southern blot) are helpful for diagnosis and monitoring of EBV-related diseases, depending on the clinical setting and the types of samples available for testing. *In situ* hybridization for EBER-RNA on biopsy samples, and more recently EBV viral load testing of blood samples, provides an accurate measure of the clinical status in patients. So far, the methods of first choice in routine EBV diagnostics are the immunofluorescence assay (IFA; still the "gold standard" method), and different enzyme immunoassay (EIA) techniques, including solid-phase ELISAs and Western blot analysis. While IFA or EIA is often used for screening, Western blot analysis is mainly performed for confirmation. Today, a number of manufacturers provide commercially available

EBV-specific IFAs and EIAs. Recently, *in situ* hybridization of EBER-RNA has become standard for EBV diagnosis in tumor cells, while qPCR procedures are used for EBV typing. Investigations are underway to better define the utility of these assays across the full spectrum of EBV-associated diseases. Additionally, gene expression profiling and array technology will likely improve our ability to sub-classify these diseases and predict responses to therapy. Some of the most common EBV-associated malignancies are listed as follows (Robertson, 2005).

Burkitt's Lymphoma

The association between Epstein–Barr virus and Burkitt's lymphoma (BL) has long been established (Bellan et al., 2003). This is a tumor of the jaw and face found in children of equatorial Africa which has rare occurrence elsewhere. The exact cause of this tumor is unclear, but there is probably a genetic factor as well as a malarial co-factor. Only 5% of BLs in the U.S. are associated with EBV, whereas in endemic areas such as eastern Brazil or Africa, nearly 90% of pediatric BLs carry EBV. The tumor cells show evidence of EBV DNA and tumor antigens, and the patients carry a much higher level of anti-EBV antibodies than other members of the population. Tumor cells are monoclonal and show a characteristic translocation between chromosomes 8 and 14 that places the c-myc oncogene under the control of the immunoglobulin heavy or light chain promoters, resulting in the up-regulation of c-myc oncogene in these cells (Damania, 2007). Burkitt's lymphoma is also associated with HIV infection, or occurs in post-organ transplant patients undergoing immunosuppressive therapies to prevent graft rejection. Burkitt's lymphoma may be one of the diseases associated with the initial manifestation of AIDS (Bellan et al., 2003; Damania, 2007).

Nasopharyngeal Cancer (NPC)

This is a tumor of epithelial cells of the upper respiratory tract that has highest incidence in the Cantonese of southern China, Hong Kong, Singapore, and Taiwan regions. There is a medium level of incidence in the Inuit populations in north America and in some populations in north Africa. The occurrence of NPC in the rest of the world is low, indicating that there may be a genetic predisposition for the development of EBV cancers in these populations, or there may be an involvement of environmental cofactors (Damania, 2007; Pipas and a., 2009).

Hodgkin's Lymphoma

Hodgkin's lymphoma, formerly known as Hodgkin's disease, is a B-cell lymphoproliferative disease of the lymphatic system in which 1% of the tumor population is comprised of Hodgkin/Reed–Sternberg (HRS) cells, which are derived from germinal center B cells. The HRS cells are multinucleated giant cells that have distinct nucleoli (Arvin et al., 2007; Damania, 2007). The first sign of Hodgkin's lymphoma is often a swollen lymph node, which appears without a known cause. The disease can spread to the nearby lymph nodes and later may spread to the spleen, liver, bone marrow, and other organs. There are three types of Hodgkin's lymphoma: lymphocyte-depleted, nodular sclerosis, and mixed-cellularity. Each of these differs in their association with EBV infection; 20% of nodular sclerosis Hodgkin's lymphomas are associated with EBV, whereas 100% and 70% of lymphocyte-depleted Hodgkin's lymphomas and mixed-cellularity Hodgkin's lymphomas, respectively, are associated with EBV infection (Damania, 2007). Immunosuppressed individuals, either due to HIV infection or immunosuppressive drugs in solid organ transplant patients, are at higher risk than the general population (Damania, 2007).

Infectious Mononucleosis

Infectious mononucleosis (also called "the kissing disease") occurs through the exchange of saliva containing EBV from infected individuals. Most people are exposed to the virus during their childhood, which is asymptomatic. However, the infected person sheds the virus from time to time throughout life. Infection acquired in adolescents and young adults leads to the development of infectious mononucleosis after 1–2 months of infection. The disease is characterized by malaise, lymphadenopathy, fever, and enlarged spleen and liver (Arvin et al., 2007; Pipas and a., 2009). The severity of disease often depends on age, with younger patients resolving the disease more quickly. Although infectious mononucleosis is usually benign, there can be complications. These include neurological disorders such as meningitis, encephalitis, myelitis, and Guillain–Barrè syndrome (Arvin et al., 2007). Secondary infections, autoimmune hemolytic anemia, thrombocytopenia, agranulocytosis, and aplastic anemia may also occur. In infectious mononucleosis, infected B cells are transformed, which proliferates and activates suppressor CD8 T cells. These T cells differ from normal T cells in appearance and are known as Downey cells (Pipas & a., 2009). This T-cell response results in enlarged lymph nodes as well as an enlarged liver and spleen. The activation of T-cells limits the proliferation of B cells and the disease resolves. Uncontrolled viral replication can lead to a severe syndrome with B-cell lymphoproliferation, leukopenia, and lymphoma. In patients with T-cell deficiency, X-linked lymphoproliferative disorder may occur. Patients with HIV infection or post-organ transplant patients who are under immunosuppressive therapies are at high risk to develop lymphoproliferative disorder (Pipas and a., 2009).

X-Linked Lymphoproliferative Disease

X-linked lymphoproliferative (XLP) syndrome is a rare immunodeficiency disease that is characterized by a susceptibility for fatal or near-fatal Epstein–Barr virus (EBV)-induced infectious mononucleosis in childhood, and a markedly increased risk for lymphoma or other lymphoproliferative diseases (Pipas and a., 2009). There is a mutation on the X chromosome that has been found to be associated with a T and NK cells lymphoproliferative disorder. The mutation is denoted as Xq25 on the long arm of the X chromosome. This mutation creates a deletion in the SH2D1A gene that codes for SLAM-associated protein (SAP). The SAP protein is important in the signaling events that activate T- and NK-cells and result in the modulation of IFN-γ. Persons with X-linked lymphoproliferative disorder have an impaired immune response to Epstein–Barr virus (EBV) infection, which often leads to death from bone marrow failure, irreversible hepatitis, and malignant lymphoma (Pipas and a., 2009).

Research Methods and Protocols

EBV infection during childhood is usually asymptomatic, however, it establishes a lifelong latent infection (Mesri et al., 2010; Robertson, 2005). EBV transforms peripheral human B-lymphocytes into indefinitely proliferating lymphoblastoid cell lines (LCLs) that allow for genetic manipulation of the virus. Lymphoblastoid cell lines (LCLs) are generated by Epstein–Barr virus (EBV) transformation of the B-lymphocytes within the peripheral blood lymphocyte (PBL) population (Robertson, 2005). as shown in Flow Chart 7.2: In step (1), latently EBV infected cells are grown in culture. (2) These cells are treated with 1 mM sodium butyrate (NaB) and 20 ng/mL of 12-O-tetradecanoylphorbol 13-acetate (TPA) to induce reactivation. (3) Culture supernatant is collected and filtered through a 0.4 μm filter to harvest the secreted virus followed by ultracentrifugation to concentrate the virus. (4) Purified EBV is then used for infecting human peripheral blood mononuclear cell (PBMCs). (5) Transformed B cells presenting EBV antigens are then selected. (6) The transformed B cells are further treated with interleukin (IL-2) for clonal expansion. (7) The selected EBV lymphoblastoid cell lines (EBV-LCLs) are analyzed for specificity and cryopreserved for further research. These lymphoblastoid cell lines (LCLs), derived from B-lymphocytes, are extensively used for in vitro research on gene expression studies and surrogate models to study genotype–phenotype relationships in humans (Robertson, 2005). Conventionally, the bacterial artificial chromosome (BAC) system is efficiently used to generate herpesvirus mutants by homologous recombination. In the BAC system, the entire EBV genome is cloned as a plasmid and propagated in *Escherichia coli*, and any mutation can be rapidly and precisely introduced into viral genes (Robertson, 2005). The EBV genome was first cloned as a bacterial artificial chromosome from EBV strain B95-5 derived from infectious mononucleosis (Arvin et al., 2007; Robertson, 2005). The establishment of this system enabled studies on epithelial cell background as well as on viral infection on other

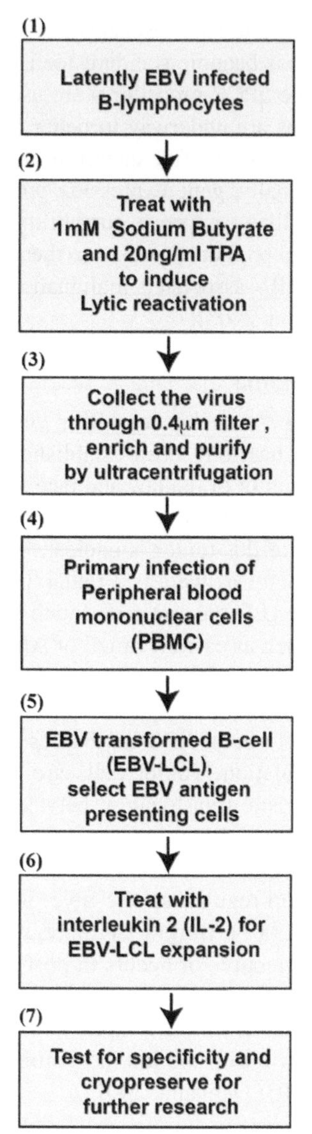

FLOW CHART 7.2 LCL generation by EBV transformation of B-lymphocytes.

B-lymphocytes like Ramos and Raji (Bellan et al., 2003; Robertson, 2005). These virus-producing cell lines provided the required tools to study the latent and lytic virus replication, virus production, and regulatory role of viral and cellular proteins in establishing virus latency, transformation, and tumorigenesis (Mesri et al., 2010; Robertson, 2005). In vitro molecular biology methods that reveal gene expression levels in LCLs are fairly comparable to the naturally occurring gene expression in primary B cells (Robertson, 2005).

CURRENT RESEARCH PERSPECTIVES

Understanding the different mechanisms by which DNA tumor viruses alter signal transduction pathways in malignant cells is important for detecting novel targets for cancer therapies (Saha et al., 2010). In addition to viral factors, host factors, genetic makeup, and geographical and environmental factors have also been found to interfere with the normal activities of signaling pathways leading to virus-induced cancers (Damania, 2007). Conventionally, for patients with AIDS-KS, highly active antiretroviral therapy is effectively used to reduce HIV infection together with up to a 90% reduction in Kaposi's sarcoma occurrence (Cai et al., 2010; Mesri et al., 2010). Furthermore, antiviral medications, including acyclovir, ganciclovir, and foscarnet are used effectively against EBV and KSHV (Cai et al., 2010; Mesri et al., 2010). Although our knowledge of viral oncoproteins and their associated cellular factors that are involved in carcinogenesis has greatly advanced, we are still lacking specific drugs that inhibit these viral proteins (Damania, 2007). Today, most anti-tumor therapies against virus-induced cancers target cellular proteins that have a role in these processes rather than the viral proteins. Several companies have efforts underway to develop small molecule drugs that target host proteins involved in cell cycle regulation, inflammatory response, proteasomes, and signal transduction pathways (Saha et al., 2010). Small molecule drugs such as nutlin-3a that target the p53-Mdm2 interaction, and Bortezomib (Velcade), a proteasome inhibitor, can be used against EBV as well as KSHV-mediated cancer cell lines. Bortezomib has been shown to inhibit proliferation and to induce apoptosis in KSHV-infected PEL cells (Saha et al., 2010). Additionally, PI3K inhibitor (LY294002), mTOR inhibitor (rapamycin), and IL-12 that negatively regulate the viral G-protein-coupled receptor (vGPCR) pathway have shown promising results against Kaposi's sarcoma (Saha et al., 2010). Other strategies like inhibition of matrix metalloproteinases (MMPs) and inhibitors for VEGF are also found to be effective against KS and HBV-associated hepatocellular carcinoma (HCC) (Saha et al., 2010). Interestingly, natural phenolic compounds like resveratrol (found in plants such as grapes) have been shown to be potent antiviral compounds against numerous viruses, including EBV, HSV, HIV, and influenza (Saha et al., 2010). Use of molecular drugs against cellular factors very often creates many undesirable off-target effects. Thus, developing vaccines or therapeutic drugs that specifically target DNA tumor viral proteins will be essential for effective protection against the virus with reduced drug cytotoxicity (Damania, 2007). Currently, promising vaccines are available only against human papillomavirus (HPV) and human hepatitis B (HBV), (Chang, 2009; Lowy and Schiller, 2006). and there are no vaccines available against EBV, KSHV, and the human polyomaviruses (Damania, 2007; Moore and Chang, 2010).

ETHICAL ISSUES

EBV infection and tumorigenesis is a complex process and is heavily affected by genetic constitution, socio-economic background, and geographical locations and subpopulation levels (Pipas and a., 2009). Generally, the prognosis and selection of the most appropriate treatment are assessed using both patient-related and standard tumor-related characteristics. The high mortality rate among patients with EBV-associated tumors is partly due to delays in diagnosis that result from the complexity of its initial clinical presentation. EBV-specific quantitative PCR is mostly employed to detect EBV from infected peripheral blood lymphocytes or biopsy samples (Robertson, 2005). A better understanding of viral gene expression in the context of EBV infections from diverse populations will prove useful in diagnosing and treating EBV-associated malignancies. Clinical research on patient samples or animal models involves an array of ethical issues and should be in accordance with World Health Organization (WHO) criteria. According to this, all clinical research involving patient samples should be approved and continuously monitored by a university or institutional ethical committee. Additionally, EBV has strict host specificity, which infects only humans; therefore, it is not practical to study the dynamics of protein expression during EBV infection and pathogenesis. However, an alternative method has been developed to study in vivo infection of EBV by generating a murine model known as "humanized mice." These mice have the potential to maintain human hematopoiesis, including human CD4+ leukocytes that can thereby support persistent EBV infection in vivo. Similar to clinical research, studies involving animal models also require approval from the institutional ethical committee. Above all, participating patients or their guardians should provide informed consent according to the institutional guidelines. Institutional ethical committees annually review the protocols and progress made in clinical research and have the authority to disapprove of a study if it is deviating from the original guidelines (Robertson, 2005).

TRANSLATIONAL SIGNIFICANCE

Early diagnosis and better treatment methods are critical for the control of EBV-associated malignancies. Currently, there are no vaccines available for EBV. Use of molecular drugs against cellular factors very often creates undesirable off-target effects. Thus, developing vaccines or therapeutic drugs that specifically target EBV viral proteins will be essential for effective protection against the virus. A better understanding of various molecular pathways that regulate EBV infection and tumoriegenesis are important for identifying novel targets for cancer therapy.

WORLD WIDE WEB RESOURCES

http://www.virology.net/

All the Virology on the WWW seeks to be the best single site for virology information on the Internet. It has a collection of all the virology-related web sites that might be of interest to our fellow virologists, and others interested in learning more about viruses.

http://www.virology.net/Big_virology/BVhomepage.html

This site talks about viral taxonomy and has a collection of virus pictures.

http://www.ncbi.nlm.nih.gov/books

NCBI Bookshelf provides free access to books and documents in life sciences and healthcare. A vital node in the data-rich resource network at NCBI, Bookshelf enables users to easily browse, retrieve, and read content, and spurs discovery of related information.

REFERENCES

Arvin, A., Campadelli-Fiume, G., Mocarski, E., Moore, E. S., Roizman, B., Whitley, R., & Yamanishi, K. (2007). *Human Herpesviruses*.

Bellan, C., Lazzi, S., De Falco, G., Nyongo, A., Giordano, A., & Leoncini, L. (2003). Burkitt's lymphoma: new insights into molecular pathogenesis. *Journal Clinical Pathology, 56*, 188–192.

Bhatia, S., Afanasiev, O., & Nghiem, P. (2011). Immunobiology of Merkel cell carcinoma: Implications for immunotherapy of a polyomavirus-associated cancer. *Current Oncology Reports, 13*, 488–497.

Blake, N. (2010). Immune evasion by gammaherpesvirus genome maintenance proteins. *The Journals of General Virology, 91*, 829–846.

Bouvard, V., Baan, R., Straif, K., Grosse, Y., Secretan, B., El Ghissassi, F., Benbrahim-Tallaa, L., Guha, N., Freeman, C., Galichet, L., Cogliano, V., & WHO International Agency for Research on Cancer Monograph Working Group (2009). A review of human carcinogens–Part B: biological agents. *The Lancet Oncology, 10*, 321–322.

Cai, Q., Verma, S. C., Lu, J., & Robertson, E. S. (2010). Molecular biology of Kaposi's sarcoma-associated herpesvirus and related oncogenesis. *Advances in Virus Research, 78*, 87–142.

Chang, M. H. (2009). Cancer prevention by vaccination against hepatitis B. *Recent Results in Cancer Research, 181*, 85–94.

Chang, Y., Cesarman, E. C., Pessin, M. S., Lee, F., Culpepper, J., Knowles, D. M., & Moore, P. S. (1994). Identification of herpesvirus-like DNA sequences in AIDS-associated Kaposi's sarcoma. *Science, 266*, 1865–1869.

Damania, B. (2007). DNA tumor viruses and human cancer. *TRENDS in Microbiology, 15*, 38–44.

Epstein, M. A. (1981). What we could do with an EBV vaccine. *Lancet, 1*, 759–761.

Feng, H., Shuda, M., Chang, Y., & Moore, P. S. (2008). Clonal integration of a polyomavirus in human Merkel cell carcinoma. *Science, 319*, 1096–1100.

Folkman, J. (2002). Role of angiogenesis in tumor growth and metastasis. *Seminars in Oncology, 29*, 15–18.

Goedert, J. J. (2001). *Infectious Causes of Cancer: Targets for Intervention*. Humana Press Inc.

Hahn, W. C., Counter, C. M., Lundberg, A. S., Beijersbergen, R. L., Brooks, M. W., & Weinberg, R. A. (1999). Creation of human tumour cells with defined genetic elements. *Nature, 400*, 464–468.

Hanahan, D., & Weinberg, R. A. (2000). The Hallmarks of Cancer. *Cell, 100*, 57–70.

Hausen, H. z (2001). Oncogenic DNA viruses. *Oncogene 20*.

Ignatovich, I. A., Dizhe, E. B., Akif'ev, B. N., Burov, S. V., Boiarchuk, E. A., & Perevozchikov, A. P. (2002). [Delivery of suicide thymidine kinase gene of herpes virus in the complex with cationic peptide into human hepatoma cells in vitro]. *Tsitologiia, 44*, 455–462.

Kao, J. -H. (2011). Molecular epidemiology of hepatitis B virus. *The Korean Journal of Internal Medicine, 26*, 255–261.

Lowy, D. R., & Schiller, J. T. (2006). Prophylactic human papillomavirus vaccines. *The Journal of Clinical Investigation, 116*, 1167–1173.

Martin, D., & Gutkind, J. S. (2008). Human tumor-associated viruses and new insights into the molecular mechanisms of cancer. *Oncogenesis, 2*(27 Suppl), S31–42.

Matsuo, T., & Itami, M. (2002). Seropositivity of human herpesvirus-8 in patients with uveitis. *Ocular Immunology and Inflammation, 10*, 197–199.

Mesri, E. A., Cesarman, E., & Boshoff, C. (2010). Kaposi's sarcoma and its associated herpesvirus. *Nature Reviews Cancer, 10*, 707–719.

Mitxelena, J., Gomez-Ullate, P., Aguirre, A., Rubio, G., Lampreabe, I., & Diaz-Perez, J. L. (2003). Kaposi's sarcoma in renal transplant patients: experience at the Cruces Hospital in Bilbao. *International Journals of Dermatology, 42*, 18–22.

Moore, P. S., & Chang, Y. (2010). Why do viruses cause cancer? Highlights of the first century of human tumour virology. *Nature Reviews Cancer, 10*, 878–889.

Narisawa-Saito, M. K. ,T. (2007). Basic mechanisms of high-risk human papillomavirus-induced carcinogenesis: roles of E6 and E7 proteins. *Cancer Science, 98*, 1505–1511.

Parkin, D. M. (2006). The global health burden of infection-associated cancers in the year 2002. *International Journal of Cancer, 118*, 3030–3044.

Pipas, B. D., & a, J. M. (2009). *DNA Tumor Viruses, 794*.

Robertson, E. S. (2005). *Epstein–Barr Virus*. Caister Academic Press.

Saha, A., Kaul, R., Murakami, M., & Robertson, E. S. (2010). Tumor viruses and cancer biology Modulating signaling pathways for therapeutic intervention. *Cancer Biology & Therapy, 10*, 1–18.

Shepard, C. W., Simard, E. P., Finelli, L., Fiore, A. E., & Bell, B. P. (2006). Hepatitis B virus infection: Epidemiology and vaccination. *Epidemiologic Reviews, 28*, 112–125.

Silva, R. L. d (2011). Polyoma BK virus: an emerging opportunistic infectious agent of the human central nervous system. *The Brazilian Journal of Infectious Diseases, 15*, 276–284.

Stevenson, P. G. (2004). Immune evasion by gamma-herpesviruses. *Current Opinion in Immunology, 16*, 456–462.

Vaughn, C. P., & Elenitoba-Johnson, K. S. (2001). Intrinsic deoxyguanosine quenching of fluorescein-labeled hybridization probes: a simple method for real-time PCR detection and genotyping. *Laboratory Investigation, 81*, 1575–1577.

Verma, S. C. ,L.K., & Robertson, E. (2007). Structure and function of latency-associated nuclear antigen. *Current Topics in Microbiology and Immunology, 312*, 101–136.

Yi-Bin, Chen, Rahemtullah, A., & Hochberg., Ephraim (2007). Primary effusion lymphoma. *The Oncologist, 12*, 569–576.

FURTHER READING

S Robertson, Erle (Ed.), (2005). *Epstein–Barr virus*. Caister Academic Press. ISBN-1-904455-03-4.

Human Herpesviruses-Biology, Therapy, and Immunoprophylaxis.Ann Arvin, 2007 Edited by Ann Arvin, Gabriella Campadelli-Fiume, Edward Mocarski, Patrick S. Moore, Bernard Roizman, Richard Whitley, and Koichi Yamanishi. Cambridge: Cambridge University Press; 2007. ISBN-13: 978-0-521-82714-0.

Medical Microbiology. (1996). In Samuel Baron (Ed.), *University of Texas Medical Branch at Galveston, Galveston, Texas* (4th ed.). Galveston: University of Texas Medical Branch at Galveston. ISBN-10: 0-9631172-1-1.

GLOSSARY

Cytopathic Effect Refers to degenerative changes in cells, especially in tissue culture.

Deregulation The act of freeing from regulation.

Hit-and-Run Viruses Viruses that can initiate cancers or play a role in their development, but then disappear from the host.

Immune Surveillance The process by which the immune system continually recognizes and removes malignant cells.

Intravasation The invasion of cancer cells through the basal membrane into a blood or lymphatic vessel.

Knockdown Gene knockdown refers to techniques by which the expression of one or more of an organism's genes is reduced.

Oncogene A gene that (in certain circumstances) transforms a normal cell into a tumor cell.

Onco-Proteins Gene products that cause the transformation of normal cells into cancerous tumor cells.

Secondary Tumor Metastasis is one of the hallmarks of malignancy. This new tumor is known as a metastatic or secondary tumor.

ABBREVIATIONS

AIDS Acquired Immunodeficiency Syndrome
BAC Bacterial Artificial Chromosome
BART BamHI A Rightward Transcripts
EBV Epstein–Barr Virus
EBNA EBV-Encoded Latent Antigen
HBV Hepatitis B Virus
HCV Hepatitis C Virus
HIV Human Immunodeficiency Virus
HPV Human Papilloma Virus
HTLV-1 Human TL Virus
KSHV Kaposi's Sarcoma-Associated Herpesvirus
LANA Latency-Associated Nuclear Antigen
LCL Lymphoblastoid Cell Lines (LCLs)
MCPyV Merkel Cell Polyomavirus
vGPCR Viral G-Protein-Coupled Receptor
TNFR Tumor Necrosis Factor Receptor
WHO World Health Organization

LONG ANSWER QUESTIONS

1. What is virus transformation? How do DNA tumor viruses transform cells?
2. How do small DNA tumor viruses differ from large DNA tumor viruses?
3. What are the different DNA tumor viruses that infect humans?
4. What are viral oncoproteins? How do they contribute to tumorigenesis?
5. A person with AIDS is diagnosed with acute B-cell lymphoma: What could be the causative agent and why?

SHORT ANSWER QUESTIONS

1. Which of the DNA tumor viruses can cause B-cell lymphoma in humans?
2. Which tumor virus has been associated with cervical carcinomas?
3. What is the necessary prerequisite for defining an infection as latent?
4. Which DNA tumor virus has been associated with hepatocellular carcinomas?
5. Why can excessive exposure to the sun cause Merkel cell carcinoma (MCC)?

ANSWERS TO SHORT ANSWER QUESTIONS

1. DNA tumor viruses that belong to the families gammaherpesvirus, Epstein–Barr virus, and Kaposi's sarcoma-associated herpesvirus (KSHV) primarily cause B-cell lymphoma in humans.
2. Infection by human papilloma virus is the most common sexually transmitted disease, and can lead to the development of cervical cancer.
3. During latency the virus remains dormant and persists in a highly ordered chromatin state as an episome whose propagation is dependent on its ability to hijack the replicative machinery of its host. During latency the virus encodes only a few latency-associated proteins that are known to be tumorigenic. Among DNA tumor viruses, latency is well studied in herpesvirus family members EBV and KSHV.
4. Hepatitis B virus (HBV), a member of the Hepadnavirus family, causes an infectious inflammatory disease of the liver. Patients with chronic hepatitis B are at high risk of developing hepatocellular carcinoma (HCC).
5. Merkel cell carcinoma (MCC) occurs most often on the sun-exposed face, head, and neck. MCPyV is the first polyomavirus that has been shown to integrate into human genomic DNA. The virus then undergoes two sequential mutational events that result in persistent T-Ag expression; this plays a key role in turning asymptomatic viral infection into an aggressive MCC. Excessive exposure to sun can induce this mutational event, leading to MCC.

Chapter 8

Animal Models for Human Disease

Mohammad Reza Khorramizadeh* and Farshid Saadat†

*Endocrinology and Metabolic Research Institute, Tehran University of Medical Sciences, Tehran, Iran and Department of Medical Biotechnology, School of Advanced Technologies in Medicine, Tehran University of Medical Sciences, Tehran, Iran, †Department of Immunology, School of Medicine, Guilan University of Medical Sciences, Rasht, Iran

Chapter Outline

SUMMARY

A useful animal model for disease must be similar in its pathology to disease conditions in humans. Experimental animal models of rheumatoid arthritis and multiple sclerosis are useful for a better understanding of disease mechanisms and for evaluating therapeutic efficacy of new and emerging drugs.

WHAT YOU CAN EXPECT TO KNOW

This chapter introduces the subject of animal models used in different human disease studies. The benefits of animal models are that one can study the mechanisms of diseases as well as test new and emerging drugs for their therapeutic efficacy. Rheumatoid arthritis (RA) is one of the autoimmune disorders for which different animal models are available, and each and every model has its merits and demerits. We will discuss the importance and induction of RA by collagen. Collagen-induced arthritis in animal models reflects characteristic features of RA patients.

Multiple sclerosis (MS) is another debilitating disease that affects the central nervous system (CNS) of humans. The animal model used to study MS is known as Experimental Allergic Encephalomyelitis (EAE). We have given the details of how to induce EAE and also how to apply this animal model. Although various ethical issues are involved with the development and use of animal models to study human disease, the importance of animal models can neither be ignored nor denied.

Animal Biotechnology. http://dx.doi.org/10.1016/B978-0-12-416002-6.00008-0

HISTORY AND METHODS

INTRODUCTION

As human beings, our body is comprised in such a manner that cells cannot be considered as a separate entity. Physiologically, homeostasis is the reason that these components live and perform their functions within that environment. Disruption of this process leads to fatal conditions and is considered a disease. To investigate the mechanism of disease and to find the means to reverse adverse conditions, various strategies are used (e.g. cell-based assays and tissue culture studies). Although these models can provide useful information, they fail to address various physiological conditions and the complex interactions among different cell types of tissues and organs.

Ideally a useful animal model for any disease has to have pathology similar to the disease condition in humans. Use of animals in research has a long history that dates back to the fourth century B.C. In the 1600s, William Harvey used animals to describe the blood circulatory system. Many scientists, such as Louis Pasteur and Emil von Behring, have used animal models for experimental purposes to prove their hypotheses.

Animal models are good for understanding disease mechanisms and treatment and for overcoming the limitations of clinical trials that use human subjects. For example, experimental animal models for diseases like rheumatoid arthritis or multiple sclerosis have been successfully employed to screen new bioengineered, chemical, or herbal therapeutics that might have the potential for treatment of human patients. So far, more than 40,000 studies have been reported in the NCBI database; they use animal models for different diseases. Animal model studies have been the main reason for a better understanding of disease mechanisms. Animal models of disease can be divided into two categories: (1) spontaneous disease models, and (2) induced disease models. In the case of induced disease models, induction can occur by various agents, both chemical and biological. This chapter discusses some of the most important animal models.

RHEUMATOID ARTHRITIS

Rheumatoid arthritis (RA) is an autoimmune disorder with progressive occurrence that preferentially affects peripheral joints. In spite of the fact that RA is severe and crippling and affects large numbers of people, very little knowledge about its etiology and pathogenesis is available in the literature.

Epidemiology and Etiology

Rheumatoid arthritis affects about 1% of the population. The ratio of the prevalence of RA in males and females is 1 to 2.5. RA can occur at any age, but it's mainly reported to affect the 40–70-year-old age group. No doubt the incidence has been reported to increase with age. The etiology of RA is unknown, but it has been predicted that genetic and environmental factors play an important role in the onset of RA. Recent advances have identified genetic susceptibility markers both within and outside of the major histocompatibility complex (MHC). Human leukocyte antigen (HLA) genes located on chromosome 6p have been found to have a strong association with rheumatoid arthritis. Individuals carrying HLA-DR4 and HLA-DR1 alleles have been shown to have a higher risk of RA. A positive correlation has been suggested for the role of HLA in terms of the severity of RA rather than onset of the disease. Although, the data regarding this conclusion is inconsistent, some of the studies have shown associations between tumor necrosis factor (TNF) alleles and rheumatoid arthritis. Other genes like those for corticotrophin-releasing hormone, interferon (IFN)-γ, and interleukin-10 (IL-10) have also been implied for RA. It can be concluded that the role of genetic components in RA is modest at the best (Viatte et al., 2013).

Epigenetics is another important factor that contributes to RA. In the case of identical twins, RA has not been shown to have 100% concordance; therefore, the role of non-genetic factors has also been implicated in the etiology of RA (Meda et al., 2011). Throughout the world, rheumatoid arthritis is more common in women than men. This indicates that hormones may play an important role in the development of the disease. Pregnancy has also been considered as a risk factor for rheumatoid arthritis. Studies show that onset of RA is rare during pregnancy, but risk increases after delivery. Smoking is associated with increased incidences of RA, especially in men. On the contrary, populations that consume a diet high in omega-3 fatty acids have been reported to be protected from rheumatoid arthritis. From experimental models in animals, a large number of infectious agents such as viruses and bacteria have also been suggested to trigger or contribute to the development of rheumatoid arthritis. However, no relationship between infectious agents and the development of RA has been found.

Pathogenesis

An inflamed synovium is central to the pathophysiology of rheumatoid arthritis. Histologically RA shows pronounced angiogenesis, cellular hyperplasia, an influx of inflammatory leucocytes, and changes in the expression of cell–surface adhesion molecules, proteinases, proteinase inhibitors, and many cytokines. Synovial changes in rheumatoid arthritis vary with disease progression. In the first weeks of the disease, tissue edema and fibrin deposition are prominent and can manifest clinically as joint swelling and pain. Within a short period, the synovial lining becomes hyperplastic, commonly becoming ten or more cells deep and consisting of type A (macrophage-like) and type B (fibroblast-like) synoviocytes that produce glycosaminoglycans (e.g. hyaluronan, as reported to be present in synovial tissue and synovial fluid). The sub-lining also undergoes alterations for its cellularity, both in cell type and cell numbers, with prominent infiltration of mononuclear cells, including T cells, B cells, macrophages, and plasma cells.

The abundance and activation of macrophages at the inflamed synovial membrane correlates significantly with the severity of the disease. Activated macrophages overexpress major histocompatibility complex (MHC) class II molecules and produce pro-inflammatory or regulatory cytokines and growth factors (IL-1, IL-6, IL-10, IL-13, IL-15, IL-18, TNF-α, and granulocyte macrophage colony stimulating factor (GM-CSF)), chemokines (IL-8, macrophage inflammatory protein 1 (MIP-1), monocyte chemoattractant protein 1 (MCP-1)), metalloproteinases, and neopterin. These biomolecules are routinely detected in inflamed joints. Most of the T cells infiltrating the rheumatoid synovium express CD45RO and CD4, which is an indication that the T-cell subset present in the synovium is memory helper T cells. Surprisingly, 10–15% of the T cells present in the case of the synovium have granzymes A and perforins. This 10–15% of cells present in the synovium represents cytotoxic T-cell subsets. Therefore, it can be concluded that CD8-expressing cells are infrequent in the synovium. In the synovial fluid of rheumatoid arthritis patients, CD4 and CD8 T cells are equally represented. TCRα/ß is expressed on most of the T cells while only a minority of cells show TCRγ/δ expression. It has, however, been found that the expression of TCRγ/δ is increased in the synovium of patients with active RA. Synovial-vessel endothelial cells transform into high endothelial venules early during the course of disease. High endothelial venules are specialized post-capillary venules usually present in secondary lymphoid tissue or inflamed non-lymphoid tissues; these venules facilitate the transit of leucocytes from the bloodstream into tissues.

The formation of locally invasive synovial tissue (i.e. pannus) is a characteristic feature of rheumatoid arthritis. Pannus is involved in the erosion of joints in rheumatoid arthritis. Pannus is histologically distinct from other regions of the synovium and shows phases of progression. Initially, there is penetration of cartilage by synovial pannus, which is composed of mononuclear cells and fibroblasts, with a high-level expression of matrix metalloproteinases (MMPs) by synovial lining cells. In later phases of the disease, cellular pannus can be replaced by fibrous pannus comprised of a minimally vascularised layer of pannus cells and collagen overlying cartilage. The tissue derivation of pannus cells has not been fully elucidated, although they are thought to arise from fibroblast-like cells (type B synoviocytes). *In vitro* work shows that these fibroblast-like synoviocytes have anchorage-independent proliferation and loss of contact inhibition, which is a phenotype usually found in transformed cells. However, the molecular pathogenic mechanisms driving pannus formation still remains poorly understood.

Clinical Manifestations

The range of presentations of rheumatoid arthritis is broad, but disease onset is insidious in most cases, and several months can elapse before a firm diagnosis can be ascertained. The predominant symptoms are pain, stiffness, and swelling of peripheral joints. Although articular symptoms are often dominant, rheumatoid arthritis is a systemic disease. Active rheumatoid arthritis is associated with a number of extra-articular manifestations, including: fever, weight loss, malaise, anemia, osteoporosis, lymphoadenopathy, and so on.

The clinical course of the disorder is extremely variable, ranging from mild, self-limiting arthritis to rapidly progressive multisystem inflammation with profound morbidity and mortality. Analyses of clinical course and laboratory and radiological abnormalities have defined as negative prognostic factors for progressive joint destruction; unfortunately, none of these are reliable enough to allow therapeutic decision-making. Frequent assessment of disease symptoms and responses to therapy is crucial for successful and long-term management of rheumatoid arthritis. Joint destruction from synovitis can occur rapidly and early in the course of the disorder; radiographic evidence is present in more than 70% of patients within the initial 2 years. More sensitive techniques such as magnetic resonance imaging (MRI) can identify substantial synovial hypertrophy, bone edema, and early erosive changes as early as 4 months after the onset of disease. These radiographic changes predate misalignment and functional disability by years; by the time physical deformity is evident, substantial irreversible articular damage has commonly occurred. Furthermore, biopsy analysis of clinically symptom-less knee joints in patients with early rheumatoid arthritis show active synovitis, highlighting the poor correlation between clinical assessment and disease progression, and the rapid development of polyarticular synovitis.

Treatment

The past decade has seen a major transformation in the treatment of rheumatoid arthritis in terms of approach and choice of drugs. The previous therapeutic approach generally involved initial conservative management with nonsteroidal anti-inflammatory drugs (NSAIDs) for several years; disease-modifying anti-rheumatic drugs (DMARDs) were withheld until clear evidence of erosion was seen. DMARDs were then added individually in slow succession as the disease progressed. This form of treatment has been supplanted by early initiation of DMARDs and combination DMARD therapy in patients with the potential for progressive disease. The idea of early intervention with DMARDs has been validated in several randomized trials. DMARDs contain medications from different classes of drugs including methotrexate, gold salts, hydroxychloroquine, sulfasalazine, ciclosporin, and azathioprine. DMARDs are often partly effective and poorly tolerated for long-term therapy. In meta-analyses of dropout rates from clinical trials, 20–40% of patients discontinued use of DMARDs assessed as monotherapy during the duration of the trial; even in clinical practice, the median duration of DMARD monotherapy was less than 2 years for non-methotrexate agents. Although there are many reasons for lack of long-term adherence to treatment, poor efficacy, delayed onset of action, and toxic effects are major limitations. Additionally, DMARDs therapy requires patients to undergo frequent monitoring of blood and physical examinations for toxic effects of treatment protocol. Results from clinical trials showed that DMARD therapy decreased markers of inflammation such as erythrocyte sedimentation rate and swollen joint counts, and that improved symptoms in a selected subset of patients; however, most patients continued to show progression of irreversible joint destruction on radiography. These findings illustrate the consequences of progressive disease, and have shown the need for development of new and more effective therapies based on the therapeutic principles used for oncology; it means that treatment protocols for RA patients require use of several therapeutic agents from different classes to be used in combination. Recent studies have shown that combination therapy of biological DMARDs like TNF-α inhibitors with methotrexate have clear-cut benefits with tolerable toxic effects. Treatment with agents that can block TNF-α function has proved to be highly effective against RA. Further studies reported down-regulation of synovial GM-CSF, IL-6, and IL-8, suggesting that TNF-α supports production of other pro-inflammatory cytokines. However, the mechanisms behind the clinical effect of the TNF-α-blocking treatment are not fully understood. In an animal model, TNF-α-blocking agents such as etanercept (a soluble TNF-α receptor) and infliximab (a monoclonal antibody) reduce the expression of vascular adhesion molecules and inhibit the spontaneous production of IL-1 and IL-6. Patients with a new onset of symptoms and those with disease of several years' duration and who had failed previous DMARD therapy all benefited. These results suggest that patients in many stages of disease progression can benefit from combination therapy (Chiu et al., 2012).

Experimental Models

In order to study the pathogenesis of RA one can use different animal models. There are many experimental models that resemble RA in different respects. Since RA is a heterogeneous disease there is probably a need for different animal models that each reflect a characteristic feature of a particular subgroup of RA patients or illustrate a particular aspect of the disease.

Spontaneous Models

Despite the fact that RA is not a spontaneously developing disease, spontaneously developing models for arthritis may be useful to study the role of genetics in the development of the disease. Transgenic mice constitutively expressing TNF-α have been shown to develop arthritis spontaneously (Keffer et al., 1991). Transgenic mice expressing a TCR specific for bovine pancreas ribonuclease develop spontaneous arthritis that is mediated by antibodies (Korganow et al., 1999). This model is particularly interesting because it demonstrates that T cells specific for a ubiquitous antigen may induce an organ-specific autoimmune disease. Expression of the gene product causes an up-regulation of several cytokines (IL-1, IL-6, TGF-β1, IFN-γ, and IL-2) and subsequent development of arthritis (Iwakura et al., 1995). There are some other spontaneous models for arthritis in non-transgenic mice (Bouvet et al., 1990).

Induced Models

Arthritis can induce by complete Freunds' adjuvant (CFA). Pearson (1956) described this model for the first time. Subsequently, it was demonstrated that other adjuvants, such as IFA, pristane, or squalene, could also induce arthritis (Carlson et al., 2000). Microbially derived products such as lipopolysaccharide (LPS), muramyle dipeptide (MDP), and trehalosedimycolate (TDM) can also induce arthritis when given with mineral oil (Lorentzen et al., 1999; Kohashi et al., 1980).

Collagen-induced arthritis (CIA) is normally induced by immunization of susceptible mouse (e.g. DBA/1) or rat (DA, Lewis) strains at the base of the tail. The inoculum used for immunization contains both adjuvant and collagen type II. The adjuvant has to be sufficiently strong to cause tissue destruction as well as induction of a strong pro-inflammatory immune response (Holmdahl and Kvick, 1992; Kleinau et al., 1995). Susceptibility to CIA is dependent on both MHC (class II region) and non-MHC genes (Lorentzen and Klareskog, 1996). Antibodies against collagen II are essential for the development of CIA. This fact has been demonstrated by passive transfer of anti-CII antibodies, which results in synovitis (Svensson et al., 1998). T cells are also important for CIA development during early stages of disease progression (Goldschmidt and Holmdahl, 1994). The dependence of both T and B cell responses has also been demonstrated in the same model (Seki et al., 1988).

Pathology of CIA

Inflamed joints in CIA are infiltrated by inflammatory cells that accumulate in the synovial membrane and fluid, similar to RA. The most frequent cell type in synovial fluid is granulocyte. There is also a great infiltration of leukocytes into the synovial membrane. These cells have signs of an activated phenotype of RA since MHC class II molecules are expressed (Klareskog and Johnell, 1988). In addition, there is intense production of macrophage-derived cytokines in inflamed joints (e.g. TNF-α and IL-1β) (Mussener et al., 1997, Ulfgren et al., 2000). A small number of T-cells are encountered, and some of these T cells have IL-2 receptor α chain up-regulated. The disease shows a thickened synovial membrane that subsequently forms a pannus on the cartilage surface (Holmdahl et al., 1998, 1991). In both CIA and RA, cartilage and bone destruction occurs mainly at the cartilage–pannus junction. There are some features of the pathology of CIA that differ from what is usually observed in RA (e.g. extra-articular manifestation). Although the compatibility of the CIA model to human RA has been argued, many pathological features of CIA are similar to rheumatoid arthritis.

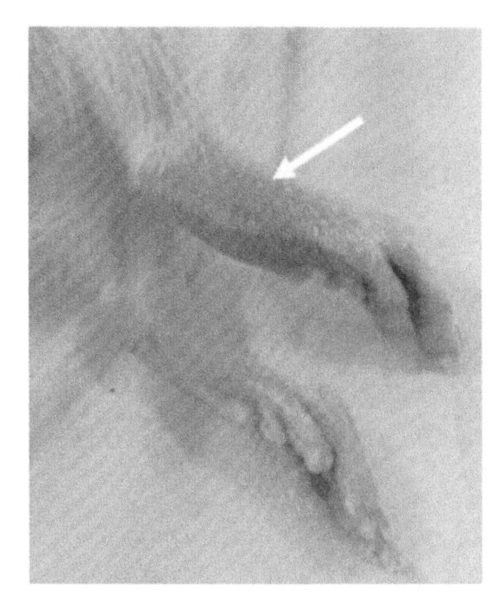

FIGURE 8.1 Inflammatory edema (white arrow).

Methodology and Protocols

Experimental collagen-induced arthritis was initiated by injecting bovine collagen type II at the base portion of the tail of the animal (Saadat et al., 2005). Male Lewis rats weighing about 160–180g were used. After induction of CIA, animals were divided randomly into four or more groups based on experimental design. At least four different groups were needed, including: a control group without arthritis, animals with collagen-induced arthritis, CIA animals with treatment, and CIA animals treated with methotrexate as a positive control.

Sample preparation: bovine collagen type II (CII) was dissolved in 0.1 M acetic acid at a concentration of 2 mg/mL by stirring overnight at 4°C (dissolved CII can be stored at −70°C if it has to be used at a later time). Before injecting the animals, CII was emulsified with an equal volume of complete Frennd's adjuvant (CFA). For induction of CIA, on Day 1 rats were injected intradermally at the base of the tail with 100 μL of emulsion (containing 100 μg of CII). After 12–16 days animals showed the development of inflammation at peripheral joints (Figure 8.1). On Day 21 a booster injection of CII in CFA was administered. This model was used to evaluate the anti-RA effect by giving intraperitoneally injections of test materials (e.g. chemical or herbal extracts). Methotrexate was a control used to evaluate the effect of the test compound and to compare the efficacy of the new compound with methotrexate. In this model the test compound was given from Day 25, where the frequency, route of administration, and dose could be selected as needed. The end point and days for evaluation of different parameters were selected; one of the most common points was Day 35 (Flow Chart 8.1). The paws and knees were then removed for histopathological assay.

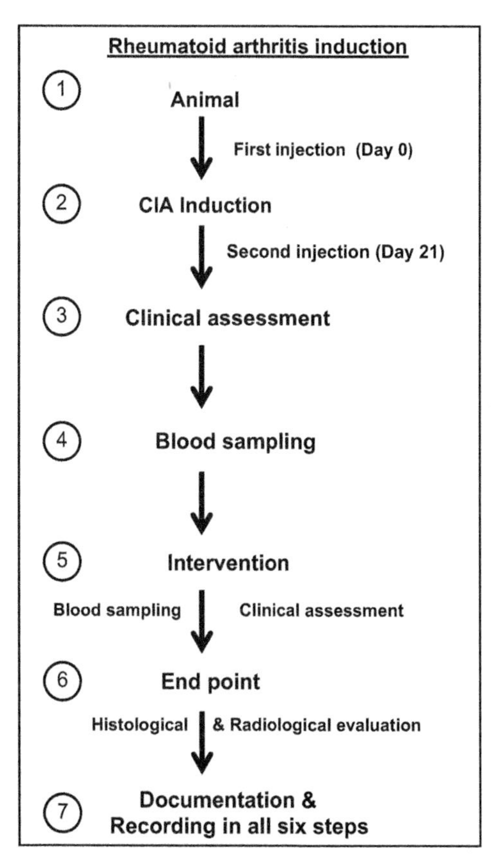

FLOW CHART 8.1 CIA induction for rheumatoid athritis.

Clinical Assessment of CIA

The visual observation can be done by using the macroscopic system as given in Table 8.1 below. The scaling to

TABLE 8.1 Scaling to Record the Signs of Collagen Induced Arthritis (CIA)

Scale	Observed Symptoms
0	No signs of arthritis
1	Swelling and/or redness of the paw or 1 digit
2	Involvement of 2 joints
3	Involvement of > 2 joints
4	Severe arthritis of the entire paw and digits

TABLE 8.2 Scoring for Histological Evaluation of Collagen Induced Arthritis (CIA)

Scale	Histological Evaluations
0	No signs of inflammation
1	Infiltration of inflammatory cells in the joints
2	Swelling and tissue edema in the joints
3	Bone erosion in the joints

record the observation should be from 0 to 4 for each paw (Szabó et al., 1998).

Histological Assessment

On Day 35, animals were anesthetized with sodium pentobarbital (45 mg/kg intraperitoneally) and euthanized. Blood was collected by intracardiac puncture, and paws and knees were removed, trimmed, and fixed in 10% buffered formalin, decalcified and then embedded in paraffin, sectioned at 5 μm, and stained with hematoxylin and eosin for histological examination. Joint damage was assessed based on synovial hypertrophy, pannus formation, inflammatory cell infiltration, and cartilage and subchondral bone destruction. Joint erosion was graded on a scale of 0–3 for each limb (Table 8.2) according to the severity of damage (Figure 8.2).

Radiographic Evaluation

Radiological scoring was performed by an investigator who was blind to the treatment protocol (on Day 35). Radiographical analysis of affected joints in control rats typically showed soft tissue swelling, joint space narrowing, reduced lucency due to demineralization, and areas of recalcification indicative of new bone formation. A score was assigned to each joint on the basis of the information in Table 8.3.

Scores were 0–3 per joint (0, normal; 3, maximum joint destruction).

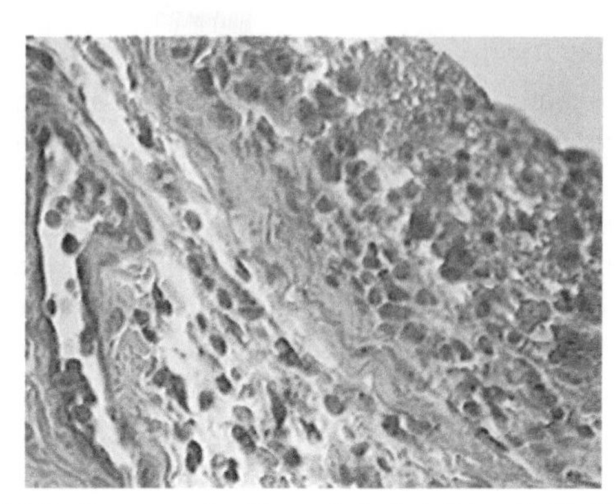

FIGURE 8.2 Representative histopathological slides of a hind limb joint of a healthy Lewis rat.

TABLE 8.3 Scoring for Radiological Evaluation of Collagen Induced Arthritis (CIA)

Scale	Radiological Evaluations
0	No alteration of joint
1	Swelling of soft tissue
2	Joint space narrowing
3	New bone formation and bone destruction

MULTIPLE SCLEROSIS

Multiple sclerosis (MS) is a chronic inflammatory demyelinating disorder of the central nervous system (CNS) that affects ~ 2.5 million individuals worldwide (National Multiple Sclerosis Society, 2008). Similar to the affected population in other autoimmune diseases, twice as many women as men have MS. Multiple sclerosis, like many other diseases, has existed as long as human life. In the 1860s, the first report by Dr. Jean-Martin Charcot certified MS as a disease. A patient of his, who suffered an unusual symptom, died. After dissection, brain lesions were discovered. He called the disease *sclerose en plaques.* Myelin was subsequently discovered, although its exact role was not recognized. About one century of research resulted in the discovery of MS as an autoimmune disease. Since this finding, extensive studies on MS have revealed some aspects of disease pathogenesis and etiology. Steroids and disease-modifying agents were used. However, this debilitating disease is not completely understood and MS remains an incurable neurological disorder.

Epidemiology and Etiology

Multiple sclerosis is the most common inflammatory demyelinating disease of the CNS in Europe and North America. The prevalence of MS in North America and Europe is ~ 80–100 per 100,000 people. However, prevalence of

MS is not globally uniform, geographically decreases in latitudes, and has been observed in only ~ 1–2 per 100,000 individuals in Africa and Asia. The etiology of MS is unknown. Both genetic involvement and environmental factors have been indicated in MS. The only consistent correlation of involvement of the MHC locus is the MHC class II allele HLA–DR2, which reflects a linkage with MS. Some other factors like dietary components (e.g. milk), pathogens like human herpes virus 6 (HHV–6), measles virus, Epstein–Barr virus, and chlamydia have been implied as etiological factors. However, the association between any of these agents with MS is debatable.

Pathogenesis

The presence of CNS inflammation is a hallmark of MS. This inflammatory process greatly increases in the CNS by activation and deregulation of different cell types of the immune system. Activation and entry of myelin-specific lymphocytes into the CNS cause damage to oligodendrocytes, leading to demyelination. Most of the cells from the immune system can contribute towards demyelination, but the main process of demylation is mediated by antibody and complement. So far, it has been noticed that antibody and complement are responsible for lesions in 40–50% of MS patients. In addition to antibody and complement, $CD8^+$ T cells play an important role in MS pathogenesis. $CD8^+$ T cells make up the largest percentage of lymphocytes found in the brain of MS patients. All neuroectodermal cells in MS lesions express MHC class I molecules, making them an excellent target for $CD8^+$ T cells. In addition to their pro-inflammatory properties, $CD8^+$ T cells can also suppress the immune system and down-regulate inflammation. However, experiments with perforin, an important regulator of cytotoxic damage to immune cells, have made it clear that $CD8^+$ T cells present at MS lesions cause cytotoxicity, which could be the main source for demyelination and axonal damage. Bystander $CD4^+$ T cells do not contribute to the demyelinating process, but once $CD4^+$ T cells move into the CNS and become activated against myelin antigen, these $CD4^+$ T cells could be contributing directly towards demyelitaion of CNS. Moreover, CD4 TH17 effector T cells are postulated to play a crucial role in the pathogenesis of MS (Bettelli et al., 2006). After demyelination, remyelination is possible, which could further damage the CNS. The ratio of demyelination to remyelination determines whether a patient will develop secondary progressive MS (SPMS) or relapse remitting MS (RRMS). If remyelination occurs before axonal damage, irreversible physiological damage can be prevented. None of the FDA-approved therapies target oligodendrocytes to stimulate remyelination, but it is a very interesting possibility for future therapeutic intervention (Smriti et al., 2007).

Clinical Manifestations

The majority of symptoms associated with MS can be directly attributed to inflammation, edema, demyelination, and/or axonal damage within the brain, spinal cord, and optic nerves. Clinical motor manifestations include: weakness, stiffness, and/or pain in arms or legs, abnormal reflex activity, and spasticity. Often, the earliest symptoms of MS are somatosensory, including numbness and tingling. In MS, cerebral involvement is often accompanied by symptoms such as ataxia and intention tremor. Many individuals with MS complain of increased urinary frequency, urgency, and incontinence. Bladder and bowel disturbances remain among the most disabling and embarrassing symptoms experienced by MS patients. Sexual symptoms are also very common among both men and women. Fatigue, sleep disturbances, depression, and deficits in cognitive functioning are also common. The clinical course of MS is often highly variable and is generally characterized by relapses or exacerbations and deterioration of neurologic function, which entitle relapsing remitting MS. The features of relapsing remitting MS are defined as "episodes of acute worsening of neurologic function followed by a variable degree of recovery, with a stable course between attacks." Approximately 80–85% of patients are initially diagnosed with RRMS that evolves from an isolated demyelinating attack, which is characterized by multifocal inflammation along with varying degrees of axonal injury. A patient may experience disease progression with or without relapses and minor remissions; that clinical condition is defined as secondary progressive MS. It has been seen that 75% of RRMA patients will eventually develop the SPMS state. Primary progressive MS (PPMS) affects ~ 10–15% of MS patients. PPMS is defined as "disease progression from onset, with occasional plateaus and temporary minor improvements in clinical condition." The duration of MS varies significantly among MS patients. Some patients will live with MS for several decades, while about 10% will develop an acute, fulminant form of MS. Patients with an acute and fulminant form of MS show a rapid deterioration in their clinical signs and symptoms that have fatal consequences; these patients usually die within one to three years after the onset of disease. In general, the clinical spectrums among MS patients represent a benign disease and a low relapse rate and these may never develop into secondary progressive disease. The heterogeneity of the clinical course of MS is shown to have similar variation in its pathology.

MS lesions were recently segregated into four distinct subtypes. The general pathology of MS, the formation of demyelinating lesions in the CNS associated with infiltrating CD3+ T cells, activated macrophages, and microglia-containing myelin debris, and infiltrating B cells, is common to all forms of the disease. It is thought that MS lesions are mediated by soluble factors such as TNF–α and immunoglobulin deposition on the myelin sheath, and the local activation of the complement cascade. The diagnostic criteria for clinically definite MS (CDMS) include factors such as clinical history, MRI imaging, and CSF abnormalities. At present, there are no identifiable biomarkers that can predict the clinical subtype of MS. Similarly, there are no factors that can assist in predicting whether a patient diagnosed with MS will develop either a progressive or benign version of the disease. The clinical and pathological heterogeneity in MS has made it important to either develop or identify reliable biomarkers. Several cytokines, immunoglobulins, MMPs, markers of axonal/neuronal injury, and apoptotic markers have been suggested to have potential as biomarkers, but these biomarkers need validation by rigorous durability trials.

Treatment

In MS, treatment strategies can be either acute or long term. During a relapse, the goal of acute treatment is to reverse neurological disability as well as to delay further neurological dysfunction, so that normal function can be restored. This type of treatment for MS patients is in contrast to the goals of long-term treatments. The main objective of long-term treatment for MS patients is to decrease relapses (both severity and frequency), which could lend support to stopping progression of disability. Patients experiencing a relapse, such as optic neuritis or transverse myelitis, are often administered high-dose corticosteroid first–line therapy. During progressive phases of the disease, patients may be prescribed immunosuppressive agents such as cyclophosphamide or mitoxantrone because the progressive phase is often accompanied by worsening inflammatory demyelination and axonal degeneration.

In 1993, IFN-β was the first agent to demonstrate significant clinical efficacy among patients suffering with RRMS. Although the exact disease-modifying effects of IFN-β in MS are unknown, several immunomodulatory mechanisms have been suggested. Presently, two forms of IFN-β, including IFN-βla (Avonex® and Rebif®) and IFN-βlb (Betaseron®), have been prescribed. Glatiramer acetate (Copaxone®) is a synthetic mixture of polypeptides that has been approved to treat RRMS. Similar to IFN-β, Glatiramer acetate is found to be not effective for progressive forms of MS. Natalizumab (Tysabri®) is an alpha-4 integrin antagonist and is the first drug of an entirely new class of immune-directed therapies that has been approved by the FDA to treat relapsing MS. Natalizumab is a humanized recombinant monoclonal antibody that blocks leukocyte migration into the CNS by binding to α-4 integrins; these are components of the Very Late Antigen-4 (VLA-4) complex constitutively expressed on the leukocyte surface. In monotherapy trials, natalizumab has been reported to reduce the risk for sustained progression of disability as well as decrease the frequency of relapses. Based on the current literature, natalizumab appears to be one of the most effective agents to prevent relapses as well as to stop disease progression. Currently, numerous other monoclonal antibodies are under investigation as potential therapies for MS; for example, anti-CD52 (alemtuzumab), anti-CD25 (daclizumab), and anti-CD20 (rituximab). More investigations and clinical trials should be designed to assess the effectiveness of several drugs and approaches that can target both inflammatory and degenerative components of MS. These kinds of approaches may offer hope for individuals who are suffering from this debilitating disease (Smriti et al., 2007).

Experimental Models

To gain ideas about MS mechanisms, a number of models have been developed. These experimental models fall into two categories: spontaneous models and induced models. Each model reflects characteristic features of MS patients and has its own merits and demerits.

Spontaneous Models

Myelin basic protein mutant (taiep rat), proteolipid protein mutants (Rumpshaker and Jimpy mice), as well as gene-knockout animals (the myelin-associated glycoprotein (MAG) knockout), show dysmyelination, altered neurotransmission, and in some instances, clinical disease. These models have frequently been used to study myelination.

Induced Models

With chemically induced lesions, viral and autoimmune models are developed to show some evidence of demyelination, which is considered a pathological hallmark of MS. Direct injection of ethidium bromide or lysolecithin into the CNS produces demyelination. These induced models are usually effectively repaired once macrophages clear the myelin debris. For this reason, these models are rarely used at the present time. A number of viruses, including Semliki Forest Virus and Theiler's Murine Encephalomyelitis Virus, have been found to induce disease by neurotrophic infection of the CNS, specifically oligodendrocytes. Finally, experimental allergic encephalomyelitis (EAE) has received the most attention as a model for MS; this animal model is routinely used for testing different therapeutic strategies. EAE exhibits many clinical and histological features of MS, and is caused by autoimmunity induced against antigens that are either expressed naturally or artificially in CNS (Denic et al., 2011).

Methodology and Protocol

The method for EAE induction and preparation of antigens to induce EAE in C57BL/6 mice was adapted from the method described by Kafami et al. (2010).

It is important for the successful induction of EAE to follow standard precautions for the use of animals. Female C57BL/6 mice that are 4–6 weeks old are used for the induction of EAE. Animals must adhere to the normal laboratory animal maintenance guide.

Protocol

Animals were immunized with the Hooke kits (Hooke labs, EK-0115, Lawrence, MA, USA). It is recommended to follow the manufacturer's instructions. A mesh was dampened in ether and put in a desiccator. The mouse was kept in the desiccator and observed until breathing slowed down to ascertain whether the mouse had been anesthetized. The mouse was removed from the anesthetic chamber and laid on its side. Two syringes were filled with 1 mL of myelin oligodendrocyte glycoprotein (MOG) emulsion with complete Freund's adjuvant. Each animal was given an injection of 200 μL/animal. The needle was gently inserted into the subcutaneous space at the base of the tail, and 200 μL of emulsion was injected into the site. Since it was difficult to give the mouse a 200 μL injection, every mouse was given a 100 μL injection at two different sites on the same day. Immediately, and after 24 hours from the first injection, each mouse was given an intraperitoneal injection of pertussis toxin (100 μL/animal, ip). The animal was observed until complete recovery and it could move without a floppy gate. This procedure was repeated for all animals. After 2–3 days the flanks were bulging in response to the subcutaneous injection (Flow Chart 8.2).

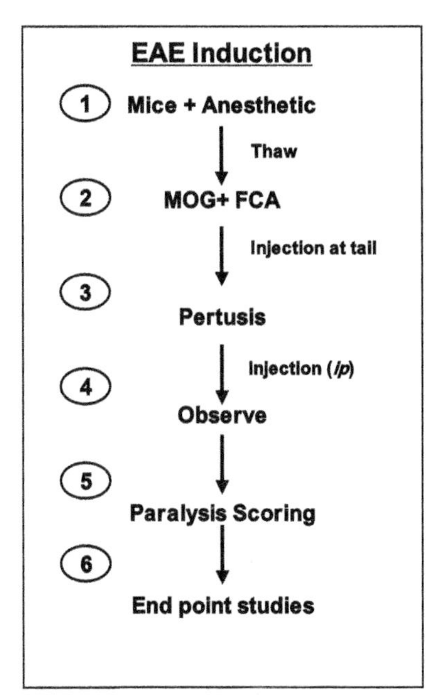

FLOW CHART 8.2 Important steps for the induction of experimental allergic encephalomyelitis (EAE).

Clinical Evaluation

One day before immunization, and from the 7th to the 35th day post-immunization, the animals were evaluated on a daily basis for signs of EAE following the 10-point score system (Table 8.4).

Three different clinical parameters were analyzed to compare the course of EAE (Figure 8.3): (1) Severity of disease as the cumulative disease index (CDI) was the mean of the clinical scores of the animals; (2) Disease onset, calculated as the mean of the first day animals showed the signs of the disease in experimental animals; and (3) Peak of disease score, which represented the mean of the highest clinical score of disease for all animals in each group. Tonicity of the tail and the distal part of the tail was ascertained by touching the tip of the tail. If the distal part of the tail was flaccid, the animal was removed from the base and observed to see if its tail remained erect or fell down (examined with the touch of a finger).

After ascertaining tonicity of the tail, the gait of the animal was observed by keeping it in an open area (like a tabletop) and allowing it to walk. After checking the gait, the hind limb was observed by grabbing its tail. After that, the paralysis score was recorded for unilateral paralysis. By holding the animal in the palm of the hand, it was easy to evaluate the type of paralysis (unilateral or bilateral). It was noted whether the mouse rolled spontaneously in its cage or was dead with complete paralysis.

Histology

After 35 days, animals that had an EAE score of 5 that did not change for 3 more days were euthanized by chloral

TABLE 8.4 Scoring Criteria for Paralysis in Case of Experimental Allergic Encephalomyelitis (EAE)

Scale	Clinical Evaluations
0	No clinical disease
0.5	Partial tail paralysis
1.0	Complete tail paralysis
1.5	Complete tail paralysis and discrete hind limb weakness
2.0	Complete tail paralysis and strong hind limb weakness
2.5	Unilateral hind limb paralysis
3.0	Complete hind limb paralysis
3.5	Hind limb paralysis and forelimb weakness
4.0	Complete paralysis (tetraplegia)
5.0	Moribund or dead

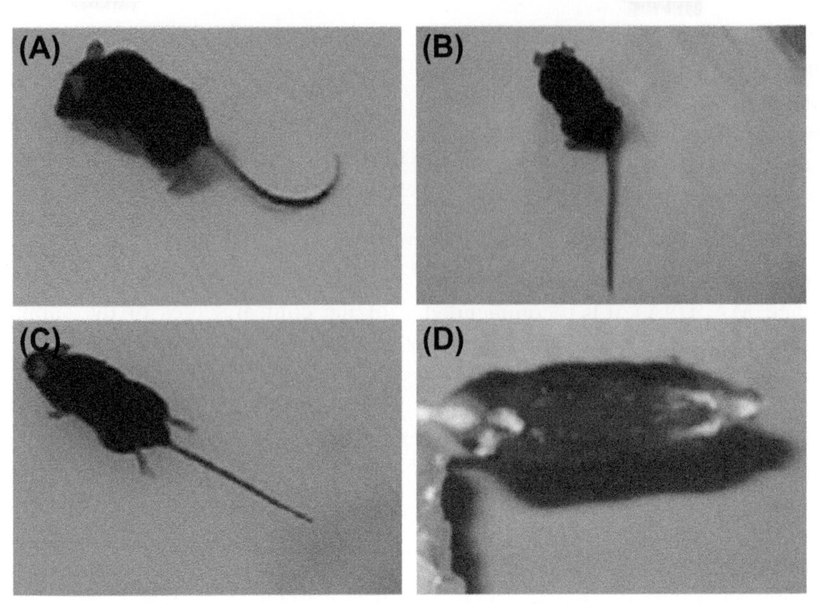

FIGURE 8.3 Clinical signs of EAE: (A) Loop tail. (B) Flaccid tail and paralyzed limbs. (C) Flaccid tail and paralyzed hind and front limbs. (D) Flaccid tail and weakness of hind limbs.

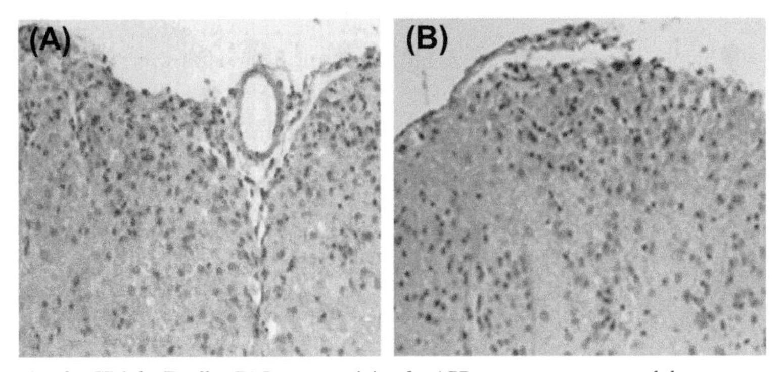

FIGURE 8.4 (A) Immunostaining for CD3 for T cells. (B) Immunostaining for APP to assess acute axonal damage.

hydrate injection (0.3 mL, ip). For histopathlogical evaluations, different tissues were harvested after dissecting the animals. Animals were placed appropriately in the dissection tray. A midline incision was made on the abdomen; the diaphragm was opened while ribbons were cut to expose the beating heart. The needle was inserted into the left ventricle of the heart while a phosphate-buffered saline (PBS) tap was allowed to fill the heart for 2 seconds. The right aorta was cut with small scissors to allow the PBS and PFA to circulate to exit. PBS allowed perfusion until the liver turned from red to yellow (~ 2–3 minutes). The best sign was when liquid flowed out of the incised left aorta and turned from red to clear. Another indicator was when PBS entered the pulmonary system and emerged through the nose of the animal. Then the PBS tap was closed and the tap was turned on for 4% paraformaldehyde (PFA, pH 7.4 at 37°C) to allow PFA to flow and perfuse the circulatory system for 3 minutes. Perfusion was evaluated by involuntary hind limb movement and tail shivering. When the mouse became stiff, it was time to stop PFA perfusion.

After perfusion was complete with PFA, the system was washed with PBS to remove residual PFA. After perfusion, the various tissues of interest were harvested and stored in fresh 4% PFA for 3 days at 4°C. Then these tissues were washed with PBS and the PFA-fixed tissue could be stored in PBS for a few months. These tissues were then available for sectioning and staining (Figure 8.4).

Immunohistochemistry

For immunohistochemistry the three sections showing the highest infiltrations were studied. An area ≥ 1.5×10^7 μm^2 from the brain/spinal cord was selected and analyzed under 200 × magnification to assess the average number of positive cells per mm^2 and to quantify it on a computerized imaging system (BX51 microscope [Olympus, Hamburg, Germany] with AnalySIS software [Special SIS Docu; Soft Imaging

System]) by planimetry. The inflammatory index had to be calculated as a percentage determined by dividing the number of visual fields with > 10 CD3 T cells by the total number of visual fields examined. Detection of amyloid precursor protein (APP) was performed for acute axonal damage.

ELISA

To assess the content of circulating pro-inflammatory cytokines like IL-6, IL-4, IL-12, IL-10, TNF-α, and IFN-γ, enzyme-linked immunosorbant assay (ELISA) was employed. To evaluate the levels of different cytokines, blood was collected into tubes by a retro-orbital plexus method. The collected blood was kept in the tube to clot. After clotting of the blood serum, it was separated and stored at −20°C. These serum samples were then used for the evaluation of different cytokines using ELISA.

Real-Time RT-PCR

In order to quantify the mRNA of different pro-inflammatory cytokines such as TNF-α and IFN-γ, anti-inflammatory cytokines like IL-10, myelin-deteriorating matrix metalloproteinase MMP-9, and the content of myelin basic protein (MBP 3-4), samples from animals had to be analyzed by real-time PCR. Animals were sacrificed with lethal injection and perfused with cold PBS. Then the limbs and muscles were removed with scissors and the skin removed from these organs. A transverse cut was made at the base of the skull and vertebral column to separate them. The nasal bridge was broken with a small scalpel and the eyeballs removed. Very thin forceps were used under the skull bones to break it into pieces from the frontal to occipital lobes. The bony connection under the cerebellum was broken to expose the cerebellum. The broken bones of the skull needed to be removed. The nerve root connection with the brain was cut. The brain was removed and stored in liquid nitrogen. For removal of the spinal cord, an oblique cut was made from the lateral side of the spinal cord (started from the cervical part) to the furthest part of the vertebral column (both sides). The spinal cord was then exposed by cutting the boney flap. The steps above were repeated to get to the coda aquina. The spinal cord was taken out by cutting its adhesion to the base. It was then stored in liquid nitrogen.

The frozen tissue sample was used for RNA extraction. First, the sample was homogenized by pushing and rotating it with a sterile glass homogenizer. Next, the homogenate sample was left on the bench top at room temperature (15–25°C) for 5 minutes to promote dissociation of nucleoprotein complexes. Then 200 μL of chloroform were added to the tube and the tube was shaken vigorously for 15 s. The tube containing the homogenate was placed on the bench top at room temperature for 2–3 min and then centrifuged again at 15,000 rpm for 15 min at 4°C. After centrifugation, the sample separated into 3 phases: an upper, colorless, aqueous phase containing RNA; a white interphase; and a lower, red, organic phase. The upper, aqueous phase was transferred to a new sterile eppendorf tube. One volume (usually 600 μL) of 70% ethanol was added to the tube containing the aqueous phase and mixed thoroughly by vortexing. Visible precipitates after addition of ethanol could then be noticed. Up to 700 μL of the sample were processed for total RNA extraction by using an RNeasy Mini spin column (Roche, Germany) according to the kit instructions. After RNA extraction, RNA was quantified spectrophotometrically and the purity of RNA was ascertained by taking out a ration between the OD at 260 and 280 nm. Quantitative real-time reverse transcriptase PCR was performed to analyze the levels of mRNA of different cytokines using cytokine-specific primers. The first step was to perform cDNA synthesis by using a cDNA synthesis kit (TaKaRa, Japan), which was followed by a Syber Green I real-time PCR master mix kit TaKaRa Japan). A house keeping gene (like the β-actin gene) was included in the study to compare the results.

ETHICAL ISSUES

The use of laboratory animals in research is of major ethical concern. Much of the argument revolves around moral values. Today there is a wide spectrum of views on animal rights. This has prompted the establishment of guidelines on the care and use of experimental animal models. The guidelines endorse some essential principles for the care and use of animals for scientific projects. The basis of these principles is to replace animals with other methods such as mathematical models, computer simulations, and *in vitro* biological systems, thus reducing the number of animals used in order to obtain valid results without unnecessary duplication, and finally, refining projects by selecting appropriate species and techniques to minimize pain or distress to animals using appropriate sedation or anesthesia.

As a researcher, one must always assume that procedures that cause pain to humans will cause pain in such situations in animals. Surgical procedures should be performed on anaesthetized animals. It should be kept in mind that if the animal would suffer severe pain during a procedure, or if at the endpoint cannot be alleviated swiftly, the animals must be killed humanely.

The transportation, housing, feeding, and handling of animals are also important. Housing facilities should be compatible with the needs of the species and equipped to achieve a high standard of animal care. The place should be designed to facilitate control of environmental factors. Cages should be comfortable and should fulfill behavioral requirements such as free movement and activity, bedding, contact with others of the same species, lighting, temperature, air quality, appropriate day/night cycles, and protection from excessive noise.

The population density of animals within cages should also be considered from an ethical standpoint.

This statement refers to the need for the reader to operate in accordance with the guidelines at her/his academy.

TRANSLATIONAL SIGNIFICANCE

The concept of translational research is to try to convert the results derived in animal models into a new understanding of disease mechanisms and therapeutics in human beings. It is a bridge from experimental models to clinical medicine. Over recent years the importance of this kind or research has progressively increased. Consequently, translational research is considered a key component to finding practical applications, especially within medicine.

With the improvement of technologies, significant progress has been made in producing various types of engineered experimental animal models based on a better understanding of the molecular and genetic principles of disease. As a result, any interventions in experimental models are more practical and repeatable when compared to patient-oriented research.

Various risk factors that are linked to, or even responsible for, differences in clinical results should also be considered as significant for the development of experimental models; this will enhance the translational value of experimental models. These risk factors can be categorized into genetic factors, acquired factors, and health conditions, which can be studied in models in a controlled manner. In medicine, the performance of successful translational research requires data from hospitals.

WORLD WIDE WEB RESOURCES

http://www.niams.nih.gov/

National Institutes of Health (NIH) site for the National Institute of Arthritis and Musculoskeletal and Skin Diseases.

http://www.ncbi.nlm.nih.gov/pubmed/

PubMed comprises over 22 million citations for biomedical literature from MEDLINE, life science journals, and online books. PubMed citations and abstracts include the fields of biomedicine and health, and cover portions of the life sciences, behavioral sciences, chemical sciences, and bioengineering. PubMed also provides access to additional relevant web sites and links to other NCBI molecular biology resources.

http://www.nlm.nih.gov/medlineplus/multiplesclerosis.html

MedlinePlus is the National Institutes of Health (NIH) site for patients and their families and friends. Produced by the National Library of Medicine, it brings you information about diseases, conditions, and wellness issues in easy-to-understand language. MedlinePlus offers reliable, up-to-date health information, anytime, anywhere, for free.

http://www.ebi.ac.uk/ipd/imgt/hla/

The IMGT/HLA Database provides a specialist database for sequences of the human major histocompatibility complex (HLA) and includes the official sequences for the WHO Nomenclature Committee for Factors of the HLA System. The IMGT/HLA Database is part of the international ImMunoGeneTics project.

REFERENCES

Bettelli, E., Carrier, Y., Gao, W., Korn, T., Strom, T. B., Oukka, M., Weiner, H. L., & Kuchroo, V. K. (2006). Reciprocal developmental pathways for the generation of pathogenic effector TH17 and regulatory T cells. *Nature*, *441*, 235–238.

Bouvet, J. P., Couderc, J., Bouthillier, Y., Franc, B., Ducailar, A., & Mouton, D. (1990). Spontaneous rheumatoid-like arthritis in a line of mice sensitive to collagen-induced arthritis. *Arthritis Rheum*, *33*(11), 1716–1722.

Carlson, B. C., Jansson, A. M., Larsson, A., Bucht, A., & Lorentzen, J. C. (2000). The endogenous adjuvant squalene can induce a chronic T-cell-mediated arthritis in rats. *Am J Pathol*, *156*(6), 2057–2065.

Chiu, Y., Ostor, A. J. K., Hammond, A., Sokoll, K., Anderson, M., Buch, M., Ehrenstein, M. R., Gordon, P., Steer, S., & Bruce, I. N. (2012). Access to the next wave of biologic therapies (Abatacept and Tocilizumab) for the treatment of rheumatoid arthritis in England and Wales. *Clin Rheumatol*, *31*, 1005–1012.

Denic, A., Aaron, J., Johnson, B., Allan, J., Bieber, C., Arthur, E., Warrington, C., Moses Rodriguez, C., & Istvan Pirko, C. (2011). The relevance of animal models in multiple sclerosis research. *Pathophysiology*, *18*, 21–29.

Goldschmidt, T. J., & Holmdahl, R. (1994). Therapeutic effects of monoclonal antibodies to alpha beta TCR but not to CD4 on collagen-induced arthritis in the rat. *Cell Immunol*, *154*(1), 240–248.

Holmdahl, R., Jonsson, R., Larsson, P., & Klareskog, L. (1988). Early appearance of activated CD4+ T lymphocytes and class II antigen-expressing cells in joints of DBA/1 mice immunized with type II collagen. *Lab Invest*, *58*(1), 53–60.

Holmdahl, R., & Kvick, C. (1992). Vaccination and genetic experiments demonstrate that adjuvant-oil-induced arthritis and homologous type II collagen-induced arthritis in the same rat strain are different diseases. *Clin Exp Immunol*, *88*(1), 96–100.

Holmdahl, R., Tarkowski, A., & Jonsson, R. (1991). Involvement of macrophages and dendritic cells in synovial inflammation of collagen induced arthritis in DBA/1 mice and spontaneous arthritis in MRL/lpr mice. *Autoimmunity*, *8*(4), 271–280.

Iwakura, Y., Saijo, S., Kioka, Y., Nakayama-Yamada, J., Itagaki, K., Tosu, M., Asano, M., Kanai, Y., & Kakimoto, K. (1995). Autoimmunity induction by human T-cell leukemia virus type 1 in transgenic mice that develop chronic inflammatory arthropathy resembling rheumatoid arthritis in humans. *J Immunol*, *155*(3), 1588–1598.

Kafami, L., Raza, M., Razavi, A., Mirshafiey, A., Movahedian, M., & Khorramizadeh, M. R. (2010). Intermittent feeding attenuates clinical course of experimental autoimmune encephalomyelitis in C57BL/6 mice. *AJMB*, *2*(1), 47–52.

Keffer, J., Probert, L., Cazlaris, H., Georgopoulos, S., Kaslaris, E., Kioussis, D., & Kollias, G. (1991). Transgenic mice expressing human tumour necrosis factor: a predictive genetic model of arthritis. *EMBO J, 10*(13), 4025–4031.

Klareskog, L., & Johnell, O. (1988). Induced expression of class II transplantation antigens in the cartilage-pannus junction in RA: chronic synovitis as a model system for aberrant T-lymphocyte activation. *Br J Rheumatol, 27*(Suppl. 2), 141–149.

Kleinau, S., Lorentzen, J., & Klareskog, L. (1995). Role of adjuvants in turning autoimmunity into autoimmune disease. *Scand J Rheumatol, 101*(Suppl), 179–181.

Kohashi, O., Tanaka, A., Kotani, S., Shiba, T., Kusumoto, S., Yokogawa, K., Kawata, S., & Ozawa, A. (1980). Arthritis-inducing ability of a synthetic adjuvant, N-acetylmuramyl peptides, and bacterial disaccharide peptides related to different oil vehicles and their composition. *Infect Immun, 29*(1), 70–75.

Korganow, A. S., Ji, H., Mangialaio, S., Duchatelle, V., Pelanda, R., Martin, T., Degott, C., Kikutani, H., Rajewsky, K., Pasquali, J. L., Benoist, C., & Mathis, D. (1999). From systemic T-cell self-reactivity to organ-specific autoimmune disease via immunoglobulins. *Immunity, 10*(4), 451–461.

Lorentzen, J. C., & Klareskog, L. (1996). Susceptibility of DA rats to arthritis induced with adjuvant oil or rat collagen is determined by genes both within and outside the major histocompatibility complex. *Scand J Immunol, 44*(6), 592–598.

Lorentzen, J. C. (1999). Identification of arthritogenic adjuvants of self and foreign origin. *Scand J Immunol, 49*(1), 45–50.

Meda, F., Folci, M., Baccarelli, A., & Selmi, C. (2011). The epigenetics of autoimmunity. *Cell Mol Immunol, 8*(3), 226–236.

Mussener, A., Litton, M. J., Lindroos, E., & Klareskog, L. (1997). Cytokine production in synovial tissue of mice with collagen-induced arthritis (CIA). *Clin Exp Immunol,107(3)*, 485–493.

National Academy of Sciences. Guide for the Care and Use of Laboratory Animals. 7. 1996. Washington D.C., National Research Council, Institute for Laboratory Animal Research, NAS.Ref Type: Catalog National Multiple Sclerosis Society 2008, *About MS: Who Gets MS?*

Pearson, C. M. (1956). Development of arthritis, periarthritis and periostitis in rats given adjuvants. *Proc Soc Exp Biol Med, 91*(1), 95–101.

Saadat, F., Cuzzocrea, S., Di Paola, R., Pezeshki, M., Khorramizadeh, M. R., Sedaghat, M., Razavi, A., & Mirshafiey, A. (2005 Aug). Effect of pyrimethamine in experimental rheumatoid arthritis. *Med Sci Monit* (8), 11. BR293-9.

Seki, N., Sudo, Y., Yoshioka, T., Sugihara, S., Fujitsu, T., Sakuma, S., Ogawa, T., Hamaoka, T., Senoh, H., & Fujiwara, H. (1988). Type II collagen-induced murine arthritis. I. Induction and perpetuation of arthritis require synergy between humoral and cell-mediated immunity. *J Immunol, 140*(5), 1477–1484.

Smriti, M., Agrawal, & Wee Yong, V. (2007). Immunopathogenesis of multiple sclerosis. *International Review of Neurobiology, 79*, 99–126.

Svensson, L., Jirholt, J., Holmdahl, R., & Jansson, L. (1998). B cell-deficient mice do not develop type II collagen-induced arthritis (CIA). *Clin Exp Immunol,111(3)*, 521–526.

Szabó, C., Virag, L., Cuzzocrea, S., Scott, G. S., Hake, P., O'Connor, M., et al. (1998). Protection against peroxynitrite-induced fibroblast injury and arthritis development by inhibition of poly(ADP-ribose) synthetase. *Proc Natl Acad Sci U S A, 9*, 3867–3872.

Ulfgren, A. K., Grondal, L., Lindblad, S., Khademi, M., Johnell, O., Klareskog, L., & Andersson, U. (2000). Interindividual and intra-articular variation of proinflammatory cytokines in patients with rheumatoid arthritis: potential implications for treatment. *Ann Rheum Dis, 59*(6), 439–447.

Viatte, S., Plant, D., & Raychaudhuri, S. (2013 Feb 5). Genetics and epigenetics of rheumatoid arthritis. *Nat Rev Rheumatol.* http://dx.doi.org/10.1038/nrrheum.2012.237. [Epub ahead of print].

FURTHER READING

Aghdami, N., Gharibdoost, F., & Moazzeni, S. M. (2008). Experimental autoimmune encephalomyelitis induced by antigen pulsed dendritic cells in the C57BL/6 mouse: Influence of injection route. *Exp Anim, 57*, 45–55.

Alonso, A., & Hernan, M. A. (2008). Temporal trends in the incidence of multiple sclerosis: a systematic review. *Neurol, 71*, 129–135.

Ascherio, A., & Munch, M. (2000). Epstein–Barr virus and multiple sclerosis. *Epidemiology, 11*(2), 220–224.

Burgoon, M. P., Williamson, R. A., Owens, G. P., Ghausi, O., Bastidas, R. B., Burton, D. R., & Gilden, D. H. (1999). Cloning the antibody response in humans with inflammatory CNS disease: isolation of measles virus-specific antibodies from phage display libraries of a subacute sclerosing panencephalitis brain. *J Neuroimmunol, 94*(1–2), 204–211.

Firestein, G. S., Budd, R. C., Gabriel, S. E., McInnes, I. B., & O'Dell, J. R. (2013). *Kelley's Textbook of Rheumatology* 19103–2899, Philadelphia, PA: Saunders, An Imprint of Elsevier. ISBN: 978-1-4377-1738-9. VOLUME II.

Giridharan, N. V., Kumar, V., & Muthuswamy, V. (May 2000). *Use of Animals in Scientific Research.* New Delhi: Indian Council of Medical Research Ministry of Health & Family Welfare.

Kannan, K., Ortmann, R. A., & Kimpel, D. (2005). Animal models of rheumatoid arthritis and their relevance to human disease. *Pathophysiology, 12*, 167–181.

Mix, E., Meyer-Rienecker, H., Hartung, H.-P., & Zettl, U. K. (2010). Animal models of multiple sclerosis – Potentials and limitations. *Progress in Neurobiology, 92*, 386–404.

Sato, F., Omura, S., Martinez, N. E., & Tsunoda, I. (2011). Animal Models of Multiple Sclerosis. *Neuroinflammation*, 55–79.

GLOSSARY

Adhesion Molecule A cell surface molecule (e.g. selectin, integrin, member of the Ig superfamily) whose function is to promote adhesive interactions with other cells or the extracellular matrix. These molecules play crucial roles in cell migration and cellular activation in innate and adaptive immune responses.

Adjuvant A substance such as complete Freund's adjuvant (CFA) that enhances T- and B-cell activation, mainly by promoting the accumulation and activation of antigen-presenting cells at the site of antigen exposure. Adjuvants stimulate expression of T-cell-activating co-stimulators and cytokines by antigen-presenting cells, and may also prolong the expression of peptide-MHC complexes on the surface of these cells.

Autoimmune Disease A disease caused by a breakdown of self-tolerance such that the adaptive immune system responds to self-antigens and mediates cell and tissue damage. Autoimmune diseases can be organ specific (e.g. thyroiditis or diabetes) or systemic (e.g. systemic lupus erythematosus).

CD Molecules Cell surface molecules expressed on various cell types in the immune system that are designated by the "cluster of differentiation (CD) number."

Disease-Modifying Anti-Rheumatic Drugs (DMARDs) DMARDs contain medications from different classes including methotrexate, gold salts, hydroxychloroquine, sulfasalazine, ciclosporin, and azathioprine. DMARDs were often only partly effective and poorly tolerated in long-term therapy of autoimmune diseases.

Enzyme-Linked Immunosorbent Assay (ELISA) A method of quantifying an antigen immobilized on a solid surface by use of a specific antibody with a covalently coupled enzyme. The amount of antibody that binds the antigen is proportional to the amount of antigen present and is determined by spectrophotometrically measuring the conversion of a clear substrate to a colored product by the coupled enzyme.

Experimental Autoimmune Encephalomyelitis (EAE) An animal model of multiple sclerosis, an autoimmune demyelinating disease of the central nervous system. EAE is induced in rodents by immunization with components of the myelin sheath (e.g. myelin basic protein) of nerves, mixed with an adjuvant. The disease is mediated in large part by cytokine secreting CD4+ T cells specific for the myelin sheath proteins.

Granulocyte–Monocyte Colony-Stimulating Factor (GM-CSF) A cytokine made by activated T cells, macrophages, endothelial cells, and stromal fibroblasts that acts on bone marrow to increase the production of neutrophils and monocytes. GM-CSF is also a macrophage-activating factor and promotes the differentiation of Langerhans cells into mature dendritic cells.

Granuloma A nodule of inflammatory tissue composed of clusters of activated macrophages and T lymphocytes, often with associated necrosis and fibrosis. Granulomatous inflammation is a form of chronic delayed-type hypersensitivity, often in response to persistent microbes, or in response to particulate antigens that are not readily phagocytosed.

Granzyme A serine protease enzyme found in the granules of CTLs and NK cells that is released by exocytosis, enters target cells, and proteolytically cleaves and activates caspases and induces target cell apoptosis.

Homeostasis In the adaptive immune system, the maintenance of a constant number and diverse repertoire of lymphocytes, despite the emergence of new lymphocytes and the tremendous expansion of individual clones that may occur during responses to immunogenic antigens. Homeostasis is achieved by several regulated pathways of lymphocyte death and inactivation.

Human Leukocyte Antigens (HLA) MHC molecules expressed on the surface of human cells. Human MHC molecules were first identified as alloantigens on the surface of white blood cells (leukocytes) that bound serum antibodies from individuals previously exposed to other individuals' cells.

Interferons A subgroup of cytokines originally named for their ability to interfere with viral infections, but that have other important immunomodulatory functions. Type I interferons include interferon-α and interferon-β, whose main functions are antiviral; type II interferon, also called interferon-γ, activates macrophages and various other cell types.

Interleukins Any of a large number of cytokines named with a numerical suffix roughly sequentially in order of discovery or molecular characterization (e.g. interleukin-1, interleukin-2). Some cytokines were originally named for their biological activities and do not have an interleukin designation.

Lipopolysaccharide (LPS) A component of the cell wall of Gram-negative bacteria that is released from dying bacteria and stimulates many innate immune responses, including the secretion of cytokines, induction of microbicidal activities of macrophages, and expression of leukocyte adhesion molecules on endothelium. LPS contains both lipid components and carbohydrate moieties.

Major Histocompatibility Complex (MHC) Molecule A heterodimeric membrane protein encoded in the MHC locus that serves as a peptide display molecule for recognition by T lymphocytes. Two structurally distinct types of MHC molecules exist. Class I MHC molecules are present on most nucleated cells, bind peptides derived from cytosolic proteins, and are recognized by $CD8^+$ T cells. Class II MHC molecules are restricted largely to dendritic cells, macrophages, and B lymphocytes, bind peptides derived from endocytosed proteins, and are recognized by CD4+ T cells.

Matrix Metalloproteinase (MMP) MMPs are a family of highly conserved endopeptidases dependent on Zn^{2+} ions for activity. MMPs can collectively cleave most extracellular matrix. At present, 25 vertebrate MMPs and 22 human homologs have been identified and characterized. MMPs participate in many physiological processes, such as embryonic development, organ morphogenesis, blastocyst implantation, ovulation, nerve growth, cervical dilatation, post-partum uterine involution, mammary development, endometrial cycling, hair follicle cycling, angiogenesis, inflammatory cell function, apoptosis, tooth eruption, bone remodeling and wound healing.

Myelin Oligodendrocyte Glycoprotein (MOG) MOG is a CNS-specific type I membrane glycoprotein of the immunoglobulin superfamily expressed mainly on the outermost layer of the myelin sheath, making it an ideal target for antibody-mediated demyelination. It is highly immunogenic, and unlike other myelin proteins used to induce EAE, is unique in inducing both an encephalitogenic T-cell response and a demyelinating response in EAE.

Multiple Sclerosis (MS) A chronic inflammatory demyelinating disorder of the central nervous system. The majority of symptoms associated with MS can be directly attributed to inflammation, edema, demyelination, and/or axonal damage within the brain, spinal cord, and optic nerves.

Nitric Oxide (NO) A biologic effector molecule with a broad range of activities that in macrophages functions as a potent microbicidal agent to kill ingested organisms.

Pannus Formation of locally invasive synovial tissue is a characteristic feature of rheumatoid arthritis.

Perforin A protein that is homologous to the C9 complement protein and is present in the granules of CTLs and NK cells. When perforin is released from the granules of activated CTLs or NK cells, it promotes entry of granzymes into the target cell, leading to apoptotic death of the cell.

Rheumatoid Arthritis (RA) An autoimmune disease characterized primarily by inflammatory damage to joints and sometimes inflammation of blood vessels, lungs, and other tissues. $CD4^+$ T cells, activated B lymphocytes, and plasma cells are found in the inflamed joint lining (synovium), and numerous pro-inflammatory cytokines, including IL-1 and TNF, are present in the synovial (joint) fluid.

Reverse Transcriptase (RT) An enzyme encoded by retroviruses, such as HIV, that synthesizes a DNA copy of the viral genome from the RNA genomic template. Purified reverse transcriptase is used widely in molecular biology research for purposes of cloning complementary DNAs encoding a gene of interest from messenger RNA.

T_H1 Cells Subset of CD4+ helper T cells whose principal function is to stimulate phagocyte-mediated defense against infections via secretion of a group of cytokines, including IFN-γ.

T_H2 Cells Subset of CD4+ helper T cells whose principal functions are to stimulate IgE and eosinophil/mast cell-mediated immune reactions via a particular set of cytokines, including IL-4 and IL-5.

T_H17 Cells Subset of CD4+ helper T cells that are protective against certain bacterial infections and also mediate pathogenic responses in autoimmune diseases.

Tumor Necrosis Factor (TNF) A cytokine produced mainly by activated mononuclear phagocytes that stimulates the recruitment of neutrophils to sites of inflammation.

TNF-α Blocking Agents A group of biological disease-modifying anti-rheumatic drugs such as Etanercept® (a soluble TNF-α receptor) and infliximab (a monoclonal antibody).

Very Late Antigen (VLA) The set of integrins that shares a common beta-1 chain.

ABBREVIATIONS

APP Amyloid Precursor Protein
CFA Complete Freund's Adjuvant
CD Cluster Of Differentiation
CDI Cumulative Disease Index
CDMS Clinically Definite MS
CNS Central Nervous System
CSF Cerebrospinal Fluid
DMARDs Disease-Modifying Anti-Rheumatic Drugs
DNA Deoxyribonucleic Acid
EAE Experimental Allergic Encephalomyelitis
EDTA Ethylene Diamide Tetraacetic Acid
ELISA Enzyme-Linked Immunosorbent Assay
FDA Food and Drug Administration
GM-CSF Granulocyte–Macrophage Colony-Stimulating Factor
H&E Hematoxyline and Eosin
HLA Human Leukocyte Antigen
HHV–6 Human Herpes Virus 6
IFN Interferon
IL Interleukin
LPS Lipopolysaccharide
MBP Myelin Basic Protein
mg Milligram
MHC Major Histocompatibility Complex
mL Milliliter
mm Millimeter
MAG Myelin-Associated Glycoprotein
MMP Matrix Metalloproteinase
MOG Myelin Oligodendrocyte Glycoprotein
MCP-1 Monocyte Chemoattractant Protein
MDP Muramyle Dipeptide
MIP-1 Macrophage Inflammatory Protein
MMPs Matrix Metalloproteinases
MRI Magnetic Resonance Imaging
MS Multiple Sclerosis
ng Nanogram
NO Nitric Oxide
NSAIDs Non-Steroidal Anti-Inflammatory Drugs
OD Optical Density
PBS Phosphate Buffered Saline
PCR Polymerase Chain Reaction
PLP Proteolipid Protein
PP–MS Primary Progressive Multiple Sclerosis
PR–MS Progressive Relapsing Multiple Sclerosis
RA Rheumatoid Arthritis
RNA Ribonucleic Acid
RPM Revolutions Per Minute
RR–MS Relapsing–Remitting Multiple Sclerosis
RT Reverse Transcriptase
RT-PCR Real-Time Polymerase Chain Reaction
SPMS Secondary Progressive MS
TNF Tumor Necrosis Factor
TDM Trehalosedimycolate
μg Microgram
μL Microliter
μm Micrometer
VLA-4 Very Late Antigen-4

LONG ANSWER QUESTIONS

1. Describe the significance of animal models in biotechnology.
2. How is CIA induced, and how can this model be used to test drugs against RA?
3. Discuss the different types of animal models for studying the pathogenesis of rheumatoid arthritis.
4. Why is the presence of inflammation in the CNS considered a hallmark of multiple sclerosis?
5. Explain the various methods for evaluation of experimental models of multiple sclerosis.

SHORT ANSWER QUESTIONS

1. Why are animal models used for the pathogenesis of rheumatoid arthritis?
2. Give an example that shows the impact of epigenetics in the initiation of RA?
3. Which types of evaluation should be performed after "collagen induced arthritis" is aroused?
4. What are the "intervening factors" in experimental models of multiple sclerosis?
5. After activation of the immune system, which type of lymphocytes enters into the central nervous system (CNS)?

ANSWERS TO SHORT ANSWER QUESTIONS

1. There are many experimental models that resemble RA in different respects. Since RA is a heterogeneous disease, there is a need for different animal models that each reflect a characteristic feature of a particular subgroup of RA patients or illustrates a particular aspect of the disease.
2. The epigenetics of RA are also responsible for the initiation of RA. Since the concordance of rheumatoid arthritis in identical twins is not 100%, other non-genetic factors also play a role in the disease etiology.
3. Daily clinical assessment according to a macroscopic scoring system, histological processing, and assessment of arthritis damage, and radiographic evaluation by an investigator blinded to the treatment protocol on Day 35.
4. Age, weight, and possible infectious diseases in animals should be considered as the intervening factors.
5. Following activation, myelin-specific lymphocytes enter into the CNS and oligodendrocytes are damaged.

Chapter 9

HIV and Antiretroviral Drugs

Ashish Swarup Verma, Iqram Govind Singh, Ruby Bansal and Anchal Singh
Amity Institute of Biotechnology, Amity University Uttar Pradesh, NOIDA (UP), India. Crosslay Wellness Program, Pushpanjali Crosslay Hospital, Ghaziabad (UP), India

Chapter Outline

SUMMARY

Acquired Immune Deficiency Syndrome (AIDS) is the terminal stage of Human Immunodeficiency Virus (HIV) infections. HIV is one of the dreaded infectious diseases of the late 20th century. This chapter provides information about the history, discovery, epidemiology and replication of HIV. Anti-HIV drugs and novel anti-HIV drug targets for development of new drugs is also discussed.

WHAT YOU CAN EXPECT TO KNOW

AIDS is considered as the final chapter in the life of HIV-infected patients. Even though we have many choices for antiretroviral drugs, the fact is, "*HIV can be controlled but cannot be cured.*" HIV is a retrovirus and its genome consists of nine genes along with a Long Terminal Repeat (LTR), which produces 15 different proteins during replication.

Animal Biotechnology. http://dx.doi.org/10.1016/B978-0-12-416002-6.00009-2

A better understanding of HIV replication has been the cause for the development of anti-HIV drugs that inhibit different stages of HIV replication. At present, we can block HIV replication by inhibiting enzymes like reverse transcriptase, protease, and integrase. Successful treatment strategies are ones that use a combination of different drugs to control HIV replication. Some other steps of HIV infections are being explored for the development of a new group of drugs. We have also discussed classification of different clinical stages of HIV patients, which is useful and crucial to make decisions for the right choice of drugs. Lately, bone marrow transplantation is on the horizon and may open new avenues to treat HIV. Methods to evaluate drug toxicity as well as anti-HIV effects of drugs are provided to aid better understanding. There is no doubt that improved antiretrovirals have increased the life-span of HIV patients post-HIV infection, which has brought up another new health concern: NeuroAIDS.

HISTORY AND METHODS

INTRODUCTION

Infectious diseases and humanity are closely intertwined with each other. Infections have been a part of human life since the inception of humanity. As human beings we are vulnerable to infections since the time of birth, and in certain circumstances one can be infected even before birth (pre-natal infections). Infections and human life are closely linked, therefore, the significance of infection in our lives can neither be denied nor ignored.

Vaccines and antibiotics have enormous implications for human welfare because they have decrease morbidity and mortality among the human population. The discovery of antibiotics and vaccines started to give us the idea that "*we are winning the war against microbes/infections.*" But is this true?

On one hand, the 20th century became famous for eradication of some important diseases like small pox, polio, etc., while on the other hand we have seen a sudden eruption of entirely new diseases like SARS, HIV, Ebola, etc. Out of these, HIV has received tremendous attention from various quarters of human society, including government organizations and non-government organizations, religious groups, clinicians, scientists, and even the media. HIV is an infectious disease that has received unexpectedly more than the desired attention. Nobody knows the answer to this intense public response seen for HIV infection, but it may be due to the shock that the initial cases reported were from one of the most resourceful nations, the United States.

DISCOVERY AND ORIGIN OF HIV

It has been over 30 years since unique cases of infection were reported for the first time. In 1981 HIV infections were not known, and those initial patients were reported to be suffering from a devastation of the immune system. The compromised immune system led to a fatal clinical condition now known as Acquired Immune Deficiency Syndrome (AIDS). There is no doubt that AIDS is the ultimate chapter of life in HIV patients. AIDS is clinically defined as a condition with $CD4^+$ T-lymphocyte counts of less than 200 cells per ml of blood along with the presence of AIDS-defining illness like HAD, HIV wasting syndrome, and AIDS-defining cancers like Kaposi's Sarcoma, etc. It is a general belief that HIV is a single type of virus, however, HIV is actually of two different types, HIV-1 and HIV-2. HIV-1 is more prevalent with a global presence and fatal consequences, while HIV-2 is geographically isolated, confined mostly to the African continent, and is less pathogenic, which means patients who are infected with HIV-2 can survive a lot longer compared to those infected with HIV-1. From this point in the chapter we will use HIV to mean HIV-1.

HISTORY OF HIV AND AIDS

The initial cases of HIV infection were reported in five gay men from Los Angeles, California. These five patients were suffering from *Pneumocystis carinii* pneumonia (PCP). This report was published in MMWR, a publication from The Centers for Disease Control and Prevention (CDC) (Gottlieb et al., 1981). PCP is a disease that is normally not reported amongst the general population; PCP infections are commonly observed only in immune compromised/immune-deficient patients. In 1981, these PCP patients were not reported to be infected with HIV or suffering from AIDS because until that time, medical science was neither aware of HIV and AIDS, nor had these words even been coined. As a matter of fact, the world community is indebted to a drug technician, Ms. Sandra Ford at the CDC, who brought this unusual observation to the attention of her bosses. She noticed an unusually high number of requests for a unique drug, pentamidine, which is used to treat PCP. She said in an interview to Newsweek.

A doctor was treating a gay man in his 20s who had pneumonia. Two weeks later he called to ask for a refill of a rare drug that I handled. This was unusual – nobody ever asked for a refill. Patients usually were cured in one 10-day treatment or they died.

Realizing the gravity of the situation, the CDC formed a "Task Force on Kaposi's Sarcoma and Opportunistic Infections (KSOI)." By this time very little information about this new deadly disease was available to anyone, therefore people were skeptical about the mode of transmission. This disease did not have any official name, so initially it was called lymphadenopathy or KSOI. It was observed that this disease was closely associated with the gay community, therefore people started calling it Gay Compromise Syndrome (GCS), Gay Related Immune Deficiency (GRID), Acquired Immunodeficiency Disease (AID), Gay Cancer, and Community Acquired Immuno Dysfunction (CAID). The growing epidemic led to the development of various organizations to deal with this deadly new disease both in the U.S. and U.K. First and foremost, these organizations started advising gay men to follow safe sex practices. By the end of 1982, a 20-month old child who had received multiple blood transfusions died due to infections related to AIDS. The death of this child raised the alarm for blood banks: as blood banks needed a safe blood supply. With time, it was noticed that the disease was reported from all walks of life, including non-homosexuals. Thus, names for the disease like GRID and GCS became completely irrelevant. Now the situation was more confusing among clinicians and doctors, and they needed to find an appropriate name for this deadly disease. Ultimately, on July 24, 1982, a meeting was convened in Washington, D.C., and the name "AIDS" was coined.

KSOI, which was formed to look after this new disease, started following and monitoring cases throughout the U.S. What was most surprising was that within a few months after the formation of KSOI, reports of cases of this disease started pouring in from all over the U.S. Finally, the case of a hemophiliac from Haiti with AIDS was reported, which led to the speculation that the disease might have originated in Haiti. As time passed it became clear that AIDS was affecting a wider group of people than was previously thought. By 1982, evidence started to accumulate that AIDS also existed among European Nations and on the African Continent. The number of cases reported in the U.S. was alarmingly high. By 1983, in the U.S. alone, 3,064 cases of AIDS had been reported, out of which 1,292 patients had died.

Active research was going on in France and the U.S. to identify the causative agent of AIDS. Dr. Luc Montagnier from the Pasteur Institute in Paris was the first one to report the causative organism for AIDS. In 1983 his group named this virus as Lymphadenopathy Associated Virus (LAV). Montagnier's work was published in the May 20, 1983 issue of *Science*, and was jointly authored by the 12 members of the team (Barre-Sinousi et al., 1983). In 2008 Luc Montagnier was finally awarded the Nobel Prize in Physiology or Medicine for his contribution to the discovery of HIV. Unfortunately, Montagnier's publication did not receive the required attention from the scientific community. Almost a year later, in the May 4, 1984 issue of *Science*, Dr. Robert Gallo from the National Cancer Institute (NCI), Bethesda, Maryland, published his work. His team named the virus Human T-Cell Lymphotrophic Virus-1 (HTLV-I). Gallo's group consisted of 13 members (Gallo et al., 1984). Dr. Gallo's publication received a great deal of attention from the scientific community. Later, in 1985, both groups published genetic sequences of their samples, and they reported > 90% similarity in sequences. The discovery of HIV was marred with numerous controversies. The Pasteur Institute in Paris, France filed a patent 4 months before Dr. Gallo's group filed their patent. However, the U.S. Patent and Trademark Office granted the patent to Dr. Gallo for blood testing of HIV. This chain of events led to a dispute between Dr. Gallo and Dr. Montagnier: Who discovered HIV? This dispute received international attention and was finally resolved by a meeting between Ronald Regan, then President of United States, and Jacques Chirac, the French Prime Minister. During this meeting it was decided that Dr. Luc Montagnier and Dr. Robert Gallo had made equal contributions towards the discovery of HIV.

The first case of HIV transmission from mother to child by breastfeeding was reported in 1985. In the mean time, the two different names for the same virus (HTLV-I and LAV) started to create confusion; therefore, it became necessary to find a common name for this organism. In 1986, the International Committee on Taxonomy of Viruses decided to drop both names (LAV and HTLV-I) and adopt a new one: Human Immunodeficiency Virus (HIV).

In 1987, WHO established special programs on AIDS, and AIDS earned the reputation as the only disease to be debated in the United Nations General Assembly. In 1987, azidothymidine (Azt) was approved by the Food and Drug Administration (FDA) for treatment of HIV. In 1988, December 1st was declared as World AIDS Day, and in 1991 a Red Ribbon was adopted as the symbol for AIDS awareness. In 1996, a joint United Nations Program on HIV became operational; it was called UNAIDS. In 1996, during the 11th International AIDS Conference in Vancouver, Canada, a report on the efficacy of antiretroviral therapy was presented. In 2001, the U.N. Secretary General granted funds to fight HIV at the global level. Later, there was the G8 Summit which has declared for Universal Access of Antiretroviral Treatment (2010).

GLOBAL DISEASE BURDEN

Since the constitution of UNAIDS by the United Nations, UNAIDS publishes annual reports about the status of HIV and AIDS. These reports provide comprehensive information about HIV and AIDS at both global and national levels. The UNAIDS Annual Report is considered to be the most authentic document for epidemiological facts. As per the UNAIDS report (2012), ~34 million people are living with HIV with an expected range of ~31.4–35.9 million. Although there is no gender bias with regards to HIV and AIDS, the most saddening truth of the UNAIDS Report is that ~3.3 million children are also infected with HIV; children in this category are up to the age of 15 years. Out of different regions of the world, the African subcontinent is the worst affected by HIV infections with ~23.5 million. The range is actually ~22.1–24.8 million, or ~2/3rd of the total people infected with HIV in the world. African countries still have the highest incidences of new HIV infections. The highest number of deaths due to AIDS is on this continent. This is indicative of the fact that even though there are serious efforts to control HIV infection at global and international levels, these measures are not as effective as they should be. The Australian continent is least affected with HIV infections and AIDS. Globally ~7,400 new individuals get infected with HIV every day, while ~5,500 people die every day due to HIV and AIDS-related causes. The difference in death and new infection rates leads to the accumulation of ~1,900 new HIV seropositives to the existing pool of HIV infected people. On an annual basis, these figures could add up to ~700,000 people per year, which is close to the population of any of the capital cities of an average European nation (barring a few large cities). Still, the majority of HIV seropositives and AIDS patients are from developing nations or under-developed nations. Such a high prevalence of HIV in these nations is indicative of multiple factors like poor implementation of surveillance programs, poor health care, poor living conditions, lack of resources, social conditions, etc. There is no doubt that more serious efforts are required at local, national, and international levels to control this epidemic in those nations that are worst affected. Despite these devastating statistics, a closer look at new data offers new insights and hope:

1. Since 1996 the prevalence of HIV infection has not seen a significant increase in numbers.
2. There is no doubt that long-term HIV survivors and new infections are major contributing factors towards the accumulation in existing pool of HIV patients.
3. There are certain nations that have shown a decline in the number of HIV patients.
4. There is a decrease in AIDS-related deaths, which is attributable to improved antiretroviral drugs, an improvement in health care facilities, and access to better antiretroviral drugs for HIV-seropositives.
5. There is no doubt that new HIV infections are declining on a global level, although the decline in HIV-seropositive individuals varies from one nation to the next.

CLINICAL STAGES OF HIV

HIV infections are prevalent across the globe. For effective management of HIV patients, and to apply appropriate treatment strategies, there is a need for classification of different clinical stages of HIV infection.

With time, various diagnostic tests were developed, along with improved drugs to treat HIV infection. A combination of drugs and diagnostic tests has been of great help in generating a reasonable amount of data regarding clinical conditions of HIV patients, as well as for the interpretation of laboratory data. Due to the non-availability of any overarching guidelines, every clinician, group of clinicians, and laboratory scientist was interpreting data to the best of their ability to treat patients. This ambiguous interpretation turned out to be a major impediment to develop the best possible treatments across the board. Non-availability of coherent guidelines became a major concern for various issues related to HIV research, awareness and treatment programs such as: (1) HIV surveillance programs, (2) epidemiology, (3) health status of patients, (4) design of new diagnostic tools, (5) recommendation of different diagnostic tests, (6) appropriate treatment strategies, (7) maintenance of clinical data, (8) consultation and advice from other clinicians for the same HIV patient(s), and (9) comparison of prevalence from one geographical location to the next, etc.

The only answer to overcome this bottleneck was to classify clinical conditions into different categories based on existing clinical and laboratory data. This task was taken up for the first time was by The Centers for Disease Control and Prevention (CDC) in Atlanta, Georgia. In 1986 they developed the first ever classification of various clinical conditions among HIV patients. This classification was tremendously useful towards clinical management of HIV patients. The CDC keeps revising this classification on the basis of improved understanding of clinical conditions and the results of diagnostic data. Later, the World Health Organization (WHO) also developed a new classification for HIV infection that is considered to be a reference standard around the world. Although various other classification systems do exist in the literature, the most widely accepted and followed systems are those developed by the CDC and WHO. These two classification systems have the same purpose, but they cater to the needs of HIV patients under different circumstances.

The CDC initially developed a classification system to address the need for better understanding of HIV infections and to develop strategies to control it in the U.S. (CDC, 1992). The criteria for the CDC's classification relies more on diagnostic end-points (i.e. CD4 counts and viral loads). It has been clearly demonstrated that these two clinical parameters are inversely related to each other. Undoubtedly, the CDC system has certain important end-points that are helpful to understand, as well as useful for the improvement of treatment strategies for HIV patients on the basis of diagnostic tests.

It is unfortunate that all HIV-infected people do not have access to the resources available in countries like the U.S. In developing countries, most HIV patients do not have access to either the necessary facilities to monitor disease progression via expensive diagnostic tests nor the expensive and effective drugs necessary for treatment. WHO gave consideration to both of these facts when addressing the problem for those individuals living in less fortunate conditions. This was the major motivating factor for WHO to develop a classification system in 1990 that was based on clinical signs and symptoms.

Classification of Clinical Stages

The WHO clinical staging system for HIV/AIDS was developed in 1990. The WHO system emphasizes clinical parameters as the guideline for clinical decision-making, and is useful in resource-limited settings. This method of clinical staging can be effectively used even without information about CD4 counts or other laboratory diagnostic tests, and it is important to realize that CD4 counts are not the prerequisite for initiating antiretroviral therapy. In 2005, WHO revised its classification system for different clinical stages of HIV infection (WHO Guidelines, 2005). WHO classification has delineated four stages for HIV patients, designated Stages I–IV. A detailed description of these four different stages is given in Table 9.1.

TABLE 9.1 WHO Classification of Clinical Stages of HIV Infections

Stage	Name	Signs and Symptoms
Stage I	Primary HIV infection	Asymptomatic
	(Seroconversion)	Acute retroviral syndrome
		CD4 > 500 cells/mm^3
Stage II	Asymptomatic Phase	Moderate weight loss
		Recurrent respiratory tract infections
		Herpes zoster
		Angular cheilitis
		Recurrent oral ulcerations
		Papular pruritic eruptions
		Seborrhoeic dermatitis
		Fungal nail infections of fingers
		CD4 > 350-499 cells/mm^3
Stage III	Generalized Lympadenopathy	Severe weight loss
		Unexplained chronic diarrhoea
		Unexplained persistent fever
		Oral candidiasis
		Oral hairy leukoplakia
		Pulmonary tuberculosis (TB)
		Acute stomatitis, gingivitis or periodontitis
		CD4 > 200-349 cells/mm^3
Stage IV	Symptomatic Phase	HIV wasting syndrome
		Pneumocystis pneumonia
		Chronic herpes simplex infection
		Oesophageal candidiasis
		Extrapulmonary TB
		Kaposi's sarcoma
		CNS toxoplasmosis
		HIV encephalopathy, etc.
		CD4 < 200 cells/mm^3

Stage-I: Primary HIV Infection or Seroconversion Stage

Within 2–4 weeks after initial exposure of HIV, primary infection represents acute HIV infection. During this stage, large numbers of viruses are produced in peripheral blood of the infected host, which leads to the activation of immune response. This causes production of protective antibodies as well as a cytotoxic T-cell response. This stage is known as "Seroconversion." During Stage-I, the most common symptoms are quite similar to symptoms of mild influenza (e.g. fever, diarrhea, sore throat, headaches, etc.); more than 65% of the cases of HIV infection are presented with these symptoms during early-stage infection. Diagnosis of Stage-I for HIV infection is based on detection of either HIV antigens or anti-HIV antibodies in blood samples. Confirmatory tests are usually recommended to those individuals who are negative in preliminary tests at the seroconversion stage (Stage-I) but considered to be high-risk cases as per their personal history.

Stage-II: Asymptomatic Stage

The next stage of HIV infection is known as the asymptomatic stage. This stage usually lasts for ~10 years post-HIV exposure. During the asymptomatic phase, viral replication slows down significantly, but replication does not stop. At Stage-II, HIV infection drops to low levels in peripheral blood, but antibodies against HIV can be detected on a regular basis. The CD4 T-cell counts are usually > 500 cells/mm^3. Some patients can show CD4 T-cell count < 500 cells/mm^3.

Stage-II can be further extended with use of the right antiretroviral drugs. At present, antiretroviral drugs can prolong this stage up to 20 years or more post-HIV infection. The main objective of any antiretroviral treatment is to keep viral replication to a minimum so that immune-status deterioration can be minimized.

Stage-III: Persistent Generalized Lymphadenopathy

There are no specific signs or symptoms for this clinical stage. Stage-III comes after the asymptomatic phase. In general, persistent generalized lymphadenopathy is observed which lasts for 3 months or more. Swollen lymph nodes are commonly observed, with a size of > 1 cm in diameter. These patients otherwise look healthy, but non-specific lymphadenopathy persists in these HIV-seropositives, and lymph node biopsy is not routinely recommended for HIV patients showing Stage-III.

Stage-IV: Symptomatic Stage

Symptomatic HIV infections are mainly presented with higher incidences of various opportunistic infections and AIDS-associated cancers. This stage of HIV infection is often characterized with multi-system diseases and infections that affect various systems of the body. A rapid decline of immune-status is a hallmark of this stage, and is due to a significant increase in HIV replication. Increased HIV replication results in rapid disease progression. Some constitutional symptoms like fever, malaise, etc., appear at this stage; these can be treated easily. It is difficult to control HIV replication at this stage. The choices for therapeutic interventions for these patients are limited and not very helpful. These symptoms usually signal the terminal stages of illness. HIV patients at Stage-IV require counseling about the final outcome of the disease; it is important that the patient be well informed. There is always the possibility of variations in the spectrum of opportunistic infections, which can vary from one geographical location to another (Walker, 2006).

WHO and the CDC update their classifications of the clinical stages of HIV infections on a routine basis. With time, new clinical symptoms are also emerging among long-term HIV seropositives. A new and emerging health concern among long-term HIV-seropositives is NeuroAIDS, which is discussed later in this chapter.

MOLECULAR BIOLOGY OF HIV

HIV is roughly spherical in shape and measures about 120 nm in diameter. Taxonomically, HIV belongs to the *Lentivirus* genus and the *Retroviridae* family. Being a member of the *Retroviridae* family, the genetic material in HIV is a single-stranded, positive-sense RNA. HIV does not have any DNA in its genome (Figure 9.1). To utilize RNA as a genetic material, HIV has the reverse transcriptase (RT) enzyme, which helps to transfer genetic information to its new progeny. The HIV genome is ~9.8 kb with 9 genes (Figure 9.2), excluding long-terminal repeats (LTRs), and produces 15 viral proteins (Abbas et al., 2008; Frankel and Young 1998).

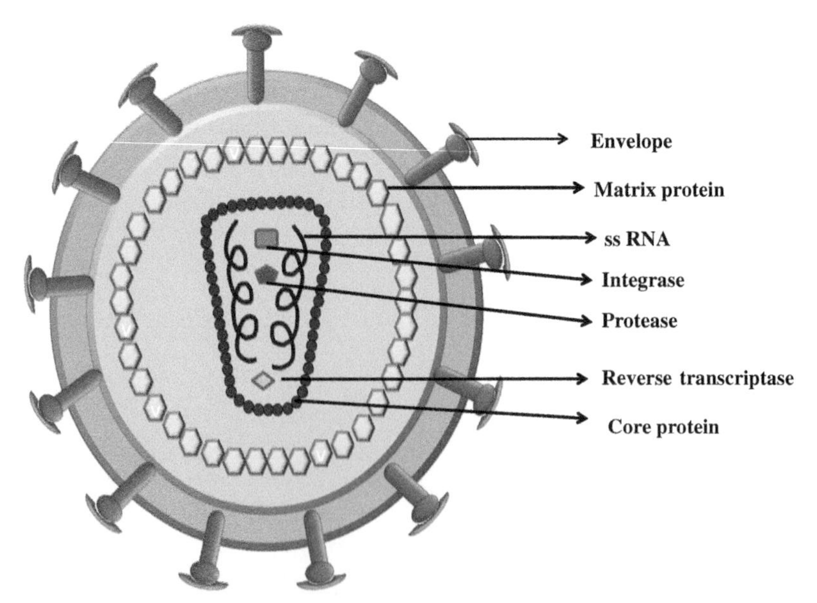

FIGURE 9.1 Structure of HIV. Graphical representation of cross-section of HIV. Envelope is the outermost layer; it consists of a lipid bilayer. The envelope layer is comprised of gp120 and gp41. The layer next to the envelope is the matrix protein. The matrix layer is followed by the core protein. At the center of the virion, two molecules of single-stranded RNA (ssRNA) and other enzymes are present. These enzymes are protease, integrase, and reverse transcriptase. Reverse transcriptase also contains RNase H. The location of each individual protein and RNA is shown in the figure. (Polymerase is not shown; it contains integrase, protease, reverse transcriptase, and RNAse H).

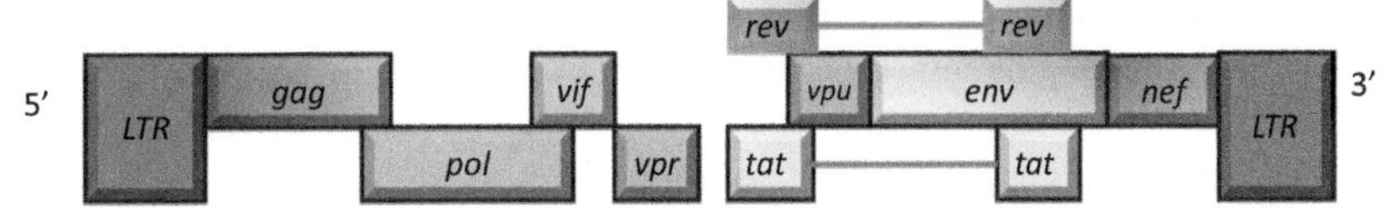

FIGURE 9.2 HIV genome. Schematic representation of the HIV genome. The genome is 9.8 kB in size, and consists of 9 genes, which are flanked by LTRs on either side of the genome. These 9 genes finally produce 15 proteins: *env*, envelope; *gag*, group specific antigen; *LTR*, Long Terminal Repeat; *nef*, negative factor; *pol*, polymerase; *rev*, regulator of expression of viral proteins; *tat*, transactivator of transcription; *vif*, viral infectivity factor; *vpr*, viral protein R; *vpu*, viral protein U.

These 15 proteins can be divided into sub-categories: (i) Structural proteins (Gag, Env, Pol), (ii) Regulatory proteins (Tat, Rev), and (iii) Accessory proteins (Nef, Vif, Vpr, Vpu).

LTRs are present at both ends of the HIV genome and help in the integration of the pro-virus into host-cell DNA. Synthesis of new virions starts from the LTR because transcriptional factors of host cells bind to the LTR region of the viral genome.

Envelope (Env)

"Env" is derived from "envelope." The genomic location of *env* is nucleotides 5771–8341; it is 2,570 nucleotides in size. The total size of the Env precursor protein is 160 kDa. The envelope is a glycoprotein and is synthesized as a polyprotein precursor. The Env precursor is known as gp160, and is processed by a cellular protease that results in production of two proteins: (i) surface Env protein, also known as glycoprotein gp120; and (ii) a transmembrane glycoprotein called gp41. The gp120 glycoprotein contains determinants that interact with T-cell receptors, and gp41 interacts with co-receptors; gp41 consists of three domains: (i) a domain essential for membrane fusion known as the ectodomain; (ii) a transmembrane protein that serves as an anchor; and (iii) a cytoplasmic tail. The gp120 glycoprotein primarily binds with the target cell due to co-receptor interaction, which induces conformational changes in gp41. Conformational changes in gp41 promote formation of a gp120/gp41 glycoprotein complex, leading to fusion of the virus to the host cell membrane. At the heart of the fusion reaction is the ectodomain of gp41; this region contains a highly hydrophobic N-terminus ("fusion peptide") and two heptad repeat motifs referred as N-helix and C-helix.

Group Specific Antigen (Gag)

Gag produces a protein of 55 kDa that contains 500 AA. The *Gag* gene encodes for Pr55Gag, which is a polyprotein precursor. Pr55Gag is cleaved, leading to formation of four proteins: p6, p7, p17, and p24, as well as two spacer peptides, p1 and p2. p17 is a matrix protein (MA), p24 is a capsid protein (CA), p7 is a nucleocapsid (NA), and p6 is known as Vpr-binding protein. The N-terminal of MA is mainly responsible for targeting and binding to plasma membrane.

Long-Terminal Repeats (LTR)

LTR is present on either side of the viral genome. It harbors *cis*-acting elements, which are required for RNA synthesis, and is the initiation site for transcription of the viral genome. *LTR* consists of three regions: U3 (unique, 3′ end), R (repeated), and U5 (unique, 5′ end). Various elements present in U3 help in direct binding of RNA polymerase II (pol II) to DNA templates. Newly synthesized viral RNA falls into three major classes: (i) unspliced RNAs, which function as precursors for Gag and Gag-Pol polyprotein; (ii) partially spliced mRNAs (~5 Kb), which encode Env, Vif, Vpr, and Vpu proteins; and (iii) small but multiple-spliced mRNAs (1.7–2.0 Kb), which encode for Rev, Tat, and Nef.

Negative Factor (Nef)

"Nef" evolved from "negative factor." Its genomic location is from 8343–8963 nucleotides, and the total length of the *nef* gene is 620 nucleotides. Nef is an accessory protein to HIV and contains about 206 AA with a molecular mass of 27 kDa. Nef is expressed during the early stages of replication in host cells. Nef down-regulates expression of cell receptors like CD4, CD8, CCR5, CXCR4, etc. These are the reasons Nef is considered an important protein for *in vivo* pathogenesis. The functions of the Nef protein, which have been observed in *in vitro* studies, are as follows: (i) It perturbs endocytosis; (ii) It modulates signal transduction pathways in infected cells; (iii) It enhances viral infectivity; and (iv) It supports fusion of HIV-1 to target cells.

Nef seems to play a significant role in alterations of CNS functioning because Nef is present in higher levels in astrocytes, causing alteration in their growth. Nef can alter electrophysiology of neurons and induce inflammatory mediators from monocytes.

Polymerase (Pol)

The name "Pol" is derived from "polymerase," and its genomic location is from 1839–4642 nucleotides; it consists of 2,803 nucleotides. Pol protein is 112 kDa and consists of 935 AA. The *pol*-encoded enzymes are initially synthesized as part of a large polyprotein whose synthesis results from a rare frame-shifting event. The individual *pol*-encoded enzymes, viral proteases (PR or p10), reverse transcriptase

(RT or p64), and integrase (IN or p32) are cleaved from the polyprotein precursor by viral proteases. Integrase protein promotes insertion of linear but double-stranded pro-viral DNA into the chromosome of the host cell. Integration is an absolutely necessary step for viral replication because integrase mutant viruses fail to spread infections.

Regulator of Expression of Viral Proteins (Rev)

The name "Rev" is derived from "Regulator of Expression of Viral Proteins." Rev is encoded by two exons, both of which are essential to producing functional proteins. The genomic location of *rev* is 5516–5591 and 7925–8199; it is comprised of 75 and 274 nucleotides respectively. Rev is a regulatory protein and is essential for regulation of viral replication. The molecular mass of Rev protein is 19 kDa, and it contains 116 AA. Rev down-regulates post-transcriptional splicing of viral mRNAs. Rev contains two functional domains: (i) an arginine-rich domain that binds with viral RNA and supports nuclear localization, and (ii) a leucine-rich domain that is hydrophobic and mediates nuclear export of viral RNA.

Transactivator of Transcription (Tat)

The word "Tat" is derived from "transactivator of transcription." Tat is a regulatory protein essential for regulation of viral replication. *tat* comprises 259 nucleotide sequences, and its genomic location is 5377–5591. Tat is an 86–110 AA long protein with a molecular mass of 16 kDa. Tat protein consists of several functional domains. Secreted Tat may be taken up by neighboring cells, therefore Tat affects both infected and non-infected cells. Tat induces apoptosis of neurons both *in vivo* and *in vitro* via oxidative stress pathways.

Viral Infectivity Factor (Vif)

Vif is an accessory protein. The name is derived from "viral infectivity factor." The genomic location of *vif* is nucleotides 4587–5165, and it is comprised of 578 nucleotides. Vif has a major role in the production of infection-competent new virions from infected cells. The molecular weight of the Vif protein is 23 kDa, and it is made up of 192 AA. Vif is expressed during late stages of HIV replication, and is localized in the cytoplasm of infected cells. Vif appears to act during viral assembly in virus-producing cells, or subsequently in virion maturation to produce virions competent for reverse transcription in the target cell.

Viral Protein U (Vpu)

Vpu produces a protein of 16 kDa that consists of 82 AA. Viral protein U is a type-1 integral membrane phosphoprotein unique to HIV. The genomic location of *vpu* is from nucleotide 5608–5856, and it consists of 248 nucleotides. Vpu enhances pathogenesis *in vivo,* even though it is not an essential protein. The Vpu protein usually gets expressed during replication in the host cell. Vpu can't be detected in virions because it does not get packaged into viral particles. It performs two major functions during HIV replication (i) It enhances the release of viral particles, and (ii) It promotes CD4 degradation. The outcome of Vpu-induced CD4 degradation is to liberate gp160 from Env/CD4 complexes in the endoplasmic reticulum.

Viral Protein R (Vpr)

Vpr consists of 96 AA. Therefore, the predicted molecular mass of Vpr is 12.7 kDa. The genomic location of Vpr is 5105–5396 and it contains 296 nucleotides. Vpr is an accessory protein of HIV, and is not essential for viral replication. Truncation of ORF from Vpr has resulted in the production of slow-replicating viral progeny. This is the rationale for the name Vpr: "viral protein, regulatory." Vpr is an accessory protein and is therefore not crucial for replication, but it has been implied in various biological functions like transcription of new viral genomes, apoptosis induction, disruption of cell-cycle control, induction of defects in mitosis, nuclear transport of the pre-integration complex (PIC), facilitation of reverse transcription, suppression of immune activation, as well as reduction of the HIV mutation rate. Vpr is capable of breaching cell membranes. This property supports entry of extracellular Vpr into uninfected cells.

REPLICATION: STEPS AND DRUG TARGETS

Knowledge of HIV replication is essential to develop better antiretroviral drugs and effective treatment strategies. Unless we know the HIV replication steps, we will not be able to find the drug targets. In other words, we can say that a better understanding of HIV replication has given us new drug targets that can be studied and exploited for development of new drugs. The latest concept that is under investigation is to explore the possibilities of blocking the entry of HIV into host cells. It is a reasonably good choice that can be used in combination with drugs rather than using it as the sole strategy to stop HIV infection (Male et al., 2006).

HIV, being a retrovirus, does not have DNA as a genetic material to pass its genetic information to the next viral generation. To overcome this problem, HIV has a unique enzyme called "reverse transcriptase." RNA can be reverse transcribed into DNA with the help of reverse transcriptase to complete the process of viral replication. HIV, like any other virus, is simply a bag of nucleoproteins. HIV cannot replicate by itself because it lacks the synthetic machinery for replication. For replication of HIV, it needs the synthetic machinery of a host or a host's cells. Replication of the virus will only start after the cells get infected with HIV;

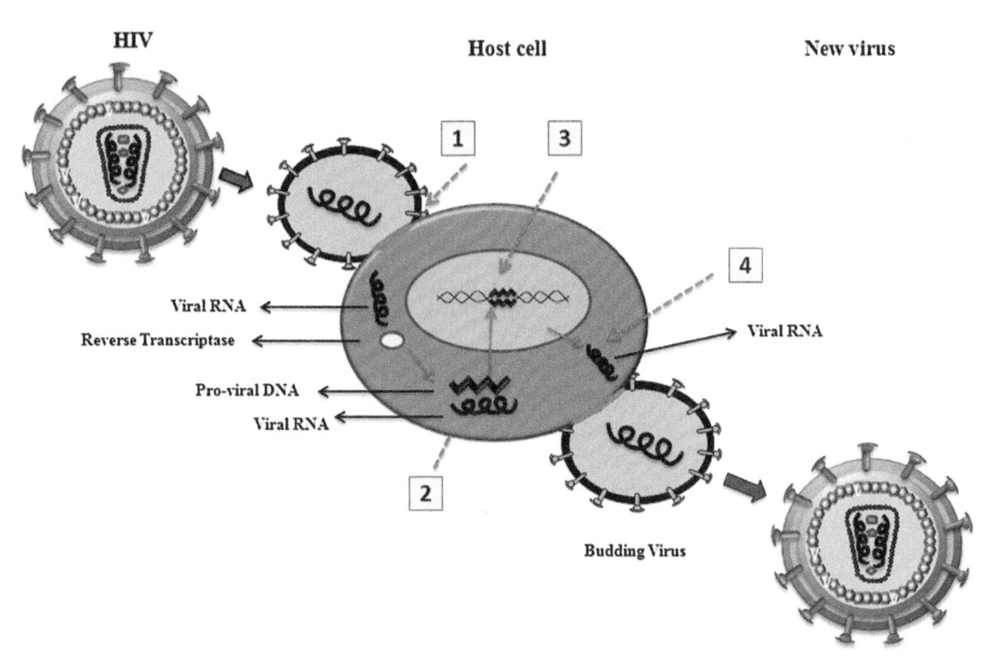

FIGURE 9.3 HIV regulation steps and drug targets. Schematic representation of HIV replication along with the stages where different groups of antiretrovirals work. During infections, HIV attaches to the cell surface of target cells and fuses with the cells to release viral RNA and other proteins. Reverse transcriptase produces pro-viral RNA in the cytoplasm. Pro-viral RNA moves to the nucleus and integrates with the host cell genome with the activity of the integrase. After integration of pro-viral DNA into the host DNA, it gives rise to mRNA, which finally translates into different proteins required for synthesis of new virions. These proteins get cleaved by proteases to get assembled into new virions; new virions are released into circulation due to budding from the cells. Steps for drug targets are mentioned in numerals in blocks: Step 1 is the target for fusion inhibitors, Step 2 is the target for reverse transcriptase inhibitors, Step 3 is the target for integrase inhibitors, and Step 4 is the target for protease inhibitors. ~: Viral RNA; ~: Pro-viral DNA.

the process of viral replication starts with the attachment of viral particle until the time of release of new virions from infected cells (Freed, 2001). The process of HIV replication can be divided into the following main steps: (1) Viral attachment and entry, (2) Reverse transcription, (3) Integration of pro-viral DNA, (4) Transcription and translation, and (5) Completion and release (Figure 9.3).

(1) Viral Attachment and Entry

This is the first step of viral replication. In this step HIV has to come closer to the cells that have specific receptors for HIV. CD4 is the main receptor for HIV attachment to the cells and the target cells for HIV infection are T-lymphocytes. CD4 is a 58 kD glycoprotein present on the surface of a unique T-cell sub-population known as T-helper cells. Apart from T cells, CD4 receptors are also expressed on other cells like monocytes, macrophages, dendritic cells, microglial cells, and several other cell types. The first step for HIV infection is the attachment of HIV for its entry into cells.

The process of HIV entry starts with the interaction of two envelope proteins, namely gp120 and gp41. The gp120 viral protein binds with the CD4 receptor with high affinity. Binding of gp120 induces a conformational change in gp41, which unfolds and moves toward the cell surface. The gp41 protein binds with co-receptors for HIV (i.e. either CXCR4 or CCR5). Attachment of both gp120 and gp41 to their specific receptors leads to fusion of the viral envelop with the plasma membrane of the target cells. Fusion of the viral envelope to the cell membrane facilitates delivery of the viral genome and proteins into the host cells. This is the most common pathway for the entry of HIV into cells, and is a lucrative target in designing drugs. This is an area of interest for drug development at the current time. Drugs that can block entry of HIV into cells are not the most effective means to stopping HIV infections. Therefore, initial research was focused on developing antiretroviral drugs targeting other crucial steps to inhibit HIV replication.

(2) Reverse Transcription

After fusion of the virus with cells, delivery of the viral genome into the host cell takes place. In the cytoplasm of host cells, reverse transcriptase starts working by copying viral RNA into complementary DNA (cDNA). This newly synthesized DNA is known as "pro-viral DNA." The major limitation of this process is that the RT enzyme does not have proofreading capabilities, making this process highly error prone. This is the reason that HIV develops resistance to drugs. For scientists, this turned out to be the most useful target to block HIV replication because it is the first step in producing new virions after viral entry. If one can block this step, then production of new viral particles could be stopped. This is the reason various classes of drugs have been developed to block this step (NRTI, NtRTI, NNRTI). These drugs are known to inhibit the activity of RT. Another benefit associated with these drugs is that alternative analogs can be used to control development of resistance against any one of these drugs.

(3) Integration of Pro-viral DNA
Once the pro-viral DNA is synthesized, it has to be integrated into the host DNA. The newly synthesized pro-viral DNA integrates with the host's cell DNA; this process takes place due to the activity of integrase. Integrase is an enzyme of viral origin. The nine genes of HIV are flanked by Long Terminal Repeats (LTRs). LTRs are essential for integration of pro-viral DNA into host DNA. If one can stop or block the integration of pro-viral DNA into the host, it can be a great strategy to stop viral replication. New drugs have been developed to block this stage of the replication cycle. Drugs that can block the integration step are known as integrase inhibitors (e.g. Isentress, which was approved by the FDA in 2007).

(4) Transcription and Translation
After integration of viral DNA into the host cell, transcription and translation of viral genes starts. In some circumstances, the pro-viral DNA gets integrated but the virus does not undergo active replication because it remains latent. This phenomenon is known as "latent infection." Production of the infective virus requires activation of various cellular transcription factors of the host cells. Host cellular transcriptional factors, upon activation, lead to transcription of viral RNA with the help of cellular RNA polymerases. Splicing of transcripts takes place in a specific manner to produce mRNAs, which finally leads to the synthesis of different viral proteins. Tat and Rev are two proteins that are produced at this stage, and they support viral replication in activated T-cells. At a later stage, Env and Gag proteins are produced. The full-length RNA binds with gag protein and gets packaged into new virus particles.

(5) Assembly and Release
This is the last and final step of HIV replication. Env glycoprotein passes through the cytoplasmic endoplasmic reticulum and gets transported to the Golgi complex. In the Golgi complex, Env protein is cleaved by proteases and processed into two glycoproteins, gp120 and gp41. These glycoproteins move toward the host cell's plasma membrane, where gp41 anchors gp120 to the cell membrane of infected cells. Gag and gag-pol polyproteins help in budding of viruses. HIV protease cleaves polyproteins into a functional HIV protein and enzymes. Cleavage of these proteins completes the maturation of the newly infective HIV. Protease inhibitors can block this step (e.g. Lopinavir, Ritonavir, etc.).

ANTIRETROVIRAL DRUGS

Antiretroviral drugs are the most commonly known for treating HIV infections. The truth is that they can be used to treat any retrovirus infection. At the start of HIV, there was no drug available to treat HIV infection. Azidothymidine was the first drug approved by the FDA (in 1985) to treat HIV infection. With time, and a better understanding of the mechanism and various steps of HIV replication helped scientists to target important steps in HIV replication (which could be blocked by different drugs). Various drugs have been developed to block these steps of HIV replication and to treat HIV infection. Treatment with antiretroviral drugs is commonly known as Antiretroviral Treatment (ART) (Olender et al., 2012; Boehringer- Ingelheim, 2005). Various antiretroviral drugs are classified into the following categories: (1) NRTIs, (2) NtRTIs, (3) NNRTIs, (4) PIs, and (5) InIs (Table 9.2).

1. Nucleoside Reverse Transcriptase Inhibitors (NRTIs)
Drugs of this category block the activity of reverse transcriptase. Conceptually these drugs are considered to be "False Building Blocks." They are called as False Building Blocks because they serve as alternative substrates for RT. During synthesis of pro-viral DNA, these drugs compete with physiological nucleosides. These drugs have a minor modification of the azido group, which is attached to the ribose sugar. The azido group leads to termination of DNA synthesis and blocks further HIV replication. Some of the common drugs of this group are AZT, ddI, ddC, d4T, 3TC, and FTC. These drugs are analogs of different nucleosides. AZT was the first drug used for HIV treatment, and the success of AZT led to the development of numerous other drugs of the same category. AZT and d4T are thymidine analogs, FTC and 3TC are cytidine analogs, ddI is an adenosine analog, and similarly, ABV is a guanosine analog.
As we know, RT does not have proofreading abilities, which is the reason that HIV resistance is commonly observed against the drugs of this group. During the course of treatment, the change in nucleoside analogs has proven to be effective in preventing drug resistance.

2. Nucleotide Reverse Transcriptase Inhibitors (NtRTIs)
This is another version of the NRTIs. The difference with NtRTIs is that these contain a nucleotide rather than a nucleoside. Nucleoside analogs must be converted to nucleotide analogs to show their inhibitory activity. NtRTIs have the added advantage that they skip the step of conversion of nucleoside into nucleotide; therefore these drugs are less toxic compared to nucleoside inhibitors. The mode of action of NtRTIs is similar to NRTIs, but NtRTIs have some serious side effects. TDF is an example of this category. A combination of TDF with ddI should be avoided for the treatment of HIV patients. NtRTIs are present in the market with various combinations.

TABLE 9.2 Antiretroviral Drugs

Class	Chemical Name	Generic Name	Trade Name
E.I or	T-20	Enfuvirtide	Fuzeon
F.I	MVC	Maraviroc	Selzentry
In.I	RGV	Raltegravir	Isentress
NRTI	ABC	Abacavir	Ziagen
	AZT	Azidothymidine	Azetidine
	ddI	Didanosine	Videx
	3TC	Lamivudine	Epivir
	d4T	Stavudine	Zerit
	ddC	Zalcitabine	Hivid
	ZDV	Zidovudine	Retrovir
NNRTI	DLV	Delavirdine	Rescriptor
	EFV	Efavirenz	Sustiva
	ETR	Etravirine	Intelence
	NVP	Nevirapine	Viramune
NtRTI	TDF	Tenofovir	Viread
PI	ATV	Atazanavir	Reyataz
	APV	Amprenavir	Agenerase
	IDV	Indinavir	Crixivan
	LPV	Lopinavir	Kaletra
	NFV	Nelfinavir	Viracept
	RTV	Ritonavir	Norvir
	SQV	Saquinavir	Invirase
	TPV	Tipranavir	Aptivus

Abbreviations: E.I., Entry Inhibitors; F.I., Fusion Inhibitors; In.I, Integrase Inhibitors; NNRTI, Non-Nucleoside Reverse Transcriptase Inhibitors; NRTI, Nucleoside Reverse Transcriptase Inhibitors; NtRTI, Nucleotide Reverse Transcriptase Inhibitors, PI, Protease Inhibitors.

3. Non-Nucleoside Reverse Transcriptase Inhibitors (NNRTIs)
   This is another group of drugs that blocks or targets the activity of reverse transcriptase. The mode of action of NNRTIs is different than for NRTIs. NNRTIs bind directly to RT. Binding of NNRTIs with reverse transcriptase makes RT unable to catalyze pro-viral DNA synthesis. NNRTIs bind with RT at a site close to the binding site of the nucleoside, therefore, NNRTIs block the binding site and reduce or inhibit nucleoside binding. These do not require any further activation for their actions (e.g. Nevirapine, Delavirdine, Efavirenz, etc.).
4. Protease Inhibitors (PIs), Entry Inhibitors (EIs) and Fusion Inhibitors (FIs)
   These are a different group of antiretroviral drugs which are used to treat HIV infections. These drugs block different steps of HIV infections and replication. Examples of each drug is mentioned in Table 9.2.
5. Integrase Inhibitors (InIs)
   As the name suggests these drugs inhibit activity of integrase. The FDA approved a drug in this category in 2007 for HIV treatment (e.g. Raltegravir).

HIV RESISTANCE AND ANTIRETROVIRAL TREATMENT

Resistance against drugs is a very common problem with HIV infections. After a certain period of time, resistance becomes such an issue that different treatment modalities have to be tried to treat patients who fail to respond to a regular course of treatment (Tang and Shafer, 2012). Since the focus of this chapter is not on the clinical aspects, a brief of three different treatment modalities to treat HIV seropositive with drug resistance is discussed.

Highly Active Antiretroviral Treatment (HAART)

This is a new name for combinational therapy for HIV infections. HAART therapy uses a combination of 3–4 antiretroviral drugs of different classes. Drugs selected for the HAART protocol are from the NRTI, NNRTI, PI, and InI categories. The strategy to treat HIV patients using a HAART regimen was decided by a panel from the National Institutes of Health (NIH) and other international organizations. HAART is designed to reduce the complexity of dose scheduling and problems related to the adherence of complex drug scheduling. A physician has to weigh the potential risks and benefits of drugs to a patient before the selection of drugs for a HAART regimen. HAART treatment is an effective way to deal with drug resistance. To overcome the problem of HIV mutation, HAART treatment targets different stages of HIV replication simultaneously so that the development of resistance against drugs is reduced to a great extent.

Salvage Therapy

Salvage therapy is also known as Mega-HAART. This therapy is used when HAART does not give the expected results. In other words, salvage therapy is a last resort to treat HIV infection in those patients who are resistant or not responding to other treatment strategies. At present, up to nine drugs can be administered in different combinations under this therapy. This is a very expensive treatment for HIV patients and it also has serious side effects.

Drug Holiday

A drug holiday is another treatment strategy for HIV patients. The rationale for this protocol is to stop treatment for a certain period of time so that HIV can exist without any selection pressure of a drug, and as a result, HIV will become more susceptible to anti-HIV drugs. If antiretroviral drugs are administered after the interruption, they may be effective in treating HIV that is resistant to drugs.

NEW TYPES OF ANTIRETROVIRALS

Since the discovery of HIV and the miseries associated with HIV infections, the scientific community has worked to prevent HIV infections by finding newer anti-HIV drugs. Tireless researches have led to the identification of novel drug targets that can prevent HIV entry into cells.

The initial process for HIV infection is the attachment of the virus to target cells. During infection, the viral envelope glycoproteins (i.e. gp120 and gp41) interact with CD4, a primary receptor for HIV infection, and chemokines receptors (CCR5 or CXCR4), which are also present on target cells (Berger et al., 1999). During HIV infection gp120 binds to CD4 and undergoes conformational changes. These conformational changes in gp120 are transmitted to gp41. These changes support gp120 interactions with co-receptors and gp41. Recently, one more host cell surface protein (i.e. protein disulfide isomerase, PDI) has been reported to play an important role in HIV entry. PDI belongs to a family of proteins that catalyzes formation, reduction, and isomerization of disulphide bonds. Intracellular PDI also assists in the folding of nascent polypeptides chains. PDI molecules that are attached to the cell surface are continuously shed and rapidly replaced by new PDIs.

Cell surface PDI has been shown to bind with CD4 receptor. Upon HIV infection, gp120–PDI–CD4 (a ternary complex) is formed. The conformational change that occurs in gp120 after binding to CD4 is because of reduction in some of its disulfide bonds, which is catalyzed by PDI of the gp120–PDI–CD4 complex (Fenouillet et al., 2001; Ryser and Fluckiger, 2005).

The knowledge gathered about gp120, gp41, CD4, and chemokine receptors suggests that this process can be divided into the following steps: (1) Interaction of gp120 with CD4, (2) Interaction of co-receptor (CXCR4/CCR5) with gp120, and (3) Activation of gp41.

Targeting Steps 1 and 2 has led to the development of various new inhibitors of HIV entry that are under clinical investigation (Table 9.3).

TABLE 9.3 Inhibitors of HIV Entry

Compound	Mechanism	Manufacturer	Identity
(A) gp-120–CD4 Binding			
BMS-378806	gp120-binding	BMS	Indole derivative
BMS-488043	-do-	BMS	2nd generation compound
Pro 542	-do-	Progenics	CD4–IgG4
TNX-355	CD4-binding	T/B	Anti CD4
(B) gp-120–Co-receptor Binding			
SCH-C*	CCR5 binding	SP	OPPA
SCH-D	-do-	SP	Not disclosed
Pro-140	-do-	Progenics	MoAb
Tak-779*	-do-	Takeda	Quaternary ammonium anilide
Tak-220	-do-	Takeda	2nd generation compound
GW873140	-do-	GSK	Piperazine derivative
UK-427, 857	-do-	Pfizer	–
AMD3100*	CXCR4 binding	AnorMed	Bicyclam
KRH-2731	-do-	Kureha	Arginine derivative
(C) Fusion Inhibitors			
Fuzeon	CHR mimic	Trimeris	Peptide

Abbreviations: B, Biogen; BMS, Bristol Myers Squibb; GSK, Glaxo Smith Kline; MoAB, Monoclonal Antibodies; OPPA, Oxymino-piperidino-piperidine-amide; SP, Schering-Plough; T, Tanox.
*Discontinued.

1. Inhibitors of gp120 and CD4 Interaction
   Targeting the first step of HIV infection (i.e. the interaction between gp120 and CD4) seems to be the most rationale step to prevent infection. BBS-378806 (BMS-806), an indole derivative developed by Bristol Meyers Squibb (BMS), inhibits the step mentioned above. BMS-806 binds at a site in gp120 that is in the vicinity to the site where the disulfide bonds (which undergo PDI-mediated reduction) are located. BMS-806 has been successfully tested against a large number of clinical isolates, even at nanomolar (nm) concentrations. However, BMS-806 did not turn out to be a very successful option because several HIV mutants were found to be replicating successfully, even in the presence of BMS-806. Another inhibitor of protein disulfide isomerase, PRO542, is under phase II clinical trials; it binds to gp120. The peptide CD4-M33 acts as a mimic for the CD4 domain and binds to gp120 effectively, at least in *in vitro* conditions (Jacobson et al., 2004; Martin et al., 2003). TNX355, a monoclonal antibody known to bind with CD4, is also under investigation for its anti-HIV

effects. The major limitation of these inhibitors is their bioavailability when administered via oral route.

2. Inhibitors of gp120 and Co-Receptor Interaction
CXCR4 and CCR5 are chemokine receptors that act as co-receptors for HIV entry into target cells. The natural ligands for CXCR4 are RANTES, MIP-1α, and MIP-1β, while the natural ligand for CCR5 is SDF-1α. The knowledge of availability of natural ligands for CCR5 and CXCR4 has provided a new direction for the development of ligands for chemokine receptors, the goal being to impair entry of HIV into target cells by blocking co-receptors. UK427,857 prevents binding of gp120 to CCR5; it is effective at low nanomolar concentrations. TAK-220 binds to CCR5 and CXCR4, and has been found to be effective to inhibit HIV entry. Both of these inhibitors are under further investigation.
AMD-3100 (a bicyclam) has been identified to work against CXCR4. AMD-3100 has been effective to suppress HIV infections, but has been reported to have serious side effects, so it was discontinued and replaced by another analog (i.e. AMD-070). AMD-070 has been reported to be effective at low nanomolar concentrations, can be orally administered, and is under clinical investigation.

3. Inhibitors of gp41 Activation
The third step is formation of the active state of gp41, which is targeted by T-20, a small 36 AA peptide. T-20 is reported to reduce viral RNA in the blood of HIV seropositives. It is FDA approved under the name Enfuvirtide, but it is not orally bio-available yet.

The role of protein disulfide isomerase (PDI) in HIV entry into target cells is an area of extensive research, with the goal of targeting PDI using drugs or inhibitors to prevent HIV infections. Various attempts to knock out PDI in mammalian cells have resulted in lethal mutations. Therefore, the only choice was to develop strategies that exclusively target the cell surface PDI. Unfortunately, there is no drug available on the market that can target cell surface PDI.

Another approach to control HIV infection could be by targeting the interaction between PDI and CD4. If an antibody can be raised that can block the interaction between PDI and CD4, then entry of HIV into its target cells can be blocked. Since PDI and CD4 both are proteins of host origin, it is easy to use them as drug targets as they are less prone to mutations. Binding between gp120 and CD4 can also be targeted for anti-HIV treatment by preventing PDI from reducing gp120 disulfide bonds. Under these circumstances, gp120 will not undergo conformational changes, so HIV entry into target cells will be prevented.

The first disulfide reduction catalyzed by PDI may result in further oxidation and reduction in gp120, like a cascade reaction. It is also possible that in the near future several potential drug targets can be identified, and better drugs could be designed to prevent HIV infections.

METHODOLOGY AND PRINCIPLES

In addition to being highly infectious, HIV was also a totally new and unknown infection to the world. The credit for the discovery of HIV goes to Luc Montaignier. The most complicated problem associated with researching HIV was the risk associated to work with it. In certain circumstances, risks for HIV infection are high. Clinicians as well as emergency responders like policemen, firemen, and paramedical staff are at high-risk of exposure due to their workplace circumstances. Under laboratory conditions the situation is a bit different as most of the time people are working under known circumstances and risk. Still, laboratory staff is always working with concentrated HIV stock or a very high virus titer. In both laboratory and clinical conditions, it is very important to follow safety procedures.

Laboratory personnel deal with HIV on a daily basis and should adhere to safety precautions whenever they perform experiments like growing HIV stocks, infecting new cells, quantifying viral infectivity in cultures, measuring end results like p24, viral load, etc. Safety precautions include use of disposable lab ware, not using sharps unless essential, not using glassware, wearing masks while working with live HIV, using double gloves that are puncture proof, and wearing disposable lab coats. All procedures should be performed under Bio-Safety Level-3 (BSL-3) conditions. None of the materials infected with HIV should be taken outside of the laboratory without inactivating HIV and following specifically outlined safety procedures.

In case of antiretroviral drug testing, HIV stock is needed, so the virus has to be grown in the laboratory. A system for monitoring HIV infection is also needed. Testing viability of cells in the presence of HIV and after adding antiretroviral drugs is the most common method. There are various methods for testing antiretroviral drugs, but the choice of method depends upon cell type and HIV (Peters et al., 2013; Zhu, 2005). Apart from this, assays are also needed to monitor end points to evaluate a drug's anti-HIV effect. Some of the most commonly used end points are determination of p24 levels, viral load (viral RNA), reverse transcriptase enzyme activity, etc. The following methods are explained here: growing HIV stock, monitoring antiretroviral drug toxicity, and evaluating anti-HIV effects of antiretroviral drugs. Also included is a brief overview of assays for antiretroviral drugs.

Growing HIV Stock

Two situations are possible to grow HIV stock. One can grow viral stock that has been obtained from patients for further study. In this case larger volumes of HIV stock is needed; usually this stock of HIV is grown in primary cells like peripheral blood mononuclear cells (PBMCs) from healthy donors.

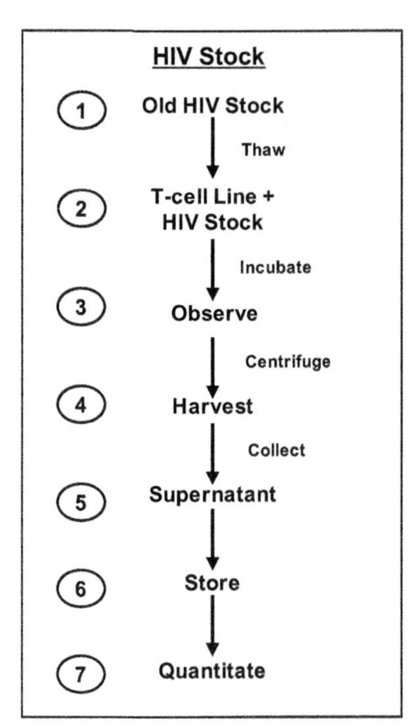

FLOW CHART 9.1 Schematic protocol to grow new HIV stock.

The other situation is to grow viral stock from a laboratory strain of HIV. Laboratories involved with HIV research require HIV stock on a routine basis for various experiments. These viral stocks can be grown from previous stocks by culturing HIV with permissive cells. HIV-infected cells can be grown for a limited period of time and HIV will replicate and be released into the culture supernatant. New stock of HIV can be harvested for further experiments. To make sure that viral stock has the same concentration every time, one has to adhere to endpoints. These endpoints could be Multiplicity of Infection (MOI), p24, RT enzymes, or viral load. Endpoints can be chosen on the basis of the facilities and expertise available in the laboratory, as well as by further application of HIV stock in the laboratory (Flow Chart 9.1). To grow viral stock, one should be familiar with the principle and rationale of the procedures and steps involved. Before starting with an HIV culture, one must consider the following points:

Principles

1. Source of HIV
   This should be taken from the previous stock or the culture growing in the laboratory. It is always a good practice to use a fixed concentration of virus or known concentration of virus to develop new viral stock. This helps in growing stock of the same quality in terms of concentration of virus in new stock.
2. HIV Permissive Cells
   Since HIV is known to grow only in $CD4^+$ cells, so it is always advisable to use cells that are $CD4^+$. In general, under laboratory conditions HIV stock is grown using T-cell lines. The reason to use T-cell lines is that they are unlimited source of cells that can be grown as (and when) needed. Some of the common T-cell lines used to grow HIV stocks are MT-4, HUT_{78}, H-9, etc. Growth of HIV can be monitored with giant cell or syncytium formation in cell cultures after infecting with HIV.
3. Infection of Cell
   It is always a good idea to use overnight grown cells for the initiation of new HIV stock. HIV stock can be added into a fixed number of cells. HIV infection can be checked for syncytium formation at regular intervals. After a certain duration of incubation, one can harvest HIV from culture supernatant. If a more concentrated stock of HIV is needed, then new cells can be added to the same culture. Addition of new cells compensates for cells that have died due to HIV-induced cell death.
4. Harvesting of Viral Stock
   After a certain period of incubation of HIV and cells, new stock of HIV can be harvested by centrifuging culture. After centrifugation, cells will pellet at the bottom of the tube and new HIV will be present in the supernatant. The cell pellet has to be discarded while the supernatant must be collected. At this time supernatant should be divided into small aliquots as needed. These aliquots can be frozen and thawed as per laboratory requirements.
5. Quantitation of Viral Stock
   One can use any of the assays like p24, RT assay, or MOI to quantify HIV titer. Detailed methods can be obtained from any standard laboratory manual regarding HIV.

Crucial Steps

1. Old Viral Stock
   Works as the source of HIV to infect new cells.
2. Mixing of Cells with HIV
   Since we used CD4 positive cells, they will be infected with HIV and allow HIV to replicate and to produce new virions.
3. Harvesting
   After incubation of HIV with cells, HIV will be released into the culture supernatant. HIV can be collected by centrifugation. Centrifugation removes the cells, and supernatant will have new HIV virions.
4. Quantitation
   This is done to measure HIV particles in culture supernatant.

Assays for Antiretroviral Drugs

Antiretroviral drugs are drugs that can work against retroviruses. In general, the term "antiretroviral drug" is used for anti-HIV drugs. Since the discovery of HIV, there has been a desperate need to develop easy and convenient methods to evaluate antiretroviral drugs. There is no suitable animal model for HIV, so it is impossible to test the activity of antiretroviral drugs in animals. There are various assays that have been developed to test antiretroviral drugs *in vitro*. Before testing antiretroviral drugs for their anti-HIV effect, it is important to study the toxicity of the drug. Toxicity as well as the antiretroviral effect of a drug can be conveniently studied under *in vitro* conditions using different cell types. Protocols for these two considerations are discussed here.

Monitoring Antiretroviral Drug Toxicity

This assay is based on the application of MTT dye, which works with the action of mitochondrial enzyme. When MTT is added to cells, it gets converted into a blue-colored product. The blue-colored product can be dissolved, and its OD measured. OD is directly proportional to the viability of cells in a linear range. Only viable cells will convert MTT and give a blue color, whereas dead cells will not be able to convert MTT. These are the criteria for the application of MTT assay to evaluate drug toxicity. The MTT assay is versatile in its applicability; it can be easily used even in a 96-well format. The use of a 96-well format has the added advantage that a large number of drugs, and different concentrations of drugs with lesser numbers of cells, can be tested at the same time (Flow Chart 9.2). To set up an experiment to study the drug's toxicity, the following procedure has to be performed: (1) culturing cells, (2) plating of cells, (3) preparation of drug dilution or concentrations, (4) addition of drugs, (5) addition of MTT, (6) stopping the MTT reaction, and (7) evaluation of data.

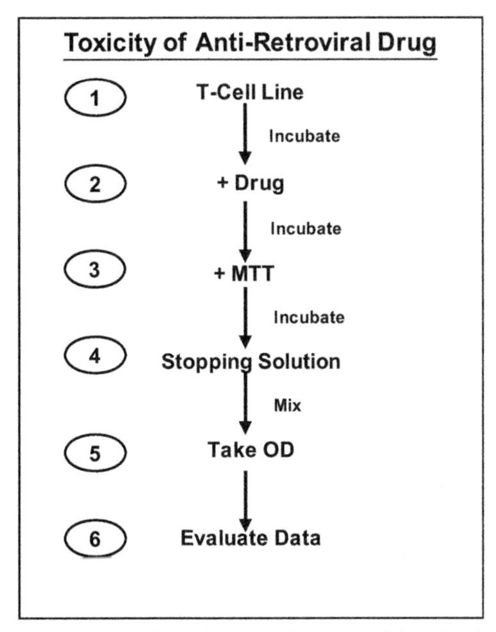

FLOW CHART 9.2 Schematic protocol to test toxicity of antiretroviral drugs using MTT assay.

Principles

1. Culturing Cells
   For this experiment one can use primary cells like PBMC or tumor cell lines. One should select cells that are targets for HIV infection. Toxicity of the drug may vary in different cell types. For anti-HIV effect it is always useful to select T-cell lines because they can be infected with HIV, hence results of drug toxicity and anti-HIV effects of the drug can be easily compared. If someone is using T-cell lines, then it is better to use overnight grown cells for this assay. As needed, cells can be grown either in large or small volumes.
2. Plating of Cells
   The next step is to plate cells in a 96-well plate. Before plating cells it is essential to know the numbers of cells required for each well so that they will give a reasonable OD that can be read. Then the volume of cell culture media in the well has to be adjusted so that the wells can also accommodate other components like drugs, stopping solution, etc. A fixed number of cells in each well will give a comparison of the toxic effect caused by the drug being tested.
3. Preparation of Drug Dilution
   One should check drug solubility before starting the experiment. Some drugs are water-soluble and some are soluble in organic solvent. When drugs are soluble in organic solvent, the solvent should be prepared with the minimal possible volume of organic solvent otherwise the organic solvent itself may contribute towards cell cytotoxicity. When using organic solvent to dissolve the drug, it is always advisable to use control wells for the solvent too, so that the effect of the solvent can be observed. The important point is to be careful when working with drugs which are not soluble in water.
4. Addition of Drugs
   Determine the drug concentration in such a manner that after addition of the drug to the cells, each well will have the desired effective concentration of drug. The usual method for doing this is to use a 2×, 5×, 10×, or 20× concentration of drugs. After addition of drugs, the plate is incubated.
5. Addition of MTT
   After the desired time interval, drug toxicity can be evaluated by adding MTT solution to each well. Then MTT has to be incubated with the cells so that live cells can metabolize MTT to give a blue color; this is done because it is an enzymatic reaction.
6. Stopping the Reaction
   The MTT reaction is stopped by adding stopping solution after the desired period of incubation. Addition of stopping solution will prevent further reaction because stopping solution causes cell death or lysis of cells.

7. Data Recording and Calculation
   Once the reaction is terminated, OD of the plate has to be read at specified a wavelength. On the basis of OD, one can find out the concentration of drugs that are toxic to the cells.

Special Note

The concentration of drugs that is toxic to cells should usually be avoided when evaluating their anti-HIV effect.

Crucial Steps

1. Culture of Cells
   The desired number of cells must be grown for the experiment because they are the target cells.
2. Cell Plating
   As per the experimental protocols, a fixed numbers of cells must be plated into the required number of wells with the required volume of culture medium.
3. Preparation of Drug Dilution
   The required concentration of drugs must be prepared by keeping in mind the effective concentration in their respective wells.
4. Addition of Drugs
   At this stage, either different drugs or different concentrations of the same drug in specified volumes must be added to each well.
5. MTT reaction
   After incubation of the cells with the drug for the specified time, MTT dye must be added so that viable cells can give the blue color.
6. Stopping the Reaction
   Stopping solution is added to stop the MTT reaction.
7. Data Recording and Calculation
   Take an OD of the plate in an ELISA reader and calculate toxicity.

Evaluating Anti-HIV Effects of Antiretroviral Drugs

This assay is simply based on principle of MTT assay for evaluation of drug toxicity. One should select cells that can be productively infected with HIV and HIV infection should induce cell death. So the cell death caused by HIV infection can be evaluated using MTT. Once we add antiretroviral drugs, and if those drugs are effective against HIV then there is a decrease in cell death due to anti-HIV effect of non-toxic concentrations of drug/s. This is one of the most convenient and cheapest assay to screen of antiretroviral drugs. Due to the convenience, this method can be easily employed for screening of large numbers of drugs and even to evaluate different analogs of same drugs to compare their efficacy (Flow Chart 9.3).

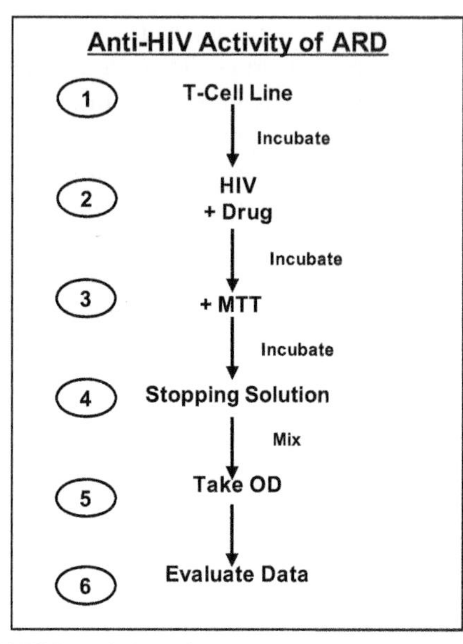

FLOW CHART 9.3 Schematic protocol to test anti-HIV activity of antiretroviral drugs using MTT assay.

It is advisable to follow all biosafety guidelines when working with infectious HIV. To set-up an experiment to study anti-HIV effect of drug/s, following steps have to be performed: (1) growing cells of choice, (2) plating of cells, (3) thawing of HIV stock, (4) preparation of drug dilution or concentration, (5) addition of drugs, (6) addition of MTT, (7) stopping the MTT reaction, and (8) evaluation of data.

Principles

1. Growing cells
   For this experiment it is usually recommended that T-cell lines be used that can be infected with HIV; cell death should occur after HIV infection. If a T-cell line is used, then it is better to use cells grown overnight for this assay.
2. Plating of Cells
   The next step is to plate cells grown overnight in 96-well plates. Before plating cells it is essential to know the number of cells required for each well so that a reasonable OD can be obtained during measurement.
3. Thawing of HIV Stock
   Take out HIV stock with a known titer. Thaw it and mix it into each well in an appropriate volume. Cells infected with HIV will show cell death.
4. Preparation of Drug Dilution
   After evaluation of drug toxicity, one should use only a non-toxic concentrations of drug to evaluate the anti-HIV effects of the drug that is being tested. It is advisable to try at least five different concentrations of the drug in order to test its anti-HIV effects.

5. **Addition of Drugs**
   Adjust the drug concentration so that after addition of other components the final drug concentrations are reached in their respective wells. Usually one can use 2×, 5×, 10×, and 20× concentrations of drugs. After addition, drug plates have to be incubated in CO_2 incubator.
6. **Addition of MTT**
   To find the effect of different concentrations of drug, the MTT solution is added to each well. Mixing can be done with either a plate mixer or multichannel pipette. Then incubate the cells for the required period of time so that the live cells can metabolize MTT.
7. **Stopping the Reaction**
   The MTT reaction can be stopped by adding the stopping solution. After adding the stopping solution one can mix the crystals formed after MTT reaction by mixing the solution.
8. **Data Recording and Calculation**
   When the reaction is stopped, the whole plate must be read for OD at the specified wavelength in the ELISA reader. On the basis of OD, one can find out the concentration of drug that is showing an anti-HIV effect.

Special Note

Use an appropriate control for the blank, drug, and HIV only. One can use XTT or other dyes for this assay as they more convenient.

Crucial Steps

1. Growing of Cells
   At this stage a desired number of cells must be grown that is permissive to HIV infections.
2. Cell Plating
   As per the experimental protocol, cells have to be plated in wells at a required volume of medium.
3. Thawing of HIV Stock
   A known HIV viral stock has to be thawed to infect cells in culture.
4. Infection of Culture
   A required volume of HIV stock has to be added into each well to infect cells.
5. Drug Preparation
   The required concentrations of drug has to be prepared by keeping in mind the effective concentration in the respective wells.
6. Evaluation of Drug
   At this stage, different drugs and their different concentrations at specified volumes must be added to respective wells.
7. MTT Reaction
   After incubation of cells with drugs for a specified time, MTT dye must be added.
8. Stopping the Reaction
   Add the stopping solution to stop MTT reaction.
9. Observation
   Take an OD of the plate and calculate toxicity.

NEUROAIDS: AN EMERGING HEALTH CONCERN

Since 1981, HIV researchers have seen enormous development towards the diagnosis of infection and improvement in antiretroviral drugs. AZT was the only drug approved in 1985. At present, HIV patients have numerous drug options and treatment modalities. An improvement in antiretroviral drugs and treatment strategies has made it possible to prolong the asymptomatic stage of HIV patients (Stage-II) to ~20 years or more post-HIV exposure. No doubt this has improved the quality of life of patients.

Early last decade, a sudden increase in neuropsychiatric complications was observed among long-term HIV seropositives. In the general population, the prevalence of neuropsychiatric diseases is 10–15%. Therefore, it was surprising to observe a prevalence of 50% or more neuropsychiatric complications among HIV seropositives (McCombe et al., 2009; Power et al., 2009). Initially, various names were given to these neuropsychiatric complications, such as HIV-associated dementia (HAD), HIV-associated encelopathy (HIVE), HIV-associated minor cognitive and motor disorders (MCMD), etc. (McArthur et al., 2005). Now all of these neuropsychiatric disorders have been grouped under the name "NeuroAIDS." In the past 10 years or more, NeuroAIDS has turned out to be another health issue among long-term HIV seropositives. Some of the common neuropsychiatric complications are listed in Table 9.4.

TABLE 9.4 NeuroAIDS: Common Neuropsychiatric Disorders

Addiction
Anxiety
Depression
Epilepsy
Mania
Mood disorders
Neurocognitive impairment
Neuropathic pain
Physical disability
Seizures

These neuropsychiatric complications in HIV-seropositives affect both the central nervous system (CNS) and peripheral nervous system (PNS). The CNS is considered to be one of the most protected organ systems in the body. It is protected by the blood–brain barrier (BBB). The main function of the BBB is to regulate the entry of any biomolecules or agents into the CNS. It is intriguing to note that most of the cells present in the CNS do not get infected with HIV; this is due to the fact that they do not express the receptors required for HIV infection. All the evidences suggests that neurons are non-permissible to HIV infections, as they too do not have receptors for HIV infectivity. The question then becomes: How are these cells affected by HIV infection? One of the best rationales for HIV infection of the CNS is the "Trojan Horse Hypothesis." According to this hypothesis, HIV enters the brain along with the infected cells from the systemic circulation (e.g. monocytes or T cells). Once HIV enters the brain, it can remain latent for a long time, and at some point in time it can start active replication, which leads to the production of various biomolecules. In turn, these viral biomolecules can induce numerous pro-inflammatory and inflammatory cytokines in the local milieu. No doubt these cytokines are produced as a part of the protective immune response, but in CNS they can cause neuronal death via apoptosis. Neurons do not regenerate, so neuronal death results in permanent damage.

Neuronal damage/death could be responsible for neuropsychiatric complications among HIV-seropositives. NeuroAIDS is expected to become a major health concern for HIV patients for the following reasons: (1) the number of patients is increasing on a daily basis, (2) symptoms of NeuroAIDS hit right at the prime age of a human life (i.e. between the ages of 35 to 45), (3) a continuous supply of antiretroviral medications is costly, (4) NeuroAIDS patients need additional medication, (5) NeuroAIDS symptoms usually render these patients less productive, which can result in a substantial loss of individual and/or family income, and (6) the eventual need for a caretaker or caregiver can (again) translate into a serious financial burden.

Although the exact reason for the onset of NeuroAIDS is not yet known, there are various possibilities under active consideration: (1) Is HIV itself responsible for NeuroAIDS? (2) Is NeuroAIDS a secondary complication that has emerged due to long-term antiretroviral treatment? (3) Is there low or negligible penetrance of ART into the brain? (4) Are low levels of persistent and chronic HIV infection contributing towards the development of NeuroAIDS? or (5) Is it a combined effect of all these possibilities? The present understanding of NeuroAIDS and its causes remains unclear.

Unfortunately, there is no good model either for *in vitro* or *in vivo* studies to understand the mechanism of NeuroAIDS. It is difficult to develop and test the efficacy of drugs to treat NeuroAIDS due to the non-availability of suitable animal models. Because of these circumstances, and by realizing the quantum of expected problems in near future, it is important to focus on a better understanding of the pathogenesis and drug targets for NeuroAIDS (Verma et al., 2010, 2012).

BONE MARROW TRANSPLANTATION: A PROBABLE CURE FOR HIV

There is no doubt that a cure for HIV must still to be found so that this dreaded disease can be brought under control. Although there is not yet any single strategy to cure HIV, some clues from HIV patients themselves have offered possibilities for possible treatments in future. There is only one patient who has been cured using this approach. This man is famously known as Berlin Man or Berlin Patient.

During HIV epidemic, it was noticed that certain individuals were classified as high-risk individuals as per the criteria for HIV infectivity. These individuals had been exposed to HIV several times, and surprisingly they did not get infected, even though they were never on any antiretroviral treatment. This observation led to the conclusion that these individuals must have "natural resistance" against HIV infections (Shearer and Clerici, 1996). Further studies have revealed that the major reason for natural resistance against HIV infection is the requirement of co-receptors for infection; under normal circumstances these co-receptors serve as receptors for chemokines. Initially it was believed that CD4 receptors, which are present on T cells and monocytes, are required for HIV infection. The co-receptors that have been identified as essential for HIV infection are CXCR4 and CCR5. On the basis of infectivity through these receptors, HIV has been classified into two different strains. One is known as the X-4 strain (which shows a preference for infectivity through CXCR-4), and the other is known as the R5 strain (which shows a preference for CCR5 receptors).

Extensive research showed that people who are naturally resistant to HIV infection have a mutation in the CCR5 gene. The mutation is a deletion of a 32-bp sequence from the gene sequences of CCR5, which is the reason it is known as a Δ32 deletion. The story got more complicated with more detailed studies when it was found that these mutations could be either homozygous or heterozygous. Homozygous mutants (i.e. CCR5Δ32/Δ32) are fully resistant to HIV infections, while individuals carrying heterozygous deletions have shown very slow progression of the disease after exposure to HIV. So far, it has been observed that this deletion exists only in Caucasians. Even among Caucasians, the frequency of this mutation is only about 1–3%. Hence, the possibility that it could be used as a common therapeutic approach is not a viable option. Nevertheless, this fact has not discouraged scientists and clinicians from making use of this natural phenomenon.

Berlin Patient earned a reputation as being the first case to test the above-mentioned fact that bone marrow transplantation with a resistant mutation could cure HIV infection. "Berlin Patient" is a 40-year old male who was HIV positive for more than 10 years. He was under a HAART regimen, and his HIV infection was under proper control due to the right choice of antiretroviral treatment. This man developed acute myelogenous leukemia (AML), and he failed to respond to AML treatment. The only option for clinicians was to treat this patient by allogeneic transplantation (Hütter et al., 2009). The surgeon treating this patient took an extra precaution to transplant this patient with bone marrow from a donor carrying CCR5Δ32/Δ32 mutations. When the patient was given this allogeneic transplantation, he did not show any signs of HIV positivity, even though antiretroviral treatment was stopped. The probable reason for HIV negativity in this patient was due to the absence of the co-receptors required for HIV infection. This appears to be a new hope for curing HIV, but it has two major limitations: (1) this mutation is rare, and (2) one has to weigh the substantial risks and benefits of bone marrow transplantation and antiretroviral treatment among HIV seropositives.

ETHICAL ISSUES

HIV does not have an ideal animal model, which is a major handicap for HIV research. The majority of data for HIV have been generated either by *in vitro* studies or directly from human patients. It is next to impossible to get human volunteers for a disease that does not have a cure. This leaves so many unanswered questions. Due to the unavailability of a suitable animal model for HIV, the only option for drug testing is to rely on *in vitro* assays or HIV patients. If a patient is already undergoing antiretroviral treatment, is it ethical to change a treatment that is already working for the patient? Similar is the case for HIV vaccine trials.

A few major ethical issues related to HIV infection are discussed below:

1. What are the legal repercussions of an HIV-infected individual knowingly infecting someone else? How can it be prevented?
2. What if a person is infected with HIV, and due to social stigmas, he/she does not inform his/her spouse? This kind of irresponsible behavior can be disastrous to a partner and raises many ethical and legal concerns. What if a pregnancy is involved? Where does it stop?
3. Transmission of HIV due to unethical professional practices is also of concern. These negligent practices can be mistakes by dentists, surgeons, phlebotomist, blood banks, etc., and their mistakes can make their patients sick for life.
4. Another concern is data obtained from partially suitable animal models like monkeys and chimpanzees that could not be applied with full confidence to human subjects too.
5. Sometimes, HIV infection is transferred from victims of crimes to emergency responders. This can lead to serious ethical and legal issues.
6. From a workplace standpoint, what if someone is HIV seropositive and is denied employment? This raises questions regarding the ethical practices of the employer.

Unless a cure is found for HIV, these ethical concerns will remain. In addition, various new ethical issues are bound to arise with time, and they will need to be addressed in order to resolve any such dilemmas.

TRANSLATIONAL SIGNIFICANCE

HIV turned out to be one of the most dreaded diseases of the 20th century. HIV is not only dangerous when contracted, but extremely infectious. The ultimate outcome of HIV infection is AIDS, which is considered to be the final chapter in the life of an HIV patient. Because most of the initial reports of HIV infection came from homosexual populations, much of the early media attention erroneously focused on socio-religious and political issues rather than the disease itself. The major obstacle in HIV research is that there is still no ideal animal model for HIV infection. Various animal models have been tried, including mice, rats, monkeys, chimpanzees, etc., but all of them have one limitation or another. This is the reason that all data about HIV infection and treatment is either based on *in vitro* studies or data from clinics.

Most of the time *in vitro* data cannot be applied as such for *in vivo* models, and certainly not for human studies. Clinical data often suggest important aspects of a disease, but individual variations can make it difficult to apply the same conclusions from one set of individuals to another. The HIV story got another twist when it was found that certain individuals do not get infected with HIV at all due to genetic variation in the receptor expression on cells of the immune system. This variation is attributed to a mutation that can be either homozygous or heterozygous. All these observations have translational significance, which must be explored vigorously to make them practical and useful. As far as the translational value for HIV research is concerned, it is just beyond imagination. Anything that is done with reference to HIV infections is important for its translational value. Even a better understanding of the mechanisms of replication and infection will help in developing a better animal model. An animal model will help in understanding the natural history of HIV, which in turn will be helpful in designing better strategies to treat HIV infections.

In fact, understanding of the steps involved in viral replication has been used to develop new antiretroviral drugs that target different steps/stages of HIV replication. This is the reason that at present we have various types of antiretrovirals like NRTIs, NNRTIs, PIs, InIs, etc. Progress in understanding the mode of action of different antiretroviral drugs has led to the development of HAART and Drug Holiday regimens. Next-generation HIV drugs are under development that will provide better treatment for HIV infections by preventing viral attachment to host cells. No doubt in-depth information about HIV replication and infection, and better treatment strategies, can bring a halt to HIV infections. This information will also be helpful in developing effective prophylactic vaccines.

It will be because of translational research that one day we may be able to say with confidence that like small pox and polio, "*HIV is now eradicated.*"

WORLD WIDE WEB RESOURCES

The World Wide Web has become an important resource for instant and easily accessible information, 24/7. Google searches (http://www.google.com) using three different keywords returned the following results: "HIV," 176 million hits; "AIDS," 312 million hits; and a combination of "HIV" and "AIDS," 308 million hits. The WWW is an excellent place to look for information about different aspects of HIV and AIDS, including history, discovery, clinical stages, clinical parameters, life cycle, routes of infection, myths, treatment types, prevalence, epidemiology, etc. Different search engines will undoubtedly return different depths of information.

Another source of information is the National Center for Biotechnology Information (NCBI) (http://www.ncbi.nlm.nih.gov). This site is maintained by the National Institutes of Health, Bethesda, Maryland. It is an excellent source for the compilation and maintenance of scientifically authenticated information. "HIV" as a key word returned more than 250,000 records, while "AIDS" gave more than 191,000 records; a combination of the two showed more than 105,000 records. This is also a good source for retrieving protein sequences, nucleotide sequences, sequence alignments, primers already in use, as well as tools for designing primers and models to predict signal transduction pathways.

Apart from these sources, Los Alamos National Laboratory, USA (www.hiv.lanl.gov) is a good source for data regarding HIV databases.

There are also various hospitals and clinics that have websites dedicated to HIV information for the general public, along with listings of clinicians treating HIV infections.

For recent developments and scientific information, one can check the websites of different journals published in this area (e.g. JAMA, NEJM, Science, Nature, JAIDS, etc.).

UNAIDS gives valuable information about HIV and AIDS (www.unaids.org). It is one of the most authenticated and updated sources of information about epidemiological data at global levels, and includes information from many different nations.

Another important source for HIV information is The Body (http://www.thebody.com), a resource with easy-to-understand explanations.

There are various other good resources available on the web, including books on HIV and AIDS. Some of them are available for free, like *HIV-2011* (http://www.hivbook.com), while others require a subscription.

As one can see, there is no shortage of web resources for HIV and AIDS, and more are being added every day.

ACKNOWLEDGMENTS

The authors would like to thank Prof. K.C. Upadhyaya, Director, AIB, AUUP, NOIDA, for providing the necessary resources to complete this book chapter. Mr. Dinesh Kumar's word processing efforts and graphical work are also duly appreciated. In addition, the authors would like to express their thanks to Priyadarshini Mallick, Sushmita Chaudhary, and Ashima Agarwal for performing extensive literature searches and providing the other materials that have brought the chapter to this stage.

REFERENCES

Abbas, K. A., Lichtman, A. H., & Pillai, S. (2008). *Cellular and Molecular Immunology* (6th ed.). India: Elsevier Press.

Barre-Sinousi, F., Chermann, J. C., Rey, F., Nugeyre, M. T., Chamaret, S., Gruest, J., Dauget, C., Axler-Blin, C., Vezinet-Brun, F., Rouzioux, C., Rozenbaum, W., & Montagnier, L. (1983). Isolation of a T-lymphocytic retrovirus from a patient at risk for Acquired Immune Deficiency Syndrome. *Science*, *220*, 868–871.

Berger, E. A., Murphy, P. M., & Farber, J. M. (1999). Chemokine receptors as HIV-1 co-receptors: roles in viral entry, tropism, and disease. *Annual Review of Immunology*, *17*, 657–700.

Boehringer-Ingelheim Pharmaceuticals Incorporated, (2005). *NNRTI Mode of Action (animation)*. www.boehringer-ingelheim.com.

Centers for Disease Control and Prevention (1992). 1993 Revised classification system for HIV infection and expanded surveillance definition for AIDS among adolescents and adults. *Morbidity and Mortality Weekly Report*, *41*, 1–19.

Fenouillet, E., Barbouche, R., Courageot, J., & Miquelis, R. (2001). The catalytic activity of Protein Disulfide Isomerase is involved in Human Immunodeficiency Virus envelope-mediated membrane fusion after CD4 cell binding. *Journal of Infectious Diseases*, *183*, 744–752.

Frankel, A. D., & Young, J. A. (1998). HIV-1: fifteen protein and an RNA. *Annual Review of Biochemistry*, *67*, 1–25.

Freed, E. O. (2001). HIV-1 Replication. *Somatic Cell and Molecular Genetics*, *26*, 13–36.

Gallo, R. C., Salahuddin, S. Z., Popovic, M., Shearer, G. M., Kaplan, M., Haynes, B. F., Palker, T. J., Redfileld, R., Oleske, J., Safai, B., White, G., Foster, P., & Markham, P. D. (1984). Frequent detection and isolation of cytopathic retroviruses (HTLV-III) from patients with AIDS and at risk for AIDS. *Science*, *224*, 500–503.

Gottlieb, M. S., Schroff, R., Schanker, H. M., Weisman, J. D., Fan, P. T., Wolf, R. A., & Saxon, A. (1981). *Pneumocystis carinii* pneumonia and mucosal candidiasis in previously healthy homosexual men: Evidence of a new acquired cellular immunodeficiency. *New England Journal of Medicine*, *305*, 1425–1431.

Hütter, G., Nowak, D., Mossner, M., Ganepota, S., Mubig, A., Allers, K., Schneider, T., Hofmann, J., Kücherer, C., Blau, O., Blau, I. W., Hofmann, W. K., & Thiel, E. (2009). Long-term control of HIV by CCR5 delta32/delta32 stem-cell transplantation. *New England Journal of Medicine*, *360*, 692–698.

Jacobson, J. M., Israel, R. J., Lowy, I., Ostrow, N. A., Vassilatos, L. S., Barish, M., Tran, D. N., Sullivan, B. M., Ketas, T. J., O'Neill, T. J., Nagashima, K. A., Huang, W., Petropoulos, C. J., Moore, J. P., Maddon, P. J., & Olson, W. C. (2004). Treatment of advanced Human Immunodeficiency Virus type 1 disease with the viral entry inhibitor PRO 542. *Antimicrobial Agents and Chemotherapy*, *48*, 423–429.

Male, D., Brostoff, J., Roth, D. B., & Roitt, I. (2006). *Immunology* (7th Ed.). Canada: Mosby Elsevier Press.

Martin, L., Stricher, F., Missé, D., Sironi, F., Pugnière, M., Barthe, P., Prado-Gotor, R., Freulon, I., Magne, X., Roumestand, C., Ménez, A., Lusso, P., Veas, F., & Vita, C. (2003). Rational design of a CD4 mimic that inhibits HIV-1 entry and exposes cryptic neutralization epitopes. *Nature Biotechnology*, *21*, 71–76.

McArthur, J. C., Brew, B. J., & Nath, A. (2005). Neurological complications of HIV infection. *Lancet Neurology*, *4*, 543–555.

McCombe, J. A., Noorbakhsh, F., Buchholz, C., Trew, M., & Power, C. (2009). NeuroAIDS: A watershed for mental health and nervous system disorders. *Journal of Psychiatry and Neurosciences*, *34*, 83–85.

Olender, S., Wilkin, T. J., Taylor, B. S., & Hammer, S. M. (2012). Advances in antiretroviral therapy. *Topics in Antiviral Medicine*, *20*, 61–68.

Peters, P. J., Richards, K., & Clapham, P. (Eds.), (2013). Human immunodeficiency viruses: Propagation, quantitation and storage. *Current Procotols in Microbiology:* (Vol. 28). USA: Wiley Press.

Power, C., Boisse, L., Rornke, & John, G. M. (2009). NeuroAIDS: An evolving epidemic. *Canadian Journal of Neurosciences*, *36*, 285–295.

Ryser, H. J. P., & Fluckiger, R. (2005). Progress in targeting HIV-1 entry. *Drug Discovery Today*, *10*, 1085–1094.

Shearer, G. M., & Clerici, M. (1996). Protective immunity against HIV infections: has nature done experiment for us? *Immunology Today*, *17*, 21–24.

Tang, M. W., & Shafer, R. W. (2012). HIV-1 Antiretroviral resistance: scientific principles and clinical applications. *Drug*, *80*, 71, 9. e1–25.

UNAIDS Report. (2012). *Report on the Global AIDS Epidemic*.

Verma, A. S., Singh, U. P., Dwivedi, P. D., & Singh, A. (2010). Contribution of CNS Cells in NeuroAIDS. *Journal of Pharmacy and Bioallied Sciences*, *2*, 300–306.

Verma, A. S., Singh, U. P., Mallick, P., Dwivedi, P. D., & Singh, A. (2012). NeuroAIDS and omics of HIV *vpr*. In D. Barh, K. Blum & M. A. Madigan (Eds.), *Omics Biomedical Perspectives and Applications* (pp. 477–511). USA: CRC Press.

Walker, B. D. (2006). AIDS and Secondary Immunodeficiency. In D. Male, J. Brostoff, D. B. Roth & I. Roitt (Eds.), *Immunology* (7th edition, pp. 311–324). Canada: Mosby Elsevier.

WHO Guidelines, 2005, www.who.int/hiv/pub

Zhu, T. (Ed.), (2005). *Method In Molecular Biology. Human Retrovirus Protocols: Virology and Molecular Biology:* (Vol. 304). USA: Humana Press.

FURTHER READING

Alfred, M., Behrens, G., Braun, P., & Bredeek, U. F. 2011. *HIV-2011*. www.hivbook.com.

Barnett, T., & Whiteside, A. (2006). *AIDS in the Twenty-First Century: Disease and Globalization*. USA: Palgrave Macmillan.

Lever, A. M. L. (1996). *The Molecular Biology of HIV/AIDS*. USA: J. Wiley & Sons.

Levy, J. A. (1998). *HIV and the pathogenesis of AIDS*. Washington. DC: American Society for Microbiology.

Sax, P. E., Cohen, C. J., & Kuritzkes, D. R. (2012). *HIV Essentials*. USA: Jones and Bartett.

Stine, G. (2011). *AIDS Update 2012* (Edition 21st). USA: McGraw Hill Companies, Inc.

Weeks, B. S., & Alcamo, I. E. (2010). *AIDS: The Biological Basis*. USA: Jones and Bartlett Learning.

GLOSSARY

Pro-Viral DNA Reverse transcriptase synthesizes DNA from viral RNA. This viral DNA is known as pro-viral DNA.

Antiretroviral Drugs Drugs used to treat HIV infection. These drugs can also be used to treat other retroviral infections (e.g. Azidothymidine).

NeuroAIDS Neuropsychiatric complications in HIV patients are grouped under the term NeuroAIDS (e.g. anxiety, mood disorders, etc.).

Opportunistic Infection An infection by a microorganism that normally does not cause disease but becomes pathogenic when the body's immune system is impaired and unable to fight off infection (e.g. *Pneumocystis carinii*).

Protease Inhibitors Drugs that inhibit the activity of proteases. These drugs are also used to treat HIV infection (e.g. Indinavir).

Integrase Inhibitors Drugs that inhibit the activity of integrase in the case of HIV infections (e.g. Raltegravir).

Reverse Transcriptase An enzyme that reverse transcribes RNA into DNA. It is found in retroviruses (e.g. HIV).

Asymptomatic Stage One of the clinical stages of HIV infection that does not show any significant signs and/or symptoms of HIV infection. HIV replication is low and controlled during this stage.

ABBREVIATIONS

AIDS Acquired Immune Deficiency Syndrome
AML Acute Mylogenous Leukemia
ART Antiretroviral Treatment
BBB Blood–Brain Barrier
CNS Central Nervous System
Env Envelope
Gag Group-Specific Antigen
HIV Human Immunodeficiency Virus
InI Integrase Inhibitor
LTR Long Terminal Repeat
Nef Negative Factor
NNRTI Non-Nucleoside Reverse Transcriptase Inhibitor
NRTI Nucleoside Reverse Transcriptase Inhibition
NtRTI Nucleotide Reverse Transcriptase Inhibitor
PDI Protein Disulfide Isomerase
PNS Peripheral Nervous System
PI Protease Inhibitors
Pol Polymerase
Rev Regulator of Viral Expression
Tat Transactivator of Transcription
Vif Viral Infectivity Factor
Vpr Viral Protein R
Vpu Viral Protein U

LONG ANSWER QUESTIONS

1. Explain HIV replication in detail.
2. What is NeuroAIDS? What are the implications of NeuroAIDS?
3. Discuss different classes of antiretroviral drugs and their mode of action, giving at least one example of each.
4. What is PDI? What is the mechanism by which PDI can inhibit HIV infection?
5. How can antiretroviral drugs be tested *in vitro?*
6. Discuss different genes of HIV and their role in HIV replication.
7. Give a brief account of HIV and AIDS history.
8. Discuss different routes for transmission of HIV infection.
9. Discuss the WHO system for classification of clinical stages of HIV infections.
10. Describe how to grow HIV stock.

SHORT ANSWER QUESTIONS

1. Name different genes of HIV.
2. Define pro-viral DNA.
3. Define NtRTI drugs and how they are different from NRTIs?
4. What is HAART?
5. What are opportunistic infections?
6. What are some other names that were once used for AIDS?
7. Name the receptor and co-receptors essential for HIV infections.
8. What is Salvage Therapy for HIV patients?
9. What is the asymptomatic stage of HIV infection?
10. What is the seroconversion stage?

ANSWERS TO SHORT ANSWER QUESTIONS

1. The HIV genome has nine genes, which are flanked by Long Terminal Repeats (LTRs). These genes are *gag, env, nef, pol, rev, tat, vif, vpr,* and *vpu.*
2. During reverse transcription, viral RNA gets transcribed into complementary DNA (cDNA) due to the action of reverse transcriptase enzyme. This cDNA is known as pro-viral DNA.
3. The full name of NtRTI is nucleotide reverse transcriptase inhibitor. It is a group of antiretroviral drugs that consists of nucleotide analogs instead of nucleosides. Their action does not require activation; this is the difference between NtRTIs and NRTIs.
4. The full name of HAART is Highly Active Antiretroviral Treatment. It is a kind of combinational therapy used to treat HIV infections. This treatment contains 3–4 drugs from different groups of antiretroviral drugs.
5. Opportunistic infections are infections that are rare in the general population. Opportunistic infections commonly occur in severely immunocompromised hosts. AIDS patients show opportunistic infections like *Pneumocystis carinii*, which usually causes pneumonia.
6. The term "AIDS" was coined by the Centers for Disease Control and Prevention (CDC), Atlanta, Georgia, in 1982. Before the term AIDS was coined, this disease had different names like Gay Compromise Syndrome (GCS), Gay Related Immuno-Deficiency (GRID), Acquired Immunodeficiency Disease (AID), Gay Cancer.
7. For HIV, CD4 is the receptor, which is found mainly on a subset of T cells known as T-helper cells. HIV also needs a co-receptor for infection, which can be either CXCR4 or CCR5.
8. As the name suggests, Salvage Therapy is designed to salvage the HIV-infected patient. Salvage Therapy is also known as Mega-HAART Therapy. This therapy is used among those HIV patients who do not respond to (or are resistant to) various antiretroviral treatments. It is an expensive treatment for HIV patients and has serious side effects.
9. Stage-II of HIV infection is known as the asymptomatic phase. In this stage, HIV replication is low and the patient does not show any significant signs and symptoms of HIV infection. A patient can remain in this stage for more than 10 years. This stage can be extended for more than 20 years with proper antiretroviral treatment.
10. Initial HIV infection (i.e. Stage-I) is classified as the "Seroconversion Stage." During this stage, the patient shows mild symptoms of the disease. Patients also produce antibodies against HIV in their serum. It can last up to 6 months.

Chapter 10

Animal Biotechnology as a Tool to Understand and Fight Aging

Pawan Kumar Maurya
Center for Reproductive Medicine, College of Medicine, Taipei Medical University, Taipei, Taiwan

Chapter Outline

SUMMARY

Aging is a biological reality and has its own dynamics that are beyond human control. It is accompanied by loss of normal function, the onset of age-related diseases, and ultimately, death. Human life expectancy is increasing due to advancement in new technologies, but age-related health problems continue to seriously compromise quality of life. The essential aim of aging research is not only to extend life span, but to also maintain good health and quality of life. Anti-aging medicines are now a multimillion-dollar industry. We are in the process of understanding the biochemical/molecular events that occur during the aging process. Dietary flavonoids are now emerging as anti-aging compounds; flavonoids have been demonstrated to have various health-associated properties. Thus, the development of preventive measures or interventions that slows down aging should be worked out to improve the quality of life for millions and reduce the costs associated with health care.

Animal Biotechnology. http://dx.doi.org/10.1016/B978-0-12-416002-6.00010-9

WHAT YOU CAN EXPECT TO KNOW

This chapter provides an excellent introduction to an overview of aging, theories of aging, reactive oxygen species, and use of various animal experimental models for the study of human aging. Translational significance of aging, strategies that will help in delaying the aging process, and ethical issues are also discussed.

HISTORY AND METHODS

INTRODUCTION

Aging is both an opportunity and a challenge. Increased life expectancy is a common feature of most countries today. At the same time, new technologies make ever more costly procedures available to health-care systems. Aging is a biological reality and has its own dynamics, which are beyond human control. Aging is defined when two criteria are met. First, the probability of death at any point in time increases with the age of the organism. This statistical definition applies from yeast to mammals and reflects the progressive nature of an organism. Second, characteristic changes in phenotype occur in all individuals over time due to limiting processes. Biologically, aging is the accumulation process of diverse detrimental changes in the cells and tissues with advancing age, resulting in an increase in the risks of disease and death (Harman 2006). There are many theories that attempt to explain the process of aging. The oxidative stress hypothesis offers the best mechanistic elucidation of the aging process and other age -related phenomena. (Figure 10.1)

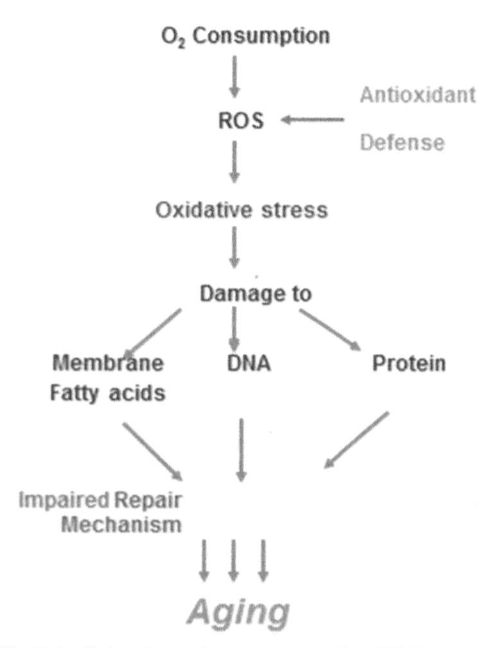

FIGURE 10.1 Role of reactive oxygen species (ROS) in process of aging.

Theories of Aging

There are more than 300 theories, none of which could qualify as being the definitive theory of aging; all of them could be, at best, labeled as "hypothesis" or "aspect theories" (Rattan 2006). Over the years, gerontologists have become resigned to the futility of formulating a unified theory of aging that can encompass its evolutionary, biological, and sociological aspects. Most significantly, it has been clearly shown that the phenotype and the rate of progression of aging are highly variable in different species, in organisms within a species, in organs and tissues within an organism, in cell types within a tissue, in sub-cellular compartments within a cell type, and in macromolecules within a cell. These observations necessarily lead to the conclusion that aging has no universal cause, phenotype, or consequence, except death.

Evolutionary Theories

The evolutionary theory describes aging as an emergent phenomenon that takes place primarily in protected environments, which allow survival beyond the natural lifespan in the wild. The natural life span of a species is termed "essential life span" (ELS) or the "warranty period." Species that undergo fast maturation and have early onset of reproduction with large reproductive potential generally have a short ELS, where as slow maturation, late onset of reproduction, and small reproduction potential of a species are concurrent with a long ELS.

Evolutionary theories argue that aging results from a decline in the force of natural selection. Because evolution acts primarily to maximize reproductive fitness in an individual, longevity is a trait to be selected only if it is beneficial for fitness. Life span is, therefore, the result of selective pressures and may have a large degree of plasticity within an individual species, as well as among species. The evolutionary theory was first formulated in the 1940s based on the observation that Huntington's disease, a dominant lethal mutation, remained in the population even though it should be strongly selected against. The late age of onset for Huntington's disease (30–40 yr) allows a carrier to reproduce before dying, thereby allowing the disease to avoid the force of natural selection. This observation inspired the Mutation Accumulation Theory of Aging, which suggests that detrimental, late-acting mutations may accumulate in the population and ultimately lead to pathology and senescence. Currently, there is scant experimental evidence for this theory of aging.

However, the basic concept that aging results from a lack of selection enjoys a wealth of experimental support. Long-lived *Drosophila* strains can be bred by selecting the offspring of older adults, demonstrating that life span can be altered directly by selective pressure. Life span is species-specific because it is largely a function of survivability and reproductive strategy in a competitive environment. Consequently, organisms that die primarily from predation and environmental hazards will evolve a life span optimized for their own particular environment. This idea was tested in a natural environment by comparing mainland opossums (that are subject to predation) to a population of opossums living on an island free of predators. The Evolutionary Theory predicts that the protected island opossums would have the opportunity to evolve a longer life span, if it were beneficial to fitness. Indeed, island opossums do live longer and age more slowly than their mainland counterparts. The observation that organisms can age in a natural environment indicates that although extending life span can be beneficial to fitness, other considerations might necessitate sacrificing longevity for reproductive fitness. This basic idea of the Disposable Soma Theory of aging argues that the somatic organism is effectively maintained only for reproductive success; afterward it is disposable. Inherent in this theory is the idea that somatic maintenance (in other words, longevity) has a cost; the balance of resources invested in longevity vs. reproductive fitness determines the life span.

Molecular Theories

The Gene Regulation Theory of aging proposes that senescence results from changes in gene expression. Although it is clear that many genes show changes in expression with age, it is unlikely that selection could act on genes that promote senescence directly. Rather, life span is influenced by the selection of genes that promote longevity. Recently, DNA microarrays have been used to assay genome-wide transcriptional changes with age in several model organisms.

Despite vigorous research, there is yet no agreement on the biochemical mechanism responsible for the loss of replicative potential of diploid culture cells. The aging process is not programmed, but is, rather, the absence of selection for maintenance. The life span of an organism is the sum of deleterious changes and counteracting repair and maintenance mechanisms that respond to damage. Aging is defined when the longevity and embryonic development are met. The probability of death at any point in time increases with age of the organism.

The importance of specific kinds of genome instability in aging is becoming increasingly apparent. The accumulation of genomic changes (i.e. point mutations), loss of repeated DNA sequence (such as ribosomal DNA rearrangements), and changes in chromosome number have been proposed as cause of aging. Age-dependent changes have also been observed in the HPRT (hypoxanthine phosphoribosyltransferase) and HLA-A (human leukocyte antigens) genes of peripheral blood lymphocytes in humans. Hypoxanthine-guanine phosphoribosyltransferase (HGPRT) is an enzyme encoded by the HPRT gene. HGPRT catalyzes the conversion of guanine to guanosine monophosphate (GMP) and hypoxanthin to ionosine ionophosphate (IMP). HGPRT has an important role in purine metabolism. HLA-A is a component of certain major histocompatibility complex I (MHC I) cell surface receptor isoforms, which are present on the surface of all nucleated cells and platelets. These play an important role in the immune system. The low frequency of these genomic changes, even in old individuals, casts doubt on their importance in aging. Aging is manifested both by the limited number of cell divisions and the change in phenotype undergone by mother cells prior to senescence. The pace of aging in mother cells is dictated by changes occurring in the ribosomal DNA. One of the most exciting developments in aging research is the identification of an insulin-like signaling pathway that regulates life span in worms, flies, and mice. Life span extension results from the activation of a conserved transcription factor in response to a reduction in insulin-like signaling, indicating that gene expression can regulate life span.

Studies of human centenarians and their relatives have identified a significant genetic aspect of the ability to survive to exceptional ages. A recent study supports the idea that exceptional longevity has a genetic component by identifying a locus on chromosome 4 that may contain gene(s) that promote longevity (Puca, *et al.* 2001). Genetic analysis of human longevity is especially important given that genetic aspects of aging are studied primarily in short-lived model organisms. The main molecular theories of aging are (a) Gene regulation theory; (b) Codon restriction theory; (c) Error catastrophe theory; (d) Somatic mutation theory; and (e) Dysdifferentiation theory.

Cellular Theories

Telomeres, the repeated DNA sequences at the ends of linear chromosomes, are unable to be fully replicated by DNA polymerases. Telomeres consist of the six-base repeating sequence TTAGGG. With each cell division, some of the telomere is lost. But the number of times that most dividing cells can divide is limited. Thus, they will shorten with cell division unless maintained via telomerases, a ribonucleopreotein enzyme that can add telomeric repeat sequences to chromosome ends. For humans, the length of the remaining telomere is usually an indicator of how many divisions a dividing cell has left. Higher levels of oxidative stress increase the rate of telomere shortening. It is proposed that telomere shortening could be a molecular clock that signals the eventual growth arrest. Strong support for this has been recently provided by demonstration that reactivation of telomerases in certain cultured human cells can extend their life span beyond their normal limits. There are several findings that relate telomere shortening to aging in vivo. There appears to be a rough correlation between telomere length and age in human soma. The Cellular Senescence Theory of aging was formulated in 1965 when cell senescence was described as the process that limits the number of cell divisions normal human cells can undergo in culture.

Free Radical Theory of Aging (FRTA)

The Free Radical Theory of Aging was first proposed in 1956 (Harman 1956). It is one of the best-known theories and remains controversial to this day. All organisms live in an environment that contains reactive oxygen species (ROS); mitochondrial respiration, the basis of energy production in all eukaryotes, generates ROS by leaking intermediates from the electron transport chain. The Free Radical Theory is further divided into several hypotheses focusing on the exclusive role of particular organelles and types of damaged molecules in the aging process. One such hypothesis argues that mutations in mitochondrial DNA accelerate free radical damage by introducing altered enzyme components into the electron transport chain. Faulty electron transport results in elevated free radical leakage and ultimately more mitochondrial DNA mutation and exacerbated oxidant production. This "vicious cycle" of mutation and oxidant production eventually leads to cellular catastrophe, organ failure, and senescence. Another hypothesis argues that free radicals cause aging when oxidized proteins accumulate in cells. An age-dependent reduction in the ability to degrade oxidatively modified proteins may contribute to the build-up of damaged, dysfunctional molecules in the cell.

System-Based Theories

In system-based theories, the aging process is related to the decline of the organ systems essential for:

1. The control and maintenance of other systems within an organism, and
2. The ability of organisms to communicate and adapt to the environment in which they live.

In humans, all systems may be considered indispensable for survival. However, the nervous, endocrine, and immune systems play a key role by their ubiquitous actions in coordinating all other systems, and in their interactive and defensive responsiveness to external and internal stimuli.

Neuroendocrine theory proposes that aging is due to changes in neural and endocrine functions that are crucial for coordinating communication and responsiveness of all body systems with the external environment, programming physiological responses to environmental stimuli, and maintaining an optimal functional state for reproduction and survival (while responding to environmental demands). These changes, often detrimental in nature, not only selectively affect the neurons and hormones that regulate evolutionarily significant functions such as reproduction, growth, and development, but also affect those that regulate survival through adaptation to stress. Thus the life span, as one of the cyclic body functions regulated by biological clocks, would undergo a continuum of sequential stages driven by nervous and endocrine signals. Alterations of the biological clock (e.g. decreased responsiveness to the stimuli driving the clock or excessive or insufficient coordination of responses) would disrupt the clock and the corresponding adjustments. An important component of this theory is the perception of the hypothalamo–pituitary–adrenal (HPA) axis as the master regulator, the pacemaker that signals the onset and termination of each life stage. One of the major functions of the HPA axis is to muster the physiological adjustments necessary for preservation and maintenance of the internal homeostasis, despite the continuing changes in the environment.

With aging, a reduction in sympathetic responsiveness is characterized by: (1) a decreased number of catecholamine receptors in peripheral target tissues; (2) a decline of heat shock proteins that increase stress resistance in many animal species, including humans; and (3) a decreased capability of catecholamine to induce these heat shock proteins. The hormones of the adrenal cortex are glucocorticoids (for the regulation of lipid, protein, and carbohydrate metabolism), mineralocorticoids (for that of water and electrolytes), and sex hormones. Among the latter is dehydroepiandrosterone, which decreases with aging; dehydroepiandrosterone replacement therapy has been advocated in humans, despite unconvincing results. Glucocorticoids, as well as other steroid hormones, are regulated by positive and negative feedback between the target hormones and their central control by the pituitary and hypothalamus. With aging – and in response to continuing and severe stress – not only feedback mechanisms may be impaired, but also glucocorticoids themselves may become toxic to neural cells, thus disrupting feedback control and hormonal cyclicity.

Caloric Restriction Theory

Evidence that calorie restriction (CR) retards aging and extends median and maximal life span was first presented in the 1930s (McCay *et al.* 1935). Since then, similar observations have been made in a variety of species including rats, mice, fish, flies, worms, and yeast. Although not yet definitive, results from the ongoing calorie-restriction studies in monkeys also suggest that the mortality rate in calorie-restricted animals will be lower than that in control subjects. Furthermore, studies shows that calorie-restricted monkeys have lower body temperatures and insulin concentrations than do control monkeys and both of those variables are biomarkers for longevity in rodents (Heilbronn and Ravussin, 2003). Calorie-restricted monkeys also have higher concentrations of dehydroepiandrosterone sulfate. The importance of dehydroepiandrosterone sulfate is not yet known, but it is suspected to be a marker of longevity in humans, although this is not observed consistently.

In humans, a major goal of research into aging has been the discovery of ways to reduce morbidity and delay mortality in the elderly. The absence of adequate information on the effects of CR in humans reflects the difficulties involved in conducting long-term calorie-restriction studies, including ethical and methodological considerations. There is also evidence that DNA damage is reduced by CR, possibly as a result of increased DNA repair capacity. If normal feeding is resumed, these animals remain fertile far longer than other animals. CR diverts energy from growth and reproduction towards somatic maintenance and thus may explain the life prolonging effect of CR.

PRINCIPLE

The basic principle that governs aging is the generation of reactive oxygen species (ROS). Although the fundamental mechanisms of aging are still poorly understood, a growing body of evidence points toward the oxidative damage caused by ROS as one of the primary determinants of aging. Aerobic cells produce ROS as a byproduct of their metabolic processes. The body's defense system tries to neutralize the ROS, but if the body's antioxidant defense system is overwhelmed, ROS will cause oxidative stress that will damage proteins, lipids and nucleic acids. A certain amount of oxidative damage takes place even under normal conditions; however the rate of this damage increases during the aging process as the efficiency of anti-oxidative and repair mechanisms decrease.

Reactive Oxygen Species – Causative Agent of Aging

A free radical exists when there are one or more unpaired electrons in atomic or molecular orbitals. Free radicals are generally unstable, highly reactive, and energized molecules. ROS can be classified into oxygen-centered radicals and oxygen-centered non radicals (Table 10.1).

Oxygen-centered radicals are superoxide anion ($\cdot O_2^-$), hydroxyl radical ($\cdot OH$), alkoxyl radical ($RO\cdot$), and peroxyl radical ($ROO\cdot$). Oxygen-centered non radicals are hydrogen peroxide (H_2O_2) and singlet oxygen ($^1O_2$). Other reactive species are nitrogen species such as nitric oxide ($NO\cdot$), nitric dioxide ($NOO\cdot$), and peroxynitrite ($OONO^-$). ROS in biological systems can be formed by prooxidative enzyme systems, lipid oxidation, irradiation, inflammation, smoking, air pollutants, and glycoxidation. Clinical studies reported that ROS are associated with many age-related degenerative diseases, including atherosclerosis, vasospasms, cancers, trauma, stroke, asthma, hyperoxia, arthritis, heart attack, age pigments, dermatitis, cataractogenesis, retinal damage, hepatitis, liver injury, and hypertension (Kumar *et al.* 2012). Free radicals have been implicated in the activation of nuclear transcription factors, gene expression, and a defense mechanism to target tumor cells and microbial infections. Superoxide anion may serve as a cell growth regulator. Singlet oxygen can attack various pathogens and induce physiological inflammatory responses. Nitric oxide is one of the most wide-spread signaling molecules that participate in every cellular and organ function in the body. Nitric oxide acts as a neurotransmitter and an important mediator of the immune response, which is also altered as a function of age (Maurya and Rizvi, 2009).

TABLE 10.1 Reactive Oxygen Species (ROS)

Oxygen Centered Radicals	Oxygen Centered Non-Radicals
$O_2\cdot$	H_2O_2
$\cdot OH$	$^1O_2$
$HOO\cdot$	
$ROO\cdot$	

Free radicals (oxidants) come from two major sources: (a) endogenous and (b) exogenous. Endogenous free radicals are produced in the body by four different mechanisms:

1. From the normal metabolism of oxygen-requiring nutrients. Mitochondria – the intracellular powerhouses that produce the universal energy molecule, adenosine triphosphate (ATP) – normally consume oxygen in this process and convert it to water. However, unwanted by-products (such as the superoxide anion, hydrogen peroxide and the hydroxyl radical) are inevitably produced, due to incomplete reduction of the oxygen molecule. It has been estimated that more than 20 billion molecules of oxidants per day are produced by each cell during normal metabolism. Imagine what happens with inefficient cell metabolism!
2. White blood cells destroy parasites, bacteria, and viruses by using oxidants such as nitric oxide, superoxide and hydrogen peroxide. Consequently, chronic infections result in prolonged phagocytic activity and increased exposure of body tissues to the oxidants.
3. Other cellular components called peroxisomes produce hydrogen peroxide as a byproduct of the degradation of fatty acids and other molecules. In contrast to the mitochondria that oxidize fatty acids to produce ATP and water, peroxisomes oxidize fatty acids to produce heat and hydrogen peroxide, which can then be degraded by the catalase. Under certain conditions, some of the hydrogen peroxide escapes to wreak havoc into other compartments in the cell.

4. An enzyme in the cells called cytochrome P450 is one of the body's primary defenses against toxic chemicals ingested with food. However, the induction of these enzymes to prevent damage by toxic foreign chemicals like drugs and pesticides also results in the production of oxidant by-products.

Superoxide Anion ($\cdot O_2^-$)

Superoxide anion is a reduced form of molecular oxygen created by receiving one electron. $\cdot O_2^-$ is an initial free radical formed from mitochondrial electron transport system. The $\cdot O_2^-$ plays an important role in the formation of other ROS in living systems. The superoxide anion can react with nitric oxide (NO·) and form peroxynitrite ($ONOO^-$), which can generate toxic compounds such as hydroxyl radical and nitric dioxide.

Hydroxyl Radical (·OH)

Hydroxyl radical is the most reactive free radical and can be formed from $\cdot O_2^-$ and H_2O_2 in the presence of metal ions such as copper or iron. Hydroxyl radicals have the highest 1-electron reduction potential and are primarily responsible for the cytotoxic effect in aerobic organism. Hydroxyl radicals react with lipids, polypeptides, proteins, and nucleic acids, especially thiamine and guanosine. They also add readily to unsaturated compounds. When a hydroxyl radical reacts with aromatic compounds, it can add on across a double bond, resulting in hydroxycyclohexadienyl radical. The resulting radical can undergo further reactions, such as reaction with oxygen, to give peroxyl radical, or decompose by water elimination to phenoxyl type radicals.

Hydrogen Peroxide (H_2O_2)

H_2O_2 can be generated through a dismutation reaction from superoxide anion by superoxide dismutase (SOD).

$$2\,O_2^{\,-} + 2\,H^+ \rightarrow H_2O_2 + O_2$$

Enzymes such as amino acid oxidase and xanthine oxidase also produce H_2O_2 from superoxide anion. H_2O_2 is highly diffusible and crosses the plasma membrane easily and is the least reactive molecule among the ROS. It is a weak oxidizing and reducing agent, and is thus regarded as being poorly reactive. It can generate the hydroxyl radical in the presence of metal ions and superoxide anion. It can produce singlet oxygen through reaction with superoxide anion or with HOCl or chloroamines in living systems. It can also degrade certain heme proteins, such as hemoglobin, to release iron ions.

Singlet Oxygen

Singlet oxygen is a non radical which is in an excited state. Singlet oxygen has been known to be involved in cholesterol oxidation. Oxidation and degradation of cholesterol by singlet oxygen was observed to be accelerated by the co-presence of fatty acid methyl ester.

Peroxyl and Alkoxyl Radicals

Peroxyl radicals (ROO·) are formed by a direct reaction of oxygen with alkyl radicals (R·). Decomposition of alkyl peroxides (ROOH) also results in peroxyl (ROO·) and alkoxyl (RO·) radicals. Irradiation of UV light or the presence of transition metal ions can cause hemolysis of peroxides to produce peroxyl and alkoxyl radicals. Peroxyl and alkoxyl radicals are good oxidizing agents. They can abstract hydrogen from other molecules with lower standard reduction potential; this reaction is frequently observed in the propagation stage of lipid peroxidation.

Nitric Oxide and Nitric Dioxide

Nitric oxide (NO·) is a free radical with a single unpaired electron. NO· itself is not a very reactive free radical, but the overproduction of NO· is involved in ischemia reperfusion, and in neurodegenerative and chronic inflammatory diseases such as rheumatoid arthritis and inflammatory bowel disease. When NO· is exposed in human blood plasma, it can deplete the concentration of ascorbic acid and uric acid, and initiate lipid peroxidation. Nitric dioxide adds to double bonds and abstract labile hydrogen atoms initiating lipid peroxidation and production of free radicals.

Peroxynitrite

Reaction of NO· and superoxide anion can generate peroxynitrite, a cytotoxic species that causes tissue injury and oxidizes low-density lipoprotein (LDL). Peroxynitrite appears to be an important tissue-damaging species generated at the sites of inflammation and has been shown to be involved in various neurodegenerative disorders and several kidney diseases. Peroxynitrite ($OONO^-$) can cause direct protein oxidation and DNA base oxidation and modification acting as a "hydroxyl radical-like" oxidant. The significance of peroxynitrite as a biological oxidant comes from its high diffusibility across cell membranes. Nitrotyrosine, which can be formed from peroxynitrite-mediated reactions with amino acids, has been found in age-associated tissues.

Enzymatic Formation

Pro-oxidative enzymes, including NADPH–oxidase, NO–synthase or the cytochrome P–450 chain, can generate reactive oxygen species. Lipoxygenase generates free radicals. Lipoxygenase needs free polyunsaturated fatty acids (PUFA), which are not present in healthy tissue. Membrane-bound phospholipase produces PUFA and lysolecithins. Lysolecithins change the cell membrane structures, and free PUFA are oxidized to form lipid hydroperoxides. Once Fe (II) is oxidized to Fe (III), lipoxygenase can convert polyunsaturated fatty acids (PUFA) into hydroperoxides. These enzymes can oxidize arachidonic acid, a PUFA rich in the central nervous system, into hydroperoxyeicosatetraenoic acid (HPETE). HPETE is then converted into leukotrienes which regulate the immune responses. The production of leukotrienes is accompanied by prostaglandins and histamine, which act as inflammatory mediators. 15–Lipoxygenase has been identified within atherosclerotic lesions, which suggests that this enzyme may be involved in the in vivo formation of oxidized lipids.

METHODOLOGY: MEASUREMENT OF FREE RADICALS AND METHODS TO MONITOR AGING

During its life span, an organism is confronted with oxidative stress on one side from intrinsic origins (like the mitochondrial power generation leaking ROS/reactive nitrogen species (RNS)) and on the other side from extrinsic origins (such as UV light, smoking etc.) Various methods on diverse systems have been documented to study oxidative stress and aging. However, most of the methods are influenced by several factors such as life style, nutrition and types of models. A commonly used alternate approach measures markers of free radicals rather than the actual radical. Markers of oxidative stress are measured using a variety of different assays.

Protein Oxidation/Protein Carbonyl Content

Proteins are the building blocks of the body and likely to be major targets, as a result of their abundance in cells, plasma, and most tissues, and their rapid rates of reaction with ROS / RNS. ROS can lead to oxidation of amino acid residue side chains, formation of protein-protein cross-linkages, and oxidation of the protein backbone, resulting in protein fragmentation. Oxidative stress leads to damage of proteins, resulting in loss of specific protein function; since proteins have unique biological functions, there are often unique functional consequences resulting from their modification. Most of enzymes are made up of proteins. Any damage in the enzyme will lead to disturbance in all metabolic activities. It is estimated that almost every third protein in a cell of an older animal is dysfunctional as an enzyme or structural protein, due to oxidative damage (Pandey and Rizvi, 2010). The measurement of protein oxidation is therefore an important factor for the prediction of the aging process.

Carbonyl group content was determined following the method as described by Renzick and Packer (1994), using the 2,4-dinitrophenylhydrazine (DNPH) assay with slight modification. Approximately 1 mg of erythrocyte membrane protein was precipitated with 20 % TCA (tricloroacetic acid) (1:1vol/vol) in an Eppendorff tube and vortexed for 30 sec. After centrifugation, the clear supernatant was discarded and the pellet was resuspended in 0.5 mL of 10 mM DNPH in 2 M HCl, and allowed to stand at room temperature for 60 minutes, vortexing at 5 min intervals, to facilitate the reaction of DNPH with pellet proteins. The protein was precipitated again with 20 % TCA and then the precipitated protein (pellet) was washed three times with 1.0 mL of 1:1(vol/vol) ethanol:ethyl acetate mixture. Finally, the pellets were dissolved in 0.7 mL of 6 M guanidine hydrochloride at 37°C. After centrifugation for 5 minutes at 6,000 × g to precipitate the insoluble material, the clear supernatant was read against a complimentary blank at their maximum absorbance of 365 nm. A parallel blank was also run with the same procedure using 2M HCl alone instead of 2,4-DNPH reagent. Carbonyl group content is expressed in nanomloes per milligram of protein using a molar absorbance coefficient of 22,000 mol L^{-1} cm-1 (Flow Chart 10.1).

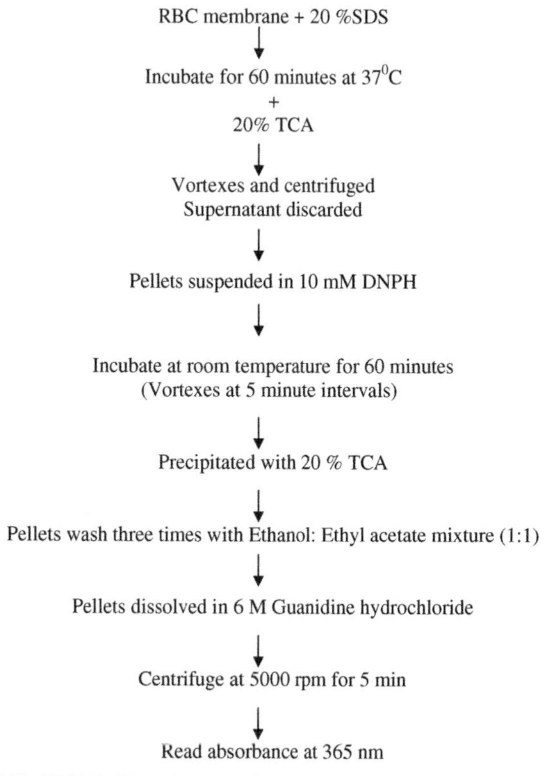

FLOW CHART 10.1 Protocol for the estimation of protein carbonyl content.

Antioxidant Capacity

Antioxidant capacity is the primary measurement to evaluate the state and potential of oxidative stress in aging and other age related diseases. Since imbalance between antioxidant and oxidants generates the condition of oxidative stress, estimation of the reducing power/antioxidant capacity the first step in the prediction of oxidative stress in the aging process. There are several methods to measure total antioxidant capacity in vitro. These methods are based on quenching of free radicals such as 1,1-diphenyl-2-picrylhydrazyl free radical (DPPH·), 2,2-azobis-3-ethylbenzthiazoline- 6-sulfonic acid (ABTS) by antioxidants, inhibition of lipid peroxidation, etc. The FRAP (ferric reducing ability of plasma) assay is superior because it is not dependent on the enzymatic/non-enzymatic method to generate free radicals prior to evaluation of the anti-radical activity of plasma. The FRAP assay offers a putative index of antioxidants, or reducing potential of biological fluids, and is simple, convenient, quick, and reproducible. Total antioxidant capacity (FRAP value) decreases as a function of human age (Rizvi *et al.*, 2006).

Ferric Reducing Ability of Plasma (FRAP) values can be determined by the following method (Flow Chart 10.2) (Benzie and Strain, 1996).

COMMON LABORATORY ANIMAL EXPERIMENTAL MODELS FOR AGING RESEARCH

Animal species phylogenetically close to humans may be used as animal models for the study of human aging. Many species with close genetic homology may serve as translational models to study aging and even age related diseases. Mice and fish are effective models for studying the mechanism of aging, and help in better understanding of genetic and physiological basis of longevity.

Mouse

Mice have relatively short life spans and share 99% of their genes with humans. Various genetic engineering technologies are available that can easily manipulate the genes of mice, which helps in the understanding genetics of human aging. Mice are used to test diets and compounds/drugs for their ability to delay aging and extend longevity in a mammalian model. Caloric restriction studies have demonstrated extended life spans in mice.

Fish

Fish have been used as a gerontological model because of their many characteristic features. Investigators have cited a number of advantages for studying aging including the availability of large cohorts of offspring from single matings, the ectothermic nature of fish, and their reasonably short life span relative to many mammalian species. Other features include the low costs for breeding and maintenance, and the ability to manipulate life span by both temperature reduction and food restriction.

In particular, guppies have proved to be an invaluable model for evolutionary analyses of aging, killifish are short-lived and may be exploitable for life span manipulation

Protocol for the estimation of total antioxidant potential (FRAP assay)

3ml of FRAP reagent + 100µl of plasma

(FRAP reagent- acetate buffer (300mM, pH 3.6) + 2,4,6- tri[2-pyridyl]-s-triazine (10mM in 40mM HCl) solution+ $FeCl_3.6H_2O$ (20mmol/liter) solution in 10:1:1 ratio respectively)

↓

Mixed vigorously

↓

The absorbance was read at 593nm at the interval of 30 seconds for 4 minutes

↓

Aqueous solution of known Fe^{2+} concentration in the range of 100-1000µmol/ liter was used for calibration

↓

Using the regression equation the FRAP values (µmol Fe (II) per liter) of the plasma is calculated

FLOW CHART 10.2 Protocol for the estimation of total antioxidant potential (FRAP assay).

studies, and zebra fish come with a formidable armament of associated biological tools from their widespread use as a model of vertebrate development. These fish are well suited for the investigation of basic processes implicated in aging, such as insulin signaling, oxidative stress, and comparative studies of species with widely divergent longevities (Gerhard, 2007).

Other Models

Many other model systems are used to study human aging, including:(1) human cells; (2) uni-cellular organisms, such as the yeast; (3) roundworm; and (4) fruit flies.

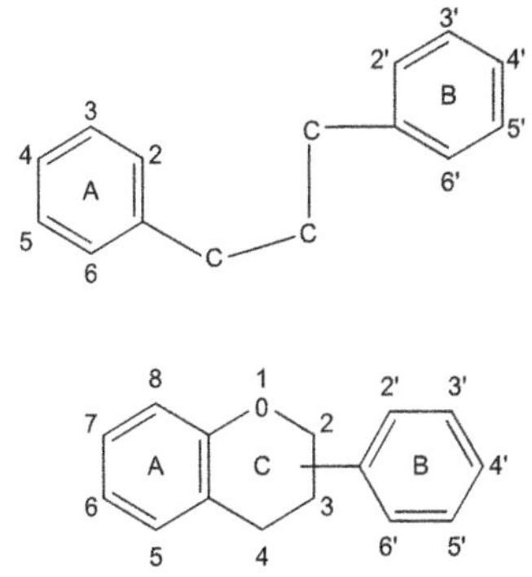

FIGURE 10.2 Basic structure of polyphenols.

POLYPHENOLS AS AN AGENT TO FIGHT AGING

Polyphenols in food plants are a versatile group of phytochemicals. It has been reported that phenolic compounds have antioxidant, antimutagenic, and free radical scavenging abilities. Epidemiological studies showed that increased consumption of phenolic compounds reduced the risk of cardiovascular disease and certain types of cancer and helped in delaying aging process (Maurya and Prakash, 2011). Moderate consumption of red wine, which contains high content of polyphenols, has been associated with the low risk of coronary heart disease. Fruits and vegetables contain diverse phytochemicals of which large fractions are polyphenols. Polyphenols that have antioxidant properties react with the ROS and form products with much lower reactivity.

Flavonoids

The flavonoids are a large and complex group of compounds that occur throughout the plant kingdom that provide flavor, color, and antifungal/antibacterial activity; they also contribute to many aspects of plant physiology. Most plant tissue can synthesize flavonoids. The flavanoids are diphenylpropane derivatives that include flavonols, flavonones, antocynidines, flavones and flavonols. More than 4,000 flavonoids have been found in plants, fruits, and vegetables. There is great interest in these phenolic compounds because of their potential role as anti-aging and cancer chemopreventive agents. This beneficial effect is considered to be mainly due to their antioxidant and chelating activities. However, some flavonoids, such as quercetin, have also been reported to be mutagenic and co-carcinogenic. The ability of flavonoids to scavenge free radicals and block lipid peroxidation raises the possibility that they may act as protective factors against cardiovascular disease and hypertension (Kumar *et al.*, 2010), and there is epidemiological evidence consistent with this hypothesis. Any physiological significance of dietary flavonoids depends upon their availability for absorption and their subsequent interaction with target tissues, but little is known about their transport across the intestine. In general, flavanoid glycosides are resistant to processing, cooking and digestion.

The flavonoids are polyphenolic compounds possessing a basic structure of 15 carbon atoms, arranged in two benzene rings joined by a linear, three-carbon chain. The chemical structures of flavonoids are based on this C_{15} skeleton, which forms a chromane ring bearing a second aromatic ring B at position 2, 3 or 4 (Figure 10.2).

Various sub groups of flavonoids are classified according to the substitution patterns of ring C. Both the oxidation state of the heterocyclic ring C and the position of ring B are important in classification. Multiple combinations of hydroxyl groups, sugars, and oxygen and methyl groups attached to these structures create various types of flavonoids. The A and C rings collectively are often termed as the flavonoid nucleus. (Figure 10.3)

Flavanoids have many biological effects countering inflammatory, bacterial, viral, microbial, hormonal, carcinogenic, neoplastic, and allergic disorders and have been reported for in both *in vitro* and *in vivo* systems (Middleton 2004). Flavonoids exert antioxidant effects by neutralizing all types of oxidizing radicals – including the superoxide (Robak *et al.* 1996) and hydroxyl radicals – and by chelation. (A chelator binds to metal ions in our bodies to prevent them from being available for oxidation.) Flavonoids can also act as powerful chain-breaking antioxidants due to the electron-donating capacity of their phenolic groups.

Tea as a Source of Anti-Aging Compounds

Tea is a natural beverage brewed from the leaves of an evergreen plant called *Camellia sinensis.* The *Camelia sinensis* is a very versatile plant that can grow under almost any conditions. Thus, tea is grown around the world from the Indian sub-continent (in India, Nepal, and Sri Lanka), to China, Japan, Indonesia, Vietnam, to the African subcontinent (in Kenya), to Latin America (in Argentina). As can be imagined, the quality of tea varies dramatically from region

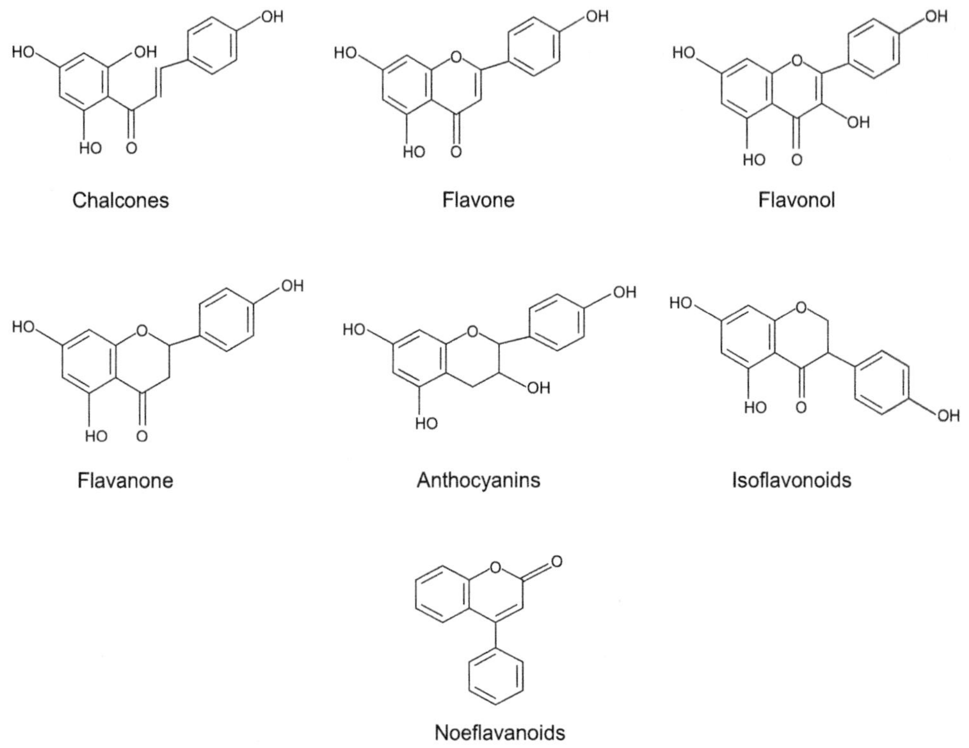

Chalcones Flavone Flavonol

Flavanone Anthocyanins Isoflavonoids

Noeflavanoids

FIGURE 10.3 Structure of different sub groups of flavonoids.

to region, with most of the variations originating in the variation in the climactic conditions of the regions where the tea is grown, and not from the differences in the tea bush itself.

Types of Teas

Green Teas

Green tea is high in an important class of beneficial substances: polyphenols. Polyphenols, acting as antioxidants, reduce the formation of many types of cancer. It seems that flavonoids slow or even halt the oxidation process that allows cholesterol to harden and build on artery walls. In this way, flavonoids have the ability to lower cholesterol and reduce the incidence of heart disease. Green tea is also high in fluoride, the same stuff found in toothpaste.

Black Teas

Tea in general contains fluoride, and all tea aids in digestion and bolsters the immune system. Black tea has both flavonoids and polyphenols, just to a lesser degree than green tea. Recent studies from the country of Denmark and from Harvard Medical School both state that drinking several cups of tea a day (either green or black) can reduce heart attacks and heart disease by as much as 60%.

Herbal Teas

Chamomile contains a mild and gentle sedative. It is recommend to people who want a tea that will calm their nerves or help them sleep, but chamomile also soothes the stomach and eases gas pains. After an illness, chamomile will even re-induce an appetite. In women, chamomile acts as an anti-spasmodic. For generations, women have used it to ease menstrual cramps. It's gentle and natural.

Cinnamon is a strong spice whose chief health benefits have to do with soothing or correcting the stomach and intestines. It helps in releasing gas. This not only helps in digestion and in settling the stomach, but it can also help in treating mild diarrhea.

Cloves have been used by herbalists for many centuries. By inhaling the strong odor of cloves, nausea is curbed and the stomach is settled. Drinking clove tea aids in digestion and curbs flatulence.

Echinacea is one of the most recognized and accepted of herbal remedies. It increases resistance to illness and helps in fighting off infections. Echinacea is a catalyst in producing white blood cells, which are the cells responsible for fighting off illness and infection.

Ginger was used in ancient times by Chinese healers. Ancient (and present day) Chinese medicine focuses on achieving balance. If something appears out of balance, add or subtract to re-balance. If, for example, something seems cold, warm it up – this is exactly what ginger does. It warms the insides. It has been used for stomach cramps and to ward off colds and flu. It is used for dieting, as ginger speeds up the metabolism. Ginger is used to ward off nausea, especially nausea due to motion sickness. Drinking a cup of ginger tea before a nausea-inducing motion experience will prevent sickness.

Ginkgo helps to alleviate hypertension. Perhaps this in itself is enough to slow aging. Ginkgo helps with circulation. It increases circulation to all parts of the body.

Ginseng is another ancient Chinese medicines. Long touted as a "restorer of vitality," taken regularly, ginseng increases energy while releasing stress. The real trick is to use ginseng consistently over a period of time. Doing so increases energy.

Licorice teas are cooling, soothing, and coating. They are especially recommended for a cough or sore throat.

Mate has caffeine in it. Grown in the rain forests of South America, it is a highly caffeinated/energizing drink. Passed around and drunk out of a gourd, in its native lands it is used in communal ceremonies and celebrations. Yerba Mate is also high in vitamin C, which in turn is an antioxidant.

Peppermint is amongst our most popular herbal tea. Nearly everyone is familiar with it, and it's an easy choice. A lot of people already know something about the health benefits of peppermint. It aids in digestion, relieves nausea, even reduces flatulence, but there are a few benefits most don't know about. Peppermint teas can help to alleviate the pain of headaches associated with menstrual cramps. It can also help those with breathing troubles, like asthma, because the smell of peppermint opens blocked breathing passages.

Rosehip is used in the blends of many teas. It is high in many vitamins, slightly acidic, and, contains pectins, so it is fruity. It acts as a mild laxative and diuretic.

Tea Catechins

Green tea has attracted significant attention recently – both in the scientific and in consumer communities – for its health benefits for a variety of disorders, ranging from cancer to weight loss. Historically, green tea has been consumed by the Japanese and Chinese populations for centuries, and is probably the most-consumed beverage, besides water, in Asian society.

The beneficial effects of green tea are attributed to the polyphenolic compounds present in green tea, particularly the catechins, which make up 30% of the dry weight of green tea leaves. The main catechins present in green tea are (−)-epicatechin (EC), epicatechin-3-gallate (ECG), (−)-epigallocatechin (EGC) and (epigallocatechin-3-gallate (EGCG) (Mukhtar and Ahmad, 2000). EGCG, the most abundant catechin in green tea, accounts for 65% of the total catechin content. A cup of green tea may contain 100–200 mg of EGCG. Catechin and gallocatechin are present in trace amounts (Figure 10.4).

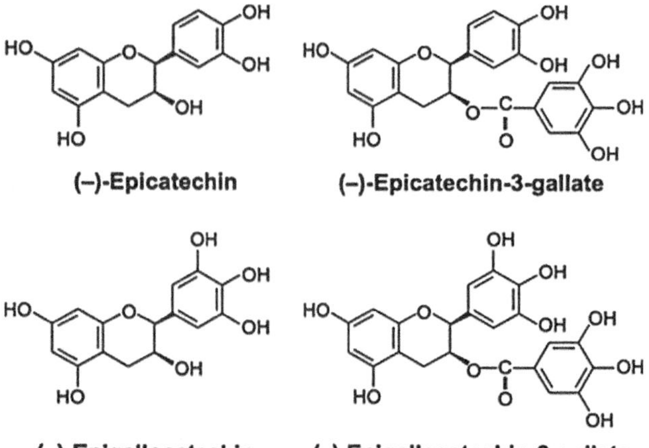

(−)-Epicatechin

(−)-Epicatechin-3-gallate

(−)-Epigallocatechin

(−)-Epigallocatechin-3-gallate

FIGURE 10.4 Structure of tea catechins.

Health Benefits of Tea

Green tea and its constituent catechins are best known for their antioxidant properties, which has led to their evaluation in a number of diseases associated with (ROS), such as cancer, and cardiovascular and neurodegenerative diseases. Several epidemiological studies as well as studies in animal models have shown that green tea can afford protection against various cancers such as those of the skin, breast, prostate and lung. In addition to the cancer chemopreventive properties, green tea and EGCG have been shown to be anti-angiogenic and anti-mutagenic. Green tea has also shown to be hypocholesterolemic and to prevent the development of atherosclerotic plaques. Among age-associated pathologies and neurodegenerative diseases, green tea has been shown to afford significant protection against Parkinson's disease, Alzheimer's disease, and ischemic damage. Green tea has also shown anti-diabetic effects in animal models of insulin resistance and has been shown to promote energy expenditure. Other health benefits attributed to green tea include antibacterial, anti-HIV, anti-aging and anti-inflammatory activity.

Molecular Mechanisms of Green Tea Effects

The health benefits of green tea are mainly attributed to its antioxidant properties and the ability of its polyphenolic catechins to scavenge ROS. These properties are due to the presence of the phenolic hydroxy groups on the B-ring in ungalloylated catechins (EC and EGC) and in the B- and D-rings of the galloylated catechins (ECG and EGCG). The presence of the 3,4,5-trihydroxy B-ring has been shown to be important for antioxidant and radical scavenging activity. The green tea catechins have been shown to be more effective antioxidants than Vitamins C and E. The metal-chelating properties of green tea catechins are also important contributors to their antioxidative activity. Recent studies have shown that misregulated iron metabolism may be a central pathological feature in Parkinson's disease and that the iron-chelating properties of EGCG are important for its protective effects in neurodegenerative diseases (Mandel et al., 2004). In addition to antioxidant effects, green tea catechins have effects on several cellular and molecular targets in signal transduction pathways associated with cell death and cell survival. These effects have been demonstrated in both neuronal cells and in tumor epithelial/endothelial cells. Green tea also inhibits angiogenesis and tumor invasion by inhibiting metalloproteinases and the vascular endothelial growth factor receptor expression and signaling in tumor and endothelial cells. In neuronal cells, however, green tea catechins serve a neuroprotective, pro-survival function. Moreover, these effects have been observed at doses far lower than those at which antitumor activities have been demonstrated.

Green Tea in Aging and Neurodegenerative Diseases

Oxidative stress is believed to be a major contributor to the pathogenesis of Parkinson's disease, especially the death of dopaminergic neurons. Recently, misregulated iron metabolism in the brain has been shown to be involved in the generation of the pathological Lewy bodies in Parkinson's disease through iron-induced aggregation of alpha-synuclein. Various studies have shown that green tea and EGCG significantly prevent these pathologies in animal models. Although there is no epidemiological evidence in human studies of the benefit of green tea for Alzheimer's disease, several studies in animal and cell culture models suggest that EGCG from green tea may affect several potential targets associated with Alzheimer's disease progression. Choi et al. (2001) showed that EGCG protects against beta-amyloid-induced neurotoxicity in cultured hippocampal neurons, an effect attributed to its antioxidant properties.

ANIMAL BIOTECHNOLOGY AS A TOOL TO UNDERSTAND AND FIGHT AGING

The aging process is an inevitable part of life for humans, and animal biotechnology plays an important role in understanding the process of aging and age-related diseases. It also provides various methods to fight aging. We are at the beginning of the biological revolution. Two centuries ago, the industrial revolution changed the way inanimate objects were manufactured. Today, the biological revolution is providing the means to create novel living organisms and combat aging and age related diseases. The increase in the human life span and the decrease in disability at older ages are a testament to the economic and social progress.

Biotechnology is a set of techniques that helps to modify living things. Biotechnology continues to deliver an impressive supply of new treatment options and interventions that will further extend healthy lives; policy makers and economists worry about the social implications of the future demand for health care. Biotechnology holds the promise of significantly improving elderly health and quality of life by alleviating the disabling conditions that plague our later years. According to both biologists and epidemiologists, the human life span continues to increase and estimates of maximal life span may be greater than initially believed.

Biotechnologies are revolutionizing the aging experience by offering earlier diagnoses, new treatments (such as regenerative and genetic interventions), and ultimately, disease prevention. Genomic studies make it possible to estimate the risk of age-related diseases. Techniques to prevent or replace lost functions are borrowing from the body's own development processes.

Aging and age-related diseases like Alzheimer's disease, Parkinson's disease and Huntington's disease may increase in coming years. However, successful treatment strategies for age-related diseases are limited, thus far. Plants have always played a major role in the treatment of human and animal diseases, and world-wide interest in the use of medicinal plants and their products is increasing, as they contain the recipe for chemical compounds of potential value in pharmaceutical products. Several Indian medicinal plants have also been proven to have anti-diabetic, anti-hypertensive and anti-aging activities. In the last decade, approaches to search for biologically active compounds have changed dramatically for a number of reasons, including advances in technology, new molecules of significant interest, changing ethical principles for organism collection, and increasing awareness of the chemical and biological potential of tropical rain forests. The pharmaceutical industry worldwide plays a major role in developing new approaches to drug discovery, aiming at faster and more efficient ways to bring new medicines to the market. Several drugs and plant products are tested in animal models to evaluate the efficacy of compounds as anti-aging. Thus we can say that animal biotechnology plays an important role in fighting aging and age-related diseases.

ETHICAL ISSUES

1. **Aging.** Aging is a biological reality and has its own dynamics, which is beyond human control. Aging is the accumulation process of diverse detrimental changes in the cells and tissues with advancing age, resulting in an increase in the risks of disease and death.
2. **Production of new animal models.** Biogerontologists employ short-lived laboratory models like yeast, free living nematodes, fruit flies, fish and laboratory mice. For *in vitro* studies, fibroblasts are the commonly studied animal cell type, originating from humans or mice. Animals that are used for aging research are typically

maintained in a non-reproductive state, physically inactive, socially isolated, and exposed to the fewest natural stressors possible. There are well-developed molecular tool kits and transgenic technologies that are used for the production of new animal models for aging research.

3. **Ethical implications for clinical and experimental research.** "Ethics refer to the moral aspects of human conduct and personal character. Often ethical dilemmas arise in relation to personal freedoms, responsibilities, and rights or obligations." (http://www.aging.pitt.edu/seniors/ethical-issues.asp). Some common situations in which ethical issues may arise in older adults:
   a. **Advance Care Planning.** Advance care planning involves advance preparation for life's unexpected emergencies. Regardless of age, advance care planning provides greater control over decisions that affect a person's future and takes into consideration the person's beliefs and preferences in the event they are unable to make decisions on their own.
   b. **Right to Privacy.** Medical treatment often involves sensitive subjects that we would rather not share with other people – even those who are very close to us. Anyone receiving treatment that does not want information shared with family, friends, or anyone else, has the right to keep that knowledge private between him/her and the health care team.
   c. **Religion and Health Care.** Religion may be a very important part of one's life. If so, religious beliefs may influence the types of medical treatment a person desires. It is crucial that a person know his/her religion's true beliefs about certain medical procedures. Many times, even the most devout people are unclear about some of their religion's rules. They may refuse medical treatment that is allowed by their faith or consent to medical treatment that is not permitted.

TRANSLATIONAL SIGNIFICANCE

The application of findings derived in basic science to the development of new understanding of disease mechanisms, diagnoses, and therapeutics in humans is known as translational research. It is the flow of ideas from basic science to clinical application. With the development of diverse new technologies, remarkable advances have occurred in the understanding of the molecular and genetic bases of aging and age-related diseases. Translational research topics in aging generally include nutrition, exercise, and metabolism of humans.

The involvement of posttranslational modifications in aging has been clearly demonstrated in recent years. Carbonylation (a hallmark of protein oxidation in general) is paradoxically decreased in histones with aging and increased by calorie restriction (CR). Acetylation of lysine 9 and phosphorylation of serine 10 in histone H3 are decreased and increased (respectively) with aging, and the acetylation level of multiple extra-nuclear proteins decreases significantly with aging. Research has shown the change was not retarded, but was increased remarkably by CR in rat liver. Based on the above findings, Nakamura *et al.* discussed possible implications of the posttranslational protein modifications in the biochemical processes underlying aging and CR-induced extension of life span (Nakamura *et al.* 2010).

WORLD WIDE WEB RESOURCES

The study of human aging is an enormous challenge. The complexity of the aging phenotype and the near impossibility of studying aging directly in humans oblige researchers to resort to models and extrapolations. Bioinformatics offer various powerful sets of tools to study aging and age related disorders. There are data-mining methods, comparative genomics to DNA microarrays, to retrieve information in large amounts of data. There are several web-based resources that provide information to study aging. The National Center for Biotechnology Information (NCBI) web page http://www.ncbi.nlm.nch.gov/ is an important resource for data mining and microarray studies. Aging-related information can be obtained from HAGR (http://genomics.senescence.info/), AGEID (http://uwaging.org/genesdb/index.php), the meta-analysis of age-related gene expression Profiles, and aging-related yeast2hybrid experiments. The above web-based resources will help in downloading promoter sequences and other genome related information. The Digital Ageing Atlas (http://human.ageing-map.org/) integrate molecular/physiological and pathological age related data. Some other useful web based resources for aging are below.

http://www.uwaging.org
http://www.antioxidants-for-health-and-longevity.com/causes-of-aging.html

REFERENCES

Benzie, I. F. F., & Strain, J. J. (1996). The ferric reducing ability of plasma (FRAP) as a measure of "Antioxidant Power": The FRAP assay. *Analytical Biochemistry*, *239*, 70–76.

Choi, Y. T., Jung, C. H., Lee, S. R., Bae, J. H., Baek, W. K., Suh, M. H., Park, J., Park, C. W., & Suh, S. I. (2001). The green tea polyphenol (–)-epigallocatechin gallate attenuates beta-amyloid-induced neurotoxicity in cultured hippocampal neurons. *Life Sciences*, *70*(5), 603–614.

Gerhard, G. S. (2007). Small laboratory fish as models for aging research. *Ageing Research Reviews*, *6*(1), 64–72.

Harman, D. (1956). Ageing: A theory based on free radical and radiation chemistry. *The Journals of Gerontology*, *11*, 298–300.

Harman, D. (2006). Free radical theory of aging: An update. *Annals of the New York Academy of Sciences*, *1067*, 1–12.

Heilbronn, L. K., & Ravussin, E. (2003). Calorie restriction and aging: review of the literature and implications for studies in humans. *The American Journal of Clinical Nutrition, 78*(3), 361–369.

Kumar, N., Kant, R., & Maurya, P. K. (2010). Concentration dependent effect of (–) epicatechin in hypertensive patients. *Phytotherapy Research, 24*(10), 1433–1436.

Kumar, N., Kant, R., Maurya, P. K., & Rizvi, S. I. (2012). Concentration dependent effect of (–) epicatechin on Na^+/K^+-ATPase and Ca^{2+}-ATPase inhibition induced by free radicals in hypertensive patients: comparison with L-ascorbic acid. *Phytotherapy Research, 26*(11), 1644–1647.

Mandel, S., Maor, G., & Youdim, M. B. (2004). Iron and alpha-synuclein in the substantia nigra of MPTP-treated mice: effect of neuroprotective drugs Rapomorphine and green tea polyphenol (–)-epigallocatechin3-gallate. *Journal of Molecular Neuroscience, 24*(3), 401–416.

Maurya, P. K., & Prakash, S. (2011). Intracellular uptake of (–) epicatechin by human erythrocytes as a function of human age. *Phytotherapy Research, 25*(6), 944–946.

Maurya, P. K., & Rizvi, S. I. (2009). Alterations in plasma nitric oxide during aging in humans. *Indian Journal of Biochemistry & Biophysics, 46*, 130–132.

McCay, C. M., Crowel, M. F., & Maynard, L. A. (1935). The effect of retarded growth upon the length of the life span and upon the ultimate body size. *The Journal of Nutrition, 10*, 63–79.

Middleton, N., Jelen, P., & Bell, G. (2004). Whole blood and mononuclear cell glutathione response to dietary whey protein supplementation in sedentary and trained male human subjects. *International Journal of Food Science, 55*, 131–141.

Mukhtar, H., & Ahmad, N. (2000). Tea polyphenols: prevention of cancer and optimizing health. *The American Journal of Clinical Nutrition, 71*, 1698S–1702S.

Nakamura, A., Kawakami, K., Kametani, F., Nakamoto, H., & Goto, S. (2010). Biological significance of protein modifications in aging and calorie restriction. *Annals of the New York Academy of Sciences, 1197*, 33–39.

Pandey, K. B., & Rizvi, S. I. (2010). Markers of oxidative stress in erythrocytes and plasma during aging in humans. *Oxidant Medical Cell Longev, 3*(1), 2–12.

Puca, A. A., Daly, M. J., Brewster, S. J., Matise, T. C., Barrett, J., Shea-Drinkwater, M., Kang, S., Joyce, E., Nicoli, J., Benson, E., Kunkel, L. M., & Perls, T. (2001). A genome-wide scan for linkage to human exceptional longevity identifies a locus on chromosome 4. *Proceedings of the National Academy of Sciences of the United States of America, 98*, 10505–10508.

Rattan, S. I. S. (2006). Theories of biological aging: Genes, proteins, and free radicals. *Free Radical Research, 40*(12), 1230–1238.

Reznick, A. Z., & Packer, L. (1994). Oxidative damage to proteins: spectrophotometric method for carbonyl assay. *Methods Enzymol, 233*, 357–363.

Rizvi, S. I., & Maurya, P. K. (2007). Markers of oxidative stress in erythrocyte during aging in humans. *Annals of the New York Academy of Sciences, 1100*, 373–382.

Rizvi, S. I., Jha, R., & Maurya, P. K. (2006). Erythrocyte plasma membrane redox system in human aging. *Rejuvenation Research, 9*(4), 470–474.

Roback, J., & Gryglewski, R. J. (1996). Bioactivity of flavonoids. *Polish Journal of Pharmacology, 48*, 555–564.

FURTHER READING

Arking, R. (2006). *Biology of Aging: Observations and Principles*. USA: Oxford University Press.

Beutler, E. (1984). *Red Cell Metabolism: A Manual of Biochemical Methods* (3 edn). New York, NY: Grune & Stratton, Inc.

Bondy, S. C., & Maiese, K. (2010). *Aging and Age-Related Disorders*. New York: Humana press.

Miwa, S., Bruce, K., & Beckman, B. (2008). *Oxidative Stress in Aging: From Model Systems to Human Diseases*. New York: Humana press.

Vassallo, N. (2008). *Polyphenols and Health: New and Recent Advances*. Hauppauge NY: Nova Science Publisher, Inc.

GLOSSARY

Aging Aging is the accumulation of changes in a person over time (e.g. human aging).

Biotechnology Biotechnology is a field of applied biology that involves the use of living organisms and bioprocesses in engineering, technology, medicine, and other fields requiring bioproducts. For example, many biotechnology companies are developing anti-aging products such as anti-aging creams. Juvista, is based on a recombinant form of human transforming growth factor-β3 (TGF-β3), which is normally present at high levels in developing embryonic skin and in embryonic wounds that heal without a scar.

Oxidative Stress Oxidative stress represents an imbalance between the production and manifestation of reactive oxygen species and a biological system's ability to readily detoxify the reactive intermediates or to repair the resulting damage. For example, lipid peroxidation, protein oxidation and DNA damage is higher in diabetic, hypertensive patients as compared to healthy, normal patients.

Polyphenols Polyphenols are a structural class of natural, synthetic, and semisynthetic organic chemicals characterized by the presence of large multiples of phenol structural units (e.g. tea catechins in green tea, resveratrol in grapes).

ABBREVIATIONS

CR Calorie Restriction
EGCG (–)-Epigallactocatechin Gallate
EGC (–)-Epigallactocatechin
ECG (–)-Epicatechin Gallate
EC (–)-Epicatechin
FRAP Ferric Reducing Ability of Plasma
H_2O_2 Hydrogen Peroxide
$^1O_2$ Singlet Oxygen
$O_2\bullet$ Superoxide Radical
•OH Hydroxyl Radical
NO Nitric Oxide
ROS Reactive Oxygen Species

LONG ANSWER QUESTIONS

1. What is aging? Why is there a decline in the regenerative potential of an organism over time?
2. When does aging begin in humans? Discuss various theories of aging, giving emphasis on the free radical theory of aging.
3. What are polyphenols? How are they classified based on their chemical structures?
4. How does biotechnology help in understanding aging and age-related phenomena?
5. What is green tea? Discuss its health benefits.

SHORT ANSWER QUESTIONS

1. What is aging?
2. Why does caloric restriction delay the onset of a number of age-related physiological and pathological changes, and increase the average and maximal life span in animals?
3. What are flavonoids?
4. How does biotechnology help to delay aging?
5. What are tea catechins?

ANSWERS TO SHORT ANSWER QUESTIONS

1. Aging is the accumulation process of diverse detrimental changes in the cells and tissues with advancing age, resulting in an increase in the risks of disease and death. Aerobic cells produce reactive oxygen species (ROS) as a byproduct of their metabolic processes. ROS cause oxidative damage to macromolecules (proteins, lipids and nucleic acids) under conditions when the antioxidant defense of the body is overwhelmed. A certain amount of oxidative damage takes place even under normal conditions, however the rate of this damage increases during the aging process as the efficiency of anti-oxidative and repair mechanisms decrease.
2. Evidence that calorie restriction (CR) retards aging and extends median and maximal life span was first presented in the 1930s. Since then, similar observations have been made in a variety of species including rats, mice, fish, flies, worms, and yeast. Although not yet definitive, results from the ongoing calorie-restriction studies in monkeys also suggest that the mortality rate in calorie-restricted animals will be lower than that in control subjects. Furthermore, calorie-restricted monkeys have lower body temperatures and insulin concentrations than do control monkeys, and both of those variables are biomarkers for longevity in rodents. Calorie-restricted monkeys also have higher concentrations of dehydroepiandrosterone sulfate. The importance of dehydroepiandrosterone sulfate is not yet known, but it is suspected to be a marker of longevity in humans, although this is not observed consistently. In humans, a major goal of research into aging has been the discovery of ways to reduce morbidity and delay mortality in the elderly. The absence of adequate information on the effects of CR in humans reflects the difficulties involved in conducting long-term calorie-restriction studies, including ethical and methodological considerations. There is also evidence that DNA damage is reduced by CR, possibly as a result of increased DNA repair capacity. If normal feeding is resumed, these animals remain fertile far longer than other animals. CR diverts energy from growth and reproduction towards somatic maintenance, and thus may explain the life prolonging effect of CR.
3. The flavonoids are a large and complex group of compounds that occurs throughout the plant kingdom, providing flavors, color, antifungal and antibacterial activity, and contributing to many aspects of plant physiology. Most plant tissue can synthesize flavonoids. The flavonoids are diphenylpropane derivatives that include flavonols, flavonones, antocynidines, flavones and flavonols. More than 4,000 flavonoids have been found in plants, fruits, and vegetables. There is great interest in these phenolic compounds because of their potential role as anti-aging and cancer chemo preventive agents. This beneficial effect is considered to be mainly due to their antioxidant and chelating activities. However, some flavonoids, such as quercetin, have also been reported to be mutagenic and co-carcinogenic. The ability of flavonoids to scavenge free radicals and block lipid peroxidation raises the possibility that they may act as protective factors against cardiovascular disease and hypertension and there is epidemiological evidence consistent with this hypothesis. Any physiological significance of dietary flavonoids depends upon their availability for absorption and their subsequent interaction with target tissues, but little is known about their transport across the intestine. In general, flavonoid glycosides are resistant to processing, cooking and digestion.
4. Biotechnologies are revolutionizing the ageing experience by offering earlier diagnoses, new treatments (such as regenerative and genetic interventions), and ultimately, disease prevention. Genomic studies make it possible to estimate the risk of age-related diseases. Techniques to prevent or replace lost functions are borrowed from the body's own development processes.
5. Most of the polyphenols in green tea are flavonols, commonly known as catechins. The main catechins present in green tea are (1) Epicatechin (EC); (2) Epicatechin-3-gallate (ECG); (3) Epigallocatechin (EGC); and (4) epigallocatechin-3-gallate (EGCG).

Section II

Animal Biotechnology: Tools and Techniques

Chapter 11

Multicellular Spheroid: 3-D Tissue Culture Model for Cancer Research

Suchit Khanna, Anant Narayan Bhatt and Bilikere S. Dwarakanath

Division of Radiation Biosciences, Institute of Nuclear Medicine and Allied Sciences, Delhi, India

Chapter Outline

SUMMARY

This section highlights the rationale, potential and implementation of multicellular tumor spheroids in cancer research. A "multicellular tumor spheroid" (MCTS) is a three-dimensional *in vitro* model system that bridges the gap between two-dimensional monolayer cell cultures and an *in vivo* tumor tissue model system. Compared to monolayer cultures, MCTS resembles tumor tissue in terms of structural and functional properties and is suitable for studying metastasis, invasion, and therapeutic screening of drugs. Spheroid models are helpful in accelerating translational research in cancer biology and tissue engineering.

WHAT YOU CAN EXPECT TO KNOW

Multi-cellular tumor spheroids (MCTS), the best described 3-D tumor models, offer an excellent *in vitro* screening system that mimics to a great extent the microenvironment prevailing in the tumor tissue, supporting studies on tumor-specific processes like angiogenesis, invasion and metastasis, as well as assessment of responses to various therapies and underlying mechanisms. Developed nearly 30 years ago as an alternative *in vitro* model to monolayer cultures, their role as a part of a high through-put, cell-based assay system in drug discovery is gaining considerable importance. Spheroids bridge the gap between monolayers and animal tumors, facilitating mechanistic studies and evaluation of anticancer therapies, particularly relevant to solid tumors. The distinct possibility of establishing MCTS from primary tumor cells and also co-culturing with different normal cells enhance the value of MCTS in cancer research and drug development. Recent developments in the generation of novel MCTS systems that allow analysis of a vast number of parameters can be conveniently employed in the

Animal Biotechnology. http://dx.doi.org/10.1016/B978-0-12-416002-6.00011-0

high throughput screening system with cell-based assays, thereby significantly shortening the time required for translational research, bridging the gap between discovery and clinical application. Thus, spheroids have the potential to enhance predictability of clinical efficacy and may minimize, if not replace, animal studies to a large extent in the near future. Together with novel and emerging tools of biotechnology, this 3-D *in vitro* model is expected to substantially reduce the cost of new drug discovery.

HISTORY AND METHODS

INTRODUCTION

Model systems play an important role in biomedicine, as they form the backbone of translational research by bridging the gap between basic concepts and discovery, with applications in the management of diseases involving diagnosis, prognosis, and therapy. In the field of oncology, they are useful in therapeutic screening, preclinical evaluation, and to even study the basic biology of tumors. It is generally observed that the degree of complexity of various models available and deployed in experimental oncology bear an inverse relationship with the level of predictability either for diagnostic purposes or therapeutic evaluation (Khaitan et al., 2006b).

In vitro cell cultures are important experimental tools in understanding the biology of neoplastic cells, as well as for the evaluation of potential therapeutic agents, and understanding of the mechanisms underlying their actions. They can easily be created and manipulated, and hence help in the systematic studies of multicellular systems. Among the *in vitro* models of tumors, 2-D models such as monolayers and suspension cultures have been used widely to study various aspects of tumor biology. However, extrapolation of findings from these models has limited value in clinics, as only a fraction of the tumor cells generally develop fully in to a cell line, and do not necessarily reflect the primary tumors from which they were derived. Most importantly, they lack three-dimensional architecture and host tissue microenvironment, which are critical features of tumors (Khaitan et al., 2006b; Vinci et al., 2012). Therefore, there has been considerable amount of effort to develop and deploy various 3-D *in vitro* culture systems (organ culture, spheroid culture) that are associated with enhanced reliability and predictability for clinical efficacy, and to minimize studies with animal models (Khaitan et al., 2006b). It is very well realized that cell-based models used in experimental studies need to recapitulate both the 3-D organization and multicellular complexity of the tumor as well as the organs to translate findings from these studies into clinical applications. Three-dimensional cultures have been utilized in biomedical research since the first half of the 20th century to gain deeper insight into the mechanisms of organogenesis and expression of malignancy. However, only a small number of 3-D model systems are sufficiently well characterized to simulate the patho-physiological cellular microenvironment in a tumor or to reconstitute a tissue-like cyto-architecture, with cell-to-cell and cell-to-matrix interactions, growth, differentiation, and therapeutic responses similar to tumors *in vivo*.

Organ culture is one of the 3-D cultures, where small pieces of tumor explants are cultured in a moist gas or air phase on the surface of a relatively large volume of stationary nutrient medium (to retain the original structural relationships and differentiation of the tissue organization), and can be used to study the interactive function and the effect of drugs and other agents (Dwarakanath et al., 1985, 1987; Lasnitzki et al., 1992). However, lack of characterized reference stock, limited availability, and high variability, as well as ethical restrictions (at least with respect to human tissue), has limited the use of organ cultures.

MULTICELLULAR TUMOR SPHEROIDS

Spheroid models are sphere-shaped cell colonies that permit growth and functional studies of diverse normal and malignant tissues. The growth of spheroids from tumor cells mimics the growth of naturally occurring human tumors, as their extracellular matrix and network of cell-to-cell and cell-to-matrix interactions are similar to *in vivo* conditions and differ from the corresponding monolayer cultures (Khaitan et al., 2006b). In spheroids as in solid tumors, cells in contact with nutrients grow quickly, while the kinetics and activity of cells further inward depend on diffusion. Various possibilities of manipulating the spheroid environment using different techniques for their generation has not only provided insights into the complexity of tumor physiology, but has also facilitated research efforts in developing novel therapeutic agents (Khaitan et al., 2006b).

This model was adopted to cancer research several decades ago by Sutherland and his co-workers, and has since then considerably contributed to our knowledge regarding various biological mechanisms of tumors, as well as in studies related to cellular response to diverse therapeutic interventions (Hirschhaeuser et al., 2010). Over the years, a wide variety of techniques have been developed for the cultivation of spheroids that span from simple culture systems using Petri dishes to novel methods like hanging drop, spinner culture, roller bottle culture, scaffold culture, microstructure based cultures, etc., that generate spheroids from a wide variety of tumor and normal cells. More recently, microfluidics-based microtechnologies have been developed for the production of uniform tumor spheroids, which are capable of generating spheroids from diverse tissue origins that can be used to model various types and stages of cancers (Hirschhaeuser et al., 2010). Schematic diagrams illustrating some of the widely used approaches, and methods for establishing homologous and heterologous spheroids are shown in Figure 11.1.

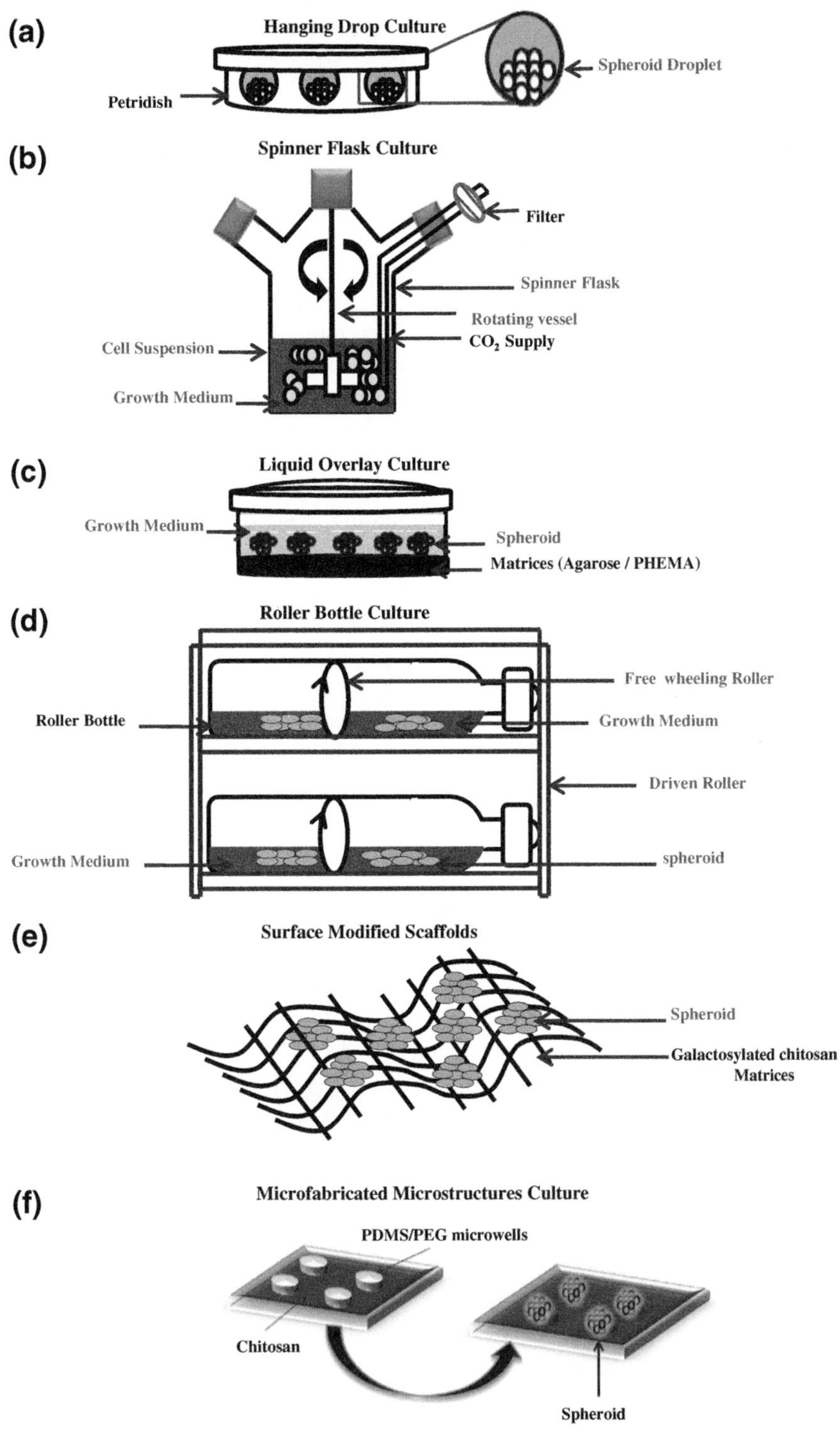

FIGURE 11.1 Different methods for the generation of multicellular tumor spheroids (MCTS).

Several well established methods can be used to assess the effects of various therapeutic agents on spheroids, including the measurement of spheroid volume and growth, which follows a Gompertz function, similar to the growth of experimental tumors in animals (Khaitan et al., 2006a, b). A combination of many analytical imaging techniques, autoradiography, the tunnel assay, bioluminescence imaging, and microelectrode-based oxygen analysis, have revealed concentric arrangement of cell proliferation, viability, and the micromilieu in large spheroids (Hirschhaeuser et al., 2010). Frozen or chemically fixed spheroid sections can be investigated for antigen expression by using various microscopy and immuno enzymatic techniques, while cells from dissociated spheroids can be used to analyze the information at the single-cell level (Khaitan et al., 2006b).

Spheroids produced from various tumor cell types have been used to study responses to various treatment modalities like radiation, chemotherapy, hyperthermia, immunotherapy, and a combination of therapeutic interventions. Spheroids, as in naturally occurring tumors, develop cell-to-cell communication, numerous communication channels including gap junctions, desmosomes, and electrical coupling, which have provided insight into the molecular mechanisms regulating cell proliferation and differentiation in tumors (Mueller-Klieser et al., 1997). Figure 11.2 illustrates various approaches currently employed using MCTS in evaluating the influence of various tumor-associated parameters on the *in vivo* response of tumors. Co-culturing normal cells with malignant cells within a spheroid can provide information on tumor invasion and angiogenesis. Growing spheroid tumor aggregates with matrigels also provide valuable information about the metastatic property of tumor cells. Although, MCTS offers many advantages over the monolayer cell cultures by mimicking *in vivo* conditions of the tumor to a very great extent, it does not eliminate the use of animal models, as it has certain limitations. For example, pharmacokinetics, pharmacodynamics and bioavailability of drugs cannot be studied using MCTS. Furthermore, the influence of the immune system on the systemic response of the organism can only be evaluated using appropriate animal models. Therefore, MCTS models cannot totally replace the testing of complete biological mechanisms relevant to the drug development.

The concentric geometry of spheroids is widely used as a model of diffusion-limited tissue with central necrosis, which develops when diffusion of oxygen becomes limited. The modes and mechanisms of cell death in

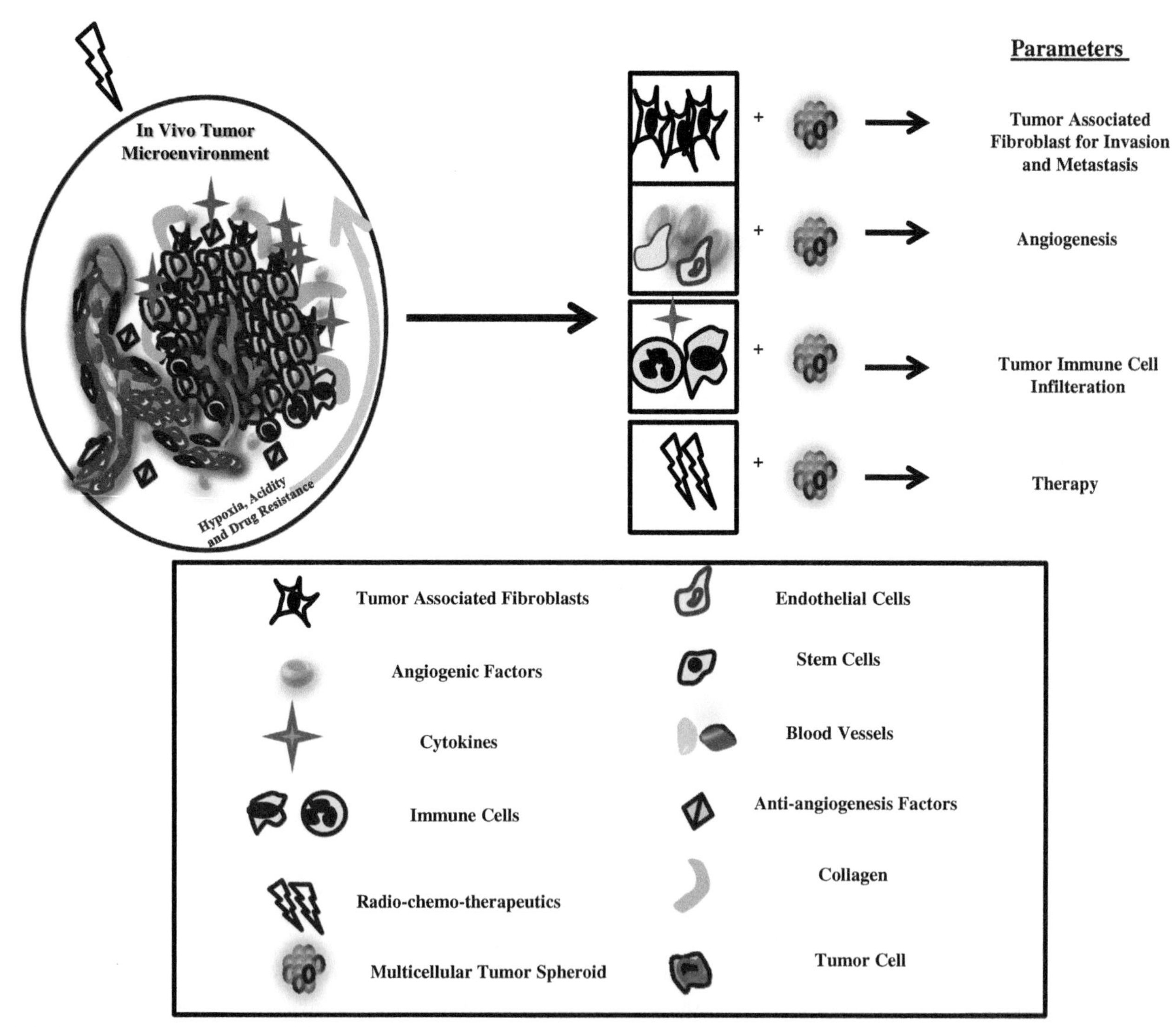

FIGURE 11.2 Approaches for studying the effects of tumor associated parameters on the *in vivo* response of tumors using multicellular tumor spheroid (MCTS).

tumor spheroids appear to be complex and involve multiple mechanisms. The three vital parameters for quantification of cellular viability in spheroids (the diameter of the spheroid, the diameter of the necrotic core, and the viable rim thickness), are different in different types of spheroids, and depend on culturing conditions. The viable rim thickness may gradually increase or decrease with expanding spheroid diameter, thereby indicating that different mechanisms are involved in the onset and expansion of the central necrosis in different spheroid types (Khaitan et al., 2006a). Multiple foci of necrotic (dead) cells confined not only to the central region and the presence of apoptotic cells have both been found in spheroids similar to *in vivo* tumors (Khaitan et al., 2006a, b). Since the mechanisms of apoptosis differs in several ways from that of necrosis, these could be independent predictors of cellular sensitivity to a particular therapeutic agent, and monitoring the level and mode of cell death that may be useful in modulating treatment or in predicting the response of tumors to treatment (Khaitan et al., 2006b; Hirshhaeuser et al., 2010).

Various parameters associated with the spheroids (such as growth, viability and cell survival, cell death, metabolic and mitochondrial status, gene expression status, levels of proteins, oxidative stress, and glutathione levels) influence the end results of *in vitro* studies aimed at understanding tumor response to various cytotoxic agents and metabolic inhibitors (Khaitan et al., 2006a). In culture, these parameters vary as a function of age and spheroid size, similar to the status in growing tumors. An exhaustive characterization in spheroids generated from a human glioma cell line clearly showed significant differences between monolayer cultures and spheroids at different ages with respect to many of the parameters (Khaitan et al., 2006a). Therefore, results obtained with any given tumor model must be interpreted carefully and extensive characterization of the model is necessary before extrapolating responses from spheroids to therapeutic responses. Furthermore, since spheroids can also be cultured from primary cells of tumors, which retain to a very great extent the biological as well as the metabolic behavior of the tumor, detailed characterization coupled with drug responses using appropriate parameters will facilitate the optimization and interpretation of predictive therapy.

Tumor spheroid cultures have been widely used in experimental radiotherapy, photodynamic therapy, hyperthermia, chemotherapy and target-specific approaches, as well as other contemporary and emerging therapies, such as anti-angiogenesis therapy, gene therapy and cell- or antibody-based immunotherapy, etc. Evidence in the literature suggests that many treatments are expected to be less effective in the 3D pathophysiological environment. Namely, 5-fluorouracil (5-FU) has higher antiproliferative effects on 2-D cultures but not in spheroids. On the other hand the hypoxia activated drug tirapazamine (TPZ) is more effective against 3D cultures (Tung et al., 2011). Therefore spheroids are most frequently considered as appropriate tools for evaluating drug candidates with enhanced tissue distribution and efficacy and can also be used for negative selection to reduce animal testing. Indeed, most therapeutic approaches were found to be less effective in 3-D than in 2D cultures. This, however, cannot be generalized, since some potential targets and signaling pathways especially, or even exclusively, play a role in the 3-D environment or milieu (Barbone et al., 2008). Consequently, the spheroid model has also been increasingly recognized as a primary tool for positive selection in innovative drug development initiatives.

Genome-wide gene expression analysis in spheroids formed by human malignant gliomas and primary porcine hepatocytes has indicated that several genes express differently during spheroid formation, and resemble the tissue of origin more closely than the monolayer. Moreover, comparative transcriptomic studies in cells from epithelial ovarian cancer, hepatocellular carcinoma, and colon cancer show that numerous genes associated with cell survival, proliferation, differentiation, and resistance to therapy are differentially expressed in cells grown as multicellular spheroids (as compared to monolayer cultures, which mainly regulate the drug response in tumor cells). The expression profiles more closely resemble the profiles of the respective tumor tissue *in vivo*, and are thus highly relevant for the establishment and testing of novel therapeutic interventions (Lin *et al.*, 2008).

HISTORICAL FACTS TOWARDS THE DEVELOPMENT OF TISSUE CULTURE TECHNOLOGY FROM 2-D AND 3-D CULTURES

In vitro 2-D cultures of animal cells were first grown in the twentieth century. In 1912, Alexis Carrel maintained chick heart cells in drops of horse plasma for the first time. After a few days, death of the explants was observed, due to exhaustion of nutrients. Cells from a given explant could be maintained indefinitely if they were periodically subdivided and fed with a sterile aqueous extract of whole chick embryos. In the early 1950s, Earle used trypsin to dissociate the cells of a whole chick embryo. When this suspension of single cells was mixed with plasma and embryo extract and placed in a sterile glass container, the cells adhered to the glass and divided to form a primary culture. The primary culture contained a variety of cell types including macrophages, muscle fibers, etc. In 1952, Gey established a continuous cell line from a human cervical carcinoma known as HeLa (Henrietta Lacks) cells. In 1961, Hayflick and Moorhead isolated human fibroblasts (WI-38), and showed that they have a finite lifespan in culture. In 1965, Harris and Watkins were able to fuse human and mouse fibroblast cells using viruses. Simultaneously, Holtfreder and Moscona pioneered the field of biomedical research by their observations on morphogenesis in spherical re-aggregated cultures of embroyonic or malignant tissues. However, it was not until the early 1970s that Sutherland and coworkers systematically investigated the response of tumour cell aggregates to antineoplastic therapy. Because the cell lines formed nearly perfect sphere-shaped aggregates, they were called "spheroids." As a consequence, a number of investigations were simulated on the studies of basic biological mechanisms (such as the regulation of proliferation, differentiation, cell death, invasion, cell–cell interaction, cell–matrix interaction, angiogenesis, or immune response (Mueller-Klieser et al., 1997)) and structural similarity to human tumors on a large scale. Thereafter, advancement in tissue culture technologies for the development and maintenance of 3-D cultures has always been helpful in therapeutic applications.

An Example Where 3-D Culture is More Beneficial over 2-D Culture

Some of the advantages and limitations of 3-D culture over 2-D cultures are summarized in Table 11.1. For example, studies with thyrocytes (thyroid epithelium cells) by Manuchamp demonstrate that the loss of follicular organization in conditions of the monolayer culture results in the loss of polarization of thyrocytes (Manuchamp et al., 1998). This leads to considerable decrease in the ability to capture and transport iodine because there are decreased intercellular contacts and connections to the basal membrane. Similar modifications were observed for the cells of other organs and tissues, including mammary gland, kidneys, etc. Therefore, cultivation in three-dimensional conditions was suggested to solve these issues.

TECHNIQUES FOR THE GENERATION OF SPHEROIDS

Success in the generation of spheroids has been possible due to availability of simple and reproducible techniques on spheroid-based applications. While generating a multi-cellular spheroid, it is essential to prevent cells from being attached to the culture ware substratum. Table 11.2 shows the advantages and limitations of numerous techniques that have been described earlier for the generation of spheroids, and widely differ on the basis of spheroid size, cell specificity, efficiency in production, influence on cellular physiology, convenience, and suitability for subsequent applications. Some of the methods for spheroid generation are briefly discussed below.

Hanging-Drop Method

The hanging-drop method was primarily developed for culturing the *in vitro* aggregation of embryonic cells for studying morphogenesis and tissue formation. The technique does not require the coating of plates for spheroid formation, but is only useful for short-term culturing. This is a very simple method in which roughly 4,000 cells in

TABLE 11.1 Advantages and Limitations of 2-D and 3-D Culture Models

Cell Culture	Merits	Limitations
2-D Culture (Monolayer and Suspension Culture)	• Absolute control on cell environment. • Cell observation, measurement, and manipulation are easier.	• Cells lose their histological organization, polarity and differentiation. • Lack of host-tissue microenvironment. • Altered gene expression and growth characteristics due to a deficiency in cell-cell and cell-matrix interactions. • Increased drug sensitivity. • Use of suspension culture is also limited by the sensitivity of some cell lines to shear stress.
3-D Culture (Organ and Spheroid Culture)	• Sufficiently well characterized to simulate the pathophysiological cellular microenvironment. • Show enhanced reliability and predictability of clinical efficacy and minimize studies with animal models. • Reconstitute a tissue-like cyto-architecture with cell–cell and cell–matrix interactions, growth, differentiation and therapeutic responses similar to tissue *in vivo*. • Gene expression profiles of 3-D model reflect clinical expression profiles of tumors.	• Lack of vasculature, host-immune interactions. • Diffusional transport limitations: O_2 and other essential nutrients may not reach all of the cells; accumulation of toxic waste products within scaffold space.

volume of 40–50 μL are placed as a hanging drop onto the underside of the lid of a tissue culture dish. When the lid is inverted, the drops are held in place by surface tension and the microgravity environment in each drop concentrates the cells, and thereafter these drops are incubated under physiological conditions until they form true 3-D spheroids, in which cells are in direct contact with each other and with extra cellular matrix components at the free liquid-air interface (Kelm et al., 2003). The method requires no specialized equipment and is useful for generating MCS of defined sizes and cell numbers. In addition, spheroids generated by this method can be either embedded in Matrigel to study angiogenesis or various viability assays like MTT (Khaitan et al., 2006b). This method has also been applied to the co-cultivation of mixed cell populations, including the co-cultivation of endothelial cells and tumor cells as a model of early tumor angiogenesis. Efficiency of MCS formation by this method can be enhanced using cell-cell cross-linking agents such as cell matrix proteins, e.g. collagen, fibronectin, the synthetic polymer Eudragit, anti-B1 integrin monoclonal antibody, poly (lactic-co-glycolic acid), etc.

Liquid Overlay Method

In this method, bacteriological or ELISA 96-well plates are used to generate multicellular spheroids. Since bacteriological plates may not always promote MCS formation due to the lack of an appropriate surface for cell attachment,

TABLE 11.2 Advantages, Limitations and Applications of Different Methods of Spheroid Generation

Method	Applications	Advantages	Limitations	References
Hanging-Drop	Useful for studying tumor physiology , metabolism, toxicology, cellular organization and development of bioartificial tissue	Affordable Spheroids uniform in cell number, size and compositions Co-culture of different cells	Labor Intensive Difficulty in high scale production	Vinci et al., 2012 Kelm et al., 2003
Soft Agar Liquid Overlay	To manufacture 3-D aggregates for studying tumor and fibroblast interactions and their role in tumor development	Affordable No shear stress Easy to set up	Spheroids formed are non-uniform in cell number, size and shape Limited mass transfer Limited cell survival	Vinci et al., 2012, Yuhas 1977,
Rotating Wall Vessel (NASA Bioreactor)	Useful for the production of large number of spheroids	Maintain cells in under low shear stress Provide constant culture conditions Efficient mass transfer Efficient for long-term maintenance of tissue-like functions and cell viability	Not useful for drug testing Expensive and require special set up Non-uniform cell size and number	Vinci et al., 2012 Ingram et al., 1997
Spinner flask / Roller Bottle / Gyratory Shaker	To generate large number of porcine hepatocyte spheroids with a concentration higher than 500 MCS/mL	Simple to perform Efficient mass transfer Increase cell viability and allow long term culturing Co-culture of different cell types	Intermediate to high shear stress Difficult to use at large scale	Vinci et al., 2012
Microfabricated Microstructures	Useful in Generating 3-D liver or stem cell spheroids, High through-put drug screening, real time imaging of cells Co-culture of different cells	Uniform and well controlled spheroid size, cell number and size Spheroids on chip	Requires specialized facilities	Hirschhaeuser et al., 2010 Dean et al., 2007
3-D Scaffolds	Tissue engineering and bioartificial liver Serve a better *in vitro* 3-D system for screening cancer therapeutics	Provides 3-D support Easy to set up	Expensive	Fischbach et al., 2007 Glicklis et al., 2000

agarose-DMEM-coated plates have been used to generate MCS (Yuhas et al., 1977). This improvisation not only provides attachment of cells to the surface, but also offers a nutritional requirement from the agar-EBME combination for the growth of spheroids. Interestingly, other hydrophobic polymers, including poly (2-hydroxyethl methacrylate) (PHEMA), or poly-*N*-*p*-vinylbenzyl-D-lactonamide can also be used in place of agar (Tong et al., 1992). The advantage of this method is that it is inexpensive and simple to perform, and provides heterogeneous size, cell number and shape of spheroids.

Microfabricated Microstructures Method

Advancement in the semiconductor industry has given rise to microfabrication techniques that have been adopted for applications in life sciences. The techniques involve the use of soft lithography in the construction of polydimethylsiloxane (PDMS) and polyethylene glycol (PEG) based microstructures. These structures can be used for biomaterial microprinting and microfluidics. One of the greater applications of microfabricated techniques is in generating MCS with uniform size, cell composition, and geometry. The apparatus consists of a microfabricated device, namely, microwells that are loaded with suspended cells, redistributed by gravity and hydrodynamic forces, and finally assembled into aggregates based on microwell geometry. With varying diameter and geometry of micro-wells, different shapes and types of spheroids can be generated, such as rods, tori or honeycombs (Dean et al., 2007). This technique has potential for the mass production of tumor spheroids to generate 3-D liver or stem cell spheroid arrays for high throughput drug screening, and in 3-D co-culture to investigate the effect of Carcinoma Associated Fibroblasts on cancer cell invasion.

Rotatory Flask Methods

The spinner flask method is also one of the most attractive and robust approaches for the generation of large numbers of spheroids of defined size-ranges under optimum growth conditions. Here, tumor cells aggregate to form spheroids in the flask without attachment to any other substrate. Spheroids of different diameter can be generated by optimizing cell seeding density, composition of medium, culture time, and spinning rate. For example, Sakai et al. (2006), used the spinner flask method for the production of porcine hepatocyte spheroids, which has great potential in the generation of bioartificial liver (BAL). In addition, the presence of Eudragit as an artificial matrix promotes hepatocyte cell aggregation, which further enhances liver function by preventing damage to the cells due to agitation. The spinner flask method generally provides high shear stress, which has an affect on cellular physiology. To overcome this problem, the National Aeronautics and Space Administration (NASA) has developed a rotating wall vessel that provides a constant supply of medium and a constant nutrient flow rate, thus resulting in better differentiation and formation of multicellular aggregation (Ingram et al., 1997). Similarly, roller bottles and gyratory shakers are also used for the generation of large amount of spheroids to avoid effects caused by stagnant medium. This system provides low shear stress due to high surface:volume ratio, which allows gas exchange at an increased rate, and thus better constant and gentle agitation. The merits of rotatory culture methods are: they are simple to perform, they can be scaled to massive production and set up as long-term cultures, there is dynamic control on culture conditions, and co-culture of different cell types is possible. The demerits are the variation in cell size and number, and that special equipment is required.

Surface Modification-Based Methods

A variety of modified substrates are also available for the generation of spheroids. For example, PVDF surfaces coated with galactose-tethered pluronic polymer are useful for hepatocyte attachment; their spheroids generate with high cell viability and enhanced functional maintenance (Du et al., 2007). Similarly, the thermoresponsive polymer, poly-N-isopropyl acrylamide (PNI-PAAm), in conjugation with collagen as modified substratum, can be used for the culture of human dermal fibroblasts. As PNI-PAAm solubility is dependent on the critical solution temperature (LCST), changes in temperature can result in the formation of spheroids from monolayer cultures. Spheroids generated via these techniques are highly suitable for tissue reconstruction (Takezawa et al., 1990). Likewise, highly porous (sponge-like) scaffolds are useful in generating immobilized 3-D arrangements of hepatocyte spheroids by providing a favorable environment and enhancing their aggregation. The scaffolds are generally made up of biomaterials such as alginate, gelatin, hyaluronan, or an alginate/galactosylated chitosan hybrid polymer. The use of scaffolds facilitates performance of implanted hepatocytes by enabling their aggregation and re-expression of differentiated function prior to implantation (Fischbach *et al.*, 2007). These porous 3-D scaffolds are useful in liver tissue engineering and bio-artificial liver (BAL). Seeding primary epithelial cells and certain immortal epithelial cell lines (e.g., MCF-10A, DU 4475) on Matrigel promotes the generation of monoclonal spheroids, which are useful for understanding the mechanism of epithelial morphogenesis.

Chip-Based Spheroid Generation

This is a more efficient method where numerous spherical multicellular aggregates (spheroids) are generated with nearly the same diameter on a microfabricated chip. These spheroids mimic real tissue morphology. Generally, the chips are made up of silicon and elastomeric microchannels with various cavities of 100–500 micron in diameter. Cells are loaded by a microfluidic channel into a silicon microchip or microwell, and thus form a localized, single spheroid within each microstructure. This method is most suitable for various biomedical applications such as cell-based biosensors for toxicological and pharmacological examinations, and in bioartificial livers (Kunz-Schughart et al., 2004).

PROTOCOL FOR TUMOR SPHEROID GENERATION

1. Thaw tumor cells from frozen stocks and subculture for >1 and <20 passages. Use standard medium for routine culturing. Keep cultures in a humidified atmosphere, with 5% CO_2 in air at 37°C.
2. Prepare single cell suspensions by mild enzymatic dissociations using trypsin/EDTA solution. Transfer stock cultures every 3rd–4th day by seeding an appropriate number of cells in tissue culture flask, depending upon the cell type and doubling time of cells.
3. To generate spheroids of 600 μm diameter, inoculate $.01 \times 10^6$ viable cells of any cancer cell line in a 96-well plate, non-coated or coated with agarose in 0.2 mL DMEM-LGD supplemented with 5% fetal calf serum, antibiotics, and 10 mM HEPES. Clusters of cells should be observable after 24 hours of initiation, but it will take nearly 4 days for these clusters to form spheroids. Individual spheroids can be used to monitor and analyze spheroid integrity, diameter, and volume under phase-contrast microscope. The pH of the medium should be monitored daily to prevent acidosis.
4. These spheroids can then further be used to monitor the efficacy of drugs.

Application of the protocol: Multicellular tumor spheroids of human glioma cell line (BMG-1) were used to understand their behavior, particularly related to the metabolism and radiation response in their application to anticancer therapy (Flow Chart 11.1) (Khaitan et al., 2006a, c).

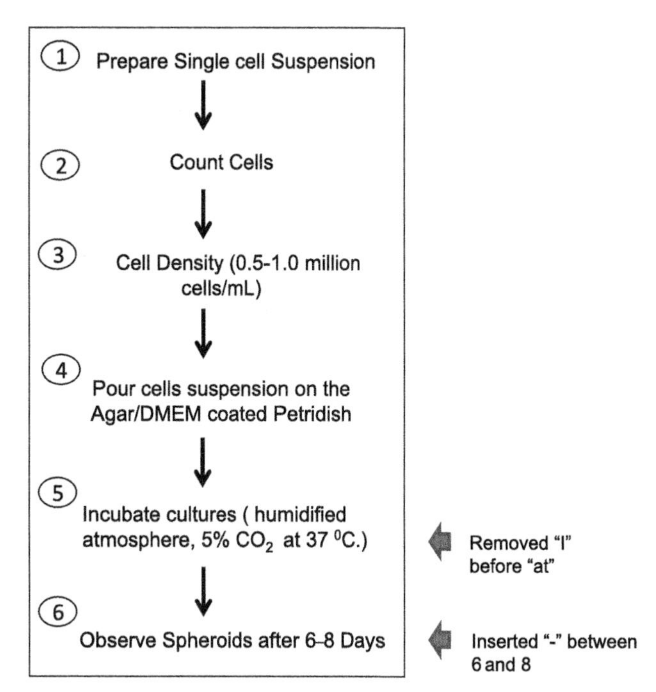

FLOW CHART 11.1 Flow chart for the preparation of spheroids by liquid overlay culture method.

DRUG TREATMENT PROTOCOL

1. For drug treatment, dilute the drugs in standard medium (e.g. DMSO, Ethanol) at 2× final concentration just before use.
2. Treat growing spheroid cultures by replacing 50% (100 μL) of the supernatant with drug-supplemented standard medium. Simultaneously, replace medium of untreated reference spheroid cultures with solvent-containing or solvent-free standard medium.
3. Incubate cultures for a definite treatment time in a humidified atmosphere with 5% CO_2 in air at 37°C.

PARAMETERS TO MONITOR DRUG EFFICACY IN 3-D CULTURES

Different parameters can be used to monitor the effect of drugs on these 3-D cultures. For example, cell viability can be determined by the use of an MTT (3-(4,5-dimethylthiozol-2-yl)-2,5-diphenyltertazolium bromide) assay, analysis of cell cycle kinetics can be done by bromo-deoxy-uridine (Brdu) labeling assay, and the percentage of apoptotic cells can be measured by a Annexin V labeling assay.

RADIATION RESPONSE OF TUMOR CELLS AND ITS MODIFICATIONS

Ionizing radiation plays an important role in the management of a majority of malignancies. Although many tumors like gliomas and several carcinomas are known to be refractory to radiotherapy with marginal benefits in survival, approaches that help in individualizing therapeutic designs based on appropriate predictive assays can improve the efficacy of these therapies. Multicellular tumor spheroids provide a useful predictive platform for the evaluation of tumor response to radiotherapy and/or chemotherapy, as well as adjuvant like radiosensitizers and chemosensitizers including metabolic modifiers and other modifiers of biological response (Khaitan et al., 2006a, b). Radiation-induced cytotoxicity is associated with two mechanistically and morphologically distinct forms of cell death that contribute to the loss of clonogenicity. Programmed cell death (apoptosis), characterized by distinct membrane alterations and DNA degradation, is induced by both extrinsic as well as intrinsic pathways. On the other hand, mitotic death, linked to the cytogenetic damage expressed as chromosomal aberrations in the metaphase and manifested as micronuclei formation in the post mitotic daughter cells, arise from residual DNA damage following induction and repair of DNA lesions.

Mitotically dead cells also undergo apoptosis as a secondary response referred to as delayed apoptosis, which may be generally observed one to two days after irradiation. The *in vivo* response of cancer cells to treatment may, in fact, be simulated in spheroids more closely than in conventional monolayer cultures, as many micro-environmental factors that contribute to the tumor cell response are better simulated in the spheroids. Since hypoxic areas in naturally occurring tumors may limit curability, one promising approach utilizes agents that label the hypoxic fraction, which has been well demonstrated using spheroids (Franko et al., 1985); many sensitizing agents have been developed using spheroids for screening and testing (Hall, 1988). However, the clinical efficacy remains to be determined. A more recent study provides a mathematical analysis of the radiation response of tumor cells, taking in to account oxygen status with the help of spheroids (Bertuzzia et al., 2010).

Radiation produces a variety of lesions in the cell. These lesions induce lethal damage in a fraction of cells (clonogenically dead cells), which lose the capability of unlimited proliferation and die, after some cell cycle delay, at a subsequent time. Thus, after irradiation, the living tumor cell population will be composed by a subpopulation of intact, viable cells and a subpopulation of live but lethally damaged, clonogenically dead cells. The death of lethally damaged cells may occur by pre-mitotic apoptosis or after one or more cell divisions (post-mitotic apoptosis). The response of spheroids to irradiation can be performed on single cells obtained after disaggregation of these spheroids, or on intact spheroids, using elimination (disintegration) of spheroidal structure and growth delay as end points. Clonogenic cell survival is difficult to perform on spheroids of human tumor cells, but the cells show tumor type dependent radiation responses and offer an approach for comparison of radiosensitivity of tumor cell lines of different histologic origin (Bertuzzia et al., 2010). Because of the oxygen concentration gradient, the potential-lethal-damage to sub-lethal-damage ratio changes with the radial distance from the spheroid boundary, leading to radio-resistance in spheroids, similar to the scenario in tumors. Therefore, radiation response using spheroids has played an important role in generating new ideas for combining radiation with drugs that can amplify the effects of radiation (radio-sensitizers, biological response modifiers, etc.) to achieve the enhanced therapeutic gain in radioresistant tumors. Since tumors with high glucose usage are generally resistant to radio- and chemotherapy, it has suggested that glycolytic inhibitors can selectively enhance the effect of radiation and chemotherapeutic drugs (Jain et al., 1996). Indeed, studies using monolayer cultures of several human and murine tumor cells have shown that the glycolytic inhibitor, 2-deoxy-D-glucose (2-DG) enhances radiation and chemotherapeutic drug induced toxicity that involve multiple mechanisms (Dwarakanath et al., 2009); however heterogeneous responses are observed in local tumor-control, tumor-bearing mice, suggesting contributions of tumor physiology and variations in host factors (Gupta et al., 2009). Interestingly, spheroids (which are relatively more radioresistant) have been found to be sensitized to a greater extent by 2-DG: a nearly 3-fold (~300%) increase was observed in the radiation-induced cell death, as compared to an increase of only 40% in monolayers under similar conditions (Khaitan et al., 2006c). Furthermore, the radiosensitization by 2-DG in spheroids was primarily due to enhancement of apoptosis, while enhanced cytogenetic damage was the major contributing factor responsible for cell death in monolayers. The differential mechanisms underlying radiosensitizing effects of 2-DG in spheroids and monolayers are consistent with the idea that these processes are regulated by different proteins and depend on different signalling pathways, which may be different in these two models.

Many studies have also shown enhanced radiosensitization by chemotherapeutics (namely, TNF, doxorubicin, misonidazole, etc.) in multi-cellular spheroids as compared to monolayer cultures (Wen-Hong et al., 2012). Thus, it is apparent that predictions and/or evaluation of newer approaches and biological response modifiers (particularly the metabolic inhibitors) are better accomplished using multicellular spheroids rather than monolayer cultures.

The multicellular spheroid model has also been found to be particularly effective in evaluating combinations of various drugs and/or radiosensitizers with fractionated irradiation regimens. Transient G_2 delay, induction of apoptosis, and late onset of DNA strand breaks following radiosensitization have been found to be more pronounced in the spheroids, similar to those used clinically. Therefore, MTS may prove to be a valuable tool in studying the response of human tumors to clinical exposure protocols, including hyper fractionation (Yuhas et al., 1984).

RESPONSE TO ANTICANCER DRUGS

Inherent as well as acquired resistance to chemotherapeutic drugs is a major obstacle that limits the success of cancer chemotherapy. Tumor response to anticancer therapeutics is influenced by many factors, including drug penetration to different parts of the tumor, the physiological status of the tumor and tumor cells (e.g. hypoxia), proliferation level (growth fraction), cell signaling status, intercellular interactions and tumor cell–extracellular matrix interaction, etc. Limited success of many chemotherapeutic drugs in clinics that have otherwise been found effective against the tumor cells has been partly due to the fact that many studies on the mechanisms of drug resistance have been carried out using monolayer cultures. Due to their three-dimensional architecture, multicellular spheroids have been found to be very useful in drug sensitivity testing (Tofilon et al., 1984), as they overcome many of the limitations posed by monolayer cultures. Several studies have demonstrated that tumor cells cultured as three-dimensional spheroids respond to drugs similarly to the *in vivo* scenario and justify the use of this *in vitro* system to develop clinically useful testing for individual human cancer patients. Since spheroids also represent the heterogeneous nature of individual tumors, they can also measure the drug sensitivities of specific cell types of cultured tumors. Disruptions of intercellular interaction and inhibition of cell adhesion show significant reduction in the resistance against many drugs in tumor spheroid (Shane et al., 2004). The usefulness of spheroids established from genetically manipulated breast cancer cell lines less sensitive to apoptosis have been recently demonstrated as a high-through-put *in vitro* system for screening drugs (Bartholoma et al., 2005).

MCTS coupled with the analysis of metabolic status has been extensively used for investigating the treatment response of tumors to chemotherapeutic drugs. More recent studies using FDG measurement of glucose utilization in spheroids of breast cancer cell lines strongly suggest that the combination of PET radiotracers and image analysis in MCTS provides a good model to evaluate the relationship between tumor volume and the uptake of metabolic tracer before and after chemotherapy (Azita et al., 2006). This feature could be used for screening as well as selecting PET-tracers for the early assessment of treatment response of tumors.

RESPONSE TO PHOTODYNAMIC THERAPY

Photodynamic therapy or PDT is a treatment that uses photosensitizing compounds along with light to kill cancer cells. While pioneering contributions of Finsen, Raab and Von Tappeiner – using a combination of light and drug administration – led to the emergence of photo-chemotherapy as a therapeutic tool, the isolation of porphyrins and their phototoxic effects on tumor tissue led to the development of modern photodetection (PD) and photodynamic therapy (PDT).

Multicellular spheroids (MCSs) represent the ideally suited model for studying the effects of PDT and to better understand tumor response to PDT without interacting with vascular systems. The use of potent photosensitizers (such as Polyvinylpyrrolidone (PVP)-hypericin and Photosan-3) has shown high therapeutic effects on spheroids of human urothelial cell carcinoma cell line as well as U-251 Mg (Terzis et al., 1997). MCSs have also proven useful for the study of complex PDT treatment regimens and combined therapies involving PDT and ionizing radiation or hyperthermia (Khaitan et al., 2006b). Computer simulations have been found to be very useful in modeling the growth of MCS and predict the effects of investigational therapies, including PDT.

RESPONSE TO ANTI-ANGIOGENESIS THERAPEUTICS

Development of tumors and their response to therapy depend on the ability of tumor cells to adopt to different environments and compete with normal cells for space and nutrients. This process of tumorigenesis involves the interaction of tumor cells with the microenvironment, where host endothelial cells (or various types of other cells) are recruited to contribute to the formation of tumor vasculature in order to provide an adequate amount of oxygen and nutrition for the growing tumor mass (termed "tumor angiogenesis"). Angiogenesis drugs act to inhibit survival of newly formed blood vessels required for tumor growth and progression. These drugs have recently shown good activity for the treatment of breast, lung, colon, and kidney cancers. However, these drugs can be toxic, and even cause death.

Hypoxia is a potent inducer of vascular endothelial growth factor (VEGF). Since MCTS develop hypoxic regions similar to *in vivo* conditions, they are good model systems to better understand mechanisms governing induction of VEGF, angiogenesis, and tumor progression, and thus provide significant therapeutic potential for anti-angiogenesis and anticancer drugs. Differential steps of tumor-induced angiogenesis have indeed been studied by a novel *in vitro* confrontation culture of avascular multicellular prostate tumor spheroids and embryoid bodies grown from pluripotent embryonic stem (ES) cells. These show that tumor-induced angiogenesis results in growth stimulation of tumor spheroids, disappearance of central necrosis, and a reduction of the pericellular oxygen pressure, coupled with an increase in the levels of HIF-1α, VEGF, heat shock protein 27 (HSP27), and P-glycoprotein (Khaitan et al., 2006b). Unequivocal induction of VEGF by hypoxia has also been confirmed using spheroids of C6 rat gliomas grown under different conditions, which was reversible following re-oxygenation (Khaitan et al., 2006b). Furthermore, under stressed conditions like glucodeprivation and low oxygen, VEGF mRNA shows more than an eight fold extended half-life as compared to non-stressed conditions. To study the induction of VEGF and GLUT-1, multicellular spheroids of glioma cells have been used in which each layer is differentially stressed by *in situ* hybridization. These findings show that two different consequences of tissue ischemia, namely, hypoxia and glucose deprivation, induce VEGF and GLUT-1 expression by similar mechanisms.

Using pre-labeled tumor cell spheroids prepared from Lewis lung carcinoma, quantitative analysis of spatio-temporal development of vasculature and its influence on the tumor growth has been carried out using microscopic examination exploiting the double labeling of tumor and vasculature (Khaitan et al., 2006b). These studies clearly demonstrate that development of tumor induced vascular network precedes the rapid growth of tumors. Tumor-mass-related stress-induced studies strongly suggest that stress related to tumor mass controls tumor growth at both the macroscopic and cellular levels, and may influence tumor progression and delivery of therapeutic agents. Stress induced VEGF activity has also been demonstrated under *in vivo* conditions using implanted neovascularizing spheroids in nude mice.

Recently, a mini tumor model in the form of a 3-D spheroid system consisting of endothelial, fibroblasts and breast tumor cell line MDA-MB-231 has been developed to study the effects of anti-angiogenesis drugs on tumor growth (Sampaio et al., 2012). It was demonstrated that this model is useful; for example, Galardin (GM6001), an inhibitor of metallo-proteases, showed inhibition of spheroid sprouting. Endostatin was found to inhibit angiogenesis through binding to integrin a5b1. Furthermore, this model has also been used to unravel new roles for mesenchymal MT1-MMP in regulating endothelial sprout formation (Reilly et al., 1997).

Recent studies with matrigel based mammalian carcinoma spheroids have clearly demonstrated that tumor growth may not be critically dependent on angiogenesis as long as a minimal intratumoral microvessel density is maintained (Jelena et al., 2006). Since there is considerable interest in targeting tumor angiogenesis for therapy, MCTS can facilitate the identification of various tumor specific factors responsible for constitutional and stress induced neovascularization, thereby aiding the development of more effective antiangiogenesis therapies.

EVALUATION OF RESPONSE TO IMMUNOTHERAPY

Tumor cells interact with various host cells like fibroblasts, endothelial cells, and infiltrating immune cells (macrophage and lymphocytes etc.) for their development. Tumor-associated immune cells play an important role in promoting or inhibiting tumor growth and progression. Since, MCTS mimics the *in vivo* environment to a large extent, co-culturing spheroids with several immune cells like monocytes, macrophages, dendritic cells, T cells or natural killer (NK) cells is helpful in studying heterologous interactions between tumor and immune cells, as well as their cellular toxicity and therapeutic effects.

Tumor microenvironment influences the differentiation of monocytes into tumor-associated macrophages (TAM). Studies using co-culture of 3-D MCS of urothelial–bladder–carcinoma cell lines J82 and RT4 with human monocytes or macrophages have been performed to determine the release of cytokines levels. The tumor spheroid of poorly differentiated J82 cells stimulates the secretion of IL-1 beta and IL-6 and also promotes the differentiation of monocytes to TAM (Konur et al., 1998). Co-cultures of urothelial bladder cell spheroids with either bacillus activated killer cells (BAK) or IL-2 (lymphokine) activated killer cells (LAK) cells have shown that BCG is an effective immunotherapy in malignant MCS tumors as compared to benign MCS, by reducing proliferation and enhancing the cytotoxicity in malignant MCS (Durek et al., 1999). Similarly, it has been observed that enhanced monocyte migration into fibroblastic tumor spheroids was due to high expression of CCL2 (monocyte chemotactic protein-1), which is under regulation of IL-6 as compared to normal fibroblasts. Moreover, *in vitro* studies using spheroid adhesion assays suggest the potential role of T-regulatory cells to adhere and transmigrate through endothelial cells in tumor-derived EC spheroids (Nummer et al., 2007). However, use of interferon-alpha (IFN-alpha) together with retinoic acid as a combinatorial immunotherapy has been shown to have a growth inhibitory effect in renal cell carcinoma (SN12C) spheroids (Rohde et al., 2004). Furthermore, as cytokine therapy, treatment of CD133+ colon cancer stem cells and spheroid cultures require use of anti-IL-4 for rendering these cells more susceptible to apoptosis through down-regulating anti-apoptotic proteins (Todaro et al., 2008).

Another novel approach is the use of these immune cells as a carrier for gene therapy. Monocytes can be used as a carrier to deliver therapeutic genes; they are pre-loaded with magnetic nanoparticles and migrate across human endothelial cell layers into a 3-D tumor spheroid in the presence of an external magnet (Muthana et al., 2008). Adenoviruses (ADV) are also an attractive candidate for the gene therapy. Modification of ADV by inserting specific genes of tumor receptors provides significant improvement in gene transfer in glioma spheroids. In addition, targeting EGFR by using specific anti-EGFR can also be used (Grill et al., 2001). Other selective approaches and new autologous cellular immunotherapies that may be more effective are currently being developed. For example, human cytotoxic T lymphocytes (CTL) are stimulated to study the interaction and assess the cytotoxicity of CTL for the tumor spheroid. Similarly, monoclonal antibodies alone or in conjugation with drugs, toxins, and radioisotopes can be used for the treatment of cancer. For example, trastuzumab (Herceptin), a monoclonal antibody that specifically targets human epidermal growth factor receptor (HER2) resulted in enhanced inhibition of proliferation of cancer cells in HER2 over-expressed spheroids. Interestingly, HER2 exist as homodimers in the 3D model as compared to heterodimers in 2D. This homodimerization leads to enhanced activation of HER2 and also results in a switch in PI3K to MAPK signaling (Pickl et al., 2009). Besides monoclonal antibodies, trifunctional antibodies can also be used to kill cancer cells. These antibodies have binding sites for two different antigens, typically CD3 and tumor antigen, and an Fc part for accessory cells (macrophages, natural killer cells and dendritic cells). The net effect is that this type of drug links T cells (via CD3) and monocytes/macrophages, natural killer cells, dendritic cells or other Fc receptor expressing cells to the tumor cells, leading to their destruction.

For example, Hirschhaeuser et al. (2010) have shown the anticancer potential of catumaxoab (a trifunctional antibody) in a co-culture of human EpCAM-positive FaDu tumor cells spheroids with human peripheral blood mononuclear cells. The results suggest that catumaxoab showed a strong dose-dependent antitumor response with decreased tumor volume, together with a massive immune cell infiltration and decreased signals for cancer cell viability and clonogenicity. However, the tumor microenvironment is frequently immunosuppressive and contributes to a state of immune ignorance, and thus results in the inability of vaccines/immunotherapy to break tolerance against tumor antigens. Factors such as lactic acid in the tumor microenvironment can inhibit the migration and differentiation of monocytes into dendritic cells and also alter their antigen presentation. High lactic acid concentrations also block the export of lactate in T cells, and thus alter their function and metabolism (Gottfried et al., 2006).

Thus, modulation of a tumor's metabolic environment can be used to enhance anti-tumor responses and thus improvisation in immunotherapy-based tumor killing.

APPLICATION OF 3-D CULTURES IN OTHER DISEASES

There are several examples where 3-D cultures have been used for various diseases. For example, Hwang et al. (2012) demonstrated that a PDMS (Polydimethylsiloxane)-based concave-patterned film could be used for designing islet spheroids with improved cellular functionality and size uniformity to cure diabetes mellitus. Similarly, primary human hepatocytes were used for the spheroid formation to study hepatitis C infection (Chong et al., 2006). Furthermore, generation of 3D human liver models by co-culturing of primary human hepatocytes (hHeps) and human adipose-derived stem cells (hADSCs) in concave micro-well based structures were used to facilitate the studies of human liver diseases like cirrhosis. Similarly, *in vitro* cardiac tissue models have principally been developed for the study of cardiac diseases and in the prediction of drug cardiotoxicity (Franchini et al., 2007). Skin, corneal, oral and vaginal tissue models are now commercially available and can be used in validated toxicological assays. These models are also useful in testing the penetration and sensitization potential of non-drug chemicals. This has been of significant interest to the cosmetics industry, which is increasingly using skin tissue models as *in vitro* substitutes for the murine local lymph node assay (LLNA), the gold standard method for testing the safety and allergenic properties of cosmetics (Carlson et al., 2008).

CONCLUSIONS

Multicellular tumor spheroids are excellent 3-D *in vitro* models that bridge the gap between monolayers and *in vivo* tumors, and facilitate mechanistic studies as well as evaluation of anticancer therapies, particularly relevant to solid tumors. With the advent of novel techniques for establishing MCTS from primary tumor cells and co-culturing with different normal cells, their value in cancer research has increased significantly. Spheroids with a heterogeneous cell population, for example, support studies on tumor-specific processes such as angiogenesis, invasion, and metastasis, as well as assessment of differential responses. Since the MCTS can be conveniently employed in the high throughput screening system with a variety of automated cell-based assays, it is expected to facilitate drug discovery and evaluation of new anticancer therapies. MCTS as a secondary *in vitro* screening system is expected to significantly shorten the time required for translational research, bridging the important and major gap between discovery and clinical studies. Thus, spheroids have the potential to enhance predictability of clinical efficacy and may minimize, if not replace, animal studies to a large extent in the near future. Together with novel and emerging biotechnological tools, this 3-D *in vitro* model is expected to substantially reduce the cost of new drug discovery, thereby making anticancer therapies more and more affordable to the public at large.

ETHICAL ISSUES

Certain Guidelines are required by the Institutional Review Board/Independent Ethics Committee (IRB/IEC) of the institutions/hospitals of the country for the collection, storage and use of biological materials (tissue/organs) in research and diagnostic investigations. Ethical guidelines are important when conducting research on stored/archived human biological tissues that had been collected for the purpose of bio-banking during routine investigation/treatment. The important ethical principles are:

1. Beneficence (doing good).
2. Non-maleficence (preventing or mitigating harm).
3. Fidelity and trust within the investigator/participant relationship.
4. Personal dignity of study participants or subjects.
5. Autonomy pertaining to informed, voluntary, and competent decision making (informed consent).
6. Privacy of personal information.

TRANSLATIONAL SIGNIFICANCE

Multicellular spheroid culture models, which have been demonstrated and established to restore the functional and micro-environmental features of *in vivo* human tumor tissue in many ways (such as the expression of antigens, pH and oxygen gradients within its microenvironment, penetration rate of growth factors and distribution of proliferating/quiescent cells within the spheroid) has contributed considerably to our knowledge of *in vivo* tumor and tissue physiology. Moreover, this model is proven to be more functionally accurate, and its gene expression profile is more similar to clinical expression profiles than the mono-layer model. Therefore, the three-dimensional multicellular tumor spheroidal (MCTS) culture has been translated as a valuable model to provide more comprehensive assessment of tumors in response to therapeutic strategies for improved clinical efficacy predictions. Indeed, the potential of sophisticated, 3-D culture systems to be incorporated into mainstream development processes for new anti-cancer therapeutic strategies is increasingly recognized; it is thought to improve the pre-animal and pre-clinical selection of both the most promising drug candidates and novel, future-oriented treatment modalities (Mueller-Klieser et al., 1997). Besides serving as a good model to test the anti-cancer drugs, this model can also be exploited significantly for understanding tumor physiology and drug penetration studies in 3-D models.

WORLD WIDE WEB RESOURCES

There are a few web sites (WWW) that provide information on 3-dimensional cultures – including multicellular spheroids – that can be easily accessed by readers. One of them is the Companion web site at www.3dcellculture.com. Designed for students; this site provides a single information source for cell culture practices using 3-D techniques. Features include:

1. Tracking of 3-D cell culture publications, with weekly citations on the site. All information can be accessed by unique Cell–Matrix–Stimulus search interface software using keyword/title/author.
2. A forum for 3-D cell culture protocols, product reviews, and other information useful in advancing 3-D culture technology.

Readers can also obtain information on various aspects of spheroids and their application in biomedical research from the websites of research institutions and universities such as University of the West of England (www2.uwe.ac.uk/services/Marketing/business/pdf/3d-cell.pdf). In addition to these academic websites, readers can also find additional information on spheroids and related products from the website of research suppliers like Invitrogen (http://www.invitrogen.com) and Sigma (www.sigmaaldrich.com) or (www.microtissues.com).

Online videos are also available for easy access of 3-D culture protocols at Jove's server (http://www.jove.com/video/2720/a-simple-hanging-drop-cell-culture-protocol-for-generation-3).

REFERENCES

Azita, M., et al. (2006). Multicellular Tumor Spheroid as a model for evaluation of [18F] FDG as biomarker for breast cancer treatment monitoring. *Cancer Cell International*, *6*(6), 1–8.

Barbone, D., et al. (2008). Mammalian target of rapamycin contributes to the acquired apoptotic resistance of human mesothelioma multicellular spheroids. *Journal of Biological chemistry*, *283*(19), 13021–13030.

Bartholoma, P., et al. (2005). More aggressive breast cancer spheroid model coupled to an electronic capillary sensor system for a high-content screening of cytotoxic agents in cancer therapy: 3-dimensional *in vitro* tumor spheroids as a screening model. *Journal of Biomolecular Screening* (10), 705–714.

Bertuzzia, A., et al. (2010). Response of tumor spheroids to radiation modeling and parameter estimation. *Bulletin of Mathematical Biology* (72), 1069–1091.

Carlson, M. W., et al. (2008). Three-dimensional tissue models of normal and diseased skin. *Current Protocols in Cell Biology*. Chapter 19, Unit 19.9.

Chong, T. W., et al. (2006). Primary human hepatocytes in spheroid formation to study hepatitis C infection. *Journal of Surgical Research* (130), 62–67.

Dean, D. M., Napolitano, A. P., Youssef, J., & Morgan, J. R. (2007). Rods, tori and honeycombs: The directed self-assembly of microtissues with prescribed microscale geometries. *FASEB Journal* (21), 4005–4012.

Du, Y., et al. (2007). Identification and characterization of a novel prespheroid 3-dimensional hepatocyte monolayer on galactosylated substratum. *Tissue Engineering* (13), 1455–1468.

Durek, C., et al. (1999). Bacillus–Calmette–Guerin (BCG) and 3D tumors: an *in vitro* model for the study of adhesion and invasion. *Journal of Urology*, *162*(2), 600–605.

Dwarakanath, B. S., & Jain, V. K. (1985). Enhancement of radiation damage by 2-deoxy-D-glucose in organ cultures of brain tumors. *Indian Journal of Medical Research* (82), 266–268.

Dwarakanath, B. S., & Jain, V. K. (1987). Modification of the radiation induced damage by 2-deoxy-D-glucose in organ cultures of human cerebral gliomas. *International Journal of Radiation Oncology Biology Physics* (13), 741–746.

Dwarakanath, B. S. (2009). Cytotoxicity, radiosensitization, and chemosensitization of tumor cells by 2-deoxy-D-glucose *in vitro*. *Journal of Cancer Research and Therapeutics* (1), S27–S31.

Fischbach, C., et al. (2007). Engineering tumors with 3-D scaffolds. *Nature Methods*, *4*(10), 855–860.

Franchini, J. L., et al. (2007). Novel tissue engineered tubular heart tissue for *in vitro* pharmaceutical toxicity testing. *Microscopy and Microanalysis*, *13*, 267–271.

Franko, A. J. (1985). Hypoxic fraction with binding of misonidazole in multicellular tumor spheroids. *Radiation Research* (103), 89–97.

Glicklis, R., Shapiro, L., Agbaria, R., Merchuk, J. C., & Cohen, S. (2000). Hepatocyte behavior within three-dimensional porous alginate scaffolds. *Biotechnology Bioengineering*, 67(3), 344–353.

Gottfried, E., Kunz-Schughart, L. A., Andreesen, R., & Kreutz, M. (2006). Brave little world: spheroids as an *in vitro* model to study tumor–immune-cell interactions. *Cell cycle*, *5*(7), 691–695.

Grill, J., et al. (2001). Combined targeting of adenoviruses to integrins and epidermal growth factor receptors increases gene transfer into primary glioma cells and spheroids. *Journal of Clinical Cancer Research*, 7(3), 641–650.

Gupta, S., et al. (2009). Enhancement of radiation and chemotherapeutic drug responses by 2-deoxy-D-glucose in animal tumors. *Journal of Cancer Research and Therapeutics*, *5*(Suppl. 1), S16–S20.

Hall, E. J. (1988). *Radiobiology for the radiologist*. Philadelphia: Lippincott.

Hirschhaeuser, F., et al. (2010). Multicellular tumor spheroids: An underestimated tool is catching up again. *Journal of Biotechnology* (148), 3–15.

Hwang, J. W., et al. (2012). Optimization of pancreatic islet spheroid using various concave patterned-films. *Macromolecular Research* (20), 1264–1270.

Ingram, M., et al. (1997). Three-dimensional growth patterns of various human tumor cell lines in simulated microgravity of a NASA bioreactor. *In Vitro Cellular & Developmental Biology – Animal* (33), 459–466.

Jain, V. K. (1996). Modification of radiation responses by 2-deoxy-D-glucose in normal and cancer cells. *Indian Journal of Nuclear Medicine* (11), 8–17.

Jelena, K., et al. (2006). Dissociation of angiogenesis and tumorigenesis in follistatin- and activin-expressing tumors. *Cancer Research* (66), 5686–5695.

Kelm, J. M., et al. (2003). Method for generation of homogeneous multicellular tumor spheroids applicable to a wide variety of cell types. *Biotechnology Bioengineering* (83), 173–180.

Khaitan, D., Chandna, S., Arya, M. B., & Dwarakanath, B. S. (2006 a). Establishment and characterization of multicellular spheroids from a human glioma cell line; Implications for tumor therapy. *Journal of Translational Medicine*, *4*(12), 4–12.

Khaitan, D., & Dwarakanath, B. S. (2006 b). Multicellular spheroids as an *in vitro* model in experimental oncology: applications in translational medicine. *Expert Opinion in Drug Discovery*, *1*(7), 663–675.

Khaitan, D., Chandna, S., Arya, M. B., & Dwarakanath, B. S. (2006 c). Differential mechanisms of radiosensitization by 2-deoxy-D-glucose in the monolayers and multicellular spheroids of a human glioma cell line. *Cancer Biology Therapy*, *5*(9), 1142–1151.

Konur, A., et al. (1998). Cytokine repertoire during maturation of monocytes to macrophages within spheroids of malignant and non-malignant urothelial cell lines. *International Journal of Cancer*, *5*(78), 648–653.

Kunz-Schughart, L. A., Freyer, J. P., Hofstaedter, F., & Ebner, R. (2004). The use of 3-D cultures for high-throughput screening: The multicellular spheroid model. *Journal of Biomolecular Screening* (9), 273–285.

Lasnitzki, I., et al. (1992). In: Freshney R. I., (ed.) *Animal Cell Culture, a Practical Approach*, Oxford: IRL Press at Oxford University Press. pp. 213–261.

Lin, R. Z., & Chang, H. Y. (2008). Recent advances in three-dimensional multicellular spheroid culture for biomedical research. Biotechnology Journal (3), 9–10. 1172-1184.

Mauchamp, J., Mirrione, A., Alquier, C., & Andre, F. (1998). Follicle-like structure and polarized monolayer: role of the extracellular matrix on thyroid cell organization in primary culture. *Biology of Cell*, *90*(5), 369–380.

Mueller-Klieser, W. (1997). Three-dimensional cell cultures: from molecular mechanisms to clinical applications. *American Journal of Physiology* (273), 1109–1123.

Muthana, M., et al. (2008). A novel magnetic approach to enhance the efficacy of cell-based gene therapies. *Gene Therapy*, *15*(12), 902–910.

Nummer, D., et al. (2007). Role of tumor endothelium in CD4+ CD25+ regulatory T-cell infiltration of human pancreatic carcinoma. *Journal of the National Cancer Institute*, *99*(15), 1188–1199.

Pickl, M., & Ries, C. H. (2009). Comparison of 3D and 2D tumor models reveals enhanced HER2 activation in 3D associated with an increased response to trastuzumab. *Oncogene*, *28*(3), 461–468.

Reilly, S. O. M., et al. (1997). Endostatin: An endogenous inhibitor of angiogenesis and tumor growth. *Cell*, *2*(88), 277–285.

Rohde, D., Brkovic, D., & Honig d'Orville, I. (2004). All-trans retinoic acid and interferon-alpha for treatment of human renal cell carcinoma multicellular tumor spheroids. *International Journal of Urology*, *1*(73), 47–53.

Sakai, Y., et al. (1996). Large scale preparation and function of porcine hepatocyte spheroids. *International Journal of Artificial Organs* (19), 294–301.

Sampaio, C. P., et al. (2012). A heterogeneous *in vitro* three dimensional model of tumor – stroma interactions regulating sprouting angiogenesis. *PLoS One*, *2*(7), 30753–30767.

Shane, K. G., Giulio, F., Ciro, I., & Robert, S. K. (2004). Antiadhesive antibodies targeting E-cadherin sensitize multicellular tumor spheroids to chemotherapy *in vitro*. *Molecular Cancer Therapy* (3), 149–159.

Takezawa, T., Mori, Y., & Yoshizato, K. (1990). Cell culture on a thermo-responsive polymer surface. *Biotechnology* (8), 854–856.

Terzis, A. J., Dietze, A., Bjerkvig, R., & Arnold, H. (1997). Effects of photodynamic therapy on glioma spheroids. *British Journal of Neurosurgery*, *3*(11), 196–205.

Todaro, M., et al. (2008). IL-4-mediated drug resistance in colon cancer stem cells. *Cell Cycle*, *3*(7), 309–313.

Tofilon, P. I., Buckley, N., & Deen, D. F. (1984). Effect of cell–cell interactions on drug sensitivity and growth of drug-sensitive and -resistant tumor cells in spheroids. *Science* (226), 862–864.

Tong, J. Z., et al. (1992). Long-term culture of adult rat hepatocyte spheroids. *Experimental Cell Research* (200), 326–332.

Tung, Y. C., et al. (2011). High-throughput 3D spheroid culture and drug testing using a 384 hanging drop array. *Analyst*, *3*(136), 473–478.

Vinci., et al. (2012). Advances in establishment and analysis of three dimensional tumor spheroid-based functional assays for target validation and drug evaluation. *Journal of Biomed-central Biology*, *10*(29), 1–20.

Wen-Hong Xu., et al. (2012). Doxorubicin-mediated radiosensitivity in multicellular spheroids from a lung cancer cell line is enhanced by composite micelle encapsulation. *International Journal of Nanomedicine* (7), 2661–2671.

Yuhas, J. M., Li, A. P., Martinez, A. O., & Ladman, A. J. (1977). A simplified method for production and growth of multicellular tumor spheroids. *Cancer Research* (37), 3639–3643.

Yuhas, J. M., Blake, S., & Weichselbaum, R. R. (1984). Quantitation of the response of human tumor spheroids to daily radiation exposures. *International Journal of Radiation Oncology Biology Physics.*, *10*(12), 2323–2327.

FURTHER READING

Acker, H., Carlsson, J., & Durand, R. (Eds.), (2011). *Recent Results in Cancer Research. Spheroids in Cancer Research: Methods and Perspectives:* (vol. 95). London: Springer.

Bjerkvig, R. (Ed.), (1992). *Spheroid Culture in Cancer Research.* Florida: CRC Press.

Cottin, S., et al. (2010). Gemcitabine intercellular diffusion mediated by gap junctions: new implications for cancer therapy. *Molecular Cancer* (9), 141.

Freshney, R. Ian (1994). *Culture of Animal Cells: A Manual of Basic Techniques and Specialized Applications.* (6th ed.). New Jersey: Wiley Blackwell.

Freshney, R. I. (2007). *Culture of Animal Cells: A Manual of Basic Techniques and Specialized Applications* (6th ed.). Oxford University Press.

Hall, E. J. (1988). *Radiobiology for the Radiologist*. Philadelphia: Lippincott.

Haycock, J. W. (Ed.), (2011). *3-D Cell Culture: Methods and Protocols*, Vol 695. Springer Protocols, New York: Humana Press.

Moorselaar, V., et al. (1990). Combined effects of tumor necrosis factor alpha and radiation in the treatment of renal cell carcinoma grown as radia spheroids. *Anticancer Research, 10*(6), 1769–1773.

Timmins, N. E., Dietmair, S., & Nielsen, L. K. (2004). Hanging-drop multicellular spheroids as a model of tumor angiogenesis. *Angiogenesis* (7), 97–103.

GLOSSARY

Gompertz Function A sigmoidal function in a mathematical model used to study the growth of tumors.

Scaffold A temporary structure of material, such as silicon, for the support of tissue/cells.

Spheroid Clump of cells grown together in tissue culture suspension.

Tori Plural of the word "torus," which is a type of 3-D circular shape.

ABBREVIATIONS

2-D 2-Dimensional
2-DG 2-Deoxyglucose
3-D 3-Dimensional
5-FU 5-Fluorouracil
ADV Adenovirus
BAK Bacillus Activated Killer Cells
BAL Bioartificial Liver
BMG-1 Human Glioma Cell Line
Brdu Bromodeoxyuridine
CCL2 Monocyte Chemotactic Protein-1
CD Cluster of Differentiation
DMSO Dimethylsulfoxide
DMEM Dulbecco's Modified Eagle Medium
DU 4475 Human Breast Cancer Cell Line
ES cells Embryonic Stem Cells
F-DG Fluordeoxyglucose
Glut-1 Glucose Transporter-1
HEPES 4-(2-Hydroxyethyl)-1-Piperazineethanesulfonic Acid
HER2 Human Epidermal Growth Factor Receptor-2
HIF-1α Hypoxia Inducing Factor-1α
Hsp27 Heat Shock Protein 27
IFN-α Interferon-α
IL-1,2,4,6 Interleukin-1, -2, -4, and -6
LAK Cells Lymphokine Activated Killer Cells
LCST Low Critical Solution Temperature
LGD Low Glucose Dextrose Media
LLNA Local Lymph Node Assay
MAPK Mitogen-Activated Protein Kinase
MCTS Multicellular Tumor Spheroid
MDA-MB-231 Human Breast Adenocarcinoma Cell Line
MTI-MMP Matrix Metalloproteinase Protein
MTT 3-(4,5-Dimethylthiozol-2-yl)-2,5-Diphenyltertazolium Bromide
PDMS Polydimethylsiloxane
PDT Photodynamic Therapy
PEG Polyethyleneglycol
PET Positron Emission Tomography
PHEMA Poly (2-Hydroxyethl Methacrylate)
PNI-PAAm Poly-N-Isopropyl Acrylamide
PVDF Polyvinyldifluoride
PVP Polyvinylpyrrolidone
SN12C Human Renal Carcinoma Cell Line
TAM Tumor Associated Macrophages
TNF-α Tumor Necrosis Factor-α
TPZ Tirapazamine
U-251Mg Human Glioblastoma Cell Line
VEGF Vascular Endothelial Growth Factor

LONG ANSWER QUESTIONS

1. How do you generate spheroids containing a precise number of cells? Write a short explanation of (a) the hanging drop method, and (b) the microfabricated microstructures based method. Why do they have an edge over other methods?
2. Describe how spheroids can be useful in studying tumor immune cell interactions and therapeutics.
3. How are spheroids useful in studying combined therapies? Briefly discuss a few therapies.
4. Discuss in brief the advantages and disadvantages of 3-D over 2-D cultures, with examples.
5. Discuss the advantages of co-culturing spheroids and how it can be useful in studying the effects of different tumor associated parameters.

SHORT ANSWER QUESTIONS

1. What are the important factors that should be considered while generating spheroids?
2. Briefly discuss the metabolic pathways involved in differentiation of immune cells.
3. Describe the major factors responsible for increased resistance of spheroids (over mono-layer) towards therapy.
4. Describe the different parameters to monitor the drug efficacy in 3-D cultures.
5. Briefly describe how cell-cell interaction alters the therapeutic response of multi-cellular tumor spheroids.

ANSWERS TO SHORT ANSWER QUESTIONS

1. Spheroid size, possible damage and influence in cellular physiology, production efficiency.
2. Lactate, growth factors, etc.
3. Hypoxia and reduced vascularization, etc.
4. MTT assay for cell viability, cell cycle kinetics by Brdu labeling, Annexin V assay for apoptosis.
5. Bystander effects.

Chapter 12

Animal Tissue Culture: Principles and Applications

Anju Verma
University of Missouri, Columbia, Missouri, USA

Chapter Outline

Animal Biotechnology. http://dx.doi.org/10.1016/B978-0-12-416002-6.00012-2

SUMMARY

Animal cell culture technology in today's scenario has become indispensable in the field of life sciences, which provides a basis to study regulation, proliferation, differentiation, and to perform genetic manipulation. It requires specific technical skills to carry out successfully. This chapter describes the essential techniques of animal cell culture as well as its applications.

WHAT YOU CAN EXPECT TO KNOW

This chapter describes the basics of animal cell culture along with the most recent applications. The primary aim is to progressively guide students through fundamental areas and to demonstrate an understanding of basic concepts of cell culture as well as how to perform cell cultures and handle cell lines. The chapter gives insights into types of cell culture, culture media and use of serum, viability assays, and the translational significance of cell culture.

HISTORY AND METHODS

INTRODUCTION

Cell culture is the process by which human, animal, or insect cells are grown in a favorable artificial environment. The cells may be derived from multicellular eukaryotes, already established cell lines, or established cell strains. In the mid-1900s animal cell culture became a common laboratory technique, but the concept of maintaining live cell lines separated from their original tissue source was discovered in the 19th century. Animal cell culture is now one of the major tools used in the life sciences in areas of research that have a potential for economic value and commercialization. The development of basic culture media has enabled scientists to work with a wide variety of cells under controlled conditions; this has played an important role in advancing our understanding of cell growth and differentiation, identification of growth factors, and understanding of mechanisms underlying the normal functions of various cell types. New technologies have also been applied to investigate high cell density bioreactor and culture conditions.

Many products of biotechnology (such as viral vaccines) are fundamentally dependent on mass culturing of animal cell lines. Although many simpler proteins are being produced using rDNA in bacterial cultures, more complex proteins that are glycosylated (carbohydrate-modified) currently have to be made in animal cells. At present, cell culture research is aimed at investigating the influence of culture conditions on viability, productivity, and the constancy of post-translational modifications such as glycosylation, which are important for biological activity of recombinant proteins. Biologicals produced by recombinant DNA (rDNA) technology in animal cell cultures include anticancer agents, enzymes, immunobiologicals (interleukins, lymphokines, monoclonal antibodies), and hormones.

Animal cell culture has found use in diverse areas, from basic to advanced research. It has provided a model system for a variety of research efforts:

1. The study of basic cell biology, cell cycle mechanisms, specialized cell function, cell–cell and cell–matrix interactions.
2. Toxicity testing to study the effects of new drugs.
3. Gene therapy for replacing non-functional genes with functional gene-carrying cells.
4. The characterization of cancer cells, the role of various chemicals, viruses, and radiation in cancer cells.
5. Production of vaccines, monoclonal antibodies, and pharmaceutical drugs.
6. Production of viruses for use in vaccine production (e.g. chicken pox, polio, rabies, hepatitis B, and measles).

Today, mammalian cell culture is a prerequisite for manufacturing biological therapeutics such as hormones, antibodies, interferons, clotting factors, and vaccines.

DEVELOPMENT OF ANIMAL CELL CULTURE

The first mammalian cell cultures date back to the early twentieth century. The cultures were originally created to study the development of cell cultures and normal physiological events such as nerve development. Ross Harrison in 1907 showed the first nerve fiber growth *in vitro*. However, it was in the 1950s that animal cell culture was performed at an industrial scale. It was with major epidemics of polio in the 1940s and 1950s and the accompanying requirement for viral vaccines that the need for cell cultures on a large scale became apparent. The polio vaccine from a de-activated virus became one of the first commercial products developed from cultured animal cells (Table 12.1).

TABLE 12.1 Historical Events in Cell Cultures

1878	Claude Bernard	Established that a physiological state of the cell similar to the live cell can be maintained even after death of the organism.
1907	Harrison	Cell entrapment and frog embryo nerve fiber growth *in vitro*.
1912	Alexis Carriel	Initiated tissue culture of chick embryo heart cells using embryo extracts as cultural media passaged for a reported period of 34 years.
1913	Steinhardt, Israeli and Lambert	Grew vaccinia virus in fragments of guinea pig corneal tissue.
1916	Rous and Jone	Used Trypsin to suspend attached cells in culture.
1927	Carrel and Rivera	First viral vaccine-against chicken pox.
1949	Enders, Weller and Robbins	Polio virus grown on human embryonic cells in culture.
1952	Gey	Establishment of continuous cell line from a human cervical carcinoma (HeLa cells).
1955	Eagle	Established nutrient requirements of cells in culture and defined culture media for growth.
1956	Little Field	HAT (hypoxanthin, aminopterin, thymidine) medium introduced for cell selection.
1961	Hayflick and Moorhead	Studied human fibroblasts (WI-38) and showed finite lifespan in culture.
1965	Ham	First serum-free HAMS's media.
1973	Kohler and Milstein	First hybridoma secreting a monoclonal antibody.
1977	Genetech	First recombinant human protein: somatistatin.
1985	Collen	Recombinant tissue plasminogen activator (TPA) in mammalian cells. Human growth hormone produced from recombinant bacteria was accepted for therapeutic use.
1986	FDA approval	First monoclonal antibody was approved by the FDA for use in humans (Orthoclone OKT3).
1986	Genetech	First recombinant protein commercialized (interferon alpha-2a).
1989	Amgen	Erythropoietin (EPO) recombinant protein produced in CHO cells available commercially.
1992	FDA approval	First genetically engineered recombinant blood-clotting factor; used in treatment of hemophilia A.
1996	Wilmut	Production of transgenic sheep (Dolly) through nuclear transfer technique.
2002	Cloneaid	Claimed to produce cloned human baby named EVE.
2004	FDA approval	First anti-angiogenic monoclonal antibody that inhibits the growth of blood vessels or angiogenesis (for cancer therapy).
2005	Birch	Reported antibody titers at an industrial scale of 5 g/L and more.
2009	Nathalie Cartier-Lacave	Combined gene therapy with blood stem cell therapy, which may be a useful tool for treating fatal brain disease.
2011	Melanie Welham, David Tosh	1M molecule treatment causes stem cells to turn into precursors of liver cells.
2012	Maria Blasco	First gene therapy successful against aging-associated decline in mice.

BASIC CONCEPT OF CELL CULTURE

Tissue culture is *in vitro* maintenance and propagation of isolated cells tissues or organs in an appropriate artificial environment. Many animal cells can be induced to grow outside of their organ or tissue of origin under defined conditions when supplemented with a medium containing nutrients and growth factors. For *in vitro* growth of cells, the culture conditions may not mimic *in vivo* conditions with respect to temperature, pH, CO_2, O_2, osmolality, and nutrition. In addition, the cultured cells require sterile conditions along with a steady supply of nutrients for growth and sophisticated incubation conditions. An important factor influencing the growth of cells in culture medium is the medium itself. At present, animal cells are cultured in natural media or artificial media depending on the needs of the experiment. The culture medium is the most important and essential step in animal tissue culture. This depends on the type of cells that need to be cultured for the purpose of cell growth differentiation or production of designed pharmaceutical products. In addition, serum-containing and serum-free media are now available that offer a varying degree of advantage to the cell culture. Sterile conditions are important in the development of cell lines.

Cells from a wide range of different tissues and organisms are now grown in the lab. Earlier, the major purpose of cell culture was to study the growth, the requirements for growth, the cell cycle, and the cell itself. At present, homogenous cultures obtained from primary cell cultures are useful tools to study the origin and biology of the cells. Organotypic and histotypic cultures that mimic the respective organs/tissues have been useful for production of artificial tissues.

How Are Cell Cultures Obtained?

There are three methods commonly used to initiate a culture from animals.

Organ Culture

Whole organs from embryos or partial adult organs are used to initiate organ culture *in vitro*. These cells in the organ culture maintain their differentiated character, their functional activity, and also retain their *in vivo* architecture. They do not grow rapidly, and cell proliferation is limited to the periphery of the explant. As these cultures cannot be propagated for long periods, a fresh explantation is required for every experiment that leads to inter-experimental variation in terms of reproducibility and homogeneity. Organ culture is useful for studying functional properties of cells (production of hormones), and for examining the effects of external agents (such as drugs and other micro or macro molecules) and products on other organs that are anatomically placed apart *in vivo*.

Primary Explant Culture

Fragments exercised from animal tissue may be maintained in a number of different ways. The tissue adheres to the surface aided by an extracellular matrix constituent, such as collagen or a plasma clot, and it can even happen spontaneously. This gives rise to cells migrating from the periphery of the explant. This culture is known as a primary explant, and migrating cells are known as outgrowth. This has been used to analyze the growth characteristics of cancer cells in comparison to their normal counterparts, especially with reference to altered growth patterns and cell morphology.

Cell Culture

This is the most commonly used method of tissue culture, and is generated by collecting the cells growing out of explants or dispersed cell suspensions (floating free in culture medium). Cells obtained either by enzymatic treatment or by mechanical means are cultured as adherent monolayers on solid substrate.

Cell culture is of three types: (1) precursor cell culture, which is undifferentiated cells committed to differentiate; (2) differentiated cell culture, which is completely differentiated cells that have lost the capacity to further differentiate; and (3) stem cell culture, which is undifferentiated cells that go on to develop into any type of cell.

Cells with a defined cell type and characteristics are selected from a culture by cloning or by other methods; this cell line becomes a cell strain.

Monolayer Cultures

The monolayer culture is an anchorage-dependent culture of usually one cell in thickness with a continuous layer of cells at the bottom of the culture vessel.

Suspension Cultures

Some of the cells are non-adhesive and can be mechanically kept in suspension, unlike most cells that grow as monolayers (e.g. cells of leukemia). This offers numerous advantages in the propagation of cells.

Cell Passage and Use of Trypsin

Passaging is the process of sub-culturing cells in order to produce a large number of cells from pre-existing ones. Sub-culturing produces a more homogeneous cell line and avoids the senescence associated with prolonged high cell density. Splitting cells involves transferring a small number of cells into each new vessel. After sub-culturing, cells may be propagated, characterized, and stored. Adherent cell cultures need to be detached from the surface of the tissue culture flasks or dishes using proteins. Proteins secreted by the cells form a tight bridge between the cell and the surface. A mixture of trypsin-EDTA is used to break proteins at specific places. Trypsin is either protein-degrading or proteolytic, it hydrolyzes pepsin-digested peptides by hydrolysis of peptide bonds. EDTA sequesters certain metal ions that can inhibit trypsin activity, and thus enhances the efficacy of trypsin. The trypsinization process and procedure to remove adherent cells is given in Flow Chart 12.1.

Quantitation

Quantitation is carried out to characterize cell growth and to establish reproducible culture conditions.

Hemocytometer

Cell counts are important for monitoring growth rates as well as for setting up new cultures with known cell numbers. The most widely used type of counting chamber is called a hemocytometer. It is used to estimate cell number. The concentration of cells in suspension is determined by placing the cells in an optically clear chamber under a microscope. The cell number within a defined area of known depth is counted and the cell concentration is determined from the count.

Electronic Counting

For high-throughput work, electronic cell counters are used to determine the concentration of each sample.

Other Quantitation

In some cases the DNA content or the protein concentration needs be determined instead of the number of cells.

Reconstruction of Three-Dimensional Structures

Cells propagated as a cell suspension or monolayer offer many advantages but lack the potential for cell-to-cell interaction and cell–matrix interaction seen in organ cultures.

Trypsinization

Rinse Monolayer Culture
↓
Add Trypsin-EDTA
↓
Incubate
↓
Stop Trypsinization
↓
Spin Cells
↓
Add Fresh Medium
↓
Count cells
↓
Seed new flask
↓
Incubate at 37°C

FLOW CHART 12.1 Trypsinization of adherent cells.

For this reason, many culture methods that start with a dispersed population of cells encourage the arrangement of these cells into organ-like structures. These types of cultures can be divided into two basic types.

Histotypic Culture

Cell–cell interactions similar to tissue-like densities can be attained by use of an appropriate extracellular matrix and soluble factors and by growing cell cultures to high cell densities. This can be achieved by (a) growing cells in a relatively large reservoir with adequate medium fitted with a filter where the cells are crowded; (b) growing the cells at high concentrations on agar or agarose or as stirred aggregates (spheroids); and (c) growing cells on the outer surface of hollow fibers where the cells are seeded on the outer surface and medium is pumped through the fibers from a reservoir.

Organotypic Culture

To simulate heterotypic cell interactions in addition to homotypic cell interactions, cells of differentiated lineages are re-combined. Co-culturing of epithelial and fibroblast cell clones from the mammary gland allow the cells to differentiate functionality under the correct hormonal environment, thus producing milk proteins.

TYPES OF CELL CULTURE

Primary Cell Culture

These cells are obtained directly from tissues and organs by mechanical or chemical disintegration, or by enzymatic digestion. These cells are induced to grow in suitable glass or plastic containers with complex media. These cultures usually have a low growth rate and are heterogeneous, however, they are still preferred over cell lines as these are more representative of the cell types in the tissues from which they are derived. The morphological structure of cells in culture is of various types: (a) epithelium type, which are polygonal in shape and appear flattened as they are attached to a substrate and form a continuous thin layer (i.e. monolayer on solid surfaces); (b) epitheloid type, which have a round outline and do not form sheets like epithelial cells and do not attach to the substrate; (c) fibroblast type, which are angular in shape and elongated and form an open network of cells rather than tightly packed cells, are bipolar or multipolar, and attach to the substrate; and (d) connective tissue type, which are derived from fibrous tissue, cartilage, and bone, and are characterized by a large amount of fibrous and amorphous extracellular materials.

Advantages and Disadvantages of Primary Cell Culture

These cultures represent the best experimental models for *in vivo* studies. They share the same karyotype as the parent and express characteristics that are not seen in cultured cells. However, they are difficult to obtain and have limited lifespans. Potential contamination by viruses and bacteria is also a major disadvantage.

Depending on the kind of cells in culture, the primary cell culture can also be divided into two types.

Anchorage-Dependent/Adherent Cells

These cells require a stable nontoxic and biologically inert surface for attachment and growth, and are difficult to grow as cell suspensions. Mouse fibroblast STO cells are anchorage cells.

Anchorage-Independent/Suspension Cells

These cells do not require a solid surface for attachment or growth. Cells can be grown continuously in liquid media. The source of cells is the governing factor for suspension cells. Blood cells are vascular in nature and are suspended in plasma and these cells can be very easily established in suspension cultures.

Secondary Cell Culture

When primary cell cultures are passaged or sub-cultured and grown for a long period of time in fresh medium, they form secondary cultures and are long-lasting (unlike cells of primary cell cultures) due to availability of fresh nutrients at regular intervals. The passaging or sub-culturing is carried out by enzymatic digestion of adherent cells. This is followed by washing and re-suspending of the required amount of cells in appropriate volumes of growth media. Secondary cell cultures are preferred as these are easy to grow and are readily available; they have been useful in virological, immunological, and toxicological research.

Advantages and Disadvantages of Secondary Cell Culture

This type of culture is useful for obtaining a large population of similar cells and can be transformed to grow indefinitely. These cell cultures maintain their cellular characteristics. The major disadvantage of this system is that the cells have a tendency to differentiate over a period of time in culture and generate aberrant cells.

CELL LINE

The primary culture, when sub-cultured, becomes a cell line or cell strain that can be finite or continuous, depending on its life span in culture. They are grouped into two types on the basis of the life span of the culture.

Finite Cell Lines

Cell lines with a limited number of cell generations and growth are called finite cell lines. The cells are slow growing (24 to 96 hours). These cells are characterized by anchorage dependence and density limitation.

Indefinite Cell Lines

Cell lines obtained from *in vitro* transformed cell lines or cancerous cells are indefinite cell lines and can be grown in monolayer or suspension form. These cells divide rapidly with a generation time of 12 to 14 hours and have a potential to be sub-cultured indefinitely. The cell lines may exhibit aneuploidy (Bhat, 2011) or heteroploidy due to an altered chromosome number. Immortalized cell lines are transformed cells with altered growth properties. HeLa cells are an example of an immortal cell line. These are human epithelial cells obtained from fatal cervical carcinoma transformed by human papilloma virus 18 (HPV18). Indefinite cell lines are easy to manipulate and maintain. However these cell lines have a tendency to change over a period of time.

Commonly Used Cell Lines

Nowadays, for the production of biologically active substances on an industrial scale, a mammalian cell culture is a prerequisite. With advancements in animal cell culture technology, a number of cell lines have evolved and are used for vaccine production, therapeutic proteins, pharmaceutical agents, and anti-cancerous agents. For the production of cell lines, human, animal, or insect cells may be used. Cell lines that are able to grow in suspension are preferred as they have a faster growth rate. Chinese hamster ovary (CHO) is the most commonly used mammalian cell line.

When selecting a cell line, a number of general parameters must be considered, such as growth characteristics, population doubling time, saturation density, plating efficiency, growth fraction, and the ability to grow in suspension. Table 12.2 shows some of the commonly used cell lines.

Advantages of Continuous Cell Lines

1. Continuous cell lines show faster cell growth and achieve higher cell densities in culture.
2. Serum-free and protein-free media for widely used cell lines may be available on the market.
3. The cell lines have a potential to be cultured in suspension in large-scale bioreactors.

The major disadvantages of these cultures are chromosomal instability, phenotypic variation in relation to the donor tissue, and a change in specific and characteristic tissue markers (Freshney, 1994).

TABLE 12.2 Commonly Used Cell Lines and Their Origins

Cell Line	Origin	Organism
H1, H9	Embryonic stem cells	Human
HEK-293	Embryonic kidney transformed with adenovirus	Human
HeLa	Epithelial cell	Human
HL 60	Human promyelocytic leukemia cells	Human
MCF-7	Breast cancer	Human
A549	Lung cancer	Human
A1 to A5-E	Amnion	Human
ND-E	Esophagus	Human
CHO	Ovary	Chinese hamster
Vero	Kidney epithelial cell	African green monkey
Cos-7	Kidney cells	African green monkey
3T3	Fibroblast	Mouse
BHK21	Fibroblast	Syrian hamster
MDCK	Epithelial cell	Dog
E14.1	Embryonic stem cells (mouse)	Mouse
COS	Kidney	Monkey
DT40	Lymphoma cell	Chick
S2	Macrophage-like cells	Drosophila
GH3	Pituitary tumor	Rat
L6	Myoblast	Rat
Sf9 and Sf21	Ovaries	Fall Army worm (*Spodoptera frugiperda*)
ZF4 and AB9 cells	Embryonic fibroblast cells	Zebrafish

GROWTH CYCLE

The cells in the culture show a characteristic growth pattern, lag phase, exponential or log phase, followed by a plateau phase. The population doubling time of the cells can be calculated during the log phase and plateau phase. This is critical and can be used to quantify the response of the cells to different culture conditions for changes in nutrient concentration and effects of hormonal or toxic components. The population doubling time describes the cell division rates within the culture and is influenced by non-growing and dying cells.

Phases of the Growth Cycle

The population doubling time, lag time, and saturation density of a particular cell line can be established and characterized for a particular cell type. A growth curve consists of a normal culture and can be divided into a lag phase, log phase, and plateau phase.

Lag Phase

This is the initial growth phase of the sub-culture and reseeding during which the cell population takes time to recover. The cell number remains relatively constant prior to rapid growth. During this phase the cell replaces elements of the glycocalyx lost during trypsinization, attaches to the substrate, and spreads out. During the spreading process the cytoskeleton reappears; its reappearance is probably an integral part of the process.

Log Phase

This is a period of exponential increase in cell number and growth of the cell population due to continuous division. The length of the log phase depends on the initial seeding density, the growth rate of the cells, and the density at which cell proliferation is inhibited by density. This phase represents the most reproducible form of the culture as the growth fraction and viability is high (usually 90–100%) and the population is at its most uniform. However, the cell culture may not be synchronized and the cells can be randomly distributed in the cell cycle.

Plateau Phase

The culture becomes confluent at the end of the log phase as growth rates during this phase are reduced, and cell proliferation can cease in some cases due to exhaustion. The cells are in contact with surrounding cells and the growth surface is occupied. At this stage, the culture enters the stationary phase and the growth fraction falls to between 0 and 10%. Also, the constitution and charge of the cell surface may be changed and there may be a relative increase in the synthesis of specialized versus structural proteins.

MONITORING CELL GROWTH

The animal cell culture can be grown for a wide variety of cell-based assays to investigate morphology, protein expression, cell growth, differentiation, apoptosis, and toxicity in different environments. Product yields can be increased if monitoring of cell growth is managed properly. A number of factors affect the maximum growth of cells in a batch reactor. Regular observation of cells in culture helps monitor cell health and the stage of growth; small changes in pH, temperature, humidity, O_2, CO_2, dissolved nutrients, etc. could have an impact on cell growth. Monitoring the rate of growth continuously also provides a record that the cells have reached their maximum density within a given time frame.

Characteristics of Cell Cultures

Animal cell cultures show specific characteristics and differ from microbial cultures. The important characteristics of the animal cell are slow growth rate, requirement of solid substrata for anchorage-dependent cells, lack of a cell wall (which leads to fragility), and sensitivity to physiochemical conditions such as pH, CO_2 levels, etc. Some of the fundamental bioprocess variables are as follows:

Temperature

Temperature is one of the most fundamental variables as it directly interferes with the growth and production processes. On a small scale, thermostatically controlled incubators can be used to control temperature. However, cell cultures grown on a large scale in bioreactors require more sensitive control of temperature. Different bioreactors use different methods to maintain the temperature of the cell culture. Temperature in a bioreactor is maintained by a heat blanket and water jacket with a temperature sensor.

pH

pH of the culture medium can be controlled by adding alkali (NaOH, KOH) or acid (HCl) solution. Addition of CO_2 gas to the bioreactor, buffering with sodium bicarbonate, or use of naturally buffering solutes help maintain the pH of the culture. A silver chloride electrochemical-type pH electrode is the most commonly used electrode in the bioreactor.

Oxygen

Dissolved oxygen is the most fundamental variable that needs to be continuously supplied to the cell culture medium. It is consumed with a carbon source in aerobic cultures (Moore et al., 1995). Diffusion through a liquid surface or membranes is one of the methods for providing dissolved oxygen to the medium.

CELL VIABILITY

The number of viable cells in the culture provides an accurate indication of the health of the cell culture (Stacey and Davis, 2007). Trypan blue and erythrosin B determine cell viability through the loss of cellular membrane integrity. Both these dyes are unable to penetrate the cell membrane when the membrane is intact, but are taken up and retained by dead cells (which lack an intact membrane). Erythrosin B stain is preferred over Trypan blue as it generates more accurate results with fewer false negatives and false positives.

Cytotoxicity

The toxic chemicals in the culture medium affect the basic functions of cells. The cytotoxicity effect can lead to the death of the cells or alterations in their metabolism. Methods to access viable cell number and cell proliferation rapidly and accurately is the important requirement in many experimental situations that involve *in vitro* and *in vivo* studies. The cell number determination can be useful for determining the growth factor activity, concentration of toxic compound, drug screening, duration of exposure, change in colony size, carcinogenic effects of chemical compounds, and effects of solvents (such as ethanol, propylene, etc.).

The assays to measure viable cells (viability assays) are as follows:

1. MTT/MTS/Resazurin assay.
2. Protease marker assay.
3. ATP assay.

The MTT assay allows simple, accurate, and reliable counting of metabolically active cells based on the conversion of pale yellow tetrazolium MTT [3-(4,5-Dimethylthiazol-2-yl)-2,5-diphenyltetrazolium bromide]. NADH in metabolically active viable cells reduces tetrazolium compounds into brightly colored formazan products or reduces resazurin into fluorescent resorufin (Figure 12.1). MTT and resazurin assays are widely used as they are inexpensive and can be used with all cell types. The protease marker assay utilizes the cell-permeant protease substrate glycylphenylalanyl-aminofluorocoumarin (GF-AFC). The substrate, which lacks an amino-terminal blocking moiety, is processed by aminopeptidases within the cytoplasm to release AFC. The amount of AFC released is proportional to the viable cell number. This assay has better sensitivity than resuzurin and the cells remain viable; thus, multiplexing is possible. The

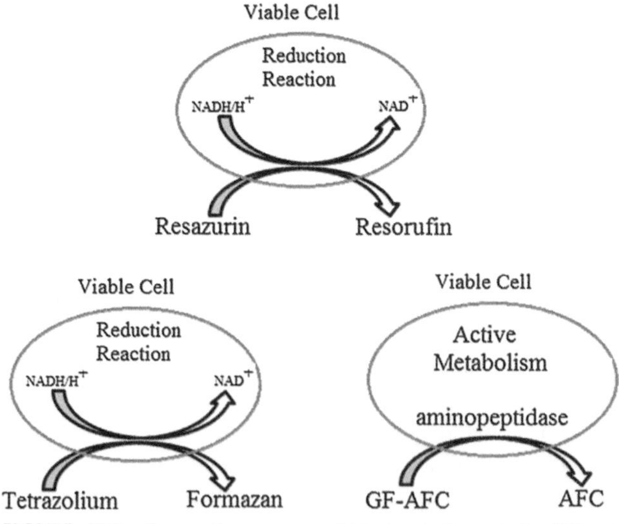

FIGURE 12.1 Schematic summary of biochemical events in different viability assays.

ATP assay is the most sensitive cell viability assay. It is measured using the beetle luciferase reaction to generate light. The MTT assay and procedure is given in Flow Chart 12.2.

Assays to detect dead cells are as follows:

1. LDH release.
2. Protease release.
3. DNA staining.

The viable cells in culture have intact outer membranes. Loss of membrane integrity defines a "dead" cell. The dead cells can be detected by measuring the activity of marker enzymes that leak out of dead cells into the culture medium or by staining the cytoplasmic or nuclear content by vital dyes that can only enter dead cells. Lactate dehydrogenase (LDH) is an enzyme that is present in all cell types. It catalyzes the oxidation of lactate to pyruvate in the presence of co-enzyme NAD^+. In the damaged cells, LDH is rapidly released. The amount of released LDH is used to assess cell death (Figure 12.2). This assay is widely used but has limited sensitivity as half-life of LDH at 37°C is 9 hours.

MTT Assay

Day 1 — **Trypsinize cells** ↓ **Count/record cells** ↓ **Cells in 96 well** ↓

Day 2 — **Treat cells with antagonist** ↓

Day 3 — **Add MTT** ↓ **Incubate** ↓ **Remove media** ↓ **Add MTT solvent** ↓ **Read Absorbance**

FLOW CHART 12.2 MTT assay.

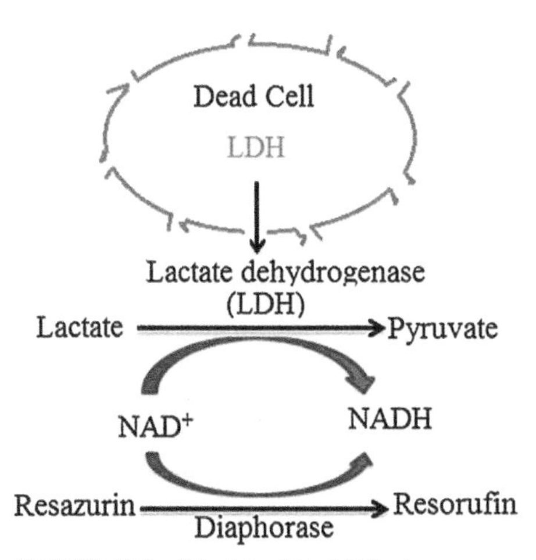

FIGURE 12.2 Principle of the LDH release assay.

The protease release assay is based on the intracellular release of proteases from the dead/compromised cell into the culture medium. The released proteases cleave the substrate to liberate aminoluciferin, which serves as a substrate for luciferase (Figure 12.3) and leads to the production of a "glowtype" signal (Cho et al., 2008).

Hayflick's Phenomenon

Hayflick limit or Hayflick's phenomena is defined as the number of times a normal cell population divides before entering the senescence phase. Macfarlane Burnet coined the term "Hayflick limit" in 1974. Hayflick (1961) demonstrated that a population of normal human fetal cells divide in culture between 40 and 60 times before stopping. There appears to be a correlation between the maximum number of passages and aging. This phenomenon is related to telomere length. Repeated mitosis leads to shortening of the telomeres on the DNA of the cell. Telomere shortening in humans eventually makes cell division impossible, and correlates with aging. This explains the decrease in passaging of cells harvested from older individuals.

CULTURE MEDIA

One of the most important factors in animal cell culture is the medium composition. *In vitro* growth and maintenance of animal cells require appropriate nutritional, hormonal, and stromal factors that resemble their milieu *in vivo* as closely as possible. Important environmental factors are the medium in which the cells are surrounded, the substratum upon which the cells grow, temperature, oxygen and carbon dioxide concentration, pH, and osmolality. In addition, the cell requires chemical substances that cannot be synthesized by the cells themselves. Any successful medium is comprised of isotonic, low-molecular-weight compounds known as basal medium, provides inorganic salts, an energy source, amino acids, and various supplements.

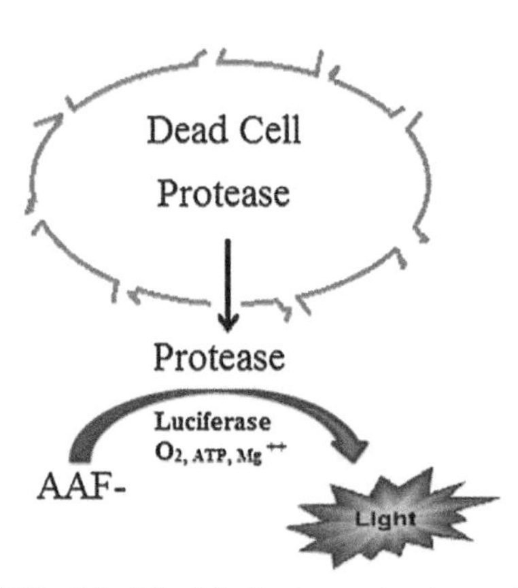

FIGURE 12.3 Principle of the luminescent protease release assay.

Basic Components in Culture Media

The ten basic components that make up most of the animal cell culture media are as follows: inorganic salts (Ca^{2+}, Mg^{2+}, Na^{+}, K^{+}); nitrogen source (amino acids); energy sources (glucose, fructose); vitamins; fat and fat soluble component (fatty acids, cholesterols); nucleic acid precursors; growth factors and hormones; antibiotics; pH and buffering systems; and oxygen and carbon dioxide concentrations.

Complete formulation of media that supports growth and maintenance of a mammalian cell culture is very complex. For this reason, the first culture medium used for cell culture was based on biological fluids such as plasma, lymph serum, and embryonic extracts. The nutritional requirements of cells can vary at different stages of the culture cycle. Different cell types have highly specific requirements, and the most suitable medium for each cell type must be determined experimentally. Media may be classified into two categories: (a) natural media and (b) artificial media.

Natural Media

Natural media consist of naturally occurring biological fluids sufficient for the growth and proliferation of animals cells and tissues. This media useful for promoting cell growth are of the following three types:

1. Coagulant or Clots: Plasma separated from heparinized blood from chickens or other animals is commercially available in the form of liquid plasma.
2. Biological Fluids: This includes body fluids such as plasma, serum lymph, amniotic fluid, pleural fluid, insect hemolymph, and fetal calf serum. These fluids are used as cell culture media after testing for toxicity and sterility.
3. Tissue Extract: Extracts of liver, spleen, bone marrow, and leucocytes are used as cell culture media. Chicken embryo extract is the most common tissue extract used in some culture media.

Artificial Media

The media contains partly or fully defined components that are prepared artificially by adding several nutrients (organic and inorganic). It contains a balanced salt solution with specific pH and osmotic pressure designed for immediate survival of cells. Artificial media supplemented with serum or with suitable formulations of organic compounds supports prolonged survival of the cell culture.

The artificial media may be grouped into the following four classes: serum-containing media, serum-free media, chemically defined media, and protein-free media.

Serum

The clear yellowish fluid obtained after fibrin and cells are removed from blood is known as serum. It is an undefined media supplement of extremely complex mixture of small and large molecules, and contains amino acids, growth factors, vitamins, proteins, hormones, lipids, and minerals, among other components (Table 12.3).

TABLE 12.3 Serum Components, Their Composition, and Role in Animal Cell Culture

Component	Probable Function
Protein and Polypeptide	
Albumin	Buffering agent and transport of lipids
Fetuin	Cell attachment
Fibronectin	Cell attachment
Transferrin	Cell attachment
Growth Factors	
Epidermal growth factor (EGF)	Mitogen
Fibroblast growth factor (FGF)	Mitogen
Amino Acids	Cellular proliferation
Lipids	Membrane synthesis
Glucose	Cellular proliferation
Hormones	
Insulin	For glucose and amino acid uptake
Transferrin	Incorporation of iron by cells
Minerals	
Iron	Enzymatic co-factors and constituents
Zinc	Enzymatic co-factors and constituents
Selenium	Enzymatic co-factors and constituents

(Modified from Freshney et al., 2005)

Advantages of Serum in Cell Culture Medium

1. It has basic nutrients present either in soluble or in protein-bound form.
2. It provides several hormones such as insulin and transferrin. Insulin is essential for the growth of nearly all cells in culture and transferrin acts as an iron binder.
3. It contains numerous growth factors such as platelet-derived growth factor (PDGF), transforming growth factor beta (TGF-B), epidermal growth factor (EGF), and chondronectin. These factors stimulate cell growth and support specialized functions of cells.
4. It supplies protein, which helps in attachment of cells to the culture surface (e.g. fibronectin).
5. It provides binding proteins such as albumin and transferrin, which helps transport molecules in cells.
6. It provides minerals like Ca^{2+}, Mg^{2+}, Fe^{2+}, K^{+} Na^{+}, Zn^{2+}, etc., which promote cell attachment.
7. It increases the viscosity of the medium, which provides protection against mechanical damage during agitation and aeration of suspension cultures.
8. It provides appropriate osmotic pressure.

Disadvantages of Serum-Containing Medium

1. Expensive: Fetal calf serum is expensive and difficult to obtain in large quantities.
2. Variation: Batch-to-batch variation occurs in serum and there is no uniformity in composition of serum. This can affect growth and yields and can give inconsistent results.
3. Contamination: Serum medium carries a high risk of contamination with virus, fungi, and mycoplasma.
4. Cytotoxic and Inhibiting Factors: The serum itself may be cytotoxic and may contain inhibiting factors, which in turn may inhibit cultured cell growth and proliferation. The enzyme polyamine oxidase in serum reacts with polyamines such as spermine and spermidine to form cytotoxic polyamino-aldehyde.
5. Downstream Processing: The presence of serum in culture media may interfere with isolation and purification of culture products. Additional steps may be required to isolate cell culture products.

Serum-Free Media

The use of serum in culture media presents a safety hazard and source of unwanted contamination for the production of biopharmaceuticals. As a number of cell lines can be grown in serum-free media supplemented with certain components of bovine fetal serum, the development of this type of medium with a defined composition has intensified in the last few decades. Eagle (1959) developed a "minimal essential medium" composed of balanced salts, glucose, amino acids, and vitamins. In the last 50 years considerable work has been carried out to develop more efficient culture media to meet the specific requirements of specific cell lines.

Advantages of Serum-Free Culture Media

1. Serum-free media are simplified and the composition is better defined.
2. They can be designed specifically for a cell type. It is possible to create different media and to switch from growth-enhancing media to differentiation-inducing media by altering the combination and types of growth factors and inducers.
3. They decrease variability from batch to batch and improve reproduction between cultures.
4. Downstream processing of products from cell cultures in serum-free media is easier.
5. They reduce the risk of microbial contamination (mycoplasma, viruses, and prions).
6. Serum-free media are easily available and ready to use. They are also cost-effective compared to serum-containing media.

Disadvantages of Serum-Free Media

1. Growth rate and saturation density attained are lower than those compared to serum-containing media.
2. Serum-free media prove to be more expensive as supplementing with hormone and growth factors increases the cost enormously.
3. Different media are required for different cell types as each species has its own characteristic requirements.
4. Critical control of pH and temperature, and ultra-purity of reagent and water are required as compared to serum-containing media.

Chemically Defined Media

These media contain pure inorganic and organic constituents along with protein additions like epidermal growth factors, insulin, vitamins, amino acids, fatty acids, and cholesterol.

Protein-Free Media

These media contain non-protein constituents necessary for the cell culture. The formulations of DME, MEM, RPMI-1640, ProCHO TM, CDM-HD are examples of protein-free media. They promote superior cell growth and facilitate downstream purification of expressed products.

CHARACTERIZATION OF CELL LINES

The characterization of cell lines is important to ensure the quality of cell-derived biopharmaceutical products. It helps in determining the cell source with regard to its identity and the presence of other cell lines, molecular contaminants, and endogenous agents. The characterization of mammalian cell lines is species-specific and can vary depending on the history of the cell line and type of media components used for culturing.

Mammalian cell line characterization can be done in four ways:

1. Identity testing.
2. Purity testing.
3. Stability testing.
4. Virological safety testing.

Identity Testing

Identity testing can be carried out by isoenzyme analysis. The banding pattern of the intracellular enzyme (which is species-specific) can be determined by using agarose gels. DNA fingerprinting and karyotyping, and DNA and RNA sequencing are alternative methods for identity testing.

Karyotyping

Karyotyping is important as it determines any gross chromosomal changes in the cell line. The growth conditions and sub-culturing of a cell line may lead to alteration in the karyotype; for example, HeLa cells were the first human epithelial cancer cell line established in long-term culture, and they have a hyper-triploid chromosome number (3n1).

Purity Testing

Bacterial and fungal contamination of cell lines occurs due to impure techniques and source material. The occurrence of contaminants can be tested by a direct inoculation method on two different media. Mycoplasma infection is contamination of cell cultures/cell lines with mycoplasmas, and it represents a serious problem. Detection by microscopy is not adequate and requires additional testing by fluorescent staining PCR, ELISA assay, autoradiography, immune-staining, or microbiological assay.

Stability Testing

Characterization and testing of cell substrate (cell line derived from human or animal source) is one of the most important components in the control of biological products. It helps to confirm the identity, purity, and suitability of the cell substrate for manufacturing use. The substrate stability should be examined at a minimum of two time points during cultivation for production. In addition, genetic stability can be tested by genomic or transcript sequencing, restriction map analysis, and copy number determination (FDA guidelines, 2012).

Viral Testing Assays

Virus testing of cell substrate should be designed to detect a spectrum of viruses. Appropriate screening tests should be carried out based on the cultivation history of cell lines. Development of characteristic cytopathogenic effect (CPE) provides an early indication of viral contamination. Some of the viruses of special concern in cell production work are human immunodeficiency virus, human papilloma virus, hepatitis virus, human herpes virus, hantavirus, simian virus, sendai virus, and bovine viral diarrhea virus. For detection of viruses causing immunodeficiency diseases and hepatitis, detection of sequences by PCR testing is adequate. Cells exposed to be serum or bovine serum albumin require a bovine virus test. Some of the viral testing assays are XC plaque assays, S+L-focus assay, reverse transcription assay. XC plaque assay is utilized to detect infectious ecotropic murine retroviruses (E-MLV). S+L-focus assay is used to test cells for the presence of infectious xenotropic and amphotropic murine retroviruses that are capable of interacting with both murine and non-murine cells. RT (real-time) assays like FPERT (real-time fluorescent product-enhanced reverse transcriptase assay) and QPERT (quantitative real-time for fluorescent product-enhanced reverse transcript assay) detect the conversion of an RNA template to cDNA due to the presence of the RT template when retrovirus infection is present in the cell line.

ADVANTAGES OF ANIMAL CELL CULTURE

1. Physiochemical and physiological condition: Role and effect of pH, temperature, O_2/CO_2 concentration, and osmotic pressure of the culture media can be altered to study their effects on the cell culture (Freshney, 2010).
2. Metabolism of cell: To study cell metabolism and investigate the physiology and biochemistry of cells.
3. Cytotoxic assay: Effect of various compounds or drugs on specific cell types such as liver cells can be studied.
4. Homogenous cultures: These cultures help study the biology and origin of the cells.
5. Valuable biological data from large-scale cell cultures: Specific proteins can be synthesized in large quantities from genetically modified cells in large-scale cultures.
6. Consistency of results: Reproducibility of the results that can be obtained by the use of a single type/clonal population.
7. Identification of cell type: Specific cell types can be detected by the presence of markers such as molecules or by karyotyping.
8. Ethics: Ethical, moral, and legal questions for utilizing animals in experiments can be avoided.

DISADVANTAGES OF ANIMAL CELL CULTURE

1. Expenditure and expertise: This is a specialized technique that requires aseptic conditions, trained personnel, and costly equipment.
2. Dedifferentiation: Cell characteristics can change after a period of continuous growth of cells in cultures, leading to differentiated properties compared to the original strain.
3. Low amount of product: The miniscule amount of monoclonal antibody and recombinant protein produced followed by downstream processing for extracting pure products increases expenses tremendously.
4. Contamination: Mycoplasma and viral infection are difficult to detect and are highly contagious.
5. Instability: Aneuploidy chromosomal constitution in continuous cell lines leads to instability.

In addition, this system cannot replace the complex live animal for testing the response of chemicals or the impact of vaccines or toxins.

ETHICAL ISSUES

Despite considerable progress in the development of cell culture techniques, the potential biohazards of working with animal and human tissues presents a number of ethical problems, including issues of procurement, handling, and ultimate use of material. In most countries biomedical research is strictly regulated. Legislation varies considerably in different countries. Research ethics committees, animal ethics committees for animal-based research, and institutional research boards for human subjects have a major role in research governance.

Some guidelines for the use of experimental or donor animals include assurances of proper conditions for housing animals and minimal pain or discomfort to any animal that is put to death or operated upon. These guidelines apply to higher vertebrates and not to lower vertebrates such as fish or other invertebrates.

USE OF FETAL BOVINE SERUM IN ANIMAL CULTURE OF MEDIA

Fetal bovine serum (FBS)-supplemented media are commonly used in animal cell cultures. In recent years, FBS production methods have come under scrutiny because of animal welfare concerns. FBS is harvested from bovine fetuses taken from pregnant cows during slaughter. The common method of harvesting the fetus is by cardiac puncture without any anesthesia. This practice of harvesting FBS is inhumane as it exposes the fetus to pain and/or discomfort. In addition to moral concerns, numerous scientific and technical problems exist with regard to the use of FBS in cell culture. Efforts are now being made to reduce the use of FBS and replace it with synthetic alternatives.

In the case of human tissues, some considerations that need to be addressed are as follows (Freshney, 2005):

1. Consent: Patient's and/or relative's approval of tissue use.
2. Project Summary: Explanation of the project, including the purpose, outcome, and medical benefits of the research.
3. Permission Requests: Paperwork regarding possible use of the tissues.
4. Ownership: Establishment of ownership with regard to cell lines and their derivatives.
5. Patent Issues: Commercial use of the tissues.

TRANSLATIONAL SIGNIFICANCE

In biomedical research the use of animal and human cell cultures has become beneficial for diverse applications. It provides indispensable tools for producing a number of products, including biopharmaceuticals, monoclonal antibodies and products for gene therapy. In addition, animal cell cultures provide adequate test systems for studying biochemical pathways, intra- and intercellular responses, pathological mechanisms, and virus production. Some of the applications of animal cell culture are discussed below.

ANTI-VIRAL VACCINES

Animal cell culture technology has played an important role in the development of viral vaccine production. The establishment of cell culture technology in the 1950s and the consequent replacement of live animals for the development of antigens has led to considerable progress in bioprocess technology. With the advent of DNA technology, molecular manipulation of viruses has led to the development of a recombinant vaccine against hepatitis B virus (HBV) and several others potential vaccines that are in the final phase of clinical trials. Table 12.4 lists recombinant hepatitis B vaccines in eukaryotic cells.

Viral Particles Production by Cell Culture

Viral particle production by cell culture differs from the production of molecules like proteins, enzymes, and toxins by bacteria or animal cells. The product formation may not be related to the development or growth of a cell, and may occur through secondary metabolic pathways, unlike virus production, which does not result from secondary metabolic

TABLE 12.4 Recombinant Hepatitis B Vaccines in Eukaryotic Cells

Trade Name	Manufacture	Recombinant Protein	Expression Host
Sci B-Vac™	SciGen	HBsAgS, M and L protein	Mammalian cells (CHO)
GenHevac B®	Pasteur-Merieux Aventis	HBsAg S and M protein	Mammalian cells (CHO)
Enivac HB	Panacea Biotec Ltd	HBsAg S protein	Yeast (*P. pastors*)
Revac-B™	Bharat Biotech International Ltd	HBsAg S protein	Yeast (*P. pastors*)

TABLE 12.5 Cell Lines Used for Viral Vaccine Production

Cell Lines	Origin
BHK-21	Kidney (hamster)
Vero	Kidney (African green monkey)
MDCK	Kidney (cocker spaniel)
CHO	Chinese hamster
Hela	Epithelial cells (human)

pathway. Virus production occurs after the viral infection directs cell machinery to perform viral particle production.

Two stages are involved in viral production:

1. Cell culture system: This requires the development of an efficient system for conversion of the culture medium substrate in the cell mass.
2. Virus production: This phase differs from the infection phase and has different nutritional and metabolic requirements. A number of immortalized cell lines are used for the industrial production of viral vaccines. Table 12.5 gives the cell lines used for vaccines.

Production of Virus-Like Particles (VLPs)

Most of the existing classical vaccines for viral disease are either altered or chemically inactive live viruses. However, incomplete inactivation of a virus or reversion of an attenuated strain can risk infection in vaccinated individuals. Viruses with segmented genomes with a high degree of genetic exchange can undergo re-assortment or recombination of genetic material with viruses of different serotypes in the vaccinated host, which can result in production of new variants of the virus. Moreover, some live virus vaccines are teratogenic; for example, Smithburn neurotropic strain (SNS) (Smithburn, 1949) and MP12-attenuated (Caplen et al., 1985) vaccine strains of the Rift Valley fever virus. A new type of vaccine that does not present the typical side effects of an attenuated or inactivated viral vaccine has been made possible with the development of recombinant DNA technology. Virus-like particles are highly effective as they mimic the overall structure of the virus, however, these particles lack the infectious genetic material. Capsid proteins can aggregate to form core-like particles (CPLs) in the absence of nucleic acids. These spontaneously assembled particles are structurally similar to authenticate viruses and are able to stimulate B-cell-mediated immune responses. In addition, VLPs stimulate a CD4-proliferative response and cytotoxic T lymphocyte response (Jeoung et al., 2011).

VLPs resemble and mimic virus structure and are able to elicit a strong immune response without causing harm. The major advantage of VLPs is their simplicity and non-pathogenic nature. They are replication-deficient as they lack any viral genetic information, thus eliminating the need for inactivation of the virus. This is important as inactivation treatments lead to epitope modifications (Cruz et al., 2002). As the structural morphology of VLPs is similar to the virus, the conformational epitopes presented to the immune system are the same as for the native virus particles. The immune response/antibody reactivity in the case of VLPs is significantly improved as VLPs present conformation epitopes more similar to the native virus. VLPs also induce a strong B-cell response. For broader and more efficient protection, it is possible to adapt one or more antigens to the multimeric protein structure. Another advantage offered by VLPs is that they significantly reduce vaccine costs as these can elicit a protective response at lower doses of antigen.

Vaccines Based on VLPs

The FDA has approved VLP-based vaccines for hepatitis B (HBV) and HPV. The HBV vaccine was approved in 1986 and the HPV one in 2006 (Justin et al., 2011). To generate immunogenic VLPs, the S gene is cloned and expressed in a eukaryotic expression host such as yeast or mammalian cells (e.g. CHO cell line). The mammalian cell culture allows easy recovery because the cells are able to secrete the antigen HBsAg. The two companies producing CHO-based vaccines are the French-based Pasteur-Merieux Aventis (Gene Hevac B) and the Israeli-based SciGen (Sci-B-Vac™). The Gene Hevac B vaccine contains the HBsAg S protein and M protein, whereas Sci-B-Vac contains the M and L proteins.

HPV Vaccine

Viruses of the Papillomaviridae family are known to induce lesions and warts and also cause cervical cancer. Fifteen strains of Papillomaviridae are known to cause cervical cancer. HPV-16 is considered a high-risk HPV type as the risk of cancer may be higher than for other high-risk HPV types. The two virally encoded proteins of HPV are L1 and L2. L1 is the main capsid protein that forms the outer shell of the virus. L2 is found in the interior of the viral particle and is less abundant. The recombinant L1 VLP is able to induce neutralizing antibodies in animals. Gardasil (the first HPV vaccine) was approved by the FDA in 2006. This vaccine is manufactured by Merck and Co., Inc. Ceravarix®, another HPV vaccine (manufactured by Glaxo Smithkline), was approved by the FDA in 2009. It uses the Trichoplusia ni (Hi-5) insect cell line infected with L1 recombinant baculovirus (Jiang et al., 1998; Wang et al., 2000).

A number other VLP-based vaccines are in clinical trials. These include the anti-influenza A M2-HBcAg VLP vaccine

TABLE 12.6 VLP Production In Mammalian and Baculo Cell Lines of Viruses Infecting Humans and Other Animals

Virus	Expression System	Cells	Recombination Protein
AAV	Human cells	HEK293	VP1, VP2, VP3
Avian influenza	B/IC (Baculovirus Insect Cell)	SF9	HA, NA, and MI
Ebola	Human cells	HEK293T	VP40 and glycoprotein
HBV	Human cells	CHO	Sag (S protein, pre S1, pre S2)
H and V	Human cells	BHK	Pr 160gag-pol
Rota virus	B/IC	Sf9	VP2, VP6, VP7
SARS	B/IC	Sf9	SP, EP and MP
SV40	B/IC	Sf9	VP1
SIV	B/IC	Sf9	Pr55 gag envelope protein

(Modified from Mello et al., 2008)

(Clarke et al., 1987), two antimalarial vaccine nicotine-Qβ VLPs (Maurer et al., 2005), and an anti-AngIIQβ VLP. VLP production in mammalian cell lines and Baculo cell lines of viruses infecting humans and other animals is summarized in Table 12.6.

RECOMBINANT THERAPEUTIC PROTEINS

Proteins play a major role in carrying out biochemical reactions, transporting small molecules within a cell or from one organ to another, formation of receptors and channels in membranes, and providing frameworks for scaffolding. The number of functionally distinct proteins in humans far exceeds the number of genes as a result of post-translational modifications. These modifications include glycosylation, phosphorylation, ubiquitination, nitrosylation, methylation, acetylation, and lipidation. Changes in protein structure as a result of mutation or other abnormalities often lead to a disease condition. Protein therapeutics offer tremendous opportunities for alleviating disease. The first therapeutic from recombinant mammalian cells was human tissue plasminogen, which obtained market approval in 1986. At present, 60–70% of all the recombinant therapeutic proteins are produced in mammalian cells.

Main Therapeutic Proteins

The main therapeutic proteins can be divided into seven groups (Walsh, 2003):

1. Cytokines
2. Hematopoietic growth factors
3. Growth factors
4. Hormones
5. Blood products
6. Enzymes
7. Antibodies

Most of the proteins have complex structures and undergo chemical modification to insure full biological activity. Protein post-translation modifications (PTM) can happen in several ways. The most widely recognized form of PTM is glycosylation, which involves extensive sequence processing and trimming in the Golgi apparatus and endoplasmic reticulum (ER). Eukaryotic cells are capable of carrying out this type of modification, and are thus preferred in biopharmaceutical processes. Hamster, BHK, and CHO cells are often the host cells of choice as glycosylation patterns generated from these cells are more similar to human patterns. Table 12.7 lists various therapeutic proteins produced in animal cell lines.

Cytokines

Cytokines are proteins of the immune system that play a central role in immune response. Cytokines are produced as a result of immune stimulus by various white blood cells. Interferons (IFNs) were the first family of cytokines to be discovered and used as biopharmaceuticals.

Applications of IFNs

IFNα is used for treatment of hepatitis, and more recently has been approved for leukemia and other types of cancers. IFNβis used for treatment of multiple sclerosis and is marketed under the names Avonex®, Belaseron®, and Rebif®. IFNγ is used for the treatment of chronic granulomatous

TABLE 12.7 Various Therapeutic Proteins Produced in Animal Cell Lines

Therapeutic Protein	Potential Application	Cell Line	Product Name
Immunoglobulin GI	Colorectal cancer	CHO	Avastin®
Immunoglobulin GI	Rheumatoid arthritis	CHO	Humira®
Immunoglobulin G2a	Non-Hodgkins lymphoma	Hybridoma CHO	Bexxar®
Immunoglobulin GI	Rheumatoid arthritis, Crohn's disease	Sp2/0	Remicade®
Immunoglobulin GI	Renal transplantation (prophylaxis of organ rejection)	Murine Myeloma	Simulect®
Immunoglobulin GI	Respiratory tract disease	NSO	Synagis®
α-Galactosidase	Fabry disease	Human fibroblasts	Replagal
Activated protein C	Severe sepsis	HEK-293	Xigris
DNAse I	Cystic fibrosis	CHO	Pulmozyme
tPA	Acute myocardial infraction	CHO	Activase®
FSH	Female infertility	CHO	Puregon®
Epoein x	Anemia	CHO	Epogen®
Epoein B	Anemia	CHO	Neorecormon®
Factor VIII	Hemophilia A	BHK	Kogenate® FS
Factor IX	Hemophilia B	CHO	BeneFIX
Factor VIIa	Hemophilia A+B	BHK	Novo Seven

(Modified from Mello and Castilho, 2008)

disease (CGD). Interleukin is another kind of cytokine that helps regulate cell growth, differentiation, and motility, and is used as a biopharmaceutical. The recombinant form of IL-2 is used for the treatment of renal cell carcinoma.

Growth Factors

Growth factors are proteins that bind to receptors on the surface of cells to activate the cells for proliferation and or differentiation. The different types of growth factors are transforming growth factor (TGF), insulin-like growth factor (IGF), and epidermal growth (EGF). The primary sources of platelet-derived growth factor (PDGF) are platelets, endothelial cells, and the placenta. Two isoforms of this protein are present in the human body and both of these have one glycosylation site and three disulfide bonds. Examples of growth factors used as biopharmaceuticals are the following:

1. Osigraft®/Eptotermin alfa (bone morphogenetic protein) is used for treatment of tibia fractures, is grown commercially in CHO cells, and was first approved in 2001 in Europe.
2. InductOS®/Dibotermin (bone morphogenetic protein) is used for tibia fractures and in spinal surgery; it is also commercially grown in CHO cells. This product was first approved in Europe in 2002.

Hormones

Insulin, glucagon, gonadotropins, and growth hormones are the most well known therapeutic hormones. The first biopharmaceuticals that obtained approval by regulatory agencies were insulin and recombinant human growth hormones. These were produced in microbial cells. The commercial recombinant forms of the gonadotropin family of hormones are Gonal-F®, Luveris®, Puregon®, and Ovitrelle. All these are produced using CHO cells, and are used for treating female infertility.

Therapeutic Enzymes

A number of recombinant therapeutic enzymes are expressed in mammalian cells. Tissue plasminogen activator (tPA) is a thrombolytic agent involved in dissolving blood clots. Recombinant tPA is commercially is known as Alteplase™ and Tenectplase™, which are used for treatment of acute myocardial infraction.

Fabry disease, a genetic metabolic disorder, is characterized by lack of enzyme α-galactosidase A. Fabrazyme®

(approved in 2001) is a recombinant α-galactosidase A, and is produced by genetically modified CHO cells.

Blood Coagulation Factors

Hemophilia A is caused by lack of blood clotting factor VIII, hemophilia B is caused by deficiency of factor IX, and hemophilia C by lack of factor XI. Factor VIII and IX are proteins. The first recombinant factor VII products were Recombinate and Kogenate®, which were expressed in CHO and BKH cells, respectively. Recombinant factor FIX is commercially sold as BeneFIX®, and is produced in recombinant CHO cells.

Antibodies

Therapeutic antibodies are used in the treatment of cancer, cardiovascular disease, infections, and autoimmune diseases. In 2004, the antibody Avasin® (Bevacizeimab) was approved for treatment of metastatic colorectal cancer. This antibody acts as an inhibitor of vascular endothelial growth factor (VEGF). Zenapax®, another commercially available antibody, is used during prophylaxis for preventing the rejection of transplanted organs. This is commercially grown in the NSO cell line and was approved for human use in 1997.

GENE THERAPY

Importance of Cell Culture in Gene Therapy

Gene therapy involves the insertion, removal, or alteration of a therapeutic or working gene copy to cure a disease or defect, or to slow the progression of a disease, thereby improving the quality of life. The human genome map was the first major step toward a new way of addressing human health and illness. Gene therapy holds great promise, however, the task of transferring genetic material into the cell remains an enormous technical challenge and requires *ex vivo* cell cultivation and adaptation from the lab to a clinically relevant state. The development of animal cell culture technology is imperative for advances in gene therapy.

Monogenic diseases caused by single gene defects (such as cystic fibrosis, hemophilia, muscular dystrophy, and sickle cell anemia) are the primary targets of human gene therapy.

The first step in gene therapy is to identify the faulty gene. This is followed by gene isolation and generation of a construct for correct expression. Integration of the gene followed by delivery of the genetic material *in vivo* or *ex vivo* is crucial to the success of gene therapy. In *in vivo* therapy, the genetic material is introduced directly into the individual at a specific site, and in *ex vivo* treatment, the target cells are treated outside the patient's body. These cells are then expanded and transferred back to the individual at a specific site. The *ex vivo* technique involves gene therapy in the cultured cells, which are expanded and subsequently transferred to the targeted tissue.

Clinical Studies

A number of clinical studies and trials for gene therapy have already been approved and are being conducted worldwide. From 1989 up to the present, about 500 clinical studies have been reported; 70% of these studies are intended for cancer treatment.

The first product designed for gene therapy was Gendicine, a medication produced by Shenzhen Sibiono Genetech, China. Gendicine is used for head and neck carcinoma treatment. The tumor 4 suppressing gene p53 in recombinant adenovirus expresses protein p53, which leads to tumor control and elimination. SBN-cel is a cell line that was subcloned from the human embryonic kidney (HEK) cell line 293, and has been used for the production of Gendicine.

BIOPESTICIDES

In recent years biopesticides have gained importance due to increased concerns about agrochemicals and their residues in the environment and food. Biopesticides provide an effective means for the control of insects and plant disease, and they are environmentally safe. The biological control of insect pests by another living organism (in order to suppress the use of pesticides) is an age-old practice. Presently, a number of biological controls are being used as biopesticides. With the high cost of chemical-based pesticides and the development of resistance to multiple chemical pesticides, baculoviruses are one of the most promising biocontrols for insect pests, and have been increasingly used effectively against caterpillars worldwide. However, the major impediment in the development of baculoviruses as biopesticides is the high cost and small volumes of *in vitro* methods. Development of an *in vitro* production process for large quantities of baculoviruses at comparable costs to chemical pesticides will help provide insect control that is safe, efficacious, cost effective, and environmentally safe.

Baculovirus Production in Animal Cell Culture

A number of factors are important for a successful commercial production of bioinsecticides:

1. Large-scale production of viruses at competitive costs.
2. Economic production of viruses (i.e. low cost for the media and running the culture).
3. Effective cell line with high virus per cell productivity.
4. With passage of the virus into cells, there is a loss of virulence and an increased risk of mutant formation; this should be avoided.

5. The quality of the polyhedral produced in the cell culture should be comparable to those obtained from caterpillars.

The insect baculovirus–cell system offers a number of advantages. It produces recombinant proteins that are functional and are immunologically active, as it is able to make post-translational modifications. The recombinant system uses a powerful promoter polyhedron.

Cell Lines for Biopesticide Production

The most commonly used cell lines in biopesticide production are the Sf21 and Sf9 cell lines, which are derived from ovarian tissues of the fall army worm (*Spodoptera frugiperda*). Sf9 cells show a faster growth rate and higher cell density than Sf21 cells, and are preferred. High Five cell lines (designated BTI-Tn-5BI-4) established from *Trichoplusia ni* embryonic tissue are also being used.

Viral Mutant Formation in Cell Culture

The continuous culturing of cells for virus production leads to virus instability and the so-called passage effect. This can result in a decrease of virulence and polyhedral production, and a variety of mutations. All these changes affect commercial production *in vitro*. Two types of mutations are commonly seen in continuous passaging of cell cultures for viral productions: (1) DIPs-defective interfering particles, and (2) Fp-few polyhedral mutations.

Fp mutations are characterized by (1) reduced polyhedral, (2) enhanced production of BV, and (3) lack of occluded virions in polyhedra. All these factors reduce the infectivity of the target pest.

Spontaneously generated Fp mutants have been reported in AcMNPV *(Autographa California* nucleopolyhedroviruses) (Wood, 1980), *Galleria mellonella* nucleopolyhedroviruses (GmMNPV) (Fraser and Hink, 1982), and *Helicoverpa armigera* nucleopolyhedroviruses (HaSNPV) (Chankraborty and Reid, 1999).

DIP mutations are the formation of defective infective particles (DIPs). They occur due to serial passaging for long periods, which results in a decrease in the filtering of infectious virus. DIPs have been reported in a number of animal virus systems and in baculovirus systems. DIP formation can be avoided by low multiplicity of infection (MOI). This minimizes the probability of the defective virus entering the cell with an intact helper virion.

MONOCLONAL ANTIBODIES

The majority of antibodies available on the market today are produced in animal cell cultures (VanDijk and Van de Winkle, 2001). Animal cells are preferred because they are capable of glycosylation and structural conformation, which is essential for a drug to be productive. Hybridoma technology has been the most widely used method for small- and large-scale production of monoclonal antibodies (mAB). However, these antibodies have limited therapeutic applications since they produce an adverse immune response on repeated use.

A number of cell lines are now being used for the production of recombinant antibodies. The Chinese hamster ovary (CHO) lines are the most commonly used. Other cell lines used are marine myelomas NSO, Sp 2/0, human embryonic kidney (HEK- 93), and baby hamster kidney (BHK).

A number of factors influence the production of mABs. For a high concentration of mAB production, the cell line should have high productivity. For high protein productivity it is important that the selected cell line be productive in order to avoid large reaction volumes and the high cost of protein purification. Cell lines with the capacity to grow without anchorage offer an advantage in terms of scaling up the process; it is much simpler than with those designed for the growth culture of anchorage-dependent cells. Sp2/0 and NSO cell lines can grow naturally in suspension; other cells such as CHO and BHK can be easily adapted to this form of cultivation.

STEM CELLS

Stem cells are unspecified cells that have the potential to differentiate into other kinds of cells or tissues and become specialized cells. The two characteristics that define stem cells are their ability of self-regenerate and to differentiate into any other cells or tissues. These cells have the capability to renew themselves to form cells of more specialized function. In recent years, stem cell research has been hailed as a major breakthrough in the field of medicine. This property of turning a cell into any other specialized-function cell has made researchers believe that stem cells could be utilized to make fully functional, healthy organs to replace damaged or diseased organs.

Culturing Embryonic Stem Cells in the Laboratory

Human embryonic stem cells are grown on nutrient broth. These cells are traditionally cultured on mouse embryonic fibroblast feeder layers (MEF), which allows continuous growth in an undifferentiated stage. The mouse cells at the bottom of the culture dish provide a sticky surface to which the cells can attach. In addition, the feeder cells release nutrients into the culture medium. Researchers have now devised animal-free culture systems for human embryonic stem cells (hESCs), and have used human embryonic fibroblasts and adult fallopian tube epithelial cells as feeder layers (in addition to serum-free mediums).

More recently, methods to sub-culture embryonic cells without the feeder layer have been developed. Martigel from BD Biosciences has been used to coat the culture plate (Hassan et al., 2012) for effective attachment and differentiation of both normal and transformed anchorage-dependent epithelioid and other cell types. This is a gelatinous protein mixture isolated from mouse tumor cells.

WORLD WIDE WEB RESOURCES

1. http://www.fda.gov
   http://www.fda.gov/downloads/biologicsbloodvaccines/guidancecomplianceregulatoryinformation/guidances/vaccines/ucm202439.pdf
   http://www.fda.gov/BiologicsBloodVaccines/Vaccines/ApprovedProducts/ucm205541.htm
   The Food and Drug Administration (FDA or USFDA) protects and promotes public health through the regulation of all foods (except meats and poultry), the nation's blood supply, and other biologics (such as vaccines and transplant tissues). Drugs must be tested, manufactured, and labeled according to FDA standards before they can be sold or prescribed.
2. http://www.promega.com
   http://www.promega.com/~/media/files/products%20and%20services/na/webinars/mechanism%20of%20toxicitywebinar2.pdf?la=en
   Promega manufactures enzymes and other products for biotechnology and molecular biology.
3. http://www.who.int
   http://www.who.int/biologicals/publications/trs/areas/vaccines/cells/WHO_TRS_878_A1Animalcells.pdf
   The World Health Organization (WHO) is a specialized agency that is concerned with international public health. It is affiliated with the United Nations and headquartered in Geneva, Switzerland. WHO ensures that more people, especially those living in dire poverty, have access to equitable, affordable care so that they can lead healthy, happy, and productive lives.
4. http://amgenscholars.com
   http://amgenscholars.com/images/uploads/contentImages/biotechnology-timeline.pdf
   Amgen Scholars provides hundreds of undergraduate students with the opportunity to engage in a hands-on summer research experience at some of the world's leading institutions.
5. http://monographs.iarc.fr
   http://monographs.iarc.fr/ENG/Monographs/vol90/mono90-6.pdf
   The IARC Monographs identify environmental factors that can increase the risk of human cancer. These include chemicals, complex mixtures, occupational exposures, physical agents, biological agents, and lifestyle factors.
6. www.iptonline.com
   http://www.iptonline.com/articles/public/IPTFIVE76NP.pdf
   IPTonline publishes "The Pharmaceutical Technology Journal," which is designed to provide information on the latest ideas, cutting-edge technologies, and innovations shaping the future of pharmaceutical research, development, and manufacturing.
7. http://www.aceabio.com
   http://www.aceabio.com/UserFiles/doc/literature/xcell_appnotes/RTCA_AppNote07_ACEA_LoRes.pdf
   ACEA Biosciences, Inc. (ACEA) is a privately owned biotechnology company. ACEA's mission is to transform cell-based assays by providing innovative and cutting-edge products and solutions to the research and drug discovery community.

REFERENCES

Caplen, H., Peters, B. J., & Bishop, D. H. L. (1985). Mutagen-directed attenuation of Rift Valley fever virus as a method for vaccine development. *Journal of General Virology*, *66*, 2271–2277.

Chakraborty, S., & Reid, S. (1999). Serial passage of a Helicoverpa armigera nucleopolyhedrovirus in Helicoverpa zea cell cultures. *Journal of Invertebrate pathology*, *73*, 303–308.

Cho, M. H., Niles, A., Huang, R., Inglese, J., Austin, C. P., Riss, T., & Xia, M. (2008). A bioluminescent cytotoxicity assay for assessment of membrane integrity using a proteolytic biomarker. *Toxicology In Vitro*, *22*, 1099–1106.

Clarke, B. E., Newton, S. E., Carroll, A. R., Francis, M. J., Appleyard, G., Syred, A. D., Highfield, P. E., Rowlands, D. J., & Brown, F. (1987). Improved immunogenicity of a peptide epitope after fusion to hepatitis B core protein. *Nature*, *330*, 381–384.

Cruz, P. E., Maranga, L., & Carrondo, M. J. T. (2002). Integrated process optimisation: lessons from retrovirus and virus like production. *Journal of Biotechnology*, *99*, 199–214.

Eagle, H. (1959). Amino acid metabolism in mammalian cell cultures. *Science*, *130*, 432–437.

FDA guidelines, 2012. http://www.fda.gov/downloads/biologicsbloodvaccines/guidancecomplianceregulatoryinformation/guidances/vaccines/ucm202439.pdf.

Fraser, M. J., & Hink, W. F. (1982). The isolation and characterization of the MP and FP plaque variants of Galleria mellonella nuclear polyhedrosis virus. *Virology*, *117*, 366–378.

Freshney, R. I. (1994). *Culture of Animal Cells: A Manual of Basic Technique* (3rd ed.). New York: Wiley-Liss Inc.

Freshney, R. I. (2010). *Culture of Animal Cells: A Manual of Basic Technique and Specialized Applications*. New Jersey: Wiley, John & Sons, Inc.

Freshney, R. I. (2011). *Culture of Animal Cells: A Manual of Basic Technique and Specialized Applications*. New Jersey: Wiley, John & Sons, Inc.

Hassan, F., Ren, D., Zhang, W., & Gu X-X. (2012). Role of c-Jun N-terminal protein kinase 1/2 (JNK1/2) in macrophage-mediated MMP-9 production in response to *Moraxella catarrhalis* lipooligosaccharide (LOS). *PLoS ONE*, *7*, e37912.

Hayflick, L., & Moorhead, P. S. (1961). The serial cultivation of human diploid cell strains. *Experiment Cell Research*, *3*, 585–621.

Jeoung, H. Y., Lee, W. H., Jeong, W., Shin, B. H., Choi, H. W., Lee, H. S., & An, D. J. (2011). Immunogenicity and safety of the virus-like particle of the porcine encephalomyocarditis virus in pig. *Virology Journal*, *8*, 170.

Jiang, B., Barniak, V., Smith, R. P., Sharma, R., Corsaro, B., Hu, B., & Madore, H. P. (1998). Synthesis of rotavirus-like particles in insect cells: comparative and quantitative analysis. *Biotechnology and Bioengineering*, *60*, 369–374.

Justin, C., Masum, H., Perampaladas, K., Heys, J., & Singer, P. A. (2011). Indian vaccine innovation: the case of Shantha Biotechnics. *Globalization and Health*, *7*, 9.

Maurer, P., Jennings, G. T., Willers, J., Rohner, F., Lindman, Y., Roubicek, K., Renner, W. A., Müller, P., & Bachmann, M. F. (2005). A therapeutic vaccine for nicotine dependence: preclinical efficacy, and Phase I safety and immunogenicity. *European Journal of Immunology*, *35*, 2031–2040.

Mello, I. M. V.G., Meneghesso da Conceicao, M., Jorge, S. A. C., Cruz, P. E., Alves, P. M. M., Carrondo, M. J. T., & Pereira, C. A. (2008). Viral vaccines: concepts, principles, and bioprocesses. In L. R. Castilho, A. M. Moraes & E. F. P. Augusto (Eds.), *Animal Cell Technology: From Biopharmaceuticals to Gene Therapy* (pp. 435–458). New York: Taylor & Francis Group.

Moore, A., Donahue, C. J., Hooley, J., Stocks, D. L., Bauer, K. D., & Mather, J. P. (1995). Apoptosis in CHO cell batch cultures: examination by flow cytometry. *Cytotechnology*, *17*, 1–11.

Smithburn, K. C. (1949). Rift Valley fever: the neurotropic adaption of virus and experimental use of this modified virus as a vaccine. *British Journal of Experimental Pathology*, *30*, 1–16.

van Dijk, M. A., & van de Winkel, J. G. J. (2001). Human antibodies as next generation therapeutics. *Current Opinion in Chemical Biology*, *5*, 368–374.

Walsh, G. (2003). *Biopharmaceuticals – Biochemistry and Biotechnology* (2nd ed.). Chichester: John Wiley and Sons.

Wang, M. Y., Kuo, Y. Y., Lee, M. S., Doong, S. R., Ho, J. Y., & Lee, L. H. (2000). Self-assembly of the infectious bursal disease virus capsid protein, rVP2, expressed in insect cells and purification of immunogenic chimeric rVP2H particles by immobilized metal-ion affinity chromatography. *Biotechnology and Bioengineering*, *67*, 104–111.

Wood, H. A. (1980). Isolation and replication of an occlusion body-deficient mutant of the Autographa californica nuclear polyhedrosis virus. *Virology*, *105*, 338–344.

FURTHER READING

Bhat, S. M. (2011). *Animal Cell Culture Concept and Application*. Oxford: Alpha Science International Limited.

Castilho, L. R., Moraes, A. M., & Augusto, E. F. P. (2008). From Biopharmaceuticals to Gene Therapy. *Animal Cell Technology*. New York: Taylor & Francis Group.

Stacey, G. N., & Davis, J. (2007). *Medicines from Animal Cell Culture*. Chichester: John Wiley & Sons.

GLOSSARY

Antigen Any substance that causes your immune system to produce antibodies against it.

Aseptic Free from pathogenic microorganisms.

Cell culture To grow cells *in vitro*.

Cytotoxicity The degree to which an agent has specific destructive action on certain cells.

Differentiation A change in a cell causing an increase in morphological or chemical heterogeneity.

Immortalized Changing a cell type with limited life span *in vitro* into a cell type with unlimited capacity to proliferate; sometimes achieved by animal cells *in vitro* or by tumor cells.

In vitro Cell growth outside the body, in glass, as in a test tube.

In vivo Cell growth in a living organism.

Medium A buffered selection of components in which an organism naturally lives or grows.

Monolayer A single layer of adherent cells on substratum.

Passage The process of passing or maintaining cells through a series of hosts or cultures.

Primary Culture A culture initiated from an explant of cells, tissues, or organs in media conducive to their growth.

Trypsinization Use of the enzyme trypsin to remove adherence proteins from a cell surface.

ABBREVIATIONS

AcMNPV *Autographa california* Nucleopolyhedroviruses
BHK Baby Hamster Kidney
CD4 Glycoprotein on the Surface of Helper T Cells That Serves as a Receptor for HIV
CDM-HD Chemically Defined Medium
CGD Chronic Granulomatous Disease
CHO Chinese Hamster Ovary
CPE Cytopathogenic Effect
CPLs Core-Like Particles
DIP Defective Infective Particle
DME Dulbecco's Modified Eagle's Media
EDTA Ethylenediaminetetraacetic Acid
EGF Epidermal Growth
ELISA Enzyme-Linked Immunosorbent Assay
E-MLV Ecotropic Murine Retroviruses
FBS Fetal Bovine Serum
Fp Few Polyhedral Mutations
FPERT Real-Time Fluorescent Product-Enhanced Reverse Transcriptase Assay
GF-AFC Glycyl-Phenylalanyl-Amino-Fluorocoumarin
GmMNPV *Galleria mellonella* Nucleopolyhedroviruses
HaSNPV *Helicoverpa armigera* Nucleopolyhedroviruses
HBcAg Hepatitis B Core Antigen
HBV Hepatitis B Virus
HEK Human Embryonic Kidney Cells
HeLA Established Human Epithelial Cell Line Derived from Cervical Caracinoma
hESCs Human Embryonic Stem Cells
Hi-5 Cells (BTI-TN-5B1-4) Derived from the Parental *Trichopulsia ni* Cell Line
HPV Human Papillomavirus
HPV18 Human Papilloma Virus 18
IFN Interferon
IGF Insulin-Like Growth Factor
IL-2 Interleukin-2
L1 VLP HPV with L1 Major Capsid Protein
LDH Lactate Dehydrogenase
mAB Monoclonal Antibody

MEF Mouse Embryonic Fibroblast Feeder Layers
MEM Minimum Essential Media
MOI Multiplicity of Infection
MP12 Strain Invented by Serial Mutagenesis of RVF Virus with Egyptian ZH501 and ZH548 Strains
MTS 3-(4,5-Dimethylthiazol-2-yl)-5-(3-Carboxymethoxyphenyl)-2-(4-Sulfophenyl)-2H-Tetrazolium
MTT 3-(4,5-Dimethylthiazol-2-yl)-2,5-Diphenyltetrazolium Bromide
NADH Nicotinamide Adenine Dinucleotide
PDGF Platelet-Derived Growth Factor
PTM Post-Translation Modifications
QPERT Quantitative Real-Time for Fluorescent Product-Enhanced Reverse Transcript Assay
rDNA Recombinant DNA
RT Real-Time Assays
SNS Smithburn Neurotropic Strain
STO Mouse Embryonic Fibroblast Cell Line
TGF Transforming Growth Factor
TGF-B Transforming Growth Factor Beta
tPA Tissue Plasminogen Activator
VEGF Vascular Endothelial Growth Factor
VLPs Virus-Like Particles

LONG ANSWER QUESTIONS

1. What are the components of serum and how do they help the cell culture?
2. What is the role of media in animal cell culture?
3. What are the advantages and limitations of animal tissue culture?
4. How can cell viability and cytotoxicity be tested in cell culture?
5. What is the role of cell culture in gene therapy and viral vaccines?

SHORT ANSWER QUESTIONS

1. What is the Hayflick effect?
2. What is the source of cells for primary monolayer cell culture?
3. Serum is one of the basic components of cell culture media (True/False)?
4. What was the first recombinant human protein?
5. What are the different phases of the growth curve?
6. Is the VLP-based HPV vaccine approved by the FDA?

ANSWERS TO SHORT ANSWER QUESTIONS

1. Limited replication capacity of cells in culture medium.
2. Organ/tissue of live animal.
3. False.
4. Somatostatin.
5. Lag phase, log phase, and plateau phase.
6. Yes, Gardasil (the first HPV vaccine) was approved by the FDA in 2006.

Chapter 13

Concepts of Tissue Engineering

Poonam Verma* and Vipin Verma†

*Department of Biochemistry, All India Institute of Medical Sciences, New Delhi, India, †Corning Life Sciences, Gurgaon, India

Chapter Outline

SUMMARY

Tissue engineering is a multidisciplinary science aimed at improving the lost functions of injured or damaged tissue. Tissue engineering offers a new hope for balancing the need and availability of organs for transplantation. It is a unique approach that utilizes the combined knowledge of various disciplines like biology, engineering, etc., to develop new tissues and bio-artificial organs.

WHAT YOU CAN EXPECT TO KNOW

Tissue engineering is a new and evolving branch of the life sciences. It uses a multidisciplinary approach to achieve its goals. Transplantation biology has evolved to a great extent to solve the problems related to injured or lost organs. The major limitation to reaping the benefits of transplantation biology is the limited availability of organs for transplantation. It is assumed and hoped that one day it will be possible to exploit the technology of tissue engineering to develop new organs as needed.

The present chapter introduces the elementary concepts of tissue engineering. The initial part focuses on the fundamental principles of tissue engineering and introduces the materials and techniques for the fabrication of scaffolds; the latter part discusses applications in selected tissues, issues and challenges, and a future course for the technology.

Animal Biotechnology. http://dx.doi.org/10.1016/B978-0-12-416002-6.00013-4

This chapter also discusses the salient features of some of the issues related to the ethical aspects of tissue engineering as well as the translational significance of this developing discipline.

HISTORY AND METHODS

INTRODUCTION

Wear and tear of tissues and organs is a natural process that becomes rapid with age. Apart from this, in accidental cases and diseases, patients either may lose organs or may face malfunction of tissues/organs, causing a severe threat to life. Conventional treatment options of such conditions include organ transplantation, surgical repair, artificial prosthesis, mechanical devices, and drug therapy. Tissue engineering has emerged as a new concept that promises the regrowth of tissue structures using cells and natural or synthetic materials.

Tissue engineering is comprised of two words: tissue and engineering. A "tissue" is a group of different types of cells that have different phenotypes but together perform a specific function; the term "engineering" refers to the application of knowledge to design and build. Thus, in a broad sense, tissue engineering is a multidisciplinary science that deals with the application of knowledge to design and construct tissues. Tissue engineering can be defined as *an interdisciplinary field that applies the principles of engineering and the life sciences to the development of biological substitutes that restore, maintain, or improve tissue functions*. The fundamental goal of tissue engineering is to create a three-dimensional mass of cells of a specific tissue that exhibit its some or whole characteristics, and which can be used to augment the desired function of a tissue. Tissue engineering can be categorized into (a) *in vitro* construction of bioartificial tissue from donor cells, and (b) *in vivo* modification of cell growth and function. The first category applies to the replacement or augmentation of malfunctioning tissues or organs, while the second category implies *in situ* regeneration. Tissue-engineered organs not only provide clinical solutions, but also can be used as "models" to study cell–cell or cell–tissue interactions, cell migrations, drug toxicity, and a number of other biological mechanisms.

HISTORY

The roots of the concept of tissue engineering can be traced back to depictions in early paintings. One of the earliest references in an artistic work is the famous painting known as "Healing of Justinian" in which St. Cosmas and St. Damien are shown doing a transplantation of a homograft limb onto a person, possibly a soldier. A better understanding of nature led to experiments in the regeneration of a living being. Paracelsus, a German-Swiss scientist who lived during the fifteenth century, tried to create life using a recipe of different chemicals. Goethe (eighteenth century) envisioned the creation of life by means of nonliving objects. The transformation of tissue regeneration from fiction to reality was made possible by the advent of clinical and basic sciences like surgery, transplantation, cell biology, biochemistry, etc. The use of wooden legs and metallic dentures can be considered earlier biomaterials for clinical purposes. The use of skin grafts can be considered the modern phase of tissue engineering. In the eighteenth century, Johann F. Dieffenbach performed some clinical experimental work on skin transplantation. Later, H.C. Bünger successfully performed autologous skin transplantation. In the twentieth century, E. Ullman performed the first kidney transplantation in animals and J.P. Merrill in humans. W.T. Green in the 1970s tried to regenerate cartilage by utilizing bone as a scaffold with chondrocytes. Yannas and Burke used skin cells on a collagen scaffold to make a skin equivalent for burn patients. The famous article of Dr. Robert Langer and Dr. Joseph Vacanti about tissue engineering in the journal Science is considered the official beginning of this area of work.

BASIC APPROACH TO TISSUE ENGINEERING: PRINCIPLES AND METHODOLOGY

In native tissues, the cells of different lineages are organized into three-dimensional forms and are surrounded by the natural extracellular matrix (ECM) that is secreted by them. Its basic components include collagen, elastin, proteoglycans, glycosaminoglycans, glycoproteins, hyaluronic acid, etc. The ECM creates a special environment in the spaces between cells and helps bind the cells in tissues together. It is also a reservoir for enzymes, inhibitors, cytokines, and many hormones and growth factors controlling cell growth and differentiation so that each tissue formed has specific characteristics and performs specific tasks in the body. Due to the diversity, organization, and distribution of constituents in the ECM, cells exhibit differential gene expressions in specific tissues. In tissue engineering, we try to recreate the artificial environment of ECM around the cells of specific tissues. For the simplest tissue engineering experiments, one requires cells, scaffolds, media, and sometimes bioreactors to scale up.

Cells

The source of the cells has a tremendous effect on the success of tissue engineering. The cells can be autologous, allogeneic, and xenogeneic. Autologous cells obtained from the individual into whom they will be implanted require prior harvesting followed by expansion in culture. Allogeneic cells are obtained from an individual of the same species other than the recipient. Xenogeneic cells are derived from individuals of different species. Allogeneic and xenogeneic sources of the cells pose challenges like host rejection, disease transmission, and ethical issues, although they are readily available in advance of need.

Scaffolds

The artificial architectures for biosynthetic ECMs are designed so that they can direct the cells to maintain a three-dimensional organization and lead the development of new tissues with suitable function. The cells should be induced by the scaffold, and can replace it with newly synthesized cell products. A large number of natural and synthetic polymers have been tried in making scaffolds for tissue engineering. Details about the scaffolds are provided in the next section.

Media

Media supplement the nutrition to cells and also complement the scaffolds to provide a complete bio-artificial extracellular matrix. The choice of medium depends upon the type of cells and the aim of the culture. Media can be divided into (a) seeding media, (b) differentiation media, and (c) maintenance media. The seeding media is the media used to introduce the cells into the scaffolds. It may or may not contain the serum. Generally, the cells remain in seeding media from six to 24 hours. The differentiation media contains a number of growth factors, cytokines, etc., to allow the cells to differentiate into the desired tissue. The maintenance media has serum with basic components to support the cells in culture.

Bioreactors

The requirement of a large number of cells and bigger engineered tissues (cell–polymer constructs) make the use of bioreactors imperative in tissue engineering. Bioreactors provide a steady and controlled microenvironment for the development of functional tissues from cell–polymer constructs. Ideally, a bioreactor should help in the uniform distribution of cells in the scaffold. It should maintain the required nutritional and oxygen concentration. A number of bioreactor designs have been proposed, like simple static or mixed spinner flasks, rotating vessels, perfusion chambers or columns, etc. Some bioreactors are also equipped with a mechanical stimulator for providing mechanical signals to the growing tissue.

Methodology

The simplest strategy for engineering any tissue is to isolate cells from the tissue of interest, followed by making a single-cell suspension, growing these cells onto a three-dimensional synthetic extracellular matrix (commonly known as a "scaffold") with appropriate growth factors, and allowing them to form a three-dimensional mass with definite biochemical or phenotypic characteristics of that tissue (Figure 13.1). This type of cell culture is known as a *three-dimensional cell culture*. This is different from a *monolayer culture* where the cells occupy the culture vessel surface to form a cellular sheet-like structure.

The simplest workflow is to select the right polymer, right cells and media, and correct/reliable assessment of the characteristics of the tissue equivalent (Flow Chart 13.1). As given in Flow Chart 13.1, if one wishes to make a tissue equivalent of hepatocytes to model the liver, one has to fabricate the three-dimensional porous scaffolds of any suitable polymer (e.g. chitosan or alginate). The hepatocytes will be isolated from the liver biopsy or from the liver of experimental animals (e.g. rat or mouse). Hepatocytes are separated by a perfusion method. The perfusion medium or solution contains an enzyme cocktail. The most common is collagenase; alternatively one can buy a ready-made medium available from different vendors, like Gibco. The major purpose of perfusion is to loosen the cells so that hepatocyte removal, collection, separation, and isolation become convenient. Cell death is a major concern during liver perfusion, and can be optimized in various situations. Isolated and purified hepatocytes are seeded with scaffolds and incubated for a week. During incubation, cells start growing into the tissue-like structures known as spheroids. Such spheroids are characterized for the liver functions where both surface markers and metabolic functions are screened. Some of the common markers for hepatocytes are albumin, alfa-feto protein, glucose-6-phosphatase, tyrosine aminotransferase, etc. These hepatocyte markers can be confirmed for mRNA either by reverse transcriptase polymerase chain reaction or any other method that depends upon the expertise and availability of the resource. The protein levels of the same markers can be confirmed by various methods, like western blot, FACS, etc. If required, after confirmation, these cultures can be further expanded and used for the intended purpose. Toxicity testing is the one example that we would like to note here.

SCAFFOLD DESIGN

For creating an engineered tissue, choice of scaffold material is important, as it has to mimic the natural ECM. A number of properties have been recognized that are essential for the design of an ideal scaffold (Table 13.1). These include biocompatibility, an interconnecting porous structure, appropriate surface chemistry for the growth of cells, easy fabrication, mechanical strength, biodegradability, and bioresorbability; they depend on the intended application. Biocompatibility of the material is the foremost requirement

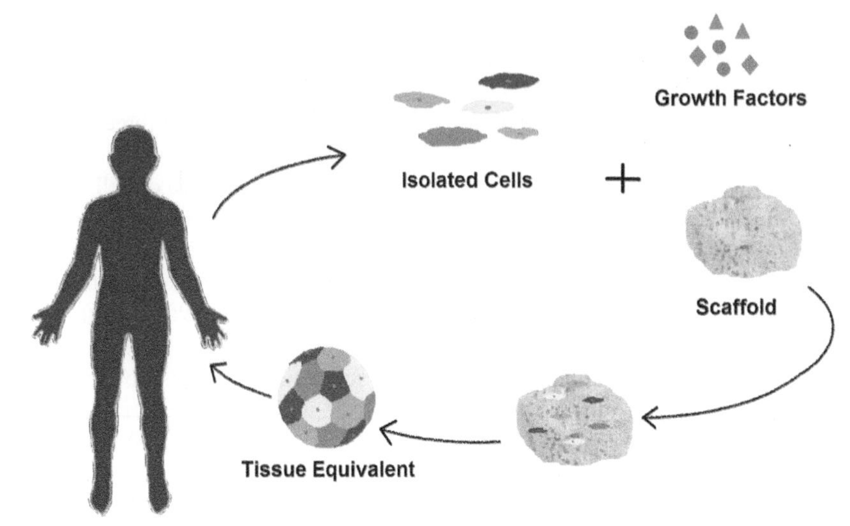

FIGURE 13.1 Basic approach to tissue engineering.

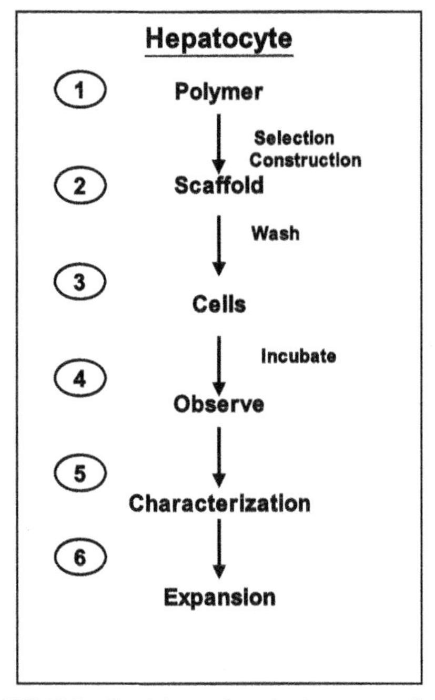

FLOW CHART 13.1 Crucial steps for using hepatocytes for tissue engineering.

as the scaffold directly interacts with the cells *in vitro* and tissues *in vivo* after implantation. Thus the material should not only be able to support the growth of the cells, but also should be immunologically the least reactive in order to be used inside the body. An interconnecting porous structure increases the surface area of the scaffold and allows more cells to be seeded on and inside it. The porous system maximizes cell-to-cell interactions and helps in the reorganization of a three-dimensional mass of the cells. In addition, the pores also allow diffusion of molecules that can be exploited to provide nutrients and gases to the cells entrapped within them. Another important consideration in scaffold design

TABLE 13.1 Considerations for Designing an Ideal Scaffold

Biocompatibility
Interconnecting porous structure
Appropriate surface chemistry
Mechanical strength
Biodegradability and bioresorbability

is surface chemistry. The surface of the scaffold material should have appropriate functional groups that can support the adherence of the cells to it. This property determines the overall architecture of the cell mass. In native tissues, a number of matrix proteins mediate cell adhesion by interacting with the appropriate cell adhesion molecules (CAMs) present on the surface of the cells. Among them, the integrin family is the most abundant and versatile. The integrin proteins interact with the RGD (R: arginine; G: glycine; D: aspartic acid) motifs of many matrix proteins, including vitronectin, fibrinogen, von Willebrand factor, collagen, laminin, etc., and are firmly anchored in the tissue. In the same manner, the materials should have cell-adhesive functional molecules to hold the cells on them. These groups can also direct the cells to grow into a specific tissue. Furthermore, the scaffold should also possess proper mechanical strength as it experiences biomechanical forces during cell adhesion and proliferation. In specific applications like vascular tissue engineering, scaffolds have to withstand hemodynamic forces due to the flow of blood. Biodegradation and bioresorbability of a material are the properties that are exploited during *in vivo* implantation of the scaffold. These properties cause the actual integration of the *in vitro* grown cells into the body, and augmentation or replacement of the injured tissue. The scaffold material is gradually degraded by the enzymes of the body and replaced by natural ECM

TABLE 13.2 Materials for Scaffolds

S. No.	Material	Example
1	Natural Polymers	Proteins - Collagen - Gelatin - Fibrin - Silk Polysaccharides - Cellulose - Chitosan - Alginate - Agarose - Hyaluronic acid
2	Synthetic Polymers	Poly (lactic acid) Poly (glycolic acid) Poly (Є-caprolactones) Poly (ortho esters) Polyethylene Polyester urethane Polysulfone Poly(tetrafluoroethylene) Poly(vinyl chloride)
3	Ceramics	Hydroxyapatite Calcium Phosphates Alumina Bioglass-ceramics

secreted by the cells. The products of the degraded scaffold material are removed from the site or are utilized as metabolites in other biochemical reactions.

MATERIALS FOR SCAFFOLDS

In native tissues, the extracellular matrix promotes the development of proper tissue structure and function, provides mechanical support for the developing tissues, localizes cells to their respective places, and regulates the expression of tissue-specific genes. In tissue engineering applications, a successful strategy involves the appropriate design and fabrication of scaffolds that can serve these functions and promote new tissue formation from cultured cells. Among all the available biomaterials (Table 13.2), polymers remain the most widely used materials for scaffolds because their properties can be changed by altering their monomers. Both synthetic and natural polymers have been used in the construction of scaffolds. Both groups have advantages as well as disadvantages. Natural polymers are abundant, are usually biodegradable, and sometimes contain groups similar to natural ECM components that can interact with the cells; however, their complexity often renders modification and purification difficult. Their batch-to-batch variation may also pose a problem. They can guide cells to grow at various stages of development.

Synthetic polymers can be formed in various compositions and their properties can be adjusted, but they generally lack biocompatibility. Therefore, choice of the material very much depends on the application. Synthetic polymers include aliphatic polyesters; for example, poly (lactic acid), poly (glycolic acid) and their copolymers, poly (ε-caprolactones), poly (ortho esters). Some important considerations that should be kept in mind while choosing the correct material for the fabrication of scaffolds according to different applications include biocompatibility, controlled degradability and bioresorbability, biomimeticity, mechanical stability, and sterilizability.

Another class of materials used for the fabrication of scaffolds is ceramics. These are the materials of metallic and non-metallic elements held together by ionic and/or covalent bonds. Ceramics have been widely used due to their biocompatibility and resemblance to the natural inorganic components of hard tissues like bone and teeth. Due to their high mechanical properties, they are used in load-bearing applications, but their disadvantage lies in their brittle nature.

SCAFFOLD FABRICATION METHODS

Preferably, the fabrication process of a scaffold should not affect the biocompatibility of the material, should allow control over porosity and pore size, may give the desired shape and size, and should not alter the physical, chemical, and biological (if any) properties of the materials. In recent years, various methods have been developed to construct scaffolds of desirable properties. These methods range from traditional techniques like solvent casting or fiber bonding, to modern computer-based design and fabrication technologies like 3D printing. Each method has its own advantages and disadvantages (Table 13.3). The choice of method depends on the applications and requirements of the tissues as no technique is ideal.

Fiber Bonding

As the name implies, the polymer fibers (e.g. polyglycolic acid) are bonded together in three-dimensional structures. The method forms highly porous scaffolds with large surface areas. Although fiber bonding forms porous scaffolds with interconnected pores, it requires a solvent, which if it remains in the scaffold, may be cytotoxic. To overcome this problem, the scaffolds are vacuum dried and sometimes involve heating at high temperature. A number of techniques are used to generate polymeric fibers. These include wet spinning, melt spinning, and electro spinning. Wet spinning is used for fiber-forming substances that have been dissolved in a solvent. In melt spinning, the fiber-forming substance is melted for extrusion through the spinneret and then directly solidified by cooling. Electro spinning

TABLE 13.3 Selected Advantages and Disadvantages of Various Fabrication Processes

S.No.	Fabrication Process	Advantages	Disadvantages
1	Fiber bonding	Porous Large surface area	Few polymers can be used Toxicity of residual solvent Poor mechanical strength
2	Solvent casting/ Particulate leaching	Control on pore size Crystalline	Few polymers can be used Toxicity of residual solvent
3	Melt molding	No use of solvent Control on pore size	Functionality of natural polymers are compromised
4	Membrane lamination	Control on pore size	Poor mechanical strength Poor interconnectivity
5	Phase separation	Incorporation of bioactive substances	No control over architecture
6	Freeze drying/ Lyophilization	Incorporation of bioactive substances	Samples are hygroscopic
7	Gas foaming	No use of solvents	No control on pore size
8	Polymer ceramic composite foam	Good mechanical properties	Residual solvent and porogen
9	Steriolithography	Accuracy for imparting small features	Only for photopolymerizable and liquid polymers
10	Selective laser sintering	No use of solvents Better compressive strength Automated	Functionality of natural polymers are compromised
11	Fused deposition modeling	No use of solvents Better compressive strength Automated	Functionality of natural polymers are compromised
12	3D Printing	Choice of materials Automated and high resolution	Poor mechanical strength Solvent toxicity
13	Pressure assisted micro syringe	Automated and high resolution	Poor mechanical strength Viscosity dependence

consists of spinning polymer solutions or melts in high electric fields. The process is based on the principle that strong electrical forces overcome weaker forces of surface tension in the charged polymer liquid. It is an inexpensive process that can be easily scaled up, employing multiple spinnerets. It is versatile in that almost any soluble polymer can be processed into nanofibers.

Solvent Casting and Particulate Leaching

In this technique the polymer is dissolved into methylene chloride or chloroform and is poured into a petri dish filled with a porogen like salt. The solvent is evaporated and the polymer with salt is kept in the water, which dissolves the salt and leaves the porous scaffold. Poly (L-lactic acid) and poly (DL-lactic-co-glycolic acid) (PLGA) scaffolds have been formed by this method. The porosity and pore size can be controlled with a suitable porogen. Scaffolds created are of defined pore size, surface, volume ratio, and crystallinity. The technique is applied to polymers dissolvable in solvents like chloroform. The residual solvent in the scaffold may denature any incorporated bioactive molecules.

Melt Molding

This technique utilizes the powdered polymer mixed with an aqueous soluble porogen, which is poured into a Teflon mold. The mold is heated above the glass transition temperature (Tg) of the polymer for an appropriate amount of time. The polymer porogen composite is removed and kept in water to dissolve the porogen.

The porous polymeric scaffold is formed to the desired shape and size. PLGA scaffolds have been constructed using this technique with gelatin as the porogen. The technique does not employ any solvent, and is carried out at a relatively low temperature; for non-amorphous polymers, high temperature is required, which prevents the incorporation of bioactive molecules. The porosity and pore size can be controlled, and the desired shape is obtained by choosing the correct mold.

Membrane Lamination

In this method, porous polymeric membranes are used in the fabrication of contour plots of specific three-dimensional shapes. The shapes of the contours are cut from these membranes. The membranes are adhered to one another (using chloroform) in the desired three-dimensional shape. This method is useful only when the original porous structure is preserved, and the boundaries between two layers should not be differentiated. The technique offers independent control of shape and porosity, but the interconnectivity of pores is limited and there are generally poor mechanical properties.

Phase Separation

In this technique, the polymer is dissolved into a suitable solvent at relatively low temperature. Phase separation is triggered by altering the physical parameters (like temperature). This causes the formation of one phase rich in polymer and other in solvent. The resulting phases are then quenched to create a two-phase solid. The solidified solvent is then removed by sublimation, and porous polymeric scaffold is obtained. This method has been developed in order to deliver drugs, proteins, or growth factors without altering their functions.The advantage of this technique lies in its ability to incorporate bioactive substances and drugs without losing their functions; on the other hand, there is no control of the internal architecture.

Gas Foaming

This technique does not use any solvent, but instead uses high-pressure gas. The polymer (PLGA) is compressed into a mold to form a solid structure. These solid structures are saturated by high-pressure CO_2 gas. Gradually the pressure is reduced, which causes pore formation in the scaffold. In another modification, a gas-foaming technique has been used in combination with a particulate leaching method. Here, PLGA was mixed with salt particles (porogen) and compressed to form solid disks. The porous structure was obtained due to gas foaming and particulate leaching.

Polymer Ceramic Composite Foam

This technique was developed to form high mechanical strength scaffolds for hard tissues like bone. In this method, ceramics like hydroxyapatite fibers are incorporated into poly (α-hydroxy ester) polymer (Thomson et al., 1998). To evenly mix polymer and ceramic, a solvent-casting technique is used. The method offers excellent mechanical properties and control over porosity, but residual organic solvent and porogen pose problems.

Solid Free Form Techniques

In the traditional methods discussed so far, there is no precise control over shape, size, thickness, porosity, and internal architecture. To address these problems, computer-based approaches have been developed to form scaffolds of predefined architecture and geometry. In these solid free form fabrication techniques, the material is taken in powder or liquid form and solidified in layers defined by a computer program. These technologies include selective laser sintering, stereolithography, fused deposition modeling, and 3D printing.

Selective LASER Sintering

This technique is solvent-free and uses heat energy to form scaffolds of predefined shapes using polymeric particles. In this method, a laser beam is directed toward a polymeric powder layer; it increases the temperature of the powder, which causes fusion of the polymeric particles. In this way, a patterned architecture is formed. It has been used to make scaffolds of polymer-coated powdered calcium phosphate.

Stereolithography

This is based on a photo-polymerization process that requires a photo-initiator. The technique makes use of a laser light beam focused on predefined regions of a liquid polymer layer, solidifying the exposed regions. This has been used to generate scaffolds of diethyl fumarate and poly (propylene fumarate). Although the technique produces precise and accurate scaffolds, its use is limited to only photopolymerizable liquid polymeric materials.

Fused Deposition Modeling

In this technique, a 3D scaffold is deposited layer-by-layer through a nozzle that is attached to a device connected to a computer. The technique is solvent-free, and the scaffolds show good compressive strengths. There is better control of the X,Y-axis than the Z-axis. Poly (ε-caprolactones) scaffolds have been constructed using this technique.

3D Printing

This method uses solvent or adhesive materials to bind polymers. An adhesive solution is deposited onto a polymeric powder bed using an inkjet printer. This method has been combined with particulate leaching methods to form porous scaffolds. The method is automated and forms highly porous scaffolds, but resolution cannot be reduced below the particle size of the polymer.

Pressure-Assisted Micro Syringe Method

In this method, the polymer is dissolved in solvent and is deposited through a syringe fitted with a 10–20 μm glass capillary needle. Thickness can be controlled by changing the syringe pressure, solution viscosity, and syringe tip diameter.

Freeze Drying

In this method, the polymer is homogenized in its solvent and water. The prepared emulsion is frozen at subzero temperatures. The frozen sample is put in a dedicated instrument (a lyophilizer), which sublimates the solvent with water in a vacuum. This not only creates highly porous scaffolds, but also, the porosity can be controlled. Another advantage is that since the whole process takes place at very low temperature, any biological molecules attached to the scaffolds do not lose their structure and function. Moreover, the prepared scaffolds can be stored at room temperature; however, since the scaffolds are hygroscopic, they have to be kept in a vacuum desiccator.

Freeze drying involves the removal of water or other solvent from a frozen product by a process called *sublimation.* Sublimation occurs when a frozen liquid goes directly to the gaseous state without passing through the liquid phase. In contrast, drying at ambient temperatures from the liquid phase usually results in changes in the product, and may be suitable only for some materials. However, in freeze-drying, the material does not go through the liquid phase, and it allows the preparation of a stable product.

EXAMPLES OF TISSUE-ENGINEERED ORGANS

Skin

Skin is the most researched area in tissue engineering, with the successful commercialization of several FDA-approved products. Examples of tissue engineered skin products are Epicel (Genzyme Tissue Repair, MA), which is an epidermal cell sheet; Integra (Integra Life Sciences, NJ); Dermagraft (Advanced Tissue Sciences, CA); and Alloderm (Lifecell Corp., TX). Some of these are dermal substitutes. Organogenesis Inc., MA, has also introduced Human Skin Equivalent (HSE) that has epidermal and dermal tissue components. The commercially available bioengineered skins are providing great support in the healing of burns. Clinically it has been observed that such skin products are effective in healing when the auto- or allografts fail.

Strategies to engineer skin tissue involve epidermal sheet fabrication, dermal replacements, or composite structures with both epidermal and dermal components. Collagen, being the major ECM component of skin, has been the most important natural polymeric scaffold for skin tissues.

Pancreas

Diabetes is a major disease that causes morbidity and mortality worldwide. According to WHO, around 280 million people will have this disease by 2025. Most of the research work in pancreatic tissue engineering has been focused on encapsulating pancreatic islet cells. The polymeric encapsulation system should be able to protect the cells from the host's immune system by providing isolation from the host's immune system. This can be achieved by controlling the pore size of the polymeric capsule. There must be proper diffusion of oxygen and nutrients across the immunoisolated capsule for the cells. The capsules should be mechanically stable and biocompatible inside the body. Early work used alginate and poly(L-lysine) polymers for encapsulating islets.

Liver

The liver is a vital organ that performs a large number of functions like metabolism and detoxification. The hepatocytes perform the functions of metabolism and detoxification. The liver has a very high regenerative capacity, and it fights continuous damage to hepatocytes due to their exposure to toxic substances. In a diseased condition, the regeneration process is compromised, leading to infection and bleeding risks, and ultimately liver failure.

The development of liver assist devices (LAD) is the major focus of liver tissue engineering. LADs consist of primary hepatocytes arranged similar to the liver on polymeric scaffolds in bioreactors. Design of the LAD should be such that it provides maximum hepatocyte viability and functionality. To meet these conditions, nutrients and gases should reach the cells, and toxins and waste should be effectively removed. LADs should be efficient in transportation of gases, nutrients, metabolites, and toxins.

Kidney

A tissue-engineered kidney should have a glomerulus and tubule to compensate for excretory functions. A synthetic membrane with suitable permeability is the key to successful development of a bio-artificial kidney. A polymeric hollow fiber design has been shown to mimic glomerular filtrations. A bio-artificial renal tubule has been designed using epithelial progenitor cells and polymeric membranes. The combination of filtration devices and tubule has paved the way to a future bio-artificial kidney.

Bone/Cartilage

The approach to engineering bone tissue is to take a suitable polymeric scaffold with native tissue, and culture osteoblasts over it. Since bones and cartilage are hard tissues, the mechanical property of the material should match the native tissue. For implantation, the scaffold should support vascular networking, and the shape of the scaffold should complement the defective site. Ideally, the body should resorb the material. Ceramics are the material of choice.

Nerves

The approach to engineering nerve tissue *in situ* is to fabricate a cylindrical structure using polymers, and entubulate cut nerves. These cylindrical structures are known as nerve conduits or guidance channels. They guide the outgrowing nervous tissue to regenerate nerve cables. They also prevent the invasion of scar tissue. Regeneration of nerves can be improved by adding supporting cells like Schwann cells.

Blood Vessels

The development of a reliable and efficient engineered blood vessel is important considering the large number of heart bypass surgeries that are performed each year. Early blood vessel constructs were built by co-culturing smooth muscle cells and endothelial cells using scaffolds of collagen.

The blood vessel should be elastic, and should have enough mechanical strength to sustain the pulsatile flow of blood. It should be non-thrombogenic, and should demonstrate patency for considerable periods of time. It should be well tolerated by the immune system. Grafts are influenced by the host tissue environment (including the hemodynamic environment), therefore, long-term resistance to hyperplasia is a major challenge (besides long-term immunocompatibility and patency).

TISSUE ENGINEERING USING STEM CELLS

Stem cells have the properties of self-renewal and differentiation. These properties have been used for engineering tissues like skin, bone, liver, pancreas, cornea, etc. In the skin tissue engineering approach, keratinocytes with epidermal stem cells were cultured to form the epidermal sheet. These sheets were kept on dermal substitutes composed of a scaffold seeded with dermal fibroblasts. Such skin tissue equivalents have the capacity to regenerate full thickness wounds.

In bone regeneration, skeletal stem cells can be isolated and cultured on three-dimensional scaffolds of ceramics that can be transplanted *in vivo* at the bone defect area. The adult mesenchymal stem cells (MSCs), which can be isolated from bone marrow, can be expanded in culture in an undifferentiated state. These MSCs have been used to engineer connective tissues. MSCs are considered good candidates for tissue engineering provided they are cultured with appropriate growth factors and cytokines that promote differentiation on suitable scaffolds. It has also been observed that MSCs secrete immunosuppressive molecules that help with immunological tolerance.

Embryonic stem cells (ESC) have both pros and cons. On one hand, they can be generated in large quantities in the laboratory in an undifferentiated state, while on the other hand their ability to differentiate efficiently is questionable. Embryonic stem cells have been used to construct synchronously contracting engineered human cardiac tissue containing endothelial vessel networks. The constructed tissue consisted of cardiomyocytes, endothelial cells, and embryonic fibroblasts. The vessels were further stabilized by mural cells that originated from embryonic fibroblasts. There are numerous articles describing the potential of stem cells in tissue engineering. Stem cells not only provide a better cellular option that can be differentiated as per specific needs, but the molecules secreted by them have been found to be helpful in integrating the transplanted tissue into the body.

ISSUES AND CHALLENGES

There are certain issues and challenges involved with the development and application of tissue engineering and its benefits to human health. The most significant challenge with tissue engineering is the shelf life of the tissue-engineered product, along with the compatibility of the tissue-engineered product with the immune system of the recipient.

The use of a non-autologous source of cells has been an issue of great concern. The presence of fetal bovine/calf serum in the medium also poses risks of infection and a strong immune response in the host body. The cells designated for transplantation should be cultured in the xenofree media with either human serum components or synthetic components. Another challenge concerns the standardization of an isolation technique for the cells that ensures no contamination, but still produces of a large number of cells without compromising their functionality and phenotype. It is still a major obstacle in tissue engineering to produce cells of the same quality from lot-to-lot. In the case of embryonic stem cells, and allogeneic and xenogeneic cells, there is also a high risk of immune rejection due to different genetic constitutions contributing foreign phenotypic expressions.

Quality control of the materials used for engineering tissue is also a matter of key concern. The material should be produced using strict manufacturing practices. The polymeric scaffolds used for making a tissue-engineering construct should be integrated into the implanted site. Their biodegradation should occur at an optimum pace, and the by-products should not produce any unwanted effects. Monitoring of the adverse effects of the by-products of biodegraded scaffold material should be closely observed, but might be challenged by the collective effects of the cells with scaffolds.

Clinical trials also have certain challenges. There is a strong possibility that the adverse events might not be detected for a long time during the clinical trials, so determining the end point might be costly. Data produced might be (to some extent) uncertain with respect to long-term effects. Post-trial follow-ups will be required for the observation of actual effects.

ETHICAL ISSUES

Tissue engineering is advancing in the form of regenerative medicine. There is no doubt that tissue engineering can offer solutions for many degenerative disorders. On one hand it will improve the quality of life of those who suffer from these diseases, but on the other hand it raises various concerns about the ethics of this new and emerging branch of science. Some of the major issues are consent from the tissue provider. In other words, would it be ethical to give one person's tissue to another without his/her consent? Even the consent discloses the identity of the person, which could lead to various social, religious, and personal issues. Should personal and religious views be considered for someone who is in desperate need of health care or treatment? There is always the possibility that transfer of tissue from one human to another could lead to additional clinical conditions, like a fatal infection.

This leads to all sorts of questions for a profound technology such as tissue engineering. Did the procedure save the patient or create more trouble? Can the donor ask for property rights? Should the donor get a share from the commercialization of the product that originated from them? What if a downtrodden individual decides to sacrifice himself or herself and exploit this technology for profit? Where to stop? When to stop? What to stop? Who should stop? Why should we stop? There are bound to be numerous ethical issues. Hopefully, with time, they can be resolved.

TRANSLATIONAL SIGNIFICANCE

Tissue engineering itself is a translational science that has innumerable applications. The information gathered through tissue engineering studies regarding the development and maintenance of any tissue or organ could be great, and could lead toward the development of new organs. There is even a possibility that some xenogenic tissue can be modified using different types of technology that make it acceptable to an entirely a different host. If this type of approach is made possible, it will most certainly be a great relief to most terminal patients in need of an organ. Tissue engineering has great potential to reverse the aging process, which would help reduce the cost of medication, and would provide the elderly a better quality of life. This technology has the potential to treat deformities from accidents, burns, gun shot wounds, etc.

WORLD WIDE WEB RESOURCES

There are numerous resources on the web for information about tissue engineering and regenerative medicine. A simple web search (e.g. with Google) yields all sorts of information about tissue engineering, everything from the basics, to different methods, protocols, and specific applications. PubMed (maintained by NCBI) can be searched for more scientific and validated information. Apart from various vendors in the field of tissue engineering (who also provide useful information on their websites), the following three sites offer excellent information on this subject.

www.tissue-engineering.net: A useful website that defines applicable terminology and provides the latest news, job postings, and various other updates in the field of tissue engineering.

http://news.discovery.com/tech/tags/tissue-engineering.htm:
Updated information about the current happenings in tissue engineering.

http://tej.sagepub.com: The *Journal of Tissue Engineering* is an open access journal.

REFERENCES

Angelova, N., & Hunkeler, D. (1999). Rationalizing the design of polymeric biomaterials. *Trends in Biotechnology*, *17*, 409–421.

Bell, E. (1995). Strategy for the selection of scaffolds for tissue engineering. *Tissue Engineering*, *1*, 163–179.

Berthiaume, F., & Yarmush, M. L. (2003). Fundamentals of tissue engineering. In B. Palsson, J. A. Hubbell, R. Plonsey & J. D. Bronzino (Eds.), *Tissue Engineering (Principles And Applications In Engineering)*. Boca Raton, FL: CRC Press. p. II-8.

Bianco, P., & Robey, P. G. (2001). Stem cells in tissue engineering. *Nature*, *414*, 118–121.

Bronzino, J. D. (2002). *The Biomedical Engineering Handbook* (Vol. 2). Boca Raton, FL: CRC Press. 109.

Caplan, A. I. (2007). Adult mesenchymal stem cells for tissue engineering versus regenerative medicine. *Journal of Cellular Physiology*, *213*(2), 341–347.

Caspi, O., Lesman, A., Basevitch, Y., Gepstein, A., Arbel, G., Habib, I. H. M., Gepstein, L., & Levenberg, S. (2007). *Circulation Research*, *100*, 263–272.

Chang, B. S., Lee, C. K., Hong, K. S., Youn, H. J., Ryu, H. S., Chung, S. S., & Park, K. W. (2000). Osteoconduction at porous hydroxyapatite with various pore configurations. *Biomaterials*, *21*, 1291–1298.

Cheung, H. Y., Lau, K. T., Lu, T. P., & Hui, D. (2007). A critical review on polymer-based bio-engineered materials for scaffold development. *Composites: Part B*, *38*, 291–300.

Chew, S. Y., Wen, Y., Dzenis, Y., & Leong, K. W. (2006). The role of electrospinning in the emerging field of nanomedicine. *Current Pharmceutical Design*, *12*(36), 4751–4770.

Chu, T. M. G., Ortan, D. G., Hollister, S. J., Feinberg, S. E., & Halloran, J. W. (2002). Mechanical and in vivo performance of hydroxyapatite implants with controlled architectures. *Biomaterials*, *23*, 1283–1293.

Ciardelli, G., Chiono, V., Vozzi, G., Pracella, M., Ahluwalia, A., Barbani, N., Cristallini, C., & Giusti, P. (2005). Blends of poly-(epsilon-caprolactone) and polysaccharides in tissue engineering applications. *Biomacromolecules*, *6*(4), 1961–1976.

Cooke, M. N., Fisher, J. P., Dean, D., Rimnac, C., & Mikos, A. G. (2003). Use of stereolithography to manufacture critical sized 3D biodegradable scaffolds for bone ingrowth. *Journal of Biomedical Materials Research*, *64 B*, 65–69.

Biomaterials. (2002). In K. C. Dee, D. A. Puleo & B. Rena (Eds.), *An Introduction To Tissue Biomaterial Interactions* (pp. 1–13). New York: Wiley-Liss.

Deng, M., Nair, L. S., Nukavarapu, S. P., Kumbar, S. G., Jiang, T., Krogman, N. R., Singh, A., Allcock, H. R., & Laurencin, C. T. (2008). Miscibility and in vitro osteocompatibility of biodegradable blends of poly[(ethyl alanato) (p-phenyl phenoxy) phosphazene] and poly(lactic acid-glycolic acid). *Biomaterials, 29*(3), 337–349.

Freed, L. E., Marquis, J. C., Nohria, A., Emmanual, J., Mikos, A. G., & Langer, R. (1993). Neocartilage formation in vitro and in vivo using cells cultured on synthetic biodegradable polymers. *Journal of Biomedical Materials Research, 27*, 11–23.

Gao, H., Gu, H., & Ping, Q. (2007). The implantable 5-fluorouracil-loaded poly(l-lactic acid) fibers prepared by wet-spinning from suspension. *Journal of Controlled Release, 118*(3), 325–332.

Green, M. M. (2000). Dynamics of cell-ECM interactions. In R. P. Lanza, R. Langer & J. Vacanti (Eds.), *Principles of Tissue Engineering* (pp. 33–55). SD, CF, USA: Academic Press.

Harris, L. D., Kim, B. S., & Mooney, D. J. (1998). Open pore biodegradable matrices formed with gas foaming. *Journal of Biomedical Materials Research, 42*(3), 396–402.

Harrison, J., Pattanawong, S., Forsythe, J. S., Gross, K. A., Nisbet, D. R., Beh, H., Scott, T. F., Trounson, A. O., & Mollard, R. (2004). Colonization and maintenance of murine embryonic stem cells on poly (alpha-hydroxyesters). *Biomaterials, 25*(20), 4963–4970.

Hutmacher, D. W. (2001). Scaffold design and fabrication technologies for engineering tissues – state of the art and future perspective. *Journal of Biomaterials Science – Polymer Edition, 11*, 107–124.

Jung, Y., Kim, S. S., Kim, Y. H., Kim, S. H., Kim, B. S., Kim, S., Choi, C. Y., & Kim, S. H. (2005). A poly (lactic acid)/calcium metaphosphate composite for bone tissue engineering. *Biomaterials, 26*(32), 6314–6322.

Kim, S. S., Utsunomiya, H., Koski, J. A., Wu, B. M., Cima, M. J., Sohn, J., Mukai, K., Griffith, L. G., & Vacanti, J. P. (1998). Survival and function of hepatocytes on a novel three-dimensional synthetic biodegradable polymer scaffold with an intrinsic network of channels. *Annals of Surgery, 228*(1), 8–13.

Langer, R. (2000). Tissue engineering. *Molecular Therapy, 1*(1), 12–14.

Langer, R., & Vacanti, J. P. (1993). Tissue engineering. *Science, 260*, 920–932.

Lee, G., Barlow, J., Fox, W., & Aufdermorte, T. (1996). *Biocompatibility of SLS Formed Calcium Phosphate Implants. Proceedings of Solid Free Form Fabrication Symposium,* Austin, TX: University of Texas. 1996, pp.15–22.

Lee, J., Tae, G., Kim, Y. H., Park, I. S., Kim, S. H., & Kim, S. H. (2008). The effect of gelatin incorporation into electrospun poly(l-lactide-co-caprolactone) fibers on mechanical properties and cytocompatibility. *Biomaterials, 29*(12), 1872–1879.

Liu, L. S., Thompson, A. Y., Heidaran, M. A., Poser, J. W., & Spiro, R. C. (1999). An osteoconductive collagen/hyaluronate matrix for bone regeneration. *Biomaterials, 20*, 1097–1108.

Lo, H., Ponticiello, M. S., & Leong, K. W. (1995). Fabrication of controlled release biodegradable foams by phase separation. *Tissue Engineering, 1*, 15–27.

Integrating cells into tissues. (2000). In H. Lodish, A. Berk, S. A. Zipursky, P. Matsudaira, D. Baltimore & J. Darnell (Eds.), *Molecular Cell Biology* (4th Ed., pp. 968–1002). New York: Freeman and Co.

Luetzow, K., Klein, F., Weigel, T., Apostel, R., Weiss, A., & Lendlein, A. (2007). Formation of poly(ε-caprolactone) scaffolds loaded with small molecules by integrated processes. *Journal of Biomechanics, 40*(1), S80–S88.

Lv, Q., Hu, K., Feng, Q., Cui, F., & Cao, C. (2007). Preparation and characterization of PLA/fibroin composite and culture of HepG2 (human hepatocellular liver carcinoma cell line) cells. *Composites Science and Technology, 67*(14), 3023–3030.

Mastrogiacomo, M., Papadimitropoulos, A., Cedola, A., Peyrin, F., Giannoni, P., Pearce, S. G., Alini, M., Giannini, C., Guagliardi, A., & Cancedda, R. (2007). Engineering of bone using bone marrow stromal cells and a silicon-stabilized tricalcium phosphate bioceramic: Evidence for a coupling between bone formation and scaffold resorption. *Biomaterials, 28*(7), 1376–1384.

Mikos, A. G., Bao, Y., Cima, L. G., Ingber, D. E., Vacanti, J. P., & Langer, R. (1993). Preparation of poly(glycolic acid) bonded fiber structures for cell attachment and transplantation. *Journal of Biomedical Materials Research, 27*, 183–189.

Mikos, A. G., Sarakinos, G., Leite, S. M., Vacanti, J. P., & Langer, R. (1993). Laminated three dimensional biodegradable foams for use in tissue engineering. *Biomaterials, 14*, 323–330.

Mikos, A. G., Thorsen, A. G., Czerwonka, L. A., Bao, Y., Langer, R., Winslow, D. N., & Vacanti, J. P. (1994). Preparation and characterization of poly(L-lactic acid) foams. *Polymer, 35*, 1068–1077.

Montjovent, M. O., Mark, S., Mathieu, L., Scaletta, C., Scherberich, A., Delabarde, C., Zambelli, P. Y., Bourban, P. E., Applegate, L. A., & Pioletti, D. P. (2008). Human fetal bone cells associated with ceramic reinforced PLA scaffolds for tissue engineering. *Bone, 42*(3), 554–564.

Nair, M. B., Babu, S. S., Varma, H. K., & John, A. (2008). A triphasic ceramic-coated porous hydroxyapatite for tissue engineering application. *ACTA Biomaterialia, 4*(1), 173–181.

Patrick, C. W., Jr., Sampath, R., & Mcintre, L. V. (2000). Fluid shear stress effects on cellular functions. In J. D. Bronzino (Ed.), *The Biomedical Engineering Handbook:* (Vol. 2, pp. 114). Boca Raton, FL: CRC Press.

Peters, M. C., & Mooney, D. J. (1998). Synthetic extracellular matrices to guide tissue formation. In Y. Ikada & S. Enomoto (Eds.), *Tissue Engineering For Therapeutic Use.* (pp. 55–65). Elsevier Science, B.V.

Pfaff, M. (1997). Recognition sites of RGD-dependent integrins. In J. A. Eble (Ed.), *Integrin-Ligand Interaction* (pp. 101–121). Heidelberg: Springer-Verlag.

Rowe, T. W. G. (1970). Freeze-drying of biological materials: some physical and engineering aspects. *Current Trends in Cryobiology*, 61–138.

Thomson, R. C., Shung, A. K., Yaszemski, M. J., & Mikos, A. G. (2000). Polymer scaffold processing. In R. P. Lanza, R. Langer & J. Vacanti (Eds.), *Principles Of Tissue Engineering* (pp. 251–262). California: Academic Press.

Thomson, R. C., Yaszemski, M. J., Powers, J. M., & Mikos, A. G. (1998). Hydroxyapatite fiber reinforced poly (alpha-hydroxy ester) foams for bone regeneration. *Biomaterials* (21), 1935–1943.

Thomson, R. C., Yaszemsky, M. J., Powers, J. M., & Mikos, A. G. (1995). Fabrication of biodegradable polymer scaffolds to engineer trabecular bone. *Journal of Biomaterials Science Polymer Edition, 7*(1), 23–38.

Trommelmans, L., Selling, J., & Dierickx, K. (2007). Ethical Issues. *Tissue Engineering. European Ethical – Legal Papers No.* (7). Leuven.

Tsang, V. L., & Bhatia, S. N. (2004). Three-dimensional tissue fabrication. *Advanced Drug Delivery Reviews, 56*(11), 1635–1647.

Vozzi, G., Flaim, C., & Ahluwalia, Bhatia S. (2003). Fabrication of PLGA scaffolds using soft lithography and microsyringe deposition. *Biomaterials, 24*(14), 2533–2540.

Wang, Y. C., Lin, M. C., Wang, D. M., & Hsieh, H. J. (2003). Fabrication of a novel porous PGA-chitosan hybrid matrix for tissue engineering. *Biomaterials, 24*(6), 1047–1057.

Weinberg and Bell. (1986). A blood vessel model constructed from collagen and cultured vascular cells. *Science*, 397–400. 231.

Whang, K., Thomas, H., & Healy, K. E. (1995). A novel method to fabricate bioabsorbable scaffolds. *Polymer, 36*, 837–841.

Wu, W., Feng, X., Mao, T., Feng, X., Ouyang, H. W., Zhao, G., & Chen, F. (2007). Engineering of human tracheal tissue with collagen-enforced poly-lactic-glycolic acid non-woven mesh: A preliminary study in nude mice. *British Journal of Oral and Maxillofacial Surgery, 45*(4), 272–278.

Yoon, B. H., Choi, W. Y., Kim, H. E., Kim, J. H., & Koh, Y. H. (2008). Aligned porous alumina ceramics with high compressive strengths for bone tissue engineering. *Scripta Materialia, 58*(7), 537–540.

Zein, I., Hutmacher, D. W., Tan, K. C., & Teoh, S. H. (2002). Fused deposition modeling of novel scaffold architectures for tissue engineering applications. *Biomaterials, 23*, 1169–1185.

FURTHER READING

Biomaterials: An Introduction. Joon Park, R. S. Lakes. Springer Press USA.

Biomaterials Science: An Introduction to Materials in Medicine. Buddy D. Ratner, Allan S. Hoffman, Frederick J. Schoen, Jack E. Lemons. Academic Press, USA.

Principles of Tissue Engineering. Robert Lanza, Robert Langer, Joseph Vacanti. Academic Press, USA.

Tissue Engineering: Engineering Principles for the Design of Replacement Organs and Tissues. W. Mark Saltzman. Oxford University Press, USA.

GLOSSARY

Biocompatibility The ability of any material to exist within a biological system without any harmful effect.

Biodegradability The ability of any material to be degraded by biological enzymes.

Bioresorbability The ability of any material to be absorbed by a biological system following its degradation.

Extra Cellular Matrix (ECM) A substance surrounding the cells in a tissue. It contains collagen, elastin, proteoglycans, glycosaminoglycans, glycoproteins, hyaluronic acid, etc. It is also a reservoir for enzymes, inhibitors, cytokines, hormones, and growth factors that control cell growth and differentiation so that each tissue formed has specific characteristics and performs specific tasks in the body.

Scaffolds The artificial architectures for biosynthetic ECMs that are designed so that they can direct the cells to maintain a three-dimensional organization and can lead the development of new tissues with suitable functions.

Tissue Engineering An interdisciplinary field that applies the principles of engineering and the life sciences to the development of biological substitutes that restore, maintain, or improve tissue functions.

Tissue A group of different types of cells that have different phenotypes, but together perform a specific function.

ABBREVIATIONS

CAMs Cell Adhesion Molecules
ECM Extracellular Matrix
ESC Embryonic Stem Cells
FDA Food and Drug Administration
LADs Liver Assist Devices
MSCs Mesenchymal Stem Cells
PLGA Poly (DL-Lactic-Co-Glycolic Acid)
RGD R (Arginine), G (Glycine), D (Aspartic Acid)
Tg Glass Transition Temperature

LONG ANSWER QUESTIONS

1. What is tissue engineering? How is it important for human health?
2. What are scaffolds? Describe the various materials used to fabricate scaffolds.
3. Describe the various fabrication techniques for making scaffolds.
4. How are stem cells useful in tissue engineering?
5. What are the current challenges to taking tissue-engineered products from the bench to the bed-side?

SHORT ANSWER QUESTIONS

1. Define the following terms:
   a. Biocompatibility.
   b. Biodegradability.
   c. Bioresorbability.
2. What are the different criteria for designing an ideal scaffold?
3. Give the name of markers for hepatocyte identification?
4. Why do we use collagenase and other enzymes to perfuse liver?
5. Write short notes on:
   a. Polymers.
   b. Ceramics.

ANSWERS TO SHORT ANSWER QUESTIONS

1a. Biocompatibility: The ability of any material to exist within a biological system without any harmful effect.

1b. Biodegradability: The ability of any material to be degraded by biological enzymes.

1c. Bioresorbability: The ability of any material to be absorbed by the biological system following degradation.

2. It should be porous with interconnecting pores, biocompatible, should have appropriate functional groups on its surface, appropriate mechanical strength, and, if required, should be biodegradable and bio-resorbable.

3. There are numerous markers that can be used for identification of hepatocytes. One can measure mRNA levels as well as protein levels of the same markers. Some of the most common markers are albumin, alpha-fetoproteins, glucose-6-phosphataase, *trans*-aminotransferse, etc.

4. The main objective of perfusion is to loosen the hepatocytes from the liver so that their identification, isolation, and culture become easier. It is always useful and helpful to use some enzyme during perfusion. Various enzymes can be included in the perfusion buffer. One of the most common enzymes used for the perfusion of the liver is collagenase.

5a. Polymers: These are the materials that have defined monomers linked together in a specific pattern.

5b. Ceramics: These are the materials of metallic and nonmetallic elements held together by ionic and/or covalent bonds.

Chapter 14

Nanotechnology and Its Applications to Animal Biotechnology

Ashok K. Adya* and Elisabetta Canetta†

*BIONTHE (Bio- and Nano-technologies for Health & Environment) Centre, Division of Biotechnology & Forensic Sciences, School of Contemporary Sciences, University of Abertay, Dundee, Scotland, UK, †Cardiff School of Biosciences, Biomedical Sciences Building, Cardiff University, Cardiff, Wales, UK

Chapter Outline

SUMMARY

Nanotechnology is advancing at a fast pace with its ramifications felt in almost every field, including animal biotechnology and life sciences. Further growth in its applications to animal nutrition, health, disease diagnosis, and drug delivery is inevitable. This chapter provides an overview of the application of nanotechnology to the study of the structure, mechanics, and biochemistry of animal cells on a nanoscale.

WHAT YOU CAN EXPECT TO KNOW

Nanotechnology is advancing at a fast pace with its ramifications felt in almost every field, ranging from materials science to food, forensic, agricultural, and life sciences, including biotechnology and medicine. Nanotechnology is already being harnessed to address many of the outstanding problems in animal biotechnology. In the next decade we expect to see further growth in its applications to animal biotechnology (e.g. animal nutrition, health, disease diagnosis, and drug delivery).

The aim of this chapter is to provide an overview and related examples of the application of nanotechnology (which concerns the manipulation of matter at the atomic and molecular scale) to the study of the changes in the structure, mechanics, chemistry, and biochemistry of animal cells that occur on a scale of nanometers (1 nm = 0.000000001 m). The different nanotechnology techniques (such as Atomic Force Microscopy, Chemical Force Microscopy, Fluid Force

Animal Biotechnology. http://dx.doi.org/10.1016/B978-0-12-416002-6.00014-6

Microscopy, Near-field Scanning Optical Microscopy, and Raman Spectroscopy and Imaging) used in biotechnology to visualize and manipulate cells, biomolecules, and proteins will be briefly explained and practical examples given and discussed. In addition, the main ethical issues related to the use of nanoparticles in animal biotechnology will be addressed, and a few examples provided. After reading this chapter, the reader should be well equipped to explore further the fascinating and ever-expanding "nanoworld" and its multi-faceted aspects. Nanotechnology is not only our nearest future, but also an already existing daily reality (especially in the materials and cosmetics industries) for us all. It is therefore paramount for future generations of scientists to know about new developments in nanotechnology and nanobiotechnology (which is the intersection of nanotechnology and biotechnology).

HISTORY AND METHODS

INTRODUCTION

The first use of the concepts in "nano-technology," even before the word existed in the dictionary, was in a talk given by physics Professor Richard Feynman at an American Physical Society meeting at Caltech on December 29, 1959. This lecture was the birth of the idea and the study of nanotechnology. Albert R. Hibbs suggested to Feynman the idea of a medical use for Feynman's micromachines. Tokyo Science University Professor Norio Taniguchi first used the term "nanotechnology" in a 1974 paper. Eric Drexler popularized the potential of molecular nanotechnology in the late 1970s and 1980s. Scanning tunneling microscopy was invented in 1981. The first book on nanotechnology, *Engines of Creation* (Eric Drexler), was published in 1986. The Atomic Force Microscope (AFM), which can be considered as being at the heart of nanotechnology, was invented by Binnig, Quate, and Gerber in 1986; 1989 saw the production of the first commercial AFM. AFM is one of the foremost tools used for imaging, measuring, and manipulating matter at the nanoscale. What Feynman and Hibbs considered a possibility is becoming a reality 54 years later.

The functionality of biosystems is governed by their nanoscale structure and by the processes that occur at this scale. For a perspective on scale, one nanometer (0.000000001 m) is one-billionth of a meter, which is the same size as a human hair split 100,000 times width-wise; the width of a DNA molecule is ~2 nm. Nanoscale processes have evolved and been optimized in nature over millions of years. For all biological and man-made systems, the first level of organization, in fact, occurs at the nanoscale, where their basic properties and functions are defined.

Nanotechnology provides the ability to work (observe, move, manipulate) at the atomic and molecular levels, atom-by-atom on a scale of ~1–100 nm to create, understand, and use new materials and devices with fundamentally new functions and properties resulting from their small scale. Nanobiotechnology, on the other hand, is a concourse of nanotechnology and biology where nanotechnology provides the tools and techniques to work from nanoscale principles to investigate, understand, and transform biological systems.

Professor Richard Feynman, a Nobel laureate, delivered his famous after-dinner speech, "*There is plenty of room at the bottom,*" at a conference in 1959. Through his charisma and genius he inspired the conceptual beginnings of nanoscience and nanotechnology (while these words were still non-existent) and laid the foundation for these fields. He talked about "maneuvering things atom by atom" and "doing chemical synthesis by mechanical manipulation." Thanks to the development of the Scanning Tunneling Microscope (STM), the Atomic Force Microscope (AFM), and other Scanning Probe Microscopy (SPM) techniques, Richard Feynman's ideas have finally been realized. Not only is it now possible to understand the functions and properties of materials and biosystems at the nanoscale, but it is also possible to perform nanomanipulatations on them for modern applications in drug delivery, nanosurgery, and nanomedicine. The nanoworld was discovered only about two decades ago, and it was driven mainly by the invention of AFM (Binning, 1986) and a variety of other methods for fabricating nanostructures.

Biological systems are in general very sensitive to their environment. For instance, a cell's environment *in vivo*, being considerably detailed and complex, can react to objects as small as a few nm, which is ~5,000 times smaller than itself. The requirement for cells to get closer to the object leads to many cell–cell adhesions with resulting biological and physiological responses/effects. When cells are taken out of the body to be cultured *in vitro*, or when external objects/devices such as prosthetics are introduced into the body, cells experience a strange new nanoworld that could be chaotic and detrimental for a cell's survival, but might offer a different or opposite signal to what the cell might normally receive. Similarly, cells are exposed to a variety of environmental and age-related stresses resulting from disease, injury, and infection. Mechanisms involved in sensing stress, and the resulting changes in cell surface morphology and physiology, are thus crucial for understanding how cells adapt and survive adverse conditions. For the above reasons alone, it becomes important to know even more about how cells react to this nanoworld (nanoenvironment) and how to control them to achieve better functionality of the biosystems.

It is now the right time to introduce the different length scales/regimes. In increasing order they are as follows: nanoscopic (~1–100 nm) < mesoscopic (~100–1,000 nm) < microscopic (< 1 μm) < macroscopic (~1 mm). While the macroscopic world (the world we see and perceive) can be scaled down through orders of magnitude to the microscopic scale with little or no change in the expected properties, it is not possible to do such scaling when we enter the nanoworld, the gateway to which is ~100–200 nm in dimension. Interfacial and quantization effects begin to play a major role at the nanoscale, and matter behaves differently at such small length scales. Physical properties of materials/biomaterials change as their size approaches nanoscale, where the fraction of atoms present at the surface of the material becomes significant. In contrast, for bulk materials larger than 1 μm, the percentage of atoms at the surface is insignificant as compared to those present in the bulk of the material. When a macroscopic device is scaled down to mesosize, it starts to reveal quantum mechanical properties.

There is evidence now in the literature to show that a variety of animal cells do respond to their nanometric environment, which could also originate from nanopatterned substratum/surfaces required to nanofabricate a wide variety of devices (bioresponsive nanomaterials) for applications in animal biotechnology and biomedicine (e.g. nanobiosensors). Although nanofabrication also falls within the realm of the present chapter dedicated to "nanotechnology and its applications to animal biotechnology," this does not constitute the major theme of this chapter. Nevertheless, for completeness we will briefly digress into this topic in subsequent sections.

While macroscopic objects can be viewed with the naked eye, magnification becomes necessary to observe microscale objects (e.g. biological specimens) with high spatial resolution (i.e. how well it is possible to distinguish the fine features of the specimen). A variety of optical/light microscopes have been developed for this purpose, but amongst other factors, the wavelength of visible light determines the resolution limit. Resolution was improved and a gain in magnification achieved by designing different types of microscopes (e.g. compound microscope, phase contrast microscope, fluorescence microscope, confocal scanning optical microscope, polarized light microscope) or by using UV radiation (shorter wavelength than visible light) at the cost of image quality. The resolution limit reached in light microscopy could be further improved by using electrons with a shorter wavelength, as in Scanning Electron Microscopy (SEM), but the requirements of ultra high vacuum and the need to use solid samples and make them conducting created severe limitations in the use of SEM for studying biological specimens without creating artifacts (e.g. due to desiccation). Additionally, SEM did not permit the study of biosystems in their native state and under physiological conditions.

Although microscopes have traditionally been tools of vital importance in the biological sciences, the real breakthrough came with the invention of Atomic Force Microscope (AFM), which is a Scanning Probe Microscopy (SPM) technique that has grown steadily since the invention of Scanning Tunneling Microscope (STM) by Binning and Rohrer, for which they won the Nobel Prize for Physics in 1986. Because AFM, as opposed to other microscopic techniques, generates images based on the measurement of tiny forces (~pN between the scanning probe/tip and the sample), it became possible to achieve molecular-level resolutions. AFM can also be performed under fluids, thus permitting samples to be imaged in near native conditions. The ability to exchange and modify the fluid during imaging further allows real-time observing/monitoring of biological processes, something that electron microscopy is still unable to offer. In the meantime, Environment Scanning Electron Microscope (ESEM) was developed, which permits the study of biological samples under low vacuum conditions. Integration of different techniques such as AFM with Raman Spectroscopy has added further impetus to nanotechnology research for applications in biotechnology and medicine because it can yield not only the morphology and nanomechanical properties of the biological specimen obtained from AFM, but also simultaneously provides chemical identification (chemical fingerprints) of the samples under investigation. Similarly, ESEM combined with Energy Dispersive Spectroscopy (EDS) can simultaneously yield information on both the morphology and elemental composition of the biological specimens under investigation.

In the following section we discuss methodologies of some relevant nanotools, nanotechniques, and nanodevices.

METHODOLOGIES

Nanotools and Nanotechniques

In the era of fast developing technology, we are currently witnessing words like nanotools and nanotechniques that are becoming more and more familiar. However, do we really know what they mean? It is obvious that these two terms are related to the word "nanotechnology," which according to the Oxford dictionary means *"the branch of technology that deals with dimensions and tolerances of less than 100 nanometers, especially the manipulation of individual atoms and molecules."*

While *nanotools* are devices, molecules, and systems that function at the nanometer scale (e.g. nanopowders used to make inks, fuel cells and batteries; nanodots employed to produce plastics, polymers, and computer memories), *nanotechniques* (which stems from the merging of the terms nanotechnology and techniques) are methods used to fabricate (e.g. top-down and bottom-up fabrication techniques), visualize, manipulate (e.g. SPM family) and to optically probe (e.g. Raman-based methods) at the nanoscale.

Top-down and *bottom-up* are two approaches used for the nanofabrication of materials. The top-down approach employs common microfabrication methods to cut, mill, and shape materials into the desired form. Examples are photolithography (i.e. use of light to transfer a photomask onto a light-sensitive coated substrate), a technique similar to the method used to make printed circuit boards routinely used in electronics, and ink-jet printing (i.e. the printer normally used to print out documents, letters, etc.). Conversely, the bottom-up approach uses the chemical and physical properties of molecules to obtain self-organized or self-assembled structures. Although nanofabrication has prevalently been used so far in solid-state physics (e.g. production of electronic components), it has also started to make an appearance in the fields of biophysics and bioengineering. Giving a detailed explanation of nanofabrication and nanotools in this chapter would be straying too far away from the main theme of this book, so no further time will be spent discussing them here.

SPM Techniques

It's appropriate to say that the SPM family is probably the pillar on which the whole nanotechnology world is built. The already well-established techniques that belong to this family are about 30 in number. Among these, the most used techniques for animal biotechnology are Atomic Force Microscopy (AFM), Chemical Force Microscopy (CFM), Fluid Force microscopy (FluidFM), and Near-field Scanning Optical Microscopy (NSOM or SNOM).

Atomic Force Microscopy (AFM): AFM is a very high-resolution (down to a fraction of one nanometer, and thus 1,000 times better than the optical diffraction limit) technique that is widely used to characterize at the nanoscale the surface of samples as different as whole cells or single molecules. In particular, in the last ten years AFM has emerged as one of the most powerful nanotechniques (Canetta and Adya, 2005; Ikai, 2010) in biology. The main reason for the huge success of AFM in biology and biotechnology lies in its ability to permit one to investigate the structure, function, properties, and interaction of biological samples in their culture media. Moreover, the extremely high resolution of the AFM allows one to observe the finest details of the samples, such as lamellopodia and cilia in cells (Deng, 2010). Another advantage of AFM is the minimal sample preparation requirements, thereby making AFM a very attractive and versatile nano-imaging technique in biotechnology. AFM images can also be processed and analyzed in order to measure at the nanoscale the surface roughness, texture, dimensions, and volumes of biological systems under study. Hence, AFM imaging does allow one not only to visualize a sample but also to measure its dimensions and determine its surface properties. Such measurements are not possible using standard light microscopes or even high-resolution electron microscopes.

We have so far talked only about the imaging capabilities of AFM. Additionally, however, AFM can also measure forces between the AFM probe (which is the heart of the AFM instrument) and the sample surface. It is therefore possible to use an AFM in force spectroscopy mode to investigate the mechanical properties of biological samples, such as adhesion, elasticity (property of a material that causes it to be restored to its original shape after applying a mechanical stress), viscosity (resistance of a fluid to deform under shear stress), and viscoelasticity (when materials change with time they respond to stress as if they are a combination of elastic solids and viscous fluids, e.g. blood is viscoelastic).

Having talked about what an AFM can do, our next priority is to understand how an AFM works. The surface of the sample is raster scanned by a probe (the so-called AFM probe) and the local interaction between the probe and the sample surface is measured (Figure 14.1). Thus, the main component (or "heart") of an AFM is the probe. The probe is called a cantilever at whose end a tip is microfabricated. The cantilever obeys the famous Hooke's law, $F = k \bullet x$, where F is the force on the cantilever, k its spring constant, and x its deflection. To measure the cantilever deflection, a laser beam is reflected off the backside of a reflective (gold- or aluminium-coated) cantilever towards a position sensitive detector (PSD, usually a four-segment photo-detector).

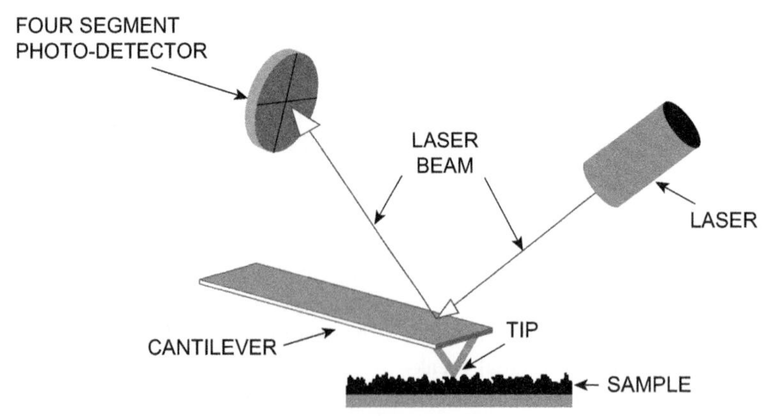

FIGURE 14.1 Operating principle of Atomic Force Microscopy (AFM).

One of the major applications of AFM in biology and biotechnology is imaging. The advantage of AFM over other high-resolution techniques such as electron microscopy is that a 3D image of the sample can be obtained. Moreover, it is possible to analyze an AFM image in order to obtain the dimensions (section analysis, i.e. length, width, and height) of the sample as well as its surface characteristics, such as surface roughness (roughness analysis) and texture.

When the AFM is used in force spectroscopy mode, then a force-distance (F-d) curve is obtained that shows the force experienced by the cantilever both when the AFM probe is brought in contact with the sample surface (*trace*, also called *approach* cycle) and separated from it (*retrace*, also called *retract* cycle) (Figure 14.2). Analyzing the F-d curves by means of theoretical models allows one to study the elastic and adhesive properties of the sample.

An AFM can also be used in force mapping mode, which combines force measurements and imaging. The output of an AFM mapping experiment is a 2D map of the elastic or adhesive properties of the sample. Figure 14.3 shows the AFM height image and the corresponding elastic map of a red blood cell.

Flow Chart 14.1 depicts the protocol for AFM imaging, AFM force spectroscopy (AFM-FS), and AFM force mapping (AFM-FM) measurements; AFM data analysis for simple AFM experiments is shown on the next page.

Chemical Force Microscopy (CFM): CFM (Noy, 2006) is a variation of AFM where the surface of the sample and the surface of the AFM probe are deliberately functionalized (see the section: "Chemical Modification of AFM Probes") in order to obtain a well-defined tip-sample interface, and therefore to measure well-defined chemical interactions.

The main difference between AFM and CFM is that with AFM the morphology of the sample is visualized by exploiting the "natural" van der Waals forces between the non-functionalized AFM tip and the sample surface, whereas CFM uses the "well-defined" interactions between

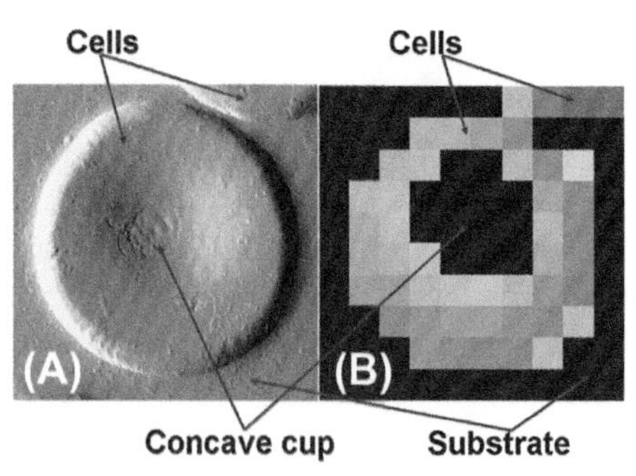

FIGURE 14.3 AFM force mapping on an individual red blood cell: (A) AFM height image of the cell on a glass slide. Note that the typical biconcave shape is clearly visible. (B) AFM elastic map of the cell shown in (a). The difference in elasticity of the glass slide (substrate) and the cell surface allows one to get an elastic map of the cell.

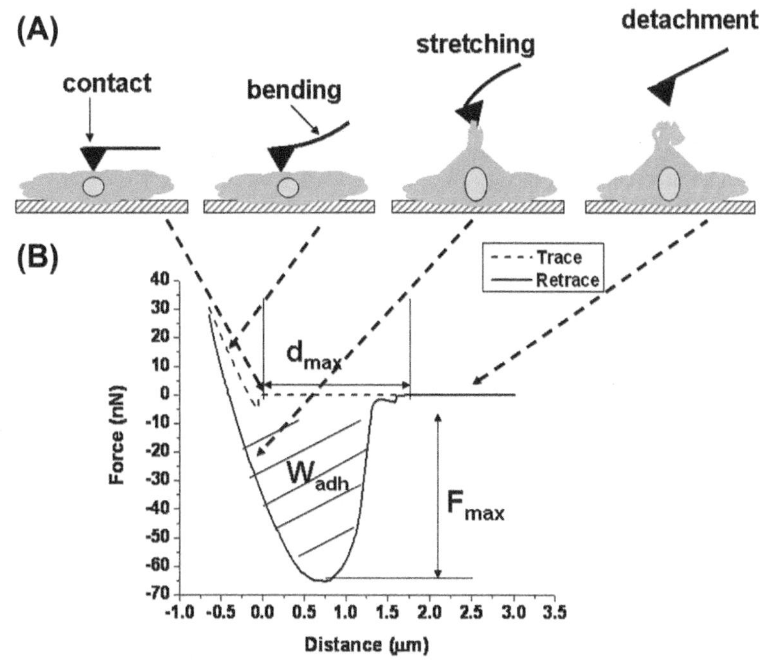

FIGURE 14.2 Sketch of the stages of an AFM force spectroscopy (AFM-FS) experiment: (A) Contact between the AFM probe and the sample surface, and cantilever bending with AFM probe indentation occurs during the trace (approach) cycle of the AFM experiment. On the retrace (retract) cycle, the sample surface is stretched until the point of detachment. (B) An example of a typical experimental *F-d* curve for a trace and retrace cycle is also presented; the stages are indicated by the arrows. The meanings of F_{max} (maximum adhesion force), d_{max} (contact point), and W_{adh} (work or energy of adhesion) are also identified. *(Adapted from Canetta and Adya, 2011.)*

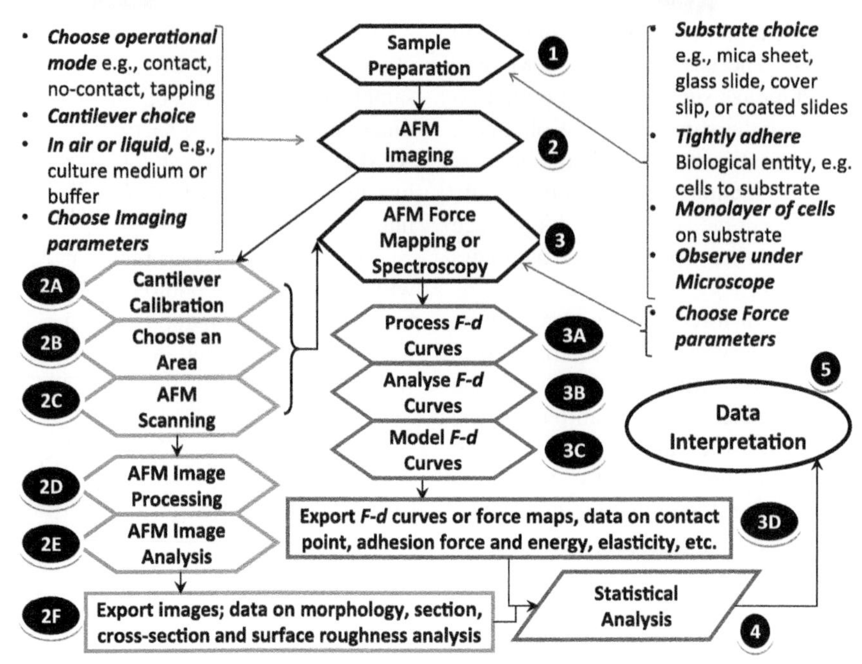

FLOW CHART 14.1 Protocol for AFM imaging, AFM force measurements, and data analysis.

the functionalized AFM tip and the sample surface (which can also be functionalized, if needed).

Fluid Force Microscopy (FluidFM): A very recent (2009) implementation of an AFM system is the so-called FluidFM (Doering, 2010). This advanced AFM combines the unique features and capabilities of nanofluidics with the high precision in positioning and the extremely high sensitivity to weak forces of an AFM (Meister, 2009). Now, the introduction of a new concept: nanofluidics. But what does nanofluidics mean, and what is it used for? Nanofluidics is the study of the behaviors and properties of fluids that are confined in a nanoscale size environment (for a review on nanofluidics, see Kirby, 2009). Because of the enhanced possibilities created by nanofluidics to control transport phenomena with unprecedented high precision, nanofluidic-based devices have been developed for cell sorting and analytical separations of DNA and proteins. Although nanofluidics is an extremely interesting and important field of research, it will not be discussed any further here in order to keep on course with the main goals of this book. Notwithstanding, it needs to be pointed out that the FluidFM system, resulting from a "marriage" of nanofluidics and AFM, has been designed to do the following: (1) "dispense and deliver" inside a living cell active agents from a solution; (2) "inject" nanoparticles and other nanoscale-size materials (e.g. fluorescein isothiocyanate, FTIC dye) inside a living cell by perforating the cell wall, and "extract" genetic material (e.g. DNA) from the cell nucleus and other cell compartments to be further manipulated; and (3) "pick and place" individual cells to move them from one place to another. This third task entails (for example) putting two cells in physical contact to investigate the phenomenon of cell–cell interaction, which can give invaluable information on how a cell adheres to another cell. This is extremely important information that can be used to better understand the phenomenon of cellular extravasation, which occurs during cancer metastasis, for example.

Near-field Scanning Optical Microscopy (NSOM or SNOM): The advantage of SNOM is that it breaks the far-field diffraction limit and uses the properties of evanescent waves (i.e. stationary waves whose intensity decays exponentially as a function of the distance from the boundary at which the wave was formed) to allow one to visualize nanoscopic structures, such as cells and biomolecules. First of all, let us understand the meaning of diffraction limit and also what it means to break it. It is well known that the resolution of a microscope is limited by the quality of the lenses and also by their alignment or misalignment. However, there is another element that affects the quality of the images obtained with a microscope and this is caused by the diffraction of the light. If a microscope can produce images with an angular resolution that is as good as the theoretical diffraction limit ($d = \lambda/2n \sin\theta$, where d is the size of a feature of the specimen that the microscope can resolve, λ is the wavelength of the light incident on the sample, n is the refractive index of the medium in which the sample is imaged, and θ is the half-angle subtended by the objective used) of the microscope itself, then the instrument is said to be *"diffraction limited."* Hence, the resolution limit (i.e. the distance of the two nearest points) of an image is $\lambda/2$. SNOM can help increase this resolution to a lateral resolution of ~20 nm and a vertical resolution of ~2–5 nm, thereby allowing the finest details of biological objects as small as biomolecules to be visualized. Moreover, SNOM is an extremely "gentle" technique because it uses a narrow light beam coming from a fiber tip to "probe" the specimen; therefore, there is no physical contact between the probe and the sample.

Although SNOM is a very powerful and non-destructive microscopy technique, its use is not as broad as that of AFM because of the many artifact issues (mainly due to tip breakage while scanning) as well as the very small working distance, the extremely shallow depth of field, and the long scan times required for large sample areas.

Raman Spectroscopy and Imaging

Raman spectroscopy (RS), and more recently, Raman imaging (RI), are extremely powerful techniques for investigating the unique "chemical fingerprints" of solid and liquid samples. RS is a vibrational spectroscopy method (Gardiner, 1989) chiefly employed to investigate the vibrational, rotational, and other low-frequency modes in a specimen. In the last 25 years RS has been widely used to probe bonds in molecules, to provide characteristic chemical information about the effects of drugs on the cell biochemistry, to understand the mechanisms of cell death and differentiation, and to discriminate between different types of cells, the most common example being the use of RS (Gremlich, 2001) to distinguish normal and cancer cells. When a sample is shined with light (i.e. a laser) of a certain wavelength, the light interacts with the sample molecules and it gets either absorbed or scattered. Most of the light will be scattered elastically (Rayleigh scattering), namely, no wavelength change occurs, and only 1 photon (quantum of light) in 10^5–10^7 photons will be scattered inelastically, thus resulting in a wavelength change (Raman scattering) (Ferraro, 2003). This change in the wavelength of the photon scattered by the sample's molecule is called "Raman shift" and it is characteristic of the molecules the incident light interacts with. Another important point of Raman scattering is that the interaction of the incident photon with the molecule of the specimen under study results in an energy exchange, and thus the energy of the scattered photon can be higher or lower than the energy of the incident photon. This change in the photon energy is directly linked to the change in the rotational or vibrational energies of the molecule with which the photon has interacted. Because these vibrational or rotational energies are specific to a chemical bond or chemical constituent of the molecule with which the incident photon has interacted, it can provide a unique "signature" of that molecule. Please remember that only 1 in 10^5–10^7 photons will be a Raman photon (the one scattered inelastically). Therefore, the intensity of the Raman scattering is significantly (~10^{-5}–10^{-7} times) weaker than that of the Rayleigh scattering. What this means is that it is not always easy to distinguish the Raman peaks from the so-called "background" generated by the environment surrounding the sample. The background could be caused, for example, by the culture medium in which the cells are kept while investigating the sample with the incident laser beam of a Raman system. To minimize such background and see more clearly the Raman peaks (which are chemical fingerprints of the sample), near-infrared laser (e.g. 785 nm) should be used with enough acquisition time (on the order of minutes) in order to produce sufficient Raman-scattering photons and intense Raman peaks.

A typical Raman setup is built in a way that a laser beam generated by a narrowband laser (i.e. a laser with a linewidth < 1 nm) is expanded through a telescope formed by two lenses to completely fill the back-aperture of the objective lens, reflected from a 45° holographic notch filter (HNF) to remove the excitation laser light, and focused on the sample by the objective lens (usually 50× or 100×). The Raman scatter from the sample is collected by the same microscope objective and filtered by the same HNF. It is then guided towards the spectrometer and acquired via a liquid nitrogen or thermally cooled CCD camera. For Raman imaging, a similar setup is used with the only difference being that the stage of the microscope is moved in the x and y directions by a motor that is synchronized with the spectroscopy CCD camera. In fact, in Raman imaging, also known as hyperspectral or chemical imaging, hundreds to thousands of Raman spectra are acquired from all over the specimen and a chemical map recreated that shows the location and amount of different components of the sample. Raman imaging finds its major applications in the study of tissues and individual cells (Figure 14.4).

The main advantages of Raman spectroscopy and Raman imaging are the very little sample preparation required and the "gentleness" of the "touch of light" on the sample. This makes RS a truly non-destructive and non-invasive technique. In addition, the ability of RS to work both *in situ* and *in vivo*, an essential factor when dealing with biological samples, has recently made this technique in great demand for animal biotechnology and life sciences. Moreover, RS provides unique information about the molecular identity of the specimen: the "chemical fingerprints," the 3D structural changes in orientation and conformation of biomolecules (e.g. collagen (Bonifacio, 2010)), the intermolecular interactions, and the dynamics of biological phenomena (e.g. cell death (Okada, 2012)). The main disadvantages of Raman spectroscopy that need to be kept in mind are, however, the very small percentage of Raman scattered photons compared to the elastically scattered photons and the strong fluorescence background. The cumulative effect of these two detriments can sometimes almost completely hide the Raman peaks. To overcome these problems, different types of Raman systems have been developed. The most broadly

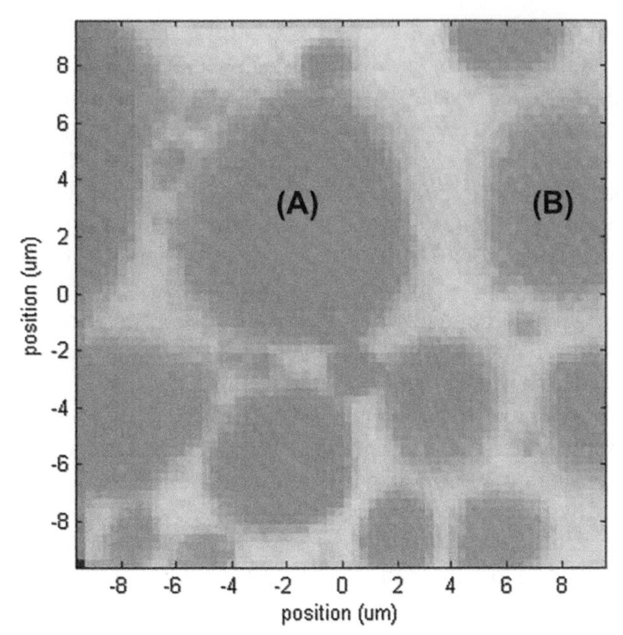

FIGURE 14.4 Raman image of a cluster of lipids in an adipocyte. The light blue color indicates the substrate, whereas the purple "blobs" are the lipid droplets. The different shadows of purple are related to the different amount of lipids inside the lipid droplets. For example, the lipid droplet labeled with (A) is richer in lipids (purple color is brighter) and of a quite homogenous composition (the intensity of the purple color is almost the same everywhere), while the lipid droplet labeled with (B) is less rich in lipids (purple color is faded) and less homogenous (different shades of purple are visible). This image is composed of 4,096 Raman spectra, making Raman imaging a quicker method to obtain large data sets compared to Raman spectroscopy.

used for biological applications are Confocal Raman Spectroscopy (CRS), Modulated Raman Spectroscopy (MRS), Surface-Enhanced Raman Spectroscopy (SERS), and Tip-Enhanced Raman Spectroscopy (TERS).

Confocal Raman Spectroscopy (CRS): CRS has the advantage over standard RS of allowing specific spatial analysis deep within transparent samples, such as living cells, by focusing the laser beam at the exact point of interest. CRS offers high depth profiling and it greatly improves the rejection of fluorescence background. The high spatial resolution and precision in laser focusing can be easily achieved by positioning a pinhole of a few microns aperture in the back focal plane of the objective lens. The pinhole will allow only the contribution from the Raman photons to go towards the spectrometer while rejecting contributions from the background.

Modulated Raman Spectroscopy (MRS): MRS (De Luca, 2010, Canetta et al., 2011) is a technique recently developed that is able to filter out the Raman spectra from the fluorescence background. The basic principle of MRS is very simple. When the laser wavelength of the excitation laser is continuously modulated, a wavelength shift of the Raman peaks occurs while the fluorescence background remains static, thereby permitting one to observe even very weak Raman features generally hidden by the fluorescence. Moreover, MRS reduces the necessary spectra accumulation times and allows real-time applications that are crucial in life science and biology.

Surface-Enhanced Raman Spectroscopy (SERS): SERS is an extremely powerful variation of standard Raman spectroscopy (RS) in which the Raman signal of the molecules is amplified by up to 10^{10}–10^{11} times by absorbing the molecules under study on rough metal surfaces (Haran, 2010), thereby allowing easy detection of single molecules and hence making this technique an essential tool in single-molecule spectroscopy.

Tip-Enhanced Raman Spectroscopy (TERS): Similar to SERS, Tip-Enhanced Raman Spectroscopy (TERS) also takes advantage of the plasmon resonance effects of the surface on which the single molecules under study are absorbed. TERS, in fact, goes even further than SERS because it uses an SPM probe tip as the enhancing surface, thereby permitting Raman scattering to occur only from a nanoscale area. TERS, therefore, allows chemical analysis and imaging at the *nanoscale*, which holds huge promise in molecular biology (Elfick, 2010).

AFM–Raman Confocal Hybrid Systems

Recently, AFM and Raman systems have been combined (integrated) together in the so-called AFM–Raman hybrid. These instruments are extremely versatile because they allow one to perform AFM and Raman experiments simultaneously on the same sample. The AFM–Raman hybrid permits researchers to perform nanoscale visualization, manipulation, and determination of mechanical properties by using AFM while allowing chemical analysis (identification) at the sub-micron scale by performing Raman spectroscopy of specimens at the same time and at the same location. Additionally, the configuration of the hybrid instrument allows one to collect TERS data and conduct SNOM experiments using just one instrument. This is extremely important in biology where the heterogeneity of the samples (e.g. cell populations) requires different experiments to be performed on the same position (e.g. the same cell) to obtain reliable and conclusive results. For example, the use of an AFM–Raman system has the potential to markedly improve the predictability of cell screening. In fact, AFM is a surface characterization technique that allows one to investigate the morphological and mechanical properties of cells and/or proteins expressed at the surface. Please remember that AFM cannot provide any information about the features of the proteins, lipids, and DNA "hidden" underneath the cell surface. Complementarity of AFM and Raman comes into play at this point because while Raman allows one to obtain information about the biochemical properties of the proteins, lipids, and DNA that lie underneath the cell membrane and inside the cell nucleus, the AFM provides surface characteristics, such as the morphology and mechanical properties of the same cell, through the AFM–Raman hybrid system.

Chemical Modification of AFM Probes

AFM is widely used to visualize at the nanoscale and to measure the nanomechanical properties of biological systems (see the next two sections). To perform these measurements, standard and unmodified AFM tips are used. Recently, however, new types of experiments (Noy, 2006, Barratin, 2008) have become possible via chemical functionalization of the AFM probes. Chemical modification of AFM tips allows one to study so-called "molecular recognition events" at a very high level of sensitivity. To make a complicated story simple, if the goal is to investigate what happens on a cell surface at the single-molecule level, then the AFM probe needs to be functionalized by "attaching" to it either ligand or receptor molecules that will bind specifically only to certain types of cell surface molecules. The major problem encountered when chemically modifying the tip is that it is not possible to control precisely its functionalization, and this "randomness" can affect the measurement of single molecule interactions.

The main applications of chemically modified AFM probes concern molecular recognition force measurements of cell–surface interactions (Li, 2011), which are extremely important in cell adhesion, cell migration, cell development, cell–cell communication and recognition, and ligand–cell membrane protein interactions; these are essential to gain a deeper knowledge of the molecular dynamics of biological processes.

Nano-Structural Features of Animal Cells and Tissues

The most straightforward application of AFM is imaging. Additionally, however, the capability of AFM to work in liquid medium allows the opportunity to image live cells in their natural physiological conditions and under controlled culture systems (e.g. keeping the cells at 37°C in a CO_2 environment and allowing continuous renewal of the culture medium via perfusion), thereby facilitating the imaging of cells in real-time for long periods of time. Furthermore, AFM images can be captured along with bright contrast, fluorescence, and confocal images with the final goal of obtaining as much information as possible from the same sample area. Recently, the combination of AFM and Raman imaging (see the section: "AFM–Raman Confocal Hybrid Systems") has allowed researchers to collect AFM images at the nanoscale and Raman chemical images on the same spot of the sample under study, thereby permitting one to obtain a complete "map" of the nanostructural and biochemical properties of the specimen.

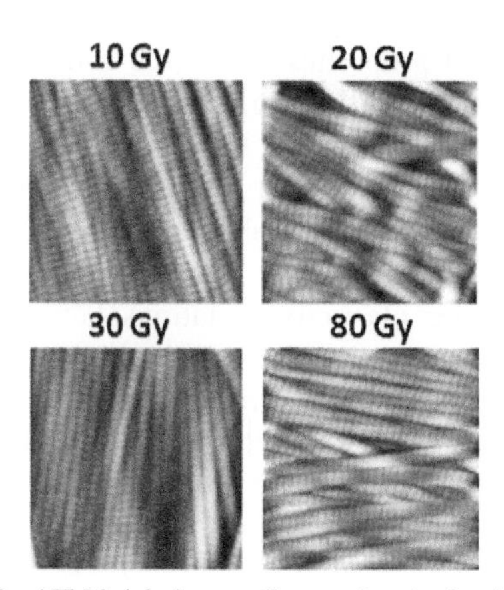

FIGURE 14.5 AFM height images (2 μm × 2 μm) of pericardial tissues exposed to 10 Gy radiation (height = 0–470 nm); 20 Gy radiation (height = 0–510 nm); 30 Gy radiation (height = 0–535 nm); and 80 Gy radiation (height = 0–650 nm).

We give here a few examples of the type of information that AFM imaging can render. AFM has been used to investigate the effects of ionizing radiation on the nanostructures of pericardium tissues (Canetta et al., 2011) with the final goal to better understand what should be the maximum cardiac doses that a breast cancer patient should receive in breast radiotherapy (Figure 14.5). Cross-sectional analysis of the AFM images showed that ionizing radiation caused an increment in the diameter of the collagen fibers that compose the pericardial tissue. Moreover, the number of fibrils affected and the extent of swelling were found to increase with the radiation dose. Another example of the use of AFM in biotechnology is stem cell profiling. Undifferentiated and differentiated stem cells were visualised (Kiss, 2011) by AFM.

AFM images showed differences in the cytoskeleton organization of undifferentiated and differentiated stem cells. Cross-section and roughness analyses of the AFM images have shown that undifferentiated cells are rounder and smaller in size, and have a smoother surface as compared to differentiated cells.

Nanomechanical Properties of Animal Cells and Tissues

AFM is widely used to investigate the nanomechanics of cells and tissues in physiological conditions. Knowing the differences in the elastic and adhesive properties of cells can be of great help in cell profiling. This is particularly true for stem cells where an appropriate differentiation protocol is still unavailable. AFM force spectroscopy was used to distinguish between undifferentiated and differentiated stem cells (Kiss, 2011) on the basis of the elasticity of their cell walls. Consistency of the mechanical properties across the cell wall of undifferentiated stem cells suggested their homogeneous nature. In contrast, typical *F-d* curves recorded at different positions on the cell wall of differentiated cells revealed different mechanical features, thereby indicating that differentiated cells have a heterogeneous nature. The observed higher elasticity variations between different cells than those on individual cells were interpreted to be arising due to significant structural differences between undifferentiated and differentiated cells. Undifferentiated cells were found to be more rigid with a higher Young's modulus (E) as compared to differentiated cells, which exhibited a lower E value.

Nanomanipulation

One of the more advanced uses of an AFM is nanomanipulation. AFM can be employed as a nanomanipulator, for example, to dissect (Rubio-Serra, 2005) metaphase chromosomes. Chromosomal dissection allows researchers to separate DNA from areas of the cell that have been cytogenetically identified in order to create genetic probes (Di Bucchianico, 2011). AFM can also be used as a low invasive cell manipulation and gene transfer system to insert pieces of DNA inside single living cells.

Nanofabrication

A variety of nanotechnology tools can be used for the chemical synthesis of nanoscale structures, nanolithography, nanoscale surface patterning, synthesis and characterization of nano materials (e.g. magnetic nanoparticles for drug delivery and cell destruction, low and high adhesion surfaces for biomedical uses such as stents, carbon nanotubes, and fluorescent quantum dots). They offer a wide range of applications, such as in drug synthesis and delivery, cell therapy and cell surgery, early disease diagnosis and prevention, medical implants, biosensors, and nanoanalyzers. It is impossible to fathom the enormity of information required to cover all these aspects in a chapter of this size. We can at best briefly introduce this topic here by including a few references for interested readers, and by giving an example or two in the next section. Different nanotools can be used; for example, nanofabrication by using scanning probe microscopy (SPM) techniques that have the additional advantage of simultaneous nanoscale imaging, surface modifications and nanolithography by using AFM, and soft lithography with self assembled monolayers. Falconnet et al. (2006) reviewed a number of techniques for cell patterning to study cellular developments during cell culture assays that provide new insights into the factors that control cell adhesion, cell proliferation, differentiation, and molecular signaling pathways.

EXAMPLES OF NANOTECHNOLOGY APPLICATIONS TO ANIMAL BIOTECHNOLOGY

This section presents some specific examples from the literature and discusses, in general, the applications of nanotechnology to animal biotechnology and biomedicine.

AFM as a Diagnostic Tool to Identify Orthopoxvirus in Animals

During the 2005 natural outbreaks of Vaccinia virus (VACV) in dairy cattle in Brazil, samples of vesicles and crusts (dried scabs) from cattle and milkers' hands were collected and subjected to further analysis (Trindade, 2007). The purified samples used were intracellular mature virus (IMV) particles, which are the predominant viral forms produced during infection. The AFM nanotechnique was employed for viral identification and characterization from clinical samples and purified viruses. The results showed that AFM can provide a rapid and biosecure tool for the diagnosis of emerging orthopoxviruses with a potential for screening bioterrorism samples. In order to minimize the risk of spread of viral diseases and to adopt effective treatment strategies, rapid detection and identification are important.

Frictional Response of Bovine Articular Cartilage

The frictional response of bovine articular cartilage was investigated at the nanoscale by comparing micro- and macroscale friction coefficients of immature bovine articular cartilage (Park, 2004). Twenty-four cylindrical osteochondral plugs in pairs from adjacent locations in six fresh 4–6 month old bovine humeral heads were harvested and divided into two groups for AFM (in physiological PBS medium) and macroscopic friction measurements. Surface roughness was acquired from surface topography, and elasticity modulus from indentation experiments with AFM. The AFM friction coefficient was found to be more representative of the equilibrium friction coefficient reported at the macroscale, which represents the frictional response in the absence of cartilage interstitial fluid pressurization. These results suggest that AFM friction measurements may be highly suited to providing greater insight into the mechanism by exploring the role of boundary lubricants in diarthrodial joint lubrication independently of fluid pressurization.

Microstructure and Nanomechanical Properties of Cortical Bone Osteons from Baboons

Recently, the effects of tissue and animal age on mechanical properties, such as elasticity modulus, stiffness, hardness, and composition, through nano-indentation and Raman experiments on bone tissues of baboons (Burket, 2011) was examined. The results demonstrated that composition and mechanical function are closely related and influenced by tissue age and animal age. Because tissue composition crucially determines the mechanical function, understanding these relationships will enhance the knowledge of normal and pathological bone function and enable the improvement of current therapies for skeletal diseases such as osteoporosis, which is known to alter the nanoscale heterogeneity of the material properties of bone tissue. Understanding aging-related material property changes at the tissue level is thus essential to predicting bone fragility, skeletal mechanical integrity, and age-related fracture.

Use of Calf Thymus DNA for Cancer Experiments

In 2011, a study of the effect of drug bioavailability of synthesized natural coumarin (SC) and nano-coumarin (NC) having potential drug value (by using DNA from calf thymus as the cellular target of therapeutic molecules) was carried out by using AFM, SEM, and a variety of other techniques and bioassays (Bhattacharyya, 2011). NC demonstrated greater efficiency of drug uptake and anti-cancer potential in melanoma cell line A375, as revealed by SEM and AFM. Nanoparticles (NP), now being increasingly used in bio-applications like therapeutics, anti-microbial agents, transfection vectors, and fluorescent labels, should be harmless to target organisms and organ systems. SC and NC showed negligible cytotoxic effects on normal skin cells and peripheral blood mononuclear cells of mice. Nanoparticles were found distributed in different tissues in mice, like heart, kidneys, liver, lungs, spleen, and brain. NC in mice brains crossed through the blood–brain barrier, but SC failed to cross the barrier, further suggesting that nano-coumarin was more active and potent than the synthetic coumarin in suppressing the expression of p53-regulated expression of Cyclin D1, survivin, Stat-3. This may have been because of the enhanced cellular uptake of the encapsulated nanoform of coumarin. These results imply that nano-coumarin may be superior to coumarin as an anti-tumor, anti-invasive, and anticarcinogenic agent. This in turn suggests it as a likely candidate for chemo-preventive drug design, because of its small size, more rapid entry into target cells, and biodegradable nature.

Characterization of Mitochondria Isolated from Normal and Ischemic Hearts in Rats

AFM has been used to observe the morphological property changes in heart mitochondria isolated from a rat myocardial infarction model (Lee, 2011). The shape parameters from AFM topography images revealed that myocardial infarction caused the swelling of mitochondria. The biophysical properties determined from AFM force spectroscopy experiments showed that the adhesion force of heart mitochondria significantly decreased by myocardial infarction and ischemic stimuli possibly induced the stiffening of mitochondria. The opening of channels on sufficient swelling can rupture the outer mitochondrial membrane and cause the release of cytochrome c, which leads to apoptotic cell death.

Polymorphism and Ultrastructural Organization of Prion Protein

Atomic force microscopy was employed (Anderson, 2006) to analyze the ultrastructure of amyloid fibrils produced from the full-length mouse prion protein (PrP). The fibrils displayed unprecedented variations in their morphologies. The results revealed extremely broad polymorphism in fibrils generated *in vitro*, which is reminiscent of high morphological diversity of scrapie-associated fibrils isolated from scrapie brains, suggesting polymorphism is peculiar for polymerization of PrP, independent of whether fibrils are formed *in vitro* or under pathological conditions *in vivo*. Prion diseases are a group of fatal neurodegenerative brain disorders found in both animals and humans, and all of them are related to misfolding and polymerization of PrP.

Ultrastructural Investigation of Animal Spermatozoa Using AFM

While sperm morphology is considered to be a prognostic factor for fertilization and pregnancy, abnormal morphology is linked to male infertility. This is why morphological changes are targeted for developing contraceptives. Maturation and capacitation are some of the post-testicular processes that mammalian spermatozoa must undergo before becoming fully competent to fertilize an egg. During their passage from the testis to the site of fertilization, mammalian spermatozoa encounter a range of fluids of very different origins and compositions that greatly influence post-testicular processes. Morphological changes in the surface structure of bovine sperm heads during acrosome reactions have been reported (Saeki, 2005) by using AFM and SEM. AFM was combined with other techniques to study the supramolecular organization of the sperm plasma membrane during maturation and capacitation of bovine, porcine, and ovine spermatozoa (Jones, 2007). For a review of AFM imaging of spermatozoa, the reader can refer to Kumar et al. (2005). AFM has also been used to study the changes in plasma membranes overlying the heads of bull, boar, goat, ram, mouse, stallion, and monkey spermatozoa during post-testicular development, after ejaculation, and after exocytosis of the acrosomal vesicle (Ellis, 2002). To study the effect of prostegerone on sperm function, we

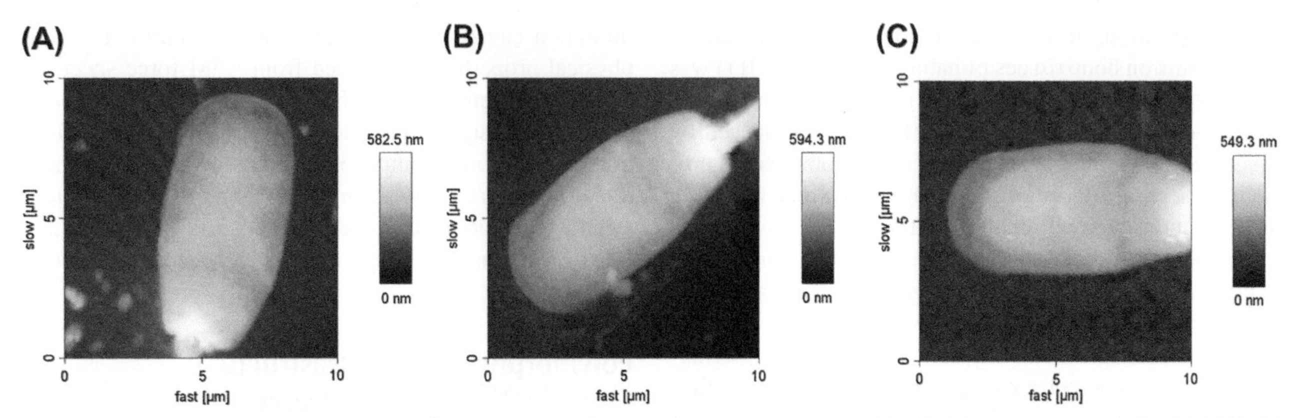

FIGURE 14.6 AFM height images (10 × 10 µm^2) of boar sperm cells (A) without prostegerone (B) with 100 nM prostegerone, and (C) with 1,000 nM prostegerone. *(Adya (2012), unpublished work.)*

recently investigated (Adya, 2012) the morphological and nanomechanical properties of boar sperm cells at different concentrations of prostegerone. Figure 14.6 shows some of the AFM height images (unpublished results) of boar sperm cells on a scale of 10 × 10 µm^2.

Multifactor Analysis of Living Animal Cells for Biotechnology and Medicine

Advances in AFM-based nanotechniques for investigating animal cells, with possible applications in animal biotechnology and medicine, have further expanded and opened new avenues for applications of nanotechnology. A particular emphasis has been to study living cells under physiological environments, to assess cell morphology and the effects of environmental and age-related stresses, to determine cell–substrate adhesion and cell–cell aggregation, to investigate the nanomechanical/biophysical properties (e.g. stiffness and elasticity) of the cell membrane, to map the cell surface by using chemically modified cantilevers, to analyze the distribution of molecular/chemical components inside the cell by using micro- and nano-surgical approaches, and to combine AFM with other techniques such as optical and confocal microscopy, and Raman spectroscopy. Most of these applications have already been discussed in the first "Methodologies" section. Raman et al. (2011) reviewed some of these.

ETHICAL ISSUES

In addition to the examples and applications given earlier, a variety of nanoparticles in different formats (e.g. nanoparticles embedded in a matrix, functionalized nanoparticles, PEG-modified iron oxide nanoparticles, polysaccharide-modified iron oxide nanoparticles, polymer-coated nanoparticles, nanostructured sol–gel silica–dopamine reservoirs, magnetic drug nanoparticles, nanoscale oil bodies, surface-enhanced Raman scattering nanoparticles, silica or iron oxide-based magnetic nanoparticles), shapes (e.g. nanorods, nanoshells, nanotubes, nanospheres, nanocages), and sizes are currently being used and further developed for targeted drug delivery with the final goal to cure cancer and other lethal diseases in mammals. Existing research has clearly demonstrated the feasibility of introducing nanoshells and nanotubes into animal systems to seek out and destroy targeted cells. Nanoparticles smaller than one micron have been used to deliver drugs and genes into animal cells. Such nanoparticles can be toxic and/or carcinogenic, and thus hazardous to mammalian health and the environment.

One major problem facing governments and many industries around the globe is the lack of information available about health hazards due to the use of different materials for the production of nanoparticles. Some national and international agencies are working continuously to check and improve new techniques for testing toxicity and hazard limits of nanoparticles before their production in companies. Nanoparticle production companies should follow the regulations, which need to be continuously updated, and check all the hazardous side effects on the environment.

Apart from the risk and environmental and health concerns, there are other ethical issues involved with nanotechnology that need to be addressed. Several international bodies are already involved in this. Scientists are now faced with ethical limits in deciding what is acceptable to do. This is true with any new developing technology. As one example of this, researchers involved in the production and testing of "nanoscale oil bodies" as oil-based carriers for antitumor drug delivery had to carry out *in vivo* studies on mice to produce a good drug as well as good therapy for the cancer. Tumor creation in mice therefore became an ethical issue for them.

TRANSLATIONAL SIGNIFICANCE

Nanotechnology itself is a translational science. Due to its multidisciplinary nature and wide-ranging scope (as discussed in earlier sections of this chapter), it drives the agenda for translational research, a "basic-to-applied" or "bench-to-bedside" approach to "translate" fundamental research results into practical applications, such as the development of new treatments for diseases in everyday practice. Scientific discoveries begin at the "bench" with basic research in which scientists study disease at a molecular or cellular level and then progress to the clinical level for the animals' wellbeing and the improvement of human health.

WORLD WIDE WEB RESOURCES

R. Feynman's lecture "There is plenty of room at the bottom"

http://www.zyvex.com/nanotech/feynman.html

This is the transcript of the famous lecture that Professor Feynman gave on December 19, 1959 at the California Institute of Technology in which he formulated the basis of nanotechnology.

Principles of Atomic Force Microscopy

http://www.weizmann.ac.il/Chemical_Research_Support/surflab/peter/afmworks/

This is an easy-to-understand introduction to the principles of AFM.

Atomic Force Microscopy (Interactive tutorial)

http://medicine.tamhsc.edu/basic-sciences/sbtm/afm/principles.php

This website gives a comprehensive introduction to the main principles of AFM, and also allows the reader to "see" how an AFM works.

Applications of Nanotechnology to Animal Biotechnology

http://www.veterinaryworld.org/Vol.2/December/Nanotechnology and its applications in Veterinary and Animal.pdf

This is a good review of the applications of nanotechnology to livestock, biotechnology, and veterinary medicine.

Nanowerk

http://www.nanowerk.com

The Nanowerk Nanotechnology Portal is one of the most comprehensive nanotechnology and nanoscience resources. It includes a nanomaterial database (nanoBASE), an exhaustive "Introduction to Nanotechnology," daily news coverage and spotlight articles (nanoNEWS), and several links related to nanotechnology (nanoLINK).

Dedicated Nanotechnology Website

http://nanotechweb.org

This is another very comprehensive website with the most up-to-date nanotechnology discoveries and news. It gives ample information on the business side of nanotechnology, and the most recent nanotechnology events, showcases, and publications. It also has a good blog and a comprehensive list of useful links to journals, research centers, professional societies, government initiatives, books, etc.

The Institute of Nanotechnology (IoN)

http://www.nano.org.uk

This site provides deep insight on the latest research activities and events related to nanotechnology in the UK.

The European Nanotechnology Gateway

http://www.ece.umn.edu/users/hjacobs/publications/Nanoforum-European Nanotechnology Gateway.htm

This site gives an exhaustive overview of the main ongoing research activities in Europe.

Understanding Nanotechnology

http://www.understandingnano.com

This is a basic, introductory website dedicated to nanotechnology. It is the most appropriate website for beginners.

ACKNOWLEDGMENTS

We are grateful to our respective institutions for providing the necessary resources to help us complete this work. One of us (AKA) gratefully acknowledges the kind hospitality extended to him by Amity University, and the sponsorship of the Royal Society of Edinburgh (RSE, UK) and the Indian National Science Academy (INSA) to facilitate his visit to India during the summer of 2011 when this work really took off. AKA is also grateful to Amity University for the kind offer of an honorary Adjunct Professorship.

REFERENCES

Adya AK (2012). Private communication (unpublished results).

Anderson, M., Bocharova, O. V., Makarava, N., Breydo, L., Salnikov, V. V., & Baskakov, I. V. (2006). Polymorphism and ultrastructural organization of Prion protein amyloid fibrils: An insight from high resolution atomic force microscopy. *Journal of Molecular Biology*, *358*, 580–596.

Barratin, R., & Voyer, N. (2008). Chemical modifications of AFM tips for the study of molecular recognition events. *Chemical Communications*, *13*, 1513–1532.

Bhattacharyya, S. S., Paul, S., De, A., Das, D., Samadder, A., Boujedaini, N., & Khuda-Bukhsh, A. R. (2011). Poly (lactide-co-glycolide) acid nanoencapsulation of a synthetic coumarin: Cytotoxicity and biodistribution in mice, in cancer cell line and interaction with calf thymus DNA as target. *Toxicology & Applied Pharmacology*, *253*, 270–281.

Binning, G., Quate, C. F., & Gerber, C. (1986). Atomic force microscopy. *Physical Review Letters*, *56*, 930–933.

Bonifacio, A., Beleites, C., Vittur, F., Marsich, E., Semeraro, S., Paoletti, S., & Sergo, V. (2010). Chemical imaging of articular cartilage sections with Raman mapping, employing uni- and multi-variate methods for data analysis. *Analyst*, *135*, 3193–3204.

Burket, J., Gourion-Arsiquaud, S., Havill, L. M., Baker, S. P., Boskey, A. L., & vanderMeulen, M. C. H. (2011). Microstructure and nanomechanical properties in osteons relate to tissue and animal age. *Journal of Biomechanics*, *44*, 277–284.

Canetta, E., & Adya, A. K. (2005). Atomic Force Microscopy: Applications to Nanobiotechnology. *Journal of Indian Chemical Society*, *82*, 1147–1172.

Canetta, E., & Adya, A. K. (2011). Nano-imaging and its applications to biomedicine, Proceedings of the International Conference on Image Analyses and Processing (ICIAP 2011). In Prof. Giuseppe Maino (Ed.), *Lecture Notes in Computer Science Series (LNCS)* (pp. 1–10). Springer.

Canetta, E., Mazilu, M., De Luca, A. C., Carruthers, A., Dholakia, K., Neilson, S., Sargeant, H., Briscoe, T., Herrington, C. S., & Riches, A. (2011). Modulated Raman Spectroscopy for enhanced identification of bladder tumour cells in urine samples. *Journal of Biomedical Optics*, *16*(3). 037002 (7 pages).

De Luca, A. C., Mazilu, M., Riches, A. C., Herrington, C. S., & Dholakia, K. (2010). Online fluorescence suppression in modulated Raman spectroscopy. *Analytical Chemistry*, *82*(2), 738–745.

Deng, Z., Lulevich, V., Liu, F., & Liu, G. (2010). Applications of Atomic Force Microscopy in Biophysical Chemistry of Cells. *The Journal of Physical Chemistry. B*, *114*, 5971–5982.

Di Bucchianico, S., Poma, A. M., Giardi, M. F., Di Leandro, L., Valle, F., Biscarini, F., & Botti, D. (2011). Atomic force microscopy nanolithography on chromosomes to generate single-cell genetic probes. *Journal of Nanobiotechnology*, *9*, 27. (7 pages).

Doering, P., Stiefel, P., Beher, P., Sarajilic, E., Bijl, D., Gabi, M., Voros, J., Vorholt, J. A., & Zambelli, T. (2010). Force-controlled spatial manipulation of viable mammalian cells and micro-organisms by means of FluidFM technology. *Applied Physics Letters*, *97*. 023701 (3 pages).

Elfick, A. D., Downes, A. R., & Mouras, R. (2010). Development of tip-enhanced optical spectroscopy for biological applications: A review. *Analytical and Bioanalytical Chemistry*, *396*, 45–52.

Ellis, D. J., Shadan, S., James, P. S., Henderson, R. M., Edwardson, J. M., Hutchings, A., & Jones, R. (2002). Post-testicular development of a novel membrane substructure within the equatorial segment of ram, bull, boar, and goat spermatozoa as viewed by atomic force microscopy. *Journal of Structural Biology.*, *138*, 187–198.

Falconnet, D., Csucs, G., Michelle Grandin, H., & Textor, M. (2006). Surface engineering approaches to micropattern surfaces for cell-based assays. *Biomaterials*, *27*, 3044–3063.

Ferraro, J. R., Nakamoto, K., & Brown, C. W. (2003). *Introductory Raman Spectroscopy*. Academic Press. (2003).

Gardiner, D. J., & Graves, P. R. (Eds.), (1989). *Practical Raman Spectroscopy*. Springer-Verlag.

Gremlich, H. U., & Yan, B. (2001). *Infrared and Raman spectroscopy of biological materials*. New York: Marcel Dekker, Inc.

Haran, G. (2010). Single molecule Raman spectroscopy: A probe of surface dynamics and plasmonic fields. *Accounts of Chemical Research*, *43*, 1135–1143.

Ikai, A. (2010). A Review on: Atomic Force Microscopy applied to Nano-Mechanics of the Cell, *Advances in Biochemical Biotechnology*, *119*, 47–61.

Jones, R., James, P. S., Howes, L., Bruckbauer, A., & Klenerman (2007). Supramolecular organization of the sperm plasma membrane during maturation and capacitation – Review. *Asian Journal of Andrology*, *4*, 438–444.

Kirby, B. (2009). *Micro- and Nanoscale Fluid Mechanics: Transport in Microfluidic Devices*. USA: Cambridge University Press.

Kiss, R., Bock, H., Pells, S., Canetta, E., Adya, A. K., Moore, A. J., De Sousa, P., & Willoughby, N. A. (2011). Elasticity of human embryonic stem cells as determined by atomic force microscopy. *Journal of Biomechanical Engineering*, *133*, 101009. (9 pages).

Kumar, S., Chaudhury, K., Sen, P., & Guha, S. K. (2005). Atomic force microscopy: A powerful tool for high-resolution imaging of spermatozoa. *Journal of Nanobiotechnology*, *3*, 9. (6 pages).

Lee, G. J., Chae, S. -J., Jeong, J. H., Lee, S. -R., H.A, S. -J., Pak, Y. K., Kim, W., & Park H-, K. (2011). Characterization of mitochondria isolated from normal and ischemic hearts in rats utilizing atomic force microscopy. *Micron*, *42*, 299–304.

Li, Y., Wang, J., Xing, C., Wang, Z., Wang, H., Zhang, B., & Tang, J. (2011). Molecular recognition force spectroscopy study of the specific lecithin and carbohydrate interaction in a living cell. *Chemical Physics and Physical Chemistry.*, *12*, 909–912.

Meister, A., Gabi, M., Behr, P., Studer, P., Voros, J., Niedermann, P., Bitterill, J., Polesel-Mari, J., Lilley, M., Heinzelmann, H., & Zambelli, T. (2009). "FluidFM: Combining atomic force microscopy and nanofluidics in a universal liquid delivery system for single cell applications and beyond. *Nano Letters*, *9*, 2501–2507.

Noy, A. (2006). Chemical force microscopy of chemical and biological interactions. *Surface and Interaction Analysis*, *38*, 1429–1441.

Okada, M., Smith, N. I., Palonpon, A. F., Endo, H., Kawata, S., Sodeoka, M., & Fujita, K. (2012). Label-free Raman observation of cytochrome c dynamics during apoptosis. *Proceedings of the National Academy of Sciences*, *109*, 28–32.

Park, S., Costa, K. D., & Ateshian, G. A. (2004). Microscale frictional response of bovine articular cartilage from atomic force microscopy. *Journal of Biomechanics*, *37*, 1679–1687.

Raman, A., Trigueros, S., Cartagena, A., Stevenson, A. P. Z., Susilo, M., Nauman, S., & Antoranz Contera, S. (2011). Mapping nanomechanical properties of live cells using multi-harmonic atomic force microscopy. *Nature Nanotechnology*, *6*, 809–814.

Rubio-Serra, F. J., Heckl, W. M., & Stark, R. W. (2005). Nanomanipulation by Atomic Force Microscopy. *Advanced Engineering Materials*, *7*, 193–196.

Saeki, K., Sumimoto, N., Nagata, Y., Kato, N., Hosoi, Y., Matsumoto, K., & Iritani, A. (2005). Fine surface structure of bovine acrosome-intact and reacted spermatozoa observed by atomic force microscopy. *The Journal of Reproduction and Development*, *51*(2), 293–298.

Trindade, G. S., Vilela, J. M. C., Ferreira, J. M. S., Aguiar, P. H. N., Leite, J. A., Guedes, M. I. M.C., Lobato, Z. I. P., Madureira, M. C., da Silva, M. I. N., da Fonseca, F. G., Kroon, E. G., & Andrade, M. S. (2007). Use of atomic force microscopy as a diagnostic tool to identify orthopoxvirus. *Journal of Virological Methods*, *141*, 198–204.

FURTHER READING

Allhoff, F. (2007). Nanoethics: the ethical and social implications of nanotechnology. In F. Allhoff & N. J. Hoboken (Eds.), Wiley-Interscience. ISBN: 9780470084168.

Bhushan, B. (2004). Springer handbook of nanotechnology. In Bharat Bhushan (Ed.), London: Springer-Verlag. ISBN: 3540012184.

Carlson, E. D. (2009). Encyclopedia of nanotechnology. In E. D. Carlson (Ed.), New York, N.Y: Nova Science Publishers, Inc. ISBN: 9781606920794.

Goodsell, David S. (2004). Bionanotechnology: lessons from nature. In N. J. Hoboken (Ed.), Wiley-Liss. ISBN: 047141719X.

Hester, R. E., & Harrison, Roy M. (2007). Nanotechnology: consequences for human health and the environment. In R. E. Hester & R. M. Harrison (Eds.), Cambridge: Royal Society of Chemistry. ISBN: 0854042164.

Hicks, T. R., & Atherton, P. D. (1997). *The nano postitioning book: moving and measuring to better than a nanometre*. Bracknell: Queensgate Instruments. ISBN: 0953065804.

Huang, Q. (2012). Nanotechnology in the food, beverage and nutraceutical industries. In Q. Huang (Ed.), Oxford: Woodhead Publishing. ISBN: 9781845697396.

Laurencin, C. T., & Nair, L. (2008). Nanotechnology and tissue engineering: the scaffold. In C. T. Laurencin & L. Nair (Eds.), Boca Raton: Taylor & Francis. ISBN: 9781420051827.

Malsch, N. H. (2005). Biomedical nanotechnology. In N. H. Malsch (Ed.), London: Taylor & Francis. ISBN: 0824725794.

Niemeyer, C. M., & Mirkin, C. A. (2004). Nanobiotechnology: concepts, applications and perspectives. In C. M. Niemeyer & C. A. Mirkin (Eds.), Weinheim: Wiley-VCH. ISBN: 3527306587.

Reisner, D. E. (2009). Bionanotechnology: global prospects. In D. E. Reisner (Ed.), London: CRC Press. ISBN: 9780849375286.

Renugopalakrishnan, V., & Lewis, R. V. (2006). Bionanotechnology: proteins to nanodevices. In V. Renugopalakrishnan & R. V. Lewis. (Eds.), Dordrecht: Springer. ISBN: 1402042191.

Shoseyov, O., & Levy, I. (2008). NanoBioTechnology: bioinspired devices and materials of the future. In O. Shoseyov & I. Levy (Eds.), Totowa, NJ: Humana Press. ISBN: 9781588298942.

Vo-Dinh, T. (2005). Protein nanotechnology: protocols, instrumentation. applications, In T. Vo-Dinh (Ed.), Totowa, N.J: Humana Press. ISBN: 1588293106.

Zhou, W., & Wang, Z. L. (2007). Scanning microscopy for nanotechnology: techniques and applications. In W. Zhou & Z. L. Wang (Eds.), New York, NY: Springer. ISBN: 9780387333250.

GLOSSARY

CCD Charged-couple device:A technology used in digital imaging.

Holographic notch filters (HNF) Holograms that provide high laser attenuation in a very narrow bandwidth.

Nanobiosensors Biosensors using nanotechnology for detection of an analyte in a complex sample matrix with rapid response and high sensitivity.

Nanodevice A manufactured device whose scale is measured in nm.

Nanofabrication Design and manufacture of devices and structures with feature sizes measured in nanometers.

Nanoparticles Particles with dimensions in the range 1 to 100 nm.

Nanotechnology The science of manipulating matter at the atomic or molecular scale.

Nanotool Any tool used in nanotechnology or having dimensions measured in nanometers (nm).

Notch filter (NF) An optical filter that allows all frequencies to pass through, except for a single frequency or a narrow band of frequencies. Narrow notch filters are used in Raman Spectroscopy.

ABBREVIATIONS

2D Two-Dimensional
3D Three-Dimensional
AFM Atomic Force Microscope
AFM-FM AFM Force Mapping
AFM-FS AFM Force Spectroscopy
CCD Charge-Coupled Device
CFM Chemical Force Microscopy
CRS Confocal Raman Spectroscopy
d_{max} Contact Point
DNA Deoxyribonucleic Acid
EDS Energy Dispersive Spectroscopy
ESEM Environment Scanning Electron Microscope
***F-d* curve** Force-Distance Curve
FluidFM Fluid Force Microscopy
F_{max} Maximum Adhesion Force
FTIC Fluorescein Isothiocyanate (Dye)
Gy International System (SI) Unit of Radiation Dose (Absorbed Energy/Mass of Tissue)
HNF Holographic Notch Filter
IMV Intracellular Mature Virus
MNPs Magnetic NPs
MRS Modulated Raman Spectroscopy
NC Nano-Coumarin
NPs Nanoparticles
NSOM Near-Field Scanning Optical Microscopy (or SNOM)
PBS Phosphate Buffered Saline
PrP Prion Protein
PSD Position Sensitive Detector
RI Raman Imaging
RS Raman Spectroscopy
SC Synthesized Natural Coumarin
SEM Scanning Electron Microscopy
SERS Surface-Enhanced Raman Spectroscopy
SPM Scanning Probe Microscopy
STM Scanning Tunneling Microscope
TERS Tip-Enhanced Raman Spectroscopy
VACV Vaccinia Virus
W_{adh} Work or Energy of Adhesion

LONG ANSWER QUESTIONS

1. Write an essay describing nanotechnology, its translational significance, ethical issues and risks involved, and explain its use in biology and biotechnology.
2. Describe the history, operating principles, and advantages and limitations of a variety of microscopic techniques ranging from optical microscopy to scanning probe microscopy.
3. Describe how nanotechnology has helped move the frontiers in animal biotechnology. Give specific examples.
4. By drawing suitable diagrams where possible, explain a variety of Scanning Probe Microscopy (SPM) and Raman Spectroscopy (RS) techniques that can be employed to investigate biological materials at the nanoscale.
5. Describe the various operational and application modes of an Atomic Force Microscope (AFM), and show how this nanotechnique has been used for a variety of applications in biotechnology. Draw and label suitable sketches (where possible) to illustrate your answer. Describe briefly a hyphenated technique of AFM.

SHORT ANSWER QUESTIONS

1. By drawing and labeling an outline sketch, describe:
   (i) The operational principle of Atomic Force Microscopy (AFM).
   (ii) The three operational modes of AFM.

   Describe the operation of Atomic Force Microscope in Force Spectroscopy (AFM-FS) application mode by sketching a Force-distance (F-d) curve and explaining it fully.

   List three different quantitative properties of a specimen that can be determined from these AFM-FS measurements.

   Explain the differences between AFM Force Spectroscopy (AFM-FS) and AFM Force Mapping (AFM-FM). What additional advantages does AFM-FM offer relative to AFM-FS?

2. Describe the basic principles of Scanning Probe Microscopy (SPM) techniques and write short notes on:
   (i) Chemical Force Microscopy (CFM).
   (ii) Fluid Force Microscopy (FluidFM).
   (iii) Scanning Near-Field Optical Microscopy (SNOM).
   What are the differences between (a) Scanning Tunnelling Microscopy (STM) and AFM, (b) Scanning Electron Microscopy (SEM) and AFM, and (c) CFM and AFM?
3. Describe the two techniques, Raman Spectroscopy (RS) and Raman Imaging (RI), and explain why they have recently been in demand.
   List the two main disadvantages of RS, and write short notes on how these detrimental effects have been overcome in:
   (i) Confocal Raman Spectroscopy (CRS).
   (ii) Modulated Raman Spectroscopy (MRS).
   (iii) Surface-Enhanced Raman Spectroscopy (SERS).
   (iv) Tip-Enhanced Raman Spectroscopy (TERS).
   Briefly describe a confocal AFM–Raman hybrid system, and explain what potential it holds in animal biotechnology.
4. What is AFM nanomanipulation? By giving a couple of examples show how it can be used in animal biotechnology. Write a brief note on the use of nanofluidics in nanobiotechnology.
5. By giving a few examples of applications of nanoparticles, show how these nanomaterials are at the leading edge of the rapidly developing field of nanobiotechnology. What are the main health and ethical issues associated with the use of nanoparticles in animal biotechnology?

ANSWERS TO SHORT ANSWER QUESTIONS

1. For a labeled outline sketch and description of the operating principle of AFM, and for the sketch of a force-distance (F-d) curve and operation of AFM-FS and AFM-FM application modes, see "Atomic Force Microscopy" in the "SPM Techniques" section. Three different quantitative properties of a specimen that can be determined from AFM-FS measurements are F_{max} (maximum adhesion force in nN), d_{max} (contact point in nm), and W_{adh} (work or energy of adhesion in J). AFM-FM has additional advantages over AFM-FS in providing two-dimensional property maps and phase distribution across a sample surface.
   AFM can be operated in three different modes, namely contact, non-contact, and intermittent contact (also known as tapping mode). Because the tip is in hard contact with the surface in *contact mode*, the stiffness of the lever (measured by its spring constant, k) has to be less than that holding atoms together, which is ~1–10 nN/nm. Most contact mode cantilevers have a spring constant < 1 N/m. In *non-contact mode*, a stiff cantilever is oscillated in the attractive regime (tip quite close to the sample but does not touch it), hence, "noncontact" mode. The forces between the tip and sample are quite low (~pN). The detection is based on measuring changes to the resonant frequency or amplitude of the cantilever. In *tapping mode*, a stiff cantilever is oscillated closer to the sample than in non-contact mode. Part of the oscillation, therefore, extends into the repulsive region so that the tip intermittently touches or "taps" the surface. Very stiff cantilevers are used because otherwise these can get "stuck" in the water contamination layer.
2. All the SPM techniques are based upon scanning a probe, typically called the *tip*, just above a surface while monitoring some interaction between the probe and the surface. To write short notes on: (i) CFM, (ii) FluidFM, and (iii) SNOM, see the appropriate heading in the "SPM Techniques" section.
   (a) The salient differences between STM and AFM are that while in STM it is the tunneling current, in AFM it is the van der Waals forces that are monitored between the tip and surface.
   (b) The salient differences between SEM and AFM are that in SEM it is the beam of electrons while in AFM it is the tip (sharp metallic tip microfabricated at the end of a cantilever or a bead, functionalized or non-functionalized) that scans line by line over the surface; also, contrast enhancing agents are often required in SEM while no contrast enhancement is required in AFM.
   (c) The salient differences between CFM and AFM are that while AFM exploits the "natural" van der Waals forces between the non-functionalized AFM tip and the sample surface, CFM uses the "well-defined" interactions between the functionalized AFM tip and the sample surface, which can also be functionalized, if required.
3. For RS and RI see the section "Raman Spectroscopy and Imaging." Requirements of very little sample preparation and the "gentleness" of the "touch of light" on the sample make them truly non-destructive and non-invasive techniques. Additionally, their ability to work both *in situ* and *in vivo*, an essential requirement when dealing with biological samples, has recently put them in great demand for animal biotechnology and life science work. The main disadvantages of RS are the small percentage of Raman scattered photons compared to the elastically scattered photons, and the strong fluorescence background. The cumulative effect of these two can sometimes almost completely hide the Raman peaks. For a description of how these detrimental effects are overcome in (i) CRS, (ii) MRS), (iii) SERS, and (iv) TERS, see the section entitled "Raman Spectroscopy

and Imaging." For the description of a confocal AFM–Raman hybrid system, see the section "AFM–Raman Confocal Hybrid Systems."

4. AFM nanomanipulation is the use of an AFM tip as a nanoscalpel or nanoneedle to perform nanopatterning (i.e. modification of surface geometry to improve cell adhesion), to inject nanoparticles and other nanoscale-size materials (e.g. fluorescent dye) inside a living cell, etc. Nanomanipulation can be used in animal biotechnology, for example, to insert pieces of DNA into single living cells for nano-genetic manipulations, to produce single genetic probes that target particular cells, and to nanodissect chromatidial and chromosomal DNA.
   Nanofluidics is the study of the behaviors and properties of fluids that are confined in a nanoscale size environment. For a brief note on nanofluidics, see the section entitled "SPM Techniques," sub-heading "Fluid Force Microscopy (FluidFM)." For a review of nanofluidics, see Kirby (2009).
5. Nanoparticles (NPs) play a very important role in the molecular diagnosis, treatment, and monitoring of therapeutic outcomes in a variety of diseases. Their nanoscale size, large surface area, and unique capabilities make them highly effective for biomedical applications, such as cancer therapy, thrombolysis, and molecular imaging. Magnetic NPs (MNPs) with functionalized surface coatings can conjugate chemotherapeutic drugs, making them useful for drug delivery, targeted therapy, magnetic resonance imaging, etc. The drug-conjugated MNPs can be rapidly released after injection or their release delayed, resulting in targeting of low doses of the drug.
   Nanoparticles can be toxic and/or carcinogenic, and thus hazardous to mammalian health. This is due to the fact that there are several entry routes in the animal body, such as the respiratory tract and skin. Once inside the animal body, nanoparticles can have potential toxic effects on the primary target sites or organs, such as the lungs and skin. Moreover, nanoparticles can start traveling inside the animal body and also have toxic effects on distant organs, such as the kidney or liver. The toxic effects of nanoparticles can even cause death of the animal.

Chapter 15

Antibodies: Monoclonal and Polyclonal

Anchal Singh, Sushmita Chaudhary, Ashima Agarwal and Ashish Swarup Verma
Amity Institute of Biotechnology, Amity University Uttar Pradesh, Noida, U.P., India

Chapter Outline

SUMMARY

Antibodies are one of the most important components of humoral immune response. They protect a host against infections. Polyclonal antibodies (PoAb) can be raised by immunizing animals. Monoclonal antibodies (MoAb) are produced by the fusion of B-lymphocytes with myeloma cells. Both polyclonal and monoclonal antibodies have wide implications for human health and welfare.

WHAT YOU CAN EXPECT TO KNOW

Antibody production is a hallmark of the adaptive immune response. Antibodies are produced either to neutralize or to eliminate antigens or pathogens. Antibodies are produced by B-lymphocytes after differentiation of B-lymphocytes into plasma cells. Polyclonal antibodies (PoAb) can be raised by immunizing animals against the antigen of choice. PoAb have various applications of analytical, diagnostic, and therapeutic significance. PoAb were used to protect against infections like diphtheria as early as 1894. The major limitations in using PoAb are the huge variations in affinity and specificity of PoAb from lot-to-lot. Animals have to be given booster injections each time before collecting PoAb. The solution to these drawbacks of PoAb was found with the discovery of monoclonal antibodies (MoAb). MoAb were discovered by Kohler and Milstein. MoAb are a perpetual source of antibodies with consistent specificity and

Animal Biotechnology. http://dx.doi.org/10.1016/B978-0-12-416002-6.00015-8

affinity. The discovery of monoclonal antibodies has not only revolutionized the area of immunodiagnostics, but the medical sciences have also benefitted enormously. MoAb have found widespread applications in clinical treatment, biomedical research, and industry.

HISTORY AND METHODS

INTRODUCTION

Antibodies belong to a group of globular proteins of serum/plasma; they are therefore also known as immunoglobulins. Antibodies react specifically with antigens, which are responsible for the production or induction of those specific antibodies. The words antibody and immunoglobulin are interchangeably used in the literature. The term immunoglobulin can be used to refer to any antibody-like molecule regardless of its antigen-binding specificity.

An antibody is a Y-shaped molecule produced by B-lymphocytes. The terminal stage of B-lymphocytes (i.e. plasma cells) are the major source of immunoglobulin production. Antibodies are the most diverse protein known so far. The most startling fact about antibodies is that these protein molecules have almost similar amino acids residues in ~90–95% of the polypeptide chain, whereas the remaining ~5–10% is comprised of a hypervariable region that shows huge variation in the amino acid residues. The variation in the variable region is up to the extent of millions of different combination of amino acids. Each antibody has two major functions: (1) antigen binding that occurs at the fragment antigen binding (Fab) portion, and (2) the effector function of antibodies, which is due to the fragment crystallizable (Fc) portion of the immunoglobulin. In the Y-shape structure of the antibody, the arms of the Y confer the versatility and specificity of response that a host can mount against antigens, while the stem region of the antibody decides the biological activities (e.g. complement-mediated lysis, phagocytosis, allergy, etc.). The biological activity always starts with binding of the antigen to the antibody. Knowledge about the presence of this protective protein in serum was known even in 1890, and was confirmed by Tiselius and Kabat in 1939. It took almost 50 years to gain an insight about immunoglobulins.

TISELIUS AND KABAT'S EXPERIMENT

Tiselius and Kabat were the first to demonstrate that immunoglobulins are globular proteins; they also demonstrated the specificity of immunoglobulins against antigens. Their conclusion was based on electrophoretic analysis of hyper-immune sera from rabbits.

Interestingly, Tiselius and Kabat used sera from hyper-immune rabbits that were immunized with ovalbumin. Tiselius and Kabat obtained four different bands after electrophoresis of unimmunized rabbit serum. These four bands belonged to four different types of proteins. During electrophoretic separation of serum, the fastest migrating protein was albumin, which was followed by alpha (α), beta (β), and gamma (γ) fractions of serum. When hyper-immune serum was electrophoresed under the same conditions, the same four bands were again observed (i.e. albumin, α, β, and γ). However, this time the γ-fraction was much higher as compared to the unimmunized serum. It was concluded that when a rabbit was immunized with an antigen like ovalbumin, it led to induction or production of protein that belonged to the γ-fraction of serum. Later, in another experiment, hyper-immune serum was absorbed with ovalbumin and then electrophoresed. Upon electrophoresis, absorbed hyper immune-sera was separated into four bands of albumin, α-, β-, and γ-globulins. Surprisingly, after absorption the electrophoretic pattern of hyper-immune serum was similar to pre-immune serum. This result led to the interpretation that the γ-globulin fraction bound to ovalbumin; hence, Tiselius and Kabat concluded that the γ-globulin fraction was specific against ovalbumin, which was used to immunize the rabbit.

This was the landmark experiment of immunology that successfully demonstrated that immunoglobulins belong to the γ-fraction of serum and that these globular proteins have specificity against antigens.

HISTORY

The protective effect of serum has been known about since the late 19th century. In 1890, Emil von Behring and Shibasaburo Kitasato developed an effective antiserum against diphtheria. They transferred serum from an animal immunized against diphtheria to animals suffering from diphtheria. This therapy cured animals suffering from diphtheria, and Behring was awarded the Nobel Prize in 1901.

The first documented record of passive immunization tried in a human being was on Christmas Night in 1891. A young boy in Berlin was cured by injection of diphtheria antitoxins. This experiment was based on the work carried out by von Behring and Kitasato at Robert Koch's Hygiene Institute at Berlin, whose findings had been published on December 4, 1890.

Since diphtheria was very common at the time and often had fatal consequences, von Behring started working towards the production of large amounts of diphtheria antiserum. This work was carried out by the Farberwerke Hoechst Company using sheep as a source for the antiserum against diphtheria. The first large-scale trial took place in 1893 with promising results. Later, sheep were replaced by horses to raise antiserum production. With use of the antiserum, mortality in Paris due to diphtheria fell from 52% to 25% (Black, 1997).

The most common problem with this passive sero-therapy was an anaphylactic reaction among treated patients. Therefore, an ammonium sulfate precipitate of the antiserum was used for the treatment, which somehow reduced the side effects of sero-therapy even though it failed to eliminate the problems related with serum therapy. The next trial was of pepsin-treated gamma-globulin, which further reduced the incidences of serum sickness. The same treatment was also applied to other infections (e.g. *Streptococcus pneumoniae* and *Neisseria meningitidis, etc.*). This method was commonly used as the only choice for treatment until the discovery of antibiotics by Alexander Fleming. In the 1920s, Michael Heidelberger and Oswald Awery showed a quantitative precipitation reaction upon interaction of antigens with antibodies.

It was realized that sero-therapy was an excellent option to treat various diseases, and in 1939 Tiselius and Kabat reported for the first time specificity of the globulin fraction in hyper-immune sera. In 1942, Merrill Chase successfully demonstrated transfer of immunity against tuberculosis by transferring white blood cells among guinea pigs. In 1944, a human immunoglobulin preparation was used to treat measles (which finally overcame the problems of serum sickness). This treatment was also demonstrated to be effective in treating patients with agammaglobulinemia, and is still in use today. In 1948, Astrid Fagreaus completed his doctoral work by demonstrating the role of plasma B cells in antibody production.

In 1959, James Gowan demonstrated the role of lymphocytes for humoral as well as cell-mediated immune response. In the meantime, various theories for the immune system and its functions were proposed by different scientists like Paul Ehrlich, Niel K. Jerne, and F. Macfarlane Burnet. Finally, Burnet's theory of "clonal selection" was widely accepted by immunologists throughout the world (Burnet, 1959). Evidence for clonal selection theory was provided by Sir Gustav Nossal by showing that one clone of a B cell produces only one antibody. In 1960, Edelman and Porter elucidated the structure of antibodies, and in 1976 Susumu Tonegawa demonstrated somatic recombination in immunoglobulin genes.

In 1975, major advancements in the field of medical science came with the discovery of monoclonal antibodies (MoAb) (Kohler and Milstein, 1975). MoAb offered a new hope for the use of antibodies again as therapeutic agents. This hope soon vanished, however, as there were so many issues related to the use of monoclonal antibodies as therapeutics. Nevertheless, advancements in the area of protein engineering renewed interest in the use of MoAb for therapeutic purposes. Protein engineering tools have helped in the development of newer types of antibodies (e.g. chimeric, humanized MoAb, and human Ab).

ELUCIDATION OF IMMUNOGLOBULIN STRUCTURE

In 1939, Tiselius and Kabat confirmed that immunoglobulins were globular proteins present in the γ-globulin fraction of serum (Tiselius, 1937; Tiselius and Kabat, 1939), but it took another ~20–30 years for scientists to elucidate the structure of immunoglobulin as it is known to the world today. The credit for elucidation of immunoglobulin structure goes to two scientists, Rodney Porter from the U.K. and Gerald Edelman from the U.S. For their landmark work in elucidating the structure of immunoglobulin, they were awarded the Nobel Prize for Medicine or Physiology in 1972. The contributions of another scientist, Nisonoff, were also important for this discovery. Porter and Edelman both were trying to deduce the structure of immunoglobulin molecules, but their approaches to this research problem were different from each other. The coming paragraphs discuss their individual experiments and how they combined their results to identify the antibody structure (Figure 15.1).

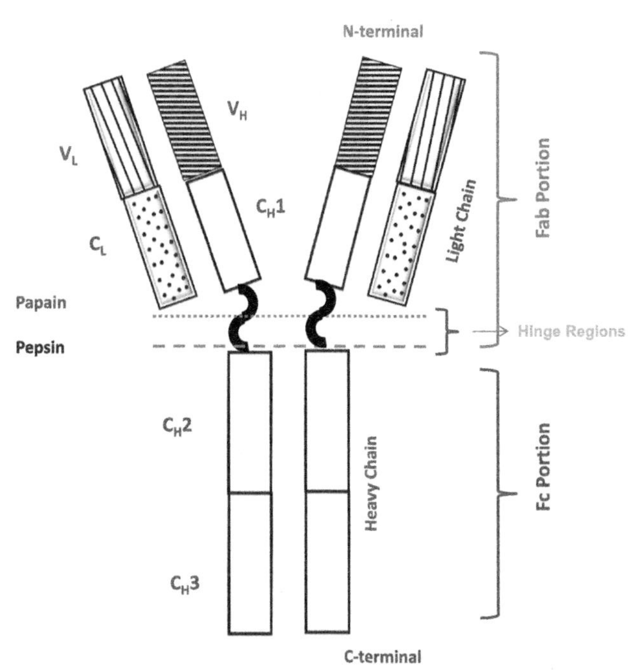

FIGURE 15.1 Structure of prototype immunoglobulins (i.e. IgG). IgG has two heavy chains and two light chains. ( ) C_H1, C_H2, and C_H3 are constant domains of heavy chains (H-chain), and ( ) C_L is a constant domain of light chain (L-Chain). ( ) V_H is the variable domain of the heavy chain and ( ) V_L is the variable domain of the light chains. The hinge region has digestion sites for papain and pepsin enzymes. The lower portion of IgG that contains only constant domains of the heavy chain is known as the fraction crystallizable (Fc) portion. Fraction antigen binding (Fab) contains C_H1 and V_H domains of the heavy chain and C_L and V_L domains of the light chain. The antibody heavy chain and light chain are held together by inter-chain disulfide bonds, but for clarity, those bonds are not shown in the figure. The hinge region of immunoglobulin lies between the C_H1 and C_H2 domains of the heavy chain. It has a site for papain digestion (........................) and pepsin digestion (_ _ _ _ _ _ _ _).

Edelman's Experiment

Edelman was using rabbit immunoglobulin-G (IgG). He treated rabbit IgG with different chemicals like DTT (a reducing agent), iodoacetamide (an alkylating agent that prevents reassociation of disrupted disulphide bonds), and a denaturing agent (a substance that disrupts noncovalent interaction) (Edelman et al.,1961; Edelman, 1973). After digestion of antibodies with different chemicals, Edelman separated digested immunoglobulins by size exclusion chromatography and found two protein peaks in equimolar ratios. The molecular masses of the two protein peaks corresponded to 50 kD and 23 kD. The protein with a molecular mass of 50 kD was designated heavy chain and the 23 kD protein was called light chain.

Conclusion of Edelman's Experiment: The initial antibody molecular weight was 150 kD, so Edelman concluded that immunoglobulins could consist of two heavy chains and two light chains that are linked to each other by disulfide bonds and noncovalent interactions.

Porter's Experiment

Porter was working on proteolytic digestion of immunoglobulins (O'Donnell et al., 1970). He was using papain enzyme to digest immunoglobulin in the presence of reducing agents like cysteine. Papain hydrolysis led to the generation of three fragments, I, II, and III. All three fragments were of the same size (50 kD), but had different electrical charges. Out of these three fragments, two were identical and maintained their ability for antigen binding. Therefore, these fragments were called fragment antigen binding (Fab). Porter knew that intact antibodies were bivalent and each Fab fragment could bind antigen, so he concluded that a Fab fragment must be univalent in nature. However, the third fragment produced due to papain digestion did not bind with antigen; it crystallized upon cold storage. Porter named this fraction the Fc fragment. The ratio between Fab and Fc fractions of immunoglobulins was found to be 2:1.

Conclusion of Porter's Experiment: On the basis of observations of Porter's experiment, it was concluded that the ratio between Fab and Fc was 2:1. Fab fragments could bind antigen whereas the Fc portion could not.

Nisonoff's Experiment

Apart from Edelman and Porter, Nisonoff was also working with immunoglobulins (Nisonoff et al., 1975). He was using pepsin (another proteolytic enzyme) to digest immunoglobulins. Pepsin digestion yielded a 100 kD fraction and numerous small fractions. The 100 kD fraction weight was double the weight of Fab as observed by Porter, so Nisonoff called this fragment $F(ab')_2$ (fragment antigen binding, divalent). The $F(ab')_2$ portion obtained from pepsin digestion could bind antigen and showed visible serological reaction (i.e. immuno-precipitation). The $F(ab')_2$ portion had both the antigen-binding sites of IgG. Amazingly, when the $F(ab')_2$ portion was treated with reducing agents, it gave two Fab-like fragments (known as Fab), as both fragments could bind antigens even after separation.

Conclusion of Nisonoff's Experiment: Pepsin digestion lead to one $F(ab')_2$ portion and smaller fragments. The $F(ab')_2$ portion could bind to antigen and was divalent.

Conclusion from Papain and Pepsin Digestion

Collectively, two enzymes (i.e. papain and pepsin) cleave in the same region of the immunoglobulin molecule. Papain cleaves at one side while pepsin cleaves immunoglobulin on the other side of the bond that holds the Fab fragments together.

The puzzle of immune structure revealed several facts:

1. Each Ig molecules has two heavy chains and two light chains.
2. Fab and Fc fragments are produced by papain digestion.
3. Immunoglobulin molecules can split into $F(ab')_2$ and smaller fragments by pepsin digestion.
4. Fab and $F(ab')_2$ can bind antigen; Fc cannot.

These landmark experiments helped to resolve the structure of immunoglobulin molecules. Still, the problem was to combine these facts to yield the final structure of immunoglobulin. The answer to this puzzle was resolved by Porter. He raised anti-sera against Fab and Fc fragments of rabbit IgG. He found that antiserum against Fc fragments reacted with H-chain and L-chain. This observation confirmed that the Fab fragment consists of both H-chain and L-chain, while the Fc portion contained only heavy chain.

IMMUNOGLOBULIN G: A PROTOTYPE FOR IMMUNOGLOBULIN

Antibodies are considered to be the workforce for humoral response, and they perform two major functions: (1) target recognition via the Fab portion, and (2) antigen elimination/effector function via the Fc portion. Antibodies are classified as polyclonal and monoclonal. Polyclonal antibodies are host proteins present in plasma and extracellular fluids, and serve as the first response against attack by any pathogen or foreign molecule. The ultimate aim of antibodies is to neutralize or eliminate any threat encountered by the immune system (Roitt and Delves, 2001). Antibodies are part of the adaptive immune response, therefore they are specific against the threat/infection; specific antibodies bind to antigens with high affinity. These two characteristics make antibodies unique tools for a variety of purposes in scientific and medical research. No other material has contributed directly or indirectly to such a wide array of scientific discoveries. The application of antibodies for numerous diagnostic assays has had a profound impact on the welfare and health of humans and animals. Polyclonal antibodies are produced *in vivo*, while biotechnological manipulations have been used to generate monoclonal antibodies.

The basic structure of immunoglobulin is made up of two different polypeptide chains. Ideally any antibody (IgG) contains two identical copies of heavy (H) chains (55 kb) and light (L) chains (28 kb). Heavy chains and light chains are held together by inter-chain disulfide bonds as well as other non-covalent interactions. In general, an IgG looks like a Y-shaped structure with molecular weight around 150 kD. The light chain contains two domains (i.e. variable and constant), while the heavy chain contains one variable domain and three or four constant domains. The Fab portion is at the N-terminal end of the heavy and light chain and is responsible for antigen–antibody binding and interaction. The Fc portion is responsible for the effector functions or biological functions of an antibody, which occurs due to binding of the Fc portion to Fc receptors present (FcR) on effector cells (Burton, 1987; Carayannopoulus and Capra, 1993; Frazer and Capra, 1999; Padlan, 1994). The Fc part is comprised of constant region domains of heavy chain, and accordingly, heavy chain binds to FcR specific for that class of antibody. The flexibility of antibodies lies in the hinge region, which is present between C_H1 and C_H2 domains (Figure 15.1).

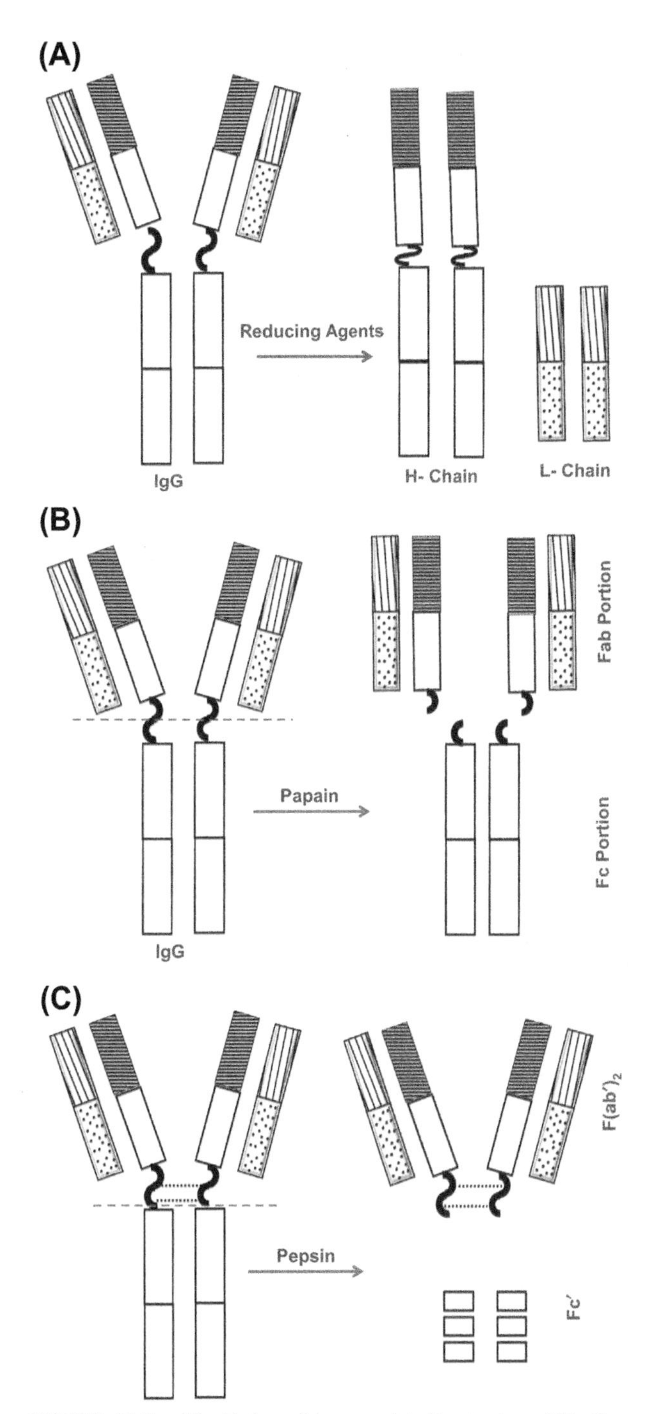

FIGURE 15.2 Elucidation of immunoglobulin structure. This figure shows the different fractions of immunoglobulin after digestion with different agents. (2A) Edelman's experiment: The figure represents digestion of immunoglobulin by reducing agents like 2-ME, which results in separation of two heavy chains (H-chain) and two light chains (L-chains). (2B) Porter's experiment: The figure represents digestion of immunoglobulin with papain. Papain digestion site is present at the upper portion of the hinge region. After digestion, it gives two Fab fragments (Fragment antigen binding) and one Fc fragment (Fragment crystallizable). (2C) Nisonoff's experiment: The figure shows digestion of immunoglobulin by pepsin. The pepsin digestion site is present at the end of the hinge region of the antibody. Disulfide bonds (▪▪▪▪▪▪▪▪▪▪▪) present in the hinge region hold the $F(ab')_2$ together, accounting for its divalent nature. Smaller fragments (called Fc′) lose their ability to bind with Fc receptors (FcR) on effector cells.

POLYCLONAL ANTIBODY *VERSUS* MONOCLONAL ANTIBODY

PoAb and MoAb are terms commonly found in immunology literature. Both of these antibodies have the same structure and function, but PoAb and MoAb differ from each other on the basis of their origin, production, and specificity. The basic difference between these two antibodies lies in the clonality of the cells that produce them. MoAb are produced by a single clone, while PoAb are produced by numerous clones together. Each of these antibodies has their own pros and cons (Lipman et al., 2005). Numerous advantages and disadvantages of PoAb and MoAb are given below for a better understanding (Table 15.1).

Polyclonal Antibodies (PoAb)

1. PoAb are also known as antiserum, so PoAb and antiserum are interchangeably used in this chapter.
2. PoAb are products of numerous B-lymphocytes.
3. Animals are the source of PoAb.
4. PoAb can be produced only *in vivo,* which means the requirement of an animal is essential.
5. PoAb are specific in nature, but show cross-reactivity due to their polyclonal nature.
6. Production as well as repeated production of PoAb requires booster injection.

TABLE 15.1 Comparison Between Polyclonal and Monoclonal Antibodies*

	Antibodies	
Characteristics	**Polyclonal**	**Monoclonal**
Adverse Effect	Serum Sickness	HAMA*
Affinity Purified Product	Not homogenous	Homogenous
Antibody Purification	May/May Not	Not essential[#]
Booster Dose	Required	Not Required
Chemically Defined	Not Well	Well
Clonality	Numerous	Single
Cross-reactivity	High to low	Low to nil
Degradability	Low	High
Denatured Antigen[$]	Detect	May/ May Not
Epitope Detection	Multiple	Single
Homogeneity (Antibody)	No	Yes
Manpower Skills	Low skill	Highly trained
Origin	Any	Mouse (Mostly)
Payload (Conjugation)	Not Easy	Easy
Production Cost		
(i) Initial	Low	High
(ii) Long-Term	High	Low
Production Requirement		
In vitro	–	+
In vivo	+	+
Specificity	Low	High
Variability[@]	High	Low

**Human Anti-Monoclonal Antibody;*
[@]Lot-to-Lot;
[#]Depends on application;
[$]Detection of antigen.

7. PoAb production does not require highly skilled manpower.
8. Sophisticated instruments are not needed to produce PoAb.
9. Sensitivity of PoAb varies from lot-to-lot, which is also attributable to their polyclonal nature.
10. Immunized animals (source of antiserum) are always vulnerable towards infections and other natural adversaries.
11. PoAb are not homogenous in nature so they cannot be easily characterized.
12. PoAb are more tolerant towards minor variation in epitopes and still recognize and bind with specific antigens. Epitope variation could be due to various factors like glycosylation, polymorphism, etc.
13. PoAb are less vulnerable to degradation.
14. PoAb can be used to detect degraded or denatured antigens (e.g. Western blotting, etc.).
15. PoAb are always a better choice to be used as capture antibodies in various immunological test (e.g. ELISA).
16. PoAb are not a good tool for affinity purification.
17. Serum sickness is the major adverse effect of PoAb.
18. Long-term production is expensive due to factors like maintenance of animal, animal deaths, etc.
19. PoAb can be used to provide passive immunity.

Monoclonal Antibodies (MoAb)

Monoclonal antibodies, as the name suggests, are products of a single clone. This single clone is selected after fusion of myeloma with B-lymphocytes. Pros and cons of MoAb are listed below.

1. MoAb are a product of a single clone of cells.
2. MoAb are produced by hybridoma cells by the fusion of B-lymphocytes with myeloma cells.
3. Initially, more time is required to produce MoAb.
4. Trained manpower is essential to produce MoAb.
5. Long-term production of MoAb is cheaper.
6. MoAb are highly specific because they are the products of a single clone.
7. The range of specificity of MoAb is very-very narrow (i.e. single epitope).
8. MoAb production requires both *in vivo* and *in vitro* systems.
9. Sensitivity of MoAb are high.
10. MoAb can be used either without purification or after purification.
11. Hybridoma cells that produce MoAb are perpetual sources of antibodies with the same specificity and sensitivity.
12. MoAb are 100% reactive towards their target.
13. Mostly MoAb are highly specific and usually do not show cross-reactivity.
14. Since they are homogenous in nature, their chemical nature can be characterized easily; therefore, MoAb are an excellent choice for conjugation to different probes.
15. When MoAb are used as therapeutic antibodies, adverse effects caused by MoAb are known as human anti-monoclonal antibody (HAMA) response.
16. MoAb are used to produce chimeric antibodies as well as humanized antibodies.
17. MoAb are an excellent choice or tool for affinity purification.

18. MoAb are well established to give highly reproducible results.
19. For the reasons that MoAb have mono-specificity and almost no cross-reactivity, they do not show any background during laboratory experiments or when used as probes.
20. Being homogenous in nature, MoAb are vulnerable to degradation because of the same susceptibility of all antibody molecules present in solution.
21. Due to mono-specificity of MoAb, they are not a suitable option for detection of denatured proteins.

NAMING MONOCLONAL ANTIBODIES

Monoclonal antibodies have found numerous applications, and this technology has seen enormous improvement in various aspects, including production and conjugation. These developments are the reason for production of new MoAb on a routine basis. As time progresses, more and more MoAb are hitting the commercial market and being applied in research laboratories. Hopefully, the development of new MoAb will continue with this uninterrupted pace. Everyone producing MoAb started naming them as per their own rationale, convenience, and other factors such as personal preferences, which led to confusion about the names of MoAb. Sometimes these names became too complicated. Sometimes the same antibody was available under different names so differentiation and identification of MoAb started getting difficult for different stakeholders. Another observation was that the same MoAb were found to be effective against more than one disease (e.g. tumors). Hence, a system to name MoAb with a systematic pattern became a necessity to avoid confusion among different stakeholders (e.g., clinicians, pharmacists, patients, etc.).

There are two main organizations that have developed guidelines for naming MoAb: (1) International Non-Proprietary Name (INN), Geneva, Switzerland; and (2) United States Adopted Names Council (USANC) Chicago, USA. The INN mode of naming MoAb is maintained under the umbrella of the World Health Organization (WHO), while the USAN system for naming MoAb is maintained by the USANC.

At present, USANC has modified its scheme for naming MoAb in such a manner that their naming system harmonizes with the naming of MoAb at the international level. As per the USANC system, MoAb should have a two-word name. However, USAN believes that while naming MoAb, this system should not be effective retrospectively to change the pre-existing names of MoAb. The rationale for USANC's decision is to avoid confusion among clinicians, pharmacists, patients, etc. with regards to the old and established names of same MoAb.

The nomenclature system uses a "stem." The stem for monoclonal antibody is "mab," while for a mixture of monoclonal antibodies, the stem is "pab." The polyclonal antibody here that is used is not the traditional polyclonal antibody, but represents a polyclonal pool of recombinant monoclonal antibodies. The naming system for monoclonal antibodies follows a pattern in which the name of each MoAb has four components: (1) prefix (1 in number), (2) infixes (2 in number), and (3) stem or suffix (1 in number). The new name, when given to a monoclonal antibody, should essentially contain following components (Table 15.2):

TABLE 15.2 List of Syllables Used to Name Monoclonal Antibody*

Meant For	Syllables	Examples
(A) Prefix		
	Variable	
(B) Infix-1		
Bacterial	-ba-/-b-	-bixumab
Bone	-so-/-s-	-somab
Cardiovascular	-ci-/-c-	-cixumab
Fungal	-fu-/-f-	-fuzumab
Immune modulator	-li-/-l-	-liximab
Interleukin as target	-ki-/-k-	-kiximab
Miscellaneous tumor	-tu-/-t-	-tuzumab
Neurons	-ne-/-n-	-nezumab
Viral	-vi-/-v-	-vizumab
(C) Infix-2		
Chimeric	-xi-	-liximab
Chimeric and humanized	-xizu-	-nexizumab
Human	-u-	-vizumab
Humanized	-zu-	-kizumab
Mouse	-o-	-bomab
Rat	-a-	-famab
Rat/mouse hybrid	-axo-	-caxomab
(D) Suffix	-mab	

**Partial list as adapted from USANC; Infix-1 represents a syllable for target or disease; Infix-2 represents syllable for source of antibody.*

$$\text{Prefix} + \text{Infix} - 1 + \text{Infix} - 2 + \text{Stem}$$

These components and their significance are discussed in the following sections.

Prefix

The prefix is the first syllable in the name of any MoAb. The main purpose of the prefix is to create a unique name for each and every MoAb. The prefix is made up of a distinct and compatible syllable. The prefixes should be selected as per the USAN list for coining names of MoAb. It is recommended to use a small prefix to keep the name of the MoAb as short as possible; in general, the syllable used as the prefix should be two or more in number, which will be helpful in differentiating new MoAb from previously assigned names of other MoAb.

Infix-1

Infix-1 in the name of an MoAb is meant to provide information about the target of an MoAb. This is the reason that infix-1 is usually directed towards the target or to the disease class against which the MoAb is effective or targeted. USANC has approved a list of syllables that are specific for the disease or for the target. If needed, an additional subclass can be added. In the case of tumors, an infix that has been recommended by USANC is either "-tu" or "-t-" and so the name of the MoAb against the tumor would be "-tuzumab/-tumab," while in case of immunomodulators, the infix could be either "-li" or "-l-" and the name of the antibodies would be written as "-liximab/-lumab/-lixizumab."

Infix-1 can be truncated to a single letter to make it easy to pronounce. This kind of situation usually arises when the infix-2 (i.e. source infix) starts with a vowel (e.g. for humanized and chimeric antibodies). Infixes indicating tumor specificity have recently been discontinued because it has been observed that some antibodies may work for more than one type of tumor. This is the reason that nowadays infixes used for tumor specificity like -col-(colon cancer), -mel- (melanoma), -got- (testes), -gov- (ovarian), and -po- (prostate) have been dropped completely or are not in use any more for naming new MoAb.

Infix-2

Infix-2 is meant to identify the source of MoAb. Certainly for therapeutic applications of MoAb, source information is crucial due to safety concerns. Sometimes MoAb are based on the sequence of one species but manufactured in an entirely different source. In these cases, infix-2 provides information about the species that serve as the source of sequences of immunoglobulins. Use of infix-2 is in harmony with the INN nomenclature system (e.g. humanized (-zu-), mouse (-o-), rat (-a-), etc.).

Additional Words

Additional words are also used while naming some monoclonal antibodies. They are typcially used as clarifying words for the nomenclature of monoclonal antibodies that have various applications and are conjugated with different types of payloads. These payloads can be of different types like radiolabels, toxins, etc. Then it is very useful to add the name of the payload along with the name of the antibody. Addition of the payload name in the MoAb name is helpful in identifying the payload attached to the MoAb. For example, in radio-labelled payloads, different words used to identify the payload have to be in the following order: (a) name of isotope, (b) symbol of element, and (c) isotope number. This is followed by the name of the monoclonal antibody: (1) Technetium Tc 99m biciromab, and (2) Indium In-111-altumomab-pentetate, etc.

Similar rules also apply to the use of additional words for toxins when conjugated as the payload to the MoAb. In that case the toxin is identified using a separate, second word ("tox") in the stem, and part of the name of the toxin has to be added to the stem (e.g. -aritox- means it contains a ricin A chain). If a drug or chemical is attached to the MoAb, then its name is incorporated into the MoAb name; for example, brentuximab vedotin, where vedotin is an anticancer drug conjugated with this antibody.

ANTIBODIES AS THERAPEUTICS: ADVERSE EFFECTS

Use of antibodies as therapeutic agents is more than a century old practice, and started with the use of antibodies/antiserum to provide passive immunity. Passive immunity provides immediate relief due to the neutralizing effect of antibodies because there is no time lapse in the production of antibodies; passive immunity does not leave immunological memory to protect against the next exposure. Unfortunately, polyclonal antibodies, when injected into a different host, lead to the generation of immunity against the antiserum. This occurs because the PoAb are recognized as foreign proteins by the recipient. That's why second exposure of the same recipient with antiserum can cause adverse effects that could be severe. An example of using PoAb as a protective agent is an antiserum against snake-bites. Antiserum is an antidote for snake venom, and anti-venom gives immediate relief from the fatal effects of snake venom. This prevents immediate death, but leaves victims suffering with other complications that can be pretty serious in certain cases. The first injection of antidote is helpful to protect the victim, but it also generates immune-memory against the anti-venom. Upon the next injection of anti-venom, numerous problems arise that could have grave consequences.

Artificial induction of passive immunity was started with the use of hyper-immune serum. The discovery of monoclonal antibodies offered a new ray of hope for improvement in transfer of passive immunity. However, application of monoclonal antibodies has its own adverse effects. In due course, various modifications of MoAb have been tried to decrease adverse effects caused by MoAb treatments. The adverse effects of therapeutic administration of antibodies are categorized into following four classes: (1) Serum Sickness, (2) HAMA, (3) HACA, and (4) HAHA (Table 15.3).

Serum Sickness

Serum sickness is a condition that arises due to the immune reaction against different proteins present in antiserum. In certain clinical conditions antiserum is administered to treat infections and other serious health risks like rabies, snake

TABLE 15.3 Antibodies as Therapeutics: A Comparison

Antibody Type	Adverse Response	Origin[@]		Human	Administration	Effector
		Constant	Variable	Immunoglobulin (%)	Frequency	Functions
Polyclonal	Serum Sickness	Animal	Animal	0	+ + +	+
Monoclonal	HAMA	Mouse	Mouse	0	+ +	+ +
Chimeric	HACA	Human	Mouse	60–65	+	+ + +
Humanized	HAHA-1	Human	Human*	90–95	+	+ + +
	HAHA-2	Human	Human*	90–95	+	+ + +

Abbreviations: HAMA, Human Anti-Monoclonal Antibody Response; HACA, Human Anti-Chimeric Antibody Response; HAHA, Human Anti-Humanized Antibody Response.
**Only CDR is of mouse origin; @Represent constant portion and variable portion of immunoglobulin;* +Low; ++Medium; +++High.

bites, etc. Immunologically, serum sickness is a delayed type of immune response. Generally, polyclonal antibodies (protective) are raised in different hosts or species (e.g. horse). Proteins present in antiserum, or even protective polyclonal antibodies themselves, are recognized as foreign or non-self entities by the immune system of the patient. The infected patient's immune system starts producing antibodies against the antiserum; this immune response could be due to formation of immune-complexes, complement-mediated immune reaction, etc.

Some of the most common symptoms of serum sickness are skin rashes, joint pain, fever, malaise, swollen lymph nodes, itching, wheezing, flushing, diarrhea, etc. Serum sickness is commonly observed within 2–3 weeks of the injection of antiserum, while in some patients, this response may appear within 3–4 days post-injection. A second exposure to antiserum leads to more severe and pronounced adverse effects as compared to the first exposure.

Generally, hyper-immune serum contains additional serum proteins; therefore, to attain an effective concentration of protective antibodies, a greater amount of hyper-immune serum has to be administered. Higher amounts of serum contain higher concentrations of other non-specific proteins, therefore, adverse effects are common and the reaction is more severe with the use of antiserum. The bottom line of this treatment modality is that patients are saved from immediate threat, but their lives become miserable due to serum sickness. Nowadays use of antiserum is restricted and limited unless it becomes a necessity.

Human Anti-Monoclonal Antibody (HAMA) Response

The discovery of monoclonal antibodies has been a great hope for the proponents of passive immunity for two reasons: (1) high specificity, and (2) a consistent source of antibodies without variability. Monoclonal antibodies used for passive immunity have two advantages: (1) lower amounts of antibodies have to be given because of their high specificity, and (2) unnecessary proteins are not present in the preparation because MoAb are purified before administration. It was fairly reasonable thinking that MoAb would resolve problems related to serum sickness caused by therapeutic use of antiserum. Initially MoAb had been a great hope among clinicians and scientists as a great modality for treatment. This belief was quickly shattered due to the HAMA response. Since MoAb are of murine origin, when MoAb are used to treat human patients, the immune systems of the patients recognize MoAb (protective) as foreign and an immune response and immune memory are generated against MoAb (protective) too. This adverse response against MoAb is known as HAMA. Of course, the HAMA response has less severe adverse effects compared to serum sickness, but the pattern of the HAMA response was similar to the pattern of serum sickness (Tjandra et al., 1990; DeNardo et al., 2003). New modifications have been developed using MoAb to make MoAb better tools to generate passive immunity.

Human Anti-Chimeric Antibody (HACA) Response

MoAb has an advantage to be a perpetual source of highly specific antibodies. This characteristic of MoAb can be further exploited by creating new modifications. Research on antibodies has delineated the fact that for antigen–antibody binding or neutralization of the antigen–pathogen, only the Fab portion of the antibody is crucial. This information was exploited to graft the Fab portion of MoAb onto human antibodies. Now, the grafted antibody had the following advantages: (1) the Fab portion maintains high specificity against the antigen, and (2) the majority of the antibody portion is of human origin. These manipulations can be achieved by the application of protein engineering, one of the latest tools that has allowed researchers to graft Fab of MoAb onto human antibodies. This modification of a human antibody is known as a "chimeric antibody." Chimeric antibodies contain only the Fab portion of mouse origin, while the rest of the chimeric antibody is of human origin. In a typical chimeric antibody, ~33% is of mouse immunoglobulin in origin and ~66% is of human immunoglobulin in origin.

With higher contents of human immunoglobulin, chimeric antibodies offer a great hope to raise passive immunity in patients. No doubt chimeric antibodies have turned out to be a better choice compared to either MoAb or PoAb as immunotherapeutic agents. At the same time, chimeric antibodies also have adverse effects because 33% of the chimeric antibody is still of mouse origin. The adverse response to chimeric antibodies is known as the Human Anti-Chimeric Antibody Response (HACA). The intensity of the HACA response is comparatively less severe compared to the HAMA response or serum sickness. The HACA response turns out to be the limitation for wide application of chimeric antibodies as therapeutic agents. The severity of the HACA response depends upon the immune status of the host (e.g. MABTHERA (Rituximab) is a chimeric antibody raised against the CD20 receptor).

Human Anti-Humanized Antibody (HAHA) Response

The HACA response again turns out to be limiting factor to using chimeric antibodies as therapeutic agents. It is a well established fact that for antigen–antibody reactions the most important part of the variable portion of the immunoglobulin molecule is the complementarity determining region (CDR). Therefore, in principle one can graft only CDRs of MoAb to human immunoglobulins; if that is possible, one can reduce the adverse effects of HACA significantly because (1) CDR maintains high specificity for the antibody to neutralize the antigen, and (2) the major portion of the hybrid immunoglobulin will be of human origin.

Protein engineering has helped in grafting CDRs of MoAb to human immunoglobulin; such a grafted antibody is known as a "humanized antibody." Humanized antibodies are similar to human antibodies, except for CDRs. Humanized antibodies contain only 6–10% mouse proteins, while the other 90–94% is human proteins. Humanized antibodies are far better than other antibodies (i.e. PoAb, MoAb, chimeric antibodies) as immunotherapeutic agents, even though just 6–10% of the mouse portion of immunoglobulin could cause adverse immune responses in some patients (HAHA) (Nechansky et al., 2010).

The severity of HAHA is significantly lower compared to adverse effects caused by serum sickness, HAMA, or HACA. HAHA response is of two types HAHA-1 and HAHA-2. The severity of HAHA-2 is less than HAHA-1, and adverse effects due to HAHA are comparatively easier to manage.

One of the best tools to provide protection in the form of passive immunity is human antibodies. The best option to solve this problem is to develop recombinant human antibodies. Hopefully, one day there will be recombinant human antibodies readily available as a means to provide passive immunity without causing any ill effects.

APPLICATIONS OF ANTIBODIES

Primarily, antibodies are designed to protect the host by neutralizing invasive entities. If antibodies can protect one animal from invaders, it was hypothesized that in a similar fashion antibodies could protect other animals from the same infection if given exogenously. With time, other applications of antibodies were also recognized and tried, but the discovery of MoAb has revolutionized the applications of antibodies (Margulies, 2005). In general, antibody applications can be divided into three main categories: (1) therapeutic, (2) analytical, and (3) preparative.

Therapeutic Applications

Antibodies are basically known to neutralize foreign molecules, toxins, pathogens, etc. These neutralizing antibodies are generated naturally by the host itself. The same rationale was used to raise anti-sera against pathogen, toxins, venoms, etc. Production of anti-venom was one of the most useful applications of PoAb to protect humans after a snakebite. No doubt this strategy saved the host from death due to venom, but the host had to suffer the long-term adverse effects of anti-venom (i.e. serum sickness). Serum sickness gets worse with the second use of anti-venom; although the host is saved from immediate death caused by the snake bite, but it can lead to life-long illness. Polyclonal antibodies have been successfully used to protect children against infectious agents like *Cornybacterium diphtheriae*. Apart from serum sickness, another major drawback of immunotherapeutic application of PoAb is variation in specificity from lot-to-lot, so it is not easy to decide the concentration or dose of antidote.

MoAb are a better choice as immunotherapeutic agents due to high specificity and the technical ability to produce same MoAb again and again without any alterations in characteristics. Therapeutic MoAb under clinical trials and on the market have been extensively reviewed in several articles (Brekke and Sandlie, 2003; Chan et al., 2009; Francis and Begent, 2003). A list of various monoclonal antibodies used as therapeutic agents is given in Table 15.4. Another advantage of MoAb is that they can be conjugated with various payloads (e.g. drugs, radio-nuclides, toxins, markers, etc.). Conjugated MoAb can be used as therapeutics for the following reasons: (1) due to their specificity they can bind or affect specific targets, and (2) when conjugated with payloads, the payload can be delivered to specific targets. This is the reason why MoAb found their therapeutic applications

TABLE 15.4 Therapeutic Antibodies Approved by the FDA*

Year	Name	Type	Target	Application
1986	Orthoclone OKT3	Mouse	CD3	Transplantation
1994	ReoPro	Chimeric	gpIIb/gpIIa	Cardiovascular Disease
1997	Rituxan	Chimeric	CD20	Non-Hodgkin's Lymphoma
1998	Zenapax	Humanized	CD25	Transplantation
1998, 1999	Remicade	Chimeric	TNF-α	Anti-arthritis
-do-	Simulect	Chimeric	CD25	Transplantation
-do-	Synagis	Humanized	RSV	Respiratory Syncytial Virus (RSV)
-do-	Herceptin	Humanized	Her-2	Metastatic Breast Cancer
2000	Mylotarg	Humanized	CD33	Acute Myeloid Leukemia (AML)
-do-	CroFab	Ovine[@]	Snake venom	Rattlesnake Antidote
2001	DigiFib	Ovine[@]	Digoxin	Digoxin Overdose
-do-	Campath	Humanized	CD52	Chronic Lymphocytic Leukemia (CLL)
2002	Zevalin	Mouse	CD20	Non-Hodgkin's Lymphoma
-do-	Xolair	Humanized	IgE-Fc	Allergic Asthma
2003	Humira	Human	TNF-α	Arthritis
-do-	Bexxar	Murine	CD20	B-Cell Non-Hodgkin's Lymphoma
-do-	Raptiva	Humanized	CD11a	Psoriasis
2004	Erbitux	Chimeric	EGFR	Breast Cancer Therapy
-do-	Avastin	Humanized	VEGF	Colorectal Cancer
-do-	Tysabri	Humanized	TNF-α	Multiple Sclerosis
2005	Lymphacide	Humanized	CD22	Non-Hodgkin's Lymphoma
-do-	Antegren	Humanized	SAM	Multiple Sclerosis
2007	HuMax-IL-15	Human	IL-15	Inflammation and Arthritis
-do-	ABT-874	Human	IL-12	Psoriasis
2008	HuMax CD4	Human	CD4	T-Cell Lymphoma

**Food and Drug Administration; this is not an exhaustive list of therapeutic antibodies. [@]Fab.*

in cancer treatment. If someone uses toxin-conjugated antibody, then it can be defined as an immuno-toxin. Similarly, using radiolabeled antibody is known as radio-immunotherapy. A new strategy has been adapted using MoAb to achieve direct cell toxicity, which is known as Antibody Directed Enzyme Prodrug Therapy (ADEPT).

Analytical Applications

Monoclonal antibodies, due to their high specificity, have turned out to be an excellent choice for various analytical applications. These applications range from analysis of laboratory data to diagnostic tests used in the clinical or pathology laboratory. Since the advent of monoclonal antibodies and development of technology to conjugate various types of probes, MoAb have revolutionized the field of immunodiagnostics. Nowadays, these tests have become so reliable that they are commonly used for diagnosis and confirmation of various diseases. Due to improvements in technology, these tests are so simple and user-friendly that even a non-technical person can perform them. Presently, immunodiagnostic tests can be directly used even in field conditions; some examples of field testing of immunodiagnostic tests are HIV tests, pregnancy tests, etc. Some of the most common examples of analytical applications of monoclonal antibodies in a laboratory environment are Western blot, radioimmunoassays, Ouchterlony's Double Diffusion (ODD), Radial Immunodiffusion (RID), Enzyme-Linked Immunosorbent Assay (ELISA), Fluorescent Associated Cell Sorting (FACS), Immunohistochemistry (IHC), Immunofluorescence, etc.

Preparative Applications

The characteristic feature of reversible reaction of antigen–antibody interaction has turned out to be a blessing for preparative applications of antibodies. This is the reason a product of choice can be isolated and purified from a complex mixture using antibodies specific for the product. The complex mixture that contains product to be purified is mixed with antibodies attached to a matrix, and the product gets bound to the antibody. The bound product (antigen–antibody complex) can be dissociated by changing the pH or ion concentration of the buffer and the product can be recovered. This process is known as immune-purification or immuno-enrichment. It is used in the laboratory for various applications like FACS, MACS, etc. This method has enormous industrial applications.

METHODOLOGY, PRINCIPLES, AND PROTOCOLS

It is a well-known fact that polyclonal antibodies and the knowledge about them laid the foundation of immunology. Later, MoAb were developed by Kohler and Milstein (it is more appropriate to use the term developed than discovered as MoAb were developed by laboratory manipulations and not produced naturally). The significance of MoAb cannot be denied because they have changed the face and scope of medical science in a number of ways. Since antibodies (either PoAb or MoAb) have enormous significance, what follows is a discussion of some of the common methods and principles involved in production of these antibodies.

Polyclonal Antibodies

The most important thing about production of polyclonal antibodies is the requirement of an animal that can produce antibodies. In general, mammals are the best choice to raise PoAb, though birds can also be good models. Usually the criteria for the selection of animal models to raise PoAb are based on two important variables: (1) application of PoAb, and (2) amount of antiserum required. If a larger volume or amount of antiserum is needed, larger animals or dairy animals are preferred. Commonly used animals for PoAb production are horses, buffalo, cows, sheep, goat, etc. If antibody is required in smaller quantities, then small animals are used to raise antiserum, such as rabbit, rat, or mouse. Mice are used mostly for investigative applications because the volume of blood or antibody that can be obtained from them is limited. Mice are a better choice when comparing differences between PoAb and MoAb raised against the same antigen, as most of the time MoAb are developed using mice as a source of B-lymphocytes. Birds are also used to raise polyclonal antibodies. One of the most commonly used avian models is the chicken. Animals are commonly employed to raise PoAb, but the procedure has the following limitations: (1) it takes time to complete immunization of animals, (2) it is expensive to maintain the herd of immunized animals, and (3) immunized animals are vulnerable to threats such as infection, which can lead to enormous losses if antibodies are used for commercial scale.

Principle

The principle for the production of polyclonal antibodies is based on the observation that an animal has to be exposed to the antigen of choice. In laboratory circumstances, or even for commercial production of PoAb, animals have to be immunized with purified antigen (Carey Henley et al., 1995; Coligan et al., 2005; Cooper and Paterson, 1995). There are various protocols for immunization of animals. Immunization requires multiple exposures of antigen at selected intervals to raise high-affinity antibodies. Affinity maturation of PoAb takes time, which is the reason for repeated antigen injections used in the immunization procedure. Repeated injections of antigen at specific time intervals induce both the primary immune response as well as the secondary immune response. It is advisable to use adjuvant along with antigen for immunizations. Adjuvant causes slow release of antigen, which supports generation of better immune response due to consistent release of antigen into the circulation. Finally, the decision to collect antibodies (serum or plasma) from immunized animals is made after the confirmation of antibody titer. A small volume of serum plasma from the immunized animal is used to confirm antibody titer. If antibody titer is not high, then booster doses of antigen are required.

It is always advisable to optimize conditions to raise PoAb (e.g. animal model, concentration of antigens, etc.). When antibodies are repeatedly required from the same immunized animal, it is advisable to give a booster injection before collecting antiserum.

Methodology and Rationale for PoAb Production

The methodology and rationale of different steps involved in PoAb production are discussed in this section. The method can be divided into steps: (1) Antigen preparation, (2) Immunization, (3) Antibody titration, and (4) Antibody isolation or purification.

Step 1: Antigen Preparation

Antigen has to be prepared for immunization so that one can produce good quality (high-affinity) antibodies. This is the reason antigen is usually injected with adjuvant. The most commonly used adjuvant is Freund's adjuvant. Freund's adjuvant contains components like mineral oil and emulsifying agents (mannide mono-oleate), which disperses the mineral oil into small droplets.

Upon mixing of antigen with Freund's adjuvants, the aqueous antigen preparation gets dispersed into small droplets that are surrounded by a film of oil, aiding slow release of antigen from the site of infection. Freund's complete adjuvant contains heat-killed mycobacteria; routinely, it is recommended that the first immunization be done with antigen mixed or emulsified with Freund's Complete Adjuvant (FCA) because heat-killed *Mycobacterium tuberculosis* is effective to induce non-specific immune response. The second and other consecutive booster injection is given with antigen mixed or emulsified with Freund's Incomplete Adjuvant (FIA). FIA contains every other component of FCA except for heat-killed *M. tuberculosis*.

Step 2: Immunization of Animals

This is the second step for generation of polyclonal antibodies. The major purpose of immunization is to expose animals with specific antigen and repeated exposure of antigen to the immune system to help in antigen recognition. Immunization causes immune activation, resulting in production of antibodies against the antigen. The amount of antigen, site of injection, and duration between booster doses are decided on the basis of the animal used for antibody production.

If a rabbit is used to raise antibodies, one has to give an injection prepared with either FCA or FIA. A subcutaneous (s.c.) route of injection is commonly advised for immunization. To immunize a rabbit, the first step is to shave the hair (fur) from the rabbit's back with a hair trimmer. After trimming the hair, the skin has to be sterilized with 70% ethanol. Then injection of the prepared antigen is given subcutaneously in the shaved skin area using a needle and syringe. If the injection volume is large, subcutaneous injections can be given at multiple sites. After the first injection of antigen, at least two more injections of booster doses have to be given to the same rabbit at an interval of 7 or 10 days. An immunization record has to be maintained, including name of antigen, concentration of antigen, volume of injections, route of injections, site of injections, dates of injection, as well as anything unusual that was noticed during the immunization process. (Note: Before immunization, it is important to check a serum sample of the rabbit to assure that serum does not have any antibody against the same antigen. It is also recommended that the rabbit be bled before each consecutive immunization.). The blood sample that is withdrawn is used to determine antibody titer as mentioned in the following text.

Step 3: Antibody Titer

During immunization and after immunization it is important to know the titer of antibodies present in the immune serum. High titer antibodies are always preferred over low titer antibodies for therapeutic, analytical, or preparative applications. Antibody titration is done by applying various assays based on antigen–antibody reactions like agglutination, hemagglutination, ELISA, Radial Immuno-Diffusion (RID), Ouchterlony's Double Diffusion (ODD), etc. Selection of the assay to quantify antibody titer is dependent upon the facilities and expertise available. One of the most common and easy methods to determine antibody titer is by agglutination test. The advantage of this test is that it is convenient to perform, requires less time to perform, and can be used for particulate antigens as well as soluble antigen. (If the antigen is particulate in nature, it can be directly used for an agglutination reaction, e.g. RBC). If the antigen is soluble, it can be performed with the use of beads (latex), Sheep Red Blood Cells (SRBC), or even Chicken RBCs (CRBC). When RBCs are used for an agglutination test, the test is known as a hemagglutination test. When soluble antigens are coated on RBC (either chicken or sheep) for the agglutination reaction, then it is known as an indirect hemagglutination test. On the basis of the aforementioned tests, one can determine titer of antibodies. Antibody titer is the decision-making factor for further booster injections. Antibody titer also suggests possible dilution of antibodies to be used for further applications.

Step 4: Purification and Identification

Titration of antibodies provides information about specificity of antibodies. Antibody titration methods can also be used to test cross-reactivity of antiserum with closely related antigens. In certain circumstances, serum or plasma containing antibodies can be used directly. Yet in some other circumstances, it is essential to use purified antibodies. There are various methods for antibody purification. Selection of the method for PoAb purification depends upon various factors like availability of resources, cost of purification, scale of purification, and downstream applications of the PoAb. The best and most convenient method for purification of polyclonal antibodies is affinity purification. Detailed protocols to raise polyclonal antibodies can be found in any standard protocol book or lab manual.

Crucial Steps for PoAb Production (Flow Chart 15.1)

1. Antigen Preparation
   Antigen has to be prepared with either FCA or FIA. Proper emulsification of antigen with adjuvant should be tested before injection (Step 1).
2. Immunization
   The animal has to be injected subcutaneously with an antigen prepared with adjuvant. After initial immunization, booster doses have to be given at specific intervals. A serum/plasma sample has to be collected every time, before injecting the antigen (Step 2 and 3).
3. Antibody Titration
   Antibody titration provides the information about affinity maturity of specific antibodies. Titration results are used as the criteria to decide booster doses and time of antibody collection. Agglutination or hemagglutination tests are used to quantify antibody titer (Step 4).
4. Purification
   If antibodies are showing high titer or expected titer, then PoAb can be used as antiserum. If unwanted protein of serum or plasma can interfere with downstream applications, then PoAb can be purified by various means (Step 5, 6, and 7).

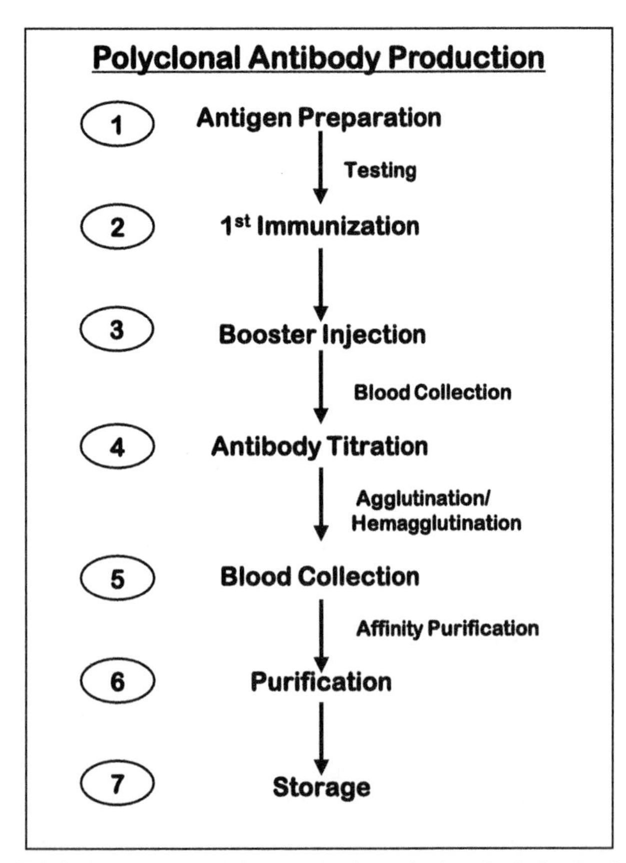

FLOW CHART 15.1 Major steps for the production of polyclonal antibodies.

Monoclonal Antibodies

MoAb are those antibodies produced from a single cell (i.e. one specific B cell). Antibodies with different specificities are present in antiserum as they are produced from numerous B cells or a clone. Undoubtedly, polyclonal antibodies are beneficial to the host, but have limited therapeutic, analytical, and preparative applications. It is easier to identify and purify antibodies of interest from a PoAb mixture; nevertheless, it is really difficult or next to impossible to identify unique B cells that produce that specific antibody. Even after tedious efforts to identify the B cell of interest, the B-lymphocyte cannot be used as a continuous source of antibody because B cells have a limited lifespan. Therefore, to have a continuous source of consistent antibodies, the only option is to have immortalized B cells. The answer for this complicated problem came with the discovery of hybridoma technology, which is the principle applied in the production of MoAb. Now the major concern is that immortalization of B cells should neither change B cells characteristics of antibody production, nor should the specificity of the antibody be changed after immortalization. MoAb came into existence in the last quarter of the 20th century with the efforts of Kohler and Milstein, who designed an experiment to immortalize an antibody-producing B-cell by fusing it with tumor cells (i.e. myeloma cells). They had fused primary B cells (antibody-producing) with myeloma cells (tumor cells) and produced a hybrid cell that retained the characteristics of both cells. The method that Kohler and Milstein used exploited good characteristics of both cell types (i.e. hybrid cells or hybridoma cells were able to retain the antibody-producing quality of B cells and the immortalization properties of myeloma cells). The word hybridoma has evolved from the union of two English words, like the union of two cells (i.e. B cells and myeloma). For the word "hybridoma," "hybrid" came from the hybridization of two cells and "oma" came from myeloma cells. These hybridoma cells became a perpetual source of antibodies having the same specificity.

Principle

The principle for the production of MoAb relies on two important factors: (1) a source of antibody-producing cells (i.e. B cells), and (2) immortalization of these cells by fusion with tumor cells. After fusion, the cell mixture has three different cell types: (1) B cells, (2) tumor cells, and (3) hybrid cells. Although it sounds very simple to generate monoclonal antibodies, this method has had some inherent limitations (i.e. selection of correct hybrid cells). The issue of selection of hybrid cells was resolved with the use of selection medium. The rationale of the selection process is discussed under the "Biochemical Pathway and Hybridoma" section of this chapter. Hybridoma cells can be grown either *in vitro* (in tissue culture) or *in vivo* (in animals) as a tumor. Hybridoma cells can be frozen at ultra-low temperatures and can be retrieved and revived from frozen stocks for monoclonal antibody production when required. Generation of monoclonal antibodies requires working both under *in vivo* and *in vitro* conditions.

Methodology and Rationale of MoAb Production

This section discusses the methodology and rationale of different steps required for the production of MoAB using mice (Coligan et al., 2005). MoAb production can be divided into five major steps: (1) Immunization of animal, (2) Fusion of cells, (3) Selection of clone, (4) Expansion of clone, and (5) Purification of monoclonal antibodies (Figure 15.3).

Step 1: Immunization of Mouse

The prerequisite for production of monoclonal antibodies is to immunize animals that can produce the antibody of choice. Immunization leads to antibody production because of the activation of B cells. After immunization, spleen is used as a source of lymphocytes. The spleen contains B-lymphocytes, which secrete antibodies against the antigen used for immunization.

For mouse immunization, the procedure followed is the same as mentioned for production of polyclonal antibodies. The essential steps to immunize a mouse are (1) antigen preparation, (2) immunization, and (3) determination of antibody titer. These steps have already been discussed in the section "Methodology and Rationale for PoAb Production."

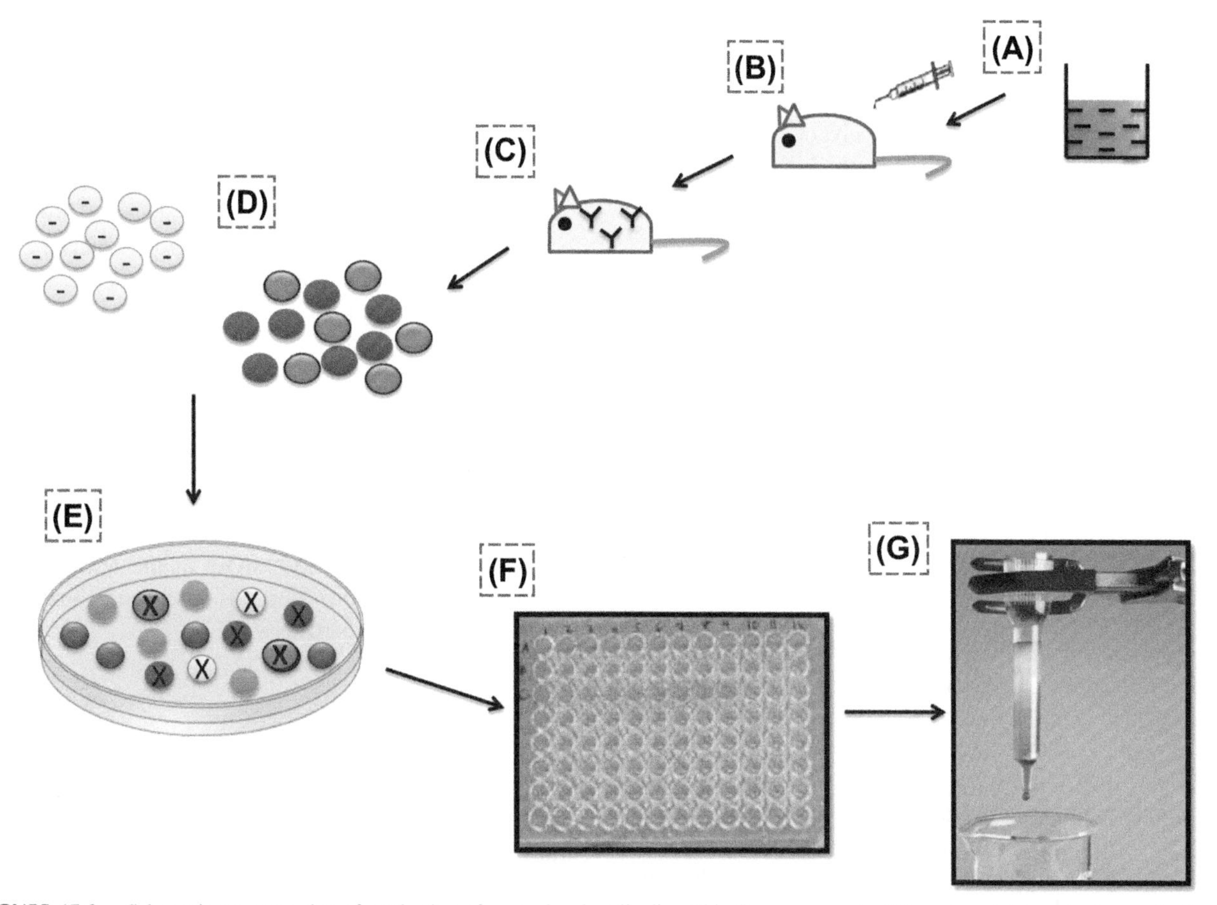

FIGURE 15.3 Schematic representation of production of monoclonal antibodies. This figure represents different crucial steps in the production of monoclonal antibodies. (A) Antigen preparations: Antigen has to be prepared with Freund's Complete Adjuvant or Freund's Incomplete Adjuvant. (B&C) Immunization: Mice have to be immunized by injecting antigen prepared with adjuvant; they also have to be given booster doses. (D) Preparation of splenocytes and fusion with myeloma: The spleen has to be removed from immunized mice and a single-cell suspension of splenocytes has to be prepared. Then splenocytes have to be fused with myeloma cells in the presence of fusogenic agents. (E) Selection of hybridoma: After fusion, a hybridoma has to be selected from the cell population mixture. Cells have to be grown in HAT selection medium so that after selection only hybridoma cells can survive, while B-lymphocytes and un-fused myeloma cells will die. (F) Screening of clone: After selection of hybridoma cells, a specific clone has to be selected. Different hybridoma cells are diluted in 96-well plates, and after a period of time each clone has to be tested for specificity against the antigen. (G) Purification of monoclonal antibodies: After selection of a specific hybridoma clone, monoclonal antibodies can be purified. If downstream application requires purification of monoclonal antibodies, then the clone can be expanded and appropriate methods can be applied for purification of monoclonal antibodies. ( ), myeloma cells; (,), lymphocytes; and (,), hybridomas. Cells marked with an "X" represent cell death in the selection medium.

Step 2: Preparation of Splenocytes

In mice, a very limited volume of peripheral blood can be drawn, which can become a limitation. Therefore, the best source for lymphocytes is spleen. Dissection of spleen and preparation of splenocytes has to be performed under aseptic conditions. The mouse has to be killed by cervical dislocation and its skin surface has to be sterilized with 70% ethanol before bringing it into the hood. The spleen has to be surgically removed under sterile conditions. Then a single-cell preparation has to be prepared by lysing RBCs using RBC lysis buffer. After RBC lysis and washing, splenocytes are ready for fusion with myeloma cells.

Step 3: Fusion of Cells

For the fusion step, splenocytes and myeloma cells are mixed together in a specific ratio. A cell suspension of mixed cells has to be treated with agents that can induce fusion. There are various agents that are used for fusion of B-lymphocytes with myeloma cells. Some of the common ones are Sendai virus, polyethylene glycol, etc. Nowadays, the most convenient and cheapest method to fuse cells is by using polyethylene glycol. After fusion, cells become quite fragile, so further steps have to be performed with utmost care. No matter which method is used, a very common problem encountered is low cell viability during the fusion.

Step 4: Selection of Hybrid Cells

When a mixture of fused cells is grown in Hypoxanthine, Aminopterin, and Thymidine (HAT) medium, the aminopterin works as a folate antagonist and blocks the *de novo* pathway for the synthesis of the nucleotide (i.e. purines and pyrimidines), so the cells that do not carry out the salvage pathway for nucleic acid synthesis eventually die in the selection medium.

In due course, primary cells (B cells) will die because they have a limited lifespan. Myeloma cells that do not get hybridized will die because of blockage of the *de novo* pathway and nonfunctional salvage pathway. Details are discussed under the section "Biochemical Pathway: Hybridoma Selection." Finally, only those cells that are fused with tumor cells will get selected; these will be the only survivors. Even the selection medium leaves a mixture of cells producing antibodies with varying specificity. To identify hybrid cells producing the antibody of interest, screening procedures are required (as discussed in the next step).

Step 5: Selection of Clones

After selection of the fused hybridoma cell, one has to select a specific clone secreting the desired antibody from the population of hybridoma cells. This experiment involves plating of cells in a 96-well plate; the experiment is designed in a manner that after dilutions, each well contains a single antibody-producing cell. After incubation, cells start dividing and the cell number increases. The time eventually comes when they start producing detectable levels of antibody against the antigen. Ideally, each well of tissue culture should produce antibodies with single specificity because cells that have grown in each well should arise from a single cell (clones). During this step, culture supernatant from each well is screened for antibody production. Appropriate protocols can be applied for screening of clones, as mentioned in the production of polyclonal antibodies. Hybridoma cells that produce antibodies against the antigen are selected. This will be the desired clone, which has to be further expanded for production of the monoclonal antibody.

Step 6: Expansion of Clone

After identification of the specific clone, it has to be expanded so that it can be used for larger antibody production. Expansion strategies are based upon need and amount of antibody required. Two methods commonly used for expansion of clones are the following:

1. *In vitro* method: In this method, as cells grow in number, they can be transferred to larger culture vessels stepwise (like 6-, 12-, 24-, 48-, and 96-well plates) so that they have enough medium and space to grow. Later they can be transferred to different sizes of tissue culture flask. It is always a good practice to freeze some of these cells at ultra-low temperatures. These frozen cells can be retrieved and revived for further production of antibodies.
2. *In vivo* method: In this method antibody-producing clones can be introduced into the peritoneal cavity of mice, where they start growing as a tumor. Antibodies will be secreted by the hybridoma (tumor) cells in ascites fluid in the peritoneal cavity of the mice. Further expansion of the clones depends upon need, availability of facilities, and expertise.

Step 7: Purification

Although it is not always essential to purify monoclonal antibodies as they are specific and produced from one clone, they can be used directly either in the form of tissue culture supernatant or in ascites form (if this does not interfere with further steps). Sometimes it might be necessary to purify monoclonal antibodies in certain conditions, like when (1) antibody concentration is low, (2) other components of tissue culture medium or ascites fluid are interfering, (3) the antibody has to be used for therapeutic purposes, (4) the antibody has to be used for diagnostic purposes, or (5) the antibody needs further manipulation (e.g. conjugation with Horse Radish Peroxidase (HRPO), alkaline phosphatase, fluorescent dye, etc.). For these kinds of applications it is always better to purify monoclonal antibodies from the tissue culture supernatant. Monoclonal antibodies can be purified by various means; the most common method for purification of antibodies is by affinity column chromatography. Detailed protocols to raise monoclonal antibodies can be found in any standard protocol book or lab manual.

Crucial Steps in Production of Monoclonal Antibodies (Flow Chart 15.2)

1. Antigen Preparation: Antigen has to be prepared with the adjuvant of choice. The most common adjuvants are either FCA or FIA. Before injecting antigen, it is advisable to test the proper emulsification of antigen with adjuvant (either FCA or FIA) (Step 1).
2. Immunization: Mice have to be immunized by injecting antigen prepared either with FCA or FIA. Usually injections are given subcutaneously, and a standard immunization protocol is followed (Step 2).

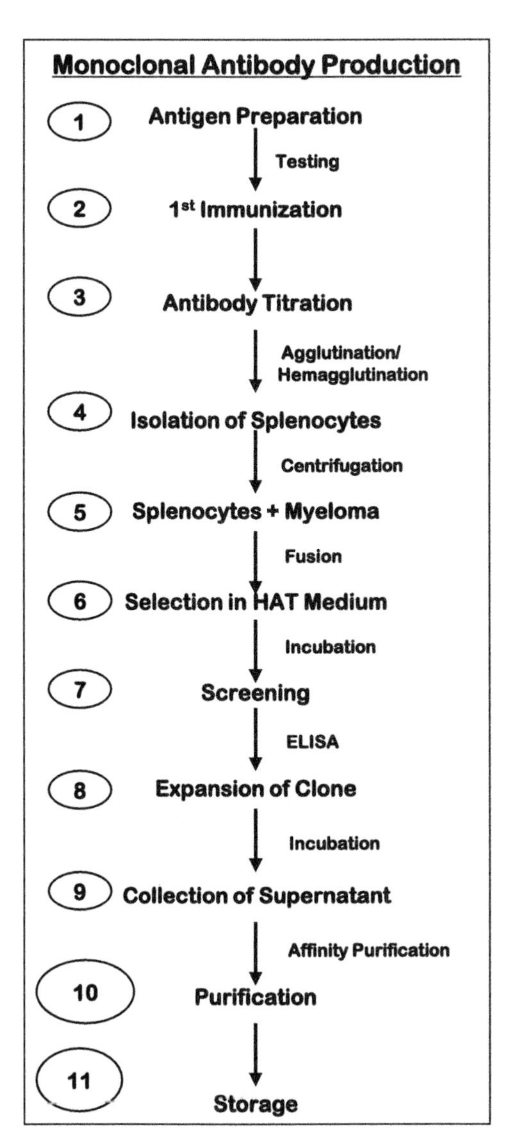

FLOW CHART 15.2 Major steps for the production of monoclonal antibodies.

3. Antibody Titration: On the basis of antibody titer, a decision is made regarding booster doses and preparation of splenocytes (Step 3).
4. Preparation of Splenocytes: In mice, splenocytes are the source of B-lymphocytes (activated). A single-cell suspension is prepared for fusion with myeloma cells (Step 4).
5. Fusion: Isolated splenocytes have to be fused with myeloma cells using fusogenic agents for the production of hybridoma cells (Step 5).
6. Selection: After fusion, three different cell types are present in the mixture (i.e. splenocytes, tumor cells, and fused cells). From this mixture of cells, only fused cells are selected with the aid of selection medium (Steps 6 and 7).
7. Expansion: After selection of fused cells, these cells have to be expanded and identified by confirming their ability to produce the antibody of interest. The unique clone has to be further expanded by different available methodologies (Steps 8 and 9).
8. Purification: Monoclonal antibodies can be purified by various methods. The most common method for purification of antibodies is by affinity column chromatography (Steps 10 and 11).

Antibody Titration

Polyclonal antibodies raised by immunization against an antigen need to be quantified for their specificity and titer. In a similar fashion, antibody titer is also required for MoAb production. The quantity of antibodies for further application can be determined by knowing the titer of the antibodies. High titer antibodies are also necessary for generation of good quality PoAb/MoAb. To ascertain titer of antibodies, one has to have an assay system to titrate antibodies either for production of polyclonal antibodies or for generation of monoclonal antibodies. Various method to titer antibodies are ELISA, ODD, RID, agglutination, hemagglutination, etc. All these methods are based on the principle of antigen–antibody binding. The choice of method to be used is dependent upon the facilities and expertise available, purpose, etc. One of the most common, convenient, and cost effective methods for this purpose is the agglutination method, which is based on the principle of antigen and antibody interactions. When antigen and antibody are present in optimum concentrations, the antigen–antibody complex forms a three-dimensional, lattice-like structure.

Agglutination reactions can be performed with particulate antigens (e.g. RBCs, bacteria, etc.). The Widal Test for *S. typhimurium* or blood group testing is also based on the principle of agglutination reactions. When the antigen is soluble, one can coat the antigen on RBCs; this reaction is called as hemagglutination.

Methodology and Rationale of Antibody Titration (Flow Chart 15.3)

1. Antigen Preparation: A soluble antigen-like bovine serum albumin (BSA) has to be dissolved in appropriate buffer. This soluble antigen is used to coat sheep RBCs (SRBCs) (Step 1).
2. Preparation of SRBCs: SRBCs have to be collected and washed. A specific concentration of SRBC has to be prepared in buffer. After making the SRBC solution, it has to be coated with tannic acid. This tannic acid-coated SRBC can be used for coating any soluble antigen (Step 2).
3. Antigen Coating: After coating SRBCs with tannic acid, they have to be mixed with antigen and incubated for a certain time period. During incubation, antigens get attached to the SRBCs and the coated SRBCs work as particulate antigens, and can be used to titrate the antibody (Step 3).
4. Antibody Titration: Hemagglutination plates (which have 96 wells) are used. In these plates one can use serial dilution of antibody. After preparing the dilutions, an equal volume of antigen-coated SRBCs are added in each well, and the plate is incubated at room temperature (Step 4).
5. Observation: Positive agglutination is seen when cells form a continuous layer in the wells. In the wells where agglutination has not occurred, SRBCs settle down at the bottom of the well, resembling a red button (Step 5).
6. Calculation: The maximum dilution showing positive agglutination, which can be observed with the naked eye or with the help of mirror, is recorded. This dilution is the antibody titer against the antigen (Step 6).

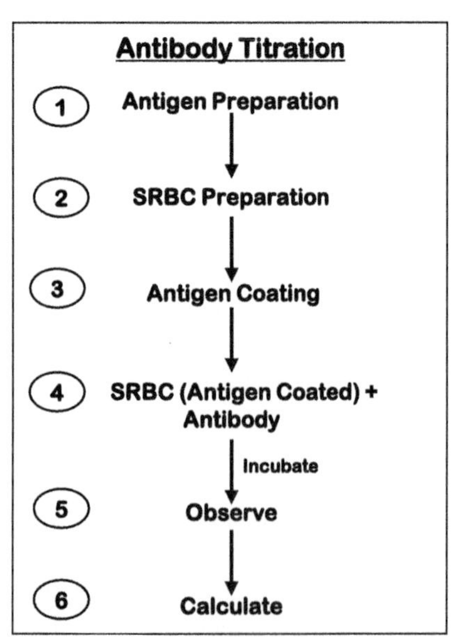

FLOW CHART 15.3 Major steps for the titration of antibodies.

BIOCHEMICAL PATHWAY: HYBRIDOMA SELECTION

Nucleotide synthesis is important and crucial for DNA synthesis and cell division. In any cell type there are two pathways for nucleotide synthesis: a *de novo* pathway and a salvage pathway. In the *de novo* pathway (new; from scratch), nucleotides are synthesized from simple precursors that cells can procure from outside sources. In the salvage pathway, synthesis of nucleotides occurs by using intermediates found in the degradation pathway of nucleotides such as nucleotides and nitrogenous bases.

It has been studied and established that cells grown *in vitro* can make use of both the pathways for nucleotide synthesis. When we use splenocytes for fusion with myeloma, the fusion cell contains two different cell types: (1) splenocytes, which are primary cells, and (2) myeloma cells, which are tumor cells. Out of these two cell types, primary cells have a limited lifespan, while tumor cells are immortal and can be grown for an unlimited period of time. To select the hybrid cells, the nucleotide biosynthesis pathway turned out to be a rescuer. Mutation in enzymes related to salvage pathways is a common method used in mammalian cell cultures. The most common target enzyme is hypoxanthine-guanine phosphoribosyl transferase (HGPRT). HGPRT is an important enzyme of the salvage pathway. Since HGPRT is a part of non-essential pathway, $HGPRT^-$ cells can grow normally and perform DNA synthesis using a *de novo* pathway. It was speculated that B cells have a finite lifespan so they will die after some time, but myeloma cells have to be eliminated from the culture medium. Therefore, myeloma cells have to be made HGPRT-deficient, and the culture should be done in a medium that favors growth of $HGPRT^+$ cells only. This way myeloma cells will not survive in the selection medium. Using the same rationale, hybridomas produced after fusion of B cells and myeloma cells will be able to survive in the selection medium. $HGPRT^-$ can be selected by growing myeloma cells in the presence of purine analogs such as 8-azoguanine (8-AG). HGPRT uses 8-AG as substrate and converts 8-AG into nucleotide monophosphate. The 8-AG-containing nucleotides are further processed during DNA synthesis and get incorporated into DNA and RNA. Substitution of 8-AG is toxic, therefore cells that are $HGPRT^+$ (or normal) will die, and only $HGPRT^-$ cells will survive when grown in the presence of 8-AG. In this way $HGPRT^-$ myeloma cells are selected. These $HGPRT^-$ myeloma cells are fused with B cells, which are $HGPRT^+$. After fusion, the mixture has different cell populations: (1) B-lymphocytes (unfused), (2) myeloma cells (unfused), and (3) fused cells (or hybrid cells). The major issue was to select only those hybrid cells that are producing the antibodies of interest. The solution to this problem came with the use of a selection medium

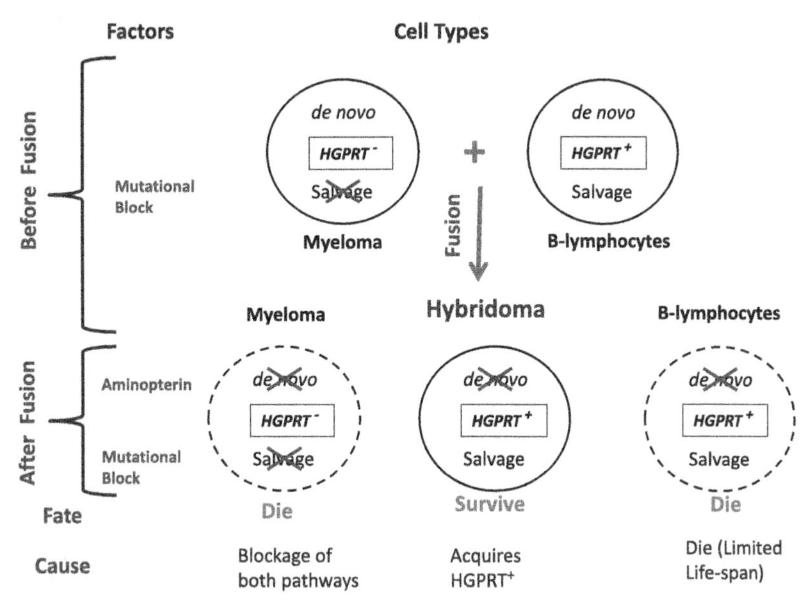

FIGURE 15.4 Biochemical pathways used for selection of hybridomas: For the production of monoclonal antibodies, myeloma cells and splenocytes are used, and after fusion, hybridoma cells are selected. Myeloma Cells that are used for the production of monoclonal antibodies are HGPRT$^-$ (hypoxanthine-guanine-phosphor-ribosyl-transferase), which is an essential enzyme of the salvage pathway. These cells are unable to carry out the salvage pathway. After fusion, hybridoma cells acquire the HGPRT$^+$ gene from B-lymphocytes. When these cells are grown with selection medium (HAT medium), aminopterin present in this selection medium blocks the *de novo* pathway, but the hybridoma can survive by synthesizing nucleotides via the salvage pathway. Myeloma cells and B-lymphocytes die in selection medium.

that allowed only hybridomas to survive. The selection medium used for MoAB production is known as HAT Medium (Hypoxanthine, Aminopterine, and Thymidine). Aminopterine blocks the *de novo* pathway, whereas hypoxanthine and thymidine are required for nucleotide biosynthesis in the salvage pathway. In selection medium, the following things occur which support selection of hybrid cells: (1) B-lymphocytes (unfused), being primary cells, die due to limited lifespan; and (2) myeloma cells that are HGPRT$^-$ die in the selection medium because they do not have a salvage pathway. Only hybridoma cells can survive because now they are HGPRT$^+$ and can use the salvage pathway for nucleotide biosynthesis (Figure 15.4).

ETHICAL ISSUES

At the present time, ethical issues have become significantly important, certainly keeping pace with antibody research and its commercialization. It has become a necessity to regulate experiments as well as industrial applications of antibodies and their modifications. Antibodies, either polyclonal or monoclonal, have to be harvested from animals. Harvesting antibodies is associated with various ethical issues regarding animal use. Animals either have to be bled or sacrificed to serve the purpose. Animal experimentation to raise antibodies involves use of Freund's Complete and Freund's Incomplete Adjuvants; these adjuvants may cause wounds at the site of injection. Therefore, dose and volume of the immunization injection have to be carefully monitored for animal health and welfare. On the one hand, animals serve as a source for antibody production, and on the other hand are good models for testing various aspects of antibodies.

These processes of animal experimentation need ethical clearance with due consideration to different aspects of antibody production (e.g. bleeding volume, bleeding site, choice of animals, etc.). Use of transgenics for antibody production raises various other ethical issues related to the use of recombinant DNA. Apart from animals, the use of human subjects is undoubtedly a serious concern for various ethical reasons, such as (1) the need for healthy volunteers to test therapeutic antibodies, (2) the trial of antibodies in terminal patients, and (3) the reliability of data obtained from such trials for further clinical utility and benefits. One of the glaring examples of antibody-related clinical trials took place in London in 2006. Six healthy individuals participated in a clinical trial for a new monoclonal antibody (TGN-1412). Within minutes of inoculation, all six volunteers were reported to have suffered multiple organ failure (Stebbings et al., 2007; Nechansky and Kircheis, 2010). These volunteers did survive, but the episode left many questions about the safest way to test antibodies in human subjects, and what the criteria should be in such a study. As time goes by, it is still difficult to find straight answers to some of these questions. New ethical issues will certainly develop over time, which will demand appropriate attention and resolutions.

TRANSLATIONAL SIGNIFICANCE

The translational value of the immune system and immunity is only limited by the imagination. It is a Herculean task to decide where to start when considering the translational significance of antibodies because anything and everything about antibodies has translational significance. Applications of immunoglobulins are growing as more and more knowledge is gained, and numerous antibody applications are being realized and recognized by various stakeholders such as scientists, clinicians, etc.

von Behring demonstrated the first translational use of antibodies when he used them to treat infectious diseases. He probably used serum without knowing the details about antibodies, except from knowledge gained by observations in day-to-day life. Later, Edward Jenner's experiment with vaccines also implicated antibodies in the induction of immunity directly or indirectly. However, the discovery of MoAb opened so many avenues for their applications. MoAb have been the main force for development of chimeric and humanized antibodies. Advancement in protein chemistry has introduced newer applications of antibodies (e.g. conjugated MoAb). Attachment of probes to MoAb has broadened the horizons for their versatile applicability in different areas of biomedical sciences. Presently, various probes can be conjugated with MoAb (e.g. alkaline phosphatase, HRPO, FITC, magnetic probes, etc.). It is difficult to imagine ELISA, FACS, immunohistochemistry, immunoelectron microscopy, or confocal microscopy without antibodies. What about purification of various products to homogeneity without immuno-affinity chromatography? With creative thinking, it will be possible to find many new applications for the specificity of antibodies.

As of now, protein engineering can be used to produce bi-specific antibodies (Presta, 2003). Employing recombinant DNA technology, it may be possible to produce multi-specific antibodies in the near future. There is always a possibility that these antibodies may either have no side effects or may have minimal and manageable side effects when used as therapeutic agents. At this juncture it is a possibility to produce completely human antibodies only by artificial means (i.e. outside the human body). These antibodies could be a boon to treat infections and other disease causative factors. Only time will tell if a magic bullet could be developed, as a solution for many illnesses, if not all.

WORLD WIDE WEB RESOURCES

The World Wide Web has become the most convenient and powerful tool for retrieving information on any subject, including scientific literature. As is well known, Google has turned out to be the most popular site for information retrieval. A search for the keyword "monoclonal antibody" at www.google.com returns more than 8.62 million hits, while "polyclonal antibody" returns 3.2 million hits. There is no doubt that Google is a gigantic resource for information regarding basic concepts on antibodies, immunoglobulin structure and function, immune response, animations, protocols for raising polyclonal and monoclonal antibodies, adjuvants, animal models for antibody production, rules and regulations, and guidelines on the use of animals for experimental studies. It is not an overstatement to say that any information related to monoclonal and polyclonal antibodies can be easily searched via Google.

Another important web resource for antibodies is the National Center for Biotechnology Information (NCBI) (http://www.ncbi.nlm.nih.gov). This website is hosted by the National Institutes of Health (NIH), Bethesda, Maryland. It is one of the most authenticated web resources for scientifically validated and updated information. One can use various search options to retrieve information on research publications, gene sequences, and protein sequences; even some of the best books are available on this site.

Another useful web resource is the ImMunoGeneTics site (http://www.imgt.org). It covers different aspects related to immunoglobulins (e.g. genetics, gene sequences, etc.).

The Antibody Resource Page (http://www.antibody resource.com) is one of the most comprehensive web pages dedicated to antibodies. This site provides information about antibody structure, function, production, research, and clinical applications, as well as great educational resources for immunology, antibody images and galleries, and names of journals and books related to the subject. This web resource also lists the suppliers of custom-made antibodies and contract service providers for antibodies, etc.

Other resources on antibodies are the web sites of Current Protocols in Immunology (http://onlinelibrary.wiley.com/book/10.1002/0471142735) and Current Protocols of Molecular Biology (http://onlinelibrary.wiley.com/book/10.1002/0471142727). These are some of the best resources for antibodies regarding principles, limitations, and benefits of a wide variety of protocols.

There is currently no way of getting around the use of animals for the production of antibodies. Therefore, there is always a need for information regarding animal welfare. For information of this nature, one can refer to the website maintained by the United States Department of Agriculture (USDA), Beltsville, Maryland. One can find numerous books and proceedings regarding antibodies at their site: http://awic.nal.usda.gov/awic/pubs/antibody/books.htm.

For monoclonal antibody naming conventions, the American Medical Association (AMA) provides good information at its site dedicated to naming biologics: http://www.ama-assn.org/ama/pub/physician-resources/medical-science/united-states-adopted-names-council/naming-guidelines/naming-biologics.page.

ACKNOWLEDGMENTS

The authors are thankful to Prof. K.C. Upadhyaya, Director, AIB, AUUP, NOIDA, (UP), India, for providing the necessary resources to complete this book chapter. Mr. Dinesh Kumar is duly acknowledged for his secretarial assistance; he was instrumental in completing the graphical work presented in this book chapter. The authors would like to express their gratitude to Ms. Priyadarshini Mallick, Mr. Deepak Kushwaha, and Mr. Ajay Yadav for their untiring efforts in performing literature searches that were essential for the completion of this book chapter.

REFERENCES

Black, C. A. (1997). A brief history of the discovery of the Immunoglobulins and origin of the modern Immunoglobulin nomenclature. *Immunology and Cell Biology*, *75*, 65–68.

Brekke, O. H., & Sandlie, I. (2003). Therapeutic antibodies for human diseases at the dawn of the twenty-first century. *Nature Reviews Drug Discovery*, *2*, 52–62.

Burnet, F. M. (1959). *The Clonal Selection Theory of Immunity*. USA: Vanderbitt University Press.

Burton, D. R. (1987). Structure and function of antibodies. In F. Calabi & M. S. Neuberger (Eds.), *Molecular Genetics of Immunoglobulins*. The Netherlands: Elsevier Science Publishers.

Carayannopoulus, L., & Capra, J. D. (1993). Immunoglobulins Structure and Function. In W. E. Paul (Ed.), *Fundamental Immunology*. USA: Raven.

Carey Hanly, W., Artwohl, J. E., & Bennett, B. T. (1995). Review of polyclonal antibody production in mammals and poultry. *Institute for Laboratory Animal Research Journal*, *37*, 93–118.

Chan, C. E., Chan, A. H. Y., Hanson, B. J., & Ooi, E. E. (2009). The use of antibodies in the treatment of infectious diseases. *Singapore Medical Journal*, *50*, 663–673.

Coligan, J. E., Kruisbeck, A. M., Margulies, D. H., Shevach, E. M., & Strober, W. (2005). *Current Protocols in Immunology*. New York: John Wiley and Sons.

Cooper, H. M., & Paterson, Y. (1995). Production of polyclonal antisera. In J. E. Coligan, A. M. Kruisbeck, D. H. Margulies, E. M. Shevach & W. Strober (Eds.), *Current Protocols in Immunology* (pp. 2.4.1–2.4.9). New York: John Wiley and Sons, Inc.

DeNardo, G. L., Bradt, B. M., Mirick, G. R., & DeNardo, S. (2003). Human antiglobulin response to foreign antibodies: therapeutic benefit? *Cancer Immunology and Immunotherapy*, *52*, 309–316.

Edelman, G. M., Benacerraf, B., Ovary, Z., & Poulik, M. D. (1961). Structural differences among antibodies of different specificities. *Proceedings of National Academy of Sciences USA*, *47*, 1751–1758.

Edelman, G. M. (1973). Antibody Structure and Molecular Immunology (Nobel prize address). *Science*, *180*, 830–840.

Francis, R. J., & Begent, R. H. J. (2003). Monoclonal antibody targeting therapy: An overview. In K. N. Syrigos & K. J. Harrington (Eds.), *Targeted Therapy for Cancer*. New York: Oxford University Press.

Frazer, J. K., & Capra, J. D. (1999). Immunoglobulins: structure and function. In W. E. Paul (Ed.), *Fundamental Immunology* (4th ed.). USA: Lippincott-Raven.

Kohler, G., & Milstein, C. (1975). Continuous cultures of fused cells secreting antibody of predefined specificity. *Nature*, *256*, 495–497.

Lipman, N. S., Jackson, L. R., Trudel, L. J., & Weis-Garcia, F. (2005). Monoclonal versus polyclonal antibodies: Distinguishing characteristics, applications and information resources. *Institute for Laboratory Animal Research Journal*, *46*, 258–267.

Margulies, D. H. (2005). Monoclonal antibodies: Producing magic bullets by somatic cell hybridization. *The Journal of Immunology*, *174*, 2451–2452.

Nechansky, A. (2010). HAHA – Nothing to laugh about. Measuring the immunogenicity (human anti-human antibody response) induced by humanized monoclonal antibodies applying ELISA and SPR technology. *Journal of Pharmaceutical and Biomedical Analysis*, *51*, 252–254.

Nechansky, A., & Kircheis, R. (2010). Immunogenicity of therapeutics: A matter of efficiency and safety. *Expert Opinion Drug Discovery*, *5*, 1–13.

Nisonoff, A., Hooper, J. E., & Spring, S. B. (1975). *The Antibody Molecule*. New York: Academic Press.

Padlan, E. (1994). Anatomy of the antibody molecule. *Molecular Immunology*, *31*, 169–217.

Pavlou, A. K., & Beslsey, M. J. (2005). The therapeutic antibodies market to 2008. *Eurpean Journal of Pharmaceutics Biopharmaceutics*, *59*, 389–396.

Presta, L. (2003). Antibody engineering for therapeutics. *Current Opinion Structural Biology*, *13*, 519–525.

Roitt, I. M., & Delves, P. J. (2001). *Roitt'S Essential Immunology*. United Kingdom: Blackwell Publishing.

Stebbings, R., Findlay, L., Edwards, C., Eastwood, D., Bird, C., North, D., Mistry, Y., Dilger, P., Liefooghe, E., Cludts, I., Fox, B., Tarrant, G., Robinson, J., Meager, T., Dolman, C., Thorpe, S. J., Bristow, A., Wadhwa, M., Thorpe, R., & Poole, S. (2007). "Cytokine storm" in the phase I trial of monoclonal antibody TGN1412: better understanding the causes to improve preclinical testing of immunotherapeutics. *Journal of Immunology*, *179*, 3325–3331.

Tiselius, A. (1937). Electrophoresis of serum globulin. *Biochemistry Journal*, *31*, 313–317.

Tiselius, A., & Kabat, E. A. (1939). An Electrophoretic study of immune sera and purified antibody preparations. *The Journal of Experimental Medicine*, *69*, 119–131.

Tjandra, J. J., Ramadi, L., & McKenzie, I. F. (1990). Development of Human Anti-Murine Antibody (HAMA) response in patients. *Immunology and Cell Biology*, *68*, 367–376.

FURTHER READING

Abbas, K. A., Lichtman, A. H., & Pillai, S. (2008). *Cellular and Molecular Immunology* (6th ed.). India: Elsevier.

Campbell, A. M. (1984). *Monoclonal Antibody Technology*. USA: Elsevier.

Committee on Methods of Producing Monoclonal Antibodies Institute for Laboratory Animal Research National Research Council (1999). *Monoclonal Antibody Production*. USA: National Academy Press.

Goding, J. W. (1996). *Monoclonal Antibodies* (3rd ed.). USA: Academic Press.

Harlow, E., & Lane, D. (1988). *Antibodies: A Laboratory Manual*. USA: Cold spring Harbor Laboratory Press.

Hay, F. C., & Westwood, O. M. R. (2003). *Practical Immunology* (4th ed.). India: Blackwell Publishing.

Male, D., Brostoff, J., Roth, D. B., & Roitt, I. (2006). *Immunology* (7th ed.). Canada: Elsevier.

GLOSSARY

Adjuvant A substance that enhances the antigenicity of an antigen. Freund's Adjuvants are the most commonly used adjuvants in the laboratory.

Antigenicity Ability of an antigen to combine/bind with the product of HM1 and/or CM1.

Antiserum The fluid component of clotted blood from an immune individual that contains antibodies against the agent used for immunizations.

Chimeric This is derived from chimera, which means a substance that is created from the proteins or genes of two species. For example, chimeric antibodies contain the amino acid sequence of a different species in another species' antibody (e.g. antibody with a human constant region and mouse variable regions).

Humanized Antibodies These antibodies have CDRs from a species other than humans, although the rest of the antibody molecule is of human immunoglobulin.

Immunization Process of artificially producing a state of immunity in an animal against an antigen (commonly known as vaccination). For laboratory purposes, immunization also means to induce antibody production against a specific antigen.

Immunogenicity The capacity of a host to induce an immune response is the characteristic of the host rather than of an antigen. This is the reason that the same antigen can cause an immune response in one host while not in another (e.g. one cannot raise an antibody against BSA in bovines, but can raise antibodies against BSA in rabbits).

Monoclonal Antibody Antibody produced by a single clone of B cells having the same antigenic specificity (e.g. Humira).

Polyclonal Antibody Antibodies that are secreted by different B-cell lineages, and therefore have different specificities.

Selection Medium Medium that favors growth of a particular cell type. For example, HAT selection medium is used for the selection of hybrid B-lymphocytes and myeloma cells.

ABBREVIATIONS

Ab Antibody/Antibodies
8-AG 8-Azo Guanine
CMI Cell Mediated Immunity
CRBC Chicken Red Blood Cells
DTT Dithiothreitol
ELISA Enzyme Linked Immuno Sorbent Assay
Fab Fragment Antigen Binding
F(ab')$_2$ Fragment Antigen Binding (divalent)
FACS Flourescent Activated Cell Sorting/Sorter
Fc Fragement Crystallizable
FCA Freund's Complete Adjuvant
FDA Food and Drug Administration
FIA Freund's Incomplete Adjuvant
FITC Fluorescein Isothiocyanate
HAT Hypoxanthine, Aminopterin, and Thymidine medium
HGPRT Hypoxanthine Guanine Phospho Ribosyl Transferase
HMI Humoral Mediated Immunity
HRPO Horseradish Peroxidase
Ig Immunoglobulin(s)
IgG Immunoglobulin G
INN International Nonproprietary Names
MACS Magnetic Activated Cell Sorting/Sorter
2-ME 2-Mercapto Ethanol
MoAb Monoclonal Antibody/Antibodies
ODD Ouchterlony Double Diffusion
PoAb Polyclonal Antibody/Antibodies
RBC Red Blood Cells
SRBC Sheep Red Blood Cells
USAN United States Adopted Names
USANC United States Adopted Names Council

LONG ANSWER QUESTIONS

1. Discuss the advantages and disadvantages of monoclonal antibodies.
2. Discuss the experiment of Porter and Edelman and how they deduced that antibodies have two heavy chains and two light chains.
3. Describe the nomenclature of monoclonal antibodies.
4. What are some applications of monoclonal antibodies as therapeutics?
5. Discuss serum sickness, HAMA, HACA, and HAHA. How are these symptoms or diseases are correlated with antibody structure?
6. What is the principle behind the selection of hybridoma cells for monoclonal antibody production?
7. How are polyclonal antibodies raised?
8. Discuss the translational significance of polyclonal and monoclonal antibodies.

SHORT ANSWER QUESTIONS

1. What is a monoclonal antibody?
2. What is serum sickness?
3. What are the different components of selection medium used for the production of monoclonal antibodies?
4. Name the cells whose salvage pathway is blocked during monoclonal antibody production.
5. Name the different fractions produced after papain digestion. Do papain-digested fragments of antibodies bind with antigens? Can any of these fragments be used for immunoprecipitation?
6. What are the benefits of the hinge region in an antibody?
7. What fractions are produced after pepsin digestion? Write one line for each fraction produced due to pepsin digestion.

ANSWERS TO SHORT ANSWER QUESTIONS

1. Monoclonal antibodies are those antibodies produced by a single clone of B-lymphocytes.
2. Serum sickness is a disease that happens due to the use of polyclonal antibodies as therapeutic agents. The adverse effect is more pronounced when the same antibody is given again. Anti-venom is one of the best examples for serum sickness.
3. Selection medium used for the production of monoclonal antibodies is known as HAT Medium. The letters in "HAT" represent its individual components: H, Hypoxanthine; A, Aminopterin; and T, Thymidine.

4. The salvage pathway is blocked in myeloma cells for the selection of hybridomas during monoclonal antibody production.
5. Papain digestion of antibodies gives Fab and Fc fragments. Papain-digested Fab fragments will bind with antigens, but neither of the two fragments will cause immunoprecipitation.
6. The hinge region of an antibody exists between the C_H1 and C_H2 domains. This region provides the flexibility for the antibody to bind with an epitope. In some antibodies, the hinge region is not present and is replaced by an extra domain of constant region. Antibodies having hinge regions are more vulnerable to degradation/digestion.
7. Pepsin treatment causes digestion of antibodies, producing $F(ab')_2$ and degraded Fc fragments. $F(ab')_2$ is divalent, binds with antigens, and is capable of immunoprecipitation. Fc fragments are extensively degraded so the Fc portion loses its ability to bind with FcR; therefore, the Fc portion produced due to pepsin digestion can not initiate the biological functions of antibodies.

Chapter 16

Molecular Markers: Tool for Genetic Analysis

Avinash Marwal, Anurag Kumar Sahu and R.K. Gaur

Department of Science, Faculty of Arts, Science and Commerce, Mody Institute of Technology and Science, Rajasthan, India

Chapter Outline

SUMMARY

The development of molecular techniques for genetic analysis has led to a great augmentation in our knowledge of animal genetics and our understanding of the structure and behavior of various animal genomes. These molecular techniques, in particular the applications of molecular markers, have been used to scrutinize DNA sequence variations in and among animal species and to create new sources of genetic variation by introducing new and favorable traits from landraces and related animal species.

WHAT YOU CAN EXPECT TO KNOW

Improvements in marker detection systems, and in the techniques used to identify markers linked to useful traits, have enabled great advances in recent years. While RFLP

Animal Biotechnology. http://dx.doi.org/10.1016/B978-0-12-416002-6.00016-X

markers have been the basis for most work in laboratory animals, valuable markers have been generated from random amplified polymorphic DNA (RAPDs) and amplified fragment length polymorphisms (AFLPs). Simple sequence repeats (SSR) or microsatellite markers have been developed more recently for the laboratory. This marker system is predicted to lead to even more rapid advances in both marker development and implementation in breeding programs. Ethical issues related to laboratory animals in research and experimentation has been debated, defended, and protested by both individuals and organizations at various levels. Responses range from personal lifestyle decisions and fervent philosophical treatises to strident arguments, violent demonstrations, and direct action. The continuum of attitudes about animals and the human relationship with animals spans the range from those who support no regulation of the human use of animals and those who advocate absolute animal liberation from all human use.

HISTORY AND METHODS

INTRODUCTION

The various genotypic classes are indistinguishable at the phenotypic level because of the dominance effect of the marker and low genome coverage. The first work on the detection of genome variation in animal livestock was based on morphological, chromosomal, and biochemical markers. Most morphological markers are sex-limited, age-dependent, and are significantly influenced by the environment. Biochemical markers show a low degree of polymorphism (Duran et al., 2009). Recent advances in molecular biology provide novel tools for addressing evolutionary, ecological, and taxonomic research questions, and the application of methods based on population genetics and statistics allows the development of animals with a high productive efficiency. Important advances to some of the economically important characteristics in several species of livestock have been achieved based on phenotypic performance; however, several limitations of these methods of improvement (based on population genetics alone) have become evident with time. Their efficiency decreases when the characteristics are difficult to measure or have low heritability (Beuzen et al., 2000). Additionally, selection has been generally limited to those characteristics that can properly be measured in large numbers of animals. Some characteristics, such as the rate of survival, are expressed too late in life to serve as useful criteria for selection. Also, traditional selection within populations has not been very efficient when the selection objective involves several characteristics with unfavorable genetic correlation, for example, milk production and the protein content of milk.

Molecular techniques allow detection of variations or polymorphisms that exist among individuals in the population for specific regions of DNA. These polymorphisms can be used to build up genetic maps and to evaluate differences between markers in the expression of particular traits in a family that might indicate a direct effect of these differences (in terms of genetic determination) on the trait. Everybody in the field of biological science has started using molecular marker technology for the purpose of so-called "gene discovery." This has resulted in the development of new types of molecular markers and various technological simplifications to resolve the problems associated with molecular marker technology (Teneva, 2009). The development of molecular biology over the past three decades has created new means for studying livestock genetics and animal breeding. Selection according to genotype has become an important tool in the breeding of farm animals. Molecular markers capable of detecting genetic variation at the DNA sequence level have removed the above mentioned limitations of morphological, chromosomal, and protein markers, and they possess unique genetic properties that make them more useful than other genetic markers. They are numerous and are distributed ubiquitously throughout the genome. DNA-based markers have many advantages over phenotypic markers in that they are highly heritable, relatively easy to assay, and are not affected by the environment (Montaldo and Meza-Herrera, 1998).

Information collected by the Food and Agriculture Organization (FAO) of the United Nations indicates that approximately 30% of the world's farm animal breeds are at risk of extinction. Conservation policies of native breeds will depend to a large extent on our knowledge of historic and genetic relationships among breeds, as well as economic and cultural factors. In the Secondary Guidelines for Development of National Farm Animal Genetic Resources Management Plans, the FAO proposed an integrated program for global management of livestock genetic resources using reference microsatellites (Oldenbrock, 2007). In India, The National Bureau of Animal Genetic Resources (NBAGR), Karnal is entrusted with characterization of important indigenous livestock breeds. India has 27 breeds of cattle, 8 of buffalo, 42 of sheep, 20 of goats, 6 of horses, and 17 of poultry. Genetic variation of the animal is the basic material, which is utilized for changing the genetic makeup or genetic potentiality of domestic species to suit our needs. Mechanization, unplanned and indiscriminate breeding among native stocks, and human bias in favor of certain breeds are directly or indirectly responsible for the dilution of Indian livestock germplasm. Hence, characterization of indigenous germplasm is essential for their conservation. The increasing availability of molecular markers in laboratory animals allows the detailed analyses and evaluation of genetic diversity, and also the detection of genes influencing economically important traits (Erhardt and Weimann, 2007).

Molecular markers have three-fold applications in gene mapping: (i) a marker allows the direct identification of the gene of interest instead of the gene product, and consequently it serves as a useful tool for screening somatic cell hybrids; (ii) use of several DNA probes and easy-to-screen techniques, a marker also helps in physical mapping of the genes using *in situ* hybridization; and (iii) molecular markers provide sufficient markers for construction of genetic maps using linkage analysis. Genetic maps are constructed on the basis of two classes of molecular markers. Type I markers that represent the evolutionary conserved coding sequences (e.g. classical RFLPs and SSLPs) are useful in comparative mapping strategies where polymorphism is not an essential prerequisite. However, these are mostly single locus and di-allelic (SLDA) and thus are not useful for linkage analysis. On the other hand, type II markers (like microsatellites markers) have higher polymorphism information content than conventional RFLPs and can be generated very easily and rapidly. Therefore, major efforts are being made to produce gene maps based on the type II markers. Further utilization of molecular markers developed from DNA sequence information, namely ASO and STMS polymorphic markers, is also helpful in the rapid progress of gene mapping (Table 16.1).

From the late 1980s, the number of publications generated has been at an unexpectedly high volume, projecting that genetic analysis in animals would be a "cake walk." Everybody in the field of biological science started using molecular marker technology for the purpose of so-called "gene discovery." The hype of this technology was so prominent and started to divide scientists working in "animal biology" into various groups. This kind of situation is not new in science. At any point in time, the typical scientific discipline tends to be seized by a particular methodology or enthusiasm and other approaches get dumped. This has resulted in the development of new types of molecular markers and various technological simplifications to resolve the problems associated with molecular marker technology. The recent PCR-based approach, gel-free visualization of PCR products, and automation at various steps, are boons to the molecular marker approaches adopted for genome mapping and genetic diversity analysis in the animal kingdom.

TABLE 16.1 Various Molecular Markers Used for Genetic Analysis of Animals

Year	Acronym	Nomenclature
1974	RFLP	Restriction Fragment Length Polymorphism
1986	ASO	Allele Specific Oligonucleotides
1988	AS-PCR	Allele Specific Polymerase Chain Reaction
1989	SSCP	Single Stranded Conformational Polymorphism
1989	STS	Sequence Tagged Site
1990	RAPD	Randomly Amplified Polymorphic DNA
1991	RLGS	Restriction Landmark Genome Scanning
1992	SSR	Simple Sequence Repeats
1994	ISSR	Inter Simple Sequence Repeats
1994	SNP	Single Nucleotide Polymorphisms
1995	AFLP	Amplified Fragment Length Polymorphism
1999	MSAP	Methylation Sensitive Amplification Polymorphism
2000	MITE	Miniature Inverted-Repeat Transposable Element
2002	SSLP	Simple Sequence Length Polymorphisms

METHODOLOGY

Molecular markers are also the prerequisite for the identification of positional and functional candidate genes responsible for quantitative traits. In this presentation, the detailed use of molecular markers for the evaluation of genetic diversity and identification of economically important traits in animal production is explained using different examples.

Restriction Fragment Length Polymorphism (RFLP)

Restriction fragment length polymorphism (RFLP) analysis is a method used to indirectly collect information on mutations occurring at specific short regions (restriction sites) (Botstein et al., 1980). RFLP enabled the detection of polymorphisms at the DNA sequence level. RFLP are first class genetic markers, allowing the construction of highly saturated linkage maps. Restriction Fragment Length Polymorphism (RFLP) is a technique in which organisms may be differentiated by analysis of patterns derived from cleavage of their DNA (Flow Chart 16.1). If two organisms differ in the distance between sites of cleavage of a particular restriction endonuclease, the length of the fragments produced will differ when the DNA is digested with a restriction enzyme. The similarity of the patterns generated can be used to differentiate species (and even strains) from one another.

Steps Involved in RFLP Analysis

1. The first step in DNA typing is extraction of the DNA from the sample, be it blood, saliva, semen, or some other biological sample.
2. The purified DNA is then cut into fragments by restriction enzymes. For example take the pattern GCGC and

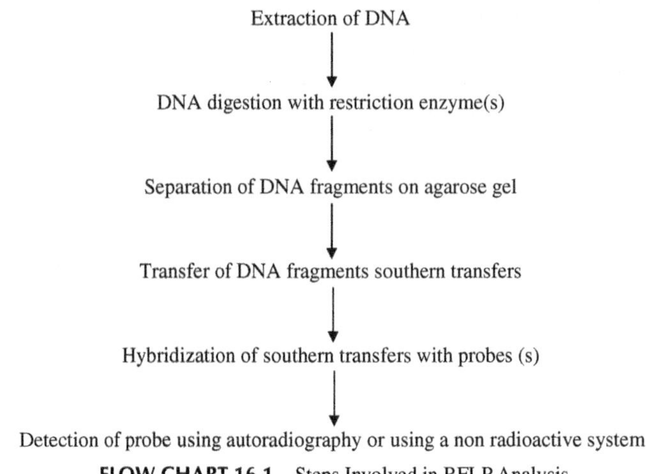
Extraction of DNA
DNA digestion with restriction enzyme(s)
Separation of DNA fragments on agarose gel
Transfer of DNA fragments southern transfers
Hybridization of southern transfers with probes (s)
Detection of probe using autoradiography or using a non radioactive system

FLOW CHART 16.1 Steps Involved in RFLP Analysis

imagine it occurs more than once in the DNA. The number of times it occurs is unique to the individual. The restriction enzyme chops the DNA in two at every place where the GCGC pattern occurs.

3. The restriction fragments have negative charge and can be separated by a technique called gel electrophoresis, which separates the pieces of DNA based on their size. The samples of DNA that have been treated with restriction enzymes are placed in separate lanes on a slab of electrophoretic gel across which is placed an electric field. The fragments migrate towards the positive electrode, the smaller fragments moving faster than the larger fragments, thus separating the DNA samples into distinct bands.
4. Put the nylon membrane onto the porous support, then the plastic mask (the mask must be slightly smaller than the gel), being very careful to slide the gel onto the mask. Close the apparatus and start the pump (you should be able to see if a vacuum is formed; if not, look for the problem).
5. If the membrane to be used is being employed for the first time, then an overnight pre-hybridization is needed, otherwise 4–5 hours should be enough. Incubate in the rotating hybridization oven at 65°C. After pre-hybridization, start hybridization by adding the boiled, labeled probe to the pre-hybridization.
6. Incubate overnight. The bands can be visualized using luminescent dyes.

Applications of RFLP

Analysis of RFLP variation in genomes was a vital tool in genome mapping and genetic disease analysis. If a researcher were to try to initially determine the chromosomal location of a particular disease gene, he/she would analyze the DNA of members of a family afflicted by the disease and look for RFLP alleles that show a similar pattern of inheritance as that of the disease. Once a disease gene was localized, RFLP analysis of other families would reveal who was at risk for the disease, or who was likely to be a carrier of the mutant genes. RFLP analysis was also the basis for early methods of genetic fingerprinting, useful in the identification of samples retrieved from crime scenes, in the determination of paternity, and in the characterization of genetic diversity or breeding patterns in animal populations. RFLP assay was carried out for rapid identification of three target species (*Siganus canaliculatus, S. corallinus* and *S. javus*) using mitochondrial gene regions to facilitate studies on species-specific spatio-temporal patterns of larval dispersal and population connectivity to aid fishery management (Ravago-Gotanco et al., 2010).

Allele-Specific Oligonucleotide (ASO)

An ASO is typically an oligonucleotide of 15–21 nucleotide bases in length. It is designed (and used) in a way that makes it specific for only one version, or allele, of the DNA being tested. An ASO is a short piece of synthetic DNA complementary to the sequence of a variable target DNA. It acts as a probe for the presence of the target in a Southern blot assay or, more commonly, in the simpler dot blot assay.

Applications of ASO

It is a common tool used in genetic testing, forensics, and molecular biology research, and is important in genotype analysis and the Human Genome Project. With an allele-specific oligonucleotide (ASO) test it was shown that the mutation co-segregated with the recessively inherited yellow coat color in the Labrador retriever. Golden retrievers also appeared to be homozygous for the mutation (Everts et al., 2000).

Allele-Specific PCR (AS-PCR)

PCR primers are chosen from an invariant part of the genome, and might be used to amplify a polymorphic area between them. In allele-specific PCR the opposite is done. At least one of the primers is chosen from a polymorphic area, with the mutations located at (or near) its 3′-end. Under stringent conditions, a mismatched primer will not initiate replication, whereas a matched primer will. The appearance of an amplification product therefore indicates the genotype (Wangkumhang et al., 2007).

Applications of AS-PCR

It is a convenient and inexpensive method for genotyping Single Nucleotide Polymorphisms (SNPs) and mutations. It is applied in many recent studies, including population genetics, molecular genetics, and pharmacogenomics. Two genetic variants of the bovine β-casein gene (A1 and B) encode a histidine residue at codon 67, resulting in potential liberation of a bioactive peptide (β-casomorphin) upon digestion. An allele-specific PCR (AS-PCR) was evaluated to distinguish between the β-casomorphin-releasing variants (A1 and B) and the non-releasing variants. AS-PCR successfully distinguished β-casein variants in 41 of 42 animals as confirmed by sequence analysis (Keating et al., 2008).

Single-Strand Conformation Polymorphism (SSCP)

This method is one of the simplest and perhaps one of the most sensitive PCR-based methods for detecting multiple mutations and polymorphisms in genes and family analysis ("fingerprinting") (Huby-Chilton et al., 2001).

Applications of SSCP

It is used as a way to discover new DNA polymorphisms apart from DNA sequencing, but is now being supplanted by sequencing techniques because of efficiency and accuracy. These days, SSCP is most applicable as a diagnostic tool in molecular biology. It can be used in genotyping to detect homozygous individuals of different allelic states, as well as heterozygous individuals who should demonstrate distinct patterns in an electrophoresis experiment. SSCP is also widely used in virology to detect variations in different strains of a virus, the idea being that a particular virus particle present in both strains will have undergone changes due to mutation, and that these changes will cause the two particles to assume different conformations, and thus be differentiable on an SSCP gel (Kubo et al., 2009). The genetic diversity of Jamunapari goats (*Capra hircus*) was investigated using an optimized non-radioactive polymerase chain reaction single-strand conformation polymorphism (PCR-SSCP) method to detect α-lactalbumin polymorphism, indicating that Jamunapari goats have high genetic variability at loci exon I of the α-lactalbumin gene (Kumar et al., 2006).

Sequence Tagged Site (STS)

A sequence-tagged site (STS) is a short region along the genome (200 to 300 bases long) whose exact sequence is found nowhere else in the genome. The uniqueness of the sequence is established by demonstrating that it can be uniquely amplified by the PCR. The DNA sequence of an STS may contain repetitive elements, sequences that appear elsewhere in the genome, but as long as the sequences at both ends of the site are unique, unique DNA primers complementary to those ends can be synthesized, the region amplified using PCR, and the specificity of the reaction demonstrated by gel electrophoresis of the amplified product.

Applications of STS

STSs are very helpful for detecting microdeletions in some genes. For example, some STSs can be used in screening by PCR to detect microdeletions in Azoospermia (AZF) genes in infertile men. Identification of genes in elephants could provide additional information for evolutionary studies and for evaluating genetic diversity in existing elephant populations. Sequence tagged sites (STSs) were identified in the Asian and the African elephant for the following genes: melatonin receptor 1a (MTNR1A), retinoic acid receptor beta (RARB), and leptin receptor (LEPR) (Burk et al., 1998).

Random Amplified Polymorphic DNA (RAPD)

Random amplified polymorphic DNA (RAPD) analysis is a PCR-based molecular marker technique that shows that the difference in the pattern of bands amplified from genetically distinct individuals behave as Mendelian genetic markers. It employs a single primer (a 10-mer) for a random amplification under specific PCR conditions. The number of amplified fragments depends on the distribution and number of annealing sites throughout the genome. In fact, amplification takes place only when primers anneal on each strand at sites not more distant than 3–4 kb. PCR random products are then detected easily on an agarose gel and the resulting banding pattern represents the DNA fingerprint. In comparison with other genetic markers, RAPD provides a more arbitrary sample of the genome and can detect an unlimited number of loci, simply by changing the base combination in the oligomer used. The most limiting property of RAPD markers is probably the dominant expression of alleles, making difficult the interpretation of multi-locus patterns. Even problems of amplification reproducibility were raised in the past. RAPDs have been widely employed for studies on taxon identification, hybridization, reproductive behavior, and population genetic structure. This technique has been employed for targeting genes of economic value (Flow Chart 16.2). RAPD markers are amplification products of anonymous DNA sequences using single, short, and arbitrary oligonucleotide primers, and thus do not require prior knowledge of a DNA sequence (Bardakci, 2001).

Steps Involved in RAPD Analysis

1. The first step is extraction of the DNA from the sample, be it blood, saliva, semen, or some other biological sample.
2. The yield of DNA per gram of tissue isolated is measured using a UV spectrophotometer at 260 nm. DNA purity is determined by estimating the ratio of absorbance at 260 nm to absorbance at 280 nm.
3. Oligonucleotide primers are used for amplification to standardize the PCR conditions. The reactions are carried out in a DNA thermocycler (Bio Rad).
4. The restriction fragments have negative charge and can be separated by a technique called gel electrophoresis, which separates the pieces of DNA based on their size. The samples of DNA that have been treated with restriction enzymes are placed in separate lanes on a slab of electrophoretic gel across which is placed an electric field. The fragments migrate towards the positive electrode, the smaller fragments moving faster than the larger fragments, thus separating the DNA samples into distinct bands.
5. Gels with amplification fragments are visualized and photographed under UV light. A medium-range DNA Ruler is used as a molecular marker for the size of the fragments.

Applications of RAPD

Inbreeding indicates the degree of homozygosity at a locus within a population. Normally inbreeding is estimated in terms of a coefficient calculated from the pedigree of an individual. If no history is available, however, there is no way to estimate the inbreeding coefficient. Sometimes data on individuals are missing, and that too can prevent the estimation of the inbreeding coefficient, which is essential for formulation of a breeding program at the farm level and for breed development. Random amplified polymorphic DNA (RAPD) analysis was carried out on 20 randomly selected animals of three Indian cattle breeds (namely Red Sindhi, Hariana, and Tharparkar) maintained at three farms: Central Cattle Breeding Farm, Chiplima, Orissa, India; Shree gaushala, Jind, Hariana, India; Central Cattle Breeding Farm, Lakhimpur-Kheri, U.P., India (Bhattacharya et al., 2003).

Restriction Landmark Genome Scanning (RLGS)

Restriction landmark genomic scanning using methylation-sensitive endonucleases (RLGS-M) is a powerful method for systematic detection of DNA methylation. This approach is based on the assumption that CpG methylation, particularly of CpG islands, might be associated with gene transcriptional regulation.

Applications of RLGS

It is a method that was used to study a mouse brain that was scanned for genomic DNA from various developmental stages to detect the transcriptionally active regions (Kawai et al., 1993).

Single Nucleotide Polymorphisms (SNP)

Single nucleotide polymorphism means a polymorphism corresponding to a difference at a single nucleotide position (substitution, deletion, or insertion). It includes detection of a single nucleotide change by a direct sequence protocol.

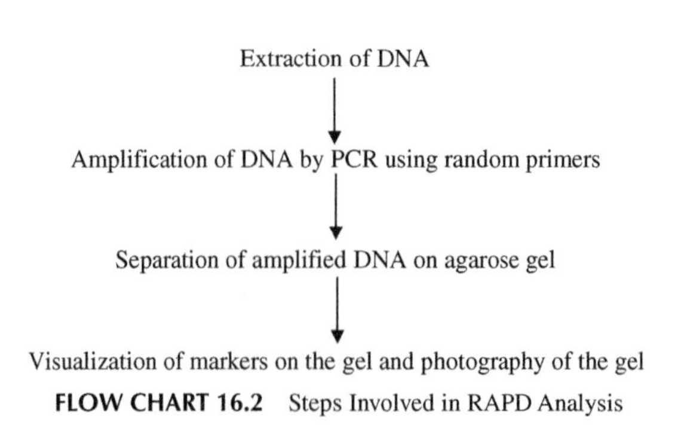
Extraction of DNA
Amplification of DNA by PCR using random primers
Separation of amplified DNA on agarose gel
Visualization of markers on the gel and photography of the gel

FLOW CHART 16.2 Steps Involved in RAPD Analysis

SNPs occur in both coding and noncoding regions of the genome. SNP detection technologies are used to scan for new polymorphisms and to determine the allele(s) of a known polymorphism in target sequences. SNP detection technologies have evolved from labor intensive, time-consuming, and expensive processes to some of the most highly automated, efficient, and relatively inexpensive methods (Kwok and Chen, 2003).

Applications of SNP

The Atlantic cod (*Gadus morhua*) is a groundfish of great economic value in fisheries and an emerging species in aquaculture. Genetic markers are needed to identify wild stocks in order to ensure sustainable management, and for marker-assisted selection and pedigree determination in aquaculture. The SNPs were tested on Atlantic cod from four different sites, comprising both North-East Arctic cod (NEAC) and Norwegian coastal cod (NCC). The average heterozygosity of the SNPs was 0.25 and the average minor allele frequency was 0.18. The SNP markers presented here are powerful tools for future genetic work related to management and aquaculture. In particular, some SNPs exhibiting high levels of population divergence have the potential to significantly enhance studies on the population structure of Atlantic cod (Moen et al., 2008).

Amplified Fragment Length Polymorphism (AFLP)

An amplified fragment length polymorphism (AFLP) is a highly sensitive method for detecting polymorphisms (Flow Chart 16.3). A different procedure combines enzymatic digestion with PCR. AFLPs are the result of digestion with restriction endonucleases, a ligation of specifically designed oligonucleotide adaptors to the ends of each fragment, and PCR amplification using primers complementary to the adaptor sequence to which an extended sequence of a few bases is added in order to reduce the number of amplified fragments. Banding patterns obtained in this way reveal useful information for a series of applications like population genetics and systematic and kinship analysis. It discriminates heterozygote from homozygote when a gel scanner is used.

Steps Involved in AFLP Analysis

1. To prepare an AFLP template, genomic DNA is isolated and digested with two restriction endonucleases simultaneously. This step generates the required substrate for ligation and subsequent amplification. The success of the AFLP technique is dependent upon complete restriction digestion; therefore, much care should be taken to isolate high quality genomic DNA intact without contaminating nucleases or inhibitors.
2. Following heat inactivation of the restriction endonucleases, the genomic DNA fragments are ligated to *EcoR I* and *Mse I* adapters to generate template DNA for amplification. These common adapter sequences flanking variable genomic DNA sequences serve as primer binding sites on these restriction fragments. Using this strategy, it is possible to amplify many DNA fragments without having prior sequence knowledge.
3. PCR is performed in two consecutive reactions. In the first reaction, called pre-amplification, genomic DNAs are amplified with AFLP primers, each having one selective nucleotide. The PCR products of the pre-amplification reaction are diluted and used as a template for the selective amplification using two AFLP primers, each containing three selective nucleotides.

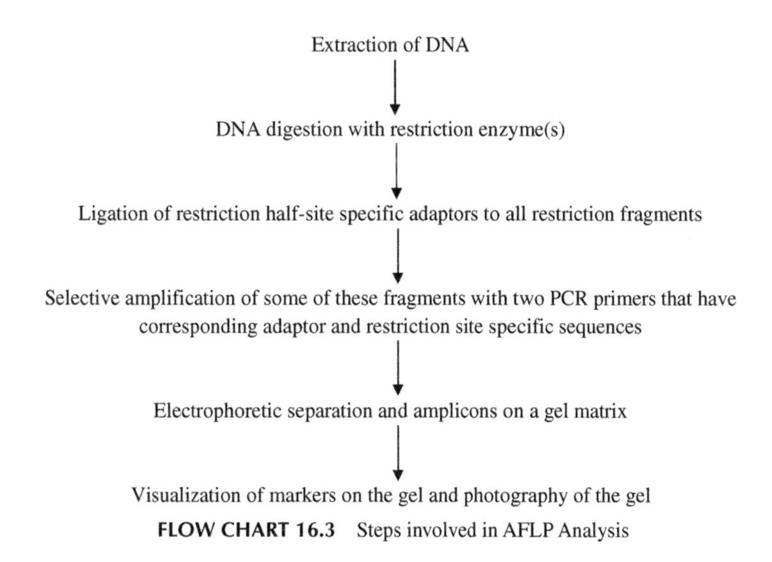

Extraction of DNA
DNA digestion with restriction enzyme(s)
Ligation of restriction half-site specific adaptors to all restriction fragments
Selective amplification of some of these fragments with two PCR primers that have corresponding adaptor and restriction site specific sequences
Electrophoretic separation and amplicons on a gel matrix
Visualization of markers on the gel and photography of the gel

FLOW CHART 16.3 Steps involved in AFLP Analysis

4. Products from the selective amplification are separated on a 5% or 6% denaturing polyacrylamide (sequencing) gel. The resultant banding pattern ("fingerprint") can be analyzed for polymorphisms either manually or using analytical software.

Applications of AFLP

AFLP is a multiplex PCR-based method in which a subset of restriction fragments are selectively amplified using oligonucleotide primers complementary to sequences that have been ligated to each end. AFLP analysis allows the reliable identification of over 50 loci in a single reaction. This technique combines the reliability of the RFLP and ease of the PCR; thus, AFLP is a new typing method for DNA of any origin or complexity. Amplified fragment length polymorphism (AFLP) was tested to assess the frequency of extra-pair parentage in a bluethroat (*Luscinia svecica namnetum*) population. Thirty-six families totaling 162 nestlings were analyzed. Using a combination of three primer pairs, an exclusion probability of 93% for the population was reached. This probability can reach 99% when considering families independently. It was revealed that extra-pair fertilizations are very common: 63.8% of all broods contain at least one extra-pair young, totaling 41.9% of all young analyzed. However, with the technique and the three primer pairs used, it was not possible to attribute the parentage exclusions to extra-pair paternity, maternity, or both. As brood parasitism has never been reported in this species, it seems likely that the exclusions are due to extra-pair males. This study shows that dominant AFLP markers can be useful for studying the mating system of taxa for which no microsatellite primers are available. This technique allows the approximate estimation of parentage exclusions despite the fact that it is not possible to know which parent has to be excluded (Questiau et al., 1999).

Methylation Sensitive Amplification Polymorphism (MSAP)

MSAP is a modification of the Amplified Fragment Length Polymorphism (AFLP) method that makes use of the differential sensitivity of a pair of iso-schizomers to cytosine methylation.

Applications of MSAP

The methylation levels of genomes were compared in swine, cattle, sheep, rat, chicken and duck, using the methylation-sensitive amplification polymorphism technique (MSAP). The results showed that the methylation levels in genomes of the species investigated were mostly 40–50% (except cattle). The methylation level varied in different species. The methylation pattern in various tissues of each species was specific; for the same species, the methylation level of the tissue genome was mostly higher than that of the blood genome. The difference of methylation level between birds and mammals was not significant, however, mammals appeared to have a lower hemi-methylation frequency and higher full methylation frequency than birds (Shao-Qing et al., 2007).

Miniature Inverted-Repeat Transposable Element (MITE)

MITE assay involves Transposon Display (TD), which is a modification of the AFLP procedure where PCR products are derived from primers anchored in a restriction site and a transposable element rather than in two restriction sites. For this candidate, primers in transposable elements are designed based on a consensus sequence generated of transposable elements. Miniature inverted-repeat TEs (MITEs) are a special type of Class 2 non-autonomous element that is present in high copy numbers in many eukaryotic genomes (Yujun and Wessler, 2010).

Applications of MITE

Amphioxus is consistently used as a model of genome evolution for understanding the invertebrate/vertebrate transition. The amphioxus genome has not undergone massive duplications like those in the vertebrates, or disruptive rearrangements like in the genome of *Ciona* (a urochordate), making it an ideal evolutionary model. Transposable elements have been linked to many genomic evolutionary changes, including increased genome size, modified gene expression, massive gene rearrangements, and possibly intron evolution. Five novel MITEs were identified by an analysis of an amphioxus DNA sequence: *LanceleTn-1, LanceleTn-2, LanceleTn-3a, LanceleTn-3b,* and *LanceleTn-4*. Several of the *LanceleTn* elements were identified in the amphioxus ParaHox cluster, and it was suggested that these had important implications for the evolution of this highly conserved gene cluster. The estimated high copy numbers of these elements implies that MITEs are probably the most abundant type of mobile element in amphioxus, and are thus likely to have been of fundamental importance in shaping the evolution of the amphioxus genome (Osborne et al., 2006).

Microsatellites

DNA microsatellite sequences are valuable genetic markers due to their dense distribution in the genome, high variability, co-dominant inheritance, and relative ease of detection. As hyper-variability is highly significant for detecting differences in a population and between individuals, microsatellite typing can reveal the degrees of polymorphism that is easy to interpret. Another complementary approach to genetic analysis by microsatellites involves the use of gene markers (functional markers) for domestic species as an indication of the extent of variation (especially at loci of economic importance), and may provide information related to the functional differences between the breeds and may have relevance in selective breeding programs to establish new breeding stocks for the purpose of maintenance of useful genetic diversity.

This candidate gene approach involves a gene with a known expression of certain proteins, and appears to provide genomic information that can be used for the genetic improvement of livestock. Microsatellites are also known as simple sequence repeats (SSRs), short tandem repeats (STRs), inter simple sequence repeats (ISSR), simple sequence tandem repeats (SSTR), variable number tandem repeats (VNTR), simple sequence length polymorphisms (SSLP), and sequence-tagged microsatellites (STMS).

Simple Sequence Repeats (SSRs)/Short Tandem Repeats (STRs)/Simple Sequence Tandem Repeats (SSTRs)

SSRs or STRs, also known as microsatellites, are repeating sequences of 2–6 base pairs of DNA. STRs are typically co-dominant. They are used as molecular markers in genetics, and for kinship, population, and other studies. They can also be used to study gene duplication or deletion. STRs are also known to be causative agents in human disease, especially neurodegenerative disorders and cancer.

Applications of SSR

To reduce analysis cost and sample consumption and to meet the demands of higher sample throughputs, PCR amplification and detection of multiple markers (multiplex STR analysis) has become a standard technique in most forensic DNA laboratories. STR multiplexing is most commonly performed using spectrally distinguishable fluorescent tags and/or non-overlapping PCR product sizes. Multiplex STR amplification in one or two PCR reactions with fluorescently labeled primers and measurement with gel or capillary electrophoresis separation and laser-induced fluorescence detection is becoming a standard method in forensic laboratories for analysis of the 13 CODIS STR loci.

SSR was used in five domestic animal species (namely buffalo, cattle, goat, sheep, and yak), and has been used as a model to investigate the relative abundance of the type of repeat motif, their distribution in coding and non-coding regions, evaluation of mitochondrial SSR data for appropriateness of the phylogenetic tree, dynamism of length and repeat motif, and extent of conservation of flanking regions across loci with respect to time (Shakyawar et al., 2009).

Inter Simple Sequence Repeats (ISSR)

Inter simple sequence repeat (ISSR)-PCR is a technique that involves the use of microsatellite sequences as primers in a polymerase chain reaction to generate multi-locus markers. It is a simple and quick method that combines most of the advantages of microsatellites (SSRs) and amplified fragment length polymorphism (AFLP) with the universality of random amplified polymorphic DNA (RAPD). ISSR markers are highly polymorphic and are useful in studies on genetic diversity, phylogeny, gene tagging, genome mapping, and evolutionary biology (Reddy 2002). Sequences amplified by ISSR-PCR can be used for DNA fingerprinting. Since an ISSR may be a conserved or non-conserved region, this technique is not useful for distinguishing individuals, but rather for phylogeographical analyses or maybe delimiting species; sequence diversity is lower than in SSR-PCR, but still higher than in actual gene sequences. In addition, microsatellite sequencing and ISSR sequencing are mutually assisting, as one produces primers for the other (Pradeep *et al.*, 2002).

Applications of ISSR

For the first time in spiders, the Inter Simple Sequence Repeat (ISSR) technique was used to study the genetic variability of Mexican populations of *Brachypelma vagans*. A non-lethal technique was used to collect samples from six populations in the Yucatan peninsula, and seven ISSR primers were tested. Four of these primers gave fragments (bands) that were sufficiently clear and reproducible to construct a binary matrix and to determine genetic variability parameters. It revealed a very high percentage of polymorphism (P 5 98.7%), the highest yet reported for tarantula spiders. The results show that the ISSR-PCR method is promising for intraspecific variation of tarantula spiders (Machkour-M'Rabet et al., 2009).

Variable Number of Tandem Repeat Markers (VNTRs)

Variable number of tandem repeat markers (VNTRs) are located in a genome where a short nucleotide is organized as a tandem repeat. These can be found on many chromosomes, and they often show variations in length. Each variant acts as an inherited allele that allows its use for identification. VNTRs are multi-allelic loci that consist of repeated core sequences (> 6 nucleotides) known as mini-satellites, are tandem repeats, and are flanked by segments of non-repetitive sequences. This allows the VNTR blocks to be extracted with restriction enzymes and analyzed by restriction fragment length polymorphism, to be amplified by polymerase chain reaction (PCR), and to have their size determined by gel electrophoresis. The versatility and efficiency of these markers in genotyping and diversity studies were tested in the laboratory; their versatility was a great advantage. They are easy to score and can be run on agarose gels, which makes them an interesting tool, especially for use in less sophisticated laboratories.

Applications of VNTRs

VNTRs are an important source of RFLP genetic markers used in linkage analysis (mapping) of genomes. They have become essential in forensic crime investigations. The technique may use PCR, size determined by gel electrophoresis, and Southern blotting to produce a pattern of bands unique to each individual. Therefore, VNTRs are being used to study genetic diversity (DNA fingerprinting) and breeding patterns in animals. VNTRs also have clinical applications like VNTR typing, the next gold standard in genotyping for early diagnosis of *M. tuberculosis* super-infection or mixed infection.

Sequence Tagged Microsatellite Site (STMS)

A sequence tagged microsatellite site is a another form of microsatellite marker, which if cloned and sequenced, can be subjected to PCR amplification; such microsatellite loci can be recovered by PCR. Microsatellite markers in the STMS format can be completely described as information in databases and can serve as common reference points that will allow the incorporation of any type of physical mapping data into the evolving map. The advantage is that band profiles can be interpreted in terms of loci and alleles, and allele sizes can be determined with high accuracy.

Applications of STMS

One of the main applications of STMS is characterizing and understanding animal genetic variation. The use of STMS markers in genetic distancing of breeds is gaining momentum. The ever-increasing knowledge of mammalian genetic structure and the development of convenient ways of measuring that structure have opened up a range of new possibilities in the areas of animal and product identification.

Simple Sequence Length Polymorphisms (SSLP)

Simple Sequence Length Polymorphisms (SSLPs) are used as genetic markers with Polymerase Chain Reaction (PCR). An SSLP is a type of polymorphism: a difference in DNA sequence between individuals. SSLPs are repeated sequences over varying base lengths in intergenic regions of DNA. Variance in the length of SSLPs can be used to understand genetic variance between two individuals in a certain species (Rosenberg *et al.*, 2002).

Applications of SSLP

Genetic monitoring is an essential component of colony management, and for the rat has been accomplished primarily by using immunological and biochemical markers. SSLPs are a faster and more economical way of monitoring inbred strains of rats. Sixty-one inbred strains of rats were characterized using primer pairs for 37 SSLPs. Each of these loci appeared to be highly polymorphic, with the number of alleles per locus ranging from 3 to 14; as a result, all 61 inbred strains tested in this study could be provided with a unique strain profile. These strain profiles were also used for estimating the degree of similarity between strains. This information may provide the rationale for selecting strains for genetic crosses or for other purposes (Otsen et al., 1995).

Example of Microsatellites

A microsatellite is a stretch of DNA with mono-, di-, tri-, or tetra-nucleotide units repeated. Microsatellites are short sequences of nucleotides (typically 1 to 5 bp) that are tandemly repeated.

(a) Repeat Units

```
AAAAAAAAAAA        =(A) 11      =mononucleotide (11bp)
GTGTGTGTGTGT       =(GT) 6      =dinucleotide (12bp)
CTGCTGCTGCTG       =(CTG) 4     =trinucleotide (12bp)
ACTCACTCACTCACTC   =(ACTC) 4    =tetranucleotide (16bp)
```

(b) Homozygous Microsatellite

```
...CGTAGCCTTGCATCCTTCTCTCTCTCTCTCTATCGGTACTACGTGG... (46 bp)
...CGTAGCCTTGCATCCTTCTCTCTCTCTCTCTATCGGTACTACGTGG... (46 bp)
5' flanking region  microsatellite locus 3' flanking region
```

(c) Heterozygous Microsatellite

```
....CGTAGCCTTGCATCCTTCTCTCTCTCTCTCTATCGGTACTACGTGG...(46bp)
CGTAGCCTTGCATCCTTCTCTCTCTCTCTCTCTCTATCGGTACTACGTGG..(50 bp)
5' flanking region  microsatellite locus  3' flanking region
```

The most common way to detect microsatellites is to design PCR primers that are unique to one locus in the genome and the base pair on either side of the repeated portion (Figure 16.1). Therefore, a single pair of PCR primers will work for every individual in the species and produce different sized products for each of the different length microsatellites.

The PCR products are then separated by either gel or capillary electrophoresis. Either way, the investigator can determine the size of the PCR product and thus how many times the dinucleotide "CA" was repeated for each allele (Figure 16.2). It would be nice if microsatellite data

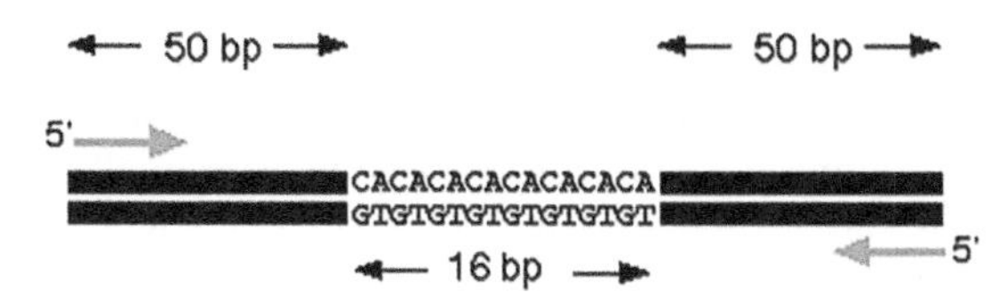

FIGURE 16.1 Detection of microsatellites from genomic DNA: Two PCR primers (forward and reverse gray arrows) are designed to flank the microsatellite region. If there were zero repeats, the PCR product would be 100 bp in length. Therefore, by determining the size of each PCR product (in this case 116 bp), one can calculate how many CA repeats are present in each microsatellite (there are 8 CA repeats in this example).

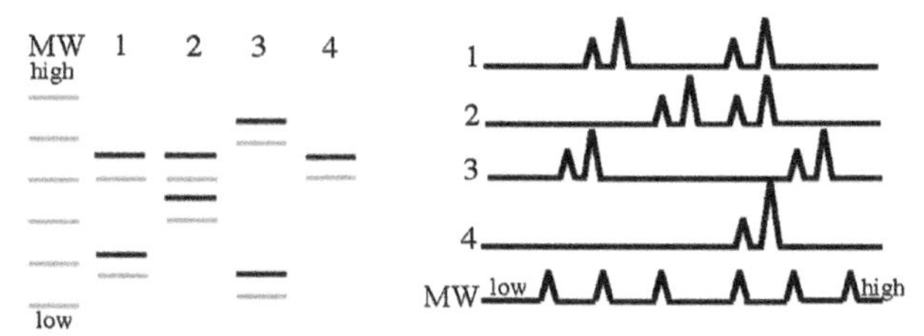

FIGURE 16.2 Stylized examples of microsatellite data: **Left Half:** Four sets of data were produced by gel electrophoresis, including major (black) and stutter (gray) bands (MW: molecular weight standards). **Right Half:** These data were produced by analysis on an automated capillary electrophoresis-based DNA sequencer. The data are line graphs with the location of each peak on the X-axis representing a different sized PCR product; the height of each peak indicates the amount of PCR product. Major bands produce higher peaks than the stutter peaks.

produced only two bands, but often there are minor bands in addition to the major bands; these are called stutter bands and they usually differ from the major bands by two nucleotides.

Advantages of Microsatellite Markers

They require low quantities of template DNA (10–100 ng). They are found in large numbers and are relatively evenly spaced throughout the genome or are randomly distributed throughout the genome. They have a high level of polymorphism and follow a typical Mendelian inheritance, which usually expresses in a co-dominant fashion, and are often multi-allelic, giving mean heterozygosity of more than 70%. Microsatellites are unaffected by environmental factors, and generally do not have pleiotropic effects on quantitative trait loci (QTL).

Disadvantages of Microsatellite Markers

They have high development costs, and heterozygotes may be misclassified as homozygotes when null-alleles occur due to mutations in the primer annealing sites. The underlying mutation model (infinite allele model or step-wise mutation model) is largely unknown. Sometimes stutter bands may complicate accurate scoring of polymorphisms. Homoplasy due to different forward and backward mutations can underestimate genetic divergence. They are time-consuming and expensive to develop.

Some General Applications of Microsatellites

Microsatellites have several applications like parentage determination, genetic distance estimation of twin zygosity and freemartinism, sexing of pre-implantation embryos, and identification of disease carriers. Applications are briefly discussed in the following section.

Parentage determination: Since the breeding value of an animal is generally estimated using the information available from its relatives, the knowledge of correct parentage is therefore a prerequisite. Parentage testing using molecular markers yields much higher exclusion probability (> 90%) than testing with blood groups (70–90%) or other biochemical markers (40–60%). Highly polymorphic DNA fingerprinting markers are quite useful for this purpose. Recently, DNA fingerprinting with oligoprobes (OAT18 and ONS1) has been successfully used for determining the parentage of an IVF buffalo calf. With the advent of PCR-based microsatellite assays, a large number of microsatellite panels have been reported that are useful for parentage testing in different livestock species.

Determination of twin zygosity and freemartinism: Correct knowledge of zygosity twins, particularly in monotocus animals, is important. Monozygotic twins provide a means for epidemiological and genetic studies, and also help in transplant matching. Individual-specific DNA fingerprinting techniques have the potential for determination of twin zygosity and demonstration of spontaneous XX/XY chimerism.

Identification of disease carrier: Many of the most serious incurable diseases result not from infections with bacteria or viruses, but from defects in genomes of the host. Certain allelic variations in the host genome lead to susceptibility or resistance to a particular disease. DNA polymorphisms occurring within a gene help to understand the molecular mechanism and genetic control of several genetic and metabolic disorders, and allow the identification of heterozygous carrier animals which are otherwise phenotypically indistinguishable from normal individuals.

ETHICAL ISSUES

The use of genetic biomarkers in epidemiological studies raises specific social and ethical issues related to the selection of molecular markers and methods of analysis, obtaining participants, the storage of biological samples and their linkage with individual data, the disclosure of information, and the publication of results. Several of these issues are similar to those associated with the use of any type of biomarker in epidemiology. Other problems are specifically related to the use of genetic material and the perception that genetic information raises special concerns regarding privacy, risk of abuse, and psychosocial impact (Hainaut and Vähäkangas, 1999). Cloning raises concerns both from ethical and practical points of view. Whether it is acceptable to clone humans is a difficult issue to deal with. Besides the low success rates seen in animals, the chance of abnormalities suggests that more information is required concerning the consequences of such practices before they would become routine in humans and animals. Advantages for animal breeding programs derived from cloning with no use of transgenesis are likely to be small (Van Vleck, 1999).

Most countries involved in biomedical research now have in place regulations governing the use of experimental or other donor animals as providers of tissues. These apply to higher animals assumed to have sufficient brain capacity and organization to feel pain and distress, and generally do not apply to lower vertebrates such as fish or invertebrates. Usually a more highly evolved animal is assumed to be sentient after halfway through embryonic development; restrictions apply to the method by which the animal is killed or operated upon (if it is to remain alive) such that the animal suffers minimal pain or discomfort. Restrictions also apply to the way the animal is housed and maintained, whether in an animal house during husbandry, under experimental conditions, or in a veterinary hospital under clinical conditions. In each case, control is usually exercised locally by an animal ethics committee and nationally by a governmental or professionally appointed body.

The current status of opinion and debate regarding ethical issues can be divided into three broad categories of relevance to animal biotechnology. The first is scientific integrity, where the focus has been on scientific fraud and the integrity of the research process. The second concerns possible harms or risks to parties affected either directly by research (including animals themselves) or through the eventual commercialization or development of products from animal biotechnology. The final category concerns a responsibility to serve as a guardian of the public interest with respect to application and development of technologies derived from new genetic sciences. It is plausible to see the scientific community as a whole having such a fiduciary obligation to the broader public with regards to the technical complexity of the issues and owing to public funding and institutional support for scientific research. The overall conclusion is that in the latter two categories especially, there is an urgent need for new participation in the deliberative consideration of ethical issues by working scientists (Thompson, 2008).

The European Group on Ethics (EGE) is an independent, pluralist, and multidisciplinary body that advises the European Commission on ethical aspects of science and new technologies in connection with the preparation and implementation of community legislation or policies. There are two ways in which farm animals in society are viewed: (1) There is the 18th century view of animals as living creatures with functional value; they are seen in terms of their purpose, which is determined by society. (2) Farm animals can also be viewed as sentient beings with inherent, intrinsic value (European Group of Ethics in Science and New Technologies (EGE), http://ec.europa.eu/bepa/expert-groups/ege/index_en.htm). The functional view of farm animal cloning includes commercial production of large numbers of high-value, elite animals, or using cloning as a method to disseminate genetically engineered animals. Using cloning for greater productivity will inevitably perpetuate many of the serious welfare problems already widespread in high-tech farming (e.g. lameness in broiler chickens, pigs, and dairy cows). Cloning and genetic engineering of farm animals is taking us in the wrong direction, toward perpetuating factory farming when all other societal trends point toward sustainable farming and respect for animals as sentient beings. The very aims of cloning are therefore an ethical issue. The practice of cloning raises ethical and welfare concerns, detailed as follows.

Invasive Medical Interventions are performed on donor animals (for oocyte extraction) and on surrogate mothers. Oocyte extraction for pigs and sheep is usually surgical, with all the accompanying stresses of recovery.

Suffering Caused to Surrogate Mothers. Pregnancy is typically prolonged, and cloned calves and lambs may be 25% heavier than normal. Higher birth weights lead to painful births and often the need for cesarean section.

Abnormal Fetal Development and Late Pregnancy Mortality lead to frequent deaths at various stages of development. Death in the second half of gestation is common.

Postnatal Mortality. The viability of cloned offspring at delivery and up to weaning is reduced compared to normal births. Surviving newborn clones have altered neonatal metabolism and physiology, and a high proportion of them die before weaning. Typical complications include gastroenteritis, umbilical infections, defects in the cardiovascular, musculoskeletal, and neurological systems, and susceptibility to lung infections and digestive disorders. These animals have short lives of suffering.

Health Problems During Life. Clones may have a greater propensity in later life for respiratory problems and immune system deficiencies compared with normal animals. Many clones have died or have had to be put down at a young age. Underlying weaknesses in cloned animals may not be fully revealed until the animals are stressed in some manner. A study undertaken at the U.S. Department of Agriculture (published in October 2005) suggested that clones may be born with crippled immune systems.

Inefficiency and Waste of Life. This includes embryos, fetuses, and mature animals that are killed as part of the procedures. A recent paper from New Zealand refers to the process as still inefficient and highly prone to epigenetic errors.

The Legal Situation in the EU. There are two important pieces of European Union law that should be considered with reference to cloned farm animals. The Protocol on Improved Protection and Respect for the Welfare of Animals in the Amsterdam Treaty of 1997 recognizes that animals are "sentient beings" and calls on the EU to "pay full respect to the welfare requirements of animals." The EU directive on the protection of animals kept for farming purposes (Directive 98/58/EC) says: "Natural or artificial breeding or breeding procedures which cause, or are likely to cause, suffering or injury to any of the animals concerned must not be practiced." The directive also says, "No animal shall be kept for farming purposes unless it can reasonably be expected, on the basis of its genotype or phenotype, that it can be kept without detrimental effect on its health or welfare." In other words, if the animal (in this case the cloned animal) is likely to suffer because of its inbuilt genetic or physiological weaknesses, then it shouldn't be on a farm. In view of the widespread suffering caused to cloned animals themselves, to the females from whose eggs they are extracted, and to the surrogate females who have the cloned embryos inserted and who usually face cesareans, Compassion in World Farming calls on the EGE to recommend a ban on the cloning of animals for food.

TRANSLATIONAL SIGNIFICANCE

A wide range of molecular marker technologies is now available for genetic studies that give us information about the kinships between breeds and the domestication process. On the other hand, molecular markers already provide new opportunities to speed up selection of routinely measured traits, to select for new traits that are costly and/or difficult to record in farm animals, and to improve animal production and productivity. Amongst these, the favorite markers for these studies are mainly microsatellites; RAPD, AFLP, ISSR, and SSR marker systems are also emerging as lead technologies. The RAPD marker system is not considered as convenient because of its inconsistency. However, RAPD assays are still used for DNA fingerprinting, along with other dominant markers such as AFLP and ISSR markers. SSR markers remain the markers of choice for genome mapping and genetic diversity analysis.

Several variations of the above mentioned marker systems are also available. These new generations of markers (namely, MITE) are in the early phase of usage and are not routinely employed in molecular marker technology laboratories. Despite an array of molecular markers available to researchers, it is still important to choose the right marker for the right problem. There is clearly potential to enhance the rates of genetic improvement by using molecular information. The full realization of this potential will require an investment in infrastructure and knowledge, and will therefore be limited in the first step mainly due to nationally and internationally important species and breeds within species. Moreover, the technology must be carefully targeted to provide optimal returns to breeding organizations and farmers.

WORLD WIDE WEB RESOURCES

The issue of animal cloning has received a great deal of attention in public discourse. Bioethicists, policy makers, and the media have been quick to identify the key ethical issues involved, and to argue (almost unanimously) on such attempts. Meanwhile, scientists have proceeded with extensive research agendas for the cloning of animals. Despite this research, there has been little public discussion of the ethical issues raised by animal cloning projects. Polling data show that the public is decidedly against the cloning of animals. To understand the public's reaction and fill the void of reasoned debate about the issue, it is necessary to review the possible objections to animal cloning and assess the merits of the anti-animal cloning stance. Some objections to animal cloning (e.g. the impact of cloning on the population of unwanted animals) can be easily addressed, while others (e.g. the health of cloned animals) require more serious attention by policy makers and the public. For more than 40 years, Congress has entrusted the Animal and Plant Health Inspection Service (APHIS) with the stewardship of animals covered under the Animal Welfare and Horse Protection Acts. APHIS continues to uphold that trust, giving protection to millions of animals each year, nationwide. APHIS provides leadership for determining the standards of humane care and treatment of animals. APHIS implements those standards and achieves compliance through inspection, education, cooperative efforts, and enforcement.

Similarly, The Bioethics Research Library at Georgetown University is an interdisciplinary and multi-format collection on ethical issues related to health care, biomedical research, biotechnology, and the environment. There are a number of organizations worldwide that are actively working for the humane treatment of animals. The Universities Federation for Animal Welfare (UFAW) is an independent registered charity that works to develop and promote improvements in the welfare of all animals through scientific and educational activity worldwide. The Great Ape Project (GAP) is an international movement that aims to defend the rights of non-human great primates (e.g. chimpanzees, gorillas, orangutans, and bonobos), our closest relatives in the animal kingdom. The mission of the Association for Assessment and Accreditation of Laboratory Animal Care International (AAALAC) is to "enhance the quality of research, teaching, and testing by promoting humane, responsible animal care and use." The organization awards accreditation to institutions that "meet or exceed AAALAC standards" regarding animal care.

Moreover, the Canadian Council on Animal Care (CCAC) is an autonomous and independent body created in 1968 to oversee the ethical use of animals in science in Canada. The CCAC is registered as a non-profit organization, and is financed primarily by the Canadian Institutes of Health Research (CIHR) and the Natural Sciences and Engineering Research Council of Canada (NSERC), with additional contributions from federal science-based departments and agencies, and private institutions participating in its programs. It is governed by a council of representatives from 22 national organizations that are permanent member organizations, and up to three limited-term member organizations.

The CCAC acts as a quasi-regulatory body and sets standards (its guidelines documents and policy statements) on animal care and use in science that apply across Canada. It is accountable to the general public and is responsible for the dissemination of information on the use of animals in science to Canadians. In addition to guidelines, documents, and policy statements, the CCAC develops comprehensive annual statistics on the number of animals used in science, and produces an annual report to disseminate information to its constituents and the general public.

The following are web resources related to animal welfare and ethical issues:

http://repository.upenn.edu
http://www.aphis.usda.gov
http://bioethics.georgetown.edu
http://www.ufaw.org.uk
http://www.greatapeproject.org/newsletters/
http://www.ccac.ca
http://www.awionline.org
http://www.vetmed.ucdavis.edu/whatsnew/article.cfm?id=2528
http://www.labanimal.com/laban/index.html
http://ec.europa.eu/bepa/european-group-ethics/publications/proceedings-ege-roundtables

REFERENCES

Bardakci, F. (2001). Random Amplified Polymorphic DNA (RAPD) Markers. *Turk J Biol*, *25*, 185–196.

Beuzen, N. D., Stear, M. J., & Chang, K. C. (2000). Molecular markers and their use in animal breeding. *The Veterinary Journal*, *160*, 42–52.

Bhattacharya, T. K., Kumar, P., Joshi, J. D., & Kumar, S. (2003). Estimation of inbreeding in cattle using RAPD markers. *Journal of Dairy Research*, *70*, 127–129.

Botstein, D., White, R. L., Skolnick, M., & Davis, R. W. (1980). Construction of a genetic linkage map in man using restriction fragment length polymorphisms. *Am J Human Genet*, *32*, 314–331.

Burk, N. E., Messer, L. A., Ernst, C. W., & Rothschild, M. F. (1998). Identification of sequence tagged sites in the Asian and African elephant. *Animal Biotechnology*, *9*(2), 155–160.

Duran, C., Appleby, N., Edwards, D., & Batley, J. (2009). Molecular genetic markers: Discovery, applications, data storage and visualisation. *Current Bioinformatics*, *4*, 16–27. 1574-8936/09.

Erhardt, G., & Weimann, C. (2007). Use of molecular markers for evaluation of genetic diversity and in animal production. *Arch. Latinoam. Prod. Anim*, *15*(Supl. 1). 2007.

Everts, R. E., Rothuizen, J., & van Oost, B. A. (2000). Identification of a premature stop codon in the melanocyte-stimulating hormone receptor gene (MC1R) in Labrador and Golden Retrievers with a yellow coat colour. *Animal Genetics*, *31*(3), 194–199.

Hainaut, P., & Vähäkangas, K. (1999). Genetic analysis of metabolic polymorphisms in molecular epidemiological studies: social and ethical implications. *IARC Scientific Publications*, *148*, 395–402.

Huby-Chilton, F., Beveridge, I., & Gasser, R. B. (2001). Single-strand conformation polymorphism analysis of genetic variation in Labiostrongylus longispicularis from kangaroos. *Electrophoresis*, *22*(10), 1925–1929.

Kawai, J., Hirotsune, S., Hirose, K., Fushiki, S., Watanabe, S., & Hayashizaki, Y. (1993). Methylation profiles of genomic DNA of mouse developmental brain detected by restriction landmark genomic scanning (RLGS) method. *Nucleic Acids Research*, *21*(No. 24).

Keating, A. F., Smith, T. J., Ross, R. P., & Cairns, M. T. (2008). A note on the evaluation of a beta-casein variant in bovine breeds by allele-specific PCR and relevance to β-casomorphin. *Irish Journal of Agricultural and Food Research*, *47*, 99–104.

Kubo, K. S., Stuart, R. M., Freitas-Astúa, J., Antonioli-Luizon, R., Locali-Fabris, E. C., Coletta-Filho, H. D., Machado, M. A., & Kitajima, E. W. (2009). Evaluation of the genetic variability of orchid fleck virus by single-strand conformational polymorphism analysis and nucleotide sequencing of a fragment from the nucleocapsid gene. *Archives of Virology*, 2009.

Kumar, D., Gupta, N., Ahlawat, S. P. S., Satyanarayana, R., Sunder, S., & Gupta, S. C. (2006). Single strand confirmation polymorphism (SSCP) detection in exon I of the α-lactalbumin gene of Indian Jamunapari milk goats (*Capra hircus*). *Genetics and Molecular Biology*, *29*(2), 287–289.

Kwok, P. Y., & Chen, X. (2003). Detection of single nucleotide polymorphisms. *Current Issues in Molecular Biology*, *5*, 43–60.

Machkour-M'Rabet, S., He´naut, Y., Dor, A., Pe´rez-Lachaud, G., Pe´lissier, C., Gers, C., & Legal, L. (2009). ISSR (Inter Simple Sequence Repeats) as molecular markers to study genetic diversity in tarantulas (Araneae, Mygalomorphae). *The Journal of Arachnology*, *37*, 10–14.

Moen, T., Hayes, B., Nilsen, F., Delghandi, M., Fjalestad, K. T., Fevolden, S. E., Berg, P. R., & Lien, S. (2008). Identification and characterisation of novel SNP markers in Atlantic cod: Evidence for directional selection. *BMC Genetics*, *9*, 18.

Montaldo, H. H., & Meza-Herrera, C. A. (1998). Use of molecular markers and major genes in the genetic improvement of Livestock. *Electronic Journal of Biotechnology*, *1*(2).

Oldenbroek, K. (2007). *Utilisation and Conservation of Farm Animal Genetic Resources*. Wageningen Academic Publishers. Page 64.

Osborne, P. W., Luke, G. N., Holland, P. W. H., & Ferrier, D. E. K. (2006). Identification and characterisation of five novel Miniature Inverted-repeat Transposable Elements (MITEs) in amphioxus (*Branchiostoma floridae*). *Int J Biol Sci*, *2*(2), 54–60.

Otsen, M., Bieman, M. D., Winer, E. S., Jacob, H. J., Szpirer, J., Szpirer, C., Bender, K., & Van Zutphen, L. F. (1995). Use of simple sequence length polymorphisms for genetic characterization of rat inbred strains. *Mamm Genome*, *6*(9), 595–601.

Pradeep, M., Sarla, R. N., & Siddiq, E. A. (2002). Inter simple sequence repeat (ISSR) polymorphism and its application in plant breeding. *Euphytica*, *128*, 9–17.

Questiau, S., Eybert, M. C., & Taberlet, P. (1999). Amplified fragment length polymorphism (AFLP) markers reveal extra-pair parentage in a bird species: the bluethroat (Luscinia svecica). *Molecular Ecology*, *8*, 1331–1339.

Ravago-Gotanco, R. G., Manglicmot, M. T., & Pante, M. J. R. (2010). Multiplex PCR and RFLP approaches for identification of rabbitfish (Siganus) species using mitochondrial gene regions. *Molecular Ecology Resources*, *10*, 741–743.

Reddy, M. P., Sarla, N., & Siddiq, E. A. (2002). Inter simple sequence repeat (ISSR) polymorphism and its application in plant breeding. *Euphytica*, *128*, 9–17.

Rosenberg, N. A., Pritchard, J. K., Weber, J. L., Cann, H. M., Kidd, K. K., Zhivotovsky, L. A., & Feldman, M. W. (2002). Genetic structure of human populations. *Science*, *298*(5602), 2381–2385.

Shao-Qing, T., Yuan, Z., Qing, X., Dong-Xiao, S., & Ying, Y. (2007). Comparison of methylation level of genomes among different animal species and various tissues. *Chinese Journal of Agricultural Biotechnology*, *4*, 75–79.

Shakyawar, S. K., Joshi, B. K., & Kumar, D. (2009). SSR repeat dynamics in mitochondrial genomes of five domestic animal species. *Bioinformation*, *4*(4), 158–163.

Teneva, A. (2009). Molecular markers in animal genome analysis. *Biotechnology in Animal Husbandry 25 (5–6)*, 1267–1284.

Thompson, P. B. (2008). (2008). Current ethical issues in animal biotechnology. *Reprod Fertility, and Development*, *20*(1), 67–73.

Van Vleck, L. D. (1999). Implications of cloning for breed improvement strategies: Are traditional methods of animal improvement obsolete? *Journal of Animal Science*, *77*(Suppl. 2), 111–121.

Wangkumhang, P., Chaichoompu, K., Ngamphiw, C., Ruangrit, U., Chanprasert, J., Assawamakin, A., & Tongsima, S. (2007). A web-based allele specific PCR assay designing tool for detecting SNPs and mutations. *BMC Genomics* (8), WASP. 275.

Yujun, H., & Wessler, S. R. (2010). A program for discovering miniature inverted-repeat transposable elements from genomic sequences. *Nucleic Acids Research*, *38*(22). MITE-Hunter.

FURTHER READING

Avise, J. C. (2004). *Molecular Markers, Natural History, and Evolution* (2nd ed.).

Freshney, R. I. (2010). *Culture of animal cells: a manual of basic technique and specialized applications* (6th ed.).

Gupta, P. K. (2008). *Molecular Biology and Genetic Engineering. Rastogi Publication*.

Lewin B (2004). Gene VIII.

Turnpenny, P., & Ellard, S. (2005). *Emery's Elements of Medical Genetics* (12th ed.). London: Elsevier.

GLOSSARY

Allele One of a number of alternative forms of the same gene or genetic locus (generally a group of genes). It is the alternative form of a gene for a character that produces different effects. The form "allel" is also used, an abbreviation of allelomorph. Sometimes, different alleles can result in different observable phenotypic traits, such as different pigmentation. However, many variations at the genetic level result in little or no observable variation.

Biochemical Markers Genes that encode proteins that can be extracted and observed (e.g. isozymes and storage proteins).

Deletions Also called gene deletion, deficiency, or deletion mutation; designated as Δ. A mutation (genetic aberration) in which a part of a chromosome or a sequence of DNA is missing.

Ethics Also known as moral philosophy, it is a branch of philosophy that involves systematizing, defending, and recommending concepts of right and wrong behavior (e.g. ethical issues related to animal research).

Freemartinism The normal outcome of mixed-sex twins in all cattle species that have been studied; it also occasionally occurs in other mammals, including sheep, goats, and pigs. A freemartin or free-martin (sometimes martin heifer) is an infertile female mammal that has masculinized behavior and non-functioning ovaries. Genetically the animal is chimeric (karyotyping of a sample of cells shows XX/XY chromosomes). Externally the animal appears female, but various aspects of female reproductive development are altered due to the acquisition of anti-Müllerian hormones from the male twin.

Homoplasy Occurs when characters are similar, but are not derived from a common ancestor. Homoplasy often results from convergent evolution.

Inbreeding Reproduction from the mating of parents that are close genetic relatives. It results in increased homozygosity, which can increase the chances of offspring being affected by recessive or deleterious traits.

Isoschizomers Pairs of restriction enzymes specific to the same recognition sequence. For example, Sph I (CGTAC/G) and Bbu I (CGTAC/G) are isoschizomers of each other. The first enzyme discovered that recognizes a given sequence is known as the prototype; all subsequently identified enzymes that recognize that sequence are isoschizomers.

Laboratory Animals Animals used for experimental purposes in the laboratory (e.g. vertebrate animals from Zebra fish to non-human primates).

Livestock Domesticated animals raised in an agricultural setting to produce commodities (e.g. food and fiber) and perform labor.

Locus In genetics and genetic computation, a locus (plural loci) is the specific location of a gene or DNA sequence on a chromosome. A variant of the DNA sequence at a given locus is called an allele. The ordered list of loci known for a particular genome is called a genetic map. Gene mapping is the procession of determining the locus for a particular biological trait.

Molecular Markers Allow the detection of variations or polymorphisms that exist among individuals in the population for specific regions of DNA (e.g. RFLP, AFLP, SNP, etc.).

Morphological Markers The first marker loci available had an obvious impact on the morphology of plants. Genes that affect form, coloration, male sterility, or resistance (among others) have been analyzed in many plant species. Examples of this type of marker include the presence or absence of awn, leaf sheath coloration, height, grain color, aroma (rice), etc. In well-characterized crops like maize, tomato, pea, barley, or wheat, tens or even hundreds of such genes have been assigned to different chromosomes.

Pharmacogenomics The branch of pharmacology that deals with the influence of genetic variation on drug response in patients by correlating gene expression or single-nucleotide polymorphisms with a drug's efficacy or toxicity. Pharmacogenomics aims to develop rational means to optimize drug therapy with respect to patients' genotypes to ensure maximum efficacy with minimal adverse effects.

Pleiotropy Occurs when one gene influences multiple phenotypic traits. Consequently, a mutation in a pleiotropic gene can have an effect on some or all traits simultaneously. This can become a problem when selection of one trait favors one specific version of the gene (allele), while selection of the other trait favors another allele.

Polymorphism Occurs when two or more clearly different phenotypes exist in the same population of a species (i.e. more than one form or morph). The term is also used somewhat differently by molecular biologists to describe certain point mutations in the genotype, such as SNPs.

Population Genetics The study of allele frequency distribution and change under the influence of the four main evolutionary processes: natural selection, genetic drift, mutation, and gene flow. It also takes into account the factors of recombination, population subdivision, and population structure. It attempts to explain such phenomena as adaptation and speciation.

Restriction Enzymes Enzymes that cut DNA at specific recognition nucleotide sequences known as restriction sites. They are commonly classified into three types that differ in their structure and whether they cut their DNA substrate at their recognition site, or if the recognition and cleavage sites are separate from one another (e.g. EcoRI, BamHI, HindIII, etc.).

Taxon A population or group of populations of organisms that are usually inferred to be phylogenetically related and which have characteristics in common that differentiate the unit (e.g. geographic population, genus, family, order) from other such units. A taxon encompasses all included taxa of lower rank and individual organisms.

Traits A distinct variant of an organism's phenotypic character that may be inherited, environmentally determined, or a combination of the two. For example, eye color is a character or abstraction of an attribute, while blue, brown, and hazel are traits.

Transposon Display A strategy that allows simultaneous detection of individual elements. For example, sequences flanking dTph1 elements are amplified by means of a ligation-mediated PCR. The resulting fragments are locus-specific and can be analyzed by polyacrylamide gel electrophoresis.

ABBREVIATIONS

AAALAC Association for Assessment and Accreditation of Laboratory Animal Care International
AFLP Amplified Fragment Length Polymorphism
APHIS Animal and Plant Health Inspection Service
ASO Allele-Specific Oligonucleotide
AS-PCR Allele-Specific PCR
AZF Azoospermia
CCAC Canadian Council on Animal Care
CIHR Canadian Institutes of Health Research
EGE European Group on Ethics
EU European Union
FAO Food and Agriculture Organization
ISSR Inter Simple Sequence Repeats
LEPR Leptin Receptor
MITE Miniature Inverted-repeat Transposable Element
MSAP Methylation Sensitive Amplification Polymorphism
MTNR1A Melatonin Receptor 1A
NBAGR National Bureau of Animal Genetic Resources
NCC Norwegian Coastal Cod
NEAC North-East Arctic Cod
NSERC Natural Sciences and Engineering Research Council of Canada
QTL Quantitative Trait Loci
RARB Retinoic Acid Receptor Beta
RAPD Random Amplified Polymorphic DNA
RFLP Restriction Fragment Length Polymorphism
RLGS Restriction Landmark Genome Scanning
SLDA Single Locus and Di-Allelic
SNP Single Nucleotide Polymorphism
SSCP Single-Strand Conformation Polymorphism
SSLP Simple Sequence Length Polymorphisms
SSR Simple Sequence Repeats
SSTR Simple Sequence Tandem Repeats
STMS Sequence Tagged Microsatellites
STS Sequence Tagged Site
STR Short Tandem Repeats
TD Transposon Display
UFAW Universities Federation for Animal Welfare
VNTR Variable Number Tandem Repeats

LONG ANSWER QUESTIONS

1. What is a molecular marker? Explain the different types of molecular markers.
2. What ethical issues are raised when molecular markers are used for the study of animals, and how can they be resolved?
3. What are microsatellite markers and why are they more useful than any other marker for diversity analysis?
4. What are the general applications of molecular markers?
5. How can animal research be studied using various World Wide Web resources?

SHORT ANSWER QUESTIONS

1. What are the ideal features of molecular markers?
2. Describe the brief history of the evolution of molecular markers.
3. What is an ethical issue?
4. Briefly explain the advantages and disadvantages of molecular markers.
5. How are molecular markers important in animal breeding?

ANSWERS TO SHORT ANSWER QUESTIONS

1. Molecular markers allow detection of variations or polymorphisms that exist among individuals in the population for specific regions of DNA (e.g. RFLP, AFLP, SNP, etc.). Molecular markers have three-fold applications in gene mapping: (1) A marker allows the direct identification of the gene of interest instead of the gene product, and consequently, it serves as a useful tool for screening somatic cell hybrids; (2) Use in several DNA probes and easy-to-screen techniques, a marker also helps in the physical mapping of

the genes using *in situ* hybridization. (3) Molecular markers provide sufficient markers for construction of genetic maps using linkage analysis.

2. Genetic maps are constructed on the basis of two classes of molecular markers. Type I markers that represent the evolutionary conserved coding sequences (e.g. classical RFLPs and SSLPs) are useful in comparative mapping strategies where a polymorphism is not an essential prerequisite. However, these are mostly single locus and di-allelic (SLDA), and thus are not useful for linkage analysis. On the other hand, type II markers (like microsatellite markers) have higher polymorphism information content than conventional RFLPs and can be generated easily and rapidly. Therefore, major efforts are being made to produce gene maps based on type II markers. Further utilization of molecular markers developed from DNA sequence information (namely ASO and STMS polymorphic markers) is also helpful in the rapid progress of gene mapping (Table 16.1).
3. Ethics, also known as moral philosophy, is a branch of philosophy that involves systematizing, defending, and recommending concepts of right and wrong behavior (e.g. ethical issues related to animal research). The use of genetic biomarkers in epidemiological studies raises specific social and ethical issues related to the selection of molecular markers and methods of analysis, obtaining participants, the storage of biological samples and their linkage with individual data, the disclosure of information, and the publication of results. Several of these issues are similar to those associated with the use of any type of biomarker in epidemiology. Other problems are specifically related to the use of genetic material and the perception that genetic information raises special concerns regarding privacy, risk of abuse, and psychosocial impact. Cloning raises concerns both from ethical and practical points of view.
4. The development of molecular biology during the past three decades has created new means for studying livestock genetics and animal breeding. Selection according to genotype has become an important tool in the breeding of farm animals. Molecular markers capable of detecting genetic variations at the DNA sequence level have removed the limitations of morphological, chromosome, and protein markers, and they also possess unique genetic properties that make them more useful than other genetic markers. They are numerous and distributed ubiquitously throughout the genome. DNA-based markers have many advantages over phenotypic markers in that they are highly heritable, relatively easy to assay, and are not affected by the environment. The disadvantage is that they have high development costs and heterozygotes may be misclassified as homozygotes when null-alleles occur (due to mutation in the primer annealing sites).
5. Information collected by the Food and Agriculture Organization (FAO) of the United Nations indicates that approximately 30% of the world's farm animal breeds are at risk of extinction. Conservation policies for native breeds will depend to a large extent on our knowledge of historic and genetic relationships among breeds, as well as on economic and cultural factors. Genetic variation of an animal is the basic material, which is utilized for changing the genetic makeup or genetic potential of domestic species to suit our needs. Mechanization, unplanned and indiscriminate breeding among native stocks, and human bias in favor of certain breeds are directly or indirectly responsible for the dilution of Indian livestock germplasm. Hence, characterization of indigenous germplasm is essential for their conservation. The increasing availability of molecular markers in laboratory animals allows the detailed analyses and evaluation of genetic diversity, and also the detection of genes influencing economically important traits.

Chapter 17

Gene Expression: Analysis and Quantitation

Denys V. Volgin

Department of Animal Biology, School of Veterinary Medicine, University of Pennsylvania, Philadelphia, Pennsylvania

Chapter Outline

SUMMARY

Quantitation of gene expression within a particular cell, tissue, or the whole organism is an important tool for animal biotechnology. Using the brain as a model system, we discuss methodological approaches to identification, collection, and processing of samples, including combined detection of mRNA and protein at the regional and single-cell levels.

WHAT YOU CAN EXPECT TO KNOW

The quantitative analysis of gene expression is an important tool for animal biotechnology and other life sciences. It involves methods that allow one to measure the level at which a target gene is expressed within a particular cell, tissue, or the whole organism. When performed at the single-cell levels, it allows one to dissect molecular regulatory mechanisms of selected cells, thus helping understand how gene expression may affect cellular phenotypes and functions. If the mammalian brain is selected as a model system, its various cells can be identified using distinct criteria, such as anatomic location, morphology, projections, and expression of neurotransmitters or other markers. In many experimental situations, it is more feasible to quantify mRNA levels as an indirect measure of gene expression. However, in studies comparing gene expression between groups, such as developmental studies or studies of the effect of various treatments, the mRNA quantitation may not suffice because changes in mRNA are not necessarily associated with proportional changes of the corresponding protein. In order to facilitate interpretation of the mRNA data, corresponding proteins should also be quantified. In addition, investigation of the

Animal Biotechnology. http://dx.doi.org/10.1016/B978-0-12-416002-6.00017-1

expression location could provide important information for the developmental studies and analysis of protein functions. In this chapter, the results of rat studies of human disease are used to discuss the methodological approaches to identification, collection, and processing of samples from heterogeneous brain regions subjected to the quantitative analysis of gene expression, with particular emphasis on the combined detection of mRNA and protein at both the regional and single-cell level.

HISTORY AND METHODS

INTRODUCTION

Gene expression is a fundamental life process providing a bridge between information encoded within a gene and a final functional gene product, such as a protein or non-coding RNA (ncRNA). For protein expression, it is a multi-stage process that includes transcription, mRNA splicing, translation, and post-translational protein modification. This process can be modulated at any stage, thus controlling the quantity and spatiotemporal parameters of the functional protein appearance. It is vital for maintaining normal cellular structure and function, and is also the basis for developmental changes, such as differentiation and morphogenesis. The ability to regulate gene expression allows cells to deliver a functional protein whenever it's needed for their normal functioning or survival. This mechanism underlies various physiological and pathological processes, including cellular adaptations to novel environments, maintenance of homeostasis, and recovery from damages.

The ability to perform analysis of gene expression in a quantitative manner has become increasingly important for animal biotechnology and other life sciences. Such an analysis usually refers to the techniques that allow one to measure the level at which a target gene is expressed within a particular cell, tissue, or the whole organism. For a basic biomedical scientist, it helps achieve understanding of the relationship between gene expression profiles and cellular or organism phenotypes. For animal biotechnology, quantitative analysis of gene expression also serves as an indispensable tool in the generation of transgenic or knockout animals. In multicellular animals, cells of the same origin are joined into tissues that in turn assemble in an organ in order to serve a common physiological function. When compared to various *in vitro* systems involving animal cells, such as tissue cultures, animal organs are characterized by various levels of cellular heterogeneity that makes quantitative analysis of gene expression and regulation a challenging task.

The brain may be viewed as a model system of a highly complex organ of vertebrate animals. It functions as the centralized control center of the animal nervous system that works like a computer. The brain receives coded information from the environment both inside and outside the animal body, has the capacity to store and processes this information, and generates multiple outputs. Within the brain, neural cells are commonly organized as layered structures (such as the cerebral cortex) or relatively distinct regional clusters, or nuclei, varying in size and shape. This chapter focuses on a number of key methodological approaches to identification, collection, and processing of samples from heterogeneous brain regions, with particular emphasis on the combined detection of messenger RNA (mRNA) and protein at the regional and single-cell levels. Following a discussion of general methodological aspects, specific examples of studies performed in our laboratory will be provided.

PRINCIPLES

Quantitation of protein-coding gene expression is typically performed by measuring the protein amounts. In addition, investigation of the expression location (gene product localization) could provide important information for developmental studies of complex organs, such as the brain, and analysis of protein function at regional and single-cell levels. However, in many experimental situations, it is more feasible to quantify levels of a functional product precursor (mRNA) as an indirect measure of the gene expression level.

Direct quantitative measurement of tissue mRNA levels can be achieved by Northern blotting. Its standard methodology includes electrophoretic separation of an RNA sample on an agarose gel, hybridization to a radioactive or chemiluminescent probe targeting the sequence of interest, and subsequent autoradiographic detection of the labeled RNA followed by densitometric measurement of the band strength in a gel image. This approach to quantitation requires relatively large samples of total RNA (5–50 μg) corresponding to a 2.5 to 25 mg tissue sample. Samples of this size obtained from brain tissues will contain a variety of cells, including neurons from the adjacent distinct regions of the brain, and glial and microvascular cells. Such heterogeneous samples may be used in the studies aimed at detection of brain-wide or larger regional changes in gene expression. However, both the basal level of expression and magnitude of changes introduced by the experimental conditions need to be quite large to overcome the relatively low detection sensitivity of this approach. In addition, the interpretation of data resulting from these studies may be quite limited due to the inability to pinpoint the exact location and cellular origin of the detected changes. To overcome the latter limitation, a complementary technique called *in situ* hybridization can be employed, but only a limited number of genes can be detected using a combination of differently labeled probes.

The sensitivity of mRNA detection can be significantly improved by using amplification of the reverse-transcribed RNA. In this technique, total RNA extracted from a tissue sample is subjected to reverse transcription (RT) followed by quantitative PCR (qPCR). As a result of RT, 10–50 μg of total RNA should yield 50–250 ng of single-stranded DNA (cDNA). The subsequent cDNA template amplification is performed in a quantitative manner using real-time PCR cyclers. Real-time PCR offers many advantages when compared to the standard PCR technique. The amplification within each sample can be monitored during the reaction using the fluorescent signal emitted by labeled hybridization probes or dyes as the DNA amplification progresses. This makes it possible to determine the PCR threshold cycle (c_T) at which the fluorescent signal reaches certain threshold levels. The c_T can be used as a quantitative measure of the starting amount of target sequence for any given sample. The reaction can be calibrated using a standard curve generated from serial template dilutions (Figure 17.1). This can allow one to obtain the absolute quantitation of the starting number of mRNA or cDNA copies (e.g. in numbers of copies per volume unit of tissue homogenate or weight unit of total RNA extracted from the sample, or copies per cell). The real-time methodology also enables verification of the specificity of the reaction prior to gel electrophoresis by melting the products after the PCR completion. This is achieved by slowly increasing heating of the products during which the fluorescent signal reflecting the amount of double-stranded DNA decreases as a function of temperature. Specific PCR amplification produces a distinct peak at the temperature specific for the target sequence on the negative derivative of the melting curve. This allows one to expedite optimization of the PCR conditions. The RT-qPCR is extremely sensitive because it is theoretically able to detect a single mRNA copy. Provided that the PCR primers and the reaction conditions have been tested and optimized, it is possible to achieve quantitative detection of selected mRNA species in reverse-transcribed RNA samples of individual CNS cells (Volgin et al., 2002).

In studies comparing gene expression between groups, such as developmental studies or studies of the effect of various treatments, interpretation of mRNA data may be limited because changes in mRNA are not necessarily associated with proportional changes of the corresponding protein (Raol et al., 2005). For example, an increased level of target mRNA may coexist with no changes or a change in the opposite direction at the protein levels because of an increased posttranscriptional degradation, or because mRNA transcription increases as a compensatory response to posttranscriptionally altered protein levels. The efficiency of mRNA screening by RT-qPCR has led to the accumulation of a wealth of data, but often there is no information about functional products of these genes. Therefore, in order to strengthen findings and facilitate interpretation of any new mRNA data, corresponding proteins should also be quantified by using various methods of immunolabeling. The immunoblotting approaches (Western blot or dot blot) are well suited for the densitometric quantitation of the protein of interest. However, it requires destruction of tissue samples, which makes impossible precise localization of proteins to a particular population of brain cells. This limitation can be circumvented by using immunohistochemical labeling of proteins in cells of a fixed tissue section. Both immunoblotting and immunohistochemistry approaches use the same general principles of antibody reaction and its visualization. An important difference is that preparation of the tissue sample for immunohistochemical labeling becomes crucial in order to preserve the tissue structure

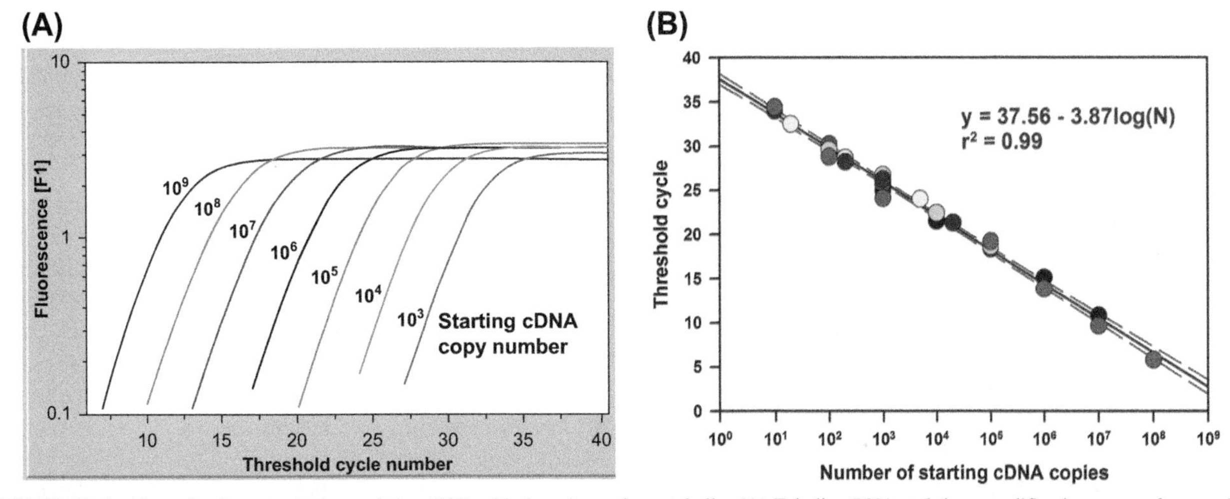

FIGURE 17.1 Example of a quantitative real-time PCR with the primers for α-tubulin. (A) Tubulin cDNA real-time amplification curves for samples containing standard dilutions of the purified target cDNA. (B) Tubulin qPCR calibration curve generated by amplifying known copy numbers of the target cDNA; circles of different colors correspond to separate calibration runs conducted during a 2-year period; dashed lines show 95% confidence intervals for the regression line across all data points.

and the ability of target protein epitopes to bind antibodies. This involves proper methods of tissue harvesting and fixation, usually achieved by using paraformaldehyde. To collect brain tissues suitable for immunohistochemistry from *in vivo* experimental conditions, the animals often need to be transcardially perfused with cold-buffered saline in order to remove blood cells, which is followed by perfusion with a fixative. Subsequently, the brain is extracted and subjected to additional fixation before it can be sectioned. In general, significant effort may be required for the optimization experiments, in which issues like excessive background labeling or insufficient target protein staining and autofluorescence related to the fixation procedure have to be systematically addressed. If the assay is not fully optimized, these issues may significantly limit the ability to quantify the immunostaining density in digital microscopic images of the sections. An additional method of quantitation available for the immunohistochemical detection is based on the counting of immunolabeling cells in well-defined regions of the section or brain nuclei. Quantitative or semi-quantitative immunohistochemical localization of the target protein often provides important complementary data when combined with mRNA studies (Volgin et al., 2003).

METHODOLOGY

Quantitation of mRNA Levels Using RT-PCR

The unmatched sensitivity of this methodological approach makes it extremely fit for mRNA-level quantitation in small tissue samples ("micropunches") or even individual cells. Extracting RNA samples from brain tissue micropunches containing only several hundred neurons gives a unique opportunity to target small, well-defined brain regions (Figure 17.2) (Comer et al., 1997; Okabe et al., 1997; Volgin and Kubin, 2007; Volgin, 2008). The amounts of cDNA generated from the micropunch samples following a successful reverse transcription is usually sufficient to quantify dozens or even hundreds of mRNA species. This can be achieved by multiple PCR reactions with different sets of specific primers (Volgin, 2008). Therefore, using RT-qPCR with the material obtained from the tissue micropunch samples enables one to solve the localization issue at the regional level and perform multiple gene expression assays, thereby making this approach very useful in regional gene expression profiling or high-throughput studies.

A further refinement of this technique led to development of a single-cell RT-PCR protocol that allows one to achieve an even higher degree of spatial selectivity (Lambolez et al., 1992; Phillips and Lipski, 2000; Sucher and Deitcher, 1995; Volgin et al., 2004b). Samples of single brain cells of interest can be collected from various *in vitro* preparations, such as slices of the brain (Sucher and Deitcher, 1995), cells isolated through acute dissociation (Phillips and Lipski, 2000; Volgin et al., 2001; Volgin et al., 2002; Volgin et al., 2004b; Volgin et al., 2008), or primary cell cultures (Chiang et al., 1994). Additionally, samples could be captured from fixed tissue using a laser dissection technique (Fink et al., 1998; Schütze et al., 1998). The collection of cells for the subsequent studies can be done reliably and efficiently when they are assembled in homogenous groups (e.g. nuclei). However, when the neurons of interest are scattered in brain regions containing multiple cell types, it requires additional effort to sort through the diversity of cellular populations. In such cases, single-cell gene profiling studies can easily become relatively inefficient and laborious. Various strategies aimed at overcoming difficulties due to cellular diversity can be employed. Cells can be selected by screening of their electrophysiological and pharmacological properties before harvesting (Sucher and Deitcher, 1995). Alternatively, neurons can be retrograde-labeled from a selected brain region using fluorescent markers (Comer et al., 1999; Volgin et al., 2001; Volgin et al., 2002; Volgin et al., 2004a; Volgin et al., 2008; Volgin et al., 2009). With the introduction of these additional procedures that also require additional equipment or surgeries, the yield of successfully isolated cells per animal can be quite limited. Cells can be selected on the basis of immunofluorescent labeling of a protein marker expressed selectively in the cells of interest. Labeled cells from fixed and stained sections can be then efficiently collected using a laser-capture microdissection technique. One limitation of this approach is that when sections are made very thin, only a fraction of mRNA from each target cell can be captured for further processing, whereas by increasing the thickness one increases a risk of capturing parts of neighboring cells (Kamme et al., 2003). In addition, the efficiency of mRNA detection can be decreased because of fixative presence, especially formaldehyde (Goldsworthy et al., 1999). To overcome some of these limitations, a method was developed that combines acute brain cell dissociation with immunocytochemical cell identification (Figures 17.3 and 17.4A–C). In this method, cells acutely dissociated from the brain slices are immunolabeled *in vitro* using fluorescent antibodies, and individual labeled cells are subsequently harvested and subjected to single-cell RT-PCR. Similar to any other mRNA processing and detection technique, this method may theoretically lead to some alterations in mRNA levels, but this was not obvious from previous mRNA detection studies performed using unfixed cell material (Volgin et al., 2003; Volgin et al., 2001; Volgin et al., 2002). With this approach, there was some reduction in the yield of cells with well-preserved morphology compared to the experiments in which dissociated brain cells are harvested immediately after they settle in a cell culture dish. This may be related to the additional time required for immunolabeling and/or multiple changes of fluid in the dish. However, this is more of a concern with neurons obtained from mature animals, whereas most cells dissociated from early postnatal animals survive the procedure

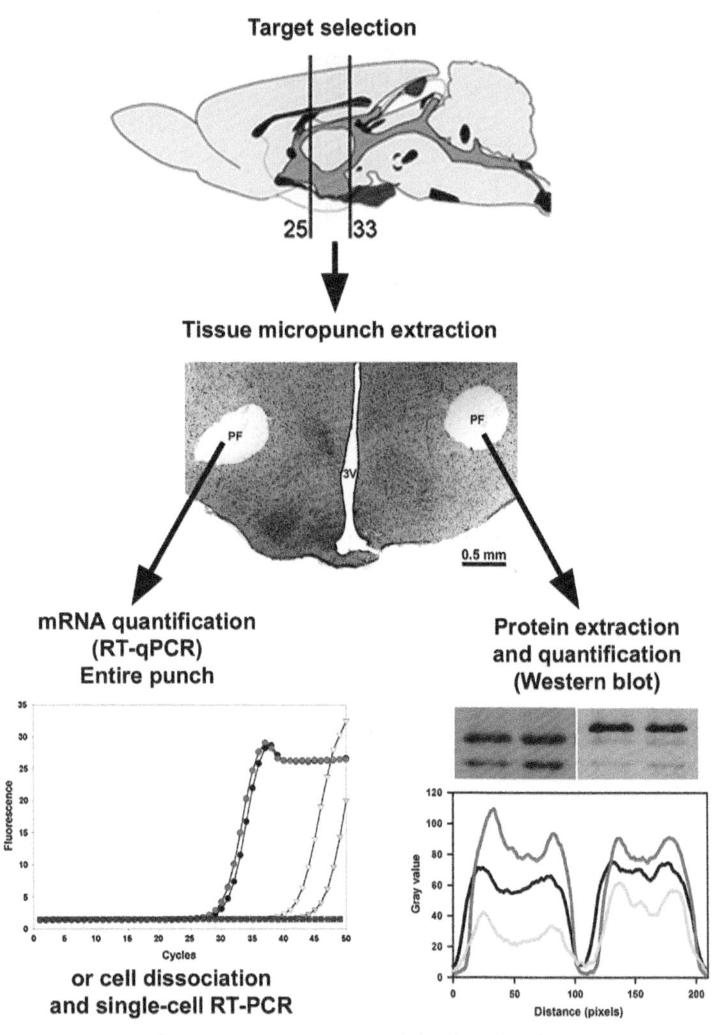

FIGURE 17.2 Main steps of the brain tissue sampling procedure. First, the anatomical location of the target brain region has to be determined. Then, thin brain slices are cut on a vibratome, and the brain region of interest is located within each slice under a dissecting microscope (details of the slice anatomy are explained in Figure 17.4A). Finally, circular micropunches are cut out from the region of interest, and tissue samples are processed for mRNA and protein quantitation. Alternatively, a tissue micropunch can be subjected to acute cell dissociation followed by cell harvesting for the single-cell RT-PCR.

much better. When compared to other techniques, immunocytochemistry-based identification of acutely dissociated neurons allows one to sample total RNA from the entire cell not treated with a fixative and requires no prior animal tracer injection surgery or additional costly equipment. This method can be potentially combined with *in vitro* pharmacological studies of labeled target cell populations prior to harvesting. The single-cell samples can be subsequently subjected to RT-PCR, microarray hybridization, or used to generate libraries for next generation sequencing, such as RNA-Seq. Although next-generation sequencing is still expensive and time-consuming when compared to RT-qPCR, it offers many additional advantages. For example, it makes it possible to distinguish mRNA splice variants and can be used to quantify mRNA expression in organisms for which limited or no sequence information is currently available.

The reverse-transcribed total RNA from single cell samples usually has very small amounts of specific cDNA species generated from the mRNAs of interest. In addition, there is often a need to split the material obtained from each cell and use only a fraction of the total cell cDNA. This is practiced in co-expression studies aimed at assessment of multiple mRNA species in the same cell (Volgin et al., 2003; Volgin et al., 2001). To allow quantitation and gel visualization of the PCR-amplified product, a sufficiently high number of qPCR cycles (40 or more) may be required. In this case, the PCR conditions should be thoroughly optimized. Such conditions should ideally enable the detection of a single copy of the target cDNA in order to allow interpretation of negative qPCR results as the evidence of absence of the mRNA of interest. In some cases, this may result in successful mRNA quantitation at the single-cell level, even if qPCR is performed with a fraction of the sample obtained from a single neuron (Volgin et al., 2002). However, such a PCR optimization often requires significant time and effort.

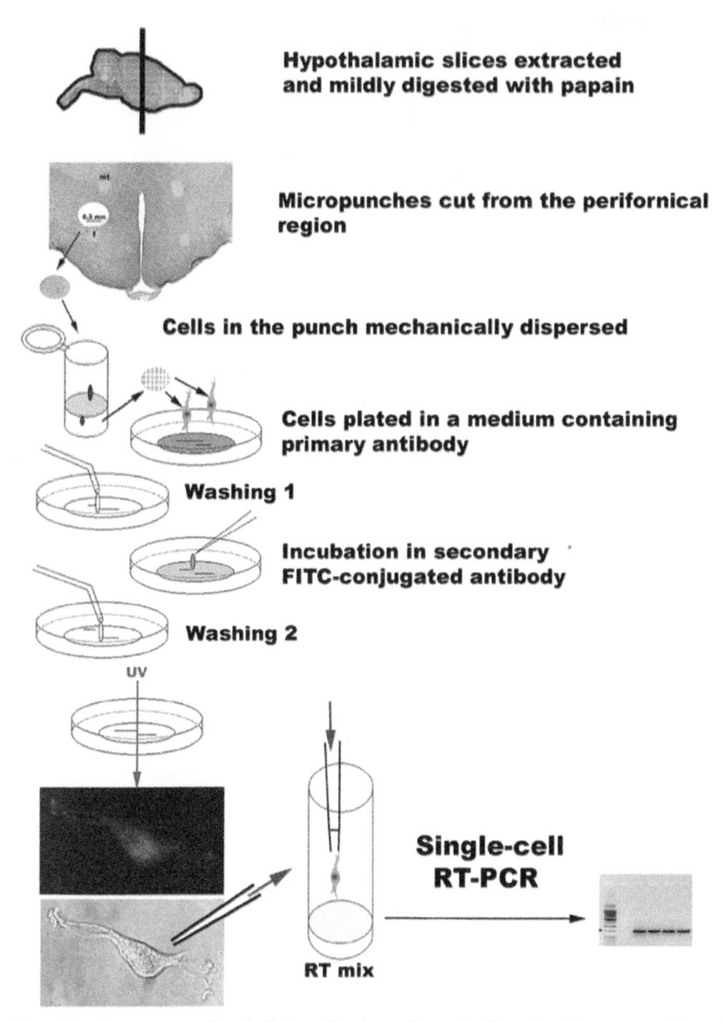
Hypothalamic slices extracted
and mildly digested with papain
Micropunches cut from the perifornical
region
Cells in the punch mechanically dispersed
Cells plated in a medium containing
primary antibody
Washing 1
Incubation in secondary
FITC-conjugated antibody
Washing 2
UV
Single-cell
RT-PCR
RT mix

FIGURE 17.3 Experimental design of the immunocytochemical identification of acutely dissociated neurons subjected to single-cell RT-PCR (all steps are self-explanatory).

One solution for this problem is using a two-round PCR amplification with either "nested" or "semi-nested" design. The first PCR round utilizes a set of specific "external" primers and should lead to initial amplification of larger cDNA sequences. To avoid accumulation of non-specific PCR products, the concentration of external primers in the reaction mix is usually reduced. In the second round, a fraction of the first-round PCR product is subjected to a second PCR with another set of specific primers, the "internal" primers designed to amplify a shorter segment enclosed within the first-round specific PCR product. The first round can be done as a multiplex PCR (with multiple primer sets). This allows one to obtain multiple pre-amplified cDNA products, which can be then detected in multiple second-round PCRs with primers recognizing their specific targets (Volgin et al., 2008). This approach makes it possible to detect dozens of target mRNA species in a single-cell sample, which makes it an efficient and sensitive mRNA profiling method. However, the direct quantitative assessment of starting cDNA amounts is not feasible in this case because of the unknown efficiency of the first round of amplification. An alternative method of relative quantitation that can be used with the two-round single-cell PCR is based on the statistical comparison of the proportions of cells expressing various mRNAs of interest (Volgin et al., 2003; Volgin et al., 2008).

Quantitation of Protein Levels

The most common approach to protein quantitation is Western blotting against the protein of interest. For this method, small samples of brain tissue, such as tissue micropunches taken from the region of interest, can be used (Figures 17.2 and 17.5A). Proteins have to be first extracted from the brain cells. The procedure includes treatment of tissue samples that helps achieve solubilization, denaturation, and elimination of protein–protein interactions by breaking the cross-linking bonds. In addition, non-protein components have to be removed before proceeding to the next step. After the extraction step, the proteins are separated using

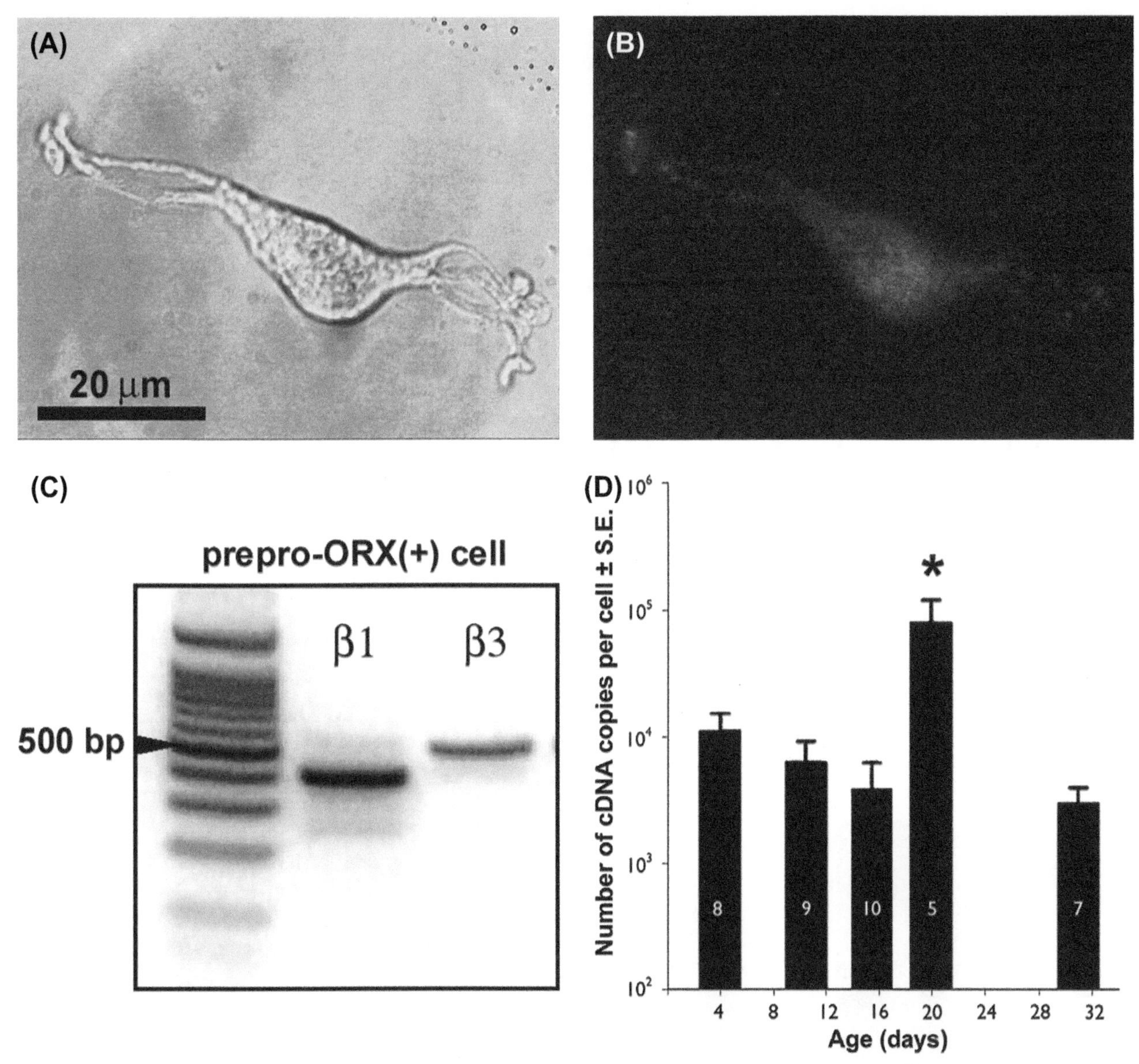

FIGURE 17.4 Example of immunocytochemically identified neuron expressing melanin-concentrating hormone (MCH) acutely dissociated from the hypothalamic PF region observed under phase contrast microscopy (A) and then under fluorescent illumination with an FITC filter (B) (Volgin et al., 2004b). (C) Example of agarose gel display of GelRed-stained PCR products for distinct $GABA_A$ receptor subunits from another single neuronal sample identified as expressing prepro-orexin. (D) Developmental changes in the orexin 2 receptor mRNA in XII motor neurons (Volgin et al., 2002). The average numbers of the the orexin 2 receptor cDNA copies per cell is determined using quantitative PCR; the numbers superimposed over the bars show the numbers of cells studied. * indicates a significant difference from rats at any other age ($p < 0.05$).

polyacrylamide gel electrophoresis (PAGE). Following the separation, proteins are transferred onto a membrane usually made from nitrocellulose or polyvinylidene difluoride. The membrane is then incubated with a blocking buffer that may contain milk or normal serum or even purified proteins in order to prevent any nonspecific binding of antibodies to the membrane surface. The transferred proteins are then probed with an antibody to the target protein in either a direct or indirect manner. In the direct detection technique, the primary antibody against the protein of interest is labeled with an enzyme or fluorescent dye for imaging and quantitation. In the indirect detection approach, the membrane is first incubated with a primary antibody. After removing unbound primary antibody, the secondary antibody against a species-specific site of the primary antibody is added. The secondary antibody can be linked to biotin, fluorescent probes such as fluorescein, or enzyme conjugates such as horseradish peroxidase. The most sensitive detection technique utilizes a chemiluminescent substrate that, when incubated with the enzyme, such as horseradish peroxidase, produces light in proportion to the amount of secondary antibody. An image of the sites at which antibodies are bound to the

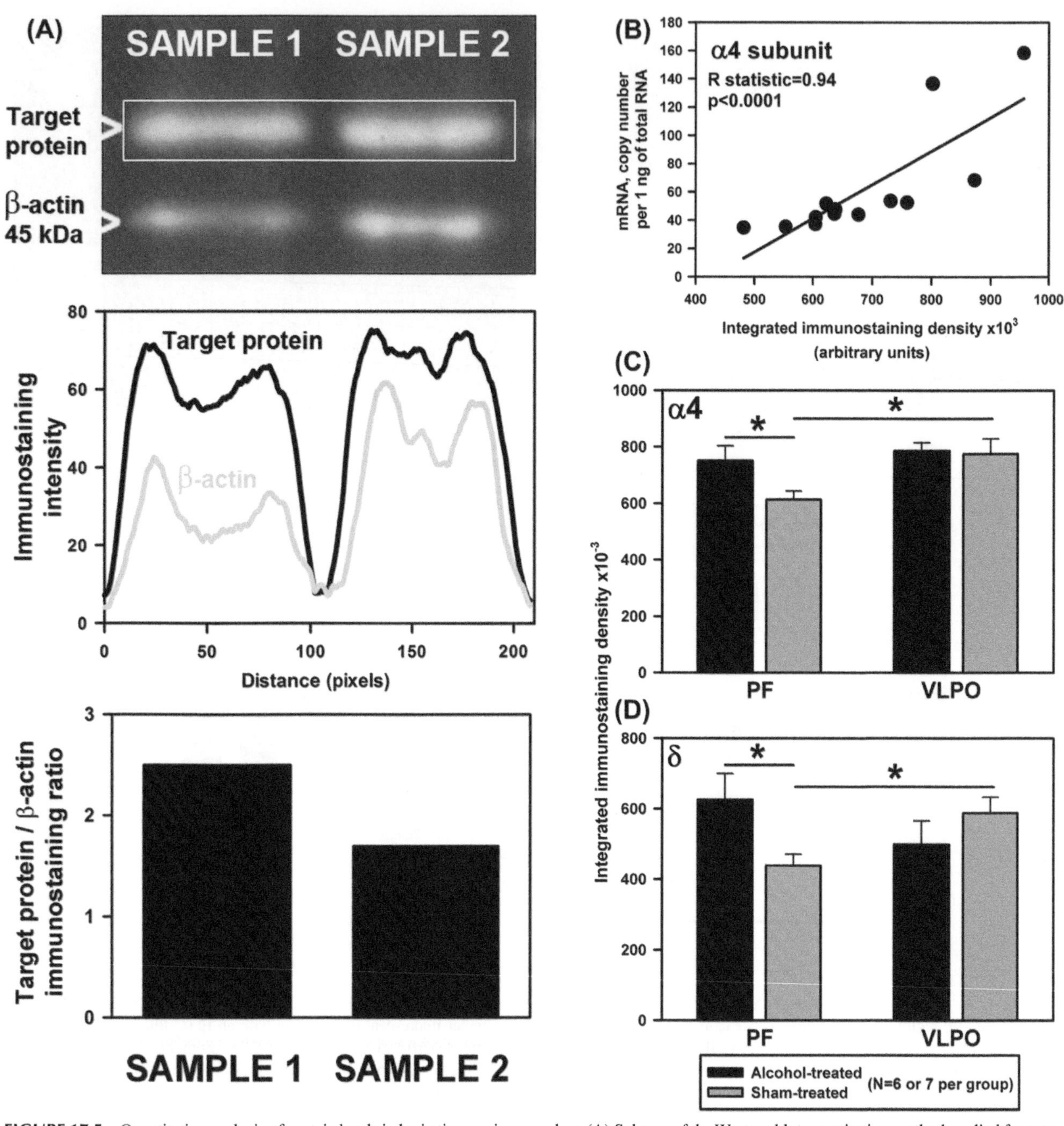

FIGURE 17.5 Quantitative analysis of protein levels in brain tissue micropunches. (A) Scheme of the Western blot quantitation method applied for comparison between two protein samples extracted from brain tissue micropunches. Plots of the distribution of their immunostaining intensity are built, the areas under the curves are measured, and the target protein-to-β-actin ratio is used to quantify the variation in the target protein levels. (B) Positive correlation between integrated density of α4 subunit dot blot immunostaining in hypothalamic micropunch tissue samples extracted from one side of the slices containing the PF or VLPO region and α4 subunit mRNA levels in samples taken from the opposite side of each slice (data from 13 pairs of samples from 7 sham-treated rats). The dot-blot immunostaining was used to determine that early developmental exposure to alcohol leads to increased immunoreactivity for α4 (C) and δ (D) subunits of $GABA_A$ receptor in the PF region in 29–30-day-old rats (Volgin, 2008). * indicates a significant difference ($p < 0.05$).

blot membrane can be then captured using X-ray film or digital imaging equipment. Another approach allows one to directly detect fluorescently tagged antibodies using a fluorescence imaging system. This method makes it possible to use more than one fluorophore and to detect multiple targets on the same membrane.

Regardless of what detection chemistry or system is used, the intensity of the signal should be proportional to the amount of the protein on the membrane. Distinct protein bands can be quantified densitometrically in digital images obtained directly from imaging equipment or scanned X-ray films (Figure 17.5A). When equal amounts of total protein

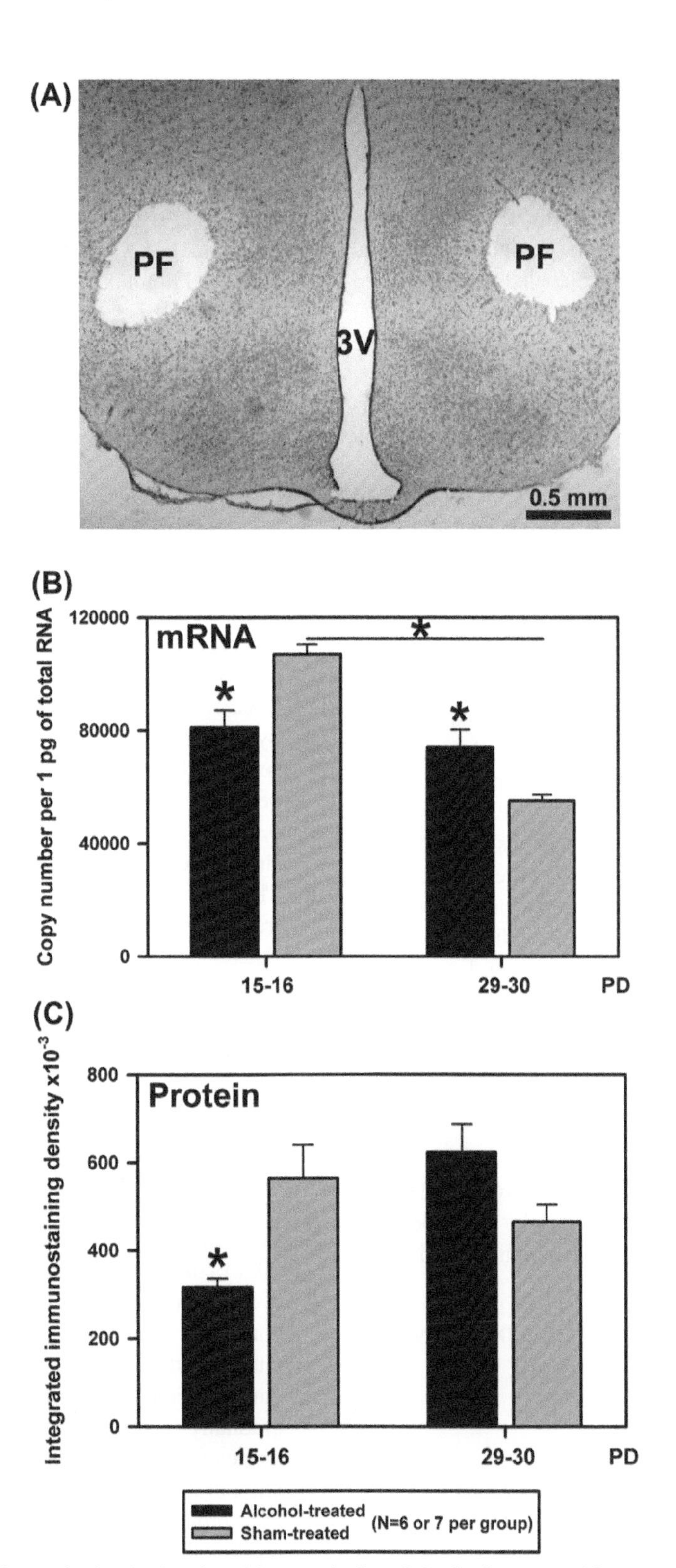

FIGURE 17.6 (A) An example of Neutral red-stained section of the posterior hypothalamic slice extracted from a 15-day-old rat showing the position of tissue micropunches taken bilaterally from the PF region (PF, 3V, third ventricle). (B,C) Developmental and alcohol-induced changes in GAD expression in the PF region (PD, postnatal day). Early developmental alcohol exposure leads to decreases in GAD67 mRNA (B) and GAD protein (C) on PD 15-16. On PD 29-30, GAD67 mRNA is higher in alcohol-treated than in control rats (B), and GAD protein in alcohol-treated rats also tends to increase (C). Data in B are expressed as the mean integrated density of grayscale images obtained from DAB-Ni-stained dot-blots processed using GAD65/67 antibodies. * indicates a significant difference between the groups ($p < 0.05$).

per each sample are used, direct comparison between the samples can be achieved. However, when protein samples are obtained from a very small tissue sample, such as a tissue micropunch, the ability to bring all samples to the same concentration of total protein is often limited. In this case, an additional protein marker, generally a product of so called housekeeping genes, can be immunolabeled and visualized on the same membrane. Subsequently, density measurement for the target protein band can be normalized relative to the density of a housekeeping gene protein in each sample.

The Western blot technique allows one to obtain information on the size of the protein in addition to its identity provided through the reaction with the specific antibody. Provided that a very selective antibody is available, and the immunolabeling conditions are well tested and optimized, a simplified immunoassay called dot-blot can be used to achieve a rapid, high-throughput protein quantitation in tissue micropunch samples. The starting amount of tissue required for the assay is much smaller than in any other immunoblotting technique, which makes it possible to quantify multiple proteins and mRNAs obtained from symmetrically located tissue micropunches of the same size (Volgin, 2008). This approach basically omits the PAGE separation step of the Western blot technique. A small drop (0.5 μl) of solubilized protein sample is absorbed directly onto a nitrocellulose membrane, which is then sequentially incubated in primary antibodies against the protein of interest, biotinylated secondary antibodies, avidin–horseradish peroxidase complex, and then visualized with diaminobenzidine with nickel ammonium sulfate. The membranes are then dried and scanned, and immunostaining density for each "dot" is quantified.

PROTOCOLS

These protocols are designed for small rodents (rats and mice). However, they may be optimized for samples from the brain or another organ extracted from any other animal species.

Protocol 1. Parallel Quantitation of mRNA and Protein from Defined Rodent Brain Region

Tissue Extraction

- Animals are deeply anesthetized (4% isoflurane), decapitated, and the whole brain or its structure of interest is rapidly removed and immersed in an ice-cold artificial cerebrospinal fluid (ACSF) with pH 7.4 and controlled osmolarity.
- The brain chunk is trimmed while immersed in the same ice-cold medium, and transverse sections (400 –700 μm thick) are cut from the region of interest with a tissue slicer (Vibratome).
- Slices containing the target area are placed in a dissection dish filled with the ice-cold ACSF and inspected under a dissecting microscope. Subsequently, 500–700 μm circular punches are cut bilaterally (in pairs) from the region of interest using a custom-cut syringe needle (Figures 17.2 and 17.6A). One of the samples from each pair is subjected to total RNA extraction and RT-qPCR, and the other sample is frozen on dry ice and stored at −80°C for the subsequent protein immunodetection.
- Each slice is then fixed in formalin and cut into 30 μm thick sections that are subsequently stained with Neutral red to verify the anatomic location of the micropunch (Figure 17.6A).

RNA Extraction and RT-qPCR

- Total RNA is extracted from each tissue micropunch sample using a commercially available, column-based kit (e.g. the RNeasy® Mini Kit from QIAGEN Inc., Valencia, CA), re-dissolved in RNase-free water, and quantified by densitometry. The parameters of RNA quality are set as follows. The A260/A280 absorbance ratio should be between 1.25 and 2.2, the A320 should be less than 0.1, and the concentration of total RNA from one punch should measure between 2 and 15 μg/mL.
- The desired amount of the extract is treated with RNase-free DNase I and reverse-transcribed using the SuperScript II reverse transcriptase (Invitrogen, Carlsbad, CA) or similar enzyme according to the manufacturer's instructions. The reaction can be performed in a conventional PCR cycler. Following the RT, the samples are chilled on ice and stored at −20 C° until qPCR is ready to be done.
- Subsequent qPCR reactions are monitored in real-time using the LightCycler® system (Roche Diagnostics, Indianapolis, IN). Fixed aliquots of each cDNA sample (e.g. 1 μl) are used for qPCR with the specific primer set for each gene selected for the study. The reactions include 30 s of initial denaturation at 95°C followed by 20–40 cycles consisting of a 0–1 s spike at 95°C and 25 s of combined annealing-elongation at 68°C, and are concluded with 30 s of final elongation. At the end of the amplification, the products are subjected to melting by gradually increasing heating to 95°C. The position and shape of the melting curve peaks provide an initial evaluation of the quality of each reaction and specificity of the PCR product. The products are then cooled and evaluated on the ethidium bromide-stained agarose gel to confirm that they are of the expected size.
- To ensure specificity of the PCR primers, all candidate primer sets are tested against the sequence data provided by the GenBank genetic database of the U.S. National Institutes of Health in order to identify a set with a minimal potential for primer dimerization and maximal target selectivity. Subsequently, the PCR conditions for each

set of primers are optimized by performing test PCRs with 100 times diluted cDNA obtained from brain tissue micropunches. Primers are deemed acceptable when the following criteria are fulfilled:

- The PCR product amplified after 35 cycles of PCR appears on the gel as a distinct band with the expected molecular weight.
- The melting curve for the PCR product has a single sharp peak at the temperature expected for the target sequence.
- In 35 cycle-long PCR runs, no specific PCR products are amplified from control samples of either non-reverse transcribed RNA or samples with no cDNA added.
- The efficiency of the PCR amplification, measured as the slope of the linear portions of log-converted fluorescence versus cycle number curves, is at least 1.6 per cycle.

qPCR Calibration Using the External Standards

- The target cDNA is amplified from a brain tissue sample using the same primer set, and the PCR product is separated on the ethidium bromide-stained 2% agarose gel. The distinct band specific for the target sequence is excised from the gel using a sterile scalpel blade and placed into a sterile PCR tube. Usually, the amount of PCR product amplified from one round of the LightCycler real-time PCR to the point of the reaction's saturation is sufficient to obtain enough cDNA for the subsequent gel purification. However, it may be necessary to use more starting material for the gel to ensure a sufficient amount of the target cDNA.
- The PCR product is extracted from the gel and purified by using QIAquick Gel Extraction Kit (QIAGEN) according to the manufacturer's instruction. The concentration of the purified standard cDNA is determined by measuring the absorbance at 260 nm. The number of cDNA copies (copies/μlμL) of the standard can be calculated using the following formula: 6×10^{23} [copies/mol] × Concentration [g/μlμL]/Molecular weight [g/mol], where average molecular weight of cDNA can be simplistically calculated as the number of base pairs multiplied by 660.
- A "sham" reverse transcription reaction is performed in the buffer that includes all the usual reagents, with the exception of mRNA. These reagents are carried over from the RT reaction in the actual PCR reactions decreasing their efficiency; these conditions should also be replicated in the calibration procedure. Therefore, the product of this sham RT needs to be added to the standard dilutions of target cDNA.
- The standard serial dilutions are prepared starting from 10^7 copies per sample and ending with the projected 10^{-1} dilution serving as a negative control, and qPCR are performed with all serial dilutions in 2–3 replications to obtain a threshold cycle number for every standard dilution (Figure 17.1A).
- The threshold cycle data are then plotted (Figure 17.1B) as a function of serial dilutions using statistical software (e.g. SigmaPlot) or the LightCycler software; the Y-intercept and the slope of the curve are determined using the linear regression. Using these data, one can calculate the number of copies of the target sequence in an unknown sample using the following formula: $N = 10^{((\text{Y-intercept} - cT)/\text{slope})}$, where c_T is the threshold cycle number for an unknown sample, and N is the copy number for this sample.

Protein Quantitation: Dot Blot and Western Blot

- Each micropunch sample is sonicated in the solubilizing buffer containing 7 M of urea, 2 M of thiourea, 0.25 mM Tris base, 4.0% CHAPS, and 1.0% NP-40. The resulting crude extract is subsequently used for both dot blot and Western blot procedures.
- For the dot blot immunodetection, 0.5 μl μL of the crude extract is absorbed onto a nitrocellulose membrane.
- The membrane is blocked in Tris-buffered saline (TBS, pH 7.4) containing nonfat dry milk and Tween 20. The blocked membranes were then incubated overnight in TBS containing primary antibody against the protein of interest followed by sequential incubation in biotinylated secondary antibodies and the avidin–horseradish peroxidase complex, and then visualized with diaminobenzidine with nickel ammonium sulfate.
- The membranes are dried at room temperature, and images of the membranes are digitized. The immunostaining density is quantified using ImageJ software (National Institutes of Health).
- For the Western blot procedure, the soluble fraction of the homogenate is loaded onto 4–15% precast polyacrylamide Tris-HCl minigels (Ready gel, Bio-Rad, Hercules, CA) in the loading buffer containing Tris-HCl, DTT, sodium dodecyl sulfate (SDS), Bromophenol blue, and glycerol.
- Proteins are separated by SDS-PAGE and transferred to nitrocellulose membranes using the Mini-PROTEAN 3 system (Bio-Rad).
- Membranes are blocked and processed with the primary antibodies obtained from the same animal source targeting the protein of interest and β-actin, a housekeeping gene product, as described above for the dot blot technique. Primary antibody binding sites are then revealed by sequential incubation with the secondary, animal source-specific, horseradish peroxidase-conjugated antibody, and the SuperSignal West Dura chemiluminescent substrate (Pierce/Thermo Scientific, Rockford, IL). The chemiluminescent signal is detected using the autoradiography film.

- Distinct protein bands are quantified densitometrically in film images using ImageJ software and the amount of the target protein is expressed relative to the density of the band for β-actin in each sample (Figure 17.5A).

Protocol 2. Single-cell RT-PCR

Acutely dissociated brain cells can be identified for their harvesting and subsequent single-cell RT-PCR based on either anatomical or functional criteria. The anatomical criteria can be fulfilled by selecting neurons retrograde-labeled from their projection sites. For example, the hypoglossal (XII) motor neurons innervate the tongue muscles, and can therefore be retrograde-labeled from these muscles using fluorescent latex microspheres ("beads") that are transported from the neuromuscular junction to the neuronal cell body. Alternatively, the dissociated brain cells can be identified *in vitro* using immunocytochemical labeling of their functional protein markers prior to single-cell RT-PCR.

Brain Cells Labeling and Cell Dissociation

- For the retrograde labeling of XII motoneurons, animals are anesthetized (3.5% isoflurane) and 6–10 μL of 25% rhodamine-dextran (Molecular Probes/Invitrogen) are injected into multiple sites of the tongue muscle; animals are allowed to recover for the 3–6 days necessary to achieve successful retrograde transport of the tracer.
- The decapitation and brain extraction procedures are performed as described in Protocol 1. The transverse medullary slices containing XII motor nucleus are placed in custom-made baskets immersed in oxygenated PIPES buffer inside a chamber with double walls between which water with the desired temperatures is circulated.
- The slices are subjected to enzymatic digestion at 35°C in PIPES buffer containing papain, L-cysteine, EDTA, and β-mercaptoethanol.
- Following digestion, the slices are washed and placed in oxygenated, HEPES-based standard external solution at room temperature.
- Circular punches, 400 μm thick, are cut out from the XII nucleus and triturated in an Eppendorf tube filled with the standard external solution using a series of fire-polished Pasteur pipettes with decreasing tip diameters. The dispersed cells are plated in custom-made Petri dishes with poly-L-lysine-coated glass bottoms made of coverslip glass. After a 30-min cell-settling period, the dishes are slowly perfused with the standard external solution.
- The micropunch locations are verified as described in Protocol 1.
- If the alternative method of cell identification based on immunocytochemical labeling of their protein markers is used, the brain cells are dissociated from intact animals following the steps described above. After the trituration step, 0.2 mL of 1–2% blocking normal serum appropriate for the selected antibodies in the standard external solution is added to the cell suspension and incubated for 10 min at room temperature.
- 0.2 mL of mixture containing diluted primary antibodies and normal serum is added to the cells. The antibody and serum concentrations can be selected on the basis of preliminary experiments with the same antibodies used for immunohistochemical labeling of paraformaldehyde-fixed sections containing the same brain region. Following the addition of the primary antibody and serum, the cell suspension is gently mixed and plated in poly-L-lysine-coated Petri dishes.
- Cells are allowed to settle for 45 min at room temperature, and then the dish is perfused with 1–2% oxygenated serum in the standard external solution at a flow rate of 0.2–0.3 mL/min in order to achieve the primary antibody washout.
- The medium from above the plated cells is gently removed with a pipette and the dish is refilled with the standard external solution containing appropriate normal serum and secondary fluorescein-tagged antibody. The cells are incubated with the secondary antibody for 30–60 min.
- The labeling procedure is concluded with the perfusion with serum-containing standard external solution followed by oxygenated standard external solution with no serum.

Cell Harvesting and Single-Cell RT-PCR

- Individual retrograde-labeled or immunolabeled neurons are identified using an inverted microscope equipped with phase contrast objectives, UV illuminator, and an FITC filter (Figure 17.4A and B).
- Under visual control through a CCD camera, identified neurons are attached by suction to the tips of glass pipettes filled with 8 μL of buffer containing 120 mM KCl, 1 mM $MgCl_2$, and 10 mM HEPES (pH 8.3).
- The entire content of the pipette is discharged into a PCR tube containing 3 μL of the reverse transcription mix (RNase inhibitor, random hexamers, dNTPs, $MgCl_2$, and DTT). The single-cell samples are stored at −20°C until submitted to reverse transcription on the same day.
- The samples are ultrasonicated for 3 min on ice. After subjecting the samples to DNAse I treatment, the reverse transcription is done for 60 min at 42°C with SuperScript II reverse transcriptase (Invitrogen). The resulting cDNA can be the divided into aliquots that are used in separate PCRs.
- The cDNA is amplified in either one-round qPCR or two-round, nested or semi-nested PCR. For the latter approach, the first round of amplification is done using a standard thermal cycler and the second round on a real-time cycler

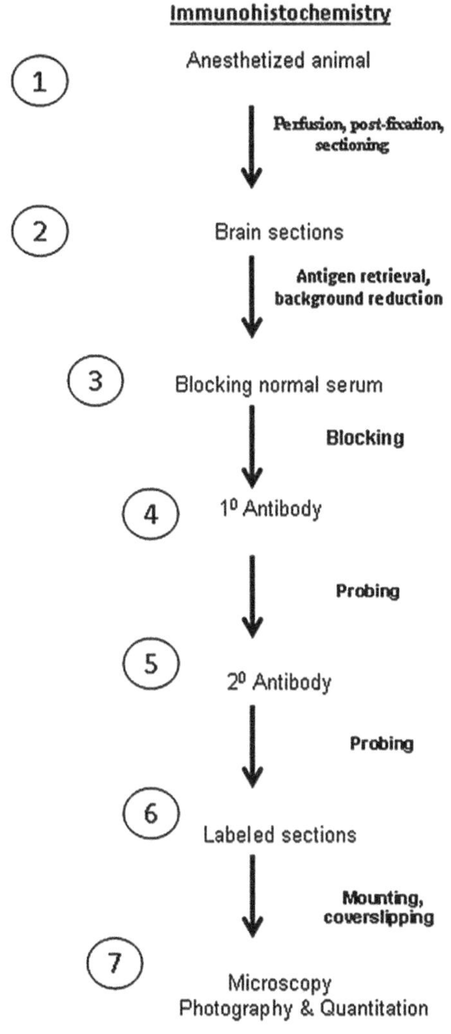

FLOW CHART 17.1

1. During the first step, deeply anesthetized animals are transcardially perfused with a fixative in order to remove all blood cells and ensure the preservation of tissue architecture and cell morphology. The brains are removed, post-fixed in the same fixative, cryoprotected, and then sectioned in the transverse plane using a cryostat into 25–35 μm thick sections.
2. This is the second step, when free-floating sections are incubated in 1% sodium borohydride to help unmask the antigens and reduce any fixative-induced autofluorescence. In addition, H_2O_2 in methanol can be used at this step in order to suppress endogenous peroxidases that may produce background signals.
3. The third step of the procedure is to block all epitopes within the tissue sections to prevent the nonspecific binding of the antibodies. This can be achieved by using normal serum that carries antibodies that bind to nonspecific sites. It is important to use serum from the animal species that the secondary antibody was made in, as opposed to the species of the primary antibody.
4. Incubation with the primary antibody is the fourth step. The primary antibody recognizes the target antigen or protein. This step should be followed by washing to remove extra antibody.
5. During the fifth step, sections are incubated with the secondary antibodies, which are either biotinylated or labeled with a fluorescent tag. With biotinylated antibodies, the sites of binding are then visualized using the avidin–biotin complex kit conjugated to either horseradish peroxidase or a fluorescent tag. This helps amplify the signal obtained by the reaction between the antigen and primary antibody.
6. This is the sixth step, when the labeled sections are serially mounted on the microscope slides, dehydrated, defatted, and cover-slipped.
7. During the last step, the labeled sections are ready to be examined under the microscope and documented by the means of photography available in the laboratory. The density of the immunolabeling and background within the structures of interest can be quantified in digital images and subjected to statistical analysis.

(Light Cycler; Roche Diagnostics). The real-time amplification protocol, qPCR calibration, the criteria for primer specificity, reaction quality control, and interpretation of negative PCRs are described in Protocol 1.

- To control for genomic DNA amplification, non-reverse transcribed single-cell samples are submitted to PCRs with primer sets used in the study. To control for the presence of the investigated mRNAs in the culture medium, one sample of the fluid from above the plated cells is collected during each cell collection session and submitted to the same RT-PCR procedures as single-cell samples.

Protocol 3. Semi-Quantitative Immunohistochemistry (Flow Chart 17.1)

- Animals are deeply anesthetized (pentobarbital, 100 mg/kg) and transcardially perfused with phosphate-buffered saline (PBS; pH 7.4) containing heparin followed by 4% paraformaldehyde and 5% dimethyl sulfoxide.
- The brains are removed, post-fixed in the same fixative overnight, washed in PBS and cryoprotected in phosphate-buffered sucrose (30%) for 2–4 days, and then sectioned in the transverse plane (25–35 μm) on a cryostat into 3–5 series. One series is mounted, stained with cresyl violet, dehydrated, and cover-slipped to provide the reference for the other series that will be subjected to immunohistochemical procedures.
- The remaining free-floating sections are subjected to incubation in 1% $NaBH_4$ (sodium borohydride; 15–30 min), methanol (70%) and H_2O_2 (0.3%) (15 min), blocking normal serum, and then primary antibody against the marker of interest. Incubations with primary antibody last 48–72 h at 4°C, after which the sections are washed and incubated with the appropriate secondary antibody. The secondary antibodies are either biotinylated or labeled with a fluorescent tag. With biotinylated antibodies, the sites of binding are subsequently visualized using the avidin–biotin complex kit conjugated to either horseradish peroxidase or a fluorescent tag.
- For dual labeling, the sections are incubated in the second set of primary and secondary antibodies selected to be compatible with the antibodies and sera used in the first round. It is possible to use either two different fluorescent tags (e.g. FITC and Cy3), or, when appropriate, a two-color horseradish peroxidase reaction. In the latter case, the first round of labeling uses heavy metals to intensify the horseradish peroxidase reaction and yield a black

product, whereas the second labeling does not include heavy metals and yields a brown product. At the end of these procedures, the sections are serially mounted, dehydrated, defatted, and cover-slipped.

- Mounted sections are observed using an upright microscope and digitally photographed. The density of the immunostaining and background is quantified using ImageJ software (National Institutes of Health) from digital images taken under constant illumination and magnification. The staining density can be determined by outlining the nucleus or region of interest, and then measuring the average grayscale intensity on a pixel-by-pixel basis. The background staining is measured by positioning a copy of the outline over an area of the same section having the lowest immunolabeling intensity. The difference between the two measurements is used to characterize the specific immunolabeling.

EXAMPLES

Prolonged Upregulation of Gene Expression for the Hypothalamic $GABA_A$ Receptors in a Rodent Model of Prenatal Exposure to Alcohol

Prenatal exposure to alcohol is associated with lasting abnormalities of sleep and wake behavior and motor development, but the underlying molecular mechanisms remain to be investigated. We hypothesized that prenatal alcohol exposure alters development of inhibitory gamma-aminobutyric acid (GABA) signaling in the brain regions important for the control of sleep and motor activity. We tested our hypothesis by assessing changes in gene expression for GABA enzymatic precursor, glutamate decarboxylase (GAD), and the selected subunits of $GABA_A$ receptors expressed in these regions in a rat model of prenatal exposure to alcohol (Volgin, 2008). On postnatal days 15–16 or 29–30, 500 μm thick hypothalamic slices were obtained from rats exposed to alcohol or those receiving a sham treatment and sacrificed at a constant circadian time (3–4 PM). Two pairs of tissue micropunches were cut bilaterally, one from the wake- and motor activity-promoting perifornical (PF) region of the posterior hypothalamus (Figure 17.6A) and the other from the sleep-promoting ventrolateral preoptic (VLPO) region of the anterior hypothalamus. One of the samples from each pair was subjected to RNA extraction and RT-qPCR, and the other sample was used for protein immunodetection, as described in Section 1 of the Methodology. In samples obtained from each region, we quantified mRNA levels for the GAD67 isoform. In the PF region of the control rats, GAD67 mRNA significantly decreased between postnatal days 15 and 30 in the PF region (Figure 17.6B). In contrast, in the rats subjected to alcohol exposure, this normal developmental change was reversed. On postnatal days 15–16, the GAD67 mRNA level was lower in the PF region in alcohol-exposed rats, whereas on postnatal days 29–30, it was higher than in control rats. The latter effect was region-specific because it did not occur in the VLPO region (not shown). These mRNA data were followed up on with protein quantitation using dot-blot immunodetection. Consistent with the findings at the mRNA level, also found were proportional changes in GAD immunoreactivity in the PF region (Figure 17.6C).

Also compared were the mRNA levels for eleven subunits of $GABA_A$ receptor (α1-5, β1-3, long and short splicing variants of γ2 subunit ($\gamma 2_{L,S}$), δ, and ε) that are expressed in the hypothalamus (Pirker et al., 2000) among the rats of different postnatal age groups and treatments. To date, two types of changes have been detected: those related to normal development and those induced by alcohol exposure. Among the developmental changes, the mRNA levels for four subunits, α4, α5, β3 and δ, decreased between postnatal days 15–16 and 29–30. The developmental changes for the α1-3, β1,2, $\gamma 2_{L,S}$, and ε subunits of $GABA_A$ receptor were not significant. The exposure to alcohol resulted in significantly altered developmental mRNA patterns for the α4, β3, and δ subunits of the $GABA_A$ receptor. The mRNA levels for these subunits in alcohol-exposed rats were elevated on postnatal days 29–30 compared to the control rats, with the increase in α4 and δ subunit mRNA becoming significant on postnatal days 15–16. The mRNA levels of $\gamma 2_{L,S}$, which did not exhibit age-related changes in control rats, also increased on postnatal days 29–30 in alcohol-exposed rats. In the VLPO region, the δ subunit mRNA increased following alcohol exposure on postnatal days 29–30, whereas all other subunits were not affected (data not shown).

As with GAD, there was a follow-up with quantitation of protein levels for selected $GABA_A$ receptor subunits. Primary antibodies were used against the α4 and δ subunits and the dot blot technique. The integrated density of dot staining for both subunit proteins was significantly correlated with the corresponding mRNA levels (Figure 17.5B). On postnatal days 29–30, immunoreactivity for both subunits was significantly increased in the PF region of alcohol-exposed rats when compared to control animals of the same age. In contrast, the levels of these subunits were not significantly changed in the VLPO region (Figure 17.5C,D). These data are consistent with the results of the mRNA study, and further support the hypothesis that alcohol exposure results in elevated $GABA_A$ receptor-mediated inhibition in the PF region.

The picture that emerges from these studies is that alcohol exposure elicits a deficiency in GABA biosynthesis due to either a decrease in GAD expression or a loss of GABAergic cells in both the PF and VLPO regions. This may result in reduced inhibition in the wake-promoting PF region, leading to sleep deficits. A compensatory reaction to these effects could be an increased expression of $GABA_A$

receptors on the PF cells. Indeed, it was found that the mRNA levels of the α4 and δ subunits started to increase early (postnatal days 15–16), and mRNA levels of some other subunits also increased by postnatal days 29–30. Most of these changes were region-specific in that they occurred in the PF region and were not evident in the VLPO region. Most previous studies reported an increase in the α4 subunit expression following acute or chronic alcohol exposure, both *in vivo* and *in vitro* (reviewed in Worst and Vrana, 2005). Importantly, earlier studies demonstrated that the α4-β3-δ combination can create a $GABA_A$ receptor subtype that is extremely sensitive to alcohol at low ("physiological") concentrations (Wallner et al., 2003). Results confirmed the importance of this $GABA_A$ receptor subtype in mediating effects of alcohol on the brain and suggested that $GABA_A$ receptors composed of the α4, β3, and δ subunits are major targets for early developmental alcohol exposure in the hypothalamic PF region.

Hypothalamic Orexin System and mRNA Expression Profiling at the Single-Cell Level

In current studies, the primary focus has been on the hypothalamic PF region; it has been particularly captivating because of its important role in the maintenance of wakefulness and motivated behavior, including alcohol- and drug-seeking behavior. Lesions of this brain region result in a behavioral state characterized by sleepiness and depressed brain activity (van Economo, 1930). The region contains multiple cell types including wake- and motor activity-related orexin-synthesizing neurons, local GABAergic interneurons that may have sleep- or wake-related activities, and melanin-concentrating hormone (MCH)-synthesizing neurons that contribute to homeostatic regulation of sleep. The excitatory orexin neuropeptides (orexin A and orexin B) are synthesized by a unique group of neurons in the posterior hypothalamus. These neurons and their orexins were implicated in the regulation of sleep, motor activity, energy homeostasis, and motivated behaviors, whereas lack of orexin signaling were strongly linked to narcolepsy/cataplexy, a debilitating disorder characterized by hypersomnolence and episodes of a sudden loss of motor tone (cataplexy) (reviewed in Aston-Jones et al., 2010; Ohno and Sakurai, 2008). Orexins are produced from their precursor, prepro-orexin, and act on two orexin receptors (type 1 and 2).

To assess mRNA expression profiles in cells of different phenotypes located in the PF region, a protocol was developed that allows one to recognize neuronal phenotypes prior to cell selection for single-cell RT-PCR studies. The protocol was developed using hypothalamic orexin- and MCH-synthesizing neurons (Volgin et al., 2004b). Cells acutely dissociated from the PF region were *in vitro* immunolabeled using anti-prepro-orexin or anti-MCH antibodies (Figure 17.4A,B), and individual labeled cells were subjected to RT-PCR. MCH mRNA was detected in 68% of cells immunocytochemically identified as MCH-containing, and 66% immunocytochemically identified as orexin-containing had prepro-orexin mRNA. These data show that the methodology allows one to select desired cells with at least 66% confidence from the PF region prior to subjecting them to the single-cell mRNA profiling. Using this protocol, a pilot study was conducted with $GABA_A$ receptor subunit mRNA profiling in single cells dissociated from the PF region. Cells were collected after immunohistochemical staining and then tested for the presence of mRNAs for three α, three β, and the ε subunit of $GABA_A$ receptor. Significant differences were found in the complements of $GABA_A$ receptor mRNAs expressed in cells of different phenotypes. Among 20 cells immunostained for orexin and positive for prepro-orexin mRNA, 6 had mRNAs for at least two β subunits (Figure 17.4C) and none was positive for GAD67 mRNA. In contrast, among 20 cells immunostained for MCH and MCH mRNA-positive, none contained mRNA for more than one β subunit, but 20% were also positive for GAD67 mRNA (significantly different from orexin cells). The latter confirmed earlier reports that some MCH cells are GABAergic (Verret et al., 2003). Among the cells negative for either orexin or MCH, it was found that 19 were positive for GAD67 mRNA. Such cells tended to differ from either orexin or MCH cells in that only one out of 14 tested had mRNA for the ε subunit, whereas 20% of orexin and MCH expressed mRNA for this subunit. These data demonstrate that the protocol enables finding significant differences in $GABA_A$ receptor mRNA expression profiles among different neurons located in a heterogeneous brain region, such as the PF region.

The orexin 2 receptor is expressed in rat XII motor nucleus, which prompted assessment of developmental changes in the expression of the orexin 2 receptor mRNA in the XII motor neurons. In retrograde-labeled and acutely dissociated XII motor neurons subjected to RT-qPCR, the number of reverse-transcribed mRNA copies per cell was significantly higher around postnatal day 20 than at any other age studied (postnatal days 4–31) (Figure 17.4D) (Volgin et al., 2002). This suggests that orexin 2 receptor mRNA synthesis increases during the critical period for the development of rapid eye movement (REM) sleep, which also corresponds to the period of the onset of narcolepsy/cataplexy symptoms in humans.

ETHICAL ISSUES

As with other animal models of human disorders, there are certain ethical criteria of research that need to be fulfilled. First, meaningful alternatives to the experimental procedures that may cause more than very brief pain or distress to the animals have to be considered. If no such alternative procedures exist or can be developed, methods that refine an animal model by eliminating or reducing pain or distress that animals undergo should be used. These methods include anesthesia and euthanasia, the method of which has to be consistent with the recommendations of the American Veterinary Medical Association Guidelines on Euthanasia. In practice, this means that all planned animal procedures have to be reviewed and approved by an appropriate oversight committee, namely the Institutional Animal Care and Use Committee. In addition, all personnel involved in the experiments should participate in training courses on the use of vertebrate animals in research. All these measures are aimed at ensuring that animal discomfort, pain, and injury are limited to the extent which cannot be avoided in the conduct of scientifically sound animal research.

TRANSLATIONAL SIGNIFICANCE

One area in which brain gene expression studies are expected to advance our understanding of the cellular, molecular, and epigenetic mechanisms of neurobehavioral abnormalities resulting from a suboptimal prenatal environment could be characterization of mechanisms underlying the altered regulation of sleep following prenatal exposure to alcohol, a socially relevant and common prenatal insult. Prenatal alcohol exposure is an established cause of abnormal sleep in infants that correlates with subsequent abnormalities of neurocognitive development and is a key predictor of disrupted sleep in older children. Numerous earlier studies established the role of insufficient sleep as a cause of neurobehavioral morbidity, including cognitive deficits, in the general population. In addition, disrupted sleep has detrimental consequences for cardiovascular and endocrine functions, and the quality of life. Data indicate that the long-term impact of this condition can be especially profound when it occurs during development. Therefore, it is possible that sleep deficits caused by alcohol exposure may be at least a significant exacerbating factor for cognitive and behavioral abnormalities associated with fetal alcohol spectrum disorders (FASD). To date, however, the cellular and molecular mechanisms of the resulting sleep deficiency remain unknown. Considering the nature of these studies, especially those aiming to dissect the molecular mechanisms of sleep abnormalities, they need to be conducted using an experimental animal model. There is a vast amount of literature on the chemical neuroanatomy of the rat brain, and extensive genomic information has been derived from rodents, especially rats. Consequently, most recent studies of the neurophysiology and neuropharmacology of sleep and long-term neurodevelopmental effects of prenatal alcohol exposure are done on rats. Animal models of this condition allow one to control experimental conditions and provide invaluable insight into the mechanisms of alcohol-induced brain damage and link them to altered behavioral functions. The information gained from these studies points to many similarities between the underlying mechanisms in all mammals, including humans. Once the molecular targets of prenatal exposure to alcohol, such as specific subtypes of GABA receptors expressed in sleep-regulating brain cells of a particular phenotype, are identified, the next step would be to determine the roles of cells expressing these target genes in the development of the multifaceted symptoms of FASD. A separate line of investigation may involve studies of the transcriptional and post-transcriptional mechanisms that mediate the changes in neurons of distinct phenotypes located in the hypothalamic PF region. For example, there is evidence that alcohol activates neuronal mRNA expression of the α4 subunit of $GABA_A$ receptors *in vitro* via heat shock factor 1 (Pignataro et al., 2007). Activation of this transcriptional pathway initiated by prenatal exposure to alcohol may play a role in FASD-related abnormalities of brain development. Since many pathways relevant for our research are phylogenetically conserved, the tools for studying these mechanisms may include simpler vertebrate model animals, such as zebrafish. Since FASD are often under-diagnosed, research on molecular markers is crucial for identification and reduction of health risks, and could provide a mechanistic basis for development of novel diagnostic and therapeutic approaches.

WORLD WIDE WEB RESOURCES

GenBank® (http://www.ncbi.nlm.nih.gov/genbank/) is the NIH database of all publicly available DNA sequences. The GenBank database enables the scientific community to gain access to the most up-to-date and comprehensive DNA sequence information. There are no restrictions on the use or distribution of GenBank data. All PCR primers originated in our laboratory and used in the experiments described in this chapter have been designed and tested using the GenBank sequence data.

ImageJ (http://rsbweb.nih.gov/ij/index.html) is public domain image processing software developed at the National Institutes of Health. Its options for image acquisition, analysis, and processing can be customized using a built-in editor and a Java compiler. The ImageJ source code is freely available. The program allows one to edit, analyze, process, save, and print color (8-bit) and grayscale, 16-bit integer and 32-bit floating-point images, and can work with multiple image formats. ImageJ is able to measure areas and pixel value statistics of the user's selections. The program can do other measurements, such as distances and angles, and can create density histograms and line profile plots. It also supports many other standard image processing functions. We use ImageJ software for quantitative analysis of the immunostaining density in protein studies involving Western or dot blots and immunohistochemistry with fixed brain sections.

Allen Brain Atlas (http://www.brain-map.org) is an excellent collection of online public resources integrating extensive gene expression and brain anatomy data, complete with a novel suite of search and viewing tools. This portal has a library of short video tutorials that illustrate different aspects of using its resources. All gene expression data including *in situ* hybridization, histological staining, microarray, and the whole of transcriptome sequencing data have been indexed and can be searched. The Allen Mouse Brain Atlas is of particular interest. It creates a cellular-resolution, genome-wide map of gene expression in the mouse brain. The complementary Allen Reference Atlas allows users to directly match gene expression patterns to various neuroanatomical structures.

REFERENCES

Aston-Jones, G., Smith, R. J., Sartor, G. C., Moorman, D. E., Massi, L., Tahsili-Fahadan, P., & Richardson, K. A. (2010). Lateral hypothalamic orexin/hypocretin neurons: A role in reward-seeking and addiction. *Brain Research, 1314*, 74–90.

Chiang, L. W., Schweizer, F. E., Tsien, R. W., & Schulman, H. (1994). Nitric oxide synthase expression in single hippocampal neurons. *Brain Research Molecular Brain Research, 27*, 183–188.

Comer, A. M., Yip, S., & Lipski, J. (1997). Detection of weakly expressed genes in the rostral ventrolateral medulla of the rat using micropunch and reverse transcription-polymerase chain reaction techniques. *Clinical and Experimental Pharmacology &Physiology, 24*, 755–759.

Comer, A. M., Gibbons, H. M., Qi, J., Kawai, Y., Win, J., & Lipski, J. (1999). Detection of mRNA species in bulbospinal neurons isolated from the rostral ventrolateral medulla using single-cell RT-PCR. *Brain Research. Brain Research Protocols, 4*, 367–377.

Economo, C. v. (1930). Sleep as a problem of localization. *The Journal of Nervous and Mental Disease, 71*, 249–259.

Fink, L., Seeger, W., Ermert, L., Hänze, J., Stahl, U., Grimminger, F., Kummer, W., & Bohle, R. M. (1998). Real-time quantitative RT-PCR after laser-assisted cell picking. *Nature Medicine, 4*, 1329–1333.

Goldsworthy, S. M., Stockton, P. S., Trempus, C. S., Foley, J. F., & Maronpot, R. R. (1999). Effects of fixation on RNA extraction and amplification from laser capture microdissected tissue. *Molecular Carcinogenesis, 25*, 86–91.

Kamme, F., Salunga, R., Yu, J., Tran, D.T., Zhu, J., Luo, L., Bittner, A., Guo, H.Q., Miller, N., Wan, J., & Erlander, M. (2003). Single-cell microarray analysis in hippocampus CA1: demonstration and validation of cellular heterogeneity. *Journal of Neuroscience, 23*, 3607–3615.

Lambolez, B., Audinat, E., Bochet, P., Crépel, F., & Rossier, J. (1992). AMPA receptor subunits expressed by single Purkinje cells. *Neuron, 9*, 247–258.

Ohno, K., & Sakurai, T. (2008). Orexin neuronal circuitry: role in the regulation of sleep and wakefulness. *Frontiers in Neuroendocrinology, 29*, 70–87.

Okabe, S., Mackiewicz, M., & Kubin, L. (1997). Serotonin receptor mRNA expression in the hypoglossal motor nucleus. *Respiratory Physiology, 110*, 151–160.

Phillips, J. K., & Lipski, J. (2000). Single-cell RT-PCR as a tool to study gene expression in central and peripheral autonomic neurones. *Autonomic Neuroscience, 86*, 1–12.

Pignataro, L., Miller, A., Ma, L., Midha, S., Protiva, P., Herrera, D., & Harrison, N. (2007). Alcohol regulates gene expression in neurons via activation of heat shock factor 1. *Journal of Neuroscience, 27*, 12957–12966.

Pirker, S., Schwarzer, C., Wieselthaler, A., Sieghart, W., & Sperk, G. (2000). GABA(A) receptors: Immunocytochemical distribution of 13 subunits in the adult rat brain. *Neuroscience, 101*, 815–850.

Raol, Y. H., Zhang, G., Budreck, E. C., & Brooks-Kayal, A. R. (2005). Long-term effects of diazepam and phenobarbital treatment during development on GABA receptors, transporters and glutamic acid decarboxylase. *Neuroscience, 132*, 399–407.

Schütze, K., Pösl, H., & Lahr, G. (1998). Laser micromanipulation systems as universal tools in cellular and molecular biology and in medicine. *Cell and Molecular Biology (Noisy-le-Grand), 44*, 735–746.

Sucher, N. J., & Deitcher, D. L. (1995). PCR and patch-clamp analysis of single neurons. *Neuron, 14*, 1095–1100.

Verret, L., Goutagny, R., Fort, P., Cagnon, L., Salvert, D., Léger, L., Boissard, R., Salin, P., Peyron, C., & Luppi, P. (2003). A role of melanin-concentrating hormone producing neurons in the central regulation of paradoxical sleep. *BMC Neuroscience, 4*, 19.

Volgin, D. V., Mackiewicz, M., & Kubin, L. (2001). α_{1B} receptors are the main postsynaptic mediators of adrenergic excitation in brainstem motoneurons, a single-cell RT-PCR study. *Journal of Chemical Neuroanatomy, 22*, 157–166.

Volgin, D. V., Saghir, M., & Kubin, L. (2002). Developmental changes in the orexin 2 receptor mRNA in hypoglossal motoneurons. *Neuroreport, 13*, 433–436.

Volgin, D., Fay, R., & Kubin, L. (2003). Postnatal development of serotonin 1B, 2A and 2C receptors in brainstem motoneurons. *European Journal of Neuroscience, 17*, 1179–1188.

Volgin, D. V., Fenik, V. B., Fay, R., Okabe, S., Davies, R. O., & Kubin, L. (2004a). Serotonergic receptors and effects in hypoglossal and laryngeal motoneurons – Semi-quantitative studies in neonatal and adult rats. In J. Champagnat, M. Denavit-Saubie G. Fortin, et al. (Eds.), *Post-Genomic Perspectives in Modeling and Control of Breathing. Advances in Experimental Medicine and Biology:* (Vol. 551, pp. 183–188). New York: Kluwer Academic/Plenum Publ.

Volgin, D. V., Swan, J., & Kubin, L. (2004b). Single-cell RT-PCR gene expression profiling of acutely dissociated and immunocytochemically identified central neurons. *Journal of Neuroscience Methods, 136*, 229–236.

Volgin, D. V., & Kubin, L. (2007). Regionally selective effects of GABA on hypothalamic GABA(A) receptor mRNA *in vitro*. *Biochemical and Biophysical Research Communications, 353*, 726–732.

Volgin, D. V. (2008). Perinatal alcohol exposure leads to prolonged upregulation of hypothalamic GABA(A) receptors and increases behavioral sensitivity to gaboxadol. *Neuroscience Letters, 439*, 182–186.

Volgin, D. V., Rukhadze, I., & Kubin, L. (2008). Hypoglossal premotor neurons of the intermediate medullary reticular region express cholinergic markers. *Journal of Applied Physiology, 105*, 1576–1584.

Volgin, D. V., Malinowska, M., & Kubin, L. (2009). Dorsomedial pontine neurons with descending projections to the medullary reticular formation express orexin-1 and adrenergic α_{2A} receptor mRNA. *Neuroscience Letters, 459*, 115–118.

Wallner, M., Hanchar, H., & Olsen, R. (2003). Ethanol enhances alpha 4 beta 3 delta and alpha 6 beta 3 delta gamma-aminobutyric acid type A receptors at low concentrations known to affect humans. *Proceedings of the National Academy of Sciences of the United States of America, 100*, 15218–15223.

Worst, T., & Vrana, K. (2005). Alcohol and gene expression in the central nervous system. *Alcohol Alcohol, 40*, 63–75.

FURTHER READING

Bartlett, J. M. S., & Stirling, D. (2003). A short history of the polymerase chain reaction. *PCR Protocols, 226*, 3–6.

Kennedy, S. (2011). *PCR troubleshooting and optimization: The essential guide*. Caister Academic Press. p. 235.

Nicholas, H. (2010). *Single-cell gene-expression analysis: quantitative RT-PCR*. VDM Publishing. p. 232.

Protein Blotting and Detection: Methods and Protocols. Humana Press. p. 588. http://www.springer.com/new+9626+forthcoming+titles+(default)/book/978-1-934115-73-2.

Swanson, L. W. (2012). *Brain Architecture, Understanding the Basic Plan* (2nd ed.). New York: Oxford University Press. p. 311.

GLOSSARY

Anesthesia Temporary blocked sensation, such as the pain feeling, caused by administration of drugs in order to allow animals to undergo surgery without pain and distress.

Autofluorescence Natural light emission by biological macromolecules and cellular structures following light absorption.

Chemiluminescence Emission of light resulting from a chemical reaction.

Epitope Part of an antigen that is recognized by antibodies.

Euthanasia (animal) The act of humanely putting an animal to death, often performed under deep anesthesia in order to minimize pain and distress.

Gamma-Aminobutyric Acid (GABA) Main inhibitory neurotransmitter of the mammalian brain that controls excitability of neural cells.

Glia (Glial Cells) Non-neuronal cells of the brain and other parts of the nervous system that chiefly provide support and protection for neurons.

Hypoglossal (XII) Motor Neurons Neural cells that innervate the tongue muscle.

Hypothalamus Part of the brain that contains multiple small nuclei with a variety of functions, including regulation of sleep and circadian rhythms, hunger, metabolism, and body temperature.

Medulla (Medulla Oblongata) Part of the brain stem that regulates various vital physiological functions such as breathing, digestion, heart and blood vessels, swallowing, and sneezing.

Melanin-Concentrating Hormone (MCH) Neuropeptide synthesized in the hypothalamus that is involved in the regulation of energy metabolism, feeding behavior, mood, and sleep.

Micropunch Small tissue sample extracted by "punching" of biological tissue with a custom-cut syringe needle.

Motor Neurons Neural cells of the central nervous system (CNS) that control muscles via their axons projected outside the CNS.

Nucleus (Brain) Structure of the brain consisting of a relatively distinct cluster of neural cells.

Orexin Europeptide synthesized in the hypothalamus that regulates arousal, motor activity, wakefulness, and appetite.

Rapid Eye Movement (REM) Sleep Stage of normal sleep during which the rapid and random movement of the eyes occurs.

Receptor (Neuronal) A protein embedded in the cell surface plasma membrane to which specific signaling molecules (neurotransmitters) may bind.

Retrograde Labeling Tracing neural connections from the synapses to the cell bodies.

Ultrasonication Disintegration of cellular structures by ultrasound.

ABBREVIATIONS

ACSF Artificial Cerebrospinal Fluid
cDNA Single Stranded DNA
CCD Charge-Coupled Device
CHAPS 3-[(3-Cholamidopropyl)Dimethylammonio]-1-Propanesulfonate
CNS Central Nervous System
c_T PCR Threshold Cycle
DTT Dithiothreitol
EDTA Ethylenediaminetetraacetic Acid
FASD Fetal Alcohol Spectrum Disorders
FITC Fluorescein Isothiocyanate
GABA Gamma-Aminobutyric Acid
GAD Glutamate Decarboxylase
HEPES (4-(2-Hydroxyethyl)-1-Piperazineethanesulfonic Acid)
MCH Melanin-Concentrating Hormone
mRNA Messenger RNA
ncRNA Non-Coding RNA
PAGE Polyacrylamide Gel Electrophoresis
PBS Phosphate-Buffered Saline
PF Perifornical Region of the Hypothalamus
PIPES Piperazine-N,N′-Bis(2-Ethanesulfonic Acid)
PCR Polymerase Chain Reaction
qPCR Quantitative PCR
REM Sleep Rapid Eye Movement Sleep
RT Reverse Transcription
SDS Sodium Dodecyl Sulfate
TBS Tris-Buffered Saline
VLPO Ventrolateral Preoptic Region of the Hypothalamus

LONG ANSWER QUESTIONS

1. RT-PCR is an extremely sensitive technique because it is theoretically able to detect a single mRNA copy. Describe potential side effects of the high sensitivity for single-cell RT-PCR detection.
2. Describe the main advantages of mRNA quantitation using RT-PCR.
3. Explain the purpose of each step of qPCR calibration using external standards.
4. Genetically modified mice lacking the brain peptide orexin or its receptor in the brain could be used as a model of the human disease narcolepsy for evaluating new narcolepsy therapies. Describe all potential approaches that would allow you to determine whether orexin gene expression is indeed absent in the mouse brain.
5. How would you design an experiment that would quantify expression of a gene that encodes multiple proteins?

SHORT ANSWER QUESTIONS

1. Why does studying gene expression in the entire animal brain have its limitations?
2. Why is it important to quantify both mRNA and protein for the gene of interest?
3. What is the key difference between the main approaches to identify brain cells for gene expression studies?
4. What are the main issues that may interfere with the quantitative analysis of proteins detected by immunohistochemistry?
5. What are the main advantages and limitations of immunoblotting methods when compared to immunohistochemistry?

ANSWERS TO SHORT ANSWER QUESTIONS

1. Because animal brain is very heterogeneous: it has multiple distinct structures built of various cell types.
2. In studies comparing gene expression between different groups of animals, such as developmental studies or studies of the effect of various treatments, the mRNA quantitation alone may not provide enough information about the gene expression because changes in mRNA are not necessarily associated with proportional changes of the corresponding protein.
3. Retrograde labeling is performed by means of survival surgery on anesthetized animals receiving an injection of a retrograde tracer. In contrast, immunolabeling of dissociated neurons is done *in vitro* after the animal has been sacrificed.
4. Excessive background immunolabeling, weak target protein staining, and autofluorescence related to the fixation procedure.
5. Immunoblotting enables better densitometric quantitation of multiple target proteins in the same sample, but requires destruction of tissue samples. This makes impossible precise localization of proteins to a particular population of brain cells, which could be achieved by immunohistochemistry.

Chapter 18

Ribotyping: A Tool for Molecular Taxonomy

S.K. Kashyap[*], S. Maherchandani[*] and Naveen Kumar[†]

*Department of Vet Microbiology & Biotechnology, Rajasthan University of Veterinary & Animal Sciences, Bikaner, Rajasthan, India, †Central Institute for Research on Goats, Indian Council of Agricultural Research, Makhdoom, District-Mathura, UP, India

Chapter Outline

SUMMARY

Ribotyping is the identification and classification of bacteria based on polymorphisms in taxon-specific "signature labels" of universal and highly conserved ribosomal RNA molecules or their genes. It began in the 1980s and evolved into various forms. It is considered a relatively stable and dependable system for molecular taxonomy, and is widely accepted by international agencies.

WHAT YOU CAN EXPECT TO KNOW

Microbial taxonomy is a dynamic and continuously evolving system, probably because of the fact that microorganisms multiply asexually and with very short generation time. They also have far greater adaptability and capability to undergo variation in response to their environment as compared with biological systems of higher order (i.e. eukaryotes). Before the advent of molecular techniques, microbial systematics was largely based on studying their phenotypic traits such as morphological, biochemical, antigenic properties, and bacteriophage susceptibility. Some of the methods based on these properties of microorganisms are still used by many laboratories for their preliminary identification, but need be supplemented with molecular techniques for deriving definitive taxonomic conclusions. Although various molecular biological techniques are being practiced for

Animal Biotechnology. http://dx.doi.org/10.1016/B978-0-12-416002-6.00018-3

this purpose, ribotyping (i.e. identification and classification based on ribosomal RNA or their genes) evolved in the 1980s, and is still considered to be a relatively stable and dependable system that is widely accepted by international agencies. Ribosomal RNAs are the core molecule of ubiquitous ribosomes. Their genes have been under the least evolutionary selection pressure, providing highly conserved regions and some less conserved sequences, which provide taxon-specific "signature labels," the basis of taxonomy. Conventional ribotyping was based on the determination of homology in 16S rRNA sequences by detecting restriction polymorphism and hybridization on membranous support, which is now being supplemented with more convenient PCR-ribotyping (i.e. determination of polymorphism in 16S and 23S rRNA intergenic space) and sequencing of whole ribosomal operons. Fluorescent *in situ* hybridization for the detection of rRNA of non-viral organisms in their natural habitat has also become fairly popular owing to its simplicity, cost effectiveness, and lack of need for culturing.

HISTORY AND METHODS

INTRODUCTION

Taxonomy is a field of science that involves description, identification, and nomenclature of an organism, which can be used in its classification. The purpose of taxonomy is to give proper signatures to an organism so that relatives of a particular taxon can be identified in a systematic manner. The information thus gathered can be used for systematics (i.e. studying their evolutionary and adaptability aspects). One of the approaches involved in systematics is the cladistic approach, where clades are formed, each consisting of an ancestor organism with all its descendants. The basic assumption behind the cladistic approach is that members of a group share a common evolutionary history and are thus more "closely related" to one another than to other groups of organisms. Another approach is evolutionary or synthetic systematics, which emphasizes evolutionary relationships. This approach suggests that the degree of genetic differences between lineages should also be used in addition to their evolutionary similarities when developing taxonomic classifications. The third approach is phenetics, or numerical taxonomy, where importance is given to multiple traits instead of relying on only a few traits. Similarities or differences between the characteristics of organisms are calculated, and clusters are formed. The clusters so formed may not necessarily reflect evolutionary relatedness. In other words, systematics is either studying the relationships between the organisms or characterizing and discriminating between different levels of taxa down to the species level and even beyond.

Microbiology and its systematics took a long time to evolve due to its limitations, as it could not have an organismal approach where phenotypic traits played an important role. From the initial approaches of studying biochemistry or physiology for systematics and taxonomy, it advanced with time to incorporate most of the modern techniques for its taxonomy. It is strongly felt that microbiological systematics is useful not only in studying microbial properties and phylogeny, but also in diagnostics where an unknown organism can be identified up to the species and sub-species levels, and also to infer some practical information to explore its medical, industrial, or environmental use. The methods and markers used for the characterization of a biological unit need to have universality (i.e. applicability to all members), reliability, and feasibility, and should also be able to reveal substitutions taking place in it over time. Of the various approaches used for this purpose, studying ribosomal RNA and its genes (rDNA) have gained wide popularity. Ribosomal RNA is universally present, large, has a high degree of sequence conservation, has an important role in the translation of genomic messages, and has stable rDNA towards horizontal gene displacement. For these reasons, this operon is an important semantophoretic molecule for studying systematics and developing phylogenies. In addition to having taxonomic applications, this technique has opened up many new fields, like metagenomic studies in a particular environment. These include unicellular eukaryotes, variability in their distribution under a given condition, and discrimination of microbial strains collected from different sources. The information can be further used in devising strategies to plan eradication or control of diseases, or in industries or environmental conservation, etc., depending upon the type of organism. The aim of this chapter is to give a brief overview of microbiological systematics, its history, and how ribotyping as a tool justifies all the approaches of systematics with its practical utilities.

HISTORICAL DEVELOPMENTS IN BACTERIAL TAXONOMY

Origination of taxonomy was one of the major events in the biological sciences in the 18th century. It was then that Carl von Linné (Carolus Linnaeus) described binomial nomenclature (taxonomy), which was accepted worldwide. The science of scientific naming started in the late sixteenth century as plants and animals were assigned long, polynomial, descriptive names based on particular characteristics. Latin, being a scholarly language, was used for this naming. As more and more species were discovered, the polynomials got longer and increasingly unwieldy. Linnaeus simplified the naming system by designating one Latin name to indicate genus and the other name for the special epithet. These binomial names were easy to remember, and people soon started using them. Eventually these binomials replaced the polynomial names completely. Linnaeus' plant taxonomy was based on the number and arrangement of stamens and stigma. Later on, the system of classification became more organized, and John Ray's practice of using morphological evidence from all parts of the plant in all stages of its development was followed. One of the important facts was that the complex morphologies and variations were used systematically to build phylogenies.

This advancement in taxonomy provided an organizational structure to biological science, and it became the most fundamental building block for information sharing on biological resources. It was an impelling force for the development of zoology and botany, as construction of phylogenies and evolutionary studies could be performed. Studies on structural and functional relationships, and on diversities and their correlation with ecologies and evolution became feasible. Later, taxonomists devised rules for nomenclature, and subsequently, botanical and zoological codes for nomenclature were developed. Even Darwin's theory of evolution did not change the Linnaean classification; it merely provided a theoretical justification.

Microbiology as a separate field evolved in the late nineteenth century with the first description of a bacterial species. As early microbiologists were botanists, they adopted its binomial nomenclature. During the late nineteenth century and early twentieth century, the scientific community realized the importance of microorganisms. These minute creatures could be found in all the ecological niches. Many new microorganisms were identified, and the scientists working on them assigned novel names; however, there was no systematic and universal approach of nomenclature, which made exchange of knowledge difficult. This peculiar need in bacteriology led to the formation of a nomenclature committee at the first international microbiological congress of the International Society for Microbiology held in Paris in 1930. Later, in subsequent meetings, the rules for bacteriological codes were established and refined. This committee developed criteria for classifying microbial species, and prepared a list of species types. During this period, repositories for type and reference of culture collections were also developed. These events led to the significant efforts by bacteriologists to develop systematics.

Since microbial morphologies did not have enough structural variability to have much phylogenetic significance (as seen in two other organismal sciences, botany and zoology), more focus was placed on the biochemical aspects of the organisms. However, comparative biochemistry cannot answer the evolutionary purists as efficiently as comparative anatomy can, which might be one of the reasons for the slow development of microbial taxonomy. Also, for taxonomy it is imperative to select specimens for comparison to other individuals. The species types of higher plants, animals, or insects could be kept in satisfactory shape and size by proper preservation, but in bacteria, pure cultures had to be maintained indefinitely without a significant change in characteristics. Furthermore, versatility in terms of adapting to the environment (which is more pronounced in bacteria as compared to higher forms) was found to pose more complexity. Still, efforts by those early microbiologists started giving structure to microbial taxonomies. Microbial systematics was open to the development of biological sciences and other newly developed technologies that had relevance in taxonomy, and natural relatedness schemes were adopted. During the 1960s, one of the major events that gave a boost to microbial taxonomy was the work of Sokal and Sneath (1963). They proposed a taxonomic system using numeric algorithms like cluster analysis rather than using subjective evaluation of their properties; they called their new system numerical taxonomy. This historical work is still followed, and has helped significantly in developing phylogenies. With the introduction of numerical phenetic analysis and molecular techniques, the inter- and intra-relatedness of species could be determined objectively. The methods being used ranged from phenotypic characterization, biochemical characterization, and DNA–DNA re-association, to base composition analysis. Another important event that occurred in the late 1970s was when Carl Woese devised a system of classification above the species level; he compared ribosomal RNA (rRNA) molecules (Fox et al., 1977). This newer view showed prokaryotes to comprise not one, but two unrelated major groups, *Archaea* and *Eubacteria*. The emergence of molecular approaches laid the foundation for different approaches in bacterial systematics. This led to differences amongst the bacterial systematists, and two schools were formed, one whose members held to traditional ideas of taxonomy, and the other who embraced molecular approaches to taxonomy. Later, in 1987, an ad hoc committee on the reconciliation of approaches to bacterial systematics concluded that "bacterial taxonomy, which began as a largely intuitive process, has become increasingly objective with the advent of numerical taxonomy and techniques for the measurement of evolutionary divergence in the structure of semantides, i.e. large, information-bearing molecules such as nucleic acids and proteins" . . . and "an ideal taxonomy would involve one system" (Wayne et al., 1987).

TYPING METHODS USED FOR BACTERIAL SYSTEMATICS

Taxonomy requires two sources of information to be investigated as extensively as possible: genomic information and phenotype. Genomic information is obtained from nucleic acids, either through sequencing or through DNA–DNA similarity, G+C mol.%, and ribotypes. Phenotype is the visible physical and biochemical expression of genotypes that result from the interaction of genotype and environment. Although many different types of techniques have been devised for analyzing microorganisms, each has its advantages and limitations as to the extent to which it can be used for taxonomy. An ideal technique would be cost effective, less time consuming, automatable, and would have good taxonomic resolution power. On the other hand, the ultimate purpose of identification needs to be defined. It may be limited to identification and to assignment of a definite taxon, or to understanding disease/infection, where the focus is on devising strategy for treatment and control (as for clinical isolates).

Phenotypic Typing Methods

Classification was initially based on morphology, biochemical characteristics, and growth conditions of the organisms. These investigations required the use of pure cultures and the characteristics studied were compared with the properties of reference strains. The similarities or differences between the reference strains and the isolate under question were the basis of bacterial identification and nomenclature (for a review see Rosselló-Mora and Amann, 2001).

- ***Morphology of Bacteria*:** The morphology of a bacterium includes the cellular and colony characteristics of the organism. The cellular characteristics such as shape, Gram reaction, type and location of flagella and its motility, help in preliminary identifications. The colony characteristics include color, dimensions, and form. Bacteria show characteristic types of growth on solid media under appropriate cultural conditions, and the colony morphology can be used in presumptive identification. These colonies may vary in size and nature (circular, wavy, rhizoid, etc.), elevation (flat, raised, convex, etc.), and opacity (transparent, opaque, translucent). The color of the colony or the changes that they bring about in their surroundings is also an indicator used for the identification of bacteria. Some bacteria also produce pigments; this feature is also used as an aid in identifying the organism.
- ***Biochemical Characteristics of Bacteria*:** The biochemical features include both primary and secondary biochemical tests for identification of microorganisms, such as growth on selective or differential media, growth in the presence of various substances such as antimicrobial agents, carbohydrate utilization tests, or presence or absence of certain enzymes. These tests are commonly used, and characteristics based on these tests are widely described in Bergey's Manual of Determinative Bacteriology (Garrity, 2001). The limitation with these tests is that laboratory to laboratory variability may be seen; furthermore, there are no rules controlling the range of diversity among the strains to be examined. The set of strains should be as large as possible, and should include cultures of historical, pathological, or environmental importance. It is important to include reference strains (type strains) whose identity has been established for comparative purposes. Moreover, routine tests should represent a broad spectrum of the biological activities of the organisms.
- ***Serotyping of Bacteria*:** Bacterial species and types can be identified by specific antigen–antibody reactions. Antigens are substances that induce the production of antibodies in a foreign species. Bacteria and bacterial components serve as excellent antigens. The test includes production of antibodies in an animal host and testing of the antiserum by either the agglutination or precipitation test. These conventional tests are now supplemented with monoclonal antibody-based, highly sensitive ELISA, and a wide range of other serological tests.
- ***Phage Typing of Bacteria*:** This method is used for typing bacteria by testing the susceptibility of the culture to lysis by each of a set of type-specific lytic bacteriophages. The phage type of the culture is identified based on its pattern of susceptibility to the different phage strains. Phage typing allows subdivision of a serological entity, as in *Salmonella typhi,* which is divisible into more than 100 different phage types.

In addition to these conventional phenotypic typing methods, there are currently many spectrophotometric methods used in the characterization of bacteria that require specialized instrumentation. The most popular methods are the following:

- ***Fourier Transform Infrared Spectroscopy (FTIR)*:** This technique is based on the observation of vibrational properties of chemical bonds when excited by an infrared beam; it enables one to assess the overall molecular composition of microbial cells in a non-destructive manner, reflected in the specific spectral fingerprints for different microorganisms. This vibrational spectroscopy technique is useful in providing rapid, relevant information of intact microbial cells, and allows rapid discrimination, classification, and identification down to the strain level, as well as monitoring of metabolite production. This method generates little waste, and thus has lower run costs.
- ***Pyrolysis Mass Spectrophotometry*:** This is a rapid and high-resolution method for the analysis of otherwise non-volatile material, and has been widely applied for discriminating between closely related microbial strains. This technique involves thermal decomposition of materials at very high temperatures. Large molecules are cleaved at their weakest points and produce smaller, more volatile fragments. These fragments can be separated by gas chromatography. This technique has the ability to analyze small amounts of biological material with minimum sample preparation to obtain fingerprint data that can be used for identification and typing of the microorganism (Goodacre and Kell, 1996).

- ***Matrix-Assisted Laser Desorption/Ionization-Time of Flight Mass Spectrometry (MALDI-TOF MS):*** This is a very sensitive fingerprinting method used for the classification and identification of microorganisms, with its applications in clinical diagnostics. MALDI-TOF is a soft ionization technique in which a biomolecule is irradiated on a UV-absorbing matrix by laser pulse. The ionized biomolecules are accelerated in an electric field and enter the flight tube where different molecules are separated according to their mass-to-charge ratio, and reach the detector at different times, thus generating spectra. A database of known organisms is used to match the isolate under investigation, providing a matching score based on identified masses and their intensity. MALDI-TOF MS has the ability to measure peptides and other compounds in the presence of salts, and to analyze complex peptide mixtures; this makes it an ideal method for measuring non-purified extracts and intact bacterial cells. Different experimental factors, including sample preparation, the cell lysis method, matrix solutions, and organic solvents, affect the quality and reproducibility of bacterial MALDI-TOF MS fingerprints (De Bruyne et al., 2011).

Genotypic Typing Methods

During the early 1960s, increasing knowledge of the properties of DNA and the development of molecular biological techniques supported the idea that bacteria might best be classified by comparing their genomes. Initially, overall base compositions of DNAs (mol.% G + C values) were used. Bacteria whose mol.% G + C values differed markedly were obviously not of the same species. However, single values obtained by the analysis of DNA base compositions allowed only superficial comparisons, and a much more precise method was needed. Thus, DNA–DNA hybridization techniques were developed. A great practical advantage of this method was that it often produced more sharply defined clusters of strains than those solely circumscribed by phenotypic traits. The phylogenetic definition of a species generally included strains with approximately 70% or greater DNA–DNA relatedness and with 5°C or less ΔT_m. Here organisms tended to be either closely related or not at all. This method is still important and valid as far as bacterial systematics is concerned, with the limitation being use of only two strains at one time, one of which is essentially the reference strain.

In addition to these methods, and with advancements in the science of genomics, many newer genomic marker techniques such as amplified fragment length polymorphism (AFLP), randomly amplified polymorphic DNA (RAPD), and restriction fragment length polymorphism (RFLP) are also being used to collect information on microorganisms.

Though such techniques have been encouraged by the ad hoc committee on species definition in bacteriology, which met in Belgium in 2002, they still consider DNA–DNA re-association and 16S rDNA sequence analysis as the major methods for the definition of bacterial species (Stackebrandt et al., 2002).

BASIS OF USING rRNA AND rRNA GENES AS TAXONOMIC TOOLS

The ribosome is a structure of ancient origin and is necessarily ubiquitous. Each cell requires ribosomes for translation; their number in a given cell may range from a few hundred to 100,000, and is believed to be fixed within a particular species. Since they perform the central function of translation in a cell, the primary structures of ribosomes are sufficiently constrained. They are formed from proteins and ribosomal RNA. Ribonucleic acid is the core molecule for the translation process, forming the framework of the ribosome from a structural point of view; thus, it has much more stringent requirements. Even minor changes in a nucleotide sequence in rRNA can produce a deleterious effect on translation. Thus, during evolution rRNA had to be conserved in its basic sequences to maintain the translation apparatus with great efficiency. This may be the reason that rRNA gene sequences have been found less prone to spread horizontally between the members of similar species. The differences between rRNAs of different cells represent discontinuities that occurred among the early ancestors of today's cells, making them suitable for phylogenetic study, where the number and distribution of different rRNA genes are taken as a measure of diversity.

Organization of the Ribosomal Operon

In prokaryotes, the smaller subunit of the ribosome (30S) has 16S rRNA (~1,600 nucleotides long), and the larger subunit (50S) has 23S rRNA (~3,000 nucleotides long) and 5S rRNA (~120 nucleotides long) molecules. The genes of these three rRNAs are typically linked together into an operon, called a ribosomal operon (Figure 18.1A and 18.1B). This operon contains genes for 16S, 23S, and 5S rRNAs in order from the 5′ to 3′ direction. These genes are separated by a short segment of nucleotides called the intergenic spacer region (ISR), which may consist of genes for some tRNA molecules. The genes for rRNA are about 1.5 kb in size in all bacteria, and have clusters of highly conserved regions, variable regions, and hypervariable regions containing species, genus, and family-specific signature sequences; this facilitates determination of evolutionary interrelatedness, generation of probes for hybridization and primers for PCR, and formation of reference databases. Of the three rRNA genes, 16S rRNA gene sequences are considered to be the most conserved, and the species having 70% or greater DNA similarities usually have more than 97% sequence identity within this gene. The remaining 3% or 45-nucleotide differences are not

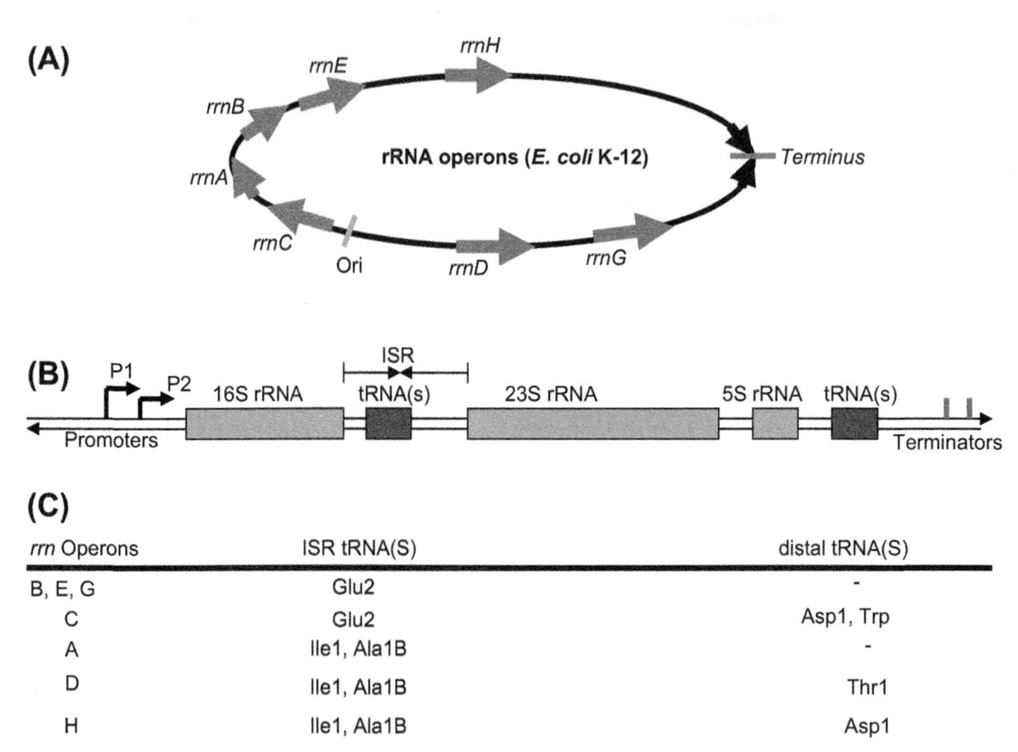

rrn Operons	ISR tRNA(S)	distal tRNA(S)
B, E, G	Glu2	-
C	Glu2	Asp1, Trp
A	Ile1, Ala1B	-
D	Ile1, Ala1B	Thr1
H	Ile1, Ala1B	Asp1

FIGURE 18.1 (A) Organization of seven different operons (*rrnA, rrnB, rrnC, rrnD, rrnE, rrnG,* and *rrnH*) present in *E. coli* K-12. (B) Organization of rRNA (rRNA and tRNA) genes and internal spacer region in a typical rRNA operon (*E. coli* K-12). (C) Type of tRNAs present in different rRNA operons of *E. coli* K-12. ISR: Internal Spacer Region; P1 and P2: Promoters.

evenly scattered along the primary structure of the molecule, but are concentrated mainly in certain hyper-variable regions, the positions of which are taxon-specific. It is in these regions that species/genus-specific sequences may be clustered. The ribosomal RNA operon is transcribed into a single preRNA molecule consisting of regions of 16S, spacer, tRNA, spacer, 23S, spacer, tRNA, spacer, and 5S rRNA sequentially, from the 5′ to 3′ direction; this is then cleaved into rRNA and tRNA molecules.

Within the chromosome of an organism there may be multiple copies of the rRNA operon. The number of copies of ribosomal RNA operons containing the genes coding for rRNA genes (16S, 23S, and 5S) and their associated internal spacer regions (ISRs) in bacterial species have been found to range from 1–15 (e.g. 7 in *Escherichia coli* and *Salmonella* and 10 in *Bacillus subtilis*). In *E. coli*, these copies are named as *rrnA, rrnB, rrnC, rrnD, rrnE, rrnH, and rrnG* (Figure 18.1A). There is some degree of variation in the length of the spacer region between 16S–23S rRNA genes in different organisms due to the number and type of tRNA genes that may be present in this region (Figure 18.1C); for example, in most of the Gram-negative bacteria, genes for $tRNA^{ala}$ and $tRNA^{ile}$ are present in this region, while in Gram-positive bacteria no tRNA gene, or $tRNA^{ala}$ or $tRNA^{ile}$ (or both) may be present. The ISRs have been suggested to reflect intra-species phylogeny. Thus, rRNA or rRNA genes can be exploited to group or type and identify bacterial organisms using different molecular tools like generating primers and probes; this process is called "ribotyping."

DIFFERENT TECHNIQUES OF RIBOTYPING

Grimont and Grimont (1986), first obtained several patterns of hybridized fragments from rRNA genes of different bacterial species, and also observed that strains showing identical patterns were highly related. This laid the foundation for utilizing ribotyping as a tool for identifying and classifying bacteria based upon differences in ribosomal RNA molecules or the genes encoding them. Ribotyping generates a highly reproducible and precise fingerprint that can be used to identify and classify bacteria. The methodology involves the fingerprinting of genomic DNA restriction fragments that contain all or part of the genes coding for the 16S and 23S rRNA (initially phylogeny based on 5S rRNA homology was also attempted, but was replaced by larger RNA molecules due to availability of more information, universality, and experimental ease). Currently, ~2,500,000 16S rRNA sequences of various species are available. This also includes a significant number of sequences retrieved by PCR from yet-uncultured microorganisms. This was the first technique used to differentiate *Archea* from *Eubacteria*. Advances in biotechnological tools has given rise to different techniques that are currently being used to resolve this operon, and extrapolating the data obtained for identification of bacteria from samples of different origins for further use as per requirements.

Conventional Ribotyping

Conventional ribotyping, a restriction fragment length polymorphism technique, is based on restriction endonuclease cleavage of total genomic DNA followed by electrophoretic separation, Southern blot transfer, and hybridization of transferred DNA fragments with a radiolabeled ribosomal operon probe (Figure 18.2). After the availability of ribosomal gene sequences and DNA analytic software, it became possible to select an appropriate restriction enzyme that cuts once within the 16S rRNA genes and once within the 23S rRNA gene, which enables detection of DNA sequence polymorphisms in ribosomal operons or immediately adjacent upstream and downstream genes flanking each ribosomal operon. The DNA is extracted from a single bacterial culture of the test organism derived from a single colony, and is subjected to restriction digestion to get its discrete sized DNA fragments. These fragments are then resolved by electrophoretic separation and transferred to a membrane (Southern blotting) (Figure 18.3A), and regions of rRNA operons are probed by hybridization with radioactively labeled probes (alternatively, other reporter systems such as fluorescent or chemiluminescent-labeled probes can be used). During species/genus-specific probe preparation, a whole 16S–23S–5S operon is first amplified by PCR; then this amplified fragment is used to generate radioactive or other labeled probes using random primers and a large DNA polymerase I (Klenow) fragment (Figure 18.3B). There are other approaches to generate probes. For example, any segment of the operon (the 16S or 23S region, or the whole operon) can be cloned separately and used as a probe. Another method employs cDNA synthesized using reverse transcriptase and a 16S or 23S rRNA template as the probe sequence. Oligonucleotides that are chemically synthesized with a "signature sequence" of the specific level of the taxon can also be labeled and used as probes. Following electrophoresis, the pattern of the rRNA gene (i.e. ribotypic bands) is then recorded, digitized, analyzed manually or with appropriate software, and stored in the database (Figure 18.3C). The variations in ribotypes among bacteria may occur due to both the position and intensity of bands complementary to rRNA genes, and are used to identify and classify the bacteria. If the entire *rrn* operon is used as the probe sequence, then the banding pattern is known as a ribotype, and if only the 16S rRNA sequence is used as the probe, the banding pattern is called a 16S ribotype. The polymorphism in ribotyping is the reflection of the conservation of rRNA operon sequences and variability of chromosomal flank sequences. In certain situations, for discrimination of strains, one, two, or more restriction endonucleases can be used simultaneously to identify a greater degree of restriction fragment length polymorphism.

Selection of Restriction Endonuclease for Ribotyping by Sequence Analysis (*In Silico*)

As of now, more than 250,000 ribosomal gene sequences are available in public databases, and with the use of DNA analytic software, it has become possible to select an appropriate restriction enzyme that cuts once within the 16S rRNA genes and once within the 23S rRNA gene, which enables detection of DNA sequence polymorphisms in ribosomal operons or immediately adjacent upstream and downstream genes flanking each ribosomal operon. *Haemophilus influenzae* strain Rd was the first available bacterial genomic sequence, and has been used as a protoype for ribotyping (Bouchet et al., 2008). *In silico*

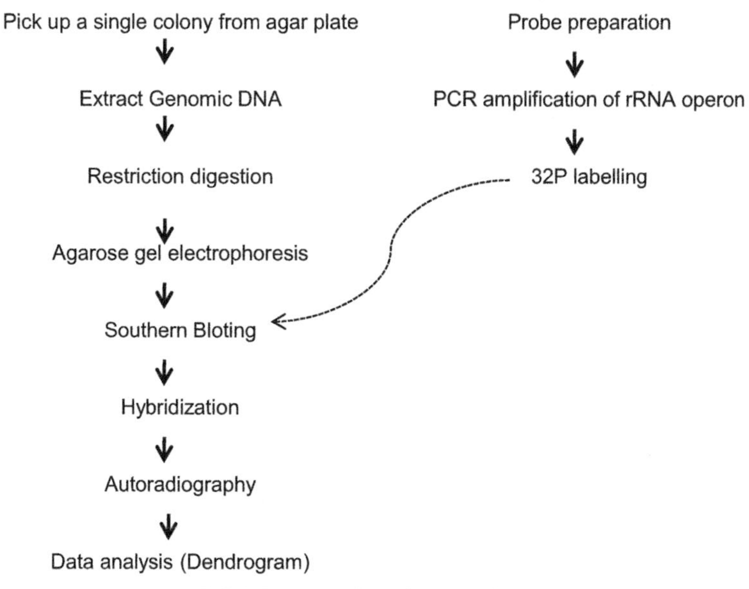

FIGURE 18.2 Ribotyping methodology.

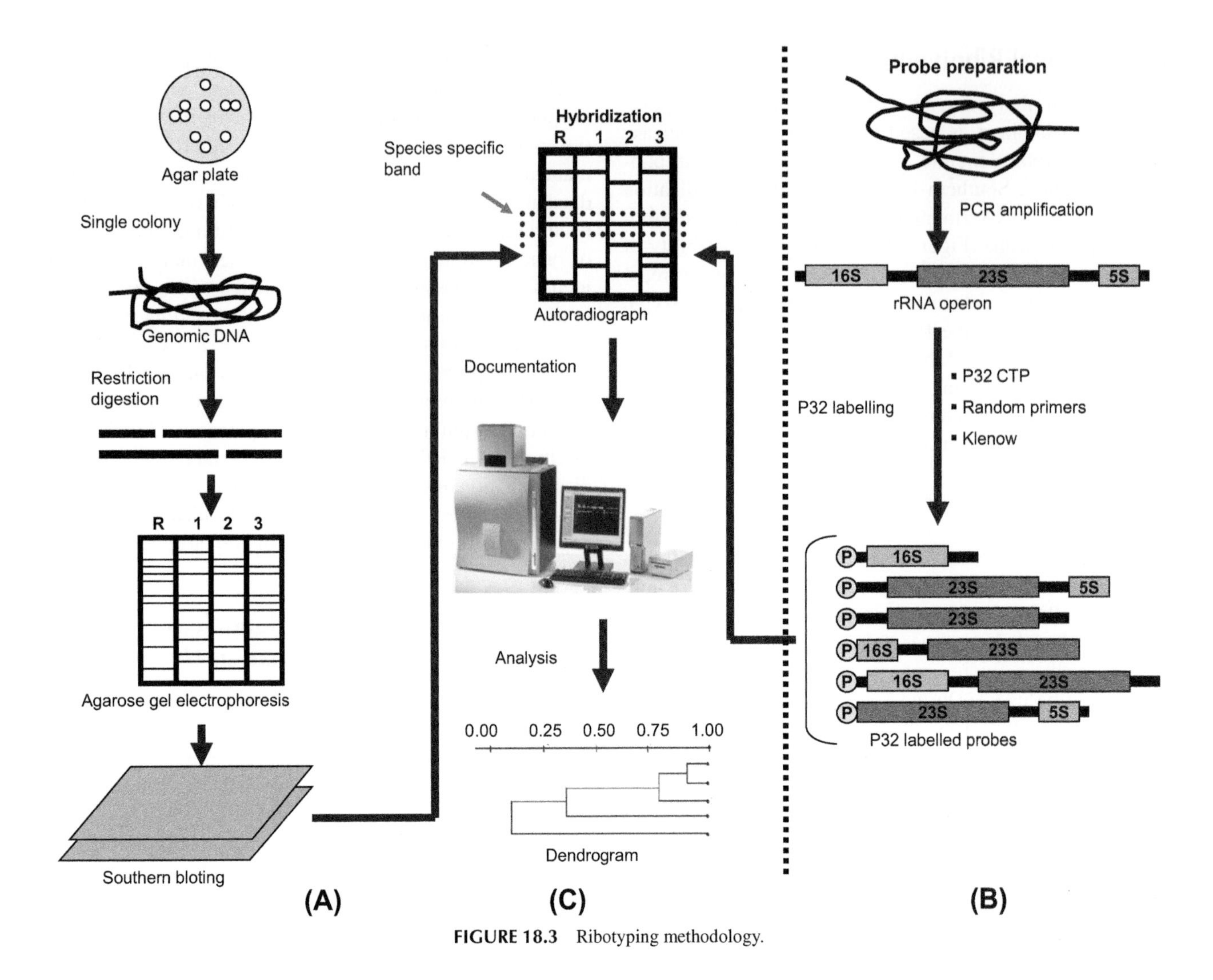

FIGURE 18.3 Ribotyping methodology.

analysis of *H. influenzae* strain Rd revealed the presence of six ribosomal operons and associated flanking sequences responsible for ribotyping polymorphisms. The initial ribotype protocol for *H. influenzae* includes the following steps:

1. Search the conserved restriction endonuclease cleavage sites within the six ribosomal operons.
2. Choose the ideal restriction enzyme that would cut once within the 16S rRNA gene and once within the 23S rRNA gene to create an internal species-specific fragment. The length, GC content, and specificity of the recognition site also determine the sensitivity of the ribotyping.
3. Scan for all publicly available 16S and 23S *H. influenzae* rRNA gene sequences (in addition to closely related species) to confirm conservation of the restriction site.
4. After selection of restriction endonuclease, the genomic DNA of the isolate to be ribotyped is digested, separated by electrophoresis, and transferred onto a nylon membrane by Southern blotting; it is then hybridized to a labeled probe containing conserved ribosomal operon sequences that enable generation of ribotype RFLP patterns.

In silico analysis of *H. influenzae* revealed that its rRNA operon has two cutting sites for *Eco*RI, one within the 16S rRNA gene and one within the 23S rRNA gene that would result in ribosomal operons plus two bands comprising the ISR located between 16S and 23S rRNA of the six rRNA operons. Interpretation of ribotyping depends in part on size variation of ISRs bands containing tRNA-encoding DNA sequences. ISR size is for the most part related to the number of tRNAs found within the ISR region (e.g. one or two in the case of *H. influenza*), and is therefore referred to as a species-specific signature band. The high degree of conservation among ribosomal operons suggests that the ribotype polymorphism is not due to the rRNA gene, but rather to flanking sequences of the rRNA genes. *In silico* analysis of 50,000 bp immediately upstream and downstream of the six ribosomal operons of *H. influenza* strain Rd revealed that rDNA flanking sequences are primarily composed of neutral housekeeping genes (encoding proteins, but not subjected to diversifying selection) evolved through point mutations (Bouchet et al., 2008). The number of variant alleles of a housekeeping gene is vast, and therefore provides the basis for ribotyping. During evolution, the bacterial genome undergoes random point mutations throughout its genome, and thus the polymorphisms in ribotyping are dependent on the rate of these point mutations in the genes flanking ribosomal operons. Change in a single base pair in a 6 bp restriction recognition site (*Eco* RI) will result in loss of the recognition sequence, and a consequent change in the ribotype RFLP fingerprint profile.

The importance of this process lies in the fact that it allows detection of point mutations, and thus can be used for strain differentiation within a species (which is important from a taxonomic point of view). The basic *in silico* approach described for *H. influenzae* for design and interpretation of a conventional ribotyping scheme has been similarly implemented for several other bacterial species (e.g. *E. coli, P. aeruginosa, B. cepacia, S. pneumoniae, N. meningitis, B. pertussis, B. parapertussis, B. broncoseptica*, etc.).

Automated Ribotyping

Automated ribotyping was first time introduced by DuPont Qualicon in 1995 for identification and characterization of bacterial isolates by generation of ribosomal RNA fingerprints (RiboPrint pattern) from bacteria. This system is reproducible, convenient, and fast. In automated ribotyping, a single colony is picked up from an agar plate, suspended in lysis buffer, transferred to a sample carrier, and loaded into the RiboPrinter. Later, inside the RiboPrinter (as with conventional ribotyping), restriction digestion of DNA (mostly *Eco* RI), separation of DNA fragments by electrophoresis on agarose minigel, transfer and capture to nylon membranes, and hybridization with a chemiluminescent-labeled DNA probe derived from *E. coli rrn* rRNA operon take place. After washing to remove unhybridized probes, the chemiluminescent patterns are then electronically imaged, processed, and compared to other DNA patterns in the RiboPrinter database. A single batch run takes place in about 8 hours, and the generated data can be distributed into the network. This system allows virtually all bacteria species to be characterized according to their specific ribotype, and to be identified as per their existing reference pattern. It is particularly useful in the epidemiological surveillance of bacterial diseases, and for rapid identification of the source of infection, which may help in formulating a strategy to prevent further spread of disease and reduce the number of victims. However, the system is expensive, the run costs are high, and a pure culture is needed; it also requires good characterization and identification in the database for effective use.

The conventional ribotyping procedure involves Southern transfer and hybridization with radioactively labeled probes, making it cumbersome; it also requires specialized laboratories with facilities for physical containment of radiation. Following the advent of new genetic tools and rapid sequencing schemes, attempts were made to develop more rapid and less labor-intensive typing schemes for microorganisms related to ribosomal RNA operons. These alternative schemes are more practical than conventional ribotyping for identification and classification of bacteria, even beyond the species level.

PCR Ribotyping

PCR ribotyping is based on amplification of the intergenic spacer region, along with some sequences of 16S and 23S rRNA genes; it also detects the length and number of polymorphisms in the bands after gel electrophoresis (Kostman et al., 1990). Conserved regions in 16S rRNA and 23S rRNA genes allow designing of primers which can anneal with the genes of a wide number of bacterial genera, enabling the amplification of intergenic spaces through PCR. In this method, the primers are chosen in such a way that they are complementary to the 3′ end of the 16S rRNA gene and the 5′ end of the 23S rRNA gene (Figure 18.4). Following PCR, the amplified product is resolved through agarose or

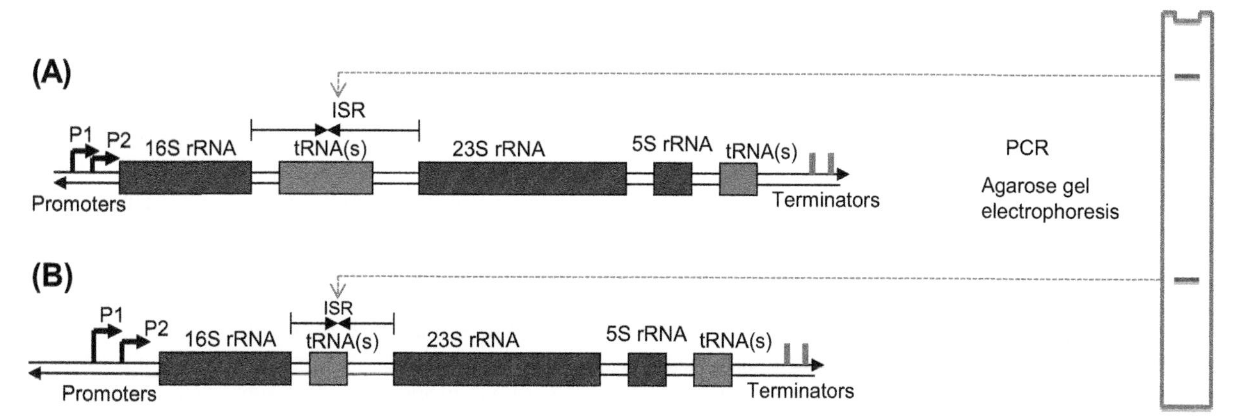

FIGURE 18.4 Length heterogeneity of internal spacer region (ISR) of rRNA operons (A and B) among bacterial species/strains results in polymorphisms that can be detected by PCR and agarose gel electrophoresis (PCR ribotyping).

polyacrylamide gel electrophoresis, and the data of size and number of bands are analyzed. The length of the spacer region can vary in different copies of the rRNA operon within the same chromosome. This is probably because of the fact that the internal spacer region is under the least selection pressure and thus may vary in length and sequence within the organisms of the same genus and species. If only one or very few bands appear after electrophoresis, techniques such as denaturing gradient gel electrophoresis, temperature-gradient electrophoresis, or SSCP can be applied to further resolve the polymorphism. After electrophoresis, more than one band may be obtained, but in a particular strain it will be constant. This technique was found to be especially suitable in clinical microbiology laboratories in epidemiological studies for determining relatedness and discrimination of strains isolated from different sources.

PCR Ribotyping and Endonuclease Subtyping

If PCR ribotyping does not result in resolving conclusive heterogeneity in the amplified bands of multiple strains of one or different species, subtyping of the strains by restriction analysis of the amplified product may generate additional data on heterogeneity in the genomic sequences, which in turn may help in discrimination of the bacterial strains of a single species. This heterogeneity results due to the high degree of sequence variability among multiple copies of the ISR, ranging from a point mutation to deletion or addition of larger segments. Such variability is mainly due to the fact that ISRs in some organisms code for one or two tRNAs. They also possess multiple alleles of the rRNA gene cluster, with considerable variation in the lengths and sequences of the ISRs both at the level of operons in the same genome and between operons of different strains within a species.

PCR Ribotyping Followed by Sequencing of ISR

Mycoplasma and *Mycobacterium* are the organisms that have only one or two copies of rRNA operons in a single organism and pose difficulty in resolution of ISR variability in terms of length and number of operons. In such situations, direct nucleotide sequencing of amplified product following PCR ribotyping of the spacer region may result in identification of variations in the sequences of genomes, thus helping in the discrimination of different strains of a species.

Since the sequences of 16S and 23S RNA genes are highly conserved, the primers first described by Kostman and associates can be used to amplify ISRs of a wide range of bacteria, making PCR ribotyping widely applicable. Moreover, this method obviates the need for the time-consuming and cumbersome Southern transfer technique used in conventional ribotyping. It is also widely used in the clinical microbiology laboratory where the focus is on expedited identification of microorganisms and determination of the source and spread of infection (to facilitate control measures).

PCR-ribotyping carried out on various bacteria has resulted in the realization that the choice of PCR primer is important in the successful detection of heterogeneity in the intergenic space, especially of the 23S rRNA gene, which is relatively less conserved than that of the 16S rRNA gene. Primers derived from certain regions of the 23S rRNA gene have been found to not be able to bind to all strains of a species, and thus fail to amplify the intergenic space through PCR.

Amplified Ribosomal DNA (rDNA) Restriction Analysis (ARDRA)

Amplified ribosomal DNA (rDNA) restriction analysis (ARDRA) is a modified method of PCR ribotyping. This simple and effective technique was first reported by Vaneechoutte and coworkers; they could distinguish nine of the 13 well-described taxa of the eubacterial family *Comamonadaceae* (Vaneechoutte et al., 1992). ARDRA is based on PCR amplification of the 16S rRNA gene, followed by restriction digestion of the amplified products. The restriction fragment patterns obtained are further used to determine the diversity or similarity between the bacteria based on the proportion of the fragments that are shared between the microorganisms under question. The advantage of this technique is that it can be used either for identification of the organism in pure cultures or in microbial communities.

Initially, DNA is extracted from the sample. This sample may have an unknown microorganism either in pure culture, where it can be cultivated, or a complex population of microorganisms. This DNA is subjected to PCR using the primers derived from the conserved regions present on both ends of the 16S rDNA gene. The entire amplicon of size of about 1,500 bp can then be either directly subjected to restriction enzyme digestion or cloned into *E. coli* cells to segregate the PCR products into individual cultures. Cloning helps yield better resolution compared to direct restriction enzyme digestion of the amplicons, especially when a complex population of organisms is being handled. As the amplicon is about 1,500 bp long, the restriction enzymes having a recognition site of more than four bases long will have less (or no) resolution. A single tetra cutter also may not give good resolution. Therefore, at least three tetra cutter restriction enzymes are used. Proper selection of the restriction enzymes is also important, as different restriction enzymes may give different resolutions of banding patterns. One study (Moyer et al., 1996) found *Hha*I, *Rsa*I, and *Bst* UI appropriate for the detection and differentiation of bacterial taxa based on criteria of RFLP size–frequency distribution patterns. These enzymes yielded the highest frequency of restriction fragment band size classes in the 250- to 550-bp range. The restriction pattern resolved after electrophoresis can further be analyzed, and phyolograms or cladograms prepared and compared with the relevant databases for identification.

This technique is useful if proper selection of restriction enzymes and electrophoresis has been done to resolve the restriction bands properly. ARDRA has proven useful only for genus differentiation and species differentiation, but not for intra-species discrimination of different bacterial isolates.

Terminal Restriction Fragment Length Polymorphism of 16S rRNA Gene

This technique was first used by Liu and coworkers (Liu *et al.*, 1997). This technique is useful in studying microbial communities in different environments. It is a PCR-based method in which the 16S rRNA gene is amplified with primers labeled at the 5′ end. The amplified products are subjected to restriction digestion using tetra cutters, and fluorescent-labeled terminal restriction fragments are then sequenced on an automated sequencer. The members of closely related phylogenetic groups often have the same terminal restriction fragment sizes.

Long PCR Ribotyping

The long PCR ribotyping method involves PCR amplification of an entire 16S–23S–5S ribosomal operon of length of up to 6 kbp, followed by restriction endonuclease digestion to detect restriction length polymorphisms. The products can be visualized in a simple manner in an ethidium bromide-stained agarose gel electrophoretogram. This method was first developed to type extremely heterogeneous (non-typable) *Haemophilus influenzae* strains, and has also been found useful in rapidly characterizing novel organisms with a high level of discrimination, reproducibility, and ease (Smith-Vaughan et al., 1995). The improved enzymatic systems for synthesis of longer DNA segments made it possible to amplify PCR products up to 15 kb. This is achieved by replacing *Taq* DNA polymerase with a combination of a high level of thermostable DNA polymerase lacking 3′–5′ exonuclease activity and a low level of thermostable DNA polymerase exhibiting 3′–5′ exonuclease activity (e.g. *Pfu* DNA polymerase) with improved base pair fidelity. It is considered more discriminatory than 16S sequencing alone because it covers the more highly variable ISR and also possesses 23S rRNA genes.

Broad-Range PCR Ribotyping

Availability of ribosomal RNA sequences of a wide range of organisms in databases has resulted in the exploration of many new possibilities. In clinical microbiology laboratories, in certain situations when organisms are visible but cannot be cultivated (especially after antibiotic therapy or if they appear to be very slow-growing), broad-range PCR-based ribotyping has yielded good results. Sequence analysis of ribosomal RNA genes of different groups of organisms has emphasized the fact that 16S rRNA genes (and also 23S rRNA genes) show highly conserved regions of varying levels of variable sites. Primers can be designed that are specific to broad groups or higher level phylogenetic taxa (i.e. *Bacteria, Archaea, Eukarya, Fungi*). Within such groups, PCR-based progressive differentiation can be achieved using primers that are targeted to detect division, family, and genus. Usually primers are directed to amplify about 300 to 900 bp regions of the rRNA gene, depending upon the quality of the template DNA; a smaller target is preferable for damaged or lower quality DNA. Appropriate positive and negative controls must be observed, and special precautions need be taken to prevent DNA contamination during sampling, in a laboratory environment, and of all reagents. Real-time PCR, if applied in place of conventional PCR, is more suitable in analyzing background false amplification.

Limitations of PCR Ribotyping

Use of the 16S–23S ISR region in studies of phylogeny, molecular evolution, or population genetics is a potentially powerful tool. However, there are certain limitations imposed by the possibility of multiple non-identical rRNA operons. These can be any of the following:

1. ISRs are of identical length but different sequence will produce identical ISR fragment patterns after gel electrophoresis.
2. Preferential amplification of some operons: the organism may contain both tRNA-containing and tRNA-lacking rRNA operons, and the operons without tRNAs may be preferentially amplified during PCR. This phenomenon may impact analyses employing restriction enzyme digests of ISR PCR products. If the ISR from one operon is amplified preferentially in one isolate, while the ISR from a different operon is preferentially amplified in another isolate, comparisons of digests of PCR products from those two isolates may be flawed. Experimental designs should therefore contain safeguards against complications imposed by the presence of multiple non-identical rRNA operons.
3. Selection of an ideal restriction endonuclease described above is not always possible.
4. rRNA genes are transcribed from the ribosomal operon as 30S rRNA precursor molecules, which are then cleaved by RNase III into 16S, 23S, and 5S rRNA molecules. rRNA molecules may likewise be fragmented due to the presence of intervening sequences (IVS) within either the 16S rRNA or the 23S rRNA. The presence of IVSs within the 16S or 23S rRNA may be misleading

in the interpretation of ribotyping RFLP patterns by (i) containing endonuclease recognition sequences of the selected restriction enzyme, and (ii) altering the size of the signature band. The appearance of an additional restriction site within the IVSs would result either in truncation of the species-specific signature band or in alteration of the flanking DNA band, depending on the position of the new cut site.

5. Split rRNA operon: After availability of whole genome sequences of some of the bacteria, split rRNA operons have been detected. Boer and Gray first reported an unusual organization of rRNA genes in the mitochondrial DNA of *Chlamydomonas reinhardtii*, where each gene was discontinuous and dispersed throughout the genome (Boer and Gray, 1988). Later, split rRNA operons were also identified in *Eubacteria* and *Archaea.* In split rRNA operons, in the most common pattern, the 23S and 5S genes form an operon, and the 16S rRNA gene is separate; there may be one or more copies of each of the genes. For example, *Helicobacter pylori* 26695 carries two 23S–5S operons, two single 16S genes, and one separate 5S gene.

Ribosomal DNA Sequence Analysis

Development of rapid and high-throughput next generation sequencing techniques (NGS) has of late resulted in the availability of rapid and unambiguous nucleotide sequence information for 16S and 23S rDNA of several bacterial species. This information is available in regularly updated public databases (GenBank: www.ncbi.nlm.nih.gov; or Ribosomal Database Project II: Cole et al., 2011) or private commercial databases for discrimination of even closely related species. The technology has enabled the determination of 16 S rRNA sequences without even culturing the organisms from clinical and environmental sources; this is known as "taxonomic profiling," which consists of amplification of a complete or selected region of the 16S rRNA region through PCR, and sequencing of the amplicon. The choice of primers determines amplification of conserved or less conserved regions of the ribosomal operon. Such "microbiome" and "metagenomic" studies earlier involved Sanger's automated capillary sequencing, which is being increasingly replaced by various next-generation sequencing platforms that enable more samples to be sequenced at a higher depth and lower cost. Amongst different NGS technologies, the Roche 454 and the Solexa system from Illumina have been the most used for metagenomic studies. The Roche 454 system involves a pyrosequencing process of clonally amplifying DNA fragments through emulsion polymerase chain reaction attached to microscopic beads in a picotitre plate leading to an average read length of 600–800 bp. The processing of the samples has been so optimized that up to 12 samples of DNA in nanogram quantities can be sequenced as single-end libraries. However, this process results in incorrect reads, especially at the homopolymeric tails. Moreover, ePCR has been shown to produce misprimed amplified products, leading to false estimates of gene numbers and reading that can be done only in one direction. The Illumina/Solexa technology is based on solid-surface PCR amplification of random DNA fragments immobilized on a surface. These are then sequenced with reversible terminators in a sequencing-by-synthesis process. The cluster density is enormous, and clustered fragments can be sequenced from both ends suitable for metagenomic studies. Sequencing technologies are evolving continuously, and many new systems are available that might prove useful for metagenomic applications, now or in the near future. The Applied Biosystems SOLiD sequencer arguably provides the lowest error rate of any current NGS sequencing technology, however, it does not achieve reliable read length beyond 50 nucleotides. Nevertheless, for assembly or mapping of metagenomic data against a reference genome, recent work has shown encouraging outcomes. Ion Torrent (and more recently Ion Proton) is another emerging technology and is based on the principle that protons released during DNA polymerization can detect nucleotide incorporation. This system promises read lengths of more than 100 bp and throughput on the order of the Roche 454 sequencing systems. Along with the evolving NGS technologies, suitable bioinformatics tools for analyzing metagenomic data have also been developed. For targeted amplicon sequencing of 16S rDNA profiles, tools such as QIIME, muther, and VAMPS were developed for comparison and analysis of microbial communities. Since the high-throughput sequencing technologies introduce errors into sequence data, there are approaches to reduce these errors by clustering the flowgrams produced by the sequencer into a smaller number of sequences likely to be present in original samples; these methods use tools such as AmpliconNoise and Denoiser, which can be applied individually or within the context of QIIME or muther tools (Kuczynski et al., 2011).

In Situ Hybridization Targeted to Detect rRNA

The technique of hybridization with labeled probes (mostly fluorescent) has also been applied for the phylogenic detection of non-viral microorganisms based on using taxon-specific (kingdom, family, genus, species, or subspecies) probes designed to anneal with rRNA *in situ* in their habitat. The technique was first applied for bacterial identification in late the 1980s using fluorescent labels (known popularly as FISH) (De Long et al., 1989) and has been further refined (review: Amann et al., 1995). Apart from being rapid and cost effective, the main advantage of this method is that the organisms can be quantitatively detected in their natural

Pick up a microbial community

Fixation and permeabilization

Hybridization using fluorescent labeled rRNA probe

Washing

Fluorescent microscopy

FIGURE 18.5 *In situ* hybridization.

habitat (i.e. aquatic, biofilms, soil, clinical samples, etc.) without the need for culturing. This technique can even be applied in mixed populations for simultaneous detection of more than one type of organism.

The (F)ISH procedure (Figures 18.5 and 18.6) consists of first designing an oligonucleotide probe targeting the hyper-variable to less variable regions of 16S or 23S rRNAs specific to various phylogenetic levels of the target organism; this uses available sequences in RNA databases and appropriate software tools for sequence comparison and probe design. The oligonucleotide probe is usually kept short in length (15 to 25 bases), and secondary structures are avoided. The oligonucleotide is labeled at the 5′ end with any of the fluorochromes (DAPI, FLUOS, TRITC, Texas Red, Cy3, or Cy5 are commonly used) emitting wavelengths of different colors. Instead of labeling with fluorochromes, oligonucleotides can be tagged with other reporter molecules such as alkaline phosphatase, horseradish peroxidase, or digoxigenin, but they require a corresponding detection system. The selected probe must be thoroughly checked for its specificity against known standards and related non-target organisms. The hybridization reaction is the most crucial step and is carried out on multi-well glass microscopic slides or membrane filters (having fixed the sample with heat, formaldehyde, or alcohol) under carefully chosen conditions of temperature (usually optimized around 46°C) and concentration of formamide. This step is followed by washing with an appropriate buffer at a specific temperature (usually 48°C). The fixation procedure of the sample varies for gram-positive and Gram-negative organisms owing to differential complexity of the cell wall structure affecting permeability of the probe. Simultaneous detection of multiple organisms can also be attempted if a combination of probes is used and they are labeled with carefully chosen fluorochromes for better visualization. In such situations, hybridization conditions of differential stringency also need to be adequately selected to allow annealing of all the target nucleic acids. Multiple probes labeled with different fluorochromes can also be applied for the detection of a single organism to improve probability of detection. Alternatively, a single oligonucleotide can be doubly labeled at the 5′ and 3′ ends with different fluorochromes to enhance the intensity and resolution (Double-labeled Oligonucleotide ProbE, DOPE-FISH).

Visualization of the fluorescent signals is conveniently achieved using an epifluorescence microscope and choosing an appropriate set of excitation and emission filters, depending upon the fluorochrome dye used. Inclusion of controls (positive and negative controls for autofluorescence of the cells) in the reaction greatly improves the reliability of the detection system. Confocal laser scanning microscopy can also be used for visualization of the fluorescence; this overcomes many of the problems encountered while using an epifluorescence microscope, especially if the specimen is thicker. Moreover, three-dimensional images can also be obtained with a Confocal Laser Scanning Microscope (CLSM), facilitating better resolution and quantitation of the target organisms. Another modification involves application of flow cytometry for the detection of fluorescence in the cells post-hybridization if the reaction has been attempted with cells in fixed and liquid suspension states. It also allows quantitation, specific sorting, and separation of desired cells in a mixed population, and increasing the concentration of target organism.

A modification of the *in situ* hybridization method aims to detect metabolically active cells in a community by growing the cells in the presence of a radioactively labeled substrate (containing ^{14}C or ^{3}H). This results in accumulation of a radioactive label inside the cells that facilitates detection by microautoradiography on a photographic emulsion. This is in turn followed by *in situ* hybridization to determine phylogeny. The technique is popularly known as MAR-FISH.

Clone-FISH

The Clone-FISH technique is a modification of the FISH technique in which the hybridization reaction is not carried out in the target organism directly. In Clone-FISH, 16S rRNA genes or their fragments of the target organisms are cloned in suitable plasmid vectors and transcribed heterogeneously into rRNA *in vivo* within a suitable *Escherichia coli* host. *E. coli* cells are then fixed and subsequently used as targets for *in situ* hybridization with the desired probe. Clone-FISH is a simple and fast technique, and is compatible with a wide variety of cloning vectors and hosts that have general utility for probe validation and screening of clone libraries. This technique is more applicable when the target organism is difficult to cultivate or is very slow growing, resulting in a very low rRNA-targeted hybridization signal and higher background noise.

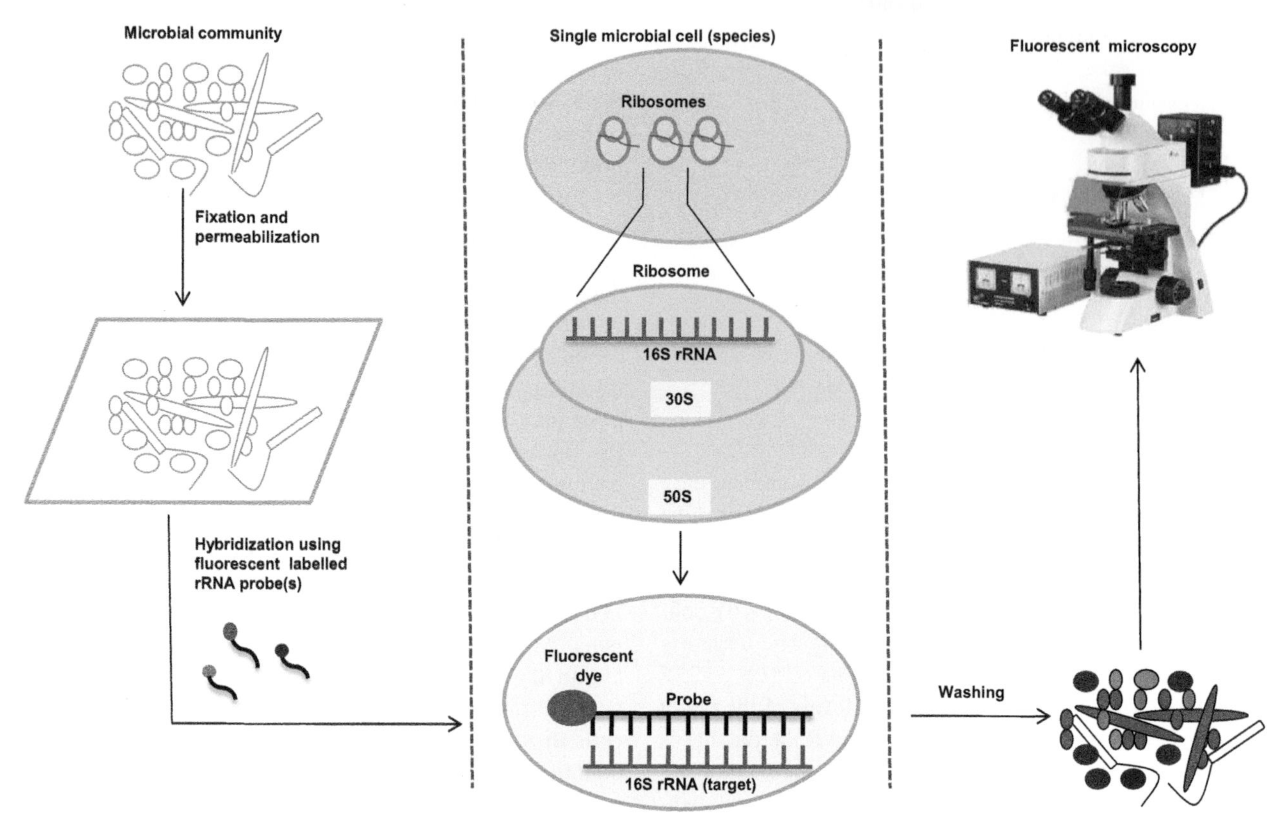

FIGURE 18.6 Steps in *in situ* hybridization.

CARD Fish

The intensity of the signals in an *in situ* hybridization reaction depends upon the stage of growth of the cell and the ribosomal contents. To increase the sensitivity of the hybridization reaction, amplification of the signals can be achieved using horseradish peroxidase-labeled probes in combination with catalyzed reported deposition (CARD) of fluorescently labeled tyramides. Tyramide is a phenolic compound, which upon activation by horseradish peroxidase (HRP) enzyme, covalently binds to adjacent electron-rich tyrosine residues in proteins (even in fixed tissue or cell preparations). The hybridization reaction involves a single oligonucleotide that is covalently cross-linked to an HRP label. This enzymatic label is stable, and after the hybridization step, amplification of the signal is achieved by radicalization of the tyramide molecule derivatives linked to fluorophores by a single HRP molecule. Precautions need be taken to stop the activity of endogenous peroxidases with the blocking agents. Even multiple targets can be detected by sequential application of the labeled probes and tyramide labeled with different chromophores. This technique has been used for detection of pelagic marine (*Bacteria, Cynobacteria* and *Acinobacteria*) and sedimentary marine (*Archaea* and *Bacteria*) as well as marine algae to study the epiphytic bacterial community.

RING (Recognition of Individual Gene) FISH

This is the method of detection of a single gene in bacteria. It involves a polynucleotide probe having only a few hundred nucleotides, but with multiple labels. With an extended hybridization time and high probe concentration, a network of probe molecules is formed that is typically characterized by halo-shaped fluorescence signals in the periphery of the cells (ring-shaped signals). The ring-shaped signals are believed to occur due to folding of the single-stranded RNA probe molecules into secondary structures, which results in the formation of a network of probes around the cell during whole-cell hybridization. This technique can be applied in combination with a conventional *in situ* hybridization reaction for identification of organisms based on 16S rRNA detection. The advantage RING-FISH is that the characteristic genes (e.g. virulence, antibiotic resistance, etc.) of the target organism can be detected along with the assignment of taxonomic level, especially in environmental samples.

Stable Isotope-Labeled Probing of rRNA and rDNA

For the identification of functionally active microorganisms in environmental samples (also in biofilms), ribosomal RNA or their genes can be detected in addition to their metabolic genes using the technique of Stable Isotope Probing (SIP). This method is based on growing the organisms in an environment containing stable isotope (nonradioactive)-labeled substrates in carefully chosen concentrations and duration of incubation, and subsequently analyzing isotope-labeled RNA (RNA-SIP) or DNA (DNA-SIP).

The commonly available isotopes for probing are compounds of ^{13}C (acetate, propionate, bicarbonate, phenol, etc.), ^{15}N (ammonium chloride), or ^{18}O (water). Labeled DNA or RNA is separated with density gradient centrifugation and subsequently 16S/23S rRNA genes are amplified by PCR or RT-PCR using the desired taxonomic level-specific primers. Isotope ratio mass spectrometry (IRMS) is used to distinguish labeled and unlabeled nucleic acids (isotope ratio). Another approach employs construction of DNA or cDNA libraries from the isolated "biomarked" nucleic acid and designing RNA probes for (F)ISH. DNA-SIP requires a longer period of incubation and is less sensitive than RNA-SIP, as RNA is synthesized at a faster rate in growing cells than DNA, and more labels are incorporated during a shorter period of incubation.

Peptide Nucleic Acid Probes

Although nucleic acid probes have been widely used for both solid- and liquid-phase hybridization, the use of peptide nucleic acid (PNA) probes is becoming more common of late owing to certain advantageous properties over nucleic acid probes. Peptide nucleic acids are nucleic acid analogs consisting of a neutral polyamide (i.e. poly-N-(2-aminoethyl)glycine) chain simulating a negatively charged poly-sugar phosphate chain of nucleic acids to which purine and pyrimidine bases are attached covalently (review: Corey, 1997). PNA probes hybridize with target nucleic acids both with DNA and RNA molecules in a similar fashion as the nucleic acid probes following conventional rules for base pairing. PNA probes are chemically and enzymatically more stable than nucleic acid probes, and the PNA-DNA/RNA hybrids have stronger binding affinity at low and medium salt concentrations. The specificity of PNA probes in terms of binding to target sequences with mismatches is also comparatively higher with lower background noise, and therefore are better suited for *in situ* hybridization reactions.

LIMITATIONS OF RIBOTYPING

The simplicity of performing and automating 16S rRNA sequencing, and availability of several large databases of reference sequences and taxonomies, have made ribotyping one of the most preferred techniques in diagnostic laboratories for identification and epidemiological studies. Its contribution to systematics has also been immense. For example, aerobic gram-negative bacteria that used acetate as a sole carbon source were placed in a single group, the pseudomonads, which comprised several genera, including Pseudomonas. However, when 16S rRNA gene sequencing was introduced, it was discovered that these organisms arose by three different evolutionary routes that are now recognized as the Alpha-, Beta-, and Gamma-proteobacteria. However, like all other techniques, it has its own limitations. The higher level of conserved sequences within rRNA operons results in lower discriminatory power. High similarity can be found in 16S rRNA gene sequences among some closely related microorganisms that lack resolution at the species level. It is well known that organisms can have identical 16S rRNA gene sequences and still belong to separate species that can be detected using other methods (Pontes et al., 2007). Another factor that leads to difficulties in interpretation is the different copy numbers of the gene in different bacterial species, which causes over or under representation of some species in rRNA target studies. Moreover, the process is dependent on the efficiency of the DNA extraction method, and is also liable to PCR biases and artifacts, such as preferential amplification of certain sequence types, generation of chimeric sequences, and false positives due to experimental contaminants. Even following electrophoresis, the varying band patterns may exhibit only slight differences in size, which can further pose difficulties in evaluating the ribotypes.

Despite these limitations, ribotyping based on 16S rRNA gene sequencing has now been widely accepted as the primary technique for the identification of cellular microorganisms, particularly after establishment of boundary conditions (i.e. if a strain shows less that 97% 16S rRNA gene homology with its highest match-described species, then it can be declared a novel species) (Stackebrandt and Goebel, 1994).

FUTURE PERSPECTIVES

Microbial taxonomy is periodically reviewed and subjected to modifications based on updating of information on newer isolates of all species across the world, and is ever-developing due to new data-generating techniques. The present system of identification and classification, predominantly based on ribotyping, though widely accepted, appears to be somewhat skewed in favor of analyzing a single trait or mono-locus housekeeping gene. Moreover, the present taxonomic system merely provides information on characterization, identification, and probable model of evolutionary relationship among its constituents. The past two decades have witnessed great advancements in the assessment of overall homology and differences amongst microorganisms. Advancements have been seen in the areas of determination of molecular structure (phenotypic) details, complete genome sequencing, (faster) analytical software tools that facilitate incorporation of multiple characteristics and housekeeping genes, and determination of both phenotypic and genotypic characteristics. Also, there is great potential for automation based on microarrays or "biochip" techniques, which may have a far-reaching impact on the identification and classification of microbes. The taxonomic system, if based on holistic properties, may also provide practical and useful information on the medical, epidemiological, and environmental significance of microorganisms.

ETHICAL ISSUES

So far, the technology for classification of microorganisms has not raised any serious ethical concerns. However, it is possible that with further expansion of this technology, there may be ethical issues that develop.

TRANSLATIONAL SIGNIFICANCE

Ribotyping of microbial isolates has helped in the taxonomic classification of organisms and building of phylogenies with a greater degree of authenticity, as discussed earlier (Grimont and Grimont, 1986). Its reproducibility and robustness have found utility in epidemiological tracking of infections, determination of clonality of strains, enabling of effective planning for prevention and disease management, and also tracking of sources of contamination in the food industry. Automated robots for ribotyping have been developed, reducing hands-on time for typing microorganisms. This technique has also been useful in studying the geographical distributions of microbiomes and in understanding the physiology of constituent uncultivable organisms.

Advancements in the biological sciences have led to the unraveling of many new phenomenon of the biosphere. Current scientific approaches have been directed towards systems biology (i.e. the study of complex interactions within biological systems). An important contribution of systems biology has been revelations about co-evolution and the close interaction between an animal and its microbiome. The role of the microbiome has been proven to be in host defense mechanisms, metabolism, reproduction, and all body functions. This intricate association between the vertebrate host and its microbiome led the National Institutes of Health to launch the Human Microbiome Project in 2008. Its goal is the identification and characterization of microorganisms associated with healthy and diseased individuals (http://commonfund.nih.gov/hmp/). Ribotyping, an important tool for giving signatures to microbes, has been one of the useful methods in deciphering this microbiome.

Another area in which ribotyping has proved to be useful is in studies related to nutrigenomics, where deciphering nutritional health effects requires an understanding of the composition of gut microflora, effects of changing diet on changes in gut microflora, and also corroboration of these changes with the health of the host.

WORLD WIDE WEB RESOURCES

http://www.ncbi.nlm.nih.gov: The web site for the National Center for Biotechnology Information provides access to molecular biology, biomedical, and genomic information.

http://www.microbial-ecology.net/probebase: probeBase is an online resource for rRNA-targeted oligonucleotide probes.

http://www.psb.ugent.be/rRNA/: The European rRNA Database provides access to all complete, or nearly complete, ribosomal RNA sequences from both the small (SSU) and large (LSU) ribosomal subunits.

http://ribosome.fandm.edu. This is a 16S and 23S ribosomal RNA mutation database that provides information on RNA of numerous organisms, and is currently under revision to include mutations in ribosomal proteins and ribosomal factors of more organisms.

http://greengenes.lbl.gov: An online, full-length (ssu) rRNA gene database. The Greengenes web application provides access to current and comprehensive 16S rRNA gene sequence alignments for browsing, blasting, probing, and downloading. The data and tools presented by Greengenes can assist researchers in choosing phylogenetically specific probes, interpreting microarray results, and aligning/annotating novel sequences.

http://ncbi.nlm.nih.gov/taxonomy: This site provides classification and nomenclature for all the organisms in public sequence databases, representing about 10% of described species on the planet.

http://commonfund.nih.gov/hmp: This site elaborates the objectives of the Human Microbiome Project (HMP), and lists other sources for genomes of many reference strains in gene banks and strain repositories.

REFERENCES

Amann, R. I., Ludwig, W., & Schleifer, K. H. (1995). Phylogenetic identification and in situ detection of individual microbial cells without cultivation. *Microbiology and Molecular Biology Reviews, 59*, 143–169.

Boer, P. H., & Gray, M. W. (1988). Scrambled ribosomal RNA gene pieces in *Chlamydomonas reinhardtii* mitochondrial DNA. *Cell, 55*, 399–411.

Bouchet, V., Huot, H., & Goldstein, R. (2008). Molecular genetic basis of ribotyping. *Clinical Microbiological Reviews, 21*, 262–272.

Cole, J. R., Wang, Q., Wang, Q.,Chai, B., & Tiedje, J. M., (2011). The Ribosomal Database Project: Sequences and software for high-throughput rRNA analysis. In: F. J. de Bruijn (ed), *Handbook of Molecular Microbial Ecology I: Metagenomics and Complementary Approaches*, Hoboken, NJ: Wiley.

Corey, D. R. (1997). Peptide nucleic acids:expanding the scope of nucleic acid recognition. *Trends in Biotechnology, 15*, 224–229.

Cowan, S. T. (1965). Principles and Practice of Bacterial Taxonomy – a Forward Look. *Journal of General Microbiology, 39*, 148–158.

De Bruyne, K., Slabbinck, B., Waegerman, W., Vauterin, P., De Baets, B., & Vandamme, P. (2011). Bacterial species identification from MALDI-TOF mass spectra through data analysis and machine learning. *Systematic and Applied Microbiology, 34*, 20–29.

De Long, E. F., Wickham, G. S., & Pace, N. R. (1989). Phylogenetic strains: ribosomal rRNA-based probes for the identification of single cells. *Science, 243*, 1360–1363.

Fox, G. E., Pechman, K. R., & Woese, C. R. (1977). Comparative cataloging of 16S ribosomal ribonucleic acid: Molecular approach to procaryotic systematics.. *International Journal of Systematic Bacteriology, 27*, 44–57.

Garrity, G. M. (2001). *Bergey's Manual of Systematic Bacteriology* (2nd ed.). New York: Springer Verlag.

Goodacre, R., & Kell, D. B. (1996). Pyrolysis mass spectrometery and its applications in biotechnology. *Current Opinion in Biotechnology, 7*, 20–28.

Grimont, F., & Grimont, P. A. D. (1986). Ribosomal ribonucleic acid gene restriction patterns as potential taxonomic tools. *Annales de l'Institut Pasteur Microbiology, 137B*, 165–175.

Kostman, J. R., Alden, M. B., Mair, M., Edlind, T. D., LiPuma, J. J., & Stull, T. L. (1995). A universal approach to bacterial molecular epidemiology by polymerase chain reaction ribotyping. *Journal of Infectious Diseases, 171*, 204–208.

Kuczynski, J., Lauber, C. L., Walters, W. A., Parfrey, L. W., Clementem, J. C., Gevers, D., & Knight, R. (2011). Experimental and analytical tools for studying the human microbiome. *Nature Reviews: Genetics, 13*, 47–58.

Liu, W. T., Marsh, T. L., Cheng, H., & Forney, L. J. (1997). Characterization of Microbial diversity by determining terminal restriction fragment length polymorphisms of genes encoding 16S rRNA. *Applied and Environmental Microbiology, 63*, 4516–4522.

Moyer, C. L., Tiedje, J. M., Dobbs, F. C., & Karl, D. M. (1996). A computer-stimulated restriction fragment length polymorphism analysis of bacterial small- subunit rRNA genes: Efficacy of selected tetrameric restriction enzymes for studies of microbial diversity in nature. *Applied and Environmental Microbiology, 62*, 2501–2507.

Pontes, D. S., Lima-Bittencourt, C. I. ·E., Chartone-Souza, E., & Amaral Nascimento, A. M. (2007). Molecular approaches: advantages and artifacts in assessing bacterial diversity. *Journal of Industrial Microbiology & Biotechnology, 34*, 463–473.

Rossellő-Mora, R., & Amann, R. (2001). The species concept for prokaryotes. *FEMS Microbiology Reviews, 25*, 39–67.

Smith-Vaughan, H. C., Sriprakash, K. S., Mathews, J. D., & Kemp, D. J. (1995). Long PCR-ribotyping of nontypeable *Haemophilus influenzae*. *Journal of Clinical Microbiology, 33*, 1192–1195.

Sokal, R. R., & Sneath, P. H. A. (1963). *Principles of Numerical Taxonomy*. San Francisco: Freeman.

Stackebrandt, E., & Goebel, B. M. (1994). Taxonomic note: a place for DNA–DNA reassociation and 16S rDNA sequence analysis in the present species definition in bacteriology. *International Journal of Systematic Bacteriology., 44*, 846–849.

Stackebrandt, E., Fredriksen, W., Garrity, G. M., Grimont, P. A. D., Kämfer, P., Maiden, M. C. J., Nesme, X., Mora, R. R., Swings, J., Trüper, H. G., Vauterin, L., Ward, A. C., & Whitman, B. (2002). Report of the ad hoc committee for the re-evaluation of the species definition in bacteriology. *International Journal of Systematics and Evolutionary Microbiology, 52*, 1043–1047.

Vaneechoutte, M., Rossau, R., De Vos, P., Gillis, M., Janssens, D., Paepe, N., De Rouck, A., Fiers, T., Claeys, G., & Kersters, K. (1992). Rapid identification of Comamonadaceae with amplified ribosomal DNA restriction analysis (ARDRA). *FEMS Microbiological Letters, 93*, 227–234.

Wayne, L. G., Brenner, D. J., Colwell, R. R., Grimont, P. A. D., Kandler, O., Krichevsky, M. I., Moore, L. H., Moore, W. E. C., Murray, R. G. E., Stackebrandt, E., Starr, M. P., & Truper, H. G. (1987). Report of the Ad Hoc Committee on Reconciliation of Approaches to Bacterial Systematics. *International Journal Of Systematic Bacteriology, 37*, 463–464.

FURTHER READING

Barbieri, M. (1981). The ribotype theory on the origin of life. *Journal of Theoetical Biology, 91*, 545–601.

Cohan, F. M. (2006). Towards a conceptual and operational union of bacterial systematics, ecology, and evolution. *Philosophical Transactions In Royal Society. B, 361*, 1985–1996.

Olsen, G. J., Lane, D. J., Giovannoni, S. J., Pace, N. R., & Stahl, D. A (1986). Microbial ecology and evolution: a ribosomal RNA approach. *Annual Reviews Microbiology, 40*, 337–365.

GLOSSARY

Genotype The genetic makeup of a cell, an organism, or an individual, usually with reference to a specific characteristic under consideration.

Metagenomics The study of metagenomes, genetic material recovered directly from environmental samples.

Microbiome A microbiome is the totality of microbes, their genetic elements (genomes), and environmental interactions in a particular environment.

Nutrigenomics A branch of nutritional genomics that is the study of the effects of foods and food constituents on gene expression.

PCR Ribotyping Amplification of the intergenic spacer region along with some sequences of 16S and 23S rRNA genes, and the detection of the length and number of polymorphisms in the bands produced after gel electrophoresis.

Phenotype The composite of an organism's observable characteristics or traits.

Pure Cultures A microbial population consisting of a single species, ideally derived from a common ancestor.

Ribotyping Tool for identifying and classifying bacteria based upon differences in ribosomal RNA molecules or the genes encoding them.

Systematics The study of the diversification of living forms, both past and present, and the relationships among living things through time.

Taxonomic Profiling Amplification of a complete or selected region of the 16S rRNA region through PCR, and sequencing of the amplicon.

Taxonomy The science of the concept and practice of grouping individuals into species, arranging species into larger groups, and giving those groups names, thus producing a classification.

ABBREVIATIONS

AFLP Amplified Fragment Length Polymorphism
ARDRA Amplified Ribosomal DNA (rDNA) Restriction Analysis
CARD FISH Catalyzed Reported Deposition
cDNA Complementary Deoxyribonucleic Acid
CLSM Confocal Laser Scanning Microscopy
DAPI 4′,6′-Diamidino-2-phenylindol
DOPE-FISH Double-Labeled Oligonucleotide Probe
ELISA Enzyme-Linked Immunosorbant Assay
ePCR Emulsion Polymerase Chain Reaction
FISH Fluorescent In Situ Hybridization
FTIR Fourier Transform Infrared Spectroscopy

G + C Guanine and Cytosine Ratio
HRP Horseradish Peroxidase
IRMS Isotope Ratio Mass Spectrometry
ISR Intergenic Spacer Region
MAR-FISH Microautoradiography and Fluorescence In Situ Hybridization
MALDI-TOF MS Matrix-Assisted Laser Desorption/Ionization-Time of Flight Mass Spectrometry
NGS Next Generation Sequencing Technologies
PCR Polymerase Chain Reaction
PNA Peptide Nucleic Acid
RAPD Random Amplification of Polymorphic DNA
rDNA Ribosomal Deoxyribonucleic Acid
RFLP Restriction Fragment Length Polymorphism
RING FISH Recognition of Individual Gene
RNA Ribonucleic Acid
rRNA Ribosomal Ribonucleic Acid
SIP Stable Isotope Probing
T_m Melting Temperature
TRITC Tetramethylrhodamine-5-(and 6-)Isothiocyanate
tRNA Transfer Ribonucleic Acid

LONG ANSWER QUESTIONS

1. Define taxonomy. How are taxonomic principles different in microorganisms and eukaryotes?
2. Give an account of different typing methods used for bacteria.
3. What is ribotyping? What are the reasons for this technique to be widely used for taxonomic classification of bacteria?
4. Describe the principle and methodology of PCR ribotyping. In what way does it improve on conventional ribotyping?
5. What are different approaches of ribotyping? Mention their relative advantages and disadvantages.

SHORT ANSWER QUESTIONS

1. How does the concept of species in bacteria differ from that of eukaryotes?
2. What is the basis of ribotype polymorphisms?
3. PCR ribotyping is based on intergenic space heterogeneity. What are the applications of PCR ribotyping?
4. What would you infer from finding a single band during PCR ribotyping?
5. What are the reasons for conventional ribotyping becoming unpopular?
6. (a) How can the intensity of the signals in (F)ISH be enhanced? (b) How can a polynucleotide probe be enhanced to have better penetration and higher binding affinity?

ANSWERS TO SHORT ANSWER QUESTIONS

1. Bacteria multiply asexually by binary fission, and thus limits to gene transfer up to the species level are difficult to ascertain.
2. Differences in the length and number of bands using the whole ribosomal operon as a hybridization probe.
3. PCR ribotyping is mainly used in the clinical microbiology laboratory for epidemiological studies and to discriminate bacterial organisms up to the species and strain level.
4. A single band in PCR ribotyping indicates a single ribosomal operon in the test organism.
5. It is a time-consuming and labor-intensive process of Southern transfer and hybridization with a labeled radioactive probe.
6. (a) By improving the penetration of probes through the cell wall via lysozyme treatment. (b) By using a double-labeled probe or more than one probe.

Chapter 19

Next Generation Sequencing and Its Applications

Anuj Kumar Gupta* and U.D. Gupta†
*C-11/Y-1, C-Block Dilshad Garden, Delhi, India, †National JALMA Institute for Leprosy & Other Mycobacterial Diseases (ICMR), Agra, UP, India

Chapter Outline

SUMMARY

Next generation sequencing (NGS) has revolutionized nearly every area of biotechnology. It has been applied to various aspects of biological science, including animal, human, and plant biotechnology. The enormous information produced by NGS assists in understanding genomic variations, disease mechanisms, and resistance, thus helping development of better diagnostics, therapies, and breeds.

Animal Biotechnology. http://dx.doi.org/10.1016/B978-0-12-416002-6.00019-5

WHAT YOU CAN EXPECT TO KNOW

This chapter describes the fundamentals of next generation sequencing technologies. After reading this chapter, one should be able to distinguish between basic concepts of Sanger sequencing and next generation sequencing. One should know the available sequencing platforms, and understand the advancements in sequencing technologies, chemistries, and their advantages and limitations and novel applications, all of which are facilitated by next generation sequencing approaches. One may also expect to know how NGS technologies work and what their translational significance is.

HISTORY AND METHODS

INTRODUCTION

For more than 30 years, Sanger sequencing by the dideoxy chain termination method has been the central approach for determining a DNA sequence and the gold standard for discussing genomic variations. During the past decade, research on genomics has been inclined towards DNA sequencing as a primary molecular research tool as genome-wide sequencing applications are now facilitated with high-throughput next generation sequencing (NGS) platforms. Virtually every area of biotechnology has been benefited from the advent of high-throughput sequencing. These NGS platforms with different sequencing chemistries and instrumentation produce sequencing data several times higher than traditional ways of sequencing in terms of number of base pairs sequenced in a massively parallel way; hence, sequencing a complete genome with desired depth now takes a matter of days or weeks in contrast to Sanger sequencing. Massively parallel sequencing has revolutionized the identification of genomic variations, disease-associated markers, and has facilitated molecular evolutionary analysis in animals and plants.

HISTORY OF DNA SEQUENCING

DNA sequencing was started with the work of Frederick Sanger at the MRC Centre, the University of Cambridge, England in 1972. Sanger developed and published a method for "DNA sequencing with chain-terminating inhibitors" in 1977 (Sanger et al., 1977a) with the use of chain-terminating dideoxynucleotide analogs that caused base-specific termination of primed DNA synthesis. The first full DNA genome to be sequenced was that of bacteriophage φX174 in 1977 (Sanger et al., 1977b). Simultaneously in 1977, Maxam and Gilbert published "DNA sequencing by chemical degradation," in which terminally labeled DNA fragments were subjected to base-specific chemical cleavage; the reaction products were then separated by gel electrophoresis (Maxam and Gilbert, 1977). The method did not gain popularity due to its technical complexity and use of hazardous chemicals. The Sanger method, on the other hand, became popular because it was more efficient and used fewer toxic chemicals. A number of sequencing trials then initiated (with the help of Sanger sequencing), including sequencing of mycoplasma, *Escherichia coli*, *Caenorhabditis elegans*, and *Saccharomyces cerevisiae*. The Institute of Genomic Research (TIGR) was the first to publish the full genome sequence of a free-living bacterium (*Haemophilus influenzae*) using a shotgun approach (by random fragmentation without the need for mapping) in 1995. In 2001, one of the greatest discoveries of biological sciences, the sequence of the human genome, was published by two independent groups led by Craig Venter of Celera Genomics, who used a shotgun sequencing approach, and Francis Collins of the National Human Genome Research Institute (NIH), who used a BAC mapping-based method.

Sanger sequencing has been used to sequence DNA regions in question, but is limited to a few genes at a time. The complexity of emerging diseases and their association with genomic changes needs a much wider picture of the genome. The demands for sequencing a large number of human individuals and other organisms, and the limitations of Sanger sequencing to produce large amounts of data efficiently and rapidly, drove the development of high-throughput next generation sequencing that could parallelize the sequencing process, thus producing thousands or millions of sequences at once.

A timeline of next generation sequencing technologies (and their platforms) and when they were introduced is shown in Figure 19.1.

GENERATION OF SEQUENCING TECHNOLOGIES

Sequencing technologies did not see substantial changes for the first 30 years of development. It's only the last decade, with accompanying changes in sequencing technology, that can be defined as a change of sequencing generations. Briefly, "generation" refers to the chemistry and technology used by the sequencing process. The first generation of sequencing was obviously defined by the Sanger and Maxam–Gilbert techniques, which were able to sequence a few hundred base pairs at a time, and could be used for individual gene sequencing. Sequencing 3 billion base pairs of the human genome would take a very long time with Sanger sequencing, as it needs to sequence about 6 million fragments of DNA of 500 bp length.

After being served by Sanger sequencing for over 30 years, the scientific community was introduced to next generation sequencing technologies, also called second-generation sequencing. The major advancement from the first generation was that second generation sequencers have been able to parallelize the sequencing reaction in a massive

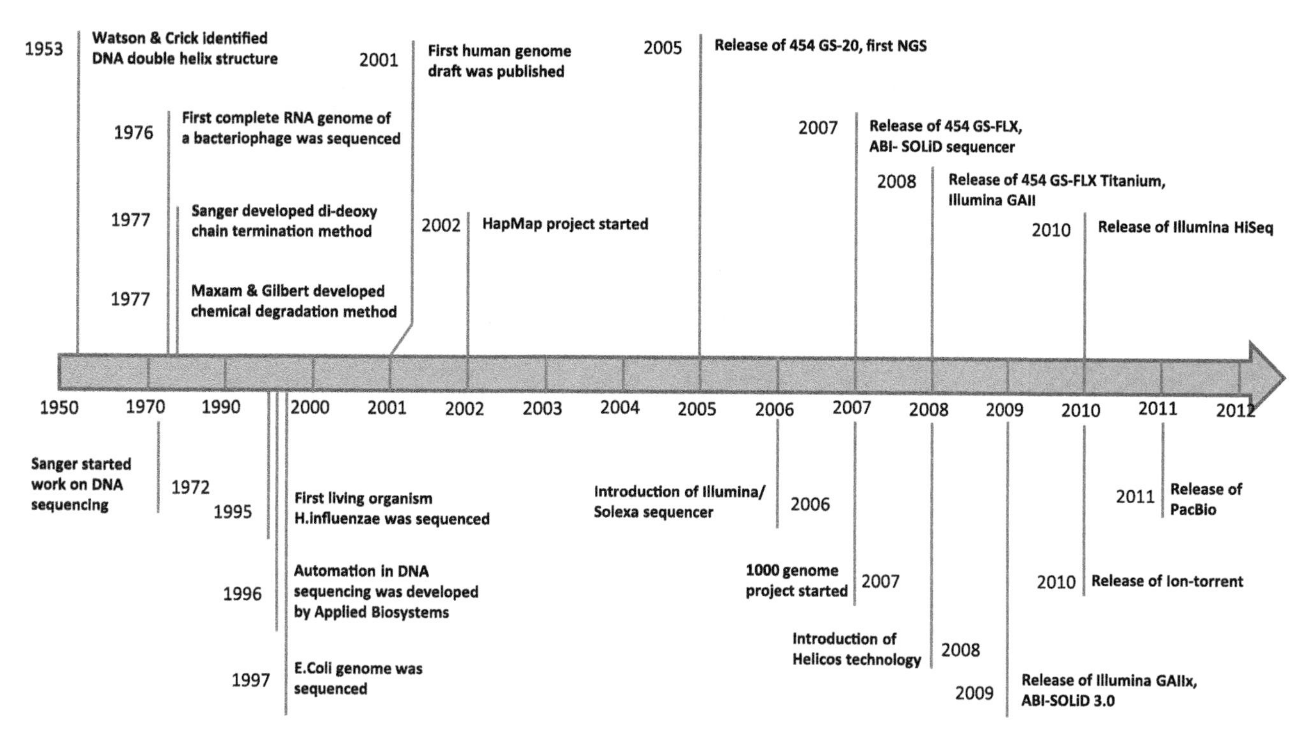

FIGURE 19.1 Timeline of introduction of DNA sequencing technologies and platforms.

manner, thus generating huge amounts of data very rapidly at a modest cost; this revolutionized the field of genomics. These include Roche 454, Illumina Solexa, and ABI-SOLiD technologies.

The term third generation sequencing (or next-next generation sequencing) is given to technologies that interrogate single molecules of DNA without amplifying them through PCR, thereby overcoming issues related to the biases introduced by PCR amplification and de-phasing. However, third generation sequencers were developed with a vision of making sequencing cheaper than second-generation sequencing. Third generation sequencers include PacBio and Helicos.

IonTorrent, however, may be placed in between second and third generation. Although it is not a single-molecule sequencing technique, what differentiates it from the second generation is that the sequencing is not based on fluorescence detection and is "post-light."

Complete Genomics, Oxford Nanopore and Plonator use different types of technologies and chemistries, not defined by third generation, and can be placed under other or fourth generation sequencing technologies.

PRINCIPLE OF SANGER SEQUENCING VS. NEXT GENERATION SEQUENCING

Sanger sequencing works on the principle of chain termination, whereby the growing DNA strand is terminated by the incorporation of dideoxy nucleotides (ddNTPs), A, T, G, and C, in different reactions. The fragments of different lengths of the amplicon are then run on a gel in different lanes to determine the sequence. In newer dye terminator sequencing, the ddNTPs are labeled with fluorescent dyes to make the fragments readable through a laser light source at unique wavelengths in capillary gels, rather than on traditional slab gels. Sequencers based on Sanger sequencing produce a read length (the length of a DNA fragment that can be sequenced at one time) of 800–1000 bp. Because at one time only one read can be sequenced in one capillary of the sequencer, the total output of the run is equal to the read length. However, sequencers with multiple capillaries enable one to sequence multiple samples at a time (e.g. 8, 16, 48, or 96, etc.).

On the other hand, next generation sequencers work on the principle of sequencing millions of DNA fragments simultaneously in a massively parallel mode to produce sequence data in megabases or gigabases. The whole genome or transcriptome of an organism is fragmented into millions of small pieces and sequenced independently in parallel. Next generation sequencing technologies lowered the cost and effort of DNA sequencing by producing huge amounts of data in massively parallel mode. It can be compared by looking at the cost per megabase (MB) of a DNA sequence generated in September 2001 (> 5,000 USD) with the same in July 2011 (< $0.10 USD). The first human genome, published in 2001, was a 13-year effort with Sanger technology and a multinational, multi-centric project. By comparison, the full genome of James Watson was sequenced over a 4-month period in 2008 with the help of next generation sequencing at the cost of less than $1 million USD.

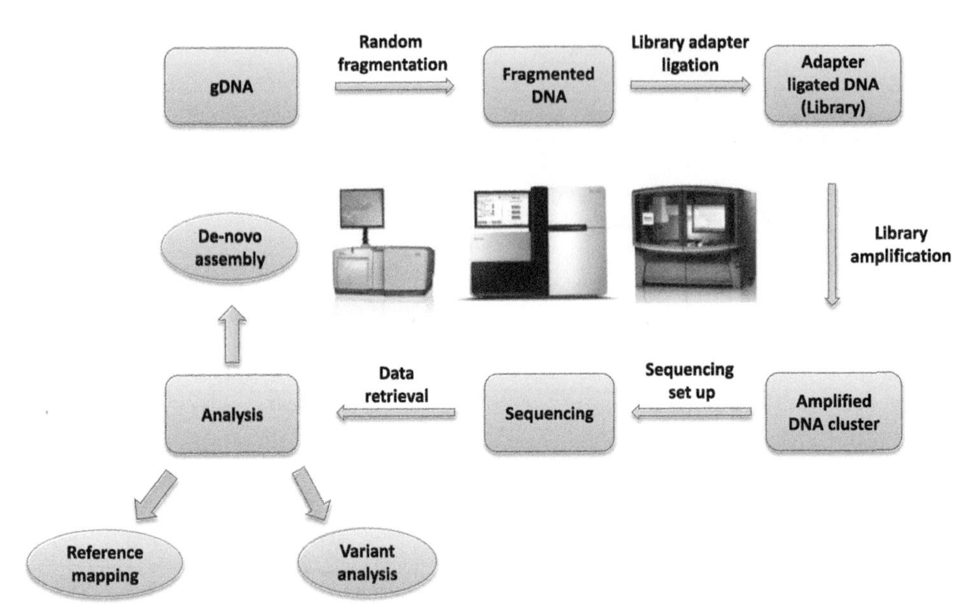

FIGURE 19.2 Basic scheme of a next generation sequencing experiment.

NEXT GENERATION/SECOND GENERATION TECHNOLOGIES

Next generation sequencing employs massively parallel sequencing of millions of DNA fragments simultaneously. This can be achieved by different technologies on different NGS platforms. Depending upon the technology used, they differ from each other in terms of read length, data produced, and data quality, and hence bioinformatics required to process and analyze that data. In this section is discussed important next generation sequencing technologies, their working principles, key features, and output data generated. The basic scheme of an NGS experiment is illustrated in Figure 19.2.

Pyrosequencing Technology

Roche-454 GS FLX: In 1993, Nyrén and his group (Nyrén et al., 1993) published a novel sequencing method based on chemiluminescent detection of pyrophosphate released during polymerase-mediated deoxynucleoside triphosphate (dNTP) incorporation, which was later commercialized by 454 Life Sciences with technical refinement (Ronaghi et al., 1996) and the use of emulsion-based PCR. Later, in 2007, Roche acquired 454. In pyrosequencing, DNA to be sequenced is fragmented and subjected to its complementary strand synthesis by DNA polymerase. As the polymerase incorporates a nucleotide in the growing chain, a pyrophosphate molecule is released. This pyrophosphate, through a series of enzymatic reactions, is converted into ATP. The ATP is used to enzymatically convert luciferin into oxyluciferin, which emits fluorescence that is recorded by the camera. By detecting this fluorescence, the incorporation of a nucleotide is confirmed. The identity of the incorporated nucleotide is known, as four dNTPs (dATP, dTTP, dCTP and dGTP) are introduced in the reaction separately in predefined cycles.

454 Sequencing uses a massively parallel pyrosequencing system capable of sequencing up to 1,000 bp of DNA in a 23-hour run on its new Genome Sequencer FLX Plus instrument. The technology works by fragmenting the DNA into approx 800–1,000 bp in length (nebulization), ligating adaptors to DNA fragments, making a library, and attaching the library to small DNA-capture beads. The beads are compartmentalized into water-in-oil emulsion microvesicles, often called micro-reactors, where clonal multiplication of single DNA molecules bound to the beads occurs during emulsion PCR. After amplification, the emulsion is disrupted, and the beads containing clonally amplified template DNA are enriched. Each amplified DNA-bound bead is placed into a tiny well on a PicoTiterPlate consisting of around 3.4 million wells. A mix of enzymes such as DNA polymerase, ATP sulfurylase, and luciferase, are also packed into the well. The PicoTiterPlate is then placed into the sequencer machine for sequencing. The GS FLX platform can produce data of approximately 450 MB with a 400–600 bp read length. The new platform GS FLX Plus is, however, able to generate approximately 700 MB of data with a read length of 700–1,000 bp.

Advantages: The advantage of 454 technology lies within its ability to sequence reads in the read length of 700–1000 bp. The longer read length is advantageous in terms of downstream bioinformatics, resulting in sequence assembly with longer contigs, higher N50 length, and less gaps, especially in *de novo* sequencing projects. Longer paired-end reads produced by the 454 platform also facilitate construction of better scaffolds.

Limitations: The 454 technology has certain limitations. The primary one is the difficulty in sequencing homopolymer repeats due to simultaneous incorporation of the same nucleotide, producing light that cannot be discriminated after a certain length (> 6 bp) with high accuracy (Mardis, 2008). Another disadvantage of the technology is the generation of relatively low bases/run (around 700 Mb) as compared to other NGS technologies. This makes it relatively expensive technology and not of priority if re-sequencing is desired with a high X depth.

Reversible Terminator Technology

Illumina Solexa: The technology of Illumina sequencing started with a concept by British scientists Shankar Balasubramanian and David Klenerman; it involved sequencing of single DNA molecules attached to microspheres. They founded Solexa in 1998, keeping single molecule sequencing in mind, but because of certain limitations, had to shift to sequencing clonally amplified DNA; the system was commercialized in 2006 as the Solexa Genome Analyzer (Voelkerding et al., 2009).

The Illumina Genome Analyzer uses flow cells consisting of optically transparent slides with eight individual lanes. Small oligonucleotide anchors are immobilized on the surfaces of these lanes. The template DNA to be sequenced is fragmented, phosphorylated at the 5′ end, and adenylated to add a single A at the 3′ end. Oligonucleotide adaptors are ligated to the DNA fragment, and the ligation is facilitated by the presence of a single T overhang on the adaptors. The adaptor-ligated oligonucleotides are complementary to the flow-cell anchors, and hence attach to the anchors. These DNA templates attached to the anchors are used to generate clusters of the same DNA fragment by amplification. A DNA fragment bends and hybridizes with its distal end to an adjacent anchor complementary to the distal end. On denaturation, both strands separate, and again bend and hybridize with their distal ends to adjacent anchors complementary to their distal ends. After multiple amplification cycles, a single DNA template makes a clonally amplified cluster with thousands of clonal molecules. Millions of clusters of different template molecules can be generated per flow cell.

For sequencing, the technology uses four fluorescently labeled nucleotides to sequence the tens of millions of clusters on the flow cell surface in parallel. In each growing chain, a single labeled deoxynucleoside triphosphate (dNTP) is added in each cycle. Due to the incorporation of the labeled nucleotide, DNA polymerization terminates, and the fluorescent dye is imaged to identify the incorporation. Then the label is enzymatically cleaved to allow incorporation of the next nucleotide (www.illumina.com).

Advantages: According to product information available from Illumina, in approximately 11–14 days, huge amounts of data are generated in the form of base pairs sequenced per run. From around 95 GB data coming out from nearly 150 bp long reads from both sides (2 x 150 bp) in the most widely cited platform Genome Analyzer IIx, the throughput has been significantly increased up to 600 GB data with 2 × 100 bp reads in newer versions of the platform (e.g. HiSeq2500 or HiSeq 2000), resulting in low cost per base. Their bench top platform MiSeq produces approximately 5 GB of data for 2 × 150 bp sequencing, or 8 GB data with 2 × 250 bp sequencing.

Large data and low cost per base renders the technology a good choice for many sequencing applications where large read length and *de novo* construction of a genome is not required (e.g. re-sequencing, ChIP sequencing, certain RNA sequencing projects, etc.).

Limitations: The major concern with the Illumina technology is that of de-phasing, which means different fragments in a cluster are sequenced with different phases; in other words, under- or over-incorporated nucleotides, because of block removal failure or other factors, result in fragments of varying lengths, which reduces precision in base calling at the 3′ ends of the fragments. De-phasing increases with increased read length. It is more common at sequences of invert repeats or GGC (Nakamura et al., 2011). Illumina technology produces reads of short length "micro-reads," hence assembly and downstream bioinformatics could be a challenge, especially for certain *de novo* sequencing. Longer run-time is also a limitation.

Sequencing by Ligation Technology

ABI SOLiD: Sequencing by ligation technology is marketed by Applied Biosystems, USA. The name SOLiD stands for Small Oligonucleotide Ligation and Detection System. This technology was developed by George Church in 2005, and was further improved and distributed by Applied Biosystems in 2007 (Voelkerding et al., 2009). The principle of this sequencing relies upon the ability of DNA ligase to detect and incorporate bases in a very specific manner. In sequencing by ligation, DNA fragments attached to beads are clonally amplified by emulsion PCR. After PCR, specific primers hybridize to the adaptor sequence of the amplified templates on the beads, which provides a free 5′ phosphate group for ligation to the fluorescently labeled probes (called interrogation probes) instead of providing a 3′ hydroxyl group as in normal polymerase-mediated extension. The interrogation probe is 8 bp in length, where the first two bases are specific, and the rest of the 6 bases are degenerate. A set of four fluorescently labeled interrogation probes, consisting of one of 16 possible 2-base combinations at the end (e.g. TT, GT, TC, CG, etc.), compete for ligation to the sequencing primer. Upon ligation, fluorescence is captured, which is corresponding to the probe ligated. For the second cycle, the "fluor" of the attached probe is removed and a

5′ phosphate group is regenerated. Multiple cycles of ligation, detection, and cleavage are performed, with the number of cycles determining the eventual read length. Following a series of ligation cycles (usually seven), the extension product is removed and the template is reset with a primer complementary to the n-1 position for a second round of ligation cycles. This process is repeated each time with a new primer with a successive offset (n-1, n-2, n-3, and so on). Thus the sequencing is divided into library preparation, emulsion PCR, bead deposition, sequencing, and primer reset. A 6–7-day long instrument run in a SOLiD 5500 system claims to generate sequence data at approximately 10–15 GB/day (total throughput 120–240 GB, 100 GB in the case of the SOLiD 4 system) with a read length of 75 bases (for mate-paired: 2 x 60 bp; for paired-end: 75 bp × 35 bp) and a sequence consensus accuracy of 99.99% (Voelkerding et al., 2009).

Advantages: The advantage of this technology is generation of sequencing data of comparatively higher accuracy than other sequencing methods. One of the reasons behind the high accuracy is sequencing with successive offset primer less by one bp so that each nucleotide of the template is sequenced twice; therefore, in order to miscall a SNP, two adjacent colors must be miscalled, which does not frequently happen.

Limitations: Among the limitations of this technology are that less data are output than with Illumina, and shorter read length, requiring close genome sequencing for mapping. Even the time taken for a whole run is about 6–7 day to complete, especially for bigger genomes.

THIRD GENERATION SEQUENCING TECHNOLOGIES

Single Molecule Real-Time Sequencing

Pacific Biosciences: The sequencing technology of Pacific Biosciences works on the Single Molecule Real-Time (SMRT) sequencing technology, and enables the observation of DNA synthesis as it occurs in real time. This is possible with the use of Zero Mode Waveguide (ZMW), specially designed micro-holes used. Around 75,000 ZMWs are fabricated per SMRT cell, enabling the potential detection of approximately 75,000 single molecule sequencing reactions in parallel.

A ZMW is a nano-hole made in a 100 nm metal film on a glass surface. Due to the small size of the ZMW, laser light of wavelength of approximately 600 nm cannot pass completely through the ZMW, and exponentially decays as it enters into it. Therefore, by applying a laser through the glass into the ZMW, only the bottom 30 nm of the ZMW become illuminated. A single DNA polymerase molecule is anchored to the bottom glass surface of the ZMW. Nucleotides, each labeled with a different colored fluorophore, are then flooded above the ZMWs. Labeled nucleotides travel down into the ZMW within microseconds, reaching the DNA polymerase, then diffuse back up and exit the hole. As laser light cannot penetrate up through the holes, it does not excite fluorescently labeled nucleotides present on the upper side of the holes. Thus the labeled nucleotides above the ZMWs are dark. Only when they diffuse through the bottom 30 nm of the ZMW do they fluoresce. Correct incorporation of nucleotide into the growing strand, results in higher signal intensity for incorporated versus unincorporated nucleotides, which are then detected. Thus, the a single nucleotide incorporation can be detected inside the ZMW.

Advantages: Pacific Bioscience sequencing technology, being a single molecule, real-time sequencing technology does not require the PCR amplification steps. This avoids the usual amplification bias in the sequenced DNA fragments. With its PacBio RS instrument, this technology generates read lengths of around 1,000–3,000 bp, with an average of 1,200 bp. Such a long read length tries to overcome the short read limitations of repeat handling and incomplete assembly, and facilitates mapping and assembly of the sequenced reads in a much better way. The time taken from sample preparation to sequencing results is also shorter, and can take less than one day. The technology has the ability to observe and capture kinetic information about polymerase activity.

Limitations: PacBio technology generates around 70 to 140 MB of data per SMRT cell, depending upon a GC content of the template DNA (www.pacificbiosciences.com) that is significantly less compared to the other technologies. However, due to significantly less data output, it is at present a struggling technology. However, short run times can overcome the problem, provided the cost remains limited for repeat runs for a desired depth. It also suffers from random insertion–deletion errors (Mardis, 2011).

True Single Molecule Sequencing (tSMS)

Helicos Biosciences: Helicos sequencing technology also operates on the principle of single molecule sequencing. By directly sequencing single molecules of DNA or RNA on the HeliScope platform, this technology, called Helicos' True Single Molecule Sequencing (tSMS™) and Direct RNA Sequencing (DRS™), increases the sequencing speed at a low cost. tSMS and DRS enable the simultaneous sequencing of large numbers of strands of single DNA or RNA molecules in which labeled bases are sequentially added to the nucleic acid templates captured on a flow cell.

Billions of single molecules of template DNA are captured on a specific proprietary surface. To this, polymerase and one of the fluorescently labeled nucleotides (C, G, A or T) is added, which is incorporated into the growing complementary strands on all the templates in a sequence-specific manner. After a wash step, the incorporated nucleotides are imaged and their positions recorded. In the next step, the fluorescent group is removed in a highly efficient cleavage process, leaving behind the incorporated nucleotide. This process continues through each of the other three bases in multiple cycles, providing a 25–55 bp read (average 35 bp) from each of those individual templates. From 600 million to 1 billion DNA strands, a total of 21 to 35 GB of sequence data is generated per run with 99.995% accuracy.

Advantages: Relying upon the single molecule sequencing principle and generating big data with low cost makes Helicos advantageous over other amplified sequencing techniques. Single molecule technology enables sequencing of RNA molecules from single cells directly without reverse transcription or amplification, without cDNA synthesis, and without PCR biases or amplification-induced errors.

Limitations: Generation of small read lengths compared to other next generation sequencing technologies could be a major limitation of Helicos technology. An important drawback to single molecule sequencing is a higher raw error rate. This can be overcome with repetitive sequencing, but increases the cost per base for a given accuracy rate.

Ion Semiconductor Sequencing

Ion Torrent: Ion Torrent technology works on the principle of detection of hydrogen ion release during incorporation of new nucleotides into the growing DNA template. In nature, when a nucleotide is incorporated into a strand of DNA by a polymerase, a hydrogen ion is released as a by-product. Ion Torrent, with its Ion Personal Genome Machine (PGM™) sequencer, uses a high-density array of micro-machined wells to perform nucleotide incorporation in a massively parallel manner. Each well holds a different DNA template. Beneath the wells is an ion-sensitive layer followed by a proprietary ion-sensor. The ion changes the pH of the solution, which is detected by an ion sensor. If there are two identical bases on the DNA strand, the output voltage is doubled, and the chip records two identical bases called without scanning, camera, and light. Instead of detecting light as in 454 pyrosequencing, Ion Torrent technology creates a direct connection between the chemical and digital events. Hydrogen ions are detected on ion-semiconductor sequencing chips. These ion semiconductor chips are designed and manufactured like any other semiconductor chips used in electronic devices. These are cut in the form of wafers from a silicon boule. The transistors and circuits are then pattern-transferred and subsequently etched onto the wafers using photolithography. This process is repeated 20 times or more creating a multi-layer system of circuits. Ion torrent PGM (Personal Genome Machine) generates a total data output of around 10–1,000 MB, depending upon the type of ion semiconductor sequencing chip used. However, in September 2012, Ion Torrent launched their bigger system, the Ion Proton. It uses larger chips with higher densities, and thus can be suitable for exome and whole-genome sequencing. Although Ion Proton is capable of generating much larger outputs (around 10 GB), it is substantially more expensive.

Advantages: Ion Torrent generates read lengths of around 200 bp, which are used to fill gaps in the assembly produced by other technologies. Although the read length is much shorter than Roche-454 and PacBio, due to lower costs, Ion Torrent can be a reasonable choice in some cases. The short run time of this technique also facilitates multiple runs for generation of more data in a given time.

Limitations: The read length of 200 bp lies in between short and long read length NGS technologies. Whereas short read technologies are facilitated by huge data generation, Ion lags behind in total data output. In this way, Ion Torrent has to prove itself as a standalone sequencing technique for *de novo* sequencing projects of big genomes.

OTHER/4TH GENERATION SEQUENCING TECHNOLOGIES

Oxford Nanopore, Polonator, and Complete Genomics sequencing can be placed under either “other” or “fourth generation” sequencing. These technologies use different types of sequencing chemistries, which are not satisfied by usual definitions of second or third generation. However, Polonator chemistry resembles, and in fact was a basis for development of, a second generation sequencing technique, ABI-SOLiD; its future idea of being a modular, cheap, and open system, makes it different from others, and puts it in the category of other/fourth generation technologies. While Complete Genomics is so far optimized only for human sequencing, these “other” technologies must prove their success in comparison to existing platforms.

Nanopore Sequencing

Oxford Nanopore: Nanopore sequencing is a method for DNA sequencing that has been under development since 1995. In February 2012, Oxford Nanopore Technologies, UK presented initial data of the technology. A nanopore is simply a small hole on the order of 1 nm in internal diameter that is made up of certain transmembrane cellular proteins. Nanopore sequencing works on the principle of minute changes in electric current across the nanopore immersed in a conducting fluid with voltage applied when a moving nucleotide (or DNA strand) passes through it. Every nucleotide on the DNA molecule, while passing through the nanopore, obstructs the nanopore to a different, characteristic degree, and the amount of change in current is characteristic for each different nucleotide. DNA may be forced to pass through the hole one base at a time, like thread through the eye of a needle. The change in the current can be directly read, and the sequence of the passing DNA can be determined by detecting changes in the current generated specific to the base passed. Alternatively, a specific nanopore can be designed to produce current changes when a specific nucleotide passes through it.

Although the technology has not been in routine use for sequencing, according to the scientists who developed it, it can sequence the entire 5.4-kilobase genome of the virus in one continuous read. However, it initially aims to deliver reads of 100 kilobases. That is still much longer than the fragments processed by other technologies. The initial system will have a feature to read DNA at a rate of hundreds of kilobases per second.

Advantages: The potential advantages of the nanopore system are that it could deliver real-time seqeuncing of single molecules at low cost; it is also expected to read very long DNA molecules in a single read. Nanopore technology is expected to provide sequencing at a very low cost, $25 to $40 per gigabase of sequence. That would mean a cost of just a few thousand dollars to sequence a human genome to the standard 30-fold coverage. Being a very fast sequencing process it could be a future choice of sequencing for nearly all applications.

Limitations: So far nanopore sequencing is still in the proof-of-concept stage and not yet commercially available, parallelized, or routinized. At this stage there are many concerns with this technology that have to be taken into account before its actual commercial use. One of these could be how to improve its resolution so that it can detect single bases with the rapid movement of DNA through the hole.

Polony-Based Sequencing Technology

Polonator Technology: This technology was developed by Dr. Church at Harvard Medical School. The vision behind the development of this technology was to keep costs down to provide an affordable system for the masses, with high quality components. It was designed to be modular, easily upgradable, and usable for a variety of applications. It was also imagined as an open system so that scientists could help with advances in hardware software, and applications.

Polonator sequencing is basically polony-based sequencing by ligation. Two flow cells are mounted within the instrument ; one undergoes the biochemistry while the other is imaged. Each flow cell has 18 individual wells, with a total of over 1 billion streptavidin-coated polystyrene beads. The reaction takes place with DNA on the surface of beads. Depending on library titration and technique, the typical output is 8 to 10 million mappable reads per lane, or about 150 million reads per dual flow cell run. With a read length of 26–28 bases, the run output is about 4 to 5 Gb. Alternatively, library beads can be enriched by removing the unamplified beads, increasing the output up to 8–10 Gb.

Advantages: The Polonator is supposed to be the least expensive sequencing technology commercially available. All software features will be freely downloadable, which will facilitate its optimization for use by others.

Limitations: Polonator technology generates a read length of only 26 bp (13 + 13 bp paired end). The duration of a run is currently about four days. In general, the best 92% of the mappable reads have a mean accuracy of above 98%.

DNA Nanoball Sequencing

Complete Genomics: There is another sequencing technology developed and commercialized by Complete Genomics, specifically optimized for human re-sequencing applications. The technology is based on a proprietary DNA array and ligation-based read technology, and claims to provide an accuracy of 99.999%, facilitating human sequencing service to scientific community.

Complete Genomics claims to deliver very accurate and low cost human sequencing, as it has brought together diverse technologies to create a comprehensive solution for large-scale sequencing of human genomes. There is a combination of technological advancements at each step of the workflow (i.e. libraries, arrays, sequencing assay, instruments, and software). They use in-house developed high-throughput sequencing instruments.

Complete genomics technology has two primary components: DNA nanoball, or DNB™ arrays and combinatorial probe-anchor ligation, or cPAL™, reads. DNA is packed efficiently on a silicon chip, making the patterned DNB arrays. A highly accurate cPAL read technology helps to read the DNA fragments using small concentrations of low-cost reagents. It is claimed that this unique combination of proprietary DNB and cPAL technologies is superior in both quality and cost to other commercially available approaches of sequencing a whole human genome at a very low consumable cost and consensus error rate. The technology of DNA Nanoball sequencing involves shearing DNA that is to be sequenced into small fragments and circularizing the fragments using an adaptor sequence. The circular fragments are replicated by a rolling circle mechanism, resulting in many single-stranded copies of each fragment. The DNA copies concatenate head to tail in a long strand, and form the compacted DNA nanoball. The nanoballs are then adsorbed onto a microarray flow-cell in a highly ordered pattern, allowing a very high density of DNA nanoballs to be sequenced. A number of fluorescent probes are then ligated to the DNA at specific nucleotide locations in the nanoball, starting unchained sequencing reactions. The color of the fluorescence at each interrogated position is recorded to detect the fluorescence, and a base call is determined. The read length generated is 70 bases per DNB, 35 from each end. It is claimed that the level and uniformity of base-call accuracy achieved gives a 35-base read that has equivalent mapping power of the longer reads from other methods.

Advantages: The technology is capable of packing a very high density of DNA nanoballs to be sequenced in an orderly manner, maximizing the number of reads per flow cell and a non-progressive cPAL, minimizing read error. The Complete Genomics technology gained special attention as it claims to provide unprecedented specificity and sensitivity of sequencing. It is advantageous for those seeking human genome sequencing for a larger number of samples at a comparable cost.

Limitations: The limitation of Complete Genomics technology lies within two things. First, it is specially optimized for humans only and is not open for other organisms. Secondly, a 35-bp read length is supposed to have few, though not all, limitations of a short read technology, such as problems in mapping for highly repetitive DNA. However, claiming extraordinary accuracy, this limitation may be somewhat minimized. The use of multiple rounds of PCR can also introduce PCR bias in the sequences.

A comparison of important next generation sequencing platforms is summarized in Table 19.1.

DOWNSTREAM BIOINFORMATICS

With the help of massively parallel sequencing, huge date is generated in the form of ATGC sequences of DNA. These sequences or reads have different lengths depending upon the technology used and samples to be sequenced. These reads need to be joined together or analyzed to generate information with biological significance. The reads are primarily handled downstream of next generation sequencing, primarily in two ways.

De Novo Assembly

Assembly is the process of using various computer programs to align multiple sequencing reads that are overlapping with one-another to reconstruct a long DNA fragment, may be the whole genome, as DNA sequencing cannot read whole genome in one go, but it reads small pieces of different length, depending on the technology used. *De novo* assembly refers to assembling reads to build a new sequence, where no existing sequence of the same or related organism/species is available. More and longer reads allow better sequence overlapping for easy assembly (Figure 19.3). Shorter sequences are faster to align, but they complicate the assembly as shorter reads are more difficult to use with repeats or near identical repeats. A number of free and commercial programs are available which can be used to assemble sequence data depending upon type of platform, read length, quality scores, and abundance or total generated data.

Mapping

Mapping pertains to assembling reads against an already existing backbone sequence of the same or related organism/species to build a sequence that is similar, but not necessarily identical, to the backbone sequence. Mapping is generally

TABLE 19.1 Comparison of Important Next Generation Sequencing Platforms

	Roche 454 GS FLX Plus	Illumina Solexa HiSeq2500	ABI SOLiD 5500xl	Ion Torrent	Polonator	Pacific Bio	Helicos	Oxford Nanopore
Sequencing methods	Pyrosequencing	Reversible Dye Terminators	Sequencing by ligation	H^+ Detection	Polony-based ligation	ZMW – Single molecule	Heliscope – Single molecule	Nanopore with DNA transistor
Read Lengths	700–1000 bp	2x100 bp	75 bp	200 bp	26–28 bp	Av 1200, upto 3 kb	25–55 (av 35 bp)	Not determined can be upto 100 kb
Sequencing run time	23 hrs	11–14 days	6–7 days	2 hrs	4 Days	Less than a day	Less than a day	Not known, can be fast
Data generated	700 MB / run	600 GB / run	10–15 GB/day	10–1000 MB (depends upon chip used)	4–5 GB	70–140 MB / cell	21–35 GB / run	Not known, but expected very large
Advantages	-Longer read length -Small data files	-Huge data	-High quality data	-Low cost -Very fast	-Least expensive -Modular	-Longer read than 454 -Fast	-Big data among SMS	-Very cheap
Concerns	-Less data -Homopolymer	-Short reads -Dephasing -Long run time	-Short reads -Long run time	-Less data -Small read	-Very short reads -Low accuracy	-Random indel errors	-Small reads - Higher raw error rate	-Practicability to be proven

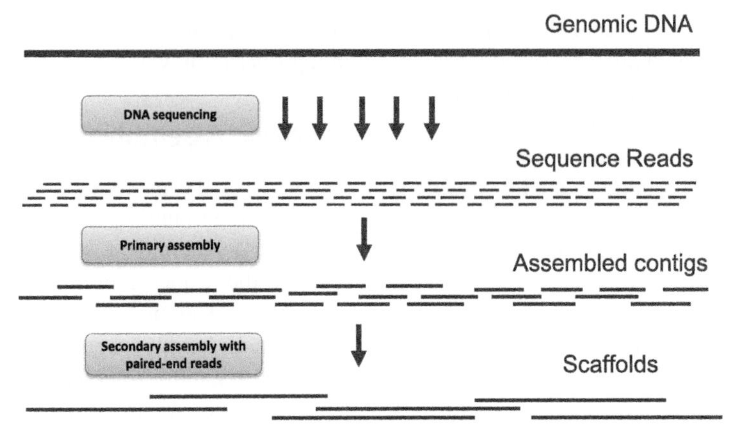

FIGURE 19.3 Schematic diagram of *de novo* sequence assembly using shotgun and paired-end read data.

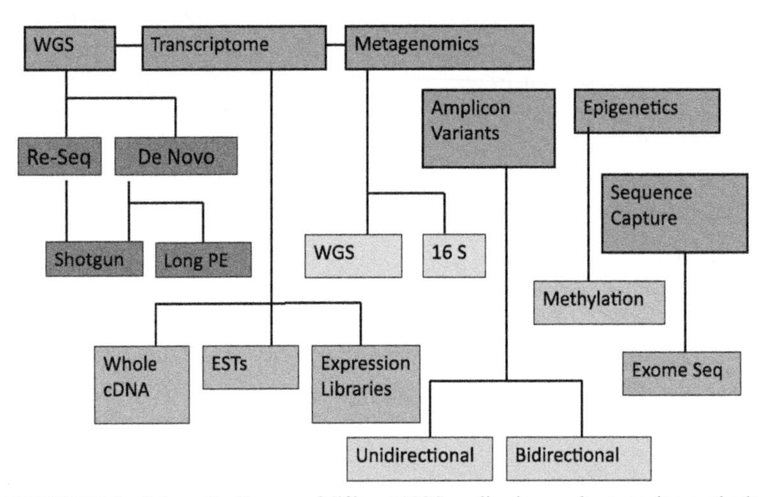

FIGURE 19.4 Schematic diagram of different NGS applications and sequencing methods.

used to figure out genomic variation in the sequenced DNA compared to the reference sequence, such as SNPs, insertions, deletions, and structural variations.

GENERAL PRINCIPLES OF NGS METHODS IN VARIOUS APPLICATIONS

Next generation sequencing can be used in a number of ways, depending upon the application. However, in every type of sequencing, the major changes occur in the sample processing and library preparation steps. Once the library is prepared, a particular sequencing platform uses the same chemistry to sequence the fragments. Major types of next generation sequencing applications and the sequencing methods used for them are shown in Figure 19.4, and an interconnected workflow for different protocols is shown in Flow Chart 19.1.

Whole Genome *De Novo* Sequencing

Before the first completely sequenced genome of a free living organism, *Hemophilus influenzae* in 1995, DNA sequencing was used to sequence individual genes, clones, DNA fragments, and engineered products. The first draft of a human genome was published in 2001 by two different Sanger sequencing approaches simultaneously. It took around five years to complete the sequencing and its assembly analysis. Since the arrival of massively parellel NGS, genomic research has been benefited from the success of sequencing new genomes for the first time (no reference genome already available) with much less effort and time. Whole genome sequencing is now the method of choice for genomic studies and a number of organisms, both eukaryotic and prokaryotic, have been sequenced at the whole genome level.

Sequence assembly refers to aligning and merging fragments of a much longer DNA sequence in order to reconstruct the original sequence. Whole genome *de novo* sequencing is basically sequencing fragments of DNA and aligning sequence reads or assembling them to make a full-length genome without referring to any previous information of available sequences for the same species. *De novo* sequencing is a challenging task and needs more sequence data and

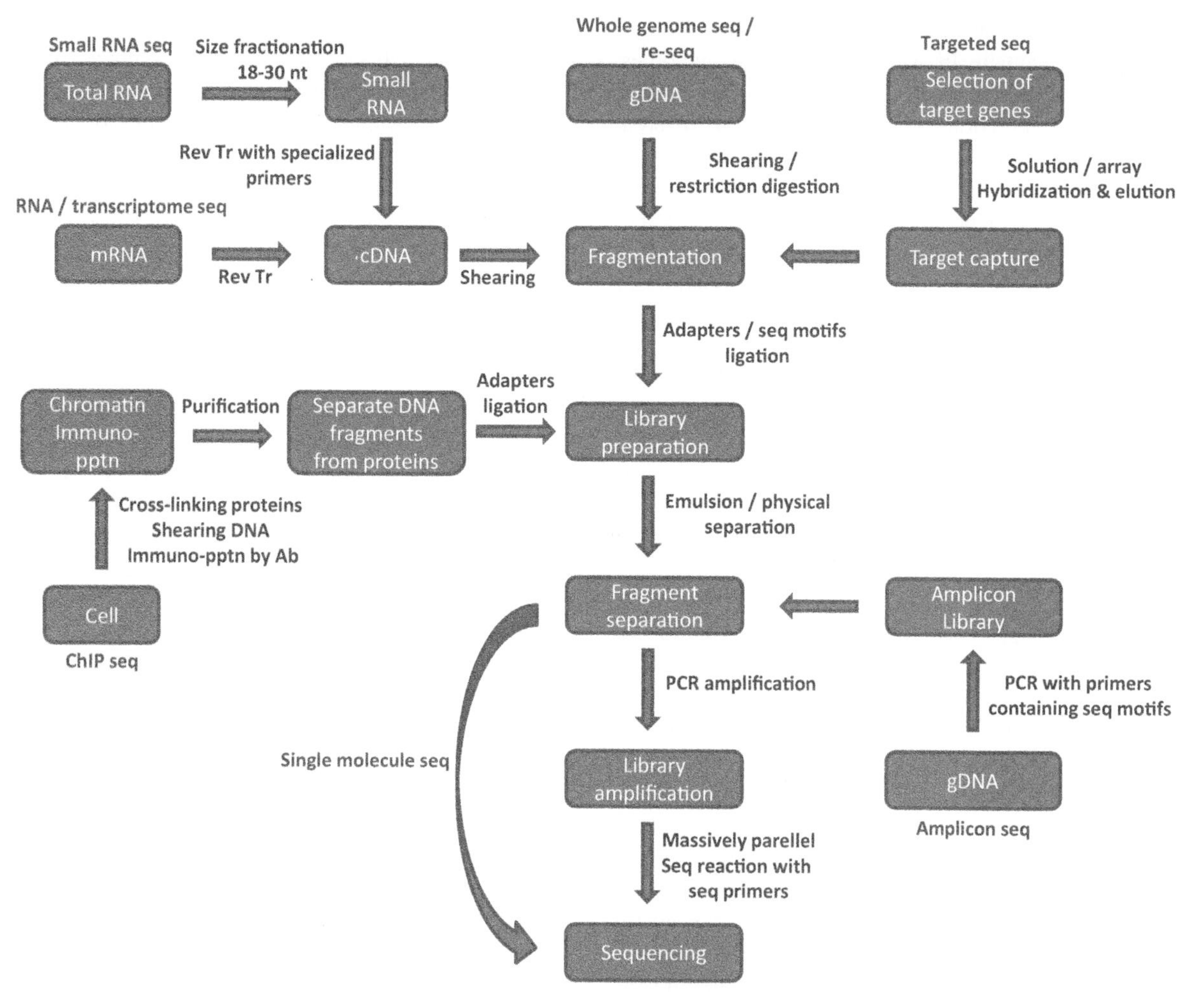

FLOW CHART 19.1 Scheme of different protocols used for different sequencing applications and their interconnections.

relatively longer read lengths so that reads can be aligned faster and a consensus sequence can easily be made. With shorter read lengths, an assembly is difficult to make due to the presence of non-matching ends and repetitive sequences; longer reads have the advantage of handling repetitive sequences more efficiently. However, even long reads cannot construct a full genome without gaps. Initial assembly of reads in the form of contigs can be further extended with the addition of paired-end read data (Figure 19.3). *De novo* sequencing assemblies face problems of handling terabytes of sequencing data, which is computer intensive and requires processing on computing clusters.

For *de novo* sequencing, the starting material is genomic DNA. DNA is fragmented, generally with a size in the range of the read length capability of the sequencing platform. Platform-specific sequencing motifs/adaptors are then attached to the fragments, purified, and quantified. DNA fragments with appropriate adaptors are collectively called libraries, the quality of which are one of the key determining factors for the success of sequencing. The library fragments are separated individually, amplified to produce several clones of the same fragment, and sent to the sequencing reactions in massively parellel format. Single molecule sequencing technologies are different in that library fragments are not amplified to avoid PCR bias, and are sequenced as single molecules (Flow Chart 19.1).

Whole Genome Re-Sequencing

With the help of next generation sequencing, reference genome sequences for many organisms are available in existing databases. Once the initial sequence for an organism's genome is available, it is possible to perform comparative sequencing or re-sequencing to identify polymorphisms, mutations, and structural variations between different or related organisms. Researchers are interested in identifying and characterizing sequence variations and understanding their biological consequences. Re-sequencing of genes or other genomic regions of interest is a key step in the detection of mutations associated with various phenotypes, strains, and diseases. Re-sequencing techniques can be used to test known mutations (genotyping) as well as to scan any mutation in a given region (variation analysis). Taking advantage of available complete genomes, researchers need to study genomic variations/alterations, such as insertions, deletions, SNPs, CNVs, etc., in different strains/varieties at different time points of growth, during drug stress, environmental stress, different phases of cell division, etc. For sequencing the whole genome of a species for which complete genome sequence information is already available, re-sequencing analysis is comparatively easy. The sequence reads generated can be easily mapped onto the sequences available in the database, and the variations can be mapped as well. For re-sequencing, platforms that can produce more data are more beneficial than those that generate longer reads with less data. The protocol for re-sequencing experiments is similar to that of *de novo* sequencing.

Targeted Re-Sequencing

Targeted re-sequencing refers to sequencing regions of interest in the genome rather than sequencing the whole genome. The most popular one is sequencing the exon part of the genome to get rid of unwanted introns. Targeted re-sequencing can be used to examine all of the exons in the genome, a part of the genome, specific gene families that are supposed to constitute known drug targets or regions and are thought to be involved in disease or pharmacogenetic effects (through genome-wide association studies). In the sequence enrichment step, there is a critical need for cost-effective solutions that can target specific genomic regions with high specificity and sensitivity, enabling the detection of both rare and common variants. Oligonucleotide microarrays and solution-based hybridization methods that have probes specific to targeted regions/exons have been used to capture target regions from a pool of genomic fragments (e.g. Agilent, Nimblegen, Illumina, etc.). Capture probes can be immobilized on a solid surface in the form of microarray probes, or can be used in solution. In microarray-based approaches, genomic DNA is fragmented and oligonucleotide linkers containing universal PCR priming sites are ligated to the fragments. The fragments with linkers are denatured, hybridized to an array, and eluted. The enriched DNA can be amplified by PCR before sequencing by next generation sequencers.

Solution-based technologies use oligonucleotides up to 170 bases in length, containing PCR priming sequences with a restriction enzyme recognition site. The oligonucleotide library is amplified by the PCR, digested with restriction enzymes, and ligated to adaptors containing the polymerase promoter site. *In vitro* transcription is performed to generate single-stranded, biotinylated cRNA capture sequences. The system utilizes 120-mer biotinylated cRNA to capture regions of interest, enriching them out of an NGS genomic fragment library (e.g. Agilent Technologies). Other groups have captured specific genomic regions by using other solution-based hybridization methods, such as molecular inversion probes (MIPs). In this technique the probes are single-stranded DNA containing sequences complementary to the target sequences in the genome. The probes hybridize to, and capture, the genomic targets. MIP probes are designed to have two genomic target complementary segments separated by a linker. When the probe hybridizes to the target, it undergoes an inversion and circularizes. The technology has been used extensively in the HapMap project for large-scale SNP genotyping, for detecting gene copy alterations, and to identify biomarkers for different diseases such as cancer.

An alternative target enrichment approach developed by Rain Dance Technologies uses a novel microfluidics technology. In this technology, PCR primer pairs for the targeted regions, genomic DNA, and PCR reagents are segregated individually in water in emulsion droplets. These droplets are merged and passed through an electrical impulse that causes them to coalesce. It generates PCR reactions at the rate of 10 million reactions per hour. The coalesced droplets are amplified by the PCR, and the amplicons, are pooled and processed for NGS.

Transcriptome Sequencing

Transcriptome pertains to RNA transcribed from the particular genome under investigation in a given condition at a particular time. The RNA content of a cell provides direct knowledge of gene regulation and protein content information. Transcriptome sequencing, also called RNA sequencing, refers to the sequencing of cDNA to get information about a sample's total RNA content at a given time in a given condition under study. This requires the conversion of mRNA into cDNA before the sequencing reaction. Different methods for in-depth characterization of transcriptomes and quantification of transcript levels have emerged as important tools for understanding cellular physiology and disease biology.

The technique has been benefited from next generation sequencing, which provides deep coverage at a base-level resolution. It provides comprehensive RNA sequencing with the information on differential expression of genes, differently spliced transcripts and gene alleles, post-transcriptional modifications, non-coding RNAs, alternative splicing, single-nucleotide polymorphism (SNP), and gene fusions. It is now possible to detect and identify nearly every transcribed molecule from short microRNA to long 5′- and 3′-untranslated regions, and even full-length mRNAs. Transcriptome sequencing has been used to discover novel gene sequences in transcriptomes. Other uses include identification of cancer-related Single Nucleotide Polymorphisms (SNPs) in transcribed sequences in human cancer transcriptomes, and confirmation of predicted genes in bacteria (Mane et al., 2009, and references therein).

Transcriptome sequencing has begun to be utilized in various clinical diagnostic applications. Where current methods require RNA to be converted to cDNA by reverse transcription prior to sequencing, and has been shown to introduce biases and artifacts that may interfere with proper characterization and quantification of transcripts, Helicos developed Direct RNA Sequencing (DRS™), which enabled a virtually unbiased view and quantification of transcriptomes. DRS can sequence RNA molecules directly in a massively parallel manner without RNA conversion to cDNA or other biasing sample manipulations such as ligation and amplification.

Amplicon Sequencing

The sequencing of PCR products is used to analyze genetic variations among different biological species and strains. Next generation sequencing, being a massively parallel sequencing technique, has facilitated ultra-deep sequencing of amplicons, making possible identification of rare variants present in very small frequencies. Amplicons are generated by PCR with primers already containing appropriate sequencing motifs, and thus amplicons themselves are the sequencing library. As in amplicon sequencing, where each amplified molecule within a mixture of amplicons can be sequenced individually, the technology is capable of detecting rare variants with detection limits of 0.5% and even lower. Amplicon sequencing is suitable for studying highly variant samples such as the hypervariable regions of antibodies, as well as samples with low variance, such as population samples for SNP detection. It is used in various projects like population diversity screenings within different individuals, identification of rare mutations/SNP-associated with various diseases, metagenome analysis (with 16S or 18S rDNA sequencing), the study of methylation patterns, etc.

ChIP DNA Sequencing

ChIP sequencing is used to analyze DNA–protein interactions within the cell. It combines chromatin immunoprecipitation (ChIP) with massively parallel DNA sequencing to identify the binding sites of DNA-associated proteins. The protocol involves chromatin immunoprecipitation to isolate DNA bound to proteins, followed by separation of DNA from proteins by purification. Purified DNA is then processed for sequencing, similar to *de novo* sequencing, but optimizes for low quantities of DNA. It is used primarily to determine how transcription factors and other DNA-binding *trans*-acting proteins bind to DNA, and how they influence cell function. Determining how proteins interact with DNA to regulate gene expression, DNA structure, and other cellular activities is essential for fully understanding many biological processes and disease states. ChIP-seq technology is currently seen primarily as an alternative to the ChIP-on-chip technique, which requires a hybridization microarray. ChIP-on-chip has been the most commonly used technique so far, and has been utilized to study these protein–DNA relationships. ChIP-sequencing can be used to precisely map global binding sites for a particular protein of interest.

Small RNA Sequencing

Small RNAs are typically 18–40 nucleotides in length, and have been shown to play a critical role in regulating gene expression in many organisms, affecting developmental timing, cell fate, tumor progression, neurogenesis, and other key cellular processes. Animals, plants, and fungi contain several distinct classes of small RNA, including microRNA (miRNA), short interfering RNA (siRNA), piwi-interacting RNA (piRNA), etc. The investigation and quantification of small RNAs in a high-throughput manner may provide valuable insight into the mechanisms of gene regulation and the involvement of non-coding RNA in important diseases such as cancer. With the enormous depth reached by next generation sequencing technologies, comprehensive information of the small RNAs of a cell can be obtained in various clinically important samples. It allows researchers to examine tissue-specific gene expression patterns, disease associations, miRNAs isoforms, and to discover previously uncharacterized small RNAs that have critical roles in the cell. Small RNA sequencing involves isolation of small RNA by size fractionation techniques followed by cDNA preparation with specialized primers capable of converting such small molecules into cDNA. cDNA is then processed for next generation sequencing.

ANIMAL BIOTECHNOLOGY AND THE CATTLE GENOME

Animal biotechnology has a long history, from traditional breeding techniques dating back to 5,000 years B.C. These techniques involve crossing different varieties of animals to produce greater genetic variety, and later selective breeding in the offspring to produce the maximum number of desirable traits. Today animal biotechnology is based on the concepts of genetic engineering, and has been facilitated by high-throughput technologies such as high-density DNA chips and next generation sequencing.

Two independent research groups simultaneously completed sequencing the human genome in 2001. This was one of the remarkable achievements of biological science and molecular biology. Human genome sequencing opened the gates for genome sequencing of other mammalian species. The decreasing cost and time of genome sequencing at a rapid rate due to technological and methodological developments enabled researchers to plan sequencing of other important organisms at a genome-wide level. In this continuation, an assembled draft of the bovine genome sequence was released in 2007, and later in 2009, the complete sequence of the bovine genome was published (Elsik et al., 2009). The complete sequence of the cattle genome resulted in the discovery of a large number of genetic variants in the form of SNPs, and was used as a powerful tool for identifying small genetic variations associated with livestock. Following this, low coverage sequencing of animals representing six other breeds of the cattle was also done to identify about 100,000 genetic polymorphisms (SNP) between individuals. A lot of sequencing data was generated over the next few years by livestock research scientists for full-length cDNAs, Expressed Sequence Tags (ESTs), highly specific bacterial artificial chromosome (BAC) end sequence markers, dense genetic linkage maps, physical maps, and cell-based resources (CSIRO).

The success story of large and complex genome sequencing added the chapter of draft sequencing of the giant panda genome, solely from paired-end next generation sequence reads, which proved suitability of using short sequence reads for the assembly of large and complex eukaryotic genomes (Li et al., 2010). The emergence of next generation sequencing also prompted the International Sheep Genomics Consortium (ISGC) to plan a Sheep Reference Genome Project in 2009. Next generation sequencing has enabled the scientific community to form various genome sequence consortiums to sequence reference genomes of important animals useful for humans.

APPLICATIONS OF NGS IN ANIMAL BIOTECHNOLOGY

Next generation sequencing has potential in every sub-field of animal biotechnology, from revolutionizing traditional animal breeding applications to the identification of very small genomic variations. Recent molecular biological approaches have benefited from the development of next generation sequencing technologies, and are supposed to help in animal research as well as translational applications (Figure 19.5).

Evolutionary Research

Evolutionary biology is a field concerned with the study of the evolutionary processes that have given rise to the diversity of life on Earth. Understanding the mechanisms of how organisms adapt to changing environments is a central topic in modern evolutionary ecology. Many fundamental questions that are important to our understanding of adaptation genetics remain unanswered because of a lack of sufficient information about molecular events that occur during the process of evolution. Recently, high-throughput technologies facilitated the study of evolutionary biology with the

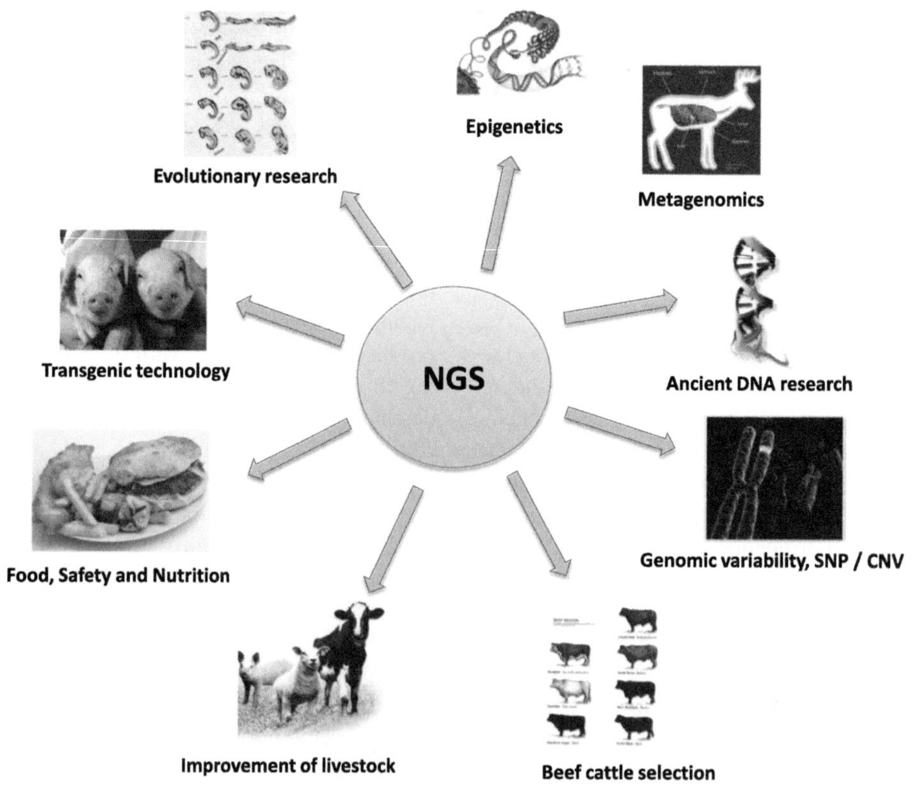

FIGURE 19.5 Different applications of next generation sequencing in animal biotechnology.

development of next generation sequencing. NGS technologies make it easier and more possible to identify genetic loci responsible for adaptation. NGS offers the opportunity to perform genomics studies on many additional ecologically interesting species, unlike earlier techniques where there was a requirement of a closely related genetic model organism to link evolutionary data with molecular events. NGS helps to find loci of small effect, because genotyping is becoming faster and cheaper, it is possible to carry out improved mapping studies with more individuals and markers (Stapley et al., 2010). NGS provides a cost effective way to perform preliminary genome-based analyses, and allows examination of some fundamental developmental and evolutionary processes of a species in the absence of a closely related genome (Subramanian et al., 2010).

Epigenetics

Epigenetics is the study of heritable changes in gene expression or cellular phenotype caused by mechanisms other than changes in the underlying DNA sequence. It is a field of rising importance in genetics and genomics. In other terms, it is the study of epigenetic factors that influence molecular properties by binding to specific DNA sequences. Examples of such changes are DNA methylation and histone modification, both of which are known to regulate gene expression without altering the associated DNA sequence. Some examples of these modifications include imprinting, X-inactivation, gene silencing, and embryonic reprogramming (Sellner et al., 2007). Epigenetic effects such as methylation involve the addition of methyl groups to certain cytosine residues, and if this occurs in the promoters of genes, transcription machinery can be blocked from binding to the DNA. The highly methylated regions in DNA tend to be less transcriptionally active, which may have an impact on the phenotypes of animals.

The study of epigenetics is important for animal breeding because it may help in finding part of the missing causality and missing heritability of complex traits and diseases. It impacts many economically important traits, from growth and development to more efficient reproduction and breeding strategies. Methylated DNA patterns are modified along the life of an individual by environmental forces, and some environments are more likely to increase certain methylation patterns; these patterns would contribute to the phenotypic variation between individuals. Furthermore, the environment may affect the methylation pattern of three generations cohabiting under the same specific circumstances at a given time during pregnancy: the productive female, the fetus, and the fetus' germ cells. Hence, what happens to an animal during its lifetime may have consequences in future generations (Gonzalez-Recio, 2011). Tissue-specific epigenetic studies in bovine genomes have the potential to select or induce favorable effects, and this information can be included in further breeding programs.

Recently, the identification of methylated DNA has become possible using some high-throughput technologies based on next generation sequencing, which provide information on DNA methylation on a genome-wide basis. The next generation sequencing technologies can easily elucidate whether nucleotides are methylated, which allows rapid understanding of these effects and their influence on phenotypes in beef cattle.

Metagenome Sequencing

Metagenomics is the study of mixed genomes. It is the study of genetic material recovered directly from samples, where a mixed population of various microorganisms is expected to be present. Traditionally, metagenome sequencing relies upon cultivating the microbial cultures from test samples and sequencing of specific cloned genes, often the 16S rRNA gene, to identify the microorganism present in the natural sample and to produce a profile of diversity. In the culturing process, a number of microorganisms usually missed out, either due to their non-culturable nature or due to overgrowing co-cultures. Furthermore, amplification of the 16S rRNA gene from a natural sample also misses out on representation of many organisms. It is now facilitated by massively parallel next generation sequencing technologies. Because of its ability to reveal the previously hidden diversity of microscopic life, metagenomics by these high-throughput techniques offers a powerful tool to see the microbial world in a mixed microbial habitat.

A novel application of this technology that is becoming increasingly important and of significant concern in animals (especially livestock) is the sequencing of gut microbiomes. Most work in the past has been done in humans and mice and has focused on profiling the 16S ribosomal RNA (rRNA) gene to identify the microbes present in the gut using traditional Sanger sequencing. However, metagenomic research has recently been accelerated due to next generation sequencing technologies.

Animals harbor several flora in their gut, which interact with the host and affect many biological processes by utilizing and converting a number of important nutrients. The study of gut microbiomes and their interactions with the genotype of the host is important because a substantial genetic diversity in the species present within the gut microbiome is expected. The gut microorganisms are supposed to have a profound impact on energy consumption/generation and utilization of important nutrients from food material.

Ancient DNA

In less than a year from its development, next generation sequencing increased the amount of DNA sequence data from extinct organisms by several orders of magnitude, which provided information to understand the origins of species. Although newer techniques aided in studying ancient DNA by increasing sensitivity and specificity, next generation sequencing has revolutionized ancient DNA research to a great extent.

Ancient DNA is usually highly fragmented with average fragment lengths ranging from 51.3 bp for some Neanderthal DNA to 142 and 164 bp, respectively, for DNA from permafrost mammoth hair. More than 20 studies have already made use of NGS to obtain sequence data from ancient remains (Knapp and Hofreiter et al., 2010 and references therein). Ancient DNA isolated from fossils usually contains very different levels of contamination and is difficult to sequence. However, large data generated by the various NGS technologies together with the short read length makes them an ideal tool for ancient DNA research. This is evidenced by the massive increase of available sequence data from long-dead organisms (since the invention of NGS) (Knapp and Hofreiter, 2010). Due to the small amount of endogenous DNA and high background contamination, shotgun sequencing of ancient DNA is not of much use; however, targeted sequencing with different capture methods is promising. Next generation-sequencing techniques have helped ancient DNA research, and thus the understanding of the evolution of organisms.

Genomic Variability and SNP/CNV Discovery

Once the genome of a species is sequenced, it facilitates identification of genomic variation within different individuals by sequencing and comparing the data of individual genomes with already available reference genomes. After successful alignment of the fragments of one or more individuals to a reference genome, different single nucleotide polymorphisms (SNPs) are identified and individuals are assigned a genotype depending upon the genomic variation they harbor. SNPs may be associated with diseases (or susceptibility to diseases) depending upon their position in the essential genes. Massively parallel sequencing strategies generate large amounts of sequencing data, produce a high depth of sequenced fragments, and help find SNPs present, even in low frequencies.

Copy number variations (CNVs) are gains and losses of a genomic DNA sequence, usually > 50 bp between two individuals of a species (Mills et al., 2011). They are alterations of the DNA of a genome that result in the cell having an abnormal number of copies of one or more sections of the DNA. Since they cover a bigger portion of the genomic sequence, they affect a wide range of phenotypic traits than SNPs, which are more frequent in the genome. CNVs thus have potentially greater effects on the phenotype of the organism than SNPs. CNVs can affect the gene structure and dosage, which may result in altered gene regulation.

Several common CNVs have been identified and shown to play important roles in normal phenotypic variability and disease susceptibility in humans and other higher organisms where they are known to be associated with diseases such as autism, schizophrenia, neuroblastoma, Crohn's disease, and severe obesity-related disorders. Recently, interest in CNV detection has extended into domesticated animals (Bikhart et al., 2012 and references therein). CNVs have been discovered by cytogenetic techniques such as fluorescent *in situ* hybridization, comparative genomic hybridization, array comparative genomic hybridization, and by SNP arrays. The microarray-based methods limit their use as only a relative copy number (CN) increase or decrease can be reported with respect to the individual reference genome. Recent advances in next generation sequencing of DNA have further enabled the systematic identification of CNVs at a higher resolution and sensitivity.

CNVs can be passed to the next generation of animals, have higher rates of accumulating mutations, and may be associated with animal health under recent selection. *Bos taurus indicus* are better adapted to warm climates and demonstrate superior resistance to tick infestation than *Bos taurus taurus* breeds due to copy number variation in associated resistance genes in their genomes (Porto-Neto et al., 2011). Similarly, milk production traits, along with other important phenotypic traits, show distinctive patterns due to CNVs in beef and dairy cattle breeds (Bikhart et al., 2012 and references therein).

Beef Cattle Selection

Beef cattle are raised for meat production (as compared to dairy cattle, which are used for milk production). Traditionally, marker-assisted selection is used for the accurate selection of specific DNA variations that have been associated with a measurable difference or effect on complex traits. Recent advancements in sequencing and genotyping technologies have enabled a rapid evolution in methods for beef cattle selection from restriction fragment length polymorphism (RFLP) markers that were low-throughput and time-consuming to the new high-density single nucleotide polymorphism (SNP) assays and next generation sequencing; in comparison, marker genotypes are easily and inexpensively generated. With the rapid development of molecular technologies, new tools have become available for beef producers to efficiently produce high quality beef for today's consumer. Technologies such as next generation sequencing help to shorten the generation interval, to identify causal mutations, and to provide information on gene expression; this strengthens our understanding of epigenetic changes and the effect of gut microbiomes on cattle phenotypes.

Rapid, accurate, and relatively low cost sequencing of genomes of individual animals has the potential to revolutionize selection in beef cattle. Massively parallel sequencing data provide information about novel as well as known polymorphisms within an individual. The discovery of mutations that actually cause variation within traits will become increasingly important, and their knowledge will allow testing across breeds, which will drastically reduce the number of loci that need to be tested to explain variations within a trait (Rolf et al., 2010).

Animal Breeding and Improvement of Livestock Productivity and Health

Animal breeding is a branch of animal science that addresses the evaluation of genetic value in terms of estimated breeding value (EBV) of domestic livestock. Animals have been selected for breeding with superior EBVs in growth rate, and egg, meat, milk, or wool production, as well as other important desirable traits. This has revolutionized agricultural livestock production throughout the world. Traditional breeding has been done with a lack of molecular information about the genes actually responsible for quantitative traits. The efficiency of these traditional methods remains limited in the case of traits that have low heritability (and cannot be correctly measured in a large number of animals), such as meat quality, internal parasite resistance, slow genetic progress, etc. (Eggen et al., 2012). NGS is transforming animal breeding due to the low cost involved in whole-genome sequencing studies that facilitate identification of genetic markers at a genome-wide level. Next generation sequencing has revolutionized the planning and implementation of livestock breeding programs. For the livestock industry, these high-throughput technologies are expected to increase the efficiency and productivity of animal breeding, whereas for consumers it is supposed to enhance security and the quality of animal products (Eggen et al., 2012).

Food, Safety, and Nutrition

The safety of food produced through animals for human consumption remains a key issue among food and nutrition biologists. The food produced through animal biotechnology includes products with genetic modifications, and there is concern about their effect on human health. The main areas of concern are allergens, bioactivity, and the toxicity of unintended expression products. As the food allergens introduced from the expression of genetically modified proteins from genetically modified animals is a big concern, the difficulty is how to accurately anticipate these before human consumption.

Next generation sequencing strategies, being more economically feasible with real-time results, can be directly applied to improve poultry production and enhance food safety. By determination of the gut microbes, genes involved in metabolic pathways, the presence of plasmids, and screening for functions such as antibiotic resistance or nutrient production can be carried out. These gut microbial flora of poultry and animals can be sequenced to determine the effect on human health and diseases (D'Souza and Hanning, 2012).

Transgenics

Transgenic animals are used as experimental models to perform phenotyping, and for testing in biomedical research. They are important model systems for establishing the mutational fingerprint for various human diseases, including cancer. Genetically modified animals are becoming more important to the discovery and development of treatments for many serious diseases. Transgenic mice are often used to study cellular and tissue-specific responses to disease. By modifying the DNA sequence or transferring DNA to an animal, certain proteins can be developed that can be used in medical treatment. Stable expression systems of human proteins have been developed in many animals, including sheep, pigs, and rats, for their commercial production. However, the mutation detection assays by conventional DNA sequencing analysis used for these transgenic systems only allow low-throughput detection of mutational fingerprints in phenotypically expressed individual mutants, which is costly, time consuming, and extremely laborious (Besaratinia et al., 2012).

APPLICATIONS OF NGS IN HUMAN HEALTH

Next generation sequencing has changed the way of observing human disease mechanisms, and has impacted both basic and clinical research. At one end, the basic research sector is fairly driven by either direct use of NGS to sequence novel variations, or by the information generated through NGS, which is used with traditional experiments. Clinical research involves high-throughput genetic testing with higher resolution and clinically relevant genetic follow-ups. A few key areas where NGS has made a significant impact are summarized in the following sections.

Cancer Research

The area of cancer research has been especially revolutionized by next generation sequencing. As cancer research has traditionally been complicated by the fact that there is no clear-cut mechanism for all types of cancers, we need to analyze a large number of genetic variations in the human genome that can be associated with cancer phenotypes. A large number of cancer cases (as well as healthy individuals) need to be studied for a comparison of their genetic make-ups with several genetic targets. With the introduction of NGS, this area has primarily been benefited, as several genomes can be sequenced simultaneously within few weeks. Targeted DNA sequencing, especially exome sequencing, allows even higher throughput at a reduced cost per sample. This has facilitated analyzing genomic variations, such as millions of SNPs to be associated with a particular phenotype in genome-wide association studies (GWAS). Worldwide collaborations for cataloging mutations in multiple cancer types are underway. In April 2008, leading teams of international scientists constituted the International Cancer Genome Consortium (ICGC) with the aim of generating high-quality genomic data from 53 tumor types over the next decade, which will help in new discoveries related to diagnostics, prognostics, and therapeutics.

Genetic Disorders

Recent efforts have demonstrated a significant opportunity for the use of NGS in the diagnosis of genetic disorders, making NGS a practical, attractive, and economically feasible technology for clinical applications. NGS has significantly improved the identification of disease-associated mutations and genetic alterations. NGS facilitates researchers with the required capacity to analyze large panels of genes for each individual. In the last few years, technological advancements in NGS, especially target enrichment methods, have resulted in the identification of variations responsible for more than 40 rare disorders. These include numerous diseases such as Schinzel–Giedion Syndrome, Sensenbrenner Syndrome, Neonatal Diabetes Mellitus, Miller Syndrome, Kabuki Syndrome, Fowler Syndrome, Hereditary Deafness, Parkinson's disease, etc. Identification of genetic causative agents in mental disorders has also been improved by the use of NGS, such as identification of mutations responsible for hyperphosphatasia mental retardation syndrome.

Human Microbiome

Microbial organisms have a close association with the human body, both beneficial and harmful. These organisms are primarily bacteria, but also include Archaea, yeasts, single-celled eukaryotes, helminths, and viruses. Many of these organisms have not been successfully cultured, identified, or otherwise characterized. A few important sites of micobial colonization in the human body include the colon, stomach, vagina, skin, esophagous, hair, nose, and mouth (oral cavity). In 2008, the Human Microbiome Project (HMP) was initiated by the United States National Institutes of Health to identify and characterize the microorganisms found in association with both healthy and diseased individuals.

Next generation sequencing technologies have become a valuable tool in the study and analysis of microbial communities in diverse environments, including the human body. It is useful in identifying microbial colonization, polymicrobial infections, and microbial communities with better resolution. With the help of NGS, one can identify new species of colonized microbes through a metagenome sequencing approach, either by employing 16S rRNA gene amplicon sequencing from a mixed population, or by a whole-genome sequencing approach. In the latter case, NGS technologies producing a long read have an advantage, as it is better assemble longer reads to form a suitable genome from reads of diverse organisms.

Pre- and Post-Natal Diagnosis

One of the valuable applications of next generation sequencing technology is molecular genetic testing in pre- and post-natal diagnostics. Traditionally, invasive methods have been used to draw samples and detect chromosomal abnormalities; these typically come with a high risk for both mother and fetus, but can provide definite genetic information about the fetus. The NGS-based detection of fetal aneuploidy in high-risk pregnancies is promising, and has successfully been used for the detection of chromosomal aneuploidy in fetal DNA from cell-free DNA fragments in maternal plasma (Fan et al., 2008; Ashoor et al., 2012). The advantage of NGS over traditional pre-natal diagnosis is that it allows non-invasive testing of fetal abnormalities with high sensitivity and specificity.

Infectious Diseases

Because of the constantly decreasing costs of DNA sequencing, next generation technologies are continuously becoming an integral part of genetic and infectious disease research and diagnostics. They have primarily been used in viral research for ultra-deep whole viral genome sequencing for influenza viruses, and for the detection of human immunodeficiency virus (HIV) genome variability and evolution within the host, and human hepatitis C virus quasispecies. NGS has also been used for monitoring antiviral drug-resistant mutations.

NGS has emerged as an extremely powerful tool for studying bacterial genomics, viral dynamics, host–response, and other aspects in infectious disease biology that were previously inaccessible. Due to the capability of NGS to sequence individual molecules in a massively parallel fashion, it is possible to segregate and sequence individual genomes, which is useful in studying the evolution of infectious agents through the history, monitoring emergence of drug resistance, and discovering novel viruses in a population.

Personalized Medicine

Every individual is different in their genetic make-up. Therefore, its susceptibility to different diseases, infections, and disorders are also different. Moreover, every individual needs different treatment and management of illness. One should know what his/her genetic make-up is in order to decide accurate and proper care (i.e. know more about one's genes and their impact). Our personal genome can provide information about increased risks for a selection of hereditary diseases. Next generation sequencing technologies can provide information on all of the different types of disease-causing alterations in individuals in the short turn-around time required to screen patients for either clinical trials, or for diagnostics in clinical settings. It is beneficial in identifying and developing panels for biomarkers, which are individual genes associated with a particular type of health condition.

FUTURE PERSPECTIVES

Next generation sequencing, being relatively inexpensive technology in view of generating sequence information at the genome-wide level, has revolutionized nearly each and every area of biotechnology. Traditional applications in medical biotechnology such as novel pathogen identification, biomarker discovery, SNP association with diseases, stem cell biology, etc., and of animal biotechnology like transgenics, production of probiotics and prebiotics, enzyme technology, animal cloning, gamete and embryo production, artificial insemination, and gene–gene and gene–environment interactions related to environmental conditions, could be studied quantitatively using modern bioinformatics tools.

In the future, sequencing of individual genomes of interest under different living, nutritional, or treatment conditions will benefit the medical and agricultural communities by providing guidance for disease control and prevention. Sequencing the genome of animals is supposed to enable scientists to more accurately identify the genetic markers useful in economically important traits. While the medical community is supposed to be benefited from the information generated by NGS in terms of better diagnostics and therapeutics, the information is also supposed to help the agricultural community to breed healthier dairy cattle that produce more and higher quality milk, as well as beef cattle that produce better quality beef. The purpose of sequencing agricultural animals has moved far beyond the original goals of serving as a model for studying human health issues. At present, sequencing of animals has goals of (but not limited to) studying traits of economic interest to raise livestock production, studying the effect of domestication, selecting high-fertility breeds, and understanding genotypic and phenotypic changes due to environmental factors like nutrition.

Biotechnology is entering the post-genome era. NGS technologies are capable of helping scientists and clinicians study genomes of individuals faster than ever before. Sequencing microorganisms and parasites in the organs of agricultural animals can also help veterinary scientists develop new vaccines and therapeutics (Bai et al., 2012).

CHALLENGES

Sequencing whole genomes by these massively parallel-sequencing technologies generates huge amounts of data that must be properly managed, stored, and analyzed. As the reagent costs of sequencing decrease with the development of better reagents and protocols, a number of sequencing projects are running simultaneously and generating enormous data in the form of millions and billions of DNA base pairs. Even though sequencing projects of numerous animals and plants are running, and many scientists have conducted high-end sequencing of various animal and plant samples, the lack of computing skills, required hardware, storage, and network capabilities necessary to manage the massive data sets generated by larger scale whole-genome sequencing studies is often felt. As sequencing studies are continuously increasing in number, the cost and complexity of data analysis and management is emerging as the primary limiting factor among researchers.

ETHICAL ISSUES

Next generation sequencing allows simultaneous sequencing of enormous amounts of DNA. Taking advantage of the tremendous power of technology, NGS has started to be used for understanding human genetic diseases. However, important ethical and social issues need to be addressed before it becomes routine for clinical diagnostic applications. The transition of next generation sequencing technology into the clinic is one of the most important aspects, but will also have significant ethical issues regarding the ownership and privacy of patient genetic information. Clinical genetic investigators may also face issues such as return of genetic information of a patient after being analyzed by NGS. For example, in 2010, members of NHGRI's ClinSeq study discovered that a patient had passed away; this person's exome had already been sequenced and analyzed. In such cases the question arises whether the results of the genetic studies should be returned to the patient, and if so, what information should be returned and to whom? The NHGRI team concluded that certain results, such as variants with well-established clinical significance, should be returned to the family because these results were relevant to the family's health and could be used in face-to-face genetic counseling sessions.

Ethical issues also cover decisions to conceal the identity of an individual from publicly accessible de-identified data. It would also be essential to confirm that would-be participants be aware of the risks before they decide whether to participate in a study, and whether participants or their family members should be informed of incidental genotype findings, especially if such findings would have some adverse effect on family health (Devarakonda et al., 2012).

TRANSLATIONAL SIGNIFICANCE

Translational research pertains to the process of translating scientific discoveries into practical (clinical) applications. Next generation sequencing is playing a transformational role in cancer discovery and genetic disorder research. Providing new insights into disease mechanisms and metabolic and signaling pathways are examples of progress made through NGS that was not previously unfeasible. Related information is being used to improve diagnostics, and to develop more effective and more personalized treatments for disease and patient care.

Furthermore, targeted next generation sequencing holds great potential for speedy, wide-ranging mutational analysis to unravel complex tumor signature variations and improve the cost-effectiveness of sequencing by focusing on the portions of the genome that are relevant for the question of study. With decreasing costs and improving technology, NGS has the potential to translate enormous amounts of raw data into useful information in almost all aspects of research into health, development, and disease.

WORLD WIDE WEB RESOURCES

Following are websites with relevant information about different types of next generation sequencing platforms and technologies. These websites have enormous information about the technology, chemistry, and applications of various NGS platforms, as most of these are from technology developers and manufacturers themselves. However, comparison of significant research outputs, and applicability and relevance to clinical applications, should be made through various publications from independent research groups, which are supposed to be unbiased and impartial with regards to application and usability of individual technology. The aforementioned websites are as follows:

1. www.seqanswers.com: This website addresses several issues related to workflow and protocol that are not necessarily provided in manuals or other webpages. NGS users create an account, and then share the issues they face while performing the actual experiments as a thread. The issue is then discussed by a number of NGS users who share their unbiased experiences and try to resolve the particular issue.
2. www.genomeweb.com: This website is useful for getting recent updates about different available technologies and platforms. The website also addresses other genomics techniques such as arrays, clinical genomics, PCR, bioinformatics, etc. Under the tab "sequencing," visitors get the most recent information about new products for novel applications in NGS that have been introduced by different NGS firms. Although the website also has news related to marketing updates, it is useful to understand what new technology is being introduced and is available for users.
3. http://en.wikipedia.org: Wikipedia is a good encyclopedia for everything. It also contains a lot of information about nearly every type and application of NGS, especially for those who are new to the technology and want to learn the basics behind a particular technique, its principle, history, workflow, etc.

Following websites have been referred for this chapter:

4. www.454.com
5. http://my454.com/products/technology.asp
6. http://www.illumina.com
7. www.appliedbiosystems.com
8. www.pacificbiosciences.com
9. http://en.wikipedia.org/wiki/Helicos_single_molecule_fluorescent_sequencing
10. www.iontorrent.com
11. http://www.nature.com/news/nanopore-genome-sequencer-makes-its-debut-1.10051
12. http://www.polonator.org
13. www.completegenomics.com
14. www.raindancetech.com
15. http://www.nimblegen.com
16. http://www.agilent.com
17. http://www.csiropedia.csiro.au/display/CSIROpedia/Cattle+genome+project
18. http://www.dnasequencing.org/history-of-dna
19. http://massgenomics.org
20. www.sanger.ac.uk
21. www.icgc.org
22. www.seqanswers.com
23. www.genomeweb.com/sequencing

REFERENCES

Ashoor, G., Syngelaki, A., Wagner, M., et al. (2012). Chromosome-selective sequencing of maternal plasma cell-free DNA for first-trimester detection of trisomy 21 and trisomy 18. *American Journal of Obstetrics and Gynecology*, *206*. 322.e1–5.

Bai, Y., Sartor, M., & Cavalcoli, J. (2012). Current status and future perspectives for sequencing livestock genomes. *Journal Animal Science Biotechnology*, *3*, 8–14.

Besaratinia, A., Li, H., Yoon, J. I., et al. (2012). A high-throughput next generation sequencing-based method for detecting the mutational fingerprint of carcinogens. *Nuclear Acids Research*. doi: 10.1093/nar/gks610.

Bickhart, D. M., Hou, Y., Schroeder, S. G., et al. (2012). Copy number variation of individual cattle genomes using next generation sequencing. *Genome Research*, *22*, 778–790.

Devarakonda, S., Govindan, R., & Hammerman, P. S. (2012). Cancer gene sequencing: Ethical challenges and promises. Virtual mentor. *American Medical Association Journal Ethics*, *14*, 868–872.

D'Souza, D., & Hanning, I. (2012). Advances in unraveling the DNA code: An introduction to next generation sequencing; In: Next generation Sequencing Tools: Applications for Food Safety and Poultry Production Symposium. *Poultry Science*, *91*, 48–49.

Eggen, A. (2012). The development and application of genomic selection as a new breeding paradigm. *Animal Frontiers*. doi:10.2527/af.2011-2027.

Elsik, C. G., Tellam, R. L., Worley, K. C., et al. (2009). The genome sequence of taurine cattle: a window to ruminant biology and evolution. *Science*, *324*, 522–528.

Fan, H. C., Blumenfeld, Y. J., Chitkara, U., et al. (2008). Noninvasive diagnosis of fetal aneuploidy by shotgun sequencing DNA from maternal blood. *Proceedings of the National Academy of Sciences of the United States of America*, *105*, 16266–166271.

González-Recio, O. (2011). Epigenetics: A new challenge in the post-genomic era of livestock. *Frontiers Genetics*, *2*. http://dx.doi.org/10.3389/fgene.2011.00106.

Knapp, M., & Hofreiter, M. (2010). Next generation sequencing of ancient DNA: requirements, strategies and perspectives. *Genes*, *1*, 227–243.

Li, R., Fan, W., Tian, G., et al. (2010). The sequence and de novo assembly of the giant panda genome. *Nature*, *463*, 311–317.

Mane, S. P., Evans, C., Cooper, K. L., et al. (2009). Transcriptome sequencing of the Microarray Quality Control (MAQC) RNA reference samples using next generation sequencing. *BMC Genomics*, *10*, 264–275.

Mardis, E. R. (2008). Next generation DNA sequencing methods. *Annual Review Of Genomics And Human Genetics*, *9*, 387–402.

Mardis, E. R. (2011). A decade's perspective on DNA sequencing technology. *Nature*, *470*, 198–203.

Maxam, A. M., & Gilbert, W. (1977). A new method for sequencing DNA. *Proceedings of the National Academy of Sciences of the United States of America*, *74*, 560–564.

Mills, R. E., Walter, K., Stewart, C., et al. (2011). Mapping copy number variation by population-scale genome sequencing. *Nature*, *470*, 59–65.

Nakamura, K., Oshima, T., Morimoto, T., et al. (2011). Sequence-specific error profile of Illumina sequencers. *Nucleic Acids Research*, *39*, e90.

Nyrén, P., Pettersson, B., & Uhlén, M. (1993). Solid phase DNA minisequencing by an enzymatic luminometric inorganic pyrophosphate detection assay. *Analytical Biochemistry*, *208*, 171–175.

Porto-Neto, L. R., Jonsson, N. N., D'Occhio, M. J., et al. (2011). Molecular genetic approaches for identifying the basis of variation in resistance to tick infestation in cattle. *Veterinary Parasitology*, *180*, 165–172.

Rolf MM, McKay SD, McClure MC et al. (2010). How the next generation of genetic technologies will impact beef cattle selection. Beef Improvement Federation Research Symposium and Annual Meeting. June 28- Jul y 1, 2010. Missouri.

Ronaghi, M., Karamohamed, S., Pettersson, B., et al. (1996). Real-time DNA sequencing using detection of pyrophosphate release. *Analytical Biochemistry*, *242*, 84–89.

Sanger, F., Nicklen, S., & Coulson, A. R. (1977a). DNA sequencing with chain-terminating inhibitors. *Proceedings of the National Academy of Sciences of the United States of America*, *74*, 5463–5467.

Sanger, F., Air, G. M., Barrell, B. G., et al. (1977b). Nucleotide sequence of bacteriophage phi X174 DNA. *Nature*, *265*, 687–695.

Sellner, E. M., Kim, J. W., McClure, M. C., et al. (2007). Board-invited review: Applications of genomic information in livestock. *Journal Of Animal Science*, *85*, 3148–3158.

Stapley, J., Reger, J., Feulner, P. G. D., et al. (2010). Adaptation genomics: the next generation. *Trends Ecology Evolution*, *25*, 705–712.

Subramanian, S., Huynen, L., Millar, C. D., et al. (2010). Next generation sequencing and analysis of a conserved transcriptome of New Zealand's kiwi. *BMC Evolutionary Biology*, *10*, 387–398.

Voelkerding, K. V., Dames, S. A., & Durtschi, J. D. (2009). Next generation sequencing: From basic research to diagnostics. *Clinical Chemistry*, *55*, 641–658.

FURTHER READING

Janitz, M. (2008). *Next Generation Genome Sequencing: Towards Personalized Medicine*. Wiley-vch Verlag Gmbh.

Kwon, Y. M., & Ricke, S. C. (2011). *High-Throughput Next Generation Sequencing: Methods and Applications. Vol. 733 of Methods in Molecular Biology*. Humana Press.

McDowall, J. S. (2013). *Next Generation Sequencing: Platforms, Resources and Data Analysis. Chapman and Hall/CRC Mathematical and Computational Biology Series*. Taylor & Francis Group publishers.

Rodriguez-Ezpeleta, N., Hackenberg, M., & Aransay, A. M. (2012). *Bioinformatics for High Throughput Sequencing*. Springer Science + Business Media.

GLOSSARY

Adaptors Adaptors are short DNA sequences that are attached to the unknown DNA fragment so that DNA with known sequences flank the unknown DNA for priming (e.g. sequencing adaptors).

Assembly Assembly refers to merging small fragments of a DNA sequence to reconstruct the original, much longer sequence. This is needed as the DNA sequencing technology/instrument cannot read whole DNA in one go, but reads the DNA in small pieces of between 20 and 1,000 bases, depending on the technology used (e.g. human genome assembly from shotgun reads).

***De Novo* Assembly** *De novo* assembly refers to assembling short reads to create a novel full-length DNA sequence with no prior reference sequence available (e.g., *de novo* assembly of plant genomes).

Di-Deoxy Nucleotides Di-deoxy nucleotides are chain-terminating nucleotides used in the Sanger method for DNA sequencing. The absence of the 3′-hydroxyl group means that, after being added to a growing nucleotide chain, no further nucleotides can be added to them as no phosphodiester bond can be created.

DNA Sequencing DNA sequencing is the process of determining the order of nucleotides within a DNA molecule. It includes a method or technology that is used to determine the order of the four bases, adenine (A), guanine (G), cytosine (C), and thymine (T), in a strand of DNA.

Emulsion PCR An emulsion is a mixture of two or more liquids that are normally immiscible. Emulsion PCR isolates individual DNA molecules in aqueous droplets within an oil phase along with other PCR reagents for amplification. A PCR reaction within each droplet then makes clonal copies of the DNA molecule followed by immobilization on beads for later sequencing (e.g. emulsion PCR used in 454 sequencing).

Genome The genome is the total nucleotide content of an organism, generally DNA, but in some viruses, RNA. The genome includes both the genes and the non-coding sequences of the DNA/RNA.

Genomic Variation Genomic variations pertain to changes in the sequence of the genome as compared to the reference genome of the same species. It can be single nucleotide changes, insertions, deletions of a segment of DNA, or large chromosomal changes.

GWAS A genome-wide association (GWAS) study is an examination of common genetic variants in different individuals of a population to analyze which variant, especially SNP, is associated with a trait or disease.

High-Throughput A high-throughput process is a scaled-up, parallelized, and automated process with rapid and greater output as compared to conventional methods.

Mapping Mapping is aligning reads against an existing genome sequence to figure out genomic variations. Mapping assembly is building a sequence from reads with the help of an existing genome sequence.

Nebulization In terms of sequencing processes, nebulization of DNA means physical shearing of DNA to produce smaller fragments of desired size range with the help of pressurized gas (e.g. nitrogen).

NGS Platform An NGS platform is a particular NGS technology with an instrument and reagent that does the sequencing (e.g. Roche 454, Illumina-HiSeq, ABI-SOLiD, etc.)

Polony Polony is a contraction of "polymerase colony," a small colony of DNA. Polonies are discrete clonal amplifications of a single DNA molecule.

Read Length Read length is the length of a DNA fragment or piece that a particular sequencing technology/instrument can sequence (read). Later these reads are aligned to form a larger DNA construct.

ABBREVIATIONS

BAC Bacterial Artificial Chromosome
cDNA Complementary DNA
ChIP Chromatin Immunoprecipitation
CNV Copy Number Variation
cPAL Combinatorial Probe Anchor Ligation
EBV Estimated Breeding Value
EST Expressed Sequence Tags
GB Gigabases
GWAS Genome-Wide Association Studies
Indels Insertions and Deletions
MB Mega Bases
NGS Next Generation Sequencing
rDNA Ribosomal DNA
SMRT Single Molecule Real Time (Sequencing)
SNP Single Nucleotide Polymorphism
ZMW Zero Mode Waveguide

LONG ANSWER QUESTIONS

1. Define next generation sequencing and its applications in animal biotechnology.
2. What are different generations of DNA sequencing? Describe their principle and applications.
3. Describe the workflow of next generation sequencing in different applications.
4. Describe use of next generation sequencing in human health and its translational significance.
5. Describe various next generation sequencing technologies, their advantages, limitations, and specific uses.

SHORT ANSWER QUESTIONS

1. What is next generation sequencing, and how is it different from Sanger sequencing?
2. Give a brief description about different generations of next generation sequencing.
3. Give a general workflow of next generation sequencing, indicating differences for various applications.
4. What is sequence assembly? Define *de novo* and mapping assemblies.
5. What is targeted sequencing? Briefly describe its use in medical research.

ANSWERS TO SHORT ANSWER QUESTIONS

1. Next generation sequencing (NGS) is a term given to sequencing technologies post-Sanger sequencing. NGS can produce huge amounts of sequencing data at incredibly low cost. While Sanger sequencing works on the principle of chain termination, whereby the growing DNA strand is terminated by the incorporation of dideoxy nucleotides (ddNTPs), A, T, G, and C, in different reactions, next generation sequencing works on the principle of sequencing millions of DNA fragments simultaneously in a massively parallel mode. Sequencers based on Sanger sequencing produce total sequencing data output of a few hundred bp at a time; NGS can produce sequence data in megabase or gigabase quantities.
2. The term "generation" refers to the chemistry and technology used by the sequencing process. First generation sequencing was primarily the Sanger and Maxam–Gilbert sequencing techniques, which were able to sequence a few hundred base pairs at a time, and were used for individual gene sequencing. Sequencing 3 billion base pairs of the human genome would take a long time with Sanger sequencing, as it needs to sequence about 6 million fragments of DNA with a 500 bp length.
   Next generation sequencing technologies, also called second generation sequencing, have parallelized the sequencing reaction in a massive manner, thus generating huge amounts of data very rapidly at modest cost, thus revolutionizing the field of genomics. They include Roche 454, Illumina Solexa, and ABI-SOLiD technologies.
   The term third generation sequencing (or next-next generation sequencing) is given to technologies that interrogate single molecules of DNA without amplifying them through PCR, thereby overcoming issues related to the biases introduced by PCR amplification and de-phasing. However, the third generation sequencers were developed with a vision of making sequencing cheaper than second generation sequencing. Third generation sequencers include those of PacBio and Helicos.

3. Next generation sequencing can be used in a number of ways depending upon the application. In every type of sequencing, the major changes occur in the sample processing and library preparation steps. Once the library is prepared, a particular sequencing platform uses the same chemistry to sequence the fragments. Major types of sequencing used for different types of applications are shown in Figure 19.4. An outline of the protocols used in various applications is summarized in Flow Chart 19.1.
4. Assembly is the process of using various computer programs to align multiple sequencing reads that are overlapping one another to reconstruct the long DNA fragment (and maybe the whole genome), as DNA sequencing cannot read whole genomes in one go, but reads small pieces of different length, depending on the technology used. *De novo* assembly refers to assembling reads to build a new sequence, where no existing sequence of the same or related organism/species is available. Mapping pertains to assembling reads against an already existing backbone sequence of the same or related organism/species to build a sequence that is similar but not necessarily identical to the backbone sequence. Mapping is generally used to figure out genomic variation in the sequenced DNA compared to the reference sequence such as SNPs, insertions, deletions, and structural variation SNPs.
5. Targeted sequencing refers to sequencing regions of interest in the genome rather than sequencing the whole genome. The most popular one is sequencing the exon part of the genome to get rid of unwanted introns. Targeted re-sequencing can be used to examine all the exons in the genome, a part of the genome, and specific gene families that are supposed to constitute known drug targets or regions and are thought to be involved in disease or pharmacogenetic effects; it is done through genome-wide association studies.

Chapter 20

Biomolecular Display Technology: A New Tool for Drug Discovery

Madhu Biyani[*,††], Koichi Nishigaki[††] and Manish Biyani[*,†]

[*]*Department of Biotechnology, Biyani Group of Colleges, Jaipur, India,* [†]*Department of Bioengineering, The University of Tokyo, Tokyo, Japan,* [††]*Department of Functional Materials Science, Saitama University, Saitama, Japan*

Chapter Outline

SUMMARY

The identification of molecules with desired function is of great significance in biology and medicine. Display technologies represent a new tool for drug discovery that facilitate screening of novel biomolecules against any target of choice. This chapter reviews the development of *in vitro* display technologies and their application in drug discovery, focusing on challenges and perspectives for rapid and efficient modern drug discovery processes.

WHAT YOU CAN EXPECT TO KNOW

Discovering and developing safe and effective drugs is one of the most challenging scientific endeavors. However, the current decrease in the number of approved drugs produced by the pharmaceutical industry and the increasing number of new, emerging, and re-emerging diseases, demands a technology-based paradigm shift towards developing a new generation of therapeutics drugs based on understanding mechanisms at the molecular level. Most drug development fails because of unacceptable toxic side effects of a drug and a limited understanding of the binding of an arbitrary drug to an arbitrary target (diseased protein). A new generation of so-called biomolecular drugs, especially peptide aptamers, was created to meet these demands. Peptide aptamers are combinatorial protein reagents characterized by high specificity and strong affinity to their targets, and some other advantageous properties such as a versatile selection process, ease of chemical synthesis, and small physical size, which collectively make them potential therapeutic agents. Recently, peptides have shown a significant impact on the success rate of drug discovery. This chapter focuses on technological advancements and emphasizes the significance of "display technologies" in the development of peptide aptamers as novel tools to discover targets, biologically relevant druggable sites, and potential drug compounds.

Animal Biotechnology. http://dx.doi.org/10.1016/B978-0-12-416002-6.00020-1

HISTORY AND METHODS

INTRODUCTION

The global war against diseases appeared to be on the road to victory when Dr. Edward Jenner discovered vaccination in 1796, one of the greatest discoveries in medicine. The next revolution was a period of accidental discoveries in the middle of the 20th century where the discovery of a drug was happened upon primarily by observing the therapeutic effects of a compound and then isolating it. Aspirin and penicillin are two well-known examples of this type of fortuitous discovery. A further revolution occurred in the latter half of the 20th century when drug discovery was driven by the introduction of force screening of large libraries of chemical compounds based on some preliminary knowledge. Statins, which lower cholesterol levels in patients with heart disease, are one example. Recently, however, two of the most challenging issues in the healthcare field have been discovered. First is the advent of new and emerging infectious diseases and the re-emergence of old diseases, as well as the rapid spread of pathogens resistant to drugs and of disease-carrying insects resistant to insecticides. Swine flu is the most recent example. It is thought to be a mutation, or, more specifically, a re-assortment of four known strains of influenza A virus subtype H1N1. The second issue is "tailoring care" to the individual patient by treating the disease specifically and not by following "one-treatment-fits-all" formulae. Sickle-cell anemia is the best example, where patients have a similar genetic lesion in the beta-globin gene, but their phenotypic diversity ranges from life threatening to symptom free. We are now entering an era of personalized medicine that had already become evident in 2003 when the International Human Genome Project completed a blueprint of the human genome and confirmed the fact that no two individuals are exactly alike, especially in terms of medicine. Therefore, a paradigm shift in medicine is essentially required, with an increased interest in the discovery of new drug candidates. However, the current productivity crisis in the pharmaceutical industry has seen fewer drugs introduced onto the market than in the late twentieth century (Kola and Landis, 2004; Pammolli et al., 2011). The major reasons identified for this were that most drug candidates had a lack of efficacy and too-high toxicity; this was mainly caused by insufficient validation of therapeutic targets and/or insufficient specificity. Therefore, a clear technology-based shift is required to address the major causes of failure and to offer a powerful new drug discovery approach that will enable a seamless process from target identification and/or validation to identification of hit drug candidates.

Completion of Human Genome Project in 2003 provided an unprecedented opportunity to elucidate the genetic basis of human disease at the molecular level. The information on genes and their link to diseases are stored in several comprehensive public databases (George et al., 2008). Among these, the GeneCards database has identified 3,204 genes that are associated with human diseases (total number of human genes is estimated near 25,000). Out of these, 523 genes are reported to encode disease-related proteins and are druggable (druggablity is the likelihood of being able to modulate a target with a drug) (Russ and Lampel, 2005). Nearly 215 of these disease genes are included in the DrugBank database with drugs available, and only 89 of the disease genes from this dataset have Food and Drug Administration (FDA)-approved drugs available (see Figure 20.1) (Sakharkar et al., 2007). Therefore, with nearly 85% of the human disease genes representing druggable targets with no drugs available, it is important to develop a powerful approach that will have a significant impact on the success rate of drug discovery.

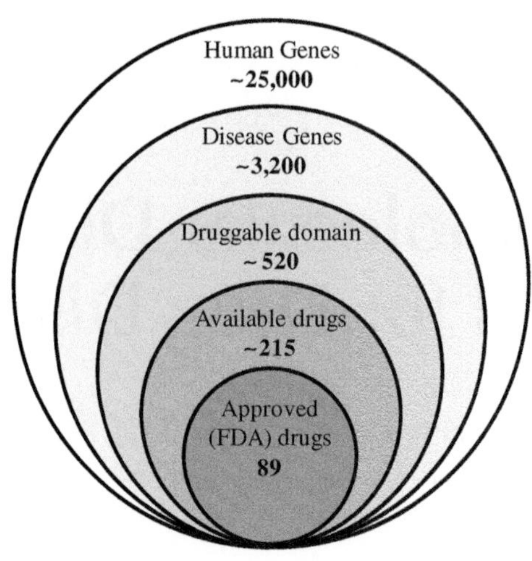

FIGURE 20.1 Cartoon representation the current status and need of a paradigm shift in modern drug research and development.

PRINCIPLE

The discovery and development of a new drug has become one of the most challenging scientific endeavors. How an individual new drug is discovered and brought to the public is a very long, complicated process that typically costs close to $1 billion and takes an average of 10 to 15 years (Figure 20.2).

The drug discovery process starts by identifying cellular and genetic chemicals in the body that play a role in specific diseases, called "target identification." After choosing a potential target, the "target validation" step is followed to insure that it actually is involved in the disease and can be acted upon by a compound (drug). Next, compounds are identified that have various interactions with these targets and are screened based on their comparative association with a desired change in the behavior of diseased cells. This is followed by "lead identification." A lead compound

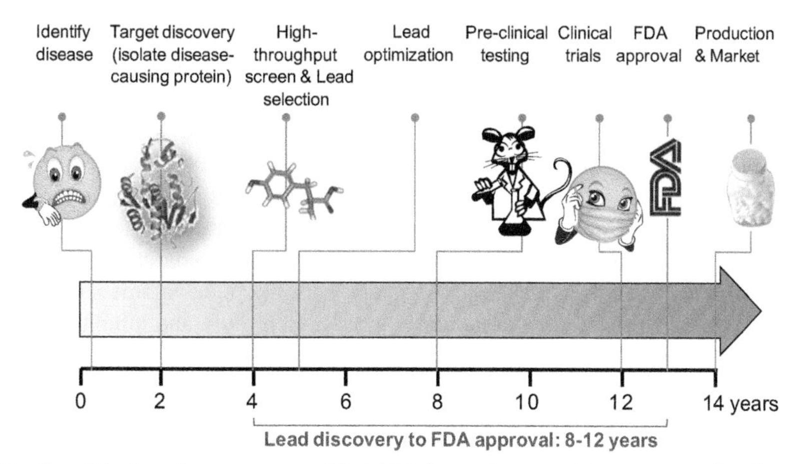

FIGURE 20.2 Stages and timeline of the drug discovery process. The initial phase, where promising targets are first discovered and linked to a disease, is shown here as taking 1–2 years, but it can require several years.

is one that is believed to have potential to treat disease. Testing and optimization are then done on each of the lead compounds by performing *in vitro* (testing on cells in a test tube) and *in vivo* (testing on living organisms) confirmation, followed by clinical evaluation. In the pre-clinical step, the lead compound is tested extensively to ensure it will be safe to administer to humans. Testing at this stage can take from one to five years and must provide information about the pharmaceutical composition of the lead compound (drug), its safety, how the drug will be formulated and manufactured, and how it will be administered to the first human subjects. In the final step, clinical testing of the drug is done by administration to healthy volunteers or patients. This usually consists of Phase-I, Phase-II and Phase-III, where in each successive phase, increasing numbers of patients are tested. After the conclusion of successful pre-clinical and clinical testing, the drug is passed to the FDA for market authorization. In general, out of several thousand new compounds identified during the discovery process, only a few are considered safe for clinical testing in patients, and only one of these compounds on average is ultimately approved as a marketed drug for treatment.

Drug discovery is plagued by time-consuming, costly processes and extremely high rates of failure. The FDA has estimated that just a 10% improvement in the ability to predict drug failures before clinical trials could save $100 million (USD) in development costs per drug. The high failure rate in principle originates during the lead generation and optimization steps. It may not be possible to speed up clinical trials, but bottlenecks in lead discovery can be tackled. What is important in the lead discovery process is that understanding principles is much more critical than calculating accuracy. For example, instead of measuring binding affinity to a few decimal places, calculating relative trends in binding affinity could be more meaningful. Therefore, explaining "why" (for example, why a small change in a drug causes a large change in its activity or why one enantiomer causes side effects while another does not) can reduce failure rates and drive better drug development. Needless to say, selecting the most efficient compounds and introducing high-throughput screening systems will dramatically change this picture.

NECESSITY: SMALL MOLECULE VS. BIOMOLECULAR (BIOLOGICS) DRUGS

The systematic search for a drug was first advocated by Paul Ehrlich, the founder of chemotherapy, who received the Nobel Prize for Medicine in 1908 for his work on the magic bullet concept (where a compound is used to target a particular biomolecule of interest). This biomolecule is essentially an intracellular or membrane-bound protein identified as a contributor to a disease state. Targeting these proteins with chemical agents led to the discovery of organic small molecules for the treatment of various types of diseases, including infectious diseases and cancer. However, entry to the market of newly approved drugs has been decreasing recently due to the lack of efficacy and very high toxicity of proposed drugs. Since most biological processes are mediated by proteins, owing to their remarkable capability of molecular recognition and specific interactions with other molecules, it is natural that protein-based biopharmaceuticals have emerged as a new generation of therapeutic drug. Among them, monoclonal antibodies such as "*Trastuzumab*" or "*Centuximab*" have gained remarkable success in the biomedical area. However, several disadvantages have become apparent with the increasing application of antibodies, including low bioavailability due to large size, limited shelf life, difficultly in producing them biologically, viral or bacterial contamination during the manufacturing process, and immunogenic effects. To reduce the immune response they inherently trigger, efforts have been devoted to designing a new generation of antibodies by switching from murine antibodies to chimeric (human 60%, murine 40%, e.g. *Centuximab*; human 90%, murine 10%, e.g. *Trastuzumab*; human 100%, e.g. *Panitumumab*). Meanwhile, insight into molecular biology has led to the invention of the Systematic Evolution of Ligands by Exponential Enrichment (SELEX) process for *in vitro* selection of high-affinity oligonucleotides (Gold and Tuerk, 1990; Ellington and Szostak, 1990). A number of DNA and RNA aptamers (short biomolecules such as DNA/RNA/peptide chains capable of identifying a target molecule with high affinity and specificity) have been generated that can bind to multiple targets in a similar mode to that of antibodies. The first RNA aptamer-based drug (*Pegaptanib*) was approved for therapeutic use by the FDA (Gragoudas et al., 2004). Peptide aptamers have recently also emerged as an attractive alternative to antibody therapy. Unlike antibodies, peptide aptamer sequences are short, easy to synthesize, chemically stable, and less immunogenic. Although the most clinically used drugs are small molecules or proteins, each of these differ significantly in their properties (Figure 20.3). Small molecules, because of their size, are advantageous for their accessibility (cell permeability and tissue penetrations), whereas proteins provide high binding affinity and target specificity. On the other hand, a peptide (as shown in the middle of Figure 20.3) can potentially combine favorable properties of both small molecules and proteins. Peptides can bind as tightly and specifically as antibodies while being small enough to enter cells and tissues.

Recent technological advances have allowed the development of peptide aptamers that can bind to protein targets with high affinity and inhibit function with high specificity (Crawford et al., 2003). In contrast to the inhibitory abilities of peptide aptamers, some peptide aptamers could be screened to activate the function of their cognate target proteins (Nouvion et al., 2007). Peptide aptamers have big advantages over small molecules in terms of specificity and affinity for targets, and over antibodies in terms of size; they therefore have great potential for further exploration as future therapeutic drugs. Recent advancements in high-throughput screening systems and *in vitro* molecular selection technologies allow the identification of novel and potentially commercially relevant peptides. These will be discussed in following sections.

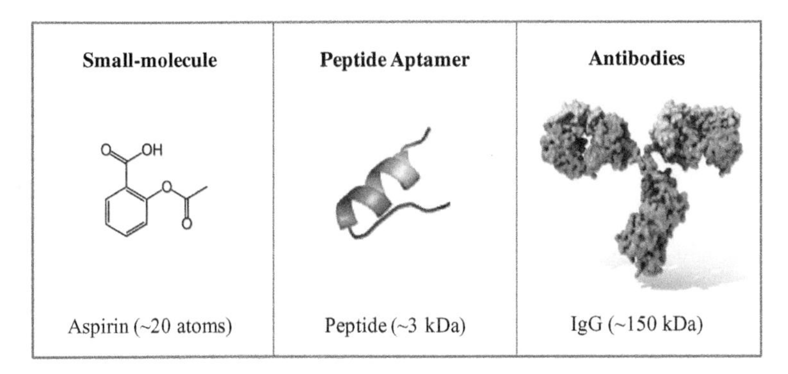

	Antibody	Peptide aptamer
Size	30 ~180 Kda	1~ 5 Kda
Dissociation constant (K_d)	10 nM	10 nM or can be improved
Library size	10^9	$10^{12\sim15}$
Cost	Expensive	Inexpensive
Quality control	Unstable	Stable
Immunogenicity	High	Less
Chemical modification	Difficult	Possible (unnatural amino acid)
Optimization	Difficult	Easy
IP Problems	Unsolved	Solved
Shelf life	Limited and sensitive to temperature	Long and stable
Production	Difficult to synthesize, requires animals, suffer from batch to batch variation	Produced by chemical synthesis (no animal) resulting in little or no batch to batch variation and extremely simple and robust

FIGURE 20.3 Comparison between major types of therapeutics.

METHODOLOGY: BIOMOLECULAR DISPLAY TECHNOLOGIES

The drug discovery process for developing new pharmaceutical compounds has traditionally depended upon an empirical approach to screening. However, because of rising R&D costs, short product lifecycles, and the public's ever-increasing demand for new, high-quality treatments, pharmaceutical companies are encountering pressure to develop new drugs faster, more cost-effectively, and in greater numbers than ever before. These pressures have dramatically transformed the process by which new drugs are discovered and developed, and have motivated pharmaceutical companies to collaborate with biotechnological companies for solutions to streamline the drug discovery process. Therefore, the discovery of highly functional and biologically active molecules is required for future therapeutics.

Biomolecular display technologies that allow the construction of a large and diverse pool (millions to trillions) of biomolecules, their display for property selection, and rapid characterization (decoding) of their structures, are particularly useful for accessing and identifying novel potential candidates for drug discovery. These display technologies basically mimic the process of natural evolution, and in general rely on the common principle of coupling (linkage) of individual biomolecules (DNA/RNA) with their encoded products (peptides or proteins). A typical display module in biomolecular display technologies consists of three major components: displayed entity, linker, and corresponding genetic code (Figure 20.4A).

A number of display technologies have been developed to provide essential tools to enrich biomolecular drugs, including nucleic acids and proteins (peptides) with desired functions, based on evolutionary molecular engineering and combinatorial chemistry (Li, 2000). Over the past two decades, many display formats have been developed that use different types of displayed entities, linkage formats, and coding strategies. These all can be systematically classified into two major categories based on their expression system, cell-based type or cell-free type. (Figure 20.4B).

Phage display is one of the first methods, and was described by George P. Smith in 1985 by displaying a foreign peptide on the coat of a filamentous phage (Smith, 1985). Since then, phage display and two other bacterial

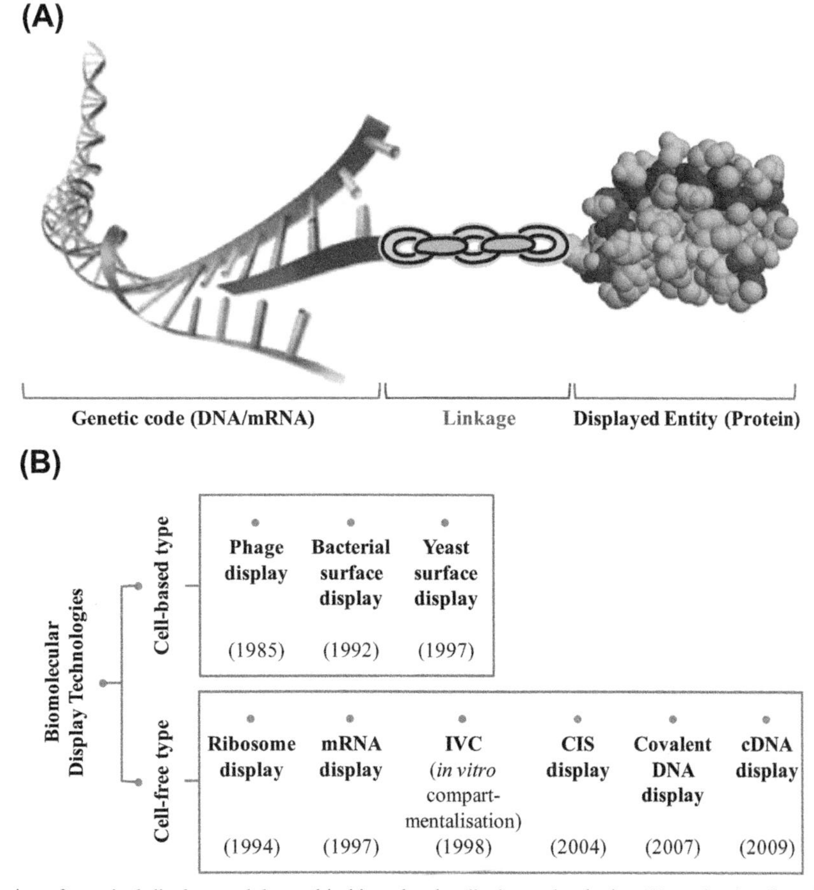

FIGURE 20.4 An illustration of a typical display module used in biomolecular display technologies (A), and a timeline of the different biomolecular display technologies developed to date (B).

and yeast surface display technologies have been introduced and widely adopted (Francisco et al., 1992; Boder and Wittrup, 1997). However, cell-based type display technology has its limitations, including slow speed (few days to weeks), limitation of library size due to the transformation efficiency of the host (approximately 10^8), and toxicity to host cells caused by expression of exogenous proteins. This led to the generation of cell-free type display technologies, which avoid the need for transformation, enable more of the sequence landscape to be displayed, and increase the probability of higher-affinity hits. The first cell-free type display technology, ribosome display, was developed in 1994 using *E. coli* lysate for the display of peptides onto the ribosomes (Mattheakis et al., 1994). During the cell-free expression, the ribosomes were stalled on the mRNA template and the nascent peptide remained in a complex, which could then be recovered by EDTA. Next, a related technology called mRNA display (or *in vitro* virus) was reported; it functions by the formation of a covalent linkage between the mRNA template and the expressed protein via puromycin. It provides the advantages of speed and stability over ribosome display (Nemoto et al., 1997; Roberts and Szostak, 1997). In 1998, Tawfik and Griffiths introduced man-made cell-like compartments using water-in-oil droplets, termed *in vitro* compartmentalization (IVC), which provides an alternative way of linking phenotype and genotype to mimic the natural compartments of living organisms (Tawfik and Griffiths, 1998). Later, some other display technologies were introduced to demonstrate their potential for unprecedented library size and stability issues: CIS display (Odegrip et al., 2004), covalent DNA display (Bertschinger et al., 2007), and cDNA display (Yamaguchi et al., 2009).

The aim of this chapter is to focus on recent advancements in display technologies that emphasize the development of peptide aptamers and the prospective advantages of these methods over antibody-based target disease diagnosis and therapy. The three major display technologies, including phage display, ribosome display, and mRNA display, are schematically shown in Figure 20.5 and discussed in the following sections.

Phage Display

Phage display is one of the first methods to introduce an indirect linkage (physical) of a protein with its DNA sequence, and is the most widely adopted technology so far for the discovery of novel drugs (Smith and Petrenko, 1997). In order to provide the linkage of protein and the encoded nucleic acid, it uses a bacterial virus. First, a gene library is genetically fused to a filamentous bacteriophage so that on co-infection by the phage, newly emerging phage particles encase individual gene sequences while displaying the corresponding gene-encoded polypeptides on their outer surface (Figure 20.5). Next, the recombinant phage DNA is transformed into *E. coli* cells, resulting in large numbers of bacterial clones, each containing the coding sequence for a unique library member. The expressed protein is incorporated into the viral coat

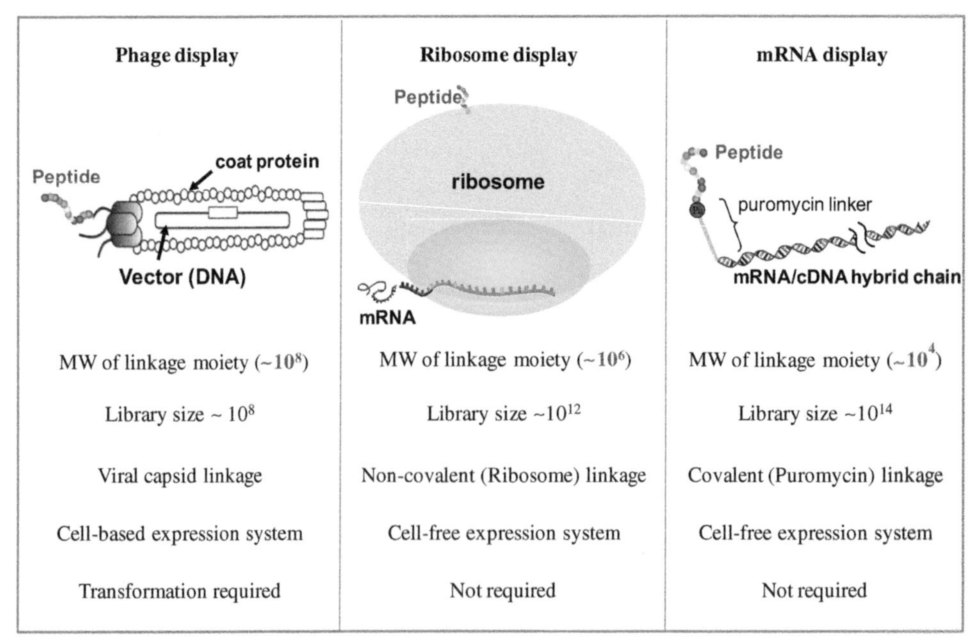

FIGURE 20.5 Illustration and comparison of the most common biomolecular display technologies. In phage display, an indirect linkage (physical) between the gene and gene product is provided by the viral capsid. In ribosome display, a non-covalent linkage is achieved by producing ternary complexes of RNA, ribosomes, and associated nascent peptides. In the mRNA display system, a covalent linkage is generated through a puromycin molecule attached to the encoding mRNA via a short DNA linker molecule.

and the cloned library DNA is packed within the virion. For the selection of potential candidates, the diverse pools of recombinant phage particles are incubated with an immobilized target molecule. Non-binding particles are washed away and those library candidates that bind specifically to the target are eluted and amplified by infection into fresh *E. coli* cultures. In this way, potential library members are enriched from very large libraries based on their binding affinity to target molecules (antibodies, enzymes, cell-surface receptors). Phage display technology provides the flexibility that selection can be performed under both *in vivo* and *in vitro* conditions.

Phage display technology consists of the following steps (Flow Chart 20.1):

1. The filamentous phage M13 is commonly used for vector construction.
2. The library is generally created or obtained through various methods as described in the flow chart. A diverse library size of from 10^{10}–10^{11} can easily be generated. The DNA library is inserted into the N-terminus of either the major coat protein (Gp8) or a minor tip protein (Gp3) of the filamentous phage. Then the target protein or the peptide is displayed as a fusion protein with Gp8 or Gp3 on the surface of the filamentous phage.

Step 1 – Construction of vector
(Most commonly used vector Ff M13)

Step 2 – Create and obtain randomized peptide display library
- Oligo-directed mutagenesis
- *In vitro* recombination
- Oligo cassette
- Recursive PCR
- Oligo-splinted assembly
- Error-prone PCR
- PCR with randomized primers
- Premade peptide libraries, gene fragmented libraries, cDNA libraries, and antibody V-region libraries

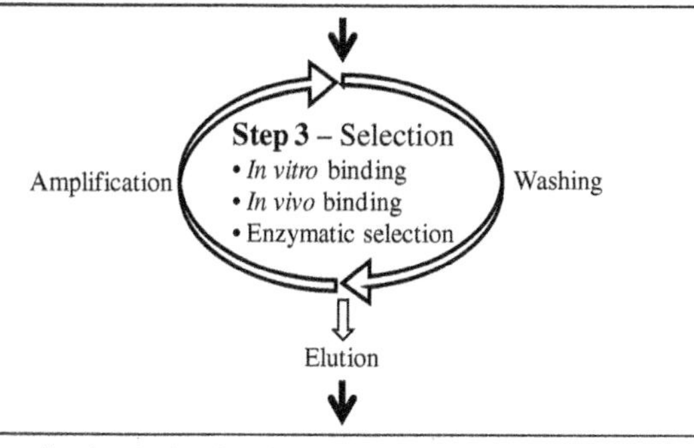

Elution

Step 4 – Analysis of selected clones
- Sequencing
- PCR fingerprinting
- Binding by phage ELISA etc.

FLOW CHART 20.1 Phage display technology

3. The phage-displayed library is selected by *in vitro* binding (the most common selection method) to a target and washing; then the retained phages are eluted. This process is also referred to as "sorting" or "biopanning." The eluted phages are used to re-infect *E. coli* for the preparation of new phage (amplified to allow a further round of selection).
4. The selected clones are identified, and the properties of displayed proteins determined. The most common technique for identifying selected clones is simply to sequence them; an alternative approach is PCR fingerprinting (the insert is amplified and digested with a frequent-cutting restriction enzyme; the pattern of bands is specific for each clone). Phage ELISA is the most popular and common technique for selection of clones.

Selection of Peptides

Phage display has been successfully applied for the isolation of peptides that bind protein targets with high affinity and specificity. Examples include: peptides that bind angiotensin converting enzyme-2 (ACE2) with a Kd ranged from 140-3 nM (Huang et al., 2003); peptides that bind each of three members of the inhibitor of apoptosis (IAP) family with the Kd ranged from 160–440 nM (Franklin et al., 2003); peptides that bind to protein kinase Cα (PKCα), an intracellular target, only under activation conditions (Ashraf et al., 2003); peptide HTMYYHHYQHHL that binds to the VEGF receptor kinase domain-containing receptor (KDR) and slows the growth of breast carcinoma BICR-H1 tumors in mice (Hetian et al., 2002); peptide that binds to recombinant human ErbB-2 tyrosine kinase receptor, which is implicated in many human malignancies, with Kd ranged from 3 nM to 5 μM (Huang et al., 2003). Most often, one or more families of related peptides are found. The common motif of one family can then be built into a secondary library and higher-affinity peptides selected. In most cases, these peptides have stronger affinities in the range of 1–10 nM. Usually, several families of binders are found to not compete for binding to the target, presumably because they bind different sites. However, when two of these molecules that bind at non-overlapping sites are joined with a suitable linker, such as several units of PEG, the affinity of the heterodimer is typically much greater than the affinity of either component. An added advantage of such a heterodimeric molecule is that it covers a larger area of the active site on the target and is more likely to interfere with protein–protein interactions. The phage display method has been the most widely used display method so far for the discovery of diagnostic and therapeutic molecules; however, the technology has limitations. The major drawbacks of this technology are the limitation of library size by the transformation efficiency of bacteria (generally 10^{7-10} members) and the toxicity to bacterial cells caused by exogenous protein expressions.

Ribosome Display

Ribosome display is the first entirely *in vitro* display technology (Mattheakis et al., 1994). It was developed by generating a protein–ribosome–mRNA ternary complex for selection in a cell-free translation system (Figure 20.5). The key idea is to translate a library of mRNA molecules that has no stop codon using cell-free system. As the translation progresses, the ribosome will run and stall to the very end of the mRNA molecule; it does not release the encoded protein since the last amino acid of the protein is still connected to the petidyl-tRNA. Thus, the ribosome that translates mRNA without a stop codon will be trapped in a form where the protein has emerged from the ribosome and the mRNA is also still connected to the ribosome. The resultant mRNA–ribosome–protein ternary complexes are then used for affinity selection on an immobilized target. mRNA of bound complexes are recovered after washing steps, reverse transcribed, and amplified and identified.

As this system is performed entirely *in vitro*, there are two main advantages over cell-based methods. First, the diversity of the library is not limited by the transformation efficiency of bacterial cells, but only by the number of ribosomes (up to 10^{14}/mL) and the number of mRNA molecules (a huge potential diversity of above 10^{12} members can be generated easily by PCR) available in the reaction test tube. Thus, the larger accessible library renders ribosome display a superior technology for selecting rare sequences with high-affinity properties. Second, random mutations can be introduced easily during PCR steps since no library needs to be transformed after any diversification step. Thus, further diversity can be continuously introduced into the DNA pools after selection, making this technology an efficient route for facile, directed evolution of biomolecules.

The methodology of ribosome display/mRNA display technology involves the following steps (Flow Chart 20.2):

1. A random DNA library containing T7 Promoter, ribosome binding site, and stem-loop and encoded polypeptide of interest is constructed and amplified by PCR.
2. The constructed DNA library, which does not carry the stop codon, is transcribed into mRNA and purified mRNA (without ligation in ribosome display and with ligation in mRNA display as shown in flow chart) and is used for the cell-free translation reaction.
3. This step is specifically for the mRNA display method only: the resultant mRNA is covalently linked to a puromycin-attached linker-DNA by a hybridization and T4 RNA ligation reaction.
4. mRNA is used as a template for *in vitro* translation to generate the ribosome display or the mRNA display protein fusion (*for enhancing the stability of ribosome complexes, the concentration of Mg^{++} is increased and the purified mRNA-protein fusion is stabilized by converting mRNA into cDNA using a reverse transcription reaction).

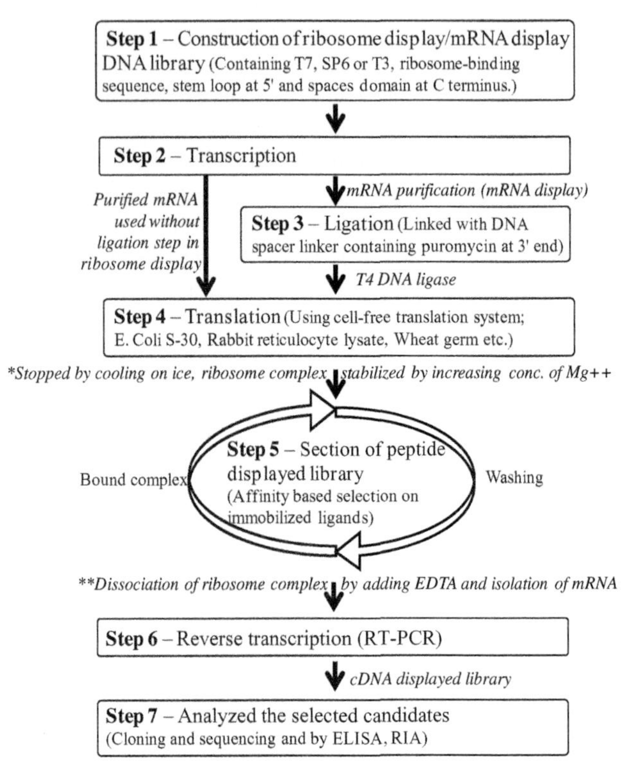

FLOW CHART 20.2 Ribosome display/mRNA display technology

5. The desired ribosome complexes of the cDNA/mRNA-protein library are followed by an affinity-based selection reaction. The library is allowed to bind with the immobilized target protein and those library components that bind the target weakly or non-specifically are removed by washing steps (**the bound ribosome complexes can be dissociated by EDTA or whole complexes can be specifically eluted with the target and then mRNA is isolated). The recovered library is then enriched for target-binding sequences.
6. Isolated mRNA is reverse transcribed to cDNA (in ribosome display). The cDNA (in both ribosome display and mRNA display) is then amplified by PCR and used for the next round of selection. In general, three to six rounds of selection are required to select the peptide with low nanomolar or sub-nano molar affinity to the target, depending on library complexity and the target protein. To select even higher affinities (pM), special strategies (off-rate) are required.
7. The selected candidates are identified by cloning and sequencing, and the binding affinity to the target protein is determined by common methods like ELISA or SPR.

Selection of Peptides

Ribosome display has been successfully applied for *in vitro* antibody selection, evolution, and humanization. Examples include: a 10-mer peptide that binds with dynophin B with

0.29 nM affinity (Mattheakis et al., 1994); a 15-mer streptavidin that binds peptides with an affinity of about 4 nM (Lamla and Erdmann, 2003); a 20-mer peptide that binds with prostate-specific antigen (PSA), a tumor marker (Gersuk et al., 1997). Despite encouraging results from ribosome display selection, it has suffered from mechanistic challenges such as a larger size of ribosome moieties, which hinder the specific binding of displayed peptides to the target protein. Another issue is the inherent instability of the resulting protein–ribosome–mRNA ternary complexes, which makes it difficult to keep this ternary complex intact during the selection step and thus restricts the choice of experimental conditions.

mRNA Display

A more robust cell-free method was devised for the display system, mRNA display or *in vitro* virus (Nemoto et al., 1997; Roberts and Szostak, 1997). Like ribosome display, mRNA display uses a complex between mRNA and the polypeptide encoded by that mRNA as the basic unit of selection. What makes mRNA display more robust than ribosome display is the covalent nature of the linkage between the mRNA and the protein in the mRNA–protein complex. This is achieved by bonding the two macromolecules through a small adaptor molecule (typically puromycin), which is ligated in advance at the end of the mRNA molecule through a short linker DNA molecule (Figure 20.5). Puromycin serves as a chemically stable small molecule mimic of aminoacyl tRNA. So, in the case of a no-stop codon, and when the ribosome reaches the end of the mRNA template, puromycin enters the peptidyltransferase site to form a covalent bond with the nascent peptide chain and makes a mRNA–protein fusion (Flow Chart 20.2).

Selection of Peptides

Streptavidin-binding peptides that contain at least one copy of histidine-proline-glutamine (HPQ) were successfully selected by mRNA display with a Kd of 2.5 nM, which in contrast were much stronger than the peptide selected by phage display with a Kd of 13-72 μM (Wilson et al., 2001). A disulfide-constrained library based on EETI-II, a knottin trypsin inhibitor, was constructed by randomizing the six residues of the trypsin-binding site, and mRNA display was used to select new trypsin-binding peptides. The selected peptides were highly homologous or identical to wild-type EETI-II, and their dissociation constants from trypsin ranged from 16 (for the wild type) to 82 μM (Baggio et al., 2002). Recently, peptide aptamers have been selected that bind with high affinity and specificity to the internal ribosomal entry site (IRES) of hepatitis C virus (HCV) mRNA (Litovchick and Szostak, 2008).

The mRNA display system is apparently the most widely used *in vitro* selection method for a few reasons, such as being *in vitro* translation, which enables both an expanded range of experimental conditions and unnatural amino acids to be incorporated by the exploitation of suppressor codons, and being of a greater diversity in the molecular library. However, there are still a few challenges that remain to be solved. In mRNA display, the purification of protein–puromycin–mRNA adducts from the ribosome presents a topological puzzle. After translation, the protein folds outside of the ribosomal tunnel to a globular domain. At the other end of the tunnel, the puromycin–mRNA reagent reacts with the polypeptide. Thus, a folded domain sits at one end of the tunnel, while the long mRNA is connected to the peptide at the other end. Whereas the purification is performed under conditions expected to dissociate the ribosome, no direct evidence is yet available that an "opening" of the tunnel takes place. Alternative explanations are that (1) the mRNA passes through the protein exit tunnel, or (2) the protein denatures and goes "backwards" through the tunnel. If such denaturation of the displayed protein is required, that might limit the application of mRNA display to proteins with robust refolding properties.

Other Display Systems

Several other strategies linking the displayed protein directly to the encoding DNA, including CIS display, covalent DNA display, cDNA display, have been described. These technologies are in principle analogous to the mRNA display, but using the DNA molecule instead of the mRNA molecule could be advantageous due to the higher stability of the DNA molecules in comparison to labile mRNA molecules. It is not evident that one selection system is advantageous over the others since they all have their advantages and limitations (Figure 20.5). Thus, there is a high demand for an efficient, sophisticated, and combined *in vitro* evolution method that can address the following shortcomings to generate novel functional peptide aptamers for drug discovery:

1. The high diversity and quality of the library are the vital strengths for the success of an *in vitro* selection method. Various methods have been reported, including the most popular error prone PCR and DNA shuffling methods for the construction of the primary library and subsequent libraries. It is yet to be decided how to choose a mutation method for making suitable and efficient primary and subsequent libraries. Therefore, a method is required that has the ability to generate and combine a wide range of sequence diversity.
2. All *in vitro* selection methods discussed above are advantageous especially in finding the peptide aptamers of high affinity from huge molecular diversity. Ideally, for the discovery of therapeutic agents, a method can be more advantageous if it can select the molecules based not only on their binding, but also on their functional abilities.

3. The various methods developed so far have been for identification of mainly enzyme inhibitors. However, there is no such general method that has been developed to identify activators, despite their emerging importance in drug discovery.
4. A general platform that can be widely applied for any type of protein target is desirable, especially in this age of personalized medicines when drug diversity is required.

A General Method for Discovery of Functional Peptide Aptamers

To tackle all the challenges discussed in the above sections, Nishigaki's group recently developed a method called evolutionary Rapid Panning Analysis System (eRAPANSY) (Kitamura et al., 2009) for acquiring protease-inhibiting peptide aptamers. This method consists of primary library construction and selection, and then construction and enrichment of the secondary library using the primary library selection products. It enables one to obtain cathepsin E-inhibiting peptides with higher affinity and activity. What is highlighted here is that in the course of evolution the "*information*" acquired in the preceding rounds of selection should be preserved as a kind of module to be used as the building blocks for successive rounds of selection. It is similar to circuit modules in electronics that can be used as a unit (preserving specific functions), or words composed of letters in linguistics that can constitute different sentences of different meanings (function). Each elementary unit, like resistors and condensers in electronics, or letters and symbols in linguistics, does not convey any meaning, but the circuit modules and the words do, and they can construct a higher-ordered structure of novel functions. Therefore, such an attempt to construct artificial modules in protein engineering is conducted in eRAPANSY. In further work, eRAPANSY was advanced as a general approach by devising a paired peptide method, called a "progressive library method (PLM)," for acquiring highly functional peptide aptamers (Figure 20.6) (Kitamura et al., 2012).

The whole process of PLM consists of three successive library constructions (using basic techniques such as block shuffling for the primary library construction (YLBS), the module-restructuring for the secondary library construction (ASAC), the module (domain) pairing for the third library construction (p&p) and selections (cDNA display for affinity-based selection and SF-link for function-based selection). Thus, the whole process in PLM is highly integrative, sophisticated, and closely resembles the process of natural evolution of proteins. Proteases involved in various human diseases are an attractive target for discovery of therapeutic candidates. Intriguingly, some proteases (e.g. cathepsin E) are found to have anti-tumorigenic roles (Kawakubo et al., 2007). The PLM method was successfully applied to screen protease-regulating peptide aptamers (Biyani et al., 2011). In addition, PLM was also used to develop a novel peptide aptamer-based technology called "pep-ELISA" (as a promising substitution for antibody-based methods), and successfully demonstrated sufficient sensitivity (10 μg/ml) for the detection of cathepsin-E in cancer tissues and plasma (Kitamura et al., 2011). Therefore, we expect that PLM holds great promise in the field of drug discovery (including diagnostics and therapeutics) and in protein science due to its powerful ability to identify *de novo* functional structures of proteins and peptide-to-peptide interactions (Flow Chart 20.3).

The methodology of the Progressive Library Method (PLM) is mainly developed for the *in vitro* evolution of peptide aptamers using cDNA-display (Yamaguchi et al., 2009) and the SF-link method (Naimuddin et al., 2007). The overall strategy of PLM is comprised of three successive library constructions and selections as described below:

1. Primary library construction (YLBS) and selection: The primary library is essential in the identification of bioactive molecules with novel functions. Primary library selection is equal to module finding. The primary library is constructed using Y-ligation-based block shuffling (YLBS) (Kitamura et al., 2002). Briefly, the whole cycle of YLBS is composed of hybridization of two sequences (5′-half and 3′-half), T4 RNA ligase ligation of two variable sequences (equal to blocks), PCR amplification of the ligated DNA, regeneration of 5′- and 3′-half precursor DNAs by restriction cleavage, and recovery of single-stranded DNAs. By repeating these steps, the number of ligated blocks and the diversity of resultant products increase exponentially. Then the cDNA-display and SF-link method are applied to the library to generate a cDNA-encoded peptide library (used for the affinity-based selection) and a cDNA-displayed peptide library tagged with an enzyme–substrate sequence (used for the function-based selection). In affinity-based selection, the selected library products contain all of the inhibitory, activating, and function-free neutral binding peptides, while in the function-based selection, peptides can be differentiated by an inhibitory and an activating peptide by employing the inverse mode of operation using the SF-link method.
2. Secondary library construction (All-Steps-All-Combinations, ASAC) and selection: Secondary library construction and selection allows further refinement of the molecules selected from the primary library. Secondary library selection is equal to module shuffling. The secondary library is constructed by employing the YLBS method with slight modifications. The peptide sequences obtained by the primary library selection were

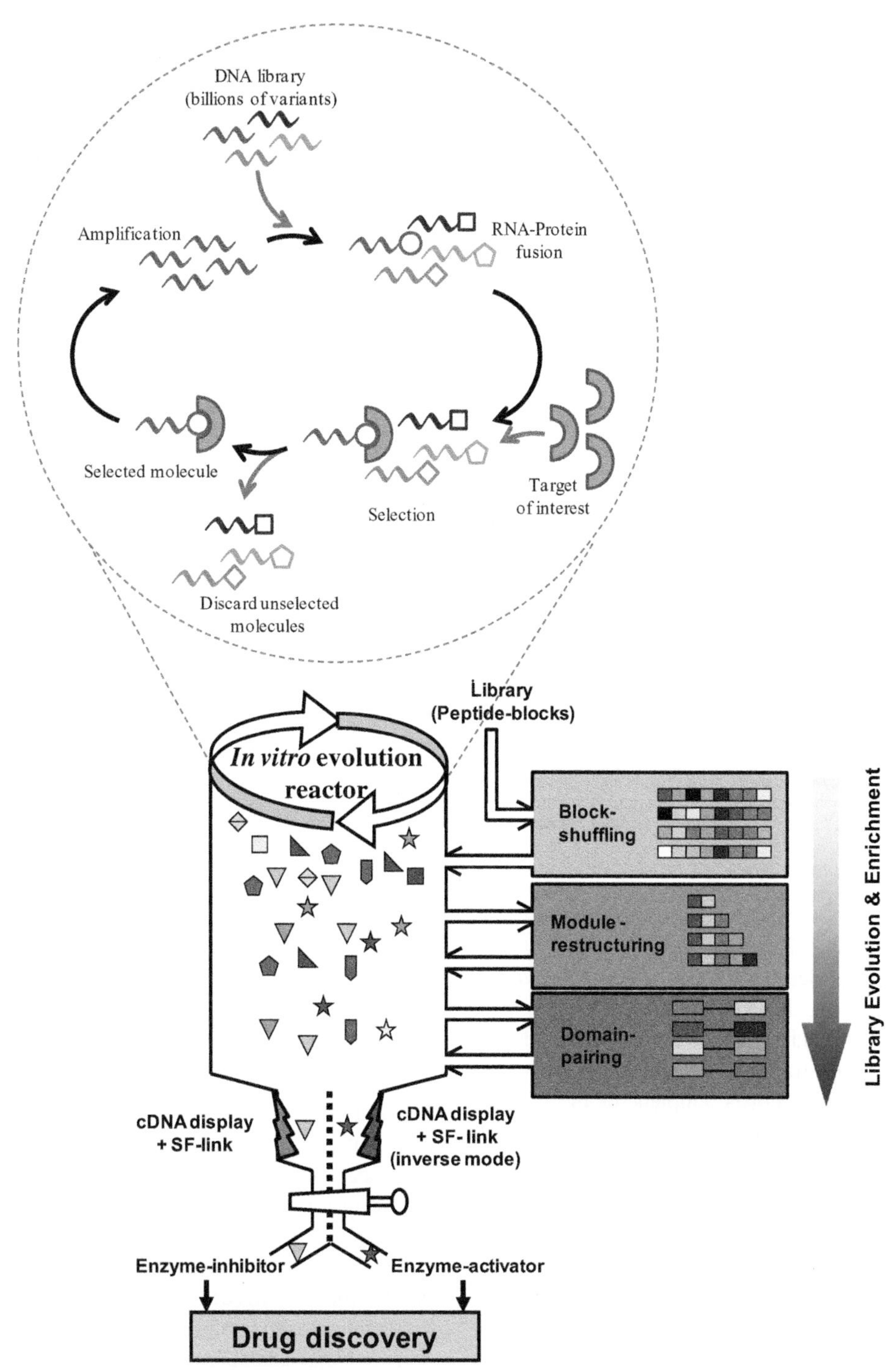

FIGURE 20.6 Schematic drawing of a general method for the *in vitro* evolution of functional peptide aptamers using PLM method.

cluster-analyzed and used to design the blocks to be constructed (tetramer). Using these blocks, YLBS-shuffling is performed. The resulting library contained all of the blocks arbitrarily shuffled with a different number of blocks (2–8 blocks), and therefore it was called an all-steps-all-combinations (ASAC) library. Then cDNA-display and SF-link methods are applied to the library as described above for the primary library and used for selections.

3. Third library construction (pair peptide library, p&p) and selection: The third library plays an important role in enhancing affinity and function at the highest level. Third library selection is equal to module pairing. The third library was constructed by combining two peptides selected from the secondary library. The YLBS method was employed to obtain paired peptides with a linker sequence separating them. In brief, a set of peptide

Step 1 – Primary library construction and selection

a. Construction of the DNA library using YLBS method
b. Convert the library into their cDNA display and SF- link format
c. Perform cDNA display selection cycle (for affinity-based selection)
d. Perform cDNA displayed SF-link selection cycle (function-based selection)
e. Analysis of selected clone peptides by sequencing, SPR, ELISA and *in vitro* /cell-based activity assay

Step 2 – Identification of peptide aptamers and used for the secondary library construction

Step 3 – Secondary library construction and selection

a. Construction of secondary DNA library using ASAC method followed by the same steps as described in Step 1

Step 4 – Identification of more functional peptide aptamers with higher affinity and used for the third library construction

Step 5 – Third library construction and selection

a. Construction of third DNA library using paired peptide method followed by the same steps as described in Step 1.

Step 6 – Identification of further more functional peptide aptamers with the highest affinity and evaluation.

FLOW CHART 20.3 Progressive library method (PLM)

blocks (5′-halves) were combined arbitrarily with a set of linker blocks and then elongated by combing with another set of peptides (3′-halves), which consisted of the same elements as those of the 5′-halves. The diversity of the constructed library was nominally ~10^4; it was then sequenced to check the conformation of the library. The same protocol is adopted for the affinity- and function-based selections of the paired peptide library as described for the primary and the secondary library selections.

CONCLUSION AND FUTURE PERSPECTIVES

Biomolecular display technologies have great potential for providing biologically active molecules for various applications, such as discovery of novel therapeutics, diagnostics, medical imaging molecules, and research reagents. These technologies also have the potential to provide the crucial interface between modern therapeutic target discovery using genomics, proteomics, and bioinformatics methods, and the development of new and potent drugs. For example, molecular probes generated from molecular display technologies for proteins related to a disease can be useful for rapid validation of those proteins as viable therapeutic targets. Therefore, these technologies have received more interest from pharmaceutical companies for rapid generation of such molecules.

Cell-free type display technologies, namely ribosome and mRNA/cDNA display, have several features that should make them more amenable to standardization and automation than phage display. They comprise a fast selection cycle, allowing processing of huge libraries, and are not limited by cellular transformation. Despite the several advantages of cell-free type display technologies, there is still a demand for a sophisticated, integrated, and general method that can become the *in vitro* display technology of choice for many applications. Antibodies and DNA/RNA aptamers obtained from molecular display technologies have shown great progress for use in the therapeutic field, as have analytical methodologies such as affinity chromatography, capillary electrophoresis, mass spectrometry, and biosensors for diagnostic assays. However, a peptide aptamer tool is still in an early stage of evolution. As reported above, the good success (from the clinical stage all the way to the pharmaceutical market; see Table 20.1) and high number of recently published papers on peptide aptamer selection using molecular display technologies represents the promise that peptide aptamers can be suitable molecules for therapeutic and diagnostic applications in the near future.

ETHICAL ISSUES

So far there are no anticipated ethical issues related to this technology. However, the possibility of ethical issues arising due to advancements and applications of this technology are entirely possible. Certainly these kinds of issues will have to be dealt with as (and when) they arise.

TRANSLATIONAL SIGNIFICANCE

Drug discovery is a wide translational area involving contributions from academic institutions, biotech companies, and large pharmaceutical corporations. With the increasing adoption of technological advancements, novel integrated approaches are emerging to generate new therapies. Biomolecular display technologies have the potential to provide the must-needed interface between modern genomics/proteomics-based therapeutic challenges and the screening of potential lead candidates for development of next-generation novel drug compounds. During the past two decades, several display technologies that mimic the process of natural evolution have been developed and applied to tackle several key issues in the process of drug discovery, including the diversity and abundance of drug candidate libraries, speed and cost effectiveness, automation, and the rapid optimization of lead candidates. Furthermore, a greater collaboration between academia, biotech companies, and pharmaceutical corporations can bring speedy progress to the development of new drugs for the market.

TABLE 20.1 Several Therapeutic Peptides That Have Reached the Market

Name	Length	Target	Indication	Company
Hematide	Dimeric	Erythropoietin	Treatment of chronic kidney-related anemia.	Affymax
POT-4	cyclic	C3 convertase	Treatment of age-related macular degeneration (AMD).	Potentia Pharmaceuticals Inc
Acthrel	41 aa	ACTH	Diagnosis of ACTH-dependent Cushing's Syndrome.	Ferring Pharms
Sarenin	8 aa	Angiotensin II receptor antagonist	Treatment of hypertension.	Norwich-Eaton Pharms; Procter & Gamble
Byetta	39 aa	Antidiabetic agents	Glycemic control in patients with type 2 diabetes mellitus.	Amylin Pharms
Fuzeon	36 aa	Anti-HIV	Treatment of AIDS/HIV-1 infection.	Roche
Acticalcin	32 aa	Calcitonins	Treatment of postmenopausal osteoporosis.	AstraZeneca
Angiomax	20 aa	Cardiovascular	Anticoagulant in patients with unstable angina undergoing PTCA.	Nycomed Pharma
Takus	10 aa	Cholecystokinin diethylamine	Diagnosis of functional state of gallbladder and pancreas.	Pharmacia and Upjohn
Geref	29 aa	GHRH and analogue	Growth hormone deficiency, diagnosis evaluation of pituitary function.	Serono Labs
Bigonist	9 aa	GnRH and analogs	Treatment of advanced prostate cancer.	Sanofi-Aventis
Antocin	9 aa	Oxytocin, antagonist	Delaying the birth in case of threat of premature birth.	Ferring Pharms
NeoTect	10 aa	Somatostatin	Diagnosis of lung tumors.	Amersham H.
Pitressin	9 aa	Vasopressin	Central diabetes insipidus.	Monarch
Naaxia	2 aa	Spaglumat	Allergic rhinitis and conjunctivitis.	Laboratoire Thea

(Adapted and updated from Vlieghe et al., 2010.)

REFERENCES

Ashraf, S. S., Anderson, E., Duke, K., Hamilton, P. T., & Fredericks, Z. (2003). *The Journal of Peptide Research*, *61*, 263–273.

Baggio, R., Burgstaller, P., Hale, S. P., Putney, A. R., Lane, M., Lipovsek, D., Wright, M. C., Roberts, R. W., Liu, R., Szostak, J. W., & Wagner, R. W. (2002). Identification of epitope-like consensus motifs using mRNA display. *Journal of Molecular Recognition*, *15*, 126–134.

Bertschinger, J., Grabulovski, D., & Neri, D. (2007). Selection of single domain binding proteins by covalent DNA display. *Protein Engineering, Design & Selection*, *20*, 57–68.

Biyani, M., Futakami, M., Kitamura, K., Kawakubo, T., Suzuki, M., Yamamoto, K., & Nishigaki, K. (2011). *In vitro* selection of cathepsin E-activity enhancing peptide aptamers at neutral pH. *International Journal of Peptides*, 834525.

Boder, E. T., & Wittrup, K. D. (1997). Yeast surface display for screening combinatorial polypeptide libraries. *Nature Biotechnology*, *15*, 553–557.

Crawford, M., & Woodman, R. (2003). Ko Ferrigno P. Peptide aptamers: tools for biology and drug discovery. *Briefings in Functional Genomics Proteomics*, *2*, 72–79.

Ellington, A. D., & Szostak, J. W. (1990). In vitro selection of RNA molecules that bind specific ligands. *Nature*, *346*, 818–822.

Francisco, J. A., Earhart, C. F., & Georgiou, G. (1992). Transport and anchoring of beta-lactamase to the external surface of Escherichia coli. *Proceedings of the National Academy of Sciences of the United States of America*, *89*, 2713–2717.

Franklin, M. C., Kadkhodayan, S., Ackerly, H., Alexandru, D., Distefano, M. D., Elliott, L. O., Flygare, J. A., Mausisa, G., Okawa, D. C., Ong, D., Vucic, D., Deshayes, K., & Fairbrother, W. J. (2003). *Biochemistry*, *42*, 8223–8231.

George, R. A., Smith, T. D., Callaghan, S., Hardman, L., Pierides, C., Horaitis, O., Wouters, M. A., & Cotton, R. G. (2008). General mutation databases: analysis and review. *Journal of Medical Genetics*, *45*, 65–70.

Gersuk, G. M., Corey, M. J., Corey, E., Stray, J. E., Kawasaki, G. H., & Vessella, R. L. (1997). High-affinity peptide ligands to prostate-specific antigen identified by polysome selection. *Biochemical and Biophysical Research Communications, 232*, 578–582.

Gold, L., & Tuerk, C. (1990). Systematic evolution of ligands by exponential enrichment: RNA ligands to bacteriophage T4 DNA polymerase. *Science, 249*, 505–510.

Gragoudas, E. S., Adamis, A. P., Cunningham, E. T., Jr., Feinsod, M., & Guyer, D. R. (2004). Pegaptanib for neovascular age-related macular degeneration. VEGF Inhibition Study in Ocular Neovascularization Clinical Trial Group. *The New England Journal of Medicine, 351*, 2805–2816.

Hetian, L., Ping, A., Shumei, S., Xiaoying, L., Luowen, H., Jian, W., Lin, M., Meisheng, L., Junshan, Y., & Chengchao, S. (2002). *The Journal of Biological Chemistry, 277*, 43137–43142.

Huang, L., Sexton, D. J., Skogerson, K., Devlin, M., Smith, R., Sanyal, I., Parry, T., Kent, R., Enright, J., Wu, Q. L., Conley, G., DeOliveira, D., Morganelli, L., Ducar, M., Wescott, C. R., & Ladner, R. C. (2003). Novel peptide inhibitors of angiotensin-converting enzyme 2. *The Journal of Biological Chemistry, 278*, 15532–15540.

Kawakubo, T., Okamoto, K., Iwata, J., Shin, M., Okamoto, Y., Yasukochi, A., Nakayama, K. I., Kadowaki, T., Tsukuba, T., & Yamamoto, K. (2007). Cathepsin E prevents tumor growth and metastasis by catalyzing the proteolytic release of soluble TRAIL from tumor cell surface. *Cancer Research, 67*, 10869–10878.

Kitamura, K., Kinoshita, Y., Narasaki, S., Nemoto, N., Husimi, Y., & Nishigaki, K. (2002). Construction of block-shuffled libraries of DNA for evolutionary protein engineering: Y-ligation-based block shuffling. *Protein Engineering, 15*, 843–853.

Kitamura, K., Yoshida, C., Kinoshita, Y., Kadowaki, T., Takahashi, Y., Tayama, T., Kawakubo, T., Naimuddin, M., Salimullah, M., Nemoto, N., Hanada, K., Husimi, Y., Yamamoto, K., & Nishigaki, K. (2009). Development of systemic in vitro evolution and its application to generation of peptide-aptamer-based inhibitors of cathepsin E. *Journal of Molecular Biology, 387*, 1186–1198.

Kitamura, K., Biyani, M., Futakami, M., Suzuki, M., Kawakubo, T., Yamamoto, K., & Nishigaki, K. (2011). Peptide aptamer-based ELISA-like system for detection of cathepsin E in tissues and plasma. *Journal of Molecular Biomarkers Diagnostic*, 2, 104. (http://dx.doi.org/10.4172/2155-9929.1000104.)

Kitamura, K., Komatsu, M., Biyani, M., Futakami, M., Kawakubo, T., Yamamoto, K., & Nishigaki, K. (2012). Proven in vitro evolution of protease cathepsin E inhibitors and activators by a paired peptides method. *Journal of Peptide Science*. http://dx.doi.org/10.1002/psc.2453.

Kola, I., & Landis, J. (2004). Can the pharmaceutical industry reduce attrition rates? *Nature Reviews. Drug Discovery, 3*, 711–715.

Lamla, T., & Erdmann, V. A. (2003). Searching sequence space for high-affinity binding peptides using ribosome display. *Journal of Molecular Biology, 329*, 381–388.

Li, M. (2000). Applications of display technology in protein analysis. *Nature Biotechnology, 18*, 1251–1256.

Litovchick, A., & Szostak, J. W. (2008). Selection of cyclic peptide aptamers to HCV IRES RNA using mRNA display. *Proceedings of the National Academy of Sciences of the United States of America, 105*, 15293–15298.

Mattheakis, L. C., Bhatt, R. R., & Dower, W. J. (1994). An in vitro polysome display system for identifying ligands from very large peptide libraries. *Proceedings of the National Academy of Sciences of the United States of America, 91*, 9022–9026.

Naimuddin, M., Kitamura, K., Kinoshita, Y., Honda-Takahashi, Y., Murakami, M., Ito, M., Yamamoto, K., Hanada, K., Husimi, Y., & Nishigaki, K. (2007). Selection-by-function: efficient enrichment of cathepsin E inhibitors from a DNA library. *Journal of Molecular Recognition, 20*, 58–68.

Nemoto, N., Miyamoto-Sato, E., Husimi, Y., & Yanagawa, H. (1997). In vitro virus: bonding of mRNA bearing puromycin at the 3′-terminal end to the C-terminal end of its encoded protein on the ribosome in vitro. *FEBS Letters, 414*, 405–408.

Nouvion, A. L., Thibaut, J., Lohez, O. D., Venet, S., Colas, P., Gillet, G., & Lalle, P. (2007). Modulation of Nr-13 antideath activity by peptide aptamers. *Oncogene, 26*, 701–710.

Odegrip, R., Coomber, D., Eldridge, B., Hederer, R., Kuhlman, P. A., et al. (2004). CIS display: In vitro selection of peptides from libraries of protein-DNA complexes. *Proceedings of the National Academy of Sciences of the United States of America, 101*, 2806–2810.

Pammolli, F., Magazzini, L., & Riccaboni, M. (2011). The productivity crisis in pharmaceutical R&D. *Nature Reviews. Drug Discovery, 10*, 428–438.

Roberts, R. W., & Szostak, J. W. (1997). RNA–peptide fusions for the in vitro selection of peptides and proteins. *Proceedings of the National Academy of Sciences of the United States of America, 94*, 12297–12302.

Russ, A. P., & Lampel, S. (2005). The druggable genome: an update. *Drug Discovery Today, 10*, 1607–1610.

Sakharkar, M. K., Sakharkar, K. R., & Pervaiz, S. (2007). Druggability of human disease genes. *The International Journal of Biochemistry & Cell Biology, 39*, 1156–1164.

Smith, G. P. (1985). Filamentous fusion phage: novel expression vectors that display cloned antigens on the virion surface. *Science, 228*, 1315–1317.

Smith, G. P., & Petrenko, V. A. (1997). Phage display. *Chemistry Reviews, 97*, 391–410.

Tawfik, D. S., & Griffiths, A. D. (1998). Man-made cell-like compartments for molecular evolution. *Nature Biotechnology, 16*, 652–656.

Vlieghe, P., Lisowski, V., Martinez, J., & Khrestchatisky, M. (2010). Synthetic therapeutic peptides: science and market. *Drug Discovery Today, 15*, 40–56.

Wilson, D. S., Keefe, A. D., & Szostak, J. W. (2001). The use of mRNA display to select high-affinity protein-binding peptides. *Proceedings of the National Academy of Sciences of the United States of America, 98*, 3750–3755.

Yamaguchi, J., Naimuddin, M., Biyani, M., Sasaki, T., Machida, M., Kubo, T., Funatsu, T., Husimi, Y., & Nemoto, N. (2009). cDNA display: a novel screening method for functional disulfide-rich peptides by solid-phase synthesis and stabilization of mRNA-protein fusions. *Nucleic Acids Research*, 37. e108.

FURTHER READING

Douthwaite, Julie A., & Ronald, H. (2012). In Ribosome Display and Related Technologies: Methods and Protocols. *Methods in Molecular Biology* (Vol. 805). Jackson: Humana Press.

Nakata, T. (2002). Total synthesis of macrolides. In S. Omura (Ed.), *Macrolide Antibiotics: Chemistry, Biology and Practice* (pp. 220–232). Academic Press.

Ng, Rick (2009). *Drug Discovery: Large Molecule Drugs. 'Drugs, from Discovery to Approval'* (2nd ed.). . 93–135.

Sidhu, Sachdev S. (Ed.), (2005). *Phage Display in Biotechnology and Drug Discovery*. Boca Raton: CRC Press.

Smith, Charles G., & O'Donnel, James T. (Eds.), (2006). *The Process of New New Drug Discovery and Development* (2nd ed.). Boca Raton: CRC Press.

Spilker, Bert (2009). Biotechnology. In *Guide to Drug Development, A Comprehensive Review and Assessment* (pp. 119–131). Wolters Kluwer Press.

GLOSSARY

Affinity Descriptive, qualitative term that indicates the relative tendency of one molecular entity to associate or interact with another.

Biomolecular Drug (Biologics) A drug made from living organisms and their products, such as a serum, toxin, antitoxin, or analogous product applicable to the prevention, treatment, or cure of diseases.

Combinatorial Library Set of compounds prepared by combinatorial chemistry. It may consist of a collection of pools, or sub-libraries.

Directed Evolution A method used in protein engineering to harness the power of natural selection to evolve proteins or nucleic acids with desirable properties not found in nature.

Druggable Site The portion of a genome that can be targeted by a drug.

Enzyme-Linked Immunosorbent Assay (ELISA) Heterogeneous assay in which an antibody linked to an enzyme is used to detect the quantity of antigen present in a sample.

Epitope Any part of a molecule that acts as an antigenic determinant. A macromolecule can contain many differerent eptiopes, each capable of stimulating production of a different specific antibody.

Functional Assay Assay in which the biological or physiological activity of the target is measured.

In Vitro In the test tube; refers to a study in the laboratory using isolated organs, tissues, cells, or biochemical systems.

***In Vitro* Selection** Selection for phenotypes (traits) in test tubes expressed at the cellular or callus level that usually possess genetic changes that control the trait.

In Vivo In the living body; refers to a study performed on a living organism.

Lead Identification A process targeted toward the generation of at least one compound series that meets the requirements for progression to lead optimization.

Lead Optimization A process in which the drug-like properties of an initial lead or lead series are improved.

Library A set of compounds (samples) produced through combinatorial chemistry or other means that expands around a single core structure.

Peptide Aptamer A new class of biomolecules with a peptide moiety of randomized sequence that are selected for their ability to bind to a given target protein under intracellular conditions.

Small Molecule Drug A drug made of organic molecules of small molecular weight (less than 1,000 Daltons) that possesses a biological activity against a protein or molecular target responsible for causing the disease (e.g. acetylsalicylic acid, aspirin).

ABBREVIATIONS

ACE2 Angiotensin-Converting Enzyme-2
ASAC All-Step-All-Combination
DNA Deoxyribonucleic Acid
E. coli *Escherchia coli*
EDTA Ethylenediaminetetraacetic Acid
EETI-ii Ecballium Elaterium Trypsin Inhibitor
ELISA Enzyme-Linked Immunosorbent Assay
eRAPANSY evolutionary Rapid Panning Analysis System
ErbB-2 Erythroblastic Leukemia Viral Oncogene Homolog
FDA Food and Drug Administration
H1N1 Hemagglutinin type-1 and Neuraminidase type-1
IAP Inhibitor of Apoptosis
IRES Internal Ribosomal Entry Site
IVC In Vitro Compartmentalization
kDa Kilodalton
KDR Kinase Domain-Containing Receptor
mRNA Messenger Ribonucleic Acid
nM Nanomolar
P&P Pair and Peptide
PCR Polymerase Chain Reaction
PEG Polyethylene Glyocol
PKC Protein Kinase Cα
PLM Progressive Library Method
PSA Prostate-Specific Antigen
RNA Ribonucleic Acid
SELEX Systematic Evolution of Ligands by Exponential Enrichment
tRNA Transfer Ribonucleic Acid
VEGF Vascular Endothelial Growth Factor
YLBS Y-Ligation Block Shuffling

LONG ANSWER QUESTIONS

1. What are the various stages of the drug discovery process? Give a brief description of each stage.
2. What is the most expensive part of the drug discovery process and why? What is the solution/approach to deal with this issue based on recently developed biotechnologies?
3. What is directed molecular evolution and what is the significance of library enrichment in the process of directed molecular evolution?
4. What is biomolecular display technology? Describe the various types of available display technologies and compare their significance to each other.
5. What is the significance of structural biology in the drug discovery process? How do biomolecular structures interact with each other and with drug molecules in their environment? Explain with examples.

SHORT ANWERS QUESTIONS

1. Differentiate small molecular drug from biomolecular drugs.
2. What is mRNA display technology and how is it different from ribosome display technology?
3. What is eRAPANSY and the progressive library method?
4. What are aptamers? Describe their types with examples.

5. What are the differences between antibodies and peptide aptamer-based therapeutics?

ANSWERS TO SHORT ANSWERS QUESTIONS

1. Small molecular drugs and biomolecular drugs can be differentiated as indicated in Table 20.2.
2. mRNA display is a cell-free system for the *in vitro* selection of proteins and peptides. It uses the principle linkage between nascent proteins (phenotypes) and their corresponding mRNA (genotype) through the formation of mRNA–polypeptide complexes. mRNA display technology has the following advantages that differentiate it from the ribosome display technology.
   a. Linkage: mRNA display uses covalent mRNA–polypeptide complexes linked through puromycin, whereas ribosome display uses stalled, non-covalent ribosome–mRNA–polypeptide complexes.
   b. Library Size: mRNA display can screen a library of size near 10^{14} or more, while for ribosome display technology it is 10^{12} or less.
   c. Selection Pressure: In ribosome display, selection stringency is limited to keep the ribosome–mRN–polypeptide in a complex because of the non-covalent ribosome–mRNA–polypeptide complexes; this is not restricted in the mRNA display method.
   d. Molecular Weight of the Linkage Moiety: The polypeptides in a ribosome display system are attached to a ribosome, which has a molecular weight of ~10^6 Da. There might be some unpredictable interaction between the selection target and the ribosome, and this may lead to a loss of potential binders during the selection cycle. In contrast, the puromycin-modified DNA spacer linker used in mRNA display technology is much smaller compared to a ribosome. This linker may have less chance to interact with an immobilized selection target. Thus, mRNA display technology is more likely to give less biased results.
3. eRAPANSY (evolutionary Rapid Panning Analysis System) is a systemic *in vitro* evolution method used to generate protease-inhibiting peptide aptamers. It was introduced by Nishigaki's lab in 2009. This method consists of primary library construction and selection, and then construction and enrichment of the secondary library using the primary library selection products. This method is successfully reported to select out the high affinity and inhibiting peptide aptamers for the protease cathepsin E, and thus their significant role in cancer therapeutics. The Progressive Library Method (PLM) is an advanced version of eRAPANSY that uses a paired peptide method for the third library construction to obtain highly functional peptide aptamers.
4. Aptamers are short oligonucleic acid or peptide molecules that bind to a specific target molecule with high affinity. Aptamers are usually generated by selecting them from a large random sequence pool via *in vitro* selection methods such as systematic evolution of ligands by exponential enrichment (SELEX), and display technologies. Aptamers can be used for both basic research and clinical purposes as macromolecular drugs. These compound molecules have additional research, industrial, and clinical applications. Aptamers can be classified as:
   a. Nucleic Acid Aptamers (DNA or RNA): Nucleic acid aptamers are engineered through repeated rounds of *in vitro* selection or equivalently, SELEX, to bind to various molecular targets. Example: *Pegaptanib,* an anti-VEGF aptamer for treatment of human ocular vascular disease.
   b. Peptide Aptamers: the peptide aptamer are short amino acids sequences (10–30 amino acids). They can be selected from combinatorial peptide libraries constructed by phage display, and other display system such as ribosome display and mRNA display. Example: *Fuzeon,* an anti-HIV aptamer for treatment of AIDS infection.
5. Please see Figure 20.3.

TABLE 20.2 Small Molecular and Biomolecular Drugs

Small Molecular Drugs	Biomolecular Drugs
Organic compound in nature (e.g. Asprin)	Biological product in nature (e.g. nucleic acids, antibiotics, peptides, etc.).
Low molecular weight size (< 800 Daltons)	Smaller (e.g. 3 kDa peptide) to larger (150,000 kDa antibiotic) size.
Chemically synthesized and produced	Both, chemically and biologically synthesized and produced.
Traditional drugs	New generation drugs.
Toxicity is high	Low toxicity.

Chapter 21

In Silico Models: From Simple Networks to Complex Diseases

Debmalya Barh*, Vijender Chaitankar†, Eugenia Ch Yiannakopoulou**, Emmanuel O Salawu††, Sudhir Chowbina***, Preetam Ghosh† and Vasco Azevedo†††

*Centre for Genomics and Applied Gene Technology, Institute of Integrative Omics and Applied Biotechnology (IIOAB), Nonakuri, Purba Medinipur, India, †Department of Computer Science, Virginia Commonwealth University, Richmond, Virginia, USA, **Department of Basic Medical Lessons Faculty of Health and Caring Professions, Technological Educational Institute of Athens, Athens, Greece, ††Institute of Bioinformatics and Structural Biology, National Tsing Hua University, Hsinchu, Taiwan, PhD Bioinformatics Program, Taiwan International Graduate Program, Academia Sinica, Taipei, Taiwan, and Institute of Information Science, Academia Sinica, Taipei, Taiwan, ***Advanced Biomedical Computing Center, SAIC-Frederick, Inc., Frederick National Laboratory for Cancer Research, National Cancer Institute, Frederick, Maryland, USA, †††Laboratorio de Genetica Celular eMolecular, Departmento de Biologia Geral, Instituto de Ciencias Biologics, Universidade Federal de Minas Gerais Belo Horizonte, Minas Gerais, Brazil

Chapter Outline

SUMMARY

In this chapter, we consider *in silico* modeling of diseases starting from some simple to some complex (and mathematical) concepts. Examples and applications of *in silico* modeling for some important categories of diseases (such as for cancers, infectious diseases, and neuronal diseases) are also given.

WHAT YOU CAN EXPECT TO KNOW

Recent advances in bioinformatics and systems biology have enabled modeling and simulation of sub-cellular and cellular processes, and disease using primarily methods from dynamical systems theory. In this approach, all interactions among all components in a system are described

Animal Biotechnology. http://dx.doi.org/10.1016/B978-0-12-416002-6.00021-3

mathematically and computed models are established. These *in silico* models encode and test hypotheses about mechanisms underlying the function of cells, the pathogenesis and pathophysiology of disease, and contribute to identification of new drug targets and drug design. The development of *in silico* models is facilitated by rapidly advancing experimental and analytical tools that generate information-rich, high-throughput biological data. Bioinformatics provides tools for pattern recognition, machine learning, statistical modeling, and data extraction from databases that contribute to *in silico* modeling. Dynamical systems theory is the natural language for investigating complex biological systems that demonstrate nonlinear spatio-temporal behavior. Most *in silico* models aim to complement (and not replace) experimental research. Experimental data are needed for parameterization, calibration, and validation of *in silico* models. Typical examples in biology are models for molecular networks, where the behavior of cells is expressed in terms of quantitative changes in the levels of transcripts and gene products, as well as models of cell cycle. In medicine, *in silico* models of cancer, immunological disease, lung disease, and infectious diseases complement conventional research with *in vitro* models, animal models, and clinical trials. This chapter presents basic concepts of bioinformatics, systems biology, their applications in *in silico* modeling, and also reviews applications in biology and disease.

HISTORY AND METHODS

BIOINFORMATICS IN ANIMAL BIOTECHNOLOGY

Biotechnology will be the most promising life science frontier for the next decade. Together with informatics, biotechnology is leading revolutionary changes in our society and economy. This genomic revolution is global, and is creating new prospects in all biological sciences, including medicine, human health, disease, and nutrition, agronomy, and animal biotechnology.

Animal biotechnology is a source of innovation in production and processing, profoundly impacting the animal husbandry sector, which seeks to improve animal product quality, health, and well-being. Biotechnological research products, such as vaccines, diagnostics, *in vitro* fertilization, transgenic animals, stem cells, and a number of other therapeutic recombinant products, are now commercially available. In view of the immense potential of biotechnology in the livestock and poultry sectors, interest in animal biotechnology has increased over the years.

The fundamental requirement for modern biotechnology projects is the ability to gather, store, classify, analyze, and distribute biological information derived from genomics projects. Bioinformatics deals with methods for storing, retrieving, and analyzing biological data and protein sequences, structures, functions, pathways, and networks, and recently, *in silico* disease modeling and simulation using systems biology. Bioinformatics encompasses both conceptual and practical tools for the propagation, generation, processing, and understanding of scientific ideas and biological information.

Genomics is the scientific study of structure, function, and interrelationships of both individual genes and the genome. Lately, genomics research has played an important role in uncovering the building blocks of biology and complete genome mapping of various living organisms. This has enabled researchers to decipher fundamental cellular functions at the DNA level, such as gene regulation or protein–protein interactions, and thus to discover molecular signatures (clusters of genes, proteins, metabolites, etc.), which are characteristic of a biological process or of a specific phenotype. Bioinformatics methods and databases can be developed to provide solutions to challenges of handling massive amounts of data.

The history of animal biotechnology with bioinformatics is to make a strong research community that will build the resources and support veterinary and agricultural research. There are some technologies that were used dating back to 5,000 B.C. Many of these techniques are still being used today. For example, hybridizing animals by crossing specific strains of animals to create greater genetic varieties is still in practice. The offspring of some of these crosses are selectively bred afterward to produce the most desirable traits in those specific animals.

There has been significant interest in the complete analysis of the genome sequence of farm animals such as chickens, pigs, cattle, sheep, fish, and rabbits. The genomes of farm animals have been altered to search for preferred phenotypic traits, and then selected for better-quality animals to continue into the next generation. Access to these sequences has given rise to genome array chips and a number of web-based mapping tools and bioinformatics tools required to make sense of the data. In addition, organization of gigabytes of sequence data requires efficient bioinformatics databases. Fadiel et al. (2005) provides a nice overview of resources related to farm animal bioinformatics and genome projects.

With farm animals consuming large amounts of genetically modified crops, such as modified corn and soybean crops, it is good to question the effect this will have on their meat. Some of the benefits of this technology are that what once took many years of trial and error is now completed in just months. The meats that are being produced are coming from animals that are better nourished by the use of biotechnology. Biotechnology and conventional approaches are benefiting both poultry and livestock producers. This will give a more wholesome affordable product that will meet growing population demands.

Moreover, bioinformatics methods devoted to investigating the genomes of farm animals can bring eventual economic benefits, such as ensuring food safety and better food quality in the case of beef. Recent advances in high-throughput DNA sequencing techniques, microarray technology, and proteomics have led to effective research in bovine muscle physiology to improve beef quality, either by breeding or rearing factors. Bioinformatics is a key tool to analyze the huge datasets obtained from these techniques. The computational analysis of global gene expression profiling at the mRNA or protein level has shown that previously unsuspected genes may be associated either with muscle development or growth, and may lead to the development of new molecular indicators of tenderness. Gene expression profiling has been used to document changes in gene expression; for example, following infection by pathological organisms, during the metabolic changes imposed by lactation in dairy cows, in cloned bovine embryos, and in various other models.

Bioinformatics enrichment tools are playing an important role in facilitating the functional analysis of large gene lists from various high-throughput biological studies. Huang *et al.* discusses 68 bioinformatics enrichment tools, which helps us understand their algorithms and the details of a particular tool. However, in biology genes do not act independently, but in a highly coordinated and interdependent manner. In order to understand the biological meaning, one needs to map these genes into Gene-Ontology (GO) categories or metabolic and regulatory pathways. Different bioinformatics approaches and tools are employed for this task, starting form GO-ranking methods, pathway mappings, and biological network analysis (Werner, 2008). The awareness of these resources and methods is essential to make the best choices for particular research interests.

Knowledge of bioinformatics tools will facilitate their wide application in the field of animal biotechnology. Bioinformatics is the computational data management discipline that helps us gather, analyze, and represent this information in order to educate ourselves, understand biological processes in healthy and diseased states, and to facilitate discovery of better animal products. Continued efforts are required to develop cost-effective and efficient computational platforms that can retrieve, integrate, and interpret the knowledge behind the genome sequences. The application of bioinformatics tools for biotechnology research will have significant implications in the life sciences and for the betterment of human lives. Bioinformatics is being adopted worldwide by academic groups, companies, and national and international research groups, and it should be thought of as an important pillar of current and future biotechnology, without which rapid progress in the field would not be possible. Systems approaches in combination with genomics, proteomics, metabolomics, and kinomics data have tremendous potential for providing insights into various biological mechanisms, including the most important human diseases.

BIOINFORMATICS AND SYSTEMS BIOLOGY

We are witnessing the birth of a new era in biology. The ability to uncover the genetic code of living organisms has dramatically changed the biological and biomedical sciences approach towards research. These new approaches have also brought in newer challenges.

One such challenge is that recent and novel technologies produce biological datasets of ever-increasing size, including genomic sequences, RNA and protein abundances, their interactions with each other, and the identity and abundance of other biological molecules. The storage and compilation of such quantities of biological data is a challenge: the human genome, for example, contains 3 billion chemical units of DNA, whereas a protozoan genome has 670 billion units of DNA. Data management and interpretation requires development of newly sophisticated computational methods based on research in biology, medicine, pharmacology, and agricultural studies and using methods from computer science and mathematics – in other words, the multi-disciplinary subject of bioinformatics.

Bioinformatics enables researchers to store large datasets in a standard computer database format and provides tools and algorithms scientists use to extract integrated information from the databases and use it to create hypotheses and models. Bioinformatics is a growth area because almost every experiment now involves multiple sources of data, requiring the ability to handle those data and to draw out inferences and knowledge. After 15 years of rapid evolution, the subject is now quite ubiquitous.

Another challenge lies in deciphering the complex interactions in biological systems, known as systems biology. Systems biology can be described as a biology-based interdisciplinary field of study that focuses on complex interactions of biological systems. Those in the field claim that it represents a shift in perspective towards holism instead of reductionism.

Systems biology has great potential to facilitate development of drugs to treat specific diseases. The drugs currently on the market can target only those proteins that are known to cause disease. However, with the human genome now completely mapped, we can target the interaction of genes and proteins at a systems biology level. This will enable the pharmaceutical industry to design drugs that will only target those genes that are diseased, improving healthcare in the United States. Like two organs in one body, systems analysis and bioinformatics are separate but interdependent.

COMMON COMPUTATIONAL METHODS IN SYSTEMS BIOLOGY

Computational methods take an interdisciplinary approach, involving mathematicians, chemists, biologists, biochemists, and biomedical engineers. The robustness of datasets related to gene interaction and co-operation at a systems level requires multifaceted approaches to create a hypothesis that can be tested. Two approaches are used to understand the network interactions in systems biology, namely Experimental and Theoretical and Modeling techniques (Choi, 2007). In the following sections is a detailed overview of the different computational or bioinformatics methods in modern systems biology.

EXPERIMENTAL METHODS IN SYSTEMS BIOLOGY

Experimental methods utilize real situations to test the hypothesis of mined data sets. As such, living organisms are used whereby various aspects of genome-wide measurements and interactions are monitored. Specific examples on this point include:

PROTEIN–PROTEIN INTERACTIONS (PPIs)

Protein–protein interaction predictions are methods used to predict the outcome of pairs or groups of protein interactions. These predictions are done *in vivo*, and various methods can be used to carry out the predictions. Interaction prediction is important as it helps researchers make inferences of the outcomes of PPI. PPI can be studied by phylogenetic profiling, identifying structural patterns and homologous pairs, intracellular localization, and post-translational modifications, among others. A survey of available tools and web servers for analysis of protein–protein interactions is provided by Tuncbag et al., 2009.

TRANSCRIPTIONAL CONTROL NETWORKS

Within biological systems, several activities involving the basic units of a gene take place. Such processes as DNA replication, and RNA translation and transcription into proteins must be controlled; otherwise, the systems could yield numerous destructive or useless gene products. Transcriptional control networks, also called gene regulatory networks, are segments within the DNA that govern the rate and product of each gene.

Bioinformatics have devised methods to look for destroyed, dormant, or unresponsive control networks. The discovery of such networks helps in corrective therapy, hence the ability to control some diseases resulting from such control network breakdowns. There has also been rapid progress in the development of computational methods for the genome-wide "reverse engineering" of such networks. ARACNE is an algorithm to identify direct transcriptional interactions in mammalian cellular networks, and promises to enhance the ability to use microarray data to elucidate cellular processes and to identify molecular targets of pharmacological drugs in mammalian cellular networks. In addition to methods like ARACNE, systems biology approaches are needed that incorporate heterogeneous data sources, such as genome sequence and protein–DNA interaction data. The development of such computational modeling techniques to include diverse types of molecular biological information clearly supports the gene regulatory network inference process and enables the modeling of the dynamics of gene regulatory systems. One such technique is the template-based method to construct networks. An overview of the method is shown in Flow Chart 21.1.

The template-based transcriptional control network reconstruction method exploits the principle that orthologous proteins regulate orthologous target genes. Given a genome of interest (GoI), the first step is to select the template genome (TG) and known regulatory interactions (i.e. template network, TN) in this genome. In step 2, for every protein (P) in TN, a blast search is performed against GoI to obtain the best hit sequences (Px). In step three these Px are then used as a query to perform a blast search against TG. If the best hit using Px as a query happens to be P, then both P and Px are selected as orthologous proteins in step four. If orthologs were detected for an interacting P and target gene then the interaction is transferred in GoI in the final step. Note that this automated way of detecting orthologs can infer false positives.

SIGNAL TRANSDUCTION NETWORKS

Signal transduction is how cells communicate with each other. Signal transduction pathways involve interactions between proteins, micro- and macro-molecules, and DNA. A breakdown in signal transduction pathways could lead

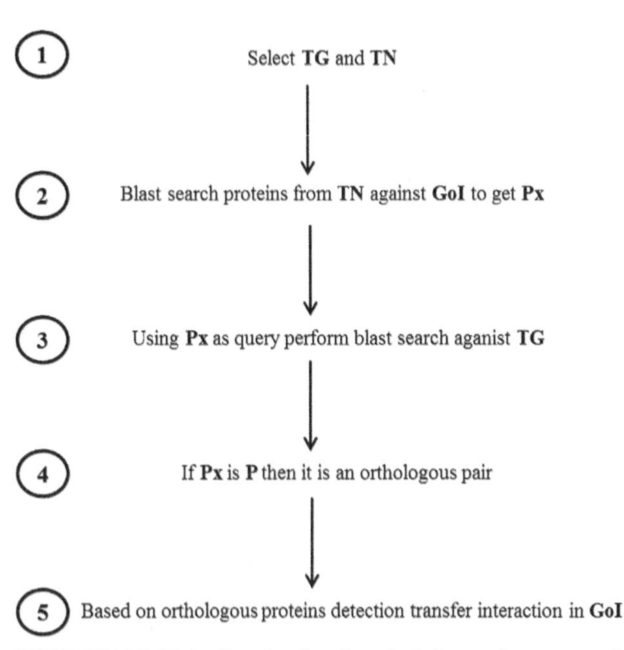

FLOW CHART 21.1 Template-based method for regulatory network reconstruction.

to detrimental consequences within the system due to lack of integrated communication. Correction of broken signal transduction pathways is a therapeutic approach researched for use in many areas of medicine.

High-throughput and multiplex techniques for quantifying signaling and cellular responses are becoming increasingly available and affordable. A high-throughput quantitative multiplex kinase assay, mass spectrometry-based proteomics, and single-cell proteomics are a few of the experimental methods used to elucidate signal transduction mechanisms of cells. These large-scale experiments are generating large data sets on protein abundance and signaling activity. Data-driven modeling approaches such as clustering, principal components analysis, and partial least squares need to be developed to derive biological hypotheses. The potential of data-driven models to study large-scale data sets quantitatively and comprehensively will make sure that these methods will emerge as standard tools for understanding signal-transduction networks.

MATHEMATICAL MODELING TECHNIQUES

The systems biology and mathematical biology fields focus on modeling biological systems. Computational systems biology aims to develop computational models of biological systems. Specifically, it focuses on developing and using efficient algorithms, data structures, visualization tools, and communication tools. A mathematical model can provide new insights into a biological model of interest and help in generating testable predictions.

Modeling or simulation can be viewed as a way of creating an artificial biological system *in vitro* whose properties can be changed or made dynamic. By externally controlling the model, new datasets can be created and implemented at a systems level to create novel insights into treating gene-related problems. In modeling and simulation, sets of differential equations and logic clauses are used to create a dynamic systems environment that can be tested.

Mathematical models of biochemical networks (signal transduction cascades, metabolic pathways, gene regulatory networks) are a central component of modern systems biology. The development of formal methods adopted from theoretical computing science is essential for the modeling and simulation of these complex networks. The computational methods that are being employed in mathematical biology and bioinformatics are the following: (a) directed graphs, (b) Bayesian networks, (c) Boolean networks and their generalizations, (d) ordinary and partial differential equations, (e) qualitative differential equations, (f) stochastic equations, and (g) rule-based formalisms. Below are a few specific examples of the applications of these methods.

Mathematical models can be used to investigate the effects of drugs under a given set of perturbations based on specific tumor properties. This integration can help in the development of tools that aid in diagnosis and prognosis, and thus improve treatment outcome in patients with cancer. For example, breast cancer, being a well-studied disease over the last decade, serves as a model disease. One can thus apply the principles of molecular biology and pathology in designing new predictive mathematical frameworks that can unravel the dynamic nature of the disease. Genetic mutations of BRCA1, BRCA2, TP53, and PTEN significantly affect disease prognosis and increase the likelihood of adverse reactions to certain therapies. These mutations enable normal cells to become self-sufficient in survival in a stepwise process. Enderling et al. (2006) modeled this mutation and expansion process by assuming that mutations in two tumor-suppressor genes are sufficient to give rise to a cancer. They modified Enderling's earlier model, which was based on an established partial differential equation model of solid tumor growth and invasion. The stepwise mutations from a normal breast stem cell to a tumor cell have been described using a model consisting of four differential equations.

Recently, Woolf et al. (2005) applied a novel graphical modeling methodology known as Bayesian network analysis to model discovery and model selection for signaling events that direct mouse embryonic stem cells (an important preliminary step in hypothesis testing) in protein signaling networks. The model predicts bidirectional dependence between the two molecules ERK and FAK. It is interesting to appreciate that the apparent complexity of these dynamic ERK–FAK interactions is quite likely responsible for the difficulty in determining clear "upstream" versus "downstream" influence relationships by means of standard molecular cell biology methods. Bayesian networks determine the relative probability of statistical dependence models of arbitrary complexity for a given set of data. This method offers further clues to apply Bayesian approaches to cancer biology problems.

Cell cycle is a process in which cells proliferate while collectively performing a series of coordinated actions. Cell-cycle models also have an impact on drug discovery. Chassagnole et al. (2006) used a mathematical model to simulate and unravel the effect of multi-target kinase inhibitors of cyclin-dependent kinases (CDKs). They quantitatively predict the cytotoxicity of a set of kinase inhibitors based on the *in vitro* IC_{50} measurement values. Finally, they assess the pharmaceutical value of these inhibitors as anti-cancer therapeutics.

In cancer, avascular tumor growth is characterized by localized, benign tumor growth where the nearby tissues consume most of the nutrients. Mathematical modeling of avascular tumor growth is important to understanding the advanced stages of cancer. Kiran et al. (2009) have developed a spatial–temporal mathematical model classified as a different zone model (DZM) for avascular tumor growth based on diffusion of nutrients and their consumption, and it includes key mechanisms in the tumor. The diffusion and nutrient consumption are represented using partial differential equations. This model predicts that onset of necrosis occurs when the concentrations of vital nutrients are below critical values, and also the overall tumor growth based on the size effects of the proliferation zone, quiescent zone, and necrotic zone.

The mathematical approaches towards modeling the three natural scales of interest (subcellular, cellular, and tissue) are discussed above. Developing models that can predict the effects across biological scales is a challenge. The long-term goal is to build a "virtual human made up of mathematical models with connections at the different biological scales (from genes to tissue to organ)."

THE CONCEPT OF MODELING

A model is an optimal mix of hypotheses, evidence, and abstraction to explain a phenomenon. Hypothesis is a tentative explanation for an observation, phenomenon, or scientific problem that can be tested by further investigation. Evidence describes information (i.e. experimental data) that helps in forming a conclusion or judgment. Abstraction is an act of filtering out the required information to focus on a specific property only. For example, archiving books based on the year of publication, irrespective of the author name, would be an example of abstraction. In this process, some detail is lost and some gained. Predictions are made through modeling that can be tested by experiment. A model may be simple (e.g. the logistic equation describing how a population of bacteria grows) or complicated. Models may be mathematical or statistical.

Mathematical models make predictions, whereas statistical models enable us to draw statistical inferences about the probable properties of a system. In other words, models can be deductive or inductive. If the prediction is necessarily true given that the model is also true, then the model is a deductive model. On the other hand, if the prediction is statistically inferred from observations, then the model is inductive. Deductive models contain a mathematical description; for example, the reaction–diffusion equation that makes predictions about reality. If these predictions do not agree with experiment, then the validity of the entire model may be questioned. Mathematical models are commonly applied in physical sciences. On the other hand, inductive models are mostly applied in the biological sciences. In biology, models are used to describe, simulate, analyze, and predict the behavior of biological systems. Modeling in biology provides a framework that enables description and understanding of biological systems through building equations that express biological knowledge. Modeling enables the simulation of the behavior of a biological system by performing *in silico* experiments (i.e. numerically solving the equations or rules that describe the model). The results of these *in silico* experiments become the input for further analysis; for example, identification of key parameters or mechanisms, interpretation of data, or comparison of the ability of different mechanisms to generate observed data.

In particular, systems biology employs an integrative approach to characterizing biological systems in which interactions among all components in a system are described mathematically to establish a computable model. These *in silico* models complement traditional *in vivo* animal models and can be applied to quantitatively study the behavior of a system of interacting components. The term "*in silico*" is poorly defined, with several researchers claiming their role in its origination (Ekins et al., 2007). Sieburg (1990) and (Danchin et al. 1991) were two of the earliest published works that used this term. Specifically, *in silico* models gained much interest in the early stages by various imaging studies (Chakroborty et al., 2003). As an example, microarray analysis that enabled measurement of genome-scale expression levels of genes provided a method to investigate regulatory networks. Years of regulatory network studies (that included microarray-based investigations) led to the development of some well-characterized regulatory networks such as *E. coli* and yeast regulatory networks. These networks are available in the GeneNetWeaver (GNW) tool. GNW is an open-source tool for *in silico* benchmark generation and performance profiling of network inference methods. Thus, the advent of high-throughput experimental tools has allowed for the simultaneous measurement of thousands of biomolecules, opening the way for *in silico* model construction of increasingly large and diverse biological systems. Integrating heterogeneous dynamic data into quantitative predictive models holds great promise for significantly increasing our ability to understand and rationally intervene in disease-perturbed biological systems. This promise – particularly with regards to personalized medicine and medical intervention – has motivated the development of new methods for systems analysis of human biology and disease. Such approaches offer the possibility of gaining new insights into the behavior of biological systems, of providing new frameworks for organizing and storing data and performing statistical analyses, suggesting new hypotheses and new experiments, and even of offering a "virtual laboratory" to supplement *in vivo* and *in vitro* work.

However, *in silico* modeling in the life sciences is far from straightforward, and suffers from a number of potential pitfalls. Thus, mathematically sophisticated but biologically useless models often arise because of a lack of biological input, leading to models that are biologically unrealistic, or that address a question of little biological importance. On the other hand, models may be biologically realistic but mathematically intractable. This problem usually arises because biologists unfamiliar with the limitations of mathematical analysis want to include every known biological effect in the model. Even if it were possible to produce such models, they would be of little use since their behavior would be as complex to investigate as the experimental situation. These problems can be avoided by formulating clear explicit biological goals before attempting to construct a model. This will ensure that the resulting model is biologically sound, can be experimentally verified, and will generate biological insight or new biological hypotheses. The aim of a model should not simply be to reproduce biological data. Indeed, often the most useful models are those that exhibit discrepancies from experiment. Such deviations will typically stimulate new experiments or hypotheses. An iterative approach has been proposed, starting with a biological problem, developing a mathematical model, and then feeding back into the biology. Once established, this collaborative loop can be traversed many times, leading to ever increasing understanding.

The ultimate goal of *in silico* modeling in biology is the detailed understanding of the function of molecular networks as they appear in metabolism, gene regulation, or signal transduction. This is achieved by using a level of mathematical abstraction that needs a minimum of biological information to capture all physiologically relevant features of a cellular network. For example, ideally, for *in silico* modeling of a molecular network, knowledge of the network structure, all reaction rates, concentrations, and spatial distributions of molecules at any time point is needed. Unfortunately, such information is unavailable even for the best-studied systems. *In silico* simulations thus always have to use a level of mathematical abstraction, which is dictated by the extent of our biological knowledge, by molecular details of the network, and by the specific questions that are addressed. Understanding the complexity of the disease and its biological significance in health can be achieved by integrating data from the different functional genomics experiments with medical, physiological, and environmental factor information, and computing mathematically. The advantage of mathematical modeling of disease lies in the fact that such models not only shed light on how a complex process works, which could be very difficult for inferring an understanding of each component of this process, but also predict what may follow as time evolves or as the characteristics of particular system components are modified. Mathematical models have generally been utilized in association with an increased understanding of what models can offer in terms of prediction and insight.

The two distinct roles of models are prediction and understanding the accuracy, transparency, and flexibility of model properties. Prediction of the models should be accurate, including all the complexities and population-level heterogeneity that have an additional use as a statistical tool. It also provides the understanding of how the disease spreads in the real world and how the complexity affects the dynamics. Model understanding aids in developing sophisticated predictive models, along with gathering more relevant epidemiological data. A model should be as simple as possible and should have balance in accuracy, transparency, and flexibility; in other words, a model should be well suited for its purpose. The model should be helpful in understanding the behavior of the disease and able to simplify the other disease condition.

IN SILICO MODELS OF CELLS

Several projects are proceeding along these lines, such as E-CELL (Tomita, 2001) and simulations of biochemical pathways. Whole cell modeling integrates information from metabolic pathways, gene regulation, and gene expression. Three elements are needed for constructing of a good cell model: precise knowledge of the phenomenon, an accurate mathematical representation, and a good simulation tool.

A cell represents a dynamic environment of interaction among nucleic acids, proteins, carbohydrates, ions, pH, temperature, pressure, and electrical signals. Many cells with similar functionality form tissue. In addition, each type of tissue uses a subset of this cellular inventory to accomplish a particular function. For example, in neurons, electro-chemical phenomena take precedence over cell division, whereas, cell division is a fundamental function of skin, lymphocytes, and bone marrow cells. Thus, an ideal virtual cell not only represents all the information, but also exhibits the potential to differentiate into neuronal or epithelial cells. The first step in creating a whole cell model is to divide the entire network into pathways, and pathways into individual reactions. Any two reactions belong to a pathway if they share a common intermediate. *In silico* modeling consists not only of decomposing events into manageable units, but also of assembling these units into a unified framework. In other words, mathematical modeling is the art of converting biology into numbers.

For whole cell modeling, a checklist of biological phenomena that call for mathematical representation is needed. Biological phenomena taken into account for *in silico* modeling of whole cells are the following:

1. DNA replication and repair
2. Translation
3. Transcription and regulation of transcription
4. Energy metabolism
5. Cell division
6. Chromatin modeling
7. Signaling pathways
8. Membrane transport (ion channels, pump, nutrients)
9. Intracellular molecular trafficking
10. Cell membrane dynamics
11. Metabolic pathways

The whole cell metabolism includes enzymatic and non-enzymatic processes. Enzymatic processes cover most of the metabolic events, while non-enzymatic processes include gene expression and regulation, signal transduction, and diffusion.

In silico modeling of whole cells not only requires precise qualitative and quantitative data, but also an appropriate mathematical representation of each event. For metabolic modeling, the data input consists of kinetics of individual reactions and also effects of cofactors, pH, and ions on the model. The key step in modeling is to choose an appropriate assumption. For example, a metabolic pathway may be a mix of forward and reverse reactions. Furthermore, inhibitors that are part of the pathway may influence some reactions. At every step, enzymatic equations are needed that best describe the process. *In silico* models are built because they are easy to understand, controllable, and can store and analyze large amounts of information. A well-built model has diagnostic and predictive abilities. A cell by itself is a complete biochemical reactor that contains all the information one needs to understand life. Whole cell modeling enables investigation of the cell cycle, physiology, spatial organization, and cell–cell communication. Sequential actions in whole cell modeling are the following:

1. Catalog all the substances that make up a cell.
2. Make a list of all the reactions, enzymes, and effectors.
3. Map the entire cellular pathway: gene regulation, expression, metabolism, etc.
4. Build a stoichiometric matrix of all the reactions vs. substances (for qualitative modeling).
5. Add rate constants, concentration of substances, and strength of inhibition.
6. Assume appropriate mathematical representations for individual reactions.
7. Simulate reactions with suitable simulation software.
8. Diagnose the system with system analysis software.
9. Perturb the system and correlate its behavior to an underlying genetic and/or biochemical factor.
10. Predict phenomenon using a hypothesis generator.

Advantages and Disadvantages of *In Silico* Modeling

In silico modeling of disease combines the advantages of both *in vivo* and *in vitro* experimentation. Unlike *in vitro* experiments, which exist in isolation, *in silico* models provide the ability to include a virtually unlimited array of parameters, which render the results more applicable to the organism as a whole. *In silico* modeling allows us to examine the workings of biological processes such as homeostasis, reproduction, evolution, etc. For example, one can explore the processes of Darwinian evolution through *in silico* modeling, which are not practical to study in real time.

In silico modeling of disease is quite challenging. Attempting to incorporate every single known interaction rapidly leads to an unmanageable model. Furthermore, parameter determination in such models can be a frightening experience. Estimates come from diverse experiments, which may be elegantly designed and well executed but can still give rise to widely differing values for parameters. Data can come from both *in vivo* and *in vitro* experiments, and results that hold in one medium may not always hold in the other. Furthermore, despite the many similarities between mammalian systems, significant differences do exist, and so results obtained from experiments using animal and human tissue may not always be consistent. Also there are many considerations that cannot be applied. For example, one cannot investigate the role of stochastic fluctuations by removing them from the system, or one cannot directly explore the process that gave rise to current organisms.

APPLICATIONS OF *IN SILICO* DISEASE MODELING IN PRACTICE

In silico modeling has been applied in cancer, systemic inflammatory response syndrome, immune diseases, neuronal diseases, and infectious diseases (among others). *In silico* models of disease can contribute to a better understanding of the pathophysiology of the disease, suggest new treatment strategies, and provide insight into the design of experimental and clinical trials for the investigation of new treatment modalities.

In Silico Models of Cancer

In silico modeling of cancer has become an interesting alternative approach to traditional cancer research. *In silico* models of cancer are expected to predict the complexity of cancer at multiple temporal and spatial resolutions, with the aim of supplementing diagnosis and treatment by helping plan more focused and effective therapy via surgical resection, standard and targeted chemotherapy, and novel treatments. *In silico* models of cancer include: (a) statistical models of cancer, such as molecular signatures of perturbed genes and molecular pathways, and statistically-inferred reaction networks; (b) models that represent biochemical, metabolic, and signaling reaction networks important in oncogenesis, including constraint-based and dynamic approaches for the reconstruction of such networks; and (c) models of the tumor microenvironment and tissue-level interactions (Edelman et al., 2010).

Statistical models of cancer can be broadly divided into those that employ unbiased statistical inference, and those that also incorporate a priori constraints of specific biological interactions from data. Statistical models of cancer biology at the genetic, chromosomal, transcriptomic, and pathway levels provide insight about molecular etiology and consequences of malignant transformation despite incomplete knowledge of underlying biological interactions. These models are able to identify molecular signatures that can inform diagnosis and treatment selection, for example with molecular targeted therapies such as Imatinib (Gleevec) (Edelman et al., 2010).

However, in order to characterize specific biomolecular mechanisms that drive oncogenesis, genetic and transcriptional activity must be considered in the context of cellular networks that ultimately drive cellular behavior. In microbial cells, network inference tools have been developed and applied for the modeling of diverse biochemical, signaling, and gene expression networks. However, due to the much larger size of the human genome compared to microbes, and the substantially increased complexity of eukaryotic genetic regulation, inference of transcriptional regulatory networks in cancer presents increased practical and theoretical challenges.

Biochemical reaction networks are constructed to represent explicitly the mechanistic relationships between genes, proteins, and the chemical inter-conversion of metabolites within a biological system. In these models, network links are based on pre-established biomolecular interactions rather than statistical associations; significant experimental characterization is thus needed to reconstruct biochemical reaction networks in human cells. These biochemical reaction networks require, at a minimum, knowledge of the stoichiometry of the participating reactions. Additional information such as thermodynamics, enzyme capacity constraints, time-series concentration profiles, and kinetic rate constants can be incorporated to compose more detailed dynamic models (Edelman et al., 2010).

Microenvironment-tissue level models of cancer apply an "engineering" approach that views tumor lesions as complex micro-structured materials, where three-dimensional tissue architecture ("morphology") and dynamics are coupled in complex ways to cell phenotype, which in turn is influenced by factors in the microenvironment. Computational approaches of *in silico* cancer research include continuum models, discrete models, and hybrid models.

Types of Cancer Models

In **continuum models**, extracellular parameters can be represented as continuously distributed variables to mathematically model cell–cell or cell–environment interactions in the context of cancers and the tumor microenvironment. Systems of partial differential equations have been used to simulate the magnitude of interaction between these factors. Continuum models are suitable for describing the individual cell migration, change of cancer cell density, diffusion of chemo-attractants, heat transfer in hyperthermia treatment for skin cancer, cell adhesion, and the molecular network of a cancer cell as an entire entity. However, these types of *in silico* models have limited ability for investigating single-cell behavior and cell–cell interaction.

On the other hand, **"discrete" models** (i.e. cellular automata models) represent cancer cells as discrete entities of defined location and scale, interacting with one another and external factors in discrete time intervals according to predefined rules. Agent-based models expand the cellular automata paradigm to include entities of divergent functionalities interacting together in a single spatial representation, including different cell types, genetic elements, and environmental factors. Agent-based models have been used for modeling three-dimensional tumor cell patterning, immune system surveillance, angiogenesis, and the kinetics of cell motility.

Hybrid models have been created which incorporate both continuum and agent-based variables in a modular approach. Hybrid models are ideal for examining direct interactions between individual cells and between the cells and their microenvironment, but they also allow us to analyze the emergent properties of complex multi-cellular systems (such as cancer). Hybrid models are often multi-scale by definition, integrating processes on different temporal and spatial scales, such as gene expression, intracellular pathways, intercellular signaling, and cell growth or migration. There are two general classes of hybrid models, those that are defined upon a lattice and those that are off-lattice.

The classification of hybrid models on these two classes depends on the number of cells these models can handle and the included details of each individual cell structure, i.e. models dealing with large cell populations but with simplified cell geometry, and those that model small colonies of fully deformable cells.

For example, a hybrid model investigated the invasion of healthy tissue by a solid tumor. The model focused on four key parameters implicated in the invasion process; tumor cells, host tissue (extracellular matrix), matrix-degradative enzymes, and oxygen. The model is actually hybrid, wherein the tumor cells were considered to be discrete (in terms of concentrations), and the remaining variables were in the continuous domain in terms of concentrations. This hybrid model can make predictions on the effects of individual-based cell interactions (both between individuals and the matrix) on tumor shape. The model of Zhang et al. (2007) incorporated a continuous model of a receptor signaling pathway, an intracellular transcriptional regulatory network, cell-cycle kinetics, and three-dimensional cell migration in an integrated, agent-based simulation of solid brain tumor development. The interactions between cellular and microenvironment states have also been considered in a multi-scale model that predicts tumor morphology and phenotypic evolution in response to such extracellular pressures.

The biological context in which cancers develop is taken into consideration in *in silico* models of the tumor microenvironment. Such complex tumor microenvironments may integrate multiple factors including extracellular biomolecules, vasculature, and the immune system. However, rarely have these methods been integrated with a large cell–cell communication network in a complex tumor microenvironment. Recently, an interesting effort of *in silico* modeling was described in which the investigators integrated all the intercellular signaling pathways known to date for human glioma and generated a dynamic cell–cell communication network associated with the glioma microenvironment. Then they applied evolutionary population dynamics and the Hill functions to interrogate this intercellular signaling network and execute an *in silico* tumor microenvironment development. The observed results revealed a profound influence of the micro-environmental factors on tumor initiation and growth, and suggested new options for glioma treatment by targeting cells or soluble mediators in the tumor microenvironment (Wu et al., 2012).

In Silico Models and Inflammatory Response Syndrome in Trauma and Infection

Trauma and infection can cause acute inflammatory responses, the degree of which may have several pathological manifestations like systemic inflammatory response syndrome (SIRS), sepsis, and multiple organ failure (MOF). However, an appropriate management of these states requires further investigation. Translating the results of basic science research to effective therapeutic regimes has been a longstanding issue due in part to the failure to account for the complex nonlinear nature of the inflammatory process wherein SIRS/MOF represent a disordered state. Hence, the *in silico* modeling approach can be a promising research direction in this area. Indeed, *in silico* modeling of inflammation has been applied in an effort to bridge the gap between basic science and clinical trials. Specifically, both agent-based modeling and equation-based modeling have been utilized (Vodovotz et al., 2008). Equation-based modeling encompasses primarily ordinary differential equations (ODE) and partial differential equations (PDE). Initial modeling studies were focused on the pathophysiology of the acute inflammatory response to stress, and these studies suggested common underlying processes generated in response to infection, injury, and shock. Later, mathematical models included the recovery phase of injury and gave insight into the link between the initial inflammatory response and subsequent healing process. The first mathematical models of wound healing dates back to the 1980s and early 1990s. These models and others developed in the 1990s investigated epidermal healing, repair of the dermal extracellular matrix, wound contraction, and wound angiogenesis. Most of these models were deterministic and formulated using differential equations. In addition, recent models have been formulated using differential equations to analyze different strategies for improved healing, including wound VACs, commercially engineered skin substitutes, and hyperbaric oxygen. In addition, agent-based models have been used in wound healing research. For example, Mi et al. (2007) developed an agent-based model to analyze different treatment strategies with wound debridement and topical administration of growth factors. Their model produced the expected results of healing when analyzing for different treatment strategies including debridement, release of PDGF, reduction in tumor necrosis factor-α, and increase of TGF-β1. The investigators suggested that a drug company should use a mathematical model to test a new drug before going through the expensive process of basic science testing, toxicology and clinical trials. Indeed, clinical trial design can be improved by prior *in silico* modeling. For example, *in silico* modeling has led to the knowledge that patients who suffered from the immune-suppressed phenotype of late-stage multiple organ failure, and were susceptible to usually trivial nosocomial infections, demonstrated sustained elevated markers of tissue damage and inflammation through two weeks of simulated time. However, anti-cytokine drug trials with treatment protocols of only one dose or one day had not incorporated this knowledge into their design, with subsequent failure of candidate treatments.

Infectious Diseases

By now the reader is expected to be familiar with the meaning and the basics of *in silico* modeling. In this section we discuss the application of *in silico* modeling in the understanding of infectious diseases and in the proposition/development of better treatments for infectious diseases. In fact, the applications of *in silico* modeling can help far beyond just the understanding of the dynamics (and sometimes, statistics) of infectious diseases, and far beyond the proposition/development of better treatments for infectious diseases. The modeling can be helpful even in the understanding of better prevention of infectious diseases.

The level of pathogen within the host defines the process of infection; such pathogen levels are determined by the growth rate of the pathogen and its interaction with the host's immune response system. Initially, no pathogen is present, but just a low-level, nonspecific immunity within the host. On infection, the parasite grows abundantly over time with the potential to transmit the infection to other susceptible individuals.

Triad of Infectious Diseases as the Source of Parameters for In Silico *Modeling of Infectious Diseases*

To comprehensively understand *in silico* modeling in the domains of infectious diseases, one should first understand the "triad of infectious diseases," and the characteristics of "infectious agent," "host," and "environment" on which the models are always based. In fact, modeling of infectious diseases is just impossible without this triad; after all, the model would be built on some parameters (also called variables in more general language), and those parameters always have their origin from the so-called "triad of infectious diseases." At this point, a good question would be: What is a "triad of infectious diseases?"

"Triad of infectious diseases" means the interactions between (1) agent, which is the disease causing organism (the pathogen); (2) host, which is the infected organism, or in the case of pre-infection, the organism to be infected is the host (thus in this case the host is the animal the agent infects); and (3) environment, which is a kind of link between the agent and the host, and is essentially an umbrella word for the entirety of the possible media through which the agent reaches the host.

Now that we have an idea on what *in silico* modeling of infectious diseases are generally based on, we will outline a better understanding of the parameters that are considered in most *in silico* disease models. To discuss the parameters in an orderly manner, we just categorize them under each of the three components of the "triad of infectious diseases," and summarize them in the next subsection. It must be emphasized at this point that (1) even though all the possible parameters for *in silico* modeling of infectious diseases can be successfully categorized under the characteristics of one of any of the three components of the "triad of infectious diseases" (agent, host, and environment), (2) the parameters discussed in the next sub-section are by no means the entirety of all the possible parameters that can be included in *in silico* modeling of infectious diseases. In fact, several parameters exist, and this section cannot possibly enumerate them all. That is why we have discussed the parameters using a categorical approach.

Parameters for In Silico *Modeling of Infectious Diseases*

Parameters Derived from Characteristics of the Agent

Some of the parameters for *in silico* modeling of infectious diseases are essentially a measure of infectivity (ability to enter the host), pathogenicity (ability to cause divergence from homeostasis/disease), virulence (degree of divergence from homeostasis caused/ability to cause death), antigenicity (ability to bind to mediators of the host's adaptive immune system), and immunogenicity (ability to trigger adaptive immune response) of the concerned infectious agent. The exact measure (and thus the units) used can vary markedly depending on the intentions for which the *in silico* infectious diseases model is built, as well as the assumptions on which the *in silico* disease model is based. From the knowledge of the agent's characteristics, one should know that unlike parameters related to the other characteristics of the agent, the parameters related to infectivity find their most important use only in the modeling of the pre-infection stage in infectious disease modeling.

Finally, some of the agent-related parameters of great importance in *in silico* modeling of infectious diseases are concentration of the agent's antigen–host antibody complex, case fatality rate, strain of the agent, other genetic information of the agent, etc.

Parameters Derived from Characteristics of the Host

The parameters originating from characteristics of the host can also be diversified and based on the intentions for which the *in silico* infectious diseases model are built and the assumptions on which the *in silico* disease model are based; however, the parameters could then be grouped and explained under the host's genotype (the allele at the host's specified genetic locus), immunity/health status (biological defenses to avoid infection), nutritional status (feeding habits/food intake characteristics), gender (often categorized as male or female), age, and behavior (the host's behaviors that affect its resistance to homeostasis disruptors).

Typical examples of host-related parameters are the alleles at some specifically targeted genetic loci; the total white blood cell counts; differential white blood cell counts, and/or much more sophisticated counts of specific blood cell types; blood levels of some specific cytokines, hormones, and/or neurotransmitters; daily calories, protein, and/or fat intake; daily amount of energy expended and/or duration of exercise; etc.

Parameters Derived from Characteristics of Environment

At first parameters originating from the environment might seem irrelevant to the *in silico* modeling of infectious diseases, but they are relevant. Even after the pre-infection stage, the environment still modulates the host–agent intersections. For example, the ability (and thus the related parameters) of the agent to multiply and/or harm the host are continually influenced by the host's environmental conditions, and in a similar way the hosts defense against the adverse effects of the agents are modulated by the host's environmental conditions. But somehow, not so many of these parameters have been included in *in silico* infectious disease models in the recent past. A few examples of these parameters are the host's ambient temperature, the host's ambient atmospheric humidity, altitude, the host's light–dark cycle, etc.

Infectious Diseases In Silico *Model Proper, a Typical Approach/Scenario*

Now that we know the parameters for *in silico* infectious disease modeling, the next reasonable question would be "What form does a typical *in silico* infectious disease model take?" So, this sub-section attempts to answer this very important question.

Let us view the *in silico* model as a system of well-integrated functional equations or formulae. Such well-integrated functional equations can be viewed or approximated as a single, albeit more complex, functional equation/formula. It is hence possible to vary any (or a combination) of the variables contained in this equation by running numerical simulations on a computer depending on the kind of prediction one wants to make. Such *in silico* models can hence investigate several (maybe close to infinite) possible data points within reasonable limits that one sets depending on the nature of the variables considered.

So the equations behind a typical infectious disease *in silico* model could take the form (Equation 21.1):

$$H = \beta \text{ (link function) } f(A) \text{ (link function) } g(E) + \varepsilon \cdots \quad (21.1)$$

where H is the output from a smaller equation that is based on host parameters; β is a constant; f and g are link functions which may be the same as or different from each other and other link functions in this system of equations; A is the output from a smaller equation that is based on agent parameters; g is a link function which may be the same or different from other link functions in this system of equations; E is the output from a smaller equation that is based on environment parameters; and Ɛ is a random error parameter.

Readers should know that we use the term "link function" to refer to any of the various possible forms of mathematical operations or functions. This means that based on the complexity of the model, a particular "link function" might be as simple as a mere addition or as complex as several combinations of operators with high degree polynomials.

H in Equation 21.1 could have resulted from a smaller model/function of the form (Equation 21.2):

$$H = \beta_h \text{ (link function) } f_{h1}(h_1) \text{ (link function) } f_{h2}(h_2) \text{ (link function)} \ldots f_{hx}(h_x) \text{ (link function) } + \varepsilon \cdots \quad (21.2)$$

where β_h is a constant; f_{h1}, f_{h2}, … f_{hx} are link functions that may be the same or different (individually) from (every) other link function in this system of equations; h_1, h_2, … h_x are a set of the host's parameters (e.g. age, gender, white blood cell count, cytokine level, etc); and Ɛ is a random error parameter.

A in Equation 21.1 could have resulted from a smaller model/function of the form (Equation 21.3):

$$A = \beta_a \text{ (link function) } f_{a1}(a_1) \text{ (link function) } f_{a2}(a_2) \text{ (link function) } \ldots f_{ax}(a_x) \text{ (link function) } + \varepsilon \cdots \quad (21.3)$$

where β_a is a constant; f_{a1}, f_{a2}, … f_{ax} are link functions that may be same or different (individually) from (every) other link function in this system of equations; a_1, a_2, … a_x are a set of the agent's parameters (e.g. case fatality rate, agent's genotype, etc); and Ɛ is a random error parameter.

In a similar way, E in Equation 21.1 could have resulted from a smaller model/function of the form (Equation 21.4):

$$E = \beta_e \text{ (link function) } f_{e1}(e_1) \text{ (link function) } f_{e2}(e_2) \text{ (link function) } \ldots f_{ex}(e_x) \text{ (link function) } + \varepsilon \cdots \quad (21.4)$$

where β_e is a constant; f_{e1}, f_{e2}, …f_{ex} are link functions which may be the same or different (individually) from (every) other link function in this system of equations; e_1, e_2, … e_x are a set of environmental parameters (e.g. host's ambient temperature, host's ambient atmospheric humidity, etc.); and Ɛ is a random error parameter.

Specific Examples of Infectious Diseases in the In Silico *Model*

Muñoz-Elías et al. (2005) documented (through their paper "Replication Dynamics of Mycobacterium Tuberculosis in Chronically Infected Mice") a successful *in silico* modeling of infectious diseases (specifically, tuberculosis). In their *in silico* modeling of tuberculosis in mice, the researchers investigated both the static and dynamic host–pathogen/agent equilibrium (i.e. mice–mycobacterium tuberculosis static and dynamic equilibrium). The rationale behind their study was that a better understanding of host–pathogen/agent interactions would make possible the development of better anti-microbial drugs for the treatment of tuberculosis (as well as provide similar understanding for the cases of other chronic infectious diseases). They modeled different types of host–pathogen/agent equilibriums (ranging from completely static equilibrium, all the way through semi-dynamic, down to completely dynamic scenarios) by varying the rate of multiplication/growth and the rate of death of the pathogen/agent (Mycobacterium tuberculosis) during the infection's chronic phase. Through their *in silico* study (which was also verified experimentally), they documented a number of remarkable findings. For example, they established that "viable bacterial counts and total bacterial counts in the lungs of chronically infected mice do not diverge over time," and they explained that "rapid degradation of dead bacteria is unlikely to account for the stability of total counts in the lungs over time because treatment of mice with isoniazid for 8 weeks led to a marked reduction in viable counts without reducing the total count.

Readers who are interested in further details on the generation of this *in silico* model for the dynamics of Mycobacterium tuberculosis infection, as well as the complete details of the parameters/variables considered, and the comprehensive findings of the study, should refer to the article of Ernesto et al. published in infection and immunity.

Another one of the many notable works in the domain of infectious disease *in silico* modeling is the study by Navratil et al. (2011). Using protein–protein interaction data that the authors obtained from available literature and public databases, they (after first curating and validating the data) computationally (*in silico*) re-examined the virus–human protein interactome. Interestingly, the authors were able to show that the onset and pathogenesis of some disease conditions (especially chronic disease conditions) often believed to be of genetic, lifestyle, or environmental origin, are, in fact, modulated by infectious agents.

Model of Bacterial and Viral Dynamics

Models have been constructed to simulate bacterial dynamics, such as growth under various nutritional and chemical conditions, chemotactic response, and interaction with host immunity. Clinically important models of bacterial dynamics relating to peritoneal dialysis, pulmonary infections, and particularly of antibiotic treatment and bacterial resisitance, have also been developed.

Baccam et al. (2006) utilized a series of mathematical models of increasing complexity that incorporated target cell limitation and the innate interferon response. The models were applied to examine influenza A virus kinetics in the upper respiratory tracts of experimentally infected adults. They showed the models to be applicable for improving the understanding of influenza In a virus infection, and estimated that during an upper respiratory tract infection, the influenza virus initially spreads rapidly with one cell, infecting (on average) about 20 others (Daun and Clermont, 2007).

Model parameter and spread of disease: Model parameters are one of the main challenges in mathematical modeling since all models do not have a physiological meaning. Sensitivity analysis and bifurcation analysis give us the opportunity to understand how model outcome and model parameters are correlated, how the sensitivity of the system is with respect to certain parameters, and the uncertainty in the model outcome yielded by the uncertainties in the parameter values. Uncertainty and sensitivity analysis was used to evaluate the input parameters play in the basic productive rate (*Ro*) of severe acute respiratory syndrome (SARS) and tuberculosis. Control of the outbreak depends on identifying the disease parameters that are likely to lead to a reduction in *R*.

Challenges in In Silico *Modeling of Infectious Diseases*

Difficulty in finding the most appropriate set of parameters for *in silico* modeling of infectious diseases is often a challenge. It is hoped this challenge will subside with the advancement in infectonomics and high-throughput technology. However, another important challenge lies in the understanding (and the provision of reasonable interpretations for) the results from all the complex interactions of parameters considered.

Neuronal Diseases

In this sub-section we focus on the application of *in silico* modeling to improve knowledge of neuronal diseases, and thus improve the applications of neurological knowledge for solving neuronal health problems. It is not an overstatement to say that one of the many aspects of life sciences where *in silico* disease modeling would have the biggest applications is in the better understanding of the pathophysiology of nervous system (neuronal) diseases. This is basically because of the inherent delicate nature of the nervous system and the usual extra need to be sure of how to proceed prior to attempting to treat neuronal disease conditions. By this we mean that the need to first model neuronal disease conditions *in silico* prior to deciding on or suggesting (for example) a treatment plan is, in fact, rising. This is not unexpected; after all, it is better to be sure of what would work (say, through *in silico* modeling) than to try what would not work.

Pathophysiology of Neuronal Diseases as the Source of Parameters for In Silico *Modeling of Neuronal Diseases*

Obtaining appropriate parameters for the *in silico* modeling of a nervous system (neuronal) disease is rooted in a good understanding of the pathophysiology of such neuronal disease. Since comprehensive details of pathophysiology of neuronal diseases is beyond the scope of this book, we only present the basic idea that would allow the reader to understand how *in silico* modeling of a nervous system (neuronal) disease can be done.

To give a generalized explanation and still concisely present the basic ideas underlying the pathophysiology of neuronal diseases, we proceed by systematically categorizing the mediators of neuronal disease pathophysiology: (1) nerve cell characteristics, (2) signaling chemicals and body electrolytes, (3) host/organism factors, and (4) environmental factors. Readers need to see all these categories as being highly integrated pathophysiologically rather than as separate entities, and also that we have only grouped them this way to make simpler the explanation of how the parameters for *in silico* modeling of neuronal diseases are generated.

When something goes wrong with (or there is a marked deviation from equilibrium in) a component of any of the four categories above, the other components (within and/or outside the same category) try hard to make adjustments so as to annul/compensate for the undesired change. For example, if the secretion of a chemical signal suddenly becomes abnormally low, the target cells for the chemical signal may develop mechanisms to use the signaling chemical more efficiently, and the degradation rate of the signaling chemical may be reduced considerably. Through these, the potentially detrimental effects of reduced secretion of the chemical signal are annulled via compensation from the other components. This is just a simple example; much more complex regulatory and homeostatic mechanisms exist in the neuronal system. Despite the robustness of those mechanisms, things still get out of hand sometimes, and disease conditions result. The exploration of what happens in (and to) each and all of the components of this giant system of disease conditions is called the pathophysiology of neuronal disease, and it this pathophysiology that "provides" parameters for the *in silico* modeling of neuronal diseases.

Parameters for In Silico *Modeling of Neuronal Diseases*

Parameters Derived from Characteristics of Nerve Cells

Some of the important parameters (that are of nerve cell origin) for a typical *in silico* modeling of a neuronal disease (say, Alzheimer's disease) are the population (or relative population) of specific neuronal cells (such as glial cells: microglia, astrocytes, etc.), motion of specific neuronal cells (e.g. microglia), amyloid production, aggregation and removal of amyloid, morphology of specific neuronal cells, status of neuronal cell receptors, generation/regeneration/degeneration rate of neuronal cells, status of ion neuronal cell channels, etc. Based on their relevance to the pathophysiology of the neuronal disease being studied, many of these parameters are often considered in the *in silico* modeling of the neuronal disease. More importantly, their spatiotemporal dynamics are often seriously considered.

Parameters Derived from Characteristics of Signaling Chemicals and Body Electrolytes

The importance of signaling chemicals and electrolytes in the nervous system makes parameters related to them very important. The secretion, uptake, degradation, and diffusion rates of various neurotransmitters and cytokines are often important parameters in the *in silico* modeling of neurodiseases. Other important parameters are the concentration gradients of the various neurotransmitters and cytokines, the availability and concentration of second messengers, and the electrolyte status/balance of the cells/systems. The spatiotemporal dynamics of all of these are also often seriously considered.

Parameters Derived from Host/Organism Factors

The parameters under host/organism factors can be highly varied depending on the intentions and the assumptions governing the *in silico* disease modeling. Nonetheless, one could basically group and list the parameters collectively under genotype (based on the allele at a specified genetic locus), nutritional status (feeding habits/food intake characteristics; e.g. daily calories, protein intake, etc.), gender (male or female), age, and behavior (host's behaviors/lifestyle that influences homeostasis and/or responses to stimuli).

Parameters Derived from Environmental Factors

A few examples of these parameters are ambient temperature, altitude, light–dark cycle, social network, type of influences from people in the network, etc.

Neuronal Disease In Silico *Model Proper (a Typical Approach/Scenario)*

Just like other *in silico* models, a neuronal disease *in silico* model is also based on what could be viewed as a single giant functional equation, which is composed of highly integrated simpler functional equations.

So the equations behind a typical neuronal disease *in silico* model could take the form (Equation 21.5):

$$\begin{aligned} N = \beta \text{ (link function) } f(C) \text{ (link function)} \\ g(S) \text{ (link function) } j(H) \\ \text{(link function) } k(E) + \varepsilon \cdots \end{aligned} \tag{21.5}$$

where N could be a parameter that is a direct measure of the disease manifestation; β is a constant; f, g, j, and k are link functions which may be the same or different from other link functions in this system of equations; C, S, H, and E are the outputs from smaller equations that are based on parameters from neuronal cell characteristics, signaling molecule and electrolyte parameters, host parameters, and environment parameters, respectively; and ε is a random error parameter.

The reader should know that each of N, C, S, H and E could have resulted from smaller equations that could take forms similar to those (Equations 21.2 to 21.4) described under *in silico* modeling of infectious diseases (previous sub-section).

Specific Examples of Neuronal Disease In Silico *Models*

Edelstein-Keshet and Spiros (2002) used *in silico* modeling to study the mechanism and/formation of Alzheimer's disease. The target of their *in silico* modeling was to explore and demystify how various parts implicated in the etiology and pathophysiology of Alzheimer's disease work together as a whole. Employing the strength of *in silico* modeling, the researchers were able to transcend the difficulty of identifying detailed disease progression scenarios, and they were able to test a wide variety of hypothetical mechanisms at various levels of detail. Readers interested in the complete details of the assumptions that govern *in silico* modeling of Alzheimer's disease, the various other aspects of the model, and more detailed accounts of the findings should look at the article by Edelstein-Keshet and Spiros.

Several other interesting studies have applied *in silico* modeling techniques to investigate various neuronal diseases. A few examples include the work of Altmann and Boyton (2004), who investigated multiple sclerosis (a very common disease resulting from demyelination in the central nervous system) using *in silico* modeling techniques; Lewis et al. (2010), who used *in silico* modeling to study the metabolic interactions between multiple cell types in Alzheimer's disease; and Raichura et al. (2006), who applied *in silico* modeling techniques to dynamically model alpha-synuclein processing in normal and Parkinson's disease states.

Alzheimer's Disease Model

A more specific example of a molecular level *in silico* Alzheimer's disease model can be found in Ghosh et al. (2010). Among the amyloid proteins, Amyloid-β (Aβ) peptides (Aβ42 and Aβ40) are known to form aggregates that deposit as senile plaques in the brains of Alzheimer's disease patients. The process of Aβ-aggregation is strongly nucleation-dependent, and is inferred by the occurrence of a "lag-phase" prior to fibril growth that shows a sigmoidal pattern. Ghosh et al. (2010) dissected the growth curve into three biophysically distinct sections to simplify modeling and to allow the data to be experimentally verifiable. Stage I is where the pre-nucleation events occur whose mechanism is largely unknown. The pre-nucleation stage is extremely important in dictating the overall aggregation process where critical events such as conformation change and concomitant aggregation take place, and it is also the most experimentally challenging to decipher. In addition to mechanistic reasons, this stage is also physiologically important as low-molecular-weight (LMW) species are implicated in AD pathology. The rate-limiting step of nucleation is followed by growth. The overall growth kinetics and structure and shape of the fibrils are mainly determined by the structure of the nucleating species. An important intermediate along the aggregation pathway, called "protofibrils," have been isolated and characterized that have propensities to both elongate (by monomer addition) as well as to laterally associate (protofibril–protofibril association) to grow into mature fibrils (Stage III in the growth curve).

Simulation of the Fibril Growth Process in Aβ42 Aggregation Ghosh et al. (2010) generated an ODE-based molecular simulation (using mass–kinetics methodology) of this fibril growth process to estimate the rate constants involved in the entire pathway. The dynamics involved in the protofibril elongation stage of the aggregation (Stage III of the process) were estimated and validated by *in vitro* biophysical analysis.

Preliminary Identification of Nucleation Mass Ghosh et al. (2010) next used the rate constants identified from Stage III to create a complete aggregation pathway simulation (combining Stages I, II, and III) to approximately identify the nucleation mass involved in Aβ-aggregation.

In order to model the Aβ-system, one needs to estimate the rate constants involved in the complete pathway and the nucleation mass itself. It is difficult to iterate through different values for each of these variables to get close to the experimental plots (fibril growth curves measured via fluorescence measurements with time) due to the large solution space; also, finding the nucleation phase cannot be done independently without estimating the rate constants alongside. However, having separately estimated the post-nucleation stage rate constants (as mentioned above) reduces the overall parameter estimation complexity.

The complete pathway simulation was used to study the lag times associated with the aggregation pathway, and hence predict possible estimates of the nucleation mass. The following strategy was used: estimate the pre-nucleation rate constants that give the *maximum lag times* for each possible estimate of the nucleation mass. This led to four distinctly different regimes of possible nucleation masses corresponding to four different pairs of rate constants for the pre-nucleation phase (Regime 1, where n = 7, 8, 9, 10, 11; Regime 2, where n = 12, 13, 14; Regime 3, where n = 15, 16, 17; and Regime 4, where n = 18, 19, 20, 21). However, it was experimentally observed that the semi-log plot of the lag times against initial concentration of Aβ is linear, and this characteristic was used to figure out what values of nucleation mass are most feasible for the Aβ42-aggregation pathway. The simulated plots show a more stable relationship between the lag times and the initial concentrations, and the best predictions for the nucleation mass were reported to be in the range 10–16.

Such molecular pathway level studies are extremely useful in understanding the pathogenesis of AD in general, and can motivate drug development exercises in the future. For example, characterization of the nucleation mass is important as it has been observed that various fatty acid interfaces can arrest the fibril growth process (by stopping the reactions beyond the pre-nucleation stage). Such in depth modeling of the aggregation pathway can suggest what concentrations of fatty acid interfaces should be used (under a given Aβ concentration in the brain) to arrest the fibril formation process leading to direct drug dosage and interval prediction for AD patients.

Possible Limitations of In Silico *Modeling of Neuronal Diseases*

Despite the fact that we have mentioned several possible parameters for *in silico* modeling of neuro-diseases, it is noteworthy that finding a set of the most reasonable parameters for the modeling is in fact a big challenge. On the other hand, understanding (and thus finding reasonable biological interpretations for) the results from the complex interaction of all parameters considered is also a big challenge. In addition, a number of assumptions that models are sometimes based on still have controversial issues. Accurately modeling spatio-temporal dynamics of neurons and neurotransmitters (and other chemicals/ligands) also constitutes a huge challenge.

CONCLUSION

Understanding the complex systems involved in a disease will make it possible to develop smarter therapeutic strategies. Treatments for existing tumors will use multiple drugs to target the pathways or perturbed networks that show an altered state of activity. In addition, models can effectively form the basis for translational research and personalized medicine.

Biological function arises as the result of processes interacting across a range of spatiotemporal scales. The ultimate goal of the applications of bioinformatics in systems biology is to aid in the development of individualized therapy protocols to minimize patient suffering while maximizing treatment effectiveness. It is now being increasingly recognized that multi-scale mathematical and computational tools are necessary if we are going to be able to fully understand these complex interactions (e.g. in cancer and heart diseases).

With the bioinformatics tools, computational theories, and mathematical models introduced in this article, readers should be able to dive into the exhilarating area of formal computational systems biology. Investigating these models and confirming their findings by experimental and clinical observations is a way to bring together molecular reductionism with quantitative holistic approaches to create an integrated mathematical view of disease progression. We hope to have shown that there are many interesting challenges yet to be solved, and that a structured and principled approach is essential for tackling them.

Systems biology is an emerging field that aims to understand biological systems at the systems level with a high degree of mathematical and statistical modeling. *In silico* modeling of infectious diseases is a rich and growing field focused on modeling the spread and containment of infections with model designs being flexible and enabling adaptation to new data types.

ETHICAL ISSUES

The advantages of avoiding animal testing have often been seen as one of the advantages offered by *in silico* modeling; the biggest advantage is that there are no ethical issues in performing *in silico* experiments as they don't require any animals or live cells. Furthermore, as the entire modeling and analysis are based on computational approaches, we can obtain the results of such analysis even within an hour. This saves huge amounts of time and reduces costs, two major factors associated with *in vitro* studies. However, a key issue that needs to be considered is whether *in silico* testing will ever be as accurate as *in vitro* or *in vivo* testing, or whether *in silico* results will always require non-simulated experimental confirmation.

TRANSLATIONAL SIGNIFICANCE

Tracqui et al. (1995) successfully developed a glioma model to show how chemo-resistant tumor sub-populations cause treatment failure. Similarly, a computational model of tumor invasion by Frieboes et al. (2006) is able to demonstrate that the growth of a tumor depends on the microenvironmental nutrient status, the pressure of the tissue, and the applied chemotherapeutic drugs. The 3D spatio-temporal simulation model of a tumor by Dionysiou et al. (2004) was able to repopulate, expand, and shrink tumor cells, thus providing a computational approach for assessment of radiotherapy outcomes. The glioblastoma model of Kirby et al. (2007) is able to predict survival outcome post-radiotherapy. Wu et al. (2012) has also developed an *in silico* glioma microenvironment that demonstrates that targeting the microenvironmental components could be a potential anti-tumor therapeutic approach.

The *in silico* model-based systems biology approach to skin sensitization (TNF-alpha production in the epidermis) and risk of skin allergy assessment has been successfully carried out; it can replace well known *in vitro* assays, such as the mouse local lymph node assay (LLNA) used for the same purpose (by Maxwell and Mackay (2008) at the Unilever Safety and Environmental Assurance Centre). Similarly, Davies et al. (2011) effectively demonstrated an *in silico* skin permeation assay based on time course data for application in skin sensitization risk assessment. Kovatchev et al. (2012) showed how the *in silico* model of alcohol dependence can provide virtual clues for classifying the physiology and behavior of patients so that personalized therapy can be developed.

Pharmacokinetics and pharmacodynamics are used to study absorption, distribution, metabolism, and excretion (ADME) of administered drugs. *In silico* models have tremendous efficacy in early estimation of various ADME properties. Quantitative structure–activity relationship (QSAR) and quantitative structure–property relationship (QSPR) models have been commonly used for several decades to predict ADME properties of a drug at early phases of development. There are several *in silico* models applied in ADME analysis, and readers are encouraged to read the review by van de Waterbeemd and Gifford (2003). GastroPlus™, developed at Simulations Plus (www.simulations-plus.com), is highly advanced, physiologically based rapid pharmacokinetic (PBPK) simulation software that can generate results within 5 seconds, thus saving huge amounts of time and cost in clinical studies. The software is an essential tool to formulation scientists for *in vitro* dose disintegration and dissolution studies. Towards next-generation treatment of spinal cord injuries, Novartis (www.novartis.com) is working to model the human spinal cord and its surrounding tissues *in silico* to check the feasibility of monoclonal antibody-based drug administration and their pharmacokinetics and pharmacodynamics study results.

The *in silico* "drug re-purposing" approach by Bisson et al. (2007) demonstrated how phenothiazine derivative antipsychotic drugs such as acetophenazine can cause endocrine side effects. Recently Aguda et al. (2011) reported a computational model for sarcoidosis dynamics that is useful for pre-clinical therapeutic studies for assessment of dose optimization of targeted drugs used to treat sarcoidosis.

Towards designing personalized therapy of larynx injury leading to acute vocal fold damage, Li et al. (2008) developed agent-based computational models.

In a further advancement, Entelos® (www.entelos.com) has developed "virtual patients," *in silico* mechanistic models of type-2 diabetes, rheumatoid arthritis, hypertension, and atherosclerosis for identification of biomarkers, drug targets, development of therapeutics, and clinical trial design, and patient stratification. Entelos' virtual Idd9 mouse (NOD mouse) can replace diabetes resistance type-1 diabetes live mice for various *in vivo* experiments.

Apart from diseases, systems level modeling of basic biological phenomena and their applications in disease have also been reported. An *in silico* model to mimic the *in vitro* rolling, activation, and adhesion of individual leukocytes has been developed by Tang et al. (2007). Developing virtual mitochondria, Cree et al. (2008) have demonstrated how the *in silico* modeling can assist in predicting the severity of mitochondrial diseases. Sobie and Wehrens, 2010 have developed *in silico* cardiac models and demonstrated how calcium mediated arrhythmias can develop. Interested readers in computational cardiology can consult with review of Trayanova, 2012.

WORLD WIDE WEB RESOURCES

Virus Pathogen Database and Analysis Resource (ViPR)

ViPR (http://www.viprbrc.org/brc/home.do?decorator=vipr) is one of the five Bioinformatics Resource Centers (BRCs) funded by the National Institute of Allergy and Infectious Diseases (NIAID). This website provides a publicly available database and a number of computational analysis tools to search and analyze data for virus pathogens. Some of the tools available at ViPR are the following:

1. GATU (Genome Annotation Transfer Utility), a tool to transfer annotations from a previously annotated reference to a new, closely related target genome.
2. PCR Premier Design, a tool for designing PCR primers.
3. A sequence format conversion tool.
4. A tool to identify short peptides in proteins.
5. A meta-driven comparative analysis tool.

As there are many different kinds of tools available the tools on the website are organized by the virus family.

The Rat Genome Database (RGD)

The Rat Genome Database (http://rgd.mcw.edu/wg/home) is funded by the National Heart, Lung, and Blood Institute (NHLBI) of the National Institutes of Health (NIH). The goal of this project is to consolidate research work from various institutes to generate and maintain a rat genomic database (and make it available to the scientific community). The website provides a variety of tools to analyze data.

Influenza Research Database (IRD)

IRD (http://www.fludb.org/brc/home.do?decorator=influenza) is one of the five Bioinformatics Resource Centers (BRCs) funded by the National Institute of Allergy and Infectious Diseases (NIAID). This website provides a publicly available database and a number of computational analysis tools to search and analyze data for influenza virus. This website provides many of the same tools that are provided at ViBR. There are numerous other tools such as Models of Infectious Disease Agent Study (MIDAS), which is an *in silico* model for assessing infectious disease dynamics. MIDAS assists in preparing, detecting, and responding to infectious disease threats.

The Wellcome Trust Sanger Institute

The Sanger Institute (http://www.sanger.ac.uk/) investigates genomes in the study of diseases that have an impact on global health. The Sanger Institute has made a significant contribution to genomic research and developing a new understanding of genomes and their role in biology. The website provides sequence genomes for various bacterial, viral, and model organisms such as zebrafish, mouse, gorilla, etc. A number of open source software tools for visualizing and analyzing data sets are available at the Sanger Institute website.

REFERENCES

Aguda, B. D., Marsh, C. B., Thacker, M., & Crouser, E. D. (2011). An *in silico* modeling approach to understanding the dynamics of sarcoidosis. *PLoS One*, *6*. (5)http://dx.doi.org/10.1371/journal.pone.0019544. e19544.

Altmann, D., & Boyton, R. (2005). Models of multiple sclerosis. Autoimmune diseases., Drug Discovery Today. *Disease Models*, *11*, 405–410.

Baccam, P., Beauchemin, C., Macken, C. A., Hayden, F. G., & Perelson, A. S. (2006). Kinetics of influenza A virus infection in humans. *Journal of Virology*, *80*, 7590–7599.

Bisson, W. H., Cheltsov, A. V., Bruey-Sedano, N., Lin, B., Chen, J., Goldberger, N., May, L. T., Christopoulos, A., Dalton, J. T., Sexton, P. M., Zhang, X. K., & Abagyan, R. (2007). Discovery of antiandrogen activity of nonsteroidal scaffolds of marketed drugs. *Proceedings of the National Academy of Sciences of the United States of America*, *104*(29), 11927–11932.

Chassagnole, C., Jackson, R. C., et al. (2006). Using a mammalian cell cycle simulation to interpret differential kinase inhibition in antitumour pharmaceutical development. *Biosystems*, *83*(2-3), 91–97.

Chakraborty, Arup K., Dustin, Michael L., & Shaw, Andrey S. (2003). *In silico* models for cellular and molecular immunology: successes, promises and challenges. *Nature immunology*, *4.10*, 933–936.

Choi, S. (2007). *Introduction to Systems Biology*. Humana Press Inc. SBN 9781597455312.

Cree, L. M., Samuels, D. C., de Sousa Lopes, S. C., Rajasimha, H. K., Wonnapinij, P., Mann, J. R., Dahl, H. H., & Chinnery, P. F. (2008). A reduction of mitochondrial DNA molecules during embryogenesis explains the rapid segregation of genotypes. *Nature genetics*, *40*(2), 249–254.

Danchin, A., et al. (1991). From data banks to data bases. *Research in microbiology*, *142.7*, 913–916.

Daun, S., & Clermont, G. (2007). *In Silico* modeling in infectious disease. *Drug Discov. Today Dis. Models*, *4*(3), 117–122.

Davies, M., Pendlington, R. U., Page, L., Roper, C. S., Sanders, D. J., Bourner, C., Pease, C. K., & MacKay, C. (2011 ; Feb). Determining epidermal disposition kinetics for use in an integrated nonanimal approach to skin sensitization risk assessment. *Toxicological Sciences*, *119*(2), 308–318. http://dx.doi.org/10.1093/toxsci/kfq326.

Dionysiou, D. D., Stamatakos, G. S., Uzunoglu, N. K., Nikita, K. S., & Marioli, A. (2004; Sep 7). A four-dimensional simulation model of tumour response to radiotherapy *in vivo*: parametric validation considering radiosensitivity, genetic profile and fractionation. *Journal of Theoretical Biology*, *230*(1), 1–20.

Edelman, L. B., Eddy, J. A., & Price, N. D. (2010). *In silico* models of cancer. *Wiley Interdiscip. Rev. Syst. Biol. Mol. 4*, 438–459.

Enderling, H., Anderson, A. R. A., et al. (2006). Mathematical modeling of radiotherapy strategies for early breast cancer. *Journal of Theoretical Biology*, *241*(1), 158–171.

Edelstein-Keshet, L., & Spiros, A. (2002). Exploring the Formation of Alzheimer's Disease Senile Plaques *in Silico*. *Journal of Theoretical Biology.*, *216*, 301–326.

Ekins, S., Mestres, J., & Testa, B. (2009). *In silico* pharmacology for drug discovery: methods for virtual ligand screening and profiling. *British Journal of Pharmacology*, *152.1*, 9–20.

Fadiel, A., Anidi, I., et al. (2005). Farm animal genomics and informatics: an update. *Nucleic Acids Research*, *33*(19), 6308–6318.

Frieboes, H. B., Zheng, X., Sun, C. H., Tromberg, B., Gatenby, R., & Cristini, V. (2006 Feb 1). An integrated computational/experimental model of tumor invasion. *Cancer Research*, *66*(3), 1597–1604.

Ghosh, P., Kumar, A., Datta, B., & Rangachari, V. (2010). Dynamics of protofibril elongation and association involved in Abeta42 peptide aggregation in Alzheimer's disease. *BMC Bioinformatics*, *6* (11 Suppl), S24.

Huang da, W., Sherman, B. T., et al. (2009). Bioinformatics enrichment tools: paths toward the comprehensive functional analysis of large gene lists. *Nucleic Acids Research*, *37*(1), 1–13.

Kiran, K. L., Jayachandran, D., et al. (2009). Mathematical modeling of avascular tumour growth based on diffusion of nutrients and its validation. *Canadian Journal of Chemical Engineering*, *87*(5), 732–740.

Kovatchev, B., Breton, M., & Johnson, B. (2012). *In silico* models of alcohol dependence and treatment. *Front Psychiatry*, *3*, 4.

Lewis, N. E., Schramm, G., Bordbar, A., Schellenberger, J., Andersen, M. P., Cheng, J. K., Patel, N., Yee, A., Lewis, R. A., Eils, R., & König, R. A., PalssonBØ. (2010). Large-scale *in silico* modeling of metabolic interactions between cell types in the human brain. *Nature Biotechnology*, *28*, 1279–1285.

Li, N. Y., Verdolini, K., Clermont, G., Mi, Q., Rubinstein, E. N., Hebda, P. A., & Vodovotz, Y. (2008; Jul 30). A Patient-Specific *in silico* Model of Inflammation and Healing Tested in Acute Vocal Fold Injury. *PLoS One*, *3*(7), e2789.

Maxwell, G., & Mackay, C. (2008 Nov). Application of a systems biology approach to skin allergy risk assessment. *Alternatives to Laboratory Animals*, *36*(5), 521–556.

Mi, Q., Rivière, B., Clermont, G., Steed, D. L., & Vodovotz, Y. (2007). Agent-based model of inflammation and wound healing: insights into diabetic foot ulcer pathology and the role of transforming growth factor-β1. *Wound Repair Regeneration.*, *15*, 671–682.

Muñoz-Elías, E. J., Timm, J., Botha, T., Chan, W. T., Gomez, J. E., & McKinney, J. D. (2005). Replication dynamics of mycobacterium tuberculosis in chronically infected mice. *Infection and Immunity*, *73*(1), 546–551.

Navratil, V., de Chassey, B., Combe, C. R., & Lotteau, V. (2011). When the human viral infectome and diseasome networks collide: towards a systems biology platform for the aetiology of human diseases. *BMC Systems Biology*, *5*, 13.

Raichura, A., Valia, S., & Gorinb, F. (2006). Dynamic modeling of alpha-synuclein aggregation for the sporadic and genetic forms of Parkinson's disease. *Neuroscience*, *142*(3), 859–870.

Sieburg, Hans B. (1990). Physiological studies *in silico*. *Studies in the Sciences of Complexity*, *12.2*, 321–342.

Sobie, A. E., & Wehrens, Z. H. T. (2010). Computational and experimental models of Ca2+-dependent arrhythmias. *Drug Discov Today Dis Models*, *6*, 5.

Tang, J., Ley, K. F., & Hunt, C. A. (2007; Feb 19). Dynamics of *in silico* leukocyte rolling, activation, and adhesion. *BMC Systematic Biology*, *1*, 14.

Tomita, M. (2001). Whole-cell simulation: a grand challenge of the 21st century. *Trends of Biotechnology*, *19*(6), 205–210.

Tracqui, P., Cruywagen, G. C., Woodward, D. E., Bartoo, G. T., Murray, J. D., & Alvord, E. C., Jr (1995 Jan). A mathematical model of glioma growth: the effect of chemotherapy on spatio-temporal growth. *Cell Prolif*, *28*(1), 17–31.

Trayanova, N. A. (2012). Computational cardiology: The heart of the matter. *ISRN Cardiol*, *2012*, 269680.

Tuncbag, N., Kar, G., Keskin, O., Gursoy, A., & Nussinov, R. (2009 May). A survey of available tools and web servers for analysis of protein–protein interactions and interfaces. *Brief Bioinform*, *10*(3), 217–232.

van de Waterbeemd, H., & Gifford, E. (2003 Mar). ADMET *in silico* modelling: towards prediction paradise? *Nat Rev Drug Discov*, *2*(3), 192–204.

Vodovotz, Y., Csete, M., Bartels, J., Chang, S., & An, G. (2008). Translational systems biology of inflammation. *PLoS Computational Biology*, *4*, e1000014.

Werner, T. (2008). Bioinformatics applications for pathway analysis of microarray data. *Current Opinion in Biotechnology*, *19*(1), 50–54.

Woolf, P. J., Prudhomme, W., et al. (2005). Bayesian analysis of signaling networks governing embryonic stem cell fate decisions. *Bioinfor matics*, *21*(6), 741–753.

Wu, Y., Lu, Y., Chen, W., Fu, J., & Fan, R. (2012 Feb). *In silico* experimentation of glioma microenvironment development and anti-tumor therapy. *PLoS Computing Biology*, *8*(2), e1002355.

Zhang, Y., Athale, C. A., & Deisboeck, T. S. (2007). Development of a three-dimensional multiscale agent-based tumor model: simulating gene-protein interaction profiles, cell phenotypes and multicellular patterns in brain cancer. *Journal of Theoretical Biology*, *244*, 96–107.

FURTHER READING

Bock, Gregory, & Goode, Jamie (Eds.), (2002). *'In Silico' Simulation of Biological Processes* (Vol. 247). Wiley.

Cronin, Mark TD., & Madden, Judith C. (2010). *Silico Toxicology: Principles and Applications*(Vol. 7). Royal Society of Chemistry.

Deisboeck, Thomas S., & Stamatakos, Georgio S. (2010). *Multiscale Cancer Modeling* (Vol. 34). CRC PressI Llc.

Flower, Darren R., Timmis, Jon, & Timmis, Jonathan (Eds.), (2007). *In Silico Immunology*. Springer.

Sharpe, Jason, John Lumsden, Charles, & Woolridge, Nicholas (2008). *In Silico: 3D Animation and Simulation of Cell Biology with Maya and MEL*. Morgan Kaufmann.

GLOSSARY

Algorithm Any well-defined computational procedure that takes some values, or set of values, as input, and produces some value, or set of values, as output.
Bioinformatics Bioinformatics is the application of statistics and computer science to the field of molecular biology.
Biotechnology The exploitation of biological processes for industrial and other purposes.
Data Structures A way to store and organize data on a computer in order to facilitate access and modifications.
Genome The complete set of genetic material of an organism.
Genomics The branch of molecular biology concerned with the structure, function, evolution, and mapping of genomes.
Gene Ontology A major bioinformatics initiative to unify the representation of gene and gene product attributes across all species.
Informatics The science of processing data for storage and retrieval; information science.
In Silico *In silico* is an expression used to mean "performed on a computer or via computer simulation."
In Vivo In microbiology *in vivo* is often used to refer to experimentation done in live isolated cells rather than in a whole organism.
In Vitro *In vitro* studies in experimental biology are those that are conducted using components of an organism that have been isolated from their usual biological surroundings in order to permit a more detailed or more convenient analysis than can be done with whole organisms.
Kinomics Kinomics is the study of kinase signaling within cellular or tissue lysates.
Oncogenesis The progression of cytological, genetic, and cellular changes that culminate in a malignant tumor.
Pathophysiology The disordered physiological processes associated with disease or injury.
Proteomics The branch of genetics that studies the full set of proteins encoded by a genome.
Sequencing The process of determining the precise order of nucleotides within a DNA molecule.
Systems Biology An inter-disciplinary field of study that focuses on complex interactions within biological systems by using a more holistic perspective.

ABBREVIATIONS

AD Alzheimer Disease
ARACNE Algorithm for the Reconstruction of Accurate Cellular Networks
BRC Bioinformatics Resource Center
CDKs Cyclin-Dependent Kinases
DNA Deoxyribonucleic acid
DZM Different Zone Model
GNW GeneNetWeaver
GO Gene Ontology
GoI Genome of Interest
IRD Influenza Research Database
LMW Low Molecular Weight
MOF Multi Organ Failure
NIAID National Institute of Allergy and Infectious Diseases
MIDAS Models of Infectious Disease Agent Study
mRNA Messenger Ribonucleic Acid
ODE Ordinary Differential Equations
P Protein
PCR Polymerase Chain Reaction
PDE Partial Differential Equations
PDGF Platelet-Derived Growth Factor
PPI Protein–Protein Interaction
RGD Rat Genome Database
RNA Ribonucleic Acid
SARS Severe Acute Respiratory Syndrome.
SIRS Systematic Inflammatory Response System
TG Template Genome
TN Template Network
ViPR Virus Pathogen Database and Analysis Resource

LONG ANSWER QUESTIONS

1. Explain the role of bioinformatics in animal biotechnology.
2. Explain the common computational methods in systems biology.
3. Explain the concept of *in silico* modeling.
4. Discuss the advantages, disadvantages, and ethical issues of *in silico* modeling.
5. What are the different application areas of *in silico* modeling? Discuss in detail how *in silico* modeling is applied in one application area.

SHORT ANSWER QUESTIONS

1. Describe the template-based methods to reconstruct Transcriptional Regulatory Networks.
2. What is the goal of *in silico* modeling?
3. What are the challenges in *in silico* modeling of infectious diseases?
4. What are the three types of cancer models discussed in the chapter?
5. Discuss the parameters considered for *in silico* modeling of infectious diseases.

ANSWERS TO SHORT ANSWER QUESTIONS

1. The template-based transcriptional control network reconstruction method exploits the principle that orthologous proteins regulate orthologous target genes. In this approach, regulatory interactions are transferred from a genome (such as a genome of a model organism or well studied organism) to the new genome.
2. The ultimate goal of *in silico* modeling in biology is the detailed understanding of the function of molecular networks as they appear in metabolism, gene regulation, or signal transduction.

3. There are two major challenges in modeling infectious diseases:
   a. Difficulty in finding the most appropriate set of parameters for the *in silico* modeling of infectious diseases is often a challenge.
   b. Understanding the results from all the complex interactions of parameters considered.
4. There are three types of cancer models. ***Continuum models:*** In these models extracellular parameters can be represented as continuously distributed variables to mathematically model cell–cell or cell–environment interactions in the context of cancers and the tumor microenvironment. ***Discrete models:*** These models represent cancer cells as discrete entities of defined location and scale, interacting with one another and external factors in discrete time intervals according to predefined rules. ***Hybrid models:*** These models incorporate both continuum and discrete variables in a modular approach.
5. There are three types of parameters considered for *in silico* modeling of infectious diseases:
   a. ***Parameters derived from characteristics of agent:*** Examples: concentration of the agent's antigen–host antibody complex; case fatality rate; strain of the agent; other genetic information of the agent; etc.
   b. ***Parameters derived from characteristics of host:*** Examples: the total white blood cell counts; differential white blood cell counts, and/or much more sophisticated counts of specific blood cell types; blood levels of some specific cytokines, hormones, and/or neurotransmitters; daily calories, protein, and/or fat intake; daily amount of energy expended and/or duration of exercise; etc.
   c. ***Parameters derived from characteristics of environment:*** Examples: host's ambient temperature; host's ambient atmospheric humidity; altitude; host's light-dark cycle; etc.

Section III

Animal Biotechnology: Applications and Concerns

Chapter 22

Transgenic Animals and their Applications

Shet Masih, Pooja Jain, Rasha El Baz and Zafar K. Khan

Department of Microbiology and Immunology, Drexel Institute for Biotechnology and Virology Research, Drexel University College of Medicine, Doylestown, Pennsylvania, USA

Chapter Outline

SUMMARY

Transgenic animals are the result of manipulation of specific genes of interest by creating animal models and bioreactors to (for example) produce vaccines and antibodies that can be used to help save millions of lives. These animals with altered genomes must therefore be used under strict ethical control.

WHAT YOU CAN EXPECT TO KNOW

Transgenic animals are animals that have been genetically transformed by splicing and inserting foreign (animal or human) genes into their chromosomes. When inserted successfully, these genes enable an animal to produce (for example) pharmaceutical proteins in its milk, urine, blood, sperm, or eggs, or to grow rejection-resistant organs for transplant. These proteins have significant therapeutic potential for the treatment of cystic fibrosis, hemophilia, osteoporosis, arthritis, and parasitic and infectious diseases such as malaria and HIV. Transgenic animals can also produce monoclonal antibodies (specifically targeted towards disease proteins) that are used in vaccine development to meet global demands.

This chapter has been framed in a manner to educate the reader about transgenic animals, methods of creating them, their effects on the environment, and associated socio-ethical concerns. Readers will also get a glimpse of the history of transgenic animal production and the implications for pharmaceutical research and industry. The overall purpose of selection and production of animal models required for human welfare has been described. In medical research, animal models are required to understand various cellular and molecular pathways; these are explained here

Animal Biotechnology. http://dx.doi.org/10.1016/B978-0-12-416002-6.00022-5

with examples. Some of the disease models that are used vigorously in the scientific community to study alterations in normal physiology of the host that lead to a diseases or disorders are also discussed. This chapter also includes a section on the engineering of animals as food sources or to produce human tissues to be used in histocompatible transplantations. Finally, ethical issues related to genetic engineering and the U.S. Food and Drug Administration's (FDA) ethical guidelines regarding the creation of transgenic animals are covered.

INTRODUCTION

The practice of carefully selecting and reproducing animals with new combinations of genes is not new. In nature, however, new gene combinations are found only in the same or similar species. Transgenic or genetically engineered animals are the result of novel gene combinations thoughtfully manipulated and implemented by scientists. Because DNA contains the universal genetic code for all living organisms, it can be easily altered and transferred between two completely unrelated organisms to produce a combination of characteristics that would not otherwise be possible.

The term "transgenic" was coined in 1981 to describe an animal in which an exogenous gene had been introduced into the genome (Gordon and Ruddle, 1981). A genetically engineered or "transgenic" animal carries a known sequence of recombinant DNA in its cells and passes the recombinant DNA on to its offspring. In the late 1980s, the term "transgenic" was extended to gene-targeting experimentation and the production of chimeric or "knockout" mice in which a gene has been selectively removed from the host genome (Beardmore, 1997). Recombinant DNA refers to fragments of DNA sequences that have been joined together in a molecular biology laboratory. The DNA to express a desired protein is constructed in such a manner that it can express the functional protein when inserted in the nucleus of a transgenic animal. The gene encoded in the resultant DNA construct is capable of producing the same kind of protein no matter which animal or microbe (or even plant) is producing it. This extra desired protein synthesis is the only difference between transgenic and non-transgenic animals; otherwise the two are identical in behavior and appearance. Some of the proteins, particularly for therapeutic purposes, have been produced in transgenic animals (Echelard et al., 2006; Lillico et al., 2007). Some of these engineered proteins provide the animals with better disease resistance than their wild-type counterparts. Other examples are proteins that provide healthier milk, meat, or eggs (Lai et al., 2006).

Transgenic animals can be produced through various methods, but in all procedures the first requirement is to generate a transgene (i.e. the DNA sequence that encodes for a particular protein along with other necessary sequences). Generally, three parts are required to construct a transgene: (1) a promoter sequence that determines which tissue should express the recombinant protein; (2) the sequence of the gene coding for amino acids of the desired recombinant protein; and (3) the sequence responsible for the termination of the expressed protein. The most common method of producing transgenic animals is microinjection, in which the constructed transgene is inserted into the male pronucleus of a freshly fertilized egg. Generally, several copies of the transgene are inserted into the male pronucleus, which is larger than the female pronucleus. The other common method for producing transgenic animals is embryonic stem (ES) cell manipulation. In this method a transgene is inserted into the stem cells of the blastocyst with the help of microinjection, chemicals, or viral transduction. The ease of screening of embryonic stem cells carrying transgenes is the main advantage leading to the high efficiency of the technique. Polymerase chain reaction (PCR) and Southern blot hybridization are two main techniques for screening the animals carrying the desired transgene.

Different categories of transgenic animals include those produced as disease models, xenotransplanters, transpharmers, food sources, and scientific research models. Through genetic engineering various disease models have been developed to mimic the human symptoms of disease. Examples of such models include the OncoMouse (mouse model for cancer study), the AIDS mouse, the Alzheimer's mouse, the HLA-A2.1/Tg mouse (to study the presentation of antigens that are normally not presented by the surface of mouse antigen-presenting cells), and Parkinson's fly. Animals that are engineered to express a desired protein in their milk are known as transpharmers; for example, mice, sheep, goats, and cows have been engineered by this method. Also, there are animals that have been engineered to produce histocompatible organs that can be implanted in humans without fear of rejection by the human body. This technique has been used for producing pigs as xenotransplanters, but the use of those organs has not been approved yet. Similarly, to meet the daily increasing demand of food, animals have been engineered that grow larger than their wild-type counterparts without requiring extra food. Two examples of transgenic food sources are Superfish and Superpig. Superfish is a promising food source, but Superpig proved futile because the animals developed multiple health issues.

Transgenic animal models are generally produced for scientific research by inserting a transgene in their DNA to study the effect of overexpression of that particular gene on the animal's physiological processes. Sometimes the gene under investigation is knocked out to determine its effect on normal body metabolism. Well-known examples of scientific research models are ANDi, the transgenic monkey; smart mouse; youth mouse; Supermouse; and influenza-resistant mouse.

Many ethical issues are associated with the production of transgenic animals. First, is it ethical to generate transgenic animals? It seems clear that altering the genome of an animal to create artistic effects (e.g. the green fluorescent rabbits) is unnecessary and cannot be recommended. If the creation of an animal would increase scientific knowledge and possibly help understand a disease condition, then alteration of the genome can be accepted by most. Although, in most transgenic experiments some animals die, the number of human lives saved as a result may be much greater than the animals' suffering. Further, it is recommended that the suffering of the transgenic animals should be reduced as much as possible. The production of all transgenic animals cannot be justified, but each experimental instance should be judged on case-to-case basis. Alzheimer's mouse, for example, feels no pain due to its transgene, and its creation is justified because studying this mouse model might lead to new Alzheimer's treatments. Conversely, the Beltsville pig or so-called Superpig experienced much suffering for a remote possibility of helping meet world food demand, so this experiment was rightly terminated. Another important example is OncoMouse, a model that straddles this ethical edge because the mouse does suffer tremendously and eventually dies of cancer, but the knowledge gained from this transgenic model is highly valuable in the fight against cancer in human beings. Therefore, the creation of OncoMouse is justified, but measures must be applied to reduce the animals' pain; or they can be sacrificed before advanced tumor development causes them unbearable suffering. Although animal physiology is not identical to human physiology, the knowledge obtained from living disease models is important and useful, thus justifying the creation of transgenic animals as disease models.

Another concern is that transgenic animals will escape and outbreed with their wild-type natural cousins. Despite this concern, transgenic animals can be reared and used with restrictive security measures in place. Established regulatory authorities such as universities' Institutional Animal Care and Use Committees (IACUC) monitor scientific research and require research scientists to justify animal use for each experiment. The creation of transgenic animals per se is opposed by many religious groups and by some voluntary and nongovernmental organizations that feel the practice interferes with nature. Most, however, have no problem if the suffering of the animals is minimized and the experiment is important scientifically.

CREATING TRANSGENIC ANIMALS

Transgenic animals can be created using various techniques (Cho et al., 2009). One of the most popular techniques is microinjection of a transgene into the male pronucleus of a freshly fertilized egg *in vitro*. A brief protocol/method for generation of transgenic mice has been described. The pronuclear-microinjection technique primarily involves five major steps that are given in Flow Chart 22.1. Another common technique is embryonic stem cell transfer. Other methods include transfer of transgenes into embryonic stem cells with the help of a chemical or virus, homologous recombination, or gene knockdown.

Construction of a Transgene

Although the genetic code is almost the same for all organisms, the fine details of gene control are different. For example, a gene from a bacterium will often not work correctly if it is introduced unmodified into an animal cell. Therefore, genetic engineers first construct a transgene containing the gene of interest plus extra DNA sequences that correctly control the function of the gene in the new animal (Figure 22.1). When constructing a transgene, scientists generally substitute the donor's promoter sequence with one that is specially designed to ensure that the gene will function in the correct tissues of the recipient animal; for example, if the gene needs to be expressed in the milk of a mammal, the promoter specific for mammary tissues is used in the transgene (Murphy et al., 1993).

Microinjection

Microinjection is the most popular technique for creating transgenic animals (Cho et al., 2009). In this technique recombinant DNA (transgene) is inserted into the male pronucleus. First, eggs are collected from super-ovulating female animals and then fertilized *in vitro*. The freshly fertilized eggs are held stable by a microtube suction device (Figure 22.2), and a solution containing 200 to 300 copies of the recombinant DNA is injected using a micropipette. The recombinant DNA is injected into the male pronucleus because it is larger than the female pronucleus (Wong et al., 2000). Ethical concerns arise because a very small percentage of animals are born with the desired transgene while a huge number of eggs are wasted in the experiment.

Embryonic Stem Cell Transfer

Another commonly used technique of creating transgenic animals is by transferring the transgene to embryonic stem (ES) cells. This technique is generally used when a particular gene must be altered in a particular tissue. A transgene is constructed to target a specific site in the genome of the animal to be created. For embryonic stem cells, a blastocyst is harvested from a female animal or by *in vitro* fertilization, and the inner cell mass is collected (Jacenko, 1997). Sometimes chemicals or steroids can be given to an animal to prevent implantation of the embryo. The transgene can be introduced into the embryonic stem cells with the help of a chemical, by a virus. Embryonic cells are used because

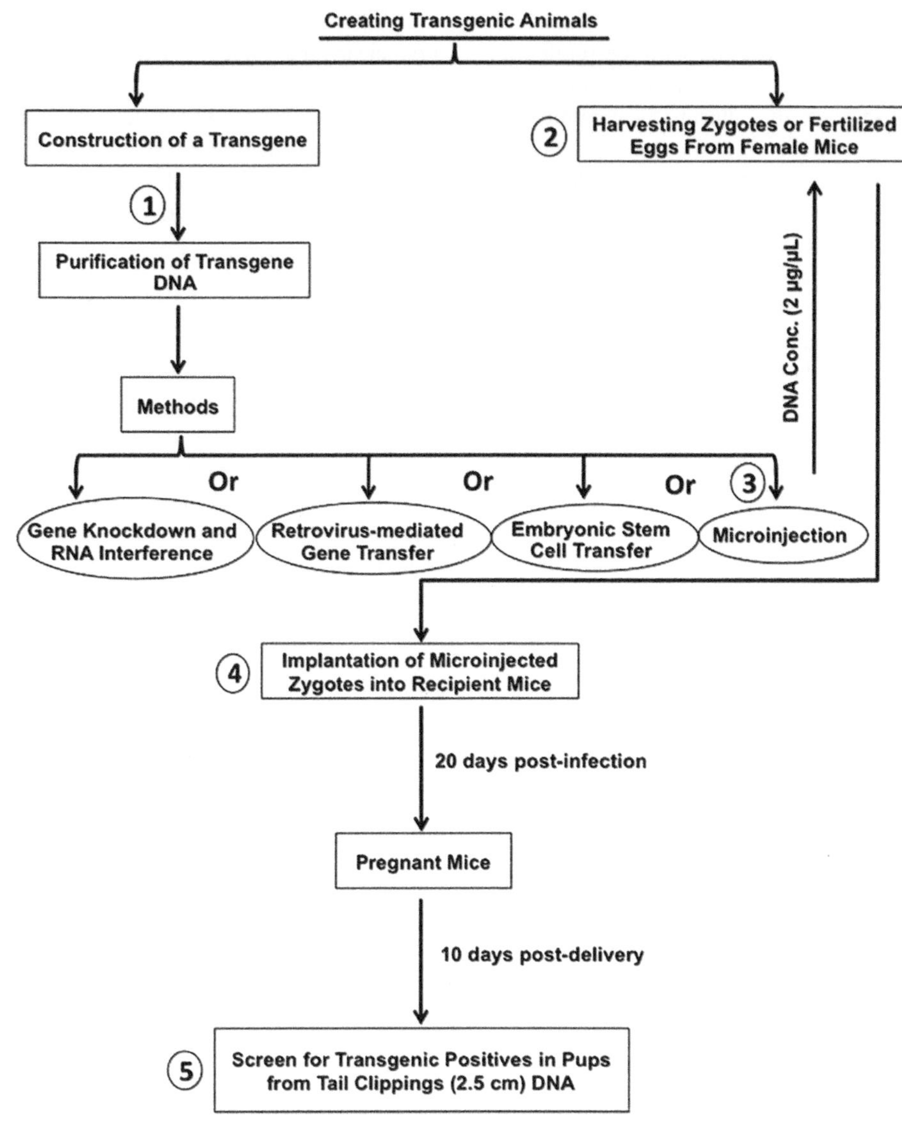

FLOW CHART 22.1

1. ***Purification of transgenic DNA constructs***. Purification of the construct is a critical step for a healthy embryo. A sucrose gradient or a gel-purification method can be adopted to obtain purified and clean DNA that is dissolved in a microinjection buffer.
2. ***Harvesting zygotes/fertilized eggs from donor female mice***. First, egg donor mice (the widely used strain FVB/N that has a large pronucleus in embryos) are superovulated by injecting (ip) pregnant mare's serum gonadotropin (PMSG) and human chorionic gonadotropin (HCG) to obtain a maximum number of zygotes for microinjection before mating. Then, isolate the ovary and oviduct through surgical procedures and release the eggs into M2 medium dishes. Finally, separate and wash the zygotes in medium and transfer the dishes at 37°C in a 5% CO_2 incubator.
3. ***Microinjection of construct.*** Purified DNA (conc. 2 ng μl^{-1}) is gently microinjected into the zygote pronucleus with a constant flow rate of 50 hPa (hectopascal) using an automated microinjector in order to initiate integration of the transgene and cell division. Transfer the zygotes in an M16 culture medium dish and maintain at 37°C in a 5% CO_2 incubator until implantation into the recipient mice.
4. ***Implantation of microinjected zygotes into recipient mice.*** Implantation of microinjected zygotes into pseudo-pregnant recipient mice is done through surgical steps. First, locate the infundibulum, then open an oviduct into which the microinjected zygotes can be carefully transferred, tuck the oviduct and ovary into the body cavity of the mice, stitch up the skin, and, finally, transfer the animals into separate cages.
5. ***Screening of pups for transgenic positives.*** The recipient mice are normally pregnant and deliver pups 20 days post-transfer of the embryos. The pups are genotyped 10 days after birth for positives (from tail clippings of ~0.5 cm) for transgenic DNA by standard procedure.

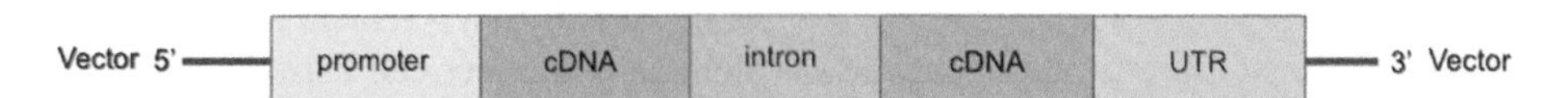

FIGURE 22.1 Transgene. The promoter (indigo) dictates in which tissue the transgene will be expressed. The transgene (orange) contains cDNA sequences encoding the protein to be expressed (e.g. a human therapeutic protein to be produced in the animal's milk). There is evidence that intron splicing plays a role in gene expression, so an intron (purple) is used (not necessarily from the gene of interest) between the cDNA sequences. The untranslated region (UTR) termination sequence (green) dictates the termination of transcription. The construct is linearized before injection (Murphy et al., 1993).

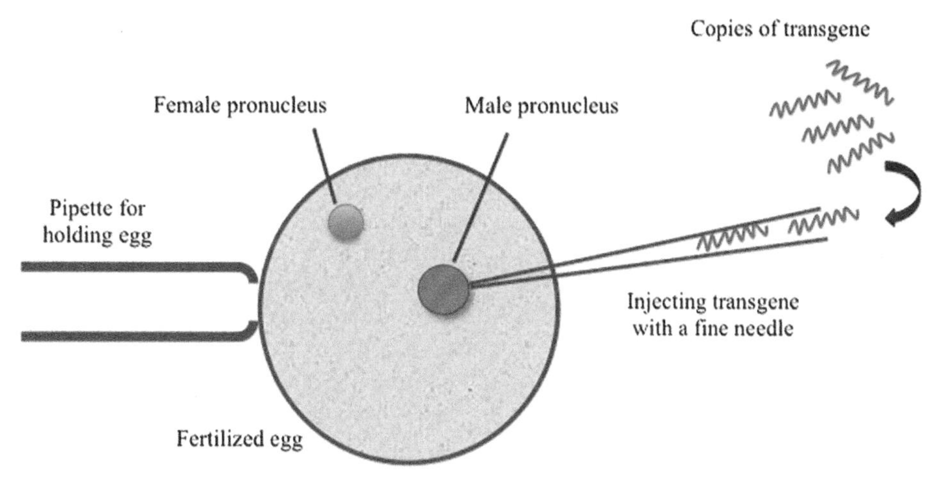

FIGURE 22.2 Microinjection. After fertilization, the egg is held stable by a microtube suction device, while a solution containing many copies of the transgene (green) is injected into the male pronucleus (blue) using a micropipette (because it is larger than the female pronucleus) (indigo) (Wong *et al.*, 2000).

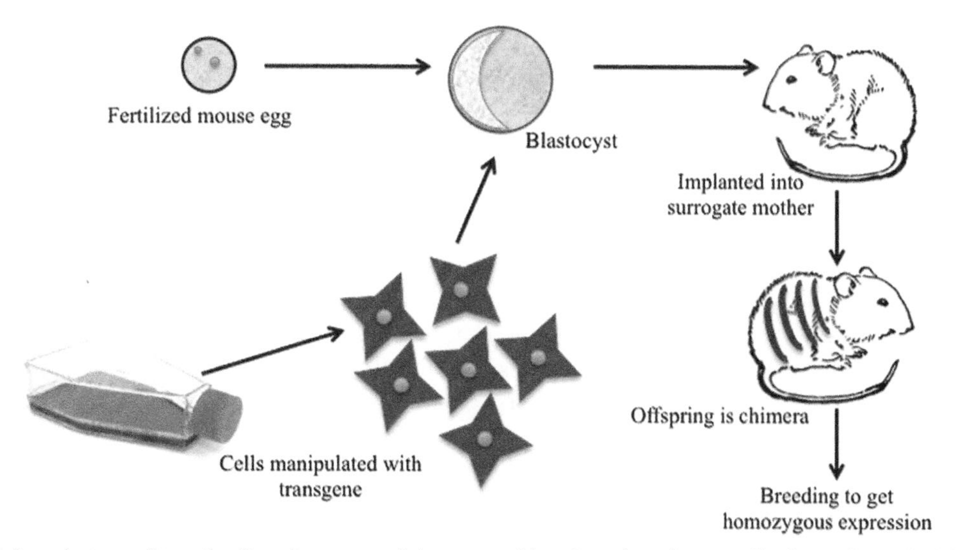

FIGURE 22.3 Embryonic stem cell transfer. Once the transgene is incorporated into the embryonic stem cells, those cells can be left to divide *in vitro*, or they can be injected into a blastocyst and implanted into a host's uterus to grow normally (Jacenko, 1997).

they are capable of differentiating into any of the three germ layers. Therefore, if the embryonic stem cells carrying the recombinant gene are injected into an embryo, that embryo, when implanted into the uterus of a host animal, will develop into an animal carrying that particular transgene (Figure 22.3). Once the transgene is inside the embryonic stem cells, with the division of the cells the transgene incorporates itself by homologous recombination. This natural technique involves crossing over between paired "sister" chromatids, which can lead to a new recombinant sister chromatid. These homologous sequences are considered when the transgene is designed in order to determine where to integrate the desired gene. The flanking sequences of that particular area are chosen and added to both flanking sides of the transgene. Thus, via this technique a transgene can be integrated at a particular position; this is not possible using general techniques of viral transduction, chemicals such as calcium phosphate or rubidium chloride, or even with microinjection, as all of these techniques insert the transgene randomly into the genome.

Retrovirus-Mediated Gene Transfer

Retroviral vectors have also been used to transfer genes of interest into animal genomes (van der Putten et al., 1985). Although, embryos can be used up to mid-gestation, 4- to 16-cell stage embryos are generally used for infection with one or more retroviruses carrying recombinant DNA (effectively transducing only mitotically active cells). Immediately following infection, the retrovirus produces a DNA copy of its RNA genome using the viral enzyme reverse transcriptase (Figure 22.4). Usually without any deletions or rearrangements, the DNA copy of the viral genome, or provirus, integrates randomly into the host cell genome. Very

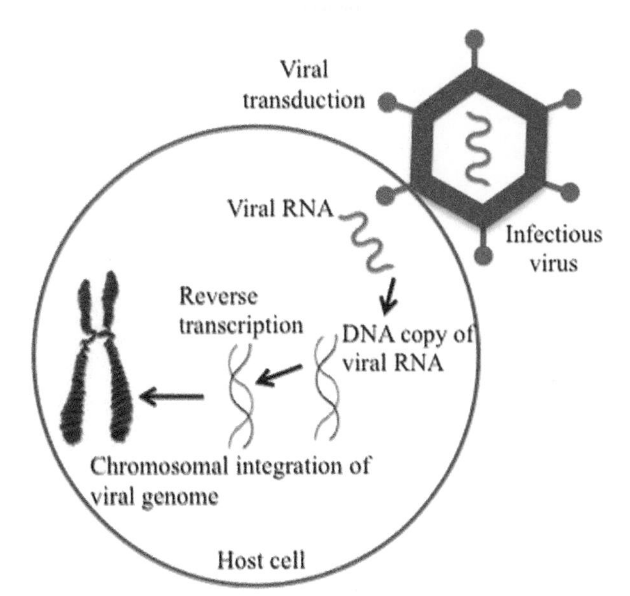

FIGURE 22.4 Retrovirus-mediated gene transfer via transduction. After viral transduction of the host cell, the viral RNA genome is reverse transcribed into DNA by reverse transcriptase (delivered into the cell along with viral RNA), and this cDNA is then integrated into the host cell genome (Han *et al.*, 1999).

high rates of gene transfer, approaching 100% efficiency, have been achieved with the use of retroviruses (Han et al., 1999). However, as with gene transfer by microinjection, integration events are random. For safety reasons, retroviruses are frequently modified by removing structural genes such as *gag*, *pol*, and *env* (which support viral particle formation). However, most retroviral lines used in transgenic animal experiments are ecotropic, meaning that they infect only the model systems (e.g. mice or rats); hence, rodent cell lines, rather than humans, could be at risk of contamination if correct precautions are not met. Disadvantages of retrovirus-mediated gene transfer include low copy number integration; the additional steps required to produce retroviruses in comparison with microinjection or embryonic stem cell-based techniques; a general limitation on the length of the foreign DNA insert (usually < 15 kb); a potential for undesired genetic recombination that might alter the replicative characteristics of the retrovirus; a high frequency of mosaicism; and finally, possible interference of retroviral sequences on recombinant gene expression.

Gene Knockdown and RNA Interference

Efforts at gain-of-function and loss-of-function modeling have usually concentrated on introducing specific mutations into the nuclear genome. RNA interference (RNAi) technology has broadened the possibilities for creating loss-of-function animal models. Short interfering RNA (siRNA) exists in a double-stranded state and inhibits endogenous genes (and/or exogenous sequences, as in viral genes) as the result of complementary sequence homology (Fire et al., 1998). From an evolutionary point of view, RNAi appears to protect the cell against foreign (e.g. viral) RNA invasion. The mechanism of RNAi is thought to involve a double-stranded (sometimes hairpin) RNA molecule that is cleaved into small fragments of ~22 nucleotides in length and assembled into a ribonucleoprotein complex referred to as the RNA-inducing silencing complex (RISC) (Reid et al., 2001). The RISC then binds to homologous mRNA and performs its endonucleolytic cleavage (Dykxhoorn et al., 2003). Recently, it has been shown that alterations in the complementary oligonucleotide (i.e. length and nucleotide composition) can have a significant effect on both the degree and the duration of "gene silencing" (Lamberton and Christian, 2003). These oligonucleotides that silence gene expression (i.e. protein synthesis) are referred to as siRNAs. Small interfering RNA has been shown to be a potent inhibitor of gene function *in vivo* (Fire et al., 1998). Mouse and rat models harboring small hairpin RNA (shRNA) transgenes, following shRNA transcription, produced lower levels of the homologous protein when compared with controls (Hasuwa et al., 2002). Gene silencing and knockdown technology has potential medical and agricultural applications, including the inhibition of viral gene transcription and of endogenous genes coding for deleterious gene products (Novina et al., 2002). In small animals such as the mouse, RNAi has several advantages over homologous recombination and embryonic stem cell-mediated gene knockout methods. RNAi can be chemically synthesized directly, thus avoiding laborious cloning steps. Therefore, this methodology is the most significant advance since nuclear transfer in effecting efficient loss-of-function modeling in mammalian species (particularly for non-murine species in which embryonic stem cell transgenesis has not been successful).

Screening for Transgenic Positives

After the transgenic animals are born, they are screened either by polymerase chain reaction or by Southern blot analysis. The expression of transgene depends on the site of integration of the transgene; some transgenes may not be expressed if they are integrated into a site that is transcriptionally inactive. A common practice therefore is to breed the animals further to obtain the optimal expression of the desired protein.

Transgenesis Versus Cloning

Transgenesis should not be confused with cloning. Cloning refers to the reproduction of an exact replica of a living organism using the DNA (without manipulation) of that organism, whereas transgenesis (genetic engineering) refers to the human manipulation of genetic material in a manner that does not occur in nature. In 1997, Dolly the sheep was born as a result of cloning experiments at Roslin Institute (Midlothian, Scotland, UK) using a somatic cell from an adult sheep.

TRANSGENIC ANIMALS AS DISEASE MODELS

Many animal models have been created for studying the mechanisms of disease or changes in physiology that characterize some disorders. A few important examples are discussed here. Many other animal disease models have been created or are on researchers' lists to be generated. Small animal models such as mouse models are favored because of the animals' relatively short gestation time and easy manipulation and housing. However, many researchers like pigs as disease models for their anatomical and physiological similarities to humans in studying inflammation, heart disease, Lou Gehrig's disease, sickle cell disease, and many other diseases.

OncoMouse

An important disease model is OncoMouse, a mouse model created to study cancer. Cancers develop in different ways from different causes; therefore, there are many ways to make a model for studying cancer. One way is to create mice that do not possess the *p53* allele, the tumor suppressor gene that is involved in most cancers and that is crucial in checking the uncontrolled growth characteristic of cancer. Removal of this allele leaves mice susceptible to many types of cancer, most frequently lymphoma (Harvey et al., 1993). OncoMouse was created in 1984 by replacement of the normal mouse *myc* gene, which is believed to regulate expression of 15% of all genes by acting as a transcription factor, with a virus tumor promoter/myc fusion recombinant gene. The mice and their offspring developed carcinomas. This first mouse was created at Harvard Medical School in Boston for DuPont (Stewart et al., 1984). The researchers applied for, and in 1988 received, a patent on the process of creating the animal (Leder and Stewart, 1984), and on the mouse itself, making OncoMouse the world's first patented animal. This caused a considerable stir in the scientific community (as will be discussed further in the "Ethics" section), because to study this cancer model, or to create a new model using Harvard's techniques of producing the OncoMouse, obtaining a license was required. DuPont argued that the patent covered any transgenic animal predisposed to cancer (Marshall, 2002). Since then, the company has allowed researchers working with the U.S. National Institutes of Health to work on the mouse for nonprofit research (Smaglik, 2000). Various experiments continue on OncoMouse that may lead to preventing and curing multiple forms of cancer.

AIDS Mouse

AIDS mouse is another good example of a disease model. AIDS mouse was created in 2001 at the University of Maryland by microinjecting the genome of HIV-1 into fertilized mouse eggs. The recombinant DNA of HIV-1 (the transgenic genome) was altered by deletion of the two genes that cause the virus to become infectious (Reid et al., 2001). Thus, HIV-1 mouse cannot transmit the virus to humans, making handling of the animal relatively safe while still allowing study of the HIV biology. This model allows researchers to study early-stage symptoms to better diagnose the disease in humans. It also allows researchers to track chronic conditions associated with AIDS and to test numerous treatments in the search for a cure for HIV disease (Kohn, 2001). Previous reports showed that chimpanzees were capable of supporting HIV replication, but no small animal developed the virus, and use of chimpanzees as a disease model is very expensive. An original female mouse that possessed the modified viral gene was bred with a healthy male mouse to produce HIV gene-bearing pups. Skin lesions were also seen in the AIDS mouse, similar to those seen in Kaposi's sarcoma, which often occurs in AIDS patients. These observations indicated that HIV may indeed be a cause of cancer (Vogel et al., 1988). Thus, AIDS mouse was a big step towards finding treatments to prevent or cure the disease.

Alzheimer's Mouse

Alzheimer's mouse is another important disease model. Alzheimer's disease has been linked to the formation of beta-amyloid plaques in the brain, places where fibers have developed tangles that can block and degrade neurons during the progression of the disease. Similar to humans with the disease, Alzheimer's mouse overproduces a protein that forms these amyloid plaques, and it displays both the symptoms and diagnostic characteristics of Alzheimer's disease (Duff et al., 1996). Alzheimer's mouse was generated in 1995 by a collaborative experiment between Worcester Polytechnic Institute and Transgenic Sciences, Inc. (which became Athena, then Exemplar Corp., then Elan Pharmaceuticals). This mouse line was generated by overexpressing a mutation that causes an aggressive early-onset form of Alzheimer's disease (Games et al., 1995). Scientists studying Alzheimer's disease wanted an animal model for some time before this breakthrough occurred. The first Alzheimer's vaccine, which almost entirely prevented the creation of amyloid plaques in young mice and even reduced the damage of the plaques already allowed to develop in older mice, was developed from research using Alzheimer's mouse (Schenk et al., 1999). This vaccine moved to human clinical trials in 2000, but was cancelled in 2001 because a minority of patients developed brain inflammation; a second-generation vaccine by the same company is now in Phase II human clinical trials and no deleterious side effects have been observed. Thus far, no vaccine has been approved by the FDA for the treatment of Alzheimer's disease in humans, but on the basis of the scientific credibility established using the mouse model, researchers may be on track to preventing and curing Alzheimer's disease entirely.

Parkinson's Fly

Studying neurological diseases without a model is very difficult, but animal models produced through transgenic approaches have yielded significant progress. A *Drosophila* fly was created at Harvard Medical School in 2000 as a model for Parkinson's disease. This fly has a mutation of the α-synuclein gene linked to inheritable Parkinson's disease. Parkinson's fly shows disease-specific characteristics that are seen in humans during the progression of the disease, such as loss of motion control and loss of dopamine. This fly carries a much simpler genome and serves as an excellent model for studying Parkinson's disease at the genetic level (Feany and Bender, 2000). Many previously unobservable characteristics in the progression of the disease have been studied with the help of this fly. The symptoms of Parkinson's generally do not become visible in humans until an estimated 60% to 80% of dopamine nerve cells have already died (Vatalaro, 2000). Parkinson's fly allows scientists to understand early-onset symptoms, which could lead to earlier diagnosis and ultimately to a treatment or cure of the disease in humans.

TRANSGENIC ANIMALS AS BIOLOGICAL MODELS

Transgenic animals are also created as biological models with the aim of increasing knowledge about genetics and expression of genes during certain natural and physiological conditions. Some of the important biological models that have been produced through transgenesis are discussed here.

ANDi (Monkey)

The first transgenic monkey, ANDi ("inserted DNA" spelled backwards), was born in 2000. ANDi is one of most important biological models (Chan et al., 2001). A harmless gene for green fluorescence protein (GFP) was inserted into ANDi's rhesus genome using an engineered virus. The eyes and fingernails of two other monkeys in the project, stillborn twins, glowed under ultraviolet light, although ANDi himself did not. The GFP gene was chosen for two reasons: first, it would have very little effect on the monkey, and detecting whether the transgene had been transmitted properly would be easy. ANDi is the only monkey of 40 fertilized eggs to be born alive and to express the gene for GFP. Therefore, ANDi proved that transgenic primates could be created and could express a foreign gene if delivered into their genome. ANDi opened the door for the creation of other primate biological models for the study of primates' natural physiology and behavior.

Doogie (Smart Mouse)

In 1999 another important biological model was created at Princeton University. The transgenic mouse "Doogie" (the smart mouse) was engineered to overexpress NR2B receptors in synaptic pathways. Overexpression of these receptors makes the mice learn faster, like juveniles, throughout their lives. When tested for learning and memory the "Doogie" mice performed better than their wild-type counterparts (Tang et al., 1999). To test the memory of a mouse, two objects are presented to the mouse in a cage for exploration. Then one object is replaced with another and the mouse is again allowed to explore the objects. If the mouse spends more time paying attention to the new object, that is a good sign that it remembers the old one. If the mouse explores each object equally, the mouse has probably forgotten the old object that it explored previously. Doogie mice perform consistently better on these tests as they grow older. In the near future, this research may lead to improvements in learning and memory in humans and other animals. The fact that overexpression of this gene improves memory confirms an old theory about how mammals think and learn (Harman, 1999). Further research on the Doogie mice can provide valuable information on human development, learning, and memory.

Supermouse

Supermouse is a transgenic biological model developed in 1982 in which the gene of a rat growth hormone is microinjected into fertilized eggs. These mice grew noticeably larger than their wild-type littermates and became the world's first expressing transgenic animals and the first ones with a noticeable phenotypic response to the transgene. Researchers hoped to use these mice to study the effects of growth hormone, accelerated animal growth, gigantism, and as a means of correcting genetic defects related to the growth pattern of animals and humans (Palmiter et al., 1982). This mouse model was also used in research aiming to create food-producing animals with accelerated growth. Correction of dwarfism is the most obvious application of these animals.

Youth Mouse

In 1997 the Weizmann Institute of Science in Rehovot, Israel, created a mouse model called "youth mouse." These mice characteristically overexpress urokinase-type plasminogen activator, primarily thought to be helpful in dissolving blood clots. These mice are smaller, eat less, and live about 20% longer than normal wild-type mice (Miskin and Masos, 1997). Overexpression of the clot dissolver extends life by preventing atherosclerosis, a process in which plaques develop in the arteries of an animal as it ages and which can lead to clots, hemorrhages, and heart attacks. Of the four lines of transgenic mice attempted, only one kind autonomously ate less and lived longer, but it also displayed infrequent muscle tremors. This transgenic mouse line, dubbed Alpha MUPA, shows the same characteristics as normal wild-type counterparts on restricted diets (Miskin et al., 1999). Therefore, "youth mouse" promises to be a useful biological model for studying development processes and aging, especially in relation to diet.

Influenza-Resistant Mouse

Influenza-resistant mice were created to study the use of genetic alterations for effecting disease resistance against (for example) influenza. These mice overproduce Mx protein, which is known to act as an antiviral agent. These mice are significantly more resistant to influenza and other orthomyxoviruses compared with their normal wild-type littermates (Staeheli et al., 1986). If these findings were applied to farm animals such as pigs and ducks, the animals' chances of infection with avian strains of influenza and other viruses, and also their chances of passing these infections to humans, might be lowered. The rate of evolution of these viruses in the animal hosts might also be lowered, thus helping humans retain an immunoprotective response against future viral outbreaks.

TRANSGENIC ANIMALS AS XENOTRANSPLANTERS

Organ transplantation is necessary in those cases when the whole-self organ fails to function (e.g. liver or kidney). However, organ transplantation can be performed only if a donor's organ is determined to be histocompatible with the patient to take over the function of the diseased or failed organ. Generally a small percentage of donated organs are found to be histocompatible with any given patient, and such matched organs are in extremely short supply. Thus, immunosuppressive drugs are frequently given to lower the patient's immune response against the transplanted organ in order to avoid graft rejection. However, this forced decrease response of the immune system leaves the patient vulnerable to opportunistic infections. Therefore, to solve the organ shortage and histocompatibility problem, xenotransplanters are being engineered to provide animal organs that are histocompatible with humans. Generally, the recipient rejects organs that come from other tissue types, even from the same species. The well-known cause of this rejection is the repertoire of antigens on the organ's surface that inform the host that the organ is non-self. Thus, the host's own immune system attacks the transplanted organ, causing a cascade of problems (most notably blood clotting) that are dangerous to the already weakened patient.

Xenotransplanters are animals engineered not to express those antigens that are recognized by the host immune system. The only animal currently chosen for xenotransplant research is the pig because its physiology closely matches that of humans, and pigs are much less expensive to study than monkeys or other primates. In the pig, a sugar called alpha-1,3-galactosyltransferase present on the surface of the cells needed to be knocked out. In 2002 at the University of Missouri, four pigs were produced whose transferase genes had been knocked out (Lai et al., 2002). A null gene was introduced into the pig embryos using the nuclear transfer technique. A second copy of the blank gene was introduced by nuclear transfer into the embryos from these adult pigs, which resulted in piglets with both copies of the gene that encode the antigen knocked out. Previously, the organs from transgenic pigs designed in this way were transplanted into baboons and no rejection was seen (Logan and Sharma, 1999). Human trials have not yet been approved for organs from transgenic xenotransplanters from fear that the pig organs will allow the crossover of viruses from animals to humans (zoonotic infection), especially to humans with weakened immune systems like patients waiting for a transplant. The most clear-cut example of zoonotic infection is influenza, which is often transmitted from pigs to humans even without organ transplantation. Human trials may begin soon, but debate is ongoing about the issue of transmission of infections.

TRANSGENIC ANIMALS AS FOOD SOURCES

Another purpose of producing transgenic animals is to create food sources to meet increasing global food requirements. Following are examples of the transgenic animals that have been created to be used as food sources.

Superpig

Growing animals more efficiently and with less food would be very beneficial to society. "Superpig," one of the food strains created, was a pig engineered to grow bigger and faster, thus producing a more efficient food source. Many of the transgenic superpigs were created by microinjection of the transgene for a growth hormone, whether porcine, ovine, bovine, or even rat (Pursel et al., 1997). The popular Beltsville pig was created in Beltsville, Maryland, under the supervision of the U.S. Department of Agriculture. Human or bovine growth hormones were expressed in these pigs (Miller et al., 1989). Unfortunately, the Beltsville pigs harbored many health problems, most commonly arthritis. Animal rights groups even claimed that the pig was impotent and had ulcers along with heart problems, lameness, kidney disease, and pneumonia. Therefore these pigs were euthanized, and biologists imposed a voluntary moratorium on performing further studies on mammals involving alteration in the expression of growth hormone.

Superfish

The creation of "Superfish" was another attempt at creating a more efficient food source. Tilapia, a species of engineered fish, was created at the Centro de Ingenieria Genética y Biotecnologia in Havana, Cuba, to overexpress its own growth hormone, which was microinjected. This animal was not transgenic, as it was not created by using another animal's gene, but it was genetically engineered. Superfish showed increased growth, but it reached an adult size that was not larger than normal tilapia (Martinez et al., 1996). Similarly, a transgenic salmon was engineered that produced growth hormone continuously, irrespective of the season (turning it off depending on the season). In this experiment the eggs of a species of usually slow-growing trout were used for microinjection with the gene of a salmon that grew very fast after many generations of selective breeding (Devlin et al., 2001). There was concern that these fish might escape into the environment, and therefore tight control was kept over transgenic fish farms (Stokstad, 2002). The biggest fear is that these fish will breed with native wild-type fish and outcompete them for food, though one of the companies involved in the farming of transgenic salmon claimed that the salmon were raised on fish pellets and therefore would not know how to forage for themselves in the wild natural environment. There is still considerable opposition to the generation and farming of these "superfish," though this transgenic fish looks like a much more likely source of food than any transgenic animal species.

TRANSGENIC ANIMALS FOR DRUG AND INDUSTRIAL PRODUCTION

Transgenic animals may better serve as important models in translating basic science breakthroughs into potential clinical applications. Moreover, the use of transgenic animals as bioreactors in the pharmaceutical industry has far-reaching implications, from protein production in various end organs (e.g. milk, blood, urine, and other tissues) to the modification of tissues and organs for transplantation. Thus, the applications of transgenic technologies, although potentially of great significance, have yet to be fully recognized.

Transgenic animal research has been used mostly in the field of human medicine. Various therapeutic proteins or peptides for the treatment of human diseases require animal cell-specific modifications to be effective, and are generally produced in mammalian cell-based bioreactors. To start a new cell culture-based manufacturing facility for a single therapeutic protein or peptide may cost more than $600 million (USD), and the resultant drug may not be affordable to most patients. It is difficult for the therapeutic-protein manufacturing industry to keep pace with the rapid increase in drug discovery and development, resulting in unmet patient needs and dramatically rising medicine costs. Genetically engineered animals may provide an important source of these protein/peptide drugs in the future because the production of recombinant proteins in the animal secretive products (e.g. the milk, blood, or eggs of transgenic animals) presents a much less expensive approach than producing therapeutic proteins in animal cells. The first human therapeutic protein, Antithrombin III (ATryn, GTC Biotherapeutics, Framingham, MA), was derived in 2006 from the milk of genetically engineered goats and was approved by the European Commission for the treatment of patients with hereditary antithrombin deficiency (Avidan et al., 2005). Serum biopharmaceutical products are also obtained from transgenic animals, such as antibodies that can be used for the treatment of infectious diseases, cancer, organ transplant rejection, and autoimmune diseases such as rheumatoid arthritis (Patel et al., 2007; Dunn et al., 2005). Presently, the main production system for such blood products is donated human blood, a system that is limited by disease concerns (e.g. HIV/AIDS), lack of qualified donors, and regulatory issues. Genetically engineered animals, such as cattle carrying human antibody genes, provide a steady supply of polyclonal antibodies for the treatment of a variety of infectious and other diseases. Much information about human diseases has also been obtained from transgenic mice models, which have become important in biological and biomedical research.

A transgenic sheep has been produced for production of α1-proteinase inhibitor (α1-PI) protein that when released in blood serum binds to the elastase secreted by neutrophils in response to certain spores, bacteria, and other antigens. Elastase released in large amounts can damage elastin in the walls of lung alveoli, leading to severe emphysema. People who have dysfunctional gene controls for α1-antitrypsin production (emphysema or cystic fibrosis) can be supported in two ways: by gene therapy using a functional α1-antitrypsin gene or by administration of high doses of α1-PI as an aerosol. Gene therapy is still debated; therefore, the only way of producing α1-PI in large quantities is by creating a transgenic animal with an α1-PI-producing gene. Pharmaceutical Protein Ltd. of Midlothian, Scotland, tried to produce this enzyme in sheep's milk. There are potential advantages of using sheep; being mammals they will produce the same kind of α1-PI as in humans. Sheep are less expensive than cows and they mature more quickly. The enzyme is formed only in the milk, which can be collected easily, and the sheep remain fit and healthy for a long time. Further, large quantities of enzyme can be produced because flocks of these sheep can be easily bred and the purified enzyme from milk will therefore be inexpensive. Once the purified enzyme is produced, it must undergo clinical trials and receive approval from the regulatory authorities before it becomes available on the market.

Several biomedical research models are being produced from transgenic animals, including livestock species produced specifically for various human afflictions such as Alzheimer's disease and ophthalmic disease, and for the possible xenotransplantation of cells, tissues, and organs (Forsberg, 2005). Transgenic animals are also helpful in studying animal diseases such as "mad cow" disease (bovine spongiform encephalopathy) and infection of the udder (mastitis) (Maga et al., 2006). Although scientists have now developed transgenic livestock for agricultural purposes, including some with enhanced production traits (Brophy et al., 2003), environmental benefits (Golovan et al., 2001), and disease resistance traits (Maga et al., 2006), no company other than Aqua Bounty (Boston, MA), which produced the growth-enhanced salmon, has announced its intent to pursue the commercialization of these agricultural applications. Economic profits are higher with the production of genetically engineered/transgenic animals for human medicine than for agricultural applications. Commercialization of agricultural applications of transgenic animals is hindered by concerns about the cost and timelines associated with the regulatory process, and by consumer acceptance issues. Also, potential investors are reluctant because public acceptance of agricultural applications of genetic engineering has generally been lower than acceptance of medical applications (e.g. recombinant insulin). Several chimeric and humanized antibodies have been marketed, and other therapeutically useful proteins are in development (Table 22.1).

TABLE 22.1 Pharmaceutically Related Products Derived from Transgenic Animals

Product	Use	Transgenic Animal	Company/ Organization
ABX-EGF	Cancer	mouse	Abgenix–Amgen
Antithrombin 3 (ATIII)	Thrombosis	goat	GTC
Butyrylcholinesterase	Biodefense	goat	Nexia
CFTR	Cystic fibrosis	sheep, mouse	PPL
Collagen I	Tissue repair	cow	Pharming
Collagen II	Rheumatoid	cow	Pharming
CTLA4Ig	Rheumatoid arthritis	goat	BMS-GTC
Factor IX	Hemophilia	pig, cow sheep	Pharming PPL
Factor VIII	Hemophilia	pig sheep	Pharming PPL
Fibrinogen	Wound healing	cow sheep	Pharming PPL
Glutamic acid decarboxylase	Type 1 diabetes	mouse, goat	GTC
Human calcitonin	Osteoporosis	rabbit	PPL
Human protein C	Thrombosis	pig, sheep	PPL
Msp-1	Malaria	mouse	GTC
Pro542	HIV	mouse, goat	GTC
Tissue plasminogen activator (tPA)	Thrombosis	mouse, goat	PPL
α-1 Anti-trypsin (AAT)	Hereditary emphysema, Cystic fibrosis	sheep	PPL
α-Fetoprotein (rhAFP)	Myasthenia gravis Multiple sclerosis Rheumatoid arthritis	goat	GTC
α-Glucosidase	Pompe disease	rabbit	Pharming
α-Lactalbumin	Anti-infection	cow	PPL

(Modified from Breekveldt and Jongerden, 1998.)

TRANSGENIC ANIMALS' IMPACT ON THE ENVIRONMENT

The possible environmental impact of transgenic animals will initially depend on their phenotype rather than on the fact that they are transgenic *per se* (Kapuscinski and Hallerman, 1990). Thus, if the animal in question has a phenotype giving it absolutely no chance of even short-term survival in the wild, then the fact that it is transgenic is of little to no concern as an environmental issue. If, conversely, the animal's phenotype does not limit, or perhaps increases, its chances for survival and dispersal in the wild, and if this animal is also capable of breeding in the wild, then this raises serious environmental concerns. The only way to ensure that transgenic animals can have no environmental impact is to make their escape or intentional release into the wild impossible, something that can never be assured. Although the long-term environmental impact of transgenic animals is hard to predict with certainty, it is generally accepted that if such impacts arise, they will be difficult or impossible to reverse.

The production and evaluation of transgenic salmon currently emphasize accelerating growth rates through engineered changes in growth hormone expression (Du et al., 1992). The rapid growth of these fish clearly suggests a competitive advantage over wild fish should transgenic individuals escape. Given that absolute body size is often a deciding factor in the outcome of competition for access to resources and/or mates, any animal that more rapidly attains a certain size is likely to displace or prey upon smaller, slower-growing individuals of the same species. Such animals may also exhibit interspecific competitive/predatory advantages due to changes in spatial and temporal distributions as a result of their faster growth. Similar outcomes are possible from fish engineered for traits such as freeze resistance, salinity tolerance, disease resistance, or other economically valuable characteristics (Maclean and Penman, 1990). Based on what is currently known about the phenotype of growth hormone-transgenic salmon, it is impossible to adequately predict the environmental outcomes should these fish escape or be released into the wild. Beyond these near-term ecological effects, the greatest concerns with transgenic fish relate to the genetic effects of interbreeding with wild populations. Even if they are not well adapted for survival in the wild, transgenic animals may have detrimental impacts on the genetic structure of wild populations by allowing the introgression of "exotic" genes into natural gene pools. Changes in the genetic makeup of a well-adapted wild population may ultimately affect the animals' abilities to withstand environmental change. Of particular concern in this regard is the so-called Trojan gene effect, whereby transgenic animals that are poorly adapted for survival in the wild but exhibit mating advantages (e.g. through faster growth and/or larger size) drive populations to extinction by successfully breeding with wild individuals, thereby reducing the fitness of their progeny (Muir and Howard, 1999). Non-transgenic farmed salmon exhibit characteristics that predispose them to such Trojan gene effects as reduced survival of progeny from mating between farmed and wild salmon (Fleming et al., 2000).

An additional environmental concern with transgenic fish is their capacity for transferring diseases and/or parasites to wild populations. This is a general concern with aquaculture (Saunders, 1991), not so much because farmed fish are an initial source of pathogens or parasites, but because their confinement allows the amplification of pathogen and parasite loads for subsequent transmission back to wild populations. Transgenic fish that are engineered for disease resistance would potentially increase the risk of becoming vectors for transferring pathogens or parasites to wild populations either through direct contact or through the release of their feces or contaminated rearing water.

The use of transgenic salmon also offers several possible environmental benefits to aquaculture. For instance, the improved food conversion efficiency of growth hormone transgenic salmon should reduce the amount of fishmeal required per unit body mass produced. Although not yet realized, the engineering of transgenic fish that can use plant meal as a protein source would have tremendous environmental benefit in this regard. Engineering disease- and parasite-resistant fish could potentially reduce the use of antibiotics and pesticides in aquaculture. Finally, if the use of transgenic fish makes on-shore, fully contained fish farming an economically viable enterprise, then this will allow the concentration and better management of fish waste products. These include organic wastes in the form of uneaten food and feces, as well as inorganic wastes such as phosphates and nitrates. The accumulation of these wastes from aquaculture can cause increased biochemical oxygen demand, eutrophication, and sedimentation problems, all of which can be eliminated in on-shore systems with recirculated water supplies.

PATENTING TRANSGENIC ANIMALS

In general, consumers have expressed support for the principle of patenting, including the patenting of genes and gene sequences (Pollara and Earnscliffe Strategy Group, 2000). When the Canadian Court of Appeals agreed (decision overturned by the Supreme Court of Canada) with the patentability of the Harvard OncoMouse, only about 50% of Canadians said they were not comfortable with the Appeals Court decision. When asked – "Is it okay for someone to have a patent on a new plant modified through the use of transgenesis?" – 66% disagreed. However, only 30% agreed that "granting a patent on an animal modified through the use of transgenesis is no different than granting a patent on a consumer product." Furthermore, 66% agreed that "we should not grant patents on a new species of guinea pig that includes human genes," and the same majority were in agreement that patents should not be granted on "a new species of chimpanzee that includes human genes" (Environics, 1998).

Harvard's OncoMouse raised general ethical issues regarding transgenic technology. It also raised two key issues for the patent system: (1) Should patents be granted at all for animals or animal varieties, particularly for higher-order animals such as mammals, even if they do otherwise meet patentability criteria (e.g. novelty, industrial applicability/usefulness, inventive step)? (2) How should moral implications be addressed in relation to specific cases (e.g. the question of suffering caused to the transgenic animal)? These issues have been resolved in different ways by the patent authorities of different countries, as the following examples illustrate.

The United States Patent Office in 1988 granted Patent #4,736,866 to Harvard College, claiming the creation of "a transgenic non-human mammal whose germ cells and somatic cells contain a recombinant activated oncogene sequence introduced into said mammal." The claim explicitly excluded humans, apparently reflecting moral and legal issues regarding patents on human beings or modification of the human genome.

The European Patent Office (EPO) considered the OncoMouse case at length and at several levels. The EPO applies the patent standards of the European Patent Convention, which contains two key relevant provisions: Article 53(a) excludes patents for inventions "the publication or exploitation of which would be contrary to *ordre public* or morality," and Article 53(b) excludes patents on "animal varieties or essentially biological processes for the production of animals." The EPO applied the utilitarian test and decided that the prohibition on patenting animal varieties did not constitute a ban on patenting animals as such. It concluded further that the OncoMouse was not an animal variety and so did not fall within that exclusion.

In order to address the *ordre public* or morality exception, the EPO developed a utilitarian balancing test. This aimed to assess the potential benefits of a claimed invention against negative aspects, in this case weighing the suffering of the OncoMouse against the expected medical benefits to humanity. Other considerations could also be taken into account in the balancing test, such as environmental risks (neutral in this case), or public unease (there was no evidence in European culture for moral disapproval of the use of mice in cancer research, thus no moral disapproval of the proposed exploitation of the invention in this case). The EPO concluded that the usefulness of the OncoMouse in furthering cancer research satisfied the likelihood of substantial medical benefit and outweighed moral concerns about suffering caused to the animal. In the original application, the claims referred to animals in general, but in the course of the proceedings, the patent was amended and finally maintained with claims limited to mice.

ETHICAL ISSUES

Production of transgenic animals remains a hot topic of debate as some people are ethically uncomfortable with the idea of genetically engineering animals. Two central ethical concerns are associated with the creation of transgenic animals. The first relates to breaching species barriers or "playing God." According to this view, life should not be regarded solely as if it were a chemical product subject to willful genetic changes and patentable for economic benefit or commercialization. The second major ethical issue is the belief that the transgenesis of animals interferes with the integrity or telos of the animal. "Telos" can be defined as "the set of needs and interests which are genetically based, and environmentally expressed, and which collectively constitute or define the form of life or way of living exhibited by that animal, and whose fulfillment or thwarting matter to that animal" (Holland and Johnson, 1998). Such concerns are not unique to genetic engineering, and traditional breeding and selection practices can also change animals in similar ways. Cows from the Belgian Blue cattle breed, for example, require cesarean delivery of their calves because they have been selected for increased birth weight resulting from the naturally occurring "double-muscle" trait and the narrowness of the cow's pelvic passageway (Vandenheede et al., 2001).

The success rate in creating transgenic animals is still low. There are more failures than successes, and the general assumption is that the higher the species of transgenic animal, the greater the cloning failure. The failures in transgenesis are animals that die before they are born or animals that are born without the transgene and are of no use for fulfilling the desired purpose. However, the number of failures is decreasing as cloning techniques improve. Comparing the prenatal deaths of a few animals against saving the lives of potentially thousands of humans suggests that the reward is very high, and improvements in the process are continuously decreasing the costs.

A very good example of a transgenic animal that does not suffer is Alzheimer's mouse. Alzheimer's mouse does poorly on maze tests, but does not feel any pain related to its condition by any standards used in laboratories. Because Alzheimer's mouse spends its days in a laboratory setting, any survival skills that would be hampered in the wild by its diminished memory do not come into play. Also, Alzheimer's mouse continues to provide significant information on Alzheimer's disease that could lead to a cure. The Alzheimer's mouse that was created in part at Worcester Polytechnic Institute (Worcester, MA) (Games et al., 1995) was used to develop a vaccine that lowers senile plaque burden in mice (Schenk et al., 1999); it is now in Phase II human clinical trials by Elan Pharmaceuticals (Dublin, Ireland). The benefits are tremendous, and the animal suffering is almost nonexistent. Thus, creation of Alzheimer's mouse was a brilliant idea.

The OncoMouse is more complicated ethically. In 2008, an estimated 7.5 million people died of cancer worldwide (American Cancer Society, 2008). Any transgenic or genetically engineered animal/mouse that can help increase knowledge about this deadly disease would have enormous medical benefits. However, the problem with the OncoMouse is that as it grows, it begins to suffer from the tumors just as humans do with this disease. Therefore, in this complex case, though the animal has enormous medical benefits, its use is associated with strong ethical contraindications. However, because most of the work on OncoMouse focuses on early development of the tumors, the mouse can be euthanized before its suffering increases unbearably. If advanced oncogenesis needs to be studied, university/institutional animal care committees could require that painkillers be used. Therefore, it is difficult to say that the creation of the OncoMouse was completely a good idea, because the mice do suffer like humans and die in pain. The need to prevent cancer and palliate its symptoms is overwhelming, however, and to save millions of human lives most scientists believe that OncoMouse is the best hope.

FDA GUIDELINES ON GENETICALLY ENGINEERED ANIMALS

The U.S. Food and Drug Administration is the lead agency responsible for the control and regulation of genetically engineered animals to be used for food, and it plans to regulate transgenic animals under the "new animal drug" provisions of the Food, Drug, and Cosmetic Act (FDCA). The new animal drug provisions ask: (1) Is the new drug safe for the animal? (2) Is the new drug effective? (3) If the drug is for an animal that is used for food, is the resulting food safe to eat? Although premarket regulatory review of genetically engineered animals is mandatory, the FDA has not yet issued general guidelines explaining what information will be required for this regulatory review. Also, the regulatory path to commercialization of genetically engineered animals remains ill defined (Pew Initiative on Food and Biotechnology, 2005). However, transgenic animal research is subject to existing FDA regulations governing animal research. All organizations or institutions receiving or applying for federal funding to carry out research using animals are required by the federal Animal Welfare Act (AWA) of 1966 to have the IACUC review research protocols involving dogs, cats, rabbits, guinea pigs, hamsters, gerbils, nonhuman primates, marine mammals, captive wildlife, and domestic livestock species used in nonagricultural research and learning. The AWA also requires the following: (1) research institutions must have a veterinary care program in place; (2) all personnel using or caring for live animals must be well qualified to do so; and (3) a mechanism must be in place at the organization for reporting of issues regarding animal care and unethical use. The AWA is administered through the U.S. Department of Agriculture (USDA) and is enforced through unannounced and random inspections by a veterinary medical officer designated by the USDA. On an international level, the Association for Assessment and Accreditation of Laboratory Animal Care (AAALAC) oversees the voluntary accreditation and assessment of research organizations committed to responsible animal care and use.

TRANSLATIONAL SIGNIFICANCE

Transgenic animals are an important part of today's consumer health world, and we may expect even more dependency in the near future as most of the recombinant proteins or monoclonal antibodies are derived from these animals. Thus, it becomes relevant to know how transgenic animals are produced and can be used for fulfilling our needs. For example, it is not possible to use human subjects to study the role of a particular gene that can be easily studied if inserted into a fertilized egg or embryo that will then eventually grow into a transgenic animal. Similarly, a particular disease condition can be generated in transgenic animals to study the effect of a specific drug that cannot be directly applied to human beings. No doubt animals do feel pain and do suffer during laboratory experiments like human beings, but this suffering is still less if compared to the relief of pain or cure from disease for millions of people worldwide. At the same time, it is important to consider the relevance of the end results, and any kind of haphazard science must not be practiced on transgenic animals. Thus, one should not blindly follow the misconceptions of allowing or denying use of transgenic animals, and should instead become aware of the actual benefits and ethical issues associated with these animals.

WORLD WIDE WEB RESOURCES

The World Wide Web has become a common source of information about any subject, including transgenic animals and various topics related to transgenics. One of the most important resources where one can get scientifically validated information is PubMed (www.ncbi.nlm.nih.gov/pubmed). Another good resource for this type of information is the National Institutes of Health (http://www.nih.gov). Apart from these two major sources, the research departments at a number of educational institutions provide good web sites with information regarding this subject, such as the following: http://www.harvard.edu (Harvard University), http://www.stanford.edu (Stanford University), http://www.mit.edu (Massachusetts Institute of Technology), and http://www.wustl.edu (Washington University St. Louis). Apart from the web resources created by academic institutions, several government organizations, like the Food and Drug Administration (FDA, http://www.fda.gov) and the U.S. Patent and Trademark Office (http://www.uspto.gov) also provide information about transgenic animals.

Although this is not an exhaustive list of resources on the subject, and new sites are constantly being created and updated on a regular basis, the following are particularly useful:

http://www.transgenelifesciences.com
http://www.med.umich.edu
http://www.hhmi.org
http://www.rnainterference.org
http://www.medicalnewstoday.com
http://www.who.int
http://www.wisc.edu
http://www.dnalc.org

ACKNOWLEDGMENTS

The authors wish to acknowledge the United States Public Health Service/National Institutes of Health grants AI 093172-01 to ZKK and AI 077414 to PJ.

REFERENCES

Avidan, M. S., Levy, J. H., Scholz, J., Delphin, E., Rosseel, P. M., et al. (2005). A phase III, double-blind, placebo-controlled, multicenter study on the efficacy of recombinant human antithrombin in heparin-resistant patients scheduled to undergo cardiac surgery necessitating cardiopulmonary bypass. *Anesthesiology, 102*, 276–284.

Beardmore, J. A. (1997). Transgenics: autotransgenics and allotransgenics. *Transgenic Research, 6*, 107–108.

Breekveldt, J., & Jongerden, J. (1998). Transgenic animals in pharmaceutical production. *Biotechnology and Development Monitor, 36*, 19–22.

Brophy, B., Smolenski, G., Wheeler, T., Wells, D., L'Huillier, P., et al. (2003). Cloned transgenic cattle produce milk with higher levels of beta-casein and kappa-casein. *Nature Biotechnology, 21*, 157–162.

Chan, A. W., Chong, K. Y., Martinovich, C., Simerly, C., & Schatten, G. (2001). Transgenic monkeys produced by retroviral gene transfer into mature oocytes. *Science, 291*, 309–312.

Cho, A., Haruyama, N., & Kulkarni, A. B. (2009). Generation of transgenic mice. (March), *Current Protocols in Cell Biology*, 1–29. chapter 19.

Devlin, R. H., Biagi, C. A., Yesaki, T. Y., Smailus, D. E., & Byatt, J. C. (2001). Growth of domesticated transgenic fish. *Nature, 409*, 781–782.

Du, S. J., Gong, Z. Y., Fletcher, G. L., Shears, M. A., King, M. J., et al. (1992). Growth enhancement in transgenic Atlantic salmon by the use of an "all fish" chimeric growth hormone gene construct. *Biotechnology (N Y), 10*, 176–181.

Duff, K., Eckman, C., Zehr, C., Yu, X., Prada, C. M., et al. (1996). Increased amyloid-beta42(43) in brains of mice expressing mutant presenilin 1. *Nature, 383*, 710–713.

Dunn, D. A., Pinkert, C. A., & Kooyman, D. L. (2005). Foundation review: Transgenic animals and their impact on the drug discovery industry. *Drug Discovery Today, 10*, 757–767.

Dykxhoorn, D. M., Novina, C. D., & Sharp, P. A. (2003). Killing the messenger: short RNAs that silence gene expression. *Nature Reveiw Molecuar Cell Biology, 4*, 457–467.

Echelard, Y., Ziomek, C. A., & Meade, H. M. (2006). Procduction of recombinant therapeutic proteins in the milk of transgenic animals. *BioPharm International, 19*, 36–46.

Environics R (1998) Renewal of the Canadian biotechnology strategy. Public opinion research. Report to the Canadian Biotechnology Strategy Task Force. Environics Research Group, Ltd., Ottawa, Ontario, Canada.

Feany, M. B., & Bender, W. W. (2000). A Drosophila model of Parkinson's disease. *Nature, 404*, 394–398.

Fire, A., Xu, S., Montgomery, M. K., Kostas, S. A., Driver, S. E., et al. (1998). Potent and specific genetic interference by double-stranded RNA in Caenorhabditis elegans. *Nature, 391*, 806–811.

Fleming, I. A., Hindar, K., Mjolnerod, I. B., Jonsson, B., Balstad, T., et al. (2000). Lifetime success and interactions of farm salmon invading a native population. *Proceedings of the Royal Society – Biological Sciences, 267*, 1517–1523.

Forsberg, E. J. (2005). Commercial applications of nuclear transfer cloning: three examples. Reproduction. *Fertility and Development, 17*, 59–68.

Games, D., Adams, D., Alessandrini, R., Barbour, R., Berthelette, P., et al. (1995). Alzheimer-type neuropathology in transgenic mice overexpressing V717F beta-amyloid precursor protein. *Nature, 373*, 523–527.

Golovan, S. P., Meidinger, R. G., Ajakaiye, A., Cottrill, M., Wiederkehr, M. Z., et al. (2001). Pigs expressing salivary phytase produce low-phosphorus manure. *Nature Biotechnology, 19*, 741–745.

Gordon, J. W., & Ruddle, F. H. (1981). Integration and stable germ line transmission of genes injected into mouse pronuclei. *Science, 214*, 1244–1246.

Han, J. J., Mhatre, A. N., Wareing, M., Pettis, R., Gao, W. Q., et al. (1999). Transgene expression in the guinea pig cochlea mediated by a lentivirus-derived gene transfer vector. *Human Gene Therapy, 10*, 1867–1873.

Harman, J. (1999). *Scientists create smart mouse: addition of single gene improves learning and memory*. Princeton: Princeton University Office of Communications. NJ. Available at http://www.princeton.edu/pr/news/99/q3/0902-smart.htm.

Harvey, M., McArthur, M. J., Montgomery, C. A., Jr., Butel, J. S., Bradley, A., et al. (1993). Spontaneous and carcinogen-induced tumorigenesis in p53-deficient mice. *Nature Genetics, 5*, 225–229.

Hasuwa, H., Kaseda, K., Einarsdottir, T., & Okabe, M. (2002). Small interfering RNA and gene silencing in transgenic mice and rats. *FEBS Letters, 532*, 227–230.

Holland, A., & Johnson, A. (1998). *Animal Biotechnology and Ethics*. London, UK: Chapman & Hall.

Jacenko, O. (1997). Strategies in generating transgenic mammals. *Methods in Molecular Biology, 62*, 399–424.

Kapuscinski, A. R., & Hallerman, E. M. (1990). Transgenic fish and public policy:Anticipating environmental impacts of transgenic fish. *Fisheries, 15*, 2–11.

Kohn, C. (2001). First HIV rat seen as best model for human studies. *Science Daily*. August 2, 2001. Available at http://www.sciencedaily.com/releases/2001/08/010806074655.htm.

Lai, L., Kolber-Simonds, D., Park, K. W., Cheong, H. T., Greenstein, J. L., et al. (2002). Production of alpha-1,3-galactosyltransferase knockout pigs by nuclear transfer cloning. *Science, 295*, 1089–1092.

Lai, L., Kang, J. X., Li, R., Wang, J., Witt, W. T., et al. (2006). Generation of cloned transgenic pigs rich in omega-3 fatty acids. *Nature Biotechnology, 24*, 435–436.

Lamberton, J. S., & Christian, A. T. (2003). Varying the nucleic acid composition of siRNA molecules dramatically varies the duration and degree of gene silencing. *Molecular Biotechnology, 24*, 111–120.

Leder, P., & Stewart, T. (1984). *Transgenic Non-Human Mammals, The Harvard Oncomouse. U.S. Patent and Trademark Office*. Patent #4,736,866 Cambridge, MA.

Lillico, S. G., Sherman, A., McGrew, M. J., Robertson, C. D., Smith, J., et al. (2007). Oviduct-specific expression of two therapeutic proteins in transgenic hens. *Proceedings of the National Academy of Sciences of the United States ofAmerica, 104*, 1771–1776.

Logan, J. S., & Sharma, A. (1999). Potential use of genetically modified pigs as organ donors for transplantation into humans. *Clinical Experimental Pharmacology and Physiology, 26*, 1020–1025.

Maclean, N., & Penman, D. (1990). The application of gene manipulation to aquaculture. *Aquaculture, 85*, 1–20.

Maga, E. A., Cullor, J. S., Smith, W., Anderson, G. B., & Murray, J. D. (2006). Human lysozyme expressed in the mammary gland of transgenic dairy goats can inhibit the growth of bacteria that cause mastitis and the cold-spoilage of milk. *Foodborne Pathogens and Disease, 3*, 384–392.

Marshall, E. (2002). Intellectual property. DuPont ups ante on use of Harvard's OncoMouse. *Science, 296*, 1212.

Martinez, R., Estrada, M. P., Berlanga, J., Guillen, I., Hernandez, O., et al. (1996). Growth enhancement in transgenic tilapia by ectopic expression of tilapia growth hormone. *Molecular Marine Biology and Biotechnology, 5*, 62–70.

Miller, K. F., Bolt, D. J., Pursel, V. G., Hammer, R. E., Pinkert, C. A., et al. (1989). Expression of human or bovine growth hormone gene with a mouse metallothionein-1 promoter in transgenic swine alters the secretion of porcine growth hormone and insulin-like growth factor-I. *Journal of Endocrinology*, *120*, 481–488.

Miskin, R., & Masos, T. (1997). Transgenic mice overexpressing urokinase-type plasminogen activator in the brain exhibit reduced food consumption, body weight and size, and increased longevity. *Journals of Gerontology Series A: Biological Sciences and Medical Sciences*, *52*, B118–124.

Miskin, R., Masos, T., Yahav, S., Shinder, D., & Globerson, A. (1999). AlphaMUPA mice: a transgenic model for increased life span. *Neurobiology of Aging*, *20*, 555–564.

Muir, W. M., & Howard, R. D. (1999). Possible ecological risks of transgenic organism release when transgenes affect mating success: sexual selection and the Trojan gene hypothesis. *Proceedings of the National Academy of Sciences of the United States of America*, *96*, 13853–13856.

Murphy, D., Carter, D. A., & Smith, D. R. (1993). Transgene design. *Methods in Molecular Biology*, *18*, 115–118.

Novina, C. D., Murray, M. F., Dykxhoorn, D. M., Beresford, P. J., Riess, J., et al. (2002). siRNA-directed inhibition of HIV-1 infection. *Nature Medicine*, *8*, 681–686.

Palmiter, R. D., Brinster, R. L., Hammer, R. E., Trumbauer, M. E., Rosenfeld, M. G., et al. (1982). Dramatic growth of mice that develop from eggs microinjected with metallothionein-growth hormone fusion genes. *Nature*, *300*, 611–615.

Patel, T. B., Pequignot, E., Parker, S. H., Leavitt, M. C., Greenberg, H. E., et al. (2007). Transgenic avian-derived recombinant human interferon-alpha2b (AVI-005) in healthy subjects: an open-label, single-dose, controlled study. *International Journal of Clinical Pharmacology and Therapeutics*, *45*, 161–168.

Pew Initiative on Food and Biotechnology. (2005). *Pew Initiative on Food and Biotechnology Exploring the Regulatory and Commercialization Issues Related to Genetically Engineered Animals*. Washington, DC.

Pollara and Earnscliffe. (2000). *Public opinion research into biotechnology issues – 3rd wave. Report to the Biotechnology Assistant Deputy Minister Coordinating Committee (BACC)*. Ottawa: Government of Canada Pollara and Earnscliffe.

Pursel, V. G., Wall, R. J., Solomon, M. B., Bolt, D. J., Murray, J. E., et al. (1997). Transfer of an ovine metallothionein-ovine growth hormone fusion gene into swine. *Journal of Animal Science*, *75*, 2208–2214.

Reid, W., Sadowska, M., Denaro, F., Rao, S., Foulke, J., Jr., et al. (2001). An HIV-1 transgenic rat that develops HIV-related pathology and immunologic dysfunction. *Proceedings of the National Academy of Sciences of the United States of America*, *98*, 9271–9276.

Saunders, R. L. (1991). Potential interaction between cultured and wild Atlantic salmon. *Aquaculture*, *98*, 51–60.

Schenk, D., Barbour, R., Dunn, W., Gordon, G., Grajeda, H., et al. (1999). Immunization with amyloid-beta attenuates Alzheimer-disease-like pathology in the PDAPP mouse. *Nature*, *400*, 173–177.

Smaglik, P. (2000). NIH cancer researchers to get free access to 'OncoMouse.' *Nature*, *403*, 350.

Staeheli, P., Haller, O., Boll, W., Lindenmann, J., & Weissmann, C. (1986). Mx protein: constitutive expression in 3T3 cells transformed with cloned Mx cDNA confers selective resistance to influenza virus. *Cell*, *44*, 147–158.

Stewart, T. A., Pattengale, P. K., & Leder, P. (1984). Spontaneous mammary adenocarcinomas in transgenic mice that carry and express MTV/myc fusion genes. *Cell*, *38*, 627–637.

Stokstad, E. (2002). Transgenic species. Engineered fish: friend or foe of the environment? *Science*, *297*, 1797–1799.

Tang, Y. P., Shimizu, E., Dube, G. R., Rampon, C., Kerchner, G. A., et al. (1999). Genetic enhancement of learning and memory in mice. *Nature*, *401*, 63–69.

van der Putten, H., Botteri, F. M., Miller, A. D., Rosenfeld, M. G., Fan, H., et al. (1985). Efficient insertion of genes into the mouse germ line via retroviral vectors. *Proceedings of the National Academy of Sciences of the United States of America*, *82*, 6148–6152.

Vandenheede, M., Nicks, B., Desiron, A., & Canart, B. (2001). Mother-young relationships in Belgian Blue cattle after a Caesarean section: characterisation and effects of parity. *Applied Animal Behaviour Science*, *72*, 281–292.

Vatalaro, M. (June 13, 2000). (2000) Fly model of Parkinson's offers hope of simpler, faster research. *NIH Record*. http://nihrecord.od.nih.gov/newsletters/06_13_2000/story05.htm.

Vogel, J., Hinrichs, S. H., Reynolds, R. K., Luciw, P. A., & Jay, G. (1988). The HIV tat gene induces dermal lesions resembling Kaposi's sarcoma in transgenic mice. *Nature*, *335*, 606–611.

Wong, R. W., Sham, M. H., Lau, Y. L., & Chan, S. Y. (2000). An efficient method of generating transgenic mice by pronuclear microinjection. *Molecular Biotechnology*, *15*, 155–159.

FURTHER READING

Ackerman, S. (2006). *ANDi: The First Genetically Engineered Monkey. Division of Comparative Medicine of the National Center for Research Resources and by the National Institute of Child Health and Human Development*. http://www.ncrr.nih.gov/newspub/apr01rpt/ANDi.asp.

Animal Aid Youth Group. (2006). *Animal Experiments*. http://www.animalaid.org.uk/youth/topics/experiments/genetics.htm.

IACUC. (2006). *General Information*. http://www.iacuc.org/.

Houdeline, L. M. (1997). *Transgenic Animals: Generation and Use*. Hardwood Academic Publisher.

Pinkert, C. A. (2002). *Transgenic Animal Technology: A Laboratory Handbook* (2nd ed.). Academic Press.

GLOSSARY

Embryonic Stem (ES) Cells Embryonic stem cells are the pluripotent cells derived from the inner mass of early-stage embryos (e.g. cells from a 4–5 days post-fertilization embryo in the case of humans).

Patent A patent is a form of intellectual property right granted by a government or a sovereign state to an inventor or their assignee for a limited period of time in exchange for the public disclosure of an invention (e.g. patent granted on Harvard OncoMouse).

RNA Interference (RNAi) RNA interference (RNAi) is a natural process that cells use to turn down, or silence, the activity of specific genes. It was first discovered in Petunia in 1998. RNAi destroys the messenger RNA that carries information coded in genes to the cell's protein-producing factories.

Transgene A transgene is an artificial gene manipulated in the molecular biology lab that incorporates all appropriate elements critical for gene expression; it is generally derived from a different species (e.g. production of α1-proteinase inhibitor (α1-PI) protein in transgenic sheep carrying a transgene of human origin).

Xenotransplantation A procedure that involves the transplantation, implantation, or infusion into a human subject (recipient) of either (a) live cells, tissues, or organs from a nonhuman animal source, or (b) human body fluids, cells, tissues, or organs that have had been produced from live non-human animal cells, tissues, or organs.

ABBREVIATIONS

AAALAC Association for Assessment and Accreditation of Laboratory Animal Care
AWA Animal Welfare Act
EPO European Patent Office
ES Cell Embryonic Stem Cell
FDCA Food, Drug, and Cosmetic Act
GFP Green Fluorescence Protein
PCR Polymerase Chain Reaction
RISC RNA-Inducing Silencing Complex
RNAi RNA Interference
shRNA Small Hairpin RNA
siRNA Short Interfering RNA

LONG ANSWER QUESTIONS

1. How are transgenic animals useful for medical research? Describe with examples.
2. What are the different methods of creating transgenic animals? Explain in detail.
3. How can transgenic animals affect the natural environment? Explain with suitable examples.
4. How can the use of transgenic animals over cell-based bioreactors be justified? Do you think transgenic animals can meet world food demands?
5. Are there any controls on the creation of transgenic animals? What is your opinion about patenting of transgenic animals?

SHORT ANSWER QUESTIONS

1. What was the first animal for which a patent was issued?
2. What is the "Trojan gene" effect?
3. What is RNAi?
4. What is the difference between cloning and transgenesis?
5. Name the agency responsible for regulation of animal experimentation in various organizations.

ANSWERS TO SHORT ANSWER QUESTIONS

1. OncoMouse was the first animal to be patented in 1988.
2. Transgenically produced animals (e.g. transgenic salmon over-expressing growth hormone gene), if they escaped into the environment, could wipe out their wild-type counterparts by competitively growing faster.
3. RNAi (RNA interference) is a technique used to inhibit the expression of protein by degrading its mRNA. This technique was first discovered in plants.
4. Cloning is the reproduction of an exact copy of a living organism using the DNA (without manipulation) of that organism, whereas transgenesis (genetic engineering) refers to the human manipulation of genetic material in a manner that does not occur in nature.
5. Institutional Animal Care and Use Committee (IACUC) is the agency responsible for regulation of animal experimentation in most of the institutions and organizations.

Chapter 23

Stem Cells: A Trek from Laboratory to Clinic to Industry

Bhudev C. Das and Abhishek Tyagi
Laboratory of Molecular Oncology, Dr. B. R. Ambedkar Center for Biomedical Research (ACBR), University of Delhi, Delhi 110007, India

Chapter Outline

Animal Biotechnology. http://dx.doi.org/10.1016/B978-0-12-416002-6.00023-7

SUMMARY

The chapter deals with the basic understanding of stem cell biology for the betterment of human health and disease, as it holds great promise for the treatment of various debilitating diseases. Stem cells can be obtained at early development stages, in cord blood, and in adult organs. Stem cells can be totipotent, pluripotent, multipotent, or unipotent. Recently, induced pluripotent stem cells (iPSCs) have opened a new avenue for therapy.

WHAT YOU CAN EXPECT TO KNOW

This chapter is intended to give the non-specialist an insight into the nuances of stem cells, their origin, functions, and utility. It distinguishes embryonic and adult stem cells, based on their characteristic features in organisms, their origin, and how they are identified. As background, the fundamental processes of embryo development are reviewed and defined. Stem cells are defined, characterized, and it is shown how they function in the intact organism during early development and later during cell regeneration in the adult as well as its role in human disease states and repair. The complexity of stem cell recovery and their manipulation into specific cells and tissue is illustrated by reviewing current experimentation on both embryonic and adult stem cells in animals and limited research on human stem cells. Also, the basics of induced pluripotent stem cells (iPSCs) and cancer stem cells (CSC) have been provided. An assessment on the use of human embryonic or adult stem cells is considered from ethical as well as religious viewpoints.

HISTORY AND METHODS

INTRODUCTION

Stem cells are one of the most exciting areas of science in modern biology. Stem cells are defined as cells with infinite proliferative capacity. They can differentiate in all types of cells, tissues, and organs of the body. The implications of stem cells can be easily appreciated in the regeneration potential of lower animals such as newts, kitchen lizards, starfish, and many others. These animals reveal a potential biological process, which could be applied to humans. The regeneration of tissues/organs is an ancient, universal, and fundamental biological process. It is even seen in Greek myth when Zeus punished Prometheus for giving fire to humans; each day eagles would eat Prometheus' liver, and each day his liver would regenerate. The identity of the specific cells that allow regeneration of some tissues was first revealed in the 1950s when experiments with bone marrow established the existence of stem cells and led to the development of bone marrow transplantation, a therapy now widely used in medicine. It is these cells that re-grow, renew, repair, and replenish the cellular requirements of the body. For the first time in history, it thus became possible for physicians to regenerate a damaged tissue/organ with a new supply of healthy cells by using the unique ability of stem cells to create many of the body's specialized cell types. These cells offer therapies for diseases that are not treatable by conventional therapies.

Stem cell research is being pursued in the hope of achieving major medical breakthroughs. Scientists are striving hard to create therapies that rebuild or replace damaged cells/organs with tissues grown from stem cells or target diseased stem cells. It offers hope for treatment of people suffering from cancer, diabetes, cardiovascular disease, spinal-cord injuries, and many other serious disorders. These magic cells serve as valuable material for drug discovery and drug testing. They are also powerful tools for better understanding of the basic biology of the human body and disease pathogenesis. Stem cell research has opened up a new direction in bringing about a revolution in the field of detection, prevention, prognostication and treatment of various chronic and untreatable diseases such as cancer.

HISTORY OF STEM CELL RESEARCH

With the advent of the microscope in the 1800s, followed by the discovery of cells, scientists became more involved in understanding the biochemistry and cell biology (i.e. how cells propagate, and how one cell gives rise to another cell and tissues. The history of stem cell research started in the late 1800s when scientists tried to fertilize mammalian eggs *in vitro,* but only slight success was achieved for want of advanced research tools. However, in 1959, in a breakthrough experiment, scientists produced rabbits using an *in vitro* fertilization technique. Another success was achieved when researchers in the mid-1960s revealed that the sexual organs of mice possess some unique cells that could give rise to various other kinds of cells. With this finding, researchers first began to think about stem cells (Table 23.1).

In the mid-1980s, scientists worked on human testicles and human blastocysts. However, a great success was achieved in 1995 and 1996 when stem cells were obtained from the embryos of rhesus monkeys. Finally in 1998, scientists at the University of Wisconsin isolated human embryo cells from a human **blastocyst**, a hollow structure made up of an outer layer of cells, a fluid cavity, and the inner mass of cells (Thompson et al., 1998). Stem cells were found in the inner mass of blastocyst; they were removed and cultured in a culture dish where the stem cells grew over time.

TABLE 23.1 Major Milestones in Stem Cell Research

1952	Frog enucleated egg experiment.
1959	First animals made by *in vitro* fertilization (IVF).
1960	Teratocarcinomas determined to originate from embryonic germ cells in mice. Embryonal carcoinoma cells (EC) identified as a kind of stem cell.
1968	The first human egg is fertilized *in vitro.*
1970	EC cells injected into mouse blastocysts make chimeric mice. Cultured SC cells are explored as models of embryonic development in mice.
1981	First mouse embryonic stem (ES) cells isolated and grown in culture.
1984-88	Pluripotent, clonal cells called embryonal carcinoma (EC) cells are developed. When exposed to retinoic acid, these cells differentiate into neuron-like cells and other cell types.
1988	First cord blood transplant.
1989	A clonal line of human embryonal carcinoma cells is derived that yields tissues from all three primary germ layers.
1994	Human embryonic stem (ES)-like cells generated.
1995	Evidence found for neural stem cells.
1998	University of Wisconsin-Madison and Johns Hopkins University isolate the first human embryonic stem cells.
2000	Singaporean and Australian scientists derive human ES cells from blastocysts.
2001	Human ES cells successfully developed into blood cells.
2002	Neural stem cells successfully developed into functional neurons.
2003	Institute of Stem Cell Research, Edinburgh, discovers key gene that keeps ES cells in a state of youthful immortality.
2003	Dolly dies after developing progressive lung disease.
2003	UK Stem Cell Bank Established.
2006	First induced pluripotent stem cells (iPS) produced from mouse cells.
2007	Development of first induced pluripotent stem cell created from human cells.

WHAT ARE STEM CELLS AND WHY ARE THEY IMPORTANT?

Every cell in the human body can be traced back to its origin in a fertilized egg that came into existence from the union of an ovum with a sperm, each with a haploid genome from mother and father, respectively. The body is made up of more than 200 different types of cells, and all of these cell types come from a pool of stem cells formed in the early embryo. During early development, as well as later in life, various types of stem cells give rise to the specialized or differentiated cells that carry out the specific functions of the body, such as skin, blood, muscle, liver, and nerve cells. Over the past two decades, scientists have been gradually deciphering the processes by which unspecialized stem cells become a variety of specialized cells in the body. This property makes stem cells most interesting for scientists seeking to create novel medical treatments that target these cells or replace lost or damaged cells. Many tissues/organs in our body continuously require replacement of aged, injured, or diseased cells. For example, turnover of blood cells in human circulation is about 10^8 cells per hour, while such turnover of epithelial cells in the small intestine is every 3–5 days and in skin it takes 3-4 weeks. Stem cells are the reserve progenitor cells that become differentiated to functional cells as required.

Thus, stem cells can be defined as a specific type of undifferentiated cells that have the capacity for unlimited self-renewal as well as the ability to differentiate into various types of specialized cells.

Stem cells are also important for living organisms for many other reasons. In the 3- to 5-day-old embryo, called a blastocyst, the inner cells that give rise to the entire body of the organism develop into many specialized cell types and organs such as the heart, lung, skin, sperm, eggs, and other tissues. In some adult tissues, such as bone marrow, muscle, and brain, a discrete population of adult stem cells generates replacements for cells that are lost due to normal wear and tear, injury, or disease.

Given their unique regenerative abilities, stem cells offer new potentials for treating diseases such as diabetes, heart disease, and many degenerative diseases. However, much work remains to be done in the laboratory and in clinics to understand how to make best use of these cells for cell-based therapies to treat diseases, particularly those that are normally untreatable.

Laboratory studies on stem cells enable scientists to learn about the cells' essential properties and what makes them different from specialized cell types (Figure 23.1). Scientists are already using stem cells in the laboratory to screen new drugs and to develop model systems to study normal growth and identify the causes of birth defects.

Research on stem cells continues to advance knowledge about how an organism develops from a single cell and how healthy cells replace damaged cells in adult organisms. Stem cell research is one of the most fascinating areas of contemporary biology, but, as with many expanding fields of scientific inquiry, research on stem cells raises questions as rapidly as it generates new discoveries.

WHAT MAKES STEM CELLS SPECIAL?

The biological key to the immense potential of stem cells rests with the special distinguishing properties of stem cell maintenance and division. A stem cell undergoes division because it wants to produce a specialized cell to fulfill the cell requirement of a particular organ system, and also wants to produce a stem cell that can function as the "Master Cell." A true stem cell possesses certain intrinsic properties that endow it with extraordinary features, such as the following:

- **Undifferentiated and Unspecialized:** *Stem cells are unspecialized and undifferentiated.* One of the fundamental properties of a stem cell is that it does not have any tissue-specific structures that allow it to perform specialized functions.
- **Self-Renewal:** Ability to give rise to identical new stem cells that are also capable of self-renewal, expansion, and differentiation, leading to maintenance of the stem cell pool.
- **Differentiation:** Stem cells can give rise to specialized cells with a particular function. When an unspecialized cell produces a specialized cell like a brain cell, muscle cell, or red-blood cell, the process is called differentiation. Examples include hematopoietic stem cells (HSCs) that give rise to all hematopoietic cells, neural stem cells (NSCs) that give rise to neurons, astrocytes and oligodenrocytes, and mesenchymal stem cells that differentiate into fibroblasts, osteoblasts, and chondroblasts. Some adult stem cells (ASCs) may give rise to only a single mature cell type, such as the corneal stem cell (Holm, 2002).

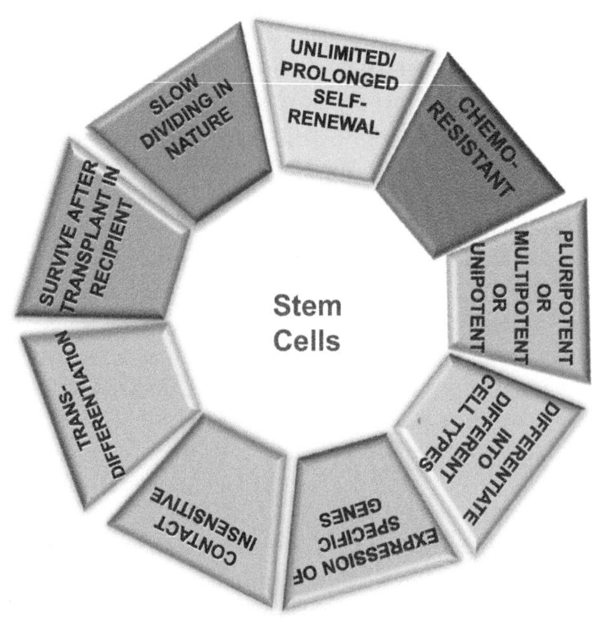

FIGURE 23.1 Properties of stem cells.

- **Homeostasis:** The ability of stem cells to both self-renew and differentiate is critical for tissue homeostasis. Only upon receipt of a stimulating signal can the stem cell become activated to divide and proliferate. Therefore, stem cell proliferation depends on dynamic niche signaling. Maintaining a balance between the proliferation signal and anti-proliferative signaling is the key to homeostatic regulation of stem cells, allowing stem cells to undergo self-renewal while supporting ongoing tissue regeneration. Any genetic mutation that leads stem cells to become independent of growth signals, or to resist anti-growth signals, will cause the stem cells to undergo uncontrolled proliferation and possible tumorigenesis.

DIFFERENTIATION POTENTIAL/POTENCY OF STEM CELLS

Special populations of mammalian cells that have unlimited self-renewal and differentiation capabilities are considered to be stem cells. Many of the terms used to define these stem cells depend on the behavior of the cells in the intact organism (*in vivo*), under specific laboratory conditions (*in vitro*), or post-transplantation *in vivo*. Cell potency specifies the differentiation potential of the stem cells to differentiate into different cell types (Table 23.2).

Totipotent Stem Cells

The fertilized egg or zygote is considered to be the first stem cell and is said to be **totipotent** (from the Latin word *totus*, meaning entire) because it has the potential to generate all the cells and tissues that make up an entire organism. As the zygote further divides and differentiates, the cells lose their regenerative potential such that until the 8-cell stage,

TABLE 23.2 Stem Cell Classes on the Basis of Potency

1.	**Totipotent**	Up to 8-cell stage produced after fusion of egg and sperm; each can give rise to embryonic as well as extra embryonic tissues (total organism).
2.	**Pluripotent**	Descendents of totipotent cells. Embryonic stem cells. Induced pluripotent stem cells.
3.	**Multipotent**	Tissue-specific family of stem cells such as hematopoietic stem cells (HSCs) and mesenchymal stem cells (MSCs).
4.	**Unipotent**	These cells can produce only one type of cells (e.g. muscle stem cells or spermatogonial stem cells).

each of them is **totipotent.** Later the embryonic cells retain **pluripotency,** which means they have the capacity to differentiate into all 210 types of body cells, excluding the extra-embryonic tissues of the placenta and umbilical cord. All these cells are descendents of a single, totipotent cell of the zygote or fertilized egg.

Pluripotent Stem Cells

The term pluripotent was used to describe stem cells that can give rise to cells derived from all three embryonic germ layers (ectoderm, mesoderm, and endoderm). These three germ layers are the embryonic source of all cells of the body. All the different kinds of specialized cells that make up the body are derived from one of these germ layers ("pluri," derived from the Latin *plures,* meaning several or many). Thus, pluripotent cells have the potential to give rise to any type of cell, a property observed in the natural course of embryonic development and under certain laboratory conditions.

Multipotent Stem Cells

These precursor cells can produce cells of only one closely related family of cells or only one primary germ layer (e.g. hematopoietic stem cells).

Unipotent Stem Cells

This is a term usually applied to a cell in adult organisms, which means these cells are capable of differentiating into only one cell lineage. "Uni" is derived from the Latin word *unus,* which means one. Adult stem cells in many differentiated, undamaged tissues are typically unipotent and give rise to just one cell type under normal conditions. This process would allow for a steady state of self-renewal for the tissue. However, if the tissue becomes damaged and the replacement of multiple cell types is required, pluripotent stem cells may become activated to repair the damage. The embryonic stem cell is pluripotent (i.e. it can give rise to cells derived from all three germ layers).

ARE STEM CELLS IMMORTAL?

Whereas most adult differentiated cells will die either by natural programming (homeostasis) or by injury, the stem cell compartment that replenishes them must persist during the whole lifespan of an organism. For this reason, these long-lived cell populations should be able to cope with two main biological phenomena: (1) no DNA damage to cause gene mutations, and (2) a normal senescence mechanism.

In long-lived organisms, tissues are more susceptible to DNA damage by environmental agents or by by-products of normal metabolism, such as reactive oxygen species. Since mutations may lead to cancer, they must be either corrected by gene repair mechanisms or destroyed by apoptosis of the cell. To avoid mutations, one of the mechanisms that seem to be used by stem cells is non-random chromosome segregation (Cairns, 1975; Cairns, 2002). This model proposes the segregation of the original ("**immortal**") DNA template to the stem cell originated by asymmetric division, whereas the newly synthesized strands (more prone to replication-induced mutation) are passed on to the daughter cells that are committed to differentiation (Potten et al., 2002).

Senescence, on the other hand, is a physiological property of normal cells. Most normal human cells are programmed for a given number of cell divisions when cultured *in vitro* **(the "Hayflick limit")** due to progressive telomere shortening that leads to a phenomenon called **replicative cell senescence**. However, immortal cells such as cancer cells and stem cells exhibit enhanced telomerase activity, which allows cells to replicate indefinitely. Understanding cellular mechanisms that control asymmetric cell kinetics may hold the key to unlocking the mystery of the "**immortal DNA strand hypothesis**" for maintaining stem cell vitality/viability.

CHARACTERISTICS OF STEM CELLS

Stem Cell Plasticity

A widely accepted definition of plasticity is yet to be established, but in general this term refers to the ability of adult stem cells to cross lineage barriers and to adopt the expression profiles and functional phenotypes of cells unique to other tissues (i.e. they differentiate into multiple cell types). For example, hematopoietic stem cells can differentiate into skeletal muscle cells, cardiac muscle cells, liver cells, or brain cells. Bone marrow stromal cells can differentiate into blood and skeletal muscle cells.

Mechanisms of Stem Cell Plasticity

There are a number of possible mechanisms that could explain these phenomena. One possibility is **trans-differentiation** of a cell directly into another cell type as a response to environmental cues. Trans-differentiation has been shown mainly *in vitro.* Some *in vivo* data also support this mechanism. Direct trans-differentiation can be clinically limited by the number of cells to be introduced into an organ without removal of existing cells. But if bone marrow cells could give rise to stem cells of another tissue, then they should in principle be able to repopulate whole organs from just a few starting cells. This model of **de-differentiation** is in agreement with the data from animal models. Genetic analysis of donor cells has brought to light another possible mechanism. The fusion of host and donor cells can give rise to mature tissue cells without trans- or de-differentiation.

The resulting heterokaryons are able to cure a lethal genetic defect and do not seem to be prone to giving rise to cancer (Kashofer and Bonnet, 2005). The ultimate proof of plasticity of stem cells should involve a direct demonstration of nuclear reprogramming from one stem cell fate to another.

Stem Cell Fate

Generally, stem cells remain quiescent and do not enter the cell cycle. This is important in sequestering a reserve pool of cells for use in times of stress, wear and tear, and during development. A stem cell may undergo apoptosis and not contribute to further development; this appears to be the norm in brain tissues where turnover of differentiated cells (neurons and glial cells) is very low. Stem cells may undergo symmetric cell divisions to self-renew or undergo terminal differentiation, or they may undergo asymmetric cell divisions to generate differentiated progeny as well as maintain a pool of stem cells (Figure 23.2). A dynamic balance between proliferation, survival, and differentiation signals ensures that an appropriate balance between stem cells, precursor cells, and differentiated cells is maintained throughout development and adult life (Rao and Mattson, 2001).

Stem Cell Quiescence

An important property of stem cells that is often overlooked, yet relates to the phenomenon of self-renewal, is that stem cells reside in the body in a state of **quiescence**. This is true for stem cells of the bone marrow and the skin. However, how these cells are induced to enter the cell cycle is not well understood. It is well understood that once these quiescent cells enter the cell cycle, they undergo a high rate of expansion and do not exit the cell cycle but undergo continuous division. Many of these cells pass through the G1 and G2 phases so quickly that these phases are virtually non-existent. Embryonic stem cells, skin stem cells, hematopoietic stem cells, and cancer cells are some of the examples which continuously divide without exiting the cell cycle (Sherr, 2000).

WHERE TO FIND STEM CELLS?

There are several major sources from which a good number of stem cells can be harvested:

Embryos

Embryonic stem cells (ESCs) are derived from the cells of the inner cell mass of the blastocyst during embryonic development. Embryonic stem cells have the capacity to differentiate into any cell type and the ability to self-replicate for numerous generations. A potential disadvantage of ESCs is their ability to proliferate endlessly unless they are controlled (Thomson et al., 1998).

Fetal Tissue

These embryonic germ cells (EGCs) are derived from cells found in the gonadal region of an aborted fetus (Shamblott et al., 1998).

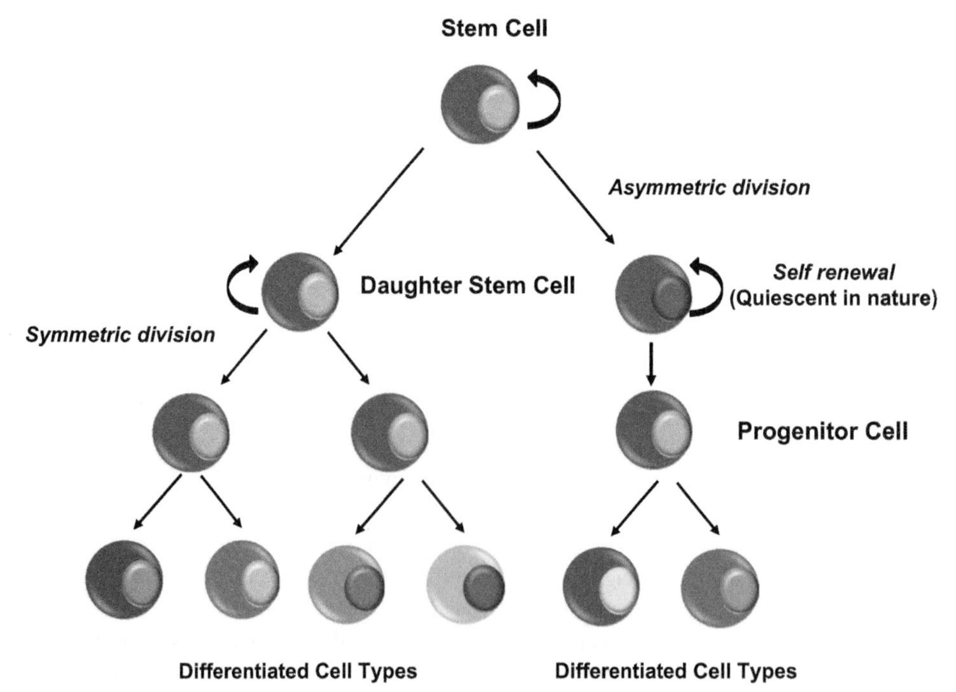

FIGURE 23.2 Asymmetric and symmetric stem cell division.

Amniotic Fluid-Derived Stem Cells (AFSCs)

AFSCs can be isolated from aspirates of amniocentesis during genetic screening. An increasing number of studies have demonstrated that AFSCs have the capacity for remarkable proliferation and differentiation into multiple lineages such as chondrocytes, adipocytes, osteoblasts, myocytes, endothelial cells, neuron-like cells, and liver cells (Barria et al., 2004).

1. **Placental Tissue:** This is a unique source of different populations of stem cells, including mesenchymal, hematopoietic, and trophoblastic stem cells, and they are possibly more primitive stem cells (Fauza, 2004).
2. **Umbilical Cord Blood Stem Cells (UCBSCs):** These stem cells are derived from the blood of the umbilical cord. There is a growing interest in the capacity of these cells for self-replication and multi-lineage differentiation. Umbilical cord blood stem cells can be differentiated into several cell types like cells of the liver, skeletal muscle, neural tissue, pancreatic cells, immune cells, and mesenchymal stem cells (Gang et al., 2004).

WHERE DO STEM CELLS COME FROM?

Stem cells are found from the early stages of human development to the end of life. All stem cells may prove useful, but each of the different types has both promise and limitations. The origin or lineage of stem cells is well understood for ES cells; their origin in adults is less clear and in some cases controversial. It is quite possible that ES cells originate before germ layer commitment, raising the possibility that this may be a mechanism for the development of multipotent stem cells, including adult stem cells. Several sources of stem cells have been discovered (e.g. embryonic, fetal, and adult stem cells), and each comes from different sources with different properties (Blau et al., 2001).

Embryonic Stem Cells (ESCs)

When a sperm fertilizes an egg it becomes a **zygote**, which can be considered the ultimate stem cell (Figure 23.3) because it can develop into any cell – not only cells of the embryo, but also cells of surrounding tissues such as the placenta. The zygote up to its 8-cell stage has the highest degree of **trans-differentiation** and is referred to as a **totipotent stem cell**, meaning that the totipotent cells are the first stage stem cells as the zygote develops into both embryonic and extra embryonic tissues (Hadjantonakis and Papaioannou, 2001). Thirty hours after fertilization, the zygote begins to divide, and by the fifth or sixth day, the cells form a **blastocyst.** A blastocyst is a pre-implantation embryo that contains all the materials necessary for the development of a complete human being. The blastocyst is mostly a hollow sphere of cells. In its interior is the **inner cell mass**, which is composed of 30–34 cells that are referred to as **pluripotent** because they can differentiate into all of the cell types of the body. These cells have somewhat less potential for

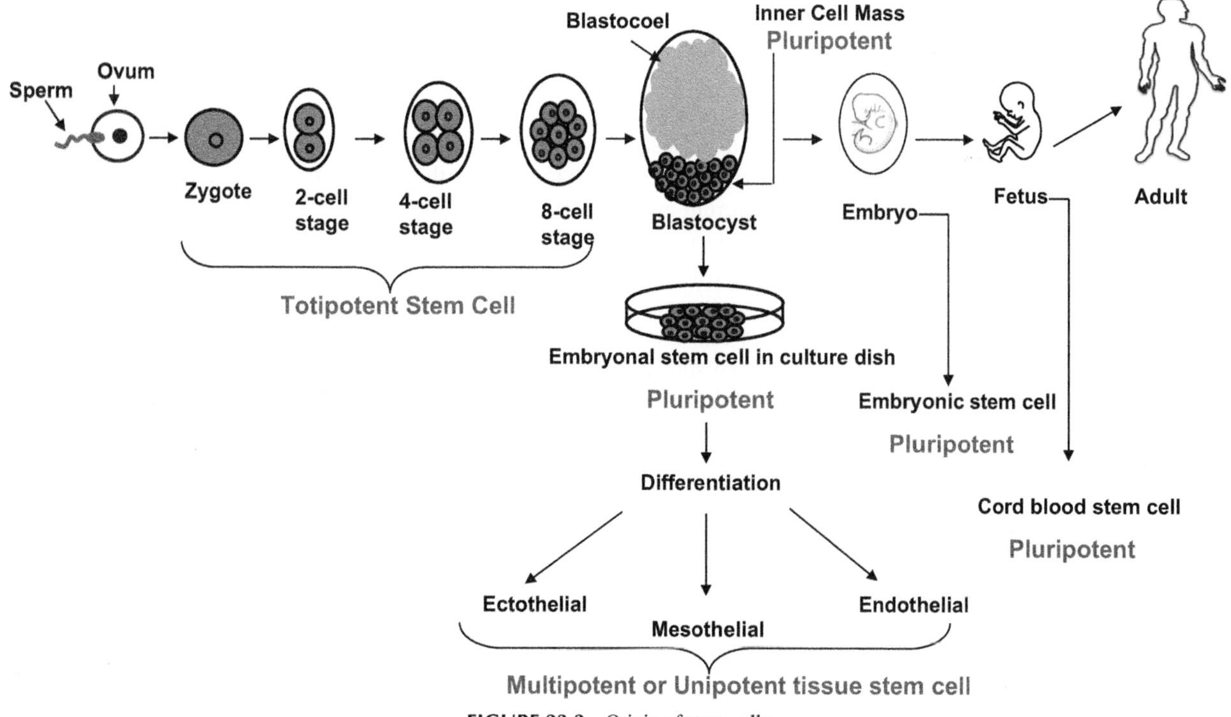

FIGURE 23.3 Origin of stem cells.

TABLE 23.3 Embryonic Stem Cells (ESCs)

1.	Derived from inner mass of cells of a blastocyst (4–6 days old fertilized human egg).
2.	Can be obtained from *in vitro* fertilized embryos or aborted embryos.
3.	Capable of differentiation into all 210 cell types and tissues of the body comprising three germ layers (ectoderm, mesoderm, and endoderm).
4.	Capacity to self-renew; pluripotent.
5.	Have potential to replace damaged or lost tissue or organs.

differentiation and are more specialized than the totipotent zygote stem cells. Those on the outer surface of the **blastocyst** develop into **placental** and **other tissues** that surround the fetus, while those inside are referred to as **inner cell mass** or **embryonic stem cells** (Tzukerman et al., 2003) (Table 23.3).

Embryonic Germ Cells (EGCs)

Five to nine weeks after fertilization, the growing embryo, now called a **fetus**, develops a region known as the **gonadal ridge**. The gonadal ridge contains the **primordial germ cells**, which will eventually develop into **eggs** or **sperm**. **Embryonic germ cells** are isolated from these primordial germ cells of the five-to-nine-week-old fetal tissue that results from elective abortions. Like ESCs, EGCs are also pluripotent (Shamblott et al., 1998). As the embryo grows, it accumulates additional embryonic stem cells in the yolk sac; as the fetus grows from eight week-old embryonic and fetal stem cells, it develops into tissues and organs. At this stage, the stem cells are more tissue-specific rather than responsible for generating all the body's different cell types. For example, fetal stem cells in the liver tend to generate liver and blood cell families. Such cells are generally designated as **multipotent**. Nevertheless, fetal cells may have an advantage over embryonic stem cells in that they may not form teratomas. Fetal liver tissue has been shown to be a rich source of stem cells (Rollini et al., 2004). Until week 12, fetal stem cells (as well as the ES cells that preceded them) have a very important property that they can be transplanted into an individual without being rejected. This is because they have little or no immunogeneic protein on their surface (Class II HLA) (O'Donoghue and Fisk, 2004). After the 12th week when fetal stem cells start expressing these immune-triggering proteins, the fetus develops the potential for tissue rejection. Accordingly, stem cells derived from these sources may have therapeutic potential only when given to the individual from whom they were derived (autologous transplantation) or to an immunologically matched donor (allogenic transplantation).

Adult Stem Cells (ASCs)

The term adult stem cell refers to the cells found in adult organs that constantly replenish the somatic cells in the tissue of their origin. A rigorous assessment of adult stem cells is required to purify a population of cells using cell surface markers, to transplant a single cell from the purified population into a syngeneic host without any intervening *in vitro* culture, and to observe self-renewal and tissue or organ regeneration for their multipotency. These cells possess a strong regenerative capability to replenish the senile or sick cells of the tissue in which they reside under pathological conditions or injury. The primary roles of ASCs in living organisms are to maintain and repair the tissue in which they are found. These cells have more limited differentiation potential, so they are called **multipotent stem cells**. ASCs are found in many tissues of the body and can be used for transplantation. Certain types of ASCs seem to have the ability to differentiate into a number of different cell types. If this differentiation of ASCs can be controlled in the laboratory, these cells may become the basis of therapies for many serious common diseases. These ASCs are very small in number in each tissue and reside in a special area of each tissue where they remain quiescent (non-dividing) for many years until they are activated by a disease signal or tissue injury (Table 23.4).

It is believed that there were primarily only two types of stem cells, embryonic and adult stem cells. There is another "new" kind of high-potential stem cell that has been developed recently called **induced pluripotent stem cells (iPSCs)**. Scientists have successfully transformed regular adult cells into stem cells using a technique called **nuclear reprogramming**. By altering the gene's somatic function in the adult cells, it is possible to reprogram the cells to act similar to the embryonic stem cells. This new technique would help avoid the controversies that come with embryonic stem cells, and would prevent immune rejection. Researchers have been able to take regular connective tissue cells and reprogram them to become heart cells. These new heart cells were injected into mice with heart failure and it improved heart function and survival time. This ingenious technique has paved the way to coax somatic cells of the body into becoming egg and sperm. It provides a potential cure for infertility as well as parenthood options for same-sex couples and those suffering from sterility.

HOW ARE STEM CELLS IDENTIFIED, ISOLATED, AND CHARACTERIZED?

The ultimate goal of stem cell research is to develop cell-based therapies to treat conditions currently untreatable by conventional therapies. To achieve this, protocols are

TABLE 23.4 Difference Between Embryonic and Adult Stem Cells

	Embryonic Stem Cells (ESCs)	Adult Stem Cells (ASCs)
1	ES cells are **pluripotent** cells capable of differentiating into any type of cells of the body (e.g. embryos).	Adult stem cells are **unipotent/multipotent** and can differentiate into the type of cells/tissue from which they are derived (e.g. bone marrow, stromal cells, brain, baby teeth, etc.).
2	Large numbers of these cells can be **easily cultured** *in vitro*.	Rarely found, and **difficult to grow** *in vitro*.
3	**Transplant rejection** occurs with ESCs.	**No transplant rejection** occurs as it uses cells from the same patient's body.
4	Possibility of infection and teratoma formation.	No such possibility.
5	May induce immune response.	Immune privileged.
6	Ethically controversial.	No ethical issues.

needed to expand, characterize, and isolate sufficient quantities of purified stem cell populations of a defined lineage. Following is brief outline of standards and criteria that may be employed when approaching the challenge of identifying, isolating, and characterizing a stem cell population.

Embryonic Stem Cell Identification

The basic characteristics of an ES cell include self-renewal, multi-lineage differentiation *in vitro* and *in vivo*, clonogenicity, a normal karyotype, extensive proliferation *in vitro* under well-defined culture conditions, and the ability to be frozen and thawed. In animal species, *in vivo* differentiation can be rigorously assessed by the ability of ES cells to contribute to all somatic lineages and produce germline chimerism. These cells must generate embryoid bodies and teratomas containing differentiated cells of the three germ cell layers. Therefore, ES cells must be shown to be positive for well known molecular markers of pluripotent cells.

Adult Stem Cell Identification

The basic characteristic of an adult stem cell is a single cell (clonal) that self-renews and generates differentiated cells. The most rigorous assessment of these characteristics is to prospectively purify a population of cells (usually by cell surface markers), transplant a single cell into an acceptable host without intervening *in vitro* culture, and observe self-renewal and tissue, organ, or lineage reconstitution.

The search for a specific molecular signature to define stem cells continues. Until that time, **stemness** remains a concept of limited utility with tremendous potential.

Methods To Identify Stem Cells

A number of methods are being used today to unravel the secrets of the stem cell. For example, confocal microscopy can study their morphology. In addition to giving information about a cell's size and shape, confocal microscopy has a three-dimensional imaging capability that can be used to examine internal cellular structures and components.

Analysis of stem cells falls into two main categories: monitoring the genetic integrity of the cells, and tracking the expression of proteins associated with potency. Genomic analysis of stem cells is necessary to ascertain whether stem cells maintained in culture for long periods have not undergone chromosomal loss, duplication, or epigenetic changes, while proteomic analysis ensures that the cells are expressing the factors necessary to maintain potency. Thus, analysis of the differentiated state through both genomic and proteomic methods ensures identification and propagation of the proper stem cell population.

Molecular markers are critical to the study of cellular differentiation and cell-fate specification. To study the transition from one cell type to another, one must conclusively identify distinct cell types. While stem cells are best defined functionally, a number of molecular markers have been used to characterize various stem cell populations. Although functions have yet to be ascertained for many of these early markers, their unique expression pattern and timing provide a useful tool to identify and isolate stem cells. Evidence regarding the role of specific markers in defining embryonic, hematopoietic, mesenchymal/stromal, and neural stem cell populations (both *in vitro* and *in vivo*) supports their significant role in characterizing stem cells (see Flow Chart 23.1). Some commonly employed methods are presented below:

By Cell Surface Marker Expression

Embryonic Stem Cell Markers

Oct-4: Oct-4 (octamer-binding transcription factor 4), also known as POU5F1 (POU domain, class 5, transcription factor 1), is a protein that in humans is encoded by the POU5F1 gene. It was originally identified as a DNA-binding protein that activates gene transcription via a *cis*-element containing an octamer motif. It is expressed in totipotent embryonic stem and germ cells. This protein is critically involved in sustaining self-renewal of undifferentiated embryonic stem cells and pluripotency. Oct-4 is not only a master regulator of pluripotency that controls lineage commitment, but is also the first and most recognized marker used for the identification of totipotent ES cells. It has been implicated in tumorigenesis of adult germ cells.

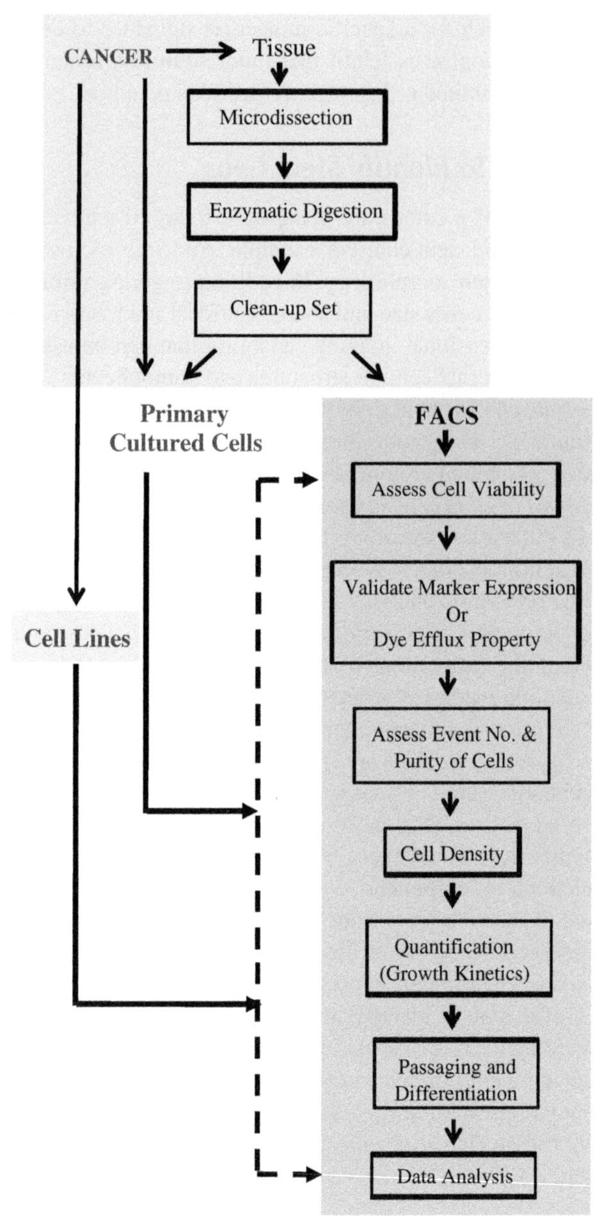

FLOW CHART 23.1 Outline for cancer stem cell isolation from cancer tissue biopsy or cancer cell lines.

Stage-Specific Embryonic Antigens (SSEAs): SSEAs were originally identified by three monoclonal antibodies (Abs) recognizing defined carbohydrate epitopes associated with lacto- and globo-series glycolipids: SSEA-1, -3, and -4. SSEA-1 is expressed on the surface of preimplantation-stage murine embryos (i.e. at the eight-cell stage) and has been found on the surface of teratocarcinoma stem cells, but not on their differentiated derivatives. Biological roles of these carbohydrate-associated molecules have been suggested in controlling cell surface interactions during development. Undifferentiated primate ES cells and human EC and ES cells express SSEA-3 and SSEA-4, but not SSEA-1.

Hematopoietic Stem Cell Markers

CD133 is a member of the penta-span transmembrane glycoproteins (5-transmembrane, 5-TM), which specifically localize to cellular protrusions. A CD133 isoform, AC133-2, has recently been cloned and identified as the original surface antigen recognized by the AC133 Ab. CD133 may provide an alternative to CD34 for HSC selection and *ex vivo* expansion. Recent studies have offered evidence that CD133 expression is not limited to primitive blood cells, but that it defines unique cell populations in non-hematopoietic tissues as well as hematopoietic stem cells, glioblastoma, neuronal and glial stem cells, and adult kidney, mammary gland, placental, digestive tract, testes, and several other cell types (Mizrak et al., 2008).

CD34 (Sialomucin) is a member of a family of single-pass transmembrane sialomucin proteins that show expression on early hematopoietic and vascular-associated tissue. It may mediate the attachment of stem cells to bone marrow extracellular matrix or stromal cells. It is considered the most critical marker for hematopoietic stem cells (HSCs) since it was found expressed on human bone marrow cells. It is also found on clonogenic progenitors, and some lineage-committed cells. Osawa first demonstrated that murine HSCs could be CD34 negative (Osawa, et al., 1996). Transplantation studies also showed repopulating activity in a CD34 cell population in fetal sheep. However, all clinical and experimental protocols, including gene therapy and HSC transplantation, are currently using CD34^{+}-enriched cell populations.

ABCG2; also known as **BCRP1 (CDw338)**, is a member of the multi-drug resistance (MDR) family of proteins that is a member of the adenosine triphosphate (ATP)-binding cassette (ABC) transporters. The physiological role of ABCG2 seems to be excretion of genotoxic substances from the body and from primitive repopulating cell populations. The human genome encodes approximately 49 ABC proteins, out of which only a fraction of them have been functionally characterized. They have been classified into seven subfamilies based on phylogenetic analysis. It was first identified in a breast cancer cell line. It belongs to the half-transporter group, and is unique as it is localized to the plasma membrane. The expression of ABCG2 appears to be highest in the mammary gland during pregnancy, as well as in variety of normal tissues, like intestine, brain, kidney, and placenta (and hematopoietic stem cells). It is therefore considered a more promising marker than CD34 for primitive HSC isolation and characterization. It is highly expressed on primitive **"Side Population" (SP)** stem cells. This SP phenotype is based on the efflux of fluorescent dyes such as Rhodamine 123, Hoechst 33342, and Dye cycle violet (DCV). Exclusive expression of ABCG2 on SP cells suggests that ABCG2 may be a potential functional marker for positive selection of pluripotent stem cells from various adult sources, and has been implicated for its functional role in stem cell biology.

Mesenchymal/Stromal Stem Cell Markers

STRO-1 is a cell surface antigen identified by STRO-1 mAb (Simmons and Torok-Storb, 1991). It is used to identify bone marrow mesenchymal stromal cells and nucleated erythroid precursors, but not committed hematopoietic progenitors. In combination with glycophorin A (GPA), STRO-1 is a useful marker for identification of mesenchymal stem cells. It has been reported that the STRO-1+/GPA-bone marrow cell subset can be used as a universal stromal feeder layer for hematopoietic stem/progenitor cells. These subsets of cells are multipotent and differentiate into fibroblasts, adipocytes, osteoblasts, chondrocytes, and vascular smooth muscle-like cells.

Neural Stem Cell Markers

PSA-NCAM (Polysialic Acid-Neural Cell Adhesion Molecule or CD56): It is a homophilic binding glycoprotein whose regulated expression along with their isoforms is critical for many neural developmental processes. A neuronal-restricted precursor identified by PSA-NCAM+ high expression can undergo self-renewal and differentiate into multiple neuronal phenotypes, whereas neonatal brain precursors are restricted to a glial fate. Increasing evidence indicates that the poly-sialic acid (PSA) portion of NCAM is known to be essential in several developmental events, including cell migration, axon growth, nerve branching, and synaptic arrangement.

Nestin: Nestin is a class VI intermediate filament protein (Michalczyk and Ziman, 2005). These intermediate filament proteins are predominantly expressed in stem cells of the central nervous system (CNS), mostly in nerve cells where they are implicated in the radial growth of the axon. However, its expression is usually transient, suggesting involvement in a major step in the neural differentiation pathway; it does not persist into adulthood. Nestin has been the most extensively used marker to identify CNS stem cells within various areas of the developing nervous system and in cultured cells *in vitro*. Nestin expression has also been discovered in non-neural stem cell populations, such as follicle stem cells and their immediate differentiated progeny, pancreatic islet progenitors, as well as hematopoietic progenitors.

Low-Affinity Nerve Growth Factor Receptor (also called the **LNGFR or p75 neurotrophin receptor**) is a type I trans-membrane protein that belongs to a large family of receptors that includes tumor necrosis factor receptors (TNFs), Fas, and approximately 25 other members. It is now known that p75NTR has surprisingly diverse effects, ranging from cell death to regulation of axon elongation. It binds to nerve growth factor (NGF), brain-derived neurotrophic factor (BDNF), neurotrophin-3 (NT-3), and neurotrophin-4 (NT-4) equally. Working in concert with Trk receptors, p75NTR recognizes neurotrophins and transmits trophic signals into the cell, suggesting its critical function during the development of the nervous system. Neural crest stem cells (NCSCs) from peripheral nerve tissues have been isolated based on their surface expression of p75NTR, which can self-renew and generate neurons and glia both *in vitro* and *in vivo*. In addition, neuroepithelial-derived p75NTR+ cells are also able to differentiate into neurons, smooth muscle, and Schwann cells in culture. Recently, p75NTR has been used as a marker to identify mesenchymal precursors as well as hepatic stellate cells.

By Dye Efflux Property

In the absence of definitive cell surface markers, an alternative approach to surface antigens is to use the **side population (SP)** phenotype. Cells subject to **Hoechst 33342** dye staining and **fluorescence-activated cell sorting (FACS)** analysis can give a profile such that those cells that actively efflux the dye appear as a distinct population of cells on the side of the dual-color emission spectra (**blue versus red**) FACS profile on a density plot; hence the name **"Side Population" (SP)** (Figure 23.4). Therefore, the SP is defined according to convention by depletion using Hoechst transporter inhibitors **reserpine** or **verapamil** or

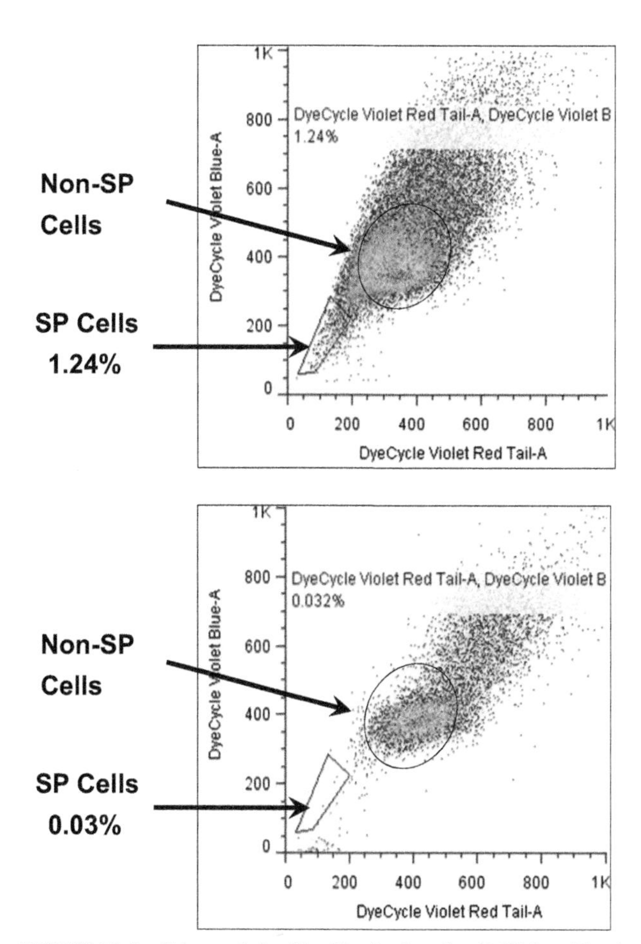

FIGURE 23.4 Side population identification based on DCV dye efflux in HPV16+ve Cervical Cancer Cell Line (SiHa).

Fumitremorgin C. This SP phenotype is attributed partly to the activity of various members of the adenosine triphosphate (ATP)-binding cassette (ABC) transporters family (such as ABCG2) that are expressed in normal stem cells and most cancer stem cells. ATP-binding cassette (ABC) transporters are a superfamily of multi-span transmembrane proteins that can pump a wide variety of endogenous and exogenous compounds out of cells, including metabolic products, lipids and sterols, and drugs, thus protecting them from cytotoxic agents. Moreover, SP cells are present in a quiescent state and possess an adhesive endoplasmic reticulum with a few ribosomes, suggestive of low metabolic activity, elements common to stem cells. This method was first used to isolate murine hematopoietic stem cells (Goodell et al., 1996). Since its initial application, Hoechst side population (SP) analysis has proven to be a valuable technique for identifying and sorting presumptive stem and early progenitor cells in a diverse range of normal and cancerous tissues across species (see the Protocols section).

On the Basis of Functional Assays

Self-Renewal Assay

Self-renewal, a hallmark feature of stem cells, is the process by which stem cells divide to make more stem cells, perpetuating the stem cell pool throughout life. A major advance in stem cell research was achieved in 1996 when undifferentiated multipotent neural stem cells were discovered to be grown and maintained as a neurosphere in suspension (Reynolds and Weiss, 1992). This anchorage-independent culture of stem cells as sphere was found to be instrumental in enriching the presumptive stem cell populations when the specific stem cell identification markers had not been defined. In brief, this assay involves culturing of stem cells dissociated as a single-cell suspension from tissues on non-adherent substrata in the presence of defined media which support growth of stem cells until they form organized floating bodies called "spheres." What is the exact nature of dividing cells within these spheres? How is the proliferation stopped when a certain number of cells in the sphere is reached? How is sphere homeostasis maintained? These are some of the interesting questions that need to be explored. One of the major disadvantages of this assay was sphere clonality, which involves formation of a secondary clonal sphere from single cells obtained from the primary sphere. The practical advantage of this method is the ability to cryopreserve the spheres for future analysis.

Slow-Cycling Population Assay

Labeling of nucleotides by 5-bromo-2-deoxyuridine (BrdU) or by (^{3}H) thymidine incorporation in proliferating cells for phenotype determination through immunoflourescence or radiolabeling are some of the conventional methods for identifying stem cell or progenitor cell populations in adult tissues. The term "slow-cycling assay" is an idiom because of the distinctive stem cell feature of slow cycling. The assumption behind this assay is that cells with slow cycling times are eminent by retention of the label, which is incorporated into DNA of cells having fast cycling times during DNA synthesis. Once administrated with a label, the cells can be monitored for a long period. Since only the slow cycling cells are likely to retain the anti BrdU antibody staining or radioactive label, they can be acknowledged as **"label retaining cells or slow cycling cells."**

Lineage Labeling Assay

Lineage labeling assay represents the most powerful and reliable tool for identifying stem cells, and was originally developed for understanding the evolution at the cellular level such as detection of cellular origin, lineage relationship, distribution pattern, and differentiation. In short, a single cell is labeled such that the label is closely transmitted to all daughters of the initial cell, resulting in a labeled clone. Labeling of cells reveals information about its state of differentiation. Lots of tracing methods, such as genetic fate mapping, development of genetic mosaics, chromosome loss, recombination, and/or replication-defective retroviral vectors utilize cell labeling. This technique represents a potentially fruitful method for identifying new stem cells with single-cell resolution.

THE STEM CELL NICHE

Several theories have been proposed to explain the methods of lineage determination: extrinsic methods (through growth factors, stroma, or external stimuli), intrinsic methods (for example, nuclear factors), or both. A "**niche**" is a subset of tissue cells and extracellular substrates that favors the existence of the stem cell in the undifferentiated state *in vivo*. Interaction with other cell types and the components of the extracellular matrix are believed to influence the survival and development of the committed cells. Research over the past decade has made it increasingly clear that the stem cell **niches provide a microenvironment** that is important in protecting and perpetuating the self-renewing, undifferentiated state of their precious resident cells. The adhesive milieu of niche allows it to retain stem cell daughters. In fact, when and how the stem cell niches develop is still not clearly understood. Also, there is considerable variation in niche design.

What Keeps Stem Cells in Their Niche?

The ability of stem cells to reside within their niches is an evolutionarily conserved phenomenon. Much of our understanding of this process comes from functional studies in *Drosophila* germ stem cells (GSCs) and mice hematopoietic stem cells (HSCs). These studies show that direct physical interactions between stem cells and their non-stem cell neighbors in the niche are critical to keeping stem cells in this specialized compartment and in maintaining stem cell character. The ability of the niche to retain its stem cells is also likely to play a role in recruiting stem cells, a process referred to as **"homing"** (Whetton and Graham, 1999). Although the molecular mechanisms of this process are not known, it seems likely that niches develop concomitantly with input from both the stem cells and the surrounding tissue. Once niches are established, they seem to be able to survive as signaling centers to attract stem cells.

How Do Stem Cells Get Activated in Their Niche?

Examining the properties of stem cells in and out of their niches has provided some interesting insights. Inside the niche, stem cells are often quiescent. Since proliferation can often be induced in tissue culture, it suggests that the niche's microenvironment is both proliferation- and differentiation-inhibitory. Also, when there is a constant flux of slowly dividing stem cells, such as when the niche becomes occupied, excess stem cells are displaced shortly after they divide, thereby physically loosening connections with the niche. Alternatively, stem cells might simply remain dormant within the niche until they become functional (e.g. in response to injury). Future research in this direction will be exciting in defining the molecular characteristics of the stem cell and its niche.

GROWING STEM CELLS IN THE LABORATORY

The ability to grow functional adult stem cells in culture creates new opportunities for drug research. Researchers are able to grow differentiated cell lines and then test new drugs on each cell type to examine possible interactions *in vitro* before applying them *in vivo*. This is critical in the development of drugs for use in clinical research because of the possibilities of species-specific interactions. It is hoped that by having these cell lines available for research, use of research animals will be reduced, and that the effects on human tissue *in vitro* will provide insight not normally available via animal testing.

With the advent of induced pluripotent stem cells (iPSCs), rather than needing to harvest embryos or eggs (which poses serious ethical issues), researchers can induce any somatic cell of the body and reprogram it to produce the desired patient-specific type of stem cell. This greatly reduces the problems associated with using animals, and allows the unlimited use of iPSCs for therapeutic purposes.

iPSCS: A STEM CELL RESEARCH BREAKTHROUGH

Stem cells are basically immortal cells that can generate any tissue type, from bone cells to brain cells; this makes them pluripotent. Scientists have been able to reprogram cells using the right cocktail of genes, enzymes, and proteins to coax the cells to express genes and factors needed to maintain the defining properties of embryonic stem cells. These genetically reprogrammed pluripotent cells are called induced pluripotent stem cells (iPSCs). This technique has made it possible to create human eggs and sperm in the laboratory that, besides other things, could increase parenthood options. An **iPSC** is any somatic cell that has been reprogrammed to exhibit pluripotent stem cell properties. Though any somatic cells have been successfully reprogrammed, fibroblast cells are most commonly used for this purpose because they can be easily extracted from a patient using a safe and non-invasive skin biopsy, and are easy to culture in a lab. The successful creation of **induced pluripotent stem cells** proved to the world that cellular differentiation is not a unidirectional process. With proper signaling, cellular differentiation can be made bidirectional. The hypothesis that cells differentiate only uni-directionally towards terminal differentiation was first challenged in 1962 by John Gurdon (Gurdon, 1962) in his pioneering work in nuclear transfer, and later with the historic birth of **Dolly** the Sheep (February of 1997) via **somatic cell nuclear transfer (SCNT).** The success of SCNT stimulated developmental biologists to begin exploring the possibility of creating pluripotent stem cells (e.g. induced pluripotent stem cells) by reprogramming fully differentiated somatic cells to an embryonic-like state. A groundbreaking discovery came in 2006 when Shinya Yamanaka demonstrated that mouse fibroblasts could be reprogrammed to **pluripotent "embryonic-like"** stem cells by over-expressing specific genetic factors. The team hypothesized that a select group of 24 pluripotency-related genes, when over-expressed in mouse somatic cells, could induce pluripotency. Of the 24 genes screened, only four were essential for reprogramming mouse fibroblasts into pluripotent stem cells, otherwise known as **induced pluripotent stem cells (iPSCs)** (Takahashi and Yamanaka, 2006). Yamanaka's team determined that the essential stemness genes are **Oct-4, SOX2, c-Myc, Klf-4**, the four genes that have important functions in the regulation of pluripotency in embryonic stem cells (Figure 23.5). One year later, the team reported that the same four factors were capable of inducing pluripotency in human somatic cells (Takahashi et al., 2007), James Thomson (Yu et al., 2007) and his team developed an **iPSC cell line** from human foreskin fibroblasts using another set of four genes: **Oct-4, SOX2, Nanog,** and **Lin28**. This discovery revolutionized the field of stem cell science and regenerative medicine because it opened the door for the development of patient-specific autologous pluripotent stem cells for clinical use.

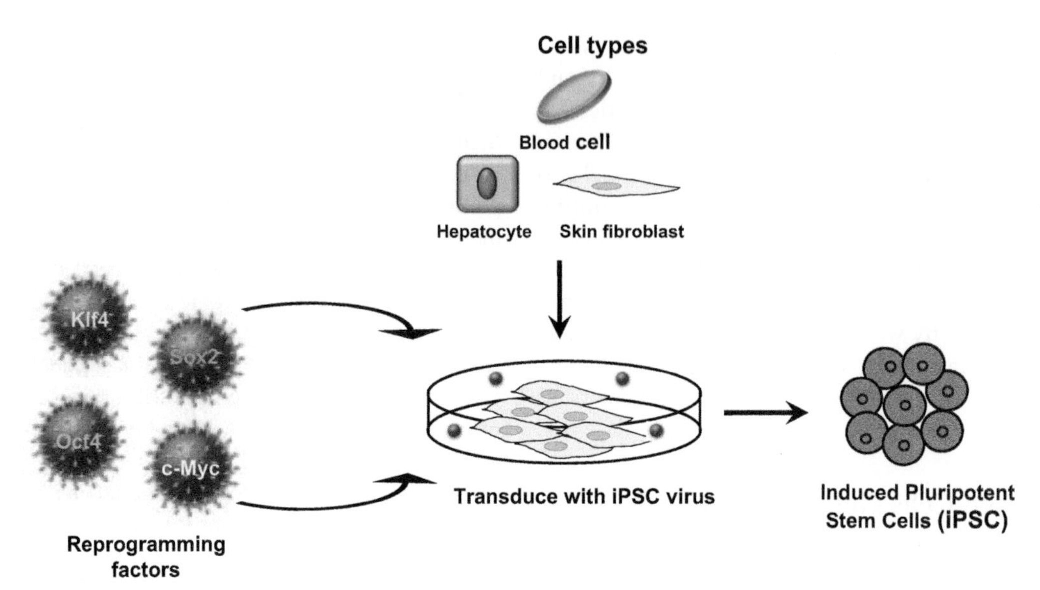

FIGURE 23.5 Generation of induced pluripotent stem cells.

A number of methods are used to artificially overexpress genes in a cell. Scientists are also exploring the use of small molecules and other technologies to reprogram cells without the use of viral vectors. The Yamanaka group utilized **retroviral vectors to deliver the transgenes** encoded for the four reprogramming factors: Oct-4, SOX2, Klf-4 and c-Myc. These four reprogramming factors function as transcription factors, meaning that they are capable of binding to DNA sequences to control the transcription of a unique set of genes. Together, Oct-4, SOX2, Klf-4, and c-Myc induce the expression of genes that are not normally expressed in the fibroblast but are expressed in pluripotent stem cells. The four transcription factors continue to drive transcription of their downstream genes, leading to the activation of other transcriptional network signaling that induces a cascade of transcriptional activity. Emergence of a colony over the period of two to three weeks is the first sign that the fibroblasts have been reprogrammed into induced pluripotent stem cells. From there, the iPSC cell colonies are isolated and expanded. They may be used to further our understanding of human development and the mechanisms involved in many devastating human diseases. iPSCs may also pave the way for infertile couples to have their own children because their egg and sperm could be created in the laboratory. This would of course raise a variety of moral and ethical concerns that would need to be dealt with.

CANCER STEM CELLS (CSCs)

Is cancer a stem cell disease? At present, there is an exciting turn of events in which the principles of stem cell biology have been applied to cancer biology. The similarity between embryonic tissue and cancer tissue was recognized as early as the 19th century. It was suggested by Recamier and Virchow that cancer arise from embryo-like cells present in the adult. This idea was formalized as the **"embryonal rest"** theory of cancer, as proposed by Cohnheim (1875) and Durante (1874). They proposed the hypothesis that stem cells "misplaced" during embryonic development were the source of tumors that formed later in life. Later studies involving tumors derived from ascites fluid in rats and leukemia cells in mice showed that a single tumor cell can give rise to a new tumor and generate heterogeneous progeny, providing strong evidence for the clonal origin of tumors (Makino, 1956).

The origin of teratocarcinomas from normal germinal stem cells is the prototype for the role of stem cells in the generation of cancer. Cancer stem cells (CSCs) could be defined as "a cell within a tumor that posses the capacity to self-renew and to cause the heterogeneous lineages of cancer cells that comprise the tumor." Normal stem cells and cancer cells share many key properties. Given the striking similarities between normal stem cells and cancer cells, such as high proliferation potential, the ability to give rise to multiple cell types, migration, expression of telomerase, similar signaling pathways operating within them, and common marker expression profiles, there is a strong possibility that cancer arises in a normal stem cell since many cellular mechanisms that are found active in cancer cells are already active in stem cells, but in a controlled fashion. Considerable research effort has been dedicated to investigating these possibilities, and our knowledge of cancer stem cells continues to rapidly evolve.

Cancer stem cells (CSCs) were first isolated from acute lymphatic leukemia, and were later isolated from several solid tumors. Most CSCs are now identified on the basis of cell surface CD (**cluster of differentiation**) markers,

TABLE 23.5 Markers Used to Identify and Isolate Stem Cells from Malignancies

CANCER	CD44	CD24	CD133	ALDH1	ESA	B1	α6	CD138	CD34	CD166	CD20
Breast	+	-	+	+	+	+	+				
Colon	+		+	+	+					+	
Prostrate	+		+	+		+	+				
Head and Neck	+			+							
Pancreatic	+	+	+	+	+						
Lung			+								
Brain			+								
Liver			+								
Melanoma	+		+			+					+
Multiple myeloma				+	+			-	+		+
Leukemia	+			+	+					+	

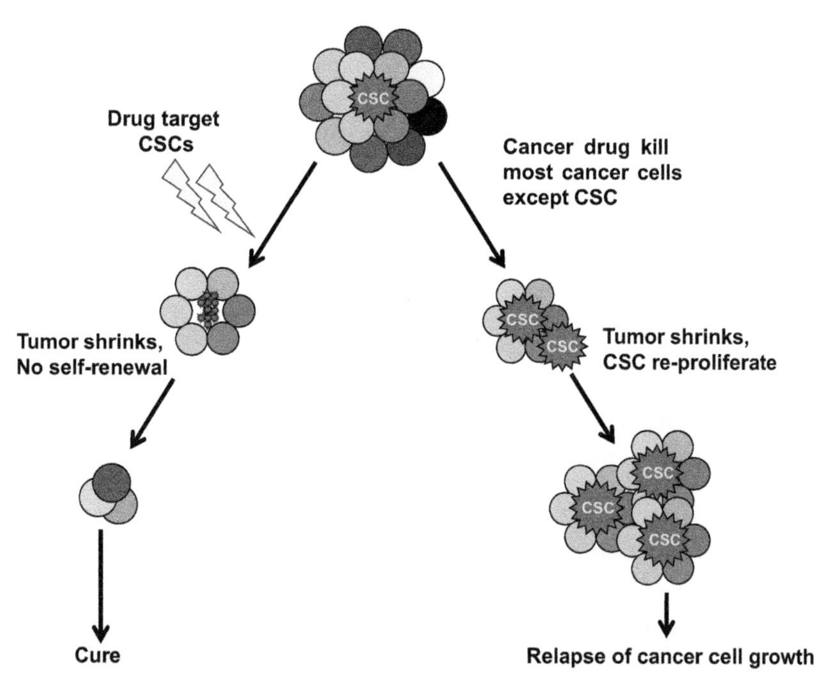

FIGURE 23.6 Outcome of cancer stem cell targeting.

which are well established for many common cancers (Table 23.5). CSCs represent only about 1% of the tumor mass, but it is only these cells that are capable of generating new tumors. Tumor cell lines such as HeLa cells contain a minor population (2–3%) of cancer stem cells. It has been established that these CSCs are capable of surviving radiation exposure and cytotoxic chemotherapeutic drug treatment (Figure 23.6). They also survive genetic insults, drug transporters, DNA repair mechanisms, and are refractory to programmed cell death. CSCs remain quiescent for a long period of time and escape standard cancer therapies. The regeneration or relapse of many cancers following chemoradiotherapy could result from the survival of cancer stem cells. Many different efforts have been combined in the search for therapeutic approaches to target these cancer stem cells.

PROTOCOLS FOR HOECHST 33342 OR DCV STAINING AND STEM CELL PURIFICATION

Hoechst or DCV staining was established in cultured cancer cell lines. Initial experiments should be performed on bone marrow in order to establish the protocol to identify the stem cell (side population, SP) from primary culture and cell lines.

Hoechst or DCV Staining

The ability to discriminate Hoechst or DCV SP cells is based on the *differential efflux* of Hoechst 33342 or DCV by a multi-drug-like transporter. This is an *active biological process*. Therefore, optimal resolution of the profile is obtained with great attention to the staining conditions. The Hoechst or DCV concentration, staining time, and staining temperature are all critical. Likewise, when the staining process is over, the cells should be maintained at 4°C in order to prohibit further dye efflux. If one adheres rigorously to the protocol below, it should be easy to find SP cells.

1. Ensure that a water bath is at precisely 37°C (check this with a thermometer). Pre-warm DMEM+ (see below).
2. Count the trypsinised cells from the T-25 flask having 50%–60% confluency and re-suspend around 10^6 cells per mL in pre-warmed DMEM+. Mix well.
3. Add Hoechst to a final concentration of 5 μg/mL (a 200× dilution of the stock) or 10 μM Dye Cycle Violet (DCV) from 5 mM stock.
4. Mix the cells well, and place them in the 37°C water bath for 90 minutes exactly. Make sure the staining tubes are well submerged in the water-bath to ensure that the temperature of the cells is maintained at 37°C. Tubes should be mixed several times during incubation.
5. After 90 minutes, spin the cells down in the COLD and re-suspend in COLD HBSS+.
6. At this point samples may be run directly on the FACS or further stained with antibodies (antibody staining of Hoechst/DCV-stained cells). All further manipulations MUST be performed at 4°C to prohibit leakage of the Hoechst or DCV dye from the cells.
7. At the end of the staining, re-suspend stained cells in cold HBSS+ containing 2 μg/mL **propidium iodide** (PI) for dead cell discrimination.

In order to confirm identification of the right cells, (1) block the population with multi-drug transporter inhibitors like verapamil of fumitremoegin C or reserpine or (2) co-stain with antibodies. Verapamil is used at 50–100 μM (purchase it from Sigma, and make a 100× stock in 95% ethanol), and is included during the entire Hoechst/DCV staining procedure, depending on the types of cells under investigation.

Materials Required

Hoechst 33342 (from Sigma) comes as a lyophilized powder and is re-suspended at 1 mg/mL in water, filter sterilized, and frozen in small aliquots.
HBSS+ (Hanks Balanced Salt Solution, Gibco) with 2% Fetal Calf Serum and 10 mM HEPES buffer (Gibco).
DMEM+ (Gibco) with 2% Fetal Calf Serum and 10 mM HEPES buffer (Gibco).
Propidium Iodide (PI) from Sigma. Final concentration should be 2 μg/mL.
DCV (from Invitrogen) comes as stock solution of 5 mM in dH_2O.

Flow Cytometry Set-Up

A UV laser is needed to excite the Hoechst dye and propidium iodide, but a **VIOLET** laser is needed for DCV. A second laser can be used to excite additional fluorochromes (e.g. FITC and PE with a 488 laser). The Hoechst dye is excited with the UV laser at 350 nm, and its fluorescence is measured with a 450/20 BP filter (Hoechst Blue) and a 675 EFLP optical filter (Hoechst Red), whereas DCV is excited with the violet laser at 407 nm and its fluorescence is measured the same as Hoechst dye. Propidium iodide (PI) fluorescence is also measured through the 675 EFLP (having been excited at 350 nm). Note that Hoechst/DCV blue is the standard analysis wavelength for DNA content analysis.

Tips for Optimal Resolution of SP Cells

Since analysis of the Hoechst/DCV dye is performed in linear mode, we have found that **good C.V.s (Coefficient of variance) are critical**; these are calibrated using beads that have a very tight distribution (e.g. DNA Check beads from BD). Also, for good C.V.s, the sample differential pressure must be as low as possible.

ETHICAL ISSUES

STEM CELL RESEARCH AT THE CROSSROADS OF RELIGION AND SOCIAL ISSUES

Stem cells are capable of generating various tissue cells that can be used for therapeutic approaches to debilitating and incurable diseases. Even though many applications of stem cells are under investigation, such research has raised high hopes and promises along with warnings and ethical and religious questions in different societies.

Many of the world's religious leaders have offered their opinions about the moral acceptability of stem cell research. While research using adult or cord-blood stem cells is relatively uncontroversial, there is no consensus on the acceptability of using human embryos in research of any kind, and therefore there is no consensus on the acceptability of acquiring human embryonic stem cell (hES) lines or using them in research. There is also no consensus on whether somatic cell nuclear transfer (SCNT) should be used to create hES lines with specific characteristics. However, from religious perspective, moral acceptability of hES research will be admissible only if it fulfils duties that are meant to respect and protect human life because it is sacred and should be used to prevent and alleviate human sufferings. Although, religious duties that protect and promote distributive justice by providing access to therapies for all people creates boundaries and limits on acceptable stem cell research. These differing opinions reflect diverse views held by, and within, the major world religions. These views are outlined in the following sections.

Greek Orthodox and Roman Catholic Churches

While the formal authorities for the Greek Orthodox and Roman Catholic Churches have come out in favor of stem cell research using adult stem cells, they have condemned hES research as immoral and illegal as it involves the willful destruction of embryos, which is considered homicide. The official position of these churches is that a human being's life begins at conception, so the human embryo has the same moral status as a full-grown individual. Since the underlying belief is in the embryo's right to life, any use of the embryo that is not for its own good is immoral, and therefore impermissible.

Protestant Churches

The United Church of Christ, a mainline Protestant denomination, has declared that they are open to embryo research, but consider that the goals of the research are of paramount importance, with considerable emphasis that the benefits from this and other medical research be distributed evenly and justly to all those in need, regardless of resources or geography. The Anglican Church, another mainline Protestant religion, is sharply divided on the ethics of hES research based on the morality of embryo research. The Protestant Church of Germany is also similarly divided, weighing the moral status of the embryo and the need to help ailing and suffering people.

Judaism

The rules of the Jewish culture are shaped by both religious text and rabbinical law. Both of these sources are relevant to the explanation of Jewish attitudes to hES research. Orthodox Jews believe that embryos do not have the same moral status as fully developed individuals. The result, therefore, is that embryos created by IVF have no special moral or legal status. Under Jewish law (*Halcha*), the fetus does not become a person (*nefesh*) until the head emerges from the womb. When the embryo is implanted, it is "as water" up to the fortieth day. After that time, and before the fetus emerges from the woman's body, it is a potential life and has great value. Since embryos used in hES research are ex-corporeal (outside the body), according to the Jewish faith it is possible to use excess IVF embryos in research. In addition to the Jewish views on the moral status of the human embryo, the Jewish religion places great emphasis on preventing and alleviating suffering. Therefore, there is a moral imperative to help those who are suffering from diseases, and to explore the potential of all types of stem cell research. This belief leads Jews to have a generally favorable view on stem cell research, including hES research.

Islamic Countries

Like other countries, some of the Islamic countries have also been involved in stem cell research for quite some time. Among the Islamic countries, Iran took the lead in hES research in 2003. In Iran, Turkey, Singapore (which has a Muslim majority), and other Islamic countries, embryo research policies are influenced by the religious belief that full human life with its attendant rights begins only after the ensoulment of the fetus. This fact, in conjunction with the importance articulated in the Quaran of preventing human suffering and illness, means that the use of surplus IVF embryos for stem cell research is relatively uncontroversial. What remains controversial in the Muslim world is creating embryos for the purpose. As with other religions, Islam and its followers have differing points of view on these issues. For example, in Egypt, a conservative religious country, the Muslim head of the Egyptian Medical Syndicate stated that embryos are early human life and should never be used in research, while the Chairman of the Islamic Law Council of North America has declared that the embryos being used in stem cell research are outside the body and therefore have no potential to become human beings. These are thus acceptable under Islamic law as sources for hES research.

Hinduism and Buddhism

Hinduism is a dominant Asian religion that varies significantly in traditions and beliefs. Closely related to Hinduism is Buddhism. There is no central Hindu or Buddhist authority to pronounce religious positions on stem cell research. There is also no Hindu or Buddhist teaching that directly addresses the morality of stem cell research. As with other major world religions, embryonic stem cell research is much more controversial. In traditional Hindu belief, conception is the beginning of a soul's rebirth from a previous life. Life in all its forms is viewed as sacred, and no harm should be caused to an embryo because it is seen as a living being. Most Buddhists have adopted the classical Hindu teaching that personhood begins at conception. Though Buddhist teachings do not directly address the issue, like Hinduism, there are two main tenets that divide Buddhists: (1) the prohibition against harming or destroying others (*ahimsa*), and (2) the pursuit of knowledge (*prajna*) and compassion (*karua*). A central belief of Hinduism and Buddhism is that an individual's soul or self is eternal. In Hinduism it is believed that the soul is passed from one living being to another in a process called reincarnation. In Buddhism, reincarnation is described as the rebirth of the self. These beliefs that the soul or the self is reborn lead to a greater acceptance of cloning technology. For this reason, although the use of embryos in stem cell research remains a divisive issue in these religions, the use of cloning technology in stem cell research is less controversial.

No matter what policy framework is adopted, whether liberal, intermediate, or restrictive, the mechanisms for assessing and regulating embryonic, stem cell, and cloning research must keep pace with the scientific discoveries that affect the research. This also holds true for the ethical issues related to differences in cultural background and religious beliefs, and finding appropriate and acceptable solutions and guidelines.

TRANSLATIONAL SIGNIFICANCE

THERAPEUTIC POTENTIAL OF STEM CELLS: A NEW AGE IN HEALTH CARE

Many degenerative diseases and neurological disorders are caused by the loss of normal cellular function in a particular organ. When cells are damaged or destroyed, they no longer produce, metabolize, or function accurately to regulate many substances essential to life. Therefore, when such specialized cells are lost due to disease or damage, other mature cells cannot fill the gap. Stem cells, in contrast, are cells at an early stage of development, and have the ability to self-renew (that is, to reproduce themselves) for indefinite periods; they give rise to a number of different kinds of mature and functionally differentiated **(specialized)** cells. There has been significant interest in the therapeutic and scientific potential of stem cells since reconstitution of the hematopoietic system was first realized by bone marrow transplantation in the 1960s. The isolation of tissue-specific, multipotent stem cells from adult organs, and the derivation of pluripotent human embryonic stem cells offer the potential for regeneration of a number of different tissues and organs susceptible to age-related degenerative conditions and traumatic injury. In the not-too-distant future it will be possible to repair heart tissue damaged by myocardial infarction, to replace neuronal cells lost in Parkinson's and Alzheimer's diseases, to transplant new insulin-producing cells for diabetics myelinating cells for individuals afflicted with multiple sclerosis, and to replace bone and cartilage lost through aging and inflammatory disease. In addition, the generation of specific populations of defined subtypes of human cells has tremendous potential to revolutionize the fields of drug discovery and investigation into the cellular basis of human disease. The newly emerging field of regenerative medicine will fundamentally alter clinical medicine and significantly influence our perceptions of aging, health, and disease, with a myriad of consequences for society at large.

Stem Cell Therapy: A Ray of Hope

In the last two decades, several important scientific discoveries have brought about a revolution in the field of medicine. In the 20th century, the effective and curative treatments of various diseases like tuberculosis and leprosy were discovered. Treatments like artificial joint replacement and bypass surgery are now within reach of the common man. On the horizon for the 21st century, stem cell therapy is a ray of hope for the treatment of intractable diseases like paralysis, Alzheimer's, cerebral palsy, multiple sclerosis, traumatic spinal cord injury, and paraplegia. This is a completely new and innovative therapy that is flourishing around the globe, and began in 2003. The discovery of the stem cell has brought about a revolution in modern medicine.

Principles of Stem Cell Therapy

Cell-based therapy is an empirical therapy. In this therapy, stem cells are induced to differentiate into the specific cell type required to repair the damaged or destroyed cells or tissues. Stem cell therapy is the use of stem cells in the treatment. Treatment can be divided in two types: autologous and allogenic.

Autologous stem cell therapy uses a patient's own stem cells, which are obtained from blood, bone marrow, etc. Allogenic stem cell therapy is designed to cure a person with the help of donated stem cells. However, in a number of diseases or disorders, the body may reject allogenic (foreign) stem cells. So far this therapy is not legally accepted in India.

There are some major changes happening in stem cell therapies that are setting a new paradigm for regenerative medicine. The opportunity to repair and regenerate tissues injured by disease and trauma is opening the way to new, optimistic treatments that need to be carefully evaluated in early clinical trials.

HOW ARE STEM CELLS USED IN CELL-BASED REGENERATIVE THERAPIES?

Stem cells hold great promise for medicine, both as a potential source of replacement cells for damaged organs, and as a scientific resource to study disease pathogenesis. Currently, stem cell-based therapy has been on-going in animal models for several diseases, including neurodegenerative disorders such as Parkinson's disease, spinal cord injury, and multiple sclerosis (Singee et al., 2007; Sharp and Keirstead, 2007). While it has been convincingly demonstrated that various embryonic and adult mouse and human SCs have differentiation capabilities, the full regenerative potential of these cells remains to be determined. Characterization of SC phenotypes is more complicated, as for the assessment of specific markers. In addition, immunological and tumorigenic concerns hamper the therapeutic use of ESCs. Adult SCs are free from ethical and immunological concerns, but are usually only available in small numbers that decrease with age. Nevertheless, many kinetic and quantitative issues must be resolved before stem cell therapy proves to be a robust and safe strategy that can be transferred to the clinic. These factors include control of differentiation, survival, and maturation of stem cells in the context of a host degenerative tissue. Furthermore, long-term and large-scale multi-center clinical studies are required to determine the precise therapeutic effect of stem cell transplantation.

CAN STEM CELLS BE USED TO FIND NEW MEDICINES?

One powerful application of human stem cells will be in the area of drug discovery and drug safety. The ability to make pure populations of different kinds of human cells suggests that different chemical compounds can be quickly tested for efficacy and safety by exposing the cells to these chemicals. Many such compounds are now tested on animals. While animal models are critically important, the use of human cells to test new medicines offers the opportunity to better determine drug effects in humans and refine both the safety and beneficial effects of new compounds. In recent years, many compounds have been withdrawn for drug toxicity, such as Grepafloxacin and Rofecoxib, leading to potential loss of huge investments in research and development. Industry collaboration in the field is critical to advancing toxicology work, as it can be used to expose drugs with dangerous side effects before they reach the market. So, while stem cells have yet to deliver on great promises for regenerative medicine, they are at least being used as important tools for research and drug testing (Figure 23.7).

POTENTIAL USES OF STEM CELLS

Stem cells have potential uses in many different areas of research and medicine, as described below. However, these applications are all likely to be several years away (Figure 23.8).

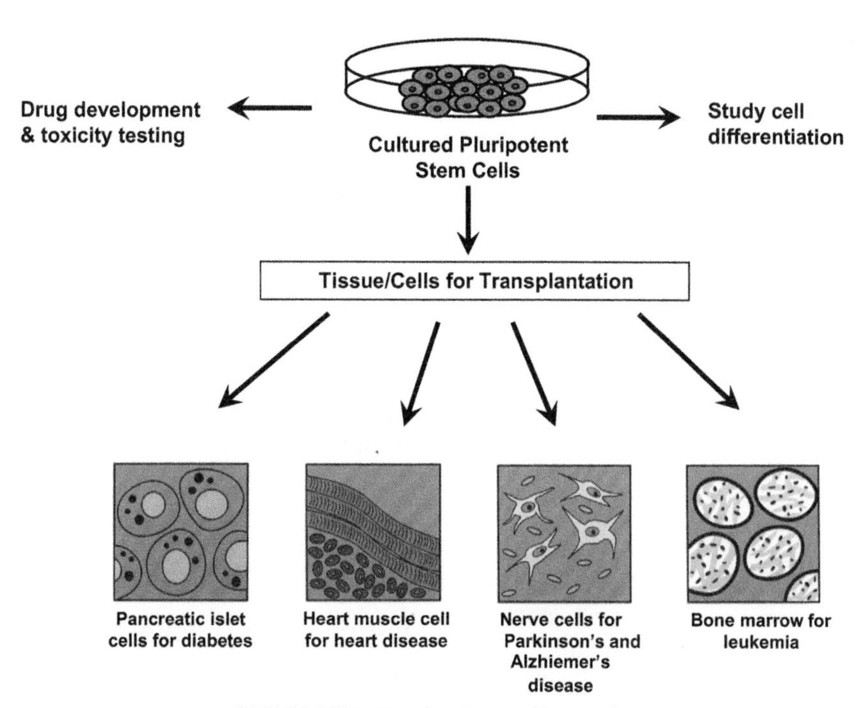

FIGURE 23.7 Promise of stem cell research.

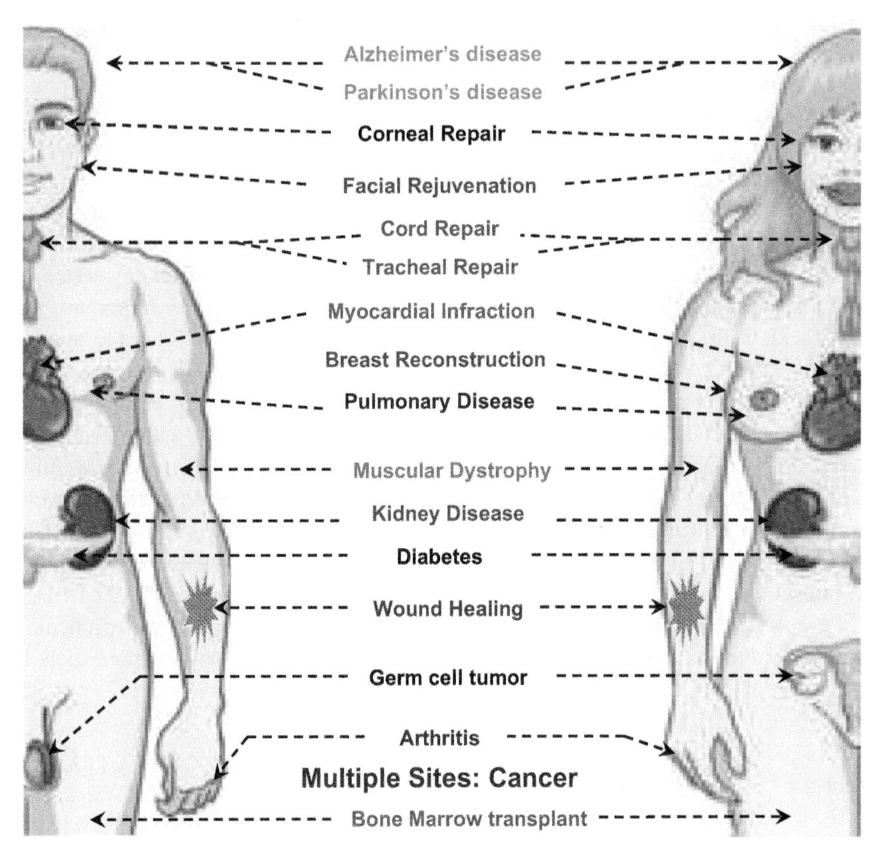

FIGURE 23.8 Potential uses of stem cells in disease.

Damaged Tissue Replacement

Human stem cells could be used in the generation of cells and tissues for cell-based therapies. This involves treating patients by transplanting specialized cells that have been grown from stem cells in the laboratory. Due to their ability to replace damaged cells in the body, stem cells could be used to treat a range of conditions, including heart failure, spinal injuries, diabetes, and Parkinson's disease. Scientists hope that transplantation and growth of appropriate stem cells in damaged tissue will regenerate the various cell types of that tissue. For example, hematopoietic stem cells could be transplanted into patients with leukemia to generate new blood cells, or neural stem cells could be able to regenerate nerve tissue damaged by spinal cord injury.

Human Development Studies

Stem cells could be used to study early events in human development to find out more about how cells differentiate and function. This could help researchers find out why some cells become cancerous, and how some genetic diseases develop. This knowledge could lead to clues about how these diseases can be prevented or cured.

New Drug Testing

Stem cells grown in the laboratory may be useful for testing drugs and chemicals before they are employed for clinical trials in humans. The cells could be directed to differentiate into the cell types that are important for screening that drug. These cells may be more likely to mimic the response of human tissue to the drug being tested than what is seen in animal models. This could make drug testing safer, cheaper, and more ethically acceptable to those who oppose the use of animals not only in pharmaceutical testing, but also all other animal experiments.

Screening Toxins

Stem cells may be useful for screening potential toxins in substances such as pesticides before they are used in the field.

Testing Gene Therapy Methods

Stem cells may prove useful during the development of new methods for gene therapy that may help people suffering from genetic diseases.

STEM CELL BANKING

Stem cell banks are increasingly seen as an essential resource for biological materials for both basic and translational research and medical treatment. Stem cell banks support transnational access to quality-controlled and ethically sourced stem cell lines from different origins. Though certain ethical and legal concerns exist with some types of stem cells, stem cell banking could do well to examine the approaches fostered by tissue banking. Like bio-banks, stem cell banks have the core objective to ensure the availability of quality and ethically approved cells or embryos for research and eventual therapies.

What is Stem Cell Banking?

A stem cell bank is a facility that stores normal stem cells without any genetic alteration or exposure to chemicals and drugs for the purpose of future use in products or medical needs. The cells are tested before, during, and upon completion of the cell banking process, or before being used for treatment. It is the most straightforward way of providing natural pluripotent stem cells either to the same individual, if it was stored for personal use, or as a pooled source for allogenic purposes.

Cord blood is one of the richest sources of hematopoietic stem cells (HSCs), is a valuable source, and is genetically unique to the baby and family. Cord blood banking is an easy, painless, and non-invasive procedure, and if ever a need arises in the future, these stem cells could be used as powerful therapies.

Why Bank Stem Cells?

Stem cell banking can serve as a lifeline that could help treat a long list of life-threatening diseases for many years to come. It gives an opportunity to have a powerful source of cells in the event of an emergent medical therapy with no immunological rejection risk. Stem cell banking is an important decision, as it is difficult to find a matching donor at the time of urgent transplant. To date, umbilical cord blood has been used in more than 12,000 transplantations for children and adults.

Stem cells have been labeled as an important biological resource that can be stored safely for future applications for diseases like Alzheimer's, diabetes, heart and liver disease, muscular dystrophy, Parkinson's disease, spinal cord injury, and stroke. Public cord blood banks store donated cord blood for potential use by transplant patients. Family cord blood banks store cord blood on behalf of the client. It offers a sense of security to the family, who can use it in the event a child or family member needs it for treatment. **Babycell,** a **Regenerative Medical Services Pvt. Ltd.**, is India's first umbilical cord blood bank for storing samples, and is an internationally accredited laboratory. The **U.S. National Stem Cell Bank (NSCB),** developed by the **WiCell Research Institute**, is a premier stem cell bank that provides hESCs to eligible scientists for use in NIH-funded research projects that meet standard quality control guidelines.

Present Scope and Future Possibilities of Stem Cell Banking

The easy availability of stem cells has allowed them to be used to treat as many as 130 different diseases, including leukemia, thalassemia, and neuro- and muscular-degenerative diseases. There are successful trials for cancer, diabetes, cardiac failure, multiple sclerosis, retinitis pigmentosa, spinal cord injuries, as well as for Alzheimer's disease and Parkinson's disease. There are more than 500 clinical trials underway at the present time. Recently, stem cell banking is also being attempted with menstrual blood cells. With the number of lifestyle disorders increasing, stem cell banking is an important option for needy patients.

Stem Cell Banks in India

Chennai-based **LifeCell** is the first Stem Cell Bank in India. **Cryocell**, a Florida-based stem cell banking facility in the U.S., has teamed up with **LifeCell. Reliance Lifesciences** in India has opened up collection centers in key locations in India, along with **Cryobank**, another pioneer in stem cell banking in India. The cost of stem cell banking differs from bank to bank.

WORLD WIDE WEB RESOURCES

Stem cell therapies offer enormous potential for the treatment of a wide range of diseases and injuries, including neurodegenerative diseases, cardiovascular disease, diabetes, arthritis, spinal cord injury, stroke, and burns. The regenerative and differentiation capacities and other potentials of stem cells make them attractive treatment modalities, but they also create challenges for the establishment of criteria to ensure development of safe and effective therapies. Although emerging regulatory procedures try to define these criteria, the absence of appropriate legislation and enforcement, desperation for cures, media hype, and the medical tourism industry exploit differences or gaps in the regulatory framework. There is a clear need for new research tools and strict regulatory guidelines to foster development of the types of stem cells that are potentially the most therapeutically useful and impactful.

Various national guidelines were made for the use of stem cell research around the globe to improve the understanding of human health and disease, and to evolve strategies to treat serious diseases. These guidelines address both ethical and scientific concerns to ensure responsible use of both good laboratory practices (GLP) and good clinical practice (GCP) in the area of stem cell research and therapy. However, stem cell therapies are not yet approved in India. Recently, the **Indian Council for Medical Research (ICMR)** and the **Department of Biotechnology** (DBT) formulated guidelines for stem cell research that exclude the use of embryonic cells for clinical treatment. While the regulatory system is still in development, there are currently different mechanisms to regulate clinical translation, variable criteria used by oversight bodies for protection of human subjects, and the ability to regulate practice of medicine separate from research. To unify and streamline the process, the National Apex Committee for Stem Cell Research and Therapy (NAC-SCRT) was created in 2009. NAC-SCRT is an interagency body created with the aim of effectively reviewing and monitoring stem cell research in India.

In the United States, the FDA's Center for Biologics Evaluation and Research, Office of Cellular, Tissue, and Gene Therapies (**CBER-OCTGT**) is charged with the oversight of stem cell products as well as other biological products. In 2005 it issued the "Tissue Rules" (21 CFR 1271), which form the basis for regulation of all human cells, tissues, and cellular and tissue-based products (HCT/Ps). Regarding stem cells specifically, in 2008 CBER-OCTGT generated guidance for ESC-based therapies as well as considerations for preclinical safety testing and patient monitoring.

Stem cell-based products present a unique regulatory challenge because standard pharmaceutical paradigms do not wholly apply, and accordingly, stem cell therapies do not neatly fit into current regulatory categories. As a result, regulatory requirements are often unclear in their application, and therefore create uncertainty. To better understand how stem cell therapies are faring in this regulatory environment, the International Society for Stem Cell Research (ISSCR), as well as the California Institute of Regenerative Medicine (CIRM), is working to establish a roadmap that protects patients and fosters the dramatic innovation in the stem cell field. These evolving frameworks will be informative for assessing risk tolerances in stem cell research to ensure development of safe and effective therapies for commercialization Table 23.6.

TABLE 23.6 Regulatory Framework of Stem Cell Research by Region

Country	Regulatory Framework	Related Websites
Argentina	Ministry of Health and Administration of Medications, Foods, and Medical Technology (ANMAT) created Instituto Nacional Central Unico Coordinador de Ablacion e Implante (INCUCAI) in 2007 as an agency relating to the use of human cells for implantation. In 2008, the Consorcio de Investigación en Celulas Madres (CICEMA) was created to foster ties between industry, academia, and the clinic.	www.msal.gov.ar www.anmat.gov.ar www.incucai.gov.ar www.cicema.org.ar/english/home_eng
China	The Minitsty of Health of the People's Republic of China (MOH) provides guidance on regulation of stem cell therapies.	www.moh.gov.cn; http://eng.sfda.gov.cn
European Union	The Advanced Therapy Medicinal Products (ATMP) Regulation was adopted by the European Medicines Agency (EMA) in 2007. Stem cell therapies are currently an area of discussion within the Committee for Advanced Therapies (CAT) created under the ATMP.	http://ec.europa.eu/health/human-use/advanced-therapies www.ema.europa.eu/ema
India	The Indian Council for Medical Research (ICMR) created the National Apex Committee for Stem Cell Research and Therapy (NAC-SCRT), which is an interagency body tasked with the oversight of the research and development of stem cell therapeutics.	www.icmr.nic.in
Japan	The Japanese Ministry of Health, Labor, and Welfare (MHLW) drafted guidelines for development and approval of novel medical products, including adult stem cells. Pluripotent cells are not yet addressed in the current version of the guidelines.	www.mhlw.go.jp/english
United States of America	The Center for Biologics Evaluation and Research (CBER), Office of Cellular, Tissue, and Gene Therapies (OCTGT) is charged with the oversight of stem cell products. In 2005, CBR-OCTGT issued guidance on cellular therapies in the "Tissue Rules" (CFR – 1271), and in 2008 provided a briefing pertaining to hESC-related therapies.	www.fda.gov/BiologicsBloodVaccines/CellularGeneTherapyProducts

Source: ISSCR Community Forum.

REFERENCES

Barria, E., Mikels, A., & Haas, M. (2004). Maintenance and self-renewal of long-term reconstituting hematopoietic stem cells supported by amniotic fluid. *Stem Cells Devoted, 13*, 548–562.

Blau, H. M., Brazelton, T. R., & Weimann, J. M. (2001). The evolving concept of a stem cell: entity or function? *Cell, 105*, 829–841.

Cairns, J. (1975). Mutation selection and the natural history of cancer. *Nature, 255*, 197–200.

Cairns, J. (2002). Somatic stem cells and the kinetics of mutagenesis and carcinogenesis. *Proceedings of the National Academy of Sciences of the United States of America, 99*, 10567–10570.

Conheim, J. (1875). Congenitales, quergestreiftes muskelsarkon der nireren. *Virchows Arch., 65*, 64.

Fauza, D. (2004). Amniotic fluid and placental stem cells. *Best Practice and Research. Clinical Obstetrics and Gynaecology, 18*, 877–891.

Gang, E. J., Jeong, J. A., Hong, S. H., Hwang, S. H., Kim, S. W., Yang, I. H., Ahn, C., Han, H., & Kim, H. (2004). Skeletal myogenic differentiation of mesenchymal stem cells isolated from human umbilical cord blood. *Stem Cells, 22*, 617–624.

Goodell, M. A., Brose, K., Paradis, G., Conner, A. S., & Mulligan, R. C. (1996). Isolation and functional properties of murine hematopoietic stem cells that are replicating in vivo. *The Journal of Experimental Medicine, 183*, 1797–1806.

Gurdon, J. B. (1962). The transplantation of nuclei between two species of Xenopus. *Developmental biology, 5*, 68–83.

Hadjantonakis, A., & Papaioannou, V. (2001). The stem cells of early embryos. *Differentiation, 68*, 159–166.

Holm, S. (2002). Going to the roots of the stem cell controversy. *Bioethics, 16*, 493–507.

Kashofer, K., & Bonnet, D. (2005). Gene therapy progress and prospects: stem cell plasticity. *Gene therapy, 12*, 1229–1234.

Makino, S. (1956). Further evidence favoring the concept of the stem cell in ascites tumors of rats. *Annals of the New York Academy of Sciences, 63*, 818–830.

Michalczyk, K., & Ziman, M. (2005). Nestin structure and predicted function in cellular cytoskeletal organisation. *Histology and Histopathology, 20*, 665–671.

Mizrak, D., Brittan, M., & Alison, M. R. (2008). CD133: molecule of the moment. *The Journal of Pathology, 214*, 3–9.

O'Donoghue, K., & Fisk, N. M. (2004). Fetal stem cells. *Best Practice & Research. Clinical Obstetrics & Gynaecology, 18*, 853–875.

Osawa, M., Hanada, K., Hamada, H., & Nakauchi, H. (1996). Long-term lymphohematopoietic reconstitution by a single CD34-low/negative hematopoietic stem cell. *Science, 273*, 242–245.

Potten, C. S., Owen, G., & Booth, D. (2002). Intestinal stem cells protect their genome by selective segregation of template DNA strands. *Journal of Cell Science, 115*, 2381–2388.

Rao, M. S., & Mattson, M. P. (2001). Stem cells and aging: expanding the possibilities. *Mechanisms of Ageing and Development, 122*, 713–734.

Recamier, J. C. A. (1829). *Recherches sur the Traitement du Cancer: par la Compression Methodique Simple ou Combinee, et sur l'Histoire General de la Meme Maladie*. Paris: Gabon.

Reynolds, B. A., & Weiss, S. (1992). Generation of neurons and astrocytes from isolated cells of the adult mammalian central nervous system. *Science, 255*, 1707–1710.

Rollini, P., Kaiser, S., Faes-van't Hull, E., Kapp, U., & Leyvraz, S. (2004). Long-term expansion of transplantable human fetal liver hematopoietic stem cells. *Blood, 103*, 1166–1170.

Shamblott, M. J., Axelman, J., Wang, S., Bugg, E. M., Littlefield, J. W., Donovan, P. J., Blumenthal, P. D., Huggins, G. R., & Gearhart, J. D. (1998). Derivation of pluripotent stem cells from cultured human primordial germ cells. *Proceedings of the National Academy of Sciences of the United States of America, 95*, 13726–13731.

Sharp, J., & Keirstead, H. S. (2007). Therapeutic applications of oligodendrocyte precursors derived from human embryonic stem cells. *Current Opinion in Biotechnology, 18*, 434–440.

Sherr, C. J. (2000). Cell cycle control and cancer. *Harvey Lectures, 96*, 73–92.

Simmons, P. J., & Torok-Storb, B. (1991). Identification of stromal cell precursors in human bone marrow by a novel monoclonal antibody, STRO-1. *Blood, 78*, 55–62.

Singec, I., Jandial, R., Crain, A., Nikkhah, G., & Snyder, E. Y. (2007). The leading edge of stem cell therapeutics. *Annual Review of Medicine, 58*, 313–328.

Takahashi, K., & Yamanaka, S. (2006). Induction of pluripotent stem cells from mouse embryonic and adult fibroblast cultures by defined factors. *Cell, 126*, 663–676.

Takahashi, K., Tanabe, K., Ohnuki, M., Narita, M., Ichisaka, T., Tomoda, K., & Yamanaka, S. (2007). Induction of pluripotent stem cells from adult human fibroblasts by defined factors. *Cell, 131*(5), 861–872.

Thomson, J. A., Itskovitz-Eldor, J., Shapiro, S. S., Waknitz, M. A., Swiergiel, J. J., Marshall, V. S., & Jones, J. M. (1998). Embryonic stem cell lines derived from human blastocysts. *Science, 282*, 1145–1147.

Tzukerman, M., Rosenberg, T., Ravel, Y., Reiter, I., Coleman, R., & Skorecki, K. (2003). An experimental platform for studying growth and invasiveness of tumor cells within teratomas derived from human embryonic stem cells. *Proceedings of the National Academy of Sciences of the United States of America, 100*, 13507–13512.

Whetton, A. D., & Graham, G. J. (1999). Homing and mobilization in the stem cell niche. *Trends in Cell Biology, 9*, 233–238.

Yu, J., Vodyanik, M. A., Smuga-Otto, K., Antosiewicz-Bourget, J., Frane, J. L., Tian, S., Nie, J., Jonsdottir, G. A., Ruotti, V., Stewart, R., Slukvin, , II, & Thomson, J. A. (2007). Induced pluripotent stem cell lines derived from human somatic cells. *Science, 318*, 1917–1920.

FURTHER READING

Lanza, R., Gearhart, J., Hogan, B., Melton, D., Pedersen, R., Thomas, E. D., Thomson, J., & West, M. (2006). *Essential of Stem Cell Biology*. London: Elsevier Academy Press.

Loring, J. F., & Peterson, S. E. (2012). *Human Stem Cell Manual: A laboratory Guide* (2 nd ed.). San Diego, CA: Academy Press.

Potten, C. S. (1997). *Stem Cells*. London: Elsevier Academy Press.

Sell, S. (2003). *Stem Cells Handbook*. New Jersey: Humana Press.

Vemuri, M. C. (2007). *Stem Cell Assays*. New Jersey: Humana Press.

GLOSSARY

Adult Stem Cells Stem cells that are present in a tissue or organ after birth; they are multipotent. (e.g. blood stem cells in bone marrow).

Allogenic Foreign to the body and during transplantation, the transplanted tissues or cells are derived from a donor who is genetically different from the recipient.

Asymmetric Division Generation of distinct cells in progeny from a single mitosis. Such division may position daughter cells in a different microenvironment or segregate into only one daughter cell.

Autologous Transplanted tissues or cells are derived from the same recipient; therefore, it is called autologous transplantation.

Blastema A population of mesenchymal stem cells that can regenerate a new limb at the amputation/cut site of an animal's limb.

Blastocyst An early embryo of a mammal (before implantation) that consists of an outer trophoblast and an inner mass of cells that form the embryonic stem cells.

Cancer Stem Cell (CSC) A very small population of slow-growing and self-renewing cells responsible for sustaining a cancer and later relapse to form the bulk of the cancer.

Cancer-Initiating Cell The term encompasses both cells of cancer origin and cancer stem cells.

Cell Line A population of cells maintained in culture that has been modified to have the capacity for indefinite cell division.

Cell Replacement Therapy Reconstitution of tissue by transplantation of functional stem-cell progeny.

Chimera Organism made up of cell populations from more than one individual.

Cleavage Division Cell division that is not associated with cell growth; thus, when one cell divides into two, it becomes half of its original size.

Clinical Trial Clinical research in which a new drug/device/method is tested in a group of patients following GMP, GCP, and ethical issues.

Clonal Analysis Study of properties of single cells, which is important for demonstration of self-renewal and potency of stem cells.

Clone An individual genetically identical to another in which there is a common ancestor or parent (e.g. identical twins).

Cord Blood The blood derived from the umbilical cord and placenta that supplies nutrients from mother to the developing fetus. This is a rich source of blood stem cells that can be easily stored for future use.

Culture Media A special fluid used to culture cells in the laboratory that contains all the nutrients cells require for growth.

Differentiation A process by which a cell changes from one type into another, loses its precursor properties, and acquires new functions of a specialized cell.

Ectoderm One of the three cell lineages in the early embryo. The ectoderm forms the outer cell layers and later develops the skin and nerve cells.

Embryo An early stage of development following fertilization of an egg with a sperm.

Embryonic Germ (EG) Stem cells that are derived from primordial germ cells and are pluripotent.

Embryonic Stem (ES) Cells These cells are pluripotent stem cells derived from the inner cell mass of a blastocyst.

Endoderm One of the three cell lineages of the early embryo that forms the innermost cells that give rise to the intestine, liver, pancreas, and lungs.

Epigenetic Regulation This is due to changes in gene function in the absence of any gene mutation. During the early stage of differentiation, certain genes are repressed or de-repressed through modification of DNA or protein, leaving only the genes that are required for the specialized functions of a differentiated cell. Epigenetic regulation maybe disturbed after artificial reprogramming in iPS cells.

Ex Vivo Outside the body (i.e. in the laboratory).

Extra-Embryonic Tissue These tissues do not form the embryo proper, but give rise to the embryonic part such as the placenta, umbilical cord, and membrane surrounding the fetus.

Feeder Cell Layer A cell layer that "feeds" other cells. Generally used for growing stem cells that remain undifferentiated. These cells are treated with X-irradiation or some drug to arrest mitosis so that they do not overgrow, but can provide nutrients for the growth of stem cells.

Flow Cytometer (Fluorescent-Activated Cell Sorter, FACS) An instrument that can count and sort fluorescence-labeled cells from a cell suspension on the basis of a cell surface marker.

Fetus Unborn offspring of mammals.

Founder/Ancestor/Precursor Cell These cells are without self-renewal ability, and contribute to tissue formation.

Lineage Priming Promiscuous expression in stem cells of genes associated with differentiation programs.

Multipotent Stem cells that can form multiple lineages and constitute an entire tissue or tissues/organ (e.g. hematopoietic stem cells).

Niche Cellular microenvironment that provides support and stimuli essential to sustaining self-renewal.

Oligopotent Stem cells that can form two or more lineages within a tissue.

Plasticity It is hypothesized that tissue stem cells may broaden potency in response to physiological demands or insults.

Pluripotent Ability to form all the body's cell lineages, including germ cells (e.g. embryonic stem cells).

Potency The extent of commitment in a cell, particularly in a stem cell.

Progenitor Cell Any dividing cell that has the capacity to differentiate.

Regenerative Medicine Reconstruction of diseased or injured tissue by activation of proliferation of endogenous cells or by cell transplantation.

Reprogramming Increase in cell potency that may occur naturally in regenerative organisms; also called de-differentiation.

Self-Renewal Cell division that generates at least one daughter cell to the mother cell with a capacity to differentiate; it is a property of stem cells.

Stem Cell Cell that can continuously produces unaltered daughters, and also has the ability to produce a lineage of cells and differentiate into specialized cells.

Stem-Cell Homeostasis Maintenance of a tissue stem-cell pool throughout the life of an organism with balanced symmetric or asymmetric self-renewal.

Stemness Hypothesis that different stem cells are regulated by common genes and mechanisms.

Unipotent Stem cells that can form a single lineage (e.g. spermatogonial stem cells).

ABBREVIATIONS

ABC ATP-Binding Cassette

ABCG2 ATP-Binding Cassette Sub-Family Member 2

Abs Antibodies

AFSC Amniotic Fluid-Derived Stem Cells
ASC Adult Stem Cells
ATP Adenosine Triphosphate
BCRP1 Breakpoint Cluster Region Pseudogene 1
BDNF Brain-Derived Neurotrophic Factor
BrdU 5-Bromo-2-Deoxyuridine
CBER-OCTGT Centre for Biologics Evaluation and Research, Office of Cellular Tissue, and Gene Therapies
CD Cluster of Differentiation
CIRM California Institute of Regenerative Medicine
c-Myc Cellular Myc Gene (Derived from Myelocytomatosis)
CNS Central Nervous System
DBT Department of Biotechnology
DCGI Drug Controller General of India
DNA Deoxyribonucleic Acid
ESC Embryonic Stem Cells
FACS Fluorescence-Activated Cell Sorting
GCP Good Clinical Practices
GLP Good Laboratory Practices
GSC Germ Stem Cells
HLA Human Leukocyte Antigen
HSC Hematopoietic Stem Cells
hES Cells Human Embryonic Stem Cells
hEG Cells Human Embryonic Germ Cells
ICM Inner Cell Mass
ICMR Indian Council for Medical Research
iPSC Induced Pluripotent Stem Cells
ISSCR International Society for Stem Cell Research
IVF *In Vitro* Fertilization
Klf-4 Krueppel-Like Factor 4
LNGFR Low-Affinity Nerve Growth Factor Receptor
MDR Multi-Drug Resistance
NAC-SCRT National Apex Committee for Stem Cell Research and Therapy
NCSC Neural Crest Stem Cells
NGF Nerve Growth Factor
NSC Neuronal Stem Cells
NSCB National Stem Cell Bank
Oct-4 Octamer-Binding Transcription Factor 4
PSC Pluripotent Stem Cell
p75NTR p75-Neurotrophin Receptor
POU5F1 POU Domain Class 5 Transcription Factor 1
PSA-NCAM Polysialic Acid-Neural Cell Adhesion Molecule
SCNT Somatic Cell Nuclear Transfer
SOX-2 SRY (Sex Determining Region Y) Box 2
SOP Standard Operative Procedures
SP Side Population
SSEA Stage-Specific Embryonic Antigens
TNF Tumor Necrosis Factor
Trk Receptor Tyrosine Kinase Receptor
UCBSC Umbilical Cord Blood Stem Cells

LONG ANSWER QUESTIONS

1. What is a stem cell? Describe in detail, how are the stem cells identified and isolated for research and clinical use?
2. What are the core genetic and epigenetic regulators of stem cells *in vivo*?
3. What are the extrinisic and environmental factors that influence stem cell renewal and differentiation, and how are they maintained?
4. How are iPSCs generated? How does one differentiate them from ESCs? Enumerate the clinical utility of iPSCs.
5. Do stem cells age? If so, how and what cause them to age? How does one justify this for cancer stem cells?

SHORT ANSWER QUESTIONS

1. What is a stem cell line?
2. What is somatic cell nuclear transfer (SCNT)?
3. What is a cancer stem cell?
4. What properties of stem cells have promise for medical treatments?
5. Differentiate between cancer cells and stem cells.
6. What are the advantages and disadvantages of embryonic and adult stem cells?

ANSWERS TO SHORT ANSWER QUESTIONS

1. It's an ongoing, living colony of stem cells in a laboratory from which cells can be obtained for research or other uses. Sometimes these are called "immortal" cell lines. Embryonic stem cells are said to be easier to grow in a stem cell line, but they also may develop genetic abnormalities and teratocarinoma.
2. Somatic cell nuclear transfer is a cloning technique. It is a way to create embryonic stem cells with the same genetic blueprint as a patient so that the cells will not be rejected by the patient's immune system. The end result of this procedure is a single cell that can grow into an embryo in the same way as an egg fertilized by a sperm normally grows into an embryo. Stem cells can be harvested from cloned embryos in the same way that they are harvested from the extra embryos from *in vitro* fertilization clinics. SCNT is also a critical tool to create disease-specific cell lines that can be studied *in vitro* to learn how complex diseases develop and to understand how certain drugs affect the progression of that disease. The technique has been proven to work in animal cells, but has not yet been proven in human cells.
3. Cancer stem cells (CSCs) are cancer cells (found within tumors or hematological cancers) that possess the capacity to self-renew and to cause the heterogeneous lineages of cancer cells that comprise the

tumor (and are therefore tumorigenic). They show drug resistance, are responsible for tumor recurrence, and are refractory to programmed cell death. The existence of cancer stem cells was first reported by Recamier in 1829.

4. The power of stem cells can be harnessed to cure degenerative diseases such as Parkinson's disease and juvenile diabetes, and to repair injuries such as trauma to the spinal cord. These diseases are often listed as targets for future stem cell therapies, but may also help in understanding those diseases.
5. Cancer cells show uncontrolled proliferation combined with increased genetic instability (reflecting clonal heterogeneity), whereas stem cells are defined as cells that have the ability to perpetuate themselves through self-renewal and to generate mature cells of a particular tissue through differentiation. Stem cells are slow-dividing quiescent cells that are tightly regulated in their niche.
6. See Table 23.7

TABLE 23.7 Advantages and Disadvantages of Embryonic and Adult Stem Cells

Embryonic Stem Cells	Adult Stem Cells
Derived from embryos and develop into all cell types of the body.	Found in organs all throughout the human body.
Primary purpose is to facilitate growth and development.	Primary purpose is to maintain and repair the body.
Difficult to differentiate uniformly and homogeneously into a target tissue.	Limited in number, finite in nature; may not live as long as ES cells in culture.
Immunogenic in nature after transplantation.	Not immunogenic and are easy to harvest.
Capable of forming tumors; promote tumor formation (tumorigenic).	Non-tumorigenic in nature; difficult to reprogram to other tissue types.

Chapter 24

Role of Cytogenetics and Molecular Genetics in Human Health and Medicine

Madhumita Roy Chowdhury* and Sudhisha Dubey†

*Genetic Unit, Department of Pediatrics, All India Institute of Medical Sciences, New Delhi, India, †Department of Genetic Medicine, Sir Ganga Ram Hospital, New Delhi, India

Chapter Outline

SUMMARY

The complexities of cytogenetics and molecular genetics have been described in easy-to-understand language. The basics of cytogenetics and molecular genetics have been explained, and examples have been used to demonstrate a correlation in a number of human diseases.

WHAT YOU CAN EXPECT TO KNOW

A chromosome is the basic unit for studies related to cytogenetics and molecular genetics. A description of chromosomes, their structure, and role in inheritance, along with numerical and structural abnormalities in humans has been provided. Different kinds of chromosomal structural abnormalities leading to chromosomal disorders have been described in the cytogenetics section. The molecular genetics section describes the structure of DNA and gene and how changes in the DNA sequences lead to manifestation of genetic disorders. Both the sections outline the basics and give an overview of how this basic information is applied in clinical perspective in the field of medical genetics. It also gives an overview of the laboratory procedures or different

Animal Biotechnology. http://dx.doi.org/10.1016/B978-0-12-416002-6.00024-9

tools applied for the detection and diagnosis of different genetic disorders. The main categories of genetic disorders – single gene, chromosomal, multigenic, and mitochondrial disorders – have been briefly explained to expose students to different patterns of inheritance and their genetic mechanism in disease, with examples.

HISTORY AND METHODS

INTRODUCTION

To understand the role of genetics in medicine, one first must know how it plays a significant role in cell division, growth, and cell differentiation, as the developmental program is implemented gradually from a zygotic stage to an adult human being. This entire process of development involves a vast array of cellular, biochemical, and molecular interactions.

The basic and fundamental laws of genetics – which revolutionized our understanding of genetics – were illustrated by a set of simple experiments in 1865 by an Austrian monk named Gregor Mendel. His breeding experiments on garden peas helped us to understand the inheritance pattern and the way different characteristics are passed on from one generation to the next. However, his work was not recognized until 16 years after his death. For his major contributions, he is known as "Father of Genetics." In 1909, Archibald Garrod illustrated the genetic basis of four congenital metabolic diseases (albinism, alkaptonuria, cysteinuria and pentosuria) and also showed that they are inherited as autosomal recessive forms.

In 1910, genetics became an independent scientific field with the fruit fly study (*Drosophila melanogaster*) by Thomas H. Morgan. Subsequently, many years of studies on *Drosophila melanogaster* by other scientists showed that genes are arranged linearly on chromosomes; this led to the theory of inheritance. DeVries in 1901 recognized that genes can change, and he coined the term "mutation" for this change. In 1927, H.J. Muller demonstrated that these mutations can be introduced by roentgen rays. Until this time, knowledge about the chemical nature of genes was not available, which had led to misconceptions about the possibility of eliminating of "bad genes" from human populations (eugenics). Close relationship between genetics and biochemistry became evident when Beadle and Tatum in 1941 observed one gene is responsible for the formation of one enzyme in fungus.

In 1928, Frederick Griffith showed that pneumococcal bacteria are capable of transferring genetic information through a process known as transformation. Later on, in 1944 Oswald T. Avery and his coworkers demonstrated that the transforming material was in fact DNA, which carried the genetic information in bacteria. A major breakthrough took place in 1953 when James D. Watson and Francis H. Crick proposed the structure of DNA as double helix (Turnpenny and Ellard, 2011).

With the development of knowledge of classical and molecular genetics, the need to explore the medical aspects of human genetics was felt. In 1949, Neel recognized that sickle cell anemia was as a hereditary disease. In the same year, Singer and Wells identified that sickle cell anemia is caused by an alteration in the normal hemoglobin (Pauling et al., 1949). The American Society of Human Genetics and the first Journal of Human Genetics (i.e. American Journal of Human Genetics) were established in the same year, and even the first textbook of human genetics appeared in 1949 (Curt Stern, Principles of Human Genetics). In 1959, 47 chromosomes were observed for the first time in a patient with Down Syndrome. Following this, molecular techniques were developed to identify genes and disease-causing mutations. Great leaps in this field have not only changed the diagnosis and management of single gene disorders, but have also provided new molecular approaches to the genetic diseases. The field of medical genetics began to develop at the end of nineteenth century and the entire human genome was sequenced in 2000. Knowledge of the genome and genetics has increased tremendously in the last few decades. This revolution in knowledge helped us to understand importance of genetics in almost every area of medicine. Genetics is now not restricted to single gene disorders, but has made remarkable contributions to other common multifactorial diseases like hypertension, psychiatric illness, and cardiovascular diseases. Thus, it is essential to introduce to students the basic concepts of cytogenetic and molecular genetics to understand the latest developments in medicine.

Most diseases have a probable genetic and environmental basis. The genetic component may be the major one, but it is notable that in up to 50% of all cases, no clear explanation can be established. The integration of cytogenetics and molecular technologies has thus opened up an exciting field in modern biology.

CYTOGENETICS: AN OVERVIEW

Cytogenetics, the study of chromosomes, was originated more than a century ago. However, it is in the last few decades that studies on human chromosomes have become a major field in the biomedical sciences. Flemming published the first drawn illustration of a chromosome in 1882 and Waldeyer used the term chromosome (Greek word for "stained body") for the first time in 1888. It took many years to identify the exact number of chromosomes in a normal diploid human cell (Gartler, 2006). In 1912, Hans von Winiwarter reported 47 chromosomes in spermatogonia and 48 in oogonia and this count remained as 48 chromosomes until Joe Hin Tjio and Albert Levan discovered the correct human chromosome count. This revolutionary finding of

Tjio's was published (with Levan as his co-author) in the Scandinavian journal *Hereditas* on January 26, 1956 (Tjio and Levan, 1956). Chromosome banding methods, for example, are today's vital tools in clinical genetics. In 1956 researchers in Sweden used aceto-orcein dye for direct chromosome staining and could count 46 chromosomes in human cells, but could only distinguish chromosomes according to their sizes and centromere positions (Trask, 2002). Based on these two criteria, human chromosomes were classified into seven groups: A through G. In the late 1960s when quinacrine mustard (QM) was used to stain human chromosomes, accurate chromosome identification could be done under a fluorescence microscope into bright and dark regions called Q-bands. Since then, numerous banding techniques have been developed, of which G-banding (Giemsa banding) is the most widely used technique for chromosome analysis. Banding techniques are extremely useful for the detection of structural changes associated with chromosomal disorders. The microscopic analysis of chromosome structure and behavior in mitosis and meiosis revealed changes in the chromosome sets of plants, animals, and man. These changes were referred as chromosomal aberrations, and research was carried out on corn by the American cytogeneticist B. McClintock in the period 1929–38.

CHROMOSOME MORPHOLOGY AND CLASSIFICATION

There are two kinds of cell division: mitosis and meiosis. Mitosis is somatic cell division responsible for the growth, proliferation and tissue differentiation of the body, whereas meiosis is responsible for the production of gametes. Because mitotic cells are easy to obtain, morphological studies are generally based on mitotic metaphase chromosomes. Chromosomes are not visible under a light microscope in non-dividing (interphase) cells. As the cell begins to divide, the threadlike chromatin material in the nucleus begins to condense; in the metaphase stage, the chromosomes are best recognizable.

A chromosome consists of two arms separated by a primary constriction called a **centromere**. The short chromosome arm is designated as p (petite) and the long arm as q (one letter after p) (according to Paris nomenclature). A centromere consists of several hundred kilobases of repetitive DNA and is responsible for movement of chromosomes during cell division.

Each chromosome consists of two identical strands known as **chromatids** or **sister chromatids,** which are visible after the s (synthetic) phase of cell cycle. Each of the two sister-chromatids contains a highly coiled double helix of DNA and is joined at centromere.

The tip of each chromosome is called telomere. Telomeres are highly conserved and consist of repeated DNA sequence TTAGGG. Telomeres maintain the structural integrity of chromosomes and help to distinguish the chromosome ends from broken DNA by sealing the ends of chromosomes, just like caps. Telomeres get shorter with each cell division and when they get too short, the cell no longer can divide and dies. This process has been associated with aging and cancers.

Chromosomes are arranged and are numbered according to their size and the position of their centromeres. A chromosome with the centromere at or near the middle is known as **metacentric**. A **submetacentric** chromosome has a centromere somewhat displaced from the middle point. **Acrocentric** chromosomes have centromeres very near to one end. **Telocentric** chromosomes, which are absent in human cells, have their centromeres at the very tip of one end (Figure 24.1).

The number of chromosomes in the somatic cell is diploid and is designated by the symbol 2N. The gametes, have the haploid number N. In humans the diploid number is 46, inheriting 23 from each parent through the sperm or egg. Homologous chromosomes form a pair with one constituent from each parent. Thus, there are 23 pairs of chromosomes in human cells. Of these, 22 pairs are known as **autosomes** and the remaining chromosome pair consists of the **sex chromosomes**, and is directly involved in sex determination. In females, the two sex chromosomes are identical (XX), whereas in males the two sex chromosomes are not

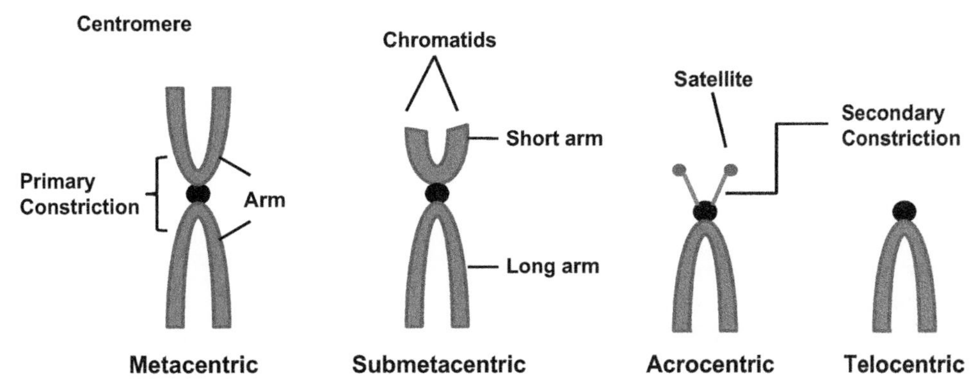

FIGURE 24.1 Diagrammatic representation showing chromosome classification according to centromere position and size.

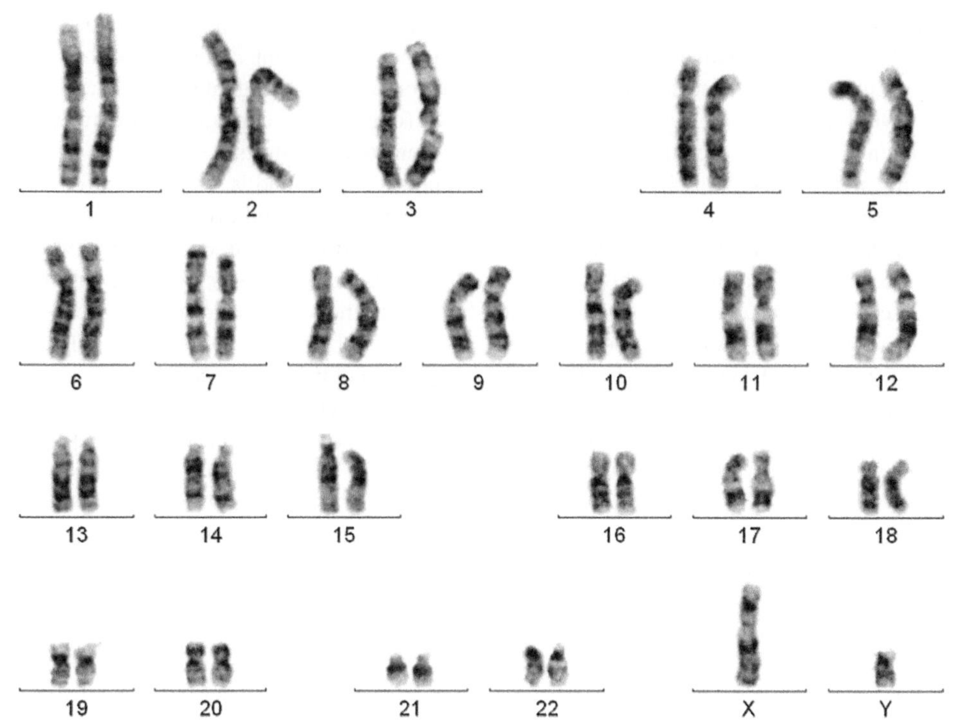

FIGURE 24.2 Karyotype showing a normal male chromosomal constitution. *(Courtesy: Cytogenetic Lab Genetic Unit, All India Institute of Medical Sciences (AIIMS), New Delhi, India.)*

identical (XY). The Y chromosome is smaller than the X chromosome (Verma and Babu, 2002).

The term **karyotype** refers to a display of the chromosomes of a cell by lining them up, beginning with the largest chromosome and with the short arm oriented toward the top (Figure 24.2). In humans, seven (A–G) groups of autosomes are recognized. Sex chromosomes (X,Y) are placed at the end. A diagram of the karyotype based on chromosome measurements in many cells is called an **ideogram.** Chromosome numbers 1–3 (A group) are metacentric, numbers 4–5 (B group) and 6–12 (C group) are submetacentric and 13–15 (D group) and 21–22 (G group) are acrocentric and have satellites and chromosomes 16–18 (E group) are again metacentric (Table 24.1).

TABLE 24.1 Grouping of Chromosomes Based on Descending Order of Size and Position of the Centromere

Sr. No	Group	Chromosome	Description
1	Group A	1–3	Large metacentric chromosomes easily identified based on their size and centromere position.
2	Group B	4–5	Large submetacentric chromosomes.
3	Group C	6–12, X	Medium-sized submetacentric chromosomes.
4	Group D	13–15	Medium-sized acrocentric chromosomes with satellites.
5	Group E	16–18	Moderately short metacentric or submetacentric chromosomes.
6	Group F	19–20	Short metacentric chromosomes.
7	Group G	21–22, Y	Short acrocentric chromosomes with satellites, Y chromosome does not bear satellites.

Nomenclature

Human chromosome nomenclature systems were developed to prevent confusion in reporting cytogenetic results. In 1995, the International System for Human Cytogenetic Nomenclature was recommended by the International Standing Committee on Human Cytogenetics Nomenclature.

A karyotype description is written from left to right without leaving any space, separating each item with a comma unless otherwise specified.The karyotype formula begins with the total number of chromosomes in the cell, followed by the notation of the sex chromosomes (first the Xs and then the Ys). A normal male is designated

TABLE 24.2 List of Some of the Commonly Used Symbols for Chromosome Nomenclature with Descriptions

Sr. No.	Chromosome Nomenclature	Symbols	Examples	Description
1	The sex chromosomes	X,Y	46,XX/ 46,XY	Normal female/normal male.
2	Deletion	del	46,XX, del (6)(p2)	Deletion of p arm of 6 at band level 2.
3	Duplication	dup	46,XY dup (6)(q21–q24)	Duplication in the long arm of chromosome 6 involving breakpoint from 6q21–6q24.
4	Inversion	inv	46,XY, inv(6)(q21–q24)	Inversion in the long arm of chromosome 6 involving breakpoint from 6q21–6q24.
5	Short arm of chromosome	p	p10	Short arm of chromosome 10.
6	Long arm of chromosome	q	q10	Long arm of chromosome 10.
7	Satellite	s	15ps	Satellites on the short arm of chromosome 15.
8	Translocation	t	46,XY,t(14;21)	Translocation involving chromosome 14 and 21.
9	Addition of the whole chromosome	(+)	47,XY,+21	Trisomy 21 (three copies of chromosome 21).
10	Deletion of the whole chromosome	(–)	45,XX,–22	Loss of chromosome 22.

as"46,XY" and a normal female as "46,XX". An extra or a missing chromosome is designated with a "+" or "–" sign, respectively, before the number of chromosome. Thus, a male with trisomy (extra chromosome) for chromosome 21 is 47,XY,+21, and a female with a monosomy for 22 is 45,XX,-22. Addition or deletion of a chromosome segment is denoted with a plus "+" or minus "–" sign after the symbol of the chromosome arm, respectively.

For example, a female having a deletion of short arm of chromosome 5 (cri du chat syndrome) is written as 46,XX,5p–. The formula for a male with a translocation between chromosome 14 and 21 is 46,XY,t (Southern and Mellor, 1975; Watson and Crick, 1953).

While describing a human chromosome complement and its abnormalities /aberrations, sex chromosome aberrations are specified first (X chromosome abnormalities are presented before those involving Y), followed by abnormalities of the autosomes listed in numerical order, irrespective of aberration type. Table 24.2 lists some of the commonly used symbols for chromosome nomenclature.

CHROMOSOMAL DISORDERS

These account for approximately 6% of all recognized congenital abnormalities, and these can be divided into numerical and structural aneuploidy, with a third category consisting of different chromosome constitutions in two or more cell lines. Aneuploidy is a term that is used when the chromosome number in the cells is not the typical number that it should be for a particular species. This gives rise to chromosome abnormalities such as an extra chromosome or loss of one or more chromosomes, and leads to a particular chromosomal disorder.

The presence of an extra chromosome is referred to as trisomy. Of the 22 autosomes in man, only three occur regularly as trisomies in live-born infants: trisomy 21 (Down syndrome), trisomy 18 (Edward syndrome), and trisomy 13 (Patau syndrome). Other trisomies are not observed in live-born infants because they are lethal in early embryonic life, and not compatible with life at birth. The presence of an additional sex chromosome (X or Y) has only mild phenotypic effects. Trisomy 21 is usually caused by failure of separation of one of the pairs of homologous chromosomes during anaphase of maternal meiosis I. This failure of the bivalent to separate is called "non-disjunction." Less often, trisomy can be due to non-disjunction occurring during meiosis II when a pair of sister chromatids fail to separate.

Down syndrome is the most common chromosomal disorder having an incidence of 1 in 800 to 1000 live births. It is also one of the most common causes of mental retardation. These patients have typical clinical features that are often easily identified even in a newborn child. Some of the distinct features of Down syndrome can be seen in Figure 24.3a, and different characteristics are mentioned below:

1. Epicanthic folds (a fold of skin of the upper eyelid that partially covers the inner corner of the eye).
2. Flat facial profile/flat nasal bridge.
3. Folded or dysplastic ears.
4. Low-set, small ears.
5. Brachycephalic (a medical term used to define a condition where the head is disproportionately wide giving it a short and broad appearance).
6. Open mouth.
7. Protruding tongue.

8. Furrowed tongue.
9. Short neck.
10. Excessive skin at nape of the neck.

The absence of a single chromosome is referred to as monosomy. Monosomy for an autosome is almost always incompatible with survival to term. Monosomy X (karyotype 45,XO) is another chromosomal disorder, also known as Turner syndrome, representing about 5% of conceptions. It affects about 1 in every 2500 females. The most common features of Turner syndrome can be observed in Figure 24.3b, while the characteristic features has been mentioned below:

1. Short stature.
2. Swelling of the hands and feet.
3. Broad chest and widely spaced nipples.
4. Low hairline.
5. Low-set ears.
6. Webbed necks (extra folds of skin on the neck).
7. Girls with Turner syndrome have gonadal dysfunction (non-working ovaries), which results in amenorrhea (absence of menstrual cycle) and sterility.

Polyploid cells contain multiples of the haploid number of chromosomes such as 69 (triploidy), or 92 (tetraploidy). It has been observed that triploidies are often the cause of spontaneous miscarriages.

Structural Abnormalities

Structural chromosome rearrangements result from chromosome breakage with subsequent re-union in a different configuration. They can be balanced or unbalanced. In balanced rearrangements, the chromosome complement is complete with no loss or gain of genetic material. Thus such rearrangements are generally harmless with the exception of rare cases in which one of the breakpoints

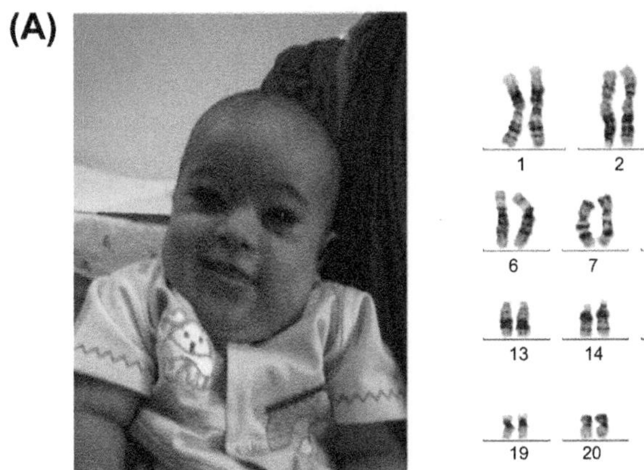

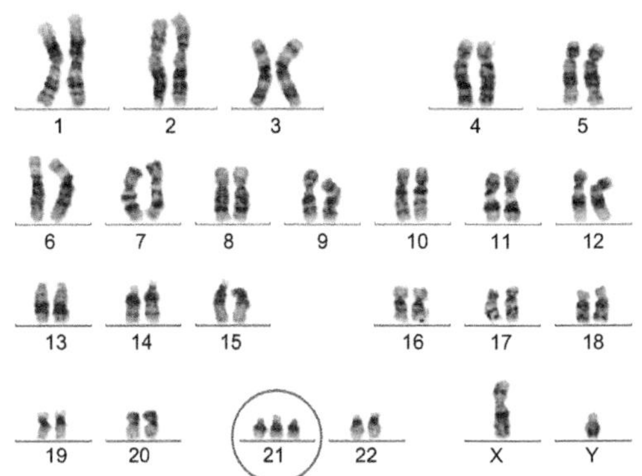

FIGURE 24.3A A Down syndrome patient showing karyotype 47,XY,+21. Courtesy: Cytogenetic Lab Genetic Unit, All India Institute of Medical Sciences (AIIMS), New Delhi, India.

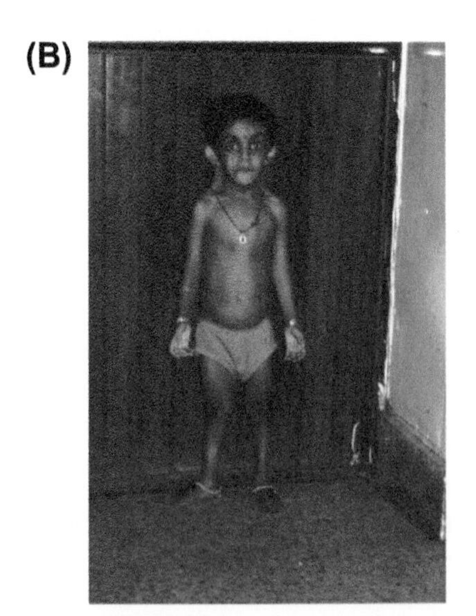

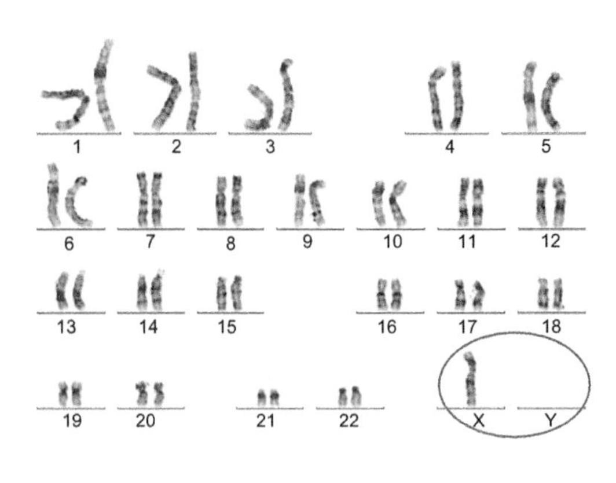

FIGURE 24.3B Clinical picture of a Turner syndrome patient and karyotype showing 45,XO. Courtesy: Cytogenetic Lab Genetic Unit, All India Institute of Medical Sciences (AIIMS), New Delhi, India.

damages an important functional gene. However carriers of balanced rearrangements are often at risk of having children with an unbalanced chromosomal complement. In an unbalanced rearrangement, there is either loss or gain of chromosomal material and the clinical effects are usually very serious. A **translocation** refers to the transfer of genetic material from one chromosome to another (Figure 24.4a). In a reciprocal translocation, two non-homologous chromosomes break and exchange fragments. Since they still have a balanced complement of chromosomes, they generally have normal phenotype. A Robertsonian translocation occurs in acrocentric chromosomes, namely 13,14,15,21 and 22. During a Robertsonian translocation, any two acrocentric chromosomes break at their centromeres and the long arms fuse to form a single chromosome with a single centromere. The short arms also fuse together and are usually lost within a few cell divisions due to the absence of centromeres. Since short arm regions do not have any essential genes, a Robertsonian translocation carrier will have no health problems but will have 45 chromosomes.

A **deletion** involves loss of part of a chromosome. A deletion can happen in every chromosome and be any size (Figure 24.4b). The consequences of a deletion depend on the size of the missing segment and the genes located on it.

Ring chromosomes usually occur when a chromosome breaks in two places and the ends of the chromosome arms fuse together to form a circular structure (Figure 24.4c) and the deleted genetic material gets lost during cell division.

An **isochromosome** is an abnormal chromosome with two identical arms, either two short (p) arms or two long (q) arms. This is sometimes seen in some females with Turner syndrome or in tumor cells (Figure 24.4d).

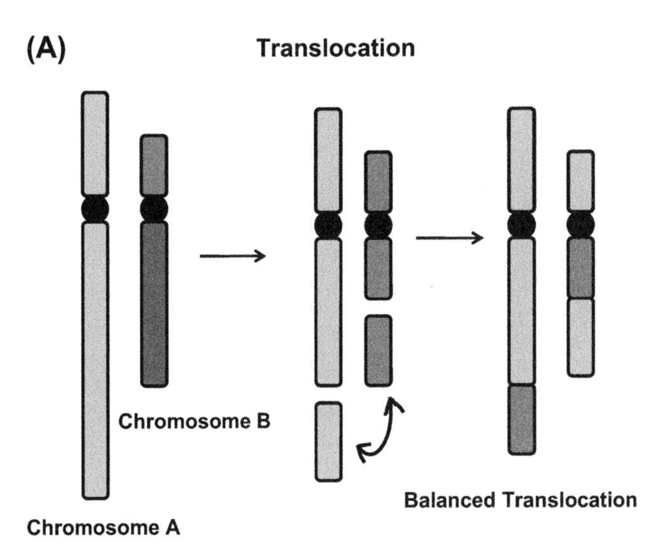

FIGURE 24.4A Diagrammatic representation of a balanced translocation.

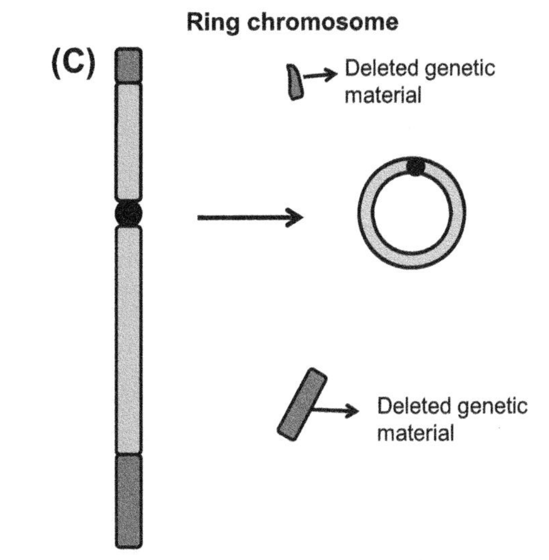

FIGURE 24.4C Diagrammatic representation of a ring chromosome.

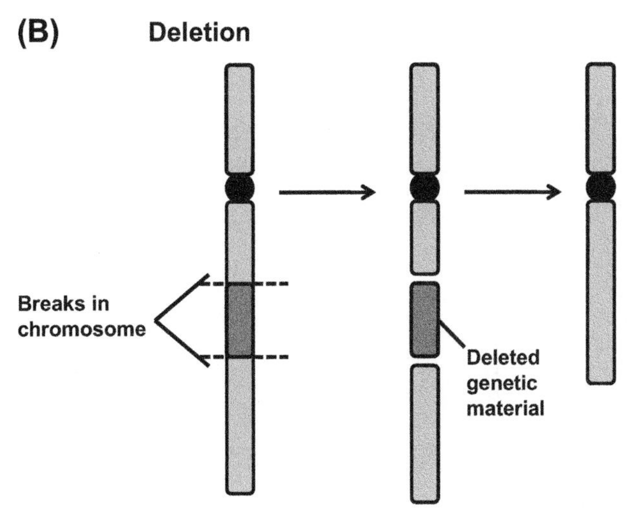

FIGURE 24.4B Diagrammatic representation showing deletion of part of a chromosome.

An **inversion** is a chromosome rearrangement where a single chromosome undergoes breakage and is then reversed and rearranged within itself. Inversions are of two types: paracentric and pericentric. **Paracentric inversions** do not include the centromere and both breaks occur on same arm of the chromosome (Figure 24.4e). **Pericentric inversions** (Figure 24.4f) include the centromere and there is a break point in each arm (p and q arm).

The terms "chimera" and "mosaic" are both used to describe people with two sets of DNA in their cells. **Chimerism** is caused by the fusing of more than one zygote, and results in a person with more than one genetic identity. **Mosaicism** refers to DNA differences that arise from only one zygote. Sometimes genetic disorders affect some cells and not others. For example, this can happen with the chromosomal disorder Down syndrome. People with Down syndrome have three copies of chromosome 21 in all cells. However, in some cases, only some cells have the extra chromosome 21, while other cells have two copies of chromosome 21; this condition is known as mosaic Down syndrome.

Mosaic disorders sometimes manifest with milder features than disorders affecting all cells, depending on the percentage of mosaicism, which varies in different patients. This is because the unaffected cells are still able to produce proteins normally and are therefore able to compensate.

Chromosome Breakage and Fragile Sites

There is another group of genetic disorders known as chromosomal breakage syndromes. The characteristic features of these disorders are increased frequency of breaks and interchanges that occur either spontaneously or when exposed to various DNA-damaging agents. The cultured cells from affected individuals exhibit elevated rates of chromosomal breakage or instability, leading to chromosomal rearrangements. Defects in the repair system lead to permanent DNA damage and thus the affected individuals are more prone to cancers. A few examples of chromosomal breakage disorders are ataxia telangiectasia, Fanconi's anemia, and Bloom's syndrome.

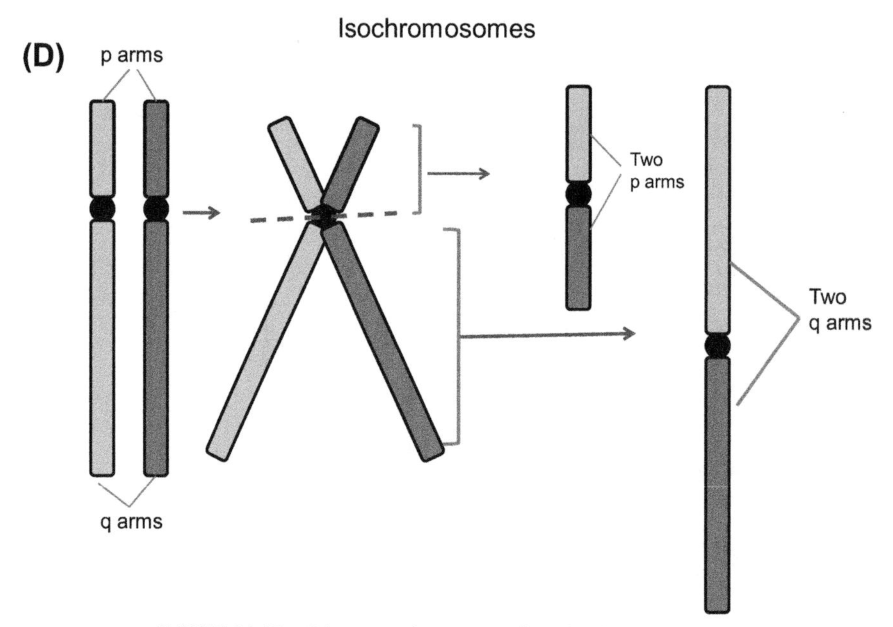

FIGURE 24.4D Diagrammatic representation of an isochromosome.

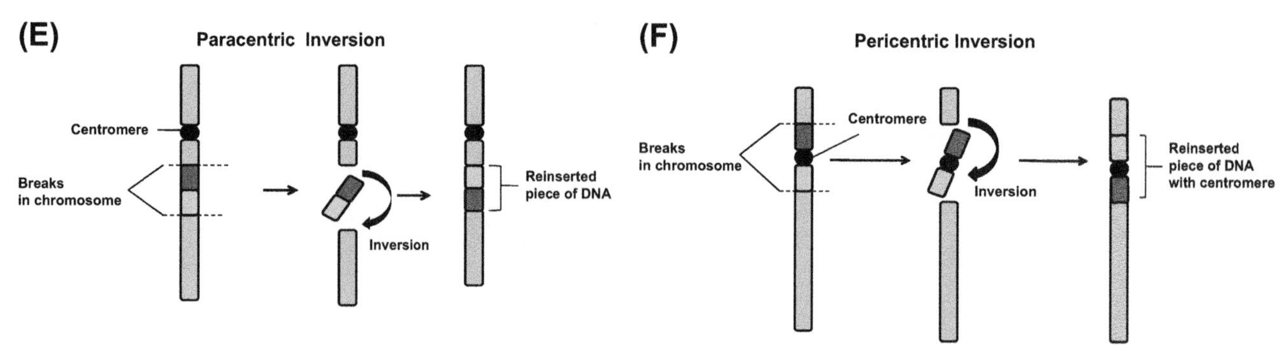

FIGURE 24.4E AND 24.4F Diagrammatic representation showing paracentric and pericentric inversion.

Chromosomal fragile sites are specific genomic regions exhibiting gaps and breaks on metaphase chromosomes. Common fragile sites are presumably present in all individuals and in all populations consisting of AT (adenine/thymidine)-rich regions and are prone to chromosomal rearrangements. These common fragile sites are of interest in cancer genetics because they are frequently affected in cancer and they can be found in healthy individuals. A well-known example is FRA3B, which is located at chromosome region 3p14.2. The FHIT gene, encompassing this FRA3B region, is a tumor suppressor gene and allelic losses at FRA3B have been observed in many types of cancer. FRA16D is another common chromosomal fragile site, located at 16q23.2. Homozygous deletion of the FRA16D locus has been observed in adenocarcinomas of stomach, colon, lung and ovary. In about 5% of the population, rare fragile sites are found which are either inherited or apparently *de-novo* and are characterized by repeat expansion composed of two or three nucleotide repeats. Fragile X syndrome is one of the most common examples of rare fragile sites present in X chromosome at Xq27.3 and causes mental impairment.

METHODOLOGY: APPLICATION OF DIFFERENT CYTOGENETIC TECHNIQUES IN DIAGNOSIS OF GENETIC DISORDERS

Cytogenetic and molecular techniques have wide applications in various kinds of cancers and in screening of various congenital anomalies. Cytogenetic approaches to study chromosomes and their association to human disease have improved greatly over the past several decades. Traditionally cytogenetics relied on cell culture, which is labor intensive. Moreover, deletions and duplications smaller than 4 million bases (4 MB or 4×10^6 base pairs) cannot be detected by routine cytogenetics. Such limitations were overcome by the advancement in technology when FISH was introduced. From 1980 onwards, a combination of cytogenetic and molecular genetic techniques started becoming popular and over the years considerable progress has been made from Fluorescence *in situ* hybridization (FISH) to array-Comparative Genomic hybridization (aCGH) (Fan, 2002). Now these modern molecular cytogenetic approaches have wider scope like:

- Examine cells from any type of tissue, even tumor cells by Flow cytometry.
- Precisely labeling the chromosomal location of any gene using different colored probes using FISH technique.

Identification of Chromosomes and Karyotyping

Cells in culture or *in vitro* are a useful model for studying the activity of cells in the whole organism or *in vivo*. Peripheral blood lymphocyte cultures are easy to obtain and generate abundant metaphases and the simplicity of the cell culture technique makes this the most convenient approach to study human chromosomes for both clinical and research purposes. After initiation of culture in a growth medium containing supplements, antibiotics, fetal calf serum, and mitogen-like phytohemagglutinin, addition of a spindle inhibitor like colchicin or colcemid after 72 hours, arrest the cells in metaphase stage. A hypotonic solution (0.075M KCl) is added, causing cells to swell and ensuring that the chromosomes are adequately dispersed within the lymphocytes. The cells are fixed before spreading on a glass slide, using methanol and acetic acid in the ratio of 3:1 (Rooney and Czepulkowski, 1997). Giemsa banding is one of the most commonly used techniques to identify chromosomes, both normal and abnormal. Giemsa reagent is a DNA stain that consists of a mixture of dyes including the basic amino-phenothiazine dyes (azure A, azure B, azure C, thionin, and methylene blue) and the acidic dye eosin. Prior to staining, an enzyme treatment with trypsin is needed to digest the proteins. This technique produces patterns of light-staining (G-light) regions and dark-staining (G-dark) regions. The pattern is unique to each chromosome, and therefore serves as a landmark for chromosome identification. The light stained regions are called **euchromatin** and contain actively expressed genes, and dark stained regions are called **heterochromatin** and contain inactive unexpressed DNA (Swansbury, 2003). A flow chart below describes the procedure of lymphocyte culture (Flow Chart 24.1).

Simultaneous to these advances in cytogenetics, the field of molecular biology was also making significant progress. Cytogenetics took advantage of this additional knowledge and technologies and a new area called molecular cytogenetics emerged, which is mainly represented by

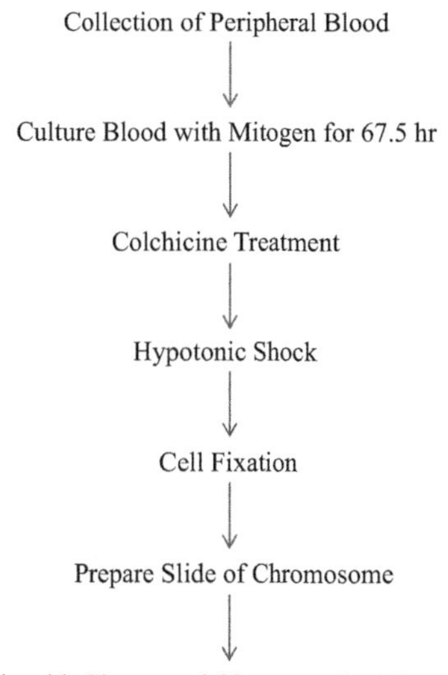
Collection of Peripheral Blood
Culture Blood with Mitogen for 67.5 hr
Colchicine Treatment
Hypotonic Shock
Cell Fixation
Prepare Slide of Chromosome
Stain with Giemsa and Observe under Microscope

FLOW CHART 24.1 Flow chart showing human lymphocyte culture.

the techniques of FISH (fluorescent *in situ* hybridization) and other, more advanced, methods such as spectral karyotyping and array-CGH, as discussed below.

Fluorescence *In Situ* Hybridization (FISH)

FISH is based on the principle of hybridization (i.e. ability of a single stranded DNA to hybridize with its complementary sequence to form a double stranded DNA). In FISH, the DNA probes which are short sequences of single-stranded DNA are tagged with fluorescent dye and applied to cell preparations on a slide under conditions that allow for the probe to attach itself to the complementary sequence in the specimen if it is present. The site of hybridization is then visualized under fluorescence microscope. It is a powerful technique used to detect chromosomal aneuploidies such as loss of a chromosomal region, a whole chromosome, or trisomies (for example Down Syndrome). This technique is also useful in identification of genetic abberations, such as the fusion of BCR and ABL genes in breast cancer, and to monitor the progression of an aberration that can help in both the diagnosis or suggesting prognostic outcomes in cancer genetics. A schematic diagram (Flow Chart 24.2) explains the FISH technique.

Spectral Karyotyping (SKY) and Multicolor FISH (M-FISH)

Spectral karyotyping and multicolor FISH are both advanced molecular cytogenetic techniques for chromosome analysis

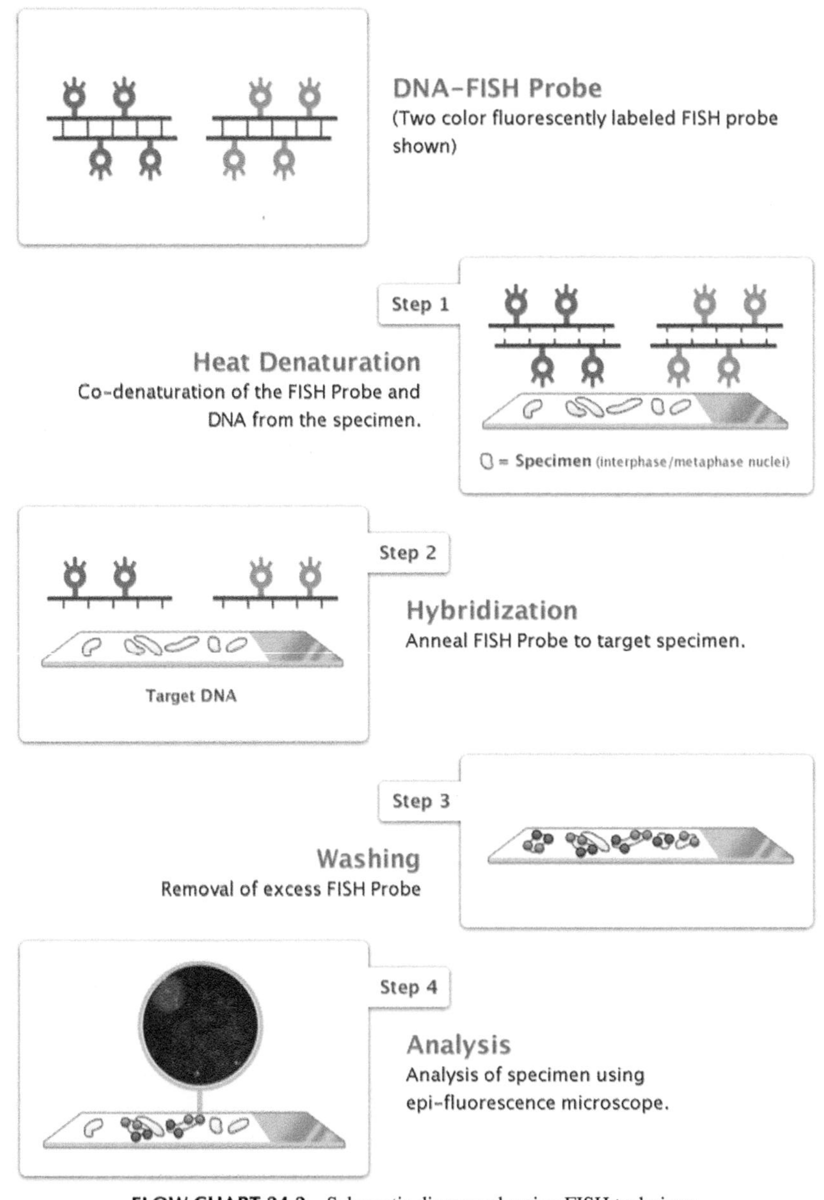
DNA-FISH Probe
(Two color fluorescently labeled FISH probe shown)
Step 1
Heat Denaturation
Co-denaturation of the FISH Probe and DNA from the specimen.
= Specimen (interphase/metaphase nuclei)
Step 2
Hybridization
Anneal FISH Probe to target specimen.
Target DNA
Step 3
Washing
Removal of excess FISH Probe
Step 4
Analysis
Analysis of specimen using epi-fluorescence microscope.

FLOW CHART 24.2 Schematic diagram showing FISH technique.

that are based on the principle of FISH. Both these techniques allow visualization of all the chromosomes simultaneously by labeling them with a combination of different colors that are spectrally distinguishable fluorochromes, but different methods are used for detecting and discriminating the different combinations of fluorescence after *in situ* hybridization. In SKY the images are captured by charge-coupled device (CCD) imaging and analyzed by using an interferometer attached to a epifluorescence microscope. Image processing software then assigns a pseudo color to each spectrally different combination, allowing the visualization of the individually colored chromosomes (Figure 24.5). In M-FISH each homologous pair of chromosomes is uniquely labeled with five fluorochromes set which are spectrally distinct in different combinations. The images are captured by band-pass filter sets and defined emission spectra are measured by dedicated M-FISH software.

These techniques have further evolved and using this same principle of hybridization one can screen the whole genome simultaneously through comparative genomic hybridization (CGH), or through an array-based method using comparative genomic hybridization (aCGH). These advanced techniques are now used to screen genomic deletion/duplication syndromes, patients with intellectual disabilities, and various cancers. These approaches are powerful tools for screening genomes for chromosomal aberrations and in cancer genetics.

Comparative Genomic Hybridization (CGH) and array-CGH (aCGH)

There are certain genetic disorders that result from unbalanced chromosomal abnormalities due to either gain or loss of genetic material; these are known as microdeletion and microduplication syndromes. It is difficult to detect these micro-copy number gains and losses using traditional techniques, which rely on the examination of a single target and prior knowledge of the region under investigation; whereas CGH can be used to quickly scan an entire genome for imbalances. This technique was first used to detect copy number changes in solid tumors. In this technique, DNA of a test and a control sample are differentially labeled and then competitively hybridized to metaphase chromosomes (reference DNA). The intensity of the fluorescent signal of the test DNA relative to that of the reference DNA is measured and linearly plotted across each chromosome, to identify copy number changes. But a limitation of the CGH technique is that it can only detect copy losses/gains which are at least 5–10 Mb in length. To overcome this limitation, another technique was developed, which combines the traditional CGH and microarray techniques, and can detect copy number changes at a level of 5–10 kilobases of DNA sequences (Speicher and Carter, 2006). In the array-CGH technique, instead of using metaphase chromosomes, thousands of short sequences of DNA are used as targets for analysis (also known as probes), which are printed on a glass slide called a chip. DNA of the test sample and a control sample are then differentially labeled with fluorescent dyes; commonly used colors are red and green which are mixed together and hybridized on to the glass slide. The fragments of DNA hybridize with their complementary strand on the array and it are then scanned in a machine called a microarray scanner, which measures the amount of red and green fluorescence on each probe. The ratios of the red to green fluorescent dyes are then calculated by the software to identify the copy number variation in the genome (Mei et al., 2000). Array CGH is becoming the technique of choice to screen chromosome abnormalities like aneuploidy, chromosomal rearrangements, and genomic deletion/duplication disorders (Figure 24.6).

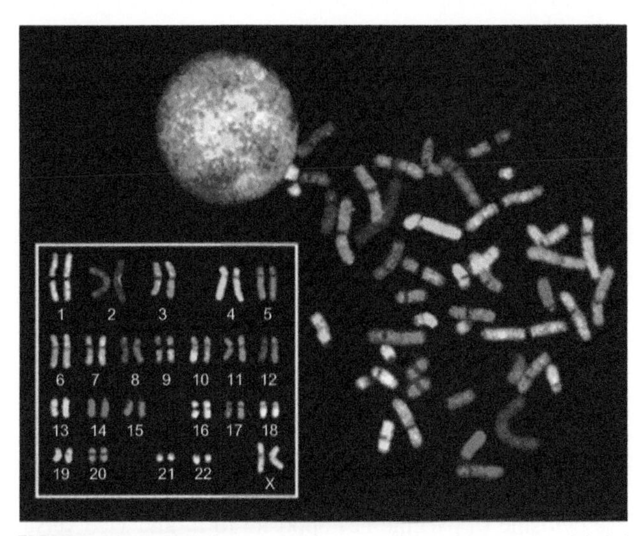

FIGURE 24.5 SKY image showing metaphase chromosomes labeled with different fluorochromes. Courtesy: en.Wikipedia.org.

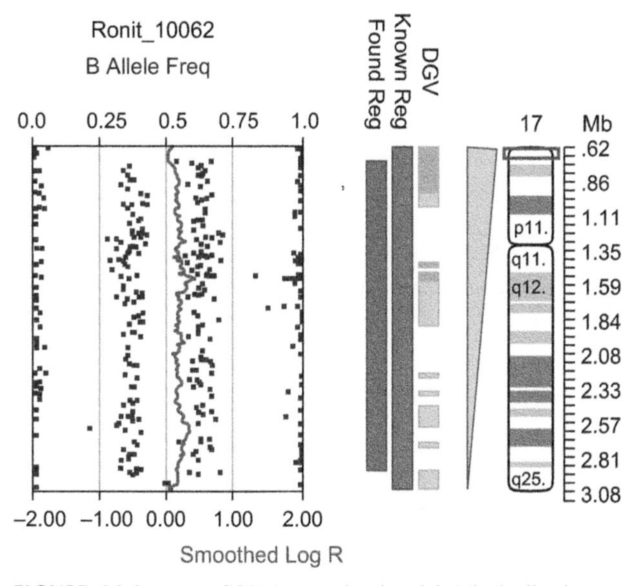

FIGURE 24.6 Array-CGH image showing 2.2 Mb duplication on chromosome 17q13.3.

Principle

The basic principle of all cytogenetic techniques is to minutely examine and scrutinize all 46 chromosomes to look for any kind of chromosomal change, either numerical or structural. Karyotyping is still considered to be the gold standard technique the world over. Later techniques started combining principles of molecular genetics with better image analysis and current bioinformatics tools. This was done in an effort to overcome some of the limitations of common methods in the study of chromosomes.

MOLECULAR GENETICS: AN OVERVIEW

Molecular genetics is the field of biology and genetics that studies the structure and function of genes at a molecular level and how they are inherited from one generation to the next. Molecular information is used to determine the patterns of descent, and to understand the genetic errors or mutations that can cause certain types of diseases (Epstein, 2003; Strachen and Read, 2004).

The history of molecular genetics dates back to 1869, when Friedrich Miescher extracted a viscous white substance from the nucleus of a cell, which was slightly acidic and rich in phosphorous and nitrogen. He named it as "nuclein" because it was found in the nucleus, but he did not know its true nature. Phoebus Levene, a Russian-American biochemist identified components inside this material (i.e. ribose in 1909 and deoxyribose in 1929). He also showed that these components were linked together in the order phosphate–sugar–base to form units, and coined the term "nucleotide" for these units. He also discovered the ribose sugar in RNA, the deoxyribose sugar in DNA. Oswald Avery in 1944 discovered that genetic information is stored in DNA and is transmitted from one generation to the next. This was a revolutionary concept, as it was opposing the prevailing concept that proteins were the hereditary material.

In 1950s, Erwin Chargaff studied DNA in many different organisms and found that the proportion of Adenine in a DNA molecule is equal with to that Thymine, and the proportion of Guanine is equal to that of Cytosine. This is known as Chargaff's rule which was further used by Crick and Watson to unravel the structure of DNA (Turnpenny and Ellard, 2011).

Hereditary Material

A cell is the basic fundamental unit of life. Each cell of the body has a darkly stained nucleus surrounded by cytoplasm. The nucleus contains the hereditary material in the form of chromosomes. These chromosomes are made up of tightly coiled very long DNA molecules that contains a series of genes. A gene in a simplest term can be defined as a segment of a DNA molecule formed by millions of nucleotides joined together in a long chain of nitrogenous bases held in position by hydrogen bonds between them and joined to each other by sugar and phosphate molecule. A gene determines the cell properties, both structure and functions, which are unique to each cell by the information contained in these sequences of DNA.

The Structure of DNA

Deoxyribonucleic nucleic Acid (DNA) is a highly complex macro or rather mega biomolecule that is essential for all known forms of life. The long chain molecule is formed of repeating units called nucleotides, so it is also known as the polynucleotide molecule. It consists of two polynucleotide antiparallel strands which are spirally coiled around each other along their lengths. The model for the structure of the DNA molecule was proposed by Watson and Crick in 1953, who won the Nobel Prize for Medicine in 1962.

Chemical Components of DNA: The highly complex DNA molecule is composed of only three types of chemical components. These are (i) deoxyribose sugar, (ii) a phosphate, and (iii) nitrogen-containing organic bases.

Deoxyribose Sugar: It is a pentose sugar (with 5 carbon atoms) having a pentagonal ring structure. Since it is called deoxyribose sugar, the nucleic acid is called deoxyribonucleic acid (DNA).

Phosphate: The phosphate in the DNA, phosphoric acid (H_3PO_4) forms the sugar-phosphate backbone of each strand.

The Nitrogen-Containing Organic Bases: These are heterocyclic compounds containing nitrogen in their rings and therefore called nitrogenous bases. DNA contains four different bases called adenine (A), guanine (G) cytosine (C), and thymine (T). These are grouped into two classes on the basis of their chemical structure: (i) Purines (with a double ring structure) and (ii) Pyrimidines (with a single ring structure).

The DNA molecule is a double helix. The molecule is formed by two antiparallel polynucleotide strands which are spirally coiled round each other in a right-handed helix. The two strands are held together by hydrogen bonds. Each strand is a long polynucleotide of deoxyribonucleotides and the backbone of the strand is formed by alternately arranged deoxyribose sugar and phosphate molecules which are joined by the phosphodiester linkages. The two strands are complementary to each other with regards to the arrangement of the bases in the two strands. For example, where adenine (a purine) occurs in one strand, thymine (a pyrimidine) is present in the corresponding position in the opposite strand, and *vice versa*. Similarly, wherever guanine (a purine) is present in one strand, the other strand has cytosine (a pyrimidine) opposite to it, and vice versa. Thus, in the double helix, purines and pyrimidines exist in base pairs (i.e. A with T, and G with C) joined by hydrogen bonds (Figure 24.7a). As a result, if the base sequence of one strand of DNA is known, the base sequence of its complementary strand can be easily deduced. In each strand, one end of the strand has one free phosphate group on carbon-5 of the sugar molecule, and is called the C-5 (or 5′) end. The other end of the strand has a free -OH on carbon-3 of the sugar molecule and is called the C-3 (or 3') end of the strand (Watson and Crick, 1953).

Before cell division occurs, the DNA material in the original or parent cell is duplicated so that after cell division, each new daughter cells get the required amount of DNA material. The process of DNA duplication is called **replication**. The most common method of DNA replication is termed as **semi-conservative** since each new cell contains one strand of original DNA and one newly synthesized strand of DNA. The original polynucleotide strand of DNA serves as a template for the formation of a new strand.

A chromosome consists of a single, very long DNA molecule that contains genetic information, and a complete set of chromosomal DNA constitutes the genome. Genomes vary widely in size: the smallest known genome (of a bacterium) contains about 600,000 DNA base pairs, while human and mouse genomes have some 3 billion base pairs.

The portion of DNA that codes for the synthesis of a specific protein is called a **gene.** The human genome is estimated to contain 20,000–25,000 genes. Each gene has a promoter region at the 5 prime end and untranslated regions or "UTRs" in both the 5′ and 3 prime regions. A gene is divided into coding regions known as **exons** and non-coding regions called **introns** (Figure 24.7b). In eukaryotes, DNA is mostly non-coding and consists of repetitive sequences.

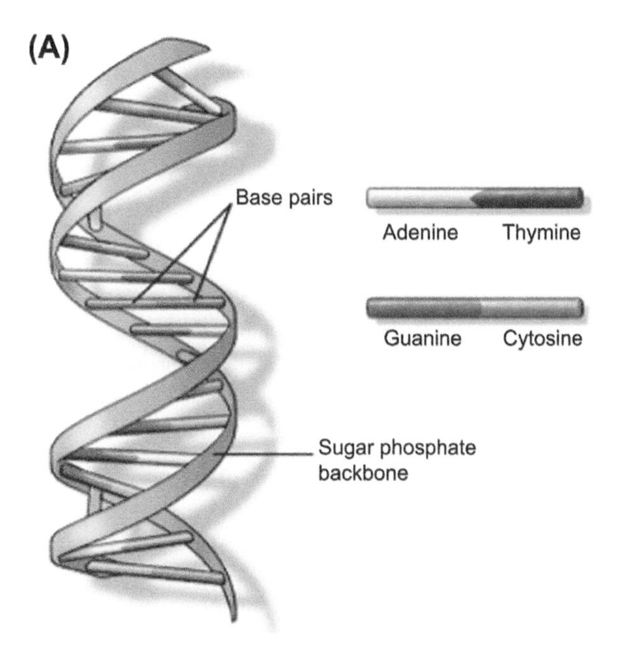

FIGURE 24.7A Double helix structure of DNA. *(Courtesy: U.S. National Library of Medicine.)*

The position of a gene on a chromosome is called its **locus** and the corresponding loci on a homologous pair of chromosomes carry genes for the same character. One of the several alternative forms of a gene at a given locus is called **allele**. The genetic make-up of a person is called the **genotype**, and the clinically manifest characters are known as the **phenotype.** Further, each protein is a long polypeptide chain molecule formed by joining amino acid molecules. The genetic code consists of 64 triplets of nucleotides. These triplets are called **codons** and each codon encodes for one of the 20 amino acids used in the synthesis of proteins.

The information in DNA is transferred to a messenger RNA (mRNA) molecule by a process called **transcription**. RNA synthesis occurs in the 5′–3′ direction and its sequence corresponds to that of the DNA strand, which is known as the sense strand.

The genetic code can be expressed as either RNA codons or DNA codons. RNA codons occur in messenger RNA (mRNA), which are actually "read" during the synthesis of polypeptides; this process is known as **translation**. A code contains signals for both starting and stopping the translation. In a nucleotide sequence, AUG is the start codon which codes for the amino acid methionine, and translation continues as the subsequent triplets are read in the same reading frame until a stop codon is encountered. Stop codons are also called "termination" or "non-sense" codons as they terminate the protein synthesis. There are three stop codons, UAA (known as ochre), UAG (amber), and UGA (opal). The nucleotide sequence between the start and stop codons is known as an **open reading frame** (ORF).

There are two classes of genes that can have an effect on how other genes function; they are called modifying genes and regulator genes. **Modifying genes** alter depending on how certain other genes are expressed in the phenotype. For instance, there is a dominant cataract gene, which will produce varying degrees of vision impairment depending on the presence of a specific allele for a companion modifying gene. However, cataracts also can be promoted by other conditions like diabetes and environmental factors, such as excessive ultraviolet radiation and alcoholism. **Regulator Genes** can either initiate or block the expression of other genes. Some genes are incompletely penetrant (i.e. their effect does not normally occur unless certain environmental factors are present). For instance, one may inherit the genes that are responsible for type 2 diabetes but never get the disease unless one becomes greatly overweight or persistently stressed psychologically.

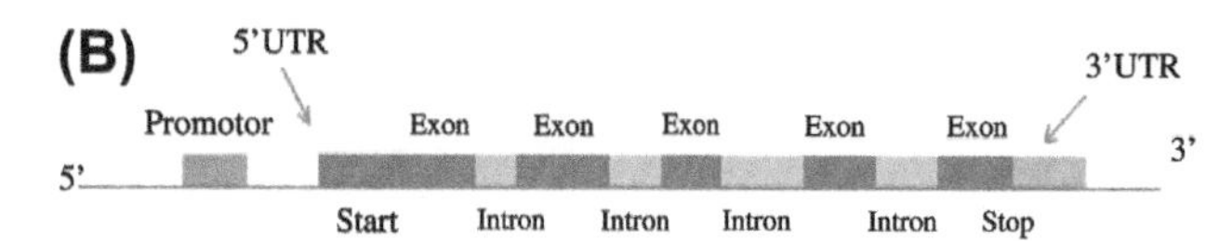

FIGURE 24.7B Structure of a typical human gene.

Mutation: A mutation is defined as any change in the nucleotide sequence of DNA. Mutations are broadly categorized into two types, fixed mutations and dynamic mutations. Fixed mutations are again subdivided into two main groups, Synonymous and Non-synonymous mutations (Haldane, 1935).

Missense Mutation: This occurs when there is a single base pair substitution, which results in different amino acids, leading to an altered protein.

For example, in Scheme 24.1 the upper panel shows a DNA sequence that codes for a segment of protein with sequence as LPHRYTNSRG. Substitution of G to A in 9th codon AGA that codes for R (Arginine), will result in AAA codon that codes for K (Lysine), thereby altering the sequence of amino acids, resulting in an altered protein as shown in the lower panel.

Synonymous Mutation: A mutation is known as synonymous if it does not alter the amino acid in the polypeptide chain. It is also known as silent mutation (i.e. it does not affect the phenotype).

Non-Synonymous Mutation: Any change in the nucleotide sequence that results in alteration of amino acids in the polypeptide chain is termed as non synonymous mutation. It effects the phenotype, depending on the importance of amino acids in the functioning of the protein.When there is a replacement of purine to purine or pyrimidine to pyrimidine (A-to-G or C-to-T), it is called **transition** and when the change is from purine to pyrimidine or pyrimidine to purine (A-to-C or G-to-T), it is known as **transversion**.Non synonymous mutations are less frequently observed than the synonymous mutations. Non-synonymous mutations are of three types: missense, non-sense and frameshift mutation.

Non-Sense Mutation: Mutations resulting in replacement or substitution that causes premature termination of translation of a peptide chain leading to loss of function/activity. For example, Scheme 24.2 shows that substitution of T to G at 5th codon TAT that codes for Y (tryptophan), results in codon TAG, which does not code for any amino acid, (and therefore terminates the peptide chain).

Frameshift Mutation: A mutation that causes an insertion or deletion of nucleotides (not multiple of three) that results in disruption of the reading frame of a polypeptide chain. This leads to inactivation of polypeptide chain/protein. When there is a deletion or insertion of one or two nucleotides, it results in the interruption of the reading frame of a protein. But if the insertion or deletion is of three or multiple of three nucleotides, then the reading frame of the protein is not changed.For example, Scheme 24.3 shows that deletion of T in the 2nd codon CCT that codes for P (proline) resulting in disruption of the reading frame, leading to an

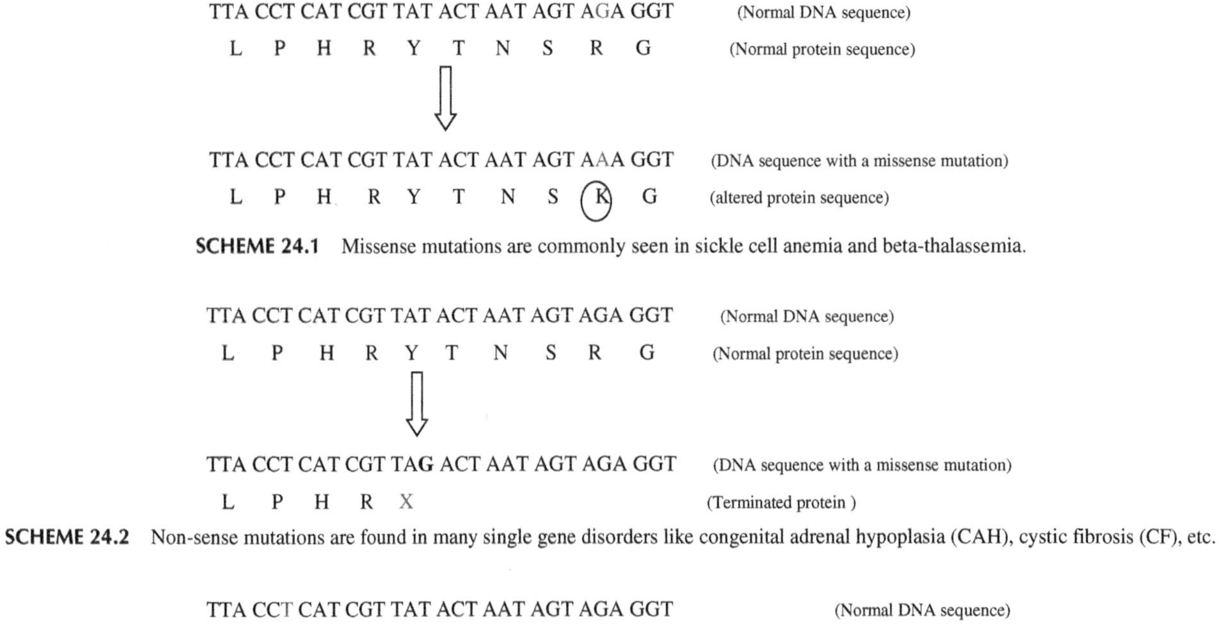

SCHEME 24.1 Missense mutations are commonly seen in sickle cell anemia and beta-thalassemia.

SCHEME 24.2 Non-sense mutations are found in many single gene disorders like congenital adrenal hypoplasia (CAH), cystic fibrosis (CF), etc.

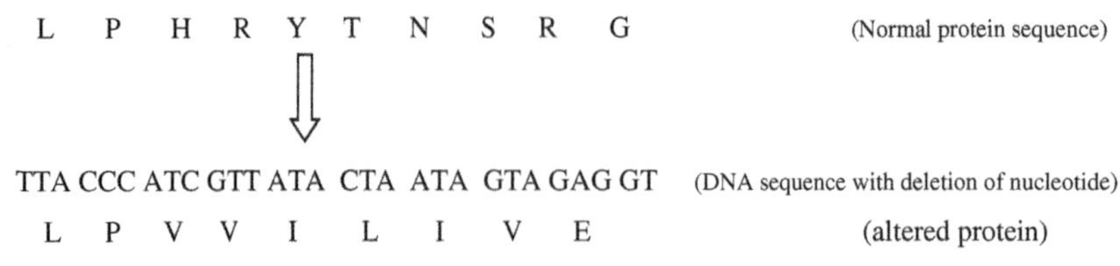

SCHEME 24.3 Deletion of one or a few nucleotide or large deletions are commonly seen in patients of Duchene muscular dystrophy (DMD) and alpha thalassemias.

alerted protein with different amino acids. Addition of one or two nucleotides will also lead to similar disruption of the protein, thus losing its significance/activity.

Scheme 24.4 shows that insertion of the three nucleotides ATG after the 2nd codon in the DNA sequence results in addition of M (methionine) at 3rd position in the protein sequence. Alhough there is a change in the polypeptide chain, the reading frame is not changed. Thus when there is a insertion of three or multiple of three nucleotides, the reading frame is not changed. Similarly, deletion of three or multiples of three nucleotides will also not change the reading frame of a protein.

Dynamic mutations: Triplet repeats are typically short sequences that are repeated many times and are commonly found in normal DNA sequences. These are highly polymorphic in nature; in other words, they are found in varying sizes in individuals. These repeats transmit normally below a certain level, and become unstable above a critical level during mitosis and meiosis, leading to triplet repeat disorders. Common examples are of triplet disorders are Fragile X mental retardation (caused by CCG repeats), Huntington's disease (caused by CAG repeats), Myotonic dystrophy (caused by CTG repeats).

For example, normal individuals have less than 45 CCG repeats in the 5′ untranslated region of the *FMR1* gene while in an affected individual with Fragile X syndrome, this number may exceed over 200 repeats, leading to silencing of the *FMR1* gene, and thus stopping the formation of FMRP protein.

Numerous diseases are clinically similar (i.e. having similar phenotype) but may have different causes. This phenomenon is called **genetic heterogeneity**. **Allelic heterogeneity** is caused due to different mutant alleles at one gene locus and **locus heterogeneity** results due to mutants at different gene loci. Genetic heterogeneity and locus heterogeneity are often used interchangeably in practice, but locus heterogeneity is only used for the involvement of different loci in the causation of a disease/phenotype individually. Genetic heterogeneity may also be used to mean a combined effect of different loci in the development of a (complex) disease (e.g. in diabetes, multiple loci are simultaneously involved in the development of diabetes).

Genetic polymorphism is a specific term that describes frequent variation at a specific locus in a genome, resulting in diversity within a population.

Genetic disorders are mainly classified into chromosomal disorders, single gene disorders, multifactorial disorders, and acquired genetic diseases. Chromosomal disorders have already been discussed in the cytogenetic section; the following sections discuss the other three main categories of genetic disorders.

Single-Gene Disorders

Over 10,000 traits or disorders in humans exhibit single gene, unifactorial, or Mendelian inheritance. However, characteristics such as height, weight and many common familial disorders, such as diabetes, hypertension, etc., do not usually follow a simple pattern of inheritance. A trait or disorder that is determined by a gene on an autosome is said to show autosomal inheritance, whereas a trait or disorder determined by a gene on one of the sex chromosomes is said to show sex-linked inheritance.

In single gene disorders, the gene responsible for the disease can be traced through families, and their occurrence in next generations can be predicted. An **autosomal dominant** trait is one that manifests in the heterozygous state (i.e. in a person having both an abnormal or mutant allele and the normal allele). Autosomal dominant disorders are generally milder than recessive disorders; a few examples are neurofibromatosis, achondroplasia, Marfan syndrome. **Autosomal recessive** disorders only manifest when the mutant allele is present in a double dose, i.e. homozygosity. Individuals heterozygous for a recessive mutant allele have a normal phenotype but genotypically are carriers. The disorder affects both males and females in equal proportions. Examples of autosomal recessive disorders are beta-thalassemia, spinal muscular atrophy, galactosemia, etc.

Sex-linked disorders can be either X-linked or Y-linked. In **X-linked recessive** mode of inheritance, a mutation in the X chromosome causes the phenotype to be expressed in males who are hemizygous for the gene mutation (i.e. they have only one X chromosome) and in females who are homozygous for the gene mutation (i.e. they have a copy of the gene mutation on each of their two X chromosomes). Carrier females who have only one copy of the mutation, so do not usually express the phenotype, although differences in X-chromosome inactivation can lead to varying degrees of clinical expression in carrier females. Examples

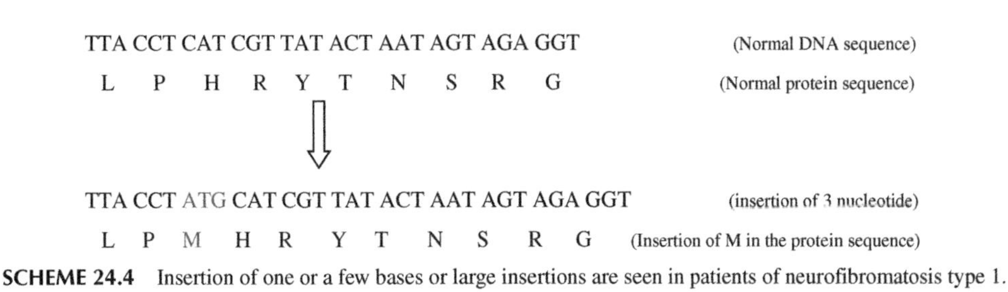

SCHEME 24.4 Insertion of one or a few bases or large insertions are seen in patients of neurofibromatosis type 1.

of X-linked recessive disorders are hemophilia, Duchene muscular dystrophy, etc. **X-linked dominant** inheritance although uncommon, includes disorders that manifest in heterozygous females as well as in males; examples include hypophosphatemic vitamin D-resistant rickets, orofaciodigital syndrome, etc.

Multigenic and Multifactorial Disorders

In contrast to the monogenic or single gene disorders, multigenic disorders are caused by less severe mutations in more than one gene. Mutation in any one of these genes alone may not cause the disease, but all these mutations present together can result in a diseased condition.

A trait is called multifactorial if multiple genes are assumed to interact with environmental factors. Most diseases develop as the result of combination of small variation in genes that can predispose an individual to a disease. Such inheritance is observed in common disorders such as hypertension, diabetes mellitus, neural tube defect, ischemic heart disease, etc. It is very important to identify susceptibility genes for common disorders so that individuals at high risk for these diseases can be identified early and preventive options can be made available to them.

Mitochondrial Disorder

In addition to a nuclear genome, our cells also contain a mitochondrial genome; each mitochondrion maintains dozens of copies of its own circular genome, and most human cells contain numerous mitochondria. Mitochondria are organelles, which are found in every cell of the human body except red blood cells. When a cell divides, its mitochondria are distributed to the two daughter cells. However, mitochondrial segregation occurs randomly and is not as organized as the highly regulated process of mitotic chromosome segregation. Therefore, cells will receive similar, but not identical, mitochondrial DNA populations. Mitochondria rely on their own set of genes, as well as on nuclear-encoded genes, in order to carry out their function as the ATP-generating (adenosine triphosphate) powerhouses of the cell. Therefore, mitochondrial mutations can lead to profound effects on cellular metabolism and function, especially in tissues that have high-energy demands, such as that of the brain, skeletal muscle, cardiac muscle, and retina. Mitochondrial inheritance is often called "maternal inheritance" because a child inherits the great majority of their mitochondria from their mother through the egg. Mitochondrial DNA contains 37 genes that are all essential for normal function of the mitochondria. Many genetic conditions are due to changes in particular mitochondrial genes, for example, MELAS (mitochondrial encephalopathy with lactic acidosis and stroke like episodes), MERRF (myoclonic epilepsy with ragged red fibers), etc.

Acquired Somatic Genetic Disorders

Somatic genetic disorders, in broad terms, are disorders in which defects are restricted to specific somatic cells. Somatic genetic defects may not be present in the zygote, but may appear during somatic cell divisions (mitosis); in single gene, multifactorial, and chromosomal disorders, the abnormality is present in all cells, including the germ cells. Such cancers are seen in patients without a family history. Since, in somatic cell disorders, the change in the genetic structure is neither inherited nor passed to offspring, such changes are called "acquired mutations." In fact, most cancers are caused due to somatic cell mutations.

METHODOLOGY: APPLICATION OF DIFFERENT MOLECULAR TECHNIQUES FOR DIAGNOSIS OF GENETIC DISORDERS

Completion of the human genome project has led us to decipher information encoded in the human genome sequence. (International Human Genome Sequencing Consortium, 2001) Simultaneously, the expansion of our knowledge on the molecular basis of human inherited diseases has brought up the need to provide diagnosis at the molecular level. This has further led to the development of various molecular diagnostic techniques to detect errors in DNA.

The first molecular diagnostics tests to reach the market were for infectious diseases, and they have now become an important tool for oncology, pharmacogenomics, genetic testing, cardiovascular disorders, neurological diseases and other disease areas. Various techniques for genetic diagnosis are being used to confirm the clinical diagnosis for a number of heritable disorders and infectious diseases. The DNA sample isolated from a patient's cell can be used to screen for innumerable diseases. With the advancement of techniques and increasing knowledge of pathophysiology of a number of diseases, diagnosis can be made well in advance for better management.

Molecular diagnostic technologies are now being used for the following applications:

Clinical Diagnostic Testing: Application of genetics in medicine and health has been around for over 50 years. Many common genetic diseases like Thalassemia, Duchenne Muscular Dystrophy (DMD), Hemophilia, Fragile X syndrome, etc, are being diagnosed by molecular tests. These methods not only diagnose but are also utilized for identification of individuals who are at increased risk of developing certain disorders later in life (e.g. Huntington's disease). In some diseases, such as breast cancer, screening for mutations in *BRCA1* and *BRCA2* genes helps in counseling family members who are at risk of developing breast cancer.

Pharmacogenomics: There are many genes responsible for metabolism of the drugs. Due to polymorphism (genetic variations) in drug metabolizing gene/genes, a person can have a different response to a particular drug than anticipated. These polymorphisms can be detected by PCR based genetic tests. Based on these polymorphisms, it is possible to identify patients who will respond to a specific drug. Accordingly, dose and timing of the drug can be calculated. This is what is known as personalized medicine. Such studies are very useful in reducing adverse drug reactions, as it is a major cause of morbidity and mortality. Some of the best known examples of drugs that show response based on genotype are isoniazid, succinylecholine, primaquine, coumarine anticoagulants, and alcohol.

Populations Screening: Population screening involves identification of carrier status (i.e. a person who inherits a genetic trait or mutation but does not show the symptoms of the disease, but they are able to pass the mutant gene in the next generation). This involves taking a small amount of blood from an individual. DNA is extracted and genetic tests are performed for confirmation of the carrier status, and thus, counseling for prenatal diagnosis. Carrier screening programs have proven effective and are being conducted where conditions like thalassemia, Tay–Sachs disease, and sickle cell disorder are fairly common. Such programs include screening of the general population or high-risk communities, antenatal screening, and cascade screening. Testing may also be useful for consanguineous couples since it increases the chance of having a child with a recessive disorder.

If certain disorders are detected early in life and are treated in a timely manner, it can increase the chances of successfully managing the disease. For example, congenital hypothyroidism (CH) affects 1 in 3000 to 4000 live births, making it the most common congenital endocrine disorder (Toublanc, 1992). If untreated early in life, CH can lead to mental retardation and abnormal growth. If early and adequate thyroxine replacement is given to these affected children, growth retardation and intellectual impairment can be avoided. So now many state governments have made provisions to screen all newborns for CH under their newborn screening programs.

Monitoring Response to Therapy: PCR has potential application in detecting residual disease or recurrence after treatment for disorders such as leukemia. PCR-based tests also help in management of these patients by determining timing, continuation, and duration of chemotherapy with relapse.

Identity Testing: Molecular tests have been very helpful in unraveling the identity of an individual. Individuals can be distinguished from one another by variable DNA sequences called DNA markers. These DNA markers are present in different regions of the genome. Each person has a particular length of these markers, making these markers specific for an individual, and hence act as DNA "fingerprints." These DNA markers can be amplified by PCR and then separated by gel electrophoresis. Length of these markers is compared between different individuals. This method can be used to determine the biological relationships between family members.These markers are also exploited for diagnosis of many diseases where it is difficult to pinpoint the exact defect in the gene or where the gene in question is very large in size. In addition, presence of maternal cells in fetal cell samples obtained for prenatal analysis, sample mix-ups, or labeling errors can be detected by fingerprinting. DNA fingerprinting can also be used to evaluate the purity of tissue samples obtained for other kinds of tests.

Histocompatibility Testing: Each person has a small, relatively unique set of HLAs that they inherit from their parents. Children, inherit half of their HLAs from the mother and half HLAs from the father. It is unlikely that two unrelated people will have the same HLA makeup, although identical twins may match each other. Histocompatibility testing is done by identifying HLA genes and antigens in a person by molecular methods and then matching up the same with donors and recipients of organ and bone marrow transplants. Maximum HLA matching of donor and recipient ensures the acceptance of blood or an organ.

To perform any genetic test, DNA is needed; it can be extracted from virtually any part of the human body, the most common sources being blood and soft tissue samples. DNA can also be extracted from semen, saliva, hair roots, etc. To characterize any portion of DNA, it needs to be cloned. In order to do this, the DNA segment of interest must be isolated by cutting the DNA with a restriction enzyme and ligating it into a vector (such as a plasmid) that can then be reproduced in bacteria. This cloning procedure is used to generate DNA libraries (a collection of DNA fragments that is stored and propagated in a population of bacteria or viruses through the process of cloning). Once cloned, DNA fragments can be characterized by various techniques like Southern blotting, by polymerase chain reaction (PCR), DNA sequencing, or by gene expression studies. Fluorescence-in-situ-hybridization (FISH) is another technique by which large deletions, translocations and inversions that are not large enough to be picked up by light microscopy, can be easily screened.

Southern Blotting

Southern blotting is a technique named after its inventor, the British biologist Edwin Southern, to detect a specific DNA sequence in DNA samples. Nucleic hybridization is used for the identification of a specific DNA segment within a genomic DNA. This technique involves mixing of denatured

DNA from two sources and then, under appropriate conditions, allowing complementary base pairing of homologous sequences. DNA from one of the sources is labeled (i.e. a DNA probe), and then identification of the target DNA from the other source is possible. In this method, restriction endonucleases are used to cut high-molecular-weight DNA strands into smaller fragments that are then electrophoresed on an agarose gel to separate them by size; they are then transferred to a membrane. The DNA fragments are denatured with alkali to make them single-stranded. These DNA fragments are transferred on to a nitrocellulose filter that binds the DNA permanently. This is called a Southern blot. This blot is then hybridized with a radioisotope-labeled probe of complementary single-stranded DNA from the gene of interest. The probe hybridizes with the complementary DNA on the Southern blot and can be visualized by autoradiography. Bands on the radiogram represent those DNA fragments that contain sequences complementary to the probe sequence. Size or length of the DNA can be estimated by the marker that is run on the gel along with the DNA sample (Southern and Mellor, 1975).

A Southern blot helps in finding out whether a particular gene is present in the genomic DNA or not and whether the size of the DNA fragment carrying the gene is normal, has decreased or increased. Hence, the presence or absence of the gene, large deletions and duplications (or even large gene arrangements) can be detected by Southern blotting.

Some disorders are caused by large deletion and duplication of the whole gene. For example in Congenital Adrenal Hyperplasia (CAH), deletion of *CYP21A2* gene is found in 25–30% of CAH patients. In such cases, the Southern blot helps in detecting gene deletion, duplication or gene rearrangements that is commonly seen in these patients.

Polymerase Chain Reaction (PCR)

PCR is an alternate technique developed in 1983 by Kary Mullis for the amplification of DNA (Mullis and Faloona, 1987). In this method, DNA is amplified in a test tube environment and requires the knowledge of the sequence of the DNA to be amplified. PCR is clearly the method of choice due to its exquisite sensitivity and specificity to detect mutations. Today, many conventional PCR methods are being replaced by real-time PCR, which allows more rapid detection and quantification of the PCR product, as well as detection of different strains of the pathogen in a patient sample.

Other newer techniques include quantitative fluorescent PCR (QF PCR) based on analysis of polymorphic small tandem repeat (STR) markers, and multiplex ligation-dependent probe amplification (MLPA), allowing relative quantification of about 40 different DNA sequences in a single reaction.

DNA Sequencing

This process is used to determine the nucleotide sequence of a given DNA fragment. With this basic information, one can locate regulatory and gene sequences, make comparisons between homologous genes across species, and identify mutations. In 1974, an American team and an English team independently developed two methods. The American team, lead by Maxam and Gilbert, developed a "chemical cleavage protocol," while Frederick Sanger, designed a procedure analogous to the natural process of DNA replication. Even though both teams shared the 1980 Nobel Prize, Sanger's method became the method of choice because of its expediency and simplicity.

This method uses the chain termination method, and extension is initiated at a specific site on the template DNA by using a short oligonucleotide "primer" complementary to the template at that region. The oligonucleotide primer is extended using an enzyme, DNA polymerase, that replicates DNA. In addition to the normal dinucleotides (dNTPs) found in DNA, dideoxynucleotides (ddNTPs) are added to prevent the addition of further nucleotides. Dideoxynucleotides are essentially the same as nucleotides but they contain a hydrogen group on the 3′ carbon instead of a hydroxyl group (OH). The DNA chain is terminated because a phosphodiester bond cannot form between the dideoxynucleotide and the next incoming nucleotide, resulting in a series of related DNA fragments that are terminated only at positions where that particular nucleotide is used. The fragments are then size-separated by polyacrylamide gel electrophoresis and the gel is then exposed to either UV light or X-ray, depending on the method used for labeling the DNA (Sanger et al., 1992).

With the advancements in technology, automated sequencing has been developed so that more DNA can be sequenced in a shorter period of time. This automated procedure, developed by Russell in 2002, is based on the same principles of Sanger's method in which reactions are performed in a single tube containing all four ddNTP's, each labeled with a different color dye (Figure 24.8).

With the advent of new technology, highly sensitive, specific and high throughput methods are now available to study DNA. Microarray is a technique that holds promise and has an exceptional sensitivity and the capacity to view genomic DNA with high resolution or to detect several pathogens simultaneously. Scientists are using this technique to measure the expression levels of large numbers of genes simultaneously or to genotype multiple regions of a genome. New platforms have been developed for sequencing large numbers of DNA fragments simultaneously in parallel, known as next generation sequencing or massively parallel sequencing. These techniques are high throughput and a whole human genome can be sequenced in a few days.

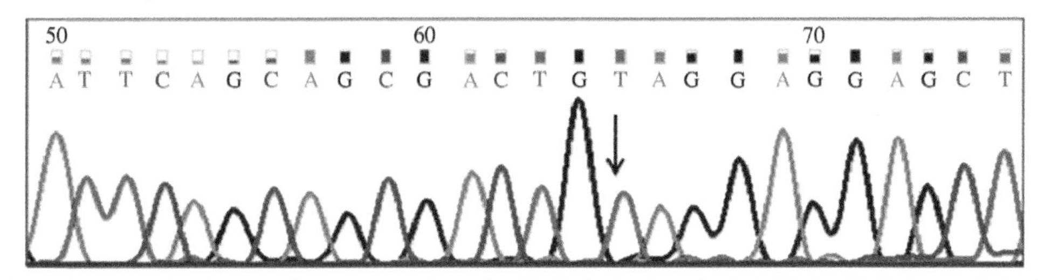

FIGURE 24.8 Electropherogram showing mutation. An arrow shows C to T substitution in codon 318 (CAG to TAG) that results in change of Gutamine (Q) coded by CAG to stop codon TAG, denoted as X.

(A)

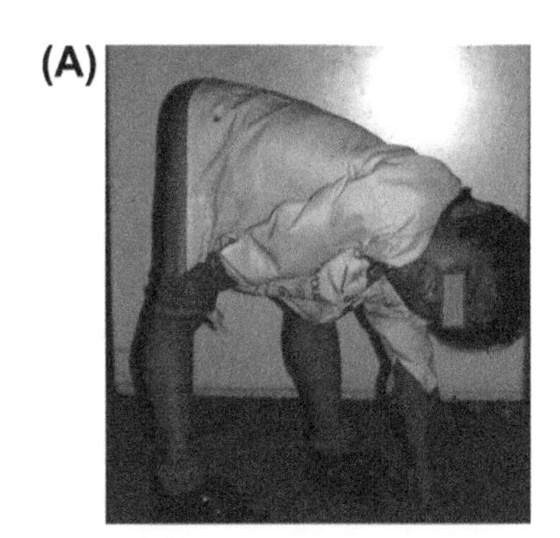

FIGURE 24.9A A patient affected with DMD.

(B)

Exon - pm
Exon - 19
Exon - 3
Exon – 8
Exon - 13

FIGURE 24.9B Multiplex PCR amplification showing presence of 5 different exons after gel electrophoresis.

Principle

The principle of all molecular genetics techniques is to elucidate the structure and function of genes. Molecular genetics techniques along with certain biochemical methods are involved with the direct study of DNA and RNA. This field started with the invention of recombinant DNA technology and the amplification of a specific region in a gene of interest, and today the entire genome can be scanned for small molecular variants (called "single nucleotide polymorphisms," or SNPs), so as to identify the variants' functions and their possible involvement in a given human disease or disorder.

The following case study should help readers to understand how the theoretical concepts of genetics are applied using the various techniques discussed above for the diagnosis of a genetic disorder.

Case Study

A couple with a 5-year old male child with symptoms like slowly progressive muscle weakness in lower legs, difficulty in climbing stairs, and getting up from a squat position, visits a doctor. This boy had an awkward gait due to weakness in his lower legs (Figure 24.9a). His maternal uncle (mother's brother) had similar symptoms and died at about 18 years of age.

Based on family history, clinical symptoms, and biochemical and histopathological tests, the doctor makes a diagnosis of Duchenne Muscular Dystrophy (DMD). DMD is an X linked disorder that affects males only. It is caused by deletion of a part of the dystrophin gene, which can be detected by the absence of the band due to non-amplification of the part of the gene corresponding to the deleted region. The dystrophin gene has 79 exons; 65% of the deletions can be identified by screening 25 exons, which are hot spots in the dystrophin gene. These 25 exons can be amplified by a multiplex PCR (amplification of 5–6 exons simultaneously in a single tube). The products are separated by agarose gel electrophoresis (Figure 24.9b). Each band in the gel corresponds to an exon. If there is deletion of any exon, the corresponding band will be absent.

Multiplex PCR was carried out in this boy, but deletion was not found in the 25 exons tested. Therefore, another advanced technique called Multiplex Ligation dependant Probe Amplification (MLPA) was used to detect all possible deletions in the rest of the exons in the dystrophin gene (Figure 24.9c). This is a high throughput method and results can be obtained in two days. No deletions were detected by MLPA technique either. There was a possibility that the child might be harboring a point mutation that could not be picked up by either multiplex PCR or MLPA, so the entire dystrohin gene was amplified into small fragments and sequenced to look for point mutations in each exon (Figure 24.9d). One mutation was found in exon 54 of the dystrophin gene. Thus, diagnosis of DMD was confirmed and family was counseled.

This case study highlights that sometimes one method is not sufficient to confirm a simple diagnosis, and hence, a combination of different techniques are needed to come to a definitive conclusion.

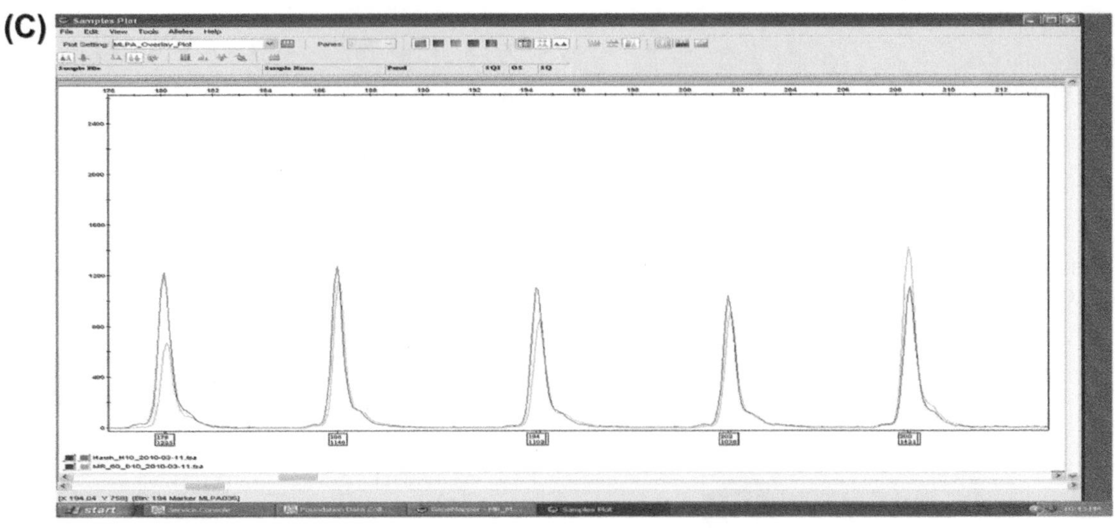

FIGURE 24.9C MLPA picture showing different peaks for different exons.

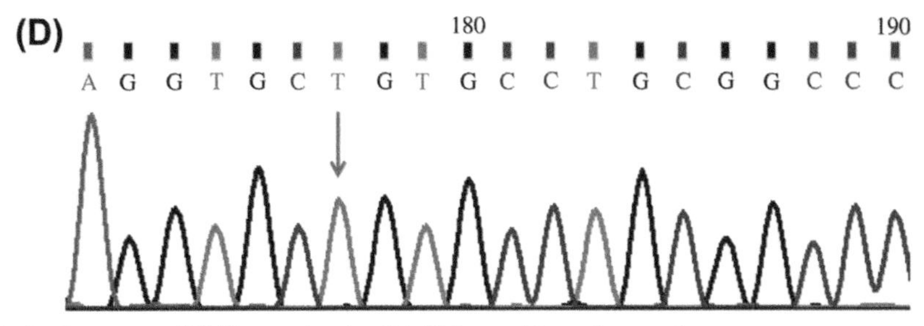

FIGURE 24.9D Partial electropherogram of DMD gene showing C to T change (shown by arrow).

ETHICAL ISSUES

Genetics is an area of medicine with enormous medical, social, ethical, and legal implications. In India, institutions like the Indian Council of Medical Research (ICMR) have issued statements that provide guidelines to physicians and address some of these concerns regarding genetic counseling and testing during pregnancy. In recent years, with the advent of *in vitro* fertilization (IVF), genetic testing has been extended to pre-implantation genetic diagnosis of embryos (PGD). This can be a useful tool in cases where one or both genetic parents have a known genetic abnormality and testing is performed on an embryo to determine if it also carries a genetic abnormality. It is an alternative option of preventing heritable genetic disease, thereby eliminating the dilemma of pregnancy termination following unfavorable prenatal diagnosis. With the development in genetic technologies, it has opened doors for new approaches to health promotion, prevention, diagnosis and treatment of both rare and common diseases. Now the focus has shifted from single gene diseases to a new field of research called genomics, in which all the genes present in the human genome are studied, as well as their interaction with each other and the influence of the environment on their functioning. This evolution has led to a more personalized approach to treat common diseases like cancer, diabetes, and heart diseases, and is referred to as "personalized medicine." Although this new approach holds great promise it, also raises issues about confidentiality and consent. The readers must be aware about the ethical aspects of genetic research, genetic testing, privacy, and disclosure of genetic information, and freedom of reproductive choices. The World Health Organization (WHO) in 1998 made a proposal – "Proposed International Guidelines on Ethical Issues in Medical Genetics and Genetic Services" – for the scientific and medical community. The entire content of the proposal has not yet reached a consensus among all nations, since laws of each nation differ with respect to some particular issues such abortion after prenatal diagnosis, choices about alternatives in assisted reproduction, and the status of the use of human embryos in genetic research. The following lines from IARC Sci Publ 154:131,2001 will highlight the sensitive issues and ethical dilemma of genetic screening:

"Many lives could be saved by screening individuals at risk and targeting preventive behavior to them but there will always be risk of making healthy people 'sick' through detecting presence of predisposing genes and potential for stigmatization and discrimination by society, insurance companies and employers." In India, ethical issues in genetics and genomics are governed by the following guidelines laid down in:

- Statement of Specific Principles for Human Genetics Research ICMR Ethical Guidelines for Biomedical Research on Human Subjects 2000, pages 49-66 and
- Ethical Policies on Human Genome, Genetic Research and Services issued by the Department of Biotechnology, GOI, January 2002.

TRANSLATIONAL SIGNIFICANCE

The term "translational research" is now being used often in scientific gatherings at academic centers and government health departments, in biomedical industries, and in private health research organizations. Translational research emphasizes the practical application of knowledge derived in basic science to the development of new understanding of disease mechanisms, diagnoses, and therapeutics in humans to improve health. In other words, it endeavors to move "from bench to bedside" or from laboratory experiments through clinical trials to develop therapeutics for direct patient applications.

WORLD WIDE WEB RESOURCES

Online tools provide the benefits of getting information at the click of a button. Remember that this technology is just a means of delivering information, so it is important to also read books and journals. Although the Internet is one of the greatest recent inventions, it can still be an unreliable resource, so it is always advisable to consult other sources to acquire in-depth knowledge about a subject.

The following are some useful websites:

1. GeneEd
   http://geneed.nlm.nih.gov/
   Developed by the National Library of Medicine in collaboration with the National Human Genome Research Institute, GeneEd explores topics such as cell biology, DNA, genes and chromosomes.
2. DNA from the Beginning
   http://www.dnaftb.org
   An animated primer on the basics of DNA, genes, and heredity.
3. Genetic Science Learning Center
   http://gslc.genetics.utah.edu/
   From the Eccles Institute of Human Genetics at the University of Utah, a website created to help people understand how genetics affects their lives and society.
4. Understanding Genetics
   http://genetics.thetech.org/
   This website provides a diverse selection of relevant resources, including reviews of basic concepts in genetics, activities, articles on genetics in the news, and a library of answers to your genetics questions. This website is produced by the Tech Museum of Innovation.
5. On-Line Abstract and Journal Articles
   http://www.ncbi.nlm.nih.gov/
   This is the website of the National Centre for Biotechnology Information. It contains links to other useful websites. The OMIM pages on this site have information regarding Mendelian inheritance in humans and information about all the known genes.
6. Genetic Disorders and Their Frequency in India.
   http://www.igdd.iicb.res.in/IGDD/home.aspx
   This is the website of the Indian Institute of Chemical Biology.

REFERENCES

Epstein, R. J. (2003). Human molecular biology: an introduction to the molecular basis of health and disease. *Journal of Society of Medicine, 96*(5), 248–249.

Fan, F. (2002). *Molecular Cytogenetics-Protocol and Applications.* Humana Press.

Gartler, S. M. (2006). The chromosome number in humans: *Nature Reviews Genetics, 7*, 655–660.

Haldane, J. B. S. (1935). The rate of spontaneous mutation of a human gene. *Journal of Genetics, 31*, 317–326.

International Human Genome Sequencing Consortium. (2001). Initial sequencing and analysis of human genome. *Nature, 409*, 860–921.

Mei, R., Galipeau, P. C., Prass, C., Berno, A., Ghandour, G., Patil, N., Wolff, R. K., Chee, M. S., Reid, B. J., & Lockhart, D. J. (2000). Genome wide detection of allelic imbalance using human SNPs and high density DNA arrays. *Genome Research, 10*. 1126–1113.

Mullis, K. B., & Faloona, F. A. (1987). Specific synthesis of DNA in vitro via polymeras catalyzed chain reaction. *Methods Enzymole, 155*, 335–350.

Pauling, L., Itano, H. A., Singer, S. J., & Wells, I. C. (1949). Sickle-cell anaemia, a molecular disease. *Science, 110*, 543–548.

Rooney, D. E., & Czepulkowski, B. H. (1997). *Human Chromosome Preparation. Essential Techniques.* Chichester: John Wiley.

Sanger, F., Nicklen, S., & Coulson, A. R. (1992). DNA sequencing with chain terminating inhibitors [classical article 1997]. *Biotechnology, 24*, 104–108.

Southern, & Mellor, Edwin (1975). Detection of specific sequences among DNA fragments separated by gel electrophoresis. *Journal of Molecular Biology, 98*(3), 503–517.

Speicher, M. R., & Carter, N. P. (2006). the new cytogenetics: blurring the boundaries with molecular biology. *Nature Reviews Genetics, 6*, 782–792.

Strachan, E. T., & Read, A. P. (2004). *Human Molecular Genetics* (3rd ed.). London: Garland Science.

Swansbury, J. (2003). Introduction to the analysis of the human G-banded karyotype. *Methods in Molecular Biology, 220*, 259–269.

Tjio, J. H., & Levan, A. (1956). The chromosome number of man. *Hereditas*, *42*, 1–6.

Toublanc, J. (1992). Comparison of epidemiological data on congenital hypothyroidism in Europe with those of other parts of the world. *Hormone Research (Basel)*, *38*, 230–235.

Trask, B. (2002). Human Cytogenetics: 46 chromosomes, 46 years and counting. *Nature Reviews. Genetics*.

Turnpenny, P., & Ellard, S. (2011). *Emery's Elements of Medical Genetics* (14th ed.). Livingstone: Churchill Livingstone.

Verma, R. S., & Babu, A. (2002). *Human Chromosome* (2nd ed.). McGraw-Hill.

Verma, I. C., & Bijarnia (2002). The burden of genetic disorders in India and a framework for community control. *Community Genet*, *5*, 192–196.

Watson, J. D., & Crick, F. H. C. (1953). Molecular structure of nucleic acid-a structure for deoxyribosenucleic acid. *Nature*, *171*, 737–738.

FURTHER READING

Glick, Bernard R., Pasternak, Jack J., & Patten, Cheryl L. (2010). *Molecular Biotechnology: Principles and Applications of Recombinant DNA*. ASM Press.

Hartl, Daniel L., & Jones - Genetics, Elizabeth W. (2009). *Analysis of Genes and Genomes*. Jones & Bartlett Learning Publishers.

Passarge, E. (2006). *Color Atlas of Genetics* (3rd ed.). Thieme Publication.

Klug, William S., & Cummings, Michael R. (1993). *Elizabeth Arnot Savage. Essentials of Genetics*. Macmillan Publishing Company.

Klug, William S., & Cummings, Michael R. (1986). *Concepts of Genetics*. Merrill Pub. Co.

GLOSSARY

Cytogenetics The study of chromosomes.

Exon Coding regions of a gene.

Idiogram A diagram of the karyotype based on chromosome measurements in many cells.

Intron Non-coding regions of a gene.

Karyotype A display of the chromosomes of a cell by lining them up, beginning with the largest chromosome and with the short arm oriented toward the top.

ORF The nucleotide sequence between the start and stop codons.

Start Codon A code that contains signals for starting the translation. For example, in a nucleotide sequence, AUG is the start codon which codes for the amino acid methionine.

Stop Codon Codon that terminates the protein synthesis in a nucleotide sequence.

ABBREVIATIONS

A Adenine
C Cytosine
G Guanine
T Thymidine
aCGH Array-Comparative Genomic Hybridization
DNA Deoxyribonucleic Acid
FISH Fluorescence *In Situ* Hybridization
HLA Human Leucocyte Antigen
mRNA Messenger RNA
ORF Open Reading Frame
PCR Polymerization Chain Reaction

LONG ANSWER QUESTIONS

1. Briefly describe the structure of a human gene.
2. What are the different kinds of mutations?
3. Describe the different kinds of inheritance patterns.
4. What are the structural abnormalities that are commonly observed in chromosomal disorders?
5. Write a brief description of the different techniques that can be used to detect chromosomal abnormalities (both numerical and structural).

SHORT ANSWER QUESTIONS

1. How are human chromosomes classified?
2. Write the chromosome nomenclature for a male having three copies of chromosome 18.
3. How would you write the chromosome complement in a female showing translocation of the long arm of chromosome 8 on the short arm of chromosome 11?
4. How are transversions different from transitions?
5. What would be the sequence of mRNA transcribed from the following DNA strand ATGCGCCATTGTGTC?

ANSWERS TO SHORT ANSWER QUESTIONS

1. Each chromosome has one short arm and one long arm, separated by a centromere. Each chromosome differs in length and also in the position of the centromere. Based on length and the position of the centromere, chromosomes are classified into A-G groups.
2. 46,XY+18
3. 46,XX,t(8q;11p)
4. Any change in the nucleotide sequence of DNA is termed a mutation. Such mutations/errors can occur during cell division (e.g. replacing adenine for guanine or cytosine for adenine, and so on). When there is a replacement of purine-to-purine or pyrimidine-to-pyrimidine (A-to-G or C-to-T), it is called transition and when the change is from purine to pyrimidine or pyrimidine to purine (A-to-C or G-to-T), it is known as transversion.
5. GACACAAUGGCGCAU

Chapter 25

Antibodies and Their Applications

Fahim Halim Khan

Department of Biochemistry, Faculty of Life Sciences, Aligarh Muslim University, Aligarh, India

Chapter Outline

SUMMARY

With the application of hybridoma technology, advances in the diagnosis and therapeutics of clinically important diseases have started to appear. Monoclonal antibodies have started to displace antiquated polyclonal antibodies, and new recombinant products such as chimeric and humanized versions of antibody constructs (scFv, dsFv, diabodies, and bispecific antibodies) are likely to become future immunological reagents that are going to revolutionize diagnostics and therapeutics for life-threatening diseases.

WHAT YOU CAN EXPECT TO KNOW

Antibodies against any protein can be made in the laboratory by injecting a test animal (e.g. mouse or rabbit) with a pure sample of the protein. With the development of modern biotechnological methodology, new and novel methods have been developed to produce tailor-made antibodies against desired antigens. These include hybridoma technology that makes monoclonal antibodies, their chimeric and human versions that are less immunogenic, development of phage display for antibody screening, small recombinant antibody constructs such as single chain variable fragment (scFv), disulfide-linked variable fragment (dsFv), diabody (bivalent antibody construct), and bispecific antibodies that are preferred for imaging tissues, a variety of clinical applications. Advances in biotechnology and molecular biology have provided methods for re-engineering mouse antibodies to completely replace the rodent antibody sequence with functionally equivalent human amino acid sequences to form human antibodies from transgenic mice. There is no doubt that in years to come antibodies will become indispensable as chemical and research reagents.

Animal Biotechnology. http://dx.doi.org/10.1016/B978-0-12-416002-6.00025-0

HISTORY AND METHODS

IMMUNODIAGNOSTICS: ROLE OF ANTIBODIES

Introduction

We are constantly exposed to a myriad of bacteria, viruses, fungi, and parasites every single day of our lives. Most of the time, we never know this because we are equipped with a defense arsenal, the immune system. This remarkable defense of the human body is constituted by a vast army of cells and associated organs. The immune system is capable of defense against pathogens that may not have been encountered before.

The immune system forms two layers of defense in the human body: non-specific or innate immunity, and specific or adaptive immunity. Innate immunity is constituted by mechanical barrier (e.g. skin, mucous membrane), chemical barrier (e.g. stomach acid), non-specific defense processes (fever, inflammation), and a vast array of pattern recognition receptors (PRRs). PRRs include receptor-like Toll-like receptors, NOD molecules, and many others. PRRs recognize a vast array of conserved microbial motifs (called Pathogen-Associated Molecular Patterns or PAMPs) present on microbes. PAMPs are absent from host cells. PRRs bind PAMPs expressed on microbial surface and clear the microbes from the host body. These PRRs can be present in both a soluble form in the human body (e.g. pentraxins) or localized on surface of the cell (Toll-like receptor).

In contrast to innate immunity, adaptive immunity is more concerned with specific aspects of immune response. Adaptive immunity is constituted primarily by antibodies, B cells and T cells. It has two important qualities: (1) It exhibits high specificity (i.e. it is specifically directed toward specific antigens/pathogens, (2) Memory (i.e. it has a unique ability to remember pathogens). This implies that if the pathogen ever tries to infect the host again, the immune system will respond to infection immediately and protect the host. The adaptive immune system is able to protect us because it has the ability to produce billions of distinct antibodies along with billions of T-cell receptors, each of which recognizes a distinct antigen present on the microbial surface, thus initiating the destruction of the invading pathogen.

History

In 1890 von Behring and Kitasato reported the existence of "antitoxin" serum factor that could protect against lethal doses of toxins in humans. This was the beginning of an era where the significance of antibodies was understood and their use in diagnostic and analytical reagents started getting exploited. With the advent of hybridoma technology by Milstein and Köhler in 1975, the way to produce custom-built antibodies *in vitro* was paved. The hallmark of monoclonal antibodies is their specific binding to a particular antigen, which enables them to find a target precisely *in vivo*. In order to make mouse monoclonal antibodies less immunogenic, Boulianne, Morrison, and coworkers (1984) engineered recombinant antibodies (chimeric antibody created by coupling the animal antigen binding variable domains to human constant domains). Going one step further on the path from fully mouse to fully human antibodies was Greg Winter and coworkers (Jones et al., 1986), who grafted mouse CDRs on human variable and constant regions to form humanized antibodies. The late 1980s onwards saw the development of bispecific antibody constructs with different specificities in their arms, and small genetically engineered antibody fragments such as scFv, dsFv, Fab, diabodies that found their applications in the medical, pharmaceutical, food, and environmental industries.

Antigens and Antibodies

Antigens (or immunogens) can be defined as those foreign molecules that can elicit an immune response. Broadly speaking, proteins and carbohydrates are usually antigenic, while lipids and DNA are usually weakly immunogenic/antigenic or non-immunogenic. Antigens as a whole are not immunogenic; they have small portions on their surface that provoke an immune response. These small restricted portions are called **antigenic determinants or epitopes**.

Antigens (such as proteins and carbohydrates) are large molecules that usually have several antigenic determinants on their surface. Since these biomolecules are present on almost every invading bacteria, virus, and other pathogen, these microbes tend to be strongly antigenic and generate a strong immune response.

Once microbes enter the host body, the immune system gets activated and protects the host through both innate and specific or adaptive immunity. Protection initiated by innate immunity is beyond the scope of this book, and hence will not be discussed here. Adaptive immunity can be antibody-mediated or T-cell mediated. B cells, upon stimulation, produce millions of antibodies that are released in the blood. These antibodies bind to the invading pathogens and clear them from the circulation. This form of immunity is mediated by antibodies (and B cells) and is called humoral immunity. The other form of adaptive immunity is called cell-mediated immunity and is mediated by T cells. T cells work in a similar manner as antibodies. T cells are equipped with specific receptors (similar to antibodies) that recognize antigens. However, T cells do not recognize antigens "alone" like antibodies. They recognize antigens only when they are associated with cell surface antigen presenting molecule: Class I and Class II major histocompatibility complexes

(MHCs). In other words, T-cell receptors recognize antigen + MHC associated complexes while antibodies/B-cells recognize antigens "alone." T cells can be subdivided into two types of antigen specific cells: cytoxic T cells (Tc cells) and helper T cells (T_H cells). Class I MHC + antigen can bind and activate only T_C cells while class II MHC + antigen can stimulate T_H cells. Upon activation, T_C cells secrete small molecules called **perforins** that perforate and kill infected target cells. Stimulation of T_H cells causes release of small messenger molecule **cytokines** that summon other immune cells to the infected site. This entire operation, whether it involves antibodies or T cells, mops up the invading pathogen and clears the host system of infection.

Antibodies are unique in their affinity and exquisite specificity for binding antigens, a quality that has made them one of the most useful molecules for biotechnology applications. An important significance of antibody-based applications is that over a third of the proteins currently undergoing clinical testing in the U.S. are antibodies. Before we move on to select applications of antibodies with possible relevance to biotechnology, let us briefly review the structure and functions of antibodies.

A typical antibody molecule consists of four polypeptides. Two identical polypeptide chains of ~450 amino acids (called heavy chains) and two identical chains of ~250 amino acids (called light chains). These four protein subunits are arranged in a Y formation. Disulfide bonds between light chain-heavy chain and between two heavy chains hold the antibody molecule together. Each light and heavy chain consists of a constant region and a variable region. Starting from the N-terminal, first 110 or so amino acids constitute the variable region. This variable region is variable in sequence and composition among antibodies of different specificities. The rest of the polypeptides in both the light and heavy chains constitute the constant region. This region shows a similar sequence and almost the same composition among different antibodies. The variable region is located at the tips of the arms of the Y-shaped molecule, and the constant region is located at the stem of the antibody. The variable region of light and heavy chains constitutes the antigen-binding site. There are three regions within variable regions (of both light and heavy chains) that show a greater degree of variability than the rest of the region. These regions (three in each chain) constitute six hypervariable regions or complementarity determining regions (CDRs). CDRs are involved in binding the antigens/antigenic determinants.

Based on the minor differences in the constant region of the heavy chain, antibodies are categorized into 5 different classes: IgG, IgM, IgA, IgD, and IgE. The structures and few important functions of antibodies are depicted in Table 25.1. From a biotechnology perspective, the most important class of antibodies is IgG, followed by IgM and IgA. The IgG antibody is the most abundant antibody in the blood. About 80% of the total serum immunoglobulin are IgG. IgG is the only antibody that can cross the placenta (i.e. move from mother to fetus). It is also the main antibody that is involved in opsonization (coating of microbes so that they are easily phagocytosed) and complement activation. IgM provides the first line of defense against invading pathogens as it is produced first in the primary immune response. Being pentameric and equipped with ten binding sites, the IgM antibody is a very effective agglutinator. IgA antibodies are the second most common antibody found in blood. They protect the exposed surface of the body against a plethora of pathogenic organisms. IgE also protects external mucosal surfaces from the assault of pathogens, but is more known for its role in allergic reaction. IgD antibodies are primarily expressed on B-lymphocytes and are probably involved in lymphocyte activation. IgD cannot bind mast cells or activate the complement system. Its concentration in serum is the lowest among all the classes of antibodies (Khan, 2009).

Polyclonal and Monoclonal Antibodies

In humans, the immune system is capable of creating millions of antibodies from which suitable antigen-binding antibodies are selected. Envious of this unsurpassed powerful system for making such exquisite binding sites, scientists have been investigating for decades various strategies to build tailor-made binding sites. The first breakthrough came in 1975 with the development of innovative hybridoma technology. Monoclonal antibody technology–hybridoma technology remains one of the core technologies of biotechnology and hundreds of diagnostically relevant hybridoma have been developed to date.

Principle

Before going into the details of this technology, let's briefly discuss how antibodies are formed when a pathogen enters a host body. A pathogen (or an antigen), due to its large size, have several antigenic determinants. Each antigenic determinant stimulates a B cell of single specificity. Since there are many antigenic determinants on an antigen, several different B cells (having different specificity) are stimulated. Upon stimulation, these B cells start secreting antibodies. Although antibodies are formed against a single antigen, they are directed towards different antigenic determinants of this antigen. Hence, they have different microspecificities, and different affinity. These antibodies, which are directed towards different antigenic determinants of one antigen, and are the product of different clones of cells, are called **polyclonal antibodies**. Polyclonal antibodies have different microspecificities and affinities towards an antigen, or more specifically, antigenic determinant. Polyclonal antibody preparations have been used for several decades to

TABLE 25.1 Class and Function of Antibodies

Class	Subclass	Light Chain	Heavy Chain	Structure	Important Function
IgG	IgG1, IgG2, IgG3, IgG4	λ or κ	γ		Complement activation, toxin neutralization, opsonization
IgA	IgA1, IgA2	λ or κ	α		Prevents viral/bacterial attachment on external surface
IgM	None	λ or κ	μ		Agglutination, killing microbes by complement activation
IgD	None	λ or κ	δ		Lymphocyte activation
IgE	None	λ or κ	ε		Counters parasitic infections, involved in allergic reactions

induce passive immunization against a variety of pathogens and their toxins, including botulinum antitoxin, tetanus antitoxin, diphtheria antitoxin, and viper venom antisera (all of them raised in horse). These antibody preparations are used prophylactically or therapeutically for preventing or treating a number of diseases.

Polyclonal antibodies constitute a minor component in the complex mixture of serum proteins in which they are present. Therefore, polyclonal antidodies lack the degree of purity and specificity required for many of the current immunological techniques where an increased assay sensitivity is required. Theoretically, if a single clone of B-cells can be isolated from an animal and grown in culture, it should provide a continuous source of antibody that has single specificity. Unfortunately, B cells live for only a few days as they survive poorly once they are taken out of the body. The solution to this problem was invented by Köhler and Milstein in 1975. They fused a rodent's mortal antibody-producing cell with an immortal tumor cell (which does not have the ability to produce antibody) to form an immortal antibody-producing cell. These cells were then isolated and cultured to become a source of antibody of single specificity. These antibodies that are directed towards a single antigenic determinant and secreted by a single clone of cells are called monoclonal antibodies. Once a pure culture of immortal hybrid cells that secretes monoclonal antibodies has been isolated, it can be grown in a culture dish or grown as ascite tumors in a mouse; it could also be frozen and reused later (Figure 25.1). These hybridomas will grow and divide and will mass produce antibodies of a single type (i.e. monoclonals).

Hybridoma Technology and Methodology

Let's discuss this technique in a little detail. A lab animal (such as the mouse) is immunized with an antigen. The spleen of the lab animal is removed and the antigen-stimulated B-cells are then isolated. These antigen stimulated B-cells, which are mortal antibody-producing cells, are then fused with immortal non-antibody producing cells. These non-antibody-producing immortal cells are special cells that have been developed by Milstein and Köhler. They are myeloma cells (cancerous B cells) that do not have the ability to produce antibodies, but can divide indefinitely. Isolated B cells are then fused with immortal myeloma cells (with the help of polyethylene glycol (PEG) or inactivated sendii virus) that fuse juxtaposed lipid bilayers of both the cells to form hybrid cells or

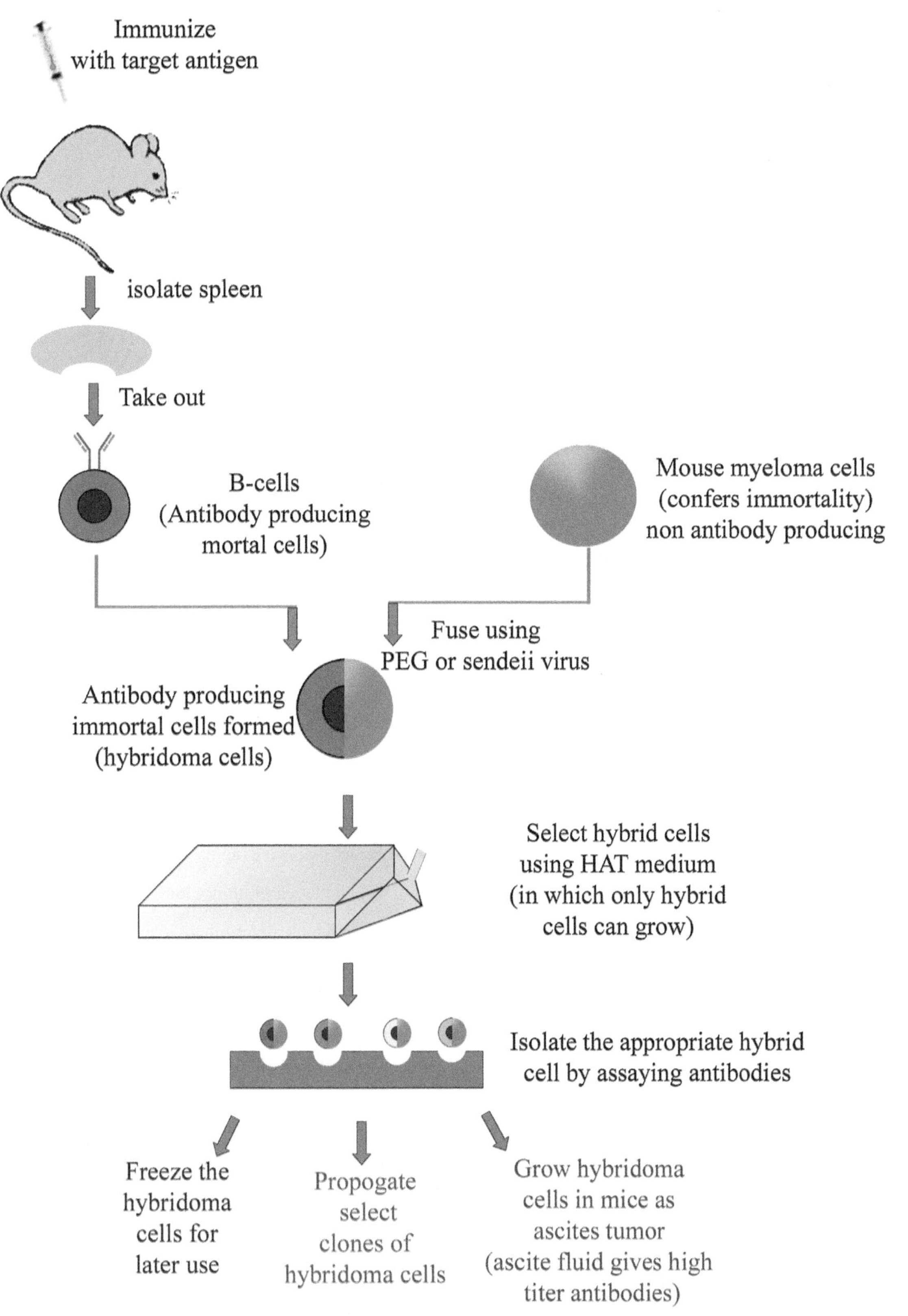
Immunize
with target antigen
isolate spleen
Take out
B-cells
(Antibody producing
mortal cells)
Mouse myeloma cells
(confers immortality)
non antibody producing
Fuse using
PEG or sendeii virus
Antibody producing
immortal cells formed
(hybridoma cells)
Select hybrid cells
using HAT medium
(in which only hybrid
cells can grow)
Isolate the appropriate hybrid
cell by assaying antibodies
Freeze the
hybridoma
cells for
later use
Propogate
select
clones of
hybridoma cells
Grow hybridoma
cells in mice as
ascites tumor
(ascite fluid gives high
titer antibodies)

FIGURE 25.1 Production of monoclonal antibodies (a simplified overview).

hybridoma cells. These hybridoma cells are then grown in medium containing aminopterin and hypoxanthine. Myeloma cell lines lack the enzyme *hypoxanthine guanine phosphoribosyl transferase* (HGPRT), a key enzyme of purine synthesis by the salvage pathway. The other *de novo* pathway is functional for these cells, but can easily be inhibited by aminopterin. In a medium containing aminopterin, only those cells that have obtained the HGPRT gene from B cells (and hence have a functional salvage pathway) and an immortality trait from myeloma cells multiply; all others die. Hence, only those cells that carry the immortality of myeloma cells and the antibody-producing capacity of B cells survive; all the other cells (including unfused cells) are eliminated. The multiplying and expanding cells are cloned by growing single cells as colonies. The antibodies produced by them are screened for antigen binding activity.

Important facts about polyclonal antibodies:

- Polyclonal antibodies recognize multiple epitopes on any one antigen, hence serum obtained after immunization with antigen will contain a heterogeneous mixture of antibodies of different affinities and microspecificities.
- They are constituted mainly of the IgG class of antibodies.
- They are inexpensive to produce as the technology/skills required for their production is much cheaper.
- The time required for producing polyclonals is usually short.

Advantages of using polyclonals:

- Polyclonals recognize multiple epitopes on any antigen.
- Polyclonals antibodies bind target proteins on multiple epitopes and hence make the protein more visible. Multiple epitopes generally provide better detection.
- Polyclonal antibodies are better suited for easy detection of denatured proteins as they are more tolerant of minor changes in the antigen (e.g., heterogeneity of glycosylation) or change in the exposure of antigenic determinants upon slight denaturation.
- Polyclonals are better suited for identifying proteins of high homology to the immunogen as there are chances that in a polyclonal mixture some antibodies will react with highly homologous, common antigenic determinants.

Disadvantages of using polyclonal antibodies:

- They are prone to batch variability (i.e. the antibody population produced against the same antigen in the same animal may not be exactly the same).
- Highly purified antigens should be used otherwise it might lead to production of crossreactive antibodies.
- Polyclonal antibodies are not useful for probing specific domains of antigens because the antiserum will contain a "mixture" of antibodies that will usually recognize many domains. Crossreactivity is one of the major problems of polyclonal antibodies as they contain antibody subpopulations that recognize different antigenic determinants on the same antigen.

Important facts about monoclonal antibodies:

- They are directed against only one epitope (antigenic determinant) on an antigen.
- Monoclonals have only one antibody subtype. They all belong to one class and subclass and have the same affinity and specificity.
- High technology and training is required for producing and maintaining monoclonals.
- The time scale required for producing monoclonals is quite long.

Advantages of using monoclonal antibodies:

- Once hybridomas producing monoclonals are made, they are a constant and renewable source of antibodies and all batches of monoclonals are identical. Monoclonals are produced by a colony of cells derived from a single hybridoma. Each cell contains identical genetic material and produces homogenous antibodies that recognize a single antigenic determinant.
- Monoclonals detect one epitope/antigenic determinant on any antigen. This offers the following advantages:
- Monoclonals specifically detect one target epitope, hence they are less likely to cross-react with other proteins.
- Because of their exquisite specificity, monoclonal antibodies are excellent as the primary antibody in an assay or for detecting antigens in tissue. They give significantly less background staining than polyclonal antibodies.
- Monoclonals give highly reproducible results.

Disadvantages of monoclonals:

- If an antigenic determinant is lost through chemical treatment of the antigen, monoclonals will not be able to detect it. Polyclonal antibodies can probably detect such an antigen as it contains antibodies that can bind more than one antigenic determinant.
- High cost and training are required for producing monoclonals.

Application of Monoclonal Antibodies

Monoclonal antibody production technology has revolutionized the world of biotechnology. Continuous advances in recombinant DNA technology over the years have provided numerous ways to design monoclonals that are more robust and efficient as compared with their original murine versions. Monoclonal antibodies have been used as diagnostics, therapeutics, research reagents, and drug targeting agents for various diseases. Hybridoma technology, along with recombinant DNA technology, have successfully led to the development of chimeric antibodies, humanized antibodies, antibody constructs, and bispecific antibodies; all these have enormous potential for diagnostic and therapeutic uses.

With the development of monoclonal antibody production technology and the availability of tailor-made antibodies that have desirable qualities (such as exquisite specificity and high affinity), a perfect antibody reagent became available to scientists and clinicians. Monoclonals slowly started replacing polyclonals as the preferred reagent for both diagnostics and therapeutics (Vitetta and Ghetie, 2006).

Immunosorbent Chromatography

The classic use of antibodies involved their use as solid-phase binding reagents. Use of monoclonal antibodies enabled binding and purification of desired proteins that displayed a single antigenic determinant. A small example: there are four hormones – chorionic gonadotropin, lutropin, thyrotropin, and follitropin – that have two chains, α and β. The α chain is common to all, and the β chain is unique to each one. When polyclonal antibodies are formed against any one of them, it cross-reacts with all of them, as they contain antibodies to the α chain (which is common to all of them). However, when monoclonals are formed against the β chain of any hormone, it detects only that particular hormone and not the other three. Monoclonal antibodies can be coupled onto solid support, such as agarose gel or sephadex matrix, and can be used as an immunosorbent column that can be used to isolate a single protein when a mixture of proteins (such as serum) is passed through it. Since monoclonal antibodies are specific for a single epitope, usually a single protein gets bound to the solid-phase monoclonal antibodies and the rest of the unbound proteins get eluted and are discarded. This target protein, which is now highly purified, is eluted and used. Use of monoclonals has allowed single-step purification of novel proteins such as human interferon and human clotting factor. Use of polyclonals for similar purification would have resulted in purification of several proteins (instead of one), as polyclonal antibodies would have cross-reacted with other proteins present in the mixture.

Blood Typing Reagent

Monoclonals have been used as preferred blood typing agents for several decades. The large volume requirements for high-quality ABO and Rh typing reagents is now made possible with the availability of monoclonal antibodies. Superior detection reagents can now be prepared from blends of at least two antibodies (i.e. anti-A and anti-B monoclonals) to optimize the intensity of agglutination for slide tests, thus making them easily visible as well as able to detect weaker sub-blood groups such as Ax and Bw. This was practical unthinkable with polyclonal antibodies.

More Sensitive in RIA

Monoclonals are several-fold more sensitive than polyclonals in detecting an antigen in serum or other body fluids. Concentration of Transcortin (cortisol-bindng protein) was determined by radioimmunoassay using both polyclonal and monoclonal antibodies. It was found that monoclonal antibodies detected cortisol at several hundred-fold lower concentration as compared to polyclonal antibodies.

Positive and Negative Selection of Cells

Positive and negative selection of cells is one of the very important classical applications of monoclonal antibodies. In positive selection, monoclonals bind to the target cell (that we want to isolate) and separate it from rest of the contaminating cells. In negative selection all the unwanted/contaminating cells are bound to monoclonal antibodies, and these antibody-coated cells are then removed from the system and leave only the cells of interest. A brief explanation follows for these cell selection processes.

Negative Selection

CD4+ T cells can be isolated from blood by depletion of all non-T cells (B cells, macrophages, and natural killer cells); cells that remain in solution afterwards are CD4+ T cells. Blood cells are incubated with a mixture of monoclonal antibodies to coat unwanted cells. The cells are targeted for removal by monoclonal antibodies that bind cell surface antigens present on contaminating cells and not on CD4+ cells such as CD14, CD16, CD36, and CD123. Moreover, these antibodies are bound on dextran-coated magnetic beads. Subsequent exposure to a strong magnetic field removes the unwanted bead-coated cells, leaving behind the desired cell population.

Positive Selection

Positive selection can be best illustrated with the following example. Metastatic carcinoma (cancer of epithelial origin) usually expresses cytokeratin 7/8 protein, which can be recognized by anti-cytokeratin antibody-1. Such malignant cells are detected by positive selection of such cells in peripheral blood, bone marrow, and lymphoid tissue of patients with metastatic carcinomas. Anti-cytokeratin antibody is coated on magnetic microbeads and incubated with disseminated cells of metastatic carcinomas. These antibodies bind and isolate/enrich tumor cells that display cytokeratin 7/8. It is reported that an enrichment of up to 10,000 times of desired cells can be achieved by this method. The desired cells are separated from the rest of the cells using a magnetic cell separator.

Diagnostic and Therapeutic Application

Ever since the first report of the successful use of monoclonal antibody for the treatment of human B-cell lymphoma in 1986 (muromonab-CD3/orthoclone OKT3), hundreds of monoclonal antibodies have been incorporated into standard therapeutic and diagnostic applications for various diseases. These monoclonal antibodies have proved to be an immensely useful scientific research and diagnostic tool.

Owing to their unrivalled specificity, monoclonal antibodies have also been in the spotlight as potential therapeutics for a variety of diseases, including cancer, infectious disease, cardiovascular disease, and transplantation disease. Major applications of monoclonal antibodies can be broadly divided into two main categories: (1) Diagnostic applications, and (2) Therapeutic applications.

During the past several decades, monoclonals have found tremendous applications in diagnostics, therapeutics, and drug targeting. The diagnosis of any infectious disease often requires the demonstration of the pathogen or a specific antibody against the pathogen or its toxin. In some infectious diseases, such as reproductive or respiratory infections, the pathogen can be demonstrated throughout the course of infection, while for others, it is visible for a short period of time. Specific antibody-based tests identify the pathogens associated with the disease or toxins they secrete or any other protein they might inadvertently release. Monoclonal antibodies recognizing unique antigenic determinants on pathogens/toxins can be developed. MAb can recognize a single antigenic determinant characteristic of a pathogen. This restricted reactivity allows for precise identification of the organism of interest, which is the major advantage of monoclonals over polyclonal antisera. In the case of a pathogen having a subtype defined by unique antigenic differences, specific monoclonal antibodies can be used, whereas conventional antisera need laborious absorption to remove cross-reactive antibodies. Specific monoclonal antibodies of diagnostic value have been tailor-made and successfully used to detect pathogens such as Leishmania, Trichomonas, Trypanosoma, and several other pathogens. With enhanced sensitivity and exquisite specificity of monoclonal antibodies, diagnostic test systems that detect a number of animal viruses have also been developed. Animal viruses such as HIV, herpes virus, rotavirus, rabies virus, human papilloma virus, and chikengunya virus are now being detected by immunodiagnostics.

Both the diagnostic and therapeutic properties of monoclonal antibodies exploit the unrivalled specificity of monoclonals. Once the correct target antigen has been identified, monoclonal antibodies can be raised against an antigen/antigenic determinant. This monoclonal antibody can be used for diagnostic or therapeutic purposes. Once injected, this antibody will build to a specific target cell and not to the multitude of other cells available. Monoclonal antibodies can be used alone (naked monoclonal antibodies) to trigger the desired response, or can be linked with toxic payloads such as radioisotopes or toxins or drugs/prodrugs (conjugated antibodies) to have the desired effect (Figure 25.2).

The proper usage of monoclonal antibody preparations is dependent on the fact that the target cell expresses antigen/antigenic determinants that are unique to that particular cell and not to neighboring cells. This is common among cancerous cells. Transformation of a normal cell to a cancerous state is normally associated with expression of unique surface antigens on tumor cells that appear foreign to the host immune system. Such antigens (called tumor antigens) are usually not expressed at all on normal cells, or expressed at very low concentration. Targeting of such antigens by naked monoclonal antibodies can have the desired effect. In other words, they can be used to kill the target cell by antibody-dependent cell-mediated cytotoxicity (ADCC). For example, in B-cell cancer of mouse, monoclonal antibodies targeted towards CD20 and CD10 antigens induced ADCC or complement-mediated cell lysis. Some antitumor effects of rituximab (a monoclonal antibody used in the treatment of non-Hodgkin's lymphoma) are used because of complement-mediated cell lysis. Another example of the use of naked monoclonal antibody is alemtuzumab (Campath), which is used to treat patients with chronic lymphocytic leukemia (CLL), cutaneous T-cell lymphoma (CTCL), and T-cell lymphoma. It is also administered in some other conditions for immunsuppression for bone marrow transplantation, kidney transplantation, and islet cell transplantation. Alemtuzumab binds to CD52 antigen, a protein present on the surface of mature B and T cells. After treatment with this antibody, these CD52-bearing cells are targeted for destruction.

***Diagnostic applications in cancer*:** For diagnostic purposes, most monoclonal antibodies are linked with a gamma emitter so that radiation can pass through several layers of cells for detection purposes. The most commonly used radioisotope is Technetium (T_c^{99m}). Some monoclonal antibodies commonly used for diagnostic purposes include CEA-Scan (mouse monoclonal antibodies that detect human carcinoembryonic antigen (CEA) expressed in high levels during colorectal cancer), Indimacis 125 (mouse monoclonal antibodies that detect tumor antigen CA-125 expressed in ovarian adenocarcinoma), and several others that are currently used in diagnostic imaging.

***Therapeutic applications in cancer*:** Monoclonal antibodies can also be used to deliver cytotoxic payloads directly to tumors by conjugating them with radioactive isotopes or toxic chemicals, drugs, or prodrugs (Table 25.2). Monoclonal antibodies attached to a

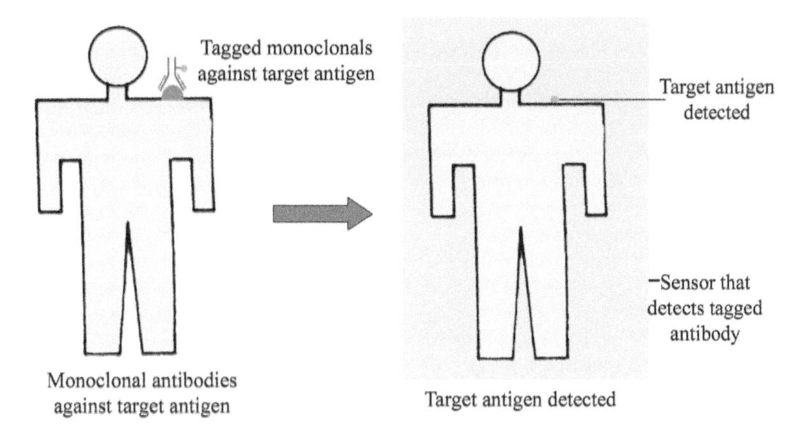

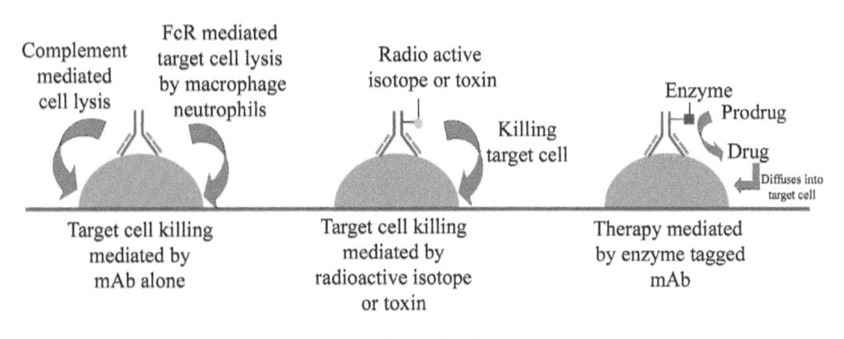

FIGURE 25.2 Major applications of monoclonal antibodies.

radioactive substance, drug, or toxin are called conjugated monoclonal antibodies. The monoclonals are used as a targeting device to take these payloads directly to the cancer cells. Conjugated monoclonals circulate in the body until they can find and bind the target antigen. They then deliver the toxic payload where it is needed most. This lessens the damage to normal bystander cells in other (or neighboring) parts of the body.

Conjugated monoclonal antibodies can be divided into several different types depending on what they are linked:

- Monoclonals linked with chemotherapeutic drugs are referred to as chemo-labeled. For example, one of the chemo-labeled antibodies approved by the FDA is brentuximab vedotin (Adcetris), which targets the CD30 antigen found on B cells and T cells. The antibody is attached to a chemotherapy drug called monomethyl auristatin E (MMAE). It is used to treat Hodgkin's lymphoma and anaplastic large cell lymphoma.
- Monoclonals linked with radioisotopes are referred to as radiolabeled. Ibritumomab tiuxetan (Zevalin) and tositumomab (Bexxar) are examples of radiolabeled monoclonals. Both of these antibodies are directed against the CD20 antigen, but they each have a different radioactive particle attached. They deliver their toxic payloads directly to cancerous B cells and are used in treating some types of non-Hodgkin's lymphoma.
- Monoclonals attached to toxins are called immunotoxins. Immunotoxins have cytotoxic poisons (toxins) attached to them, which makes them similar in many ways to chemo-labeled and radiolabelled monoclonal antibodies. At this time no immunotoxins are approved to treat cancer, although several of them are being explored.

Radioisotopes commonly linked to monoclonal antibodies for cytotoxic effect include strong emitters such as iodine-131 and 125, and yttrium-90. For selective delivery of drugs to target cells, several types of drugs and prodrugs have also been linked to monoclonal antibodies and targeted selectively to tumor sites. Chemotherapeutic drugs such as adriamycin, auristatins, methotrexate, doxorubicin, and prodrugs such as etoposide, have been linked to mAb and selectively delivered to tumor sites with moderate success.

Choice of Linkers

The large size of intact antibody–drug-linked conjugates prevents them from crossing plasma membranes (such as lysosomal membranes) once they are endocytosed. We

TABLE 25.2 Antibodies Approved as Drugs by the Food and Drug Administration (FDA)*

Product	Directed Against (Antigen)	Therapeutic Area	Trade Name
	Murine Monoclonal		
Muromonab-CD3	CD3	Arthritis inflammation	Orthoclone OKT3
Arcitumomab	Carcinoembryonic antigen	Metastatic colorectal cancer (detection)	CEA Scan
Tositumomab	C2O	Oncological disease	Bexxar
Ibritumomab Tiuxetan	C2O	Non-Hodgkin's lymphoma	Zevalin
Imciromab	Cardiac myosin	Cardiovascular disease	Myoscint
Capromab pentetate	Tumov surface antigen PSMA	Prostate adeno carcinoma (detection)	Prostascint
Chimeric			
Basiliximab	C25 (IL-2-receptor)	Prophylaxis of acute organ rejection	Simulect
Infliximab	TNF-α	Crohn's disease	Remicade
Rituximab	CD 20	Non-Hodgkin's lymphoma	Rituxan
Antibody Fragments			
Nofetumomab (Murine Fab)	Carcinoma associated antigen	Small cell lung cancer (detection)	Verluma
Igovomab (Murine $(Fab)_2$ fragment	Tumor antigen–CA	Ovarian carcinoma	Indimacis 125
Sulesomab (Murine Fab)	Granulocyte antigen NCA-90	Osteomyelitis (detection)	Leukoscan
Abciximab (Chimeric Fab)	Platelet surface receptor	Prevention of blood clot	Reopro
Humanized			
Daclizumab	CD25 (IL-2receptor)	Prevention of acute kidney transplant rejection	Zenapax
Gemtuzuumab	CD 33	Acute myeloid leukemia	Mylotarg
Trastuzmab	Human epidermal growth factor receptor (Her-2)	Metastatic breast cancer	Herceptin
Ranibizumab	Vascular endothelial growth factor (VEGF) receptor	Eye diseases	Lucentis
Palivizumab	Respiratory syncytial virus F Protein	Respiratory tract disease	Synagis
Human Phage Display/ Synthetic Antibody			
Adalimunab	TNF-α	Immune disorders, Crohn's disease	Humira
Xenomouse			
Panitumumab	Epidermal growth factor receptor (EGFR)	Metastatic colorectal cancer	Vectibix

**This table gives only a few representative examples.*

know that drugs conjugated to antibodies can exert their specific action only after dissociation from the antibody and subsequent diffusion to their targets in the cytoplasm. Studies have revealed that, upon internalization, the conjugates were delivered to lysosomes, where their antibody moiety gets completely digested and releases protein-free derivatives of the cytotoxic drug.

With this in mind, scientists have designed various linkers that are relatively stable outside the cells (i.e. in the blood), but labile once the antibody–drug-linked conjugate is inside the cell. For example, anthracyclin drugs (such as doxorubicin and calicheamicin) were conjugated to antibodies via hydrazone-based linkers that are relatively stable at neutral pH of the blood, but are hydrolyzed in the acidic environment of phagolysosomes. Similarly, another cytotoxic class of drug, toxoids, were coupled to antibodies via pH-labile ester bonds that break at low pH. Similar drugs, auristatins, were attached via disulfide linkage. These can get cleaved by disulfide exchange with thiols such as reduced glutathione, which is about 1,000-fold higher concentration inside cells than in blood. Cytotoxic drugs such dolastatins and cyclopropylindole have been conjugated via peptide linkers containing a valine–lysine sequence. These are stable in blood but get degraded in lysosomes, presumably by enzymes such as cathepsin B or β-glucuronidase.

Chimeric and Humanized Antibodies

Ideally, infusion of murine monoclonal antibodies for therapeutic purposes should generate an immediate immune response. These monoclonals have been designed for specific antigens on target cells. Surprisingly, initial clinical trials involving administration of monoclonal antibodies into human subjects proved quite disappointing (Hwang and Foote, 2005). There were three main reasons for this:

- Monoclonal antibodies were initially produced in mouse, and hence these antibodies were treated as foreign by the human system. Introduction of these murine monoclonals elicited a strong immune response that entailed cleaning of monoclonals before they reached the target site. Such an immune response that occurs in the human body after the administration of mouse antibodies is also known as the human anti-mouse antibody (HAMA) response. A single injection of murine monoclonal elicited an immune response in about 55–80% of all patients; and human anti-mouse antibodies are usually detected within two weeks of antibody administration. Such an immune response in human subjects usually destroys subsequent administration of therapeutic monoclonals, severely limiting its efficacy.
- The human effector functions (complement activation and binding to Fc receptors) were poorly recruited by the Fc regions of murine monoclonal antibodies.
- Murine monoclonals displayed a shorter half-life in serum when administered to humans.

Methodology

Genetic engineering partly solved these problems by providing an alternative route. Since almost all of the above problems were associated with Fc regions of murine monoclonals, the mouse Fc region was replaced by their human counterparts and its Fab region (which recognizes antigen) was left intact. This technique of recombinant DNA technology was invented by Morrisen et al. (1984). The strategy entailed production of a new form of monoclonal antibody consisting of mouse variable (Fab) regions and human constant (Fc) regions. Such monoclonals are referred to as **chimeric antibodies**. Briefly, production of chimeric antibodies is similar to production of monoclonals, as both involve fusion of mouse antibody-producing cells with immortal myeloma cells that cannot produce antibody. For synthesizing chimeric antibodies, mouse B-cell DNA is isolated and its constant region is replaced by its human counterpart. The mouse variable region is left intact. This chimeric mouse–human gene of the antibody is then introduced into mouse myeloma cell lines for the production of chimeric monoclonal antibodies. Such chimeric antibodies have been shown to be less immunogenic (chimeric antibodies were immunogenic in about 1% of human patients), have an extended serum half-life (about 250 hrs as compared to 35 hrs for murine monoclonals), and exhibit increased Fc-mediated effector functions as chimeric antibodies carry a human Fc region. It has been estimated that the constant region contributes 90% of the immunogenicity of chimeric antibodies, with the variable region contributing the remaining 10%.

The entire Fab region of an antibody is not essential for binding of the antigen. Regions within the variable region of light and heavy chains that bind the antigenic determinant are referred to as **complementarily determining regions (CDRs)** or **hypervariable** regions. Overall, each antigen-binding site (one in each arm) has six CDRs, three from the light chain and three from the heavy chains. Further humanization of the chimeric antibody was undertaken to reduce the immunogenicity of monoclonals to the minimum. This involves cutting out the nucleotide sequences of these six CDRs from the mouse antibody and splicing them into human antibody genes (i.e. in both light and heavy chains). Such hybrid antibodies that carry only murine CDRs in an otherwise human antibody are referred to as **humanized antibodies**. This technique was perfected by Jones et al. in 1986. This strategy, termed CDR grafting, decreases the amount of murine sequence in a monoclonal from 30% (in chimeric antibody) to 3% (in humanized antibody), and greatly reduces the risk of HAMA. Humanization has overcome major hurdles that limited the therapeutic effectiveness of murine monoclonals.

Phage Display Technology: Screening Recombinant Antibody Libraries

A phage display antibody library is a collection of bacteriophage particles that display antibody (fragments) protruding from their surface. Antibody fragments are cloned into bacteriophage such that antigen binding variable regions are displayed on the phage surface, providing a phenotype that can be used to select the phage as well as identify the genotype of recombinant antibody clone (Hoogenboom, 2005). The phage particles that are normally chosen are M13, or fd phage vector. Both vectors are bacteriophages and contain outer coats made up of protein. The antibody (fragments) genes are so ligated into the phage DNA that they are expressed on the coat proteins. The antibody fragment is ligated/fused with gene III protein of M13 (which is expressed as five copies on the surface of phage) or gene III protein of fd phage virion (which is expressed about four copies per virion). The phage particles are then incubated with their host *E. coli*, which facilitates phage replication. Bacteriophage DNA then directs the synthesis of fusion gene products containing the protein of interest in the coat. The entire phage library can then be screened in order to identify the antibody fragment of interest as described below.

Methodology

In 1985, Smith described for the first that time that a foreign DNA sequence could be cloned into filamentous bacteriophage and that such cloned sequence can be expressed on the surface of the phage particle. McCafferty et al. (1987) used this technique for the first time for expression of antibody fragments. To clone antibody fragments, V_H and V_L mRNA are isolated from hybridoma or spleen or lymph cells. mRNA is reverse transcribed into cDNA and amplified by polymerase chain reaction. V_H and V_L genes are then fused with a short linker gene that codes for $(Gly_4Ser)_3$ sequence. This V_H–linker–V_L gene construct codes for an antigen-binding fragment that is referred to as a single chain variable fragment (scFv). This gene construct is then fused with gene III protein of M13 or fd phage virion. The phages are then incubated (transformed) with *E. coli*. Phage particles transform *E. coli* and replicate inside the bacterial cell. The fusion gene product is expressed on surface coat proteins of newly formed phage particles.

The phage particles are then screened to identify those phages that code for the protein of interest. This selection process is called **biopanning** or **affinity screening**. In biopanning, the library of phages displaying the antibody fragments (or its corresponding ligand) is passed over immobilized antigen (or any other target molecule) bound to a column/bead/membrane. All the phages that display desired antibody fragments are retained on the solid support. The bound phage is then subsequently eluted either by changing the pH of the elution buffer or by adding competing ligands in the eluting buffer (Figure 25.3). The eluted phage particle can then be incubated again with *E. coli* and this process may be repeated to isolate those antibody fragments that bind with highest specificity or affinity. In 2002, adalimumab, a phage display derived antibody, became the first marketed fully human monoclonal antibody product (Leader and Golan, 2008). This fully human antibody was

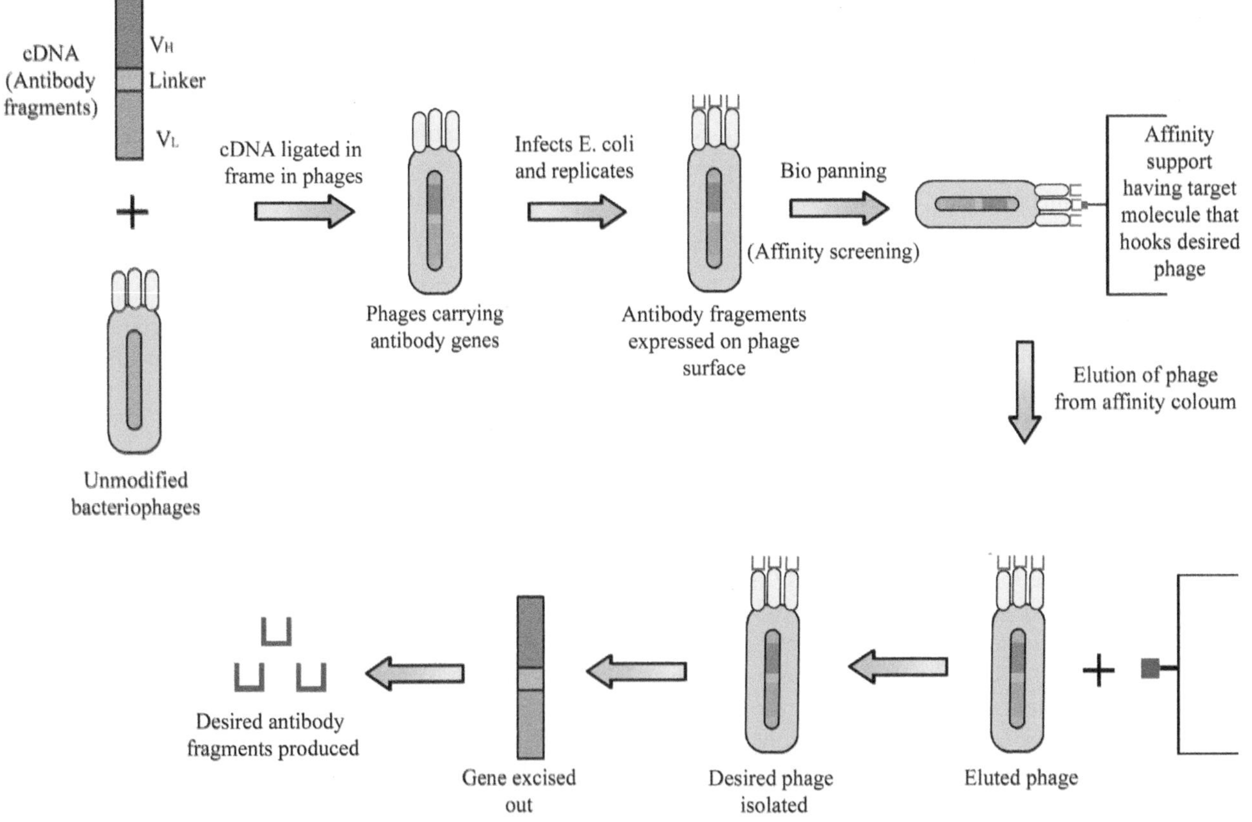

FIGURE 25.3 Phage display technology for antibody screening.

not rapidly cleared by human patients' immune systems and displayed a half-life of 2–3 weeks.

Recombinant antibody library types can be of two types: (1) Immune antibody libraries, and (2) Non-immune antibody libraries. Immune libraries are those that contain genes/cDNA coding for antibody/antigen-binding fragment derived from B-lymphocytes that have been previously immunized with target antigen. Such libraries have a V_H/V_L gene pool cloned from B cell obtained from immunized animals or naturally infected or immunized humans. Non-immune antibody libraries are produced just as immune antibody libraries but using B cells from non-immunized donors. These include healthy un-immunized humans or animals not immunized with target antigen. Needless to mention, non-immune libraries will generate a small number of positive clones for any antigen screened.

Antibody Constructs

In addition to single chain variable fragment (scFv), other commonly used antibody constructs include disulfide linked variable fragment (dsFv), antigen binding fragment (Fab), diabody, and bispecific antibodies. In dsFv, V_H and V_L chains are expressed as separate polypeptides in bacteria. These two chains are covalently assembled by engineering an interchain disulfide bond to the dsFv antibody. Antigen-binding fragments that do not have this specially constructed interchain disulfide bond have also been produced (called Fv fragments), but were found to be prone to aggregation and dissociation.

Fab antibody fragment constructs are similar to dsFv fragments but have an extra constant domain in both the light and heavy chains. Fabs consist of two polypeptide chains expressed in bacteria. One contains the polypeptide-coding region of the light chain variable domain plus the constant domain, and the other contains the heavy chain variable domain plus one constant domain. Just as in dsFv, light and heavy chains are linked by a disulfide bond. It should be noted that scFv contains two disulfide bonds whereas Fabs contain five disulfide bonds. Since Fabs are relatively large constructs and they have a large interface of light chain and heavy chain along with the presence of disulfide bonds, they show increased stability. Even though Fabs require the association of two chains, they often occur quite efficiently in bacteria.

Recombinant antibody fragments such as scFv, dsFv, Fab, can be expressed efficiently in microbes, particularly in *E. coli*. Since disulfide bonds form in an oxidizing environment and proteins are synthesized in cytosol (which is highly reducing), these proteins are exported from the cytosol into the periplasmic space. The periplasmic space is the compartment between inner and outer membranes of the bacterial cell. This space is highly oxidizing in nature. Export into this space is easily accomplished by attaching one of the well-characterized leader peptides to the N-terminus of the antibody fragment. Proteins expressed in the periplasm can be recovered from osmotic shock or from total cell lysate. Moreover, prolonged high-level expression of antibodies at 37°C renders the outer membrane of *E. coli* permeable and proteins and protein fragments can be recovered from the culture media.

Diabody

Recombinant DNA technology was able to generate small bivalent antibody constructs (diabody) consisting of two V_H and two V_L domains. In a diabody, each V_L domain is connected with one V_H domain in scFv by a short linker. This construct is similar to scFv fragments with the difference that the peptide linker is shorter (range from 3 amino acids to ~10 amino acids) than that which is present in the scFv fragment. The shortening of the linker does not allow bending of the polypeptide chain and formation of the antigen-binding site from the V_L and V_H domain from the same "scFv" fragments. Hence, two such scFv chains associate or dimerize to form two antigen-binding sites. Each antigen-binding site of the diabody is constructed by a V_H domain of one "scFv" fragment and a V_L domain of the other "scFv" fragment. The resulting diabody has two antigen-binding sites pointing in opposite directions. If two of the same scFv fragments are allowed to dimerize, a bivalent antibody construct results, which is similar to the antibody but lacks an Fc region. If two different "scFv" fragments are used, the resulting diabody will form two different antigen-binding sites simultaneously. Such diabodies are referred to as bispecific antibodies or bispecific diabodies (e.g. diabody–vcF4 that is anti-CD30 and is conjugated with the potent anti-tumor drug monomethyl auristatin).

Human Antibodies from Transgenic Mouse

Advances in molecular biology involving the manipulation of gene sequences *in vitro* and the expressions of these manipulated sequences in bacterial, fungal, and mammalian cell culture provided methods for re-engineering mouse antibodies to partially replace the rodent antibody sequence with a functionally equivalent human amino acid sequence, thus reducing the overall immunogenicity without destroying the recognition properties of the original antibody. These methods for re-engineering rodent monoclonal antibodies via chimerization and humanization proved to be quite successful. Shortly after that, new technologies were developed that allowed the production of novel monoclonal antibodies from human immunoglobulin gene sequences. These new technologies can be broadly divided into two categories (a) re-assembling recombinant human antibodies from specially constructed gene libraries such as phage display antibody libraries, and (b) synthesizing human antibodies in transgenic animals comprising a human germline configuration of immunoglobulin genes.

Methodology

In 1985, Alt et al. for the first time suggested that transgenic technology could be useful for generating new human monoclonals starting from an un-rearranged, germline-configuration transgene. In 1994, two groups reported generation of mice that carried a human immunoglobulin germline sequence and disrupted mouse immunoglobulin genes. These two groups, one led by Longberg (1994) and the other by Green (1994), disrupted the endogenous mouse heavy chain gene and light chain gene and introduced a human transgene encoding heavy chain and light chain into a mouse embryonic stem cell. The egg was kept in culture during the first few days of embryonic development. The engineered embryos were then implanted into the womb of a female mouse. The implanted embryos developed normal (but transgenic) animals. Once born, those animals that carry the transgene and maintain it stably were called founder animals.

Founder mice (carrying human antibody transgene) appeared to carry out VDJ recombination, class switching, and somatic hypermutation of the human germline antibody genes in a normal fashion, thus producing high-affinity antibodies with completely human sequences. These animals expressed both IgM and IgG. Despite the fact that human immunoglobulin transgenic mice expressed hybrid B-cell receptors (i.e. human immunoglobulin, mouse Ig, mouse Ig, and other signaling molecules), B cells developed and matured as normal B cells. The transgenic mice could be injected with the desired antigen and the resulting antibody isolated. Moreover, monoclonal antibodies could also be prepared by standard hybridoma technology. Half-lives of human antibodies recovered from such transgenic mice were similar to antibodies recovered from humans. For example, in monkeys, the half-life of transgenic mouse-derived human antibodies was found to be 9.5–11 days, similar to the half-life of human IgG (9.6 days), and different from murine IgG, which had a half-life of 1.9 days. In 2006, panitumumab (which inhibits epidermal growth factor receptor, EGFR) became the first marketed monoclonal antibody from a transgenic mouse platform. Since then, around 70 different therapeutic drugs derived from either phage display or transgenic mouse platforms have entered human clinical testing.

Bispecific Antibodies

Bispecific antibodies are those antibodies or antibody-constructs that have dual-specificity in their binding arms. Naturally occurring antibodies are monospecific, and have the same specificity in both their antigen-binding arms. Bispecific antibodies usually do not occur in nature, but are constructed by recombinant DNA or cell-fusion technologies. The classic approach to constructing and producing bispecific antibodies is Quadroma technology. This technology involves fusion of two hybridoma cell lines that produce the desired monospecific antibodies. These two hybridoma cell lines are fused to generate a Quadroma cell that expresses monoclonal antibodies with the desired specificities of the bispecific antibody. Bispecific antibodies carry two different antigen-binding sites. The specificity of one arm is different from the other. Such pairing of two different halves of the antibody has one other effect on antibody structure. The Fc portion of the bispecific antibody does not resemble any of the constituting monospecific antibodies (i.e. it is different from Fc portions of either of the parent monospecific antibodies). The "new" Fc portion may have desired properties or may generate an undesirable effect. To avoid an undesired effect, bispecific Fab fragments (or specifically $F(ab)'_2$) can be prepared by enzymatic digestion, or, better still, by recombinant DNA technology.

The vast majority of bispecific antibodies were designed to redirect cytotoxic effector cells (T_c cells, NK cells, neutrophils) towards targets cells (such as tumor cells). Those bispecific antibodies were developed in which one arm of the bispecific antibodies recognized target cells such as tumor cells and the other arm activated effector cells.

This resulted in bringing tumor cells in close proximity to activated effector cells that ultimately caused elimination of tumor cells causing cancer. A number of bispecific antibody constructs are currently being made and tested in animal models or are undergoing clinical trials. The best characterized molecules found on effector cells against which bispecific antibodies have been made include molecules such as CD64 (IgG receptor) and CD89 (IgG receptor), which are both present on neutrophils, monocytes, and macrophages; CD16 (IgG receptor) on NK cells; and CD-3 (signal transduction unit), which is present on T-cells. One arm of the bispecific antibody binds these molecules on effector cells and the other arm binds to the tumor cell triggering lysis of target tumor cells. A recombinant (CD3 × CD19) bispecific antibody was highly effective against CD19-positive lymphoma cells in clinical trials. Another CD16 x CD30 bispecific antibody is also being explored for Hodgkin's lymphomas in animal models.

Bispecific antibodies can also be used for specific targeting of tumor cells by toxins, chemotherapeutic drugs, or radioisotopes, as shown by *in vitro* and *in vivo* studies. Therapeutics based on bispecific antibodies have not yet yielded the anticipated clinical success. However, bispecific formats are still being actively pursued in preclinical and early clinical development of bispecific antibodies. The critical selection of the antigen for tumor targeting, as well as the complexity and cost of producing bispecific antibodies, are the main issues that need to be explored before the full potential of bispecific therapeutics can be realized.

ETHICAL ISSUES

Antibodies have been the paradigm of binding proteins with desired specificities for more than one century, and during the past decade their recombinant or humanized versions have entered clinical applications with remarkable success. Today, more than 20 different antibodies have been approved in Europe and the U.S., providing considerable market potential for the pharmaceutical and biotech industry. There are several reasons for the remarkable success of antibodies as a class of biological drugs. First, they can rather easily be generated against a wide range of target molecules either by classical immunization of animals followed by protocols for monoclonal antibody preparation, or, more recently, via *in vitro* selection from cloned or synthetic gene libraries. However, the use of mice to produce monoclonal antibodies has been marred with controversy. The production of monoclonal antibodies using animals (such as mice) entails procedures that have the potential to cause considerable pain and distress to animals. Not only this use of monoclonal antibodies, but their derived products have perils as these products are of animal origin and are therefore recognized and dealt with as foreign entities in human trials. Some important ethical issues related to these are discussed as follows.

Risk to animals:

- Monoclonal production involves the use of mice. These mice are used to produce both antibodies in abdominal fluid (ascites) and to grow abdominal tumor cells (hybridomas). Both these procedures inflict pain, discomfort, abdominal discomfort, and indigestion in animals. The removal of ascites fluid may cause shock in the animals, which results from rapid fluid loss.
- In order to achieve an enhanced immune response in the mouse, adjuvants are used. Adjuvants release the antigen into the mouse over a long period of time, which often results in painful lesions at the site of injection. The use of adjuvants, such as Freund's Complete Adjuvant (FCA) or Freund's Incomplete Adjuvant (IFA), is quite common. Freund's adjuvant creates a severe inflammatory response in animals that enhances the immune response, however, it also causes tissue necrosis, ulceration, self-trauma, hunching, decreased appetite, and weight loss. Freund's Complete Adjuvant has actually been banned in the Netherlands and the United Kingdom due to these reasons.
- Pristane is a chemical that is injected into a mouse's peritoneal cavity; it creates an environment conducive to rapid growth of fluid-producing tumors. It induces granulomatous reactions and prevents peritoneal fluid drainage. Consequently, pristane increases monoclonal antibody-rich fluid yield. However, its usage has been associated with weight loss, hunched appearance, and inactivity and distress in mice.
- During the production of monoclonals, the mouse's spleen is removed to provide a source of antibody-producing cells. This involves killing the animal that has already undergone quite a lot of discomfort.

Risks to participants in human trials:

Apart from the ethical issues involving animals, there are concerns regarding the safety and efficacy of monoclonals and their derived products that are administered as drugs in "First in" human trials. In March 2006, six healthy volunteers took part in a Phase I clinical study conducted for a CD28 super-agonist antibody (TGN1412) in London. Even though the dose of antibody was 500 times smaller than what was found safe in animal studies, within minutes all six volunteers suffered multiple organ failure, probably as a result of CD28 antibody-induced activation and proliferation of regulatory $CD4^+$, $CD25^+$, and T cells (that play an important role in autoimmune diseases) that started attacking the body tissues. All the volunteers survived, but it raised serious issues about the conduct of drug trials.

Armed with current scientific knowledge about monoclonal antibodies and antibody constructs, scientists and ethical committees around the globe must tackle the issues raised above and make informed decisions about their production and use. As a society, we must balance the advantages that a new medicine brings with the dangers that its production and use might induce. Only then can we make balanced decisions at an individual, local, national, and global level about the ethical use of pharmaceutical drugs, including monoclonal antibodies and their constructs.

TRANSLATIONAL SIGNIFICANCE

The effective use of accumulated knowledge across various antibody development programs remains a major challenge. Translation of these strategies for monoclonal antibodies that takes science from the bench to the bedside also remains a major challenge. This conversion requires balanced integration of relevant knowledge with respect to target antigen properties, antibody design criteria, their *in vivo* and *in vitro* efficacy, and a conscience as to how it can be used for betterment of the human population. Monoclonal and polyclonal antibodies have a broad spectrum of effects and uses in the everyday lives of citizens and scientist across the globe. Table 25.3 lists some important immunodiagnostic kits that are being routinely used today. ELISA is one of the most common immunotechniques

TABLE 25.3 Some Commonly Used Immunodiagnostic Kits

Diagnostic Kit Purpose	Immune Reaction/Immuno Techniques Employed	Antibody Type Used	Commercial Product Available
Blood Group typing	Agglutination	Polyclonal/monoclonal	Alba sera, Alba clone
Detection of			
•Analyte	Immunoaffinity	Polyclonal / monoclonal	Randox colum for detecting salbutamol
•Toxins	ELISA	Polyclonal	Alfastar for detecting alfatoxins
•Virus	Immunoaffinity capture	Polyclonal	Abcap antibodies to detect enteric virus in water
•Protozoa	ELISA	Monoclonal	Epituub *G, Lambilia* detection kit
•Bacteria	Agglutination	Monoclonal	Bengals screen kit to detects *vibrio cholera*
Cell Isolation	Negative selection of cells	Polyclonal	CD4+ isolation kit (easy sep kit)
	Positive selection of cells	Polyclonal	Isolation of any blood cells (easy sep kit)

(1) Dissolve antigen in buffer

↓

(2) Coat ELISA plate with antigen

(3) ↓ Wash

(4) Block remaining protein binding sites

↓

(5) Add tests antiserum (containing 1° antibodies)

↓ Wash

(6) Add conjugated 2° antibodies

↓ Probe

(7) Record and quantify

FLOW CHART 25.1 Enzyme-Linked Immunosorbent Assay (ELISA)

employed in these kits, and is briefly discussed below (and in Flow Chart 25.1):

1. Dissolve the antigen in an appropriate buffer. The range is usually 1–10 μg/ml, though it has to be optimized for each antigen.
2. The antigen is then coated or immobilized onto a microtiter plate by incubating the antigen overnight in the wells of the plate.
3. Washing is performed to remove unbound antigen.
4. Once unbound antigen is removed, a concentrated solution of non-interacting proteins, such as bovine serum albumin (BSA) or casein, is added to all wells. This step is known as blocking because these proteins block non-specific adsorption of other proteins to the binding sites on the plate.
5. This is followed by application of primary (1°) antibody onto wells of the ELISA plate. Primary antibody is directed towards the antigen that is to be detected. After this step, unbound antibodies are removed by washing.
6. The microtiter plate is then incubated with secondary (2°) antibody, which can be monoclonal or polyclonal in nature and has a probe attached to it. The probe can be an enzyme probe, florescent probe, radioactive probe, etc. The secondary antibody binds the primary antibodies and helps to amplify the signal obtained by the reaction between the antigen and primary antibodies. The plate is then washed to remove excess unbound secondary antibody–probe conjugates from the plate.
7. The next step is dependent upon the probe attached to the secondary antibody. If the secondary antibody is attached with an enzyme probe, then the membrane is incubated with the appropriate enzyme substrate to produce the colored reaction product, which can be detected spectrophotometrically. If, however, the probe is fluorescent or radioactive, then an appropriate imaging and documentation system is required.

WORLD WIDE WEB RESOURCES

In this technological age, there is a flood of information from the Internet. These web resources have made huge amounts of information available at the click of a button. However, such information has its own pros and cons. Listed below are a few important ones:

PROS

- The information is available to anyone, anytime.
- Quick access: Just go to Google and click.
- Sites can provide additional information as links on the same topic on the same site.
- The site can be frequently updated (even daily) to provide the latest information.
- The author doesn't need to be present to provide someone information.
- Information is easy to gather and print.
- It is easy to use the information to create your own resources.

CONS

- A computer with Internet access is required. This is still not common in third world countries.
- Sites may not be authentic. Try using sites such as PubMed (http://www.ncbi.nlm.nih.gov/pubmed), or sites that end with .edu or .gov, or other reputed pages. These should provide reliable information.
- Sometimes scientific information is available on a paid site, which is usually expensive.
- There is no opportunity to ask questions or have things explained in greater detail, though some forums have provisions to post questions.
- Last but not the least, a computer may crash or a site may be removed from the Internet at anytime by its creator.
- Note: Avoid using personal pages to gather scientific information unless they are reputed. Personal pages may provide inaccurate information, as anyone can update any fact on a personal page.

WEBSITES

1. Mike Clark's homepage: www.path.cam.ac.uk. This is a wonderful site for antibody structure, function, and different uses of antibodies.
2. Antibody resource page: http://www.antibodyresource.com. Battery of easy resource sites ranging from structure to therapeutic aspects of antibodies to immunology journals.

REFERENCES

Alt, F. W., Blackwell, T. K., & Yancopoulos, G. D. (1985). Immunoglobulin genes in transgenic mice. *Trends in Genetics*, *1*, 231–236.

Green, L. L., et al. (1994). Antigen specific human monoclonal antibodies from mice engineered with human Ig heavy and light chain YACs. *Nature Genetics*, *7*, 13–21.

Hoogenboom, H. R. (2005). Selecting and screening recombinant antibody libraries. *Nature Biotechnology*, *23*, 1105–1115.

Hwang, W. Y., & Foote, J. (2005). Immunogenicity of engineered antibodies. *Methods*, *36*, 3–10.

Jones, P. T., Dear, P. H., Foote, J., Neuberger, M. S., & Winter, G. (1986). Replacing the complementating determining regions in a human antibody with those from a mouse. *Nature*, *321*, 522–525.

Khan, F. H. (2009). *The Elements of Immunology* (1st ed). New Delhi: Pearson Education.

Köhler, G., & Milstein, C. (1975). Continuous cultures of fused cells secreting antibody of predefined specificity. *Nature*, *256*, 495–497.

Leader, B. B. Q., & Golan, D. E. (2008). Protein therapeutics: a summary and pharmacological classification. *Nature Reviews in Drug Discovery*, *7*, 21–39.

Longberg, N., et al. (1994). Antigen specific human antibodies from mice comprising four distinct genetic modifications. *Nature*, *368*, 856–859.

McCafferty, J., Griffiths, A. D., Winter, G., & Chiswell, D. J. (1990). Phage antibodies: filamentous phage displaying antibody variable domains. *Nature*, *348*, 552–554.

Morrison, S. L., Johnson, M. J., Herzenberg, L. A., & Oi, V. T. (1984). Chimeric human antibody molecules: mouse antigen binding domains with human constant region domains. *Proceedings of the National Academy of Science USA*, *81*, 6851–6855.

Vitetta, E. S., & Ghetie, V. F. (2006). Immunology, considering therapeutic antibodies. *Science*, *313*(5785), 308–309.

FURTHER READING

Atlas of Immunology (2nd ed.) by Julius M. Cruse, Robert E. Lewis. (2003). Boca Raton: CRC Press.

Encyclopaedia of Immunology (2nd ed.) Editor-in-chief, P. J. Delves, consultant editor, I. M. Roitt. (1998). Academic Press.

Encyclopaedia of Molecular Biology by John Kendrew. (1994). (Ed) Wiley-Blackwell.

Encyclopaedia of Microbiology (3rd ed.) by Editor-in-Chief: Moselio Schaechter. (2009). Elsevier.

Encyclopaedia of Molecular Cell Biology and Molecular Medicine, (2nd ed.) by Robert Meyers. (2004). (Ed) Wiley-Blackwell.

GLOSSARY

Antigen (Immunogen) Foreign molecules (usually) that can elicit an immune response. Proteins and carbohydrates are usually antigenic, while lipids and DNA are usually weakly immunogenic/antigenic or non-immunogenic (e.g. bovine serum albumin, capsular polysaccharide of bacteria).

Bispecific Antibodies Antibody or antibody construct that has dual specificity in the binding arms. Bispecific antibodies usually do not occur in nature, but are constructed by recombinant DNA or cell fusion technologies (e.g. recombinant (CD3 × CD19) bispecific antibody).

Monoclonal Antibodies Antibodies directed against a single antigenic determinant and produced by a single clone of cells (e.g. CEA-Scan, a mouse monoclonal antibody that detects human carcinoembryonic antigens).

Diabody An antibody generated by a recombinant DNA technology antibody construct consisting of two variable heavy chain region (V_H) and two variable light chain region (V_L) domains (e.g. C6.5, which targets human growth receptor 2).

Epitopes (Antigenic Determinants) Small portions of antigens that provoke an immune response.

Humanized Antibodies Hybrid antibodies that carry only murine Complementary Determining Regions (CDRs) in an otherwise human antibody are referred to as humanized antibodies (e.g. Trastuzmab, an antibody that is directed against human epidermal growth factor receptor, Her-2).

Hypervariable Regions or Complementarity Determining Regions (CDRs) The regions within the variable region of light and heavy chains that bind antigenic determinants.

Phage Display Technology A biotechnological technique that allows expression of exogenous peptides on the surface of filamentous bacteriophage (e.g. adalimumab, a phage display-derived antibody).

Polyclonal Antibodies Antibodies that are directed toward different antigenic determinants of one antigen and are a product of different clones of cells. Although antibodies are formed against a single antigen, they are directed towards different antigenic determinants of this antigen.

ABBREVIATIONS

ADCC Antibody-Dependent Cell-Mediated Cytotoxicity
CDR Complementarity-Determining Regions
CEA Carcino-Embryonic Antigen
dsFv Disulfide-Linked Variable Fragment
EGFR Epidermal Growth Factor Receptor
Fab Antigen-Binding Fragment
HAMA Human Anti-Mouse Antibody
HGPRT Hypoxanthine Guanine Phosphoribosyl Transferase
MHC Major Histocompatibility Complex
NK Cell Natural Killer Cell
PAMPs Pathogen-Associated Molecular Patterns
PRR Pattern Recognition Receptor
scFv Single Chain Variable Fragment

LONG ANSWER QUESTIONS

1. What are monoclonal antibodies? How are they different from polyclonal antibodies? Discuss some of their important diagnostic and therapeutic applications.
2. What are bispecific antibodies? How are they made? Discuss their important potential applications in medical biology.
3. What is phage display technology? Discuss, how can it be used to screen recombinant antibody fragments via a biopanning method.
4. Briefly discuss recombinant antibody fragments such as scFv, dsFv, and Fab. How are they different from diabodies?
5. Differentiate between chimeric and humanized antibodies. What are their advantages over murine monoclonal antibodies?

SHORT ANSWER QUESTIONS

1. Human anti-mouse antibody (HAMA) response that occurs in human after administration of murine antibodies is minimal in which type of monoclonal antibody? Why?
2. What is the role of aminopterin in HAT medium used to screen hybridoma cells?
3. What is the difference between scFv and a diabody?
4. Can a diabody act as a bispecific antibody?
5. List two main differences between monoclonal and polyclonal antibodies.

ANSWERS TO SHORT ANSWER QUESTIONS

1. Since HAMA is primarily directed against the murine part in monoclonal antibodies, it will be lowest where the murine component is the least, which is the case in humanized antibodies. Therefore, humanized versions of monoclonal antibodies will have a minimal HAMA response.
2. Aminopterin is a powerful inhibitor of the enzyme dihydrofolate reductase, which is the key enzyme in *de novo* synthesis of DNA. Mutant tumor cells have only one (*de novo*) pathway working in them for DNA (purine) synthesis, so adding aminopterin is done to kill any un-fused tumor cells that are still in culture. Hybridoma cells (fusion products of tumor cells and antibody-producing cells) survive because another pathway (salvage pathway) is working in B cells and provides purines for DNA synthesis.
3. In a single chain variable fragment (scFv), V_H and V_L domains are linked together by a short peptide linker that is usually 15–20 amino acids long, while in diabodies two scFv are assembled together. Moreover, monomers in diabodies have short linkers that are usually five amino acids long.
4. Bispecific antibodies are those antibodies or antibody constructs that have dual specificity in their binding arms. Naturally occurring antibodies have the same specificity in both arms. Yet a diabody can act as a bispecific antibody if it is so constructed that it has two different specificities in its two different "arms."
5. The first difference between these antibodies is that monoclonal antibodies are directed against single antigenic determinants while polyclonal antibodies are directed against several antigenic determinants on the same antigen. Second, monoclonal antibodies are derived from cells that are exact replicas or clones of each other and have the same parent cell. Polyclonal antibodies are derived from different B cells.

Chapter 26

Vaccines: Present Status and Applications

Dinesh K. Yadav, Neelam Yadav and Satyendra Mohan Paul Khurana
Amity Institute of Biotechnology, Amity University, Haryana, India

Chapter Outline

SUMMARY

Vaccine development and usage over the years has significantly reduced the number of infections and diseases. Improved knowledge of immune protection and a big leap in genetic engineering has allowed the induction of a variety of new vaccines through the manipulation of DNA, RNA, proteins, and sugars. Creation of attenuated mutants, expression of potential antigens in live vectors, and purification and direct synthesis of antigens in new systems have immensely improved vaccine technology. Both infectious and non-infectious diseases are now within the realm of vaccinology. The profusion of new vaccines has enabled the targeting of new populations for vaccination as well as the cure and removal of infectious agents from their natural reservoirs. Still, as with ancient infections like malaria and new infections like HIV, a potent vaccine is elusive, which poses a big challenge to the scientific world.

WHAT YOU CAN EXPECT TO KNOW

After studying this chapter the reader is expected to have an understanding of the historical prospects of vaccine development, an underlying concept of vaccine types, their merits and demerits, modern vaccines, ethical issues related to vaccines, and their future challenges.

HISTORY AND METHODS

INTRODUCTION

The immune system is a remarkably versatile defense system that has evolved to protect animals from invading pathogenic microorganisms. It is able to generate an enormous variety of cells and molecules capable of specifically recognizing and eliminating an apparently unlimited variety of foreign invaders. These cells and

Animal Biotechnology. http://dx.doi.org/10.1016/B978-0-12-416002-6.00026-2

molecules act together in a dynamic network to defend the host. Immune recognition is remarkable for its specificity. The immune system is able to recognize subtle chemical differences that distinguish one foreign pathogen from another. Furthermore, the system is able to discriminate between foreign molecules and the body's own cells and proteins. The term immunity is derived from the Latin words "*immunis*" and "*immunitas*," meaning "to exempt." The first recorded deliberate attempts to induce immunity were performed by the Chinese and Turks in the fifteenth century. Various reports suggest that the dried crusts derived from smallpox pustules were either inhaled into the nostrils or inserted into small cuts in the skin, a technique called ***variolation***. Buddhist monks used to drink snake venom to develop immunity to snake bites. In 1718, Lady Mary Wortley Montagu observed the positive effects of variolation on the native population and attempted the technique on her own children. English physician Edward Jenner is considered the founder of vaccinology.

The variolation technique was significantly improved and tested by Jenner. He observed that milkmaids who had contracted the mild cowpox disease were subsequently immune to the fatal smallpox disease. Jenner reasoned that introduction of fluid from a cowpox pustule into people (i.e. inoculation) might protect them from smallpox. He inoculated an eight-year-old boy with fluid from a cowpox pustule and later deliberately infected the child with smallpox. As predicted, the child did not develop smallpox symptoms. This was experimentally supported by Louis Pasteur, who succeeded in growing the bacterium thought to cause fowl cholera in culture and then showed that chickens variolated with the cultured bacterium developed cholera. When these variolated chickens were infected with potent live bacterium, chickens became infected with cholera, but they recovered and were completely protected from the disease. Pasteur hypothesized and proved that aging had weakened the virulence of the pathogen and that such an attenuated strain might be administered to protect against the disease. He called this attenuated strain a **vaccine** (from the Latin *vacca*, meaning "cow") in honor of Jenner's work with cowpox inoculation. Although Pasteur proved that vaccination worked, he did not know the reason behind its protective function.

Following are some key milestones in vaccine development:

1885: Louis Pasteur, known for his animal vaccines, injected a rabies vaccine into two people, which caused a controversy. Few people at the time were comfortable with the idea of introducing a deadly, live virus into a human being.
1896: Vaccine for cholera and typhoid were developed using killed versions of bacteria.
1897: A killed vaccine for the plague was developed.
1923: A powerful toxin from diphtheria bacterium was chemically inactivated and used as a "toxoid" to kill bacteria. Before the toxoid vaccine, there were as many as 200,000 cases each year and 15,000 deaths.
1926: A killed vaccine for pertussis ("whooping cough") was developed using the whole pertussis organism.
1927: A tetanus "toxoid" was developed. Before the tetanus vaccine, there were about 600 cases a year in the U.S. with 180 deaths (today there are about 70 cases with 15 deaths). By the late 1940s tetanus was combined with diphtheria and pertussis as the children's vaccine "DTP."
1954: Jonas Salk developed a killed polio virus, which decreased paralysis cases from 20,000 in 1952 to 1,600 in 1960.
1961: Alfred Sabin developed an oral polio vaccine using a live virus; it was easy to administer and was successful at eliminating the spread of polio.
1963: A safe and effective measles vaccine was developed, reducing the number of cases from four million in 1962 (with 3,000 deaths) to 309 cases in 1995 (with no deaths).
1964: A killed rabies vaccine was developed, but required up to 30 painful shots in the abdomen. By 1980, a newer version required only five shots in the arm to protect against this deadly disease.
1967: A vaccine for mumps was licensed, reducing the incidence from about 200,000 cases annually (with 20 to 30 deaths) to about 600 cases (with no deaths).
1970: Several strains of Rubella were weakened (attenuated) to create a vaccine. Between 1964 and 1965 there were about 12 million cases, leading to birth defects in 20,000 children. By 1971, measles, mumps, and rubella vaccines were combined into a single shot known as MMR.
1970s and 1980s: Meningoccocal, pneumococcal, and haemophilus influenza type b (Hib) vaccines were developed using a part of the bacteria cover to provide a safe antigen for the body to react to (subunit vaccine). These vaccines helped protect against life-threatening meningitis, blood infections, and some pneumonia diseases.
1986: A vaccine for hepatitis B was licensed with an antigen that is cloned rather than grown.
1990: A killed vaccine for hepatitis A was developed.
1995: A varicella (chicken pox) vaccine was licensed for use in children.
1996: The first "DTaP" vaccine was approved using only part of the pertussis organism. Combined with diphtheria and tetanus vaccines, it reduced annual pertussis deaths significantly after the DTP vaccination.
2000: Premature death related to influenza was estimated at 20,000 people annually. Influenza vaccine use reached 70 million doses.

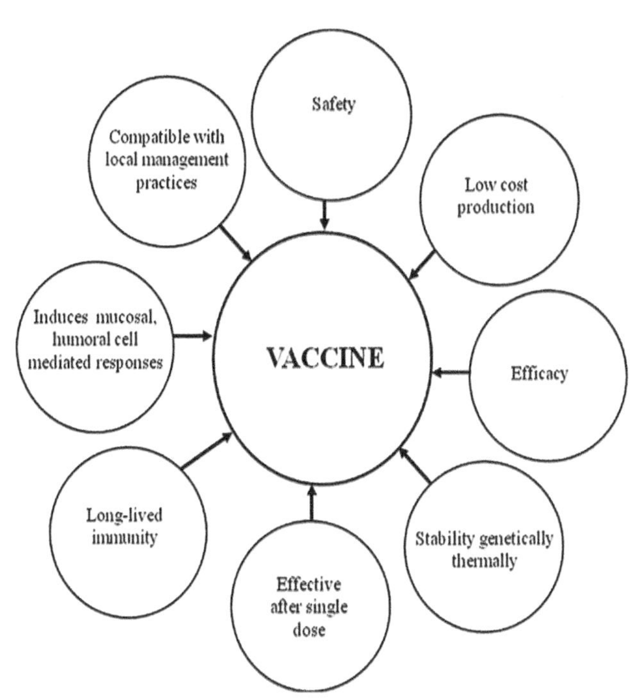

FIGURE 26.1 Properties of a vaccine.

Infectious diseases contribute to more than 45% of the total deaths in developing countries (http://www.who.int/healthinfo/global_burden_disease). Vaccination is the most effective means of preventing infectious diseases. World Health Organization (WHO) information shows that the mortality of more than 15 million children across the globe is due to infectious diseases, especially in developing countries. Mass immunization programs could save these lives. The success of any mass immunization program largely depends upon the availability of cost-effective and immunoprotective vaccines against these dreadful communicable diseases. There are many issues associated with the production of vaccines (Figure 26.1).

TYPES OF VACCINES

Vaccines work by teaching our immune system to recognize and remember the foreign pathogenic bacteria or viruses. After vaccination, our immune system stores the antigen-specific memory cells. If in the future there is recurrent exposure to the actual disease, our immune system then opsonises the bacteria or viruses much more quickly. Vaccines are made by manipulating germs or parts of germs.

Traditional Vaccines

The development of vaccinations against harmful pathogenic microorganisms represents an important advancement in the history of modern medicine. In the past, traditional vaccination has relied on two specific types of microbiological preparations to produce material for immunization and generation of a protective immune response. These two categories involve either (1) living infectious material that has been manufactured in a weaker state and therefore inhibits the vaccine from causing disease, (2) or inert, inactivated, or subunit preparations.

The decision of scientists about the appropriation of the vaccine type was based on the disease-causing agent and the natural behavior/course of the disease. Whole virus vaccines, either live or killed, constitute the vast majority of vaccines in use at the present time. However, recent advances in molecular biology have provided alternative methods for the development of improved vaccines.

There are many approaches to design vaccines against a microbe. These choices are typically based on fundamental information about the microbe, such as how it infects host cells and how the immune system responds to it. In addition, there are practical considerations, such as regions of the world where the vaccine would be used. The following are some of the options that researchers might pursue:

- Live, attenuated vaccines
- Inactivated vaccines
- Subunit vaccines
- Toxoid vaccines
- Conjugate vaccines

Two additional types of vaccines currently being researched are:

- DNA vaccines
- Recombinant vector vaccines

Live, Attenuated Vaccines

Methodology

Live, attenuated vaccines contain a version of the living pathogenic microbe that has been attenuated or weakened in the lab so that it has lost its significant pathogenicity (Flow Chart 26.1). This is accomplished by serial passage through a foreign host (tissue culture, embryonated eggs, or live animals for multiple generations). This extended passaging introduces one or more mutations into the new host. The mutated pathogen is significantly different from the original pathogenic form, so it can't cause disease in the original host but can effectively induce the immune response. Live, attenuated virus vaccines are prepared from attenuated strains that are almost or completely devoid of pathogenicity but are capable of inducing a protective immune response. They multiply in the human host and provide continuous antigenic stimulation over a period of time. Several methods have been used to attenuate viruses for vaccine production.

Live, attenuated vaccines stimulate protective immune responses when they replicate in the host. The viral proteins

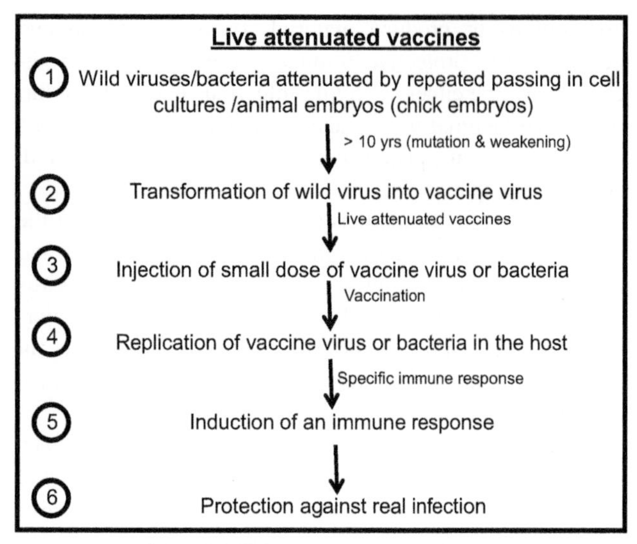

Live attenuated vaccines
1 Wild viruses/bacteria attenuated by repeated passing in cell cultures /animal embryos (chick embryos)
> 10 yrs (mutation & weakening)
2 Transformation of wild virus into vaccine virus
Live attenuated vaccines
3 Injection of small dose of vaccine virus or bacteria
Vaccination
4 Replication of vaccine virus or bacteria in the host
Specific immune response
5 Induction of an immune response
6 Protection against real infection

FLOW CHART 26.1 Live attenuated vaccines

produced within the host are released into the extracellular space surrounding the infected cells and are then acquired, internalized, and digested by scavenger cells, the antigen-presenting cells (APCs) that circulate throughout the body. These APCs include macrophages, dendritic cells, and B cells, which work together to expand immune response. The APCs recirculate and display fragments of the processed antigen on their cell surface, attached to MHC class II antigens. This complex of processed foreign antigen peptide and host MHC class II antigens form part of the specific signal with which APCs (along with the MHC peptide complex) triggers the activation of T-helper lymphocytes. The second part of the activation signal comes from the APCs themselves, which display on their cell surface co-stimulatory molecules along with MHC-antigen complexes. Both drive T-cell expansion and activation through interaction with their respective ligands, the T-cell receptor complex (TCR), and the co-stimulatory receptors CD28/CTLA4 present on the T cell surface. Activated T cells secrete molecules that act as powerful activators of immune cells. Also, as viral proteins are produced within the host cells, they are processed through proteasome degradation. Small parts of these processed intracellular proteins associate with cytosolic host cell MHC class I and display on the cell surface. These complexes together are recognized by a second class of T cells, killer or cytotoxic cells. This recognition, along with other stimulation by APCs and production of cytokine-stimulated T cells, is responsible for the development of mature cytotoxic T cells (CTL) capable of destroying infected cells. In most instances live infection induces lifelong immunity. Evidence is available that favors the fundamental role of cytokines in the differentiation of memory cells. Helper T (Th) cell-regulated B-cell immunity progresses in an ordered cascade of cellular development that culminates in the production of antigen-specific memory B cells. The recognition of peptide MHC class II complexes on activated antigen-presenting cells is critical for effective TH cell selection, clonal expansion, and effector TH cell function development. Cognate effector Th-cell–B-cell interactions then promote the development of either short-lived plasma cells (PCs) or germinal centers (GCs). These GCs expand, diversify, and select high-affinity variants of antigen-specific B cells for entry into the long-lived memory B-cell compartment. Upon antigen re-challenge, memory B cells rapidly expand and differentiate into PCs under the cognate control of memory TH cells. Although live, attenuated preparations are the vaccines of choice, they do pose the risk of reversion to their pathogenic form, and can cause infection.

Use of a related virus from another animal: The earliest example was the use of cowpox to prevent smallpox.

Administration of pathogenic or partially attenuated virus by an unnatural route: The virulence of the virus is often reduced when administered by an unnatural route. This principle is used in the immunization of military recruits against adult respiratory distress syndrome using enterically coated, live adenovirus type 4, 7, and 21.

Passage of the virus in an "unnatural host" or host cell: Often, the attenuated form of the organism (or virus) is obtained by serial passage or culture of the active organism in culture media or cells. In these cases, the molecular basis of attenuation is unknown. The major vaccines used in man and animals have all been derived this way. After repeated passages, the virus is administered to the natural host. The initial passages are made in healthy animals or in primary cell cultures. There are several examples of this approach: the 17D strain of yellow fever was developed by passage in mice and then in chick embryos. Polioviruses were passaged in monkey kidney cells and measles in chick embryo fibroblasts. Human diploid cells are now widely used (such as the WI-38 and MRC-5). The molecular basis for host range mutation, which is the basis of attenuation of pathogens, is only now being understood.

Live, attenuated vaccines are relatively easy to create for certain viruses. Vaccines against measles, mumps, and chickenpox, for example, are made by this method. Live, attenuated vaccines are more difficult to create for bacteria. Bacteria have thousands of genes and thus are much harder to control. Scientists working on a live vaccine for a bacterium, however, might be able to use recombinant DNA technology to remove several key genes. This approach has been used to create a vaccine against *Vibrio cholerae* that causes cholera. However, the live cholera vaccine has not been licensed in the United States.

Development of temperature sensitive mutants: This method may be used in conjunction with the above method.

Live, attenuated vaccines cannot be administered to individuals with weakened or damaged immune systems. To maintain potency, live, attenuated vaccines require refrigeration and protection from light.

Examples of currently available live, attenuated vaccines against viral infections include measles, mumps, rubella (MMR), cowpox, yellow fever, influenza (FluMist®) intranasal vaccine), and oral polio vaccine. Live, attenuated bacterial vaccines include tuberculosis, BCG, and oral typhoid vaccine.

Today, it is likely that regulatory agencies would require an understanding of the basis of attenuation. Therefore, development of any new attenuated form of mycobacterium for use as a candidate vaccine is likely to include the introduction of one or more specific mutations into the genome of the pathogen. Likely candidates include mutations that interfere with the synthesis of an amino acid or nucleic acid component essential for the growth of the organism.

Some examples of such live vaccines have already been prepared and evaluated in preclinical trials.

Advantages of Live, Attenuated Vaccines

Since a live, attenuated vaccine is the closest thing to a natural infection, these vaccines are good "teachers" of the immune system. The attenuated vaccines elicit strong immunoprotective cellular and antibody responses, and often confer lifelong immunity with only one or two doses.

Disadvantages of Live, Attenuated Vaccines

Despite the advantages of live, attenuated vaccines, there are some disadvantages as well. It is the nature of living things to change or mutate, and the organisms used in live, attenuated vaccines are no different. The major disadvantage of attenuated vaccines is that secondary mutations can lead to reversion to virulence and can thus cause disease. There is another possibility of interference by related viruses, as is suspected in the case of oral polio vaccine in developing countries. Also, not everyone can safely receive live, attenuated vaccines. People with immune-compromised, damaged, or weakened immune systems (due to chemotherapy, HIV infection, or even pregnancy) cannot be given live vaccines. Another limitation is that live, attenuated vaccines usually need the cold chain to stay potent, and skilled health care workers to handle them. It adds extra cost that hampers mass immunization programs in developing countries that lack widespread refrigeration and have limited access to skilled health care workers; hence, a live vaccine may not be the best choice. The possibility of contaminating viruses in cultured cells aggravates the vulnerability of live, attenuated vaccines.

Inactivated Whole Virus Vaccines

Considering the disadvantages of attenuated vaccines, another approach was used to develop vaccines. In this approach the causative whole pathogen is killed or inactivated.

Methodology

Heat, chemicals (such as formaldehyde or β-propiolactane), or radiation are typically used to inactivate antigens (Flow Chart 26.2). Chemical treatment destroys a pathogen's ability to replicate, but keeps the immunogenic structure intact in its natural form. It is crucial to ensure the integrity of the structure of antigenic epitopes of surface antigens. Thereby, inactivated, whole-pathogen vaccines provide protection by directly generating T_H and humoral immune response against the pathogenic immunogen. In the absence of cellular production of the antigen, these vaccines are usually devoid of the ability to induce significant T cytotoxic responses. In addition, these vaccines are not actually produced in the host by replication of pathogen, and therefore, to be effective, non-replicating virus vaccines must contain much more antigen than live vaccines (which are able to replicate in the host). The immunity induced by inactivated killed vaccines frequently decreases during the life of the host and may require additional boosters to achieve lifelong immunity. However, killed vaccines offer some important advantages over live vaccines: they are produced outside the host, and they can be designed to contain only the specific antigenic target of the pathogen that is involved in the development of protective immunity (and thus exclude all other viral components).

Examples of currently available inactivated vaccines are limited to inactivated whole viral vaccines against influenza, polio, rabies, and hepatitis A. Whole inactivated bacterial vaccines include pertussis, typhoid, cholera, and plague. "Fractional" vaccines include subunits (hepatitis B, influenza, acellular pertussis), and toxoids (diphtheria, tetanus). See Table 26.1 for examples.

Advantages of Inactivated Whole Virus Vaccines

Such vaccines are more stable and safer than live vaccines. The dead microbes can't mutate back to their

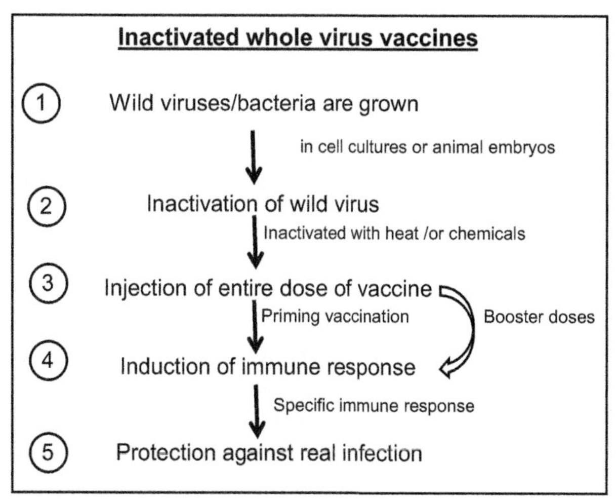

FLOW CHART 26.2 Inactivated whole virus vaccines

TABLE 26.1 Bacterial and Viral Vaccines Currently Used

Bacterial Vaccines		
Type	**Current**	**Under Trial**
Live, attenuated	B C G	*V. cholerae* *S. typhi*
Inactivated	*V. cholerae* *B. pertussis* *S. typhi*	*V. cholerae* plus subunit A *M. leprae*
Subunit	*H. influenzae* *N. meningitidis* *S. pneumoniae*	*S. typhi* *H. influenza* DT conjugate
Toxoid	Tetanus, diphtheria	
Viral Vaccines		
Type	**Current**	**Under Trial**
Live, attenuated	Vaccinia Measles Yellow Fever Mumps Polio A Adeno Aa Rubella Varicelle zoster	Cytomegalo Hepatitis A Influenza Dengue RotaA Parainfluenza Japanese Encephalitis Polio A
Inactivated	Polio Influenza Rabies Japanese encephalitis	Hepatitis A
Subunit	Hepatitis B, influenza	

(Adopted from Encyclopedia of Immunology. Edited by I.M. Roitt & P.J. Delves, Academic Press, San Diego, 1992.)

disease-causing state. Inactivated vaccines usually don't require refrigeration, and they can be easily stored and transported in a freeze-dried form, which makes them cheap and easily accessible to people in developing countries.

Disadvantages of Inactivated Whole Virus Vaccines

Most inactivated vaccines stimulate a weaker immune response than do live vaccines, so it likely requires multiple booster doses to maintain a protective immunity level. This can be a drawback in areas where people don't have regular access to health care and can't get booster shots on time. Excessive treatment can destroy immunogenicity, whereas insufficient treatment can leave infectious virus capable of causing disease. In addition, there is an increased risk of allergic reactions due to the presence of large amounts of unrelated structural antigens of microbes. Table 26.2 shows the comparison of live, attenuated vaccine and inactivated whole virus vaccine.

TABLE 26.2 Comparison Between Live, Attenuated Vaccine and Inactivated Whole Virus Vaccine

Features	Live	Dead
Dose	Low	High
No. of doses	Single	Multiple
Need for adjuvant	No	Yes
Duration of immunity	Many years	Short
Antibody response	IgG	IgA IgG
Cell-mediated immunity	Good	Poor
Reversion to virulence	Possible	Not possible

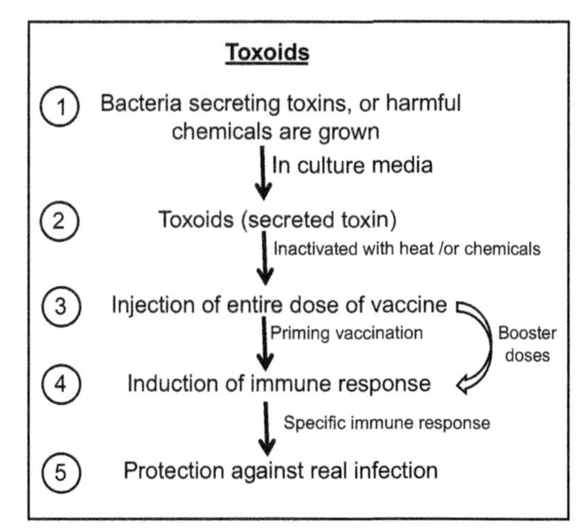

FLOW CHART 26.3 Toxoids

Toxoid Vaccines

Toxids are bacterial toxins, usually the exotoxins, secreted by pathogens that produce many of the disease symptoms after infection.

Methodology

Toxoid vaccines (e.g. vaccines for diphtheria and tetanus) are made by purifying the bacterial exotoxin (Flow Chart 26.3). Toxicity of purified exotoxins is then suppressed or inactivated either by heat or with formaldehyde (while maintaining immunogenicity) to form **toxoids**. Vaccination with toxoids induces anti-toxoid antibodies that are able to bind with the toxin and neutralize its deleterious effects. However, procedures for the production of toxoid vaccines ought to be strictly controlled to achieve detoxification/inactivation without excessive modification of the antigenic epitope structure. This technique is reserved for diseases in which the secreted toxins are the main cause of the illness. Such "detoxified" toxins are harmless/safe and are used as vaccines.

When the immune system receives a vaccine containing a harmless toxoid, it learns how to fight off the natural toxin.

The immune system produces antibodies that opsonize the bacterial toxins. Vaccines against diphtheria and tetanus are the best examples of toxoid vaccines.

Subunit Vaccines

Originally, non-replicating vaccines were derived from crude preparations of a virus from animal tissue. As the technology for growing viruses to high titers in cell cultures advanced, it became practicable to purify viruses and viral antigens. Advances in biotechnology have made it possible to identify the peptide sites encompassing the major antigenic sites of viral antigens. Hence, instead of using the entire microbe for immunization, a subunit component of the antigens that can best stimulate the immune system is used as the vaccine. However, increasing purification may lead to a decrease in immunogenicity of the subunit vaccines. This decrease necessitates the coupling of subunit vaccines to a carrier molecule called an adjuvant. Subunit vaccines can contain from 1 to 20 antigens. Since only the specific antigenic determinants (the parts of the antigen that antibodies or T cells recognize and bind to) of the antigens are used for this type of vaccine, the risk of adverse effects is significantly lowered and there is no chance of reversal of virulence. Examples of purified subunit vaccines include the vaccines for influenza virus *Haemophilis influenzae* A and B (HiA & HiB) and hepatitis B antigen (HBsAg).

Methodology

In the course of development of a subunit vaccine, identification of an immunoprotective antigen and its antigenic epitopes is the prerequisite (Flow Chart 26.4). Subunit vaccines can accommodate 1 to 20 or more antigens. Although identification of the best simulating antigens to induce the immune system is a tricky, time-consuming process, once identified, they can be made by either of the following procedures:

- The microbe can be grown in the laboratory, and then chemicals can be used to break it apart. The important antigens can then be purified and used as subunit vaccines.
- The antigen molecules can be manufactured from the microbe using recombinant DNA technology. Vaccines produced this way are called "recombinant subunit vaccines."

There are four genetically engineered vaccines currently available:

1. Hepatitis B vaccines are produced by insertion of a segment of the hepatitis B virus gene into the gene of a yeast cell. The recombinant yeast cells produce pure hepatitis B surface antigen.
2. Human papillomavirus vaccines are produced by inserting genes for a viral coat protein into either yeast (similar to hepatitis B vaccines) or into insect cell lines. Viral-like particles are produced in the yeast/insect cell lines, and when used as vaccines, induce a protective immune response.
3. A live typhoid vaccine (Ty21a) is the *Salmonella typhi* bacteria that has been genetically modified to not cause illness.
4. A live, attenuated influenza vaccine (LAIV) has been engineered to replicate effectively in the mucosa of the nasopharynx but not in the lungs.

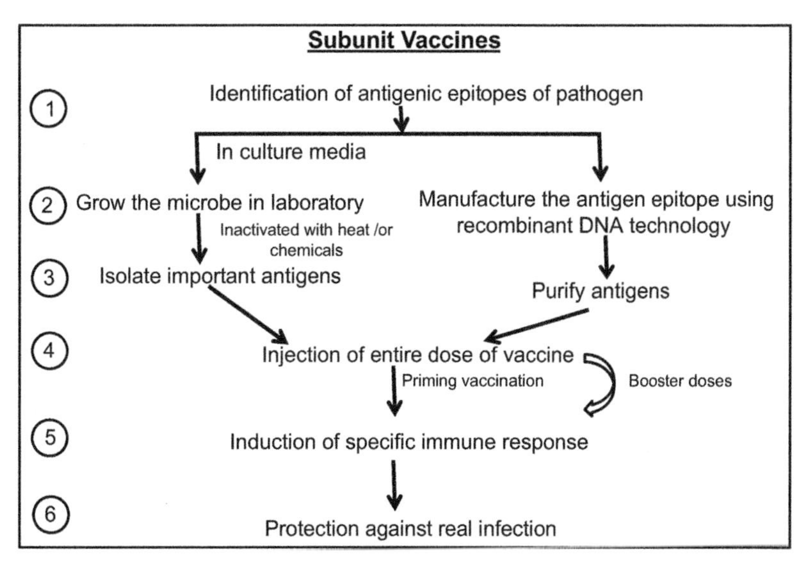

FLOW CHART 26.4 Subunit vaccines

Conjugate Vaccines

If a bacterium possesses an outer coating of sugar molecules called polysaccharides, as many harmful bacteria do, researchers can try developing a conjugate vaccine for it (Flow Chart 26.5). Polysaccharide coatings disguise bacterial antigens so that the immature immune systems of infants and younger children can't recognize or respond to them. Therefore, conjugate vaccines have been developed by attaching the polysaccharide to a stronger protein. When the immature immune system responds to the linked protein, it also responds to the polysaccharide and defends against the disease-causing bacterium. Conjugate vaccine examples include an influenza vaccine for *Haemophilus Influenzae* B (HiB) and a pneumoccocal vaccine (Prevnar®).

DNA Vaccines

DNA vaccination is a relatively recent development in vaccine technology. It involves the direct introduction of a plasmid into the appropriate tissue containing the complete expression cassette independently encoding the antigens against which the immune response is sought (Koprowski and Weiner, 1998). DNA immunization is a novel technique used to efficiently stimulate humoral and cellular immune responses to protein antigens. Direct gene transfer of plasmid DNA into mouse muscle without the need for any special delivery system was demonstrated (Wolff et al., 1990). Plasmid DNA encoding reporter genes induced protein expression within the muscle cells and provided evidence that naked DNA could be delivered *in vivo* to direct protein expression. DNA vaccines usually consist of plasmid vectors (derived from bacteria) that contain heterologous genes (transgenes) inserted under the control of a eukaryotic promoter, thus allowing protein expression in mammalian cells (Davis, 1997).

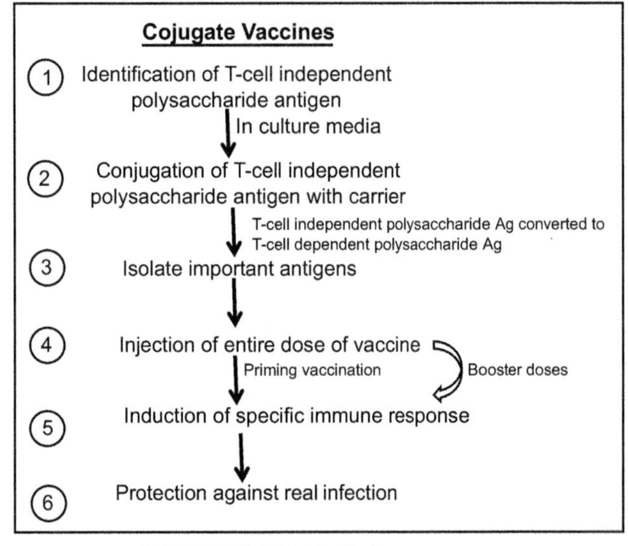

FLOW CHART 26.5 Conjugate vaccines

The use of genetic material to deliver genes for therapeutic purposes has been practiced for many years. Experiments outlining the transfer of DNA into cells of living animals were reported as early as 1950. Later experiments using purified genetic material only further confirmed that the direct DNA gene injection in the absence of viral vectors results in the expression of the inoculated genes in the host. There have been additional experiments that extend these findings to recombinant DNA molecules, further illustrating the idea that purified nucleic acids can be directly delivered into a host and proteins produced. Genetic immunization with complete expression cassettes containing human growth hormone gene produced detectable levels of the growth hormone in mice. Immunized mice developed antibodies against the human growth hormone. It described the ability of inoculated genes to be individual immunogens (Koprowski and Weiner, 1998).

DNA-based immunization has become a novel approach to vaccine development. Direct injection of naked plasmid DNA induces strong immune responses to the antigen encoded by the gene. Once the plasmid DNA construct is injected, the host cells take up the foreign DNA, expressing the viral gene and producing the corresponding viral protein inside the cell. This form of antigen presentation and processing induced both MHC class I- and class II-restricted cellular and humoral immune responses (Encke et al., 1999).

Still in the experimental stages, these vaccines show great promise, and several types are being tested in humans. DNA vaccines take immunization to a new technological level. These vaccines dispense with both the whole organism and its parts and get right down to the essentials: the microbe's genetic material. In particular, DNA vaccines use the genes that code for the all-important antigens.

Methodology

Construction of DNA Vaccines

DNA vaccines are composed of bacterial plasmids. Expression plasmids used in DNA-based vaccination normally contain two units: (1) the antigen expression unit composed of promoter/enhancer sequences followed by antigen-encoding and polyadenylation sequences, and (2) the production unit composed of bacterial sequences necessary for plasmid amplification and selection. The construction of bacterial plasmids with vaccine inserts is accomplished using recombinant DNA technology (Flow Chart 26.6). Once constructed, the vaccine plasmid is transformed into bacteria, where bacterial growth produces multiple plasmid copies. The plasmid DNA is then purified from the bacteria by separating the circular plasmid from the much larger bacterial DNA and other bacterial impurities. This purified plasmid DNA is used as the vaccine (Figure 26.2).

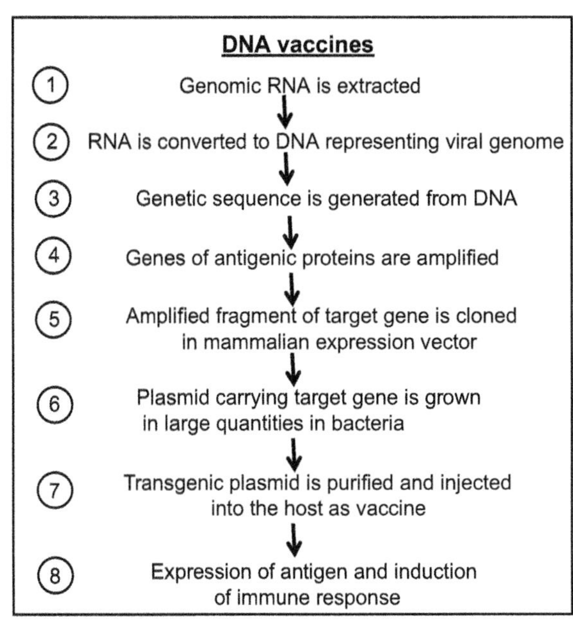

DNA vaccines
1 Genomic RNA is extracted
2 RNA is converted to DNA representing viral genome
3 Genetic sequence is generated from DNA
4 Genes of antigenic proteins are amplified
5 Amplified fragment of target gene is cloned in mammalian expression vector
6 Plasmid carrying target gene is grown in large quantities in bacteria
7 Transgenic plasmid is purified and injected into the host as vaccine
8 Expression of antigen and induction of immune response

FLOW CHART 26.6 DNA vaccines

A DNA vaccine against a microbe would evoke a strong antibody response to the free-floating antigen secreted by cells, and the vaccine would also stimulate a strong immune response against the microbial antigens displayed on cell surfaces. The DNA vaccine couldn't cause the disease because it wouldn't contain the microbe, just copies of a few of its genes. In addition, DNA vaccines are relatively easy and inexpensive to design and produce.

So-called naked DNA vaccines consist of DNA that is administered directly into the body. These vaccines can be administered with a needle and syringe or with a needleless device that uses high-pressure gas to shoot microscopic gold particles coated with DNA directly into cells. Sometimes, the DNA is mixed with molecules that facilitate its uptake by the body's cells. Naked DNA vaccines being tested in humans include those against the influenza and herpes viruses.

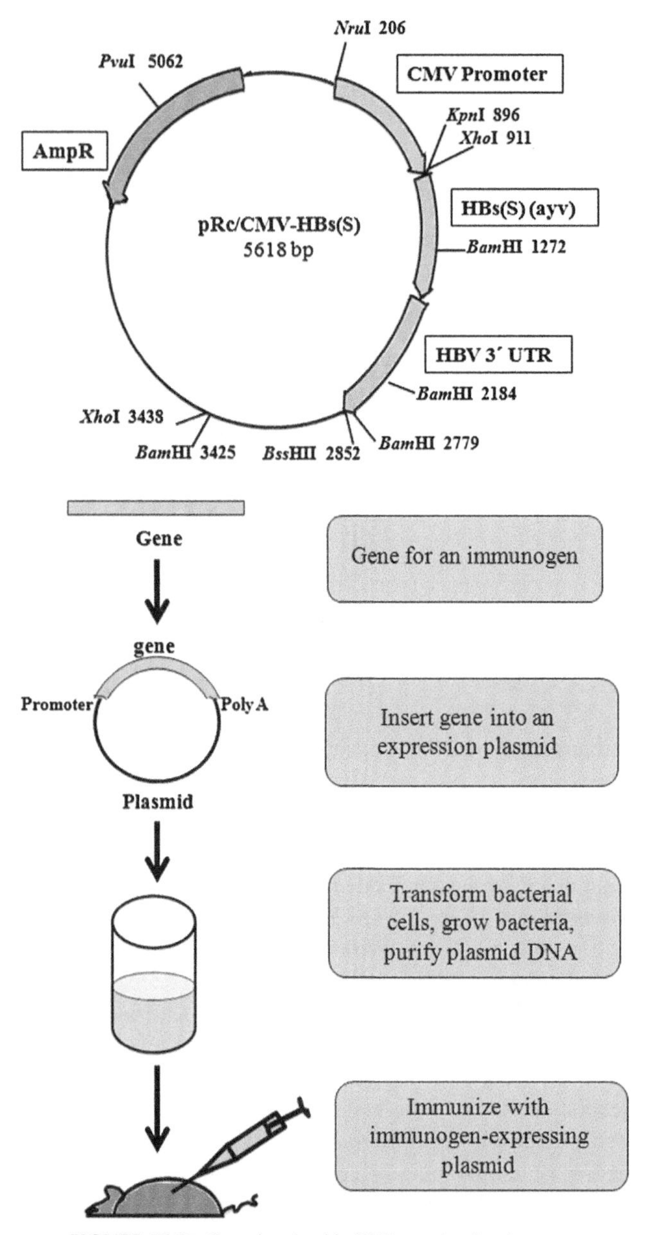

FIGURE 26.2 Steps involved in DNA vaccine development.

Mechanisms

A plasmid vector that expresses the protein of interest (e.g. viral protein) under the control of an appropriate promoter is injected under the skin or muscle of the host (Figure 26.3). After uptake of the plasmid, the protein is produced endogenously and processed intracellularly into small antigenic peptides by the host proteases. The peptides then enter the lumen of the endoplasmic reticulum (ER) by membrane-associated transporters. In the ER, peptides bind to MHC class I molecules. These peptides are presented on the cell surface in the context of the MHC class I. Subsequently, CD8+ cytotoxic T cells (CTL) are stimulated and they evoke cell-mediated immunity. CTLs inhibit viruses through both cytolysis of infected cells and non-cytolysis mechanisms such as cytokine production (Encke et al., 1999).

The foreign protein can also be presented by the MHC class II pathway by APCs (which elicits helper T-cell (CD4+) responses). These CD4+ cells are able to recognize the peptides formed from exogenous proteins that were endocytosed or phagocytosed by APC and degraded to peptide fragments and loaded onto MHC class II molecules. Depending on the type of CD4+ cell that binds to the complex, B cells are stimulated and antibody production is stimulated. This is the same manner in which traditional vaccines work (Schirmbeck and Reimann, 2001).

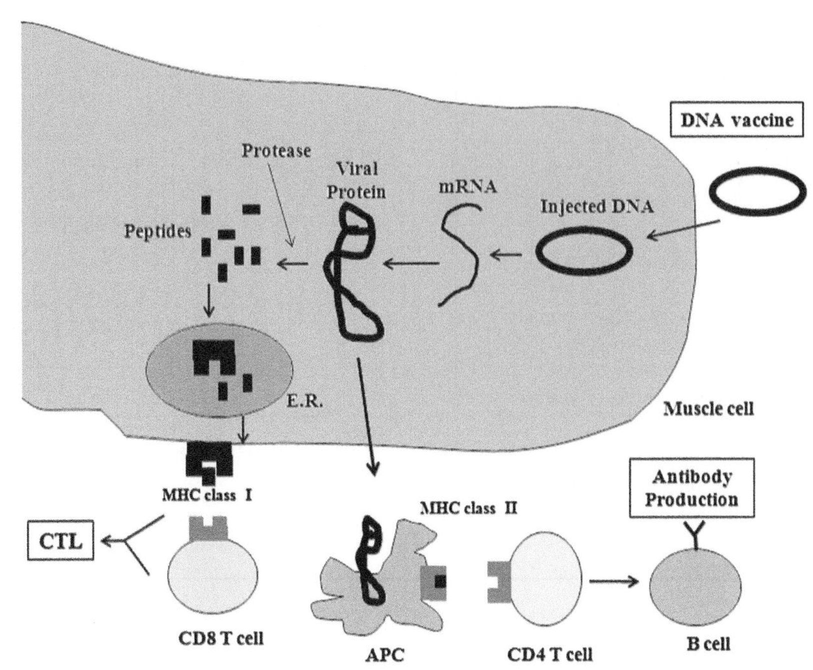

FIGURE 26.3 Action mechanism of DNA vaccine.

Advantages of DNA Vaccines

DNA immunization offers many advantages over traditional forms of vaccination. It is able to induce the expression of antigens that resemble native viral epitopes more closely than standard vaccines do since live attenuated and killed vaccines are often altered in their protein structure and antigenicity. Plasmid vectors can be constructed and produced quickly, and the coding sequence can be manipulated in many ways. DNA vaccines encoding several antigens or proteins can be delivered to the host in a single dose, only requiring a microgram of plasmids to induce immune responses. Rapid, large-scale production is available at costs considerably lower than traditional vaccines, and they are also very temperature-stable, making storage and transport much easier. Another important advantage of genetic vaccines is their therapeutic potential for ongoing chronic viral infections. DNA vaccination may provide an important tool for stimulating an immune response in HBV, HCV, and HIV patients. The continuous expression of the viral antigen caused by gene vaccination in an environment containing many APCs may promote successful therapeutic immune responses that cannot be obtained by other traditional vaccines (Encke et al., 1999). This subject has generated a lot of interest in the last five years.

Disadvantages of DNA Vaccines

Although DNA can be used to raise immune responses against pathogenic proteins, certain microbes have outer capsids made up of polysaccharides. This limits the extent of the usage of DNA vaccines because they cannot be the substitute for polysaccharide-based subunit vaccines.

Future of DNA Vaccines

It has recently been discovered that the transfection of myocytes can be amplified by pre-treatment with local anesthetics or with cardiotoxin, which induce local tissue damage and initiate myoblast regeneration. Gaining a full understanding of this mechanism of DNA uptake could prove helpful in improving applications for gene therapy and gene vaccination. Both improved expression and better engineering of the DNA plasmid may enhance antibody response to the gene products and expand the applications of the gene vaccines.

Recombinant Vector Vaccines

Recombinant vector vaccines are experimental vaccines similar to DNA vaccines, but they use an attenuated virus or bacterium to introduce microbial DNA to cells of the body. "Vector" refers to the virus or bacterium used as the carrier.

In nature, viruses latch on to cells and inject their genetic material into them. In the lab, scientists have taken advantage of this process. They have figured out how to take the roomy genomes of certain harmless or attenuated viruses and insert portions of the genetic material from other microbes into them. The carrier viruses then ferry that microbial DNA to cells. Recombinant vector vaccines closely mimic a natural infection and therefore stimulate the immune system.

Attenuated bacteria can also be used as vectors. In this case, the inserted genetic material causes the bacteria to display the antigens of other microbes on its surface. In effect, the harmless bacterium mimics a harmful microbe, provoking an immune response.

Researchers are working on both bacterial and viral-based recombinant vector vaccines for HIV, rabies, and measles.

MOLECULAR FARMING USING PLANTS AS BIO-REACTORS

Plants have been used as herbal drugs for millennia; they also play an important role in modern medicine. Recent advances in molecular biology techniques have helped in the development of new strategies for the production of subunit vaccines comprising proteins derived from pathogenic viruses, bacteria, or parasites. Although mammals, their tissues and cell lines are currently utilized for the commercial production of vaccines, these systems are expensive and their scale-up is not easy (Larrick and Thomas, 2001; Houdebine, 2009). Plants have emerged as a promising system to express and manufacture a wide range of functionally active pharmaceutical proteins (Daniell et al., 2009) of high value to the health industry with advantages over traditional bioreactors (Figure 26.4). Technological advancements have played a vital role in establishing the use of plants as "surrogate production organisms" or bioreactors. Figure 26.5 shows the general procedure to develop and characterize the plant-made injectible, oral, or edible vaccine (Davoodi-Semiromi et al., 2009). One or more immunoprotective antigens of pathogens can be produced in plants (Ashraf et al., 2005) by the expression of gene(s) encoding the protein(s). In recent years, plant-based novel production systems aimed at developing "edible" or "oral" vaccines have been discussed in detail (Ma et al., 2003; 2005; Koprowski, 2005; Lal et al., 2007; Mishra et al., 2006; Houdebine, 2009; Rybicki, 2010). Compared to traditional vaccines, edible vaccines offer simplicity of use, lower cost, convenient storage, economic delivery, and mucosal immune response.

The original concept of edible vaccines implied that transgenic fruits or vegetables expressing an antigen from a virus or bacteria could be eaten raw without any prior processing, and could act as a vaccine for launching a sufficient protective immune response against a particular disease. Currently, it is widely accepted that this original concept was rather naive, mainly for two reasons. First, different fruits from the same plant express different levels of antigens; therefore it is crucial to make plant-derived vaccines by using a pool of fruits with homogeneous antigen concentrations (vaccine dose). In general, at least a minimum processing, pooling, and freeze-drying of fruits from the same or different plants is required before incorporation into formulations or capsules for oral vaccination. Second, it is important to ensure complete separation of fruits or vegetables intended for human or animal consumption from fruits or vegetables intended for pharmaceutical purposes like vaccines. Compared to traditional vaccines, edible vaccines offer simplicity of use, lower cost, convenient storage, economic delivery, and mucosal immune response.

Hence, the approach to "edible vaccines" has been replaced by the idea of "plant-derived vaccine antigens."

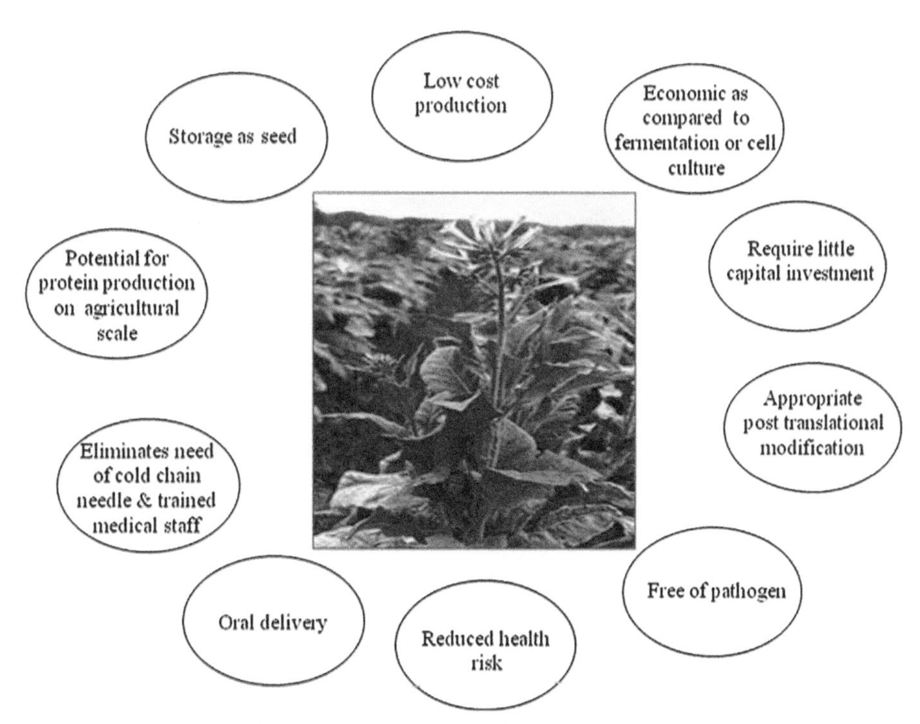

FIGURE 26.4 Advantages of using plants as bioreactors.

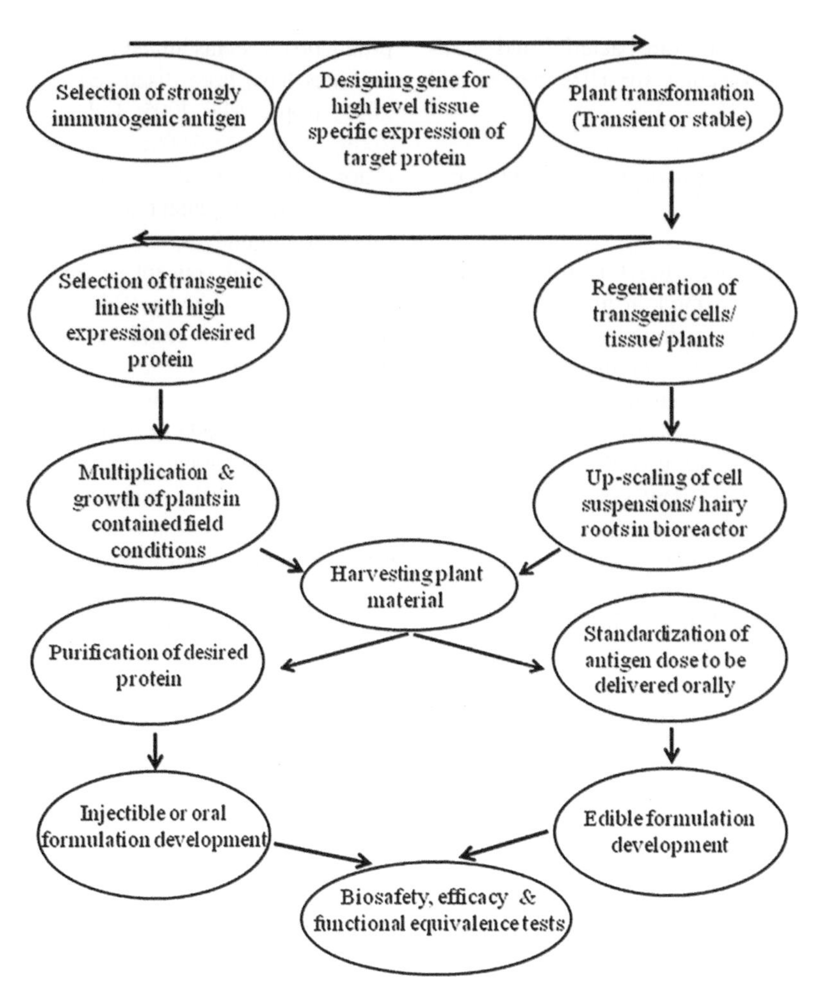

FIGURE 26.5 Steps involved in the production of plant-derived vaccines.

Antigen expression in plant tissue opens an important alternative to meet the global demand for cheaper, safer, and quality vaccines. Topographic compartmentalization of vaccine antigens in plant cells, tissues, or organs is a kind of encapsulation and provides a protective covering by protecting them from proteolytic degradation *in vivo*. Such edible encapsulation protects the antigen from degradation in mucosal and gut systems and allows its efficient absorption. This can facilitate transportation without expensive infrastructure at the manufacturing site where dose standardization and packaging are undertaken.

ROLE OF ADJUVANTS

Adjuvants are important in enhancing the immune response to antigens. The addition of adjuvants to vaccines sustains and directs the immunogenicity and modulates appropriate immune responses. This reduces the amount of antigen required and improves the efficacy of vaccines. Traditional live vaccines do not require the addition of adjuvants, while modern recombinant vaccines, particularly highly purified or synthetic antigens, require adjuvants to induce a protective and long-lasting immune response. Although aluminium salts are the most commonly used adjuvant for human vaccines, they are weak and have complex mechanisms that favor induction of antibodies rather than cellular immunity. New forms of vaccine adjuvants that have been proposed for various vaccines feature oil-based emulsions, bacterial products (*Vibrio cholerae* toxin B subunit, *E. coli* heat-labile enterotoxin B subunit or CpG nucleotides), viral products (virus-like particles), plant products (saponin derivatives), biodegradable particles (liposomes), molecular adjuvants, and synthetic adjuvants (Reed et al., 2009). Adjuvant mechanisms include depot effects, recruitment of innate immunity-specific targeting mechanisms, and carrier functions that hold the antigen in an appropriate conformation. The safety of proposed adjuvants is a primary consideration. Therefore, it is often necessary to devise methods to reduce or eliminate the reactive effects of an adjuvant while preserving the efficacy. The most effective use of adjuvants for certain types of vaccines, particularly for stimulating mucosal immunity, may be to combine the adjuvant with a particular mode of delivery, such as oral, intranasal, or transcutaneous immunization. Carriers that combine both the adjuvant and the antigen in a single formulation can serve as the basis for creation of important formulations for improved vaccines. The enhancement of immunogenicity of an antigenic protein is an important aspect if vaccine antigens, expressed at a modest level in edible plant parts, have to succeed in mounting sufficiently high immune response after passage through the mucosal and gut route. Use of improved adjuvants and better targeting to the immune system might compensate in part for low delivery of antigen. One of the targeting strategies involves linking antigens to molecules that bind well to immune system components such as M cells in the intestinal lining. M cells take up antigens that enter the small intestine and pass them to other cells of the immune system. If white blood cells (helper T lymphocytes) recognize the fragments as foreign, they induce B lymphocytes (B cells) to secrete neutralizing antibodies and initiate other strategies against the perceived enemy. The cholera toxin B subunit (CTB) and *E. coli* heat-labile enterotoxin B subunit (LTB) are potent mucosal immunogens and adjuvants. They both bind directly to the GM1-ganglioside receptor molecules on M cells by fusing antigens from other pathogens to any of these subunits (Cuatrecasas, 1973; Mishra et al., 2006). This ushers foreign antigens into the M cells. By fusing antigens to this subunit, it is possible to improve uptake of antigens by M cells and enhance immune responses. The carrier molecules also serve to modulate immune response against watery diarrhea (Yasuda et al., 2003). Tiwari et al. (2009) reported the use of cholera toxin B subunit as an adjuvant, fused with rabies glycoprotein antigen. This fusion construct was expressed in peanut seeds using a seed-specific promotor. The transgenic peanut seeds showed a high expression of the functional fusion of cholera toxin B subunit–rabies glycoprotein. Work is in progress to examine if the transgenic peanut seeds give active protection against rabies virus infection.

Adjuvants are used to potentiate the immune response:

1. Functions to localize and slowly release antigen at or near the site of administration.
2. Functions to activate APCs to achieve effective antigen processing or presentation.

Materials that have been used as adjuvants include:

1. Aluminium salts.
2. Mineral oils.
3. Mycobacterial products (e.g. Freund's complete and incomplete adjuvants).

IMMUNE-STIMULATING COMPLEXES (ISCOMs)

The immune-stimulating complex (ISCOM) is an efficient mucosal delivery system for respiratory virus envelope antigens. ISCOM allows selective incorporation of viral envelope proteins by hydrophobic interaction into a defined supra-molecular structure of Quillaja saponins. In this a mixture of the plant glycoside, saponin, cholesterol, and phosphatidylcholine provide a vehicle for presentation of several copies of the protein on a cage-like structure. Such a multimeric presentation mimics the natural situation of antigens on microorganisms. These in-built Quillaja saponins have strong adjuvant activity. Antigens are arranged in an accessible, multimeric, physically well-defined cage-like liposomal structure of 40 nm diameter. Recent studies have also shown that the immunogenicities of envelope proteins of respiratory viruses and the B subunit of cholera toxin, when incorporated in ISCOM, are greatly enhanced after mucosal administration, leading to potent mucosal IgA and systemic immune responses. Furthermore, it was reported that ISCOMs containing antigens from *Echinococcus granulosus* efficiently induced humoral and cell-mediated immune responses against carbohydrate antigens by intranasal administration. It shows great promise for the presentation of genetically engineered proteins.

Similar considerations apply to the presentation of peptides. It has been shown that by building the peptide into a framework of lysine residues so that eight copies instead of one copy are present, the immune response induced was of much greater magnitude. A novel approach involves the presentation of the peptide in a polymeric form combined with T-cell epitopes. The sequence coding for the foot and mouth disease virus peptide was expressed as part of a fusion protein with the gene coding for the hepatitis B core protein. The hybrid protein, which forms spherical particles 22 nm in diameter, elicited levels of neutralizing antibodies against foot and mouth disease virus that were at least a hundred times greater than those produced by the monomeric peptide.

PROTOCOL FOR THE DEVELOPMENT OF VACCINES

Following is the protocol explaining the development of a candidate vaccine using traditional or modern recombinant DNA technology:

1. The need for a vaccine against a specific infection or disease is established.
2. The causative agent of the disease is identified, isolated, and grown in the laboratory under *in vivo*/*in vitro* culture conditions.

3. Traditional vaccines: Pathogens are attenuated or chemically inactivated in the laboratory.
4. Recombinant vaccines: The immunoresponsive antigen that promotes immunoprotective antibodies is identified. The candidate immunoprotective antigen is genetically or proteomically modified and proliferated in a suitable expression system.
5. Appropriate methods of purification are used to obtain a traditional or recombinant immunoprotective candidate vaccine.
6. Depending on the type of disease, an animal model (e.g. mice, ferrets, rabbits, Rhesus monkeys, and chimpanzees) is chosen to test the candidate vaccine for its efficacy in protection against live infections or disease.
7. Methods are determined to measure the immune response against the candidate vaccine and to identify the immune response necessary to reach an appropriate level of protection.
8. The clinical trial protocols necessary to test the candidate vaccine in human volunteers and establish its success in the final target are developed. In Phase 1 trials, the putative vaccine is tested for safety, immunogenicity, and dosage efficacy. In Phase 2 trials, investigators test safety and immunogenicity with a larger group of volunteers. Phase 3 trials study the candidate vaccine's efficacy and safety among an even larger group, often in the thousands. After successful Phase 3 trials, the vaccine is approved and recommended for safe immunization programs.

TABLE 26.3 Infectious Agents That Pose Significant Human Health Problems

	Licensed Vaccine Available?
SEXUALLY-TRANSMITTED DISEASES	
Human immunodeficiency virus (HIV)	No
Human papilloma virus (HPV)	No
Herpes simplex virus (HSV)	No
Chlamydia trachomatis	No
Neisseria gonorrhoeae	No
Treponema pallidum	No
RESPIRATORY AGENTS	
Respiratory syncytial virus (RSV)	No
Parainfluenza virus	No
Human metapneumovirus (HMPV)	No
Group of streptococci	No
Chlamydia pneumonia	No
Meningococcous B	No
ENTERIC AGENTS	
Salmonella species	No
Shigella species	No
Escherichia coli (ETEC, EHEC, EPEC)	No
Helicobacter pylori	No
Noroviruses	No
VECTOR BORNE AGENTS	
Plasmodium falciparum	No
Dengue fever virus	No
Hantaviruses	No
Borrelia burgdorferi	No
Schistosoma species	No
Leishmania species	No
Hookworm (multiple genera)	No
OTHERS	
Hepatitis C, E viruses	No
Cytomegalovirus	No
Group B *Streptococcus*	No

FUTURE CHALLENGES IN VACCINE DEVELOPMENT

The discovery of vaccination by Edward Jenner ranks as one of the most important medical advances of all time. Millions of lives are saved each year by vaccination against a range of diseases. Current vaccination methods essentially work on the same principles established by Edward Jenner more than 200 years ago; and has a huge impact on public health. However, the detailed immunological relationship of protective immunity imparted by successful conventional techniques against most infections is not clearly understood.

Foremost Infectious Disease Problems

Despite the best efforts of researchers, medical professionals, and public health officials, a number of infectious diseases continue to pose significant public health problems. Diseases of concern in the developed world are strikingly different from diseases that developing nations contend with, in part because of the success of vaccination efforts in wealthier nations. Also, some diseases are limited geographically for ecological reasons, such as the need for a suitable vector or appropriate habitat for survival. The agents listed in Table 26.3 are noteworthy for their high infection rates, for exhibiting high morbidity or mortality, for their economic impact in both developed and developing nations, and for whether vaccines are available for them.

Infectious Disease Threats

A number of infectious agents that are relatively rare today are poised for an upsurge in incidence because of either "natural" or terrorism-related causes. The natural threats are led by the influenza strain H5N1, which, like many other natural threats, is a zoonotic organism, an infectious agent that crosses over from animals to humans. Originally found in poultry, the virus has sickened humans in Asia twice in the past decade. Zoonoses may quickly become a serious problem for human health if the agents adapt to the human environment by mutation, recombination, or reassortment, by acquisition of new genes or plasmid, by phage interchange of genetic material, or by geographical advance of their vectors.

Human pathogens that could be employed for terrorist purposes must be easily deliverable, but they need not be extremely virulent. Infectious diseases must only be severe enough to instill panic and disrupt civil life in order to be of use to a terrorist group. Many agents could be enhanced through engineering to either make the disease they cause more severe or to enable the agent to escape vaccines or treatment.

ETHICAL ISSUES

Vaccines are responsible for many global public health successes. Even so, vaccinations have also long been the subject of various ethical controversies. There are real risks associated with vaccination programs (Caplan and Schwartz, 2008). In the past, live attenuated vaccines against respiratory syncytial virus resulted in more severe disease and increased mortality in infants.

In 1976 the vaccination program against swine influenza in the U.S. was stopped because it was thought to be associated with a concomitant increase in Guillain–Barré syndrome. This association could not be confirmed after seasonal influenza vaccination. The whole-cell pertussis vaccine was previously thought to be associated with an increased risk of post vaccination encephalopathy, which lacked evidence for such casual association (Ray et al., 2006).

The key ethical debates related to vaccine regulation, development, and use generally revolve around (1) mandates, (2) research and testing, (3) informed consent, and (4) access disparities.

MANDATES

Ethical debates and objections to school mandates and other mandates arise because some individuals and communities disagree with the mandates, and/or have religious or philosophical beliefs that conflict with vaccination. For example, in an effort to protect the greatest number of people, public health vaccine regulations can infringe upon individual autonomy and liberty. Tension results if individuals want to exercise their right to protect themselves and/or their children by refusing vaccination, if they do not accept existing medical or safety evidence, or if their ideological beliefs do not support vaccination. Many scientific and medical research studies have found that individuals who exercise religious and/or philosophical exemptions are at greater risk of contracting infections, which puts them and their communities at risk (Feikin et al., 2000; Salmon and Omar, 2006). Thus, medical and public health advocates often struggle to balance the ethics of respecting individual beliefs and protecting the community's health.

VACCINE RESEARCH AND TESTING

Ethical discussions also surround the research and testing of vaccines, including discussions about vaccine development, study design, population, and trial location. Licensed vaccines go through many years of research and must pass rigorous safety and efficacy standards. The vaccine development and research process involves experts from diverse disciplines, including scientific and social fields, public health, epidemiology, immunology, statistics, as well as experts from the pharmaceutical industry. These stakeholders may have conflicting priorities and motives, which contributes to various ethical discussions. Additionally, it is important to understand a vaccine's safety and efficacy in various populations; testing a vaccine in vulnerable populations, such as children, also raises ethical concerns. Researchers must balance the need to protect children's safety with the need to adequately understand how a vaccine will perform and affect children when administered. Similarly, it is important to understand how vaccines affect people in developing countries. Yet, conducting vaccine research in developing countries raises a list of ethical concerns, such as how to provide necessary screening or treatment if diseases are detected, how to meaningfully involve local communities in the research design process, how to ensure the trial and vaccine can be supervised by local ethics review panels, and how to ensure that participants understand consent (Snider, 2000). Ethical discussions are a key component of HIV vaccine research and development because HIV vaccines pose numerous unique ethical challenges (Celada et al., 2011). For example, the AIDS stigma may put vaccine trial participants at psychological risk if they encounter discrimination. In addition, researchers have to figure out how to provide appropriate and adequate medical care and protection from stigma for participants who are screened and found to be HIV positive. Researchers have to consider that if participants misunderstand the trial, they may think that they are protected from the virus and put themselves at risk. The complexity of these issues places ethics analyses at the forefront of HIV vaccine research.

INFORMED CONSENT

Ethical debates also surround vaccine implementation and delivery, such as those with informed consent. Although federal guidelines do not require consent before vaccination, the National Childhood Vaccine Injury Act of 1986 requires that doctors give vaccine recipients, or their parents or legal representatives, a Vaccine Information Statement (VIS). The VIS provides basic information about vaccine risks and benefits and is designed to provide the information a patient or parent needs to make an informed decision.

ACCESS ISSUES

Many vaccine-related ethical debates center on the argument that access to vaccinations depends to some extent on socioeconomic and other demographic factors. Implicit in these discussions is the concern that all lives be treated with equal value and that everyone be offered the opportunity of protection by vaccination. Global disparities signal the need for continued efforts to address these concerns.

TRANSLATIONAL SIGNIFICANCE

New science, new technologies, and an ever more sophisticated understanding of immunology have yet to make a significant impact on world vaccination. Novel and improved vaccines for threatening agents are currently under development. The struggle to develop a vaccine for HIV has been ongoing since the virus was first discovered in 1982, and though a great deal of progress has been made, an effective and safe vaccine is still elusive. Nearly all vaccine designs and vectors have been tried in the effort to build an HIV vaccine. Some of the promising formulations today include a prime-boost vaccine (so named because of the two-pronged strategy of administering a primer vaccine and following up later with a booster vaccine) that is being tested in Thailand, and a vaccine based on three recombination adenovirus constructs, each expressing a different HIV gene.

The excellent work of the academic community in the development of early vaccines and in antigen discoveries has been exciting, but the wide variety and complexity of vaccine products and their production require technologies, expertise, production infrastructure, and regulatory insights that are as yet not widely available. Identification of trained individuals in translational research and vaccine development will help carry innovations from the laboratory to the market place.

There is a growing demand for vaccine safety, fueled in part by anti-vaccination groups. As disease recedes, the need for vaccination becomes less evident to the public, and more people opt out of vaccination and depend instead on mass immunization of people around them. Of course, mass immunization fails if too many people refuse to be vaccinated. Also, there have been some serious safety issues associated with vaccines, such as cases of paralysis after the oral polio vaccine, and disseminated infections after Bacille Calmette–Guerin. For this reason, older vaccines need to be re-examined for their safety, and, if necessary, improved; this was done for the whole-cell pertusis vaccine, which is now made in cell culture instead of in the brain. In the near future, Jenner's vaccinia will be replaced by further attenuated vaccinia, and Bacille Calmette–Guerin will be replaced by engineered vaccines for tuberculosis. Indeed, one of the advantages of newer molecular technologies is improved safety. As risk–benefit ratios become more controversial, when the presence of disease declines, it will be important to reduce vaccine-associated reactions to a minimum.

WORLD WIDE WEB RESOURCES

http://www.dnavaccine.com/

This interesting web site contains references on the production and administration of a wide variety of vaccines.

http://www.cdc.gov/vaccines

This website provides information related to vaccines, immunization against preventable diseases, vaccine safety, and vaccine side effects.

http://www.vaccines.org

This website provides access to up-to-the-minute news about vaccines and an annotated database of vaccine resources on the Internet.

REFERENCES

Ashraf, S., Singh, P. K., Yadav, D. K., Shahnawaz, M. D., Mishra, S., Sawant, S. V., & Tuli, R. (2005). High level expression of surface glycoprotein of rabies virus in tobacco leaves and its immunoprotective activity in mice. *Journal of Biotechnology, 119*, 1–14.

Caplan, A. L., & Schwartz, J. L. (2008). Ethics. In S. A. Plotkin, O. A. Walter & P. A. Offit (Eds.), *Vaccines*. Philadelphia: Saunders.

Celada, M. T., Merchant, R. C., Waxman, M. J., & Sherwin, A. M. (2011). An ethical evaluation of the 2006 centers for disease control and prevention recommendations for HIV testing in heath care settings. *The American Journal of Bioethics, 11*, 31–40.

Cuatrecasas, P. (1973). Gangliosides and membrane receptors for cholera toxin. *Biochemistry, 28*, 3558–3566.

Daniell, H., Singh, N. D., Mason, H., & Streatfield, S. J. (2009). Plant-made vaccine antigens and biopharmaceuticals. *Trends in Plant Science, 14*(12), 669–679.

Davis, H. L. (1997). Plasmid DNA expression systems for the purpose of immunisation. *Current Opinion in Biotechnology, 8*, 635–640.

Davoodi-Semiromi, A., Samson, N., & Daniell, H. (2009). The green vaccine: A global strategy to combat infectious and autoimmune diseases. *Human Vaccines, 5*(7), 488–493.

Encke, J., Putlitz, J. Z., & Wands, J. R. (1999). DNA Vaccines. *Intervirology, 42*, 117–124.

Feikin, D. R., Lezotte, D. C., Hamman, R. F., Salmon, D. A., Chen, R. T., & Hoffman, R. E. (2000). Individual and community risks of measles and pertussis associated with personal exemptions to immunization. *Journal of American Medical Association, 284*, 3145–3150.

Houdebine, L. M. (2009). Production of pharmaceutical proteins by transgenic animals. *Comparative Immunology, Microbiology and Infecttion Disease, 32*, 107–121.

Koprowski, H. (2005). Vaccines and sera through plant biotechnology. *Vaccine, 23*, 1757–1763.

Koprowski, H., & Weiner, D. B. (1998). *DNA Vaccination/ Genetic Vaccination*. Heidelberg: Spriner-Verlag. p. 198.

Lal, P., Ramachandran, V. G., Goyal, R., & Sharma, R. (2007). Edible vaccines: current status and future. *Indian Journal of Microbiology, 25*, 93–102.

Larrick, W., & Thomas, D. W. (2001). Producing proteins in transgenic plants and animals. *Current Opinion in Biotechnology, 12*, 411–418.

Ma, J. K. -C., Barros, E., Bock, R., Christou, P., Dale, P. J., Dix, P. J., Fischer, R., Irwin, J., Mahoney, R., Pezzotti, M., Schillberg, S., Sparrow, P., Stoger, E., & Twyman, R. M. (2005). Molecular farming for new drugs and vaccines. Current perspectives on the production of pharmaceuticals in transgenic plants. *EMBO Reports, 6*, 593–599.

Ma, J. K. -C., Drake, P. M. W., & Christou, P. (2003). The production of recombinant pharmaceutical proteins in plants. *Nature Reviews Genetics, 4*, 794–805.

Mishra, S., Yadav, D. K., & Tuli, R. (2006). Ubiquitin fusion enhances cholera toxin B subunit expression in transgenic plants and the plant-expressed protein binds GM1 receptors more efficiently. *Journal of Biotechnology, 127*, 95–108.

Ray, P., Hayward, J., Michelson, D., Lewis, E., Schwalbe, J., Black, S., Shinefield, H., Marcy, M., Huff, K., Ward, J., Mullooly, J., Chen, R., Davis, R., & The vaccine safety datalink group (2006). Encephalopathy after whole cell pertussis or measles vaccination: lack of evidence for a causal association in a retrospective case-control study. *The Pediatric Infectious Disease Journal, 25*, 768–773.

Reed, S. G., Bertholet, S., Coler, R. N., & Friede, M. (2009). New horizons in adjuvants for vaccine Development. *Trends in Immunology, 30*, 23–32.

Roitt, I. M., & Delves, P. J. (Eds.), (1992). *Encyclopedia of Immunology*. San Diego: Academic Press.

Rybicki, E. P. (2010). Plant-made vaccines for humans and animals. *Plant Biotechnology Journal, 8*, 620–637.

Salmon, D. A., & Omar, S. B. (2006). Individual freedoms versus collective responsibility: Immunization decision making in the face of occasionally repeating values. *Emerging Themes in Epidemiology, 3*, 1–3.

Schirmbeck, R., & Reimann, J. (2001). Revealing the potential of DNA-based vaccination: Lessons learned from the Hepatitis B virus surface antigen. *Biological Chemistry, 382*, 543–552.

Snider, D. E. (2000). Ethical issues in tuberculosis vaccine trials. *Clinical Infectious Diseases, 30 Suppl.*(3), S271–S275.

Tiwari, S., Mishra, D. K., Roy, S., Singh, A., Singh, P. K., & Tuli, R. (2009). High level expression of a functionally active cholera toxin B: rabies glycoprotein fusion protein in tobacco seeds. *Plant Cell Reports, 28*, 1827–1836.

Wolff, J. A., Malone, R. W., Williams, P., Chong, W., Acsadi, G., Jani, A., & Felgner, P. L. (1990). Direct gene transfer into mouse muscle in vivo. *Science, 247*, 1465–1468.

Yasuda, Y., Isaka, M., Taniguchi, T., Zhao, Y., Matano, K., Matsui, H., Morokuma, K., Maeyama, J., Ohkuma, K., Goto, N., & Tochkubo, K. (2003). Frequent nasal administrations of recombinant cholera toxin B subunit (r-CTB)-containing tetanus and diphtheria toxoid vaccines induced antigen-specific serum and mucosal immune responses in the presence of anti-rCTB antibodies. *Vaccine, 21*, 2954–2963.

FURTHER READING

Kuby Immunology, Goldsby RA, Kindt TJ, Obsorne BA. WH Freeman.
Basic Immunology, Abbas AK, Lichtman AH. Saunders W.B. Company
Fundamentals of Immunology, Paul WE. Lippincott Williams and Wilkins
Immunology, Anderson WL. Fence Creek Publishing (Blackwell).
Robbins, A., & November, Freeman P. (1988). Obstacles to developing vaccines for the Third World. *Scientific American, 259*(5), 126–133.

GLOSSARY

Attenuated vaccine A weakened pathogen used as vaccine to stimulate protective immunity (e.g. BCG vaccine, MMR, cowpox, oral polio vaccine etc.).

DNA vaccine A vaccine in which a purified DNA carries a sequence encoding the expression of the subunit antigen of interest under the control of eukaryotic promoter (e.g. DNA expressing rabies glycoprotein was used as a vaccine for eliciting protective immunity). For example, avidin, an egg protein with important properties used as a diagnostic reagent in transgenic maize. Many DNA vaccines are in clinical and pre-clinical trials, including vaccines for influenza, malaria, tuberculosis, Ebola, herpes, HIV, hepatitis etc.

Molecular farming Application of biotechnology to agricultural plants in order to use them as bioreactors for the production of valuable plant-made pharmaceuticals.

Subunit vaccine A preparation of pathogenic constituent (instead of whole pathogen) that can be administered to stimulate protective immunity (e.g. purified hepatitis B surface antigen (HbsAg) heterologously expressed in yeast cells and used as a vaccine to elicit protective immunity).

Vaccine A preparation of a pathogenic agent or its constituent that can be administered to stimulate protective immunity (e.g. cow pox and chicken pox vaccines etc.).

Variolation Smearing of a skin tear with cowpox skin crust to confer immunity to smallpox.

ABBREVIATIONS

AIDS Acquired Immune Deficiency Syndrome
APCs Antigen-Presenting Cells
CTLs Cytotoxic T Lymphocytes
DNA Deoxyribonucleic Acid
GCs Germinal Centers
HBV Hepatitis B Virus
HiA *Haemophilis influenza A*
HiB *Haemophilis influenza B*
HBsAg Hepatitis B Antigen
HIV Human Immunodeficiency Virus
ISCOMs Immune-Stimulating Complexes
LAIV Live Attenuated Influenza Vaccine
LTB Heat-Labile Enterotoxin B Subunit
MHC Major Histocompatibility Complex
PCs Plasma Cells
RNA Ribonucleic Acid
TCR T-cell Receptors
Th Helper T cells
Ty21a Live Typhoid Vaccine
VIS Vaccine Information Statement

LONG ANSWER QUESTIONS

1. What should be properties of an ideal vaccine?
2. Explain the advantages and disadvantages of an attenuated vaccine.
3. Why we cannot follow the classical paradigm to develop a vaccine against HIV infection?

4. What is the major concern with DNA vaccine?
5. Why attenuated vaccines are more likely to induce cell-mediated immunity than killed vaccines

SHORT ANSWER QUESTIONS

1. Is it possible to make vaccine against all diseases?
2. Why DNA vaccines cannot be prepared against all antigens?
3. What is the basic advantage of edible vaccines?
4. Explain whether HIV virus be used to develop a killed vaccine.

ANSWERS TO SHORT ANSWER QUESTIONS

1. It is not possible to develop vaccine against all disease because the antigenicity of the causative agent/molecule is extremely important to induce the humoral and cell mediated immunity. Thus the chemical nature of antigen determines the possibility of making the vaccine against the disease.
2. DNA vaccines can only be prepared against the proteinacious antigens. As DNA molecule can only determine the translation of proteins. Hence DNA vaccine cannot be prepared against sugar and lipid antigens.
3. Basic advantage of edible vaccines is the low cost of vaccination programs. It significantly reduces the need of cold chain, injection, adjuvants and skilled worker to administer the vaccine.
4. Inactivated or attenuated HIV virus loses its antigenicity, hence the use of killed HIV virus cannot be used as a vaccine. Furhter vaccines protect against disease, but not against infections and a very high rate of mutation in HIV virus evades the developed immunity.

Chapter 27

Safety Assessment of Food Derived from Genetically Modified Crops

Premendra D. Dwivedi, Mukul Das, Sandeep Kumar and Alok Kumar Verma
Food, Drug and Chemical Toxicology Group, CSIR-Indian Institute of Toxicology Research, Lucknow, U.P., India.

Chapter Outline

SUMMARY

GM foods are essential for meeting the food demands of the world's ever-increasing population. Various approaches are available for assessing the safety of GM foods. These include the use of bioinformatics, examining the stability of protein in the gastrointestinal tract, and allergenicity testing using animal models. This chapter explores the safety related concerns of GM foods.

WHAT YOU CAN EXPECT TO KNOW

These days, crops are being genetically modified by artificial insertion of genes to meet the requirements of an ever-growing population. These transgenic variants have improved agronomic characteristics (e.g. insect resistance, herbicide resistance, disease resistance, and drought tolerance), but the safety of all new foods (including GM) crops should be ensured prior to their release onto the market. Questions of safety, including toxicity, allergenicity, and environmental impact of GM crops and foods, are frequently posed. Avoidance of allergens is the best preventive approach to dealing with allergies as there is no permanent cure. Therefore, it becomes essential to assess the allergenic potential of genetically modified (GM) crops before their commercial release. Basic safety assessment concerns include the question of whether a newly expressed protein's allergic potential is lesser than, equal to, or higher than that of its native crop. A step-by-step approach has been used to ensure the safety of GM foods. This chapter includes the basic information regarding the protease and

Animal Biotechnology. http://dx.doi.org/10.1016/B978-0-12-416002-6.00027-4

thermal-resistant properties shown by allergens, and mechanistic aspects involved therein. Simulated Gastric Fluid, Simulated Intestinal Fluid, and Thermal Treatment assays are described in detail to enhance the knowledge of *in vitro* methods of allergenicity assessment.

HISTORY AND METHODS

INTRODUCTION

Genetically modified (GM) crops are needed to meet the requirement of the ever-growing world population. With the continued increase in population, the major challenge is how to manage food for everyone. To some extent, GM foods may fulfill this requirement. By the use of the latest molecular biology techniques, desirable traits in plants can be introduced by artificial insertion of genes from unrelated species, or sometimes from an entirely different kingdom. Development of GM crops began in the late 1980s with the advancement in biotechnological techniques for directly altering the DNA of the genome, and rearrangement of DNA by using methods such as electroporation or infection with recombinant vectors (e.g. *Agrobacterium tumefaciens*). These GM crops had new traits introduced to them as compared to native crops (e.g. insects resistance, herbicides, disease resistance, drought tolerance, and improved nutritional content). Conventional plant breeding methods were time consuming and imprecise. However, desired traits could be inserted into a plant with higher accuracy using genetic engineering methods (e.g. insertion of the Bt gene into corn, which offers insect resistance). This gene, isolated from a bacterium named *Bacillus thuringiensis*, produces a protein that can kill insect larvae. The first GM crops, tomato and soybean, were evaluated for risk assessment and approved by the United States Food and Drug Administration (USFDA). It should be ensured that GM foods are safe before their release onto the market. There is a chance that an inserted gene could result in the translation of a protein that has the ability to provoke an allergenic response that would sensitize consumers. Furthermore, it is also possible that an inserted gene could elicit allergenic potential by cross reactivity in an already sensitized population. To save sensitive consumers from unwanted exposure to allergens, appropriate preventive measures should be taken. Humans are exposed to a variety of allergens present in the environment and food. Pollens, fungi, insects, and a variety of food products of animal or plant origin, may be harmful to the exposed, sensitized group. It is often believed that the GM crops/foods may cause additional problems if effective measures are not taken. Also, people should be aware of the ill effects of allergens, as no effective medical treatments are currently available for treating this health concern. Therefore, it is essential to assess the allergenic potential of GM crops and foods prior to commercialization.

The 15th anniversary (1996–2010) of the commercialization of GM crops concluded recently. Accumulated hectares utilized for irrigation of the GM crops during this time exceeded 1 billion, which is equivalent to the total area of the United States or China, clearly signifying that biotech crops have put down strong roots. A record 87-fold increase in hectares occurred between 1996 and 2010, which makes GM crops the fastest adopted crop technology in the history of modern agriculture. In 2011, commercially cultivated GM crops were Alfalfa (*Medicago sativa*), Argentine Canola (*Brassica napus*), Bean (*Phaseolus vulgaris*), Carnation (*Dianthus caryophyllus*), Chicory (*Cichorium intybus*), Cotton (*Gossypium hirsutum L.*), Creeping Bentgrass (*Agrostis stolonifera*), Flax or Linseed (*Linum usitatissumum L.*), Maize (*Zea mays L.*), Melon (*Cucumis melo*), Papaya (*Carica papaya*), Petunia (*Petunia*), Plum (*Prunus domestica*), Polish canola (*Brassica rapa*), Poplar (*Populus nigra*), Potato (*Solanum tuberosum L.*), Rice (*Oryza sativa L.*), Rose (*Rosa hybrida*), Soybean (*Glycine max L.*), Squash (Cucurbita pepo), Sugar beet (*Beta vulgaris*), Sweet pepper (*Capsicum annuum*), Tobacco (*Nicotiana tabacum L.*), Tomato (*Lycopersicon esculentum*), and Wheat (*Triticum aestivum*). About 45 countries are moving ahead with the development of various GM crops, including: Argentina, Australia, Bolivia, Brazil, Burkina Faso, Canada, Chile, China, Colombia, Costa Rica, Czech Republic, Egypt, Arab Republic, El Salvador, Germany, Honduras, India, Iran, Islamic Republic, Japan, Korea, Malaysia, Mexico, Myanmar, The Netherlands, New Zealand, Pakistan, Paraguay, Philippines, Poland, Portugal, Romania, Russian Federation, Singapore, Slovak Republic, South Africa, Spain, Sweden, Switzerland, Taiwan, Thailand, Turkey, the United Kingdom, the United States, and Uruguay (Source: http://www.isaaa.org/gmapprovaldatabase/default.asp). Herbicide and drought tolerance, insect resistance, improved nutritional characteristics of foods or feeds, and altered fatty acid profiles are major choices of developers in most GM crops nowadays.

The first international and national provisions for the safety assessment and regulation of GM crop-derived foods were started by the Organization for Economic Co-operation and Development (OECD, 1986), and the first regulatory approval of a GM crop came after approximately one decade (in 1995). In 1996, the International Food Biotechnology Council (IFBC) and the International Life Sciences Institute (ILSI) jointly developed a decision-tree approach (Metcalfe et al., 1996) that is widely accepted by regulatory authorities all over the world. In GM foods, the inserted proteins need to be assessed before their insertion, and it is very important to ensure that the products of novel genes introduced into GM crop are not harmful. It is also important to ensure that the process of transformation does not cause any unintended change in the characteristics and levels of expression of endogenous allergenic proteins. The safety assessment focuses on the new gene products and whole foods derived from the GM crop. Both intended and potentially unintended effects of the genetic modification should be taken into account. The assessment of GM crops involves steps like characterization of the parent crop, characterization of the donor organisms from which any recombinant DNA sequences are derived, the transformation process and the introduced recombinant DNA sequences, safety assessment of the introduced gene products (proteins as well as metabolites), and food safety assessment of whole food derived from edible parts. Recently, GM brinjal in India was suspended (for an indefinite time period) prior to its intended release due to safety-related issues (Kumar et al., 2011).

During safety assessment of GM foods, allergenicity is one of the most important issues. Food allergy is an increasing global health concern. It is an immune provocation in susceptible individuals, triggered by certain food proteins, including proteins derived from GM foods. Sensitization develops when a susceptible individual is exposed to an amount of protein sufficient to induce an immune response. Subsequent exposure to the same or similar allergenic protein in sensitized individuals can provoke an adverse reaction. Allergic reactions may be mild and local, but sometimes can be severe, systemic, and fatal. Severe allergic reactions with a rapid onset of symptoms are known as anaphylaxis reactions. The susceptibility of any individual is dependent on several factors, including genetic predisposition and environmental factors.

In a food, every protein may not be responsible for provoking immunological reactions, but certain proteins that can induce allergic complications are known as allergens. Every allergen is an antigen, but not every antigen is an allergen. In an allergenic protein, certain regions have immune reactive capacity; these regions are known as epitopes. It has been reported that certain biochemical characteristics are shared by many (but not necessarily all) food allergens; one such characteristic is the relative stability and resistance to the denaturation of proteins. Pepsin resistance is thought to be an important property of allergens because when any portion of the protein remains intact, the chance of an immune response is higher. In the United States, each year about 30,000 people come to hospital emergency departments due to food-induced anaphylaxis, and nearly 200 people die (Sampson, 2003). In the United Kingdom, millions of people suffer from food-induced allergic reactions. The Food and Agriculture Organization (FAO) of the United Nations and the World Health Organization (WHO) expert consultation committee on the allergenicity of foods derived from biotechnology documented an approach for assessing the allergenic potential of novel proteins in transgenic crops (Codex Alimentarius, 2003).

RATIONALE FOR THE ALLERGENICITY ASSESSMENT OF GM FOODS

The rationale behind the allergenicity assessment of GM foods is to compare the allergenicity of a GM food with an equivalent non-GM food variety. It focuses on whether a newly expressed protein has allergic potential or not. To ensure the safety of the inserted protein, the GM crop, if found to have increased allergic potential compared to its non-GM counterpart, is not approved by regulatory agencies. GM food with allergenic potential may cause life-threatening allergic reactions in the already sensitized group as there may be sequence homology between known allergens and the novel inserted protein. Various methods have been used in a step-by-step manner to ensure the safety of GM foods. Among these, a few important points to ponder are:

- The source of the introduced protein: It should be preferably from a non-allergenic source.
- A measure of the abundance of the novel protein in the food.
- Comparison of amino acid sequences of newly expressed protein with known allergens present in allergen databases like SDAP and AllergenOnline.
- Stability of the protein in the gastrointestinal tract (GI tract).
- Thermal stability of the novel proteins.
- Various animal models like mouse, rat, guinea pig, dog, and pig have been developed and are being used for allergenicity assessment of GM foods.
- Serum IgE binding test to evaluate the presence of specific IgE antibodies against the protein of interest in serum.

Safety assessment approaches regarding GM foods differ from country to country, but the basic principle for allergenicity assessment is based on a common approach that is internationally acceptable. Several international organizations like FAO/WHO, OECD, and ILSI have provided guidelines for the safety assessment of GM foods, a prerequisite for their release onto the market.

Out of several approaches used for the assessment of allergenicity of the novel protein, stability to digestion in the GI tract is considered a primary requirement. It should be performed for all newly inserted or expressed proteins. Stability in the gastrointestinal tract is the first step towards the evaluation of the allergenicity of food proteins.

MECHANISM OF FOOD PROTEIN-INDUCED ALLERGENICITY

Proteins are an integral part of food, and they provide nutritional support for the body. Food proteins may or may not be allergenic. Allergenic proteins are thought to commonly induce immunological responses via the GI tract, and due to this, digestibility and gut permeability are important factors when assessing the allergenic potential of novel proteins. Protein processing starts in the mouth where it is mixed with saliva that contains the enzyme "ptyalin." No significant enzyme is found in the mouth for protein digestion. The main digestion of protein takes place in the stomach and small intestine. The environment of both the compartments is different. Food proteins move into the GI tract, where it gets digested with pepsin in the stomach and broken down into smaller peptide fragments. These fragments are then subjected to digestion in the intestine, and finally amino acids or small peptides get absorbed.

It is believed that intact proteins or peptides of higher molecular weight have the ability to sensitize the individual as well as release allergic mediators. Some portion of food remains undigested, and this may contain allergenic proteins as well. The intestinal absorption of food allergens and the immune responses to them are interrelated; the nature of the allergen can decide the type of immune responses generated. The intestinal epithelium, which is joined together with its neighbors via tight junctions and mucus produced by goblet cells, may act as a barrier, and can limit the permeation of macromolecules, ingested pathogens, parasites, toxins, and anti-nutrients by tight cell junctions; however, proteins have been reported to cross the intestinal barrier in an intact form.

The GI absorption of intact food allergens such as ovalbumin, peanut allergens, and Gly m Bd 30K from soybean using *in vivo* models has been well documented. Many food allergens are proteins having intramolecular disulfide bonds (2S) that make them stable and resistant to pepsin digestion (Kumar et al., 2012). Higher stability of proteins in the hostile GI tract for a period of time could be sufficient to elicit an immune provocation. The stable proteins or peptide fragments are internalized, processed, and presented by antigen-presenting cells (APC) like macrophages, dendritic cells (DCs), and B cells present in the outer layer of the intestine (i.e. lamina propria). In the intestine, the DCs are the most commonly found antigen-presenting cells. DCs in the intestine are generally found in semi-mature forms, and once these cells come in contact with food allergens, they get activated, become mature, and finally move to the Peyer's patches. In the Peyer's patches, $CD4^+$ Th2 cells recognize these allergens. These Th2 cells secrete cytokines like IL-4 and IL-13, which along with CD40 legend, help in class-switching to immunoglobulin E (IgE). The newly formed IgE antibodies move to mast cells and bind the Fc epsilon receptor 1 (FcεR1). This is known as priming of IgE on mast cells or basophil cells (Kumar et al., 2012). After subsequent secondary exposure to the same (or even similar) allergens, the allergens get linked between two IgE molecules (IgE–allergen–IgE). This phenomenon is also known as cross-linking of allergen. The cross-linking of these allergens initiates a cascade of reactions involving kinases like Lyn and Syk that ultimately ends in degranulation of mast cells. The degranulation results in the secretion of allergic mediators like histamine, mast cell proteases, prostaglandins, leukotrienes, serotonin, β-hexosaminidase, cytokines, etc. These mediators cause several disorders in the body like breathlessness, sneezing, dilation of blood capillaries. If the reaction is severe, then it can be life-threatening and the situation is known as anaphylaxis. A brief outline of pepsin-resistant protein-induced food allergies is given in Table 27.1.

SIMULATED GASTRIC FLUID (SGF) ASSAY

During the allergic assessment of GM foods, the resistance of proteins to proteolytic digestion has been detailed in current safety assessment guidelines from multiple studies. It is reported that proteins with resistance to pepsin digestion in SGF can induce allergenic reactions. The assumption behind the SGF assay had simple reasoning that nutritionally desirable proteins should be rapidly digested, and therefore have less opportunity to exert adverse health effects when consumed. This assumption appears to have been confirmed in several cases, like peanuts, eggs, milk, soybeans, red gram, green gram, red kidney beans, and chickpeas. Therefore, the stability of a transgenic protein to pepsin digestion under acidic conditions (pH 1.2 to 2.0) is generally assumed to be a simple but effective test for assessing the allergenic risk for transgenic proteins (Astwood et al., 1996). Along with other positive evidence, there are reports indicating no link between the stability of a protein in SGF and its allergenicity (Fu et al., 2002).

Therefore, it can be inferred that correlation between allergenicity and digestive stability is not absolute in each and every case. Since resistance to degradation by acid proteases is used to make regulatory decisions, further studies are needed to explore this correlation.Taken together, SGF assays may not equally mimic mammalian *in vivo* digestion conditions, but the stability of a transgenic protein or a digestion fragment in SGF may be related to its allergic potential, as pepsin resistance can provide sufficient time to interact with the GI tract's immune component and induction of allergic reactions.

How the SGF Assay Works

The adult human GI tract is a long tube with a size of approximately 9 meters running through the body from the oral to the anal aperture. The comparative digestibility of allergenic and non-allergenic proteins is highly significant when stability to digestion is used as the basis to predict the allergenic potential of novel proteins (Thomas et al., 2004). Pepsin is an aspartic protease generated from the autocleavage of pepsinogen under acidic conditions in the stomach. Protein degradation by pepsin or pepsinolysis is

TABLE 27.1 Stability* of Food Allergens in SGF and SIF

S.N.	Protein Group	Protein Source	SGF Stability (min)	SIF Stability (min)
	Allergenic Proteins			
1.	β-Lactoglobulin	Cow's Milk	120	5
2.	BSA	Cow's Milk	0	120
3.	A-Lactalbumin	Cow's Milk	0	15
4.	Lactoperoxidase	Cow's Milk	0	120
5.	Ovalbumin Egg	Egg	5	5
6.	Ovomucoid	Egg	0	60
7.	Conalbumin	Egg	0	120
8.	Lysozyme	Egg	60	120
9.	*Ara h* 1	Peanut	5	15
10.	*Ara h* 2	Peanut	0.5	0.5
11.	Peanut Lectin	Peanut	5	120
12.	Soybean Lectin	Soybean	5	120
13.	Trypsin Inhibitor	Soybean	120	120
14.	Patatin	Potato Tuber	0	0.5
15.	Papain	Papaya	0	120
16.	Bromelain	Pineapple	0	120
	Non-Allergenic Proteins			
17.	α-Lactalbumin	Human Milk	0	60
18.	Zein	Corn	120	0.5
19.	Trypsin Inhibitor	Bovine Pancreas	120	120
20.	Red Kidney Bean Lectin	Red Kidney Bean	15	120
21.	Pea Lectin	Pea	5	120
22.	Lentil Lectin	Lentil	0.5	120
23.	Lima Bean Lectin	Lima Bean	5	120
24.	Jack Bean Lectin	Jack Bean	15	120
25.	Cytochrome c	Bovine Heart	0	60
26.	Rubisco	Spinach Leaf	0	120
27.	Phosphofructokinase	Potato Tuber	0	5
28.	Sucrose Synthetase	Wheat Kernel	0	0.5

**Stability was measured as the last time period (in minutes) that the protein could be seen in the SDS–PAGE gel.*
(**Source:** Fu et al., J Agric Food Chem, 2002, 50, 7154–7160).

generally a very rapid reaction, but this is affected by the secondary or tertiary structure of protein substrate. The most effective pH for optimum activity for pepsinolysis ranges between 1.8 and 3.2, but pepsin is generally irreversibly denatured at pH 6 to 7. Due to the fact that a pH greater than 6 can irreversibly denature pepsin activity, the SGF reaction can be stopped by neutralizing aliquots of the solution at different incubation periods using bases like $NaHCO_3$ and NaOH. These incubation mixtures can be analyzed to track the digestion of substrate proteins using

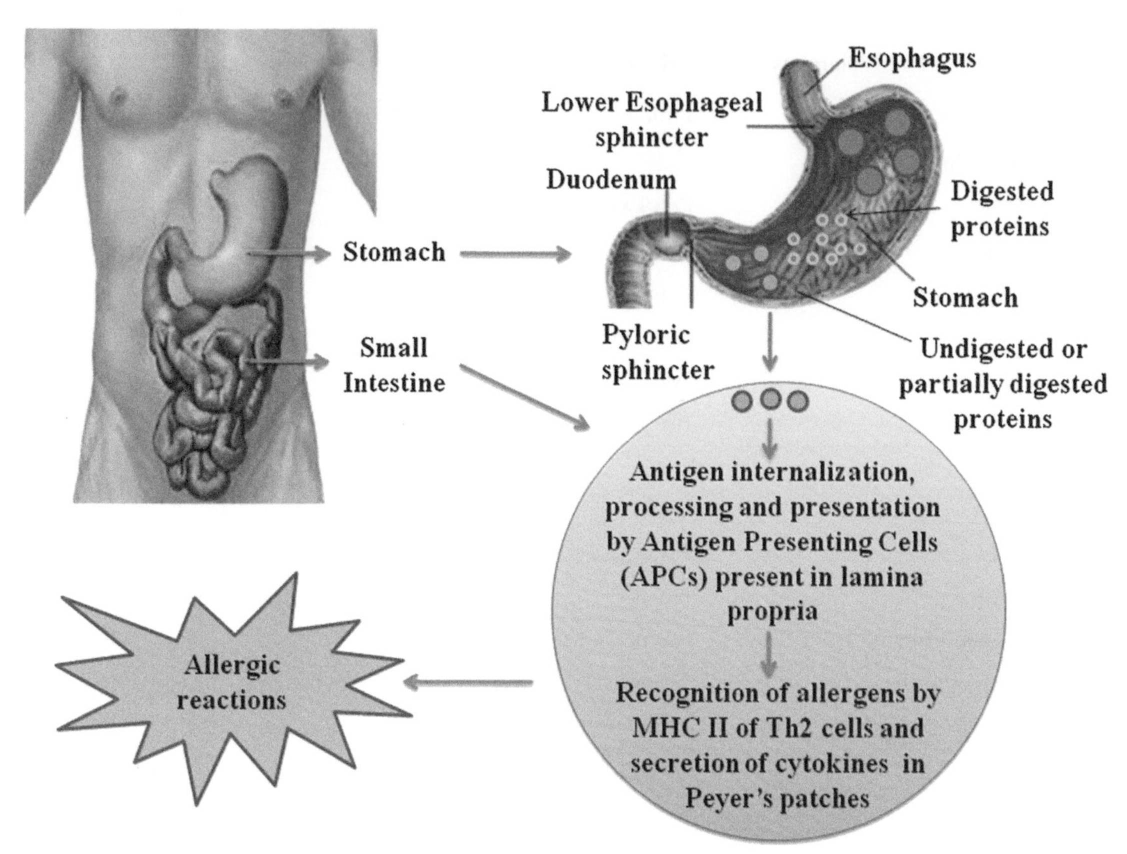

FIGURE 27.1 The basic outline of enzymatic action of pepsin on allergen.

SDS–PAGE. The food allergen first encounters proteolytic enzymes in the stomach, and then the fragmented peptides move into the intestine from the stomach (or get digested completely). The basic outline of enzymatic action of pepsin on allergen is given in Figure 27.1. Pepsin has broad substrate specificity, and preferentially cleaves proteins at amino acid leucine, phenylalanine, and tyrosine from the C- terminus (Schnell and Herman, 2009). A pictorial demonstration of pepsin action during an SGF assay is given in Figure 27.2.

Components of SGF

The SGF is a set of reagents held under specific conditions: 0.32% pepsin, pH 1.2, and 37°C. These conditions are designed to stimulate human gastric conditions present in the stomach (Astwood et al., 1996). Differences in pepsin concentration, pH, protein–substrate concentration, and detection procedures (resolution on SDS–PAGE gel, loading quantity, and protein staining methods) are considered as major regulatory factors in the SGF assay. WHO/FAO recommend the protocol described by Astwood et al. (1996). Another multi-laboratory protocol for the SGF assay was reported (Thomas et al., 2004). The laboratories involved were the ILSI; Health and Environmental Sciences Institute, Washington, D.C., USA; Sanquin Research, Amsterdam, The Netherlands; Monsanto Co., St. Louis, MO, USA; The Dow Chemical Co., Midland, MI, USA; Syngenta Central Toxicology Laboratory, Alderley Park, UK; Bayer CropScience, Sophia Antipolis, France; U.S. Food and Drug Administration National Center for Food Safety and Technology, Summit Argo, IL, USA; DuPont Co., Newark, DE, USA; University of Nebraska, Lincoln, NE, USA; Bayer Crop Science, Research Triangle Park, NC, USA; Syngenta Biotechnology Inc., Research Triangle Park, NC, USA; and the National Institute of Health Sciences, Tokyo, Japan. This study explored the fact that when the SGF assay was performed by different researchers with aliquots of the same reagents under similar test conditions, a panel of scientists could identify a similar time point for protein bands to become undetectable on SDS–PAGE gels.

A fully validated SGF assay should be reproducible, robust, and relevant, and must be largely insensitive to factors that are likely to vary among laboratories. Kinetic data analysis can be used as an interpretation tool for SGF assays. Herman et al. (2006) incorporated kinetic concepts into SGF studies rather than using a single time point when a protein band was no longer visible on an SDS–PAGE gel or Western blot. Protein bands on SDS–PAGE gels were quantified by densitometry over a digestion time point, and the pattern of protein degradation was modeled using a negative exponential equation (pseudo-first-order decline)

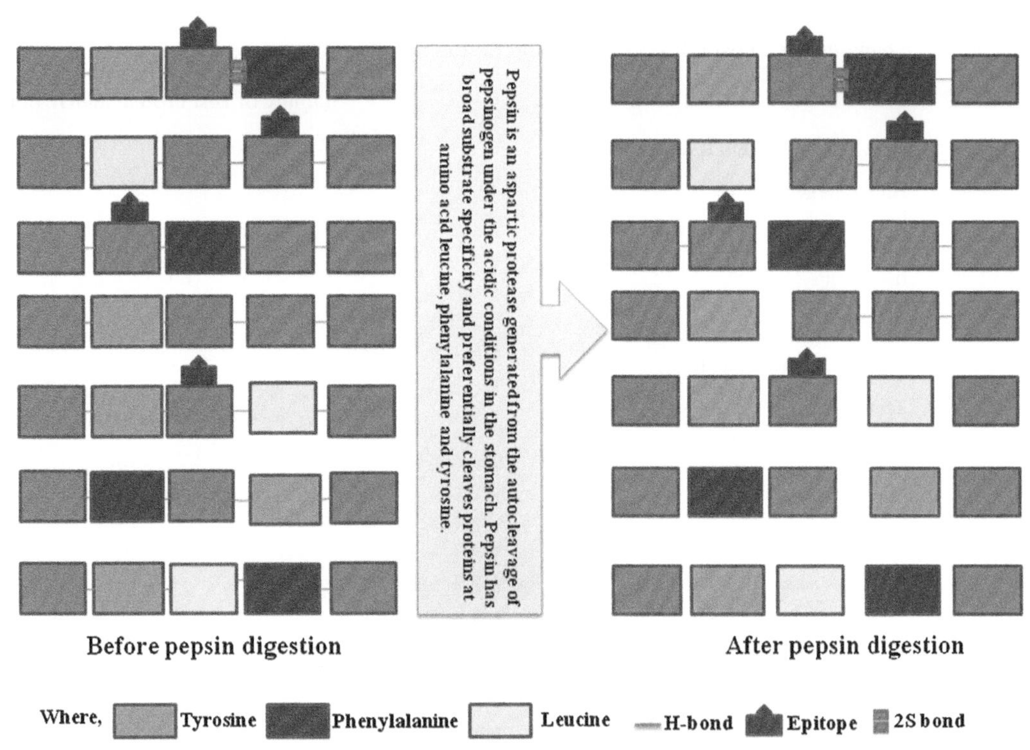
Pepsin is an aspartic protease generated from the autocleavage of pepsinogen under the acidic conditions in the stomach. Pepsin has broad substrate specificity and preferentially cleaves proteins at amino acid leucine, phenylalanine and tyrosine.
Before pepsin digestion
After pepsin digestion
Where,
Tyrosine
Phenylalanine
Leucine
H-bond
Epitope
2S bond

FIGURE 27.2 Mechanism of action of pepsin on polypeptides.

(Herman et al., 2006). Thus, a kinetic approach to analysis of SGF results uses multiple data points and relative protein decline to overcome some of the shortcomings associated with observing the first time point at which a protein is no longer visible. Researchers have focused on the most persistent protein fragment when assessing allergenic risk. Variation between pepsin lots, pepsin concentration, and substrate concentration did not substantially affect estimated degradation, although low purity pepsin lots had moderately lower catalytic power.

General Protocol of the SGF Assay

The first detailed study regarding the SGF assay was done in the late nineties by Astwood and his group. Since, then many modifications have been suggested to enhance its utility, but in each and every case, the basics of the assay are more or less similar. The pepsin activity recommended by several groups ranged between 5,000–20,000 units/mg of test protein. The pepsin concentration (3.2 mg/mL), the ratio of the pepsin to protein (3:1), and pH (1.2 and 2.0), along with different time point incubation (at 37°C), was similar everywhere (Flow Chart 27.1). The nature of the substrate sometimes influences the optimum activity of pepsin, with a relatively broad pH range between 1.2 and 3.5. Nevertheless, in a study performed on codfish allergens, all proteins were degraded to small fragments within 1 min and lost their IgE-binding capacity when the digestion was performed at physiologic conditions (pH 2.5); however, a small pH shift from 2.5 to 2.75 hampered the pepsin digestion of several allergens including the major fish allergen parvalbumin Gad c 1. It was also indicated that food allergens are not necessarily more stable in SGF as compared to non-allergenic proteins. In the majority of cases, it was found to be true. Many proteins with unproven allergenicity exhibit high stability (Table 27.2).

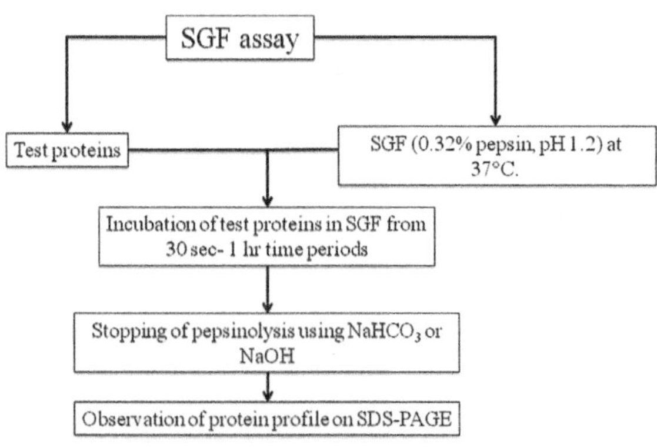
SGF assay
Test proteins
SGF (0.32% pepsin, pH 1.2) at 37°C.
Incubation of test proteins in SGF from 30 sec- 1 hr time periods
Stopping of pepsinolysis using NaHCO$_3$ or NaOH
Observation of protein profile on SDS-PAGE

FLOW CHART 27.1 SGF Assay Protocol

Factors Relevant to Gastrointestinal Digestion of Allergens

The SGF was designed to mimic the mammalian GI system. However, several other factors can play a role, like the

TABLE 27.2 Effect of Changing the Pepsin/Test Protein Ratio on the Stability* Observed in an SGF Assay

		Stability of Test Protein in SGF (min)		
S.N.	Pepsin/Test Protein Ratio (w/w)	10:1	1:1	1:10
	Food Allergens			
1.	β-lactoglobulin B	120	120	120
2.	ovalbumin	5	60	120
3.	papain	0	0	0
	Non-Allergenic Proteins			
4.	Zein	60	60	120
5.	Pea lectin	5	120	120
6.	Cytochrome c	0	0.5	0.5
7.	Sucrose synthetase	0	0	0

*Stability was measured as the last time period (in minutes) that the protein could be seen in the SDS–PAGE gel
(**Source:** Fu et al., J Agric Food Chem, 2002, 50, 7154–7160.)

buffering effect of food ingredients, mechanical breakdown of food tissue, range of stomach pH, additions of surfactants (phospholipids) and gastric lipase in physiological amounts, peristalsis, as well as possible emulsification of lipids and gastric emptying etc.

Supportive and Negative Evidence of SGF

Several independent researchers have reported supporting data regarding the utility of the SGF assay. Although the SGF assay is not exactly similar to *in vivo* digestion, its results are very close to the results of *in vivo* digestion of the mammalian system (Natalija et al., 2010). More interestingly, it has also been reported that the proteins stable to pepsin digestion have shown Western blotting on at least one-dimensional SDS–PAGE (Misra et al., 2011). Though there are several supporting papers on SGF assays, a number of subsequent reports have indicated an elusive link between protein stability in SGF and allergenicity.

Prediction of allergenicity using SGF assays has been supported in a number of reports, and is a general requirement as a part of the allergenicity assessment of transgenic proteins expressed in food; however, the predictive power of this assay has also raised several questions. There are many reasons behind the elusive predictive capability of the SGF assay, including lack of consideration of the prevalence of the allergen in food, effects of food processing, and food–matrix interactions. The food–matrix interactions may play an important role because components of food may sequester certain proteins away from the acid and pepsin in gastric fluid. The purified kiwi allergen, Act c 2, was digested quickly in SGF, but was protected from digestion by fruit pectin both *in vitro* and *in vivo*. Similarly, in transgenic corn expressing the *Escherichia coli* heat-labile enterotoxin facilitated the association of this protein with starch granules that protected it against digestion in SGF. Thus, evaluating the allergenicity of purified proteins in the SGF assay may not always lead to desired outcomes.

Although the value of comparing the stability of proteins in SGF for the purpose of evaluating the allergenic potential of novel food proteins is dubious, such comparisons are routinely used for this purpose. The resistance to *in vivo* digestion of a food protein increases its potential for causing an allergic reaction in susceptible individuals. Some peptide fragments of digested proteins can be recognizable by allergen-specific T cells. However, the amount of food protein and the conditions that can trigger the allergic reaction are largely unknown.

SIMULATED INTESTINAL FLUID (SIF) ASSAY

General Protocol of SIF Assay

There is a general belief that allergenic proteins remain unaffected in the proteolytic environment of the human GI system, and can be absorbed through the intestinal mucosa (Taylor et al., 1987). Therefore, a novel protein should be subjected to digestion in SGF and SIF. For the SIF assay, pancreatin is used in the preparation of SIF.

SIF is prepared according to a standard protocol as described in the United States Pharmacopoeia (1995). Briefly, 1 g/100 mL of pancreatin is dissolved in 0.05 M KH_2PO_4, pH 7.5. Aliquots of SIF are placed in

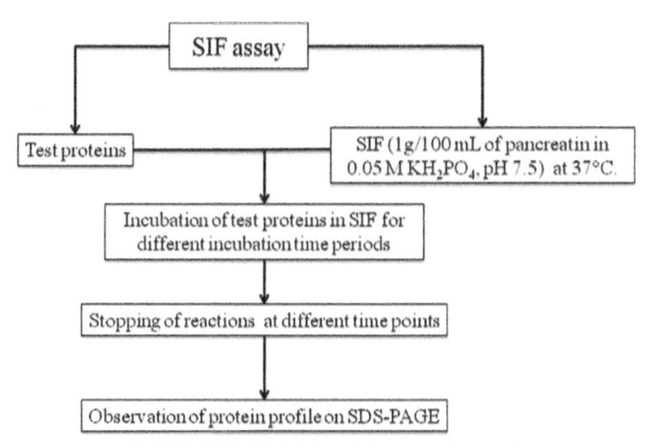

FLOW CHART 27.2 SIF Assay Protocol

microcentrifuge tubes and incubated at 37°C for 10 min in a water bath. The test protein at a concentration of 5 mg/mL (in 0.05 M KH_2PO4, pH 7.5) should be added to each of the microcentrifuge tubes to start the reaction. At different time intervals such as 0, 0.5, 5, 15, 60, and 120 min, Laemmli buffer should be added to each tube. Samples are boiled for 10 min in a water bath. The SDS–PAGE analysis, in combination with densitometry, is performed for comparing the degradation at different time points (Flow Chart 27.2).

Effect of Assay Conditions on Protein Stability in the SIF Assay

Assay conditions affect the relative digestibility of protein in SIF. The relative amount of enzyme and test protein used in an SIF assay affects the results for a particular protein. Changes in the ratio of enzymes to protein or change in pH affects the digestibility of the protein (Table 27.3). The ratio of enzyme-to-test protein (by weight) varies greatly in studies: from a 1:250 ratio of trypsin-to-soybean β-conglycinin (Kamata et al., 1982), a 5,000:1 ratio of pancreatin-to-neomycin phosphotransferase II (Fuchs et al., 1993), a 1:10 or 1:100 ratio of trypsin-to-phaseolin (Nielsen et al., 1988), and a 25:1 ratio of chymotrypsin/trypsin-to-insecticidal crystal protein Cry1 Ab expressed in transgenic tomatoes (Noteborn et al., 1995). SIF can have a significant impact on the evaluation of the allergenic potential of a food protein derived from GM crops if the ratio of enzyme-to-test protein varies.

Supportive and Negative Evidences of the SIF Assay

The stability of food proteins under *in vitro* digestive conditions is widely used as a criterion for the prediction of protein allergenicity by the agricultural biotechnology industry. Several guidelines have considered SGF or SIF digestibility as a predictive tool for estimation of allergenic potential of proteins. However, there is a need to establish standardized assay conditions that are globally accepted. A comparative study of the digestibility of food allergens and non-allergenic proteins in SGF and SIF was performed by Fu et al. (2002) to provide experimental evidence for the hypothesis. Nineteen allergenic and non-allergenic food proteins were subjected to digestion in SGF and in SIF for 2 h and 45 min, respectively. However, results did not indicate that food allergens are more stable to digestion *in vitro* than non-allergenic proteins. Such studies raise the doubts regarding digestibility assays. Despite some criticism, the digestibility assay is commonly used as a criterion for allergenicity assessment of GM foods.

Some major allergens like conalbumin are stable in SIF up to 120 min. Plant lectins (soybean, peanut) also have stability in SIF for 120 min. Bromelain and papain, well known allergens of the papain superfamily showed higher stability in SIF. Ovomucoid and lysozyme, other known allergens from egg, show higher stability in SIF, up to 60 min and 120 min, respectively. Cow's milk allergen and lactoperoxidase ae also stable in SIF for 120 min. There are several examples of allergens that have higher stability in SIF, as given in Table 27.1 (Fu et al., 2002). Therefore, *in vitro* digestion assay can be used to estimate whether a novel protein has allergenicity or not. Sometimes small fragments of degraded protein in the GI tract are sufficient to provoke allergenic responses.

Contradictory Results of the SIF Digestibility of Food Proteins

The stability of food proteins in SIF is varied, ranging from 0 to 120 min in the case of known food allergens, and also in the case of the proteins with unproven allergenicity. Few known protein allergens rapidly degrade in SIF. β-Lactoglobulin B is stable in SGF for 120 min, but gets digested within 5 min in SIF. Ara h 2, a major allergen of peanuts, is degraded within 0.5 min in SGF and in SIF.

Allergens like ovalbumin, conalbumin, and papain are stable in SIF, although instant degradation of these proteins occurs in SGF. A few non-allergens like RUBISCO and cytochrome *c* are also resistant to digestion in SIF for 120 min. Some allergens like α-casein, shrimp tropomyosin, Ara h 2, and patatin are labile to digestion in SIF, whereas plant lectin from red kidney bean, pea, and lentil (with unproven allergenicity) show stability for up to 120 min in SIF.

Digestibility of peanut and hazelnut allergens was investigated by the SGF and SIF methods (Vieths et al., 1999). They studied the digestion of both allergens in SGF for 2 h, and subsequent digestion in SIF for 45 min. Hazelnut allergens get rapidly degraded, while peanut protein allergens were found to be relatively stable. It is not necessary that food allergens with higher allergenicity are more stable in SIF as compared to allergens with lower allergenicity. For example, α-casein and Ara h 2 are rapidly degraded in SIF,

TABLE 27.3 Enzyme/Test Protein Ratios Used for SIF Assay

S.N.	Protein	Source	Enzyme Used	Enzyme/Test Protein Ratio	References
1.	Albumin	Beans	Trypsin	1:40	Marquez and Lajolo (1981)
2.	Phaseolin	Dry beans	Trypsin	1:10 or 1:100	Nielsen et al. (1988)
3.	β-Conglycinin	Soybean	Trypsin	1:250	Kamata et al. (1982)
4.	α-Lactoglobulin	Milk	Trypsin	1:100	Maynard et al. (1997)
5.	Neomycin phosphotransferase II (NPTII)	Recombinant *E. coli*	Pancretin	5000:1	Fuchs *et al.* (1993)
6.	CRYIA(b) NPTII	Recombinant *E. coli*	Chymotrypsin/ trypsin	25:1	Noteborn *et al.* (1995)

(**Source:** Fu et al., Ann NY Acad Sci, 964: 99–110, 2002).

while minor allergens like plant lectin are stable in SIF up to 120 min. These studies indicate that there is no direct correlation between stability in SIF and the allergenicity of novel proteins, indicating that digestibility of proteins in SIF does not provide a satisfactory enough result to be used as a promising method for allergenicity assessment of proteins.

In some cases, a direct correlation exists between the digestibility of a protein under *in vitro* conditions and its allergenicity. It is a widely accepted method for allergenicity assessment, but is not able to eliminate false results completely. It is sometimes difficult to say whether a given protein has allergenic potential or not. These methods fail to identify allergens that are labile to digestion in SGF and SIF. Moreover, allergens that provoke allergenicity through other routes like a respiratory route or by skin contact cannot be identified. Despite these limitations, stability to digestion is still a relevant method for the assessment of the allergenic potential of transgenic proteins inserted into foods.

THERMAL TREATMENT ASSAY

Heat-resistant proteins are not digested in the GI tract, so there are chances that these thermal-resistant proteins can elicit an allergenic response in an individual. Resistance to thermal treatment has been observed in the case of several food allergens, so it was logical to think that there is a direct correlation between heat stability and allergenicity. If a newly expressed protein shows heat-resistant properties under higher temperature, further experimental analysis is required to determine allergenicity of that protein. Therefore, a heat-resistant property can be utilized as an indicator as to whether a particular protein has immune-provoking potential or not. As in the case of GM food, a newly expressed protein or inserted protein can be checked to view its allergenicity by a thermal treatment assay in combination with other known methods. The three-dimensional structure of the majority of proteins correlates with their functional activities. As thermal treatment can bring about substantial changes in the structure of a protein, it can also modify the protein's allergenic nature. A better understanding of thermal processing-induced biochemical and immunological changes in food allergens of a crop/food may help in understanding whether thermal processing can be useful in reducing the allergenicity of the protein. During thermal processing, neoantigens sometimes form; this provides encouragement to develop new diagnostic tools.

Mechanism of Thermal Treatment Assay

Thermal exposure can occur in many ways during food preparation, including baking, cooking, roasting, grilling, drying, pasteurization, and sterilization. Many factors, such as digestibility (resistance to pepsin in the GI tract), solubility, and the ability to be absorbed intact across the intestinal tract, are responsible for the allergenicity of proteins. In general, heat treatment normally changes the structure of a protein, so it may increase or decrease the antigenicity of the protein. Prior heat treatment probably increases the digestibility of the protein, so the absorption of resulting polypeptides in the GI tract also increases; hence, the possibility of the protein eliciting an allergenic response decreases. However, in some cases, thermal processing may reduce the digestibility of a particular allergen, or some neoantigens may be formed that were not originally present in the protein (Paschke, 2009). This phenomenon may enhance the allergenic problem in sensitive patients, and these newly formed antigens or neoantigens may have the ability to sensitize a new consumer group. The major aim of thermal processing is to examine the effect of heat treatment on the structural and immunological properties of food

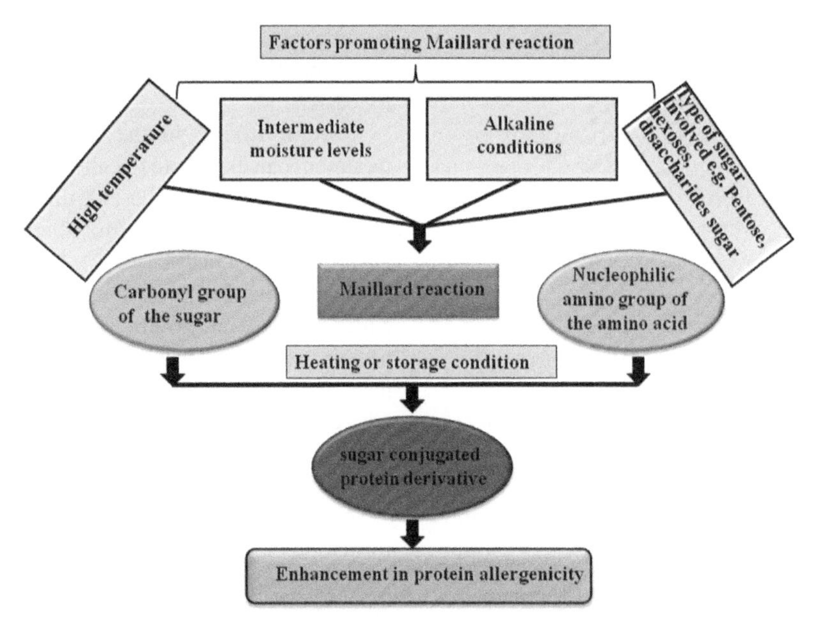

FIGURE 27.3 Maillard reaction and its relation with protein allergenicity.

allergens and identification of appropriate treatment methods for the reduction/elimination of allergenic residues in food.

One major factor for the formation of neoantigens may be the Maillard reaction (i.e. the interaction of the protein component with sugar residues upon heating that generates a sugar-conjugated protein derivative that causes enhancement in the allergenicity of the protein). A chemical reaction between the carbonyl group of the sugar reacts with the nucleophilic amino group of the amino acid, usually requiring heat. This is known as the Maillard reaction, and it leads to brown coloration during heating or storage. Several factors, like high temperature, intermediate moisture levels, and alkaline conditions, promote the Maillard reaction. The type of sugar involved also has an effect on the Maillard reaction. Pentose-reducing sugars are more reactive than hexoses, which are more reactive than disaccharides. Beyer et al. (2001) reported that in cooked peanuts, IgE-binding increased due to the Maillard reaction. A brief outline of the Maillard reaction is given in Figure 27.3.

The antibodies recognize and interact with IgE-binding epitopes presented on allergenic proteins. These IgE-binding epitopes are of two types: linear and conformational. In linear epitopes, amino acids are arranged in linear order in the polypeptide chain; in conformation epitopes, amino acids that are far apart in the primary sequence are brought together during polypeptide chain folding. Linear epitopes can be more allergenic as compared to conformational epitopes, as these are mostly resistant to heat treatment and have the ability to sensitize and provoke allergenic responses. Thermal processing mostly affects conformational epitopes, as heat treatment may break the bonds. Refolding allows formation of native conformational epitopes in most cases, but few new allergens may be formed that need further efforts to minimize the risk associated with neoantigens. Thus, thermal treatment may decrease or exacerbate or have no effect on antigenic behavior; this depends on the properties of a particular protein.

Standard Protocol for Thermal Treatment Assay

Based on the report of a joint FAO/WHO expert consultation on the allergenicity of foods derived from biotechnology (2001) and a guideline to the conduct of food safety assessment of foods derived from GM crops by the Codex Alimentarius Commission (2003), the Indian Ministry of Science and Technology (2008) prepared a standard protocol to assess the thermal stability of proteins. During the assay procedure, protein samples (conc. 1 mg/mL) are dissolved in suitable buffer and incubated at 25, 37, 55, 75, and 95°C for 30 minutes. The samples are boiled with Laemmli buffer and protein profiles obtained at different time points are compared to determine the biological activity of the protein. The stability of the protein at a particular temperature is determined from the remaining biological activity after a 30-minute incubation at that temperature (Flow Chart 27.3). Thermally treated proteins are divided into three categories: stable, partially stable, and labile. According to DBT guidelines (2008), proteins with more than 50% biological activity remaining are considered stable at that temperature. Biological activities between 50 and 10% are partially stable, and less than 10% biological activity indicates that the protein is labile. So, thermal treatment can provide a clue that a newly expressed

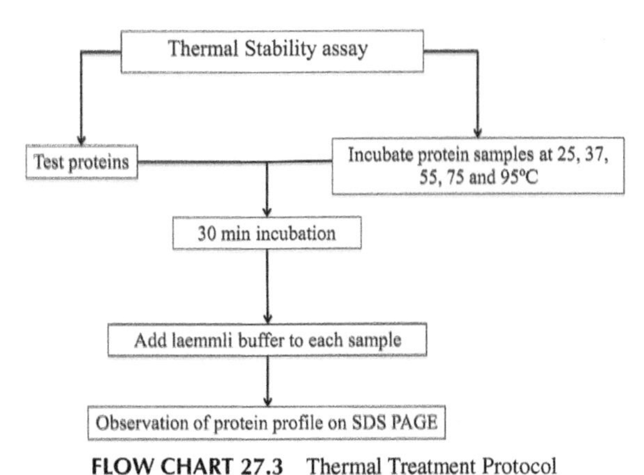

FLOW CHART 27.3 Thermal Treatment Protocol

protein is stable, partially stable, or labile, and hence have corresponding allergenic potential.

Functional Stability of Proteins and Importance of Thermal Stability Assay

Transgenic crops undergo a rigorous safety assessment procedure before commercialization due to a worldwide anti-GM movement (Kumar et al., 2011). Currently available assays cannot provide exact information regarding allergenicity of transgenic foods. For the above-mentioned reasons, the weight-of-evidence approach was introduced to assess the allergenic potential of GM crops in which thermal stability has been included as a part of the risk assessment procedure (Codex Alimentarius, 2009). It is known that cooked foods do retain their ability to cause allergic reactions. Based on these observations, it was suggested by some regulatory agencies that the heat stability test should be performed to ensure additional safety of transgenic food proteins (Codex Alimentarius Commission, 2003; Indian Ministry of Science and Technology, 2008). The functional activity of the protein can be destroyed at a particular temperature. So, thermal treatment of the protein is performed, and biological activity is measured before and after thermal treatment. It may provide information regarding the temperature and time at which activity of the protein is destroyed by heat. Measuring biological activity of the protein after exposure to heat can be one of the basic methods for risk assessment of novel proteins in transgenic crops (Indian Ministry of Science and Technology, 2008). However, it needs to be mentioned that protein denaturation does not always lead to loss of allergenicity. Thermal treatment can result in loss of epitopes originally present in their native form. It is also possible that the protein becomes more allergenic as compared to the native form because hidden epitopes may get exposed following heat treatment, resulting in enhanced allergenicity (Figure 27.4). Sometimes heat-mediated denaturation has no effect because out of the two types of epitopes (linear and conformational), only linear epitopes are more stable and have the ability to provoke allergenicity. Conformational epitopes lose their allergenicity easily.

Several studies in the literature suggest that thermal treatment can eliminate the allergenicity of allergens such as patatin protein in potatoes, chitinases in fruits, the hazelnut Cor a 1.04 allergen, Kiwi fruit allergens, and several legume allergens (Verma et al., 2012). In the case of soybean, heat treatment changes the protein's profile and immunological properties, but no effect on overall allergenicity of soy protein was found (Besler et al., 2001). Some allergens are heat stable (e.g. allergens of milk, egg, fish, and peanuts), whereas other food allergens are partially stable (e.g. soybean, cereals, celery, tree nuts, and their products) or labile (fruits of the Rosaceae family and carrots).

A major problem during the heat treatment process is the formation of new allergens, or neoallergens. These neoallergens, present in cooked food, may sensitize an individual. Neoallergens have been identified from pecans, wheat flour, roasted peanuts, lentils, almonds, cashew nuts, walnuts, soybeans, shrimp, scallops, tuna, eggs, apples, plums, milk, and potatoes. Changes in protein conformation may either have no effect on the protein allergenicity, or can modulate allergenicity (Paschke, 2009). Storage conditions are also an important factor affecting the properties of proteins, and may change their profile.

A food protein that is not normally allergenic may change its immunological properties during food storage. Codina et al. (1998) demonstrated that heat treatment can increase the allergenicity of soybean hull as two neoallergens (15.3 and 10 kDa) are formed. During transport and storage conditions, heat is generated, which may enhance the allergenicity of soybean hull. So, there are chances of allergenic conditions. For example, if novel transgenic crops are assessed for their allergenicity for all parameters except thermal treatment, it cannot be said with certainty what will be the behavior of that crop after prolonged storage. The chances of being allergenic will probably increase, as observed for many fruits and vegetables where the proteome changed and higher allergenicity was found (as in the case of apples). In other food crops, there is no change in immunoreactivity during storage (e.g. mangoes). However, comparison of the protein profile of food crops before and after thermal treatment may provide a clue regarding the allergenicity of food crops.

Mechanical processes such as stirring can also influence the allergenic properties of food proteins. Although such treatments can cause surface denaturation of food proteins, a significant effect on allergenicity has not been demonstrated. Heat-induced denaturation of proteins can reduce the allergenic potential of the food product. Protein denaturation minimizes allergenicity by changing protein conformation, which is the result of destruction of IgE-reactive conformational epitopes. Fiocchi (1995)

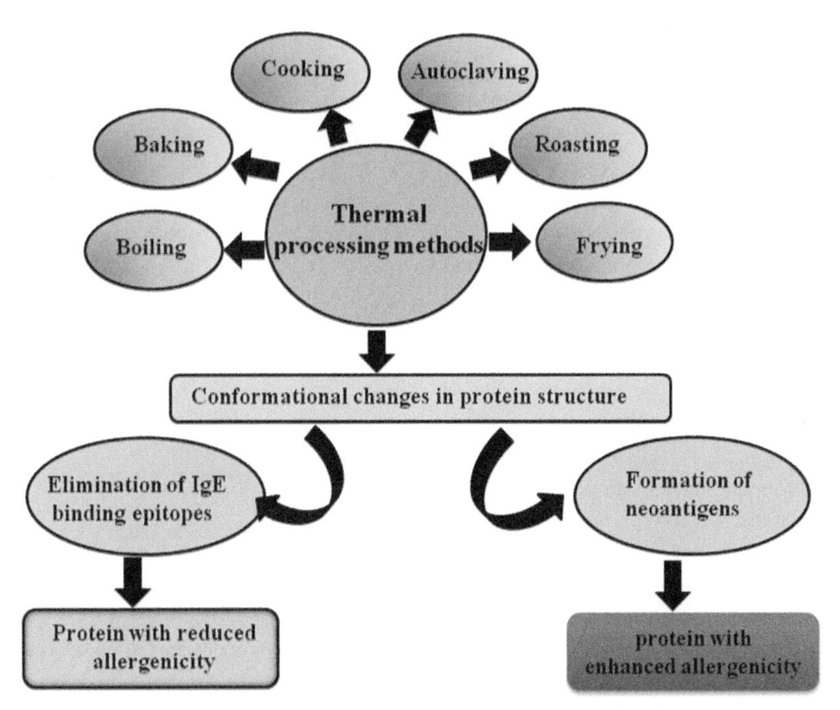

FIGURE 27.4 Thermal processing methods and their effect on protein allergenicity.

showed that thermal treatment may reduce the allergenicity of beef and purified bovine allergens. Some food allergens may be thermostable. For example, in the case of the major peanut allergen Ara h 1, the immunoreactive conformational epitopes are resistant to heat-induced denaturation. In potato allergen patatin (Sol t 1), heat treatment caused denaturation of the protein; however, the protein re-natured upon cooling. During production of canned lychee fruit, thermal treatment for longer periods does not decrease allergenicity. However, low-molecular-weight immunoreactive proteins were eliminated by heat treatment.

Contradictory Results in Thermal Treatment Procedure

Retention or enhancement in the allergenicity of some foods after cooking is an important feature of some food protein allergens. Functional assays are used to measure the heat stability of newly expressed proteins (Indian Ministry of Science and Technology, 2008), but functional stability is not related to allergenicity. A study done by Herouet et al. (2005) demonstrated that phosphinothricin acetyltransferase (PAT) protein that is safely consumed and gets inactivated at 40–45°C (15 min) or 60°C (10 min) is clearly detectable even after heat treatment at 100°C. Although the PAT protein loses its enzymatic activity, immunoreactivity is still detectable. In both heat-treated and untreated proteins, the same epitopes were recognized by IgG with the use of anti-PAT antibodies.

Conclusively, thermal treatment of novel food proteins probably has no definite predictive value in the allergenicity assessment process. The thermal treatment assay can be used more appropriately in such conditions where toxicological responses are associated with biological activity. It does not provide reliable information regarding the allergenic potential of a novel food protein. Although the thermal stability assay provides no extra information for the safety assessment of a novel protein, it can be used to demonstrate the allergenicity of processed or cooked food because, functional activity of some foods are associated with heat stability. Various reports have indicated a reduction in immunoreactive proteins following heat treatment. Sometimes, the thermal treatment approach generates neo-allergens that were originally absent in the native variety. There is still a need to explore the utility of this method in the allergenicity assessment of a novel protein. As the impact of heat treatment differs from one food constituent to another, it is difficult to demonstrate the correlation of thermal treatments on protein allergenicity. Thermal treatment can provide useful information for risk assessment of biotech crops. As this method is included in standard guidelines as a food safety approach, it must be performed for all transgenic crops with inserted gene products.

ETHICAL ISSUES

GM food safety has been a controversial issue for a long time. Safety and potential risks, as well as ethical concerns associated with GM food, are still debated. There are laws in different countries focusing attention on the use and labeling requirements of GM foods. There are also social and ethical issues regarding GM food production worldwide. A gene inserted from animal to plant may create ethical or religious problems in certain cases. For example, eating traces of genetic material from pork could be a problem for certain religious or cultural groups, although genetic material is either DNA or RNA from any source, and has only A, T(U), G, and C nucleotides. There is also fear that the distribution of GM seeds may be taken over by large multinational companies, affecting the socially equitable distribution of benefits. There is also an important issue associated with the protection of the natural environment and biological diversity, as GM foods can affect the wild varieties. However, most of these fears are unfounded, lack evidence, and can be considered more as "fear of the unknown."

TRANSLATIONAL SIGNIFICANCE

GM foods may be useful for solving the world hunger problem as they offer improved products with desirable qualities. Adoption of GM crops is increasing worldwide, but risks associated with environmental and human health still exist. It raises the issue that pre- and post-market surveillance of GM-derived food should be performed in order to avoid unintended effects. Among the various methods, SGF, SIF, and thermal stability assays are the few that are performed first for any newly expressed protein. SGF, SIF, and thermal stability assays are included in the standard guidelines for addressing the safety concerns of GM foods. Even though SGF and SIF assays have some contradictory evidence, overall these assays yield primary information regarding the probability of the allergenic potential of food proteins. The thermal assay is not a commonly used parameter for safety assessment of GM-derived foods, but can add some value to the safety assessment.

WORLD WIDE WEB RESOURCES

1. http://www.foodhaccp.com/online.html:

This is a comprehensive food safety information website, and contains periodic newsletters, journals, and updates regarding food safety concerns.

2. http://www.who.int/foodsafety/publications/biotech/20questions/en/:

This site presents useful information regarding the safety of GM foods and possible pros and cons.

3. http://www.gmo-compass.org:

This database contains information about every genetically modified plant that has been approved or is awaiting approval in the EU. Information on the food and feed produced from the respective GM plant is also available.

4. http://www.centerforfoodsafety.org/campaign/genetically-engineered-food/crops/:

The Centre for Food Safety (CSF) seeks to halt the approval, commercialization, or release of any new genetically engineered crops until they have been thoroughly tested and found to be safe for human health and the environment.

5. http://ec.europa.eu/food/food/intro/white_paper_en.htm:

This site presents white papers that describe efforts to modernize legislation into a coherent and transparent set of rules, reinforcing controls from the farm to the table, and increasing the capability of the scientific advice system so as to guarantee a high level of human health and consumer protection related to food safety.

ACKNOWLEDGMENTS

Alok Kumar Verma and Sandeep Kumar are thankful to CSIR, New Delhi for the award of their Senior Research Fellowships. This is CSIR-IITR manuscript # 3024. This work was financially supported by InDepth Project (BSC0111) of CSIR-IITR.

The authors have declared no conflict of interest.

REFERENCES

Astwood, J. D., Leach, J. N., & Fuchs, R. L. (1996). Stability of food allergens to digestion *in vitro*. *Nature Biotechnology*, *14*(10), 1269–1273.

Besler, M., Steinhart, H., & Paschke, A. (2001). Stability of food allergens and allergenicity of processed foods. *J Chromatogr B Biomed Sci Appl*, *756*(1-2), 207–228.

Beyer, K., Morrow, E., Li, X. M., Bardina, L., Bannon, G. A., Burks, A. W., & Sampson, H. A. (2001). Effects of cooking methods on peanut allergenicity. *The Journal of Allergy and Clinical Immunology*, *107*(6), 1077–1081.

Codex Alimentarius Commission, 2003. Alinorm 03/34: Joint FAO/WHO Food Standard Programme, Codex Alimentarius Commission, Twenty-Fifth Session, Rome, Italy, 30 June–5 July, 2003. Appendix III, Guideline for the conduct of food safety assessment of foods derived from recombinant-DNA plants, and Appendix IV, Annex on the assessment of possible allergenicity. pp. 47–60.

Alimentarius, Codex (2009). *Foods Derived from Modern Biotechnology*. Rome: FAO/WHO. 1–85.

Codina, R., Oehling, A. G., Jr., & Lockey, R. F. (1998). Neoallergens in heated soybean hull. *International Archives of Allergy and Immunology*, *117*(2), 120–125.

Fiocchi, A., Restani, P., Riva, E., Restelli, A. R., Biasucci, G., Galli, C. L., & Giovanni, M. (1995). Meat allergy II – effects of food processing and enzymatic digestion on the allergenicity of bovine and ovine meats. *Journal of the American College of Nutrition*, *14*, 245–250.

Fu, T. J., Upasana, R. A., & Catherine, H. J. (2002). Digestibility of Food Allergens and Non-allergenic Proteins in Simulated Gastric Fluid and Simulated Intestinal Fluids: A Comparative Study. *Journal of Agricultural and Food Chemistry*, *50*(24), 7154–7160.

Fuchs, R. L., Ream, J. E., Hammond, B. G., Naylor, M. W., Leimgruber, R. M., & Berberich, S. A. (1993). Safety assessment of the neomycin phosphotransferase II (NPTII) protein. *Bio/Technology*, *11*, 1543–1547.

Herman, R., Gao, Y., & Storer, N. (2006). Acid-induced unfolding kinetics in simulated gastric digestion of proteins. *Reg Tox Pharm*, *46*(1), 93–99.

Herouet, C., Esdaile, D. J., Mallyon, B. A., Debruyne, E., Schulz, A., Currier, T., Hendrickx, K., van der Klis, R. J., & Rouan, D. (2005).

Safety evaluation of the phosphinothricin acetyltransferase proteins encoded by the pat and bar sequences that confer tolerance to glufosinate-amonium herbicide in transgenic plants. *Reg Tox Pharm, 41*(2), 134–149.

Indian Ministry of Science and Technology. Department of Biotechnology, Government of India. (2008). Protocols for food and feed safety assessment of GE crops. *Protein Thermal Stability*, 18–21.

Kamata, Y., Otsuka, S., Sato, M., & Shibasaki, K. (1982). Limited proteolysis of soybean beta-conglycinin. *Agric Biol Chem, 46*, 2829–2834.

Kumar, S., Misra, A., Verma, A. K., Roy, R., Tripathi, A., Das, M., & Dwivedi, P. D. (2011). Bt Brinjal in India: A long way to go. *GM Crops, 2*(2), 92–98.

Kumar, S., Verma, A. K., Das, M., & Dwivedi, P. D. (2012). Molecular mechanisms of IgE mediated food allergy. *International Immunopharmacology, 13*(4), 432–439.

Marquez, U. M. L., & Lajolo, F. M. (1981). Composition and digestability of albumins, globulins, and glutinins from *Phaseolus Vulgaris. Journal of Agriculture and Food Chemistry, 29*, 1068–1074.

Maynard, F., Jost, R., & Wal, J. M. (1997). Human IgE binding capacity of tryptic peptides from bovine α-lactalbumin. *Int. Arch. Allergy Immunol. 113*, 478–488.

Metcalfe, D. D., Astwood, J. D., Townsend, R., Sampson, H. A., Taylor, S. L., & Fuchs, R. L. (1996). Assessment of the allergenic potential of foods derived from genetically engineered crop plants. *Critical Reviews in Food Science and Nutrition, 36*, S165–S186.

Misra, A., Kumar, R., Mishra, V., Chaudhari, B. P., Raisuddin, S., Das, M., & Dwivedi, P. D. (2011). Potential allergens of green gram (Vigna radiata L. Millsp) identified as members of cupin superfamily and seed albumin. *Clinical and Experimental Allergy, 41*(8), 1157–1168.

Nielsen, S. S., Deshpande, S. S., Hermodson, M. A., & Scott, M. P. (1988). Comparative digestibility of legume storage proteins. *Journal of Agricultural and Food Chemistry, 36*(5), 896–902.

Noteborn, H. P. J.M., Bienenmann-Ploum, M. E., Van Den Berg, J. H. J., et al. (1995). Safety assessment of the Bacillus thuringiensis insecticidal crystal protein CRYIA(b) expressed in transgenic tomatoes. In Genetically Modified Foods, Safety Aspects, K.-H., Engel, G. R. Takeoka & R. Teranishi (Eds.), *ACS Symposium Series 605* (pp. 134–147). Washington, DC: American Chemical Society.

OECD Recombinant DNA Safety Considerations: Safety Considerations for Industrial, Agricultural and Environmental Applications of Organisms Derived by Recombinant DNA Techniques (aka 'Blue Book'). (1986). Paris: OECD.

Paschke, A. (2009). Aspects of food processing and its effect on allergen structure. *Mol. Nutr. Food. Res., 53*(8), 959–962.

Polovic, N., Obradovic, A., Spasic, M., Plecas-Solarovic, B., Gavrovic-Jankulovic, M., & Velickovic, T. C. (2010). In vivo digestion of a thaumatin-like kiwifruit protein in rats. *Food Digestion* (1–2), 5–13.

Safety assessment of foods derived from genetically modified microorganisms, a joint FAO/WHO expert consultation on foods derived from biotechnology. (24 to 28 September 2001). Geneva, Switzerland.

Sampson, H. A. (2003). Anaphylaxis and emergency treatment. *Pediatrics, 111*(6 Pt 3), 1601–1608.

Schnell, S., & Herman, R. A. (2009). Should digestion assays be used to estimate persistence of potential allergens in tests for safety of novel food proteins? *Clin Mol Allergy, 7*, 1.

Taylor, S. L., Lemanske, R. F., Jr, Bush, R. K., & Busse, W. W. (1987). Food allergens: structure and immunologic properties. *Annals of Allergy, 59*(5 Pt 2), 93–99.

Thomas, K., Aalbers, M., Bannon, G. A., Bartels, M., Dearman, R. J., Esdaile, D. J., Fu, T. J., Glatt, C. M., Hadfield, N., Hatzos, C., Hefle, S. L., Heylings, J. R., Goodman, R. E., Henry, B., Herouet, C., Holsapple, M., Ladics, G. S., Landry, T. D., MacIntosh, S. C., Rice, E. A., Privalle, L. S., Steiner, H. Y., Teshima, R., Van, R. R., Woolhiser, M., & Zawodny, J. (2004). A multi-laboratory evaluation of a common in vitro pepsin digestion assay protocol used in assessing the safety of novel proteins. *Regulatory Toxicology and Pharmacology, 39*, 87–98.

United States Pharmacopoeia. (1995). Simulated gastric fluid, TS. In *The National Formulary 18; Board of Trustees* (pp. 2053). Rockville, MD: United States Pharmacopoeia Convention, Inc.

Verma, A. K., Kumar, S., Das, M., & Dwivedi, P. D. (2012). Impact of thermal processing on legume allergens. *Plant Foods for Human Nutrition, 67*(4), 430–441.

Vieths, S., Reindl, J., Muller, U., et al. (1999). Digestibility of peanut and hazelnut allergens investigated by a simple in vitro procedure. *Eur Food Res Technol, 209*, 379–388.

FURTHER READING

Biotechnology and Safety Assessment. (2002). *Thomas & Fuchs* (3rd Ed.). Academic Press. ISBN: 9780126887211.

The Food Safety Information Handbook. (2001). *Cynthia A. Roberts*. Oryx Press. ISBN 1-57356-305-6.

Genetically Modified Foods. (2002). *Debating Biotechnology*, Michael Ruse, David Castle Prometheus Books. ISBN 1-57392-996-4.

Genetically Modified Organisms in Agriculture. (2001). *Economics and Politics 1st Edition, G Nelson*. Academic Press. ISBN: 9780125154222.

Hull & Tzotzos & Head Genetically Modified Plants. (2009). *Assessing Safety and Managing Risk* (1st Ed.). Academic Press. ISBN: 9780123741066.

Paschke, A. (2009). Aspects of food processing and its effect on allergen Structure. *Molecular Nutrition & Food Research, 53*(8), 959–962.

GLOSSARY

Allergens Protein responsible for induction of an allergy; has an epitope as an immunogenic site.

Allergy Immune provocation in susceptible individuals by certain harmless environmental substances.

GM Crops Crops derived from the insertion of desired genes for high yield, salt tolerance, disease resistance, and other improvements.

GM Food Food derived from GM crops.

Safety Assessment Testing for adverse effects induced by GM foods on biotic and abiotic systems.

ABBREVIATIONS

APC Antigen Presenting Cells
DCs Dendritic Cells
FAO Food and Agriculture Organization
GI Tract Gastrointestinal Tract
GM Crops Genetically Modified Crops
IFBC International Food Biotechnology Council
ILSI International Life Sciences Institute
OECD Organization for Economic Co-operation and Development
SGF Simulated Gastric Fluid
SIF Simulated Intestinal Fluid

USFDA United States Food and Drug Administration
WHO World Health Organization

LONG ANSWER QUESTIONS

1. Explain the basic mechanism and methodology of the SGF assay.
2. Describe the methodology of the SIF assay and its mechanism.
3. What are the principle, basic mechanism, and methodology of the thermal stability assay?
4. How can thermal treatment affect the allergenicity potential of foods? Describe the mechanistic aspects of the Maillard reaction.
5. What are the major lacunae in SGF, SIF, and thermal stability assays?

SHORT ANSWER QUESTIONS

1. What are GM foods?
2. What are the advantages of GM food production?
3. What are the safety-assessment needs of GM foods, and how are they different from traditional food crops?
4. Give three *in vitro* assays for allergenicity assessment of GM foods.
5. What are the components of Simulated Gastric Fluids?
6. List the ingredients in Simulated Intestinal Fluids.

ANSWERS TO SHORT ANSWER QUESTIONS

1. Genetically modified (GM) foods are foods derived from organisms whose genetic material (DNA) has been modified in a way that does not occur naturally (e.g. through the introduction of a gene from a different organism).
2. Food derived from genetically modified crops (GM food) is essential for fulfilling the food demands of our fast-growing population. These GM crops have improved agronomic characteristics (e.g. insect resistance, herbicide resistance, disease resistance, and drought tolerance property). Furthermore, more nutritious crops with higher yields (as compared to their natural counterparts) can be produced by this technology.
3. Safety concerns, including toxicity, allergenicity, and the environmental impact of GM crops and foods, are frequently raised. The safety assessment of GM foods generally investigates: (a) toxicity assessment, (b) the potential to elicit an allergic reaction (allergenicity), (c) the stability of the inserted gene, (d) the nutritional effects associated with genetic modification, and (e) any unintended effects that could result from the gene insertion. Traditional food crops are bred using plant-breeding methods, and they have genes only from the same species.
4. Three commonly used *in vitro* methods for allergenicity assessment include simulated gastric fluid assay, simulated intestinal fluid assay, and thermal treatment assay.
5. Simulated gastric fluid is a set of reagents held under specific conditions (0.32% pepsin, pH 1.2, and temperature 37°C) that is designed to simulate human gastric conditions present in the stomach.
6. For the SIF assay, simulated intestinal fluid is prepared according to standard protocol as described in the United States Pharmacopoeia (1995). Briefly, 1 g/100 mL of pancreatin is dissolved in 0.05 M KH_2PO4 at pH 7.5 and a temperature of 37°C. Test protein at a concentration of 5 mg/mL is used for the assay.

Chapter 28

Nanotechnology and Detection of Microbial Pathogens

Rishi Shanker*, Gulshan Singh*, Anurag Jyoti*, Premendra Dhar Dwivedi† and Surinder Pal Singh**

*Nanotherapeutics & Nanomaterial Toxicology Group, CSIR-Indian Institute of Toxicology Research, Lucknow, U.P., India, †Food Toxicology, CSIR-Indian Institute of Toxicology Research, Lucknow, U.P., India, **CSIR-National Physical Laboratory, New Delhi, India

Chapter Outline

SUMMARY

The optical characteristics and functionalization of gold nanoparticles with DNA can be exploited to generate conjugated DNA probes for the detection of pathogens. Gold nanoparticle probes with low-cost instrumentation (or a "spot and read" system) can be a viable alternative for on-site detection and monitoring of pathogens.

WHAT YOU CAN EXPECT TO KNOW

The consumption of unsafe water and food in developing countries is one of the major causes of infectious disease outbreaks. The existing methods for the detection of pathogens prevalent in water and food samples are expensive, time consuming, and highly diverse. A majority of these pathogens often escape detection by conventional

Animal Biotechnology. http://dx.doi.org/10.1016/B978-0-12-416002-6.00028-6

methods. The detection of target pathogens requires the development of innovative, simple, rapid, sensitive, and highly specific methods to overcome existing drawbacks for the management of infectious disease outbreaks. The recent advancements in nanotechnology have led to the development of nanoparticle-based facile assays for specific detection of the bioanalytes of clinical interest. Gold nanoparticles (GNPs) with unique optical properties and high surface area are being extensively used for facile detection of bioanalytes of interest in the samples. The outstanding physicochemical properties of GNPs have proved advantageous over conventional detection methods for diagnostic purposes. The colloidal solution of GNPs exhibits intense red and blue/purple colors depending on the size, shape, and degree of aggregation of nanoparticles. The present chapter encompasses the application of nanotechnology in the field of pathogen detection, and provides insight on how nanotechnology can be exploited to overcome the problems related to existing methods.

HISTORY AND METHODS

INTRODUCTION

Rapid population growth and industrialization have led to the deterioration of the microbiological quality of water, adversely affecting human health and sustainable development. Water plays a significant role in the transmission of human diseases. Typhoid fever, infectious hepatitis, cholera, traveler's diarrhea, amebic and bacillary dysenteries, and other gastrointestinal diseases are waterborne. The occasional outbreaks of waterborne diseases point towards the need for strict monitoring and management of the "water quality" from public and private water supplies. "Water quality" is a technical term that is based upon the characteristics of water in relation to guideline values of what is best for human consumption and for all usual domestic needs, including personal hygiene. Microbial, biological, chemical, and physical aspects are important components of water quality. When referring to microbial aspects, microorganisms that are known to be pathogenic should be absent in drinking (potable) water. Potable water is water that has been either treated, cleaned, or filtered, and meets established drinking water standards as set by regulatory authorities like the World Health Organization (WHO), the Bureau of Indian Standards (BIS), the American Public Health Association (APHA), and the United States Environmental Protection Agency (USEPA). This water is expected to be realistically free from harmful bacteria and contaminants, and considered "safe" for drinking, or cooking and baking purposes. Municipal water that has been UV-irradiated, filtered, distilled, or purified, falls into the category of potable water. Hence, water treatment regimens that employ disinfection methods and the execution of bacteriological surveillance programs have resulted in decreased occurrences of water-related illness.

Management of the frequency of waterborne disease outbreaks has become a challenging task. Globally, the source of almost two-thirds of the drinking water consumed is surface water, which may be easily contaminated by sewage discharges, animal defecation, and municipal and industrial wastes. Fecal wastes from domestic animals, wildlife, and humans (to varying extents) are incorporated into the soil. These fecal wastes can also enter the water stream directly, or through poorly processed sewage effluents, by percolation of water pipelines, malfunctioning septic tanks, and seepage from sanitary landfills. Different pathogenic viruses, bacteria, and parasites may be found in the feces of domestic animals, wild animals, and humans, along with the non-pathogenic bacteria and parasites that exist in large numbers in the feces of animals as well as in soil and water. Hence, it is important to identify the etiological agent for appropriate treatment, interventions, and control.

However, the identification and monitoring of specific pathogens in low concentrations or doses in the presence of a large number of background microflora is a daunting task. The presence of "indicator organisms," generally non-pathogenic microorganisms, points towards the presence of enteric pathogens in a sample. Indicator organisms play an important role in predicting the probability of the occurrence of pathogens that are quite low in number. There are a few criteria for a microorganism to be declared an "indicator." It must be present when pathogens are present in water, absent in uncontaminated water, present in higher numbers than pathogens in contaminated water, must survive better in water than pathogens, and must be easy to analyze. Therefore, the need to identify, classify, and delineate the permissible limits of "indicators" of microbial water quality in different sectors of water have been described in this chapter.

INDICATORS OF MICROBIAL WATER QUALITY

In water quality assessment, the fecal indicator bacteria (FIB) are used to measure the sanitary quality of water for recreational, industrial, agricultural, and water supply purposes. FIB's are natural inhabitants of the gastrointestinal tract of humans and other warm-blooded animals. Generally harmless, these are released into the environment with feces, and on exposure to a variety of the environmental factors (Ashbolt et al., 2001). In general, it is believed that the fecal indicator adapted to live in the gastrointestinal tract cannot grow in natural environments. However, survival of fecal indicator bacteria in water is influenced by environmental factors like sunlight, temperature, nutrient competition with bacteria naturally inhabiting the water, predation by protozoa and other small organisms, and toxic industrial wastes. Studies have shown that FIB survive from a few hours up to several days in water, but may survive for days or months in sediments where they may be protected from sunlight and predators. The survival time of fecal indicator bacteria in water is a function of many environmental influences, and there is no common factor that applies collectively to all water bodies, or even at different seasons in a year for a single body of water. It is assumed that the mortality of pathogens and FIB are equal. Therefore, the presence of relatively high numbers of FIB in the environment indicates the likelihood of the presence of other pathogens as well.

The environmental indicators of water quality are described in Figure 28.1. Coliforms and related pathogens are broadly categorized into total coliforms (TC), fecal coliforms (FC), or thermo-tolerant coliforms and other

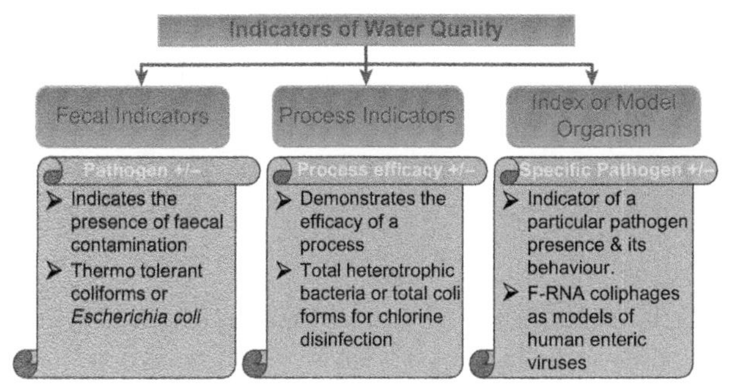

FIGURE 28.1 Types of water indicators.

indicator organisms (Figure 28.2). Studies show that 60 to 90% of total coliforms are fecal coliforms, and 90% of fecal coliforms, are *Escherichia coli.* The permissible limits of "indicators" of water quality are summarized in Table 28.1.

NEED FOR DETECTION OF WATER- AND FOOD-BORNE PATHOGENS

"Indicators" are useful to access pathogenic microorganisms, but why is detection of pathogens so urgently needed? The most frequently encountered water- and food-borne diseases (i.e. traveler's diarrhea, typhoid, and cholera) are caused by consumption of foodstuffs, including meat products and contaminated water. In the developing world, particularly southeast Asia, communities use untreated water for drinking, food preparation, and other domestic purposes, particularly in urban, suburban, and rural environments. It has been reported that the presence of water- and food-borne bacteria are a major cause of the economic burden on the food industry in developing countries (Wang et al., 2010). It is therefore necessary to detect these pathogens at an early stage to circumvent the spread of disease and epidemics. At this juncture, it is essential to know about the major players in the world of pathogenic microbes.

The pathogenic group of bacteria includes pathotypes of *Escherichia coli*, such as ETEC (Enterotoxigenic *E. coli*) and EHEC (Enterohemorrhagic *E. coli*), *Salmonella* spp., *Vibrio cholerae, Campylobacter* spp. (including antibiotic-resistant

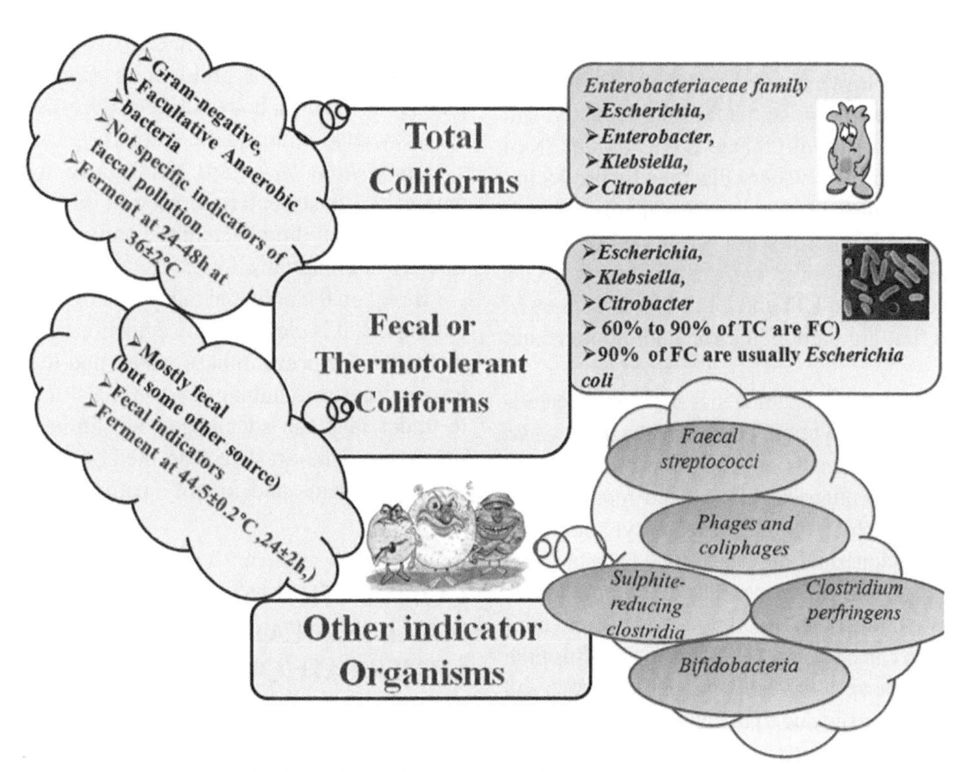

FIGURE 28.2 Different types of indicator organisms.

TABLE 28.1 Summary of Water Quality Criteria for Microbiological Indicators

Water Use	*Escherichia Coli* *CFU	*Enterococci* *CFU	*Fecal Coliforms* *CFU
Raw Drinking Water – no treatment	0/100 mL	0/100 mL	0/100 mL
Raw Drinking Water – disinfection only	≤ 10/100 mL 90th percentile	≤ 3/100 mL 90th percentile	≤ 10/100 mL 90th percentile
Raw Drinking Water – partial treatment	≤ 100/100 mL 90th percentile	≤ 25/100 mL 90th percentile	≤ 100/100 mL 90th percentile
Raw Drinking Water – complete treatment	None applicable	None applicable	None applicable
Livestock – free range animals	None applicable	None applicable	None applicable
Livestock – general livestock use	200/100 mL maximum	50/100 mL maximum	200/100 mL maximum
Livestock – closely confined (no treatment)	0/100 mL maximum	0/100 mL Maximum	0/100 mL Maximum
Livestock – closely confined (disinfection only)	≤ 10/100 mL 90th percentile	≤ 10/100 mL 90th percentile	≤ 10/100 mL 90th percentile
Livestock – closely confined (partial treatment)	≤ 100/100 mL 90th percentile	≤ 100/100 mL 90th percentile	≤ 100/100 mL 90th percentile
Livestock – closely confined (complete treatment)	None applicable	None applicable	None applicable

**CFU: Colony Forming Unit.*
(Modified from Water Quality Criteria for Microbiological Indicators (Warrington et al., 2001))

Campylobacter jejuni). Among the *E.coli* pathotype family, EHEC (foremost representative of *E. coli* 0157:H7) are the major culprits of water- and food-borne diseases in humans. These bacterial strains are characterized by the production of one or more types of cytotoxins that cause tissue damage in humans and animals. Enterotoxigenic *Escherichia coli* (ETEC), the causative agent of travelers' diarrhea, is one of the important pathogens in the farming industry, and is found in cattle and weaning piglets. ETEC strains from humans cause mild or severe watery diarrhea by producing a heat-labile enterotoxin (LTI), similar in structure to cholera toxin and heat-stable enterotoxins (ST IA and/or ST Ib). The heat-labile enterotoxins of *E. coli* are oligomeric toxins with two major serogroups, LTI and LTII. LTI is expressed by *E. coli* strains that are pathogenic for both humans and animals.

Salmonella, another typical pathogen, causes gastroenteritis and typhoid in humans. Typhoid caused by the *Salmonella* enterica serotype *Typhi* remains an important public health problem in developing countries. It has been reported that South Asian countries exhibit a high burden of typhoid fever. Furthermore, India, Indonesia, Bangladesh, and Pakistan have been identified as high infection zones (caused by *Salmonella* spp.). The severity of infections of *Salmonella* is due to their infective dose, which can be as low as 15–100 CFU. This high vulnerability to waterborne *Salmonella* infections in Asia and other developing countries is due to the lack of potable water and the dependence of a large population on natural resources for daily water requirements.

Vibrio cholerae and *Vibrio parahaemolyticus* are pathogens that cause diarrhea in humans. *V. parahaemolyticus* is an invasive organism affecting primarily the colon, whereas *V. cholerae* is non-invasive, affecting the small intestine through secretion of an enterotoxin. *Vibrio cholerae* causes a globally prevalent gastrointestinal disease, cholera, which remains a persistent problem in many countries. These pathogens occur in both marine and freshwater habitats, and are associated with aquatic animals.

Campylobacter jejuni has been associated with dysentery-like gastroenteritis, as well as with other types of infection, including bacteremic and central nervous system infections in humans.

Based on the potential risk posed by the above-mentioned pathogens, it is clear that the detection of these organisms is the key to the prevention of water- and food-borne epidemics or diseases in humans and animals. It is therefore useful to understand the advantages and limitations of the existing state-of-the-art detection methods used to detect indicator organisms and identify pathogenic variants of such microbes.

CONVENTIONAL METHODS TO DETECT FECAL INDICATOR ORGANISMS AND OTHER PATHOGENIC BACTERIA

Conventional techniques like culture-based methods have been recommended and used routinely for decades for identification and detection of pathogens. In culture-based methods, microbes grow on specific culture media at a specific temperature for a particular time period. Their characterization is based on the morphology of bacterial colonies and confirmation by biochemical tests. The most prevalent conventional methods, including culture-based methods, for pathogen detection are described in the following sections.

Most Probable Number Method

The most probable number (MPN) technique is an important technique for estimating microbial populations in soils, waters, food matrices, and agricultural products. Many soils are heterogeneous, therefore exact cell numbers of an individual organism are impossible to determine. The MPN technique is used to estimate microbial population where heterotrophic counts are difficult. This technique does not rely on quantitative assessment of individual cells; instead it relies on specific qualitative attributes of the microorganism being counted. The MPN technique estimates microbial population sizes in a liquid substrate; the method is tedious, and takes 24 to 48 hours.

Membrane Filtration Method

The membrane filter (MF) technique is used to test relatively large volumes of sample, and yields numerical results more rapidly than the MPN method. The membrane filter technique is extremely useful in monitoring drinking water and a variety of natural waters. On the basis of the MF technique, the coliform group may be comprised of all aerobic and many facultative anaerobic, Gram-negative, non-spore-forming, rod-shaped bacteria; these develop a red colony with a metallic sheen within 24 hours at 35°C on an endow-type medium containing lactose. Some members of the total coliform group may produce a dark red or nucleated colony without a metallic sheen, and are classified as typical coliform colonies after verification. Pure cultures of coliform bacteria produce a negative cytochrome oxidase (CO) and positive β-galactosidase (ONPG) reaction. Generally, all red, pink, blue, white, or colorless colonies that lack sheen are considered non-coliforms by this technique. However, the MF technique has limitations, particularly when testing waters with high turbidity or a presence of non-coliform bacteria. Hence, for such waters, or when the membrane filter technique has not been used previously, it is desirable to conduct parallel tests with the multiple-tube fermentation technique to demonstrate applicability.

Defined Substrate Methods

Media without harsh selective agents but specific enzyme substrates provide significant improvements in recoveries and identification of target bacteria such as coliforms and *E. coli*. Furthermore, the enzyme-based methods appear to

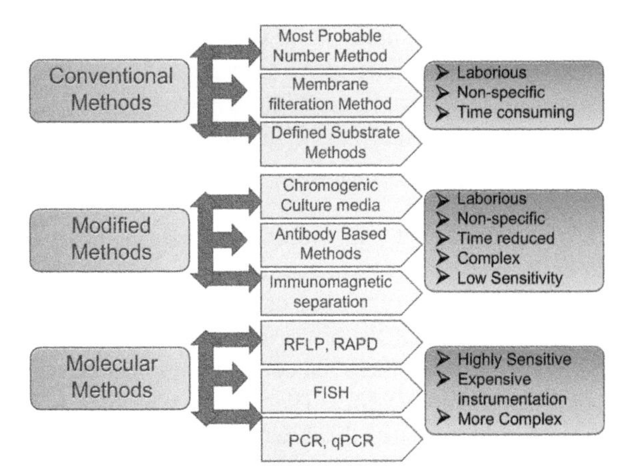

FIGURE 28.3 Different methods of pathogen detection.

pick up traditionally non-culturable Coliforms. Total coliforms are members of genera or species within the family *Enterobacteriaceae*, capable of growth at 37°C, which and that possess β-galactosidase.

This defined substrate approach has been advocated by the International Standards Organization for miniaturized MPN-based methods for Coliforms/*E. coli* and enterococci (ISO/FDIS, 1999). Certain methods in vogue are presented in Figure 28.3.

Rapid Detection Using Chromogenic Substrates

Chromogenic compounds added to conventional or newly devised media are used for the isolation of indicator bacteria. These chromogenic substances are modified either by enzymes (which are typical for the respective bacteria) or by specific bacterial metabolites. After modification, the chromogenic substance changes its color or fluorescence, thus enabling easy detection of colonies displaying the metabolic activity. In this way, these substances are used to avoid the need for isolation of pure cultures and confirmatory tests. The time required for the determination of different indicator bacteria can be reduced to between 14 to 18 hours. When necessary, the surface antigens can be selected and used as the pathogen recognition element for confirmation.

Immunological Methods

Antibody-Based Detection

Antibodies are immunoglobulins secreted by B-cells, and are recruited by the immune system to identify and neutralize foreign objects such as bacteria and viruses. Antibodies possess highly specific binding and recognition domains that can be targeted to specific surface structures of a pathogen. Immunological methods using antibodies are widely used to detect pathogens in clinical, agricultural, and environmental samples. As always with immunological techniques, the specificity of the reagents and optimization of their use is a critical issue. Although total Coliforms are a broad group and likely to be unsuitable immunological targets in environmental waters, *E. coli* could be identified from other Coliforms.

It is evident that the aforementioned methods and approaches are either laborious and time-consuming, or non-specific with a probability of false positives. Another major disadvantage of the above methods is the failure to detect non-cultivable microbial communities that are viable but non-culturable (VBNC). The VBNC state of a bacterium is its ability to cause the disease and yet fail to respond to enumeration through classical culture procedures. The VBNC state of bacterial enteropathogens poses a potential threat to human and animal health because the failure to culture such organisms in current resuscitation protocols leads to incorrect estimation.

Immunomagnetic Separation and Other Rapid Culture-Based Methods

Immunomagnetic separation (IMS) offers an alternative approach to the rapid identification of culturable and non-culturable microorganisms. The principles and application of the method are simple, but rely on suitable antibody specificity under the experimental conditions. Purified antigens are typically biotinylated and bound to streptoavidin-coated paramagnetic particles (e.g. Dynal™ beads). The raw sample is gently mixed with the immunomagnetic beads, then a magnet is used to hold the target organisms against the wall of the recovery vial, and non-bound material is poured off. If required, the process can be repeated, and the beads can be removed by simple vortexing. Target organisms can then be cultured or identified by direct means. The IMS approach may be applied to recovery of indicator bacteria from water, but is possibly more suited to replace labor-intensive methods for specific pathogens. *E. coli* O157 recovered from water samples were detected using this technique in some studies. Furthermore, *E. coli* O157 detection following IMS can be improved by electrochemiluminescence detection. However, the IMS/culture methods are also accompanied by disadvantages, such as the ability of non-specific binding, the need for physicochemical conditions such as pH and temperature, sensitivity to chemicals in the samples, the high cost of monoclonal antibody production, and limited shelf life.

MOLECULAR METHODS BASED ON GENETIC SIGNATURE OF TARGET PATHOGEN

The nucleic acid sequences are unique to all living organisms. These genetic signature sequences are the potential targets to differentiate one organism from another and to diagnose various disease-causing agents. In the post-genomic era, large numbers of microorganisms have been sequenced. In early 2013, ~18,000 prokaryotic genomes have been sequenced (NCBI Genome database, http://www.ncbi.nlm.nih.gov/genome/browse/). This sequence database has made it possible to analyze microbial pathogens at the molecular level.

The application of molecular methods has to be considered within the framework of quality management for potable water. The new methods will influence epidemiology and outbreak investigations more than the routine testing of processed drinking water. Certain molecular approaches like Restriction Fragment Length Polymorphism (RFLP), Random Amplification of Polymorphic DNA (RAPD), and Florescence *In Situ* Hybridization (FISH) are extensively used for pathogen detection, but each has limitations. In FISH-based detection, gene probes with a fluorescent marker are used, typically targeting the 16S ribosomal RNA (16S rRNA). Concentrated and fixed cells are permeabilized and mixed with the probe. The stringency of the homology between the gene probe and the target sequence are influenced by incubation temperature and the addition of chemicals. A single fluorescent molecule within a cell does not allow detection, and target sequences with multiple copies in a cell have to be selected (e.g. there are 10^2–10^4 copies of 16S rRNA in active cells). Low-nutrient environments may result in cells entering a non-replicative, viable, but non-culturable (VBNC) state for many pathogens. Such a state may give a false result that makes culture-based methods unreliable and can be overcome using molecular approaches.

Polymerase Chain Reaction Technique and Quantitative PCR

The powerful molecular technique, Polymerase Chain Reaction (PCR), allows amplification of target DNA to generate multiple copies that can be detected. PCR has been validated by the International Organization for Standardization (ISO), and is now used for testing of food-borne pathogens (Malorny et al., 2008). One problem with PCR is that the assay volume is on the order of microliters and requires that the sample be concentrated to the microliter range. The water sample has to be concentrated and purified using adequate methods, as natural water samples often contain inhibitory substances such as humic acids and iron that concentrate with the nucleic acids. Hence, it is critical to have positive and negative controls with each environmental sample PCR to check for inhibition and specificity. It may also be critical to find out whether the signal obtained from the PCR is due to naked nucleic acids, or living and dead microorganisms. The sensitivity of the PCR is often not sufficient, and post-PCR processing and analysis is needed. Additionally, PCR requires technical equipment and laboratory setup, which is not suitable for on-site diagnostics. Quantitative PCR (qPCR), also called Real-Time PCR, a fluorescence-based detection format, is more sensitive than conventional PCR. Samples can be analyzed in real time with higher specificity and sensitivity. No post-PCR processing is required. The technique has been applied to the location of non-point sources of pathogen contamination and environmental risk assessment (Singh et al., 2010).

Fluorescence labeling is a commonly used and well-established method, but has disadvantages like a lower stability of fluorescent dyes and the requirement of expensive readout systems. The fluorescence-based detection systems have many significant drawbacks, including susceptibility to photobleaching, complexity, sensitivity to contamination, cost, and reliance on relatively expensive equipment to probe their presence in an assay.

Although diverse, the existing methods have certain disadvantages that limit their use in point-of-care settings and field situations due to the requirements of sophisticated instrumentation and trained personnel. Faster, simpler, and more reliable detection methods would largely provide support to help protect consumers. Direct detection methods that provide quick, accurate, simple, and cost effective devices to be used on-site are highly desired. Advances in human and animal science are placing increasingly stringent demands on diagnostic and clinical tests to enhance sensitivity, specificity, and thresholds.

There is an extensive need for the selection of a unique system that will allow the pathogen recognition element to be accepted for an ideal detection format. It is becoming evident that new pathogen detection methodologies enabled by nanotechnology-based approaches have the potential to provide better options.

NANOTECHNOLOGY AND ITS PROMISE

The term nano, derived from the Greek word for *dwarf*, is usually combined with a noun to form words such as nanometer, nanorobot, and nanotechnology. In the last two decades, nanoscience and nanotechnology have seen a plethora of new developments in almost every field of science and technology, especially in biology and medicine. Nanotechnology has set high expectations in biological and medical sciences to solve key questions concerning bio-systems that operate at the nanoscale. It deals with the creation of functional materials, devices, and systems on the nanometer scale length of 1–100 nm. The ability to manipulate and engineer materials at the nanoscale (atomic, molecular, and macromolecular) enables one to tune the physical and chemical properties of desired materials according to specific applications.

The surface functionalization of nanomaterials by biomolecules has led to the development of new interdisciplinary research areas like biomedical nanotechnology, nanomedicine, diagnostic devices, theranostics, contrast agents, nanobiosensors, and targeted drug delivery vehicles. Functionalities can be added to nanomaterials by interfacing them with biological molecules or structures (Figure 28.4a); that makes nanomaterials useful in various technological areas as well as in both *in vivo* and *in vitro* applications. Nanotechnology has played a significant role in the development of affinity sensors (e.g. antibody–antigen interaction-based biosensors and ultrasensitive DNA hybridization detection). At the nanoscale some materials have been shown to exhibit extraordinary optical properties relative to their bulk counterparts. The light-scattering power of nanoparticles is orders of magnitude greater than fluorescent labels. Moreover, the optical signals generated from these nanoparticles are not prone to photobleaching, compared to their organic dye counterparts. Semiconductor quantum dots are one of the pioneering examples that distinctly show size-dependent emission of different colors with high quantum yield when excited with a single wavelength. The ability to simultaneously tag multiple biomolecules with these quantum dots has provided the opportunity to develop new optical diagnostic tools and the ability to observe complex cellular changes and associated events. On the other hand, noble metal nanoparticles have attracted much interest because of their unique physico-chemical properties, including large optical field enhancements that result in the strong scattering and absorption of light in the visible region due to the presence of surface plasmon resonances (SPRs). Biochemical assays based on light-scattering signals from metal nanoparticles have been widely used in the determination of the affinity interaction between DNA, proteins, and drugs, and have led to the development of easy-to-use optical sensing devices.

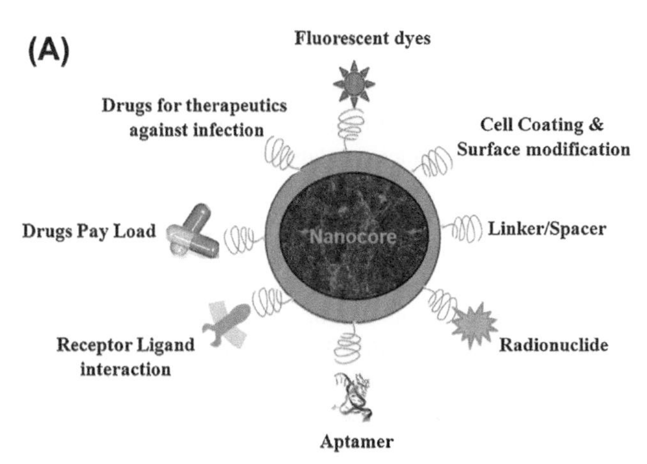

FIGURE 28.4A Functionalization of nanoparticles.

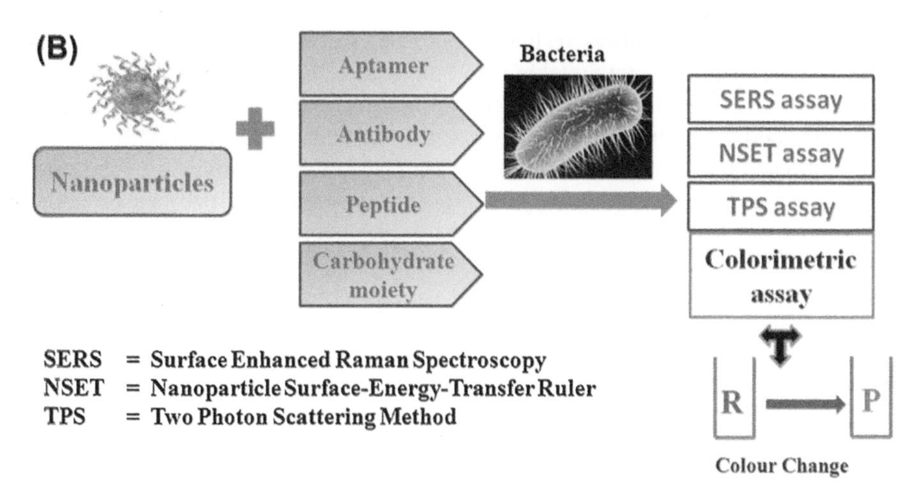

FIGURE 28.4B Nanomaterial applications in pathogen detection.

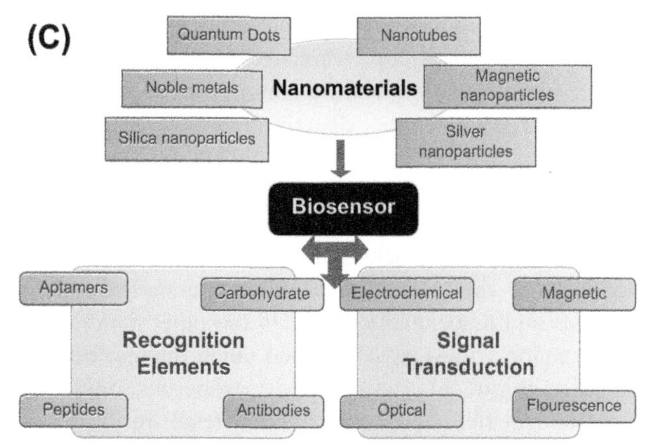

FIGURE 28.4C Nanomaterial components involved in pathogen detection. *(Modified from Vikesland et al., 2010.)*

Metallic Nanoparticles

Colloidal metal nanoparticles (e.g. gold and silver nanoparticles) have attracted tremendous research interest. Metallic nanoparticles in colloidal solution appear colored; for example, colloidal gold nanoparticles generally appear ruby red, purple, blue, and orange, depending upon their shape, size, and synthesis conditions. Similarly, titanium (Ti) and platinum (Pt) nanoparticles give blue and dark brown colors. The different colors of metal nanoparticles arise from SPRs and their confinement. Among metals, gold nanoparticles exhibit excellent biocompatibility, electronic, optical, and catalytic properties. Gold nanoparticles (GNPs) have found a distinguished place in bio-diagnostics due to their size-dependent optical properties, variety of surface coatings, and biocompatibility (Mirkin et al., 1996; Storhoff et al., 2004; Sato et al., 2007, Pandey et al., 2008). Spherical gold nanoparticles exhibit SPR-related optical absorption at 520 nm, which strongly depends on particle size and morphology.

The rich surface chemistry of gold nanoparticles allows surface modifications with various biofunctional groups, such as nucleic acids, sugars, and proteins, via the strong affinity of the gold surface with thiol ligands (Schofield et al., 2007); this creates multi-functionality to tailor the needs of biomedical applications including imaging, diagnostics, and therapy. The conjugation of nanoparticles with biomolecules (e.g. proteins and DNA) can be done either by direct covalent linkage or by non-covalent interactions. Biomolecules are often covalently linked to ligands on the nanoparticles' surface via traditional coupling strategies such as carbodiimide-mediated amidation and esterification.

The high surface-to-volume ratio of GNPs makes surface electrons sensitive to minor changes in the dielectric (refractive index) constant of the medium. Therefore, changes to the surface chemistry and environment of these particles (surface modification, aggregation, medium refractive index, etc.) lead to colorimetric changes of the dispersions that become the basis for detection of any analyte of interest. This has further facilitated the application in bio-detection via numerous methods (Boyer et al., 2002). The various synthetic methods for gold nanoparticles are discussed in the methodology section.

HISTORY

The detection of microbial pathogens remains a challenging task despite great strides made in the past three decades. The most frequently used culture-based methods have undergone diverse modifications, including specific substrates for enzymes present in target organisms. However, the issues of specificity and sensitivity directed towards low doses of organism, viable but non-culturable states, and long incubation periods, still elude the culture technique. The immunological methods using antibodies to detect pathogens in different domains of an environment, as well as in clinical settings, followed culture techniques. Immunomagnetic separation evolved to improve detection, but faced specificity and limited shelf-life problems. The understanding of nucleic acid structure and function led to the era of gene sequencing, DNA hybridization, and consequent genetic signature-based detection of organisms. A milestone in molecular biology, the polymerase chain reaction (PCR) technique developed by Kary Mullis in 1983 led to a burst in PCR use in pathogen detection between 1992 and 1999. This was followed by advancements in PCR, including real-time PCR or quantitative PCR (qPCR) based on fluorescence chemistries. This technique has been widely used for quantitative enumeration of pathogens in different domains from 1999 until today. Fluorescence-based detection is complex, and has issues of photo-bleaching, contamination, cost, and dependence on expensive instrumentation that limit applications for "on-site detection." These methodologies delineate the need for simple assays and less expensive detection system. Nanoscience provides new horizons in bio-nanotechnology for the detection of pathogens based on genetic signatures. In the last decade, nano-based approaches that exploit the unique properties of nanoparticles have shown potential for the development of novel pathogen-detection systems.

DETECTION PRINCIPLE

The efficient use of nanomaterials in biological systems relies on knowledge of the *nano–bio interface*. Gold nanoparticles have found widespread applications in the life sciences, and serve as excellent standards to understand more general features of the nano–bio interface because of their many advantages over other inorganic materials. The bulk material is chemically inert. Gold's background concentration in biological systems is low, which makes it relatively easy measure at the parts-per-billion level or lower in water. The unique optical and electronic properties of GNPs enable them to conjugate biological molecules like RNA and DNA and serve as scaffolds for nanostructures. GNP interactions with light are governed by size, environment, and physical dimensions. The fluctuating electric fields of a light ray promulgating near a colloidal nanoparticle interact with free electrons to cause the intense oscillation of electron charge that is in resonance with the frequency of visible light. These resonant oscillations are known as surface plasmons. Surface plasmon resonance for monodispersed gold nanoparticles (~30 nm) causes light absorption in the blue–green portion of the spectrum (~450 nm), while red light (~700 nm) is reflected, yielding a rich red color. The wavelength of surface plasmon resonance-related absorption shifts to longer wavelengths, with an increase in particle size. Red light is then absorbed, and blue light is reflected, resulting in solutions with a pale blue or purple color. The change in optical properties as a result of shape, size, and aggregation pave the way for the development of colorimetric and sensitive detection systems for specific bioanalytes. As an example, the DNA hybridization event using GNP is recognized by a change in color that appears as a result of DNA hybridization, which brings GNPs in close proximity.

Several proof-of-concept studies demonstrate the use of GNPs in biomedical applications like chemical sensing, biological imaging, drug delivery, and cancer treatment. A number of factors, namely size and shape of the nanoparticle, refractive index of the surrounding media, and inter-particle distance, are taken into account for use in colorimetric detection of DNA (Baptista et al., 2008). Mirkin and co-workers (1996) reported the colorimetric detection of DNA targets based on the cross-linking mechanism use of GNP probes. The two different batches of probes are designed to target the DNA. Thus, upon the addition of target DNA, a polymeric network of GNP probes is generated due to aggregation, turning the solution from red to blue (Mirkin et al., 1996). This aggregation mechanism is mainly applied to detect small-sized targets. The GNP aggregation induced by interparticle cross-linking is a relatively slower process. The relatively slow aggregation is due to the nature of the interparticle cross-linking aggregation mechanism. In general, the aggregation is driven by random collisions between nanoparticles with relatively slow Brownian motion (Sato et al., 2003).

Similarly, DNA detection based on a non-cross-linking mechanism has been well documented. Sato and co-workers described a non-cross-linking mechanism of GNP aggregation for DNA detection. The single-stranded DNA can be immobilized on GNPs above the physiological temperature. The GNP probes aggregate together at a considerably high salt concentration when the target DNA is perfectly complementary to the probe (Sato et al., 2007). In non-cross-linking aggregation systems, the van der Waals force of attraction dominate. Many parameters, including surface charge properties (e.g. charge density, the amount of associated counter ions) and entropy factors are also involved in aggregation.

Compared to interparticle cross-linking aggregation systems, the non-cross-linking aggregation mechanism has some attractive features. Aggregation induced by the non-cross-linking process is very rapid, leading to the development of faster assays. The interparticle attractive forces (van der Waals forces) dominate over the interparticle repulsive forces, which results in rapid aggregation (Sato et al., 2003). The use of GNP probes for the colorimetric detection of DNA targets represents an inexpensive and simple workable alternative to fluorescence- or radioactivity-based assays (Storhoff et al., 2004).

METHODOLOGY

Synthesis of Gold Nanoparticles

Gold nanoparticles can be synthesized using various synthetic routes. In a typical synthesis, tetrachloroauric acid

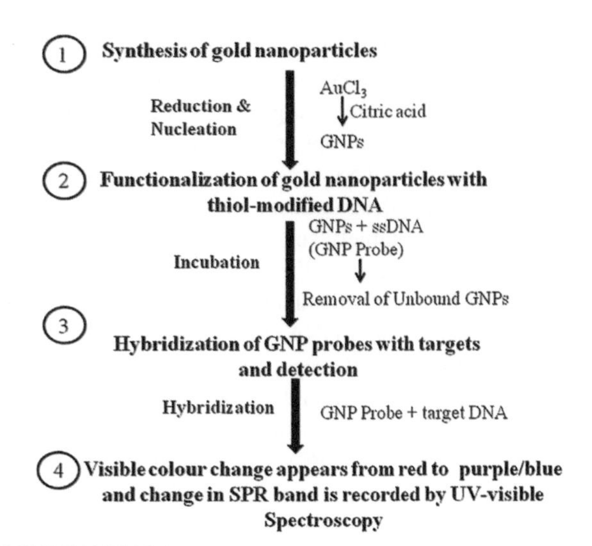
1 Synthesis of gold nanoparticles
Reduction & Nucleation
AuCl3
Citric acid
GNPs
2 Functionalization of gold nanoparticles with thiol-modified DNA
Incubation
GNPs + ssDNA
(GNP Probe)
Removal of Unbound GNPs
3 Hybridization of GNP probes with targets and detection
Hybridization
GNP Probe + target DNA
4 Visible colour change appears from red to purple/blue and change in SPR band is recorded by UV-visible Spectroscopy

FLOW CHART 28.1

1. This is the first step of synthesis of GNPs by the Turkevitch and Brust method. Citric acid reduces gold chloride (i.e. $AuCl_3$) and triggers the nucleation of Au ions to form nanoparticles, followed by its adsorption to the surface, which provides colloidal stability to nanoparticles due to its negative charge.
2. In this second step, GNPs are functionalized with ssDNA. Gold nanoparticles are incubated with thiol-modified ssDNAs for 16–20 hrs, and unbound ssDNA is removed by centrifuging the solution. The pellet of DNA-conjugated GNPs is washed and stored in 0.3 M tris-acetate NaCl (pH 8.2) at room temperature.
3. Next is hybridization of the formed GNP probes with targets and subsequent detection. After hybridization with the target, the GNP probes come in close proximity, and as a result, the particles aggregate and the solution appears purple/blue.
4. This is the last step, in which visible color change appears as a red to blue/purple change in the SPR band, and is recorded by UV-visible spectroscopy.

($HAuCl_4$) is mixed with a reducing agent, which leads to the reduction of Au ions to form nanoparticles. In the most reliable and popular method by Turkevich and Brust, citric acid first reduces gold ions from $HAuCl_4$ and triggers the nucleation to form nanoparticles, followed by its adsorption to the surface, which provides colloidal stability to nanoparticles due to its negative charges (Brust et al., 1998). This method produces monodisperse spherical gold nanoparticles with a diameter in the 10–20 nm range. (Step 1 in Flow Chart 28.1)

Computation of ssDNA Sequences for Functionalization of Gold Nanoparticles

The GNPs are bio-functionalized with the thiol-modified single-stranded DNA (ssDNA) or oligonucleotides to generate a GNP probe for target DNA detection. The bioinformatics tools are used to compute ssDNA sequences or probes complementary to the target gene sequence of the pathogen of interest. The first step towards this is the retrieval of nucleotide sequences of targeted genes of selected organisms from GenBank (www.ncbi.nlm.nih.gov). Multiple sequence alignment of the retrieved conserved region (http://www.ebi.ac.uk/Tools/msa/clustalw2/) is followed by computation of oligonucleotide probes using web-based or dedicated software such as Primer 3, Lasergene, or Beacon Designer. An analysis for cross homology and secondary structure is carried out by BLAST (http://blast.ncbi.nlm.nih.gov/Blast.cgi) and Mfold (www.bioinfo.rpi.edu/applications/mfold) servers. The computed ssDNA sequences are then synthesized in a DNA synthesizer for functionalization of GNPs.

Functionalization of Gold Nanoparticles with Thiol-Modified DNA

The computed oligonucleotides are synthesized with modification of the thiol group (–SH) at their 5′ or 3′end. These modified oligonucleotides are then attached to the GNPs through chemisorption of the thiol group onto the surface of the GNPs; GNP probes are then generated. In order to functionalize with DNA, gold nanoparticles (selected size ~20±0.2 nm diameter) are incubated with different batches (number varies with the target size) of thiol-modified ssDNAs separately for 16–20 hours with an oligonucleotide concentration of 2 μM. The unbound ssDNA is then removed by centrifugation at 16,000 × g for 15 min, and the pellet of DNA-conjugated GNPs is washed in 0.1 M Tris-acetate and NaCl (pH 8.2) buffer. The ssDNA-grafted GNP is then stored in 0.3 M Tris-acetate NaCl (pH 8.2) at room temperature. The absorption spectra of the bio-functionalized GNPs are recorded by UV-visible spectroscopy for confirmation of DNA immobilization (Steps 2–4 in Flow Chart 28.1). The basic steps of DNA detection using gold nanoparticle probes are depicted in Figure 28.5a

EXAMPLES

Enterohemorrhagic *E. coli* (EHEC) serotype O157:H7 is one of the most deadly pathogens. EHEC produces *stx1*- and *stx2*-type enterotoxins and symptoms, such as abdominal pain and watery diarrhea. Many patients develop life-threatening diseases, such as hemorrhagic colitis (HC) and hemolytic-uremic syndrome (HUS), the deadly consequences of these cytotoxins. The natural reservoirs of EHEC are domestic and wild ruminant animals, which shed the bacteria along with their feces into the environment. The products of animal origin, such as meat and milk, are at risk of contamination with EHEC originating from animals. Consumption of food containing EHEC was identified in different countries as a major route of human infection with these pathogens. Hamburger, vegetables, and fruit juices have frequently been contaminated with pathogenic *E. coli*, and have also been sources of infection. Apparently, cattle, the natural reservoir for pathogenic strains, have often been implicated in *E. coli* infections. Following are examples for detection of EHEC using GNPs.

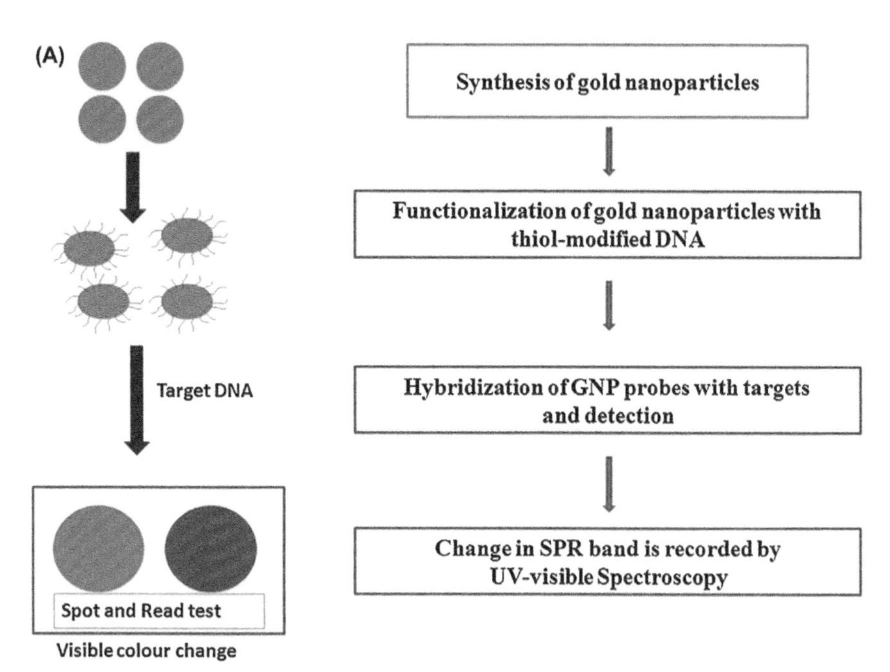

FIGURE 28.5A Methodology.

Colorimetric Detection of DNA of Shiga Toxin Producing *Escherichia coli* (Using Bio-Conjugated Gold Nanoparticles)

EHEC causes bloody diarrhea in humans through the production of shiga-like toxin. The toxin is encoded by *stx2* gene in *E. coli*. The existing methods for detection of EHEC are culturing of the bacteria on fluorogenic-substrate media, which is time-consuming. Molecular methods such as PCR and real-time PCR assays are also used, which require expensive instrumentation. In recent studies, the optical properties of gold nanoparticles (GNPs) have been exploited for detection of the nucleic acid of *Escherichia coli*. The PCR product of *stx2* gene representing EHEC signature has been targeted using gold nanoparticle probes. Gold nanoparticles of 20±0.2 nm were synthesized by citrate reduction and characterized by UV-visible spectroscopy and transmission electron microscopy. Two different batches of thiolated, single-stranded DNA (19 and 22 bp) complementary to the target are grafted onto the GNPs (Table 28.2). The hybridization of GNP probes with target DNA led to a change in color from red to purple that is visible with the naked eye (Jyoti et al., 2010). The hybridization-induced aggregation was also observed by transmission electron microscopy.

TABLE 28.2 Single-Stranded Thiol-Modified Oligonucleotide Probes and Their Complementary Sequences

Probe/ Synthetic Target	Nucleotide Sequence (5′-3′)	Length (bp)
sx2F	HS-$(C)_6$-GGAGTTCAGTGGTAATACAATG-	22
sx2R	HS-$(C)_6$-GCGTCATCGTATACACAGG-	19

(Modified From Jyoti et al., 2010.)

Colorimetric Detection of Enterotoxigenic *Escherichia coli* (ETEC) Gene Using Gold Nanoparticle Probes

In the present example, the colorimetric detection of heat-labile toxin gene *LT1* (1257 bp) of ETEC was shown. Multiple probes were used to target the gene to increase specificity towards the target as well as the aggregation of gold nanoparticles. A total of eight GNP probes were used to target the different locations on the target DNA sequence. The oligonucleotides were computed based on the conserved signature gene of ETEC. The GNPs were functionalized with thiol-modified single-stranded DNAs to facilitate hybridization with the target. After hybridization, change in the surface plasmon resonance-related band was observed by UV-visible spectroscopy. This led to a visible colorimetric change of reaction assay mixture from red (λ_{max} = 524 nm) to purple (λ_{max} = 552 nm) that is clearly visible to the naked eye. TEM was used to evaluate the aggregation and reduction in the interparticle distances of gold nanoparticles. TEM confirms the hybridization, aggregation, and reduction in the interparticle distances of the GNP probes in the presence of target DNA (Figure 28.3). In addition, the assay shows its specificity by differentiating the DNA of

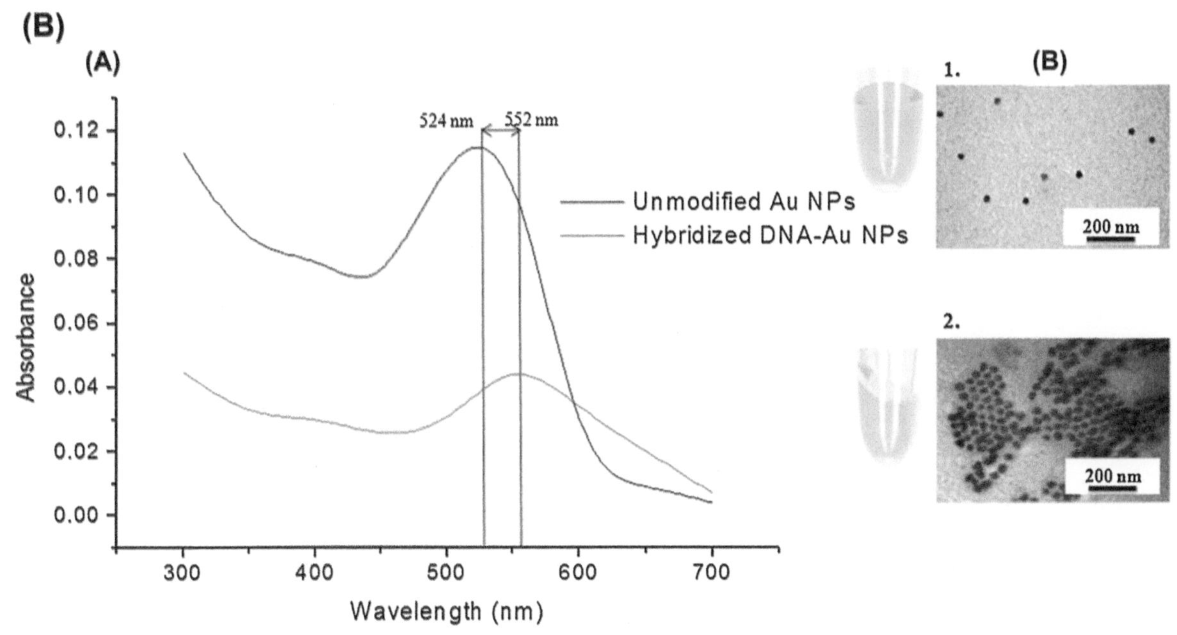

FIGURE 28.5B (A) UV–Vis spectra of unmodified GNPs (black) and hybridized DNA–GNPs (red); (B) Transmission electron micrograph of the (1) monodispersed GNPs (20±0.2 nm) and (2) hybridized polymer network corresponding to the change in solution color from red/pink to purple.

Enterohemorrhagic *E. coli* (EHEC) from ETEC, a closely related pathotype of *E. coli*.

The aggregation and the spectral shift in the plasmon band leading to a change in color observed with target DNA indicates the possibility of a simple and rapid colorimetric "spot and read" test in contrast to amplification-based fluorescence detection methods (Figure 28.5b).

Clinical Significance of Nanoparticle-Based Detection

In recent years, nano-based approaches have been exploited for detection of other pathogens of clinical significance, like mycobacterium and other pathogenic species that affect humans and animals as causative agents of tuberculosis, leprosy, and paratuberculosis. Direct detection of unamplified DNA from pathogenic mycobacteria using DNA-derivatized gold nanoparticles was reported by (Liandris et al., 2009). Different nanodiagnostics systems have been developed for the molecular diagnosis of tuberculosis-like nanoparticle-based systems, such as gold, silver, silica, and quantum dots (QDs); they have been the most widely used for TB diagnostics due to their unique physicochemical properties. For detection of multiple bacterial genomic DNA, gold nano-rod probes were used; they have high sensitivity and excellent specificity for reading the decrease in the sensitive longitudinal absorption band (Wang et al., 2012). The development of nano-enabled assays and on-site detection systems for EHEC and ETEC can perhaps help in the reduction of morbidity and mortality in children in rural ecozones. In summary, detection systems with high sensitivity and selectivity, and that are easy-to-use and rapid, are of immediate need; nanoparticle-based detection shows great promise for the development of such nanodevices.

ETHICAL ISSUES

Containment of hazards in laboratory experiments is one key to safety. Microbiologists have gained valuable experience over decades of handling extremely dangerous natural organisms, such as smallpox virus and cholera bacteria. According to *Biosafety in Microbiological and Biomedical Laboratories* (CDC, 2009), safe handling and containment of infectious microorganisms and hazardous biological materials are the mandatory principles of working with microbes. The basics of containment include the use of safety equipment, proper microbiological practices, and safeguarded facilities that protect laboratory workers, the sterile environment, and the public from exposure to infectious microorganisms that are handled and stored in the laboratory. This inclusion of risk assessment in our routine practice can prevent laboratory-associated infections. Individual workers handling the pathogenic microorganisms must understand the containment conditions under which infectious agents can be safely maneuvered and secured. Correct information and the use of appropriate techniques and equipment will enable the microbiological and biomedical communities to prevent personal, laboratory, and environmental exposures to potentially infectious agents or biohazards. Furthermore, limited understanding of the adverse impact of engineered nanomaterials or

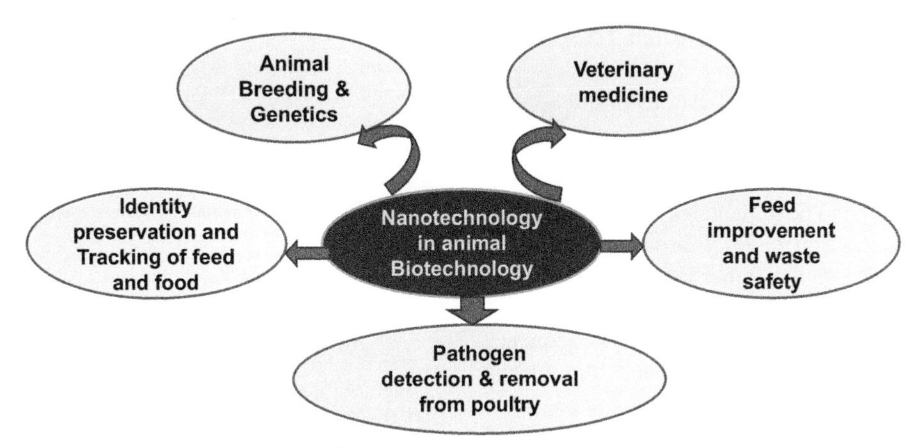

FIGURE 28.6 Applications of nanotechnology in animal biotechnology.

nanoparticles in biological systems demands safe handling via basic laboratory safety norms, and disposal as per international guidelines (Dhawan et al., 2011).

TRANSLATIONAL SIGNIFICANCE

Nanomaterial-based DNA detection and biosensors for pathogens are attracting much attention due to their comparatively high sensitivity and non-complexity (Figure 28.4b). The replacement of fluorescently labeled DNA probes with a class of metallic nanoparticle-conjugated probes appears promising because it can minimize or eliminate the necessity of using expensive and complex instrumentation. Nano-sized, multipurpose sensors are being developed to detect almost everything from physiological parameters to toxic compounds. Nanosensors can detect very small amounts of a chemical contaminant, virus, or bacteria in food systems. Carbon nanotubes are being investigated as biosensors to detect glucose, ethanol, hydrogen peroxide, immunoglobulins, and as electrochemical DNA hybridization biosensors (Figure 28.4c). Nanotechnology has the tremendous potential to revolutionize the agricultural and livestock sectors. It can provide new tools for molecular and cellular biology, biotechnology, veterinary physiology, animal genetics, and reproduction; this would greatly increase the sensitivity of detection of biological materials with quantities typically in the nano-to-picoliter range. (Figure 28.6)

Nanoparticles are being used to remove *Campylobacter* and *E. coli* from poultry products (Manuja et al., 2012). *Listeria monocytogenes*, another food-borne pathogen, was detected in spiked milk samples by magnetic nanoparticle-based immune-magnetic separation combined with real-time PCR. Gold nanoparticles have been used to detect contamination of melamine in raw milk samples by the naked eye, with no sophisticated instruments required. The method is also promising for detection of melamine contamination in other foods, such as eggs and animal feed. A fluorescent, bio-barcoded DNA assay has been developed for the rapid detection of *Salmonella enteritidis* based on two nanoparticles (Zhang et al., 2009). A hand-held chip detects animal cytochrome b genes in food or feed products. Nanosensors can detect very small amounts of chemical contaminants, viruses, or bacteria in food systems.

FUTURE APPROACHES

Nanotechnology has the potential to revolutionize biomedical science of the 21st century through a paradigm shift in diagnostics, drug discovery and delivery, vaccine development, and tissue engineering. In the domain of veterinary science, nanotechnology has improved animal reproduction, hygiene and health, and nutrition. Food science and technology has seen new nanoscience-based developments in food storage, especially in animal food like red meat and fish, by increasing shelf life and fortified nutritional value.

In microbiology, nanomaterials are paving the way as new bactericides promising to fight even multidrug resistance. Nanotechnology tools could enable the understanding of how bacteria work, while providing new opportunities to probe the dynamic and physical aspects of molecules, molecular assemblies, and intact microbial cells, whether in isolation or under *in vivo* conditions. Furthermore, developments in nanoscience are leading to new, sensitive, and faster methods for pathogen detection in the form of nanosensors. The detection of relatively large-sized double-stranded DNA using stable DNA-functionalized gold nanoparticles is a quite simple colorimetric approach, without the use of complex instrumentation. However, the developed colorimetric approaches still need research advancements in terms of sensitivity and on-site instrumentation before they will be ready for commercialization.

WORLD WIDE WEB RESOURCES

1. NCBI Genome Database: http://www.ncbi.nlm.nih.gov/genomes/lproks.cgi.
2. Multiple sequence alignment: http://www.ebi.ac.uk/Tools/msa/clustalw2.

3. BLAST (Basic Local Alignment Search Tool): http://blast.ncbi.nlm.nih.gov/Blast.cgi.
4. mfold: www.bioinfo.rpi.edu/applications/mfold.
5. Wikipedia: http://en.wikipedia.org/wiki/Surface_plasmon_resonance.

The NCBI houses genome sequencing data in GenBank, and an index of biomedical research articles in PubMed Central and PubMed, as well as other information relevant to biotechnology.

ClustalW is a program for global multiple sequence alignment. It constructs pairwise sequence alignments. This heuristic method does a pairwise progressive sequence alignment for all the sequence pairs that can be constructed from the sequence set. A dendrogram (guide tree) of the sequences is then generated according to the pairwise similarity of the sequence. Finally, a multiple sequence is constructed by aligning sequences in the order defined by the guide tree.

BLAST is an algorithm for comparing primary sequence information, such as nucleotides of DNA sequences or the amino acid sequences of different proteins. A BLAST search compares a query sequence with a library or database of sequences, and identifies library sequences that resemble the query sequence above a certain threshold. Different types of BLASTs are available according to the query sequences.

In silico is an expression used to mean "performed on computer or via computer simulation." The phrase is coined from the Latin phrases *in vivo* and *in vitro*, and is commonly used in biology; it refers to experiments done in living organisms and outside of living organisms, respectively. The *in silico* approach reduces the time and expense involved in the bench work used for experimentally testing several PCR primers to validate a protocol.

mfold is one of the earliest systems designed for molecular biology computations. The program uses mainly thermodynamic methods to predict the secondary structures of RNA and DNA.

REFERENCES

Ashbolt, N. J., Grabow, W. O. K., & Snozzi, M. (2001). Indicators of microbial water quality. Water Quality: Guidelines, Standards and Health. World Health Organization (WHO). Water Quality: Guidelines, Standards and Health. In Lorna Fewtrell & Jamie Bartram (Eds.), London UK: IWA Publishing. ISBN: 1900222 28 0.

Baptista, P., Pereira, E., Eaton, P., Doria, G., Miranda, A., Gomes, I., Quaresma, P., & Franco, R. (2008). Gold nanoparticles for the development of clinical diagnosis methods. *Analytical and Bioanalytical Chemistry*, *391*, 943–950.

Boyer, D., Tamarat, P., Maali, A., Lounis, B., & Orrit, M. (2002). Photothermal imaging of nanometer-sized metal particles among scatterers. *Science*, *297*, 1160–1163.

Brust, M., Bethell, D., Kiely, C. J., & Schiffrin, D. J. (1998). Self-assembled gold nanoparticles thin films with non-metallic optical and electronic properties. *Langmuir*, *14*(19), 5425–5429.

Centers for Disease Control and Prevention, National Institutes of Health. (2009). Biosafety in Microbiological and Biomedical Laboratories. (BMBL 5th Ed.*U.S. Department of Health and Human Services*, pp. 21–1112). Public Health Service., HHS Publication No. (CDC).

Dhawan, A., Shanker, R., Das, M., & Gupta, K. C. (2011). Guidance for Safe Handling of Nanomaterials. *Journal of Biomedical Nanotechnology*, *7*, 218–224.

ISO/FDIS. (1999). *Water Quality – Detection and enumeration of Escherichia coli and coliform bacteria in surface and waste water – Part 3. Miniaturised method (Most Probable Number) by inoculation in liquid medium*. Geneva: International Standards Organization. 9308–3.

Jyoti, A., Pandey, P., Singh, S. P., Jain, S. K., & Shanker, R. (2010). Colorimetric detection of nucleic acid signature of shiga toxin producing *Escherichia coli* using gold nanoparticles. *Journal of Nanoscience and Nanotechnology*, *10*, 4154–4158.

Liandris, E., Gazouli, M., Andreadou, M., Comor, M., Abazovic, N., Sechi, L. A., & Ikonomopoulos, J. (2009). Direct detection of unamplified DNA from pathogenic mycobacteria using DNA-derivatized gold nanoparticles. *Journal of Microbiological Methods*, *78*, 260–264.

Malorny, B., Löfström, C., Wagner, M., Krämer, N., & Hoorfar, J. (2008). Enumeration of *Salmonella* bacteria in food and feed samples by real-time PCR for quantitative microbial risk assessment. *Applied Environmental Microbiology*, *74*, 1299–1304.

Manuja, A., Kumar, B., & Singh, R. K. (2012). Nanotechnology developments:opportunities for animal health and production. *Nanotechnology Development*. DOI: 10.4081/nd.2012.e4.

Mirkin, C. A., Letsinger, R. L., Mucic, R. C., & Storhoff, J. J. (1996). A DNA-based method for rationally assembling nanoparticles into macroscopic materials. *Nature*, *382*, 607–609.

Pandey, P., Singh, S. P., Arya, S. K., Sharma, A., Datta, M., & Malhotra, B. D. (2008). Gold nanoparticle–polyaniline composite films for glucose sensing. *Journal of Nanoscience and Nanotechnology*, *8*, 3158–3163.

Sato, K., Hosokawa, K., & Maeda, M. (2003). Rapid aggregation of gold nanoparticles induced by non-cross-linking DNA hybridization. *Journal of the American Chemical Society*, *125*, 8102–8103.

Sato, K., Hosokawa, K., & Maeda, M. (2007). Colorimetric biosensors based on DNA–nanoparticle conjugates. *Analytical Science*, *23*, 17–20.

Schofield, C. L., Field, R. A., & Russell, D. A. (2007). Glyconanoparticles for the colorimetric detection of cholera toxin. *Analytical Chemistry*, *79*, 1356–1361.

Singh, G., Vajpayee, P., Ram, S., & Shanker, R. (2010). Environmental reservoirs for enterotoxigenic *Escherichia coli* in south Asian gangetic riverine system. *Environmental Science and Technology*, *44*, 6475–6480.

Storhoff, J. J., Lucas, A. D., Garimella, V., Bao, Y. P., & Muller, U. R. (2004). Homogeneous detection of unamplified genomic DNA sequences based on colorimetric scatter of gold nanoparticle probes. *Nature Biotechnology*, *22*, 883–887.

Vikesland, P. J., & Wigginton, K. R. (2010). Nanomaterial enabled biosensors for pathogen monitoring – a review. *Environment and Science and Technology*, *15*, 3656–3669.

Wang, S., Singh, A. K., Senapati, D., Neely, A., Yu, H., & Ray, P. C. (2010). Rapid colorimetric identification and targeted photothermal lysis of *Salmonella* bacteria by using bioconjugated oval-shaped gold nanoparticles. *Chemistry – A European Journal*, *16*, 5600–5606.

Wang, X., Yuan, L., Wang, J., Wang, Q., Xu, L., Juan, D., Yan, S., Zhou, Y., Fu, Q., Wanga, Y., & Zhan, L. (2012). A broad-range method to detect genomic DNA of multiple pathogenic bacteria based on the aggregation strategy of gold nanorods. *Analyst, 137*, 4267.

Warrington, P. D. (2001). *Water Quality Criteria for Microbiological Indicators Prepared pursuant to Section 2(e) of the Environment Management Act, 1981*.

Zhang, D., Carr, D. J., & Alocilja, E. C. (2009). Fluorescent bio-barcode DNA assay for the detection of Salmonella enterica serovar Enteritidis. *Biosensors and Bioelectronics.2009, 24*, 1377–1381.

FURTHER READING

Fournier-Wirth, C., & Coste, J. (2010). Nanotechnologies for pathogen detection: Future alternatives? *Biologicals, 38*(1), 9–13.

Kuzma, J. (2010). Nanotechnology in animal production – Upstream assessment, of applications. *Livestock Science, 130*, 14–24.

Lu, Y., & Liu, J. (2006). Preparation of aptamer-linked gold nanoparticle purple aggregates for colorimetric sensing of analytes. *Nature Protocols, 1*, 246–252.

Mousumi, Debnath, Bisen, Prakash S., & Prasad, Godavarthi B. K. S. (2010). *Molecular Diagnostics: Promises and Possibilities*. USA: Springer.

Sekhon, B. S. (2012). Nanoprobes and their Applications in veterinary Medicine and Animal Health. *Research Journal of Nanoscience and Nanotechnology, 2*, 1–16.

GLOSSARY

DNA Detection Identification and detection of DNA molecules using molecular techniques.

Gold Nanoparticles Suspension of nanometer-sized particles of gold in a fluid having intense red color (less than 100 nm).

Nanotechnolgy The engineering of functional systems at the molecular scale.

Thiol-Modified DNA Single-stranded DNA with thiol modification (–SH group) at either the 5′ or 3′ terminal.

Surface Plasmon Resonance (SPR) The resonant, collective oscillation of valence electrons in a solid stimulated by incident light.

ABBREVIATIONS

APHA American Public Health Association
BIS Bureau of Indian Standards
BLAST Basic Local Alignment Search Tool
CO Cytochrome Oxidase
FC Fecal Coliforms
FIB Fecal Indicator Bacteria
FISH Fluorescence *In Situ* Hybridization
GNPs Gold Nanoparticles
MF Membrane Filter
MPN Most Probable Number
NSET Nanoparticle Surface-Energy-Transfer Ruler
PCR Polymerase Chain Reaction
QDs Quantum Dots
qPCR Quantitative Polymerase Chain Reaction
RAPD Random Amplification of Polymorphic DNA
RFLP Restriction Fragment Length Polymorphism
SERS Surface-Enhanced Raman Spectroscopy
SPR Surface Plasmon Resonances
TC Total Coliforms
TEM Transmission Electron Microscopy
TPS Two-Photon Scattering Method
USEPA United States Environmental Protection Agency
VBNC Viable But Non-Culturable
WHO World Health Organization

LONG ANSWER QUESTIONS

1. Define nanotechnology.
2. What is nanobiotechnology?
3. Discuss the following applications of nanotechnology.
   a. Medical.
   b. Nanotoxicology.
   c. Nanotechnology and environment.
4. Discuss in detail the principle of using gold nanoparticles for DNA detection.
5. Describe the applications of nanomaterials for pathogen detection.

SHORT ANSWER QUESTIONS

1. What is the significance of particle size in the nanodomain?
2. Briefly talk about metal nano-particle synthesis by the colloidal route.
3. What is the principle behind the color change in a hybridization solution?
4. Differentiate the use of cross-linking and non-cross-linking patterns of GNP probes in DNA detection.
5. How does electron microscopy confirm the detection of target DNA?

ANSWERS TO SHORT ANSWER QUESTIONS

1. Nanotechnology deals with the creation of functional materials, devices, and systems on a nanometer scale length (1–100 nm). Small particles exhibit a high surface-to-volume ratio compared to their bulk counterparts, and become more reactive due to the presence of large numbers of atoms on their surface. Quantum confinement effects at a nanoscale level impart exceptional physico-chemical properties to nanoparticles, which make them promising for various technological applications. In biology, these small-size particles find immense potential for improving our understanding of cell functioning, as most biological activity happens at the nanoscale.
2. Metal nanoparticles are synthesized by reduction of metal cations using appropriate reducing agents in

aqueous and/or non-aqueous solvents. A stabilizer is also used during the synthesis to prevent aggregation of nanoparticles. The synthesis can be generalized with the following equation: $M^{+} + ne^{-} + \text{stabilizer} \rightarrow M^{\circ}$

3. After DNA hybridization, nanoparticles come closer to each other and act like an aggregate. This aggregate of nanoparticles causes a shift in light scattering that leads to a change in color.
4. The method based on GNP cross-linking involves attachment of non-complementary DNA oligonucleotides capped with thiol groups to the surfaces of two batches of GNPs. A polymer network is formed when DNA, complementary to the two grafted oligonucleotides, is added to the solution. This condensed network self-assembles the conjugated GNPs into aggregates with a concomitant change of color from red to purple. This technique is most suitable for tracking small synthetic target sequences of up to 50 bp. Another GNP aggregation system induced by non-cross-linking DNA hybridization involves immobilization of single-stranded DNA on gold nanoparticles above the physiological temperature. The nanoparticles conjugated with oligonucleotide probes aggregate together at considerably higher salt concentrations when the target DNA is perfectly complementary to the probe. This can help in tracking only the short synthetic oligonucleotides (up to 20–30 bp) targets, which, again, do not represent the sequences of a pathogenic DNA or PCR product.
5. In transmission electron microscopy (TEM), a beam of electrons is transmitted through an ultra-thin specimen, interacting with the specimen as it passes through. TEMs are capable of imaging at a significantly higher resolution, so if there is hybridization of GNP probes with target DNA, the GNP probes come in close proximity, and as a result, the particles aggregate. This aggregation can be visualized by TEM imaging.

Chapter 29

Biotechnological Exploitation of Marine Animals

Surajit Das
Department of Life Science, National Institute of Technology, Rourkela, Odisha, India

Chapter Outline

SUMMARY

The marine environment is the most widespread, untapped, least explored reservoir of flora and fauna. This chapter covers promising applications of marine biotechnology for the improvement of human welfare. Included is an exploration of marine diversity, and sustainable applications for food, pharmaceuticals, and ornamental purposes. Other aspects related to the genomics of marine life are also presented.

WHAT YOU CAN EXPECT TO KNOW

The marine world represents a largely untapped reservoir for biotechnological applications, as the least-explored, diverse microorganisms and other floral and faunal diversity reside in it. Marine organisms are of enormous scientific interest as they possess wide diversity, unique structures, metabolic pathways, reproductive systems, and sensory and defense mechanisms (due to their adaptation to the extreme environmental conditions

Animal Biotechnology. http://dx.doi.org/10.1016/B978-0-12-416002-6.00029-8

unparalleled by their terrestrial counterparts). In recent years, marine biotechnology has received huge attention due to its vast array of applications in the food industry, pharmaceuticals, the ornamental industry, and other fields. However, knowledge in marine genomics needs to be broadened for better output from the above-mentioned industries.

HISTORY AND METHODS

INTRODUCTION

The ocean is a vast realm that contains many strange and wonderful creatures. It is often considered as a source of a beauty, mystery, and variety of life, and attracts people from around the world. Marine life represents a vast source of human wealth. It provides food, medicines, and raw materials, in addition to offering recreation to millions and supporting tourism. In economic terms, it has been estimated that the ocean's living systems are worth more than $20 trillion a year.

Marine biotechnology is defined as the application of scientific and engineering principles to the processing of materials by marine biological agents to provide goods and services. Marine biotechnology explores the oceans to develop novel pharmaceutical drugs, chemical products, enzymes, and other industrial products and processes. It also plays a vital role in the advancement of biomaterials, health care diagnostics, aquaculture and seafood safety, bioremediation and biofouling (Thakur and Thakur, 2006).

MARINE BIORESOURCES AND BIOTECHNOLOGY

Historical Background

People probably started learning about marine life from the first time they saw the ocean. After all, the sea is full of good things to eat. Archaeologists have found piles of shells, the remains of ancient "clambakes," dating back to the Stone Age. While they gathered food, people learned through experience which things were good to eat and which were bad. The tomb of an Egyptian pharaoh, for example, bears a warning against eating a pufferfish, a type of poisonous fish. Coastal peoples in virtually every culture developed a store of practical knowledge about marine life and the oceans (Castro and Huber, 2003). Knowledge of the ocean and its organisms developed as people gained skills in navigating seas around the globe. However, according to Hindu mythology and Shrimad Bhagavata Purana, understanding and utilizing ocean resources was done through "Samudra Manthan" by gods and demons to yield nectar (*Amrit*) by churning the sea.

TABLE 29.1 Average Depths and Total Areas of Major Ocean Basins

	Area		Average Depth	
Ocean	Millions of km^2	Millions of $mile^2$	Meters	Feet
Pacific	166.2	64.2	4,188	13,741
Atlantic	86.5	33.4	3,736	12,258
Indian	73.4	28.3	3,872	12,704
Arctic	9.5	3.7	1,330	4,360

(Castro and Huber, 2003)

The oceans cover 71% of the Earth's surface. They are not distributed equally with respect to the equator; about two-thirds of the Earth's land area is found in the Northern Hemisphere, which is only 61% ocean, while in the Southern Hemisphere, the ocean is about 80%. The oceans are traditionally classified into four large basins (Table 29.1). The Pacific is the deepest and largest ocean, almost as large as all the others combined. The Atlantic Ocean is a little larger than the Indian Ocean, but the two are similar in average depth. The Arctic is the smallest and shallowest ocean. Connected or marginal to the main ocean basins are various shallow seas, such as the Mediterranean Sea, the Gulf of Mexico, the South China Sea, the Bay of Bengal and the Arabian Sea. Main animals found in these oceans, seas and bays have shown wide diversity in terms of biology and chemical ecology. Some of the most important characteristics of the major animal phyla of marine origin have been listed in Table 29.2.

Biotechnologically Important Marine Animals

Sponges: Sponges are animals with complex aggregations of specialized cells. These cells are largely independent on each other and do not form true tissues or organs. Sponges mostly are marine and they are structurally the simplest multicellular animals. All are sessile, living attached to the bottom or any substratum. They show an amazing variety of shapes, sizes, and colors, but with a simple body plan. Numerous tiny pores, or ostia, on the surface of the body allow water to enter and circulate through a series of canals where plankton and organic particles are filtered out and eaten. This network of canals and a relatively flexible skeletal framework give most sponges a characteristic spongy texture. Because of their unique body plan, sponges are classified under the phylum Porifera or "pore bearers."

Jellyfish: Jellyfish (also known as jellies or sea jellies) are free-swimming members of the phylum Cnidaria. Jellyfish are not actually fish, the word jellyfish is used to denote

TABLE 29.2 Important Characteristics of the Major Animal Phyla of Marine Origin

Phylum	Representative Groups	Distinguishing Features	General Habitat
Porifera (sponges)	Sponges	Collar cells (choanocytes)	Benthic
Cnidaria (cnidarians)	Jellyfishes, sea anemones, corals	Nematocysts	Benthic, pelagic
Ctenophora (comb jellies)	Comb jellies	Ciliary combs, colloblasts	Mostly pelagic
Platyhelminthes (flatworms)	Turbellarians, flukes, tapeworms	Flattened body	Mostly benthic, many parasitic
Nemertea (ribbon worms)	Ribbon worms	Long proboscis	Mostly benthic
Nematoda (nematodes)	Nematodes, roundworms	Body round in cross section	Mostly benthic, many parasitic
Annelida (segmented worms)	Polychaetes, oligochaetes, leeches	Segmentation	Mostly benthic
Sipuncula (peanut worms)	Peanut worms	Retractable, long proboscis	Benthic
Echiura (echiurans)	Echiurans	Non-retractable proboscis	Benthic
Pogonophora (beard worms)	Beard worms, vestimentiferans	No mouth or digestive system	Benthic
Mollusca (mollusks)	Snails, clams, oysters, octopuses	Foot, mantle, radula (absent in some groups)	Benthic, pelagic
Arthropoda (arthropods)	Crustaceans (crabs, shrimps), insects	Exoskeleton, jointed legs	Benthic, pelagic, some parasitic
Ectoprocta (bryozoans)	Bryozoans	Lophophore, lace-like colonies	Benthic
Phoronida (phoronids)	Phoronids	Lophophore, worm-like body	Benthic
Brachiopoda (lamp shells)	Lamp shells	Lophophore, clam-like shells	Benthic
Chaetognatha (arrow worms)	Arrow worms	Transparent body with fins	Mostly pelagic
Echinodermata (echinoderms)	Sea stars, brittle stars, sea urchins, sea cucumbers	Tube feet, five-way radial symmetry, water vascular system	Mostly benthic
Hemichordata (hemichordates)	Acorn worms	Dorsal, hollow (and ventral) nerve cords, gill slits	Benthic
Chordata (chordates)	Tunicates, vertebrates, (fishes, etc.)	Dorsal, hollow nerve cord, gill slits,	Benthic, pelagic

several different kinds of cnidarians, all of which have a basic body structure that resembles an umbrella, including scyphozoans, staurozoans (stalked jellyfish), hydrozoans, and cubozoans (box jellyfish). Jellyfish are "bloomy" by nature of their life cycles. Blooms are produced by their benthic polyps usually in the spring when sunlight and plankton increase. Most jellyfish have a second part of their life cycle, the polyp phase. Single polyps, arising from a single fertilized egg, develop into a multiple-polyp cluster and are connected to each other by strands of tissue called stolons.

Gastropods: The gastropods (class Gastropoda) are the largest and most varied group of mollusks. Snails are the most familiar gastropods, but the group includes other forms, such as limpets, abalones and nudibranchs. There are perhaps 75,000 species under garstropods, mostly marine. A typical gastropod (the term means "stomach footed") can best be described as a coiled mass of vital organs enclosed by a dorsal calcareous shell. The shell rests on a ventral creeping foot and is usually coiled. Most gastropods use their radula to scrape algae from rocks, as found in periwinkles (*Littorina*), limpets (*Fissurella, Lottia*), and abalones (*Haliotis*). Some, like mud snails (*Hydrobia*) are deposit feeders on soft bottoms. Whelks (*Nucella, Buccinum*), oyster drills (*Murex, Urosalpinx*), and cone shells (*Conus*) are carnivores.

Bivalves: Bivalves (class Bivalvia) are the clams, mussels, oysters, and similar mollusks. Bivalves have a modified molluskan body plan. The body is laterally compressed and enclosed in a two-valved shell. The gills, expanded and folded, are used not only to obtain oxygen, but also to filter and sort small food particles from the water. The inner surface of the shell is lined by the mantle, so that the whole body lies in the mantle cavity, a large space between the two halves of the mantle. Strong muscles are used to close the valves. Clams (*Macoma, Mercenaria*) use their shovel-shaped foot to burrow in sand or mud. Mussels (*Mytilus*) secrete strong byssal threads that attach them to rocks and other surfaces. Oysters (*Crassostrea*) cement their left shell to a hard surface, often the shell of another oyster. They have been swallowed by lovers of good food for thousands of years. Pearl oysters (*Pinctada*) are the source of most commercially valuable pearls. Pearls are formed when the oyster secretes shiny layers of calcium carbonate to coat irritating particles or parasites lodged between the mantle and the iridescent inner surface of the shell, which is called mother-of-pearl. Cultured pearls are obtained by carefully inserting a tiny bit of shell or plastic in the mantle. Some scallops (*Pecten*) live unattached and can swim for short distances by rapidly ejecting water from the mantle cavity and clapping the valves. The largest bivalve is the giant clam (*Tridacna*), which grows to more than 1 m (3 ft) in length.

Cephalopods: Cephalopods (class Cephalopoda) are voracious predators with specialized locomotary organs. This group includes octopuses, squids, cuttlefishes, and other fascinating creatures. All 650 living species of cephalopods are marine-based. A cephalopod (the name means "headfooted") is like a gastropod with its head pushed down toward the foot. Octopuses (*Octopus*) have eight long arms and lack a shell. They are common bottom dwellers. Including arms, the size varies from 5 cm (2 in) in the dwarf octopus *(Octopus joubini)* to a record of 9 m (30 ft) in the Pacific giant octopus.

Squids (*Loligo*) are better adapted for swimming than are octopuses. The body is elongated and covered by the mantle, which also forms two triangular fins. Squids can remain motionless in one place or move backward or forward just by changing the direction of the siphon. Eight arms and two tentacles, all with suckers, circle the mouth. The tentacles are long and retractable and have suckers only at the broadened tips. They can be swiftly shot out to catch prey. The shell is reduced to a chitinous pen that is embedded in the upper surface of the mantle. Adult size varies from tiny individuals of a few centimeters in length to 20 m (66 ft) in the giant squid (*Architeuthis*).

Cuttlefishes (*Sepia*) resemble squids in having eight arms and two tentacles, but the body is flattened and has a fin running along the sides. Cuttlefishes, which are not fish at all, have a calcified internal shell "cuttlebone" that aids in buoyancy.

Arthropods: Arthropods (phylum Arthropoda) comprise the largest phylum of animals, with more than a million known species. Of all the animals on earth, three out of four are arthropods. They have invaded all types of environments on the earth's surface, including the oceans. Marine arthropods encompass a huge variety of animals such as barnacles, shrimps, lobsters, and crabs, to cite a few. The arthropod body is segmented and bilaterally symmetrical with jointed appendages, such as legs and mouthparts.

The overwhelming majority of marine arthropods are crustaceans (subphylum Crustacea), a large and extremely diverse group that includes shrimps, crabs, lobsters, and many less familiar animals. With around 10,000 species, the decapods (the term means "ten legs") are the largest group of crustaceans. They include the shrimps, lobsters, and crabs. Decapods feature five pairs of walking legs, the first of which is heavier, and usually has claws used for feeding and defense. The carapace is well developed and encloses the part of the body known as the cephalothorax. The rest of the body is called the abdomen. Shrimps and lobsters tend to have laterally compressed bodies with distinct and elongated abdomens, the "tails" we like to eat so much. Shrimps are usually scavengers and feed on bits of dead organic matter on the bottom. Lobsters (*Homarus*) and the clawless spiny lobster (*Panulirus*) are mostly nocturnal and hide during the day in rock or coral crevices. Their feeding habits are almost like those of shrimps, however they are also known to catch live prey.

Other marine Arthropods include Horseshoe Crabs. The horseshoe crabs are the only surviving members of a group (class Merostomata) that is widely represented in fossil records. The five living species of horseshoe crabs (*Limulus*) are not true crabs but "living fossils."

Echinoderms: Sea stars (class Asteroidea, e.g. *Asterias*), brittle stars (class Ophiuroidea, e.g. *Ophiothrix spiculata*), sea urchins (class Echinoidea, e.g. *Echinometra*), sea cucumbers (class Holothuroidea, e.g. *Thelenota rubralineata*), and several other forms make up the echinoderms (phylum Echinodermata). The echinoderms display many traits that are unique among invertebrates. Adult echinoderms are radically symmetrical, like cnidarians. Echinoderms typically have a complete digestive tract, a well-developed coelom and an internal endoskeleton. It is secreted within the tissues, rather than externally like the exoskeleton of arthropods. Though sometimes it looks external, as in the spines of sea urchins, the endoskeleton is covered by a thin layer of ciliated tissue. Spines and pointed bumps give many echinoderms a spiny appearance, and hence the name Echinodermata, meaning "spiny-skinned." A network of water-filled canals make up a water vascular system that is unique to echinoderms.

Chordates: The phylum Chordata is divided into three major subphyla. All chordates possess at least during part of their lives a dorsal nerve cord, gill slits, a notochord, and a post-anal tail. Fish were the first vertebrates appeared more than 500 million years ago. The first fishes probably evolved from an invertebrate chordate not much different from the lancelets or the tadpole larvae of sea squirts that still inhabit the oceans. Fishes feed on nearly all types of marine organisms. They are the most economically important marine organisms as they are a vital source of protein for millions of people.

GENETIC ENGINEERING AND ITS APPLICATION IN AQUACULTURE

The increase in the production of aquatic organisms through the use of biotechnology over the last two decades indicates that in a few generations biotechnology may overtake conventional techniques, at least for the commercially more valuable species. In the last few years, genetics has contributed greatly to fish culture through the application of the more recent techniques developed in biotechnology and in genetic engineering. At present, the most commonly used methods in fish biotechnology are chromosome manipulation and hormonal treatments, which can be used to produce triploid, tetraploid, haploid, gynogenetic and androgenetic fish. These result in the production of individuals and lineages of sterile, monosex, or highly endogamic fish. The use of such strategies in fish culture has a practical objective the control of precocious sexual maturation in certain species; other uses are the production of larger specimens by control of the reproductive process, and the attainment of monosex lines containing only those individuals of greater commercial value. The use of new technologies, such as those involved in gene transfer in many species, can result in modified individuals of great interest to aquaculturists and play important roles in specific programs of fish production in the near future.

Transgenic Fish

The role of aquaculture in increasing fish production is well recognized and in recent years an appreciable progress has been made in the fields of selection, inbreeding, hybridization, and sex control. One important difference between fish and terrestrial animals for cultivation and genetic improvement is that, usually, fish have higher levels of genetic variation, and hence more scope for selection, than most mammals or birds (Foresti, 2000). The approach to genetic improvement of aquatic organisms that has emerged as a discipline in its own right in recent years is transgenesis, the transfer of new genes into hosts. Transgenic fish (or mollusks or crustaceans) can be defined as possessing within their chromosomal DNA, either directly or through inheritance, genetic constructs that have artificial origins.

Transgenic fish technology has great potential in the aquaculture industry. By introducing desirable genetic traits into fishes, mollusks, and crustaceans, superior transgenic strains can be produced for aquaculture. The development of transgenic fish has undergone intensive research. Transgenic fish are being developed for both academic and applied goals, allowing the production of useful model systems as well as new genetic strains with improved characteristics for aquaculture. A variety of genes have now been introduced into fish with the goal of influencing traits such as growth, improved food conversion, maturation, freezing tolerance, flesh quality, tolerance to low oxygen concentration, and disease resistance.

A foreign gene can be transferred into fish *in vivo* by introducing DNA either into embryos or directly into somatic tissues of adults. Direct delivery of DNA into fish tissues is a simple approach providing fast results and eliminating the need for screening transgenic individuals and selecting germline carriers.

Methods of Gene Transfer

Fish as models for transgenic technology are widely used, for having the following features that render fish more suited than mouse for gene transfer experiments in vertebrates:

- The fecundity of fish is much higher, increasing the number of eggs available for microinjection.
- Fertilization and incubation of fish eggs is external.
- Fish eggs are relatively large, facilitating handling.

Basic steps employed in transgenic fish production:

1 *Acquiring the gene:* The procurement of a suitable gene sequence along with regulatory flanking sequences from a gene library.
2. *Cloning of the gene:* The desired gene sequence is cloned into a plasmid or phage vector and grown up in suitable bacteria. Harvesting of the gene from the bacterial cell follows this. Cloning facilitates the availability of several million copies of the gene sequences.
3. *Gene transfer:* The methods for gene transfer include electroportation, microinjection, sperm-mediated gene transfer, microprojectile and liposome-mediated gene transfer. Among these, there are two main techniques that researchers use to transfer genetic material in fish. One is called *microinjection*, in which the genetic material is injected into newly fertilized fish eggs. However, this method is time consuming, so researchers prefer to use *electroporation.* This involves transferring the genetic material, or DNA, into fish embryos through the use of an electrical current. Electroporation is an effective, simple and rapid method that uses electrical pulses to permeabilize the cell membrane, thus allowing the entry of macromolecules into the cell. Essentially, the method is to place the fertilized fish eggs or embryos in a solution containing the DNA of interest in between two electrodes and to pulse them with electricity set at certain field strengths.

Chromosome Manipulation

Fish species are generally very tolerant of artificial manipulation of their chromosomes during early development and this property has been exploited for the production of inbred lines, monosex populations, and the control of ploidy. A variety of different techniques have been applied to obtain polyploids, gynogenetics and androgenetics, sex-reversed individuals and transgenics. Methods used include interspecific hybridization or the control of sex by the administration of sex steroids to larvae or juveniles, and even surgical or autoimmune castration. The artificial modification of the chromosome set of an organism permits the production of monosex and sterile individuals, while gynogenesis and androgenesis provide methods for the rapid production of inbred populations which can be used in cross breeding programs.

Polyploidy

Manipulation of chromosome numbers to give polyploidy individuals is common in aquaculture. The retention of the second polar body during the second meiotic division in the oocyte results in triploid individuals, which have two chromosome sets from the mother and one from the father. Unlike mammals, where chromosomal rearrangements of this magnitude are usually fatal, many fish with three sets of chromosomes survive quite readily. In addition to triploids, individuals with four sets of chromosomes (tetraploids) or those with both of their chromosome sets derived from a single parent can be produced. Triploid fish are of interest because their sterility is useful in aquaculture and fisheries' management. Sterility caused by genetic and physiological factors leads to different characteristics in male and female sexual development.

Male triploids undergo considerable changes in secondary sex characteristics and gonad development at the time of maturation, affecting the carcass appearance and reducing meat quality. Female triploids, on the other hand, normally have minimal gonad development and maintain carcass quality throughout the period of maturation of their diploid counterparts. The performance of triploid fish in production situations has been found to be comparable to that of diploid individuals, and the use of triploid females should be considered when they are to be grown past the time of normal sexual maturation.

Polyploidy of penaeid shrimps is still in infancy. However, polyploidy is currently the only known technique that can achieve the dual outcomes of reproductive sterility for genetic protection and skewing of sex ratios towards a high proportion of females, which are larger than males in all of the penaeid shrimp (Sellars et al., 2010). Three main categories of polyploidy have been studied in penaeid shrimp: meiosis I triploidy, meiosis II triploidy and mitotic tetraploidy (Figure 29.1). Meiosis I triploidy has been

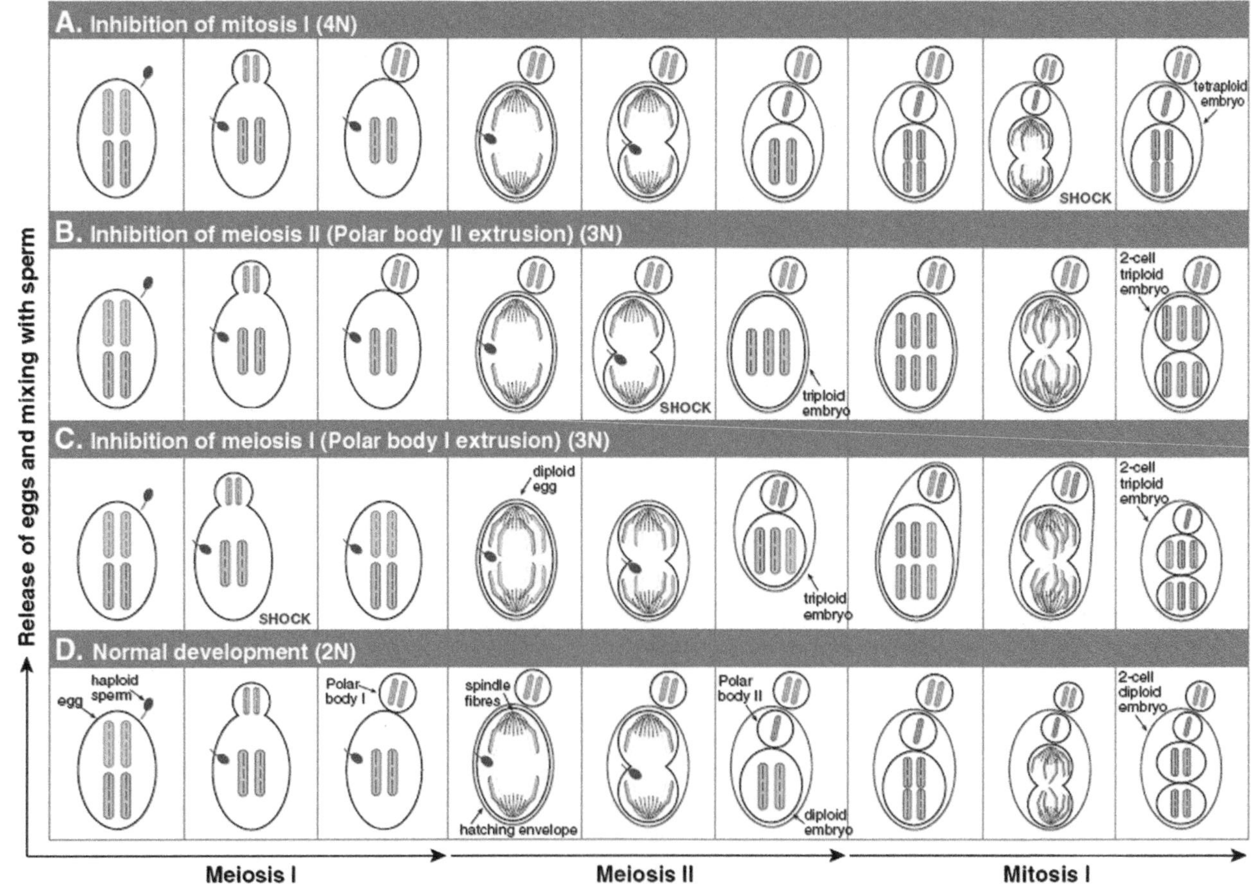

FIGURE 29.1 Process of inducing (A) Meiosis I triploidy, (B) Meiosis II triploidy and (C) Mitotic tetraploidy, and (D) Normal development in penaeid shrimp. *(Source: Sellars et al., 2010)*

studied in ***Fenneropenaeus chinensis*** and *Metapenaeus japonicus*. Meiosis II triploid larvae have been reported in *F. chinensis*, *Penaeus monodon*, *M. japonicus*, *P. indicus*, and *Litopenaeus vannamei* (see review Sellars et al., 2010). However, of these species rearing from egg to adult has only been reported (and thus successful) for *F. chinensis* and *M. japonicus*.

Gynogenesis

Gynogenesis is the production of offspring with genes from the mother only. In practice, in fish and some other organisms, it is possible to produce offspring from a mature female with no paternal genetic contribution. The technique for producing gynogenetic individuals requires the inactivation of the male genome and the diploidization of the female genetic material in the zygote, in a process induced by physical or chemical agents. Gynogenesis can be achieved easily in rainbow trout and other species of fish by fertilizing eggs with irradiated sperm and inducing polar body retention through the same type of treatments (pressure or temperature shocks) used to produce triploids. In the same way, if the eggs are irradiated and then fertilized by normal spermatozoa, androgenetic individuals can be produced under certain conditions, with only a set of parental chromosomes. These haploids are then changed to diploids by heat shock treatment before the first cell division (Figure 29.2).

Gynogenetic individuals are used to produce all-female fish populations. The reasons for inducing gynogenesis range from the production of monosex populations to the development of partially or completely inbred organisms. Partially inbred offspring (from meiotic gynogenesis) may be useful in genome studies for examining the map position of different loci in relation to the centromeres of their chromosomes. If completely inbred animals (from mitotic gynogenesis) survive to maturity, their eggs can be subjected to a second round of gynogenesis, producing true clones in large numbers. Gynogenesis is the development of embryos from eggs without genetic contribution from penetrating sperm. There are two approaches in inducing gynogenesis. They are:

1. ***Meiotic gynogenesis:*** Ovulated eggs are exposed to the biologically active but genetically "blank" milt. As a sperm cell activates an egg, intracellular mechanisms are started to eject the second polar body. At this critical point, a physiological shock such as heat, cold, or pressure should be applied to stop the loss of the second polar body. This allows the egg to proceed with normal development by utilizing the two sets of chromosomes, one set from oocyte and the other set from the second polar body. This process is called meiotic gynogenesis. Therefore, the offspring are the products of maternal inheritance. In order to preserve an endangered Chinese paddlefish *Psephurus gladius*, meiotic gynogenesis has been employed by using ultra-violet irradiated *Acipenser schrenckii* sperm (Zou et al., 2011). Meiotic gynogenesis has also been reported in European sea bass, *Dicentrarchus labrax* to obtain triploidy (Peruzzi and Chatain, 2000). Meiotic gynogenesis has short-term disadvantages but have long-term advantages. Because females produce eggs without undergoing meiosis, the eggs become clones. This resembles parthenogenesis, which is much more beneficial for an organism.
2. ***Mitotic gynogenesis:*** This is the second method of gynogenesis. In this method, the irradiated sperm is used to activate the egg cell, but the polar bodies are

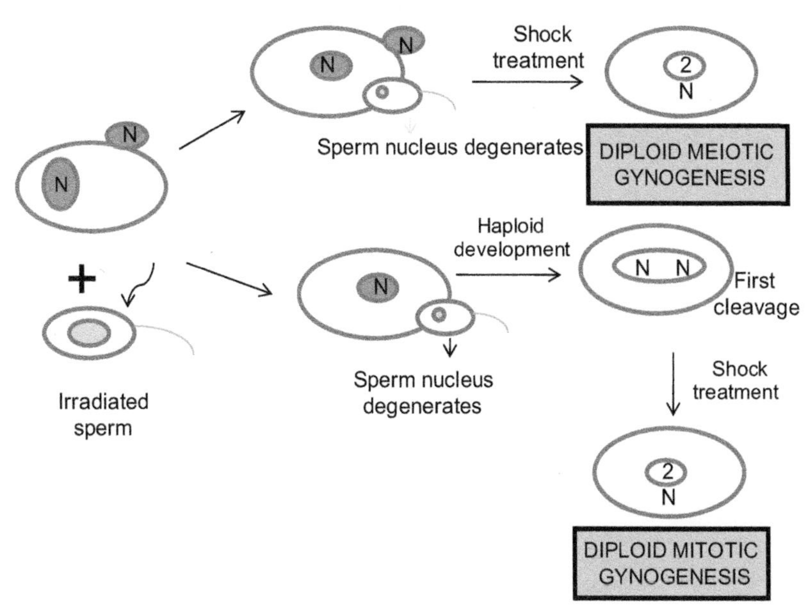

FIGURE 29.2 Gynogenesis induction method: Diploid meiotic and mitotic gynogenesis. *(Padhi and Mandal, 2000)*

allowed to expel out. This results in an activated egg, with a unique 1N set of chromosomes, provided by the maternal parent. When the haploid egg begins its development through the usual process of mitosis, the normal mitotic process is interrupted by physiological treatment (through application of heat, cold, or pressure) and transformed into viable 2N organism. In case of ornamental fishes like *Cyprinus carpio*, to obtain identical color patterns of commercial interest, mitotic gynogenesis has been employed using microsatellite DNA as markers (Alsaqufi et al., 2012). Mitotic gynogenesis is of upmost importance as it results in fully homozygous offspring (F = 1) as it is achieved by inhibition of first meiotic cleavage after duplication of the genome. By the use of this reproduction method, genetically identical fish (clonal lines) are obtained after two generations. However, mitotic gynogenesis is more difficult to induce experimentally, and there is a greater chance of causing inbreeding depression.

Androgenesis

Androgenesis is the development of embryos without genetic contribution from oocytes. The mechanism associated with androgenesis is the same as that of mitotic gynogenesis. The only difference in practice is the destruction of genetic material of the mature oocyte and not the sperm. Thus, the offspring are the products of potential inheritance alone (Figure 29.3). There are many androgenic clones that occur in nature, and they can also be generated artificially. There are many reports of the androgenesis procedure via heterospecific insemination in fishes, including *Cyprinus carpio*, *C. auratus*, *C. idella*, *Puntius conchonius*, *Pangasius schwanenfeldii*. In this regard, artificial androgenesis has been limited to commercially important food and ornamental fishes. *C. carpio* is the universal sperm donor and *O. mykiss* is the universal recipient. The major advantage of this technique is the tracing of density and distribution of parental genome from an early embryonic stage, as well as confirmation of parental origin of haploid androgenotes that scrumble at the embryonic stage. In the case of a mammal model, both the live donor and the recipient are required, however, in the case of the fish model, only a live recipient is required and the donor can be from the sperm of either a live or post-mortem-preserved donor (Pandian and Kirankumar, 2003).

Sex Reversal

Sex reversal in fish is one of the most successful applications of genetics to aquaculture. Sex manipulation in fish is desired either to overcome the problem of overpolulation or to have a population of the desired sex with better traits. For many cultured species, the production of one sex of individuals is more profitable than the other.

The sex of the fish can be reversed by administration of one or the other androgen or estrogen through diet or water. Generally, androgens induce masculinization and estrogens feminization. Treatment at the labile phase in an appropriate dose is important for successful sex reversal. The most commonly practiced methods are:

1. ***Dietary Supplementation:*** The hormone mixed feed is the most widely used method.
2. ***Immersion Technique:*** Juvenile fishes are immersed in hormone-treated water. This method is applicable in those species where the labile phase starts soon after hatching.
3. ***Injection:*** Injection requires a lower quantity of hormone to cause sex reversal. However, it is very labor intensive.

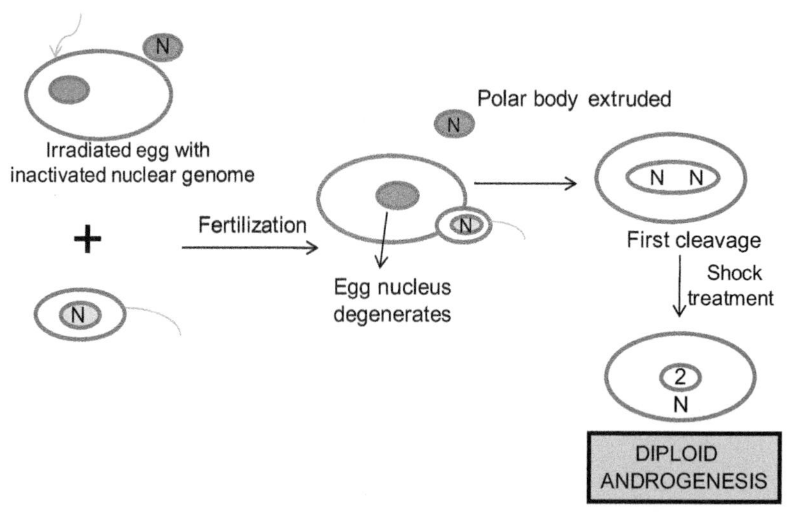

FIGURE 29.3 Androgenesis induction method. *(Padhi and Mandal, 2000)*

Cryopreservation of Gametes

Cryopreservation is a branch of cryobiology that relates to the long-term preservation and storage of biological material at very low temperature, usually −196°C, the temperature of liquid nitrogen. At this temperature cellular viability of gametes can be stored in a genetically stable form in order to make available gametes both as and when required. This method, therefore, can be successfully used as a fishery management tool. Spermatozoa could be cryopreserved more easily for their structural simplicity and small size. Cryopreservation involves three basic steps (freezing, storage, and thawing), which require extenders and cryoprotectants.

1. ***Extenders:*** Undiluted gametes are not suitable for freezing, so they must be diluted with a suitable extender. An extender is a solution consisting of inorganic and organic chemicals resembling that of blood or seminal plasma in which the viability of spermatozoa can be maintained during *in vitro* storage. The chemical formulations of the extenders used for cryopreserving spermatozoa vary widely. In general, simpler extenders, some containing only NaCl, $NaHCO_3$, and lecithin have been shown to be successful.
2. ***Cryoprotectants:*** Cryoprotectants are added to extenders to minimize the stress on cells during freezing, e.g. DMSO, glycerol, methanol, etc. (permeating cryoprotectants) and egg yolk, milk, and some proteins (non-permeating cryoprotectants). A non-permeating cryoprotectant is often used in conjunction with a permeating cryoprotectant.

MARINE ORNAMENTAL FISH TRADE AND CHROMOSOMAL MANIPULATION

Marine ornamental fishery has been a great resource and candidate species gain high prices due to their vibrant color. The estimated value of the marine ornamental trade is 200–330 million USD per year. Unlike freshwater aquaria species (where 90 per cent of fish species are currently farmed), the great majority of marine aquaria are stocked from wild-caught species (Andrews, 1990). According to data held in Global Marine Aquarium Database, a total of 1,471 species of marine ornamental fishes are traded globally (Table 29.3). Most of these species are associated with coral reefs, although a relatively high number of species are associated with other habitats, such as seagrass beds, mangroves, and mudflats. Generally the ornamental fishes are selected based on their body color (preferably attractive), body shape (unique shape compared to food fishes), and aquarium suitability. The ornamental coral reef fishery is a multi-million dollar industry that supports thousands of fisheries in developing countries and provides aquarium hobbyists with over 1,400 species of marine fishes. These days, enhanced production in marine ornamental fish culture is mainly obtained by long-term selective breeding to produce novel traits and varieties, and when species are difficult to obtain from the wild. This selective breeding supports conservation as it eliminates environmental damage and enhances the production of domesticated species (Tlusty, 2002).

The advanced molecular techniques like exploitation of genetic gains by manipulation at the zygote, gamete, chromosome, and gene levels, (such as chromosome set manipulation, intergeneric and interspecific hybridization, sperm cryopreservation, nuclear transfer and transgenesis) can also be applied to increase the yield and to overcome the age-old practices of selective breeding to obtain an efficient strain (Basrur and King, 2005). There are many advantages of practicing chromosomal manipulations in marine ornamental fish, as intergeneric hybrids grow faster than the parental species.

The chromosomal manipulation in marine ornamental fish culture can be used for two different purposes, either to induce polyploidy (especially to obtain triploid or tetraploid fish) or to reproduce fish by uniparental chromosome inheritance to obtain gynogenetic and androgenetic fish. The techniques of chromosome set manipulation can be applied to complement other techniques like interspecific hybridization, artificial speciation, genetic engineering, sex control, population control, selective breeding, pure line or clone cross-breeding, and sperm cryopreservation.

TABLE 29.3 Dominant Fishes Targeted for the Global Marine Aquarium Trade

Fish Type	Family	Volume (%)	Value (%)
Damsel/ Anemonefish	Pomacentridae	29	13.0
Angelfish	Pomacanthidae	24	46.0
Butterflyfish	Chaetodontidae	11	10.0
Wrasse	Labridae	7	12.0
Blennies/ Gobies	Blennidae/ Gobidae	5	3.0
Triggerfish/ Filefish	Balistidae/ Monacanthidae	4	2.5
Hawksfish	Cirrhitidae	2	3.0
Groupers/ Basselets	Serranidae	2	1.5
Other	33 families	15	8.0

(Bruckner, 2005)

ADVANCES IN MARICULTURE

Mariculture is a specialized branch of aquaculture, which involves the cultivation of marine organisms for food and other products in the open ocean, ponds, or raceways that are filled with seawater. This technology has rapidly progressed in the last two decades, due to new technology, greater understanding of biology of farmed species, improvements in formulated feeds, increased water quality in closed farm systems, higher demand for seafood products, farm expansion, and government interest.

Captive Rearing Technology

Captive rearing technology is of utmost importance for marine fishes, as the larvae do not readily accept rotifers as a first feeding organism. In this case, size-sorted, wild plankton are provided to them when they reach first feeding. Hence, the fish larvae can be easily developed and can grow to adults in captive rearing. The successful breeding of marine fishes under captive conditions depends on the development of healthy brood stock, knowledge of modes of their reproduction, adequate lighting, water quality, acceptable first live food organisms, healthy environment for metamorphosis, and overall husbandry techniques, which predict the possibility of captive breeding and survivability of larvae (Rowe and Hutchings, 2003).

Captive rearing of marine oysters increases their production; they are commonly consumed by humans and can be used for pearl production. Captive breeding of oysters has various advantages. For example, the oyster changes sex twice during a single season, and young oysters remain inside the parent's shell for most of the larval period. Humans have used mussels as food for a long time, as they are the excellent sources of selenium and vitamin B_{12}. However, mussels are on the verge of extinction, and necessary care should be taken (like captive rearing) to increase productivity and conserve biodiversity, either by keeping the adults alive in captivity, breeding them in captivity (although this can be difficult), or by rearing the young in captivity. In captive rearing, necessary steps should be taken to (1) honor the innate rhythm associated with breeding-cycle timing (which relates to the temperature of the environment), and (2) ensure the survival of juvenile pearl mussels by maintaining an appropriate population density (and situation for the adults) (Menon and Pillai, 2001).

Feed Technology: Microencapsulated, Micro-Coated, and Bio-Encapsulated Feeds

In recent years, there has been considerable improvement in feeding technology in fisheries, which has increased the growth of shrimp culture practices. The availability of artificially produced diets to replace cultured live food organisms could alleviate many of the problems currently limiting shrimp hatchery production by (1) reducing the level of technical skill required to operate a hatchery, (2) assuring a reliable supply of a nutritionally balanced larval feed, (3) reducing sources of contamination and larval disease, and (4) simplifying hatchery design and capital cost requirements, thereby facilitating small-scale hatchery development.

Microencapsulated feeds have increased larval survival rates to 70%. The food grains are encapsulated along with the flavors, which depend on the physical and chemical forms of the food grains. Bio-encapsulation is an easy step that fits into preexisting on-site live food production for marine fishes. It includes the adhesion of biological materials like proteins and microorganisms to be used as probiotics to increase the yield of the feeds. *Bacillus licheniformis*, *B. subtilis* and *B. circulans* are the most common probiotic organisms used in encapsulation practices.

Application of Nanotechnology in Aquaculture

Nanotechnology has a tremendous potential to revolutionize agriculture and allied fields including aquaculture and fisheries. It can provide new tools for aquaculture, fish biotechnology, fish genetics, fish reproduction, and aquatic health, etc. Nanotechnology tools like nanomaterials, nanosensors, DNA nano-vaccines, gene delivery, and smart drug delivery have the potential to solve many puzzles related to animal health, production, reproduction, and prevention and treatment of diseases. It is sensible to presume that in the upcoming years, nanotechnology research will reform science and technology and will help boost livestock production. Nanotechnology applications in the fish processing industry can be utilized to detect bacteria in packaging, color quality, and safety by increasing barrier properties. The areas related to aquaculture and fisheries where nanotechnology can be applied are:

1. ***DNA Nano-Vaccines:*** Use of nanoparticle carriers like chitosan and poly-lactide-co-glycolide acid (PLGA) for vaccine antigens, together with mild inflammatory inducers, may give a high level of protection to fishes and shellfishes, not only against bacterial diseases, but also from certain viral diseases with vaccine-induced side effects. Further, the mass vaccination of fish can be done using nanocapsules containing nano-particles. These will be resistant to digestion and degradation. These nanocapsules contain short strand DNA which when applied to water containing fishes are absorbed into fish cells. An ultrasound mechanism is used to break the capsules, which in turn releases the DNA, thus eliciting an immune response in fish due to the vaccination. Similarly, oral administration of these vaccines and

site-specific release of the active agent for vaccination will reduce the cost and effort of disease management, the application of the drug, and vaccine delivery for sustainable aquaculture (Gudding et al., 1999).

2. ***Gene Delivery:*** The development of new carrier systems for gene delivery represents an enabling technology for treating many genetic disorders. However, a critical barrier to successful gene therapy is still the formulation of an efficient and safe delivery vehicle. Non-viral delivery systems have been increasingly proposed as alternatives to viral vectors owing to their safety, stability, and ability to be produced in large quantities. Some approaches employ DNA complexes containing lipid, protein, peptide, or polymeric carriers, as well as ligands capable of targeting DNA complexes to cell-surface receptors on the target cells and ligands for directing the intracellular traffic of DNA to the nucleus. Promising results were reported in the formation of complexes between chitosan and DNA. Although chitosan increases transformation efficiency, the addition of appropriate ligands to the DNA–chitosan complex seems to achieve a more efficient gene delivery via receptor-mediated endocytosis (Kumar et al., 2008). These results suggest that chitosan has comparable efficacy without the associated toxicity of other synthetic vectors and therefore, can be an effective gene-delivery vehicle *in vivo.*

MARINE GENOMICS

Understanding the properties and functions of the genome is a fundamental task in modern bioscience. Molecular biology has a major role in many aspects of marine biotechnology. The study of the genomics of different commercially important fish is related to fishery. Genome analysis of marine microorganisms facilitates the use of genes for cell factories and bioindicator strains as well as indication of new drug targets. The marine sponge is a primitive organism in the animal kingdom and the genome analysis of such a primitive organism is of special interest in molecular evolution. Efforts in this area have also proven that the application of molecular biological techniques in ecological studies will be helpful to explore molecular biodiversity, symbiosis, and defense mechanisms.

Genetic analysis of marine life is increasing our understanding of how organisms have evolved and the roles they play in ecosystems, helping scientists analyze the health of the oceans and discover potential pharmaceuticals from the sea. Genome biology applied to marine organisms is known as marine genomics. It is a combined study of how organisms work, evolve, and adapt at the genomic level, the understanding of which can be implicated in the fields of human health and disease, food, and ornamental practices. Marine genomics can be applied to study the differences within and among the marine populations using various molecular techniques like microarrays, high-throughput sequencing, genotyping, population genetics and evolutionary analysis. In this regard, the adaptive difference in gene expression patterns and molecular mechanisms to pollution and other toxicological stress play a role in the difference among the individual marine organisms (Whitehead et al., 2011). In another approach, measurement of the development, organ-specific metabolism, oxygen consumption, and enzyme function can be combined with measures of mRNA and protein expression using microarray or proteomics (Tomanek, 2011).

In order to understand the biotechnological potential of the marine organisms, assessment of their genetic capabilities, i.e. sequencing of their genome and annotation of the genes, is required. Sequencing of phylogenetically diverse microbial genomics results in discovery of novel proteins, and the trend of discovery is linear, which demonstrates the marine environment to be a reservoir of undiscovered proteins (Angly et al., 2006). Besides prokaryotes, the abundance of marine viruses exceeds that of prokaryotes by a factor of at least ten. Therefore, marine viruses are untapped genetic sources of truly marine character, which could provide novel proteins, genetic tools, and unexpected functions. The genomics of marine eukaryotes (which comprises microalgae, macroalgae, seaweeds and protozoa) is the least explored area of genomics. However, the study of metazoan genomics is highly biased towards vertebrates, especially mammals, due to their medical and economic relevance. Until now, only a few commercially relevant marine invertebrates, such as mussels and oysters, have been sequenced, because of their importance as aquaculture species (Cunningham et al., 2006).

The discovery of advanced techniques led to the development of genomics research of marine communities. After the successful implication of metagenomics in the year 2001, it has became technically possible, due to the availability of bacterial artificial chromosomes (BACs), which enable the cloning and sequencing of long stretches of environmental DNA (Liu et al., 2009). Metagenomics works like a shotgun by taking all the genes of a community and putting them into large clone libraries to make them available for use in biotechnological applications. Current metagenomic studies target all domains of life and a broad range of environments. Meta-transcriptomics and meta-proteomics have been successfully applied in addition to marine genomics to provide exciting insights into the functioning of microbial communities. However, these approaches lack broader applications, owing to their complexity, and are of limited value for biotechnological exploitation. A recent sequencing technology has been discovered which enables 10 to 100 times faster automated sequencing to nucleic acids. These new sequencing technologies provide a read length of 50–450 nucleotides, which can generate 20–200 Mb of raw sequence data per run. Hence, the application of more and more genomic and metagenomic analyses and deep sequencing will generate large datasets from marine environments (Chan, 2009). Bioinformatics resources and tools have been developed in an attempt to maximize the capacity to analyze these vast datasets. This so-called e-infrastructure has to support advanced data acquisition, data storage, data management, data integration, data mining, data visualization, and other computing- and information-processing services over the Internet. Therefore the provision of dedicated web-based resources and e-infrastructures is essential for advanced research in marine ecology and biotechnology. Another emerging field is systems biology, which aims at a system-level understanding of biological systems. In systems biology, organisms are studied as an integrated and interacting network of genes and these interactions determine the functions of an organism, which largely depends on the mathematical tools to understand gene function relationships (Brown and Botstein, 1999).

Knowledge of metabolic pathways and their link with genomics and other "omics" aspects of marine organisms are the important basis of production of unique compounds. In order to increase the productivity of marine organisms, the metabolic pathways need to be inserted into a new host organism, which can grow easily. Metabolic engineering should be used to optimize genetic and regulatory pathways to increase the cells' production of certain compounds; these have been discovered for the prokaryotes but still it needs to be developed for the eukaryote systems. If the right targets for metabolic engineering are properly chosen, better processes can be developed. The application of these engineered cells improves the prospects for commercial production of bioactive compounds for food and pharmaceutical industries, which reduces the cost of production and makes it more sustainable. Engineered organisms are expected to become more commonly used in the future, but biosafety and consumer acceptance aspects will need to be taken into account.

DISEASE DIAGNOSIS OF CULTIVABLE ANIMALS

As aquaculture has undergone rapid expansion and intensification, disease occurrence also increased exponentially. This has put tremendous pressure on research organizations and researchers to focus their attention to develop efficient diagnostic tools. Disease control and prevention in aquaculture is largely a function of management, and accurate and timely diagnoses of diseases play a significant role in disease control strategies. It is important to detect pathogens from infected fish (clinically and sub-clinically) and in their environment. Of the various modern and sophisticated tools used for disease diagnosis, the application of antibody- and DNA-based probes for pathogen detection and identification have promising potential on the development of accurate, rapid, and sensitive diagnostic tests.

Immunodiagnostics

Immunodiagnostics is the diagnostic procedure using antigen-antibody reactions as the primary means of detection. This technique is applicable for the detection of even a small amount of biochemical substances where antibodies specific for a desired antigen can be conjugated with a radiolabel, fluorescent label, or color-forming enzyme, which is used as a probe to detect it. There are many well-known applications of it like immunoblotting, ELISA, and immunohistochemical staining of microscopic slides. The major advantages of using this technique involve its speed, accuracy, and simplicity, which lead to rapid techniques for the diagnosis of diseases, microbes, and drugs *in vivo*. In this regard, marine animals provide a useful source for natural products for immunodiagnostic applications. Keyhole limpet hemocyanin (KLH), a large, multisubunit, oxygen-carrying metalloprotein is of upmost importance; it is found in the hemolymph of the giant keyhole limpet *Megathura crenulata*. KLH is a large heterogenous glycosylated protein, each subunit containing two copper atoms, which together bind with a single oxygen molecule. Though it is a potent immunogenic, it is still safe for human applications and is used as a vaccine carrier protein for the production of antibodies for research, biotechnology, and therapeutic application. There are many advantages of using KLH as an effective carrier protein, as its large size and numerous epitopes generate a substantial immune response, and the abundance of lysine residues coupling with haptens allow a high hapten:carrier protein ratio, thus increasing the likelihood of generating hapten specific antibodies. As KLH is a limpet derivative, it reduces false positive results in immunological applications in mammalian model organisms.

Enzyme Immunoassay

The interaction of an antibody with an antigen forms the basis of all immunochemical techniques. Immunoassays are both qualitative as well as quantitative assays. A labeled antibody/antigen is used to visualize the immune reaction.

Enzyme-Linked Immunosorbent Assay (ELISA)

The Enzyme-linked Immunosorbent Assay (ELISA) has become a widely used serological technique. There are two basic methods: the direct ELISA detects antigens, and the Indirect ELISA detects antibodies. A microtiter plate with numerous shallow wells is used in both procedures. There are three enzymes used for color development with the second antibody. The earliest was alkaline phosphatase;

the most common now is horseradish peroxidase. The third is beta-galactose (from *E. coli*), which, since it is active at higher pH, is preferred for antigen absorption.

The substrates generally employed for alkaline phosphatase and beta galactose are p-nitrophenylphosphate and *o*-nitrophenylbeta-D-galactopyranoside, respectively, with color development being detected at 405 and 420 nm. In the case of horseradish peroxidase, o-phenylenediamine (OPD) or tetramethyl benzedine (TMB) are most commonly used. More sensitive ELISA detection system may be obtained by the use of flurogenic substrates for alkaline phosphatase or beta-galactosidase. Most ELISA results are read in a spectrophotometer adapted for microtiter plates. This is an excellent method for obtaining printed results for storage and it removes the subjective element. Briefly, incubate 100 μL of standard or cell extract. Then add 100 μL of detection antibody and incubate for 1 h. Further incubate with 100 μL of HRP anti-rabbit antibody for 30 min. Add 100 μL of stabilized chromogen and incubate for 30 min. Then add 100 μL of stop solution and observe for development of color. Perform all the steps at room temperature. The flow chart of this experiment is given at Flow Chart 29.1.

Dot Immunobinding Assay

The dot immunobinding assay, in which the antigen is attached to nitrocellulose paper in a series of dots, is the system of choice for screening on a limited budget. It is claimed to be equally sensitive to or more sensitive than ELISA assays. The method is similar in principle to ELISA except it uses nitrocellulose paper. The original methods involved the application of dots of the antigen to nitrocellulose sheets, followed by cutting up of the sheets so that square pieces of paper containing the dot were put in the microtiter wells for incubation with the antibodies. A variation of this involved the inversion of the microtiter plates (containing the antibodies) over the sheets with the matrix of dotted sensitive antigen; a tight seal was made for the antigen-antibody incubation. The dot-ELISA is a very sensitive assay for detecting or quantifying antigen or antibody.

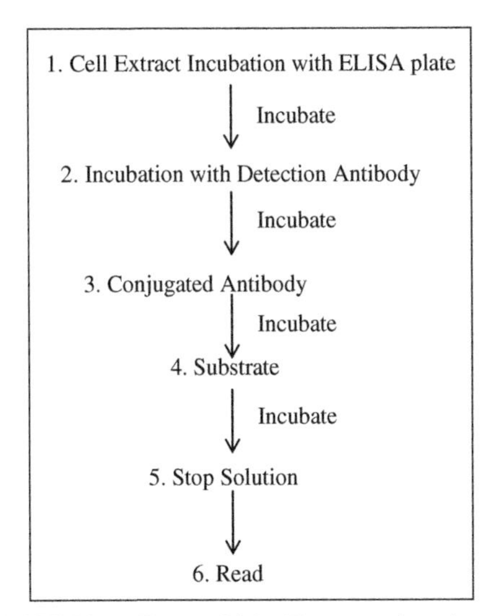

FLOW CHART 29.1 Enzyme-Linked Immunosorbent Assay (ELISA)

Western Blotting

Western blotting is a technique by which proteins can be transferred from polyacrylamide gel to a sheet of nitrocellulose in such a way that a faithful replica of the original gel pattern is obtained. A wide variety of analytical procedures can then be applied to the immobilize protein.

In this technique a sheet of nitrocellulose is placed against the surface of a SDS–PAGE protein fraction gel and a current applied across the gel (at right angles to its face); this causes the protein to move out of the gel and in to the nitrocellulose where they bind firmly by non-covalent forces. The technique has three steps: protein separation by SDS–PAGE, blotting, and immunoassay. Proteins immobilized on nitrocellulose sheets can be used to detect their respective antibodies. The sensitivity of the technique allows the detection of a specific antibody in a serum of low titer. Since antigen–antibody precipitation is not required, Fab (Fluorescent antibody) fragments and monoclonal antibodies produced by hybridomas can also be used for analysis. Thus, the technique can be employed for screening of hybridoma clones.

Latex Agglutination Test

Latex refers to the microscopic polymeric particles, which act as the base for various immunoassays and tests. Microsphere-based diagnostic tests (qualitative, yes/no results) and assays (quantitative results) are usually based upon the very specific interaction of antigen (Ag) and antibody (Ab). Sub-micron sized polystyrene (PS) microspheres are used for solid support. Ab or Ag can be absorbed to them. The "sensitized" micropsheres then act to magnify or amplify the reaction, which takes place when they are mixed with a sample containing the opposite reactant. In simple particle agglutination, a positive test results when a drop of uniformly dispersed, milky-appearing Ab-coated beads on a glass slide react with Ag in a drop of microsphere, resulting in a curdled milk-like appearance. Alternatively, Ag-coated particles are agglutinated by a positive sample of Ab.

Monoclonal Antibodies

Antibodies that are produced by hybridoma are known as monoclonal antibodies; they are highly specific and can be produced in unlimited quantities. The usefulness of monoclonal antibodies stems from three characteristics: their specificity of binding, their homogeneity, and their ability to be produced in unlimited quantities. The production of monoclonal antibodies allows the isolation of reagents with a unique chosen specificity. Because all of the antibodies produced by descendents of one hybridoma cell are identical, monoclonal antibodies are powerful reagents for testing for the presence of a desired epitope. Hybridoma cell lines also provide an unlimited supply of antibodies. Antibodies produced in mice or rats are most commonly used for diagnostic purposes, as they are more readily produced. The major advantage of monoclonal antibodies over conventional sera in this application is probably their ready availability for an indefinite period at a standard titer. Their high specificity has added greatly to the accuracy and speed of the diagnosis.

DNA-Based Diagnosis

The most recent development in diagnostics has utilized molecular biology to design a new generation of diagnostic tools, the PCR (Polymerase Chain Reaction) and Gene Probes. The key to DNA-based diagnostics is the generation of the target pathogen through recombinant DNA technology. This is done by purifying the infectious agent of interest and isolating its nucleic acid. The isolated DNA is then subjected to restriction digestion and cloning. From the selected clones, desired DNA fragment are sequenced. Once the adequate genetic information (sequence information) is generated, the information can be used in PCR. The DNA-based disease diagnosis in different animals is described below.

Shrimp

In the aquaculture of penaeid shrimps in Asia, white spot disease caused by white spot syndrome virus (WSSV) is the major cause of morbidity and mortality, resulting in huge economic losses. To date, no treatment is known to control this disease. Hence, early diagnosis followed by suitable management practices is the only alternative in tackling this disease. In the case of white spot virus, ectodermal or mesodermal tissues of the shrimp infected with WSSV can be used for viral purification. Viral DNA isolation is done using proteinase K and CTAB treatment followed by phenol-chloroform extraction and ethanol precipitation. After checking the purity of the DNA using electrophoresis, sequence information of WSSV is generated following the cloning and sequencing of the WSSV genome.

Oyster

In the *Crassostrea* protozoan parasite, *Haplosporidium nelsoni* is the etiologic agent of oyster MSX (multinucleate sphere unknown) disease. The polymerase chain reaction (PCR) technique has been widely used in pathogen detection, even for species that cannot be cultured. For the diagnosis of oyster MSX disease by molecular techniques, oligonucleotide probes for *Haplosporidium nelsoni* are used to bind to the parasites in oyster hemolymph and tissue sections. PCR–agarose gel electrophoresis and PCR–agarose gel electrophoresis with Southern blot hybridization (using radioisotopes) are used for assessing the infection of oysters by the bacterial agents, as well as the protozoan parasite. For the MSX disease, PCR–agarose gel electrophoresis with PCR re-amplification and PCR–ELISA method for direct detection of *Haplosporidium* was also found useful.

THERAPEUTICS FROM MARINE ANIMALS

Most of our medicine comes from natural resources and scientists are still exploring the organisms of tropical rain forests for potentially valuable medical products. Historical records show that human beings were aware of the venomous nature of some sea creatures for, at least, the last 4,000 years. More than 2,000 years ago, the extracts of marine organisms had been used as medicine.

Marine Natural Products of Animal Origin

Jellyfish: Several marine species turn out harmful substances that are used for defense/offense purposes; among these, cnidarians (jellyfish) are noteworthy (Mariottini and Pane, 2010). In 1961, Osamu Shimomura of Princeton University extracted green fluorescent protein (GFP) and another bioluminescent protein, called aequorin, from the large and abundant hydromedusa *Aequorea victoria*, while studying photoproteins that cause bioluminescence by this species of jellyfish. Three decades later, Douglas Prasher, a post-doctoral scientist at Woods Hole Oceanographic Institution, sequenced and cloned the gene for GFP. Martin Chalfie of Columbia University soon figured out how to use GFP as a fluorescent marker of genes inserted into other cells or organisms. Roger Tsien of the University of California, San Diego, later chemically manipulated GFP in order to get other colors of fluorescence to use as markers. In 2008, Shimomura, Chalfie, and Tsien won the Nobel Prize in Chemistry for their work with GFP. Jellyfish are also harvested for their collagen, which can be used for a variety of applications, including the treatment of rheumatoid arthritis.

GFP is a beta barrel structure, consisting of eleven β-sheets and six α helices containing the covalently bonded chromophore 4-(*p*-hydroxybenzylidene) imidazolidin-5-one (HBI) in the center. The presence of the hydrogen bonding network and electron-stacking interactions with the side chains influence the color, intensity, and photostability of GFP. The fluorescent chromophore is stable to a variety of adverse conditions like heat, extreme pH, and chemical denaturants. GFP comprises of 238 amino acids, and has an excitation peak at 395 nm and an emission peak at 508 nm. In cell and molecular biology, GFP gene is reportedly used to report of gene expression. In this context, GFP gene can be introduced into the target organism and can be maintained in the genome through breeding, injection with a viral vector, or transformation. It has been widely applied in many bacteria, yeast, fungus, fish, plants, as well as mammalian cells. Apart from gene expression studies, it has wide applications in the field of fluorescent microscopy, viability assay, and a new line of transgenic GFP has been used for gene therapy as well as regenerative medicine.

All jellyfish sting their prey using nematocysts (also called cnidocysts); these are stinging structures located in specialized cells called cnidocytes, which are characteristic of all Cnidaria. Contact with a jellyfish tentacle can trigger millions of nematocysts to pierce the skin and inject venom, yet the sting of only some jellyfish species causes an adverse reaction in humans. When a nematocyst is triggered by contact between the predator and its prey, pressure builds up rapidly inside it, up to 2,000 psi, until it bursts open. A lance inside the nematocyst pierces the victim's skin, and poison flows into the victim. Touching or being touched by a jellyfish can be very uncomfortable, sometimes requiring medical assistance; sting effects range from no effect to extreme pain to death. Because of the wide variation in response to jellyfish stings, it is wisest not to contact any jellyfish with bare skin. Even beached and dying jellyfish can still sting when touched.

Scyphozoan jellyfish stings are often uncomfortable, though not generally deadly, but some species of the class *Cubozoa*, or the Box jellyfish, such as the famous and especially toxic Irukandji, can be deadly. Stings may cause anaphylaxis, which may result in death. Hence, victims should immediately get out of the water. Medical care may include administration of an antivenom. The three goals of first aid for uncomplicated jellyfish stings are preventing injury to rescuers, deactivating the nematocysts, and removing tentacles attached to the patient. Rescuers should wear barrier clothing, such as panty hose, wet suits or full-body sting-proof suits. Deactivating the nematocysts, or stinging cells, prevents further injection of venom.

Sponge: Sponges (Phylum: Porifera) are evolutionarily ancient metazoans that have existed for 700–800 million years. They not only populate in tropical oceans, but also occur in temperate waters, and even in freshwater. Marine sponges have provided a vast resource in the search for bioactive secondary metabolites and potential drug leads (Abbas et al., 2011). These secondary metabolites in sponges play a vital role within their survival in the marine ecosystem and have gained attention in biomedical potential, pharmaceutical relevance, and various biotechnological applications (Thakur and Muller, 2004). Pharmaceutical interest in sponges was aroused in the early 1950s by the invention of unknown nucleosides: spongothymidine and spongouridine within the marine sponge *Cryptotethia crypta* (Alvarez et al., 2000). Apparently, out of the thirteen marine natural products that are currently under clinical trials as new drug candidates, twelve are derived from invertebrates (Thomas et al., 2010). Among them, Porifera remains the foremost necessary phylum, because it provides a large variety of natural products, particularly novel pharmacologically active compounds. Numerous marine sponges are giving raise to many natural products that are currently in clinical trial. L, L-Diketopiperazine called cyclo-(L-Pro-L-Phe) was isolated from South China ocean sponge *Stelletta tenuis* and showed antimicrobial activity (Li, 2009). Coriolin B isolated from the Indo-Pacific sponge *Jaspis aff. Johnstoni* exhibited sturdy inhibition of human breast and CNS cell lines (Kim et al., 2011). Sponge *Chondrosia reniformis* produced prugosenes, which are used as templates for new anti-infectives (Na et al., 2010).

Cuttlefish and Squid: Cuttlefish (*Sepia officinalis*) defend themselves by ejecting a dark ink (commonly known as sepia), which consists of a suspension of melanin granules throughout a viscous, colorless medium. At the top of the maturation process, ink gland cells of the digestive tract inside the mantle cavity degenerate and shed their contents into the ink sac; the sac acts as a reservoir of the exhausted material. Each production and ejection of the ink seems to be regulated by the glutamate/nitric oxide/cGMP-signaling pathway that is localized within the ink gland (Palumbo et al., 2000). Besides large amounts of melanin, the ink also contains proteins, glycosaminoglycans, lipids, etc. Cuttlefish ink is an ancient Chinese medicine listed in the *Compendium of Materia Medica* compiled by Shizhen Li, a renouned doctor at the time of the Ming Dynasty, and was initially employed to treat heart pain, however, modern clinical medication has proven that it is a good hemostatic medicine, which also provides significant curative effects in gynecology, surgery, and so on (Zhong et al., 2009).

Squid ink from *Loligo* spp. also possesses a wide range of biological roles. It has leukocyte-number elevating, anti-retrovirus (Rajaganapathi et al., 2000) and anti-bacterial properties. In recent years the exhaustive development of medicines from marine sources has resulted in the crucial search for effective cell-protective medicines from the ocean. One example is the detailed biochemical study of cuttlefish ink and squid ink is a major step toward the development of chemotherapeutic drugs.

Sea Snake: There are many varieties of venomous snakes in the whole world. Among them, sea snakes are unique in that they spend most of their life in the sea. Generally, most sea snakes are venomous, containing a mixture of proteins, which are diverse in function, and many highly toxic compounds (Tu, 1977). Closely related to land kraits and cobras, sea snakes have the same type of venom, only much more potent (Komori et al., 2009). This venom contains neurotoxins that act on the nerve cells of the bitten victim, paralyzing the complete respiratory system and ultimately causing death. Sca snakes are the most successful marine reptiles inhabiting the warm tropical waters of the world. There are 70 species of sea snakes belonging to 5 sub-families inhabiting the world's oceans and estuaries. In the Indian context, sea snakes are classified in to three main families (Hydrophiidae, Colubridae and Acrochordidae) and having three subfamilies (Hydrophiinae, Laticaudinae and Homalopsinae) and several genera under each family. All sea snakes are poisonous and very harmful to victims. Most sea snakes contain neurotoxins and are toxic because of the prey they eat (Komori et al., 2009).

All snake venoms are purely protein in nature and have different unique amino acid sequences and active sites, so they act as good inhibitors of the several biochemical pathways of the human body (Nagamizu et al., 2009). Snake venoms comprise a natural library of valuable bioactive substances for hemostasis and thrombosis. Snake venom cofactors are useful for clinical evaluation or sub-diagnosis of bleeding disorders, as well as for basic investigation into the molecular mechanisms of platelet plug formation induced by von Willebrand factor (VWF) and platelets. Snake venoms contain a variety of bioactive substances that influence hemostasis, thrombosis, and coagulation of mammalian blood (Andrews et al., 2004). These proteins have been used not only as specific reagents for the basic study of thrombosis and hemostasis, but are also expected to have clinical applications, e.g., as anti-thrombotic or diagnostic reagents (Marsh, 2001; 2005; Clemetson et al., 2008).

E. Conus: Cone snails (*Conus*) produce a distinctive repertoire of venom peptides that are used both as a defense mechanism and also to facilitate the immobilization and digestion of prey (Olivera and Cruz, 2001). These peptides known as *conus* act on homologous mammalian ion channels due to the degree of structural conservation exhibited by the voltage- and ligand-gated ion channels across higher eukaryotes. Moreover, mammalian ion channels exhibit diverse tissue expression patterns. This difference in tissue expression patterns was demonstrated with conotoxins that target the nicotinic acetylcholine receptor (nAChR) subtypes present at the invertebrate neuromuscular junctions, while not present in vertebrate neuromuscular junctions, are expressed in tissues relevant to pain. Thus peptides that target these ion channels may potentially be analgesic therapeutic agents in vertebrates (Olivera et al., 1987).

Conus peptides, such as the μ-conotoxins and ω-conotoxins, are currently being used as standard research tools in neuroscience. The μ-conotoxins are used for the immobilization of skeletal muscles without affecting axonal or synaptic events because of their ability to block the muscle Na^+ channel (Catterall et al., 2005). The ω-conotoxins are used as standard pharmacological reagents in voltage-gated calcium (Ca^{2+}) channel-related research and are used to block neurotransmitter release (Ichida et al., 2000; Olivera et al., 1994).

Many *Conus* spp. have been studied extensively for their peptides and activity. Peptides of *Conus* showed varied activity, and trials have also been conducted for potential treatment for severe chronic pain in patients and for chronic neuropathy (Lubbers et al., 2005). Overall, the conotoxins or conopeptides are mainly responsible for the active blocking or inhibitions and regulation of different cell transportation channels like Voltage gated ion channels, Na^+ channels, Ca^{2+} channels, K^+ channels and ligand-gated ion channels (Essack et al., 2012) (Table 29.4).

Commercial Bio-Products from Marine Organisms

By the early 1950s, Ross Nigrelli of the Osborn Laboratories of the New York Aquarium extracted a toxin from cuvierian organs of the Bahamian sea cucumber, *Actynopyga agassizi.* He named this toxin as "holothurin," which showed some antitunour activity in mice. From this humble beginning, the number of potential compounds isolated from the marine realm has virtually soared, and this number now exceeds 10,000, with hundreds of new compounds still being discovered every year. With the combined efforts of marine natural product chemists and pharmacologists, a number of promising identified molecules are already in market, clinical trials or in pre-clinical trials (Table 29.5). The antiviral compound Ara-A (active against Herpes virus) and anti-tumour compound Ara-C (effective in acute lymphoid leukemia) were obtained from sponges, and these compounds are now in clinical use. Arabinosyl cytosine (Ara-C) is currently sold by Pharmacia and Upjohn Company under the brand name Cytosar-UP. Apart from these, other products (such as blood-clotting compounds from cone snail, anti-inflammatory ointment from sea sponge, anticancer substance and disinfectants from shark, gene therapy vehicle and adhesive from shellfish's chitosan) are under development.

Enzyme inhibitors have received increasing attention as useful tools in the study of enzyme structures and reaction mechanisms. They also find applications in pharmacology and agriculture. Recently, marine organisms are increasingly recognized as a fruitful source for potential enzymes inhibitors. For example, a bryozoan, *Bugula neritina* has been the source of a family of protein kinase C (PKC) inhibitors called bryostatins, which are currently in clinical trials for cancer.

In the field of marine biotechnology, the living fossil, horse-shoe crab, is important, as its amoebocytes react with bacterial endotoxins and thus detect early infections in humans as well as traces of LPS (pyrogen) in biotechnological products. Many invertebrates, because of their simple cellular structures, provide a rich source of new information and serve as desirable non-mammalian models for research. A major area of emphasis is on genetic control of

TABLE 29.4 Amino Acid Sequences of Different Conotoxins Found in *Conus**

Peptide	Gene Family	Target	Amino Acid Sequence	Con Snails (Species)
Lt5d	T superfamily	Na+ channel	DCCPAKLLCCNP	*C. litteratus*
Lt6c	O1 superfamily	Na+ channel	WPCKVAGSPCGLVSECC GTCNVLRNRCV	*C. litteratus*
TIIIA	M superfamily	rNav1.2 rNav1.4	RHGCCKGOKGCSSRECR PQHCC	*C. tulipa*
Cal12a	O2 superfamily	Na+ channel	DVCDSLVGGHCIHNGC WCDQEAPHGNCCDTDG CTAAWWCPGTKWD	*C. californicus*
Cal12b	O2 superfamily	Na+ channel	DVCDSLVGGHCIHNGC WCDQDAPHGNCCDTDG CTAAWWCPGTKWD	*C. californicus*
BullIA	M superfamily	Nav1.4	VTDRCCKGKRECGRWC RDHSRCC	*C. bullatus*
BullIB	M superfamily	Nav1.4	VGERCCKNGKRGCGRW CRDHSRCC	*C. bullatus*
BullIC	M superfamily	Nav1.4	IVDRCCNKGNGKRGCSR WCRDHSRCC	*C. bullatus*
SIIIA	M superfamily	rNav1.2 rNav1.4	ZNCCNGGCSSKWCRDH ARCC	*C. striatus*
SIIIB	M superfamily	rNav1.2 rNav1.4	ZNCCNGGCSSKWCKGH ARCC	*C. striatus*
FVIA	O1 superfamily	(Ca^{2+} channels) N-type	CKGTGKSCSRIAYN CCTGSCRSGKC	*C. fulmen*
Sr11a	Kv1.2 Kv1.6	I2 superfamily	NQQCCWRSCCRGEC EAPCRFGP	*C. spurius*
RIIIj	Kv1.2	M superfamily	LPPCCTPPKKHCPAP ACKYKPCCKS	*C. radiatus*

*Targeted to Voltage gated ion channels, Na^+ channels, Ca^{2+} channels, K^+ channels, ligand-gated ion channels.
(Reviewed in Essack et al., 2012)

normal development and of tumour formation. In addition, many models shed light on the mechanism of nerve cells in marine invertebrates, which have direct implications in human and other mammalian systems. Sea urchins provided new information to scientists on fertilization, a fundamental biological process. Thus, marine model systems could provide new insight in to basic biological principles that will benefit further developments in medicine and industry.

Green Fluorescent Protein from Jelly Fish and Its Application

Green fluorescence protein (GFP) was first discovered in the 1960s and 1970s, along with the luminescent protein aequorin from *Aequorea victoria* (Shimomura et al., 1962). GFP fluorescence occurs when aequorin interacts with Ca^{2+} ions, which induces a blue glow, and some of this luminescent energy is transferred to the GFP, thus converting the overall color to green. The utilities of GFP came into the limelight when Douglas Prasher cloned and sequenced the wild type GFP gene in 1992. GFP is a protein composed of 238 amino acid residues, which exhibits bright green fluorescence in blue and ultraviolet light. Many marine organisms produce a similar green fluorescence, but in this case, the protein is isolated from the jellyfish *Aequorea victoria*. It is of utmost importance, as in cell and molecular biology it is frequently used as a reporter for gene expression by introducing the green fluorescence gene in many bacteria, yeast, and other fungi, fish, plants, fly, and mammalian cells to be used as biosensors (Soboleski et al., 2005).

In order to increase potential usage, different researchers have engineered different mutants of GFPs. A single point

TABLE 29.5 Marine By-Products Currently on the Market or in Clinical Phases

Product	Source	Application Area	Status
Ara-A	Marine sponge	Antiviral	Market
Ara-C	Marine sponge	Anticancer	Market
Okadaic acid	Dinoflagellate	Molecular probe	Market
Manoalide	Marine sponge	Molecular probe	Market
Vent™ DNA polymerase	Deep-sea hydrothermal vent bacteria	PCR enzyme	Market
Aequorin	Bioluminiscent jelly fish, *Aequro victoria*	Bioluminiscent calcium indicator	Market
Green fluorescent protein (GFP)	Bioluminiscent jelly fish, *Aequro victoria*	Reporter gene	Market
Phycoerythrin	Red algae	Conjugated antibodies used in ELISAs and flow cytometry	Market
Cephalosporins	*Cephalosporium* sp. Marine fungi	Antibiotic	Market
Yondelis™	Sea squirt	Cancer	Clinical phase II/III
Zinconotide	Cone snail	Chronic pain	Clinical phase III
Dolastatin	Sea slug	Cancer	Clinical phase II
Bryostatin- 1	Bryozoan	Cancer	Clinical phase II
Squalamine lactate	Shark	Cancer	Clinical phase III
Steroid	Sponge	Inflammation, Asthma	Clinical phase II

(Thakur and Thakur, 2006)

mutation improved the spectral characteristics of the GFP, resulting in increased fluorescence, photostability, a shift of the major extinction peak which matches the spectral characteristics of the commonly available FITC filter sets, thus increasing the practical utility of GFPs. Many other mutants have also been constructed for different color development, in particular blue fluorescent protein (EBFP), cyan fluorescent protein (ECFP), yellow fluorescent protein derivatives (YFP), and BFP derivatives. Genetically encoded FRET reporters sensitive to cell signaling molecules such as calcium or glutamate, protein phosphorylation state, protein complementation, receptor dimerization and other processes, provide highly specific optical readouts of cell activity in real time (Chudakov et al., 2010). GFP contains a typical β barrel structure consisting of one β-sheet with α-helix containing the covalently bonded chromophore 4-(p-hydroxybenzylidene)imidazolidin-5-one (HBI), which runs through the center. In the absence of a properly folded GFP scaffold, HBI is non-fluorescent; this exists mainly in unionized phenol form in wild type GFP, which is achieved by post the translational modification called maturation. Thus, the hydrogen-bonding network and electron-stacking interactions with these side chains influence the colour, intensity and photostability of GFP and its numerous derivatives. Protection of the chromophore fluorescence from quenching by water is achieved by the tightly packed nature of the barrel, which excludes solvent molecules (Fei and Hughes, 2001).

There is vast use of GFPs in nature in biology and other biological disciplines. Other fluorescent proteins are toxic in nature when used in living systems, but the naturally fluorescent molecules like GFPs are nontoxic when illuminated in living cells. GFPs have been widely used in labeling the spermatozoa of various organisms for identification purposes, as in the case of *Drosophila melanogaster*, where expression of GFP can be used as a marker for particular characteristics. GFPs can also be used in various structures enabling morphological distinctions. In other cases, GFP genes can be spliced into the genome of the organisms in the region of DNA that codes for target proteins which is controlled by the same regulatory sequences. Combining the several spectral variants of GFPs is a useful trick for the analysis of brain circuitry as well as the sensors of neuron membrane potential. GFPs can be a useful assay for detection of viable cells in cryobiology.

ETHICAL ISSUES

Marine biotechnology is currently facing a huge variety of ethical and legal issues. The most embarrassing situation is the demarcation of the oceanic territories of a country. This represents the first hurdle in studying and using organisms and resources found in another country's territory. If one country has a rich resource and better selection of marine life, it does not imply that any country can come in and infringe on its territory for their own research. However, scientists still struggle to come to terms with other countries for geographical boundaries. Another major ethical issue lies with the maintenance of a stable ecosystem and species. As scientists try to modify living organisms, it may threaten to damage the surrounding wildlife and habitat. In this regard, marine biotechnologists should be aware of what they are doing to make sure to keep the environment healthy and stable. Another major problem regarding the use of marine resources is its sustainability. Overexploitation of resources to meet the needs of the growing human population is a major threat. Recently, a total of 80% of the world's fisheries have been overexploited, depleted, or are in a state of collapse. Twenty-nine percent of fish and seafood species have been in a state of collapse and rest of them are projected to collapse by 2048. People take advantage of the limitation of geographical boundaries of marine environments to intrude and over-exploit resources, without hesitating to use sophisticated instruments and equipment. Strict legislation and its proper implementation may solve the issue for sustainable use of marine resources. Hence, many precautionary principles should be followed to explore the potential of marine environment, such as moving the research away from an environmentally sensitive area, or using less invasive techniques, such as computer modeling.

TRANSLATIONAL SIGNIFICANCE

The success story of marine biotechnology for field application is promising but is not yet optimal. Hence, the primary focus of marine biotechnological research should be more products oriented and to aim at solving local and global issues. So far, translational research has achieved many milestones, like the discovery of GFP proteins, a revolution in marine aquaculture for food, as well as ornamental field, microbial-enhanced oil recovery, and treatments for leukemia and cancer. However, many problems are still there, such as maintaining productivity and sustainability of the ocean, mobilization of knowledge, bioeconomy, promise of marine biotechnology for the benefit of people, treatment of many incurable diseases, treatment of pollutants, and much more. Hence, research should be targeted to direct transfer of the outcomes to field conditions for a better society. Collaborative research (in spite of geographical boundaries), and overcoming ethical issues, should be the ultimate goals for achieving the promises of marine biotechnology.

FUTURE DIRECTIONS

Blue biotechnology is the newcomer of the group of red, green and white biotechnology, standing primarily for marine biotechnology. The rich biodiversity of marine biota and their unique physiological adaptations to the harsh marine environment has coupled with new developments in biotechnology. It has opened up new and exciting vistas for the exploration of life-saving drugs, novel pharmaceuticals, industrial products and processes. Application of scientific and engineering principles to the processing of materials by marine biological agents to provide goods and services is called marine biotechnology. It deals with exploring the oceans to develop novel pharmaceutical drugs, chemical products, enzymes, and other industrial products and processes. However, the field of marine biotechnology is still in its infant stage and it faces huge challenges from technical, regulatory, political, and environmental viewpoints.

Most of our medicines come from natural resources, and, more than 2,000 years ago, the extracts of marine organisms were used as medicines. The numbers of potential compounds isolated from marine environments have exceeded 10,000 and still new compounds are discovered every year. A number of promising identified molecules are already in the market after successful clinical trials, by the combined efforts of marine natural product chemists and pharmacologists, and these precious natural products have been obtained from marine microorganisms as well as invertebrates such as sponges, mollusks, bryozoans, tunicates, etc. The success stories of marine biotechnological applications include: the commercialization of antibiotic cephalosporin from marine fungus, cytostatic cytarabine from sponge, anthelmintic insecticide kanic acid from red alga, analgesic zincototide from mollusk, etc. The antiviral compound, Ara-A, and anti-tumor compound, Ara-C, obtained from sponges are now under clinical trial, and apart from these, blood-clotting compounds from cone snail, anti-inflammatory ointments from sea sponges, anti cancer substances and disinfectants from shark, and a gene therapy vehicle and an adhesive from shell fish's chitosan are under development.

Mariculture is the most mature and highly successful example of progress in the field of marine biotechnology. Biotechnological applications to improve aquaculture are focused on species diversification, optimum food and feeding, and health and disease management, with minimum environmental impacts. Use of recombinant technologies is in progress to develop genetically modified organisms with useful features such as fast growth, resistance to pathogens, temperature and salinity tolerance, etc. Molecular biology approaches have also resulted in invention of new feedstocks and vaccines for aquaculture to increase productivity. The field of marine biotechnology needs to be developed continuously to meet the global demand of aquaculture and fish production.

WORLD WIDE WEB RESOURCES

Sustainable use of marine bioresources needs thorough knowledge, understanding, proper identification, assessment, and conservation of local marine inhabitants. Marine animals house an abundant unexploited wealth of biomolecules, which needs to be explored for bioprospecting, discovery of novel bioactive molecules, and conservation. For further details, web resources given below may be explored:

http://www.marinebiotech.org/
http://www.marinebiotech.eu/
http://www.bioresourcebiotech.com/
http://agsci.oregonstate.edu/brr/
http://www.usda.gov/wps/portal/usda/usdahome?navid=AQUACULTURE
http://www.lsuagcenter.com/en/our_offices/research_stations/Aquaculture/
http://www.euromarineconsortium.eu/fp6networks/marinegenomics
http://www.marinegenomics.org/
http://www.mendeley.com/groups/1063651/fish-marine-nanotechnology/
http://www.northwestern.edu/newscenter/stories/2013/02/from-sticky-marine-mussels-to-nanotech.html

REFERENCES

Abbas, S., Kelly, M., Bowling, J., Sims, J., Waters, A., & Hamann, M. (2011). Advancement into the Arctic region for bioactive sponge secondary metabolites. *Marine Drugs, 9*, 2423–2437.

Alsaqufi, A. S., Gomelsky, B., Schneider, K. J., & Pomper, K. W. (2012). Verification of mitotic gynogenesis in ornamental (koi) carp (*Cyprinus carpio* L.) using microsatellite DNA markers. *Aquaculture Research.* http://dx.doi.org/10.1111/j.1365-2109.2012.03242.x.

Alvarez, B., Crisp, M. D., Driver, F., Hooper, J. N. A., & Soest, R. W. M.V. (2000). Phylogenetic relationships of the family Axinellidae (Porifera: Demospongiae) using morphological and molecular data. *Zoologica Scripta, 29*, 169–198.

Andrews, C. (1990). The ornamental fish trade and fish conservation. *Journal of Fish Biology, 37*, 53–59.

Andrews, R. K., Gardiner, E. E., & Berndt, M. C. (2004). Snake venom toxins affecting platelet function. *Methods in Molecular Biology, 273*, 335–348.

Angly, F. E., Felts, B., Breitbart, M., Salamon, P., Edwards, R. A., Carlson, C., Chan, A. M., Haynes, M., Kelley, S., Liu, H., Mahaffy, J. M., Mueller, J. E., Nulton, J., Olson, R., Parsons, R., Rayhawk, S., Suttle, C. A., & Rohwer, F. (2006). The marine viromes of four oceanic regions. *PLOS Biology, 4*, 2121–2131.

Basrur, P. K., & King, W. A. (2005). Genetics then and now: breeding the best and biotechnology. *Scientific and Technical Review of the Office International des Epizooties, 24*, 31–49.

Brown, P. O., & Botstein, D. (1999). Exploring the new world of the genome with DNA microarrays. *Nature Genetics, 21*, 33–37.

Bruckner, A. W. (2005). The importance of the marine ornamental reef fish trade in the wider Caribbean. *Revista de Biologia Tropical, 53*, 127–138.

Castro, P., & Huber, M. E. (2003). *Marine Biology* (4th ed.). McGraw-Hill.

Catterall, W. A., Goldin, A. L., & Waxman, S. G. (2005). International union of pharmacology. XLVII. Nomenclature and structure–function relationships of voltage-gated sodium channels. *Pharmacological Review, 57*, 397–409.

Chan, E. Y. (2009). Next-generation sequencing methods: Impact of sequencing accuracy on SNP discovery. *Methods in Molecular Biology, 578*, 95–111.

Chudakov, D. M., Matz, M. V., Lukyanov, S., & Lukyanov, K. A. (2010). Fluorescent proteins and their applications in imaging living cells and tissues. *Physiological Reviews, 90*, 1103–1163.

Clemetson, K. J., & Clemetson, J. M. (2008). Platelet GPIb complex as a target for anti-thrombotic drug development. *Thrombosis Haemostatis, 99*, 473–479.

Cunningham, C., Hikima, J. I., & Jenny, M. J. (2006). New resources for marine genomics: Bacterial artificial chromosome libraries for the eastern and Pacific oysters (*Crassostreavirginica* and *C. gigas*). *Marine Biotechnology, 8*, 521–533.

Essack, M., Bajic, V. B., & Archer, J. A. C. (2012). Conotoxins that confer therapeutic possibilities. *Marine Drugs, 10*, 1244–1265.

Fei, Y., & Hughes, T. E. (2001). Transgenic expression of the jellyfish green fluorescent protein in the cone photoreceptors of the mouse. *Visual Neuroscience, 18*, 615–623.

Foresti, F. (2000). Biotechnology and fish culture. *Hydrobiologia, 420*, 45–47.

Guddinga, R., Lillehauga, A., & Evensen, O. (1999). Recent developments in fish vaccinology. *Veterinary Immunology and Immunopathology, 72*, 203–212.

Ichida, S., Abe, J., Zhang, Y. A., Sugihara, K., Imoto, K., Wada, T., Fujita, N., & Sohma, H. (2000). Characteristics of the inhibitory effect of calmodulin on specific [125i] omega-conotoxin GVIA binding to crude membranes from chick brain. *Neurochemistry Research, 25*, 1629–1635.

Kim, G. Y., Kim, W. J., & Choi, Y. H. (2011). Pectenotoxin-2 from marine sponges: A potential anti-cancer agent – A review. *Marine Drugs, 9*, 2176–2187.

Komori, Y., Nagamizu, M., Uchiya, K., Nikai, T., & Tu, A. T. (2009). Comparison of Sea Snake (*Hydrophiidae*) neurotoxin to Cobra (*Naja*) neurotoxin. *Toxins, 1*, 151–161.

Kumar, S. R., Ahmed, V. P. I., Parameswaran, V., Sudhakaran, R., Babu, V. S., & Hameed, A. S. S. (2008). Potential use of chitosan nanoparticles for oraldelivery of DNA vaccine in Asian sea bass (*Latescalcarifer*) to protect from *Vibrio (Listonella) anguillarum. Fish and Shellfish Immunology, 25*, 47–56.

Li, Z. (2009). Advances in marine microbial symbionts in the China Sea and related pharmaceutical metabolites. *Marine Drugs, 7*, 113–129.

Liu, H., Jiang, Y., Wang, S., Ninwichian, P., Somridhivej, B., Xu, P., Abernathy, J., Kucuktas, H., & Liu, Z. (2009). Comparative analysis of catfish BAC end sequences with the zebrafish genome. *BMC Genomics, 10*, 592.

Lubbers, N. L., Campbell, T. J., Polakowski, J. S., Bulaj, G., Layer, R. T., Moore, J., Gross, G. J., & Cox, B. F. (2005). Postischemic administration of CGX-1051, a peptide from cone snail venom, reduces infarct size in both rat and dog models of myocardial ischemia and reperfusion. *Journal of CardiovascularPharmacology, 46*, 141–146.

Mariottini, G. L., & Pane, L. (2010). Mediterranean jellyfish venom: A review on Scyphomedusae. *Marine Drugs, 8*, 1122–1152.

Marsh, N., & Williams, V. (2005). Practical applications of snake venom toxins in haemostasis. *Toxicon*, *45*, 1171–1181.

Menon, N. G., & Pillai, P. P. (2001). *Perspectives in Mariculture*. Kochi, India: The Marine Biological Association of India.

Na, M., Ding, Y., Wang, B., Tekwani, B. L., Schinazi, R. F., Franzblau, S., Kelly, M., Stone, R., Li, X. C., Ferreira, D., et al. (2010). Anti-infective discorhabdins from deep-water Alaskan sponge of the genus Latrunculia. *Journal of Natural Products*, *73*, 383–387.

Nagamizu, M., Komori, Y., Uchiya, K., Nikai, T., & Tu, A. T. (2009). Isolation and chemical characterization of a toxin isolated from the venom of the sea snake, Hydrophistorquatusaagardi. *Toxins1*, 162–172.

Olivera, B. M., & Cruz, L. J. (2001). Conotoxins, in retrospect. *Toxicon*, *39*, 7–14.

Olivera, B. M., Cruz, L. J., de Santos, V., LeCheminant, G. W., Griffin, D., Zeikus, R., McIntosh, J. M., Galyean, R., Varga, J., Gray, W. R., et al. (1987). Neuronal calcium channel antagonists: Discrimination between calcium channel subtypes using omega-conotoxin from *Conus magus* venom. *Biochemistry*, *26*, 2086–2090.

Olivera, B. M., Miljanich, G. P., Ramachandran, J., & Adams, M. E. (1994). Calcium channel diversity and neurotransmitter release: The omega-conotoxins and omega-agatoxins. *Annual Review of Biochemistry*, *63*, 823–867.

Padhi, B. K., & Mandal, R. K. (2000). *Applied Fish Genetics*. Visakahapatnam, India: Fishing Chimes.

Palumbo, A., Poli, A., Di Cosmo, A, & d'Ischia, M (2000). N-methyl-D-aspartate receptor stimulation activates tyrosinase and promotes melanin synthesis in the ink gland of the cuttlefish Sepia officinalis through the nitric oxide/cGMP signal transduction pathway. *Journal of Biological Chemistry*, *275*, 16885–16890.

Pandian, T. J., & Kirankumar, S. (2003). Androgenesis and conservation of fishes. *Current Science*, *85*, 917–931.

Peruzzi, S., & Chatain, B. (2000). Pressure and cold shock induction of meiotic gynogenesis and triploidy in the European sea bass, Dicentrarchuslabrax L.: relative efficiency of methods and parental variability. *Aquaculture*, *189*, 23–37.

Rajaganapathi, J., Thyagarajam, S. P., & Edward, J. K. (2000). Study on cephalopod's ink for anti-retroviral activity. *Indian Journal of Experimental Biology*, *38*, 519–520.

Rowe, S., & Hutchings, J. A. (2003). Mating systems and the conservation of commercially exploited marine fish. *Trends in Ecology and Evolution*, *18*, 567–572.

Sellars, M. J., Li, F., Preston, N. P., & Xiang, J. (2010). Penaeid shrimp polyploidy: Global status and future direction. *Aquaculture*, *310*, 1–7.

Shimomura, O., Johnson, F., & Saiga, Y. (1962). Extraction, purification and properties of aequorin, a bioluminescent protein from the luminous hydromedusan, Aequorea. *Journal of Cellular and Comparative Physiology*, *59*, 223–239.

Soboleski, M. R., Oaks, J., & Halford, W. P. (2005). Green fluorescent protein is a quantitative reporter of gene expression in individual eukaryotic cells. *The FASEB Journal*. http://dx.doi.org/10.1096/fj.04-3180fje.

Thakur, N. L., & Thakur, A. N. (2006). Marine biotechnology: An overview. *Indian Journal of Biotechnology*, *5*, 263–268.

Thomas, T. R. A., Kavlekar, D. P., & LokaBharathi, P. A. (2010). Marine drugs from sponge–microbe association – A review. *Marine Drugs*, *8*, 1417–1468.

Tlusty, M. (2002). The benefits and risks of aquacultural production for the aquarium trade. *Aquaculture*, *205*(3-4), 203–219.

Tomanek, L. (2011). Environmental proteomics: changes in the proteome of marine organisms in response to environmental stress, pollutants, infection, symbiosis, and development. *Annual Review of Marine Sciences*, *3*, 373–399.

Whitehead, A., Triant, D. A., Champlin, D., & Nacci, D. (2010). Comparative transcriptomicsimplicates mechanisms of evolved pollution tolerance in a killifish population. *Molecular Ecology*, *19*, 5186–5203.

Zhong, J. P., Wang, W., Shang, J. H., Pan, J. Q., Li, K., Huang, Y, & Liu, HZ (2009). Protective effects of squid ink extract towards hemopoietic injuries induced by Cyclophosphamine. *Marine Drugs*, *7*, 9–18.

Zou, Y. C., Wei, Q. W., & Pan, G. B. (2011). Induction of meiotic gynogenesis in paddlefish (Polyodonspathula) and its confirmation using microsatellite markers. *Journal of Applied Ichthyology*, *27*, 496–500.

FURTHER READING

Holmer, M., Black, K., Duarte, C. M., Marba, N., & Karakassis, I. (2008). *Aquaculture in the Ecosystem*. Springer publications.

Jefferson, T. A., Webber, M. A., Pitman, R., & Jarrett, B. (2007). *Marine Mammals of the World: A Comprehensive Guide to Their Identification*. Academic Press.

Levinton, J. S. (2008). *Marine Biology: Function, Biodiversity, Ecology*. Oxford University Press.

Miller, C. B., & Wheeler, P. A. (2012). *Biological Oceanography*. Wiley Blackwell.

Munn, C. (2003). *Marine Microbiology: Ecology & Applications: Ecology and Applications. Advanced Text*.

GLOSSARY

Abalone Common name for any group of small or large edible sea snails and marine gastropod mollusks of family Haliotidae.

Blue Biotechnology Application of biotechnology for describing marine and aquatic animals.

Clams One or more species of commonly consumed bivalves that burrow in sediments.

Cnidocysts A characteristic subcellular organelle responsible for the stings delivered by jellyfish.

Cubozoa They are the cnidarian invertebrates distinguished by their cube-shaped medusa commonly known as box jellyfish.

Hydrozoa A taxonomic class of individually very small, predatory animals, some solitary and some colonial, most living in saltwater.

Limpet The common name for a different group of sea and freshwater snails, for those that do not appear to be spirally coiled in adult stage.

Metagenomics The study of metagenomes, the genetic material recovered directly from environmental samples.

Nematocysts A capsule within specialized cells of certain coelenterates, such as jellyfish, containing a barbed, threadlike tube that delivers a paralyzing sting when propelled into attackers and prey. Also called stinging cell.

Nudibranch A member of Nudibranchia is a group of soft-bodied, marine gastropod mollusks which shed their shell after their larval stage which are noted for their often extraordinary colors and striking forms.

Ostia A small opening or orifice, as in case of a fallopian tube, vagina, maxillary sinus, etc.

Penaeid Shrimps Shrimps belonging to the family Penaeidae of marine crustacean containing economic important species like tiger prawn.

Scallops A marine bivalve mollusk of the family Pectinidae, which are cosmopolitan in nature.

Tunicates Members of the tunicata, a subphylum of the phylum Chordata which are marine filter feeders with a saclike morphology.

ABBREVIATIONS

Ab Antibody
Ag Antigen
Ara-C Arabinosyl Cytosine
BACs Bacterial Artificial Chromosomes
BFP Blue Fluorescent Protein
DMSO Dimethyl sulfoxide
DNA Deoxyribonucleic Acid
EBFP Enhanced Blue Fluorescent Protein
ECFP Enhanced Cyan Fluorescent Protein
ELISA Enzyme-Linked Immunosorbent Assay
FRET Fluorescent Resonance Energy Transfser
GFP Green Fluorescent Protein
HBI 4-(*p*-Hydroxybenzylidene) imidazolin-5-one
HRP Horseradish Peroxidase
KLH Keyhole Limpet Hemocyanin
mRNA Messenger Ribonucleic Acid
MSX Disease Multinucleate Sphere Unknown Disease
nAChR Nicotinic Acetylcholine Receptor
OPD *o*-Phenylenediamine
PKC Protein Kinase C
PCR Polymerase Chain Reaction
PLGA Poly-Lactide-Co-Glycolide Acid
PS Polystyrene
SDS-PAGE Sodium Dodecyl Sulfate/Polyacrylamide Gel Electrophoresis
TMB Tetramethyl Benzedine
VWF von Willebrand Factor
WSSV White Spot Syndrome Virus
YFP Yellow Fluorescent Protein

LONG ANSWER QUESTIONS

1. Write a detailed explanation of chromosomal manipulation in fish.
2. Write an essay on the prospects and scope of marine biotechnology.
3. Describe the tools for disease diagnosis in aquaculture.
4. Describe immunodiagnostics in aqua-farming.
5. Write a short essay on marine peptides and their pharmacological applications.

SHORT ANSWER QUESTIONS

1. What is the difference between mitotic and meiotic gynogenesis?
2. What does "extenders" mean?
3. Briefly describe the largest phylum of animals.
4. Why are fish suitable models for studying transgenic technology?
5. How is *Aequorea victoria* related to *in vitro* study designs?

ANSWERS TO SHORT ANSWER QUESTIONS

1. Meiotic gynogenesis involves the participation of oocyte and second polar body; sperm has no role in this phenomenon. However, in the case of mitotic gynogenesis, irradiated sperm is used to activate the egg where the polar bodies are expelled out.
2. An extender is a solution consisting of organic and inorganic chemicals that resembles blood or seminal plasma, which is used for the long term *in vitro* storage of spermatozoa to maintain its viability.
3. Arthropods (three out of four animals on earth) constitute the largest phylum of animals. Marine arthropods include crustaceans like shrimps, crabs, lobsters and many more familiar animals that are mostly decapod in nature.
4. There are many aspects that render fish more suitable models than mice for transgenic technology, including the higher fecundity of fish, which increases the availability of eggs for microinjection. The other advantages include the external fertilization and incubation of fish eggs, and relative large size of fish eggs, facilitating handling.
5. *Aequorea victoria*, a jellyfish, is known to produce the fluorescent protein GFP, which has larger applications in the field of research. It is used as a fluorescent dye to measure the expression level of a gene, confirm transformation experiments, is used in fluorescence microscopy, and is used in gene therapy as well as regenerative medicine.

Chapter 30

Herbal Medicine and Biotechnology for the Benefit of Human Health

Priyanka Srivastava[*†], **Mithilesh Singh**[**], **Gautami Devi**[*] **and Rakhi Chaturvedi**[*]

**Department of Biotechnology, Indian institute of Technology – Guwahati, Guwahati, Assam, India, **G.B. Plant Institute of Himalyan Environmental and Development, Sikkim Unit, Panthang, Gangtok, Sikkim, India, †Division of Biomedical Sciences, School of Bio Sciences and Technology, Vellore Institute of Technology, Vellore, Tamil Nadu, India*

Chapter Outline

SUMMARY

The present chapter discusses the importance of plants and their metabolites in herbal medicines. Various examples of biotechnological tools have been highlighted how plants can be exploited commercially without affecting their natural population. Furthermore, the chapter discusses processing plants for herbal medicine and drug discovery from natural products.

WHAT YOU CAN EXPECT TO KNOW

How do herbal medicines compare to conventional forms of medicine? What are their advantages and limitations? Besides this, what are their methods of production from plant sources, and what are the various techniques required to analyse and characterize them?

HISTORY AND METHODS

INTRODUCTION

Herbal medicines refer to the use of plant seeds, berries, roots, leaves, bark, or flowers for medicinal purposes (Figure 30.1). Medicinal plants have been a major source of drugs for thousands of years, and even today they are the basis of systematic traditional medicines in almost all countries of the world. Unani and Ayurveda systems of medicine are two of the classic and oldest examples of this category. Around 80% of the population in developing countries is completely dependent on plants for their primary health care (Bannerman et al., 1983). Even in developed countries, which are enormously advanced in terms of medicinal chemistry, over one-fourth of all prescribed pharmaceuticals originate directly or indirectly from plants

Animal Biotechnology. http://dx.doi.org/10.1016/B978-0-12-416002-6.00030-4

FIGURE 30.1 Herbal medicines. *(Courtesy: Google)*

(Newman et al., 2000). Furthermore, out of 252 drugs considered as indispensable by the World Health Organization (WHO), 11% are mainly derived from flowering plants, and 28% of synthetic drugs are obtained from natural precursors (Namdeo, 2007).

As already mentioned, herbal medicines are derived from plants. Understandably, these pharmaceuticals are produced solely from massive quantities of whole plant parts, which can lead to problems. One problem is that excessive harvesting can diminish local plant populations and erode genetic diversity. A second, but important, concern is inconsistency of the derived products in terms of quality and quantity. The latter can spell trouble in terms of safety, supply, and economic feasibility of these herbal products on a commercial scale. In order to overcome these bottlenecks, domestication and acceptance of good agricultural practices are crucial, especially for revival of diminishing plant populations. However, the conventional methods of plant propagation are lengthy and time consuming. The long cultivation periods between planting and harvesting make the entire process cumbersome and uneconomical, which in turn leads to the high cost of drugs. Moreover, wild populations are susceptible to problems of disease, drought, environmental fluctuations, low rate of fruit set, and poor seed yield, germination, and viability. This vulnerability of plants also affects batch-to-batch consistency of derived metabolites to be used as drugs or drug precursors. Clearly, there is an urgent need of alternative and complimentary methods for uniform production of herbal medicine. In this context, tools and techniques of biotechnology, like *in vitro* plant, cell, tissue, and organ culture, offer solutions in terms of mass propagation of plants in a shortened time span, occupying much less space than wild populations, and uniform production of metabolites all year round, irrespective of seasons and vagaries of climatic conditions. One clarification required at this point is the term "metabolites." The majority of the compounds used as drugs are secondary metabolites (Kubmarawa et al., 2007), whose production is largely affected by environmental fluxes.

Traditional Medicine

The World Health Organization (WHO) defines traditional medicine as being the "sum total of knowledge, skills, and practices based on the theories, beliefs and experiences that are indigenous to different cultures, which are used to maintain health, as well as to prevent, diagnose, improve, or treat physical and mental illnesses." Every early civilization used plants as their main source of medicine, and most of the world's population still relies on them. The first recorded literature on medicinal plants can be traced back to early human history, the Atharvaveda (2000 B.C.) in India. With time, the original population of an area gained knowledge which plants could be used for certain diseases or states of illness. In addition, they also gained knowledge of the harmful and poisonous plants. It is evident that the modern drug industry has been developed to a considerable degree as a result of plant-based traditional medicines.

There are a few closely related terms in use today, meanings of which should be understood clearly. *Traditional medicine* refers to the following components: acupuncture (China), Ayurveda (India), Unani (Arabic countries), traditional birth attendant's medicines, mental healer's medicines, herbal medicines, and various forms of indigenous medicines. *Complementary or alternative medicine* refers to a broad set of health care practices that are not part of a country's own tradition, and are not integrated into the dominant health care system. Traditional medicine has maintained its popularity in all regions of the developing world, and its use is rapidly spreading in industrialized countries.

Ancient System of Medicine

Ayurveda, perhaps the most ancient of all medicinal traditions, is probably older than traditional Chinese medicine. It is derived from "Ayur" meaning "life," and "Veda," meaning "knowledge." Ayurveda means the "science of life." It takes a holistic view of human beings, their health, and illness. It aims at positive health, which has been defined as a well-balanced metabolism coupled with a healthy state of being. According to Ayurveda, disease can arise from the body and/or mind due to external factors or intrinsic causes. The origin of Ayurveda is lost in prehistoric antiquity, but its characteristic concepts appear to have matured between 2,500 and 500 B.C. in ancient India. The earliest references to drugs and diseases can be found in the Rigveda and Atharvaveda.

Ayurvedic drugs have been found to perform very well against chronic ailments. Today, they are also attracting attention for diseases for which there are no (or inadequate) drugs for treatment in modern medicine, such as metabolic and degenerative disorders. Most of these diseases have multifactorial causation, and there is a growing awareness that in such circumstances, a combination of drugs, acting at a number of targets concurrently, is likely to be more effective than drugs acting at one target. Ayurvedic drugs, which are often multi-component, have a special impact on such conditions. For various reasons, Ayurveda has not included much of modern science/scientific tools. Studies of the biological activity of multicomponent Ayurvedic drugs will bring Ayurveda into the mainstream of scientific investigations.

METHODOLOGY

Investigation of Medicinal Plants

Medicinal plants have formed the basis of health care throughout the world since the earliest days of civilization, and are still widely used, and have noteworthy significance in international trade. Recognition of their clinical, pharmaceutical, and economic value is still growing, although this varies widely between countries. Plants are important for pharmacological studies and drug development, not only when bioactive compounds are used directly as therapeutic agents, but also as starting materials for the synthesis of drugs or as models for pharmacologically active compounds.

Each plant species has its own specific set of secondary metabolites. Apart from the family Poaceae, which harbors the world's worst weeds but is low in medicinal plants, many of the top twelve weed families are also the ones that are important for medicinals. The ecological and biochemical evidence suggest the preponderance of weeds in medicinal floras. Secondary compounds in plants are involved in the interaction of the plant with its environment and are important for ecological functions such as allelopathy, insect and animal attractants for pollination, seed dispersal, and for chemical defense against microbes, insects, and herbivory (Bourgaud et al., 2001). These compounds do not participate in the vital metabolic processes of the plant system, but are the ones that exhibit bioactivity, and can serve as medicinals for humans. The spectrum of chemical structures synthesized by the plant kingdom is broader than that of perhaps any other group of organisms (Rao and Ravishankar, 2002).

In the present scenario, a large proportion of the drugs used in modern medicine are either directly isolated from plants, or synthetically modified from a lead compound of natural origin. However, rarely is the drug isolated in the pure, usable form. What is initially obtained is the crude extract, which requires stepwise purification to obtain the finished product. The finished product as herbal medicine most of the time is a mixture of several compounds. When each and every component in the mixture is characterized qualitatively and quantitatively, it is called *"characterized extract,"* which is understandably more desirable than the *"uncharacterized extract."* Plant extracts are known to consist of many chemicals, and among them, a few compounds could be acting synergistically. Sometimes, isolation of the compounds from the extract may cause a decrease in desired activity, which underlines the importance of extract screening (Orhan et al., 2009).

Evidence-based studies on the efficacy and safety of traditional Indian medicines are limited. The essential ingredients in most formulations are not precisely defined. This is one of the most important challenges to scientists attempting to identify a single bioactive compound. Therefore, in-depth studies and more stringent conditions should be followed to make a herbal formulation so that the role of each and every component is known.

Drug discovery is the process by which drugs are discovered or designed. Plants have long been a very important source of drugs, and many plant species have been analyzed to see if they contain substances with therapeutic activity. Many plant drugs of folklore were investigated to determine the active ingredient in the mixture. Several reviews are available in the literature pertaining to approaches for selecting plants as candidates for drug discovery programs.

Today, many new chemotherapeutic agents are obtained synthetically, based on "rational" drug design. The study of natural products has many rewards over synthetic drug design. The former leads to materials having new structural features with novel biological activity. In this context not only do plants continue to serve as possible sources for new drugs, but chemicals derived from the various parts of these plants can also be extremely useful as lead structures for synthetic modification and optimization of bioactivity. The starting materials for about one-half of the medicines we use today come from natural sources. There is no doubt that the future of plants as sources of medicinal agents for use in investigation, prevention, and treatment of diseases is very promising.

Drug discovery from natural resources is a very tedious process. It involves identification of plant material, extraction, preliminary phytochemical screening of the crude extract, evaluation of biological activity, isolation of various bioactive compounds, and finally elucidation of structures. If the molecule is appealing, with strong pharmacological properties, then further preclinical studies are conducted on the molecules, such as toxicity, stability, and solubility studies. After undertaking these studies, if it is found that a molecule is substantially more active than the currently used drug, only then are processes developed for its economical and easy isolation from the source so that it can be readily available for therapeutic use.

In the context of isolation and screening of chemicals from plants that may possess medicinal properties, different approaches can be used. The process of obtaining bioactive substances and their chemical characterization can be schematically represented as in Flow Chart 30.1.

Extraction

Extraction involves the separation of medicinally active fractions of plant from inactive or inert components by using selective solvents through extraction procedures. The products so obtained from plants are relatively complex

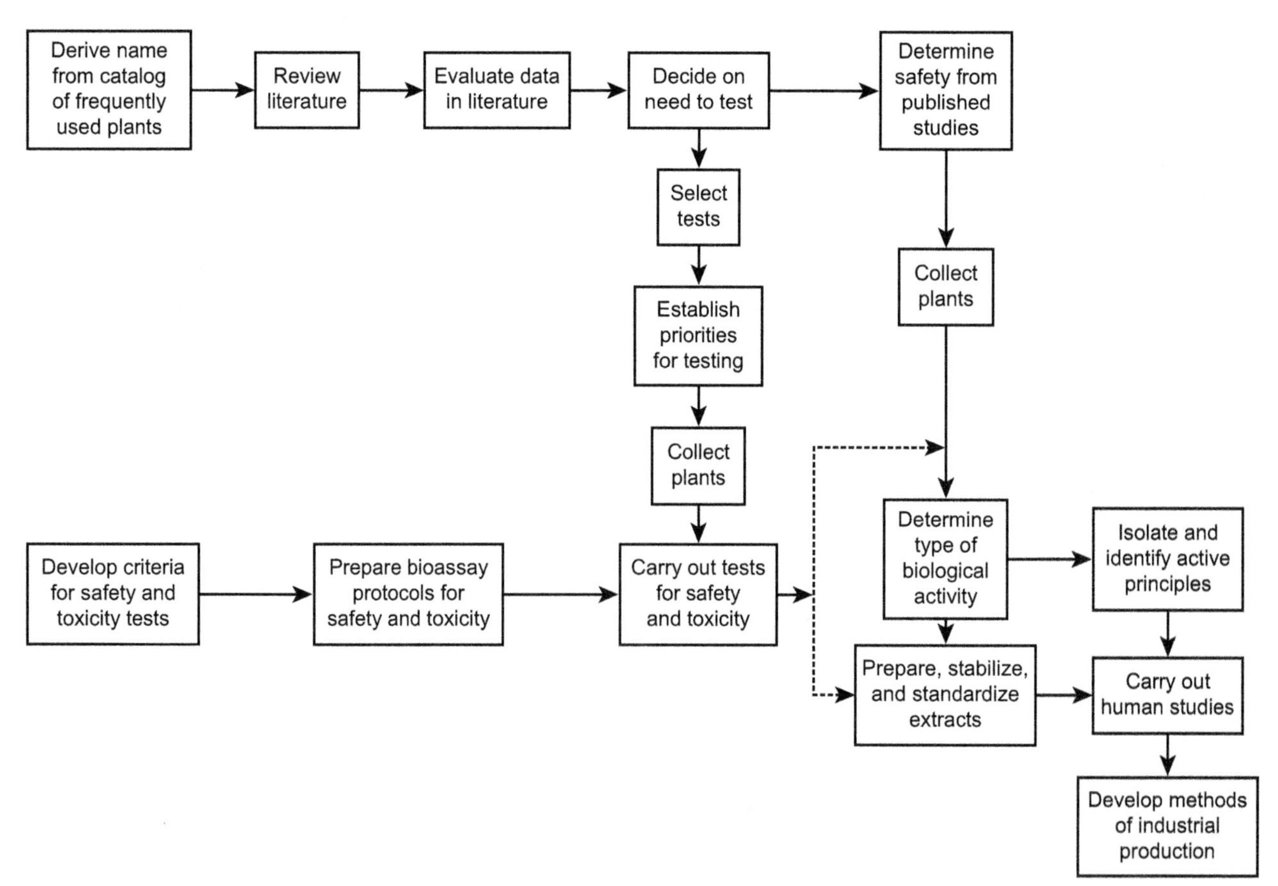

FLOW CHART 30.1 Flow chart of sequence for the study of plants used in traditional medicine. *(Adapted from Fabricant and Farnsworth, 2001.)*

mixtures of metabolites in liquid, semi-solid, or (after removing the solvent) dry powder form. This is the critical first step in the investigation of medicinal plants.

The selection of a solvent system mainly depends on the exact nature of the bioactive compounds being targeted because during the extraction process, solvents diffuse into the solid plant material and solubilize compounds of similar polarity. The extraction of hydrophilic compounds uses polar solvents, such as methanol, ethanol, or ethyl acetate. For extraction of more lipophilic compounds, dichloromethane is used. In a few cases, extraction with hexane is used to eliminate chlorophyll and oil.

As the target compounds may be non-polar to polar and thermally labile, the suitability of the methods of extraction must be well thought out. Different methods, such as sonication, heating under reflux, soxhlet extraction, and others, are commonly used for plant sample extraction. Additionally, plant extracts are also prepared by maceration or percolation of fresh green plants or dried powdered plant material in water and/or organic solvent systems.

Other modern extraction techniques include solid-phase micro-extraction, supercritical-fluid extraction, pressurized-liquid extraction, microwave-assisted extraction, solid-phase extraction, and surfactant-mediated techniques, which possess certain advantages.

Chemical Screening

This technique is also known as phytochemical screening. In this method, aqueous and organic extracts are prepared from those plant samples that are the reservoir of secondary metabolites, such as leaves, stems, roots, or bark. The plant extracts are then analyzed for the presence of secondary metabolites like alkaloids, terpenes, and flavonoids. Standard tests are available in the literature for each class of compounds to be analyzed. Following this, a simple separation technique like thin-layer chromatography (TLC) is generally used to analyze the number and type of components present in the mixture. In TLC, the extracts are loaded in a glass coated with silica gel or other adsorbent, which is then kept in a chromatographic chamber containing a suitable running solvent. This technique mainly consists of a mobile phase and a stationary phase, whereby the compounds are separated based on their polarity. Sometimes a developing solvent might also be used after the plate has been taken out of the chromatographic chamber to detect the chemicals. This approach has been used in the past, and is still being used in developing countries. Since the isolation of pure bioactive components is a long and tedious process, this procedure enables the early recognition of known metabolites in the extracts, and is thus economically viable. The tests are simple to perform, however, it is not suitable for the efficient separation of metabolites, and has low selectivity and sensitivity of detection, which makes it difficult to detect traces of components in the sample.

Biological Assays

Plant extracts have served as an important source of bioactive compounds for many drug discovery programs, and several important drugs have been isolated and identified from plants. In any isolation program in which the end product is a drug or lead compound, some type of bioassay screening or pharmacological evaluation must necessarily be used to guide the isolation process towards the pure bioactive component.

The selection of the biological assay to be adopted usually depends on the target syndrome as well as on the available information about the plant to be studied. For instance, if a plant has an ethanopharmacological history of use against a particular disease, then one would rationally use a specific bioassay technique that can predict the reputed therapeutic activity in order to isolate the lead that is responsible for that biological activity.

In the past, the extracts from plants were mainly evaluated in experimental animals, primarily mice and rats. Currently, anti-microbial assay by the disk diffusion method is in practice. However, this technique had several disadvantages. Firstly, the phytochemical extracts are highly heterogeneous due to the presence of a mixture of different bioactive components. A desired biological response may not be due to a single bioactive compound, but to a mixture of several bioactive compounds. Moreover, although several new bioassay techniques have been developed, at present these techniques are still expensive, time-consuming, and technologically complicated. The major disadvantage of bioassay techniques is the use of biological organisms, particularly mice and rats, which is not practical as these living organisms most often have to be sacrificed. Lastly, isolation, screening, and quantification of a specific bioactive compound are difficult using biological assays. Hence, this technique is losing popularity.

Isolation and Characterization of Bioactive Compounds

Due to the fact that plant extracts usually contain various types of compounds with different polarities, their separation still remains a big challenge for the process of identification and characterization of bioactive compounds. Apart from this, there are always chances of wide variations with respect to their chemical content in crude drugs/raw materials of plant origin due to varied reasons such as climatic conditions, geographical distribution, source and season of collection, and lack of scientific methods of post-harvest processing, storage, and preservation. Therefore, identification and quantification of bioactive compounds are essential prerequisites for herbal drug development (Flow Chart 30.2). Thin-layer chromatography is a powerful and simple analytical tool used for this purpose. However,

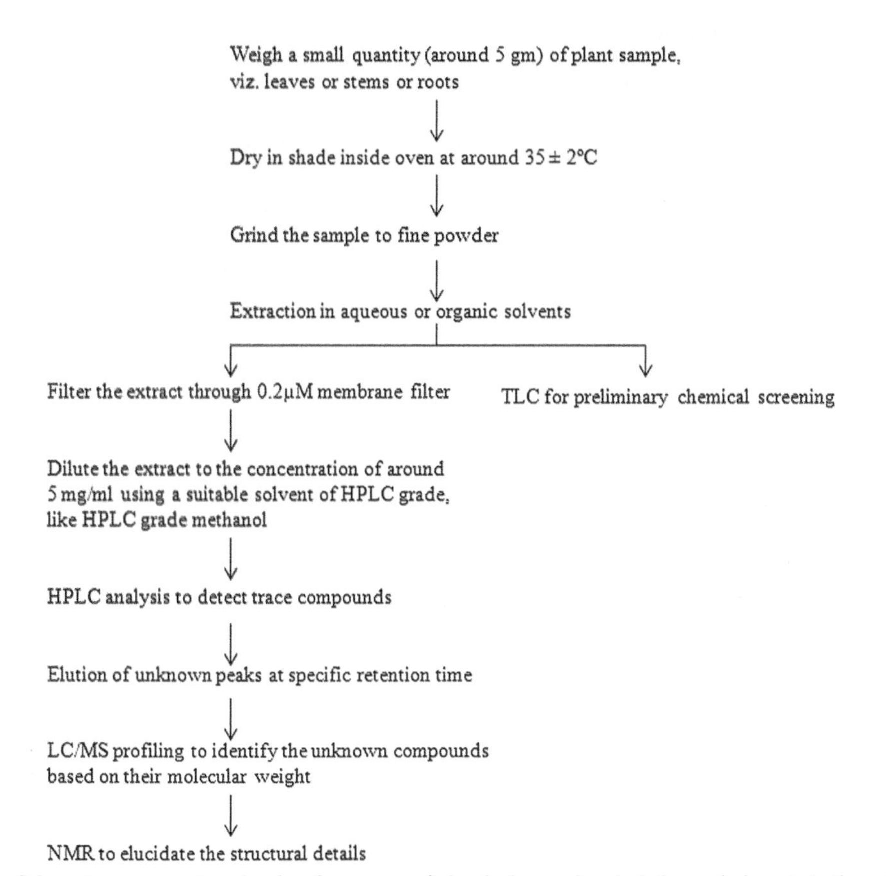
Weigh a small quantity (around 5 gm) of plant sample, viz. leaves or stems or roots
Dry in shade inside oven at around 35 ± 2°C
Grind the sample to fine powder
Extraction in aqueous or organic solvents
Filter the extract through 0.2μM membrane filter
TLC for preliminary chemical screening
Dilute the extract to the concentration of around 5 mg/ml using a suitable solvent of HPLC grade, like HPLC grade methanol
HPLC analysis to detect trace compounds
Elution of unknown peaks at specific retention time
LC/MS profiling to identify the unknown compounds based on their molecular weight
NMR to elucidate the structural details

FLOW CHART 30.2 Schematic representation showing the process of chemical screening, isolation and characterization of bioactive substances from plants.

there are situations where this tool does not give satisfactory results because of its own limitations. High-pressure liquid chromatography (HPLC), liquid chromatography/mass spectrometry (LC/MS), nuclear magnetic resonance (NMR), etc., are well-suited quantitative and qualitative analytical methods of choice to control the quality of phytopharmaceuticals.

High-pressure liquid chromatography, also called as high-performance liquid chromatography (HPLC), is an important analytical tool for the efficient localization and rapid characterization of natural products. It involves the injection of a small volume of liquid sample into a tube packed with porous particles (stationary phase), and the individual components of the sample are pulled along the packed tube (column) by a solvent (mobile phase) moved by gravity. A pump forces the liquid through the column at a specific flow rate and generates high pressure. The column packing separates the components of the sample by various physical and chemical interactions between the molecules and the packing material. The separated components get collected at the exit of the column and are detected by several techniques like UV, fluorescence detection, diode array detection, etc. Data is generated in the form of chromatograms, where individual components show peaks at specific retention times at which the component was eluted. Since, HPLC has a high resolution and is very sensitive, this technique is suitable for the detection of trace components whose concentration in the sample is very low.

The processing of a plant crude extract to provide a sample suitable for HPLC analysis, as well as the selection of solvent for sample reconstitution, can have a significant bearing on the overall success of natural product isolation and identification. The source material (e.g. dried powdered plant) will initially need to be treated in such a way as to ensure that the compound of interest is efficiently liberated into solution. This is where an efficient extraction protocol becomes important. An organic solvent may be used for extraction, and then solid material is removed by centrifugation and filtration of the extract. The filtrate is then concentrated and injected into an HPLC instrument for separation. Use of guard columns is necessary in the analysis of crude extract. Many natural product materials contain significant levels of strongly binding components such as chlorophyll and other endogenous materials that may in the long term compromise the performance of analytical columns.

Liquid chromatography coupled to mass spectrometry (LC/MS) is a newer technique, and is one of the most sensitive methods of molecular analysis. It yields information on the molecular weight and structure of the analytes. A component showing a specific retention time in HPLC can be eluted out at that particular retention time, and its mass spectral analysis can be done to get more details about its molecular weight and structure. An MS detector senses a compound eluting from the HPLC column first by ionizing it, and then by measuring its mass or by fragmenting the molecule into smaller pieces that are unique to the compound. The MS detector can sometimes directly identify the compound since every compound has its own unique mass spectrum and acts as a fingerprint for that particular compound.

Nuclear magnetic resonance (NMR) is another important analytical tool that helps in elucidation of the structural details of bioactive compounds. NMR has the ability to provide a detailed picture of molecules. Even the conformational space of molecules can be studied in great detail using this tool. This technique probes the magnetic properties of nuclei induced by their spin states. Almost every element has an isotope that is magnetically active, and their magnetic vectors align in an external field either parallel or anti-parallel to the field. There is always a small energy difference associated with the parallel and anti-parallel orientations, and the difference in energy can be visualized by irradiation with proper radio frequencies. The amount of splitting of energy levels is different for different nuclei, and is linearly dependent on the magnetic field. Therefore, different nuclei can be observed at different radio frequencies, and hence, each radio frequency becomes unique for a particular nucleus and can be easily identified.

Gas chromatography/mass spectrometry (GC/MS) is based upon the partitioning of compounds between a liquid and a gas phase. This technique is widely used for the qualitative and quantitative analysis of a large number of herbal drugs because it has high sensitivity, reproducibility, and speed of resolution. It has proved to be most valuable for the separation of volatile, non-polar, and semi-polar bioactive compounds. In GC/MS, the sample is injected into a long tubular column, the chromatography column, which has a high boiling point stationary phase, such as silicon grease. The basis of the separation is the difference in the partition coefficients of volatilized compounds between the liquid and gas phase as the plant metabolites are carried through the column by the inert carrier gas (e.g. nitrogen, helium, or argon). The time taken by the sample to pass through the length of the column is referred to as its retention time (RT). The RT for a given sample is an identifying characteristic. The detector for the GC is the mass spectrometry (MS) detector. As a sample exits the end of the GC column, it is fragmented by ionization, and the fragments are sorted by mass to form a fragmentation pattern.

BIOTECHNOLOGICAL APPROACHES FOR HERBAL DRUG PRODUCTION

Intact plants in the field or wild habitats produce high-value bioactive compounds. However, the quantity and availability of these economic products from natural resources restrict their maximized uses for the benefit of humankind. For the last few years, as the demand for bioactive compounds has increased, exploitation of medicinal plants has also increased. Hence, there is an urgent need to develop an alternative method for the large-scale production of metabolites and quality plants. In this respect, biotechnology put forward an attractive alternative to whole-plant extraction for homogeneous, controlled production, especially, when we take the commercial demand into picture. It also results in more consistent yield and quality of the products, irrespective of the seasons and the regions. Biotechnology offers an opportunity to exploit plant cells, tissues, organs, or entire organisms by growing them *in vitro* and genetically manipulating them to get desired compounds (Rao and Ravishankar, 2002). Many biotechnological strategies, such as embryogenesis, organogenesis, screening of cell lines, media optimization, and elicitation, can be carried out for enhanced production of secondary metabolites from medicinal plants. The subsequent sections briefly discuss the different *in vitro* culture techniques that can be used for herbal drug production.

Organ Cultures

The selection of an appropriate technique depends on the results that one wants. In plants where molecules of interest are localized in specialized cells, dedifferentiated cultures are not desirable. Therefore, establishment of organogenic cultures would be advantageous. Under *in vitro* conditions, redifferentiation is generally associated with an improved synthesis of secondary metabolites (Collin 2001). This is probably due to the appearance of complex cells and tissues that are metabolically more proficient. In all redifferentiated cell lines, along with the shoot-forming nodules, non-morphogenic cell masses are also present, which though non-morphogenic, might have a certain degree of differentiation at the cellular stage, and due to co-evolution, imitate the biochemistry of redifferentiated cells (Brown et al., 1986). The reports on *Artemisia annua* and *Azadirachta indica* stated that artemisinin and azadirachtin production, respectively, were very poor in dedifferentiated callus cultures, and a certain degree of redifferentiation was obligatory for compound production. Organogenesis was also found to be an essential prerequisite for steroidal saponin production in *Ruscus aculeatus*. Similar observations were made for the biosynthesis of picroside in *Picrorhiza kurroa,* wherein the metabolite did not accumulate in the dedifferentiated callus cultures, but occurred specifically in

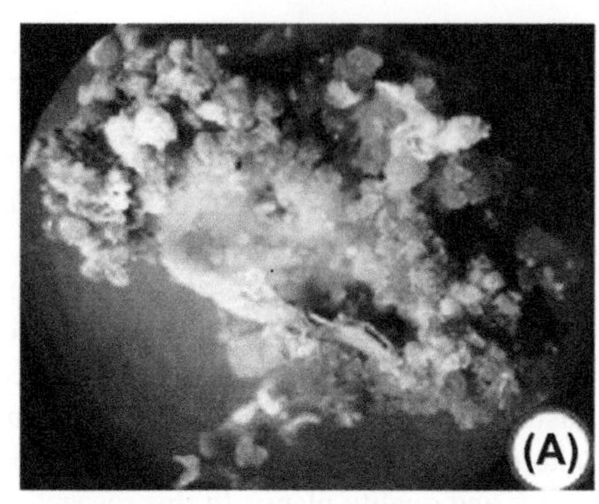

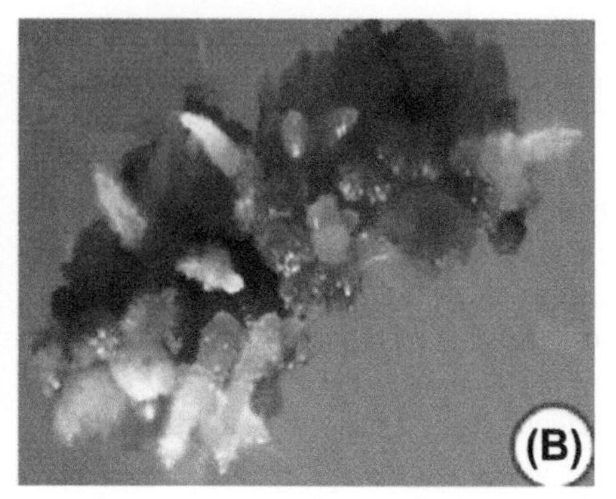

FIGURE 30 2 Neem organogenesis from leaf explants indirectly via callusing: (A) Shoot differentiation. (B) Root differentiation

the redifferentiated cultures. Berkov et al. (2010) also demonstrated that alkaloid synthesis in *Pancratium maritimum* is closely related to tissue differentiation.

Since it was observed that production of bioactive compounds is generally higher in organized plant tissues; there are attempts to regenerate whole plant organs (i.e. shoots or roots) under *in vitro* conditions, either directly from explants, or indirectly via an intervening callus phase (Figure 30.2). As expected, such regenerating cultures produce patterns of secondary metabolites that are similar to the field-grown parent plant, with the added advantage of improved production of metabolites. Another advantage of using the organized cultures is that they are relatively more stable in the production of secondary metabolites than cultures of undifferentiated cells, such as cells in callus or suspension cultures (Rao and Ravishankar, 2002).

Callus Cultures

Callus culture is the culture of dedifferentiated plant cells induced on media usually containing relatively high auxin concentrations or a combination of auxin and cytokinin under *in vitro* conditions. In plants, where sought after metabolites are present in leaves, establishing *in vitro* cultures from leaves and using them for the extraction of compounds would be an ideal alternative. Callus cultures containing the bioactive substances are collected at a specific stage (usually during the stationary phase of their growth cycle, since secondary metabolite production is greater during the stationary phase), dried, extracted, and the extract then taken for identification and quantification of the desired medicinal compound using HPLC, LC-MS, etc. The further scale-up and yield enhancement studies of the compound are performed by raising the callus in suspension, first in a shake-flask culture, and then in a suitably designed bioreactor, to maximize its production.

Suspension Cultures

A breakthrough in cell-culture methodology occurred with the successful establishment of cell lines capable of producing high yields of secondary compounds in cell suspension cultures (Zenk, 1978). During the past decades, this approach of metabolite production has attracted much academic and industrial interest. The technique of using plant cell suspension cultures for secondary metabolite production is based on the concept of biosynthetic totipotency of plant cells, which means that each cell in the culture retains the complete genetic information for production of the range of compounds found in the whole plant. Cell suspension cultures are initiated from established callus cultures by inoculating them into liquid media. The cultures are then kept in glass flasks under continual agitation on horizontal or rotating shakers; they can eventually be transferred to a specialized bioreactor. Cells in suspension cultures grow much better than in semi-solid media because of better mixing of oxygen and nutrients during shaking conditions.

Productivity of suspension cultures is critical to the practical application of this cell technology for bioactive compound production. To improve the production of secondary metabolites in *in vitro* cultures, various strategies such as the manipulation of parameters of the environment and medium, selection of high-yielding cell clones, precursor feeding, and elicitation can be opted for.

Case Study: *Lantana camara* L

This example using *Lantana camara* L. shows how plant tissues can be employed in tissue culture and further in biochemical studies.

Lantana camara L. (Sage (English) or Caturang (Hindi)) is an aromatic, evergreen shrub belonging to the family Verbenaceae. It is a reservoir of several important bioactive molecules. It has been listed as one of the important medicinal plants in the world (Sharma et al., 2000). For many years, natural products from *Lantana* have been used in the prevention and cure of many serious diseases, including cancers. The most significant bioactive molecules of this plant are shown in Figure 30.3.

For establishing tissue cultures, the first prerequisite is the selection of healthy plant material. Thus, for this study, leaves from *Lantana* plants bearing pink-yellow flowers were

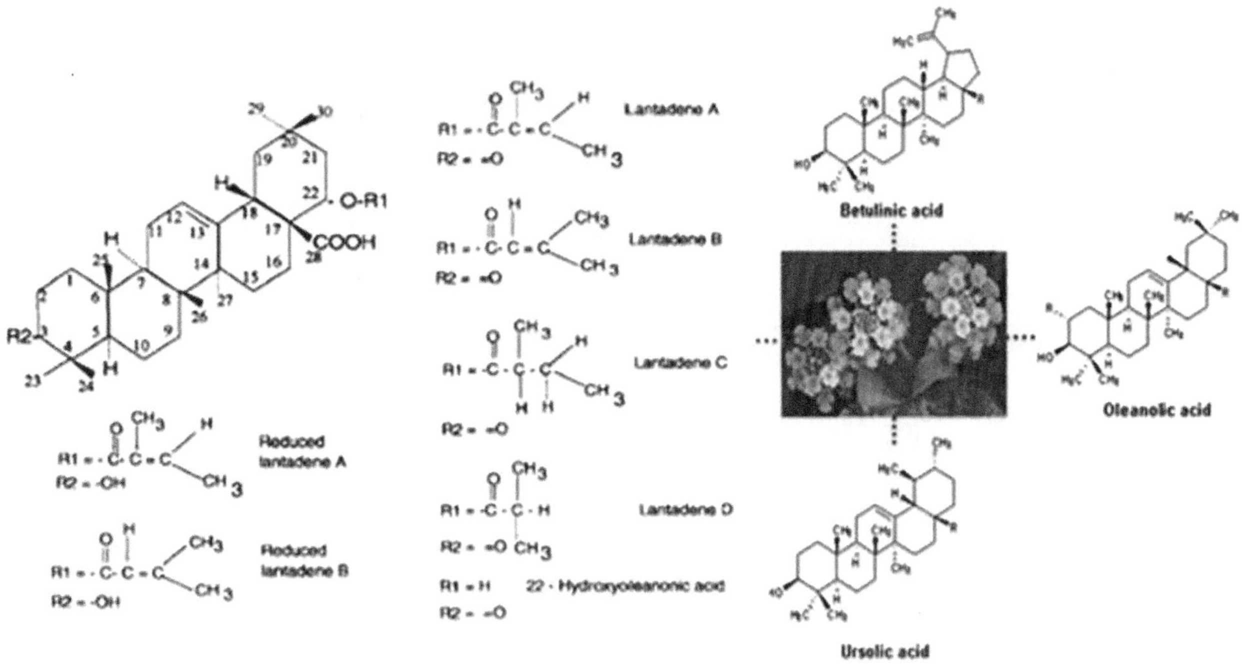

FIGURE 30.3 Bioactive compounds of *Lantana*.

picked. Leaves were disinfected using 1% (v/v) Tween-20 and 0.1% (w/v) mercuric chloride, followed by three rinses in sterile distilled water after each step. The leaf disk explants were prepared using a cork borer of 5 mm diameter. The basal media used in all the experiments related to callus induction and proliferation consisted of MS (Murashige and Skoog, 1962) medium enriched with 30 g/L sucrose and solidified with 0.8% agar (HiMedia Laboratories, Mumbai, India). The pH of the media was adjusted to 5.8 before autoclaving at 1.06 kg cm^{-2} and 121°C for 15 min. The media was supplemented with different plant growth regulators (auxins and cytokinins) at defined concentrations. Remaining steps are explicitly described in Figure 30.4.

OPPORTUNITIES AND CHALLENGES

The consumption of herbal medicines and the importance of the herbal medical industry are fast growing and widespread. According to estimates of the World Health Organization, more than 80% of the world's population depends primarily on herbal medicines. The ancient art of herbal medicine is fast developing today, and is undergoing something of a renaissance all over the world, particularly in developed countries. Most of the ingredients used in herbal medicines are taken from wild plants, and the increasing demand for medicinal plants, along with habitat loss, is putting pressure on many species. Indiscriminate harvesting from the wild has led to loss of genetic diversity, diminishing populations, local extinctions, and habitat destruction. This has raised the ire of plant conservationists.

Domestic cultivation of medicinal plants offers a viable conservation strategy, and also eliminates the problems that are generally faced in herbal extracts, such as misidentification, genetic and phenotypic variability, extract variability and instability, toxic components, and contamination. Optimized yield and uniform high-quality product can also be achieved through cultivation. However, in a rapidly shifting and fashion-prone market, the cultivator has to make the difficult decision of which particular species to grow. Therefore, the difficulty in predicting which extracts will remain marketable is another serious obstacle in bringing medicinal plants into successful commercial cultivation.

Although a large number of plant species used in herbal medicine are cultivated, a great majority of them are still utilized from the wild population. There are certain difficulties faced by growers in the cultivation of herbal plants because of low germination rates or specific ecological requirements. Lack of knowledge about the specific requirements for pollination, seed germination, and growth are the main hindrances in the cultivation of herbal plants. Fungal infection or mechanical damage frequently results in low germination rates that can be easily overcome by improved seed treatments and by ensuring optimal storage conditions. Moreover, difficult-to-grow herbal plants can be easily cultivated on a commercial scale by using controlled environments, including hydroponic systems.

Another major challenge faced in the production of herbal medicines is that the main bioactive component, which is the major ingredient in the herbal medicine, is synthesized in a very small quantity in the specific plant. This is obvious, as the bioactive components are mainly produced as secondary

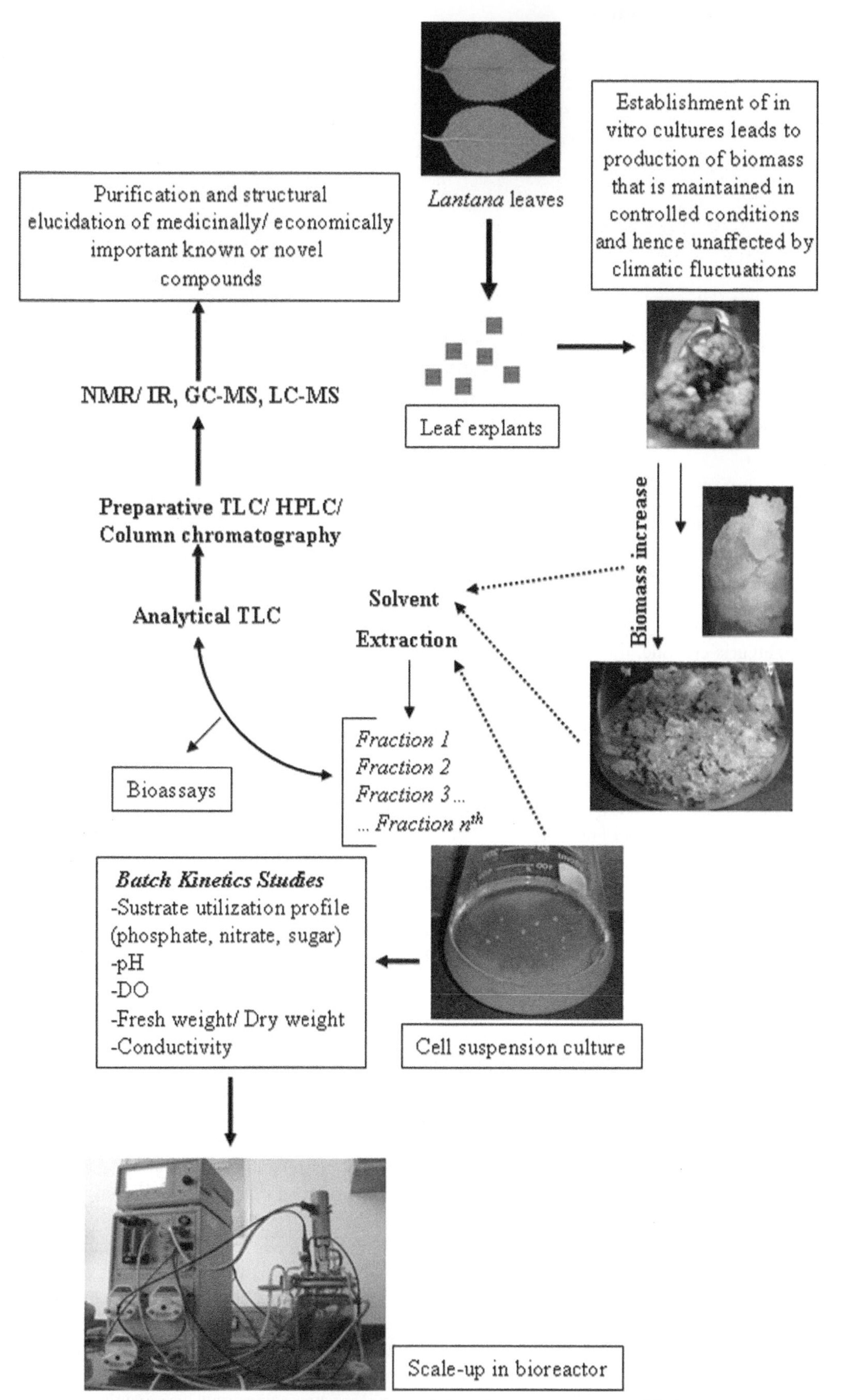

FIGURE 30.4 Isolation of bioactive compounds from *Lantana camara*, a medicinal plant.

metabolites in plant cells that are produced in small quantities. This leads to cutting down of a large number of herbal plants for producing a single drug. However, by the use of modern tissue culture techniques and genetic transformation that alters the pathways for the biosynthesis of target metabolites, today this wasteful harvesting technique can be easily overcome.

Together with supporting the use of herbal medicines, it is high time for everyone, herbalist and conservationist alike, to reduce the overexploitation of the world's wild plants. In the modern world, the trade in medicinal plants is ever-increasing, but largely unmonitored. At the moment, many harvesting practices are unsustainable, which is threatening populations of medicinal plants and their habitats, and also the livelihoods of those people engaged in their collection. It is time for the conservationists, the government, and each and every one of us to find workable global solutions.

CONCLUSIONS AND OUTLOOK

Medicinal plants are widely used by the people living in both rural and urban areas. Globalization has greatly renewed the interest in herbal medicines, and today most people prefer to take herbal medicines as an alternative therapy. This resurgence in plant remedies has mainly resulted from the following factors:

1. Herbal medicines are found to be highly effective in curing diseases.
2. Most modern drugs have one or more side effects.
3. Development of science and technology.

In addition to these factors, economic advantages also contribute to their ever-increasing popularity. Development of modern science and technology, and further studies into traditional plant medicines conducted with modern theories and techniques have greatly enriched the use of herbal medicines by absorbing new ideas and concepts from traditional plant medicine from all over the world. This has led to the tremendous expansion of the herbal medicine industry in the last few decades, and has paved the route for employment of millions of unemployed persons. Looking at all these factors, we can say that the in the not-too-distant future, traditional plant medicine will become an area of major importance in the health care system. However, efforts should be made to achieve sustainable harvesting of medicinal plants so that they are not overexploited. Also, in order to utilize the available resources of medicinal plants to their full extent, social, cultural, and economic problems, lack of well-planned and integrated strategies, and poor access to scientific information must be dealt with first.

ETHICAL ISSUES

Although approximately 80% of people today depend upon herbal medicine as a component of their primary health care, there is still concern about the safety and efficacy of herbal drugs. Despite the fact that herbal medicine can potentially contribute to the improvement of health care, many major challenges must be overcome prior to the successful incorporation of herbal remedies into medicine. Beneficence, non-malfeasance, patient autonomy, justice, and public accountability are the pillars of bioethical principles, which are religiously followed in conventional medicine. They guide the clinicians such that the patients' interests are best served. As the use of complementary medicine (including herbal medicines) becomes increasingly popular, it is becoming apparent that the same bioethical principles are applicable to these alternate forms of health care (Kemper and Cohen, 2004). Beneficence is the principle that says it is a clinician's responsibility to promote a patient's well-being; clinicians must take appropriate measures to ensure that some positive outcome will occur. Non-malfeasance is the responsibility to not hurt others. This ethical principle is almost the same as beneficence, but with important distinctions, as one's duty to prevent harm is not the same as the duty to promote well-being (Beauchamp and Childress, 2009). Patient autonomy is a foundation of conventional medicine that is pertinent to the use of herbal medicines too. In most parts of the world, consumer access to herbal medicines is controlled by prescription, thus allowing for extensive use. With self-care as one component of patient autonomy, another key element is that the patient has sound information to make an informed treatment decision (Ernst and Cohen, 2001). Time and again researchers come across cases where a patient has gathered information about herbal medicines from relatives, friends, magazines, and the Internet (Gardiner and Riley, 2007; Khader et al., 2008; Low, 2009), all of which are perceived as less reputable than official sources (Health Canada and Reid, 2005).

TRANSLATIONAL SIGNIFICANCE

Animal models are used in the study of human diseases because both animals and humans are similar in genetics, anatomy, and physiological aspects. Also, animal models are often preferable because of their easy and abundant supply and ease of manipulation. Also, for statistical analysis, a sufficient number of specimens must be used for a particular experiment. Therefore, scientists cannot conduct research on just one animal or human, and it is easier for scientists to use sufficiently large numbers of animals instead of humans to get reliable results. Only in cases of advanced clinical trials are humans used for investigations. Otherwise, animals like mice, rats, monkeys, dogs, and several fungal, bacterial, and plant species, are used as model organisms for such studies. However, even with the evident similarities between animal models and humans, only about 1% of drugs reach the last phase of clinical trials. As far as herbal medicines are concerned, the chemical constituents present in them are a part of the physiological functions of living plants, and therefore they have better compatibility with the human body. However, scientific proof of this statement is not sufficient, and this is therefore one major area where research can be carried out.

WORLD WIDE WEB RESOURCES

One of the first steps in the use of herbal medicine is to find out the best source for complete information about herbs and/or derivatives. At present, the Web is the most powerful (and perhaps most familiar) tool, but the Internet, like other resources, has its own strengths and weaknesses.

The major strength of the Internet is that it is an especially valuable research tool when looking for information that is current and frequently updated. It is also quick to access.

As far as weaknesses go, the Internet is not the best place to find established viewpoints in their original form since it is often the case that information is changed from its original source. Information on the Internet is often second-, third-, or even fourth-hand. Published books remain the safest place to get established facts and opinions, especially when looking for traditional ideas.

However, the following web sites do provide comprehensive information on herbal medicines:

http://ethnomedicinetomodern.blogspot.in/
http://www.umm.edu/altmed/articles/herbal-medicine-000351.htm
http://www.nlm.nih.gov/medlineplus/herbalmedicine.html

REFERENCES

Bannerman, R. H. (1983). The role of traditional medicine in primary health care, traditional medicine and health care coverage. *World Health Organization, Geneva*, 318–327.

Beauchamp, T. L., & Childress, J. F. (Eds.), (2009). *Principles of Biomedical Ethics* (6th edn.). New York: Oxford University Press.

Berkov, S., Pavlov, A., Georgiev, V., Weber, J., Bley, T., Viladomat, F., Bastida, J., & Codina, C. (2010). Changes in apolar metabolites during *in vitro* organogenesis of *Pancratium maritimum. Plant Physiology and Biochemistry*, *48*, 827–835.

Bourgaud, F., Gravot, A., Milesi, S., & Gontier, E. (2001). Production of plant secondary metabolites: a historical perspective. *Plant Science*, *161*, 839–851.

Brown, J. T., & Charlwood, B. V. (1986). Differentiation and monoterpene biosynthesis in plant cell cultures. In P. Morris, A. H. Scragg, A. Stafford & M. W. Fowler (Eds.), *Secondary Metabolism in Plant Cell Cultures* (pp. 68–74). Cambridge: Press Syndicate of the University of Cambridge.

Collin, H. A. (2001). Secondary product formation in plant tissue cultures. *Plant Growth Regulation*, *34*, 119–134.

Ernst, E., & Cohen, M. H. (2001). Informed consent in complementary and alternative medicine. *Archives of Internal Medicine*, *161*, 2288–2292.

Fabricant, D. S., & Farnsworth, N. R. (2001). The value of plants used in traditional medicine for drug discovery. *Environmental Health Perspectives*, *109*, 69–75.

Gardiner, P., & Riley, D. S. (2007). Herbs to homeopathy – medicinal products for children. *Pediatric Clinics of North America*, *54*, 859–874.

Canada, Health, & Reid, I. (2005). *Baseline natural health products survey among consumers*.

Kemper, K. J., & Cohen, M. (2004). Ethics meet complementary and alternative medicine: New light on old principles. *Contemporary Pediatratics*, *21*, 61–67.

Khader, Y., Sawair, F. A., Ayoub, A., Ayoub, N., Burgan, S. Z., & Amarin, Z. (2008). Knowledge and attitudes of lay public, pharmacists, and physicians toward the use of herbal products in north Jordan. *Journal of Alternative and Complementary Medicine*, *14*, 1186–1187.

Kubmarawa, D., Ajoku, G. A., Enwerem, N. M., & Okorie, D. A. (2007). Preliminary phytochemical and antimicrobial screening of 50 medicinal plants from Nigeria. *African Journal of Biotechnology*, *6*, 1690–1696.

Low, D. T. (2009). The use of botanicals during pregnancy and lactation. *Alternative Therapies in Health and Medicine*, *15*, 54–58.

Murashige, T., & Skoog, F. (1962). A revised medium for rapid growth and bioassays with tobacco cultures. *Physiol. Plant*, *15*, 473–497.

Namdeo, A. G. (2007). Plant cell elicitation for production of secondary metabolites: a review. *Pharmacology Reviews*, *1*, 69–79.

Newman, D. J., Cragg, G. M., & Snader, K. M. (2000). The influence of natural products upon drug discovery. *Natural Product Reports*, *17*, 215–234.

Orhan, I., Deliorman, O. D., & Özçelik, B. (2009). Antiviral activity and cytotoxicity of the lipophilic extracts of various edible plants and their fatty acids. *Food Chemistry*, *115*, 701–705.

Rao, R. S., & Ravishankar, G. A. (2002). Plant cell cultures: Chemical factories of secondary metabolites. *Biotechnology Advances*, *20*, 101–153.

Zenk, M. H. (1978). The impact of plant cell culture on industry. In T. A. Thorpe (ed.), *Frontiers of Plant Tissue Culture*. Calgary: IAPTC.

FURTHER READING

Hostettmann, K. (1998). Strategy for the biological and chemical evaluation of plant extracts. *Pure and Applied Chemistry*, *70*, 1–9.

Klefenz, H. (2002). *Industrial Pharmaceutical Biotechnology*. New Delhi: Business Horizons Pharmaceutical Publishers.

Kohli, J. P. S. (2009). *Dictionary of Pharmaceuticals and Biotechnology*. New Delhi: Business Horizons Pharmaceutical Publishers.

Makkar, H. P. S., Sidhuraju, P., & Becker, K. (2010). *Plant Secondary Metabolites*. New York: Humana Press.

Verpoorte, R. (2000). *Metabolic Engineering of Plant Secondary Metabolism*. Springer Verlag.

GLOSSARY

Bioactivity Specific effect on, or a reaction in, a living being upon exposure to a substance.

Biosynthetic Totipotency The inherent potentiality of a plant cell to give rise to a whole plant.

Dedifferentiation The phenomenon of a mature cell reverting to its meristematic state and forming undifferentiated callus tissue.

Plant Metabolite The intermediates and products of metabolism. Usually restricted to small molecules of a plant.

Morphogenic The development of form and structure during growth.

Redifferentiation The phenomenon of whole-plant formation from undifferentiated callus tissue.

Secondary Metabolite Organic compounds that are not directly involved in the normal growth, development, or reproduction of a plant, but often have an ecological role, such as attractant of pollinators and chemical defense against microorganisms. Humans use secondary metabolites as medicines, flavorings, and recreational drugs.

Traditional Medicine (TM) Refers to the knowledge, skills, and practices based on the theories, beliefs, and experiences, used in the maintenance of health, and in the prevention, diagnosis, improvement, or treatment of physical and mental illness.

Natural Product A chemical compound or substance produced by a living organism. A natural product often has pharmacological or biological activity for use in pharmaceutical drug discovery and drug design. A natural product can be considered as such even if it can be prepared by total synthesis.

ABBREVIATIONS

GC/MS Gas Chromatography/Mass Spectrometry
HPLC High-Performance Liquid Chromatography
LC/MS Liquid Chromatography/Mass Spectrometry
MS Mass Spectrometry
NMR Nuclear Magnetic Resonance
RT Retention Time
TLC Thin-Layer Chromatography
WHO World Health Organization

LONG ANSWER QUESTIONS

1. Write an essay on plant secondary metabolites.
2. Elucidate various steps for the study of plants in traditional medicine.
3. What is drug discovery? What are different ways for drug discovery from natural products?
4. Write a detailed account of the tools and techniques of plant tissue culture and highlight the importance of each.
5. Enlist and describe in detail important analytical techniques associated with characterization of medicinal metabolites.

SHORT ANSWER QUESTIONS

1. Define the term "secondary metabolites."
2. What is ethnobotany?
3. Differentiate between *characterized* and *uncharacterized* plant extracts.
4. Give the names of three solvents that can be used for the extraction of hydrophilic compounds?
5. Which analytical technique can be used for the separation and identification of volatile compounds?

ANSWERS TO SHORT ANSWER QUESTIONS

1. Secondary metabolites are compounds that are not directly involved in primary metabolic processes of an organism. They generally defend the organisms from environmental stresses and predators.
2. Ethnobotany is the study of how people of a particular region relate to the plants of their environment.
3. Characterized extracts are ones where each component, its concentration, and function, are known; for uncharacterized extracts, the entire components of the mixture and the role they play are not known.
4. Methanol, ethanol, and acetone.
5. GC-MS.

Chapter 31

Perspectives on the Human Genome

Aruna Kumar* and Kailash C. Upadhyaya*

*Amity Institute of Biotechnology, Amity University, Noida, Uttar Pradesh, India

Chapter Outline

SUMMARY

The chapter describes human genome projects led independently by an international collaboration of countries and a private company. The human genome sequence has opened up a whole new chapter of understanding. How this information is revolutionizing the study of biology and medicine is elaborated upon.

WHAT YOU CAN EXPECT TO KNOW

Human genome sequencing is one of the greatest endeavors of biology. Because of the efforts of publically funded human genome projects, the sequence is freely available to the public, which has contributed to significant discoveries worldwide. This chapter concentrates on understanding the architecture of the human genome in light of human genome sequencing, and how this knowledge has revolutionized the study of biology and medicine. The chapter briefly describes the human genome projects carried out by a publicly funded international consortium and a private company. The major part of the chapter elucidates various interesting findings of the human genome, and the outcomes and implications of this information. The Human Genome Project also contributed to advancement of new sequencing technologies, making personalized genome sequencing and individualized drug therapy a certainty in the near future.

HISTORY AND METHODS

INTRODUCTION

Humans suffer from a plethora of diseases caused either by an infectious agent, a nutritional deficiency, or genetic factors. Changes in an individual's genetic material can result in a genetic disorder. DNA sequences determine not only known genetic diseases such as hemophilia, sickle cell anemia, cystic fibrosis, etc., but also influence how a person responds to a drug or an infectious agent, and perhaps even our mood swings and behavior. Complex diseases such as cancer, diabetes, asthma, cardiovascular diseases, mental illness, etc. have also been shown to have some degree of genetic predisposition. The need to sequence the human genome arose from the urge to understand the genetic basis underlying diseases and the response of an individual. Understanding the human genome will help in understanding these genetic factors leading to improved methods of disease diagnosis, better prevention, and risk assessment in

Animal Biotechnology. http://dx.doi.org/10.1016/B978-0-12-416002-6.00031-6

terms of predisposition of individuals to diseases or drug toxicity (and consequently their treatment). Understanding the sequence of the human genome also has tremendous potential in biology. The Human Genome Project was initiated with this optimism that knowledge gained from understanding the human genome sequence would immensely accelerate the pace of biomedical research.

HUMAN GENOME SEQUENCING PROJECT

History

In 1985, scientists met at the University of California to explore the prospect of sequencing the entire human genome. The Human Genome Project (HGP) was initiated by the United States Department of Energy (DOE) in 1987, and was later joined by the National Institutes of Health (NIH). In 1990, James Watson was nominated as Director of the NIH component, which later became the National Human Genome Research Institute (NHGRI). In 1992, Watson resigned and the position was taken over by Francis Collins, who successfully led the project to completion. Major funding for the Human Genome Project came from the United States Department of Energy (DOE) and the National Institutes of Health (NIH). Several other countries also became associated with the project, and the International Human Genome Sequencing Consortium (IHGSC), an open collaboration involving twenty centers in six countries, was formed. Outside US, the UK Medical Research Council and the Wellcome Trust supported the genomic research in the UK. Additional contributions came from France, Japan, Germany, and China.

At the time the Human Genome Project was being undertaken by the DOE and NIH, a private company named Celera Genomics under the direction of Craig Venter also initiated sequencing of the human genome (Venter et al., 2001). The company intended to sequence the human genome within three years. This provided the drive for the HGP, as it was believed that the human genome sequence should be freely available. The HGP proposed new goals for 1998–2003. The major goal was to release a "working draft" by the end of 2001, and a complete sequence of the human genome by 2003. By the end of 1999, full-scale sequencing of the human genome began. The sequence of chromosome 22, the first human chromosome to be sequenced entirely, was published in *Nature* (Dunham et al., 1999). The second human chromosome to be sequenced by HGP scientists was chromosome 21, which is involved in Down's syndrome, Alzheimer's, and cancer (Hattori et al., 2000). By June 2000, a working draft of the human genome, covering more than 90% of the genome, was made available, followed by publication of the finished sequence, covering 99% of the genome, in 2003; this marked the 50th anniversary of elucidation of the DNA double helix structure (IHGSC, 2001; IHGSC, 2004). For more information visit www.genome.gov; www.yourgenome.org. Both public (IHGSC) and private enterprises (Celera Genomics) had completed their respective draft genome sequences. The IHGSC had used hierarchical or map-based, or BAC-based approach to achieve its goal while Celera Genomics sequencing was accomplished by a whole genome shotgun method, skipping the mapping phase (described later in the chapter). With the publication of the human genome sequence, it was expected that scientists would be able to understand the molecular basis of many genetic diseases. With the advent of next generation sequencing technology, understanding the molecular intricacies of complex human diseases is in the realm of possibility. There was an euphoria when the draft sequence of the human genome was announced in 2000 simultaeously in Washington, D.C. by the President of the United States and in London by the Prime Minister of England. In the media, this landmark achievement was compared to landing a man on the moon. This was the beginning of the "post-genomic era."

Some major bottlenecks encountered in the publicly funded Human Genome Project were non-availability of high-density genetic and physical maps. These maps are used in assembling the sequence of the whole genome of complex organisms like humans. Keeping in view the small cloning capacity of the then-current cloning vehicles, there was a need to look for better alternatives that made cloning of large inserts possible. The time taken and cost of sequencing using existing technologies at that time was high. Also, there was a need to store and analyze large amounts of sequencing data, and hence there was a need to develop better algorithms for computing (e.g. development of software packages to analyze sequence data). Practical difficulties such as cloning bias and the presence of a large quantity of repeat sequences in the human genome all provided hurdles to sequencing and assembly of human genome sequences. As a result, the first draft of the human genome had many gaps. Notwithstanding, the Human Genome Project was aided by several breakthroughs, some of which have been described as follows:

- Sanger's method of sequencing, its automation and miniaturization, use of fluorescent dNTPs, and development of capillary-based sequencing machines greatly accelerated the speed of sequencing.
- DNA-based genetic markers that helped in the construction of high-density genetic and physical maps greatly helped in assembly of YAC/BAC clones, and hence sequence assembly.
- Polymerase chain reaction, which was invented in 1983, also propelled genetic research.
- Development of large insert cloning systems such as *Escherichia coli*-based bacterial artificial chromosomes (BAC), which can carry large segments of DNA, contributed to HGP's success.
- Development of software packages such as PHRED and PHRAP for sequence analysis, etc.

Human Genome: Organization and Perspective

Sequencing of the human genome represents a significant milestone in the history of biomedical science. The efforts of the Human Genome Project have provided detailed information about the structure, organization, and function of the human genome. In order to understand human genetics and diseases, it is important to understand the structural and functional complexities of the human genome. The human cell consists of two different genomes: nuclear and mitochondrial. The size of the nuclear genome is ~3,200 Mb, and is organized into 24 (22 + X + Y) different chromosomes; when one talks about the human genome, it invariably means the nuclear genome. The mitochondrial genome is ~16,569 bp long, and is described later in this chapter.

The human genome sequences are split into 24 chromosomes. There are 22 autosomes, one X chromosome, and one Y chromosome. Each chromosome has multiple domains: (1) the centromeric region required for chromosomal separation during cell division, (2) the telomeric region required for maintaining structural integrity of chromosomal DNA during DNA replication, and (3) the chromosomal arms designated the short arm (p) and the long arm (q). Human chromosomes are composed of euchromatic and heterochromatic regions. Most genes are located in the gene-rich, transcriptionally active regions of the chromosome known as euchromatin. The other parts of the chromosome constitutes heterochromatin, which is composed of highly condensed and transcriptionally inactive regions.

Complexity of Human Genome

The human genome is a very large and complex genome. Unlike prokaryotes, which have compact genomes, the human genome is not constrained. It contains several different kinds of sequences. Only 30% of the genome contains genes or gene-related sequences. The remaining 70% of the sequence constitutes various kinds of repeats, pseudogenes, transposable elements, and many other uncharacterized sequences.

Gene Content

Knowledge of a complete set of genes and protein is integral to the study of human biology and medicine. However, this task has remained elusive, and their numbers have been fluctuating. A "gene" is defined as the part of nucleotide sequence that is necessary for the synthesis of a functional polypeptide or RNA molecule. It includes all the sequences required for the production of a particular RNA transcript, and involves distinct regulatory and coding regions. The end product of a gene could be a protein (protein-coding genes) or RNA (RNA coding genes). Evolution has defined the type and complexity of transcriptional machinery and genes. In prokaryotes, a single RNA polymerase transcribes both RNA and protein-coding genes, and has DNA-binding activity (i.e. it recognizes specific nucleotide sequences within the regulatory region of the gene). Prokaryotic protein-coding genes are often polycistronic, and their transcript is colinear with the amino acid sequence of the polypeptide it encodes. Eukaryotes on the other hand, have at least three different transcriptional machineries to transcribe different categories of genes (i.e. rRNA genes (Class I genes), protein-coding genes (Class II genes), and tRNA genes (Class III genes)). None of the eukaryotic RNA polymerases has DNA-binding activity, implying that they do not recognize specific nucleotide sequences in the regulatory region to initiate transcription. The transcription unit of eukaryotic protein coding genes can be split into exons and introns (split genes).

Estimation of the number of protein-coding genes were undertaken after completion of the human genome project. There are approximately 20–22,000 protein-coding genes in the human genome. As shown in Table 31.1, human chromosomes vary in size. They are arranged in order from longest to shortest (1 to 22), chromosomes 19 and 21 being exceptions (since the karyotyping precedes the sequencing). The largest chromosome is chromosome 1, which is ~249 Mb and contains ~2,012 known protein-coding genes. The smallest is chromosome 21 (~48 Mb) with 225 genes.

The X chromosome is ~155 Mb with 815 genes, and the Y chromosome is one of the smallest, with a size similar to chromosome 19 (~ 59 Mb); it has the least number of genes (only 45). The gene density of chromosomes varies. For example, chromosome 2 is ~243 Mb long and contains ~1,203 genes, whereas chromosome 11 (~135 Mb) contains ~1,258 genes. Chromosome 19, on the other hand, is only ~59 Mb, but carries about 1,399 genes. Thus, the highest gene density is on chromosome 19. Chromosome 17, which is about ~81 Mb, has 1,158 genes, and also has a high gene density (Table 31.1).

Many human genes have exons separated by long introns, and are located outside the heterochromatic region as expected. There is considerable variation in the overall gene and intron sizes. Usually, longer genes are the result of very long introns, and not the result of coding for longer products. Some human genes are exceedingly large. For example, the largest known human gene, "dystrophin," which is associated with Duchenne/Becker muscular dystrophy, is approximately 2.4 Mb, but the transcript is only 14,000 nucleotides long. Apparently the large size is due to intronic sequences. The gene encoding the muscle protein "titin" has the longest coding sequence (~27,000 amino acids), the longest single exon (17 kb), and also the largest number of exons (178). The gene density varies from region to region. The GC-rich regions tend to be gene-dense, with

TABLE 31.1 Ensemble Database Version 68.37

Chromosome	Length (Mb)	Protein-Coding Genes	rRNA Genes	Pseudogenes
1	249.25	2,012	66	1,130
2	243.19	1,203	40	948
3	198.02	1,040	29	719
4	191.15	718	24	698
5	180.92	849	25	676
6	171.12	1,002	26	731
7	159.14	866	24	803
8	146.36	659	28	568
9	141.21	785	19	714
10	135.53	745	32	500
11	135.01	1,258	24	775
12	133.85	1,003	27	582
13	115.17	318	16	323
14	107.35	601	10	472
15	102.53	562	13	473
16	90.35	805	32	429
17	81.19	1,158	15	300
18	78.08	268	13	59
19	59.13	1,399	13	181
20	63.02	533	15	213
21	48.13	225	5	150
22	51.30	431	5	308
X	155.27	815	22	780
Y	59.37	45	7	327

many compact genes, whereas genes with large introns are found in AT-rich regions (IHGSC, 2001).

Non-Coding Genes: As explained earlier, of about 3.2 billion-bp of the human genome sequence, less than one-third are gene or gene-related sequences. More than two-thirds of the sequence includes a variety of repeat sequences, both tandem as well as dispersed repeats. Some of the repeats are in a very high copy number, repeated more than a million times. A part of the sequence does not encode any protein, and is known to synthesize non-coding RNAs (ncRNAs). This has been one of great surprises of the post-genomic era. Certain genes encoding RNAs have been known for a while, but they have been small RNAs besides rRNAs and tRNAs that have been involved in information transfer process. There are two categories of ncRNAs: (1) microRNAs (miRNAs) involved in the phenomenon of RNA interference, and (2) a large number of sequences that encode RNA of up to several thousand nucleotides. The latter category, untranslated ncRNAs, has been implicated in a variety of biochemical processes and possibly has a regulatory role in gene expression. This realization has led to an entire field of inquiry called RNA biology.

In recent times, there has been a complete rethinking of the role of RNAs. Other untranslated RNA molecules that are involved in a variety of functions, including gene regulation, RNA processing, protein synthesis, etc., have been identified. Whole-genome or high-throughput transcripts analyses have revealed that the human genome is pervasively transcribed, and a greater part of the euchromatic region is represented in the primary transcripts. Many of these transcripts overlap with protein-coding genes, or are found in the region believed to be transcriptionally silent (The ENCODE Project Consortium, 2007). Some of ncRNAs are described as follows (data and information adapted from IHGSC, 2001):

Transfer RNAs (tRNAs) are adapters that decode the nucleotide sequence of mRNA into an amino acid sequence. tRNAs are small RNA molecules of about 80 nucleotides. They play an important role in protein synthesis. The dispersal of tRNA throughout the genome is non-random, with tRNA genes seen in clusters throughout the genome. In fact, chromosomes 1 and 6 account for half of the tRNA genes. Chromosome 6 harbors the largest tRNA gene clusters in a region of about 4 Mb. This region contains an almost sufficient set of tRNAs, representing 36 out of 49 anticodons. Only tRNAs coding for Asn, Cys, Glu, and selenocysteine are missing. Chromosome 7 carries eighteen out of 30 Cys tRNAs in a region of about 0.5-Mb while chromosome 22 and Y carries no functional tRNA genes.

Ribosomal RNAs (rRNAs) are components of ribosomes that are involved in the process of translation. Ribosomes are protein-synthesizing machinery of the cell that are made up of complexes of RNAs and proteins. Each ribosome has two subunits: large (60S) and small (40S). The large subunit is made up of proteins and 28S, 5.8S, 5S rRNAs, while the small subunit is composed of 18S rRNA complexed with proteins. Ribosomal genes are present in clusters on the short arms of five human acrocentric chromosome pairs (13, 14, 15, 21, and 22) in tandem arrays. Each repeat produces one precursor transcript that is processed to produce three types of rRNAs. The 5S rRNA genes also occur in tandem repeats. Chromosome 1 harbors the largest cluster of 5S rRNA genes. Since rRNA genes are present in arrays of tandem repeats, they are not sequenced in their entirety, and are thus underrepresented in the human genome draft sequence. Nevertheless, several rDNA-derived sequences are dispersed throughout the genome. The number of ribosomal genes on each chromosome is depicted in Table 31.1.

Small Nuclear RNAs: Several small RNA molecules have been identified that play a role in gene expression, mostly at the level of post-transcriptional processing. Some of these are described below:

Spliceosomal Small Nuclear RNAs (snRNAs): Eukaryotic pre-mRNA undergoes post-transcriptional processing (splicing), in which introns are removed to form mature mRNA. Some introns are self-splicing, and others require a spliceosome; a large ribonucleoprotein is made up of over 200 proteins and five small nuclear RNAs called U1, U2, U4, U5, and U6. These RNAs form part of a major or U2-dependent spliceosome, and process GU/AG splice sites. U11, U-12, U4atac, and U6atac, together with U5, form a minor spliceosome. A number of genes (~1,944) coding for snRNAs have been identified throughout the genome. Some of these RNAs are present in clusters; for example, U2 RNAs are clustered in tandem arrays at q21-q22. Thirty U1 RNA genes are present in a non-uniformly organized locus at 1p36.1.

Small Nucleolar RNA Genes (snoRNAs): rRNAs are synthesized in the nucleolus, and small nucleolar RNA (snoRNAs) are extensively involved in processing and modification of rRNA in the nucleolus. These RNAs modify regions of rRNA (e.g. the peptidyl transferase center and the mRNA–decoding center). In addition, many other targets, including snRNA, are being identified. A total of 1,521 snoRNA genes distributed among different chromosomes have been identified.

Besides these, an increasing number of other ncRNAs that are involved in a plethora of functions have been identified, or are being identified. Table 31.2 lists other types of ncRNAs. For more information refer to Matera et al., 2007; Ghildiyal and Zamore, 2009; Carthew and Sontheimer, 2009; Taft et al., 2010; Kaikkonen et al., 2011; Wright and Bruford, 2011; Zhang et al., 2012; Strachnan and Read; 2011) or visit www.noncode.org (NONCODE, 2005).

miRNAs are ~19–25 nucleotides long, evolutionarily conserved, small single-stranded molecules that regulate gene activity at the post-transcriptional level by a phenomenon related to **RNA interference**. The primary transcript of miRNA has a 5′ cap, 3′ polyA tail, and an inverted repeat that can base pair to form a hairpin structure. A nuclear RNase III complex (Drosha) processes this transcript to release a typical hairpin loop precursor miRNA (pre-miRNA) that moves to the cytoplasm where it is cleaved by the enzyme Dicer, a cytoplasmic RNase III, to release mature miRNA. In the cytoplasm, an effector complex called RNA-induced silencing complex (RISC, contains argonaute RNase) degrades one strand of the miRNA to leave a mature single-stranded miRNA (guide strand) bound to argonaute. This complex then selectively base pairs with target RNA that has the sequence complementary to the guide strand. The binding usually occurs at the 3′-untranslated regions (3′-UTRs) of the target mRNA sequence, but in some cases 5′-UTRs are also involved. Thus, an miRNA can target several gene transcripts. Depending on the extent of similarity, target RNA is degraded, destabilized, or prevented from being translated. A total of 1,756 miRNA genes have been identified in the human genome. Chromosome 1 has 134 miRNA genes, while Y chromosomes has the least (~15). Chromosome 19 has 110 miRNA genes. ncRNAs have occupied center stage because of their involvement in many human diseases. Many human diseases have been identified that may be due to mutation or dysfunction of ncRNAs; for example, Prader–Willi Syndrome, cancer, central nervous disorders, cardiovascular diseases, etc. (Davis-Dusenbery and Hata, 2010; Deiters, 2010; Taft et al., 2010). A full catalog of human miRNAs can be found at www.mirbase.org. Utilizing the properties of miRNAs and RNA interference mechanisms, they are being used experimentally and have great potential as therapeutic agents in terms of down-regulating the expression of specific gene(s).

TABLE 31.2 Types of Non-Coding RNAs

Name	Function
Small Cajal body RNA (scaRNA)	Modification of snRNA in Cajal bodies.
RNA ribonucleases RNase P RNase MRP	An endoribonuclease that cleaves pre-tRNA in nucleus. Cleaves rRNA in nucleolus and is involved in mtDNA replication.
Small cytoplasmic RNAs 7SL RNA	Component of signal recognition particle.
TERC (telomerase RNA component)	RNA component of the telomerase that extend telomers or ends of chromosomes.
Vault RNA	Found in ribonucleoprotein complex that is believed to be involved in drug resistance.
Long transcripts Xist	Involved in X-chromosome inactivation and dosage compensation.
Tsix RNA	An antisense to Xist.
scAluRNA	Transcribed from Alu repeats in primates.
PCGEM1 RNA	Involved in prostate cell biology and tumorigenesis.
microRNAs	Several miRNAs have been identified that regulate gene expression.
Endogenous short interfering RNA (endo-siRNA)	Involved in post-transcriptional regulation of transcripts and transposons.
Piwi-binding RNA (pi-RNA)	Involved in regulating transposon activity in germ-line cells.
Promoter-associated RNAs (PARs)	Regulate gene expression.
Transcription-initiation RNAs (tiRNA)	Regulate gene expression.
Centrosome-associated RNAs (casiRNAs)	Guide local chromatin modifications.

Genes and Diseases: Genetics plays an important role in disease occurrence and progression. There are many gene mutations that have been known to cause diseases/disorders. These may be classified as monogenic diseases (simple diseases or Mendelian diseases). As the name suggests, these types of diseases are due to mutations in a single gene. These diseases can be followed by pedigree analysis of families. They can be dominant or recessive, autosomal, or sex-linked. Examples include hemophilia, sickle cell anemia, cystic fibrosis, etc. Information about these can be accessed from the Online Mendelian Inheritance in Man (OMIM) database on the NCBI website (http://www.ncbi.nlm.nih.gov/omim). Attempts have been made to understand the underlying molecular basis of these diseases. Many of the diseases are caused by point mutations, whereas others have deletion or frameshift mutations. A class of neuromuscular diseases such as Hungtington's diseases, Fragile X chromosome, myotonic dystrophy, etc. are caused due to expansion of trinucleotide repeats. In the case of Hungtington's disease, the repeats code for the polyglutamine stretch in the respective protein product. These repeats may be present in 5′- or 3′-UTR intron or promoter region where they can cause loss or gain of function. This results in non-Mendelian-type of inheritance called anticipation, which is an increase in the probability of onset and severity of the disease as it passes through generations (Siyanova and Mirkin, 2001).

Polygenic disorders (or complex diseases) on the other hand are caused due to mutations in multiple genes. Examples include heart disease, Alzheimer's disease, diabetes, cancer, obesity, etc. Since in such complex diseases there are contributions from many genes, and they are often influenced by environmental factors, there is high degree of epigenetic modulation. These diseases are challenging with respect to decoding their molecular scenario and developing an animal model system. However, whole-genome sequencing has revealed association of certain mutations to the disease conditions.

Changes in the number or structure of chromosomes, which are the carriers of genetic material, also result in genetic disorders. For example, Down's Syndrome or trisomy of chromosome 21 occurs due to the presence of an extra chromosome 21. Similarly, Turner Syndrome (XO, 45) is due to the loss of one X chromosome, and Klinefelter Syndrome is due to the presence of an extra X chromosome (XXY, 47). Changes in chromosomal structure due to translocation (Cri du cat syndrome), inversion, or deletion (Sotos Syndrome) are also associated with abnormalities. Mutations in mitochondrial DNA also result in several diseases (described later in the text).

Biological Insights into the Human Genome: The availability of the human genome sequence has also provided several interesting insights, some of which are discussed below (adapted from IHGSC, 2001; IHGSC, 2004). Such knowledge was possible only after the sequencing of the whole genome.

Long-Range Variation in the GC Content: The presence of GC-rich and GC-poor regions in the human genome sequence had prompted the idea that these regions might have different biological characteristics, such as gene density, repeat sequences, etc. Availability of the genome sequence made possible analysis of the GC content of the human genome at a global level. The expected GC content is $41 \pm \sqrt{((41)(59)/n)}\%$. However, there are regions in the genome that show extreme variations in GC content. Long-range variation in GC content is seen throughout the genome. For example, the distal 48 Mb of chromosome 1p and the 40 Mb region of chromosome 13 have average GC content of 47.1% and 36%, respectively. Cytogenetic analyses have revealed significant association between large GC-poor regions and "dark G" (Giemsa) bands, and the lightest G-bands with high GC content.

CpG Islands: Dinucleotides CpG are extensively under-represented in the human genome. The reason for this is that most CpG dinucleotides are methylated at cytosine residues, and spontaneous deamination of methylated cytosine gives rise to thymine. As a consequence, over long evolutionary periods, CpG readily and gradually mutated to TpG dinucleotides, making them rare. However, there are "CpG" islands in the genome which are found at the 5′-end of genes and play a role in many processes, such as gene silencing, genomic imprinting, etc., thus putting these regions under functional constraint. The availability of the whole genome sequence enables one to have a global look at the CpG content and its distribution. Most of the CpG islands have 60–70% GC content. The majority of them (about 75%) are less than 850 bp. The longest CpG island is about 36,619 bp long, and is present on chromosome 10. Most chromosomes have 5–15 islands per Mb. Chromosome Y has an unusually low 2.9 islands per Mb, while chromosome 19 has 43 islands per Mb. The number of CpG islands is found to be associated with the number of genes on the chromosome.

Repeat Content of the Human Genome: The majority of the human genome is made up of repetitive DNA elements. The human genome has a much greater portion (50%) of these repeat sequences than the *Arabidopsis*, the nematode worm (*Caenorhabitis elegans*), and the fruit fly (*Drosophila*) (Table 31.3). These repeats fall into five main categories:

1. Transposable element-derived repeats (also called interspersed repeats).
2. Inactive (partially) retroposed copies of cellular genes (including protein coding genes and small structural RNAs), also referred to as processed pseudogenes.
3. Simple sequence repeats (including microsatellites and minisatellites).
4. Segmental (inter- and intra-chromosomal) duplications.
5. Tandemly repeated sequences such as at centromeres, telomeres, ribosomal gene clusters.

Transposable Element-Derived Repeats: Transposable elements are found to be a ubiquitous component of all the genomes studied so far. These are mobile DNA sequences that can move from one region to another region of the genome. On the basis of their structure and mechanism of transposition, they are classified into two main categories: Type I and Type II elements. Type I elements (retrotransposons) are thought to be derived from retroviruses and their transposition requires a reverse-transcription step that is mediated via an RNA intermediate by a mechanism called "copy and paste." They are further subdivided into two types based on the presence or absence of repeats: LTR and non-LTR types. The non-LTR retrotransposons include the LINE and SINE elements. The Type II elements are DNA transposons, and their movement is mediated by a DNA intermediate by a mechanism called "cut and paste." Transposons that can transpose independently are called autonomous, and others that require the presence of a cognate element for transposition are called non-autonomous (Slotkin and Martienssen, 2007). Many human repeat sequences are derived from transposable elements. About 45% of the genome is comprised of potentially mobile elements (Table 31.4).

TABLE 31.3 Mobile Elements in Genomes of Various Organisms

Organism	Genome Size (Mb)	Number of Genes	Fraction TE
Saccharomyces cerevisae	13.5	~6,000	~2.4
Drosophila	180	~13,600	~30
Human	3,300	~22,000	~45
Arabidopsis	125	~25,600	~14
Rice	389	~37,550	~33
Maize	2,300	~32,540	~84.2

TABLE 31.4 Summary of Mobile Elements in the Human Genome

Element	Fraction TE (%)	Copy Number
L1(LINE)	16.9	0.5×10^6
Alu (SINE)	10.6	1.1×10^6
L2 (LINE)	3.2	0.3×10^6
MIR (SINE)	2.5	0.46×10^6
LTR Elements	8.3	0.3×10^6
DNA Elements	2.8	0.3×10^6
Processed Pseudogenes	< 1.0	$1\text{-}2 \times 10^4$
Total	~45	$\sim 3 \times 10^6$

(IHGSC, 2000).

Long Interspersed Nuclear Elements (LINEs): These are the most primitive autonomous mobile elements found in the human genome. In humans, three distantly related LINE families are present: LINE1, LINE2, and LINE3, of which only LINE1 is still active. LINE machinery has resulted in reverse transcription in the human genome (including those of SINE and processed pseudogenes). These transposons are about 6 kb long, encoding two ORFs and possess a polymerase II promoter. Transposition of LINE occurs via RNA that assembles with encoded proteins and moves back to the nucleus. LINE transposition preferably occurs in AT-rich regions. The preference of LINEs for AT-rich DNA could be explained by the fact that these regions are poor in genes, and as a result these elements do not encounter a functional selection pressure. The LINE endonuclease cleaves at TTTT/A, in the DNA sequence where the LINE element is inserted into the genome.

Short Interspersed Nuclear Elements (SINEs): Unlike LINEs, SINEs are non-autonomous transposons. They cannot transpose independently, and are believed to use the LINE machinery for transposition. Nevertheless, they have a very high copy number within the mammalian genome. They share their 3′-end with the LINE element. They are 100–400 bp long, encode no protein, but possess an internal polymerase III promoter. There are three distinct monophyletic families of SINEs: the active Alu, the inactive mammalian-wide interspersed repeat (MIR), and Ther2/MIR3. The promoter of all SINEs but one is derived from tRNA genes. The Alu elements are the only active class of SINEs that have derived a promoter from signal recognition particle component 7SL. The Alu family is the most abundant sequence in the human genome, occurring more than once in 3 kb on an average. They are of comparatively recent evolutionary origin. An Alu repeat is about 280 bp long and consists of two asymmetric tandem repeats, each about 120 bp long followed by a short An/Tn sequence (Strachnan and Read; 2011). Again, unlike LINEs, which are preferentially located in AT-rich regions, Alu repeats are found in gene-rich, high GC-rich regions, indicating their positive role in genome function. The fact that Alu uses LINE machinery for transposition, but is found in a GC-rich region, indicates that its distribution is influenced by evolutionary forces. Some Alu sequences are transcribed, and it is believed that these elements may have been involved in the evolutionary dynamics of the human genome.

LTR Retrotransposons: These are also of two types, autonomous and non-autonomous. They are flanked by long terminal repeats that carry transcriptional regulatory elements. The autonomous retrotransposons contain gag and pol genes that are closely related to retroviral proteins and encode a reverse transcriptase, RNase H, protease, and integrase. Transposition occurs through reverse transcription of an RNA intermediate in the cytoplasm. The mammalian genomes also have vertebrate-specific endogenous retroviruses (ERVs), which seem to be active in the mammalian genome.

DNA Transposons: The class II transposons are similar to bacterial transposons. They have terminal inverted repeats that flank these elements and encode a transposase. This protein recognizes the terminal inverted repeats, and mediates transposition by excising the transposon out of the donor position and integrating it into the new acceptor site. The transposition may occur through a cut-and-paste mechanism in which the donor site is repaired by removal of the transposon or filled with a copy of the transposon by gap repair. They can be autonomous or non-autonomous type. The latter type lacks the transposase enzyme, and is usually a deletion derivative of the cognate autonomous elements. There are seven major classes of DNA transposons that can be further subdivided into many families with independent origins (RepBase, www.girinst.org/repbase/update/index.html). Some examples of DNA transposons are MER1-Charlie, MER2-Tigger, Mariner, PiggyBac-like, etc. DNA transposons have a short lifespan within a species, and have not been as successful as LINEs. The transposase cannot differentiate between an active and an inactive copy of a DNA transposon in the nucleus; as a result, the transposon activity declines because of accumulation of inactive copies in the genome.

Previously named as "junk DNA," selfish, or parasitic elements, these sequences actually serve a purpose of coding RNA that regulates other genetic and cellular functions. They serve as epigenetic regulators of the genome and are involved in many epigenetic mechanisms, such as imprinting, X-inactivation, position effect variegation, etc. (Slotkin and Martienssen, 2007). These sequences have played an important role in influencing the genome by causing rearrangements, chromosomal breakage, illegitimate recombination, modification, and reshuffling of existing genes, thus creating new genes and modulating the overall GC content. They also provide tools for population genetic studies.

Pseudogenes: Pseudogenes are the non-functional copies of normal genes with which they share sequence homology. They are of two types; Duplicated pseudogenes and Processed pseudogenes. Pseudogenes that are derived from duplication of genomic DNA are called duplicated pseudogenes, as they possess characteristic exon–intron structure. Such pseudogenes are derived by tandem duplication, and as a result they are located close to the functional copy. In contrast, "retro-transposed" or processed pseudogenes (PPs), are derived from reverse transcription of processed transcripts that integrate into the genomic DNA. Therefore, they are devoid of introns. They also possess polyA tracts and direct repeats at either end of the pseudogenes. Lack of a promoter usually results in an inactive gene copy. There are about 14,427 pseudogenes in the human genome. Prevalence of pseudogenes in the human genome has been problematic for genome annotation. Availability of a near-complete sequence has made the analysis of pseudogenes possible. The highly expressed housekeeping genes have multiple processed pseudogenes. For example, ribosomal proteins account for ~20% of human PPs. Protein-coding pseudogenes are more easily identified than RNA pseudogenes (Pink et al., 2011).

Pseudogenes were believed to be nonfunctional, but in the post-genomic era this view is changing. Evidence shows that many pseudogenes (2 to 20%) are transcribed into RNA. Examples include pseudogenes for tumor suppressor PTEN, glutamine synthetase, glyceraldehyde-3-phosphate dehydrogenase, etc. Many pseudogenes have been found to have tissue-specific expression, and some have an expression pattern that is different from the parent gene. Alteration in pseudogene expression has been seen under different physiological conditions, such as diabetes and cancer. Some pseudogenes have been implicated in gene regulation in mice via siRNA production (Pink et al., 2011). For more information visit www.pseudogene.org.

Simple Sequence Repeats (SSRs): SSRs are another class of repeats represented in the human genome. SSRs, which are 1–13 bases long, are called microsatellites, whereas minisatellites have longer repeat units (14–500 bases). SSRs comprise about 3% of the human genome, of which the greatest contribution comes from dinucleotide repeats. There is about one SSR per 2 kb. The most commonly present SSRs are AC and AT, followed by AG, AAT, and AAC. Trinucleotides are less frequent than dinucleotides.

Segmental Duplications: These duplications represent transfer of 1–200 kb of genomic sequence from one location to another. Segmental duplications are regions ≥ 1 kb in length, and have sequence identity of ≥ 90%. These duplications appear to have arisen recently, as indicated by high similarity in their sequences, and their absence in closely related species. There is a high proportion of large duplications in the human genome as compared to *Drosophila* and *C. elegans*. Particularly, the pericentromeric and subtelocentromeric regions of chromosomes largely have recent segmental duplications. Approximately 5.3% of the euchromatic genome is covered by segmental duplications (IHGSC, 2004). Segmental duplications are both interchromosomal and intrachromosomal. Interchromosomal duplications involve transfer of a region between non-homologous chromosomes; for example, the 9.5 kb region of the adrenoleukodystrophy locus from Xq28 is found on chromosomes 2, 10, 16, and 22. Intrachromosomal duplications occur within a chromosome or chromosome arm, and include low copy number repeats. The Y chromosome is peculiar since > 25% of its total length carries segmental duplications as large as 1.4 Mb of 99.97% similarity.

Mitochondrial Genome: As mentioned earlier, human cells have another genome that is present in the mitochondria. Mitochondria are the major energy generator of the cell, and are involved in cellular respiration. They also carry their own DNA, which is small in size. The human mitochondrial genome is 16,569 bp long. It is a circular molecule that is present in multiple copies. The two strands of mtDNA have significant differences in G and C composition. One strand is rich in guanines, and hence called the heavy (H) strand, whereas the other, or light (L) strand, has more cytosines. The transcription of each strand begins at the major promoter regions, PL and PH1, which are present in the control region (including the D-loop). The promoter PL predominantly transcribes the light chain, and PH1 transcribes the H-strand to generate polycistronic RNAs. The mature RNAs are generated by cleavage of multigenic transcripts. Replication of both the H- and L-strand starts at a specific origin; the replication of the H-strand begins at the D region. When two-thirds of the H-strand has been synthesized, the origin of replication of the L-strand, which lies between two tRNA genes, becomes exposed, at which point the L-strand is synthesized in the opposite direction. For more information see www.mitomap.org.

Because it was small in size, the mitochondrial genome was the first to be sequenced. The complete sequence of the mitochondrial genome was available in 1981 by Sanger and colleagues (Anderson et al., 1981). The mitochondrial genome codes for 37 genes that include 22 tRNAs, 2 rRNAs, and 13 protein-coding genes. The entire mitochondrial DNA (mtDNA) is transcribed, and contains very small noncoding regions. The gene density of mtDNA is 1 per 0.45 kb. All 37 mitochondrial genes lack introns. Protein-coding genes encode proteins involved in the electron transport chain and oxidative phosphorylation. Mitochondrial genetic maps are also available at www.mitomap.org. The vast majority of mitochondrial proteins are encoded by nuclear DNA. The mitochondrial genetic code has 60 codons and 4 stop codons. The stop codons include UAA and UAG (which are universal stop codons), and AGA and AGG (which specify arginine in the nuclear genetic code). The universal stop codon UGA codes for tryptophan in mitochondria, and AUA codes for methionine. Thus, the mitochondrial genome has a unique genome organization and is different from the nuclear genome.

Mutations in mtDNA have been implicated in a broad spectrum of degenerative diseases involving the central nervous system, heart, muscle, kidneys, eyes, etc., as well as aging. These mutations are inherited from mothers' side or through accumulation of mutations with age in somatic cells. Because mitochondria are inherited from the maternal side, the affected mother can have affected progeny, but affected fathers will not transmit the disease (maternal inheritance). The other characteristic of diseases that are coded by mitochondria is recurrent heteroplasmy. This is because there are several copies of mtDNA in a eukaryotic cell. Mutations may be present in all the copies (homoplasmy) or only a fraction of them (heteroplasmy). As they replicate and segregate randomly to daughter cells, their ratio changes with each replication. Depending upon the mutation, heteroplasmy, and tissue affected, there is wide diversity in disorders/diseases caused by mtDNA abnormalities. Mitochondrial myopathies are neuromuscular diseases caused due to mutation in the mitochondrial genome. Some common mitochondrial myopathies include myoclonus epilepsy, Kearns–Sayre Syndrome, mitochondrial encephalopathy, etc. It is suspected that many other diseases such as diabetes mellitus, Alzheimer's, and Parkinson's diseases, etc., are due to dysfunction of mitochondria (Crimi and Rigolio, 2008). There is also interest in mitochondrial tRNAs, as several pathological point mutations associated with neuromuscular disorders and diverse clinical phenotypes have been identified. For more information visit Mamit-tRNAdb (mamit-trna.u-strasbg.fr) and the mitomap database (www.mitomap.org).

Development of Next Generation Sequencing Technology: One of the major goals of HGP was to increase the output and reduce the cost of sequencing, and to carry out research on new technologies. Major developments have occurred in this field, which has abridged the time and cost involved in sequencing the whole genome. The cost of sequencing has been reduced from $5,292.39 per Mb in 2001 to $0.09 per Mb in 2012. This has been possible through development of "massively parallel" sequencing technologies developed independently but significantly differ from capillary sequencing methods. These sequencing technologies exclude the bacterial cloning step and use different methods to produce sequence information of millions of DNA molecules simultaneously. Examples include the Roche/454 FLX pyrosequencer, the Solexa/Illumina genome analyzer, and Applied Biosystems' SOLiD™ sequencer (Mardis, 2008, 2011). These technologies are also called next or second-generation sequencing technologies. The third-generation sequencing technologies that involve sequencing single molecule, are also being developed (review by Schadt et al 2010; Thompson and Milos, 2011). It took 13 years to obtain the first human genome sequence. The whole genome sequence of James Watson was obtained in just two months using massively parallel sequencing. This was the first genome sequenced by a next generation technology (Wheeler et al., 2008). Thus, these massively parallel sequencing technologies are reducing the time and cost involved in sequencing. Several companies now advertise to sequence the human genome for just $6,000. These sequencing technologies are also contributing to the detection of variants, including rare, copy number, or structural variants, SNPs, metagenomics, and transcriptomics studies such as gene expression, small RNA, or novel RNA discovery.

PRINCIPLE

The major objectives of the human genome project were to sequence the entire human genome, identify all 20,000–25,000 genes, and to provide high-quality data of human genome sequences as a free resource. This was by far the most complex project of biology undertaken by mankind. In addition, the human genome has greater than 50% of repeat elements with regions of genomes that are duplicated or present in tandem or dispersed across the genome. It was therefore anticipated that without the development of genetic and physical maps, it would not be feasible to construct the human genome, as the presence of repetitive DNA would lead to misassemblies. Hence, a strategy was used by IHGSC; it was known as a "hierarchical," "clone-by-clone," "map-based," "BAC"-based approach (Flow Chart 31.1) (IHGSC, 2001; www.genome.org). This approach requires construction of a genetic map, a physical map, and a comprehensive library of large insert DNA clones prior to human genome sequencing. Consequently, the initial major focus of HGP was construction of high-density genetic and physical maps. The development of DNA-based markers: Restriction Fragment Length polymorphism (RFLPs), and later PCR-based marker Simple Sequence Length Polymorphism (SSLPs) that could detect DNA polymorphism, greatly enhanced the construction of a high-density linkage map. Physical mapping was also carried out at the same time. Another PCR-based DNA marker, called a sequence-tagged site (STS), was very useful in the construction of physical maps (Olson et al., 1989). STSs are short unique sequences of known DNA that occur only once in the genome. These can be derived from expressed sequence tags (ESTs), microsatellites (SSLPs), or random genomic sequences (Brown, 2007). STS derived from SSLPs are particularly useful as they provide a direct connection between the physical and genetic maps. These molecular markers are very useful for ordering large segments of DNA. Hence, once the genetic and physical maps were made, the next step was to sequence each BAC clone by shotgun sequencing.

Prior to human genome sequencing, sequencing of more modest-sized genomes of model organisms such *as E. coli*, yeast, fruit fly, and *C. elegans* was carried out. Of these, *C. elegans* was the first animal to be completely sequenced. It has about 16.5% repeat elements in its genome. This provided the assurance that large-scale genome sequencing of complex multicellular organisms is possible. Sequencing of fruit fly and worm genomes also contributed to the development of software required to assemble and sequence annotations.

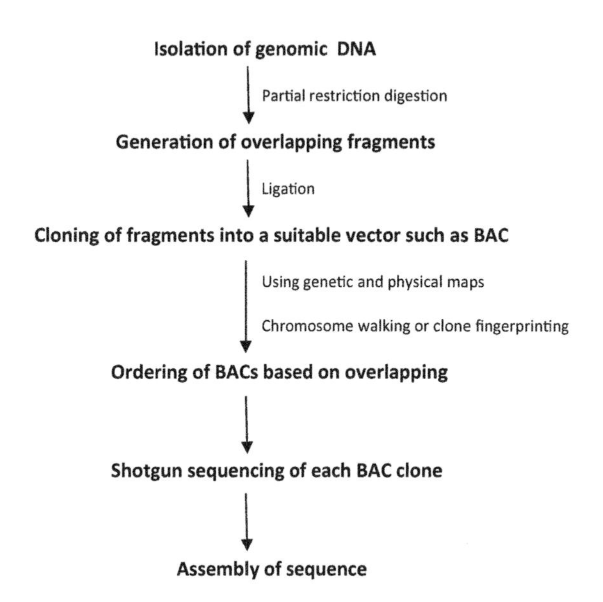
Isolation of genomic DNA
Partial restriction digestion
Generation of overlapping fragments
Ligation
Cloning of fragments into a suitable vector such as BAC
Using genetic and physical maps
Chromosome walking or clone fingerprinting
Ordering of BACs based on overlapping
Shotgun sequencing of each BAC clone
Assembly of sequence

1. The DNA is isolated from samples provided by anonymous donors
2. Next step is to generate overlapping fragments of desirable size into BAC vectors which have high cloning capacity
3. The BAC library thus obtained is used to prepare an ordered map with help of use of genetic and physical maps
4. Each BAC clone is further fragmented into smaller size and sequenced via shotgun sequencing
5. The genome sequence is assembled using computer programs

FLOW CHART 31.1 Hierarchical, or clone-by-clone, map-based, BAC-based approach (IHGSC, 2001).

An alternative strategy of whole-genome shotgun sequencing was used by Celera Genomics. The feasibility of this method was demonstrated by sequencing the *Haemophilus influenza* genome (Fleischmann et al., 1995). This strategy was first proposed by Weber and Myers, and later by Craig Venter of Celera Genomics, as a rapid way to sequence the human genome. Unlike the strategy used for the publically funded genome sequencing, this strategy involves generation of small fragments of the genome, random shotgun sequencing to ~5-fold coverage, and using software to obtain sequence assembly without prior ordering of BACs. It was argued that the human genome is too complex to be sequenced by this method. To test the hypothesis, the *Drosophila* genome was sequenced over a span of one year by this method. This project demonstrated that the chromosomal assembly is possible with high accuracy, order, and with < 10 fold coverage (Venter et al., 2001).

METHODOLOGY

In a hierarchical method (Adapted from IHGSC, 2001) (Flow Chart 31.1), DNA was obtained from anonymous human volunteers of diverse backgrounds. All samples were stripped of their labels and were given random labels such that the identity of donors for the library could not be ascertained (in accordance with U.S. Federal Regulation). The genome was fragmented into overlapping DNA molecules of high molecular weight (~100–200 kb) by partial restriction digestion. Please note sequencing a large genome such as the human genome requires vectors such as YAC or BAC, which has the capacity to clone large DNA molecules. A library of clones is thus obtained. Each clone has a contiguous stretch of genomic DNA, and it is expected that all the clones together comprise the complete genome. The DNA cloned in BACs is then ordered, covering the genome. Ordering of BACs is on the basis of the way in which they overlap, which shows that they contain overlapping segments of the genome. This can be done in several ways: chromosome walking, or clone fingerprinting and BAC end alignment. The existing genetic and physical maps were used for this purpose. Once this is done, the next step is to sequence them. The genomic fragment cloned in a BAC is too large to be sequenced. Hence, it must be sheared into smaller fragments before being sequenced by shotgun method. To increase efficiency and reduce price, it is desirable to cover the whole genome using a minimum number of clones, and only sequence them. The selected clones were subjected to shotgun sequencing. The sequencing data obtained from various centers could be processed and assembled using PHRED and PHRAP software packages. The PHRED software package made use of "base-quality control," which is helpful in monitoring raw data quality.

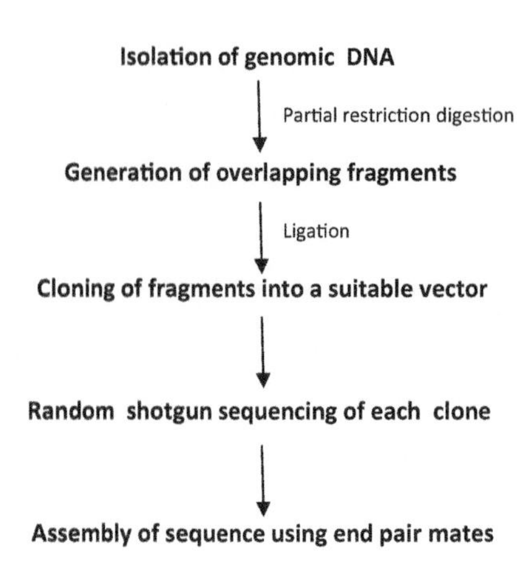
Isolation of genomic DNA
Partial restriction digestion
Generation of overlapping fragments
Ligation
Cloning of fragments into a suitable vector
Random shotgun sequencing of each clone
Assembly of sequence using end pair mates

1. The DNA is isolated from samples provided by anonymous donors
2. Next step is to generate overlapping fragments of desirable sizes
3. This is followed by cloning of DNA fragments in appropriate vectors
4. Random shotgun sequencing is carried simultaneously for all the clones
5. The genome sequence is assembled using computer programs

FLOW CHART 31.2 Whole genome shotgun sequencing approach (Venter et al., 2001).

The PHRAP computer program assembled the sequence into sequence contigs on the basis of base-quality scores, and assigned "assembly-quality scores." Assembled contigs of more than 2 kb were deposited in a public database within 24 hrs of assembly, and were freely available to the public as per the principles of HGP.

The hierarchical approach is technically demanding, slow, and costly because it requires construction of a genetic and physical map before sequencing is undertaken. There is also a possibility of rearrangements of large-insert clones, which can cause complications during genome assembly. This can be prevented using quality control measures. Nevertheless, this approach greatly reduces the challenge of assembling the finished sequence of complex and outbred organisms, like humans, who have a high percentage of repetitive DNA. Because the relative order of BACs is known, the possibility of long-range misassembly is reduced. In the whole-genome approach (described below), regions in which assembly is incorrect are not identified. This approach ensures complete coverage of the genome, and helps reduce sequencing redundancy. It was believed that this strategy would be able to identify regions that showed cloning biases, which could then be focused for additional sequencing. As the IHGSC was a collaboration of different countries, the work and responsibility could be easily disseminated among its members using this strategy (IHGSC, 2001).

Whole Genome Shotgun Sequencing: For HGP, after obtaining a certificate to protect the privacy of individuals, the DNA from five individuals – one African-American, one Asian-Chinese, one Hispanic-Mexican, and two of Caucasians – were selected for genome sequencing (Flow Chart 31.2). This method involves breaking DNA into smaller size fragments, and requires generation of a high-quality plasmid library. Clones are randomly sequenced, and a master sequence is obtained based on overlapping stretches of sequences using a computer program. The central feature of this strategy is to obtain paired-end reads (mate pairs), which are 500–600 bp in length from both ends of the inserts. The end sequences of the BACs provide a connection between the continuous sequences across the genome, which allows simultaneous mapping. As a result, the time taken by this method to produce a whole genome sequence is greatly reduced (Venter et al., 2001).

EXAMPLES WITH APPLICATIONS

The availability of a reference human genome sequence has given several new insights into biology, and has also enabled a broad range of scientific advances, some of which are discussed below:

Identification of Disease-Related Genes: Even before sequencing of the human genome began, linkage maps prepared by pedigree analyses were used to identify genes responsible for causing disease. Mapping genes using RFLP helped in defining gene positions. This helped in positional cloning of genes. Huntington disease was the first genetic disease mapped on chromosome 4 (using a DNA polymorphism). The first human disease gene to be positionally cloned was chronic granulomatous disease, followed by Duchenne muscular dystrophy, and retinoblastoma. The availability of genetic and physical maps (and later the human genome) allowed rapid identification of candidate genes *in silico*. Mendelian disorders are now readily identified in short timespans, aided by the whole-genome sequence. More than 3,000 Mendelian diseases have been cataloged (http://www.ncbi.nlm.nih.gov/omim). Several genes of biomedical importance have been identified using resources and information from HGP. Other examples include breast cancer susceptibility (BRCA2), total color-blindness (CNGB3), X-linked lymphoproliferative disease (XLP), etc (IHGSC, 2001).

Genome-Wide Association Studies: With the availability of the human genome, the next logical step was to identify genetic variants that predispose some individuals to common diseases such as cancer, diabetes, depression, heart diseases, etc. Unlike Mendelian diseases, these common diseases are caused by many genetic and environmental factors working together. To understand the genetic factors involved, one can compare the genomes of healthy and disease-carrying individuals to look for association of genetic factors with that particular disease. Genome Wide Association (GWA) studies have been defined as "any studies of common genetic variation across the entire human genome designed to identify genetic associations with observable traits" (Manolio and Collins, 2009). Such genetic analysis in humans requires variations or polymorphisms. The International HapMap Project was conceived in 2002 with the aim of providing data on human genetic variation that could be used for gene identification associated with common diseases. The International HapMap project is an extensive collaborative effort between Canada, China, Japan, Nigeria, US, and UK. Its goal is to create a database of human genetic variations, particularly single nucleotide polymorphisms (SNPs), which are present once in every 300 bp of the human genome. In addition to their abundance, SNPs are easy to genotype on large-scale, are present throughout the genome, and hence they are useful in GWA studies (http://www.hapmap.org; http://hapmap.ncbi.nlm.nih.gov/). Even before the complete human genome sequence was available, more than 1.4 million SNPs had been identified, and thereafter their numbers have been increasing. If a particular SNP or group of SNPs is more common among people carrying the disease/trait that SNP is more likely to be associated with and can be used to locate genes involved in the phenotype. Thus, HapMap made possible GWA studies, in which numerous SNPs are tested for their association with a condition or disease. These studies have provided valuable insight into the physiopathology of diseases. Such studies are now possible, and they are less time consuming and cheaper because of the availability of the human genome sequence and cost-effective genotyping technologies. For more information, an updated catalog of published GWA studies can be obtained from http://www.genome.gov/GWAStudies or Hindorff et al. Some examples of GWA collaborations are the following: Genetic Association Information Network (GAIN), which studies ADHD, bipolar 1 disorder, schizophrenia, major depressive disorder, etc.; and Cancer Genetic Markers of Susceptibility (CGEMS), which studies prostrate, breast, lung, bladder, renal, and pancreatic cancers, etc. (Manolio and Collins, 2009; Manolio et al., 2008). Such studies have implicated previously unknown mechanisms in certain conditions. For example, in age-related macular degeneration, a complement-mediated inflammation is involved. Cell-cycle regulators (CDKN2A and CDKN2B) have been shown to be associated with coronary disease, and CDKAL1 in Type 2 diabetes (Manolio, 2010). The ultimate goal of the GWA studies is to provide information so that better decisions can be made in terms of risk assessment, diagnosis, and drug therapies. Information from such studies will be useful in identifying new therapeutic targets, and in pharmacogenetic and pharmacogenomic studies (Zhang et al., 2008; Green et al., 2011). The use of genomic approaches in understanding the molecular basis of diseases is best depicted by the study of a gastrointestinal disorder: Crohn's Disease. GWAS has revealed the role of autophagy, innate immunity, and interleukin signaling in this ailment. Thus, a genomics approach has led to better understanding of diseases, and has also contributed to development of new treatments.

ETHICAL ISSUES

The bright side of whole-genome sequence information is that it has tremendous potential for use in medicine, biology, genetic counseling, etc. However, access to this information also has a potential dark side because it can impact individuals in various ethical, legal, and social ways. One of the goals of HGP was to address the ethical, legal, and social issues (ELSI) that could arise as a consequence of genomic research (Resources from www.genome.gov). With the whole genome sequences and subsequent data freely available to the public, concerns were raised. The information can be misused by schools, courts, adoption agencies, employers, and in particular, health insurers, who may discriminate on the basis of an individual's genetic makeup. For example, if a person is a carrier or has a predisposition to certain diseases that can be determined by genetic tests, he or she could be unfairly refused employment by a company. A medical insurance agency might not cover such an individual, or might charge hefty insurance premiums. This is discrimination based on a person's genetic information, and could be exploited by unscrupulous companies. Social issue concerns, such as stigmatization of an individual based on race or ethnicity and their genetic differences, could have a profound psychological impact or could create social disparity. In order to prevent misuse of genomic information, the Genetic Information Non-Discrimination Act (GINA) was passed in the U.S. (Hudson et al., 2008). This law prohibits discrimination of individuals by insurance companies and employers based on information from genetic tests.

TRANSLATIONAL SIGNIFICANCE

Development of pharmaceutical drugs to treat disorders and diseases requires understanding of suitable drug targets. Information about the complete set of human genes and proteins has increased the search for suitable drug targets. In addition, knowledge about paralogs has provided an opportunity for new therapeutic interventions. Already, several new paralogs of common drug targets have been identified *in silico* that could represent new drug target candidates. Some examples include calcitonin (CALCA), dopamine receptor, D1-α (DRD1), insulin-like growth factor-1 receptor (IGF1R), etc (IHGSC, 2001).

As mentioned previously, GWAS is helping to understand the molecular basis of complex diseases, which in turn is pinpointing or identifying new or candidate targets for drug development. In addition, Hapmap and GWAS have allowed identification of clinically important genetic polymorphisms that have been linked to differences in response to drugs by different individuals. Consequently, the field of pharmacogenomics (Pharmacology + Genomics) has evolved in the post-genome era; it is based on the fact that the response of different individuals to a drug varies due to variations in their DNA sequences. These could be variations in drug receptors, drug transporters, or drug-metabolizing enzymes (Nwanguma, 2003). Completion of HGP, one can identify people who can or cannot efficiently metabolize a drug. This information can also be used to distinguish individuals who show an adverse response to a drug. Such studies will assist in providing patients and physicians with individualized drug therapy or personalized medicine; this will make it possible to decide which drug and what dosage should be prescribed based on a patient's genetic predispositions. Clinical testing for genetic disorders is already available; this will have implications for carrier screening, prenatal diagnosis, preimplantation genetic diagnostics, pre-symptomatic testing, etc. Results based on these tests will be useful for discussing the probability of having genetic defects in progeny, and hence may be valuable in marriage counseling, *in vitro* fertilization, etc.

Personalized Genome Sequencing: In September 2007, the fully sequenced DNA of Craig Venter (Celera Genomics) was published, making it the first diploid human genome of an individual to be published (Levy et al., 2007). This marked the opening of an era of individualized genomic information. The sequencing of Watson's genome the following year in just two months (using second generation sequencing) is leading to make the era of personal genome analysis a reality (Wheeler et al., 2008).

1000 Genomes Project: An international consortium with major funding from the Wellcome Trust Sanger Institute (UK), the Beijing Genomics Institute (China), and NHGRI announced the 1000 GENOME Project; the aim is to identify DNA variations at a very high resolution. Private companies later joined this effort, making it an international public–private consortium. Genomes of 2,500 people from 27 populations around the world (e.g. Caucasians, Africans, Chinese, Koreans, Indians, etc.) are being studied using the next generation sequencing. This will significantly speed up identification of disease-related variants, and hence the process of diagnosis, treatment, and prevention of common diseases. Additional information about the project can be obtained from http://www.1000genomes.org/.

Encode Project: Short DNA sequences located in the upstream region of genes have been known to regulate the expression of genes, either by transcriptional activation of genes, upregulation or downregulation, or modulation of gene expression. There are a variety of elements known to be present throughout the genome. The goal of the pilot phase of the Encode Project (2003–2007) was to comprehensively study the structural and functional elements encoded in 1% of the human genome. The objective was to map a variety of sequences involving genes (protein-coding and non-coding exons), regulatory regions such as promoters, enhancers, termination sites, RNA transcripts, binding sites of transcription factors, epigenetic changes such as DNA methylation, chromatin accessibility, etc. All this information has further advanced our understanding of the function of the human genome. More information can be obtained from their website (http://genome.ucsc.edu/ENCODE). With the success of the pilot project, the focus is now on the remaining 99% of the genome (The ENCODE Project Consortium, 2007; The ENCODE Project Consortium, 2011). Such studies would not have been feasible on a large scale without the decoding of the human genome.

Comparative Genomics: A complete version of the human genome has also significantly enhanced comparative genomics. The human genome can now be compared to that of a mouse, rat, fly, etc., in order to understand human genetic makeup and function. Laboratory experiments on these model organisms can reveal the functions of genes, and thus their homologs in humans. Several other genomes of vertebrates and nonvertebrates have been sequenced, or are being sequenced, such as zebrafish, pufferfish, sea urchin, honeybee, platypus, dog, chicken, and bovine, as well as primates such as chimpanzee, rhesus macaque, etc. The chimp genome sequence is the first non-human primate sequence available, and provides an important tool to study our genome with respect to a closely related species. The human and chimp genomes are 99% identical, with 10 times more genetic differences than between two humans, and 60 times less than between mouse and human (resources from www.genome.gov). Such information will be useful in comparative genomics, and will allow biologists to elucidate gene functions as well as do evolutionary studies. Such studies will also identify key developments that led to vertebrate and non-vertebrate lineages.

Animal biotechnology has made significant contributions to HGP, and holds the potential to contribute much more. Animal model systems not only allow us to study the molecular basis of the pathophysiology of human diseases, but also provide systems for developing and testing new therapies. Of these, the mouse has succeeded as the best human disease model system so far (e.g. Duchenne Muscular Dystrophy, DMD; cancer; etc.). The mouse has a similar set of genes to humans, and consequently is the best model for understanding the function of genes. Knock-in, constitutive knockout, conditional knockout, and knockdown mice, in combination with RNA interference technology, provide useful tools to study gene functions (Chaible et al., 2010). More than 1,000 mouse knockout mutations are available, which will help in understanding the functions of disrupted genes, especially those of biomedical importance.

WORLD WIDE WEB RESOURCES

http://asia.ensembl.org/Homo_sapiens/Info/Index: Provides open access to HGP and other sequencing projects.

http://www.ncbi.nlm.nih.gov/omim: Maintains a catalog and information about Mendelian diseases.

http://www.genome.gov/GWAStudies: Contains a catalog of the genetic associations between diseases and genomic loci.

http://genome.gov/encode: Has a catalog of functional elements in the human genome.

http://www.1000genomes.org: A database that stores information about DNA variations.

http://hapmap.ncbi.nlm.nih.gov: A database of SNPs.

http://www.genome.gov: Provides information regarding human genome project.

http://www.noncode.org: Describes and catalogs noncoding RNA.

http://www.mitomap.org: A database focused on the mitochondrial genome.

REFERENCES

Anderson, S., Bankier, A. T., Barrell, B. G., De Bruijn, M. H. L., Coulson, A. R., Drouin, J., Eperon, I. C., Nierlich, D. P., Roe, B. A., Sanger, F., Schreier, F., Smith, A. J. H., Staden, R., & Young, I. G. (1981). Sequence and organization of the human mitochondrial genome. *Nature*, *290*, 457–465.

Carthew, R. W., & Sontheimer, E. J. (2009). Origins and mechanisms of miRNAs and siRNAs. *Cell*, *136*, 642–655.

Chaible, L. M., Corat, M. A., Abdelhay, E., & Dagli, M. L. Z. (2010). Genetically modified animals for use in research and biotechnology. *Genetics and Molecular Research*, *9*, 1469–1482.

Crimi, M., & Rigolio, R. (2008). The mitochondrial genome, a growing interest inside an organelle. *International Journal of Nanomedicine*, *3*, 51–57.

Davis-Dusenbery, B. N., & Hata, A. (2010). Mechanisms of control of microRNA biogenesis. *Journal of Biochemistry*, *148*, 381–392.

Deiters, A. (2010). Small molecule modifiers of the microRNA and RNA interference pathway. *The AAPS Journal*, *12*, 51–60.

Dunham, I., et al. (1999). The DNA sequence of human chromosome 22. *Nature*, *402*, 489–495.

The ENCODE Project Consortium. (2007). Identification and analysis of functional elements in 1% of the human genome by the ENCODE pilot project. *Nature*, *447*, 799–816.

The ENCODE Project Consortium. (2011). A User's Guide to the Encyclopedia of DNA Elements (ENCODE). *PLoS Biology*, *9*, e1001046.

Fleischman, R. D., Adams, M. D., White, O., et al. (1995). Whole-genome random sequencing and assembly of Haemophilus influenza Rd. *Science*, *269*, 496–512.

Ghildiyal, M., & Zamore, P. D. (2009). Small silencing RNAs: an expanding universe. *Nature Reviews Genetics*, *10*, 94–108.

Hattori, M., et al. (2000). The DNA sequence of human chromosome 21. *Nature*, *405*, 311–319.

Hindorff, L. A., MacArthur, J., Wise, A., Junkins, H. A., Hall, P. N., Klemm, A. K., & Manolio, T. A. A catalog of published genome-wide association studies. Available at: www.genome.gov/GWAStudies.

Hudson, K. L., Holohan, M. K., & Collins, F. S. (2008). Keeping pace with the times-the genetic information nondiscrimination act of 2008. *The New England Journal of Medicine*, *358*, 2661–2663.

IHGSC (International Human Genome Sequencing Consortium (2001). Initial sequencing and analysis of the human genome. *Nature*, *409*, 860–921.

IHGSC (International Human Genome Sequencing Consortium (2004). Finishing the euchromatic sequence of the human genome. *Nature*, *431*, 931–945.

Kaikkonen, M. U., Lam, M. T. Y., & Glass, C. K. (2011). Non-coding RNAs as regulators of gene expression and epigenetics. *Cardiovascular Research*, *90*, 430–440.

Levy, S., et al. (2007). The diploid genome sequence of an individual human. *PLoS Biology*, *5*, e254.

MITOMAP. (2011). *A human mitochondrial genome database*. http://www.mitomap.org.

Manolio, T. A., Brooks, L. D., & Collins, F. S. (2008). A HapMap harvest of insights into the genetics of common disease. *The Journal of Clinical Investigation*, *118*, 1590–1605.

Manolio, T. A., & Collins, F. A. (2009). The HapMap and Genome-Wide Association Studies in Diagnosis and Therapy. *Annual Review of Medicine*, *60*, 443–456.

Manolio, T. A. (2010). Genomewide association studies and assessment of the risk of disease. *The New England Journal of Medicine*, *363*, 166–176.

Mardis, E. R. (2008). Next-generation DNA sequencing methods. *The Annual Review of Genomics and Human Genetics*, *9*, 387–402.

Mardis, E. R. (2011). A decade's perspective on DNA sequencing technology. *Nature*, *470*, 198–203.

Matera, A. G., Terns, R. M., & Terns, M. P. (2007). Non-coding RNAs: lessons from the small nuclear and small nucleolar RNAs. *Nature Reviews Molecular Cell Biology*, *8*, 209–220.

NONCODE: An integrated knowledge database of non-coding RNAs (2005). *Nucleic Acids Research 33*, database issue D112–D115.

Nwanguma, B. C. (2003). The human genome project and the future of medical practice. *African Journal of Biotechnology*, *2*, 649–656.

Olson, M., Hood, L., Cantor, C., & Botstein, D. (1989). A common language for physical mapping of the human genome. *Science*, *245*, 1434–1435.

Pink, R. C., Wicks, K., Caley, D. P., Punch, E. K., Jacobs, L., & Carter, D. R. F. (2011). Pseudogenes: Pseudo-functional or key regulators in health and diseases. *RNA*, *17*, 792–798.

Schadt, E. E., Turner, S., & Kasarskis, A. (2010). A window into third-generation sequencing. *Human Molecular Genetics*, *19*, R227–R240.

Siyanova, E. U., & Mirkin, S. M. (2001). Expansion of trinucleotide repeats. *Molecular Biology*, *35*, 168–182.

Slotkin, R. K., & Martienssen, R. (2007). Transposable elements and the epigenetic regulation of the genome. *Nature Reviews Genetics*, *8*, 272–285.

Taft, R. J., Pang, K. C., Mercer, T. R., Dinger, M., & Mattick, J. S. (2010). Non-coding RNAs: regulators of disease. *Journal of Pathology*, *220*, 126–139.

Thompson, J. F., & Milos, P. M. (2011). The properties and applications of single-molecule DNA sequencing. *Genome Biology*, *12*, 217.

Venter, J. C., Adams, M. D., Myers, E. W., et al. (2001). The sequence of the human genome. *Science*, *291*, 1304–1351.

Wheeler, D. A., et al. (2008). The complete genome of an individual by massively parallel DNA sequencing. *Nature*, *452*, 872–876.

Wright, M. W., & Bruford, E. A. (2011). Naming 'junk': Human non-protein coding RNA (ncRNA) gene nomenclature. *Human Genomics*, *5*, 90–98.

Zhang, R., Zhang, L., & Yu, W. (2012). Genome-wide expression of non-coding RNA and global chromatin modification. *Acta Biochimica et Biophysica Sinica*, *44*, 40–47.

Zhang, W., Ratain, M. J., & Dolan, M. E. (2008). The HapMap resource is providing new insights into ourselves and its application to pharmcogenomics. *Bioinformatics and Biology Insights*, *2*, 15–23.

FURTHER READING

Brown, T. A. (2007). *Genomes* (3rd ed.). Garland Science, Taylor and Francis Group, LLC.

Crimi, M., & Rigolio, R. (2008). The mitochondrial genome, a growing interest inside an organelle. *International Journal of Nanomedicine*, *3*, 51–57.

Green, E. D., Guyer, M. S., & NHGRI (2011). Charting a course for genomic medicine from base pairs to bedside. *Nature*, *470*, 204–213.

The International SNP Map Working Group. (2001). A map of human genome sequence variation containing 1.42 million single nucleotide polymorphisms. *Nature*, *409*, 928–933.

Kogelnik, A. M., Lott, M. T., Brown, M. D., Navathe, S. B., & Wallace, D. C. (1996). MITOMAP: a human mitochondrial genome database. *Nucleic Acids Research*, *24*, 177–179.

Pevsner, J. (2009). *Bioinformatics and Functional Genomics* (2nd ed.). Wiley-Blackwell.

Primrose, S. B., & Twyman, R. M. (2006). *Principles of gene manipulation and genomics* (7rd ed.). Blackwell Publishing.

Strachan, T., & Read, A. (2011). *Human Molecular Genetics* (4th ed.). Garland Science, Taylor and Francis Group, LLC.

Tam, O. H., et al. (2008). Pseudogene-derived small interfering RNAs regulate gene expression in mouse oocytes. *Nature*, *453*, 534–538.

ABBREVIATIONS

BAC Bacterial Artificial Chromosome
DNA Deoxyribonucleic Acid
DOE Department of Energy
ELSI Ethical, Legal, and Social Issues
ERV Endogenous Retrovirus
EST Expressed Sequence Tag
GAIN Genetic Association Information Network
GINA Genetic Information Non-Discrimination Act
GWA Genome Wide Association
HGP Human Genome Project
IHGSC International Human Genome Sequencing Consortium
NIH National Institutes of Health
NHGRI National Human Genome Research Institute
RNA Ribonucleic Acid
MIR Mammalian-Wide Interspersed Repeat
miRNA MicroRNA
mtDNA Mitochondrial DNA
ncRNA Non-Coding RNA
OMIM Online Mendelian Inheritance in Man
PP Processed Pseudogene
RFLP Restriction Fragment Length Polymorphism
RISC RNA-Inducd Silencing Complex
rRNA Ribosomal RNA
snoRNA Small Nucleolar RNA
SNP Single Nucleotide Polymorphism
SSLP Simple Sequence Length Polymorphism
SSR Simple Sequence Repeat
STS Sequence-Tagged Site
tRNA Transfer RNA

GLOSSARY

Bacterial Artificial Chromosome *Escherichia coli* F-plasmid-based DNA cloning vector with capability to clone large DNA fragment.

Clone fingerprinting A technique to identify overlapping clones by generating a DNA fragment profile.

Contig A set of overlapping DNA fragments / sequences.

CpG islands Regions of genome which are rich in GC content, and usually found at the 5′ end of the genes. C in the CpG is a potential site for methylation.

Dicer Ribonuclease that plays a central role in RNA interference and gene silencing.

DNA Deoxyribonucleic acid is a genetic material.

Euchromatin The decondensed form of chromatin that stains lightly in G banding and contains transcriptionally active genes.

Finished sequence A sequence of DNA which is no less than 99.99% accurate, and ideally has no gaps. In case gaps cannot be closed, size of the gap is indicated.

Genome A complete set of genetic information of an organism. In case of diploid organism, one set constitutes the genome.

Heterochromatin The condensed form of DNA that appears as darkly stained bands in G banding

Human genome project A publically funded project that was initiated with the goal of sequencing the entire human genome and to identify all genes and other features associated with it.

Noncoding RNA A plethora of RNA molecules, which do not code for any protein and are involved in variety, possibly regulatory functions.

PCR A technique to amplify exponentially a segment of DNA in a vial.

Positional cloning Also called map-based cloning is a procedure by which a gene is cloned based on information about its position on a physical/genetic map.

Pseudogene A nonfunctional copy of a gene.

Repititive DNA A DNA sequence which is present multiple times in a genome.

Restriction Fragment Length Polymorphism (RFLP) A class of molecular markers that detects DNA polymorphism by use of restriction enzymes and change in size of restriction fragments, generated due to nucleotide substitutions in the recognition site of the restriction enzyme.

Retrotransposon Transposable elements that replicate via RNA intermediate, possibly originated from retroviruses.

RNA interference (RNAi) A term used for the process of gene silencing mediated by miRNA.

Sequence tagged site (STS) A defined unique sequence present only once in a genome.

Shotgun approach A method of sequencing in which DNA is randomly broken into fragments and sequenced.

Simple Sequence Length Polymorphism (SSLP) A type of molecular technique that detects DNA polymorphism on the basis of variation in length of repeat sequences between individuals.

Single Nucleotide Polymorphism (SNP, pronounced as snip or plural snips) A DNA sequence variation occurring due to a single nucleotide (A, T, G, or C) in chromosomes.

Transposons A mobile DNA sequences that can move from one region to another region within the genome.

Transfer RNA Adaptor molecules that decode nucleotide sequence of mRNA into amino acid sequence

Whole genome shotgun approach Strategy used by Celera Genomics to sequence human genome by random shotgun sequencing

LONG ANSWER QUESTIONS

1. Describe various strategies used for sequencing the human genome. Explain advantages and disadvantages of each.
2. What are repetitive DNA elements? Describe them with respect to the human genome.
3. Describe non-coding RNAs.
4. Elaborate on human genome organization.
5. Describe how information from HGP contributed to the welfare of human society as a whole.

SHORT ANSWER QUESTIONS

1. Write short notes on the following:
   a. miRNA.
   b. Pseudogenes.
   c. LINE.
   d. SINE.
2. Write a short note on the mitochondrial genome.
3. What are the advantages of using a hierarchical method for the sequencing of complex eukaryotic genomes?
4. List major bottlenecks encountered by the publically funded genome project, and breakthroughs that helped to achieve its goal.
5. Discuss various ethical, legal, and social issues associated with sequencing the human genome.

ANSWERS TO SHORT ANSWER QUESTIONS

1a. miRNA are a class of noncoding RNAs. They are ~19–25 nucleotides long, evolutionarily conserved, small, single-stranded molecules that regulate gene activity at the post-transcriptional level by a phenomenon related to RNA interference. The primary transcript of miRNA has a 5′-cap, 3′-polyA tail, and an inverted repeat that can base pair to form a hairpin structure. A nuclear RNase III complex (Drosha) processes this transcript to release an atypical hairpin loop precursor miRNA (pre-miRNA), which moves to the cytoplasm where it is cleaved by the enzyme Dicer, a cytoplasmic RNase III, to release mature miRNA. In the cytoplasm, an effector complex called RNA-induced silencing complex (RISC) containing argonaute RNase degrades one strand of the miRNA to leave a mature single-stranded miRNA (guide strand) bound to the argonaute. This complex then selectively base pairs with the target RNA with the sequence complementary to the guide strand. The binding usually occurs at the 3'-untranslated regions (3′-UTRs) of the target mRNA sequence, but in some cases 5′-UTRs are also involved. Thus, an miRNA can target several gene transcripts. Depending on the extent of similarity, target RNA is degraded, destabilized, or prevented from being translated. A total of 1,756 miRNA genes have been identified in the human genome. Chromosome 1 has 134 miRNA genes, while the Y chromosome has the least (~15). Chromosome 19 has 110 miRNA genes. ncRNAs seem to have occupied a center stage because of their involvement in many human diseases. Many human diseases have been identified that may be due to mutations or dysfunction of ncRNAs. For example: Prader–Willi Syndrome, cancer, central nervous disorders, cardiovascular diseases, etc. Utilizing the properties of miRNAs and RNA interference mechanisms, they have great potential as therapeutic agents in terms of down-regulating the expression of specific genes.

1b. Pseudogenes are the non-functional copies of normal genes with which they share sequence homology. They can be of two types; Duplicated pseudogenes and processed pseudogenes. Pseudogenes derived from the duplication of genomic DNA are called duplicated pseudogenes, as they possess characteristic exon–intron structure. Such pseudogenes are derived by tandem duplication, and as a result they are located close to the functional copy. In contrast, "retrotransposed," or processed pseudogenes (PPs), are derived from reverse transcription of processed transcripts that integrate into the genomic DNA. Therefore, they are devoid of introns. They also possess polyA tracts and direct repeats at either ends of the pseudogenes. Lack of a promoter usually results in an inactive gene copy. There are about 14,427 pseudogenes. The prevalence of pseudogenes in the human genome has been problematic for genome annotation. The highly expressed housekeeping genes have multiple processed pseudogenes. For example ribosomal proteins account for ~20% of human PPs. Protein-coding pseudogenes are more easily identified than RNA pseudogenes.

Pseudogenes were believed to be non-functional, but evidence has shown that many pseudogenes (2 to 20%) are transcribed into RNA. Examples include pseudogenes for tumor suppressor PTEN, glutamine synthetase, glyceraldehyde-3-phosphate dehydrogenase, etc. Many pseudogenes have been found to have tissue-specific expression, and some have an expression pattern that is different from the parent gene. Alteration in pseudogene expression has been seen under different physiological conditions, such as diabetes and cancer. Some pseudogenes have been implicated in gene regulation in mice via siRNAs production.

1c. LINEs are a type of non-LTR retrotransposon that replicate via an RNA intermediate. They are the most primitive autonomous mobile elements found in the human genome. In humans, three distantly related LINE families are present: LINE1, LINE2, and LINE3, of which only LINE1 is still active. LINE machinery has resulted in reverse transcription in the human genome (including those of SINE and processed pseudogenes). These retro transposons are about 6 kb long, encoding two ORFs and possess a polymerase II promoter. Transposition of LINE occurs via RNA that assembles with encoded proteins and moves back to the nucleus. LINE transposition preferably occurs in AT-rich regions. The preference of LINEs for AT-rich DNA could be explained by the fact that these regions are poor in genes, and as a result these elements do not encounter a functional selection pressure. The LINE endonuclease cleaves at TTTT/A in the DNA sequence where the LINE element is inserted into the genome.

1d. SINE is a type of non-LTR retrotransposon that replicate via an RNA intermediate. Unlike LINEs, SINEs are non-autonomous transposons. They cannot transpose independently, and are believed to use the LINE machinery for transposition. Nevertheless, they have a very high copy number within the mammalian genome. They share their 3′-end with the LINE element. They are 100–400 bp long, encode no protein, but possess an internal polymerase III promoter. There are three distinct monophyletic families of SINEs: the active Alu, and the inactive mammalian-wide interspersed repeat (MIR) and Ther2/MIR3. There is a promoter of all SINEs, but one is derived from tRNA genes. The Alu elements are the only active class of SINEs that have a derived promoter from the signal recognition particle component 7SL. The Alu family is the most abundant sequence in the human genome, occurring more than once in 3 kb on an average. They are of comparatively recent evolutionary origin. An Alu repeat is about 280 bp long, and consists of two asymmetric tandem repeats, each about 120 bp long followed by a short A_n/T_n sequence. Again, unlike LINEs, which are preferentially located in AT-rich regions, Alu repeats are found in gene-rich, high GC-rich regions, indicating their positive role in genome function. The fact that Alu sequences use LINE machinery for transposition (but are found in GC-rich regions), indicates that their distribution is reshaped by evolutionary forces. Some Alu sequences are transcribed, and it is believed that these elements may have been involved in the evolutionary dynamics of the human genome.

2. The human mitochondrial genome is small in size (16,569 bp). It is a circular molecule present in multiple copies. The two strands of mtDNA have significant differences in G and C composition. One strand is rich in guanines, and is hence called the heavy (H) strand, and the other, the light (L) strand, has more cytosine. The transcription of each strand begins at major promoter regions, PL and PH1, which is present in the control region, including the D-loop. The promoter PL predominantly transcribes the light chain, and PH1 transcribes the H-strand to generate polycistronic RNAs. The mature RNAs are generated by cleavage of multigenic transcripts. The replication of both H- and L-strand starts at a specific origin, the replication of the H-strand begins at the D region. When two-thirds of the H-strand has been synthesized, the origin of replication of the L-strand, which lies between two tRNA genes, becomes exposed; the L-strand is synthesized in the opposite direction.

Being small in size, the mitochondrial genome was the first to be sequenced. The complete sequence of the mitochondrial genome was made available in 1981 by Sanger and colleagues. The mitochondrial genome has 37 genes that include 22 tRNAs, 2rRNAs, and 13 protein-coding genes. The entire mitochondrial DNA (mtDNA) is transcribed, and it contains small non-coding regions. The gene density of mtDNA is 1 per 0.45 kb. All 37 mitochondrial genes lack introns. Protein-coding genes encode proteins involved in the electron transport chain and oxidative phosphorylation. In fact, a vast majority of proteins present in mitochondria are encoded by nuclear genes. The mitochondrial genetic code has 60 codons and 4 stop codons. The stop codons include UAA and UAG, which are universal stop codons, and AGA and AGG, which specify arginine in the nuclear genetic code. The universal stop codon (UGA) codes for tryptophan in mitochondria, and AUA codes for methionine. Thus, the mitochondrial genome has its unique genome organization and is different from the nuclear genome.

3. A hierarchical approach greatly reduces the challenge of assembling the finished sequence of complex and outbred organisms (like humans) that have a high percentage of repetitive DNA. Because the relative order of BAC is known, the possibility of long-range misassembly is reduced. This approach also ensures complete coverage of the genome, and helps to reduce the sequencing redundancy. In addition, it was believed that this strategy would be able to identify regions that show cloning biases, which could then be focused for additional sequencing. As the IHGSC was a collaboration of different countries, the work and responsibility could be easily disseminated among its members by this strategy.

4a. Some major bottlenecks encountered in the publicly funded Human Genome Project were the following:

- **i.** The non-availability of high-density genetic and physical maps. These maps are used in assembling the sequence of the whole genome of complex organisms like humans.
- **ii.** Keeping in view the small cloning capacity of the then current cloning vehicles, there was a need to look for better alternatives, which made cloning of large inserts possible.
- **iii.** The time taken and cost of sequencing using existing technologies at that time was very high.
- **iv.** There was a need to store and analyze large amounts of sequencing data, and hence there was a need to develop better algorithms for computations (e.g. development of software packages to analyze sequence data).
- **v.** Practical difficulties such as cloning bias and the presence of a large quantity of repeat sequences in the human genome all provided a hurdle to sequencing and assembly of the human genome sequence.

4b. The Human Genome Project was aided by several breakthroughs, some of which have been described as follows:

- **i.** Sanger's method of sequencing, and its automation and miniaturization, use of fluorescent dNTPs, and development of capillary-based sequencing machines, greatly accelerated the speed of sequencing.
- **ii.** DNA-based genetic markers that helped in the construction of high-density genetic and physical maps, greatly helped in assembly of YAC/BAC clones, and hence the sequence assembly.
- **iii.** Polymerase chain reaction, which was invented in 1983, also propelled genetic research.
- **iv.** Development of large insert cloning systems such as *Escherichia coli*-based bacterial artificial chromosomes (BAC), which can carry large segments of DNA, contributed to the success.

5. With the whole-genome sequence and subsequent data freely available to the public, concerns were raised as to how this knowledge would impact an individual or society as a whole. There was apprehension that this information would be misused by schools, courts, adoption agencies, employers, and particularly health insurers, who would discriminate on the basis of an individual's genetic makeup. For example if a person is a carrier or has predisposition to certain diseases that can be determined by genetic tests, he or she may not be hired by a company. A medical insurance agency might not cover such an individual, or might charge hefty insurance premiums. This would be discrimination based on genetic information of a person, and could be exploited by greedy companies. Social issue concerns, such as stigmatization of an individual based on race, ethnicity, or genetic differences, could have a profound psychological impact or disparity. In order to prevent misuse of genomic information, the Genetic Information Non-Discrimination Act (GINA) was passed in the U.S. (Hudson et al., 2008). This law prohibits discrimination of individuals by insurance companies and employers based on information from genetic tests.

Chapter 32

Ethical Issues in Animal Biotechnology

Abhik Gupta

Department of Ecology & Environmental Science, Assam University, Silchar, India

Chapter Outline

SUMMARY

This chapter provides a brief overview of relevant ethical theories and principles for examining the various ethical issues in animal biotechnology. These issues include intrinsic concerns such as religious objections, and extrinsic concerns such as those for the environment. Ethical issues pertaining to the creation of chimeras and the biopharming of animals are also discussed.

WHAT YOU CAN EXPECT TO KNOW

Simply ethics is a set of moral principles that guide the decisions and actions of people and enable them to determine right and wrong behavior. Ethical theories are divided into meta-ethics, normative ethics, and applied ethics. Metaethics explores the foundations of morality; normative or prescriptive ethics lists ethical behavior; and applied ethics deals with specific issues that call for debate and discussion, using the concepts and principles contained in metaethics and normative ethics. Applied ethics include diverse issues such as animal rights, environmental ethics, euthanasia, cloning, xenotransplantation, and others. New applied ethics issues are likely to emerge from time to time with the introduction of new technology and practices.

The ancient Greeks like Aristotle and Plato based their ethics on virtue, happiness, and the soul. The deontological approaches to ethics describe discharge of duties irrespective of consequences as the hallmark of an ethical way of life. On the contrary, utilitarian ethics is a form of consequentialist ethics that considers the action that brings the greatest good to the greatest number as ethical.

Animal Biotechnology. http://dx.doi.org/10.1016/B978-0-12-416002-6.00032-8

The major principles of normative ethics, medical ethics, and ethics of science and technology, include beneficence, non-maleficence, autonomy, justice, human dignity, equality, tolerance, informed consent and choice, animal rights and welfare, and environmental compatibility, among a host of others. These principles can be important tools for an ethicist to examine particular animal biotechnology innovations.

The growing interest in ethical analysis of animal biotechnology stems from the high manipulative ability of the gene-based technologies that place greater control with humans than ever before. These technologies are consequently subject to ethical analysis from religious, environmental, safety-based, animal rights-based, and welfare-based approaches.

Concerns over animal biotechnology can include both intrinsic and extrinsic concerns. Of the former, biotechnology has been accused of playing God by infringing upon the divine order of the nature of a species by transferring genes from one species to another and creating hitherto unknown forms of life. This is the most common intrinsic concern, although there is hardly any consensus among the adherents of the different religions, or even within the different sects or groups of a given religion on this concern. Another concern is the insertion of genes from prohibited animals into other animals and plants, which raises the question of their acceptability as food. Similarly, animal genes inserted into a plant could be thought of compromising the vegetable nature of the plant itself. As opposed to these religious objections, the secular intrinsic concerns allege that biotechnology is disturbing the natural species boundaries and is therefore unnatural, as it destabilizes the evolutionary order of life. However, the detractors of this concern point out that because of hybridization occurring in nature, species boundaries cannot be regarded as fixed and immutable.

Surveys to record public perceptions about animal biotechnology in Europe and the United States revealed that most people were against animal cloning for food, while they approved the use of animal biotechnology for medical purposes, subject to the framing of strong regulatory guidelines.

Among the extrinsic concerns, animal welfare in terms of both physical and mental health is judged against the yardstick of the five freedoms laid down in the Brambell Report of 1964. While taking into account many curtailments of these five freedoms in traditional animal breeding, it can be seen that animal biotechnologies such as gene knockout technology, produce a large number of surplus animals, which coupled with the current state of inefficiency and unpredictability of gene technologies, result in several health problems in livestock animals.

The concern for human health arises from two directions, namely use of genetically engineered animals or their products as food, and their use in medicine, along with their attendant ethical dilemmas. The ethical issues arising out of these concerns are evaluated in relation to a concise set of public health ethics.

The environmental concerns of animal biotechnology can be viewed in relation to ethical positions in human relationships with the environment, which could be categorized as anthropocentric (human-centered), biocentric (life-centered), and ecocentric (ecosystem-centered). Deep ecology is similar to ecocentrism, with an emphasis on self-realization. Anthropocentrism recognizes only instrumental or extrinsic value in the environment, while the others recognize intrinsic value in varying degrees. Viewed against this perspective, animal biotechnology poses certain risks to the environment. The major concern is the release of transgenic animals into the environment, where they can create various problems for their non-transgenic counterparts. Transgenic fish especially pose such hazards to a large extent. The precautionary principle should be applied to take decisions for safeguarding the environment from such hazards.

Ethical issues also arise when human cells such as pluripotent stem cells are placed in a non-human animal to create human-to-animal chimeras, which is accused of compromising human dignity and generates moral abomination by evoking the idea of bestiality. Chimeras also represent a metaphysical threat to the self-image of human beings. There is a utilitarian angle to the chimera debate in view of the medical benefits, but this ought to be regulated by not allowing chimeras with human cognitive abilities to be produced, especially in non-human primates.

The last issue that has been addressed in this chapter is that of biopharming, where specific human genes are engineered into cattle, goat, pig, sheep, or even rabbits, to produce milk containing a number of human proteins of therapeutic value. The ethical issues in animal biopharming revolve around risks to human health, gene transfer, and animal welfare, and can be analyzed by general utilitarian, public health, and animal welfare principles, as well as the precautionary principle. The message of the chapter is that ethical analysis of any new technology can help remove concerns in everybody's mind and enable its safe and wise adoption for the benefit of all.

HISTORY AND METHODS

INTRODUCTION

According to the Internet Encyclopedia of Philosophy, the field of ethics (or moral philosophy) involves systematizing, defending, and recommending concepts of right and wrong behavior. The word ethics has its root in the Greek word ethos or ethicos, which means the moral ideas or attitudes that belong to a particular group or society; in other words, it can mean its custom, habit, character, or disposition. Put simply, ethics is a set of moral principles that guide the decisions and actions of people and enable them to determine right and wrong behavior. Ethical theories are divided into three broad areas: metaethics, normative ethics, and applied ethics. Metaethics explores the foundations of morality, probing the theoretical basis and scope of moral values. In other words, it tries to define morality itself. In contrast, normative or prescriptive ethics tells us what is ethical and what is not. Applied ethics deals with specific issues that call for debate and discussion. It uses the concepts and principles contained in metaethics and normative ethics as tools to deal with issues that include animal rights, environmental ethics, euthanasia, cloning, and xenotransplantation, among many others. Issues in applied ethics may arise from time to time and generate public debate.

For example, exposure to non-ionizing radiation from mobile towers (cellphone base stations) has been shown by some studies to be harmful to human health as well as other life forms, like plants and animals. However, several other studies also exist that show it to cause no perceptible damage. Thus the scientific evidence for either its capacity to cause harm or its innocuous nature can be said to be far from unequivocal. In such a situation, would it be ethical to set up mobile phone towers in densely populated localities in urban areas? What kind of policy decision taken by the city administration and other concerned authorities or the promotional policy adopted by mobile phone companies may be considered ethical? And what should be considered an ethical stand on the setting up of mobile towers in commercial or residential areas or near schools and hospitals? One may argue that since the scientific evidence for potential harm by mobile towers to human health or the environment is not convincing, technological progress, improved and facile communication, and economic development through trade, employment generation, and the like should not be hampered (i.e. setting up towers should be allowed to continue). However, others may cite the precautionary principle, which states that "where there are threats of serious or irreversible damage, lack of full scientific certainty shall not be used as a reason for postponing cost-effective measures to prevent environmental degradation" (COMEST, 2005); they may then plead for not allowing mobile towers to be set up inside residential or other vulnerable areas until they are proved to be harmless through conclusive scientific evidence.

With the increasing emergence of issues (such as the one above) in different fields of science and technology that are not easily resolved, applied ethics has assumed importance not only in policy-making and governance, but in our day-to-day lives as well. The ethical debates around many contentious issues such as end-of-life decisions, euthanasia, cloning, and others have questioned the very basis of normative theories and spilled over into the domain of metaethics. Animal biotechnology also opens several ethical issues that need to be discussed in order to have a complete understanding of the pros and cons of different techniques that are employed, or are likely to be employed, in the near future. While identifying and discussing ethical issues, it must be borne in mind that ethics is a prescriptive, and not a descriptive, discipline. It is concerned with what ought to be and not what is. There is a tendency to group moral and ethical issues together. Despite the overlap commonly found between morals and ethics, these two fields of enquiry are distinct from each other. Most people have some moral views on various subjects, which may be similar or different (in varying degrees) from another individual's. For example, one may think it totally wrong to cut down forests or go hunting, while another person may feel it is alright to clear forests for developmental projects, but not for urban expansion, and so on. Thus, moral views come naturally to people and may depend on their upbringing and experiences, their type of education or religious training, and a host of other factors. While ethics also deals with right and wrong, it provides arguments and justifications about the stand that it takes on a given issue. Some people may instinctively think that cloning or transfer of genes is wrong, and this is part of their moral views on these issues. Ethicists, on the other hand, may also be against cloning or gene transfer, but they will present reasons and justifications to their views. Thus moral views are more based on intuitive feelings, while ethical stands rely more on logical or factual arguments and analyses (Straughan, 1999).

A BRIEF OVERVIEW OF ETHICAL THOUGHTS AND PRINCIPLES

Virtue Ethics

The ancient Greek moral theories attached great importance to virtue, happiness (eudaimonia), and the soul. Two well-known works on ethics by the Greek philosopher Aristotle (e.g. the Nicomachean Ethics and the Eudemian Ethics) contain examinations of, and discussions on, the subject of eudaimonia, which connotes happiness or flourishing, and virtue or excellence. These are the most important among the character traits that a person must have in order to lead an ethical life. Virtue was an important parameter in ancient Indian texts like the Mahabharata or the Ramayana where, for example, a warrior or a king was expected to have certain virtues in order to be considered an ideal example of his class. He was expected to sacrifice his possessions, his loved ones, and even his life, to uphold these virtues. However, as issues (controversial or otherwise) in science and technology or other areas of applied ethics were almost non-existent, there was no ethical debate or discussion in this area. Nevertheless, parts of an ancient Indian text, the Kautilya's Arthasastra, which was written or compiled between the 3rd century B.C. and the 2nd century A.D., prescribed ethically proper ways for carrying out agricultural activities and forestry. It also gave detailed instructions for how to treat elephants, which were the living war machines of that period, as well as guidelines for the welfare of other domestic and wild animals.

Deontological (Duty-Based) Ethics

Another major theory of normative ethics that provides valuable guidelines for adopting ethical positions on specific issues in applied ethics, includes the deontological approach to ethics. The deontological theories are based on the concept of duty or obligation, and are often called non-consequentialist, as they are not dependent on the consequences of a given action. For example, our duties or obligations to children, to the old and infirm, to our community, or to not commit felonies or murder, are considered ethical actions regardless of their consequences. The duty-based approach to ethics was first developed by the 17th century German philosopher Samuel Pufendorf, who classified duties under three major categories (i.e. duties to God, oneself, and others). The renowned philosopher Immanuel Kant based his deontological approach on the concept of categorical imperative, which recommends an action irrespective of its contribution to personal happiness. Thus, we are to treat another person as an end, and not as a means to an end. In other words, we should not use other people as mere instruments to achieve our own happiness. Another well-known, duty-based approach is the rights theory postulated by the 17th century British thinker John Locke, who stated that every human being enjoys certain rights given by God that include life, liberty, health, and possessions. Any action that hampers or alienates these rights is, therefore, unethical. The British philosopher W.D. Ross developed the most recent duty-based approach, where he identified some universal duties such as fidelity, reparation, gratitude, justice, beneficence, self-improvement, and non-maleficence.

Consequentialist Ethics

As the name indicates, a consequentialist ethical theory judges the rightness or wrongness of an action by the nature of its consequences. Utilitarianism is an important consequentialist ethical theory. In its classical form, postulated by the British philosopher Jeremy Bentham (1748–1832) and his follower John Stuart Mill (1806–1873), utilitarianism is concerned with maximization of good and aims to bring about "the greatest amount of good for the greatest number." While maximizing the good, classical utilitarianism treats everybody's good as the same, without any partiality, and the reason that prompts somebody to undertake an action is the same for everybody else. John Sidgwick (1838-1900) and G.E. Moore (1873-1958) were among the other thinkers who contributed richly to utilitarian ethical thinking. Consequently, utilitarianism today has many variations. It is also perhaps the most commonly used tool in the hands of policy-makers when deciding whether to build a dam across a river by submerging villages or tracts of agricultural or forest land, acquiring private land for building a highway, approving a new drug, or banning a pesticide.

PRINCIPLES

Many general principles of normative ethics, medical ethics, and the ethical guidelines for science and technology are also applicable to issues in animal biotechnology. Some such principles are defined as follows (Weed and McKeown, 2001):

1. **Rationality:** Rationality is an important prerequisite for ethics, as all morally appropriate actions should be defendable by logical reasoning.
2. **Beneficence:** Beneficence means doing good. From an ethical perspective, it has an underlying utilitarian basis that is concerned with producing the greatest possible proportion of good over evil for the greatest number of people.
3. **Non-maleficence:** It means not doing harm.
4. **Least harm:** This principle deals with a situation where neither choice is beneficial. In such a case, one ought to choose the option that is less harmful in extent or magnitude, or which causes harm to the least number of people.

Beneficence, non-maleficence, and least harm can be seen to be part of a continuum where doing good is the best option, but when that is not possible, doing any harm at all must be avoided. In the worst scenario of not being able to totally prevent harm, the least harmful option ought to be chosen.

5. **Autonomy:** This involves respect for persons who should enjoy freedom from coercion. It automatically opposes the principle of paternalism, which tends to impose upon others' will. In other words, the principle of autonomy allows people to make their own decisions.
6. **Informed consent:** A cardinal principle in medical and research ethics, informed consent involves provision of complete information, including risk involved, chances of failure, etc., about a course of treatment or a trial offered to patients or volunteers. They can then either opt for the suggested treatment/trial or decide against it.
7. **Informed choice:** This is one step beyond informed consent, where a range of options is offered to a patient/volunteer with explanations about the pros and cons of each option. The patients/volunteers can select the preferred one.

We can see here that the path of ethical progression moves from paternalism, where consent is taken for granted, to informed consent, to informed choice, with greater autonomy of the subject at each successive stage.

8. **Justice:** Justice means treating everybody with fairness, impartiality, and consistency, with equitable distribution of risks and benefits of research, health care, or other goods and services. It should be borne in mind that the word justice in an ethical vocabulary is distinct from the retributive justice of criminal law.
9. **Confidentiality:** This implies respect for privacy of information provided by patients or research subjects.
10. **Equality**: This is the duty to treat every human subject as a moral equal.
11. **Tolerance:** This calls for understanding and acceptance of diverse viewpoints that may be expressed by different subjects or stakeholders.
12. **Human dignity:** This warrants treating every human being with dignity and keeping intact their dignity as human beings.

It may be said that the five aforesaid ethical principles are closely related with some amount of overlap so that observance or violation of one is likely to affect the others to varying degrees.

13. **Animal rights:** The Merriam-Webster Dictionary defines animal rights as "rights (as to fair and humane treatment) regarded as belonging fundamentally to all animals."
14. **Animal welfare:** This pertains to the general health, happiness, and safety of an animal. Modern views of animal welfare tend to be concerned not only with the health of the animal's body, but also its feelings (Hewson, 2003).
15. **Environmentally benign nature:** This implies that an essential prerequisite for an action to be ethically appropriate is that it not be harmful to the environment.
16. **Intrinsic value:** A value in itself, regardless of its utility for others. For instance, people generally believe that other human beings have a value in themselves as conscious intelligent beings, regardless of their utility for others. Thus, human beings do not lose their value even if they are not able to work or be useful to other human beings.
17. **Extrinsic value:** Derives from the objective properties of something or somebody that has use for others by virtue of its functions. For example, a pen or pencil has value only until it is considered fit to write with, after which it loses its value.
18. **The Precautionary Principle**: According to the definition adopted in the Rio Declaration (1992), the precautionary principle (PP) states that "where there are threats of serious or irreversible damage, lack of full scientific certainty shall not be used as a reason for postponing cost-effective measures to prevent environmental degradation" (COMEST, 2005).

The aforesaid principles can serve as powerful tools in the hands of ethicists to examine the different animal biotechnological innovations.

METHODOLOGY

The methodological approach followed in this chapter to examine several important innovations in animal biotechnology with ethical analysis follows an analytical step-by-step model (Flow Chart 32.1). Available supportive data on positive and negative aspects of the innovation are extensively used. This information is first subjected to the relevant intrinsic (both religious and secular) and extrinsic (in terms of environment, health, animal welfare and any other relevant concerns) factors. This is followed by an analysis of the innovations from different normative ethics approaches and the precautionary principle, wherever applicable. The whole process is examined against the backdrop of ethical principles such as beneficence, non-maleficence, autonomy, justice, and others, depending upon their applicability in a given situation. At the end of this exercise, the pros and cons of the technology get sorted out, and a final ethical decision can be made.

APPLICATION OF ETHICS IN ANIMAL BIOTECHNOLOGY

There has been a surge in studies on the ethical concerns of biotechnology, including animal biotechnology. Why have ethical questions become an integral part of public concern about the phenomenal breakthroughs in biotechnology research in recent years? Perhaps it is because of the increasing ability of humans to chart and manipulate the genome of plants and animals and utilize this knowledge to innovate in several subject fields, such as medicine, agriculture, veterinary science, and others (i.e. humans are able to tinker with the very basis of life). A large number of people are not comfortable with such manipulations, and believe that they are tantamount to playing God, which raises questions about the ethical propriety of such research. Such beliefs that humans should not interfere with the fabric of life are usually linked to religious faith, and certain worldviews and convictions.

Other concerns include animal rights and animal welfare; possible harmful effects on human consumers, recipients or volunteers, along with infringement of their autonomy; and potential impact on the environment. Thus, of the different ethical principles discussed in the Principles

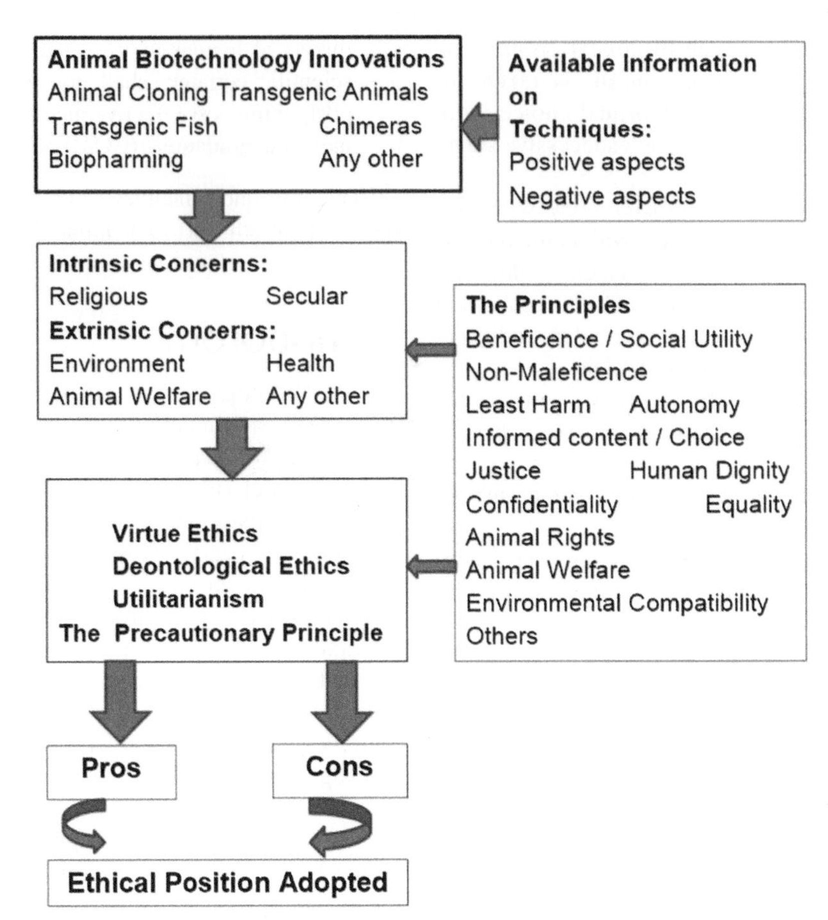

FLOW CHART 32.1 Methodology adopted in the ethical analysis of animal biotechnology innovations.

section, those like beneficence, informed consent and choice, animal rights and welfare, environmentally benign nature, and the precautionary principle (among others) are especially relevant for biotechnology in general, and animal biotechnology in particular.

Ethical questions are also coming more and more to the foreground because of the revolutionary changes in animal breeding in the last few decades. The pre-twentieth century animal breeders did not have knowledge of the underlying genetic mechanisms that controlled the expression of a given morphological character, and based their experiments on experience and observation. This approach changed in the twentieth century when knowledge of Mendelian genetics was applied to achieve greater control over breeding. Beginning in the 1960s, animal breeding experienced progressively greater control over associated processes with the application of modern biotechnological techniques such as artificial insemination (AI), which became more effective and widespread with the development of various techniques and concepts: deep-freezing semen, superovulation of females, embryo transfer, and cloning. With greater control and deeper manipulation, ethical, regulatory, and safety-related concerns were also expressed from more quarters and at increasing frequency.

ETHICAL CONCERNS IN ANIMAL BIOTECHNOLOGY

Before going into the ethical concerns in animal biotechnology, it is to be understood that ethics cannot provide conclusive proof about the rightness or wrongness of a technology. There could be arguments in favor of or against a given technology and its applications. However, ethical debates and discussions can provide a basis for individuals to reach their own decisions instead of being provided with a paternalistic prescription. Such an approach in turn can pave the way for taking appropriate policy decisions.

Intrinsic Concerns

Before discussing the ethical concerns of specific biotechnology applications, it is appropriate to introduce the ethical concerns about biotechnology in general. Some researchers have called these concerns intrinsic if they are inherent in the technologies themselves, and extrinsic if they revolve around the consequences of the technology (Straughan, 1999; Kaiser, 2005). If a given practice or technique is intrinsically wrong, then there is no further grounds for continuation of that practice or technique. For example, slavery is intrinsically unethical, and there is no excuse to perpetuate it. The major intrinsic arguments against biotechnology are summarized in the following sections.

Religious Intrinsic Critique of Biotechnology

One of the most common ethical allegations against modern biotechnology is that it amounts to playing God. Large numbers of people belonging to certain denominations of Christianity, as well as other religious faiths, believe that the divine and human realms are distinct from each other. The natural world is made up of myriad kinds of plants and animals that exhibit similarities and differences among each other in varying degrees. This arrangement is part of a divine order. The ability of modern biotechnology to transfer genes from one species to another and make replicas having characteristics selected by humans violates this divine order. The characteristics of a given species are the gifts of God, and therefore, any human fabrication of the very basis of life is tantamount to playing God. This critique of biotechnology in general, and genetic engineering in particular, delves into the domain of metaethics as it defines and delimits the rights of humans to alter the living world, which is the creation of the Almighty God. Such acts, therefore, cannot receive ethical sanction/approval.

However, there is lack of consensus among the believers in different religions, including Christianity, on such a negative view of biotechnology. There is support for the dominion over nature by humans in Christianity itself, and many believe that God has provided this opportunity to humans (the chosen ones among all the species) to work as co-creators (Straughan, 1999). Genetic changes, albeit less drastic and taking place over a long time period, have also been achieved via selective breeding throughout ages. One example is the numerous breeds of dogs that have widely different morphological characters and behavioral attributes produced through selective breeding. This raises the question of whether the time span in which genetic changes in a species can be produced (and which in the case of genetic engineering happens to be short) is an important factor for considering biotechnology as intrinsically wrong and equivalent to playing God. It is debatable whether such an argument can be considered valid. Nevertheless, gene transfer techniques bestow much more precision and control in the hand of humans over the life of animals than was available earlier. The question therefore arises whether there should be a limit to the modifications that humans should be allowed to make in the genetic composition of animals. At the same time, numerous modifications and re-designing of nature have taken place, and are continuing around us. Exotic plants and animals have been introduced into strange environments, many of which have turned into invasive species that in turn resulted in the extinction or severe decimation of native life forms. Although these changes did not take place through direct genetic manipulations, they could also be perceived as playing God, since the morphology of a species or the composition of plant or animal communities of a given area are also the works of God. Perceptions and beliefs are important issues to consider when deciding where to draw the line on tinkering with the genetic makeup of various organisms.

Another issue of a religious nature that gene transfer creates is that many religions, such as Hinduism, Sikhism, Judaism, and Islam, prohibit the consumption of certain animals, some even recommending a purely vegetarian diet. If the gene from any such prohibited animal is inserted into a plant or other animal that humans are normally allowed to consume, how does that change the status of the altered organism? How do genetically modified (GM) food items fit into the guidelines of various religious doctrines? Or if genetic material from a fish or some other animal is inserted into a fruit or vegetable, would the latter still be considered vegetarian food? No ready answers to these controversies are available, as none of the earlier methods of selective breeding of plants and animals crossed the genetic barrier between species.

Religious Critique of Human Stem Cell Research

Human embryonic stem cells (hESCs) are cells derived from the early embryo that are pluripotent and can differentiate into any cell type of the adult body. Stem cell research or therapeutic cloning, therefore, holds great promise for developing effective medical treatments for many hitherto incurable diseases such as Parkinson's and Alzheimer's diseases, diabetes, spinal cord injury, and others. However, religions such as Catholicism object to the utilization of human embryos, as it is considered a human being with full moral standing from the moment of union of the male and female gametes. Consequently, the violation of its right to life at the hands of stem cell research is not approved by the Catholic religion. Larijani and Zahedi (2004) have explained that according to Islamic Jurisprudence, ensoulment of embryos occurs after 120 days of fertilization. Therefore, embryos that have existed for less than that period of time can be used for essential research, such as curing disease. Judaism, as well as Hinduism and Buddhism, put great importance on healing and compassion, and consequently do not raise any serious objections to hESC research.

Religious Critique of Human Reproductive Cloning

Human reproductive cloning is the production of a human fetus from a single cell by asexual reproduction. When compared to embryonic stem cell research and therapeutic cloning, human reproductive cloning has attracted more serious criticism from a number of religions, as it is seen as directly challenging the authority of God, and is tantamount to playing God. Most Churches unequivocally object to human cloning. Hinduism advises caution and careful analysis before embarking on such research, while Buddhism and Judaism are relatively more tolerant to human cloning. On the contrary, the wasteful use of embryos, unnaturalness of reproduction, social harms, and contradiction of the diversity of creation, are some of the many reasons for Islam not being supportive of human reproductive cloning (Larijani and Zahedi, 2004).

Religious Critique of *In Vitro* Fertilization

In vitro fertilization (IVF) is an assisted reproductive technology (ART) that comprises a major technique for resolving infertility problems. In IVF, sperm is used to fertilize the egg outside the body. The fertilized egg is then transferred to the uterus of the patient in order to have a successful pregnancy and childbirth. The Catholic Church has expressed its moral opposition to IVF, and considered it immoral. Islam accepts IVF, provided that it is carried out between a legally married couple, but it does not approve of third party donations of sperm, eggs, or uterus.

Secular Intrinsic Objections to Biotechnology

Besides the intrinsic objections to biotechnology from a religious point of view, there are secular arguments against it as well. Biotechnology could be considered intrinsically unethical because it attempts to cross the natural dividing lines or barriers that naturally exist between species. Biotechnology, according to these critiques, is intrinsically wrong because it violates the natural integrity of species by deliberately bringing about genetic modifications.

The detractors of this view argue that the natural boundaries between species are not as fixed and immutable as they are made out to be. A certain amount of natural hybridization takes place in natural systems, and is a part of the evolutionary process of speciation. For example, gene flow between species has been shown to occur fairly frequently in insects such as heliconiinae butterflies, cicadas, sea anemones, parasites, fish, dolphins, and killer whales. An estimated 10% of animal species are known to hybridize in nature. Thus the species boundary could be visualized as a continuum (Mallet et al., 2007).

Nevertheless, this evidence of the natural transgression of species boundaries does not entirely nullify the intrinsic secular arguments against biotechnology since the drastic genetic modifications in biotechnological research, such as transfer of genetic material from plants or microbes to animals and vice versa, finds no parallel in nature, and could be treated as a gross and anthropogenic breach of species boundaries.

Public Perceptions About Genetic Modifications in Biotechnology

How do common people who are the potential recipients of biotechnological innovations respond to genetic engineering? Some surveys to gauge public response have been carried out in Europe and the U.S.; it is not known whether such surveys have been carried out in developing countries. Several Eurobarometer surveys on biotechnology have been conducted in EU countries since 1989 (Lassen *et al.*, 2006). The responses indicate that there is no outright rejection of biotechnology by Europeans, and there is even a wide acceptance of those technologies that have medical applications. For instance, in the 2010 survey, stem cell research and embryonic stem cell research received the support of 68 and 63% of the respondents, respectively. It is also noteworthy that xenotransplantation, which involves introduction of human genes into animals to produce organs for transplantation into humans, did not receive majority approval for a long time. However, in the 2010 survey, 58% of the respondents approved this technology. On the contrary, animal cloning for food products was among the least popular technology, receiving support from only 18% of the respondents. "Unnaturalness" was one of the major objections to cloned animal food and GM food crops. Thus the concern expressed is intrinsic in nature. Furthermore, even in the case of technologies receiving majority support, such as human enhancement, xenotransplantation, gene therapy, and embryonic and non-embryonic stem cell research, a considerable percentage of the public expressed their approval, subject to the administration of strict regulatory laws (Gaskell et al., 2010). In the U.S., a 2003 survey conducted by Rutgers University revealed that the majority of interviewees disapproved of the idea of animal-based genetically modified food, and considered creation of hybrid animals through genetic manipulation morally wrong (Van Eenennaam, 2006; CAST, 2010).

People are the ultimate end-users of technology. Therefore, their perceptions, even if more intuition-based than founded in logic or scientific information, have to be respected and taken into account before introducing new technologies and related products on a large scale. The ethical principles of informed consent and informed choice have to be followed, and paternalism avoided at all cost. Biotechnology and genetic engineering may not be unethical *per se*, but their imposition on an uninformed or poorly informed public is definitely unethical. The surveys conducted in Europe and the U.S. point out the need to inform and educate the people regarding the pertinent facts about gene technology so that they are able to make informed decisions.

Extrinsic Concerns

Animal Welfare

How is animal welfare best defined? Traditionally, the term animal welfare pertained to the animal's morphological and physiological status, its level of nutrition, and the quality and adequacy of shelter provided to it, etc. In other words, the criteria for welfare had to do with the physical health and the body of the animal. However, a more modern concept of welfare would ideally take into account the mental health of the animal as well. For example, a pet dog which is getting adequate food and housing and is in a good state of bodily health could still be facing cruel treatment from one of its masters, which results in its remaining in a perpetual state of anxiety. In such a case, its welfare has been violated. Thus, the definition of animal welfare is that it comprises the state of the animal's body and mind, and the extent to which its nature (genetic traits manifested in breed and temperament) is satisfied (Hewson, 2003).

While evaluating the animal welfare concerns of animal biotechnology, it is necessary to examine the nature and extent of infringement, if any, that biotechnology can inflict on animal welfare. In this context, it is necessary to list the five freedoms advocated in the Brambell Report that are the most well-known yardsticks of animal welfare. Roger Brambell drafted this document on the welfare of farm animals in 1965, and submitted it to the UK government. The guidelines subsequently developed by the Farm Animal Welfare Advisory Committee were later elaborated as the five freedoms that comprise the following:

1. Freedom from hunger and thirst by ready access to fresh water and a diet to maintain full health and vigor.
2. Freedom from discomfort by providing an appropriate environment, including shelter and a comfortable resting area.
3. Freedom from pain, injury, or disease by prevention or rapid diagnosis and treatment.
4. Freedom to express normal behavior by providing sufficient space, proper facilities, and company of the animal's own kind.
5. Freedom from fear and distress by ensuring conditions and treatment that avoid mental suffering.

It may be noted that these five freedoms take care of the animal's physical needs as well as its feelings, and provide a broad guideline for animal welfare. However, each of these five points needs to be delineated for particular farm animals. For example, the food composition and quantity for broiler chickens and dogs, and their space requirements, would be different. Again, it would also vary among different dog breeds. These five criteria could serve as a basic yardstick to find out whether animal biotechnology meets or violates the minimum basic requirements of animal welfare. It is perhaps worth mentioning here that the current practices of animal rearing that are not mediated by biotechnology also pose challenges to these five freedoms. For instance, British and French bull dogs, pugs, boxers, and several others experience breathing problems, high blood pressure, and low oxygen levels in blood due to their short muzzle, accompanied by a disproportionately large amount of muzzle soft tissue, which is characteristic of these breeds. Similarly, some dog breeds have difficulty giving birth without surgical intervention, while others have difficulty walking, and so on. Broiler poultry also have leg problems, while the double-muscled Belgian Blue cattle experience extreme calving difficulties, among a host of other problems. Questions may therefore be raised as to the ethical propriety of creating these breeds just for human entertainment and business, and subjecting them to a life afflicted by suffering and incapacitation.

While looking for ethical issues related to biotechnology (i.e. animal welfare), one has to examine the technology of genetic engineering that involves manipulation of animal genomes using recombinant DNA (rDNA) techniques. Gene knockout techniques are used in animal experimentation where a gene with unknown functions is removed to determine its role in physiological processes. Animal cloning is another technique that is going to be increasingly used in animal biotechnology. The welfare aspects of transgenic and cloned animals, therefore, need to be evaluated.

The ethical basis for the treatment of animals by humans can be summarized in three or four approaches. The Greek philosopher-naturalist, Aristotle (384–322 BCE), and the philosopher St. Thomas Aquinas (1225–1274), were of the view that since only humans possessed the ability to use reason to guide their actions, only they could be accorded moral standing. The French mathematician-philosopher René Descartes (1596–1650) held the view that animals were mere machines possessing only instrumental or extrinsic value, and were exclusively meant for service to (and use by) humans. This exploitative view is consequently called the Cartesian view. Descartes believed that animal behavior could be explained on the basis of mechanistic principles, as an animal was akin to an "automaton" (a machine or mechanical device) produced by humans, and not to a conscious being. Needless to say, such an ethic would allow experimentation without expressing any concern for animals, and is no longer acceptable. A second approach is a kind of ethical utilitarianism put forward by the renowned ethicist Peter Singer, which states that the pain and suffering experienced by an animal during a given experiment or use should be weighed against the benefits accrued from that act or experimentation before deciding on a course of action. The "rights" approach of Tom Regan rejects this kind of a trade-off ethic. Regan argued that any being that was the "experiencing subject of a life" possessed inherent or intrinsic value irrespective of its having sufficient intelligence, autonomy, or reason. It would be ridiculous to consider a human being in this manner; therefore following this principle we can also recognize intrinsic value in animals, so they deserve to be accorded moral standing. It would therefore be unethical to use animals simply as a means to an end. In this sense, the animal rights argument is very much a part, or an extension of, the human rights movement. Another pioneering researcher in animal ethics,

Bernard Rollin, pointed out that despite their good intentions of using animal experiments for curing disease or protecting people from toxicity, scientists were nevertheless responsible for a lot of animal suffering. The increasing use of animals in research necessitates the creation of a new ethic. Such an ethic would prevent cruelty and ensure for animals a life that respects their natural behavior and preferences to the best extent possible, and would remove or reduce pain without abolishing animal use in agriculture and research (Rollin, 2011).

The above discussion shows that the implications of animal biotechnology in animal welfare should be assessed for each specific type of technological application against the backdrop of the five freedoms and the various ethical approaches, like deontology, utilitarianism, and the rights approach. Some examples are provided here to illustrate this point of view.

As said earlier, gene knockout technology involves the removal of a particular gene in order to understand its role in physiological processes. One problem associated with this technology is the production of surplus animals that do not have the desired genotype. A small percentage of the embryos in genetic engineering experiments survive, and of those that survive, a mere 1–30 % carries the intended genetic alteration. With the substantial increase of such experimentation worldwide, the number of surplus animals also increases greatly, creating problems of animal welfare in terms of pain and suffering experienced by them. The present uncertainty and inefficiency of genetic engineering experiments also add to animal welfare problems. For example, unpredictable anomalies such as lameness, susceptibility to stress, and low fertility characterized the early transgenic livestock. Cloning has also remained inefficient, and cloned animals suffer several abnormalities. These include developmental abnormalities like prolonged gestation, large birth weight, reduced placenta, and abnormalities in hepatic, cardiovascular, renal, neural, and muscular tissues, resulting in pain and suffering for the newborn as well as the surrogate mother. However, with continuing refinement of genetic engineering techniques, unpredictability is expected to be reduced, leading to improvement of animal welfare (Ormandy et al., 2011). Pigs with a gene for human somatotropin or bovine somatotropin (bST) exhibited weight gain and improved ability to convert feed to flesh; but at the same time, they also experienced a number of physical problems such as lameness, lethargy, and even gastric ulcers. Similarly, cloning from blastomeres and somatic cells may result in large calves and lambs, with possibilities of more serious effects when cloning is done from somatic cells. Cloning is still an inefficient process, resulting in a high percentage of surviving eggs showing abnormalities in placental morphology, fetal development, the immune system, brain, and digestive system (reviewed in US National Research Council, 2004). This evidence, although far from adequate, points out the need to have continuous monitoring of animal welfare in transgenic and cloned animals. However, the situation is expected to improve with refinement of existing technologies.

Human Health

Nature of Risk

The concern for human health arises from two directions, namely use of genetically engineered animals or their products as food, and their use in medicine (along with their attendant ethical dilemmas). As the public surveys in Europe and the U.S. show, there is extremely low public acceptability of GM animal food, while relatively high approval of genetic medicine. Large percentages of the interviewees perceived GM/cloned animal food as low on benefits, unsafe, inequitable, and worrying. However, both the European Food Safety Authority (EFSA) and the U.S. Food and Drug Administration (FDA) observed that the reasons for not approving GM animal food mostly lacked a scientific basis. However, there are concerns that the animal organs used in xenotransplantation may transmit viruses into the human body.

Public Health Ethics and Animal Biotechnology

Schröder-Bäck (2007) formulated a set of concise principles of public health ethics that can be useful for evaluating the health risks and benefits of animal biotechnology. Based on these principles, some key questions could be asked to examine the ethical propriety of animal biotechnology (i.e. human health). The following are basic principles and questions:

1. Social Utility: This is equivalent to beneficence in medical ethics. Therefore, one could ask whether, or how much, a given animal biotechnology would be indispensable to society, or is it merely intended to advance the interests of a small group of people, such as scientists and corporations. In other words, a utilitarian analysis can be useful in adopting an ethical position regarding a particular technology.
2. Respect for Human Dignity: The utilitarian bias of social justice is somewhat counterbalanced by having respect for human dignity, which prevents the sacrifice or instrumentalization of the interests of individual human beings. Therefore, this principle could protect specific groups from possible victimization because of the adoption or rejection of a particular technology. For example, people suffering from kidney, liver, or cardiac malfunction, and their friends and relatives, would welcome advances in xenotransplantation, irrespective of the ethical implications. Thus social utility and respect for human dignity in a way complement each other and enable to us to examine both of these principles in a balanced manner.

3. Social Justice: This comprises another facet for balancing social utility. According to this principle, not only the utility, but the distributive justice, is taken care of. Thus, a particular animal biotechnology ought to be assessed with regards to its ability to look after the interests of all spectrums of people, including disadvantaged people and people in economically developed, developing, and under-developed countries or regions. This could take the form of access to a given technology by all groups of people.
4. Efficiency: This is also an important parameter to determine the appropriateness of any technology. An efficient technology would allow it to reach more people in a country or region with limited resources at its disposal.
5. Proportionality: The fifth (and last) principle demands insurance that the health benefits of animal biotechnology outweigh the possible infringement on autonomy, privacy, and all other negative impacts.

Environmental Concerns

Ethical Positions

The ethical positions that humans can adopt in their relationships with the environment could be categorized as anthropocentric (human-centered), biocentric (life-centered), and ecocentric (ecosystem-centered). Deep ecology is similar to ecocentric positions, but with a strong emphasis on Self-realization ("Self" with a capital "S") (Sessions, 1995). An anthropocentric environmental ethic grants moral standing exclusively to human beings. It considers nonhuman natural entities and nature as a whole to be only a means for human ends. Thus it recognizes intrinsic value only in human beings, with nature accorded extrinsic value only. A biocentric ethic recognizes intrinsic value in animals, especially the sentient animals, while an ecocentric ethic assigns intrinsic value to nature as a whole, including all life forms, and even entire ecosystems, and ultimately, the biosphere. Environmental risks posed by all technology (including biotechnology) need to be evaluated against these ethical positions. Even with a moderately anthropocentric stance, one cannot ignore the potential and existing threats of technology and development to natural systems, including their non-human denizens.

Concerns Posed by Transgenic Technology

The major environmental concern with transgenic animals is that they may be accidentally or deliberately released into the environment, causing environmental problems. However, in such cases the impact of the transgenic animals would be like that caused by the introduction of invasive alien species into the environment. Transgenic fish provide an example of such concerns. If a transgenic fish species has high juvenile and adult viability, high fecundity, and high fertility, then it could play the role of an invasive species, spreading its genes into the aquatic ecosystem. Thus the impact of transgenic fish release will depend on the net fitness of the species compared to that of the wild stock; if it is higher than that of the latter, then the transgene will spread through the wild population (Kaiser, 2005). Transgenic fish with greatly enhanced growth are coveted both in aquaculture and sport fishery. Transgenic coho salmon (*Oncorhynchus kisutch*), rainbow trout (*O. mykiss*), cutthroat trout (*O. clarki*), and chinook salmon (*O. tshawytsha*) have juveniles 10–15 times larger than their non-transgenic counterparts. Both juvenile and adult GM tilapia are known to be three times the size of non-GM members of the same species. These growth-enhanced fish can reach sexual maturity earlier than their wild relatives, resulting in a lasting genetic effect in the population. In a model developed with the fish Japanese medaka (*Oryzias latipes*), it was shown that transgenes could spread rapidly in the population, provided certain conditions are fulfilled (Muir and Howard, 2001). The transgenic Atlantic salmon, which is to be marketed soon, grows twice as fast as the wild individuals. It contains the DNA sequence of the code for the growth hormone of Chinook salmon, and regulatory sequences derived from Chinook salmon and ocean pout. This creates the risk of these transgenics escaping into natural ecosystems and eventually decimating or even wiping out wild populations. However, the company producing the transgenic salmon has offered assurances of full-proof culture systems far from the sea, with adequate preventive measures. Furthermore, most of the fish are reported to be sterile triploids that pose low risk to the wild populations (Marris, 2010).

Precautionary Principle

The regulatory issues emerging from human health and environmental concerns of biotechnology can be discussed under the guidelines of the UN precautionary principle.

This can be especially useful for animal biotechnology or any other emerging technologies that are characterized by a lack of full scientific certainty. However, this uncertainty cannot be an excuse for not adopting cost-effective measures to minimize health risks or prevent environmental degradation. The EU decision to impose a ban on cloned animals for food not only reflects a respect for public opinion, but also adherence to the precautionary principle, notwithstanding the fact that evidence for adverse impacts of GM animal food is far from conclusive.

SOME CHALLENGING ETHICAL ISSUES IN ANIMAL BIOTECHNOLOGY

Chimeras

In this section, the ethical implications of the production of chimeras, which are combinations of cells of different embryonic origin, including those of humans and non-human animals, are discussed. Chimeras are also defined as creatures with cells, tissues, or organs from individuals of two different species (interspecific chimeras) (Greely et al., 2007). Human-to-animal chimeras are those where human cells such as pluripotent stem cells are placed in a non-human animal. Such chimeras could be fetal or adult depending on whether the human cells were placed in an embryo or a postnatal animal. Similarly, there could be animal-to-human chimeras that come more under the heading of xenotransplantation.

Four ethical arguments can be raised against chimeras: moral taboo, species integrity, unnaturalness, and human dignity. The questions of moral taboo and playing God, as well as crossing species boundaries, have already been discussed in an earlier section. Hybrids occur in nature, and artificial hybrids of both plants and animals have been accepted by society without any ethical reservations. So the real question raised by the human-to-animal chimeras is whether the species boundary is unique and consequently inviolable or not. The other question is that if the human stem cells are planted in a non-human embryo and not allowed to develop into adults, what ethical position should be adopted? Should the termination of human-to-animal chimeric embryos receive the same ethical attention as human abortion? Again, in another possible scenario, what ethical stance should be taken if a chimeric mouse having human sperm mates with another chimeric mouse having human eggs? It is also felt that creation of interspecies creatures from human material evokes the idea of bestiality, an act of moral abomination, just as sexual intimacy between humans and animals is considered abominable and a moral taboo. Chimeras also represent a metaphysical threat to the self-image that human beings have (Robert and Baylis, 2003).

Here it could also be argued that species, besides being genetic and ecological constructs, could also be viewed as mental or cultural constructs so far as human beings are concerned. We have mental images of a cat or a monkey, and for a human being. If human-to-animal or animal-to-human chimeras violate those images, it could lead to moral-ethical problems. On the contrary, a human being with a pig's heart, liver, kidney, or neural tissue, or a cat with grafts of human tissue but otherwise human-like or cat-like characteristics (respectively) may not disrupt that mental construct of the species. Such constructs of species have been called common-sense species categories (Karpowicz, 2003). The author adopts a utilitarian argument by pointing out the fact that stem cell research could pave the way for the successful treatment of hitherto untreatable diseases such as those caused by dysfunction of retinal or neural tissues. Mouse chimeras with areas of human neural or retinal tissue could find immense use in therapies, and should be ethically defendable. Regarding the question of granting greater moral status to such chimeras, Karpowicz further argues that the mouse brain is small and has a much shorter development time than that of humans. In order to have a neuronal complexity equivalent to a human brain, a chimeric mouse would need to have a huge brain several hundred times larger than its own. Only such a chimera, which is extremely unlikely to be produced, would create moral abomination or demand greater moral status. In this context, a question arises as to whether transplantation of human stem cells into the brain of non-human primate embryos should be allowed, as it would definitely increase the risk of a future moral confusion. Johnston and Eliot (2003) are of the opinion that the ethical parameter that is most offended by chimeras is human dignity in terms of cruelty to such a creature, and treating it as a means to an end and not an end in itself, which is morally unacceptable. Greely et al. (2007) in their review of the findings of a host of other workers, concluded that chimeras compromised human dignity if the human cells transplanted into another animal were allowed to proliferate to develop into a human-like brain rendering human-like capabilities in these animals. They also reported that research involved in transplanting human neural stem cells into the brains of non-human primates should be regulated to minimize the risk of the resulting chimera having human-like cognitive capacities.

Thus the creation of human-to-animal chimeras poses certain interesting ethical questions that need to be addressed before such research could move forward to produce creatures that may have more and more human-like characteristics in their brains.

Animal Biopharming

Biopharming is a popular term that describes the manufacture of complex therapeutic proteins such as blood clotting factors, fibrinogen, and alpha-1-antitrypsin in the milk of transgenic animals such as sheep, goat, cattle, and others. Alpha-1-antitrypsin is used in the treatment of cystic fibrosis in children, thus lending great utilitarian value to such innovative technologies. While recombinant proteins like insulin and human growth hormone have been manufactured for a long time in genetically engineered bacteria and yeast, more complex proteins can only be manufactured in mammalian cells. For example, the gene coding for alpha-1-antitrypsin can also be inserted into plants and microbes, but they lack the ability to add a carbohydrate moiety that is needed to make it biologically active. The biochemical similarity of animals to humans lends them an advantage over plants and microbes. Another advantage of animal biopharming is that lactating goats, sheep, or cattle can be made to secrete the therapeutic protein in their milk, from which it can be purified. Antithrombin, an anticoagulant protein, was the first therapeutic protein produced in the milk of transgenic goats that has subsequently been approved for administration to persons with a congenital defect that prevents them from producing sufficient amounts of this protein. Spider silk has also been produced in the milk of goats in a similar fashion. A large number of other therapeutic proteins are now in the process of being produced in the milk of transgenic animals. These products include albumin, human growth hormone, collagen, fibrinogen, lactoferrin, humanized polyclonal antibodies, and others, and they are being produced in the milk of cows, goats, sheep, pigs, and rabbits (Goven et al., 2008).

Ethical issues in animal biopharming revolve around risks to human health, gene transfer, and animal welfare, and can be analyzed by general utilitarian, public health, and animal welfare principles, as well as the precautionary principle. The risk factors have been reviewed in detail in Goven et al. (2008). The information provided in the following sections is largely taken from this work and subjected to ethical analysis.

Risks to Human Health

The major potential health risks include contamination of milk with infectious agents such as prions in cattle. Prions are associated with transmissible spongioform encephalopathies (TSEs), including bovine sponigioform encephalopathy (BSE), variant Creutzfeldt–Jakob disease (vCJD), and scrapie. Although prion aggregation, activation, and transmission are complex phenomena, this is a major health hazard that can occur from biopharming. Other possible health risks include allergenic and immunogenic responses, and autoimmune reactions. These may be linked to the unpredictability of the microinjection process of producing transgenic animals. These risks, therefore, need detailed scrutiny not only from scientific experimentation and assessment, but also from the point of view of public health ethics, medical ethics, and the precautionary principle. Striking a balance between social utility/beneficence and human dignity, and social justice and efficiency, is an essential exercise that should be undertaken for a realistic analysis of health risks. While biopharming holds great promise for providing health benefits to people, it should be noted that the only biopharm product that has received marketing authorization from the European Medicines Agency (EMEA) is ATryn®, which is a form of recombinant antithrombin produced in transgenic goats. Several others are in the preclinical and research stages.

Food Chain Contamination

All animals used in biopharming have to lactate in order to produce milk, and therefore have to produce offspring. This results in the production of a large number of surplus animals. It is probable that these animals will be sold on the market and thus enter the human food chain. Similarly, the excess milk that is not required or suitable for pharmaceutical production will also enter the human food chain. This problem calls for strict regulatory mechanisms to prevent the entry of any such products legally or illegally into the market. Countries with inadequate policy frameworks and/or poor governance are therefore at higher risk from these problems. Another hazard could be in the form of these animals being marketed as processed meat or milk/milk products through clandestine channels, a possibility that cannot be entirely ruled out.

Escape of Biopharm Animals

Small animals pose a higher risk of such escape, and these may mix and mate with their relatives outside. The risks are similar to those posed by the possible escape of transgenic fish or other transgenic animals. The risks are relatively less in large, fully domesticated animals like cattle. From an ethical standpoint, careful weighing of potential benefits against risks may help to arrive at appropriate decisions. Besides utilitarian principles, this problem also raises the question of the value of animals. Should the animals be accorded only extrinsic or instrumental value, where the highest or sometimes sole priority is attached to their use-value for humans? On the contrary, should they be regarded as partners worthy of receiving intrinsic value because of their central role in producing life-saving molecules that alleviate human suffering? These questions call for intensive ethical discussion. Again, rigorous application of the precautionary principle is needed in these cases to do justice to all concerned.

Horizontal Gene Transfer

Horizontal gene transfer may take place through the presence of the programmer gene in blood, feces, and other waste materials, and could be spread by other organisms such as bacteria and blood-sucking insects. These possibilities, however remote, constitute another set of risks in biopharming. Again, the precautionary principle and the principles of beneficence, non-maleficence, and least-harm can be the major instruments in resolving this issue.

Welfare Issues of Biopharm Animals

This is one of the most important ethical concerns associated with biopharming. Various abnormalities of cloned animals cause high morbidity and mortality. Conditions such as large-calf syndrome, hydrops, or hydroallantois and musculature abnormalities occur at higher rates among cloned animals. The ethical implications of these welfare issues have already been discussed, and are applicable to biopharm animals as well.

Constitution of Ethics Committees

The formation of ethics codes and committees can be traced back to the Nuremberg Trials in 1945 and 1946, held after World War II in order to try Nazi war crimes. Several crimes were committed in the guise of scientific research that involved inhuman and unethical treatment of prisoners and detainees in concentration camps. Such experiments involved immersing the subjects in ice-cold water or exposing them to freezing temperatures; muscle, bone, and nerve transplantations without anesthesia; infecting subjects with malarial parasites and other pathogens; head injury experiments; mustard gas exposure; and so on. The subjects (who included children) were forced to participate without any informed consent. The trials led to the drafting of the Nuremberg Code, which first addressed the question of volunteer consent for protecting human subjects in research. Subsequently, the World Medical Association (WMA) framed the Declaration of Helsinki, which still serves as a standard for physicians undertaking research on human subjects. This declaration was subsequently revised and updated several times. Important principles and guidelines were also provided in the Belmont Report (1979) published by The National Commission for the Protection of Human Subjects of Biomedical and Behavioral Research, USA. This report emphasized several ethical principles such as autonomy, respect for persons, beneficence, and justice. The Declaration of Helsinki gave precedence to the wellbeing of human research subjects over every other consideration. The Declaration stated that since medical research with human subjects involved risks to the subject, the details of such research must be clearly defined and described in a research protocol. This protocol should explain how it conforms to the principles of the Declaration and should provide information on funding, sponsors, institutional affiliations, conflicts of interest (if any), and so on. This protocol was to be submitted to a research ethics committee before starting the study. These committees were required to be free from any undue influence or interference from sponsors and other interested parties so as to be able to serve as an independent body to protect the rights and ensure safety of participants in human trials. They were to take into consideration pertinent rules and regulations of the country where the research was to be carried out, along with applicable international norms and standards, but were not to withdraw or reduce any protection for research subjects contained in the Declaration. Once approved, no further changes could be made to the protocol without prior consideration and approval of the Committee. Such independent bodies are known as institutional review boards (IRBs), independent ethics committees (IECs), or ethical review boards (ERBs). It is now customary for every university, institute, hospital, or other organization where research with human subjects is conducted, to have one of these bodies for ensuring the ethical propriety of research, and for safeguarding the dignity, safety, and rights of human subjects. The goal is to obtain the prior informed consent of human subjects, assess risks, and protect individual privacies before any trial would become mandatory. An IRB or IEC must have a minimum of five voting members, at least one of whom must be a non-scientist, and one member should in no way be connected with the institution concerned.

TRANSLATIONAL SIGNIFICANCE

HUMAN THERAPEUTIC CLONING AND OTHER TECHNIQUES IN ANIMAL BIOTECHNOLOGY

Generally speaking, translational research (TR) means research aimed at converting the findings of basic science to their practical applications. More specifically, however, it means the clinical application of the outcome of basic research. TR can therefore be said to hold the key to the ultimate fate of scientific findings. Many path-breaking discoveries and inventions in every conceivable scientific discipline never find their deserving use in society because of the absence of appropriate TR. Although basic research is not aimed at solving practical problems, it can nevertheless provide clues or directions towards solutions to practical problems. Thus it could provide the knowledge platform from which practical application-oriented studies could take off. In this sense, basic and translational research could be said to represent a sort of continuum. TR is often recognized to have two distinct dimensions: one involves applying the findings of basic research or preclinical studies to the development of clinical trials (first stage or T1), followed by the adoption of best practices in the community (second stage or T2) (Rubio et al., 2010). Thus TR moves from basic research to patient-oriented research to culminate in population-based research, with society receiving the benefits and enduring the risks (if any) at this stage. However, it must be remembered that this flow is not strictly unidirectional, as there could be frequent to-and-fro movement and exchange of knowledge and experience between T1 and T2 (and even between TR as a whole) and basic research, as new issues and questions arise while applying the results at the community level. A large proportion of significant findings in basic science that hold great promise for developing cures for deadly diseases never reach the therapeutic development and application stage because of the absence of proper translational studies.

Citing the possible developments in human embryonic stem cell research as an example, use of pluripotent embryonic stem cells (ESC) has created immense possibilities for developing therapeutic solutions for hitherto incurable diseases such as neurodegenerative diseases and others.

It is to be understood that the main objective of raising ethical questions is to provide a moral–ethical framework to regulate and streamline these developments for the maximum welfare of humankind without compromising human dignity or harming other forms of life and the environment as a whole.

CONCLUSIONS

The preceding discussion of a number of important ethical concerns of animal biotechnology clearly shows that an ethical analysis of the different innovations cannot provide any readymade answers to the contentious issues raised because of their highly complicated nature. However, ignoring pertinent ethical issues can bring about detrimental consequences not only for the different stakeholders, but for the technology itself. In recent years, we have witnessed huge public outcries against several new technologies. It is a matter of argument whether such resistance has a strong scientific basis, but at the same time the paternalistic attitude often adopted by a segment of scientists and policy makers has not helped the situation. This is true not only for animal biotechnology, but for many other technologies and even developmental activities, including those in developing countries. A communication gap between the promoters and detractors of new technologies is a major cause of the resultant imbroglio. The theories and principles of ethics constitute valuable tools for creating an atmosphere of dialog among the different factions, and to ultimately arrive at the safest and wisest possible use of a given technology in a given time and place for the welfare of humans, non-human organisms, and natural systems alike.

WORLD WIDE WEB RESOURCES

http://www.unesco.org/

The United Nations Educational, Scientific, and Cultural Organization (UNESCO) website is one of the richest repositories of publications on bioethics and ethics of science and technology, including biotechnology. After logging on to the website, one can go to Social and Human Sciences, where the Themes drop-down menu leads to Bioethics and then to the Publications section, which contains books, reports, and advice on almost all conceivable ethical issues in science and technology. Furthermore, the site also contains announcements on seminars, workshops, and other events on bioethics. Information on the objectives and mode of functioning of the International Bioethics Committee (IBC), the Intergovernmental Bioethics Committee (IGBC), and the Global Ethics Observatory (GEObs), is also useful for understanding ethical perspectives in science and technology.

http://www.nhmrc.gov.au

This is the website of the Australian government's National Health and Medical Research Council. This site provides resources on the ethics of animal research and on medical ethics. The areas covered are diverse, such as perspectives on health ethics and human research ethics committees, animal research ethics, human embryos and cloning, ethical conduct in human research, research integrity, and the like.

http://ec.europa.eu

The website of the European Commission provides information on ethical issues pertaining to agriculture, the environment, and other areas. After accessing the website, one can go to the desired language (English, French, etc.), and then to Policies, Legislations, and Public consultations. In the Science and Technology sub-section in the Policies section, one can click on Ethics in Science and then on several ethics links, including those of the European Commission, as well as a large number of National Ethics Committees within and outside Europe.

www.nap.edu

The website of The National Academies Press, National Academy of Sciences, contains information on animal ethics and other issues. Free registration is possible, whereupon one can access and download a large number of documents. In the Biotechnology section, publications on translational studies, biosecurity, biosafety, and other subjects are available. Publications on the care of laboratory animals, regulations on the use of animals in neuroscience research, the use of non-human primates like chimpanzees in biomedical and behavioral research, and a host of other topics can be accessed for free download and/ or online reading.

www.scu.edu/ethics/

The website of the Markkula Center for Applied Ethics, Santa Clara University, California, is another important site for online resources on bioethics, including ethical issues in animal biotechnology. Discussions on the implications of important ethical principles in pharmacogenomics research and applications such as informed consent, confidentiality, justice, and equity can be found here. Ethical issues in translational aspects of pharmacogenomics are also presented in the context of the normative framework.

http://www.bbsrc.ac.uk

This is the website of the Biotechnology and Biological Sciences Research Council, UK. In the Science in Society section, an external link to Understanding Animal Research provides resources such as documents, images, and videos on the use of animals in research. In the Publications section, Topic-Based Publications contains write-ups on food security, industrial biotechnology, stem cell science, and related issues.

http://plato.stanford.edu

The website of the Stanford Encyclopedia of Philosophy (Principal Editor: Edward N. Zalta) is a valuable repository of information on all aspects of philosophy and ethics, including applied ethics. For example, a search using the term bioethics can lead to information on issues like human/non-human chimeras, cloning, the ethics of stem cell research, and other issues. General ethics topics like metaethics, deontology, utilitarianism, etc., can also yield substantial information in these areas.

http://www.iep.utm.edu

The Internet Encyclopedia of Philosophy contains useful definitions and explanations for concepts in ethics and philosophy. The search option can be used to access resources on ethical concepts such as deontology, metaethics, utilitarianism, virtue ethics, and the like, as well as coverage of important ethicists and philosophers.

REFERENCES

Council for Agricultural Science and Technology (CAST). (2010). *Ethical Implications of Animal Biotechnology: Considerations for Animal Welfare Decision Making*. CAST Issue Paper No. 46. Ames, Iowa: CAST.

COMEST (World Commission on the Ethics of Scientific Knowledge and Technology). (2005). *The Precautionary Principle*. Paris: UNESCO.

Gaskell, G., Stares, S., Allansdottir, A. et al., (2010) Europeans and Biotechnology in 2010 – Winds of Change? A Report to the European Commission's Directorate-General for Research. Brussels: European Commission.

Goven, J., Hunt, L., Shamy, D. & Heinemann, JA. (2008) Animal Biopharming in New Zealand: Drivers, Scenarios and Practical Implications. Constructive Conversations / Kōrero Whakaaetanga (Phase 2): Research Report No. 12. http://www.conversations.canterbury.ac.nz/documents/animalbioreport.pdf

Greely, H. T., Cho, M. K., Hogle, L. F., & Satz, D. M. (2007). Thinking about the human neural mouse. *American Journal of Bioethics*, *7*(5), 27–40.

Hewson, C. J. (2003). What is animal welfare? Common definitions and their practical consequences. *Canadian Veterinary Journal*, *44*, 496–499.

Johnston, J., & Eliot, C. (2003). Chimeras and "human dignity." *American Journal of Bioethics*, *3*(3), w6–w8.

Kaiser, M. (2005). Assessing ethics and animal welfare in animal biotechnology for farm production. *Scientific and Technical Review*, *24*(1), 75–87.

Karpowicz, P. (2003). In defense of stem cell chimeras: a response to "crossing species boundaries." *American Journal of Bioethics*, *3*(3), 17–19.

Larijani, B., & Zahedi, F. (2004). Islamic perspective on human cloning and stem cell research. *Transplantation Proceedings*, *36*, 3188–3189.

Lassen, J., Gjerris, M., & Sandøe, P. (2006). After Dolly – ethical limits to the use of biotechnology on farm animals. *Thericogenology*, *65*, 992–1004.

Mallet, J., Beltran, M., Neukirchen, W., & Linares, M. (2007). Natural hybridization in heliconiine butterflies: the species boundary as a continuum. *BMC Evolutionary Biology*, *7*, 28.

Marris, E. (2010). Transgenic fish go large. *Scientific American*. September 14, 2010.

Muir, W. M., & Howard, R. D. (2001). Fitness components and ecological risk of transgenic release: a model using Japanese medaka (*Oryzias latipes*). *The American Naturalist*, *158*(1), 1–16.

Ormandy, E. H., Dale, J., & Griffin, G. (2011). Genetic engineering of animals: Ethical issues, including welfare concerns. *Canadian Veterinary Journal*, *52*, 544–550.

Robert, J. S., & Baylis, F. (2003). Crossing species boundaries. *American Journal of Bioethics*, *3*(3), 1–13.

Rollin, B. E. (2011). Animal rights as a mainstream phenomenon. *Animals*, *1*, 102–115.

Rubio, D. M., Schoenbaum, E. E., Lee, L. S., Schteingart, D. E., Marantz, P. R., Anderson, K. E., Platt, L. D., Baez, A., & Esposito, K. (2010). Defining translational research: implications for training. *Academic Medicine*, *85*(3), 470–475.

Schröder-Bäck, P. (2007). Principles for public health ethics: a transcultural approach. *Eubios Journal of Asian and International Bioethics*, *17*, 104–108.

Sessions, G. (1995). *Deep Ecology for the 21st Century*. Boston: Shambhala.

Straughan, R. (1999). *Ethics, Morality and Animal Biotechnology*. Swindon, UK: Biological Science Research Council.

US National Research Council Committee on Identifying and Assessing Unintended Effects of Genetically Engineered Foods on Human Health. (2004). *Safety of Genetically Engineered Foods: Approaches to Assessing Unintended Health Effects*. Washington, DC. The National Academies Press.

Van Eenennaam, A. L. (2006). What is the future of animal biotechnology? *California Agriculture*, *60*(3), 3http://repositories.cdlib.org/anrcs/californiaagriculture/.

Weed, D. L., & McKeown, R. E. (2001). Ethics in epidemiology and public health I. Technical terms. *Journal of Epidemiology and Community Health*, *55*, 855–857.

FURTHER READING

Dennis, M. B., Jr. (2002). Welfare issues of genetically modified animals. *ILAR Journal*, *43*(2), 100–109.

Greene, M., Schill, K., Takahashi, S., et al. (2005). Moral issues of human non-human primate neural grafting. *Science*, *309*, 385–386.

Regan, T., & Singer, P. (1989). *Animal Rights and Human Obligations* (2nd ed). Engelwood Cliffs, NJ: Prentice Hall.

GLOSSARY

Animal Welfare A modern concept of animal welfare takes into account not only an animal's morphological and physiological status, its level of nutrition, and the quality and adequacy of its shelter, but also its mental health.

Anthropocentrism It is the human-centered ethical position in relation to the environment. It puts foremost importance on human needs and priorities.

Antithrombin Antithrombin is an anticoagulant protein. It is the first therapeutic protein produced in the milk of transgenic goats.

Applied Ethics Applied ethics deals with specific issues using the concepts and principles contained in metaethics and normative ethics as tools. It deals with wide-ranging issues like animal rights and environmental ethics, to euthanasia, cloning, and xenotransplantation, among others.

Biocentrism The biocentric or life-centered approach to environmental ethics recognizes intrinsic value in animals, especially sentient animals.

Biopharming Biopharming is a popular term that describes the manufacture of complex therapeutic proteins such as blood clotting factors, fibrinogen, and alpha-1-antitrypsin in the milk of transgenic animals such as sheep, goat, cattle, and others.

Chimera Chimeras are a combination of cells of different embryonic origin, including those of humans and non-human animals.

Deep Ecology Deep ecology puts emphasis on intrinsic relationships between man and environment, and a deep respect for all living organisms and ecosystems. The Norwegian philosopher Arne Ness is the main proponent of deep ecology.

Deontological Ethics Deontological ethics are based on the concept of duty or obligation, and are often called non-consequentialist, as they are not dependent on the consequences of a given action.

Ecocentrism An ecocentric ethic recognizes intrinsic value in nature as a whole.

Eudaimonia An important component of Aristotlean ethics, eudaimonia connotes happiness or flourishing, and virtue or excellence.

Gene Knockout Techniques These are used in animal experimentation where a gene with unknown functions is removed to determine its role in physiological processes.

Human Embryonic Stem Cells (hESCs) These are cells derived from the early embryo that are pluripotent and can differentiate into any cell type of the adult body.

Independent Ethics Committee An independent ethics committee (IEC) is also known as an institutional review board (IRB) or ethical review board (ERB). It is a committee that is authorized to examine, monitor, review, and approve (if found appropriate) biomedical and behavioral research involving humans, to ensure patient or volunteer safety.

Metaethics Metaethics studies the foundations of morality and establishes the theoretical basis and scope of moral values. In other words, it tries to define morality itself.

Normative or Prescriptive Ethics Normative or prescriptive ethics determines what is ethical and what is not.

Prion Prions are infectious pathogens that are devoid of nucleic acids and consist of a re-folded protein. They are known to cause several diseases, including bovine spongiform encephalopathy (BSE) in cattle and Creutzfeldt–Jakob disease (CJD) in humans.

Utilitarian Ethics Utilitarianism is concerned with maximization of good, and aims to bring about "the greatest amount of good for the greatest number."

ABBREVIATIONS

ART Assisted Reproductive Technology
BSE Bovine Sponigioform Encephalopathy
bST Bovine Somatotropin
CAST Council for Agricultural Science and Technology
COMEST World Commission on the Ethics of Scientific Knowledge and Technology
CJD Creutzfeldt–Jakob Disease
ERB Ethical Review Board
hESC Human Embryonic Stem Cell
IEC Independent Ethics Committee
IRB Institutional Review Board
IVF *In Vitro* Fertilization
TR Translational Research

LONG ANSWER QUESTIONS

1. Give a brief overview of the different ethical thoughts and theories. Enlist the major principles that could be of use in resolving ethical issues in science and technology.
2. Discuss the major intrinsic ethical concerns in animal biotechnology. What are your views on the accusation that animal biotechnology is "playing God." Do you think that religious beliefs should be used to impose restrictions on biotechnology research?
3. Give an account of the major extrinsic concerns in animal biotechnology.
4. Discuss the ethical issues arising out of the creation of chimeras.
5. Describe the potential benefits and risks associated with animal biopharming.

SHORT ANSWER QUESTIONS

1. Name the three broad areas of ethics.
2. Explain the term beneficence.
3. What is the most important guiding principle in animal welfare?
4. Name the five concise principles of public health ethics.
5. Distinguish among anthropocentric, biocentric, and ecocentric approaches in environmental ethics.

ANSWERS TO SHORT ANSWER QUESTIONS

1. The three broad areas of ethics are metaethics, normative ethics, and applied ethics.
2. It literally means "doing good." From an ethical perspective, it has an underlying utilitarian basis that is concerned with producing the greatest possible proportion of good over evil for the greatest number of people.
3. The five freedoms outlined in the Brambell Report (1964) provide the most useful guideline in animal welfare. These include freedom from hunger and thirst; from discomfort; from pain, injury, or disease; for expressing normal behavior; and from fear and distress.
4. These are social utility, respect for human dignity, social justice, efficiency, and proportionality.
5. These approaches are human-centered, life-centered, and ecosystem-centered, respectively. A purely anthropocentric ethic does not recognize intrinsic values in non-human organisms and ecosystems.

Index

Note: Page numbers followed by "f" denote figures; "t" tables.

A

B

G

H

J

K

O

P

T

U

V

W

X

Y

Z

Color Plates

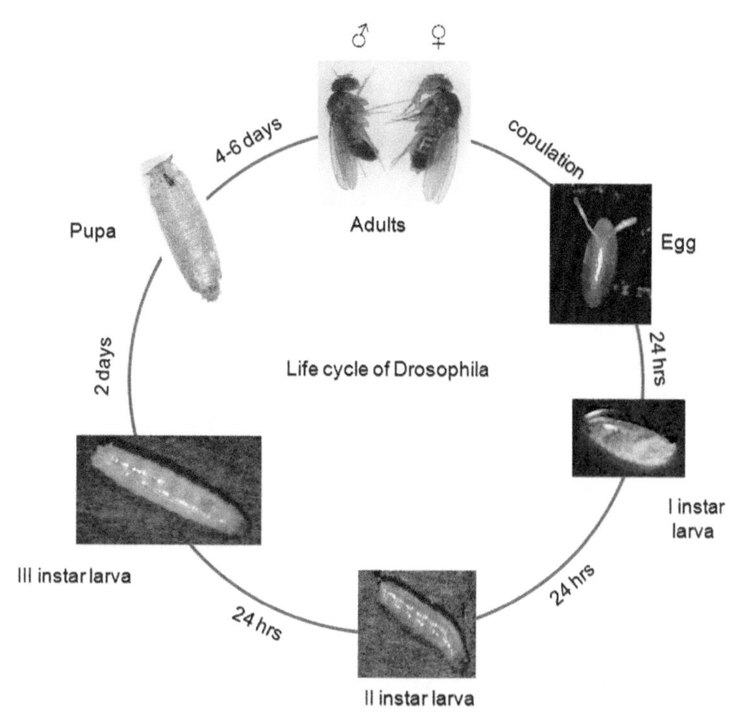

FIGURE 1.1 Life cycle stages of *Drosophila melanogaster*.

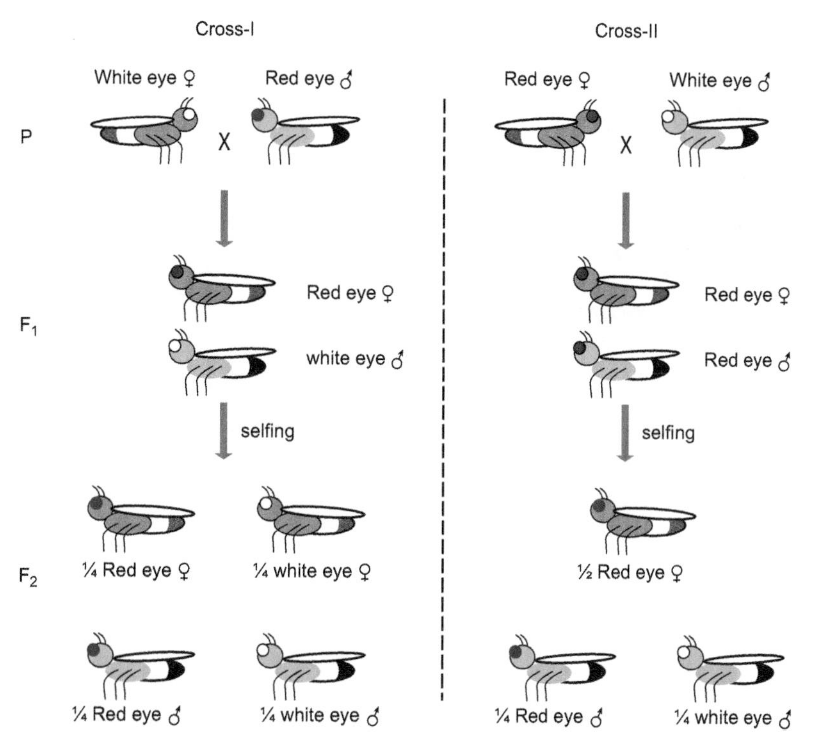

FIGURE 1.2 Schematic depiction of classes of genes associated with pattern formation in *Drosophila melanogaster*.

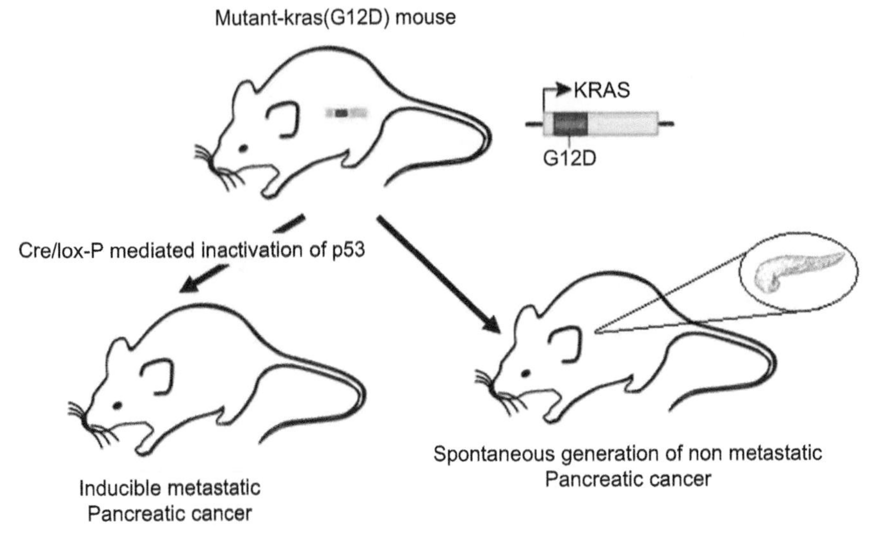

FIGURE 5.2 Generation of spontaneous tumor models for carcinogen studies.

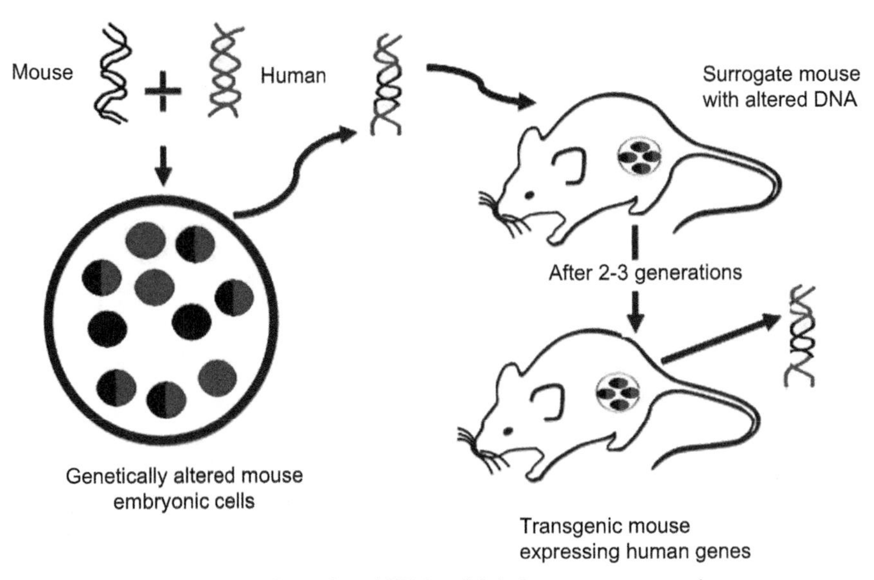

FIGURE 5.3 Generation of GEM models in immunocompetent mice.

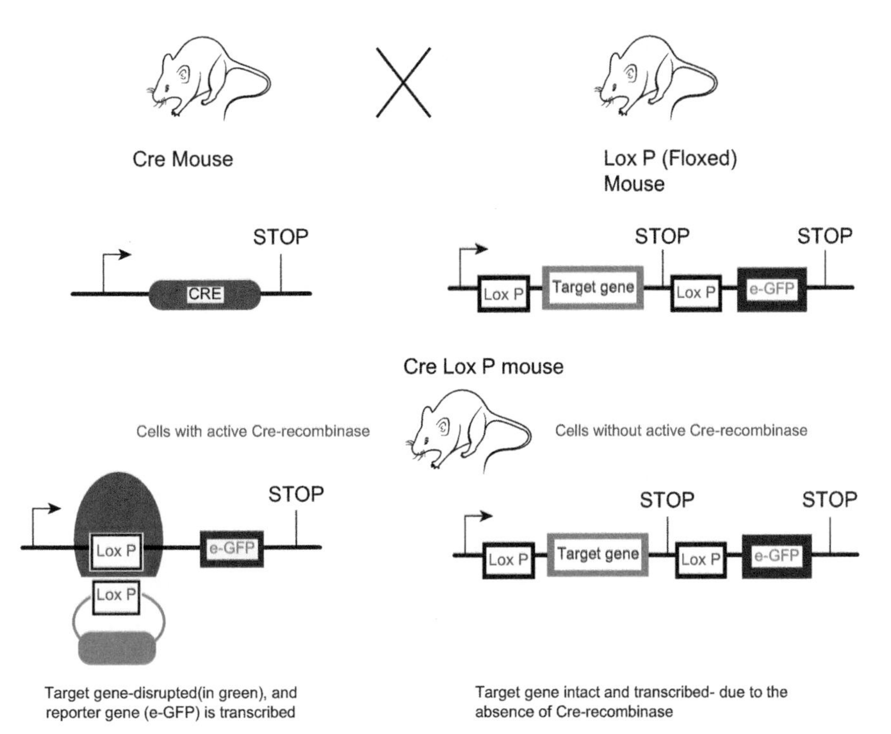

FIGURE 5.4 Generation of the Cre/Lox mouse model.

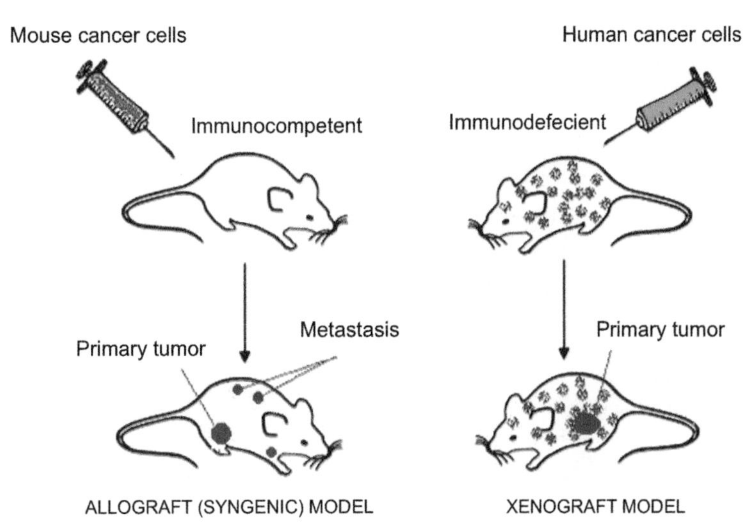

FIGURE 5.5 Generation of allograft and xenograft tumor models.

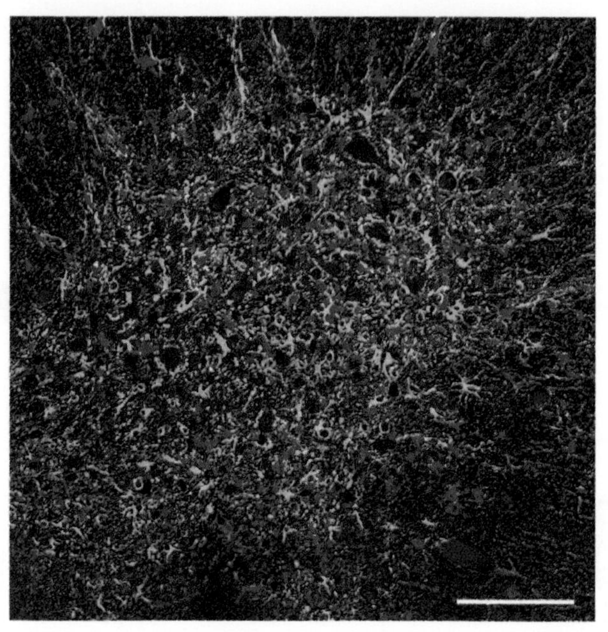

FIGURE 3.6 **Activated Microglia and Astrocytes in Lumbar Spinal Cord of Symptomatic Mutant SOD1 Mice.** Red: microglia stained with anti-Mac2 antibody, Green: astrocytes stained with anti-GFAP antibody, Blue: motor neurons stained with anti-neurofilament H antibody. Bar: 100 μm.

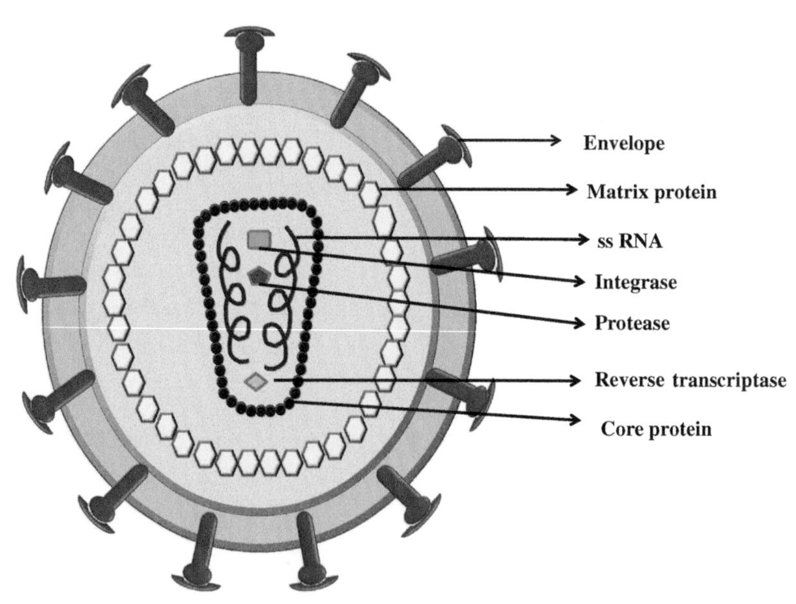

FIGURE 9.1 **Structure of HIV.** Graphical representation of cross-section of HIV. Envelope is the outermost layer; it consists of a lipid bilayer. The envelope layer is comprised of gp120 and gp41. The layer next to the envelope is the matrix protein. The matrix layer is followed by the core protein. At the center of the virion, two molecules of single-stranded RNA (ssRNA) and other enzymes are present. These enzymes are protease, integrase, and reverse transcriptase. Reverse transcriptase also contains RNase H. The location of each individual protein and RNA is shown in the figure and the molecular mass of the protein is shown in brackets. (Polymerase is not shown; it contains integrase, protease, reverse transcriptase, and RNAse H).

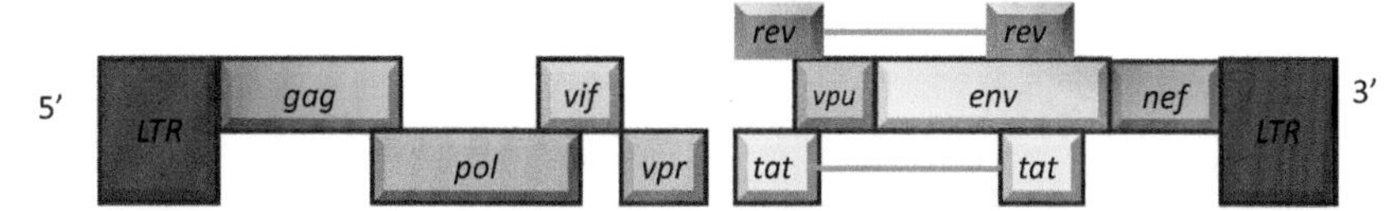

FIGURE 9.2 **HIV Genome.** Schematic representation of the HIV genome. The genome is 9.8 kB in size, and consists of 9 genes, which are flanked by LTRs on either side of the genome. These 9 genes finally produce 15 proteins: *env*, envelope; *gag*, group specific antigen; *LTR*, Long Terminal Repeat; *nef*, negative factor; *pol*, polymerase; *rev*, regulator of expression of viral proteins; *tat*, transactivator of transcription; *vif*, viral infectivity factor; *vpr*, viral protein R; *vpu*, viral protein U.

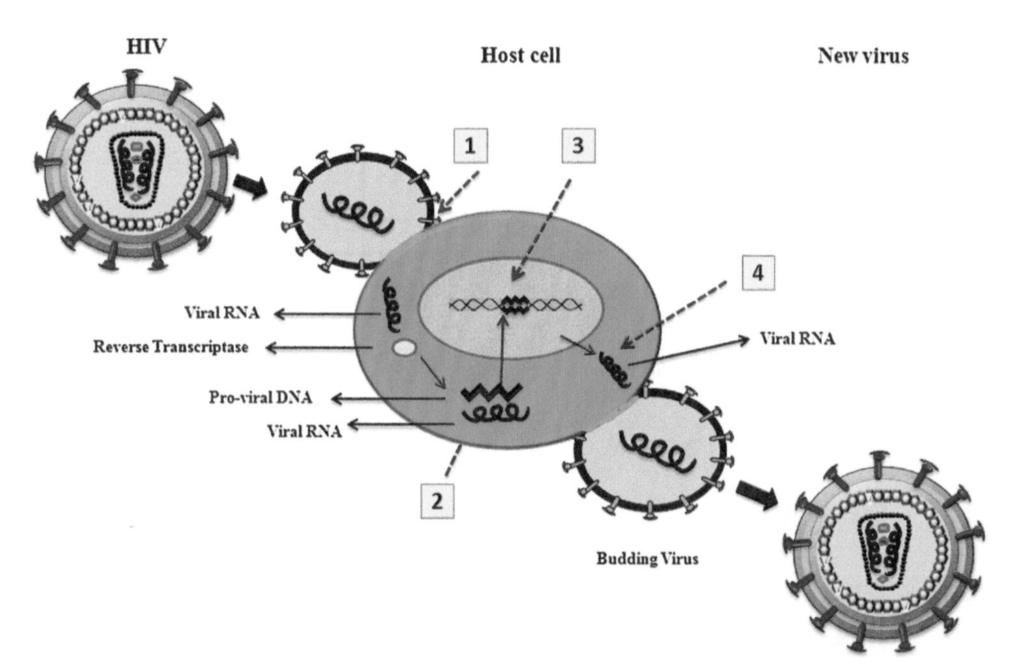

FIGURE 9.3 **HIV Replication Steps and Drug Targets.** Schematic representation of HIV replication along with the stages where different groups of antiretrovirals work. During infections, HIV attaches to the cell surface of target cells and fuses with the cells to release viral RNA and other proteins. Reverse transcriptase produces pro-viral RNA in the cytoplasm. Pro-viral RNA moves to the nucleus and integrates with the host cell genome with the activity of the integrase. After integration of pro-viral DNA into the host DNA, it gives rise to mRNA, which finally translates into different proteins required for synthesis of new virions. These proteins get cleaved by proteases to get assembled into new virions; new virions are released into circulation due to budding from the cells. Steps for drug targets are mentioned in numerals in blocks: Step 1 is the target for fusion inhibitors, Step 2 is the target for reverse transcriptase inhibitors, Step 3 is the target for integrase inhibitors, and Step 4 is the target for protease inhibitors. : Viral RNA; : Pro-viral DNA.

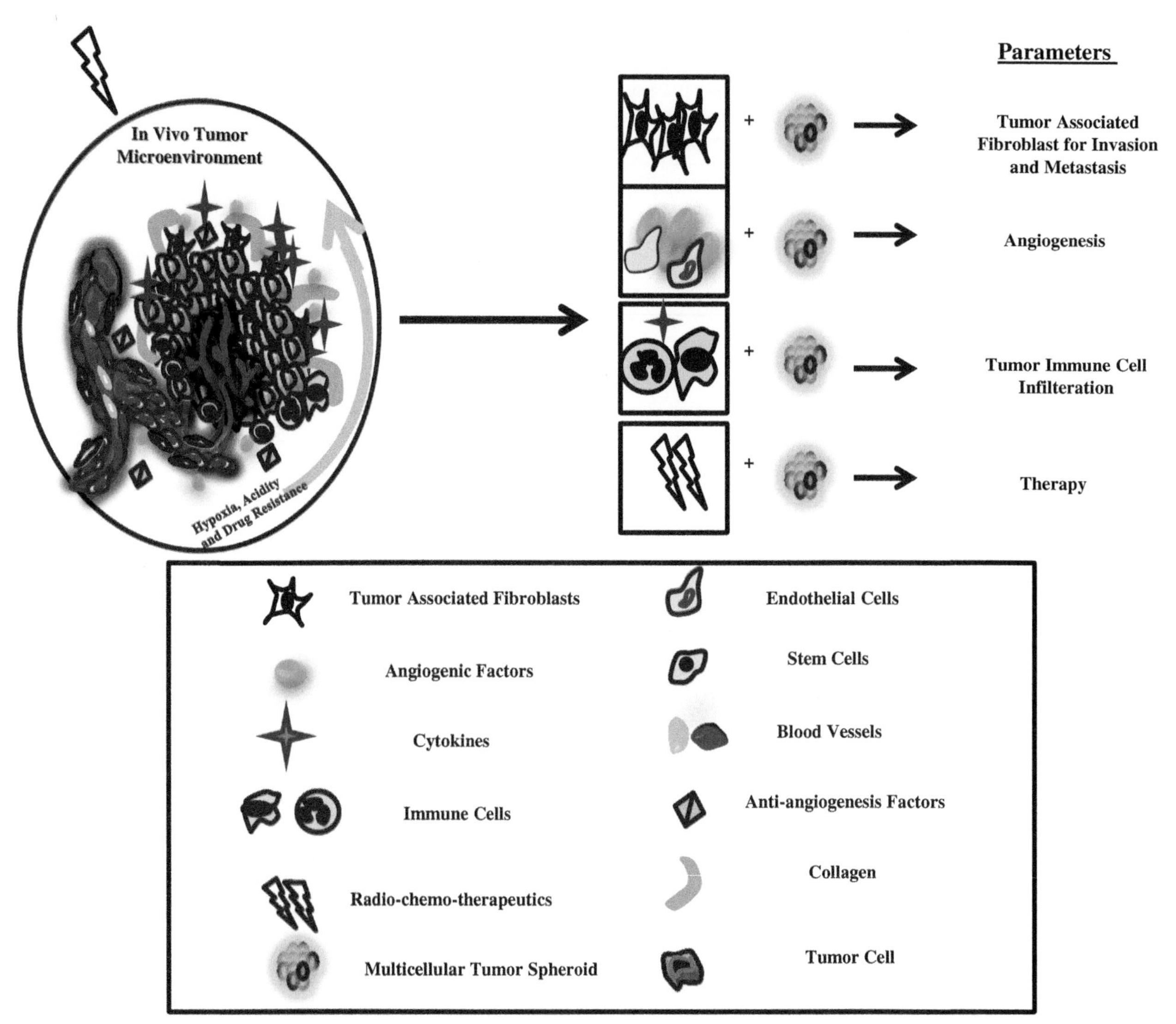

FIGURE 11.2 Approaches for studying the effects of tumor associated parameters on the *in vivo* response of tumors using multicellular tumor spheroid (MCTS).

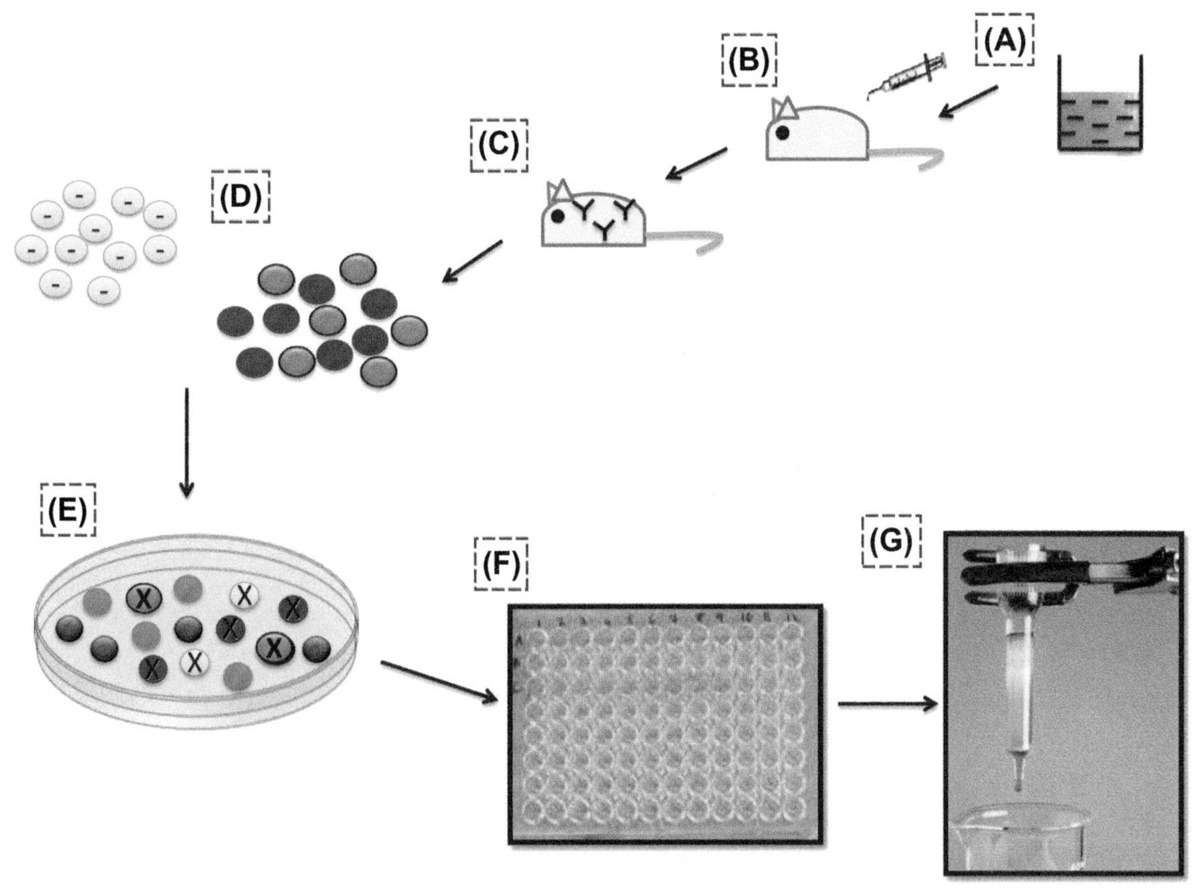

FIGURE 15.3 **Schematic Representation of Production of Monoclonal Antibodies.** This figure represents different crucial steps in the production of monoclonal antibodies. **(A) Antigen Preparations:** Antigen has to be prepared with Freund's Complete Adjuvant or Freund's Incomplete Adjuvant. **(B &C) Immunization:** Mice have to be immunized by injecting antigen prepared with adjuvant; they also have to be given booster doses. **(D) Preparation of Splenocytes and Fusion with Myeloma:** The spleen has to be removed from immunized mice and a single-cell suspension of splenocytes has to be prepared. Then splenocytes have to be fused with myeloma cells in the presence of fusogenic agents. **(E) Selection of Hybridoma:** After fusion, a hybridoma has to be selected from the cell population mixture. Cells have to be grown in HAT selection medium so that after selection only hybridoma cells can survive, while B-lymphocytes and un-fused myeloma cells will die. **(F) Screening of Clone:** After selection of hybridoma cells, a specific clone has to be selected. Different hybridoma cells are diluted in 96-well plates, and after a period of time each clone has to be tested for specificity against the antigen. **(G) Purification of Monoclonal Antibodies:** After selection of a specific hybridoma clone, monoclonal antibodies can be purified. If downstream application requires purification of monoclonal antibodies, then the clone can be expanded and appropriate methods can be applied for purification of monoclonal antibodies. ( ), myeloma cells; (,), lymphocytes; and (,), hybridomas. Cells marked with an "X" represent cell death in the selection medium.

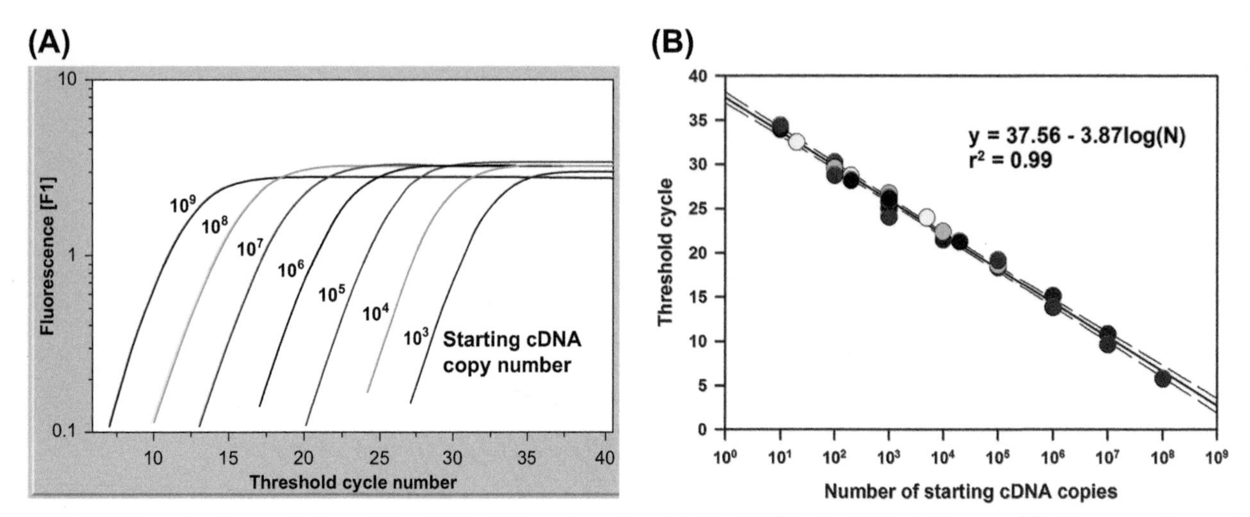

FIGURE 17.1 Example of a quantitative real-time PCR with the primers for α-tubulin. (A) Tubulin cDNA real-time amplification curves for samples containing standard dilutions of the purified target cDNA. (B) Tubulin qPCR calibration curve generated by amplifying known copy numbers of the target cDNA; circles of different colors correspond to separate calibration runs conducted during a 2-year period; dashed lines show 95% confidence intervals for the regression line across all data points.

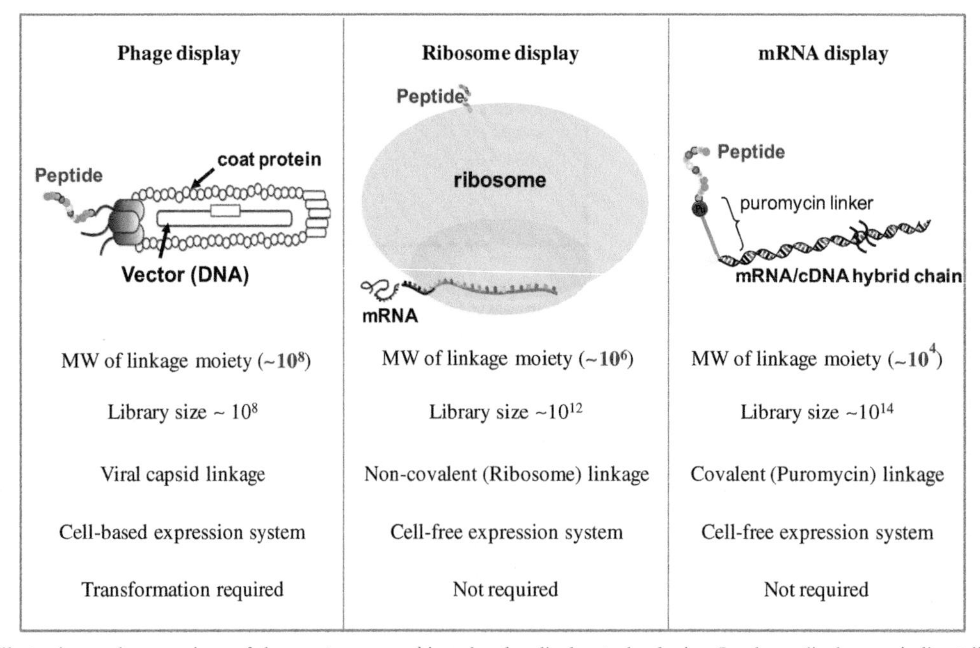

FIGURE 20.5 Illustration and comparison of the most common biomolecular display technologies. In phage display, an indirect linkage (physical) between the gene and gene product is provided by the viral capsid. In ribosome display, a non-covalent linkage is achieved by producing ternary complexes of RNA, ribosomes, and associated nascent peptides. In the mRNA display system, a covalent linkage is generated through a puromycin molecule attached to the encoding mRNA via a short DNA linker molecule.

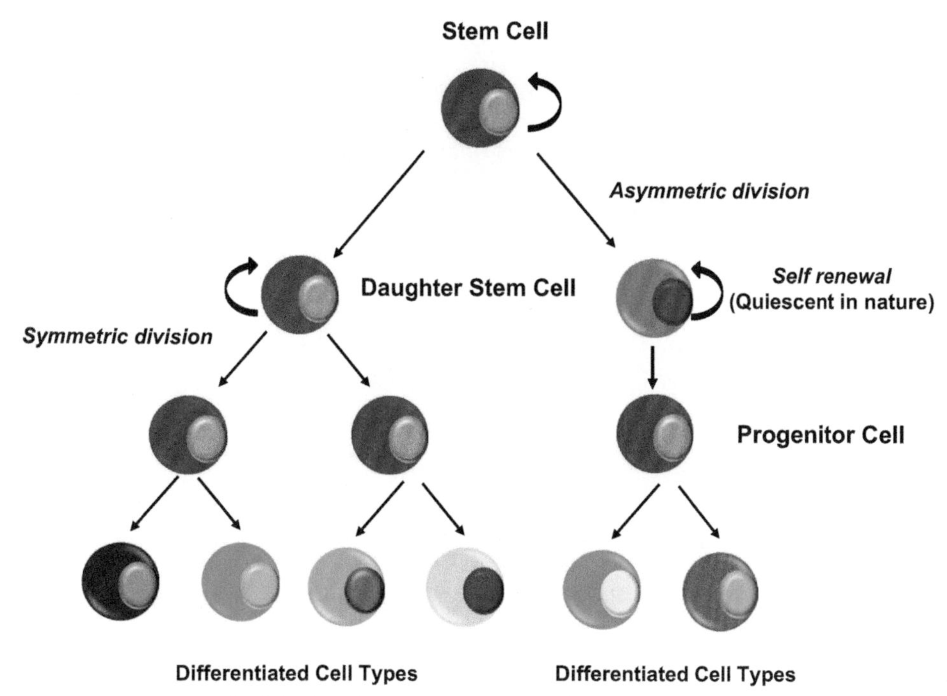

FIGURE 23.2 Asymmetric and symmetric stem cell division.

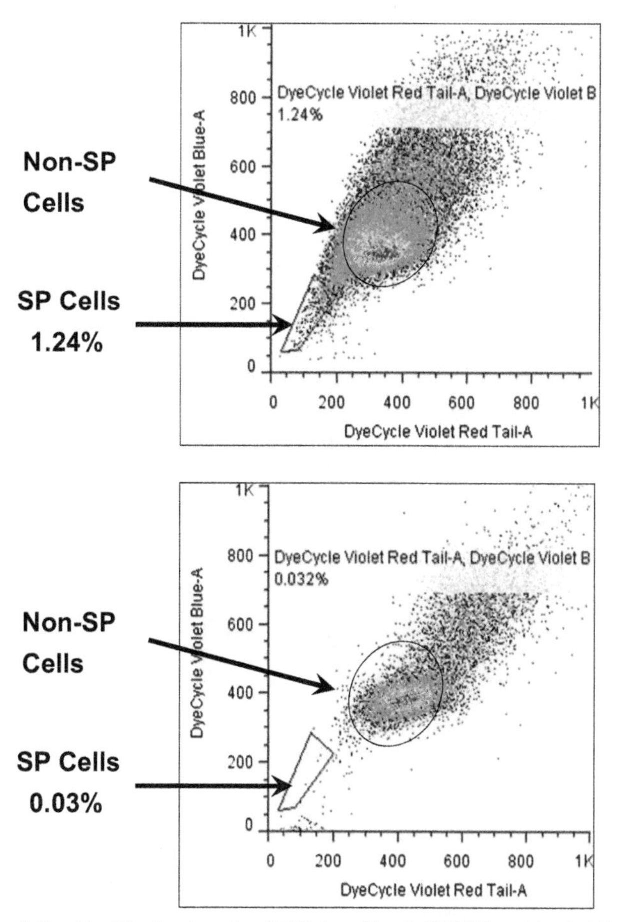

FIGURE 23.4 Side population identification based on DCV dye efflux in HPV16+ve Cervical Cancer Cell Line (SiHa).

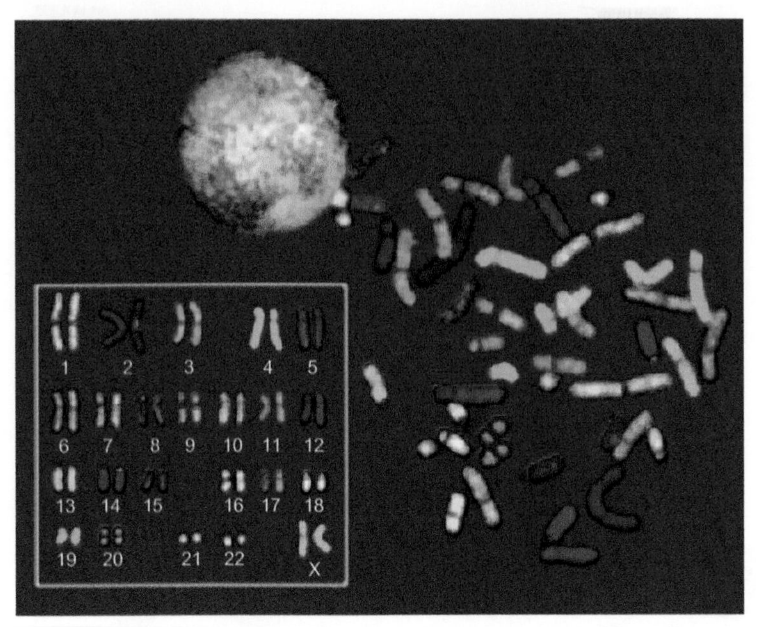

FIGURE 24.5 SKY image showing metaphase chromosomes labeled with different fluorochromes.

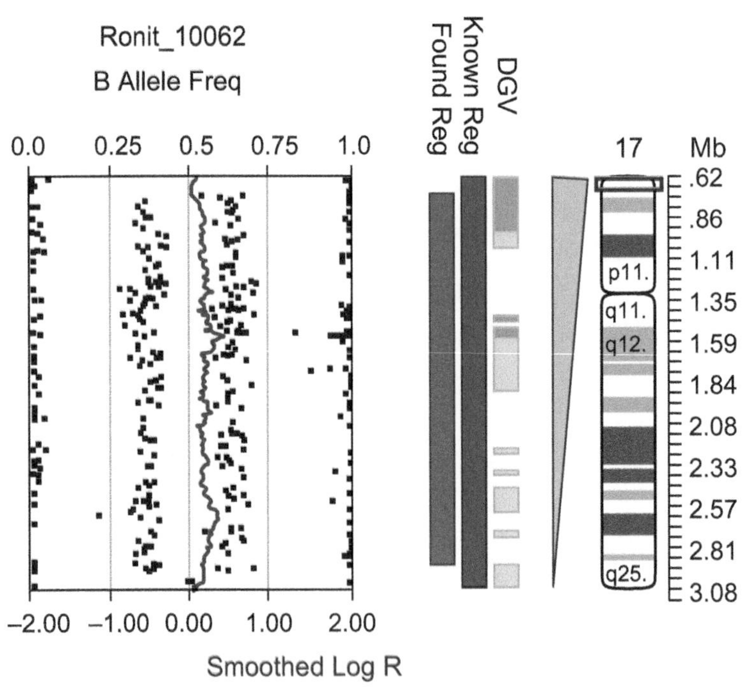

FIGURE 24.6 Array-CGH image showing 2.2 Mb duplication on chromosome 17q13.3.

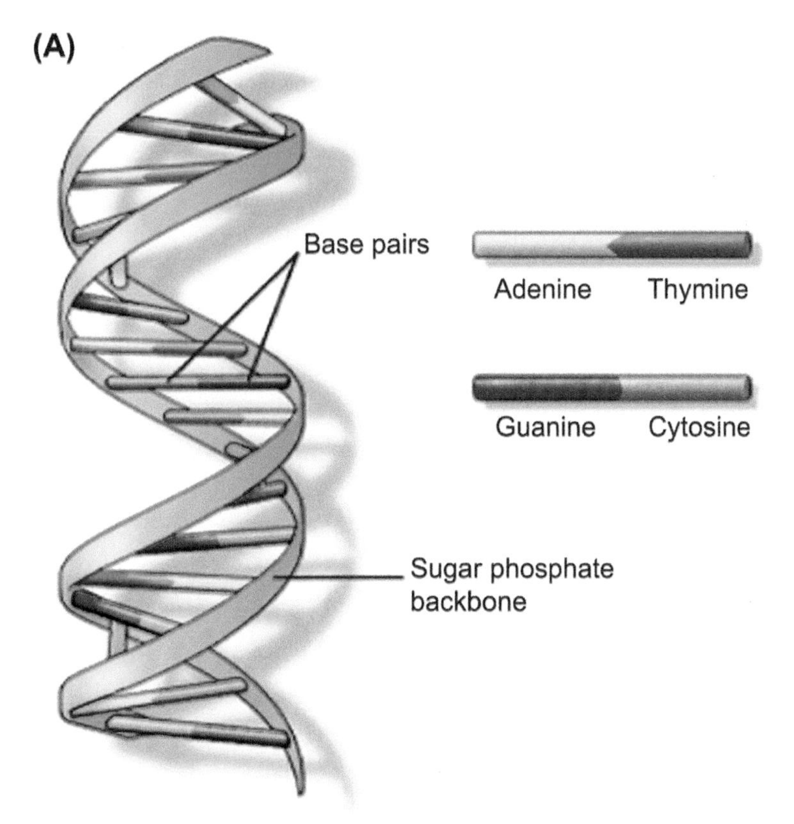

FIGURE 24.7A Double helix structure of DNA. *(Courtesy: U.S. National Library of Medicine.)*

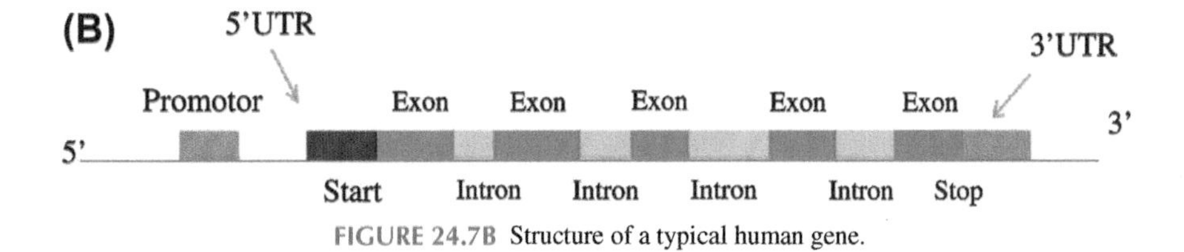

FIGURE 24.7B Structure of a typical human gene.

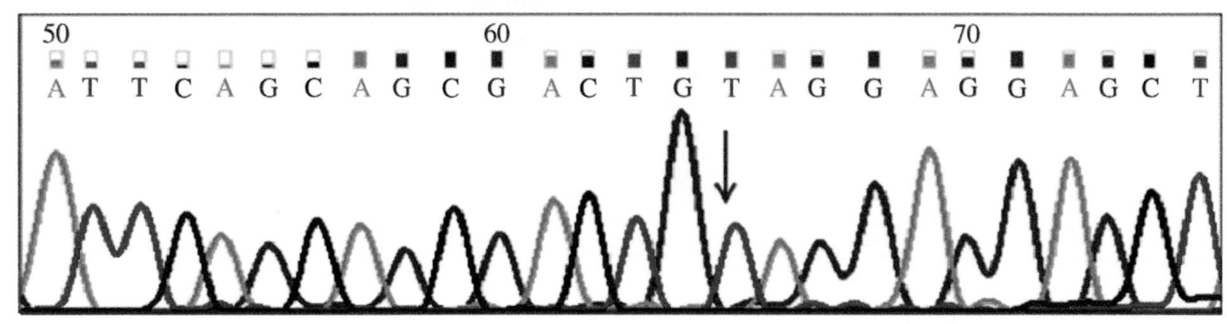

FIGURE 24.8 Electropherogram showing mutation. An arrow shows C to T substitution in codon 318 (CAG to TAG) that results in change of Gutamine (Q) coded by CAG to stop codon TAG, denoted as X.

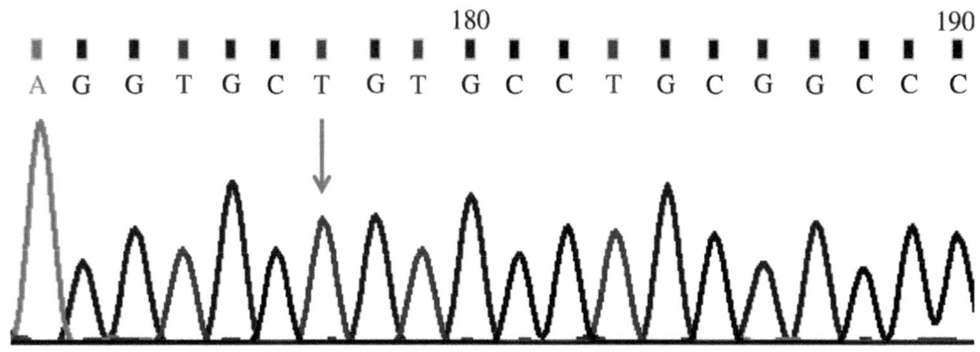

FIGURE 24.9D Partial electropherogram of DMD gene showing C to T change (shown by arrow).

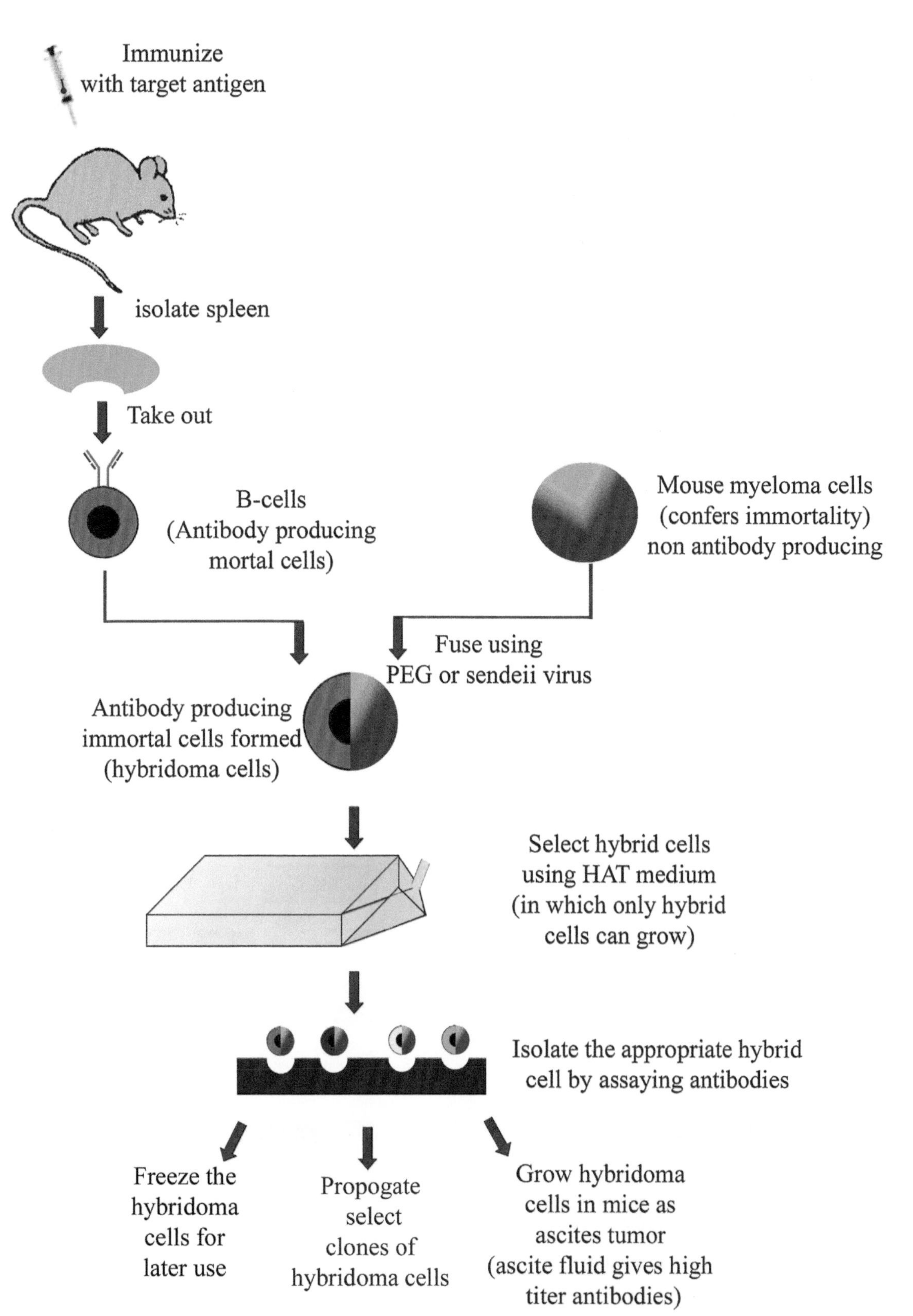
Immunize
with target antigen
isolate spleen
Take out
B-cells
(Antibody producing
mortal cells)
Mouse myeloma cells
(confers immortality)
non antibody producing
Fuse using
PEG or sendeii virus
Antibody producing
immortal cells formed
(hybridoma cells)
Select hybrid cells
using HAT medium
(in which only hybrid
cells can grow)
Isolate the appropriate hybrid
cell by assaying antibodies
Freeze the
hybridoma
cells for
later use
Propogate
select
clones of
hybridoma cells
Grow hybridoma
cells in mice as
ascites tumor
(ascite fluid gives high
titer antibodies)

FIGURE 25.1 Production of monoclonal antibodies (a simplified overview).

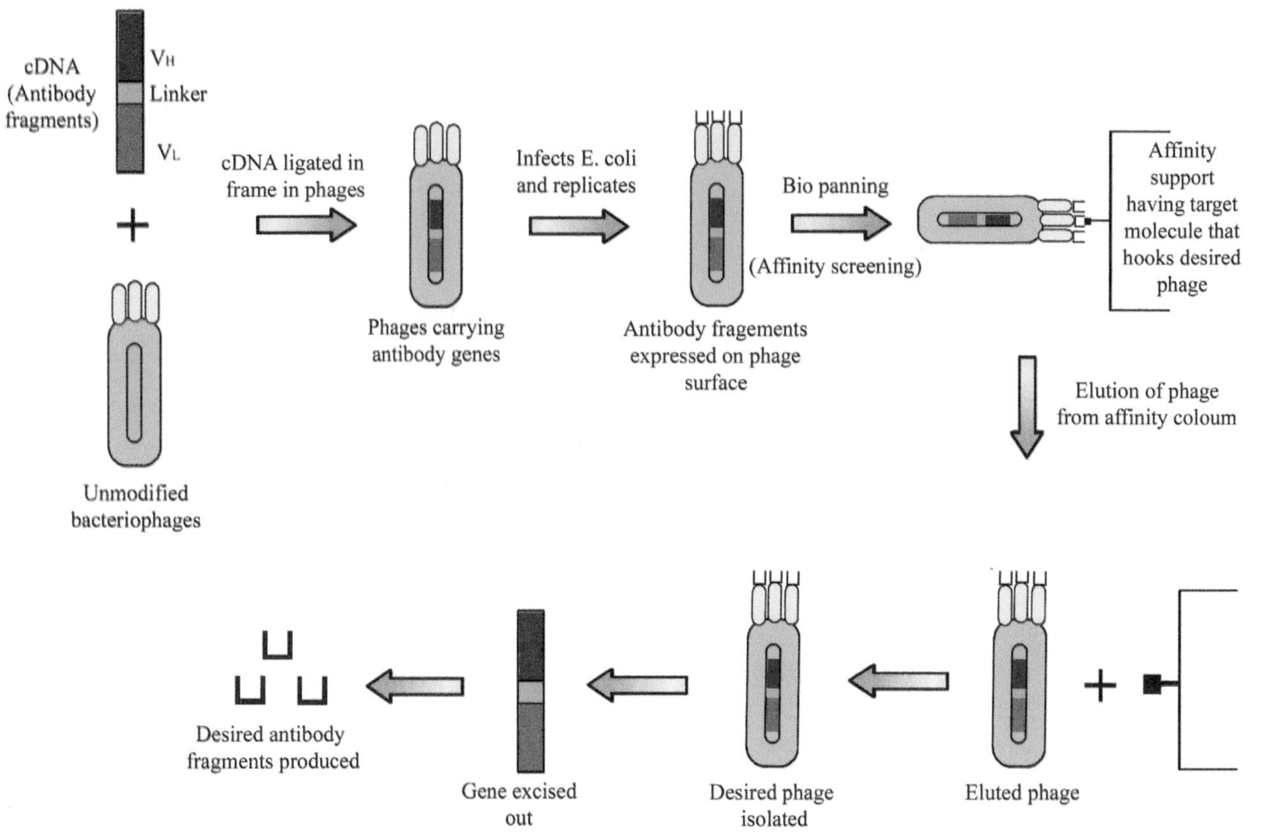

FIGURE 25.3 Phage display technology for antibody screening.

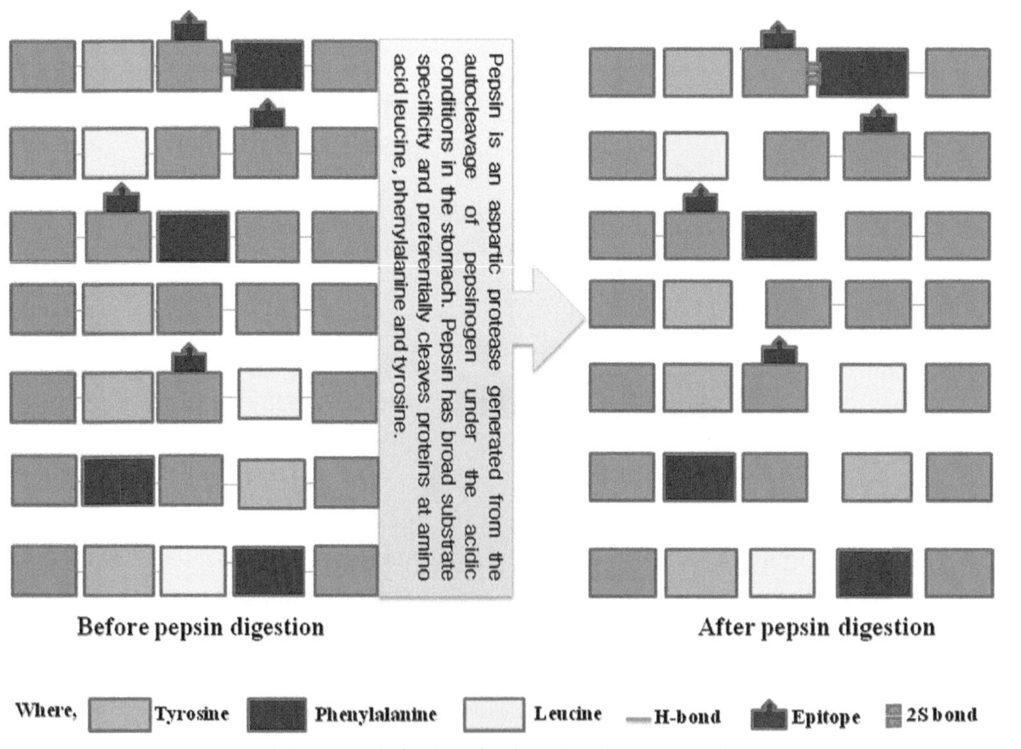

FIGURE 27.2 Mechanism of action of pepsin on polypeptides.

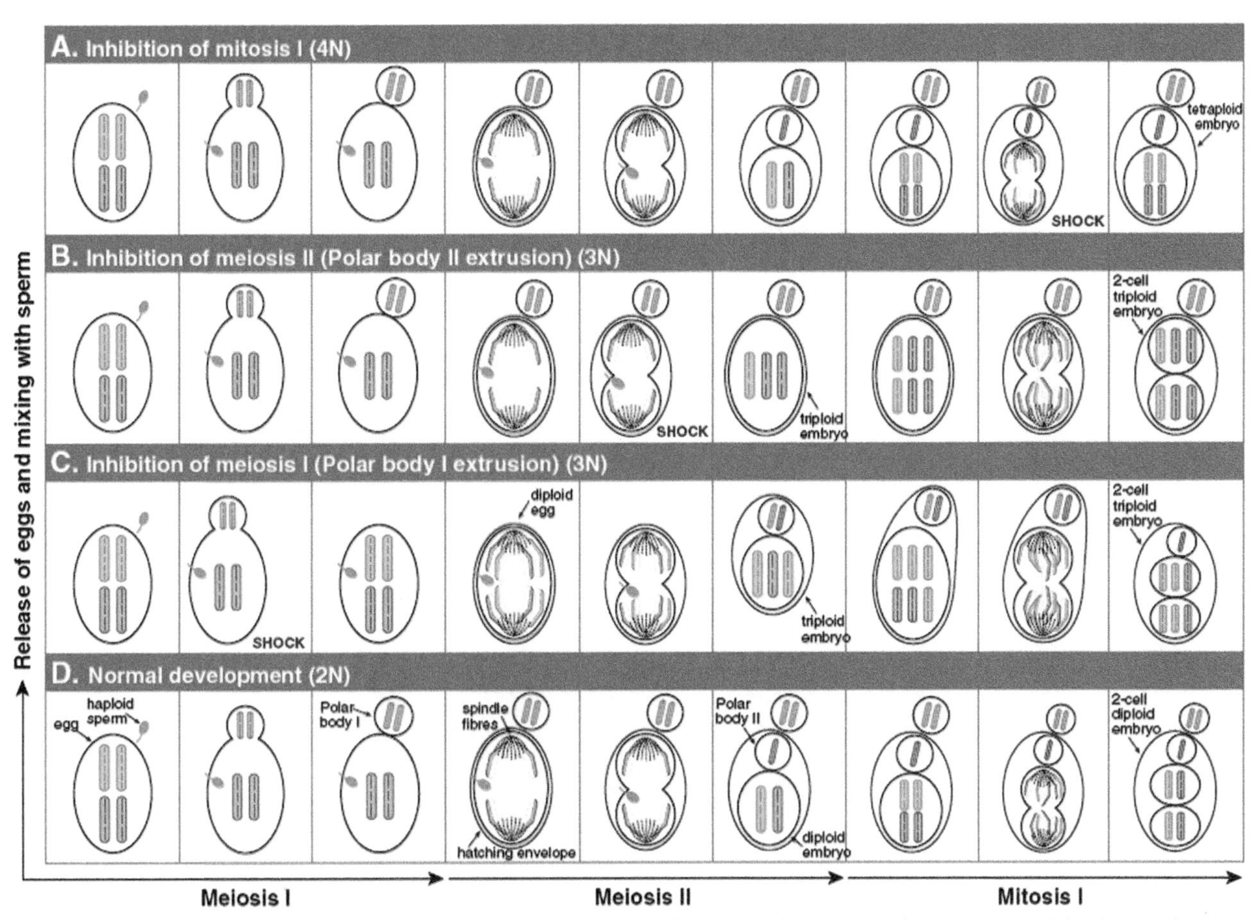

FIGURE 29.1 Process of inducing (A) Meiosis I triploidy, (B) Meiosis II triploidy and (C) Mitotic tetraploidy, and (D) Normal development in penaeid shrimp. *(Source: Sellars et al., 2010)*

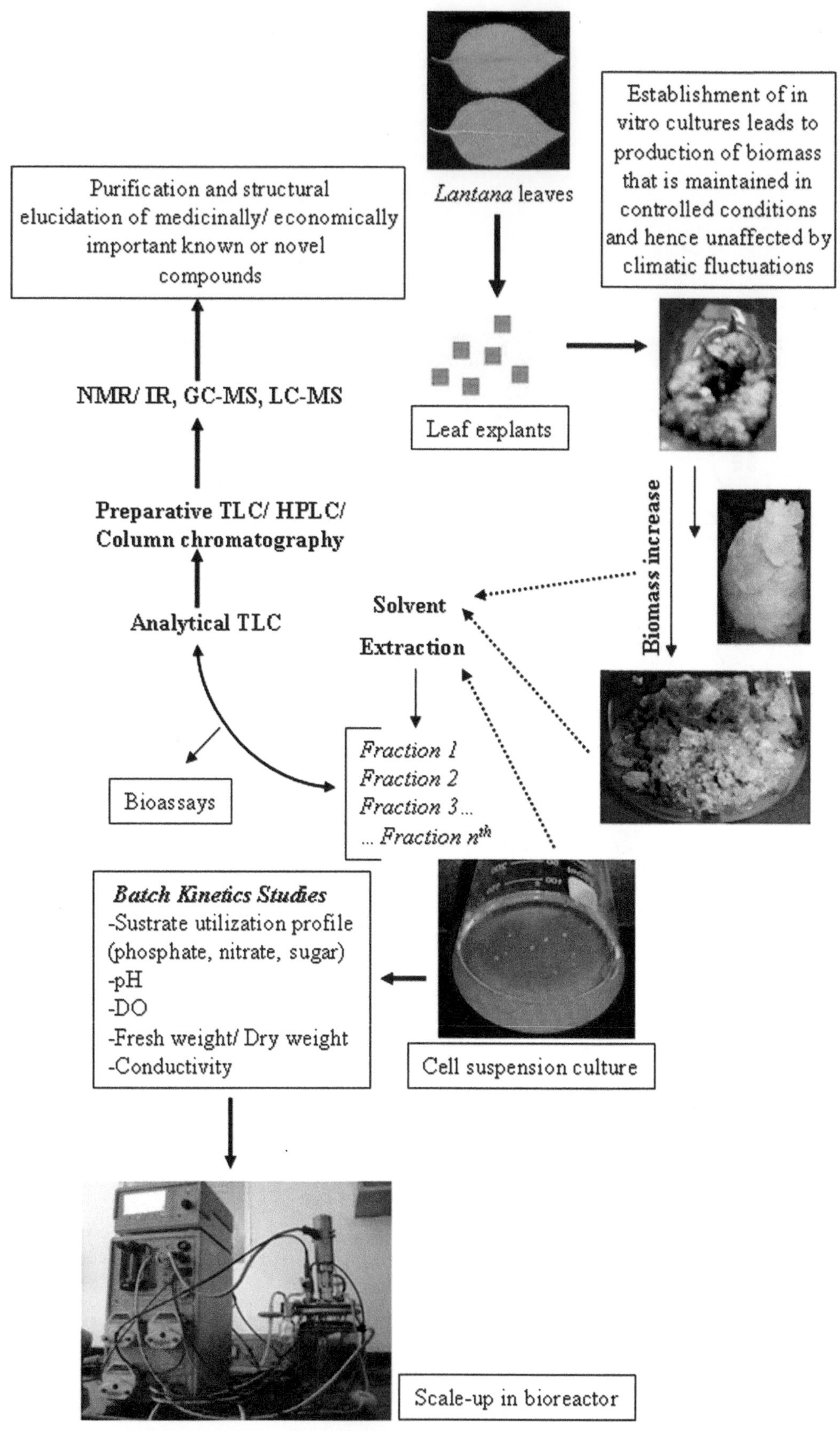

FIGURE 30.4 Isolation of bioactive compounds from *Lantana camara*, a medicinal plant.

Printed and bound by CPI Group (UK) Ltd, Croydon, CR0 4YY
08/05/2025
01865026-0004

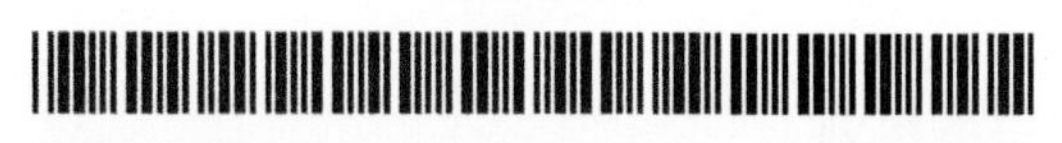